Dr. Christian M. Walz
Berlin Nov 2014

KENNEWICK MAN

Peopling of the Americas Publications

GENERAL EDITORS
Michael R. Waters and Ted Goebel

 Sponsored by the Center for the Study of the First Americans

KENNEWICK MAN

The Scientific Investigation of an Ancient American Skeleton

Edited by
Douglas W. Owsley
and *Richard L. Jantz*

Texas A&M University Press
College Station

Copyright © 2014 by Smithsonian Institution
All rights reserved
First edition
Manufactured in China by Everbest Printing Co.
 through FCI Print Group

This paper meets the requirements of ANSI/NISO Z39.48-1992 (Permanence of Paper).
Binding materials have been chosen for durability.
∞

Library of Congress Cataloging-in-Publication Data

Kennewick Man (2014)
 Kennewick Man : the scientific investigation of an ancient American skeleton / [edited by] Douglas W. Owsley and Richard L. Jantz. — First edition.
  pages cm
 "Peopling of the Americas publications" — Page preceding title page.
 Includes bibliographical references and index.
 ISBN 978-1-62349-200-7 (cloth : alk. paper) — ISBN 978-1-62349-234-2 (e-book) 1. Kennewick Man. 2. Paleo-Indians—Anthropometry—Washington (State) 3. Indians of North America—Washington (State)—Antiquities. 4. Washington (State)—Antiquities. 5. Human remains—Repatriation—Washington (State) 6. Cultural property—Repatriation—United States. 7. Archaeology—Technological innovations. I. Owsley, Douglas W., and Jantz, Richard L., editors of compilation. II. Title.
 E78.W3K45 2014
 979.7'01—dc23
                    2014005528

Contents

Introduction: The Scientific Investigation of Kennewick Man   *1*
  Douglas W. Owsley

ESTABLISHING CONTEXT   5

1 The People Who Peopled America   *7*
  Bradley T. Lepper

2 Geography, Paleoecology, and Archaeology   *30*
  James C. Chatters

3 Chronology of the Kennewick Man Skeleton   *59*
  Thomas W. Stafford Jr.

4 The Precedent-Setting Case of Kennewick Man   *90*
  Alan L. Schneider and Paula A. Barran

5 Reflections of a Former US Army Corps of Engineers Archaeologist   *108*
  Larry D. Banks

6 Curation History and Overview of the Plaintiffs' Studies   *110*
  Cleone H. Hawkinson

STUDYING SKELETAL EVIDENCE   137

7 Skeletal Inventory, Morphology, and Pathology   *139*
  Douglas W. Owsley, Aleithea A. Williams, and Karin S. Bruwelheide

8 Dentition   *187*
  Christy G. Turner II

9 Dental Microwear   *195*
  Mark F. Teaford and Sireen El Zaatari

10 Orthodontics   *207*
  John L. Hayes

11 Body Mass, Stature, and Proportions of the Skeleton   *212*
  Benjamin M. Auerbach

12 Reconstructing Habitual Activities by Biomechanical Analysis of Long Bones   *232*
  Daniel J. Wescott

13 Bones of the Hands and Feet   *249*
  D. Troy Case

14 The Natural Shocks that Flesh is Heir To   *279*
  Della Collins Cook

15 Occupational Stress Markings and Patterns of Injury   *290*
  James C. Chatters

16 Stable Isotopic Evidence for Diet and Origin   *310*
  Henry P. Schwarcz, Thomas W. Stafford Jr., Martin Knyf, Brian Chisholm, Fred J. Longstaffe, James C. Chatters, and Douglas W. Owsley

17 Taphonomic Indicators of Burial Context   *323*
  Douglas W. Owsley, Aleithea A. Williams, and Thomas W. Stafford Jr.

18 Benthic Aquatic Algae: Indicators of Recent Taphonomic History   *382*
  James N. Norris and Douglas W. Owsley

19 Postmortem Breakage as a Taphonomic Tool for Determining Burial Position   *393*
  Hugh Berryman

APPLYING TECHNOLOGY TO INTERPRETATION 415

20 Computed Tomography, Visualization, and 3D Modeling *417*
Rebecca Snyder

21 Prototype Accuracy and Reassembly *429*
David R. Hunt

22 Molding and Casting Methods *447*
Steven J. Jabo

23 The Point of the Story *452*
Dennis J. Stanford

INCORPORATING POPULATION DATA 459

24 The Ainu and Jōmon Connection *463*
C. Loring Brace, Noriko Seguchi, A. Russell Nelson, Pan Qifeng, Hideyuki Umeda, Margaret Wilson, and Mary L. Brace

25 Cranial Morphometric Evidence for Early Holocene Relationships and Population Structure *472*
Richard L. Jantz and M. Katherine Spradley

26 Two-Dimensional Geometric Morphometrics *492*
M. Katherine Spradley, Katherine E. Weisensee, and Richard L. Jantz

27 Morphological Features that Reflect Population Affinities *503*
George W. Gill

28 Identity Through Science and Art *519*
Karin S. Bruwelheide and Douglas W. Owsley

LEARNING FROM EARLY HOLOCENE CONTEMPORARIES 535

29 Evidence of Maritime Adaptation and Coastal Migration from Southeast Alaska *537*
E. James Dixon, Timothy H. Heaton, Craig M. Lee, Terence E. Fifield, Joan Brenner Coltrain, Brian M. Kemp, Douglas W. Owsley, Eric Parrish, Christy G. Turner II, Heather J. H. Edgar, Rosita Kaaháni Worl, David Glenn Smith, and G. Lang Farmer

30 A New Look at the Double Burial from Horn Shelter No. 2 *549*
Margaret A. Jodry and Douglas W. Owsley

ANTICIPATING KENNEWICK MAN'S FUTURE 605

31 Storage and Care at the Burke Museum *607*
Cleone H. Hawkinson

32 Who Was Kennewick Man? *622*
Douglas W. Owsley and Richard L. Jantz

Acknowledgments *651*

Index *657*

KENNEWICK MAN

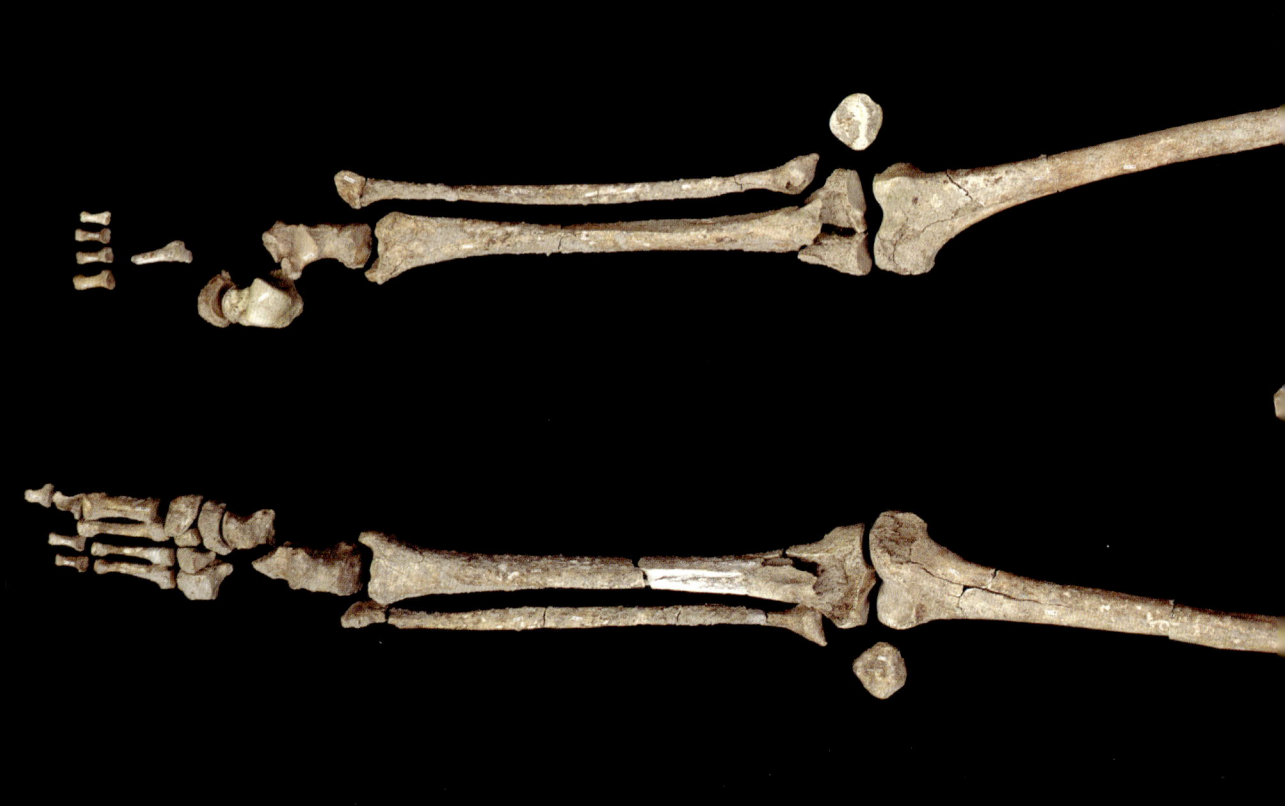

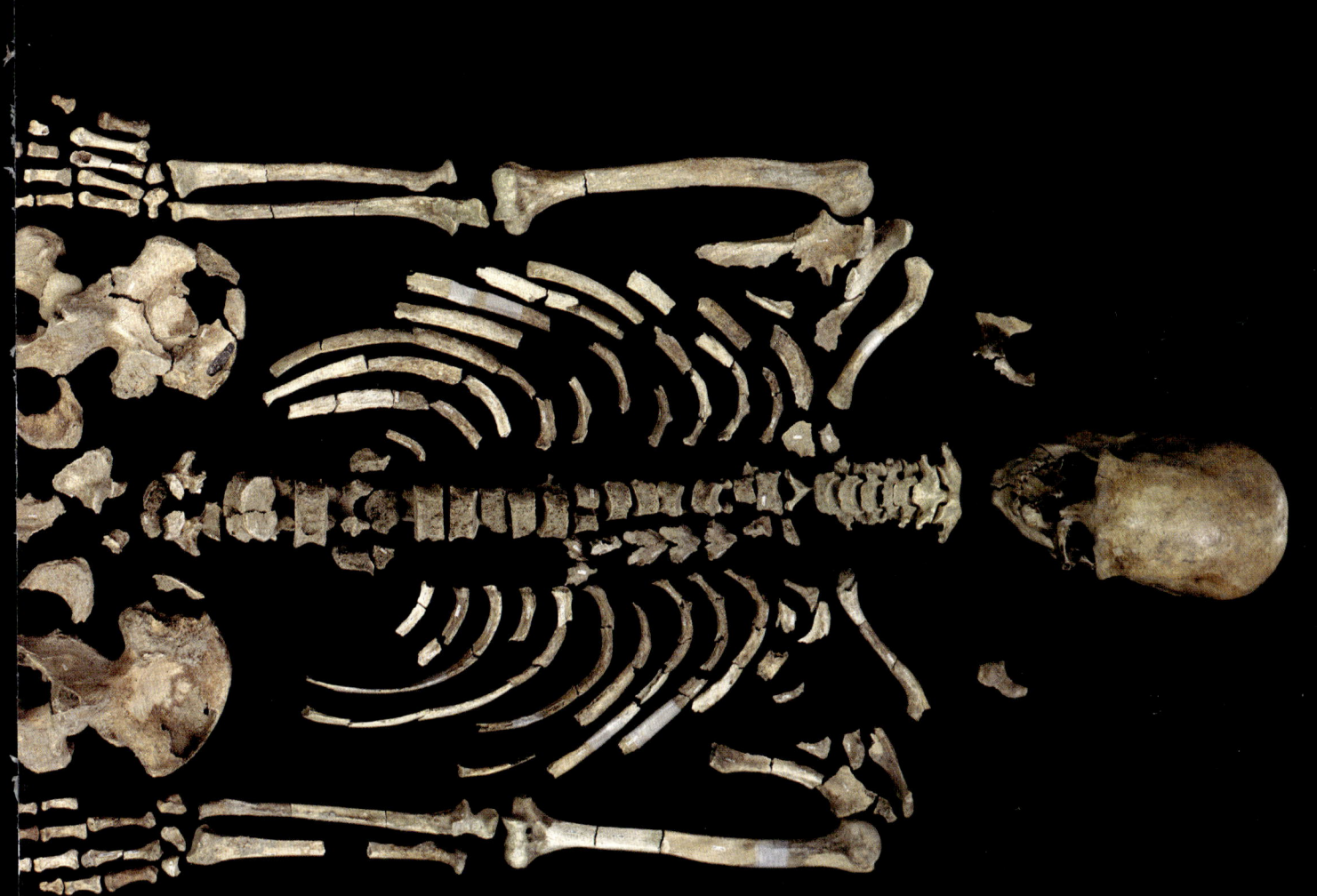

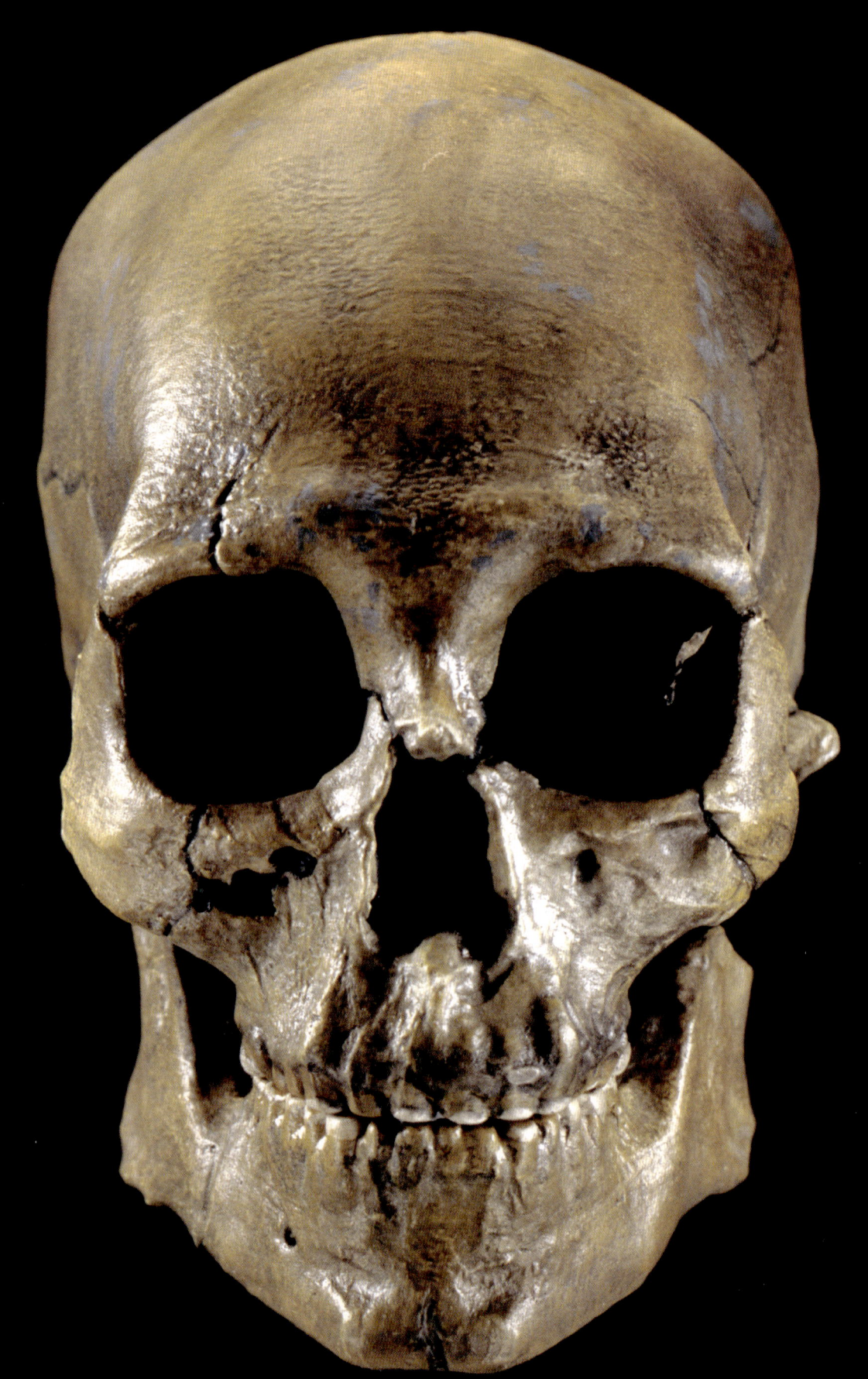

INTRODUCTION

The Scientific Investigation of Kennewick Man

Douglas W. Owsley

The study began with the discovery of about 300 bones and fragments of a skeleton whose skull spoke of uncertain ancestry and whose hip held a stone point. There were no other artifacts, not even a grave. Yet, with an understanding of how to interpret bones, investigators closely examined the fragments and reconstructed not only the individual's height, weight, and body build, but also his facial appearance. Among other details, information was gathered on his preferred foods, his main occupation, a description of his burial, and clues to his ancestors. This volume presents the results of the comprehensive scientific study of Kennewick Man, one of the most complete ancient human skeletons ever found in North America.

Although written for a rigorous, professional audience, the information, collected by experienced investigators in their fields, is structured in hopes of also reaching the interested, informed public. The chapters provide background and context, include detailed results and interpretations drawn from independent and interrelated skeletal investigations, and explain how the datasets were collected and analyzed, often within a comparative framework. Although the chapters are cross-referenced and coordinated, they were not designed to achieve correspondence on every assessment. As part of the scientific process, there can be varying interpretations; different points of view on the same data need to be presented with supporting arguments.

One point that is without debate is the rare opportunity Kennewick Man presents to learn about Paleoamericans and the conditions experienced during the early Holocene. While we may make inferences about the peopling of the Americas, the process of migration was far more complicated than appreciated even a decade ago. Recent work suggests that it began earlier than generally believed and involved movements from more than one homeland by both foot and watercraft. Human skeletal remains provide remarkable insights into this process, and bioarchaeology is one particularly informative means for investigating the past. There is still much to be learned. The contents of this volume, introduced in the following few pages, are intended to persuade others to advance the discussion, not to end it.

ESTABLISHING CONTEXT

Providing the reader with a context for Kennewick Man—both within the framework of Paleoamerican studies and this skeleton's discovery, curation, and analysis—is one goal of this volume. Lepper's review of the skeletal evidence from North America and Mexico points to the scarcity of data on late Pleistocene and early Holocene human remains. Summary burial and demographic information demonstrate that discoveries older than 8,000 radiocarbon years are not only rare but often incomplete due to recovery circumstances, poor preservation, or mortuary practices such as cremation. Lepper's chapter emphasizes the exceptional significance of the nearly complete skeleton of Kennewick Man. Chatters follows with a description of the discovery and provides an overview of how the bones were collected. Background information on the greater Columbia Basin, including its geography, hydrology, modern climate and ecology, geochronology, paleoecology, and human prehistory is also presented.

An accurate geologic age is crucial to sorting out the archaeological, cultural, and environmental contexts that form a basis for establishing Kennewick Man's contemporaries and for assessing whether he was different physically or behaviorally from them or subsequent groups. Stafford presents the chemical procedures used to purify the bone collagen, which in turn allowed for accurate dating of the remains. This process establishes the chronological foundation that places this paleoanthropological study within the context of time. Stafford's chemical analyses also provide a background for determining the effect of a marine diet on Kennewick Man's radiocarbon age.

An understanding of the circumstances of Kennewick Man's discovery and knowledge of his radiocarbon age set the stage for the legal battle over access to the remains (*Bonnichsen et al. v. US*), which was preceded by months of fruitless talks with the US Army Corps of Engineers (Corps), nearly six years in court, almost two years of an appeal that settled the case, and one year to prepare and obtain the Corps' approval of the study plan. Banks, a retired senior archaeologist with the Corps, offers a personal account of the agency's actions, which barred researchers from studying the skeleton. Attorneys Schneider and Barran describe case actions and decisions that eventually allowed scientific analyses. The lawsuit also involved clarification of the meaning of the words "Native American" and other key terms used in the Native American Graves Protection and Repatriation Act (NAGPRA). The District Court ruled that scientists have standing to challenge agencies' over-interpretation of NAGPRA. This decision has the potential to impact future interpretations of federal cultural resource laws including NAGPRA, the Archaeological Resources Protection Act, and the National Historic Preservation Act.

The curation history of the skeleton, beginning with its discovery in 1996 and ending with the plaintiffs' analyses at the Burke Museum in 2006, is presented by Hawkinson. Transfers of custody, repatriation requests, and detailed accounts of the plaintiffs' studies review the skeleton's recent history as well as efforts to secure the right to study the remains.

STUDYING SKELETAL EVIDENCE

The main objective of this volume is, of course, to present what has been learned about Kennewick Man as a result of scientific analyses. The varied, in-depth assessments of the skeleton reflect the study plan approved by the Corps and stem from two separate sessions for data collection, totaling only 16 days. During this time the many investigators, sharing the space and the bones, conducted their examinations and recorded data. Close and overlapping quarters resulted in information exchanges that led to new avenues of investigation, expanded sets of expert analyses, and an extraordinary number of photographs recording what was found.

An overview of the remains by Owsley et al. includes a complete bone and dental inventory and observations on bone morphology and pathology. It lays the foundation for subsequent chapters that focus on specific aspects of the skeleton.

Three chapters are devoted to research on Kennewick Man's dentition. Turner examines population affinity and the use of the teeth in task activities. Teaford and El Zaatari analyze patterns of enamel micro-wear related to food preparation and dietary contaminants. Hayes presents a clinical orthodontic assessment, comparing Kennewick Man's dentition to a modern population.

Bone structure, as it relates to body-build and activity patterns, is a recurring topic in several chapters. The body composition of Kennewick Man is discussed by Auerbach, who estimates stature and body mass and compares Kennewick Man to other Paleoamericans as well as to modern populations. Wescott examines external and computed tomography (CT)-derived cross-sectional properties of limb bones as a means of determining activity patterns. His biomechanical assessment compares the limb bones of Kennewick Man with late Pleistocene and Holocene groups involved in diverse subsistence strategies. The number of well-preserved hand and foot bones recovered provides an unprecedented opportunity to assess Kennewick Man's use of his hands and feet in everyday life. Case examines these small bones for asymmetry, developmental defects, and taphonomic changes, with handedness being among several conclusions.

Pathological changes and bone morphology are discussed by Chatters, who interprets patterns of traumatic injury and healing, changes in joint surfaces, and musculoskeletal markings as evidence of behavior. Cook also examines the skeleton for pathological changes. Despite congruence among some skeletal indicators of disability, her assessment differs in several respects from both Chatters and previous analyses conducted as part of the federal government's study team. Studies expressing multiple points of view are especially relevant in demonstrating the need for continued access to the remains for future analysis and interpretation.

The meticulous analytical methods used by Stafford were the basis for the carbon, nitrogen, oxygen, and phosphate isotope values reported by Schwarcz et al. Both chemical analyses chapters evaluate the implications of these findings and their interpretations with regard to the origin of Kennewick Man, an important topic revisited in the concluding chapter with the inclusion of additional food web and human comparative data. The chronological and isotopic investigations provide an increased understanding of how diagenesis changes original radiocarbon and stable isotope data and how accurate values can be obtained.

Three chapters focus on determining whether Kennewick Man was intentionally buried; they utilize information on bone preservation and the results of taphonomic assessment. This question was central to the analyses, but complicated by the recovery of scattered bones from the lake shoreline rather than the recovery of bones in situ. Evidence detailing the skeleton's original burial context is documented by Owsley et al., Norris and Owsley, and Berryman; it includes sediment deposition on Kennewick Man's bones, patterns of algal staining, and postmortem bone breakage. The first two contribute information on skeletal positioning and the length of time from exposure to recovery. The distribution and characteristics of postmortem breaks in the skeleton, and their sequence of occurrence, are used to further deduce in situ positioning and the erosion sequence. These complementary analyses demonstrate the potential for multiple lines of evidence to unravel postmortem history and provide a compelling model for taphonomic assessment of remains from disturbed contexts.

APPLYING TECHNOLOGY TO INTERPRETATION

Four chapters describe applications of emerging technologies to the Kennewick Man skeleton. Snyder explains how high-resolution, industrial-grade CT scans of the multiple pieces of the skull and right innominate were used to create stereolithographic resin models of each fragment. Hunt accepts the challenging task of assembling an accurate reconstruction of the original cranium and mandible using the translucent resin models. He describes the process of creating this proxy for the original skull, and with Jantz and Spradley, assesses the accuracy of the final model obtained by this laborious process. Over time, the photosensitive resins used to create the models described by Jabo will shrink. This instability requires the prototypes to be molded and cast in plaster, which can then be painted for visual fidelity. Virtual extraction and solid replication of the projectile point in the right ilium through CT analysis and rapid prototyping allows Stanford to describe damage to both the point's base and tip, and to tentatively identify the projectile point type. He stops short of definitively classifying the point due to limitations in viewing the actual object, but promotes the likelihood that an atlatl was used to inflict the wound. While these studies help to establish a precedence for other analyses, they also reveal limitations to their application and encourage future refinement.

INCORPORATING POPULATION DATA

Because cranial morphology is heritable and selectively neutral, it is useful for estimating the ancestral origins of unknown individuals. Craniometrics can therefore be used to test hypotheses about the peopling of the New World. In this volume, four chapters compare Kennewick Man's cranial measurements to data for Old and New World groups. Brace et al. evaluate the connection between Kennewick Man and the Paleolithic inhabitants of northeast Asia by comparing Kennewick Man to their worldwide database. This analysis examines the possibility that the Jōmon, presumed ancestors of the Ainu and direct descendants of the Paleolithic inhabitants of northeast Asia, represent a source from which the western hemisphere was initially settled. The variable set in Brace's database is focused primarily on measurements of the face and is smaller than that used by Jantz and Spradley, who identify a strong similarity between the cranial morphology of Kennewick Man and Polynesians. Spradley et al. apply two-dimensional geometric morphometrics to determine how Paleoamerican crania differ from Archaic period groups and late prehistoric/historic Native Americans. Visual appreciation of this variation is enhanced by wireframe models of crania obtained by connecting landmarks. Gill compares Kennewick Man to other populations using both metric and nonmetric approaches to the study of the skull and femur.

The translation of the results of the skeletal research to a broad audience is discussed in the chapter on the facial reconstruction of Kennewick Man. Forensic facial reconstruction methodology and anatomical artistry are combined by Bruwelheide and Owsley as they work to re-create Kennewick Man's appearance. Underlying bone structure determines the physical attributes of the head and face while the appearance of the eyes, lips, and hair were influenced by population attributes suggested by the morphometric comparisons. Personal characteristics, such as his age and tissue scarring from an injury to the forehead, were also factored into the finished bust. This reconstruction uses the results of the study plan's comprehensive analysis to achieve the most informed interpretation of his appearance.

LEARNING FROM EARLY HOLOCENE CONTEMPORARIES

Comparisons to other ancient human remains provide a frame of reference for interpreting data obtained from Kennewick Man. In this section, analysis of a ca. 10,000-year-old-skeleton from the coast of southeast Alaska by Dixon's research team exemplifies specific studies that need to be completed as part of a comprehensive study of Kennewick Man. This discovery from On Your Knees Cave includes geographical sourcing of lithic artifacts, offering insight into well-developed early Holocene trade networks or procurement strategies that required the use of watercraft. Isotope values for this man from the coast reflect dependence on marine resources and are similar to those of Kennewick Man.

Jodry and Owsley integrate information on skeletal morphology, burial setting, and associated mortuary items from a site in Texas to suggest that unique behavioral and social roles can be identified and that, unlike Kennewick Man, not all Paleoamerican males were engaged primarily in physical activities associated with hunting. This kind of analysis adds complexity and depth to Paleoamerican studies that have been primarily based on comparing lithic technologies. This chapter clearly demonstrates that continued access to ancient remains allows for new comparisons and reinterpretations of physical evidence.

ANTICIPATING KENNEWICK MAN'S FUTURE

The fate of Kennewick Man and the possibilities for future analyses are topics addressed near the end of the volume. Hawkinson raises concern about the long-term conservation of the skeleton at the Burke Museum based on ten years of data provided by the Corps. Politics and different cultural perspectives are also intimately tied to the fate of the remains, which without question represent one of North America's most significant bioarchaeological discoveries. The volume concludes with a synthesis of what has been revealed about Kennewick Man and identifies areas that still need to be investigated.

The study of this skeleton goes beyond a better understanding of a single individual from ancient America; through Kennewick Man, a greater appreciation for the complexity of the Paleoamerican story has also been acquired. The American public has shown a remarkable interest in the discovery, as well as in the scientific study of human remains, both modern and ancient. People of all ages want to learn more about this subject and to understand how information is collected from skeletal remains. In this regard, Kennewick Man has been the subject of numerous articles and books written for adults and younger readers. Kennewick Man has truly become a teacher for all ages.

On a final note, we have gained a greater appreciation for the different connotations and symbolic meanings one skeleton can have for present-day people. Kennewick Man has created a range of perspectives within and among groups internal and external to the court case, including employees of contributing federal agencies, other scientists, and Native American groups. The Kennewick Man discovery and analysis is a topic of passionate interest and debate within the field of anthropology, engendering support from some and criticism from others. Both positions were helpful. Critical commentaries helped us to better define our research objectives and clarify our analytical methodology. From long careers in physical anthropology, forensic anthropology and archaeology, we continue to be fascinated by this unparalleled opportunity to gain new knowledge. The contributors to this volume have learned a great deal about Paleoamerican life. We hope the reader does as well.

ESTABLISHING CONTEXT

Kennewick Man's significance is based on his context—within both the field of Paleoamerican studies and the court of law. His skeleton represents an invaluable source of information for this time period. Ancient skeletal remains are extremely rare in comparison to the thousands of later prehistoric and historic period skeletons. Most remains of comparable antiquity are incomplete with poor bone preservation—making the relatively complete and well-preserved skeleton of Kennewick Man that much more exceptional.

To fully understand the Kennewick Man story, it is crucial to also realize the significance of the skeleton within the context of the law. Through the Native American Graves Protection and Repatriation Act (NAGPRA), tribes were able to claim remains if they could establish a shared group relationship with the skeleton. Without confirming such a link through study, the US Army Corps of Engineers immediately decided to expeditiously repatriate Kennewick Man to tribes wanting to rebury him. The scientists, having had their many requests for access denied, felt that there was no way to prevent the loss of the skeleton except to file a lawsuit.

The results of the Kennewick Man case represent significant milestones in the interpretation of NAGPRA and the rights of scientists to study ancient remains.

The People Who Peopled America

Bradley T. Lepper

The story of the peopling of the Americas is about humans discovering and settling a truly New World. Curiously, archaeologists sometimes lose sight of the actual people involved, presumably because they so seldom encounter their physical remains. For example, Dillehay (2000:227) claimed that North and South America are "the only continents on the planet where our knowledge of an early human presence comes almost exclusively from traces of artifacts and not from human skeletal remains." Not surprisingly, perhaps, archaeologists tend to become obsessed with the elegant flint spear points and the bones of mammoth and giant bison among which they are sometimes found in lethal association. Yet, without diminishing the importance of the stones and animal bones upon which the understanding of the first Americans largely has been based (Adovasio 2003; Bonnichsen and Turnmire 1999; Dillehay 2000; Ubelaker 2006), it must be recognized that without the remains of the people themselves and the information contained in their bones, graves, and funerary offerings, the view is of a stage littered with fragments of scenery and discarded props, but with no actors.

Human skeletal remains of great antiquity are infrequently found in the Americas, but "frequency" is relative, and Dillehay (2000) has overstated the disparity between North America and the rest of the Paleolithic world. One survey of human remains from late Paleolithic Europe listed 56 individuals from the period between 15,000 and 10,000 years ago (Holt and Formicola 2008). Information is known on 74 sets of human remains from the same period in North America (Table 1.1). Europe has an area over 10 million square kilometers, whereas North America, including Canada, the United States, and Mexico, encompass about 22 million square kilometers. Therefore, in Europe there are about 5.6 documented sets of late Upper Paleolithic human remains per million square kilometers, and in North America there are about 3.4.

Clearly, the density of documented human remains is lower in North America, but since the documented European Upper Paleolithic human paleontological record encompasses about 2,000 more years than the equivalent record in North America, some of the discrepancy can be attributed to the accumulation of burials over those additional centuries. Moreover, the vastly greater extent of the late Pleistocene glaciation in North America relative to Europe meant that a greater proportion of its landmass would have been inaccessible to human occupation for much of this period. Finally, modern humans had entered Europe by perhaps 40,000 years ago, which means that by 15,000 years ago populations in many parts of Europe likely approached carrying capacity. In contrast, humans either were not yet present in the Americas, or had only recently arrived by 15,000 years ago, and it would be many centuries before population levels reached a density equivalent to that which characterized Europe at the beginning of this period. All of this suggests that human remains are not anomalously rare in the western hemisphere and that understanding of the earliest Americans is being hampered by the perception that such discoveries are rarer than they actually are.

Of course, none of this should be taken to suggest that Paleoamerican human remains are commonplace discoveries. On the contrary, human remains dating to this early period are, in absolute terms, quite rare. Although Table 1.1 lists a total of 324 possible burials, only 152 skeletons are sufficiently well preserved, complete, and adequately reported, even to be identified reliably as male or female. Many are so fragmented and incomplete that they fit comfortably in cupped hands.

One reason for the absolute, if not relative, rarity of ancient human remains is the long span of time over

TABLE 1.1 Paleoamerican human remains dated at least 8,000 RC yr. BP.

Date[1]	Site	MNI[2]	Age	Sex	Type of burial	Burial location	Grave offerings, number & type	Reference
13,270 ± 340 (SI-2488)	Meadowcroft, PA	1	—	—	accidental? firepit	rockshelter		Sciulli 1982
13,240 ± 1010 (SI-2065)	Meadowcroft, PA	1	—	—	accidental? firepit	rockshelter		Sciulli 1982
11,670 ± 60 (UCR-4000/CAMS-87301)	Eva de Naharon, Tulum Caves, Mexico	1	20–30	F	unknown	cave	0	González González et al. 2008
10,960 ± 380 (CAMS-16810)	Arlington Springs, CA	1	adult	M	unknown	stream bank	0	Chawkins 2006; Dixon 1999; Johnson et al. 2002; Powell 2005:148
10,900 ± 300 (based on associated artifacts)	Fishbone Cave, Burial 2, NV	1	25–44	F	unknown	cave	5+; hide, basketry, cordage	Dixon 1999; Orr 1956; Powell 2005; Owsley pers. comm. 2010
10,755 ± 74 (OxA-10112)	Peñon III, Mexico	1	25	F	unknown	unknown	unknown	González et al. 2003; González-Jose et al. 2005; Powell 2005:154
10,705 ± 35 (CAMS-80538)	Anzick 1, MT	1	0.6–1	M	unknown	rockshelter	>100; lithics	Lahren and Bonnichsen 1974; Waters and Stafford 2007; Wilke et al. 1991; Morrow et al. 2006; Owsley and Hunt 2001; Stafford 1994; Owsley pers. comm. 2014; Waters pers. comm. 2014
10,625 ± 95 (Beta-43055/ETH-7729)	Buhl, ID	1	17–21	F	shallow pit	gravel bar overlooking Snake River	4	Dixon 1999:126–128; Green et al. 1998; Herrmann et al. 2006
10,500	Chimalhuacán, Mexico	1	—	—	unknown	unknown	> 4	González-Jose et al. 2005
10,470 ± 490	Mostin 1978, CA	1	adult	F	—	eroding from creek bank	3; ground stone	Dixon 1999:128–129; Taylor et al. 1985:138
10,260 ± 340 (UCLA-1795A)	Mostin Burial 4, CA	1	—	—	—	eroding from creek bank	unknown	Dixon 1999
10,260 ± 190 (Gak-3998)	Warm Mineral Springs 1, FL	1	30–50	M	accidental?	sink hole	1	Clausen et al. 1975; Dixon 1999:138; Powell 2005:131
10,260 ± 190 (Gak-3998)	Warm Mineral Springs 2, FL	1	20–25	F	accidental?	sink hole	unknown	Powell 2005:131

(continued)

TABLE 1.1 (continued) Paleoamerican human remains dated at least 8,000 RC yr. BP.

Date[1]	Site	MNI[2]	Age	Sex	Type of burial	Burial location	Grave offerings, number & type	Reference
10,260 ± 190 (Gak-3998)	Warm Mineral Springs 3, FL	1	50+	F	accidental?	sink hole	unknown	Dixon 1999:138; Powell 2005:133
10,260 ± 190 (Gak-3998)	Warm Mineral Springs 4, FL	1	subadult	—	accidental?	sink hole	unknown	Powell 2005:131
10,200 ± 65 (OxA-10225)	Tlapacoya I, Mexico	1	30–35	M	—	sink hole	unknown	González et al. 2003
10,190 ± 1450 (TX-2595)	Little Salt Spring, FL	1	adult	M	—	sink hole	3+	Clausen et al. 1975; Clausen et al. 1979; Powell 2005:129; Wentz and Gifford 2007
10,020 ± 50 (CAMS-61133)	Arch Lake Woman, NM	1	17–19	F	extended in prepared burial pit	on dune crest overlooking Arch Lake basin	4: necklace (19 talc beads), unifacial stone tool, bone tool, pulverized red ochre (likely in pouch)	Owsley et al. 2010
c. 10,000	Crowfield, Ontario	1	—	—	cremation	small knoll along a gully	177 lithics; incl. 29 fluted bifaces	Deller and Ellis 1984
c. 10,000	Sloan, AR	30	—	—	extended?	prominent knoll on Crowley's Ridge overlooking Cache River valley	avg. 8 per burial; range: 1–38	Morse 1997
c. 10,000	Midland, TX	1	27–32	F	unknown, accidental?	—	0	Stewart 1955; Wormington 1957; Holliday 1997:101; Holliday and Meltzer 1996; Owsley pers. comm. 2010
10,000–8,200	Whitewater Draw 1, AZ	1	27–34	F	flexed	—	0	Waters 1986; Owsley pers. comm. 2010
10,000–8,000	Whitewater Draw 2, AZ	1	adult	—	unknown	—	0	Dixon 1999:139; Friends of America's Past 2001
10,000–5,000	Seminole Sink, TX	22	5–6	—	unknown—1 subadult cremation	small sink hole	unknown	Powell 2005:140–141

(continued)

TABLE 1.1 (*continued*) Paleoamerican human remains dated at least 8,000 RC yr. BP.

Date[1]	Site	MNI[2]	Age	Sex	Type of burial	Burial location	Grave offerings, number & type	Reference
9,710 ± 40 (CAMS-60681)	Horn Shelter 2 – A, TX (Tx-1722)	1	37–44	M	flexed, double	rockshelter	>10	Redder 1985; Redder and Fox 1988; Young 1988; Owsley pers. comm. 2010
9,690 ± 50 (CAMS-51794)	Horn Shelter 2 – B, TX (Tx-1722)	1	10–11	F	flexed, double	rockshelter	1+	Young 1988; Owsley pers. comm. 2010
9,975 ± 125 (AA-4805)	Olive Branch, IL	1	adult	F	cremation	floodplain	0	Gramly 2002; Bruwelheide and Owsley 2002
9,920 ± 250 (I-6897)	Tlapacoya XVIIIa, Mexico	1	adult	—	unknown	unknown	unknown	Dixon 1999:140; Lorenzo and Mirambell 1999; Gonzalez et al. 2003
9,880 ± 50 (CAMS-32038)	On Your Knees Cave, AK	1	20–22	M	accidental?	cave		Dixon 1999; Kemp et al. 2007; Owsley pers. comm. 2010
9,870 ± 50 (Beta-120802)	Marmes F 1, WA	1	15–20	F	cremation?	floodplain		Dixon 1999; Fryxell et al. 1968; Hicks 2004; Krantz 1979
9,870 ± 50 (Beta-120802)	Marmes F 2, WA	1	6	—	cremation?	floodplain		Dixon 1999; Fryxell et al. 1968; Hicks 2004; Krantz 1979
9,870 ± 50 (Beta-120802)	Marmes F 3, WA	1	15–25	M	cremation?	floodplain		Dixon 1999; Fryxell et al. 1968; Hicks 2004; Krantz 1979
9,870 ± 50 (Beta-120802)	Marmes F 4, WA	1	adult	—	cremation?	floodplain		Dixon 1999; Fryxell et al. 1968; Hicks 2004; Krantz 1979
9,870 ± 50 (Beta-120802)	Tlapacoya XVIIIb, Mexico	1	adult	—	cremation?	unknown	unknown	Dixon 1999:140; Gonzalez et al. 2003; Lorenzo and Mirambell 1999
9,700 ± 250 (GX-0530)	Gordon Creek, CO	1	30–34	F	flexed	arroyo bank	6; lithics, bone bead, red ochre	Breternitz et al. 1971; Dixon 1999:124–125; Powell 2005:151; Swedlund and Anderson 1999; Owsley pers. comm. 2010
9,700 ± 500 (M-130)	Graham Cave, MO	1	—	—	secondary	cave	5; point, hammerstone, antler tool, wolf canine pendant	Chapman 1975:97; Hoard et al., 2004; Logan 1952
9,670 ± 120	Cutler Ridge, FL	10	—	—	accidental? hearth	solution hole, rockshelter	NA	Dolzani 1986; Doran and Dickel 1988:371; Morrell 1997; Powell 2005:133

(*continued*)

TABLE 1.1 (continued) Paleoamerican human remains dated at least 8,000 RC yr. BP.

Date[1]	Site	MNI[2]	Age	Sex	Type of burial	Burial location	Grave offerings, number & type	Reference
9,650 ± 124 (Tx-4793)	Wilson-Leonard, TX	1	24–28	F	flexed	—	5; shark tooth fossil, groundstone chopper, hematite	Bousman et al. 2002; Collins 1998; Powell 2005:143; Owsley pers. comm. 2010
9,515 ± 155 (GX-19422-G)	Wizards Beach, NV	1	40–49	M	unknown	—	0	Edgar 1997; Tuohy and Dansie 1997; Owsley pers. comm. 2010
9,480 ± 160 (DIC-160)	Squaw Rockshelter, OH	2	25	1F	unknown	rockshelter	0	Brose 1989; Prior 1989; cf. Prufer 2001
9,470 ± 60 (UCR-3477)	Grimes Shelter 1 (9F-NSM-743), NV	1	8–10	F	unknown	rockshelter	0	Dixon 1999; Powell and Neves 1998; Tuohy and Dansie 1997; Owsley pers. comm. 2010
9,470 ± 60 (UCR-3477)	Grimes Shelter 2, NV	1	16–18	M	unknown	rockshelter	0	Dansie 1997; Owsley pers. comm. 2010
c. 9,450–8,850	Ashworth 4, KY	1	adult	F	flexed	rockshelter	1; point (?)	DiBlasi 1981
c. 9,450–8,850	Ashworth 9, KY	1	—	—	unknown	rockshelter	0	DiBlasi 1981
9,435 ± 270 (GX-4126)	Ice House Bottom 1, TN	1	adult	F	secondary, shallow pit	rockshelter	2 mammal bones	Chapman 1977
9,430 ± 40 (Beta-120803)	Marmes H1, WA	1	adult	M	cremation? hearth	cave		Dixon 1999; Fryxell et al. 1968; Hicks 2004; Krantz 1979
9,430 ± 40 (Beta-120803)	Marmes H2, WA	1	adult	—	cremation? hearth	cave		Dixon 1999; Fryxell et al. 1968; Hicks 2004; Krantz 1979
9,430 ± 40 (Beta-120803)	Marmes H3, WA	1	adult	—	cremation? hearth	cave		Dixon 1999; Fryxell et al. 1968; Hicks 2004; Krantz 1979
9,430 ± 40 (Beta-120803)	Marmes H4, WA	1	8–14	—	cremation? hearth	cave		Dixon 1999; Fryxell et al. 1968; Hicks 2004; Krantz 1979
9,430 ± 40 (Beta-120803)	Marmes H5, WA	1	8–14	—	cremation? hearth	cave		Dixon 1999; Fryxell et al. 1968; Hicks 2004; Krantz 1979
9,430 ± 40 (Beta-120803)	Marmes H6, WA	1	8–14	—	cremation? hearth	cave		Dixon 1999; Fryxell et al. 1968; Hicks 2004; Krantz 1979
9,415 ± 25 (weighted mean of multiple dates)	Spirit Cave 2, NV	1	40–44	M	flexed	cave	3+; rabbit-skin blanket, moccasins, burial mats	Edgar 1997; Eiselt 1997; Jantz and Owsley 1997; Tuohy and Dansie 1997

(continued)

TABLE 1.1 (continued) Paleoamerican human remains dated at least 8,000 RC yr. BP.

Date[1]	Site	MNI[2]	Age	Sex	Type of burial	Burial location	Grave offerings, number & type	Reference
9,270 ± 60 (UCR-3480/ CAMS-30558)	Spirit Cave 1, NV	1	adult	F	unknown	cave	0	Tuohy and Dansie 1997
8790 ± 110 (NZA-1102)	Browns Valley, MN	1	32–36	M	pit	floodplain	lanceolate points	Anfinson 1997:32; Dixon 1999:120-121; Jenks 1934, 1937; Myster and O'Connell 1997; Powell 2005:136-7; Owsley pers. comm. 2010; Stafford pers. comm. 2010
9,040 ± 200 (UCLA-1795C)	Mostin, Burial 1, CA	1	—	—	unknown	eroding from stream bank	unknown	Dixon 1999
9,040 ± 50 (UCR-3478/ CAMS-30557)	Spirit Cave 3, NV	1	18–22	F	cremation	cave	0	Tuohy and Dansie 1997; Owsley pers. comm. 2010
c. 9,000–7,000	Cueva del Tecolote, Mexico	2	—	—	unknown	cave	1	González-Jose et al. 2005
c. 9,000–8,000	Jerger 3, IN	1	—	—	cremation	sand ridge overlooking marsh	>7; lithics, marine shell	Tomak 1979
c. 9,000–8,000	Jerger 9, IN	1	—	—	cremation	sand ridge overlooking marsh	>4; lithics	Tomak 1979
c. 9,000–8,000	Jerger 13, IN	1	—	—	cremation	—	>4; lithics	Tomak 1979
9,000	Metro Balderas, Mexico	1	—	—	unknown	—	unknown	González-Jose, et al. 2008
c. 9,000–8,000	Namu, BC	1	—	—	accidental?	sea coast at mouth of salmon spawning stream	—	Dixon 1999:119
9,000 ± 82 (UCLA-1292B)	Rancho La Brea, CA	1	16–17	F	accidental?	tar pit	—	Dixon 1999:130; Kroeber 1962; Powell 2005; Owsley pers. comm. 2010
8,965 ± 30 (UCIAMS-35589)	San Miguel Man (CA-SMI-608), CA	1	33–39	M	unknown	—	unknown	Owsley pers. comm. 2010; Stafford pers. comm. 2007
8,810 ± 250 (TX-6049)	Stigenwalt, KS	1	—	—	accidental?	—	—	Hoard et al. 2004; Thies 1990:109

(continued)

TABLE 1.1 (continued) Paleoamerican human remains dated at least 8,000 RC yr. BP.

Date[1]	Site	MNI[2]	Age	Sex	Type of burial	Burial location	Grave offerings, number & type	Reference
8,610 ± 90 (mean of multiple dates)	Anzick 2, MT	1	6–8	—	unknown	rockshelter	0	Morrow et al. 2006; Owsley and Hunt 2001; Stafford 1994; Waters and Stafford 2007
c. 8,500	Reiner, WI	1	—	—	cremation	—	Eden-Scottsbluff points	Mason and Irwin 1960
8,480 ± 390 (NMC-1216)	Cummins, Ontario	1	—	—	cremation	glacial beach	0	Dawson 1983; Julig 1984
8,480 ± 110 (ISGS-236)	Koster, IL	9	5 adult, 4 subadult	2 M, 3 F	adults: flexed; subadults: 3 extended; 1 flexed	floodplain	Only 1 male (Burial 80) had offering: bone atlatl weight?	Brown and Vierra 1983:107; Hoard et al. 2004; Walthall 1999:12
c. 8,450–7,750	Ice House Bottom 2, TN	1	35–40	F	secondary, cremated, shallow basin	rockshelter	0	Chapman 1977:113
8,358 ± 21 (mean of two dates: UCIAMS-II6396; UCIAMS-II6397)	Kennewick Man, WA	1	35–39	M	extended	floodplain	0	Chapter 3
8690 ± 40 (CAMS-76640)	Chancellor 1, CA (pta 1812)	1	33–44	M	flexed double burial	—	0	Kennedy 1983; Owsley pers. comm. 2010
8,350 ± 90 8690 ± 40 (CAMS-76640)	Chancellor 2, CA (pta 1812)	1	40–54	F	flexed double burial	—	0	Kennedy 1983; Owsley pers. comm. 2010
8,250 ± 115 (S-1737)	Gore Creek, BC	1	23–39	M	flexed	wall of washout on Gore Creek	0	Cybulski et al. 1981; Dixon 1999:19
8,170 ± 100 (Beta-38554)	Hourglass Cave, CO	1	35–40	M	accidetal?	cave	0	Powell 2005:141
8,169 ± 488	Modoc 20, IL	1	35	M	semi-flexed	rockshelter	0	Fowler 1959; Neumann 1967
8,169 ± 488	Modoc 22, IL	1	13	F	flexed	rockshelter	0	Fowler 1959; Neumann 1967
8,169 ± 488	Modoc 25, IL	1	47	F	flexed	rockshelter	0	Fowler 1959; Neumann 1967
8,169 ± 488	Modoc 27, IL	1	45	F	flexed	rockshelter	0	Fowler 1959; Neumann 1967
8,169 ± 488	Modoc 28, IL	1	45–50	F	flexed	rockshelter	0	Fowler 1959; Neumann 1967
8,169 ± 488	Modoc 29, IL	1	47	M	flexed	rockshelter	0	Fowler 1959; Neumann 1967

(continued)

TABLE 1.1 (*continued*) Paleoamerican human remains dated at least 8,000 RC yr. BP.

Site	Date[1]	MNI[2]	Age	Sex	Type of burial	Burial location	Grave offerings, number & type	Reference
Stick Man, WA	8140 ± 50 (CAMS-55193); 8110 ± 50 (CAMS-55196)	1	35–49	M	unknown	unknown	unknown	Chatters et al. 2000; Owsley pers. comm. 2010
Windover (all), FL	8,120 ± 70 (TO-241)	168 (incl. #90, 102, and 154)	(101 adults; 67 sub-adults)	47 M, 47 F	95% flexed (160 flexed; 8 extended)	pond	stone tools, bone tools, wood artifacts, basketry and cordage	Dickel 2002
Windover 90, FL	8,120 ± 70 (TO-241)	1	11	F	flexed	pond	22; stemmed point, bone shaft wrench, antler point, shark tooth tool, awl	Dickel 2002
Windover 102, FL	8,120 ± 70 (TO-241)	1	30–40	M	extended	pond	0; (cranium not part of original burial)	Dickel 2002; Doran and Dickel 1988
Windover 154, FL	8,120 ± 70 (TO-241)	1	47	M	semi-flexed	pond	6; shark's tooth, chert biface knife, deer bone awl, canid bone awl, deer antler tine, butchered deer bone	Dickel et al. 1989
Kaskaskia Mine, Burial 1, IL	8,050 ± 35 (AO592)	1	12–15	?	flexed	rockshelter or cave	0	Hargrave et al. 2006
Las Palmas, Tulum Caves, Mexico	8,050 ± 130 (UGA-6828)	1	44–50	F	flexed, on surface	cave	0	González González et al. 2008
El Templo, Tulum Caves, Mexico	c. 8,000	1	25–30	M	extended, on surface	cave	0	González González et al. 2008
White River Forest, CO	c. 8,000	1	41–45	M	accidental?	near entrance to cave		Dixon 1999:125
Pelican Rapids, MN	7,850 ± 50 (CAMS-10354)	1	13–15	F	unknown	small knoll	5	Jenks 1933, 1938; Hrdlička 1937; Myster and O'Connell 1997; Fiedel 2004; Owsley pers. comm. 2010
Total: 324			159 adults/ 85 subadults	72 males/ 80 females				

[1] Dates are given in radiocarbon years before the present (RC yr. BP) when available in the reference. Lab numbers for samples are provided when known.
[2] MNI. Minimum number of individuals.

which the remains have been subjected to the variously destructive chemical and geological processes of decomposition and weathering. Moreover, deep burial under millennia of waterborne or windblown sediment will have rendered many others inaccessible. Another reason so few Paleoamericans have been recovered is the vastness of the American landscape in comparison to the relatively small numbers of people in these early populations. There also are cultural factors that may be involved, such as mortuary practices that lower the probability of bone preservation. Finally, as Dillehay (2000) suggests, strategies for finding ancient mortuary sites may be flawed.

Regardless of the explanation, the fact of the absolute rarity of documented ancient human remains in America means that each new discovery represents a vitally important addition to knowledge of the lives and deaths of the first Americans. The accidental discovery in 1996 of Kennewick Man, or the Ancient One as he is known to many American Indians, represented an opportunity to learn from the physical remains of a man who, while not in the vanguard of exploration, still is among the most ancient human remains discovered in America. The extraordinary completeness of the skeleton and the high degree of preservation make this discovery exceptional and one of the most potentially informative sets of human remains ever recovered in North America.

RECOVERED HUMAN REMAINS

The remains of human bodies encompass much of the individual's life story, from nearly the beginning to the end—and beyond. When the remains have been intentionally buried, or otherwise processed postmortem, the mortuary facilities and funerary offerings (if any) may reveal how the group to which he or she belonged responded to that death.

The intensive focus on particular individuals is unusual for archaeology, although the postprocessual critique has challenged the discipline to seek methods for telling the stories of individuals (Hodder 2000). Human remains provide the ultimate means for fulfilling this goal.

Paleoamericans are people who lived during the period from the earliest reliably documented evidence of a human presence in the western hemisphere, around 15,000 radiocarbon years before the present, up to the somewhat arbitrary date of 8,000 years ago. (The exception to this rule is the inclusion of the Pelican Rapids skeleton, Minnesota Woman, which is dated to 7,850 ± 50 years ago, because of her historical importance for Paleoamerican studies.) This designation encompasses material that traditionally would be separated into the more familiar Early and Late Paleoindian as well as Early Archaic periods. The intent is to emphasize the extreme antiquity and, often, the biological distinctiveness of these early populations, not to assert by nomenclatory fiat that these ancient people cannot be ancestral to modern American Indians (Owsley and Jantz 2001; cf. Hamilton 2008).

Paleoamericans in the literature include 324 individuals from 54 sites (Table 1.1). In addition, five sites contain 18 features inferred to be individual graves but without preserved osteological remains (Table 1.2). When these are included, the total number of reported Paleoamericans is 342 from 59 sites. Given the area involved and the time span of about 350 human generations, it is a desperately small sample on which to base conclusions about the character of the populations from which these individuals are drawn.

There is an even greater dearth of human remains from Eastern Asia, the presumed homeland of the Paleoamericans. For example, only one Paleolithic burial is documented in all of Siberia (Alekseev 1998): a double burial of a 3–4-year old and a 1-year-old from a burial pit in the floor of a dwelling at the Mal'ta site (Goebel 2001b). Late Pleistocene human remains have been recovered from the sites of Afontova Gora II and Novoselovo VI, but the fragmentary remains are from uncertain contexts (Goebel 2001a).

The samples of Paleoamerican and Paleosiberian burials certainly are heavily biased by factors of preservation and discovery. For example, 56 percent (183) of Paleoamerican human remains are from Florida. Indeed, 52 percent (168) are from the Windover site alone. Another 9 percent (30) are from the Sloan cemetery in Arkansas. The rest are scattered unevenly over 21 other American States, Canada, and Mexico. Similar biases attend the samples of human remains from Paleolithic Europe. Of the 56 late Upper Paleolithic human remains documented, 66 percent (37) are from Italy, a

TABLE 1.2 Sites inferred to represent Paleoamerican burials, without identifiable human remains, dated at least 8,000 years old.

Age	Site	MNI	Burial location	Grave offerings	Reference
10,500	Bull Brook, MA	14	kame terrace	average 3 per burial	Jordan 1960:113–114
10,000	Gorto, MI	1	small rise on a beach ridge	35 Plano points	Buckmaster and Paquette 1988
10,000	Hawkins, AR	1	prominent ridge overlooking Flambeau River	40; including 18 Dalton points	Morse 1971
9,000–8,000	Deadman Slough, WI	1	floodplain	heat-fractured bifaces; >120 fragments	Meinholz and Kuehn 1996
8,500–6,000	Pope, WI	1	—	5 Scottsbluff points, broken	Ritzenthaler 1972

country of barely 300,000 square kilometers (Holt and Formicola 2008).

Several American localities, including Bull Brook, Gorto, Hawkins, Deadman Slough, and Pope, are inferred to include burials but lack preserved skeletal remains (Table 1.2). This illustrates the taphonomic challenges involved with making generalizations about Paleoamerican burial practices. A large proportion of the documented burials include either no nonperishable funerary offerings or only a small number of objects that, if not found with human remains, would provide no clear indication that a burial had been present (Table 1.1). Therefore, many burial locations will have left virtually no readily identifiable traces.

Of interest in this regard are the impressive deposits, or caches, of Clovis artifacts, such as those found at Anzick and East Wenatchee. These have been variously interpreted as mortuary offerings, or as true "caches" of material intended for future use (Lepper 1999; Meltzer 2002), or as ceremonial offerings unrelated to mortuary practices (Gillespie 2007; Lepper and Funk 2002). Anzick is the only major deposit to be found that appears to have been associated directly with human remains, but the unprofessional manner in which the artifacts and human remains were removed limits interpretation.

Another important issue is that some proportion of the human remains does not derive from formal burials at all. Some sites may represent places where humans died without culturally appropriate mortuary treatment. These may be sites of accidental death (including, possibly, Arlington Springs, Warm Mineral Springs, Little Salt Spring, Midland, On Your Knees Cave, and Cutler Ridge) or places where people died as a result of interpersonal violence (possibly Marmes Rockshelter).

The sample also is heavily biased toward the more recent end of the Paleoamerican timeline. There are, at most, two sets of human remains that are older than 12,000 years, or less than one-half of one percent of the total sample. Human remains that are between 12,000 and 10,000 years old include 15 individuals, or 5 percent of the total. Finally, the individuals 10,000 years or younger include 307 individuals, or 95 percent of the total. For those rare sites that have been preserved, discovered, and reported, there is the issue of the uneven quality of the data in the reports. Doran (2002:26) observed that for many early skeletons, "neither the site nor the skeletal material has been adequately described and information is limited. Only a handful of these samples have been analyzed in a manner minimally acceptable to the osteological community." Swedlund and Anderson (1999:569) pointed out that the Gordon Creek Woman, "one of the earliest reported and best-documented Paleoindian burial sites," has been cited infrequently and often is omitted entirely from discussions of the first Americans. They interpret "some of the lack of interest to the fact that she is female" and so lacked both the more strongly developed diagnostic skeletal features sought by physical anthropologists, as well as funerary offerings in the form of the projectile points so admired by archaeologists specializing in Paleoamerican studies (Swedlund and Anderson 1999:570). Owsley and Jantz (2001:566), on the other hand, assert that this report has been neglected because it provides "little usable osteological data" and because the "reigning paradigm . . . identified ancient Americans as being just like recent Native Americans and they were, therefore, not of unusual research interest." I think Owsley and Jantz are correct, but there is more to the story. I suggest that the inattention to Paleoamerican skeletal remains, both male and female, by archaeologists, especially recently, relates to the social context in which this research takes place and in which the study of American Indian, or First Nations, human remains has become highly contentious (e.g., Killion 2008; Weiss 2008). Why else would archaeologists ignore such a rich source of data on Paleoamerican biology, social organization, and ceremonial life?

In spite of these issues, there is much vital information about the lives and deaths of Paleoamericans in these reports. Generalizations derived from the cumulative circumstances of each individual death, while lacking in statistical validity, can provide a place to begin to comprehend broad biological and cultural relationships across time and space.

Burial Type

The majority of Paleoamerican burials in the current sample are flexed or semiflexed interments, but of the 183 examples, 160 are from the single site of Windover. If these are excluded, then extended burials predominate with 35 occurrences. There are 21 documented examples of cremation burials, presuming the Marmes remains are correctly identified as cremated. Additionally, there are rare examples of secondary burials, remains that may have been left exposed deliberately on the ground surface within caves or rockshelters, and one woman who may have been buried deliberately in a tar pit. Obviously, the extent to which deliberate exposure was a preferred method of mortuary treatment will be nearly impossible to evaluate given the odds against the preservation of human remains left exposed to the elements, coupled with the difficulties in distinguishing such remains from an accidental death.

Age, Sex, and Status

For 80 (25 percent) individuals, neither age nor sex can be identified from the available literature (Table 1.1). Of the remaining 243, 159 are listed as adults and 85 as subadults (younger than 17 years old). The sex ratio is approximately equal with 72 males and 80 females. Females are more heavily represented in subadult burials (5:1); otherwise, given the vagaries of very small sample sizes, both sexes seem equally represented in other age groups.

Frequently, variability in both funerary offerings and in the effort involved creating the grave is used as an index of social status (O'Shea 1984). Based on the documented burials it seems, not surprisingly, that these people lived in egalitarian groups in which there was little status differentiation on the basis of either age or sex. There are burials with extraordinary funerary offerings, but there is no clear patterning in the data to indicate any consistent age or gender-specific ranking.

At the Windover site, which encompasses the largest single collection of Paleoamerican human interments, the excavators observed no significant age- or sex-based cultural differences (Dickel 2002). Adults generally were more likely to be buried with offerings (Dickel 2002), and there were a few gender-specific artifact associations. Females exclusively were buried with bone pins, antler punches, and bird-bone tubes (Dickel 2002:84), while adult males tended to have a "greater number of bone and antler artifacts." Canid teeth, sometimes hafted onto wooden handles for use as scrapers, were found only with males (Dickel 2002).

At the Sloan site, the earliest recognized cemetery in the Americas, the evidence indicated "minimal social differentiation" (Morse 1997:93). Although the meager and poorly preserved skeletal material did not allow determination of age or sex, "the composition of the inferred artifact clusters does not suggest the presence of functionally differentiated tool kits" (Morse 1997:92).

In the broader sample, the majority of the deceased were buried without any preserved funerary offerings. In contrast, two of the most lavish assemblages of grave goods were buried with subadults.

At the Anzick site in Montana, an nine-month-old child appears to have been buried with a set of remarkable Clovis artifacts, including 112 ochre-covered bifaces and 11 bone rods, which may have served as foreshafts or points. The bifaces included eight Clovis projectile points and several large, bifacial cores (Lahren and Bonnichsen 1974; Morrow et al. 2006; Owsley and Hunt 2001; Wilke et al. 1991).

At Windover, an 11-year-old female was buried with the "highest counts and greatest diversities" of offerings of any grave at the site (Dickel 2002:95). This girl was buried with 22 formal artifacts, including a bone shaft wrench, a stemmed projectile point, a barbed antler point, a shark tooth hafted in a wooden handle, awls, and a "seed cache" (Dickel 2002:94–95, 110). She was accompanied by a neonate in a globular woven bag (Dickel 2002).

The Horn Shelter 2 double burial of an adult male with a subadult female included a rich assortment of mortuary offerings. There were three turtle shells under the male's head with a fourth positioned in front of his face and a fifth beneath his pelvis. He appeared to be wearing four coyote canine tooth pendants, and talons from a Swainson's hawk were found in his mouth (Owsley et al. 2010; Redder 1985). There was a bone artifact in the turtle shell beneath his pelvis and a deer antler shaft wrench near his leg. Offerings under the head of the male included two antler tools, two sandstone abraders, a chunk of red ochre, one flint biface, and one piece of cut deer bone. More than 80 marine shell beads were placed near the heads of both individuals.

Interpersonal Violence

Keeley (1996:39) asserted that "homicide has been practiced since the appearance of modern humankind and that warfare is documented in the archaeological record of the past 10,000 years in every well-studied region" (see also Ember and Ember 1994; Milner 1995).

Hill et al. (2007:449) state that, for the Hiwi hunter-gatherers of Venezuela, the "mortality profile is characterized by notably high rates of violence and accidental trauma. The data show that . . . about 36% of all adult deaths were from warfare/homicide." Moreover, they suggest the high death rate from interpersonal violence among the Hiwi might be more characteristic of Paleolithic human demographics than previously reported lower rates for African hunting-gathering groups where there had been "interference by powerful state-level societies prior to demographic study" (Hill et al. 2007:451). Corroboration for this view is found in the accumulating evidence of violent death and trophy-taking for the Late Archaic period in eastern North America (Mensforth 2001; Smith 1993, 1995, 1997).

Another review of Paleoamerican skeletal remains conceded only that "interpersonal violence *may* have been a common occurrence among the First Americans" (Powell 2005:164). The data assembled in this review do not justify such equivocation. As exemplified powerfully in the remains of Kennewick Man, interpersonal violence clearly was part of the life experience of many Paleoamericans (Chapter 7).

Examples of Paleoamericans who were killed or grievously injured as a result of human conflict include Burial 102 from the Windover site, a 30- to 40-year-old male with an antler projectile point embedded in his pelvis. His cranium was not present in the burial, suggesting the possibility that it had been removed by his attacker as a trophy (Dickel 2002). Burial 154 at Windover was a 47-year-old male with a traumatic injury to his left eye orbit, possibly caused by a bone or antler projectile point, and a shaft fracture of the left ulna, commonly referred to as a "parry fracture" (Dickel et al. 1989). Cranial depression fractures in three adults and two subadults as well as parry fractures of ulnae are evident in other individuals at Windover (Dickel et al. 1989).

Burial 4 at the Ashworth Shelter is an adult female with a stone projectile point embedded in her third thoracic vertebrae (DiBlasi 1981).

The young adult male at Grimes Shelter has a cut mark on his left second rib with small flakes of obsidian embedded in the bone, strongly suggesting that he was the victim of interpersonal violence (Owsley, personal communication 2010).

Spirit Cave 2, also known as Spirit Cave Man, is an adult male who had suffered a fracture of the frontal bone that was not completely healed at the time of his death (Jantz and Owsley 1997). This injury was likely related to interpersonal violence (Edgar 1997).

Marmes F 3, a young adult male, has a healed injury to his left parietal. The human remains from Marmes are fragmentary and appear to have been broken, burned, and left lying on the ground surface. Fryxell et al. (1968:513) noted that the bones are "cracked in the usual manner used aboriginally to extract marrow." Krantz (1979) examined all the skeletal material from Marmes and concluded that it likely represented cannibalism, noting cut marks on at least one femur. Dixon (1999) and Hicks (2004) regard the evidence from Marmes as more consistent with cremation and ritual processing of the remains, but this does not seem to be a parsimonious interpretation of the data.

The human remains from Cutler Ridge are similar to those from Marmes. They consist of small fragments of burned human bone from a deposit also containing burned animal bones along with burned and unburned limestone pebbles and cobbles (Dolzani 1986; Powell 2005). This commingling of human and animal bone is consistent with an interpretation of cannibalism.

At Meadowcroft Rockshelter, excavators recovered two small fragments of human bone from firepits. These might represent the remains of cannibalism, but since so few bone fragments were found in the features, it is plausible to argue that the bones are the remains of bodies that were cremated in the rockshelter and then collected for secondary burial elsewhere.

Dongoske et al. (2000:180) caution that any statements regarding cannibalism by Native Americans invariably "dredges up a long history of oppression and racism." Yet when the facts are consistent with such an interpretation, it would be irresponsible to not consider cannibalism, recognizing, of course, the range of behaviors encompassed by that single, provocative word. These include ritual cannibalism, survival cannibalism, vengeance cannibalism, and subsistence cannibalism.

Some Paleoamerican skeletons show evidence of periodic dietary stress, such as Harris lines or enamel hypoplastic defects (Powell 2005). San Miguel Man, the adolescent from the Kaskaskia Mine site, and the male from On Your Knees Cave all exhibited hypoplastic defects on their teeth, indicating periods of dietary stress (Dixon 1999; Hargrave et al. 2006; Owsley, personal communication 2010). The adult male from the Horn Shelter exhibited Harris lines in his long bones, indicating he had experienced dietary stress as a subadult (Dixon 1999). The Buhl Woman exhibited 15 Harris lines in her right femur suggesting "seasonal nutritional deficiencies" (Green et al. 1998:448). Warm Mineral Springs 3, an elderly woman, exhibited "signs of dietary deficiency" (Dixon 1999:138). Such evidence is consistent with the possibility that Paleoamerican groups occasionally suffered episodes of starvation that may have caused some to engage in survival cannibalism. Revealingly, the two individuals from the Mal'ta site in Siberia also had indications of hypoplasia, indicating they suffered from malnutrition at some point in their short lives (Goebel 2001b).

To assert that cannibalism is the simplest explanation for certain patterns of Paleoamerican human bone processing and disposal is not inherently racist or ethnocentric. Paleoanthropological studies suggest cannibalism was practiced by Paleolithic groups in Europe (Villa 1992) and genetic studies indicate that cannibalism must have been fairly widespread among hominid ancestors (Mead et al. 2003). Although the practice is, today, almost universally abhorred, many people, in different times and varying circumstances, have engaged in it. To suggest that Paleoamericans sometimes practiced some form of cannibalism is only to suggest that they behaved like other humans.

Mortuary Ceremony

The formal burials of humans afford a clear window into the ceremonial lives of Paleoamericans. There are a few striking similarities with some Eurasian Upper Paleolithic burial practices. For example, the spectacular trove of artifacts deposited with the 18-month-old child at Anzick (Lahren and Bonnichsen 1974; Morrow et al. 2006; Wilke et al. 1991) is reminiscent of a double burial of two children at the Mal'ta site as well as a double burial of two juveniles at Sunghir in Russia (Formicola and Buzhilova 2004). At Mal'ta, a stone slab covered both bodies, which had been placed one on top of the other (Alekseev 1998). The burials were covered with red ochre and included a rich assortment of funerary offerings, including stone and bone tools, "a diadem made from mammoth tusk" (Derev'anko 1998:8), seven decorated ivory pendants, an ivory bracelet, and 120 flat beads (Goebel 2001b). Similarly, the double burial at the Chancellor's site, in which an adult male was buried in line with an older female with his right thumb and left middle finger removed and placed in his mouth (Kennedy 1983), is reminiscent of the Upper Paleolithic mortuary tableaux at Dolní Věstonice (Formicola et al. 2001).

At the Sloan site, some of the burials included offerings of exceptionally large Dalton points, frequently made of exotic white or black cherts, which appear to have served a ceremonial function. "These impressive artifacts did not seem to have been made for use in the 'real' world. It is more likely that they were grave offerings that had been manufactured (or resharpened) before being placed in the graves at the Sloan site" (Yerkes and Gaertner 1997:69–70). Similarly, extra large Clovis points in deposits such as the Richey cache have been interpreted as grave offerings, although at Richey there was no evidence for accompanying human remains (Gramly 1993). Moreover, Gramly (1993:6) argues that "notwithstanding the extraordinary size and finish" of the points, some "appear to have been well used" and thus, "were not just ritual objects." At other sites, such as Crowfield, Gorto, Deadman Slough, and Pope, the presumed burial offerings were broken into pieces terminating any possible "real" world usage.

Ochre, or hematite, was a part of the funerary rituals at a number of sites, including Sloan, Horn Shelter, Gordon Creek, Graham Cave, Wilson-Leonard Burial II, Jerger, and Browns Valley. Arch Lake Woman appears to have been buried with a leather pouch containing a quantity of red ochre and a unifacial flint tool (Owsley et al. 2010). Roper (1996:40–41) observed that ochre is a common component of Clovis artifact caches, which "may or may not [always] be associated with burials." Among later Paleoamerican cultures, ochre is found both at burial and camp sites but, oddly, is not generally associated with artifact caches. It is relevant to note that the only documented Upper Paleolithic burial from Siberia also included the use of red ochre (Derev'anko 1998), so the association of this material with mortuary ritual potentially has deep roots.

Places of Burial

The selection of places to bury or deposit the dead includes layers of significance encompassing practical concerns as well as spiritual and sociopolitical meanings. When a group selects a geographic location as a place of burial, the site frequently is invested with the spiritual qualities deriving from the particular religious beliefs about the dead, the social qualities of a memorial to an honored member of a family or larger social group, and the political connotations of a territorial claim. The challenge for archaeologists is how to discern these various elements of significance in the geographic contexts, modes of interment, and associated funerary offerings.

Many Paleoamerican burials are located in prominent geographic situations (Table 1.1). Rockshelters and caves frequently were selected as burial locations from the earliest (Meadowcroft Rockshelter, Naharon Cave, Fishbone Cave, Anzick) through the latest periods (Modoc Rockshelter, Ice House Bottom). Similarly, the Windover and Little Salt Spring burials were situated in "mortuary ponds" (Wentz and Gifford 2007:334). It is not known whether these sites had a particular spiritual attraction as entrances to the watery Beneath World, which figures prominently in many historic American Indian traditions (Lankford 2004), for example, or whether they were merely conveniently conspicuous places that could be relocated reliably by those who might not have had occasion to return to the site for many years.

A number of burials, such as Buhl, Arch Lake, Deadman Slough, and the Sloan cemetery, appear to have been located on topographical prominences. Once again, the significance of these features for the Paleoamericans may have been pragmatic, spiritual, political, or some combination of these.

Based on the documented archaeological record, Paleoamericans appear to have made relatively few attempts to create visible monuments or memorials that would have allowed individual graves to be relocated easily or to serve as any sort of territorial marker. Burials at Horn Shelter, Graham Cave, and Ashworth Shelter appear to have been covered with stone slabs, but as these were located within rockshelters, their visibility would have been somewhat limited. It is, of course, possible that graves were marked by wooden posts or other perishable markers, such as were found at Windover (Dickel 2002). However, in most archaeological contexts, the wood would not have been preserved and, given the ways in which most discoveries were made, postmolds, if present, likely would not have been observed.

The Windover, Little Salt Spring, and Sloan sites, at least, clearly represent formal cemeteries used repeatedly by social groups over relatively extended periods of time. To the extent that formal cemeteries are interpreted as "legitimizing corporate rights of the group to their home territory through claims of lineal descent from ancestors buried within these ritual spaces" (Walthall 1999:5), these sites indicate a surprising degree of sociopolitical complexity for this early period. An alternative interpretation is that cemeteries "may have provided a ritual node or focus for dispersed, residentially mobile, forager societies. Such places may have exerted a gravitational pull in a largely ritual dimension" (Walthall 1999:23).

CONCLUSIONS

"For many of our ancestors, skeletal analysis is one of the only ways that they are able to tell us their stories . . . it is difficult to speak with a voice made of bone. Nevertheless, while so much has been lost, these individuals have found one last way to speak to us about their lives" (Lippert 1997:126).

Archaeologist and Choctaw Indian Dorothy Lippert has written that "in the end, what we may be able to learn from the skeleton of this man [the Ancient One/Kennewick Man] is all we could learn from any one of ourselves: He was a human being, just like us" (2005:279). While the importance of that fundamental fact is not in dispute, there certainly are particular aspects to each person's life that may reveal experiences characteristic of populations at certain times and places as well as those unique to an individual. Both kinds of knowledge are essential in constructing a reasonably comprehensive biography of a person's life and times.

The North American men, women, and children represented in the documented human remains older than 8,000 years do not constitute an adequate sample from which to draw reliable generalizations and conclusions about Paleoamerican demography, social organization, and ritual practices. Given the wide variability in the sample, it appears that few generalizations are possible.

Ouzman (2008:270), citing Fiedel (2004), claims that since "over a dozen human remains" are documented from the period 10,000 to 7,000 years ago, the bones of Kennewick Man are not all that significant and, therefore, it would be more "socially productive" to give them to modern American Indians for reburial. Marks (2002) makes the same basic argument. This proposition ignores the fact that the skeleton of Kennewick Man is far better preserved and more complete than most skeletons of a comparable antiquity. Moreover, the suggestion that because several sets of Paleoamerican human remains already are available and any additional discoveries would be superfluous ignores the fact that each individual offers a unique and personal story. Each is a tessera in the late Pleistocene to early Holocene mosaic of human experience. Should scientists listen to one, or a few, of these voices made of bone and silence the others forever? More pragmatically, conclusions based on small samples simply are not reliable. And, in this temporal and geographic context, even 324 is a hopelessly small sample from which to derive valid generalizations.

The lives of Paleoamericans were at times harsh; they periodically suffered from hunger, infections, broken bones, and violence (Dixon 1999). Nevertheless, most Paleoamericans would have enjoyed seasonally abundant natural resources, a largely egalitarian social organization, and a supportive network of extended kin.

There is a surprising amount of evidence for dietary stress as a more or less regular aspect of many people's lives in the late Pleistocene and early Holocene, though it does not appear to have been any more extreme than what hunter-gatherer groups typically experience (Hill et al. 2007; Yesner 1994). Interpersonal violence appears to have been more common among Paleoamericans than expected, given their low population density, the consequent reduced probability of encountering a hostile group, and the necessity of social mechanisms to facilitate the positive interactions between groups that would ensure access to potential mating partners. Perhaps some of the intergroup violence was related to the forcible appropriation of mating partners that may not have been shared willingly. (See Wilkinson and Van Wagenen [1993] for examples of this from a later time period.)

The deaths of Paleoamericans occasioned a variety of ceremonial responses depending on the time, place, circumstances of death, and the idiosyncrasies of the people in the deceased's social group. The average lifespan for Paleoamericans seems to be more or less typical of hunter-gatherer populations, but groups must have frequently been on the razor's edge of reproductive viabililty due to a characteristically high infant mortality rate. Perhaps this explains the occasional extravagance of funerary offerings for some subadults (e.g., Anzick and Windover Burial 90). It may have been an outpouring of grief over the loss of the small band's hope for the future.

Flexed interments are more common than extended throughout the sequence. Cremations are not uncommon and can be early, as at Crowfield, or late, as at Reiner. Funerary offerings are rare and usually quite limited in quantity, but they can be lavish, especially at the earlier sites, such as Anzick, Crowfield, Gorto, and Hawkins. The mortuary program reflects little in the way of systematic status differentiation, yet the burial of children with spectacular offerings at Anzick and, less spectacularly, at Windover, suggests either the precocious beginnings of ascribed status or circumstances surrounding particular deaths that warranted special ceremonial responses.

Several Paleoamerican skeletons have been turned over to American Indian tribes for burial (e.g., Buhl, Browns Valley, On Your Knees Cave, Pelican Rapids, Marmes), and others continue to be sought by various tribes (Schneider and Bonnichsen 2005) in spite of the fact that human remains of this antiquity are not considered to qualify for inclusion under the terms of the Native American Graves Protection and Repatriation Act (NAGPRA) (Lepper 2002, 2003). Moreover, contrary to what might be inferred from the rhetoric of many activists, universal repatriation is "not unanimously advocated by American Indians and Alaskan Natives" (Hall and Wolfley 2003:28). In a survey of 508 federally recognized tribes from 32 states, there was "no Pan-Indian position on these complex and culturally and historically dependent issues"; and "many respondents believe that the study of human remains may be useful and necessary" (Hall and Wolfley 2003:31).

There have been numerous instances of American Indians working together with archaeologists and physical anthropologists to learn the stories held within ancient human remains. These stand in marked contrast to the hostility and intransigence of the tribes involved in the Kennewick Man case (Burke et al. 2008; Killion 2008; Owsley and Jantz 2002).

In 2008, the Tlingit elder Rosita Worl, president of the Sealaska Heritage Institute, told the *Alaska Daily News* reporter George Bryson, "When this 10,300-year-old person was found on Prince of Wales [On Your Knees Cave], the way it was interpreted was that we had one of our ancestors offering himself to give us knowledge." Yarrow Vaara, a student intern working at the site who also was a member of the Tlingit tribe, "regarded the human remains she was helping to excavate as possibly those of her great uncle who was teaching her about her history" (Dixon 1999:111). This statement coheres with the matrilineality of the Tlingit, where uncles have an honored role as mentors in the clan.

The investigators of the Buhl skeleton wrote that they "greatly appreciate the cooperation and tolerance of the Shoshone-Bannock tribes of Fort Hall, Idaho," who allowed study of the remains, including radiocarbon dating of bone (Green et al. 1998:452).

Such cases are promising illustrations of ways in which archaeologists and physical anthropologists are working together with American Indians to record the stories embodied in these ancient human remains. Yet, in both instances, the human remains were reburied at the conclusion of the study, likely because the Native peoples believed that, once the stories had been told, there was nothing new left to learn.

Few reliable generalizations are derivable from the available evidence of Paleoamerican human remains. This is partly due to the fact that there are so few individuals from such a vast area stretching over more than seven millennia. It also seems at least partly due to the fact that the cultural practices of the peoples residing in the various regions of North America over this extended period of time were quite diverse, resulting in haphazard sampling from an unknown number of populations across a vast and varied continent over an enormous span of time. The human remains in museum collections must be curated so that new and improved methods of analysis can be applied to them. Moreover, new discoveries must be sought to fill in the blanks and gain some sense of the variability in biology and culture across space and through time. Tragically, the political environment today is such that not only are Paleoamerican human remains in museum collections increasingly subject to "repatriation" to modern groups to whom the remains may be unrelated, but also the likelihood of new discoveries has been greatly diminished by a policy of avoidance of controversy by federal agencies and museums (Weiss 2008).

Kintigh (2008:203) observed that a "least effort approach" increasingly characterizes the interactions among federal agencies, tribes, and cultural resource management firms, such that when human remains are encountered, there is "minimal or no recording of graves" and, if the remains are recovered, there is a "speedy reburial in the absence of any demonstration of cultural affiliation." This is due to the fact that "tribes with more stridently antiscientific positions tend to be more vocal" and "agencies tend to be more responsive to more vocal tribes," which is to the detriment of both science and the American Indian groups that might wish to learn from these ancient persons who might be their ancestors.

Perhaps the stories that Kennewick Man is able to tell will help to demonstrate the importance of the respectful curation of human remains. The information presented in this volume adds immeasurably to the understanding of the life and times of a remarkable man. The fact that many results reported herein are markedly different from those obtained in previous analyses demonstrates how vital it is that human remains be available for reexamination by other scholars with ever-improving methods. Unfortunately, for many Paleoamerican human remains, the opportunity to continue conversations with them has been lost. New questions must remain unasked. And entire chapters of North American history therefore remain unwritten.

REFERENCES

Adovasio, James. 2003. *The First Americans: In Pursuit of Archaeology's Greatest Mystery*. Modern Library, New York.

Alekseev, V. 1998. The physical specificities of Paleolithic hominids in Siberia. In *The Paleolithic of Siberia: New Discoveries and Interpretations*, edited by Anatoliy P. Derev'anko, 329–335. University of Illinois Press, Urbana.

Anfinson, Scott F. 1997. *Southwestern Minnesota Archaeology: 12,000 Years in the Prairie Lake Region*. Minnesota Historical Society, St. Paul.

Bonnichsen, Robson, and Karen Turnmire, eds. 1999. *Ice Age Peoples of North America: Environments, Origins, and Adaptations of the First Americans*. Oregon State University Press, Corvallis.

Bousman, C. Britt, Michael B. Collins, Paul Goldberg, Thomas W. Stafford, Jan Guy, Barry W. Baker, D. Gentry Steele, Marvin Kay, Anne Kerr, Glen Fredlund, Phil Dering, Vance Holliday, Diane Wilson, Wulf Gose, Susan Dial, Paul Takac, Robin Balinsky, Marilyn Masson, and Joseph Powell. 2002. The Paleoindian-Archaic transition in North America: New evidence from Texas. *Antiquity* 76:980–990.

Breternitz, David A., Alan C. Swedlund, and Duane C. Anderson. 1971. An early burial from Gordon Creek, Colorado. *American Antiquity* 36(2):170–182.

Brose, David S. 1989. The Squaw Rockshelter (33CU34): A stratified Archaic deposit in Cuyahoga County. *Kirtlandia* 44:17–53.

Bruwelheide, Karin L., and Douglas W. Owsley. 2002. Appendix C. Cremated human bone from the Olive Branch site, Alexander County, Illinois. In *Olive Branch: A Very Early Archaic Site on the Mississippi River*, by Richard Michael Gramly, 245–250. American Society for Amateur Archaeology, North Andover, MA.

Brown, James, and Robert Vierra. 1983. What happened in the Middle Archaic? Introduction to an ecological approach to Koster Site archaeology. In *Archaic Hunters and Gatherers in the American Midwest*, edited by James Phillips and James Brown, 165–195. Academic Press, New York.

Bryson, George. 2008. "Saliva Samples Could Reveal Ancient Alaskan's Descendants." *Anchorage Daily News*. June 6. http://www.adn.com/life/story/428227.html.

Buckmaster, Marla M., and James R. Paquette. 1988. The Gorto site: Preliminary report on a late Paleo-Indian site in Marquette County, Michigan. *Wisconsin Archeologist* 69(3):101–123.

Burke, Heather, Claire Smith, Dorothy Lippert, Joe Watkins, and Larry Zimmerman, eds. 2008. *Kennewick Man: Perspectives on the Ancient One*. Left Coast Press, Walnut Creek, CA.

Chapman, Jefferson. 1977. *Archaic Period Research in the Lower Tennessee River Valley – 1975. Icehouse Bottom, Harrison Branch, Thirty Acre Island, Calloway Island*. Department of Anthropology Report of Investigations 18. University of Tennessee, Knoxville.

Chatters, James C. 2000. The recovery and initial analysis of an early Holocene human skeleton from Kennewick, Washington. *American Antiquity* 65(2):291–316.

Chatters, James C., Steven Hackenberger, Alan Busacca, Linda S. Cummings, Richard L. Jantz, Thomas W. Stafford, and R. Ervin Taylor. 2000. A possible second early Holocene skull from eastern Washington, USA. *Current Research in the Pleistocene* 17:93–95.

Chawkins, Steve. 2006. "Ancient Bones Belonged to a Man—Probably." *Los Angeles Times*. September 11. http://articles.latimes.com/2006/sep/11/local/me-bonesside11.

Clausen, Carl J., H. K. Brooks, and Al B. Wesolowsky. 1975. The early man site at Warm Mineral Spring, Florida. *Journal of Field Archaeology* 2(3):191–213.

Clausen, Carl J., A. D. Cohen, Cesare Emiliani, J. A. Holman, and J. J. Stipp. 1979. Little Salt Spring, Florida: A unique underwater site. *Science* 203(4381):609–614.

Collins, Michael B., ed. 1998. *Wilson-Leonard: An 11,000-Year Archeological Record of Hunter-Gatherers in Central Texas.*

5 vols. Texas Archeological Research Laboratory Studies in Archeology 31. The University of Texas at Austin and Archeology Studies Program Report 10. Environmental Affairs Division, Texas Department of Transportation, Austin.

Cybulski, Jerome S., Donald E. Howes, James C. Haggerty, and Morely Eldridge. 1981. An early human skeleton from south-central British Columbia: Dating and bioarchaeological inference. *Canadian Journal of Archaeology* 5:49–59.

Dansie, Amy. 1997. Early Holocene burials in Nevada: Overview of localities, research and legal issues. *Nevada Historical Quarterly* 40(1):4–14.

Dawson, Kenneth Cephus Arnold. 1983. Cummins Site: A Late Palaeo-Indian (Plano) Site at Thunder Bay, Ontario. *Ontario Archaeology* 39:3–31.

Deller, D. Brian, and C. J. Ellis. 1984. Crowfield: A preliminary report on a probably Paleo-Indian cremation in southern Ontario. *Archaeology of Eastern North America* 12:41–71.

Derev'anko, Anatoliy P. 1998. A short history of discoveries and the development of ideas in the Paleolithic of Siberia. In *The Paleolithic of Siberia: New Discoveries and Interpretations*, edited by Anatoliy P. Derev'anko, 5–13. University of Illinois Press, Urbana.

DiBlasi, P. J. 1981. "A New Assessment of the Archaeological Significance of the Ashworth Site (11Bu236): A Study in the Dynamics of Archaeological Investigation in Cultural Resource Management." Master's thesis, University of Louisville, KY.

Dickel, David N. 2002. Analysis of mortuary patterns. In *Windover: Multidisciplinary Investigations of an Early Archaic Florida Cemetery*, edited by Glen H. Doran, 73–96. University Press of Florida, Gainesville.

Dickel, David N., C. Gregory Aker, Billie K. Barton, and Glen H. Doran. 1989. An orbital floor and ulna fracture from the early Archaic of Florida. *Journal of Paleopathology* 2(3):165–170.

Dillehay, Thomas D. 2000. *The Settlement of the Americas: A New Prehistory*. Basic Books, New York.

Dixon, E. James. 1999. *Bones, Boats, and Bison: Archaeology and the First Colonization of Western North America*. University of New Mexico Press, Albuquerque.

Dolzani, Michael. 1986. Florida site yields human remains. *Mammoth Trumpet* 2(4):1, 3.

Dongoske, Kurt E., Debra L. Martin, and T. J. Ferguson. 2000. Critique of the claim of cannibalism at Cowboy Wash. *American Antiquity* 65(1):179–190.

Doran, Glen H. 2002. Introduction to wet sites and Windover (8BR246) investigations. In *Windover: Multidisciplinary Investigations of an Early Archaic Florida Cemetery*, edited by Glen H. Doran, 1–38. University Press of Florida, Gainesville.

Doran, Glen H., and David N. Dickel. 1988. Radiometric chronology of the Archaic Windover archaeological site (8Br246). *Florida Anthropologist* 41(3):365–380.

Edgar, Heather J. H. 1997. Paleopathology of the Wizards Beach Man (AHUR 2023) and the Spirit Cave Mummy (AHUR 2064). *Nevada Historical Society Quarterly* 40(1):57–61.

Eiselt, B. Sunday. 1997. Fish remains from the Spirit Cave paleofecal material: 9,400 year old evidence for Great Basin utilization of small fishes. *Nevada Historical Society Quarterly* 40(1):117–139.

Ember, Melvin, and Carol R. Ember. 1994. Cross-cultural studies of war and peace: Recent achievements and future possibilities. In *Studying War: Anthropological Perspectives*, edited by S. P. Reyna and R. E. Downs, 185–208. Gordon and Breach, Langhorne, PA.

Fiedel, Stuart J. 2004. The Kennewick follies: "New" theories about the peopling of the Americas. *Journal of Anthropological Research* 60(1):75–110.

Formicola, Vincenzo, and Alexandra P. Buzhilova. 2004. Double child burial from Sunghir (Russia): Pathology and inferences for Upper Paleolithic funerary practices. *American Journal of Physical Anthropology* 124(3):189–198.

Formicola, Vincenzo, Antonella Pontradolfi, and Jiří Svoboda. 2001. The upper Paleolithic triple burial of Dolní Věstonice: Pathology and funerary behavior. *American Journal of Physical Anthropology* 115(4):372–379.

Fowler, Melvin L. 1959. *Summary Report of Modoc Rock Shelter, 1952, 1953, 1955, 1956*. Illinois State Museum Report of Investigations 8. Illinois State Museum, Springfield.

Friends of America's Past. 2001. Evidence of the Past: A Map and Status of Ancient Remains. http://www.friendsofpast.org/earliest-americans/map.html.

Fryxell, Roald, Tadeusz Bielicki, Richard D. Daugherty, Carl E. Gustafson, Henry T. Irwin, and Bennie C. Keel. 1968. A human skeleton from sediments of mid-Pinedale age in southeastern Washington. *American Antiquity* 33(4):511–515.

Gillespie, Jason D. 2007. Enculturating an unknown world: Caches and Clovis landscape ideology. *Canadian Journal of Archaeology* 31(2):171–189.

Goebel, Ted. 2001a. Siberian Late Upper Paleolithic. In *Encyclopedia of Prehistory*, vol. 2: *Arctic and Subarctic*, edited by Peter N. Peregrine and Melvin Ember, 186–191. Kluwer, New York.

———. 2001b. Siberian Middle Upper Paleolithic. In *Encyclopedia of Prehistory*, vol. 2: *Arctic and Subarctic*, edited by Peter N. Peregrine and Melvin Ember, 192–196. Kluwer, New York.

González, Silvia, José Concepción Jiménez-López, Robert Hedges, David Huddart, James C. Ohman, Alan Turner, and Jose Antonio Pompa y Padilla. 2003. Earliest humans in the Americas: New evidence from Mexico. *Journal of Human Evolution* 44(3):379–387.

González González, Arturo H., Carmen Rojas Sandoval, Alejandro Terrazas Mata, Martha Benavente Sanvicente, Wolfgang Stinnesbeck, Jeronimo Aviles O., and Eugenio Acevez. 2008. The arrival of humans on the Yucatan Peninsula: Evidence from submerged caves in the state of Quintana Roo, Mexico. *Current Research in the Pleistocene* 25:1–24.

González-Jose, Rolando, Maria Cátira Bortolini, Fabrício R. Santos, and Sandro L. Bonatto. 2008. The peopling of America: Craniofacial shape variation on a continental scale and its interpretation from an interdisciplinary view. *American Journal of Physical Anthropology* 137(2):175–187.

González-Jose, Rolando, Walter Neves, Marta Mirazón Lahr, Silvia González, Héctor Pucciarelli, Miquel Hernández Martínez, and Gonzalo Correal. 2005. Late Pleistocene/Holocene craniofacial morphology in Mesoamerican Paleoindians: Implications for the peopling of the New World. *American Journal of Physical Anthropology* 128(4):772–780.

Gramly, Richard Michael. 1993. *The Richey Clovis Cache: Earliest Americans Along the Columbia River*. Persimmon Press Monographs in Archaeology, Kenmore, NY.

———. 2002. *Olive Branch: A Very Early Archaic Site on the Mississippi River*. American Society for Amateur Archaeology, North Andover, MA.

Green, Thomas J., Bruce Cochran, Todd W. Fenton, James C. Woods, Gene L. Titmus, Larry Tieszen, Mary Anne Davis, and Suzanne J. Miller. 1998. The Buhl burial: A Paleoindian woman from southern Idaho. *American Antiquity* 63(3):437–456.

Hall, Teri R., and Jeanette Wolfley. 2003. Working Together: A survey of tribal perspectives on NAGPRA: Repatriation and study of human remains. *The SAA Archaeological Record* 3(2):27–34.

Hamilton, Michelle D. 2008. Colonizing America: Paleoamericans in the New World. In *Kennewick Man: Perspectives on the Ancient One*, edited by Heather Burke, Claire Smith, Dorothy Lippert, Joe Watkins, and Larry Zimmerman, 128–137. Left Coast Press, Walnut Creek, CA.

Hargrave, Eve, Kristin Hedman, and Rebecca Wolf. 2006. *Isolated Human Burial at the Kaskaskia Mine Site (11R487), Randolph County, Illinois*. Research Report 104. Illinois Transportation Archaeological Research Program, University of Illinois, Urbana-Champaign.

Herrmann, Nicholas P., Richard L. Jantz, and Douglas W. Owsley. 2006. Buhl revisited: Three-dimensional photographic reconstruction and morphometric reevaluation. In *El Hombre Temprano en América y sus Implicaciones en el Poblamiento de la Cuenca de México*, edited by José Concepción Jiménez López, Silvia González, José Antonio Pompa y Padilla, Francisco Ortiz Pedraza, 211–220. Instituto Nacional de Antropologia e Historia, Mexico City.

Hicks, Brent A. 2004. Summary of results. In *Marmes Rockshelter: A Final Report on 11,000 Years of Cultural Use*, edited by Brent A. Hicks, 373–387. Washington State University Press, Pullman.

Hill, Kim, A. M. Hurtado, and R. S. Walker. 2007. High adult mortality among Hiwi hunter-gatherers: Implications for human evolution. *Journal of Human Evolution* 52(4):443–454.

Hoard, Robert J., William E. Banks, Rolfe D. Mandel, Michael Finnegan, and Jennifer E. Epperson. 2004. A Middle Archaic burial from east central Kansas. *American Antiquity* 69(4):717–735.

Hodder, Ian. 2000. Agency and individuals in long-term processes. In *Agency in Archaeology*, edited by Marcia-Anne Dobres and John E. Robb, 21–33. Routledge, London.

Holliday, Vance T. 1997. *Paleoindian Geoarchaeology of the Southern High Plains*. University of Texas Press, Austin.

Holliday, Vance T., and David J. Meltzer. 1996. Geoarchaeology of the Midland (Paleoindian) site, Texas. *American Antiquity* 61(4):755–771.

Holt, Brigitte M., and Vincenzo Formicola. 2008. Hunters of the Ice Age: The biology of Upper Paleolithic people. *American Journal of Physical Anthropology* 137(S47):70–99.

Hrdlička, Aleš. 1937. The Minnesota 'Man.' *American Journal of Physical Anthropology* 22(2):175–199.

Jantz, Richard L., and Douglas W. Owsley. 1997. Pathology, taphonomy, and cranial morphometrics of the Spirit Cave Mummy. *Nevada Historical Society Quarterly* 40(1):62–84.

———. 2001. Variation among early North American crania. *American Journal of Physical Anthropology* 114(2):146–155.

Jenks, Albert E. 1933. Minnesota Pleistocene *Homo*: An interim communication. *Proceedings of the National Academy of Sciences of the United States of America* 19(1):1–6.

———. 1934. The discovery of an ancient Minnesota maker of Yuma and Folsom flints. *Science* 80(2070):205.

———. 1937. *Minnesota's Browns Valley Man and Associated Burial Objects*. Memoir 49. American Anthropological Association, Menasha, WI.

———. 1938. Minnesota Man: A reply to a review by Dr. Aleš Hrdlička. *American Anthropologist* 40(2):328–336.

Johnson, John R., Thomas W. Stafford, Jr., Henry O. Ajie, and Don P. Morris. 2002. Arlington Springs revisited. *Proceedings of the Fifth California Islands Symposium*, edited by David R. Browne, Kathryn L. Mitchell, and Henry W. Chaney, 541–545. Santa Barbara Museum of Natural History, CA.

Jordan, Douglas F. 1960. "The Bull Brook Site in Relation to 'Fluted Point' Manifestations in Eastern North America." Ph.D. dissertation, Harvard University, Cambridge, MA.

Julig, Patrick. 1984. Cummins Paleo-Indian site and it paleoenvironment, Thunder Bay, Canada. *Archaeology of Eastern North America* 12:192–209.

Keeley, Lawrence H. 1996. *War Before Civilization*. Oxford University Press, New York.

Kemp, Brian M., Ripan S. Malhi, John McDonough, Deborah A. Bolnick, Jason A. Eshleman, Olga Rickards, Cristina Martinez-Labarga, John R. Johnson, Joseph G. Lorenz, E. James Dixon, Terence E. Fifield, Timothy H. Heaton, Rosita Worl, and David Glenn Smith. 2007. Genetic analysis of early Holocene skeletal remains from Alaska and its implications for the settlement of the Americas. *American Journal of Physical Anthropology* 132(4):605–621.

Kennedy, G. E. 1983. An unusual burial practice at an early California Indian site. *Journal of New World Archaeology* 5(3):4–7.

Killion, Thomas W., ed. 2008. *Opening Archaeology: Repatriation's Impact on Contemporary Research & Practice*. School for Advanced Research Press, Santa Fe.

Kintigh, Keith W. 2008. Repatriation as a force for change in Southwestern archaeology. In *Opening Archaeology: Repatriation's Impact on Contemporary Research & Practice*, edited by Thomas W. Killion, 195–207. School for Advanced Research Press, Santa Fe.

Krantz, Grover S. 1979. Oldest human remains from the Marmes site. *Northwest Anthropological Research Notes* 13(2):159–174.

Kroeber, A. L. 1962. The Rancho La Brea skull. *American Antiquity* 27(3):416–417.

Lahren, Larry, and Robson Bonnichsen. 1974. Bone foreshafts from a Clovis burial in southwestern Montana. *Science* 186(4159):147–150.

Lankford, George E. 2004. World on a string: Some cosmological components of the Southeastern Ceremonial Complex. In *Hero, Hawk, and Open Hand: American Indian Art of the Ancient Midwest and South*, edited by Richard F. Townsend and Robert V. Sharp, 206–217. Yale University Press, New Haven.

Lepper, Bradley T. 1999. Pleistocene peoples of midcontinental North America. In *Ice Age Peoples of North America: Environments, Origins, and Adaptations of the First Americans*, edited by Robson Bonnichsen and Karen L. Turnmire, 362–394. Oregon State University Press, Corvallis.

———. 2002. Judge rules scientists can study Kennewick Man. *Mammoth Trumpet* 18(1):1–3, 18–19.

———. 2003. Major decision: Kennewick Man case. *Mammoth Trumpet* 19(1):1, 3–4, 18–19.

Lepper, Bradley T., and Robert E. Funk. 2002. Paleo-Indian: East. In *Handbook of North American Indians*, vol. 3: *Environment, Origins, and Population*, edited by Douglas H. Ubelaker, 171–193. Smithsonian Institution Press, Washington, DC.

Lippert, Dorothy. 1997. In front of the mirror: Native Americans and academic archaeology. In *Native Americans and Archaeologists: Stepping Stones to Common Ground*, edited by Nina Swidler, Kurt E. Dongoske, Roger Anyon, and Alan S. Downer, 120–127. Altamira Press, Walnut Creek, CA.

———. 2005. Remembering humanity: How to include human values in a scientific endeavor. *International Journal of Cultural Property* 12(2):275–280.

Littleton, Judith. 2007. From the perspective of time: Hunter-gatherer burials in southeastern Australia. *Antiquity* 81:1013–1028.

Logan, Wilfred D. 1952. *Graham Cave: An Archaic Site in Montgomery County, Missouri*. Memoir of Missouri Archaeological Society 2. Missouri Archaeological Society, Columbia.

Lorenzo, Jose Luis, and Lorena Mirambell. 1999. The inhabitants of Mexico during the upper Pleistocene. In *Ice Age Peoples of North America: Environments, Origins, and Adaptations of the First Americans*, edited by Robson Bonnichsen and Karen Turnmire, 482–496. Oregon State University Press, Corvallis.

Marks, Jonathan. 2002. *What It Means to be 98% Chimpanzee: Apes, People, and their Genes*. University of California Press, Berkeley.

Mason, Ronald J., and Carol Irwin. 1960. An Eden-Scottsbluff burial in northeastern Wisconsin. *American Antiquity* 26(1):43–57.

Mead, Simon, Michael P. H. Stumpf, Jerome Whitfield, Jonathan A. Beck, Mark Poulter, Tracy Campbell, James B. Uphill, David Goldstein, Michael Alpers, Elizabeth M. C. Fisher, and John Collinge. 2003. Balancing selection at the prion protein gene consistent with prehistoric kurulike epidemics. *Science* 300(5619):640–643.

Meinholz, Norman M., and Steven R. Kuehn. 1996. *The Deadman Slough Site (47 Pr-46)*. Museum Archaeology Program, Archaeology Research Series Number 4. State Historical Society of Wisconsin, Madison.

Meltzer, David J. 2002. What do you do when no one's been there before? Thoughts on the exploration and colonization of new lands. In *The First Americans: The Pleistocene Colonization of the New World*, edited by Nina G. Jablonski, 27–58. Memoirs of the California Academy of Sciences No. 27. California Academy of Sciences, San Francisco.

Mensforth, Robert P. 2001. Warfare and trophy taking in the Archaic period. In *Archaic Transitions in Ohio and Kentucky Prehistory*, edited by Olaf H. Prufer, Sara E. Pedde, and Richard S. Meindl, 110–138. Kent State University Press, OH.

Milner, George R. 1995. An osteological perspective on prehistoric warfare. In *Regional Approaches to Mortuary Analysis*, edited by Lane Anderson Beck, 221–244. Plenum Press, New York.

Morell, Virginia. 1997. First Floridians found near Biscayne Bay. *Science* 275(5304):1258–1259.

Morrow, Juliet E., and Stuart J. Fiedel. 2006. New radiocarbon dates for the Clovis component of the Anzick site, Park County, Montana. In *Paleoindian Archaeology: A Hemispheric Perspective*, edited by Juliet E. Morrow and Cristóbol Gnecco, 123–138. University Press of Florida, Gainesville.

Morse, Daniel F. 1971. The Hawkins Cache: A significant Dalton find in northeast Arkansas. *Arkansas Archeologist* 12(1):9–20.

———. 1997. *Sloan: A Paleoindian Dalton Cemetery in Arkansas*. Smithsonian Institution Press, Washington, DC.

Myster, Susan M. Thurston, and Barbara O'Connell. 1997. Bioarchaeology of Iowa, Wisconsin, and Minnesota. In *Bioarchaeology of the North Central United States*, edited by Douglas W. Owsley and Jerome C. Rose, 147–239. Arkansas Archeological Survey Research Series 49. Arkansas Archeological Survey, Fayetteville.

Neumann, Holm Wolfram. 1967. *The Paleopathology of the Archaic Modoc Rock Shelter Inhabitants*. Illinois State Museum Report of Investigations 11. Illinois State Museum, Springfield.

Orr, Phil C. 1956. *Pleistocene Man in Fishbone Cave, Pershing County, Nevada*. Bulletin of the Department of Archaeology, Nevada State Museum 2:1–20.

O'Shea, John M. 1984. *Mortuary Variability: An Archaeological Investigation*. Academic Press, Orlando.

Ouzman, Sven. 2008. Law or lore? Speaking sovereignty in the Kennewick case. In *Kennewick Man: Perspectives on the Ancient One*, edited by Heather Burke, Claire Smith, Dorothy Lippert, Joe Watkins, and Larry Zimmerman, 268–270. Left Coast Press, Walnut Creek, CA.

Owsley, Douglas W., and David R. Hunt. 2001. Clovis and early Archaic period crania from the Anzick site (24PA506), Park County, Montana. *Plains Anthropologist* 46:115–124.

Owsley, Douglas W., and Richard L. Jantz. 2001. Archaeological politics and public interest in Paleoamerican studies: Lessons from Gordon Creek Woman and Kennewick Man. *American Antiquity* 66(4):565–575.

———. 2002. Kennewick Man—A kin? Too distant. In *Claiming the Stones/Naming the Bones: Cultural Property and the Negotiation of National and Ethnic Identity*, edited by Elazar Barkan and Ronald Bush, 141–161. Getty Publications, Los Angeles.

Owsley, Douglas W., Margaret A. Jodry, Thomas W. Stafford, Jr., C. Vance Haynes, Jr., and Dennis J. Stanford. 2010.

Arch Lake Woman: Physical Anthropology and Geoarchaeology. Texas A&M University Press, College Station.

Powell, Joseph F. 2005. *The First Americans: Race, Evolution, and the Origin of Native Americans.* Cambridge University Press, Cambridge.

Powell, J. F., and W. A. Neves. 1998. "Dental Diversity of Early New World Populations: Taking a Bite Out of the Tripartite Model." Paper presented at the 67th Annual Meeting of the American Association of Physical Anthropologists, Salt Lake City, UT.

Prior, Fred. 1989. Skeletal remains from Squaw Rockshelter. *Kirtlandia* 44:55–58.

Protsch, Reiner R. R. 1978. *Catalog of Fossil Hominids of North America.* Gustav Fischer, New York.

Prufer, Olaf H. 2001. The Archaic of northeastern Ohio. In *Archaic Transitions in Ohio & Kentucky Prehistory*, edited by Olaf H. Prufer, Sara E. Pedde, and Richard S. Meindl, 183–209. Kent State University Press, Kent, OH.

Redder, Albert J. 1985. Horn Shelter No. 2: The south end, a preliminary report. *Central Texas Archeologist* 10:37–65.

Redder, Albert J., and John W. Fox. 1988. Excavation and positioning of the Horn Shelter's burial and grave goods. *Central Texas Archeologist* 11:1–12.

Ritzenthaler, Robert. 1972. The Pope site: A Scottsbluff cremation? in Waupaca County. *Wisconsin Archeologist* 53(1):15–19.

Roper, Donna C. 1996. Variability in the use of ochre during the Paleoindian period. *Current Research in the Pleistocene* 13:40–42.

Sciulli, Paul W. 1982. Human remains from Meadowcroft Rockshelter, Washington County, southwestern Pennsylvania. In *Meadowcroft: Collected Papers on the Archaeology of Meadowcroft Rockshelter and the Cross Creek Drainage*, edited by Ronald C. Carlisle and J. M. Adovasio, 175–185. University of Pittsburgh, Cultural Resource Management Program, PA.

Schneider, Alan L., and Robson Bonnichsen. 2005. Where are we going? Public policy and science. In *Paleoamerican Origins: Beyond Clovis*, edited by Robson Bonnichsen, Bradley T. Lepper, Dennis Stanford, and Michael R. Waters, 297–312. Texas A&M Press, College Station.

Smith, Maria Ostendorf. 1993. A probable case of decapitation at the Late Archaic Robinson site (40SM4), Smith County, Tennessee. *Tennessee Anthropologist* 18(2):131–142.

———. 1995. Scalping in the Archaic period: Evidence from the western Tennessee valley. *Southeastern Archaeology* 14(1):60–68.

———. 1997. Osteological indications of warfare in the Archaic period of the western Tennessee valley. In *Troubled Times: Violence and Warfare in the Past*, edited by Debra L. Martin and David W. Frayer, 241–265. Gordon and Breach, Amsterdam.

Stafford, Thomas W., Jr. 1994. Accelerator C-14 dating of human fossil skeletons: Assessing accuracy and results on New World Specimens. *Method and Theory for Investigating the Peopling of the Americas*, edited by Robson Bonnichsen, and D. Gentry Steele, 45–55. Center for the Study of the First Americans, Corvallis, OR.

Stewart, T. Dale. 1955. Description of the human skeletal remains. In *The Midland Discovery*, edited by Fred Wendorf, Alex D. Krieger, and Claude C. Albritton, Jr., 77–90. University of Texas Press, Austin.

Swedlund, Alan, and Duane Anderson. 1999. Gordon Creek Woman meets Kennewick Man: New interpretations and protocols regarding the peopling of the Americas. *American Antiquity* 64(4):569–576.

Taylor, R. E., L. A. Payer, C. A. Prior, P. J. Slota, Jr., R Gillespie, J. A. J. Gowlett, R. E. M. Hedges, A. J. T. Hull, T. H. Zabel, D. J. Donahue, and R. Berger. 1985. Major revisions in the Pleistocene age assignments for North American human skeletons by C-14 accelerator mass spectrometer: None older than 11,000 years B.P. *American Antiquity* 50(1):136–140.

Thies, Randall M. 1990. *The Archeology of the Stigenwalt Site, 14LT351.* Kansas State Historical Society, Contract Archeology Series, Publication 7.

Tomak, Curtia H. 1979. Jerger: An early Archaic mortuary site in southwestern Indiana. *Proceedings of the Indiana Academy of Science* 88:62–69.

Tuohy, Donald R., and Amy J. Dansie. 1997. New information regarding Early Holocene manifestations in the Western Great Basin. *Nevada Historical Society Quarterly* 40:24–53.

Ubelaker, Douglas, ed. 2006. *Handbook of North American Indians*, vol. 3: *Environment, Origins, and Population.* Smithsonian Institution Press, Washington, DC.

Villa, Paolo. 1992. Cannibalism in prehistoric Europe. *Evolutionary Anthropology* 1(3):93–104.

Walthall, John A. 1999. Mortuary behavior and early Holocene land use in the North American midcontinent. *North American Archaeologist* 20(1):1–30.

Waters, Michael R. 1986. Sulphur Springs Woman: An early Holocene skeleton from southeastern Arizona. *American Antiquity* 51(2):361–365.

Waters, Michael R., and Thomas W. Stafford, Jr. 2007. Redefining the age of Clovis: Implications for the peopling of America. *Science* 315(5815):1122–1126.

Weiss, Elizabeth. 2008. *Reburying the Past: The Effects of Repatriation and Reburial on Scientific Inquiry*. Nova Science Publishers, New York.

Wentz, Rachel K., and John A. Gifford. 2007. Florida's deep past: The bioarchaeology of Little Salt Spring (8SO18) and its place among mortuary ponds of the Archaic. *Southeastern Archaeology* 26:330–337.

Wilke, Philip J., J. Jeffrey Flenniken, and Terry L. Oxbun. 1991. Clovis technology at the Anzick site, Montana. *Journal of California and Great Basin Anthropology* 13(2):242–272.

Wilkinson, R. G. and K. M. Van Wagenen. 1993. Violence against women: Prehistoric skeletal evidence from Michigan. *Midcontinental Journal of Archaeology* 18:190–216.

Wormington, H. M. 1957. *Ancient Man in North America*. Denver Museum of Natural History Popular Series no. 4. Denver Museum of Natural History, Denver.

Yerkes, Richard, and Linda M. Gaertner. 1997. Microwear analysis of Dalton artifacts. In *Sloan: A Paleoindian Cemetery in Arkansas*, edited by Dan F. Morse, 58–71. Smithsonian Institution Press, Washington.

Yesner, David R. 1994. Seasonality and resource "stress" among hunter-gatherers: Archaeological signatures. In *Key Issues in Hunter-Gatherer Research*, edited by Ernest S. Burch, Jr., and Linda J. Ellanna, 151–167. Berg Publishers, Oxford.

Young, Diane. 1988. An osteological analysis of the Paleoindian double burial from Horn Shelter No 2. *Central Texas Archeologist* 11:11–115.

Geography, Paleoecology, and Archaeology

James C. Chatters

The chapters in this volume are written by specialists, each of whom focused on specific observations or characteristics of the Kennewick skeleton. Inferences from these scientific studies are based on features seen in the light of each analyst's experience with other skeletal collections, ancient and modern. However, their findings must be placed in context to form a coherent narrative of Kennewick Man's life, death, and burial. This chapter provides that background information in three sections. In the first section, regional geography, geochronology, paleoecology, and human prehistory provide an environmental, geologic, and cultural setting that informs interpretations of evidence from the skeleton. The paleoecological and prehistorical aspects of this discussion focus on the period from 8,500 to 8,000 radiocarbon years ago (RC yr. BP) or 9,500 to 8,900 years ago (CAL yr. BP), which encompasses the age of the skeleton. The second section presents local conditions in which the skeleton was deposited, including geohydrology, geoarchaeology, paleoecology, and archaeology of the discovery site and its immediate surroundings. The third section describes the archaeological context of the discovery site, noting methods used to recover the remains as well as the sequence and location in which the skeletal fragments were recovered. The latter two sections complement the analysis presented in Chapter 17, by providing information about the discovery site.

REGIONAL CONTEXT
Geography
The City of Kennewick, Washington, is situated in the bowl-like Columbia Basin, a 200 kilometer-diameter feature surrounded by the Cascade Range, Okanogan Highlands, Rocky Mountains, and Blue-Ochoko Mountains (Figure 2.1). Kennewick lies in the lowest point of the basin, only 104 meters above sea level. Land rises gently to the north and east, dissected by shallow but steep-walled canyons and coulees, and marked by widely separated, anticlinal ridges. Those ridges, known as the Yakima Folds, become higher and more closely spaced to the west and northwest. Horse Heaven Hills, the southernmost of these east-west trending structures, rises immediately south of Kennewick. Perpendicular to the east end of Yakima Ridge, the next fold to the north, is a narrow anticline that extends southeastward to parallel the north edge of the Horse Heaven Hills. The summits of this anticline, known as Badger, Red, and Rattlesnake mountains, are the local high points. At more than 937 meters, Rattlesnake Mountain reaches an elevation equal to the lowest passes of the Cascade Range. Badger Mountain, which lies immediately west of Kennewick, and the nearby crest of the Horse Heaven Hills share an elevation of 480 meters.

Despite the proximity of steep mountainsides, most of the terrain around Kennewick ranges from relatively level to gently sloping (Figure 2.2). The land to the north and east is rolling gravel plain traversed by sand dunes trending southwest to northeast. To the south, once past the crests of the Horse Heaven Hills, the landscape becomes a rolling, gently southwardly sloping loess plain dissected by small, intermittent streams. Although not vertically challenging, the land is not entirely pedestrian friendly. Soft sand and loess give under the feet, while tussocks of bunchgrass and the coppice dunes that form around the stems of sagebrush and spiny hopsage create markedly uneven micro-topography.

Hydrology
Kennewick is situated at the confluence of three rivers: Columbia, Snake, and Yakima. The Yakima River, smallest of the three, has one of the largest basins on the east

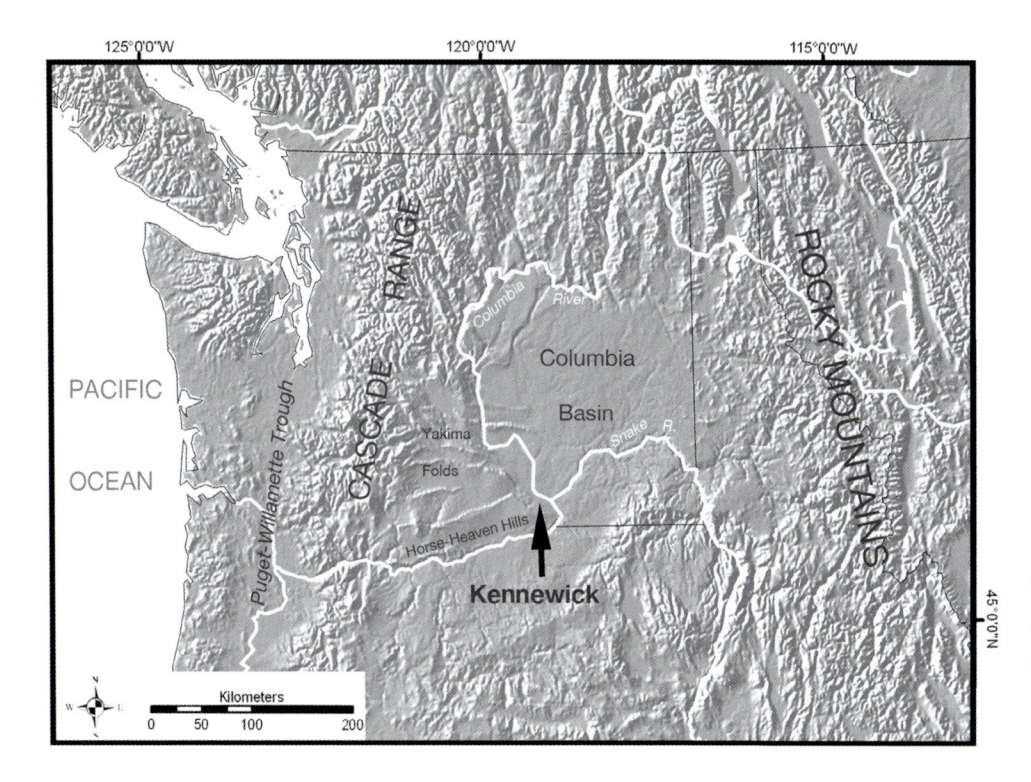

FIGURE 2.1 Relief map of the Pacific Northwest, showing the location of Kennewick, Washington, in the Columbia Basin.

flank of the Cascade Range. Entering the Columbia Basin from between Yakima Ridge and the Horse Heaven Hills, this river has a relatively low gradient over much of its length and numerous moraine-dammed lakes at its headwaters. The Snake and Columbia rivers cut canyons across the Columbia Basin. The Snake River, which enters from the east and joins the Columbia downstream of the discovery site, rises in western Wyoming and drains the Rocky Mountains from northern Nevada and Utah to the Idaho Panhandle. The Columbia is the master stream of the region. Above its confluence with the Yakima and Snake rivers, this immense stream draws snowmelt from the Cascade Range, Okanogan Highlands, and the Northern Rockies from northwest Montana and the Idaho Panhandle to central British Columbia. Before dams were constructed to control its flow, the mean daily discharge of the Columbia River between its confluences with the Snake and Yakima rivers ranged from 1,600 cubic meters per second in early fall to more than 11,000 in June (Nelson and Hauschild 1970). In those days, the water was clear and exceedingly cold even in late summer because much of it originated from snow and ice fields in the high mountains.

Modern Climate and Ecology

The Columbia Basin is one of the driest parts of the Pacific Northwest. The Cascade Range blocks most of the northeasterly airflow from the Pacific Ocean, placing central Washington in a rain shadow. Precipitation averages 20.3 centimeters annually; the majority falls as rain and snow between November and March (US Climate 2011) with a secondary peak in late spring. Summers are hot and winters are mild; the mean high temperature in July reaches 31.8 degrees C; January lows average −2.2 degrees C. Because summers are dry and warm, most small streams from the Yakima Folds and across the Columbia Basin flow only intermittently; perennial natural springs are rare.

The low, winter-dominant precipitation supports a shrub-steppe ecosystem in which the local climax species on stable landforms are big sagebrush and bluebunch wheatgrass (Daubenmire 1970). Grasses in this ecosystem are cool-climate bunching species that produce green foliage in spring and begin to die back by early summer. Forbs, many of which possess edible roots used historically by native human populations, follow a similar pattern. In sandier habitats, such as the dunes that cross the lowland deposits of glacial gravel, vegetation is sparse

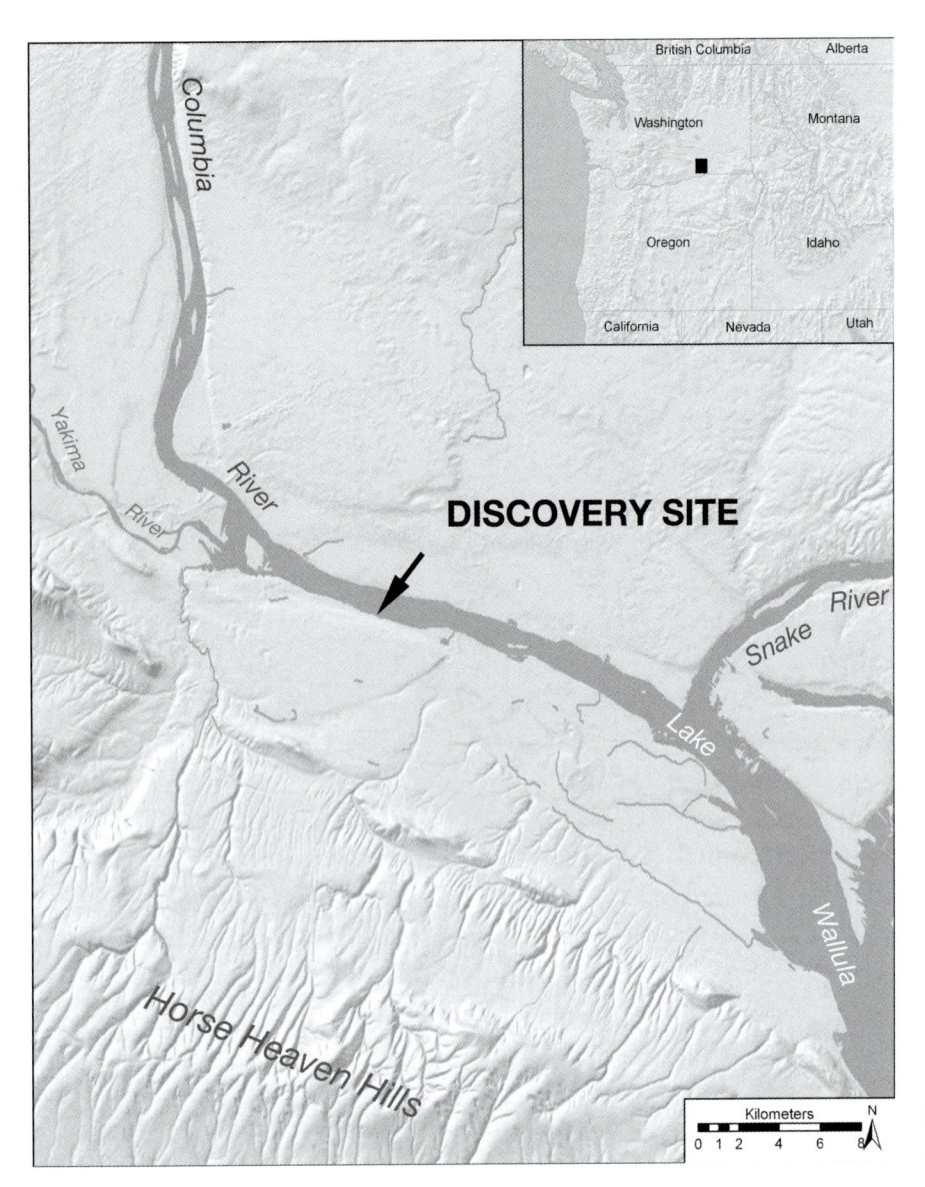

FIGURE 2.2 Geography of the discovery site.

and dominated by large-seeded grasses such as Indian ricegrass and needle-and-thread grass. Finer-textured, alkaline soils are inhabited by Great Basin wild rye, saltgrasses, and spiny hopsage, a perennial chenopod.

Trees are rare, with cottonwood and willow occurring primarily along small watercourses and shaded draws. By virtue of wide annual variations in discharge, the Columbia and Snake rivers historically supported few trees along their banks. With the exception of relict stands of western juniper found in large sand dunes northeast of the Columbia-Snake river confluence, all trees are deciduous. The nearest coniferous forests are above 800 meters elevation on the north flank of the Blue Mountains, 65 kilometers to the southeast, and above 1,000 meters elevation on the eastern slopes of the Cascade Range, 88 kilometers to the west. In unforested parts of the Columbia Basin, the proportion of grass increases to the east, such that the southeastern third of the region is predominantly a bunchgrass steppe (Franklin and Dyrness 1988).

Sparse vegetation cover means low secondary productivity in the central Columbia Basin. Big game animals, including mule deer, whitetail deer, and Rocky Mountain elk, inhabit the steppes today (Fish and Wildlife Service 2011), but data from recent archaeological sites indicate that a more diverse ungulate community existed at the time of European contact. Archaeological data from sites in the region indicate that elk were historically rare in the central Columbia Basin (Lyman 2004), but bighorn sheep occupied steep canyon walls (Lyman 2009), and pronghorn were common on the shrub steppe (Lyman 2007). Bison were occasional visitors, but dense human populations prevented the development of significant herds after around 1,800 RC yr. BP (1,740 CAL yr. BP) (Chatters 2004). Jackrabbits were abundant. In the late prehistoric faunal assemblages from Strawberry Island, which contains the most intensively investigated late prehistoric sites in the Kennewick vicinity, pronghorn and jackrabbits far outnumber any other mammalian prey (Schalk and Olson 1983).

Birds and aquatic fauna are, and before the erection of dams were, supported primarily by salmon returning to spawn in their natal streams. Salmonid fishes, which include chinook, coho, and sockeye salmon as well as steelhead trout, enter the river between April and midwinter, depending on species and strain. Fish runs were historically immense. Before the dams were constructed, fish processors harvested between 8 and 22 million kilograms of salmon annually from the Columbia River (COE 1950). Taking into account a high level of wastage and escapement, Gresh et al. (2000) estimated that spawning fish brought between 77 and 103 million kilograms of biomass into the Columbia River Basin annually before the dams were erected (present levels are only 6 to 7 percent of that amount). Dying and decomposing, the fish served as the foundation of the nutrient cycle in an otherwise nutrient-poor system. Maturing salmonid smolt and other large fish in the river basin (including three species of sucker, three of large cyprinids, burbot, and white sturgeon) either fed directly on the carcasses, or on algae or invertebrates nourished by the influx of carbon, nitrogen, and phosphorus (Bilby et al. 1996). White sturgeon were anadromous in pre-reservoir times. Two large bivalve mollusks—the western pearl mussel and Rocky Mountain ridgeback—once lived in great colonies on the riverbed and still can be found in abundance in undammed, unpolluted streams. As detritus feeders, they also probably survived in large part on salmon-supplied nutrients.

Large birds of the region are primarily migratory. Sandhill cranes feed in the steppes. Waterfowl, including a wide variety of ducks and Canada geese, rest in the Columbia River in spring and fall. Coots and a subset of the migrants nest in swamplands provided by dams and irrigation runoff. Bald eagles, which subsist on ducks and fish such as salmon and the abundant non-native carp, are common and, along with black cormorants, are year-round residents. The most abundant bird in summer is the glaucous gull, which breeds and rears its young by the tens of thousands on island rookeries. Today fruit orchards and garbage in the nearby landfill sustain the gulls. Formerly, gulls and eagles likely lived primarily on spawning and spawned-out salmon, while the cormorants depended on salmon fry and smolts on their outward migrations.

Geochronology

The Columbia Basin is underlain by flood basalts deposited during the late Miocene (15 to 7 million years). More than 3,000 meters thick, this rock mass depressed the earth's crust, contributing to the depth of the basin (Hooper 1982). During the Pliocene (5.2 to 1.8 million years), rivers from the surrounding ancestral mountain ranges filled the basin with more than 100 meters of gravel and fine sediment known as the Ringold Formation (Gustafson 1978). During the Pleistocene (1.8 million years to 12,000 CAL yr. BP), alpine glaciations advanced short distances down the east slopes of the southern Cascade Range and continental glaciers extended south of the Okanogan Highlands. During the Wisconsinan glaciation of the late Pleistocene (110,000 to 12,000 CAL yr. BP), Cordilleran ice extended as far south as the cities of Waterville and Spokane, which are both more than 100 kilometers north of Kennewick (Waitt and Thorsen 1983). East of Spokane, the ice dammed the ancestral Clark Fork River of northwestern Montana, creating a massive meltwater lake, Glacial Lake Missoula. A second lake, Glacial Lake Spokane, occupied the ancient Columbia River valley from Spokane westward, blocked by glacial ice near what is now Grand Coulee Dam.

Glacial Lake Missoula repeatedly burst through its ice dam, producing catastrophic floods that deepened river canyons and produced many of the Columbia Basin's coulees (Baker et al. 1991). Hydraulically dammed at Wallula Gap, a few kilometers south of the Snake River, the floodwaters filled the Columbia Basin, once reaching a pool level 360 meters above sea level. As they drained, the floodwaters deposited a series of sand and gravel terraces in the Kennewick vicinity. Earlier floods sliced across the Columbia Basin and coursed down the Snake River, but later flooding events followed the canyon of the Columbia River since they were no longer impeded by glacial ice on the Waterville Plateau. Lake Missoula floods ended between 12,500 and 13,000 RC yr. BP (14,800 to 15,900 CAL yr. BP) (Atwater 1986). Additional outbursts from Lake Columbia, a pool dammed by outwash from Cascade alpine glaciers, spilled down the canyon of the Columbia for another thousand years or more (Gough 1995). These last floods established the modern-day course of the Columbia River in the central Columbia Basin.

During the Holocene, the Columbia and Snake rivers underwent multiple episodes of aggradation and degradation. From earlier than 10,000 until 8,000 RC yr. BP (11,500–8,900 CAL yr. BP), the rivers aggraded their floodplains, depositing horizontal beds of fine-sandy alluvium in what is identified as the T-2 terrace (Chatters and Hoover 1992) or the Early Holocene Alluvium (Hammatt 1977). The rivers began a period of channel degradation after 8,000 RC yr. BP (8,900 CAL yr. BP), down cutting and isolating the Early Holocene Alluvium as a terrace (hereafter, the early Holocene terrace). Thereafter, sediment was added to the early Holocene terrace surface as alluvial fans produced by small, steep tributaries flowing off the Pleistocene flood terrace and through eolian redeposition of material from nearby Pleistocene landforms and later floodplain surfaces. Tephra from the climactic eruption of Mount Mazama, which dates to 6,850 RC yr. BP (7,687 CAL yr. BP) (Bacon 1983; Zdanowitz et al. 1999) and created Crater Lake, is commonly found within this eolian cap. Aggradation that began shortly after 4,300 RC yr. BP (4,860 CAL yr. BP) deposited a second terrace (Chatters and Hoover 1992). After a second degradation episode that ended circa 2,500 RC yr. BP (2,600 CAL yr. BP), the rivers established their historic floodplains.

Paleoecology

The Columbia Basin has been semiarid to arid throughout the Holocene (Chatters 1998a). The period between 8,500 and 8,000 RC yr. BP (ca. 9,500–8,900 CAL yr. BP), however, witnessed a change in the distribution of temperature and precipitation that profoundly affected the resources on which people depended. What is known about the shift is pieced together from numerous, often obscure, forms of data, each of which provides a detail important to the story. Pollen preserved in lakes and bogs records the vegetation of the region (Barnosky et al. 1987; Mehringer 1985a). The distribution and composition of plant communities compared to those of historic times provide evidence for differences in the seasonality and amount of precipitation. Carp Lake, which lies within the Ponderosa pine forest where the Horse Heaven Hills meet the Cascade Range (Barnosky 1985), and Wildcat Lake, in the grass-steppes 80 kilometers northeast of Kennewick (Mehringer 1985a, 1985b), are the pollen localities nearest the discovery site.

Information about the composition of both terrestrial and aquatic faunal communities and their habitats comes primarily from archaeological and paleontological contexts (Butler and Chatters 2003; Chatters et al. 1995; Lyman 1992, 2004a, 2004b, 2007; Lyman and Livingston 1983). Floodplain sediments and remains of freshwater mussel fauna preserved in such settings, for example, are a record of stream conditions, particularly as they pertain to the potential for the abundance of anadromous fish, particularly salmon (Chatters and Hoover 1992; Chatters et al. 1995). Clear streams with rocky bottoms are the optimal environment for salmon spawning, while turbid streams with soft bottoms are poor salmon habitats.

The relative proportion of rock fall to loess in caves and rockshelters is an indicator of winter temperature because rock cracking increases under periods of prolonged cold (Hallet 1983; Walder and Hallet 1985). Variations in regional summer paleotemperature can be inferred from elevations of timberlines in the Cascades and Rockies, which are indicated by wood and forest soils above modern timberline (Luckman and Kearney 1986; van Ryswick and Okazaki 1979). The lapse rate, or temperature change with elevation, is generally 0.6 degrees C per 100 meters of altitude (Barry and Chorley 1968).

The age of water from Columbia Basin aquifers, measured by ^{14}C dating of dissolved carbon dioxide (Silar 1969), tracks changes in both the abundance and timing of precipitation. Precipitation falling as snow is more likely to slowly melt and seep into the ground since most plants remain dormant, whereas warm season rain is more likely to either quickly run off or be taken up by actively growing plants before it can percolate below rooting depth and become part of the groundwater. Table 2.1 summarizes the paleoecological evidence measured by each of these ecological proxies for the period before and after 8,500–8,000 RC yr. BP (ca. 9,500–8,900 CAL yr. BP).

Paleoecological proxy evidence and general circulation models (Kutzbach and Geuter 1986; Kutzbach et al. 1998) both show that the early Holocene was a time of marked seasonality, with hotter summers and colder winters than today. In the Columbia Basin, greater rock spalling rates bespeak more frigid winters, and elevated timberlines reveal greater summer heat. Low rates of groundwater recharge, along with the abundance of grass at Wildcat Lake and shrub steppe around Carp Lake, indicate precipitation was skewed more toward summer than it is now. All evidence points to a more continental climate. Low near-shore biological productivity in the Northern Pacific Ocean near the Oregon-California border before 8,000 RC yr. BP (8,900 CAL yr. BP) indicates winds blew more from the continent than from the ocean during this time (Sancetta et al. 1992). The summer-dominant precipitation that favored grasslands, which extended up mountain slopes well beyond its modern range, also would have favored warm season grasses that tend to utilize the C4 photosynthetic pathway over cool season grasses that utilize the C3 pathway. In the area east of Kennewick, low rates of loess deposition in Wildcat Lake are evidence that grass cover was more dense than at any other time in the Holocene. Bison and elk, which are primarily grass consumers, were more abundant relative to other ungulates.

Palynological evidence of more extensive steppe and reduced forest cover indicate that precipitation, and hence runoff, was between 30 and 50 percent lower than it is today (Chatters 1998b; Mehringer 1985a). The lack of forest cover in surrounding mountains meant that the watershed likely experienced rapid snowmelt and had waters that warmed earlier in the spring. Rapid runoff from an open watershed would have resulted in sediment-laden streams and deposition of thick alluvium on floodplains. High frequencies of the Rocky Mountain ridgeback mussel during this time indicate that riverbeds tended to contain more sand than gravel. With warmer water, sandy beds, and lower overall runoff, those streams that continued to flow had less bed area in which salmon could spawn (Chatters et al. 1991; Neitzel et al. 1991). Lower near-shore marine productivity may have meant that anadromous fishes also found less favorable offshore habitat during the early Holocene.

To summarize, the early Holocene was a time of more-continental climate with seasonal temperature extremes and summer-dominant precipitation. This favored grasslands and the animals that grazed upon them. Dense grass cover in the southeastern Columbia Basin near Kennewick may have made this area one of the more productive terrestrial environments in the region. Salmon productivity was probably at its lowest point during the Holocene, even as terrestrial productivity was relatively high. Salmonid productivity may have been lower in the early and middle Holocene than it was historically. Those salmonids that did exist in the Columbia River system, however, had to pass through the area near Kennewick en route to upstream spawning grounds.

Between 8,500 and 8,000 RC yr. BP (9,500 and 8,900 CAL yr. BP), when Kennewick Man occupied the region, climatic conditions were shifting. Rockfall sharply declined in caves and rockshelters, indicating warmer winters, although continued high timberlines show that summers remained warm. Ponderosa pine forests began to expand downslope at the same time grasslands in the Columbia Basin were replaced by an open canopy shrub steppe. Expansion of deep-rooted pine and sagebrush, along with evidence for renewed groundwater recharge, indicate that precipitation was changing to winter dominant. Renewed upwelling off the Oregon-California coast shows that winds had taken on their modern directions.

As conifer forests and shrub steppe expanded, the grazing ungulates bison and elk declined in favor of browsers such as pronghorn and deer. The coincidence of arid conditions and the increase in nonedible pines at the expense of grasses in the highlands indicates that terrestrial animal productivity was probably lower than during the early Holocene. This would have been particularly true in the Columbia Basin, where overall vegetation density declined markedly. North of Kennewick, the absence of Mount Mazama tephra in the dunes of Hanford Reach National Monument indicates that during the middle Holocene poor vegetation cover allowed the wind to actively deflate sand dunes and keep them migrating across the landscape (Gaylord 2011).

TABLE 2.1 Evidence for paleoecological conditions in the Columbia Basin in the early and middle Holocene.

Paleoecological Proxy	Early Holocene 10,000 to 8500–8000 RC yr. BP	Middle Holocene 8500–8000 to 5400 RC yr. BP
Palynology	Carp Lake was surrounded by a sagebrush and chenopod-dominated steppe until 8500 RC yr. BP (Barnosky 1985), whereas Wildcat Lake was surrounded by grassland similar to that of today (Mehringer 1985a, b). Sites in the eastern Columbia Basin and lower slopes of the Rocky Mountains were all occupied by grassland (Barnosky et al. 1987; Chatters 1998).	Ponderosa pine forest began to move downslope around Carp Lake, but sagebrush began to invade the steppes around Wildcat Lake. The sagebrush was probably initially three-tip sage, a species now found in wetter parts of the shrub steppe, rather than the more drought-tolerant big sagebrush Mehringer (1985b). By 8000 BP, big sagebrush dominated the entire lowland of the Columbia Basin. Forests expanded in the Rocky Mountains.
Vertebrate fauna from archaeological and paleontological sites	Pleistocene megafauna were extinct by this time; modern faunal communities were established. Bison and elk were more common in proportion to pronghorn and deer than they were in the middle and later Holocene (Lyman 2004a, b). Direct evidence for salmon abundance is lacking. Fish found at Hetrick (Rudolph 1995) and Marmes Rockshelter (Butler 2004) were primarily resident, non-salmonid species.	Proportions of elk declined in relation to pronghorn after 7000 RC yr. BP. Bison declined after 8000 RC yr. BP. Small resident species such as sucker and northern pikeminnow dominate, although salmon are well represented at other sites on the Snake River (Chatters et al. 1995; Butler and Chatters 2004) and were abundant at the Five-Mile Rapids site on the lower Columbia River by 8100 RC yr. BP (Butler and O'Connor 2004).
Floodplain sediment and mussel faunas (Chatters and Hoover 1992)	Sedimentation rates were low before 9000 RC yr. BP, but rapid between then and 8000 RC yr. BP as rivers aggraded the early Holocene floodplain. Mussel faunas indicate high turbidity and fine-textured beds in trunk steams.	Eolian sedimentation on early Holocene floodplain indicates channel downcutting. Because watersheds were still largely unforested, water continued to be highly turbid with fine-textured beds in trunk streams. Water temperatures were high.
Cave and rockshelter sediments (Fryxell 1963; Chatters 1989)	Sediment is primarily roof-fall during this period, with very little interstitial loess. This stratum is often as thick or thicker than the subsequent loess, demonstrating overall greater rates of rock spalling. Winters were very cold.	Sediment is primarily loess. This indicates winters were warmer than they were in the early Holocene. Late Holocene deposits contain more rockfall, so it appears middle Holocene winters were warmer than they were during the past few millennia.
Fossil timberlines	From 9700 RC yr. BP until as late as 5000 RC yr. BP, timberlines were from 100 to 300 meters above historic levels indicating that summer temperatures were between 0.6 and 2.0°C higher (Clague and Mathewes 1989; Luckman and Kearney 1986; Van Ryswick and Okazaki 1979).	
Loess deposition in Wildcat Lake (Mehringer 1985b)	The rate of sedimentation was low, lower in fact than it was during the most recent 2500 years. Mehringer interprets this as evidence for a denser vegetation cover, as might be expected of a grass steppe.	The sedimentation rate accelerated at 8000 RC yr. BP and soon reached its Holocene maximum, indicating greater soil exposure and accessibility of loess to wind mobilization.
Ground water ages (Silar 1969)	The near-complete absence of water from this age range, bracketed as it is by dated water from before and after this period, indicates aquifer recharge was not occurring. Precipitation was more summer-dominant.	Ground water dates consistently from 8500 RC yr. BP onward, evidence that the modern winter-dominant pattern was established at about this time.

At the same time terrestrial productivity declined, salmon productivity appears to have been on the rise. Although vegetation cover indicates that precipitation, and therefore stream flow, was well below modern levels, expanding forests would have led to cleaner streams and thus some improvement in spawning habitat while changed wind directions led to improved biological productivity offshore (Sancetta et al. 1992). Excavations at Five Mile Rapids on the Columbia River, 224 kilometers downstream of Kennewick, have demonstrated that salmon were migrating in large numbers at least as early as 8,100 RC yr. BP (9,000 CAL yr. BP) (Butler and O'Connor 2004). Although fishes found in archaeological sites after this time continue to be dominated by non-salmonid residents, including suckers and northern pikeminnow, the balance of resource availability was shifting from terrestrial to aquatic as the early Holocene came to a close. Salmon are present in sites along the rivers upstream of the Columbia-Salmon confluence, but only the chinook salmon (*Oncorhynchus tschytscha*) has been identified to species (Reid and Chatters 1997).

Human Prehistory

Kennewick Man dates to a time of transition between two distinct cultural traditions. The Western Stemmed Tradition was nearing the end of its reign in the Columbia Plateau as the Old Cordilleran Tradition began its rise (Chatters 2010a; Chatters et al. 2012). What is known about the lifeways of these two traditions in the Columbia Basin region, with an emphasis on subsistence behavior, is summarized below based on Chatters et al. 2012 (see also Ames 2000; Ames et al. 1998; and Chatters and Pokotylo 1998).

Western Stemmed Tradition

During the terminal Pleistocene and early Holocene, most of temperate North America west of the Rocky Mountains was occupied by cultures of a tradition that has been variously labeled the Western Pluvial Lakes tradition (Bedwell 1973), the Western Stemmed Tradition (Bryan 1980), and the Paleoarchaic (Beck and Jones 1997; Willig and Aikins 1988). Manifestations in the Columbia Basin are known as the Phillipi phase on the middle Columbia below its confluence with the Snake River (Dumond and Minor 1983) and as the Windust phase above that confluence (Leonhardy and Rice 1970; Rice 1972).

Archaeological assemblages attributed to these phases contain a wide variety of stone and bone implements. Stone tools include large, thin, stemmed, lanceolate projectile points with distally expanding blades, as well as large end and side scrapers produced on flakes with little edge modification. Other tools include keeled scrapers or scraper-planes, denticulate scrapers, burins, gravers, drills, biface cores that also become projectile or knife preforms, cobble implements, and retouched and utilized flakes that often have multiple modified edges. Manos and milling stones occur infrequently and have often been interpreted as pigment-grinding implements rather than plant processing tools. Chipped-stone crescents and bola stones are unique to this tradition. Bolas are plumb-bob-shaped stones the size of a small chicken egg with a single circumferential groove around the long axis.

Projectile points, which were produced through bifacial percussion reduction and finished with broad, commedial pressure flakes, appear to have been initially long and wide (Butler 1965; Rice 1972), but they were often resharpened to stemmed stubs. This intensity of reworking and the presence of use-wear polish on remnants of resharpened blades indicate these implements were used as both projectiles and knives (Chatters and Prentiss 2010; Galm and Gough 2008).

Bone tools have been found in Western Stemmed Tradition sites when conditions favor bone preservation. Functionally diverse, they include atlatl spurs; wedges; unilaterally barbed bird-spear or fish-harpoon points; single piece and composite bone shafts; and tiny, eyed needles (Galm and Gough 2008; Irwin and Moody 1978; Leonhardy and Rice 1970; Rice 1972). Harpoon points, crescents, bolas, atlatl spurs, and multi-piece bone shafts all represent compound tools, indicating a complex technology with a high degree of planning depth (Torrence 1983). Planning depth is a measure of the time taken by hunter-gatherers to prepare implements for killing and capturing prey. The more concentrated prey are in time and space, the more important it is to have a technology prepared to capture them efficiently. Composite tools are an indicator that more time was spent preparing for short periods of prey capture.

Toolkit complexity did not extend to food processing

technology, which appears to have been limited. The presence of a small number of lightly utilized milling stones at the Marmes Rockshelter, Washington (Rice 1969; Hicks 2004), and the Vine site, Washington (Lewarch and Benson 1991), as well as a high frequency of such tools at the Goldendale site, Washington (Warren et al. 1963), provides possible evidence of seed processing (although Goldendale may not belong to this tradition). There is no evidence that members of this tradition developed or adopted hot rock cooking. Thermally altered rock, which would be produced by such cooking methods, is rare and usually consists of large fragments indicative of surface fires (e.g., Sappington and Schuknecht-McDaniel 2001).

Western Stemmed Tradition sites, although relatively uncommon, are found in almost every type of habitat. Remains of camps most often occur on or overlooking the Columbia and Snake rivers (Cressman et al. 1960; Leonhardy and Rice 1970; Valley 1975) and their smaller tributaries (Ames et al. 1981; Davis and Sisson 1998). They also are found in uplands adjacent to river canyons (Galm and Gough 2008), on the shores of the large lakes that surround the Columbia Basin (Miss and Hudson 1987; Salo 1987), along lowland streams and marshes well away from the major rivers (Daugherty 1956), and high into the Cascade Range and Blue Mountains (Lewarch and Benson 1991; Mack et al. 2010). Many sites show evidence of repeated reuse, indicating a seasonal round of camp movements. Lind Coulee, in the central Columbia Basin, was an often-used hunting camp (Irwin and Moody 1978). Goldendale was a seed gathering camp (Warren et al. 1963). Hornby Creek (Weisz 2006) and the Road Cut site (Cressman et al. 1960) appear to have been visited to fish and/or hunt scavenging birds, and Marmes Rockshelter may have been a winter camp. Other sites appear to have been used only once, leading Ames (1988) to suggest the settlement strategy was semi-logistical. Individual occupation events cover small areas, contain few artifacts, and exhibit a high degree of discreteness in features and artifact patterning (Chatters 1987), indicating that the social groups of the tradition were both small and mobile.

The tendency to use sites repeatedly for limited functions is reflected in individual artifacts. Some of the specialized implements in the toolkit have ecologically restricted distributions, evidence that they were designed and used for exploiting a single resource (Chatters et al. 2012). Bola stones are a good example of this patterning. With one exception (a single specimen found in the Blue Mountains (Brauner 1985)), bolas occur on river floodplains, usually below falls or rapids, or at the mouths of lakes. The largest numbers have been recovered at the mouth of Pend Orielle Lake (Weisz 2006) and at Five Mile Rapids on the Columbia River (Cressman et al. 1960). This distribution is instructive. Post-reproductive fish, juvenile out-migrants stunned by the downstream plunge, and resident fish may concentrate in quiet pools below cataracts, becoming food for bears, coyotes, and fish- and carrion-eating birds. Mouths of large lakes may act to concentrate fish in a similar way. If the grooved stones were indeed parts of bolas for hunting birds, these settings would have been effective places for their use. If the stones were weights for hook-and-line fishing, the prey would most likely have been piscivorous fish like northern pikeminnow or bull trout (Wydosky and Whitney 1979), which also congregate in such localities to feed.

Although northern pikeminnow and other non-salmonid fish have been recovered (Butler 2004; Rudolph 1995), faunal remains from archaeological sites indicate that most Western Stemmed Tradition people living on the Plateau focused their attention on big game; small game and fish were usually of secondary importance. Large ungulates—deer, elk, or bison—rank first or second in number of identified specimens in nearly all assemblages, although waterfowl, rabbits, and marmots often rank in the top four (Chatters et al. 2012). Although shellfish are abundant at Marmes Rockshelter (Ford 2004), fish are almost never present in significant amounts in assemblages definitively attributable to the Western Stemmed Tradition.

The Road Cut site at Five Mile Rapids, one of the most intensively excavated sites and the one closest to Kennewick, may be an exception, but findings from early work at the site (Cressman et al. 1960) are difficult to interpret and use with confidence. The excavators report large numbers of salmon vertebrae and bird remains from the site. Inspection of graphs representing the concentration of faunal remains seems to indicate that in the deepest levels (level 28 and below), which contain the Western Stemmed Tradition assemblage, some salmon were present along with thousands of bird bones. Most common were gulls, bald eagles, condors, and cormorants. Nearly all the recovered bola stones were obtained from these same levels, making a strong case for their use as bird-hunting implements.

Only one collection of human remains from the Columbia Basin region is attributable with certainty to the Western Stemmed Tradition—cremated bone representing up to 10 individuals found in the deepest strata of Marmes Rockshelter and its adjacent floodplain (Krantz 1979). At least six individuals are represented by the bones in the rockshelter, and four more were found on the floodplain below. Although these remains are often reported as being from a cremation hearth, the lack of evidence for in situ burning indicates bodies were cremated elsewhere and the ash transported to Marmes Rockshelter for interment (Chatters 2010b). They might have been carried from the cremation locality to Marmes Rockshelter in twined bags similar to those found containing cremated bones at Spirit Cave, a Western Stemmed Tradition site in western Nevada (Dansie 1997). Secondary interment is consistent with high residential mobility (Wallthal 1999), corroborating the evidence provided by artifact patterning and feature discreteness. One other individual, known as Stick Man, appears to have been interred on his side after having been painted with red ochre. Like Kennewick Man, Stick Man postdates the youngest Western Stemmed Tradition assemblages but predates the oldest Old Cordilleran Tradition component (Chatters et al. 2000).

The Western Stemmed Tradition occupied the Columbia Plateau from as early as 11,600 RC yr. BP (13,400 CAL yr. BP) until sometime between 8,500 and 8,000 RC yr. BP (9,500 and 8,900 CAL yr. BP) (Chatters et al. 2012). The most recent radiocarbon dates for this tradition, with the exception of one anomalous age from the Wildcat Canyon site in northern Oregon, range between 8,800 and 7,850 RC yr. BP, with the oldest coming from the Columbia Basin region and youngest from central Idaho and the most southeastern portion of Oregon (Table 2.2). In the area immediately surrounding the Kennewick Man discovery site, the youngest dates for the Western Stemmed tradition are a radiocarbon date of 8,525 ± 100 RC yr. BP (9,604–9,426 CAL yr. BP) from Marmes Rockshelter (Sheppard et al. 1987) and an optically stimulated luminescence (OSL) date of 7,200 ± 500 BC (equivalent to 8,700–9,700 CAL yr. BP) from the Beech Creek site near the crest of the Cascade Range (Mack et al. 2010).

Old Cordilleran Tradition

The Old Cordilleran Tradition (Butler 1961), thought to be a derivative of the upper Paleolithic Dyuktai culture of eastern Siberia, was ubiquitous on the Columbia Plateau and the adjacent Northwest Coast throughout the middle Holocene (Chatters 2010a; Chatters et al. 2012). The tradition first appeared along the northern Northwest Coast around 9,300 RC yr. BP (10,500 CAL yr. BP) and advanced southward and inland over a period of 1,600 years, coming to dominate the Plateau throughout the middle Holocene (Chatters et al. 2012). Although historically given a variety of local labels, it is now commonly known as the Olcott phase in western Washington (see Chatters et al. 2011) and the Cascade phase in the Columbia Plateau (Leonhardy and Rice 1970).

The Old Cordilleran Tradition is distinct from the Western Stemmed Tradition in having simpler bone and stone technologies. Stone tools include large, leaf-shaped knives and leaf-shaped projectile points, small end and side scrapers, drills, gravers, and cobble tools. The projectile points tend to be narrow and thick with serrated edges and show no sign of having been used secondarily as knives (Chatters and Prentiss 2010). Microblades are sometimes found in assemblages, most often high in the Cascade Range (Daugherty et al. 1987a, 1987b). Core tools, particularly those known as cobble choppers, as well as round cortical spalls struck from river cobbles occur much more frequently in Old Cordilleran Tradition assemblages than in the Western Stemmed Tradition. Bone implements are less common, less diverse, and appear in different forms. The most common are splinter awls, wedges, and large needles, although barbs from fishhooks and leisters have also been found. The toolkit overall contains far fewer composite tools, and those composite implements that are found were fishing oriented.

The Old Cordilleran Tradition's food processing technology was advanced relative to that of the Western Stemmed. Food-grinding tools, including manos, milling stones, and edge-ground cobbles, occur in most

TABLE 2.2 Most recent radiocarbon ages of the Western Stemmed Tradition and earliest dates on the Old Cordilleran Tradition in and near the Columbia Basin.

Site	Location	14C or OSL Date	CAL yr. BP	Reference
Most Recent Western Stemmed Tradition Dates[1,2]				
Beech Creek	W-Cent WA	7200 ± 500 BC*	9700–8700	Mack et al. 2010
Marmes Rock Shelter	SE WA	8,525 ± 100	9604–9426	Sheppard et al. 1987
Hatwai I	N Idaho	8,830 ± 310	10,249–9528	Ames et al. 1981
Cougar Mt. Cave	Cent. Oregon	8,510 ± 250	9887–9136	Ferguson and Libby 1962
Ft Rock Cave	Cent. Oregon	8,550 ± 150	9743–9318	Bedwell 1973
Harney Lake	E Oregon	8,680 ± 55	9680–9551	Gehr 1980
Dirty Shame Rock Shelter	SE Oregon	7,850 ± 120	8972–8523	Aikens et al. 1977, Hanes 1988
Shoup Rock Shelter	Cent. Idaho	8,125 ± 230	9398–8727	Swanson and Snead 1966
Redfish Overhang	Cent. Idaho	8,060 ± 285	9290–8599	Sargeant 1973
Earliest Old Cordilleran Tradition Dates				
Cascadia Cave	W-Cent. Oregon	7910 ± 280	9089–8420	Newman 1966
5-Mile Rapids	S-Cent. WA	8090 ± 90	9189–8778	Butler and O'Connor 2004
Ash Cave	S-Cent. WA	7940 ± 150	8993–8608	Butler 1962
Marmes Rock Shelter	SE WA	7840 ± 150[3]	8971–8458	Rice 1969
Thorn Thicket	SE WA	7710 ± 180	8766–8336	Sprague and Combes 1966
Plew Site	N-Cent. WA	7730 ± 120[3]	8633–8394	Draper 1986
10NP453	N Idaho	7980 ± 40	8984–8776	Ridenour 2006

[1]Davis and Sisson (1998) have a date of 8410 ± 70 RC yr. BP from Cooper's Ferry, which they rejected in favor of a much earlier date. A similar age on stylistically comparable artifacts from the Lind Coulee Site has been superseded by multiple AMS dates at around 10,000 RC yr. BP (Craven 2003).

[2]A date of 7370 ± 190 RC yr. BP is cited for the Wildcat Canyon site (Dumond and Minor 1983), but this is inconsistent with all other evidence for the region, which shows Old Cordilleran Tradition to be ubiquitous after ca. 7500 RC yr. BP. The finding is perhaps due to the assignment of disparate units to the Phillipi Phase component at this site.

[3]These dates were obtained from freshwater mussel shell, which can be an unreliable material for radiocarbon age determination.

lowland sites, where they account for a much higher percentage of the total tool inventory. Roasting pits and large quantities of small, thermally altered rocks attest to the importance of earth-oven cooking and stone-boiling. Although foods were more heavily processed during the Old Cordilleran Tradition, there is no evidence that food storage was an important component of the subsistence strategy.

In contrast to their Western Stemmed Tradition predecessors, who made widespread use of the Plateau environment, the people of the Old Cordilleran Tradition confined their activities to the major river corridors and surrounding mountain ranges (Chatters and Pokotylo 1998). Their sites are numerous along the Columbia and Snake rivers, where they consist of small concentrations of thermally altered rock, lenses of mussel shells, small collections of cobble tools and flakes, and a few formed tools and vertebrate faunal remains. Outside the river canyons, the Columbia Basin was virtually abandoned.

Artifact assemblages in the small residence camps are functionally undifferentiated (Ames 1988; Bense 1972), indicating the same suite of activities at each encampment. This fact, along with small habitations and low frequencies of site reuse, mark the Old Cordilleran practitioners of the time before 6,850 RC yr. BP (7,680 CAL yr. BP) as highly mobile foragers who tended to engage in many of the same activities at each residence camp, regardless of location. Except for probable moves between upland and lowland habitats, there is no evidence for seasonal patterning in subsistence activities.

The quasi-logistical strategy seen during the Western Stemmed Tradition had disappeared.

Food processing technologies and faunal assemblages demonstrate major changes between the end of the Western Stemmed Tradition and beginning of the Old Cordilleran. Stone-boiling, earth-oven cooking, and greater frequencies of grinding implements strongly indicate plants played a greater role in Old Cordilleran subsistence. High rates of dental caries in human skeletons from the Rocky Mountains in central Idaho show that processed carbohydrates were particularly important there, at least by the later part of the tradition. Animal exploitation changed from a focus on large mammals to an emphasis on small mammals, fish, and river mussels. Fish rank first or second in importance in 7 of 13 well-documented vertebrate assemblages predating 6,850 RC yr. BP (7,680 CAL yr. BP). Small mammals were the most important food source in another three. Large ungulates are prominent only in rockshelters, which probably served as convenient shelters for hunters. Mussels typically far outnumber all vertebrates in riverine contexts.

Human skeletal remains associated with the early (pre-6,850 RC yr. BP or 7,680 CAL yr. BP) Old Cordilleran have been found only at Marmes Rockshelter in the Columbia Basin. The remains represent two individuals whose scattered bones do not appear to have been formally interred. Later interments, including primary burials as well as secondary burials and cremations, took place in formal cemeteries. Large quantities of grave goods often accompanied the remains in what is known as the Western Idaho Burial complex (Pavesic 1985). Skeletons associated with the Old Cordilleran are morphologically distinct from those thought to have been Western Stemmed Tradition practitioners (Chatters 2010a).

The Old Cordilleran Tradition consistently appears in and east of the Cascade Range after 8,100 RC yr. BP (ca. 9,000 CAL yr. BP) (Table 2.2). The earliest date is at Five Mile Rapids, where it is associated with a salmon-dominated faunal assemblage dated to 8,090 ± 90 RC yr. BP (ca. 8,990 CAL yr. BP) (Butler and O'Connor 2004). Shortly after 8,000 RC yr. BP (8,900 CAL yr. BP), the Old Cordilleran is found in the Oregon Cascades and in the Columbia Basin as far east as northern Idaho. Ash Cave, located a short distance up the Snake River from its confluence with the Columbia, is the closest dated early Old Cordilleran component to Kennewick. Small, undated sites are scattered along the surface of the early Holocene terrace for at least 60 kilometers upstream of the Kennewick Man discovery site (Chatters 1989; Andrefsky et al. 1996).

Comparison of the Traditions

Although the practitioners of both the Old Cordilleran and Western Stemmed Traditions were highly mobile foragers, these occupants of the Columbia Basin led markedly different lives. Western Stemmed Tradition people followed an annual pattern of migrations throughout the landscape that exploited different plant and animal resources in distinct habitats by using implements specially designed for each prey type. They were primarily hunters who took their prey with both atlatls and bolas. Fishing was secondary and plants were not exploited significantly, probably due to a lack of appropriate technology for converting complex carbohydrates into food.

In contrast, Old Cordilleran people spent most of their time along major rivers, subsisting primarily on fish, freshwater mussels, and small mammals. In the high mountains they hunted larger prey, particularly deer. Although they used both lowlands and montane environments, their record shows no sign of seasonal or geographic variability in tool technologies. Their toolkit was generalized and versatile. With new methods for processing plant foods, they occupied a lower trophic level than Western Stemmed hunters, which gave their environment a higher effective carrying capacity.

Simultaneous changes in basic lithic reduction strategy, tool and processing technologies, settlement patterns, implement and basketry styles, and the physical characteristics of the people themselves show the transition from the Western Stemmed to the Old Cordilleran Tradition to have been an ethnic replacement event that was fundamentally genetic (Chatters 2010a; Chatters et al. 2012). The Old Cordilleran appears earlier west of the Cascade Range than it does in the interior, which is evidence that the newcomers first adapted to coastal environments and then followed familiar habitat up the rivers. Exactly when this expansion took place is unclear due to a lack of dated components in the Columbia Basin between 8,500 and 8,000 RC yr. BP (9,500–8,900 CAL yr. BP). The near simultaneous appearance of the Old

Cordilleran Tradition throughout the region within a century of 8,000 RC yr. BP, however, indicates that the Western Stemmed Tradition probably persevered, perhaps as a lowered population, during the interim.

Regional Context Synopsis

Kennewick Man lived in a semiarid steppe environment at a time of widespread change. At the time of his death, open sagebrush steppe in the lowlands and pine forests on the mountain slopes were beginning to replace extensive grasslands. A distinctly seasonal continental climate with cold, relatively dry winters and hot, moist summers was shifting toward a maritime climate with higher winter precipitation and warmer, more-equable seasons. As the land dried and watersheds became more conducive to salmon reproduction, the balance of resources began to shift from the terrestrial to the aquatic environment. Exactly what culture occupied the region at this time is unclear. Western Stemmed Tradition peoples, with their complex, seasonally specific technology and settlement pattern, probably remained. However, their emphasis on terrestrial resources would have become less viable as the climate shifted the balance of productivity to the rivers. Old Cordilleran peoples already occupied parts of the region west of the Cascade Range (Chatters et al. 2011; Borden 1960), even if they were not yet making incursions into the Columbia Basin.

THE CONTEXT OF THE DISCOVERY SITE

The site where Kennewick Man's remains were discovered, registered with the State of Washington as 45BN495, is situated in Columbia Park on the shore of Lake Wallula, a reservoir formed on the Columbia and lowermost Snake and Yakima rivers by McNary Dam. It is located on the Columbia River at mile 331.7, 46°14.5' north latitude and 119°10' west longitude with an elevation of 104 meters above sea level (Figure 2.3). The adjacent land is an approximately 75-meter-wide early Holocene terrace with a shoreline elevation of 106 meters. At the time of the discovery, riparian scrub of orchard grass, willow, and locust trees occupied this section of the terrace. Other parts of the terrace are maintained under manicured lawn grass. This property, owned by the US Army Corps of Engineers (Corps), lies within the city of Kennewick, Washington.

Geohydrology

In the Kennewick vicinity, the Columbia River flows from west to east along a channel that is straight, with little meandering or point bar formation (Chatters 1989). Midchannel islands exist both upstream and downstream from the discovery site, but their elevation above the early Holocene terrace and their continuity with topographic features outside the channel that rest well above peak historic flow indicate that they were produced by the glacial outburst floods that created the channel.

Corps maps of the Columbia River before it was dammed (Figure 2.3; COE 1950) show a channel 880 meters wide at peak annual flow beside the discovery site. Confined as it is between terraces of Pleistocene gravel, the Columbia River could have moved no more than 220 meters laterally from its historic position at any time during the Holocene. In March 1951, when data for the map were collected, the river filled only half its channel, with its southern bank not more than 20 meters north of the discovery site. Although the active channel could have been narrower during drier periods of the Holocene, its movements would have been restricted to the 1,100-meter span between the early Holocene terrace on the south bank and the terminal Pleistocene gravel on the north. If the early Holocene river was able to move laterally (northward) and was no wider at its peak discharge than the Columbia was in March 1951 (440 m), its near bank must still have been no farther than 660 meters from the discovery site. However, the gravel bar opposite the discovery site is continuous with and likely a contemporary of the Pleistocene gravel ridge to its west. That gravel bar would have prevented much lateral movement, limiting a lower-discharge Columbia River at flood stage to a channel approximating that recorded in March 1951. Therefore, Kennewick Man's remains are likely to have lain near the riverbank of his time.

Today, the impounded river has minimal velocity compared with its condition before it was dammed. Boat wakes and wind-driven waves are the primary erosive forces along the reservoir banks. Although the lake level is controlled somewhat by seasonal variations in Columbia River discharge, it is managed primarily by the Corps. Consequently, its water level varies less than three meters throughout the year.

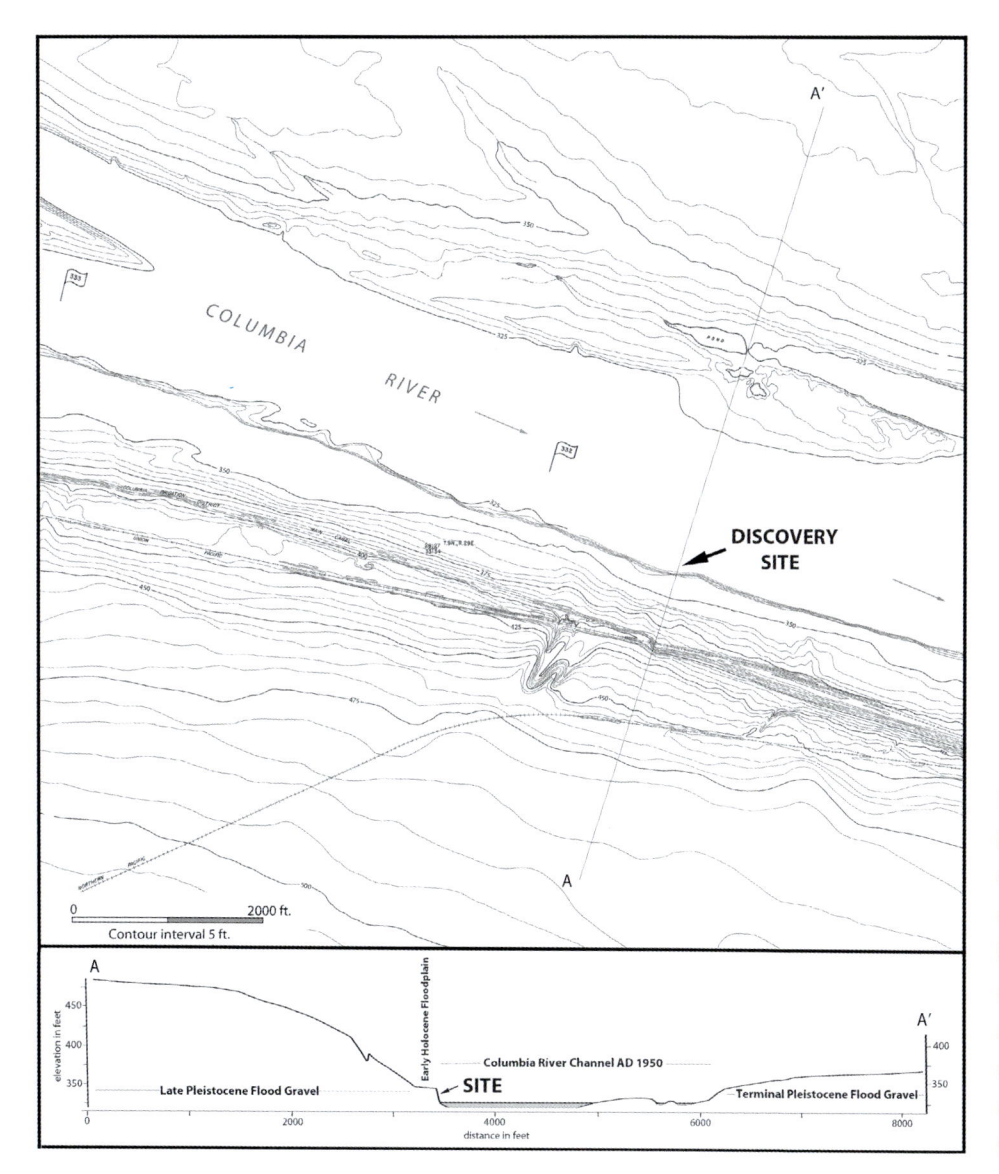

FIGURE 2.3 Topographic map and cross-section of the Columbia River channel and associated late Pleistocene and Holocene deposits at the Kennewick Man discovery site. Based on USACE (1952) with most cultural features removed for simplicity.

Geoarchaeology

The remains of Kennewick Man eroded from the early Holocene terrace on the Columbia River, about midway between the mouths of the Yakima River on the west and the Snake River on the east. Geologic investigations of the site have found two lithostratigraphic units (Figure 2.4) (Huckleberry et al. 1999). Unit II, the lower of the two, is a series of horizontal beds of fine sand and silt up to three meters deep. Individual beds within the unit exhibit a more-or-less distinct coarse-to-fine upward grading, indicating low-energy overbank deposition. Bed thicknesses range from five centimeters in the lowermost sections to as much as 50 centimeters in the upper part of the unit. Boundaries between these uppermost beds grade over one to two centimeters and are partially obscured by bioturbation. Below two meters, the boundaries remain distinct and a few beds exhibit faint, climbing ripple bed forms.

Tephra layers from eruptions of Glacier Peak and Mount Saint Helens are absent from even the deepest strata of Unit II, indicating an age of no greater than 11,200 RC yr. BP (13,100 CAL yr. BP) for the earliest beds (Mullineaux 1986). Four radiocarbon dates obtained by Corps geologists on organic carbon from this unit range from 9,010 ± 50 RC yr. BP (10,150 ± 75 CAL yr. BP) at 50 centimeters below the Unit II surface to 15,330 ± 60 RC yr. BP (18,400±300 CAL yr. BP) at 3.1 meters (Wakeley et al. 1998). While the uppermost date is stratigraphically

James C. Chatters

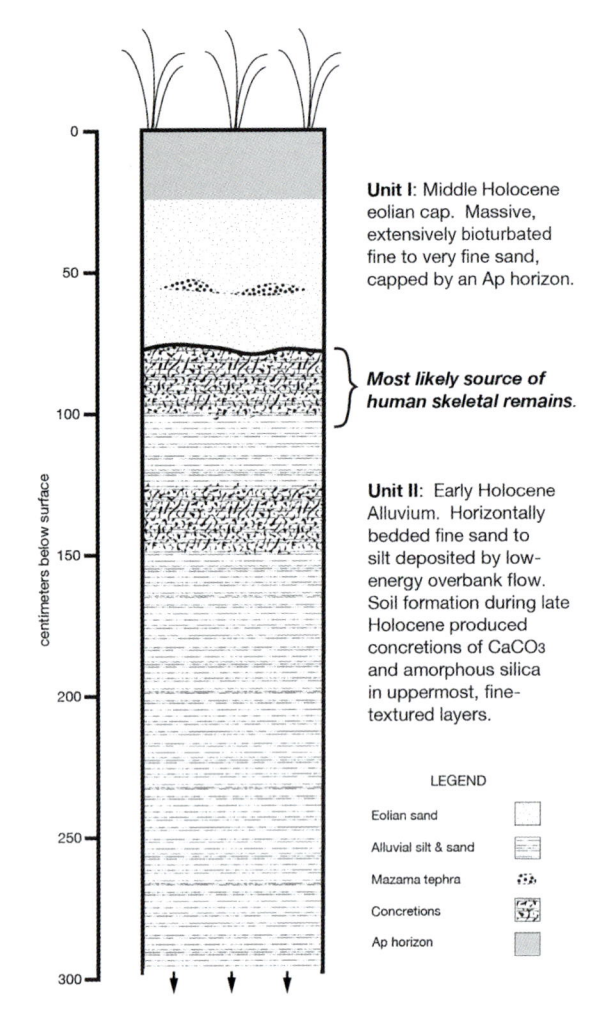

FIGURE 2.4 Schematic profile of the Holocene terrace at the discovery site.

Unit I: Middle Holocene eolian cap. Massive, extensively bioturbated fine to very fine sand, capped by an Ap horizon.

Most likely source of human skeletal remains.

Unit II: Early Holocene Alluvium. Horizontally bedded fine sand to silt deposited by low-energy overbank flow. Soil formation during late Holocene produced concretions of $CaCO_3$ and amorphous silica in uppermost, fine-textured layers.

LEGEND
- Eolian sand
- Alluvial silt & sand
- Mazama tephra
- Concretions
- Ap horizon

consistent with the ages obtained for Kennewick Man (Chapter 3; Taylor et al. 1998) and the established age of early Holocene alluvium elsewhere in the Columbia Basin (Chatters and Hoover 1992; Hammatt 1977), the other three dates are all older than the missing Glacier Peak tephra, leaving all four open to question.

Alluvial charcoal derived from older source deposits and laid down with the sediment may account for the discrepancy. A disconformity that is not physically evident cannot be ruled out but is unlikely given the glacial flooding history of this reach of the Columbia River. Glacier Peak tephra is observable in alluvial strata near the mouth of the Yakima River, where it is several meters above the surface elevation of the early Holocene terrace on the Columbia (Chatters 2000).

Unit I, which ranged from 60 to 75 centimeters thick near the discovery site, is a massive, eolian, fine to very fine quartz sand containing evidence of extensive bioturbation (Huckleberry et al. 1999). Tephra from the Mazama eruption occurs as a discontinuous layer less than five centimeters thick, near the base of the unit. Deposition of the unit began before the tephra fall in 6,850 RC yr. BP (7,687 CAL yr. BP) (Bacon 1983; Zdanowitz et al. 1999) and continued for an undetermined time. Multiple lenses of mussel shell have been found elsewhere in Unit I on the early Holocene terrace in Columbia Park (Miller 2000; Wakeley et al. 1998), indicating human occupation of the terrace after alluvial deposition ceased. Radiocarbon dates on shell from the midden closest to the discovery site produced radiocarbon shell dates of $6,230 \pm 60$ and $6,190 \pm 80$ RC yr. BP (7,140 and 6,980 CAL yr. BP) (Wakeley et al. 1998). Although shell ages often deviate by as much as 2,000 years from associated charcoal dates in the Columbia River basin (Chatters 1986), these results indicate that sand was still being deposited on the terrace after 4,000–6,000 RC yr. BP.

Huckleberry et al. (1998) concluded that the contact between Units I and II was conformable. Three lines of evidence support this inference. First, the contact between the units is planar, exhibiting at most a slight undulation where it is not obscured by bioturbation of the overlying stratum. No cut-and-fill structures could be observed in the 345-meter reach included in the study. Second, although no buried A horizon can be found at the surface of Unit II, such an absence is not uncommon in the Columbia Basin where weak buried A horizons are obscured by oxidation during the formation of soil in overlying deposits (McDonald and Busacca 1992). Finally, such a planar contact is the norm in early Holocene terrace remnants observed along the mid-Columbia and lower Snake rivers. It is highly unlikely that sediment would be uniformly stripped from all terraces in the region between the cessation of alluvial deposition and the onset of eolian deposition during the middle Holocene.

The lithostratigraphic units underwent pedogenesis (soil formation) during and primarily following the deposition of Unit I. A modern brown Ap horizon caps Unit I. Carbonate and silica concretions formed between 70 and 170 centimeters below surface. These concretions, which Huckleberry et al. (1999) interpret as casts of voids produced by roots and insect burrows, are concentrated in two silt-rich strata within one meter of the Unit II surface. One occurs at a depth of 75 to 100 centimeters and the other between 125 and 150 centimeters.

Erosion of the early Holocene terrace and deflation of sediments containing the remains of Kennewick Man occurred primarily after Lake Wallula was created. Since the dam's completion in 1954, wave action along the shoreline has caused progressive wasting of the fine-textured sediment. Study of historic photographs by the Corps (Wakeley et al. 1998) found that the shoreline had regressed southward by 10 to 15 meters. In 1997 at the discovery site, large stones and tires formed a linear scatter approximately that distance from the cutbank. Such debris is commonly placed along riverbanks and terrace escarpments in the Columbia Basin, making it probable that these items mark the historic edge of the terrace.

The postglacial geologic history of the discovery site can be summarized as follows (based on Huckleberry et al. 1998:19):

- Late glacial outburst floods formed a gravel-bordered channel before 11,200 RC yr. BP (13,100 CAL yr. BP).
- After 11,200 RC yr. BP, episodic vertical accretion in a marginal levee-flood basin environment created the early Holocene floodplain.
- Approximately 8,000 RC yr. BP (8,900 CAL yr. BP), and certainly before 6,850 RC yr. BP (7,860 CAL yr. BP), the Columbia River downcut, isolating the early Holocene terrace.
- Deposition resumed on the terrace, primarily by eolian sedimentation. Soil formation proceeded slowly.
- The surface stabilized approximately 4,000–5,000 RC yr. BP (4,500–5,700 CAL yr. BP) and the main episode of soil development ensued.
- Lake Wallula was created and shoreline erosion ultimately exposed the skeleton.

Stratigraphic Origin of the Skeleton

The skeleton was coated with carbonate cemented sediment concretions, indicating its origin in one of the uppermost silt-rich strata of Unit II. Chatters (2000) and Huckleberry and Stein (1999) conducted a series of analyses comparing sediment from the bone surfaces with samples from the stratigraphic column in an attempt to pinpoint the skeleton's source. Chatters (2000), assisted by Alan Busacca of Washington State University, Pullman, compared mean particle size of three samples of concretions from bone surfaces to a series of sediment samples taken in December 1996 from the bank exposure nearest the place where the skull had been found (between 2 and 5 m due south). Samples were from alternate 10-centimeter intervals of Unit I and alternate five centimeter intervals of Unit II between 20 and 210 centimeters below surface in a continuous profile. Sediments from the skeleton most closely matched those from 80–85 centimeters (15 cm from the Unit II surface in this profile) and 135–140 centimeters. Busacca then compared textural profiles for various samples with sediment in concretions from the bone surfaces, finding that the 80–85 centimeter sample was the best fit. Based on these comparisons, Chatters placed the skeleton between 70 and 100 centimeters below the ground surface and in the uppermost 30 centimeters of Unit II, within the uppermost silty stratum of that unit.

Huckleberry and Stein (1999) worked primarily with a less-precise series of samples taken by Corps staff from a discontinuous profile 10 meters southeast of the point where the skull was found (CPP054) and with a series of cemented and loose samples from the skeleton. They conducted granulometric, micromorphological, trace element, X-ray diffraction (XRD) and thermogravimetric (weight loss on heating) analyses on varying sets of samples, finding that only the thermogravimetric and XRD analyses provided meaningful results. Thermogravimetric analyses showed that the organic matter and carbonate content from the 80–90 centimeter depth of profile CCP054 was almost identical to that found in concretions from the skeleton. Linear regression of sample data also showed that mineralogic composition was similar between sediments from the skeleton and stratigraphic samples taken between 62 and 110 centimeters below surface, or the lowest eight centimeters of Unit I and uppermost 40 centimeters of Unit II. By XRD, the closest match was between the skeleton and 70–80 centimeters below surface, with the next-closest matches being the unfortunately thick 95–135 centimeter interval and 80–90 centimeter interval. Overlap of these findings indicates that the skeleton, if found in the vicinity of profile CPP054, came from upper Unit II at between 70 and 110 centimeters below surface. Huckleberry and Stein, however, took the conservative position of placing the bones between 70 and 120 centimeters below ground surface. Clay mineralogy of a concretion taken from the right fourth metatarsal (97A.I.25c) further supports an upper Unit II position. That concretion contained allophane, a byproduct of tephra weathering. Mount Mazama tephra in lower Unit I is the only possible source for this material. It is thus appropriate to state that the Kennewick Man remains weathered from a pedogenic carbonate horizon located in the uppermost 30–50 centimeters of Lithostratigraphic Unit II.

Paleoecological Evidence

Samples of sediment collected in 1996 from a section profiled by Chatters and Ray Tracy of the Corps Walla Walla District (Chatters 2000:Figure 2.4) were processed for pollen and phytoliths. Three samples were selected for this study, two from the uppermost concretion-rich zone of Lithostratigraphic Unit II (75–80 and 85–90 cm below surface) and one from a fine-textured layer just beneath the lower concretion-rich zone (155–160 cm). These samples were selected because they came from finer textured horizons that had a greater chance of containing intact pollen. Phytoliths and pollen were extracted from 100-gram samples by PaleoResearch Institute of Golden, Colorado, using standard techniques.

Pollen

Chatters identified pollen from the three samples using published identification keys and his personal pollen reference collection. Pollen found in the samples was sparse and often poorly preserved, which is common for pollen from nonaquatic depositional environments in the Columbia Basin. Only the 75–80 centimeter and 155–160 centimeter samples contained a high enough density of intact palynomorphs to make analysis feasible. In those two samples, multiple slides were counted in order to obtain a minimum of 750 control grains. Only 147 fossil grains could be identified in the 155–160 centimeter sample and half that many in the 75–80 centimeter sample. The 85–90 centimeter sample produced only seven identifiable grains after a count of 223 control grains, at which time its analysis was abandoned. Its limited content was similar to that of the other two samples. Results of the pollen analysis are summarized in Figure 2.5. Both samples are indicative of steppe vegetation. Arboreal taxa are poorly represented, accounting for 18 percent at 75–80 centimeters and only 7.4 percent at 155–160 centimeters. The same taxa characterize both samples. *Pinus* is most common, with Cupressaceae (probably *Juniperus*), *Alnus* (alder), and *Celtis* (hackberry) represented by one or two percent each. Given their low percentages, the pine, alder, and juniper pollen probably grew at great distances from the site. Only hackberry, an insect-pollinated species now found on rocky slopes along the Snake River, in the moister, eastern part of the Columbia Basin (St. John 1963), probably lived nearby.

The two samples differ in the types and frequencies of non-arboreal pollen. *Artemisia* (sagebrush) dominates the deeper sample at 36 percent of identified specimens. Chenopodeaceae (goosefoot family), and Poaceae (grass) follow at 22 and 18 percent, respectively. Nyctaginaceae, which is probably from a species of *Abronia* (sand verbena) is also well represented at 11 percent. Brassicaceae (mustards), longspine Asteraceae (aster family), *Sarcobatus* (greasewood), and *Eleagnus* (probably *E. commutata*, wolf willow) each occur at less than 3 percent. In contrast, Poaceae at 38 percent and *Artemisia* at 34 percent, dominate the 75–80 centimeter sample. No other non-arboreal type is represented by more than two specimens. Uncommon types are similar to those found at 155–160 centimeters, except that *Eleagnus* is absent and Caryophyllaceae (pink family) and Liliaceae (lily family) are present.

Sagebrush and chenopods dominate over grass in the 155–160 centimeter sample, indicating a community in which big sagebrush and spiny hopsage or winterfat, coexisted with Sandberg's bluegrass. Such communities are found in the driest settings of the region, with the two chenopods occurring more commonly in alkaline soils (Daubenmire 1970; Franklin and Dyrness 1988). The strong representation of what was probably sand verbena suggests a bare, sandy setting nearby. Together, this group of pollens indicates that multiple microhabitats existed on the actively aggrading floodplain, from sandier patches to finer textured, seasonally wet, alkaline swales.

The high frequency of grass and sagebrush and the

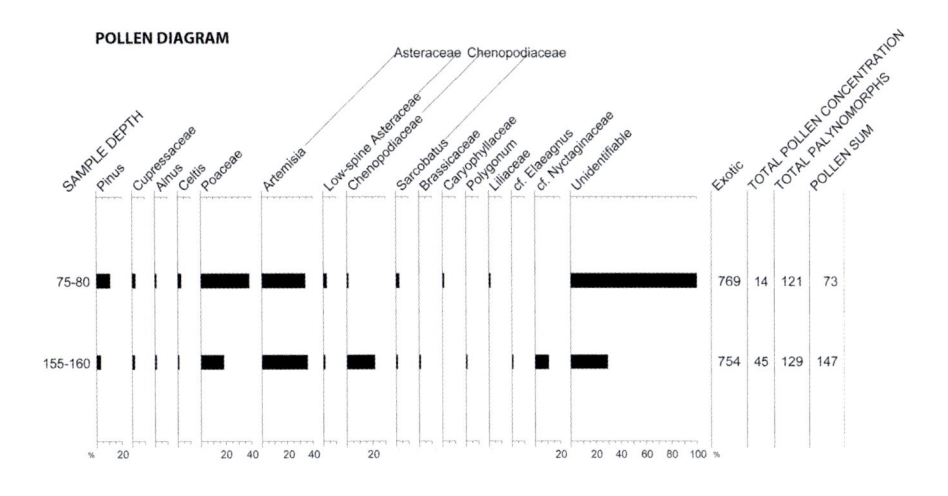

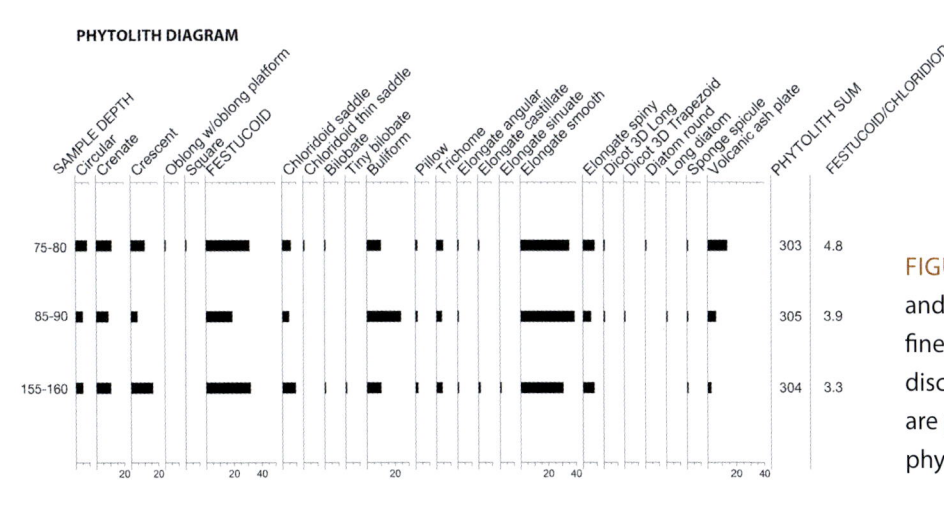

FIGURE 2.5 Diagrams of pollen and phytolith frequencies from fine-textured strata in Unit II at the discovery site. All graphed values are percentages of the pollen or phytolith sum.

near-complete absence of chenopods in the 75–80 centimeter sample suggest a grass-dominated shrub steppe, not unlike that recorded for the early Holocene at Wildcat Lake (Mehringer 1989a, 1989b). The lack of diversity in the assemblage indicates a more stable, less variable floodplain surface toward the end of Unit II deposition.

Phytoliths

Linda Scott Cummings of PRI identified the phytoliths, counting a minimum of 300 specimens from each sample (Cummings 1999). Results (Figure 2.5) are consistent with regional paleoecological conditions toward the end of the early Holocene. Most phytoliths represent grasses, of which the festucoid and chloridoid short forms are the most illuminating. Festucoid phytoliths are representative primarily of cool season grasses, although they are also found in smaller proportions in warm season (chloridoid and panicoid) grasses. Chloridoid grasses are warm season types typical of more arid environments, whereas panicoids are warm season taxa common to humid climates. Saltgrasses, which are found primarily in alkaline soils near water, are the most common chloridoid grasses in the Kennewick vicinity today (Daubenmire 1970). In the Kennewick samples, the ratio of festucoid to chloridoid forms increases from 3.2 in the lower concretion horizon to 4.4 at the surface of Unit II. This is consistent with the shift from more summer-dominant to winter-dominant precipitation and from more seasonal to less seasonal temperatures between 8,500 and 8,000 RC yr. BP (9,500 and 8,900 CAL yr. BP). It is also consistent with the pollen data, in which chenopods indicate the presence of alkaline soils near the discovery site. Buliform phytoliths, which are abundant in the 85–90 centimeter sample, are formed under moist conditions and may indicate an episode of more mesic climate or an interval of ponding on the floodplain. Occasional diatoms and sponge spicules are to be expected in sediments deposited by overbank floods.

Local Paleoecological Setting

Pollen and phytoliths from discovery-site sediments show that steppe vegetation occupied the early Holocene floodplain. Changes in proportions of festucoid and panicoid grasses between the upper and lower pedogenic carbonate horizons may have been a result of changing climate but in light of the pollen evidence for alkaline habitats, are more elegantly explained by changes in the microhabitats as the floodplain surface increased in elevation relative to the river level. Trees are conspicuously rare. The samples have a low frequency of arboreal taxa, particularly pine, which are more common in modern pollen spectra of steppe sites in the Columbia Basin. This is evidence that forests had not yet begun to expand downslope at the time these sediments were deposited. The complete absence of riparian flora, such as willow (*Salix*), currant (*Ribes*), or cottonwood (*Populus*) is evidence either that no gallery forest existed along the early Holocene Columbia River, or if a narrow band of riparian flora did exist near the river channel, it was too far away for its pollen to be deposited in the discovery site. Therefore, at the time the sediments of upper Unit II were being deposited, this site was a grassy shrub steppe located somewhat inland from the then-active river channel.

Archaeology

Site 45BN495 lies within the boundaries of the Tri-Cities Archeological District (NHRP 1983), which encompasses the Columbia River shoreline within eight kilometers upstream and downstream of the mouth of the Yakima River. The district includes 20 archaeological sites, ranging from Late Prehistoric pithouse villages and cemeteries to open camps dating to the Cascade phase of the Old Cordilleran Tradition (Hartmann 1986; Wakeley et al. 1998). It is said to contain the highest density of archaeological sites on the Middle Columbia and Lower Snake rivers (Cleveland et al. 1976). Few of these sites are located on the early Holocene floodplain within Columbia Park, but survey of that developed area has been neither intensive nor complete. At least three shell middens dating to the Cascade phase have been found (Miller 2000; Wakeley et al. 1998), but no site of the Western Stemmed Tradition has been reported in the district or within 50 kilometers of the discovery site.

When the Kennewick skeleton was collected in 1996, a thin scatter of lithic artifacts and historic debris was visible along the reservoir beach, but items in an eroding historic trash pit were the only artifacts found in situ. Artifacts collected along with the skeletal remains consisted of 18 historic artifacts, 5 lithic artifacts, and 21 pieces of faunal material (Sappington 1998). Historic artifacts include a bone-handled, pewter-inlaid knife, ceramic shards, and cut nails dating from the late nineteenth to early twentieth century, as well as shards of glass dating from the late nineteenth century until the mid-twentieth century. Lithic artifacts were two chalcedony flakes, and one flake and two flaked cobbles of basalt. None of the lithic artifacts is temporally diagnostic. Faunal remains consisted of cut and uncut elements, probably all from domestic animals, and four river-mussel shells of indeterminate antiquity.

A light scatter of lithic debitage and two discrete archaeological sites were found along the face of the early Holocene terrace during the 1997 geoarchaeological investigation (Huckleberry et al. 1998). A middle Holocene shell midden located in stratigraphic Unit I was recorded 150 meters east of the discovery site. No artifacts were found in association with it. The second site was a lithic scatter of possible early Holocene origin. Between 200 and 255 meters east of the discovery site, archaeologists found a concentration of lithic flakes and a few small fire-cracked rocks. Lithics included approximately 24 flakes of quartzite and basalt, and one each of jasper and obsidian. Two bifaces, one of them stylistically diagnostic, were found among them. The diagnostic specimen was the base of a leaf-shaped projectile point of igneous material with a nearly parallel-sided proximal end and distally diverging blade edges (Figure 2.6a). The flaking pattern is broad-commedial, a type characteristic only of Western Stemmed Tradition projectile points found in this region (Chatters and Prentiss 2010). Such projectile points are identified as Haskett (Butler 1965) and sometimes Cascade Type A (Lohse and Schou 2008). This style has been found in the Columbia Basin at the Lind Coulee (Daugherty 1956; Irwin and Moody 1978), Sentinel Gap (Galm and Gough 2008), and Hatwai sites (Ames et al. 1981). Assemblages containing Haskett points consistently date earlier than 10,000 RC yr. BP (11,500 CAL yr. BP). Attempts to determine the stratigraphic origin of this lithic scatter were unsuccessful.

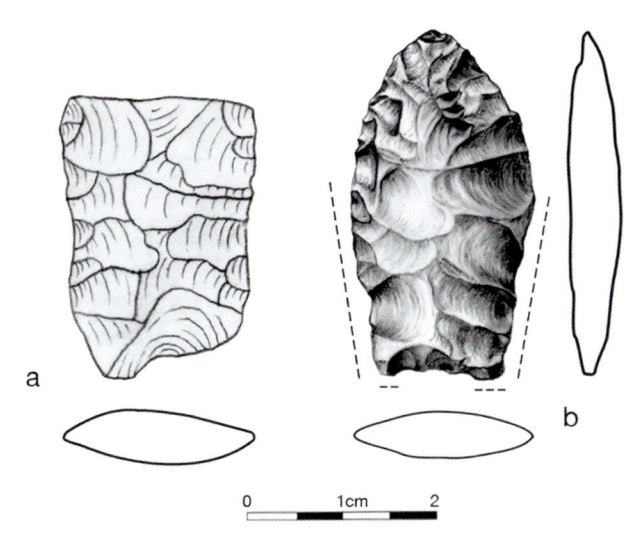

FIGURE 2.6 Projectile points found eroded from the early Holocene terrace near the discovery site: a) observed in 1997 approximately 215 meters west of where the skull was found; b) found in 1969 by Doug Twehus. Dashed lines paralleling right point indicate the presence of edge grinding.

Western Stemmed Projectile Point

In 2000, an artifact collector showed Chatters a projectile point reportedly found years earlier on the shoreline at Columbia Park. The black ignimbrite projectile point (Figure 2.6b) is 35 millimeters long, 19 millimeters wide, and 5 millimeters thick. Its ovate blade constricts slightly to form a 15 millimeter wide, nearly parallel-sided stem that narrows one millimeter to the slightly concave base. Flaking is broad-commedial over most of both faces of the stem and blade, but the distal blade shows the more random flaking pattern characteristic of resharpening. Steep beveling of flake removals at the basal margin and their invasion of broad, commedial flakes on the stem indicate that the stem was once longer and had been repaired after a bending fracture. The basal margin, stem, and proximal blade are edge-ground for a length of 17 millimeters. In outline, cross-sectional properties, flaking pattern, and edge grinding, this artifact closely matches broad-bladed, weakly stemmed projectile points reported from Strata I and II at Marmes Rockshelter (Rice 1969:Figures 23a–c), from the deepest levels of the Road Cut site (Cressman et al. 1960: Figure 46a, groups E and K), and Period I at Windust Caves (Rice 1965: Figures 12–13). Typologically, it is Windust, a style within the Western Stemmed Tradition. When its original base was intact, it may have had a form more closely resembling the Haskett type. The Windust style was produced throughout the early Holocene (Rice 1972).

The artifact was recovered from the reservoir shoreline in 1968 by Tri-Cities resident Doug Twehus. In 2000, Twehus accompanied Chatters to Columbia Park to show where he had found the projectile point. After taking bearings against bridges and landmarks on the opposite bank and walking several hundred meters along the shoreline, he identified the location of the find. He had stopped within a few meters of where the Kennewick Man skeleton had been recovered. Twehus noted, however, that the bank had eroded back several meters since he collected the artifact. Given the uncertainties of recalling the source of a find made so long ago, it is entirely possible that the projectile point came from the same lithic scatter as the other Haskett-like projectile point, found 200 meters east of the skeleton.

Context of the Skeleton

The Kennewick Man skeleton was found out of stratigraphic context and without any clear association with any other archaeological materials. The skeleton's 8,400 RC yr. BP age places it near the end of the Western Stemmed Tradition occupation and 400 years before the earliest documented Old Cordilleran Tradition occupation of the Columbia Basin. The discoveries in 1968 and 1997 of Western Stemmed Tradition-style projectile points along the nearby reservoir shore indicate that a people participating in that tradition were present within 200 meters of the skeleton's resting place at some time in the early Holocene.

DISCOVERY AND RECOVERY

Kennewick Man's remains were found in secondary depositional context and with no associated contemporary archaeological materials. Despite this fact, the position of the bones as they were found along the reservoir shoreline contributes to an understanding of their taphonomy. The remains were recovered over a period of one month in midsummer 1996, when they were under the auspices of the Benton County Coroner. An additional fragment was recovered during a brief geoarchaeological study in December 1997. The bones were found along approxi-

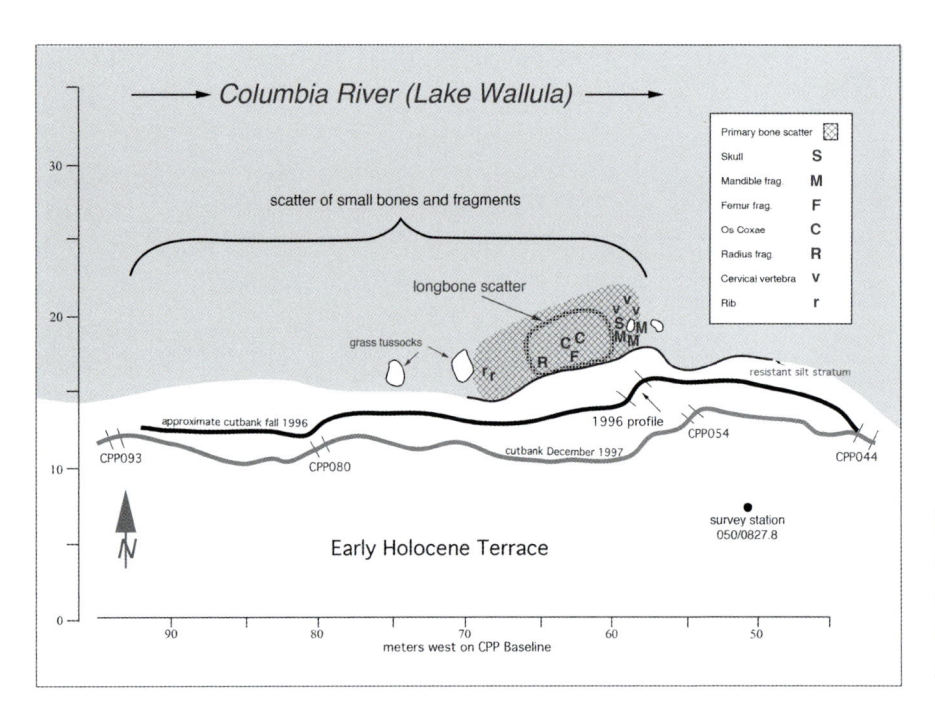

FIGURE 2.7 Plan of the discovery site, showing where skeletal remains were recovered and the location of sediment profiles documented in 1996 and 1997.

mately 30 meters of reservoir shoreline, distributed as displayed in Figure 2.7. Table 2.3 lists which bones were found during each site visit.

July 28, 1996

Two young men, Will Thomas and Dave Deacy, attending a hydroplane race at Columbia Park found fragments of a human skull exposed in mud below the surface of the Columbia River. They reported finding the cranium adjacent to and just upstream of a small tuft of grass, not more than 2 meters offshore of the eroding floodplain bank (Figure 2.7). They hid their find in nearby bushes and later, after the races ended, contacted police. After inspecting the find and cordoning off the discovery site, the police turned the cranium over to the Benton County Coroner who delivered the remains to Chatters for forensic anthropological assessment. The remains consisted of the neurocranium and maxillae in two fragments. Zygomatic, nasal, and malar bones were missing, as were the zygomatic processes of both temporals. All teeth but the third molars were present and well-cemented into the maxillae.

The evening of July 28, Chatters accompanied the coroner and Benton County Search and Rescue personnel to the find site, where the team worked by the light of powerful lanterns. The cutbank was first inspected for exposed bones and stratigraphic evidence of a grave, neither of which was evident. Near-shore reservoir mud, which was still inundated, was inspected to locate additional skeletal elements. Large fragments of the os coxae, femora, tibiae, and other elements were found in the mud several meters upstream of the tuft of grass beside which the skull had been found. Visible elements appeared scattered within 10 meters upstream of the grass tuft. To protect the remains from curious visitors, the coroner gathered up the larger fragments.

July 29, 1996

With the coroner's approval, Chatters and an intern revisited the discovery site in the afternoon. In strong daylight, the cut bank was again inspected for signs of in situ bones or evidence of a burial pit, but none was found. The scatter of early historic artifacts, including a pewter-handled knife, a few pieces of lithic debitage, and fire-cracked rock was observed on the reservoir shore. Additional skeletal elements and fragments were visible in the exposed mud below the previous day's high water mark, again within 10 meters upstream of the grass tuft. These included a fragment of mandible immediately adjacent to the grass tuft and a distal femur (see photograph in Chatters 2001) three to four meters upstream from the tuft, near the location where the os coxae had been found. Thorough inspection of wave-sorted calcite concretions and other debris that marked the previous day's reservoir

TABLE 2.3 Chronology of skeletal element collection in summer 1996.

Date	Elements collected as documented in notes of J. C. Chatters[1]
July 28	right and left malars, body of right mandible, right and left os coxae, sacrum fragment, proximal and distal left femur, proximal right femur, right humerus in two fragments, proximal left[2] tibia, atlas, axis, 2 thoracic vertebrae, proximal and distal right ulna, proximal left ulna, distal right radius, 6 rib fragments
July 29	ramus of right mandible, distal left radius, midshaft left ulna, scapula fragments, midshaft and distal right femur, midshaft left tibia, proximal right tibia, first and second metacarpals, other hand and foot bones, several unspecified vertebrae, several rib fragments
July 31	body of left mandible, 2 third molars, 3 scapula fragments, clavicle fragment, proximal left humerus, proximal right radius, left patella, 5 fragments of right and left fibulae, distal left tibia, 9 vertebrae and fragments, including cervicals 3 through 7, 17 rib fragments
August 3	3 rib heads, 1 hand bone
August 5	glenoid and acromion process of left scapula, midshaft left femur, right patella, distal right tibia, 5 fibula fragments, left calcaneus, right first metatarsal, 2 other metapodials, 8 phalanges, complete lumbar vertebra, 9 additional fragments of thoracic and lumbar vertebrae, 13 rib fragments.
August 11	rib fragments, small bones of hands and feet
August 28	3 small fragments of rib and cortical bone

[1]Does not include some fragments that were not identified until later analysis.
[2]Notes list this as right tibia, but July 29 notes indicate it was probably the left.

level yielded numerous hand and foot bones, along with fragments of ribs and other postcranial elements.

July 31, 1996

Having secured an emergency Archaeological Resources Protection Act permit from the Corps, Walla Walla District, and accompanied by Dr. Kenneth Reid, Chatters began systematically recovering bones that had been secondarily deposited in the reservoir mud. Work began near the grass tuft, where mud initially was washed through a 1/8-inch screen. This led to recovery of another mandible fragment, two third molars, and all remaining cervical vertebrae, confirming this as the resting place of the skull. Concerned that the use of a shovel risked breaking delicate bones, workers began to hydraulically excavate the soft mud, by pouring buckets of water over it to wash out the bones. Applied within 3 to 4 meters of the grass tuft, this procedure led to recovery of many additional bones, primarily from the axial skeleton and pectoral girdle. Another search of the wave-concentrated debris revealed more rib fragments and small bones.

August 5, 1996

After again searching wave-concentrated debris on August 3, and assisted by Thomas McClelland, Chatters again used the hydraulic method on the bottom mud west (upriver) of the area searched on July 31. The more than 40 fragments recovered in this manner came primarily from the lower axial skeleton and pelvic girdle. Two scapula fragments were the only pieces from the pectoral girdle found on this day. Another search of wave concentrated material produced more rib fragments.

August 11 and 28, 1996

Searches of wave-concentrated debris led to recovery of additional hand, foot, and rib bones.

December 14, 1997

One rib fragment was found on the beach by Chatters during geologic studies of the discovery site by the Corps (Huckleberry et al. 1998).

Discovery Synopsis

The skeleton was recovered from a secondary deposit of reservoir mud using a combination of surface collection, screening, and hydraulic excavation. Remains were scattered along a 30-meter length of beach, with larger elements concentrated in a 10-meter-long area. Notes taken during the recovery indicate that the skull and cervical vertebrae lay at the east end of the scatter, with remains from the pelvic girdle toward the west end, and the pectoral girdle and rest of the axial skeleton lying between. Small bones, particularly those of the hands and feet, along with rib fragments were concentrated in windrows along the upper beach, extending west (upstream). No contemporaneous archaeological materials were found in certain association with the bones.

REFERENCES

Aikens, C. Melvin, David L. Cole, and Robert Stuckenrath. 1977. Excavations at Dirty Shame Rockshelter, southeastern Oregon. *Tebiwa: Miscellaneous Papers of the Idaho State University Museum of Natural History* 4:1–29.

Ames, Kenneth M. 1988. Early Holocene forager mobility strategies on the southern Columbia Plateau. In *Early Human Occupation in Far Western North America: The Clovis-Archaic Interface*, edited by Judith A. Willig, C. Melvin Aikens, and John L. Fagan, 325–360. Nevada State Museum Anthropological Paper 21. Nevada State Museum, Carson City.

Ames, Kenneth M. 2000. Chapter 2. Review of the Archaeological Data. In *Cultural Affiliation Report* [September 2000]. National Park Service, http://www.nps.gov/archeology/kennewick/ames.htm.

Ames, Kenneth M., Don E. Dumond, Jerry R. Galm, and Rick Minor. 1998. Prehistory of the Southern Plateau. In *Handbook of North American Indians*, vol. 12: *Plateau*, edited by Deward E. Walker, 103–119. Smithsonian Institution Press, Washington, DC.

Ames, Kenneth M., James P. Green, and Margaret Pfoertner. 1981. *Hatwai (10NP143): Interim Report*. Boise State University Archaeological Report 9. Boise State University, Boise.

Andrefsky, William, Jr., Laurie L. Hale, and David A. Harder, eds. 1996. *1995 WSU Archaeological Block Survey of the Hanford 600 Area*. Project Report 29. Washington State University Center for Northwest Anthropology, Pullman.

Atwater, Brian S. 1986. Pleistocene glacial-lake deposits of the Sanpoil River Valley, northeastern Washington. *United States Geological Survey Bulletin* 1661.

Bacon, Charles R. 1983. Eruptive history of Mount Mazama and Crater Lake Caldera, Cascade Range, U.S.A. *Journal of Volcanology and Geothermal Research* 18(1–4):57–115.

Baker, Victor R., Bruce N. Bjornstad, Alan J. Busacca, Karl R. Fecht, Eugene P. Kiver, Ula L. Moody, J. G. Rigsby, Dale F. Stradling, and A. M. Tallman. 1991. Quaternary geology of the Columbia Plateau. In *Quaternary Nonglacial Geology: Coterminus U.S.*, edited by Roger B. Morrison, 215–246. Geological Society of America, Boulder, CO.

Barnosky, Cathy W. 1985. Late Quaternary vegetation in the southeastern Columbia Basin, Washington. *Quaternary Research* 23(1):109–122.

Barnosky, Cathy W., Patricia M. Anderson, and Patrick J. Bartlein. 1987. The northwestern U.S. during deglaciation: Vegetational history and paleoclimatic implications. In *North American and Adjacent Oceans During the Last Glaciation. The Geology of North America*, vol. K-3, 289–222. Geological Society of America, Boulder, CO.

Barry, Roger G., and Richard J. Chorley. 1968. *Atmosphere, Weather and Climate*. Methuen, London.

Beck, Charlotte, and George T. Jones. 1997. The terminal Pleistocene/early Holocene archaeology of the Great Basin. *Journal of World Prehistory* 11(2):161–236.

———. 2010. Clovis and Western Stemmed: Population migration and meeting of two technologies in the intermountain west. *American Antiquity* 75(1):81–116.

Bedwell, Steven F. 1973. *Fort Rock Basin: Prehistory and Environment*. University of Oregon Books, Eugene.

Bense, Judith A. 1972. "The Cascade Phase: A Study in the Effects of the Altithermal on a Cultural System." Ph.D. dissertation, Washington State University, Pullman.

Bilby, Robert E., Brian R. Fransen, and Peter A. Bisson. 1996. Incorporation of nitrogen and carbon from spawning coho salmon (*Oncorhynchus kisutch*) into the trophic system of small streams: Evidence from stable isotopes. *Canadian Journal of Fisheries and Aquatic Sciences* 53(1):164–173.

Birkeland, Peter W. 1984. *Soils and Geomorphology*. Oxford Press, New York.

Borden, Charles E. 1960. DjRi3, an early site in the Fraser Canyon, British Columbia. *National Museum of Canada Bulletin* 162:101–118.

Brauner, David R. 1985. "Early Human Occupation in the Uplands of the Southern Plateau: Archaeological Excavations at the Pilcher Creek Site, Union County, Oregon." Report on file, Department of Anthropology, Oregon State University, Corvallis.

Bryan, Alan L. 1980. The stemmed point tradition: An early technological tradition in western North America. In *Anthropological Papers in Memory of Earl H. Swanson, Jr.* edited by Lucille B. Harten, Claude N. Warren, and Donald R. Tuohy, 77–107. Special Publication of the Idaho State University Museum, Pocatello.

Butler, Virginia L. 2004. Fish remains. In *Marmes Rockshelter: A Final Report on 10,000 Years of Cultural Use*, edited by Brent A. Hicks, 320–337. Washington State University Press, Pullman.

Butler, Virginia L., and James C. Chatters. 2003. The Holocene history of salmon in the Columbia River Basin. *Geological Society of America Abstracts with Program* 35(6):607.

Butler, Virginia L., and Jim E. O'Connor. 2004. 9,000 years of salmon fishing on the Columbia River, North America. *Quaternary Research* 62(1):1–8.

Butler, B. Robert. 1961. *The Old Cordilleran Culture in the Pacific Northwest*. Idaho State College Museum Occasional Paper 5. Idaho State College Museum, Pocatello.

———. 1962. *Contributions to the Prehistory of the Columbia Plateau: A Report on Excavations in the Palouse and Craig Mountain Sections*. Idaho State College Museum Occasional Paper 9. Idaho State College Museum, Pocatello.

———. 1965. A report on investigations at an early man site near Lake Channel, southern Idaho. *Tebiwa* 9(2):1–20.

Chatters, James C., ed. 1989. "Hanford Cultural Resources Management Plan." Prepared for U.S. Department of Energy by Pacific Northwest Laboratory, Richland, WA.

Chatters, James C. 1987. Hunter-gatherer adaptations and assemblage structure. *Journal of Anthropological Archaeology* 6(4):336–375.

———. 1996. "Paleoecology and Geochronology of Hungry Horse Reservoir, Flathead National Forest, Montana: A First Approximation." Research Report P7, Applied Paleoscience, Richland, WA.

———. 1998a. Environment. In *Handbook of North American Indians*, vol. 12: *Plateau*, edited by Deward E. Walker, 29–48. Smithsonian Institution Press, Washington DC.

———. 1998b. "The Central Cascades Paleoecological Project: Palynological Investigations at Four Sites in the Gifford Pinchot and Mount Hood National Forests, Washington and Oregon." Research Report P9, Applied Paleoscience, Richland, WA.

———. 2000. The recovery and initial analysis of an early Holocene human skeleton from Kennewick, Washington. *American Antiquity* 65(2):291–316.

———. 2004. Safety in numbers: The influence of the bow and arrow on village formation on the Columbia Plateau. In *Complex Hunter-Gatherers: Evolution and Organization of Prehistoric Communities on the Plateau of Northwestern North America*, edited by William C. Prentiss and Ian Kiujt, 67–83. University of Utah Press, Salt Lake City.

———. 2010a. Peopling the Americas via multiple migrations from Beringia: Evidence from the early Holocene of the Columbia Plateau. In *Human Variation in the Americas: The Integration of Archaeology and Biological Anthropology*, edited by Benjamin M. Auerbach, 51–76. Southern Illinois University Press, Carbondale.

———. 2010b. Patterns of death and the peopling of the Americas. In *Symposio International III: El Hombre Temprano en America*, edited by José C. Jiménez López, Carlos Serrano Sánchez, Arturo González González, and Felisa J. Aguilar Arellano, 53–74. Instituto National de Antropología y Historia, Mexico City.

Chatters, James C., Virginia L. Butler, Michael J. Scott, David M. Anderson, and Duane A. Neitzel. 1995. A paleoscience approach to estimating the effects of climatic warming on salmonid fisheries of the Columbia River basin. In *Climate Change and Northern Fish Populations*, edited by Richard J. Beamish, 468–473. Canadian Special Publication of Fisheries and Aquatic Sciences 121. National Research Council of Canada, Ottawa.

Chatters, James C., Jason B. Cooper, Philippe D. LeTourneau, and Laura C. Rooke. 2011. "Understanding Olcott: Archaeological Data Recovery at Sites 45SN28 and 45SN303, Granite Falls, Snohomish County, Washington." Report for Snohomish County Department of Public Works by AMEC Earth & Environmental, Inc. Bothell, WA.

Chatters, J. C., H. A. Gard, and P. E. Minthorn. 1991. "Hanford Cultural Resources Laboratory Annual Report for Fiscal Year 1990." PNL-7853. NTIS.

Chatters, James C., Steven Hackenberger, Alan Busacca, Linda S. Cummings, Richard L. Jantz, Thomas W. Stafford, and R. Ervin Taylor. 2000. A possible second early Holocene skull from eastern Washington, USA. *Current Research in the Pleistocene* 17:93–95.

Chatters, James C., Steven Hackenberger, Anna M. Prentiss, and Jayne-Leigh Thomas. 2012. The Paleoindian to Archaic transition in the Pacific Northwest: In situ development or ethnic replacement? In *From the Pleistocene to the Holocene: Human Organization and Cultural Transformations in Prehistoric North America*, edited by C. Britt Bousman and Bradley J. Vierra, 37–66. Texas A&M Press, College Station.

Chatters, James C., and Karin A. Hoover. 1992. Response of the Columbia River fluvial system to Holocene climatic change. *Quaternary Research* 37(1):42–59.

Chatters, James C., Duane A. Neitzel, Michael J. Scott, and Steven A. Shankle. 1991. Potential effects of global climate change on Pacific northwest spring Chinook salmon

(*Oncorhynchus tschawytscha*): An exploratory case study. *Northwest Environmental Journal* 7(1):71–92.

Chatters, James C., and David Pokotylo. 1998. An introduction to Plateau prehistory. In *Handbook of North American Indians,* vol. 12: *Plateau,* edited by Deward E. Walker, 73–80. Smithsonian Institution Press, Washington DC.

Chatters, James C., and William C. Prentiss. 2005. A Darwinian macroevolutionary perspective on the development of hunter-gatherer systems in northwestern North America. *Journal of World Archaeology* 37(1):46–65.

Chatters, James C., and Anna M. Prentiss. 2010. Technological and functional analysis of lithics from 2008 excavations at the Beech Creek Site. In *Archaeological Data Recovery at the Beech Creek Site (45LE415), Gifford Pinchot National Forest, Washington,* by C. A. Mack, J. C. Chatters, and A. M. Prentiss, 49–123. Heritage Program, Gifford Pinchot National Forest, Pacific Northwest Region, USDA Forest Service, Trout Lake, WA.

Cleveland, Gregory C., Bruce D. Cochran, Julie Giniger, and Hallat H. Hammatt. 1976. *Archaeological Reconnaissance on the Mid Columbia and Lower Snake River Reservoirs for the Walla Walla District Army Corps of Engineers.* Project Report 27, Washington Archaeological Research Center, Washington State University, Pullman.

COE. 1950. Columbia River below Yakima River. In *Review Report on Columbia River and Tributaries, Northwestern United States, Letter from the Secretary of the Army,* Appendix L. House Document No. 531, 81st Congress, 2nd Session, Washington, DC.

———. 1952. *McNary Reservoir Topographic Map.* U.S. Army Corps of Engineers, Walla Walla District, Walla Walla, WA.

Craven, Sloan L. 2003. "Lithic Variation in Hafted Bifaces from the Lind Coulee Site (45GR97), Washington." Master's thesis, Washington State University, Pullman.

Cressman, L. S., David L. Cole, Wilbur A. Davis, Thomas M. Newman, and Daniel J. Scheans. 1960. Cultural sequences at The Dalles, Oregon: A contribution to Pacific Northwest prehistory. *Transactions of the American Philosophical Society* 50(10):1–108.

Cummings, Linda S. 1999. Draft report on opal phytoliths from the Kennewick Site, Washington. Letter submitted to J. C. Chatters by PaleoResearch Laboratories, Golden, CO.

Dansie, Amy. 1997. Early Holocene burials in Nevada: Overview of localities, research, and legal issues. *Nevada Historical Society Quarterly* 40(1):4–14.

Daubenmire, Rexford. 1970. Steppe vegetation of Washington. *Washington Agricultural Experiment Station Technical Bulletin* 62. Pullman, WA.

Daugherty, Richard D. 1956. Archaeology of the Lind Coulee site, Washington. *Proceedings of the American Philosophical Society* 100(3):233–278.

Daugherty, Richard D., J. Jeffrey Flenniken, and Jeanne M. Welch. 1987a. *A Data Recovery Study of Judd Peak Rockshelters (45-LE-222) in Lewis County, Washington.* Studies in Cultural Resource Management 8. USDA Forest Service, Pacific Northwest Region, Portland.

———. 1987b. *A Data Recovery Study of Layser Cave (45LE223) in Lewis County, Washington.* Studies in Cultural Resources Management 7. USDA Forest Service, Pacific Northwest Region, Portland.

Davis, Loren G., and David A. Sisson. 1998. An early stemmed point cache from the lower Salmon River Canyon of west central Idaho. *Current Research in the Pleistocene* 15:12–14.

Draper, John A. 1986. 45OK424 site report. In *The Wells Reservoir Archaeological Project*, vol. 2: *Site Report,* edited by J. C. Chatters. Central Washington Archaeological Survey, Archaeological Report 86-6. Central Washington University, Ellensburg.

Dumond, Don E., and Rick Minor. 1983. *Archaeology in the John Day Reservoir: The Wildcat Canyon Site.* Anthropological Paper 30. University of Oregon, Eugene.

Easterbrook, Don J. 1970. *Landforms of Washington.* Western Washington State College, Union Printing Company, Bellingham.

Ferguson, G. J., and W. F. Libby. 1962. UCLA radiocarbon dates. *Radiocarbon* 4:109–114.

Franklin, Jerry F., and C. T. Dyrness. 1988. *Natural Vegetation of Oregon and Washington.* Oregon State University Press, Corvallis.

Ford, Pamela J. 2004. Invertebrate fauna (shellfish). In *Marmes Rockshelter: A Final Report on 11,000 Years of Cultural Use,* edited by Brent A. Hicks, 339–345. Washington State University Press, Pullman.

Fryxell, Roald. 1963. "Late Glacial and Post Glacial Geological and Archaeological Chronology of the Columbia Plateau, Washington; An Interim Report to the National Science

Foundation, 1962–1963." Washington State University, Pullman.

Galm, Jerry R., and Stan Gough. 2008. The projectile point/knife sample from the Sentinel Gap site. In *Projectile Point Sequences in Northwestern North America,* edited by Roy L. Carlson and Martin Paul Robert Magne, 209–220. Archaeology Press, Simon Fraser University, Burnaby, BC.

Gaylord, David R. 2011. Tephrochronology of late Pleistocene and Holocene sand dune deposits, Hanford Reach National Monument, WA. *Geological Society of America Abstracts with Programs* 43(5):272.

Gehr, Keith D. 1980. "Late Pleistocene and Recent Archaeology and Geomorphology of the South Shore of Harney Lake, Oregon." Master's thesis, Portland State University, Portland.

Gough, Stan. 1995. "Description and Interpretation of Late Quaternary Sediments in the Rocky Reach of the Columbia River Valley, Douglas County, Washington." Master's thesis, Eastern Washington University, Cheney.

Gresh, Ted, Jim Lichatowich, and Peter Schoonmaker. 2000. An estimate of historic and current levels of salmon production in the northeast Pacific ecosystem: Evidence of nutrient deficiency in the freshwater systems of the Pacific Northwest. *Fisheries* 25(1):15–21.

Gustafson, Eric P. 1978. *The Vertebrate Faunas of the Pliocene Ringold Formation, South Central Washington.* Museum of Natural History Bulletin 23. University of Oregon, Eugene.

Hallet, Bernard. 1983. The breakdown of rock due to freezing: A theoretical model. In *Proceedings of the 4th International Conference on Permafrost, Fairbanks, Alaska, July 1983,* 433–438. National Academy Press, Washington, DC.

Hammatt, Hallet H. 1977. "Late Quaternary Stratigraphy and Archaeological Chronology in the Lower Granite Reservoir Area, Lower Snake River, Washington." Ph.D. dissertation, Washington State University, Pullman.

Hanes, Richard C. 1988. *Lithic Assemblages of Dirty Shame Rockshelter: Changing Traditions in the Northern Intermontane.* Anthropological Paper 40. University of Oregon, Eugene.

Hartmann, Glenn D. 1986. *Preliminary Test Excavations at Three Prehistoric Archaeological Sites in Franklin County, Washington.* Eastern Washington University Reports in Archaeology and History 100–156. Archaeological and Historical Services, Cheney, Washington.

Hicks, Brent A., ed. 2004. *Marmes Rockshelter: A Final Report on 11,000 Years of Cultural Use.* Washington State University Press, Pullman.

Hooper, Peter R. 1982. The Columbia River basalts. *Science* 215(4539):1463–1468.

Huckleberry, Gary, and Julie K. Stein. 1999. Chapter 3. Analysis of sediments associated with human remains found at Columbia Park, Kennewick, WA. In *Report on the Non-Destructive Examination, Description, and Analysis of the Human Remains from Columbia Park, Kennewick, Washington* [October 1999]. National Park Service, http://www.nps.gov/archeology/kennewick/huck_stein.htm.

Huckleberry, Gary, Thomas W. Stafford, Jr., and James C. Chatters. 1998. "Preliminary Geoarchaeological Studies at Columbia Park, Kennewick, Washington, U.S.A." Report submitted to U.S. Army Corps of Engineers, Walla Walla District, March 23, 1998.

Irwin, Ann M., and Ula Moody. 1978. *The Lind Coulee Site (45GR97).* Washington Archaeological Research Center Project Report 56. Washington State University, Pullman, WA.

Krantz, Grover S. 1979. Oldest human remains from the Marmes Site. *Northwest Anthropological Research Notes* 13(2):159–173.

Kutzbach, John E., and Peter J. Geutter. 1986. The influence of changing orbital parameters and surface boundary conditions on climate simulations for the past 18,000 years. *Journal of Atmospheric Sciences* 43(16):1726–1759.

Kutzbach, J., R. Gallimore, S. Harrison, P. Behling, R Selin, and F. Laarif. 1998. Climate and biome simulations for the past 21,000 years. *Quaternary Science Reviews* 17(6-7):473–506.

Leonhardy, Frank C., and David G. Rice. 1970. A proposed culture typology for the lower Snake River, Southeastern Washington. *Northwest Anthropological Research Notes* 4(1):161–168.

Lewarch, Dennis E., and James R. Benson. 1991. Long-term land use patterns in the southern Washington Cascade Range. *Archaeology in Washington* III:27–40.

Lohse, E. S., and C. Schou. 2008. The Southern Columbia Plateau projectile point sequence: An informatics-based approach. In *Projectile Point Sequences in Northwestern North America,* edited by Roy L. Carlson and Martin P. R. Magne, 187–208. Archaeology Press, Simon Fraser University, Burnaby, BC.

Luckman, B. H., and M. S. Kearney. 1986. Reconstruction of

Holocene changes in alpine vegetation and climate in the Maligne Range, Jasper National Park, Alberta. *Quaternary Research* 26(2):244–261.

Lyman, R. Lee. 1992. Influences of Mid-Holocene altithermal climates on mammalian faunas and human subsistence in eastern Washington. *Journal of Ethnobiology* 12(1):37–62.

———. 2004a. Prehistoric biogeography, abundance, and phenotypic plasticity of elk (*Cervus elaphus*) in Washington State. In *Zooarchaeology and Conservation Biology*, edited by R. L. Lyman and K. P. Cannon, 136–163. University of Utah Press, Salt Lake City.

———. 2004b. Late-Quaternary diminution and abundance of prehistoric bison (*Bison* sp.) in eastern Washington state, USA. *Quaternary Research* 62(1):76–85.

———. 2007. The Holocene history of pronghorn (*Antilocapra americana*) in eastern Washington State. *Northwest Science* 81(2):104–111.

———. 2009. The Holocene history of bighorn sheep (*Ovis canadensis*) in eastern Washington State, northwestern USA. *The Holocene* 19(1):143–150.

Lyman, R. Lee, and Stephanie D. Livingston. 1983. Late Quaternary mammalian zoogeography of eastern Washington. *Quaternary Research* 20(3):360–376.

Mack, Cheryl, James C. Chatters, and Anna M. Prentiss. 2010. *Archaeological Data Recovery at the Beech Creek site (45LE415), Gifford Pinchot National Forest, Washington*. Pacific Northwest Region, US Forest Service, Trout Lake, WA.

McDonald, Eric V., and Alan J. Busacca. 1992. Late Quaternary stratigraphy of loess in the Channeled Scabland and Palouse regions of Washington state. *Quaternary Research* 38(2):141–156.

Mehringer, Peter J., Jr. 1985a. Late Quaternary pollen records from the interior Pacific Northwest and northern Great Basin of the United States. In *Pollen Records of Late-Quaternary North American Sediments*, edited by Vaughn M. Bryant, Jr. and Richard George Holloway, 167–190. American Association of Stratigraphic Palynologists, Dallas.

———. 1985b. "Late Quaternary Vegetation and Climates of South-Central Washington." Report to Battelle/Pacific Northwest Laboratories, Richland, WA.

Miller, Carey L. 2000. "Edison Street Area Cultural Resources Testing in Columbia Park." Report prepared for the City of Kennewick by the Confederated Tribes of the Umatilla Indian Reservation Cultural Resources Protection Program, Mission, OR.

Minor, Rick, and Katherine A. Toepel. 1986. *Archaeological Assessment of the Bob's Point Site, (45KL219), Kickitat County, Washington*. Heritage Research Associates Report 42. Eugene, OR.

Miss, Christian J., and Loralea Hudson. 1987. *Cultural Resources Collection Analysis: Albeni Falls Project, Northern Idaho*. U.S. Army Corps of Engineers, Seattle, WA.

Mullineaux, Donald R. 1986. Summary of pre-1980 tephra-fall deposits erupted from Mount St. Helens, Washington State, U.S.A. *Bulletin of Volcanology* 48(1):17–26.

Neitzel, Duane A., Michael J. Scott, Steven A. Shankle, and James C. Chatters. 1991. The effect of climate change on stream environments: The salmonid resource of the Columbia River Basin. *Northwest Environmental Journal* 7(2):271–294.

Nelson, Jack L., and W. L. Hauschild. 1970. Accumulation of radionuclides in bed sediments of the Columbia River between the Hanford Reactors and McNary Dam. *Water Resources Research* 6(1):130–137.

Newman, Thomas M. 1966. *Cascadia Cave*. Idaho State University Museum Occasional Paper 18. Idaho State University Museum, Pocatello.

Pavesic, Max G. 1985. Cache blades and turkey tails: Piecing together the Western Idaho Archaic Burial Complex. In *Stone Tool Analysis: Essays in Honor of Don E. Crabtree*, edited by M. G. Plew, J. C. Woods, and M. G. Pavesic, 55–89. University of New Mexico, Albuquerque.

Pennefather-O'Brien, Elizabeth E., and Michael Strezewski. 2002. An initial description of the archaeology and morphology of the Clark's Fork skeletal material, Bonner County, Idaho. *North American Archaeologist* 23(2):101–115.

Prentiss, Anna M., and David S. Clarke. 2008. Lithic technological organization in an evolutionary framework: Examples from North America's Pacific Northwest region. In *Lithic Technology: Measures of Production, Use, and Curation*, edited by William Andrefsky, 257–285. Cambridge University Press, Cambridge.

Reid, Kenneth C., and James C. Chatters. 1997. "Kirkwood Bar: Passports in Time Excavations at 10IH699 in the Hells Canyon National Recreation Area, Wallowa-Whitman National Forest." Rainshadow Research Project Report 28 and Applied Paleoscience Project Report F-6, Pullman, WA.

Rice, David G. 1969. "Preliminary report, Marmes Rockshelter Archaeological Site." Washington State University, Pullman.

———. 1972. *The Windust Phase in Lower Snake River Region Prehistory*. Report of Investigations 50. Washington State University Laboratory of Anthropology, Pullman.

Rice, Harvey S. 1965. "The Cultural Sequence at Windust Caves." Master's thesis, Washington State University, Pullman.

Ridenour, D. I. 2006. "Results of Investigations at We'petes Pa'axat (10NP453), Clearwater River Region, North Central, Idaho." Master's thesis, University of Idaho, Moscow.

Rudolph, Theresa. 1995. *The Hetrick Site: 11,000 Years of Prehistory in the Wieser Valley*. Science Applications International Corp., Boise.

Salo, Lawr V. 1987. *Similkameen River Multipurpose Project Feasibility Study Cultural Resources Reconnaissance Technical Report*. U.S. Army Corps of Engineers, Seattle, WA.

Sancetta, Constance, Michell Lyle, Linda Heusser, Rainer Zahn, and J. Platt Bradbury. 1992. Late-glacial to Holocene changes in winds, upwelling, and seasonal production of the northern California current system. *Quaternary Research* 38(3):359–370.

Sappington, R. Lee, and Sarah Schukencht-McDaniel. 2001. Wewukiyepuh (10-NP-336): Contributions of an early Holocene Windust Phase site to lower Snake River prehistory. *North American Archaeologist* 22(4):353–370.

Sappington, R. Lee. 1998. "Analysis of Cultural Materials Collected in the Vicinity of 'Kennewick Man' at Columbia Park, Kennewick, Washington." Report to Walla Walla District, U.S. Army Corps of Engineers, Walla Walla, WA.

Sargeant, Kathryn E. 1973. "The Haskett Tradition: A View from Redfish Overhang." Master's thesis, Idaho State University, Pocatello.

Schalk, Randall F., and Debra L. Olson. 1983. The faunal assemblages. In *The 1978 and 1979 Excavations at Strawberry Island in the McNary Reservoir*, edited by Randall F. Schalk, 75–107. Laboratory of Archaeology and History, Project Report 1. Washington State University, Pullman.

Sheppard, John C., Peter Wigand, Carl E. Gustafson, and Meyer Rubin. 1987. A reevaluation of the Marmes Rockshelter radiocarbon chronology. *American Antiquity* 52(1):118–125.

Silar, Jan. 1969. *Groundwater Structures and Ages in the Eastern Columbia Basin, Washington*. College of Engineering Bulletin 315. Washington State University, Pullman.

Sprague, Roderick, and John D. Combes. 1966. *Excavations in the Little Goose and Lower Granite Dam Reservoirs, 1965*. Report of Investigations 37. Washington State University, Laboratory of Anthropology, Pullman.

St. John, Harold. 1963. *Flora of Southeastern Washington*. Outdoor Pictures, Escondido.

Swanson, Earl H., Jr., and Paul G. Sneed. 1966. *Birch Creek Papers No. 3: The Archaeology of the Shoup Rock Shelters in East Central Idaho*. Idaho State University Museum Occasional Paper 17. Idaho State University Museum, Pocatello.

Taylor, R. E., Donna L. Kirner, John R. Southon, and James C. Chatters. 1998. Radiocarbon dates of Kennewick Man. *Science* 280(5367):1171–1172.

Thoms, Alston V. 1989. "Northern Roots of Hunter-Gatherer Intensification: Camas and the Pacific Northwest." Ph.D. dissertation, Washington State University. University Microfilms, Ann Arbor, MI.

Torrence, Robin. 1983. Time budgeting and hunter-gatherer technology. In *Hunter-Gatherer Economy in Prehistory: A European Perspective*, edited by Geoff Bailey, 11–22. Cambridge University Press, Cambridge.

U.S. Climate Data. 2011. Climate-Kennewick-Washington. http://www.usclimatedata.com/climate.php?location=USWA0205.

U.S. Fish & Wildlife Service. 2011. Mammals of the Hanford Reach National Monument. Hanford Reach National Monument. http://www.fws.gov/hanfordreach/mammals.html.

Valley, Derek R. 1975. "Excavations at the Rock Island Overlook Site." Washington State University Laboratory of Anthropology, Pullman.

van Ryswyck, A. L., and R. Okazaki. 1979. Genesis and classification of modal subalpine and alpine soil pedons of south-central British Columbia, Canada. *Arctic and Alpine Research* 11(1):53–67.

Waitt, Richard B., Jr., and Robert M. Thorson. 1983. The Cordilleran ice sheet in Washington, Idaho, and Montana. In *Late-Quaternary Environments of the United States*, vol. 1: *The Late Pleistocene*, edited by Stephen C. Porter, 53–70. University of Minnesota Press, Minneapolis.

Wakeley, Lillian D., William L. Murphy, Joseph B. Dunbar, Andrew G. Warne, Frederick L. Briuer, and Paul R. Nickens. 1998. *Geologic, Geoarchaeologic, and Historical Investigation of the Discovery Site of Ancient Remains in*

Columbia Park, Kennewick, Washington. U.S. Army Corps of Engineers Technical Report GL-98-13. Prepared for U.S. Army Engineer District, Walla Walla. U.S. Army Corps of Engineers Waterways Experiment Station, Vicksburg, MS.

Walder, Joseph, and Bernard Hallet. 1985. A theoretical model of the fracture of rock during freezing. *Geological Society of America Bulletin* 96(3):336–346.

Wallthal, John A. 1999. Mortuary behavior and early Holocene land use in the North American midcontinent. *North American Archaeologist* 20(1):1–30.

Warren, Claude N., Alan L. Bryan, and Donald R. Tuohy. 1963. The Goldendale site and its place in Plateau prehistory. *Tebiwa* 32(2):68–185.

Weisz, Gary J. 2006. Bola stones and line weights from the Hornby Creek site, Northern Idaho. *Idaho Archaeologist* 29(2):19–24.

Willig, Judith A., and C. Melvin Aikens. 1988. The Clovis-Archaic interface in far western North America. In *Early Occupation in Far Western North America: The Clovis-Archaic Interface*, edited by Judith A. Willig, C. Melvin Aikens, and John L. Fagan, 1–40. Nevada State Museum Anthropological Papers, Carson City.

Wydosky, Richard S., and Richard R. Whitney. 1979. *Inland Fishes of Washington*. University of Washington Press, Seattle.

Zdanowicz, C. M., G. A. Xielinski, and M. S. Germani. 1999. Mount Mazama eruption: Calendrical age verified and atmospheric impact assessed. *Geology* 27(7):621–624.

Chronology of the Kennewick Man Skeleton

Thomas W. Stafford Jr.

The geologic age of Kennewick Man is the foundation for his importance in American archaeology. Thought to have been a historic period burial when first discovered, a stone projectile point in Kennewick Man's pelvis was the initial indication that the skeleton could date before European contact. Subsequently, James Chatters submitted a left fifth metacarpal for radiocarbon dating and received a result of 8410 ± 60 RC yr. Because the skull had unique features, and also because of the unexpected type of hip injury, the skeleton's geologic age needed to be verified.

In December 1997, three teams examined the stratigraphy of the discovery site to determine if the geology supported the skeleton's 8410 RC yr age (Wakeley et al. 1998). These field examinations included U. S. Army Corps of Engineers (Corps) Waterways Experiment Station (WES) geologists, scientists on behalf of the plaintiffs, and representatives from the Tribes. The work yielded the following reports and publications: Wakeley et al. 1998, Huckleberry et al. 1998, Huckleberry and Stein 1999, Chatters 2000, and Huckleberry et al. 2003. This fieldwork and additional radiocarbon dating confirmed the overall 8410 RC yr age estimate for the skeleton and that the skeleton's enclosing sediments were early Holocene alluvium.

A fundamental issue in Bonnichsen et al. v. United States et al. was whether or not the skeleton predated European contact in North America. If pre-Columbian in age, it would be considered "Native American" per the Native American Graves Protection and Repatriation Act (NAGPRA). To further investigate the skeleton's age, the National Park Service (NPS) enlisted three radiocarbon laboratories in 1998 and 1999 to measure additional ^{14}C dates directly from the skeleton. The results ranged from 5750 ± 100 to 8410 ± 40 RC yr (McManamon 2000). Because these ages were from the human bones and all ^{14}C dates significantly predated European contact in North America, the NPS determined that the chronological information demonstrated " . . . that the Kennewick skeletal remains are 'Native American as defined by NAGPRA . . . '" and that "all the dates obtained predate 6000 BP and are clearly pre-Columbian" (McManamon 2000).

During the plaintiffs' 2005 and 2006 studies at the Burke Museum, Seattle, Washington, the Corps granted permission for Stafford, one of the plaintiffs' scientists, to remove small fragments of bone from fifteen different samples that were remnants of bone previously tested and returned by radiocarbon and DNA labs. These remnants were used for the stable isotope, amino acid, and AMS ^{14}C analyses reported in this chapter and in Chapter 16.

1996–2000 RADIOCARBON STUDIES
First Radiocarbon Measurement on Kennewick Man

In 1996, Chatters submitted the left fifth metacarpal to the University of California-Riverside Radiocarbon Laboratory for ^{14}C dating (Chatters 2001:47). The age was reported as 8410 ± 60 (UCR-3476), with a δ^{13}C value of –14.9‰ (Taylor et al. 1998). The chemical fraction dated was "total amino acids," and the bone analyzed was characterized as having a collagenous composition, with 68.8 percent the protein content of a modern bone (Table 3.1).

Taylor et al. 1998 used this δ^{13}C value to estimate possible marine reservoir effects and posited that marine-derived foods, mostly salmon, comprised 70 percent of Kennewick Man's diet. They noted that end-member δ^{13}C values for 100 percent marine and 100 percent terrestrial foods were –12.8‰ and –19.6‰, respectively, and therefore Kennewick Man's –14.9‰ isotopic composition represented a 70 percent marine diet. Based on an estimated

TABLE 3.1 Previous (1996–1999) ^{14}C measurements on Kennewick Man bones.

AMS Lab Number	Skeletal Element and USACOE Catalog Number	Chemical Fraction Dated	Fraction Modern (Fm) ± 1SD	$\delta^{13}C$‰ (VPDB)	^{14}C Age, RC yr ± SD	Collagen Preservation Relative to Modern Bone (Modern = 100%)	Collagen Yield, wt. % (Modern = 20%)	Marine Reservoir-Corrected Age, RC yr ± SD	Marine Reservoir-Corrected age, CAL BP (2σ)	Comments
First ^{14}C Measurement (Taylor et al. 1998; 1999)										
UCR-3476/ CAMS-29578	Left 5th metacarpal; APS-CPS-01	Total amino acids	0.3510 ± 0.0024	–14.9[1] –15.4[2]	8410 ± 60	68.8%	~13.6[3]	7880 ± 160[1]	8340–9200[1]	"Collagenous, well preserved protein" ~70% marine reservoir correction
USACOE-Sponsored ^{14}C Measurements (McManamon 2000)										
BETA-133993	One-half of right 1st metatarsal CENWW.97.R.24(Mta) DOI 1a (9.1 g)	Alkali-washed collagen	—	–12.6	8410 ± 40	~1.5%	0.3[4]	ND	ND	"Plenty of carbon"[5] Collagen dated[6]
UCR-3807/ CAMS-60684	One-half of right 1st metatarsal CENWW.97.R.(Mta) DOI 1b	Total amino acids	0.3633 ± 0.0014	–10.8	8130 ± 40[7]	14.3%		ND	ND	"Non-collagenous"[8] "Apparent ^{14}C age"
AA-34818	Portion of left tibia, cnemial crest CENWW.97.L.20b DOI 2a	Gelatin	0.4889 ± 0.0066	–21.9	5750 ± 100	<1.6%[9]	~0.8%[10]	ND	ND	"Below acceptable protein yield; no confidence in result; minimum age"[11]
UCR-3806/ CAMS-60683	Portion of left tibia, cnemial crest CENWW.97.L.20b DOI 2b	Total amino acids	0.4216 ± 0.0015	–10.3	6940 ± 30[7]	2.3%		ND	ND	"Non-collagenous"[8] "Apparent ^{14}C age"

Abbreviations: AA – University of Arizona; NSF – Arizona AMS Facility, Tucson; Beta – Beta Analytic, Inc., Miami, Florida; CAMS – Center for Accelerator Mass Spectrometry, Lawrence Livermore National Laboratory, Livermore, California; UCR – The Radiocarbon Laboratory of the University of California, Riverside; ND – Not determined

[1]Taylor et al. 1998.
[2]Taylor 1999.
[3]Taylor (1999) reported UCR-3476 bone preservation was 68.8% of a modern bone's collagen yield, which is 20–21 wt.%.
[4]Based on 30 mg of collagen extracted from 9.1 g of bone; dated collagen was 33.7% carbon (3.2 mg carbon/9.5 mg collagen) (Hood 1999b).
[5]Hood 1999a.
[6]Hood 1999b.
[7]Reported by UCR as "Apparent ^{14}C age" (Taylor 1999).
[8]Reported by UCR as "non-collagen" (McManamon 2000; Taylor 1999).
[9]Based on Kennewick total-bone containing 0.07%N versus 4.5%N for modern bone.
[10]630 mg of bone yielded 21.8 mg of "gelatin" that contained 1.9%C (Donahue 2000a), rather than collagen's normal 41%C by weight. 630 mg of modern bone containing 20% collagen (126 mg collagen) that is 41%C would yield 52 mg of C. 630 mg of Kennewick bone yielded 0.42 mg of C, which is a 0.8% yield (0.42/52).
[11]Donahue 2000b.

marine ^{14}C offset from the Gulf of Alaska of 750 years (Stuiver et al. 1986), they calculated a marine reservoir offset of 530 years (0.70 marine diet × 750 years), which corrects the 8410 ± 60 conventional ^{14}C age to 7880 ± 160 RC years. Calendar correcting 7880 ± 160 gave a two sigma age range of 8340–9200 CAL yr BP (Taylor et al. 1998) (Table 3.1). Because no δ^{15}N values were measured on bone collagen at that time, a more accurate estimate of marine foods in the diet was not possible.

Taylor (2000; Taylor et al. 2001) analyzed additional bones in 1998 and 1999 (Table 3.2) to assess how collagen preservation varied among skeletal elements and if any bones were suitable for aDNA analyses. Among eleven bones, collagen preservation varied from 2.3 percent to 68.8 percent of modern, four had non-collagenous amino acid compositions, while seven were collagenous, and their δ^{13}C values spanned from –10.8 to –14.9‰ (Table 3.2).

The conclusions from these first analyses were as follows: the measured 8410 ± 60 RC yr age represented an age of 8340–9200 CAL BP, Kennewick Man had a diet based significantly (70 percent) on marine resources, and collagen preservation among skeletal elements varied widely, even within the same bone.

Waterways Experiment Station Radiocarbon Dating

During December 12–18, 1997, field work, Corps WES geologists examined the exposed geologic profile at the discovery site and collected samples for sedimentological analyses and radiocarbon dating using a Vibracore. Ten cores were taken at five locations along a 255 ft (77.7 m) NNE-SSW transect running along the base of the exposed geologic profile (Wakeley et al. 1998:F2).

Two shell samples, collected from each end of the discovery site's geologic profile, were dated 6090 ± 80 RC yr (Beta–113977) and 6510 ± 60 RC yr (Beta–113838), respectively (Table 3.3). The shells immediately overlay Mazama volcanic ash, or Unit III in WES geologic nomenclature. The volcanic tephra at the Kennewick Man discovery site was identified petrographically by Andrei Sarna-Wojcicki (Wakeley et al. 1998:Appendix H)

TABLE 3.2 UC-Riverside amino acid and stable isotope analyses on Kennewick Man bone[1].

Sample Number	Skeletal Element	Amino Acid Composition Relative to Modern Bone	δ^{13}C‰ (VPDB)	Comments
		1996 Analysis		
UCR-3476	Left 5th metacarpal	68.8%	–14.9	Collagenous
		1999 Analyses		
UCR-3807	Right 1st metacarpal	14.3%	–10.8	Non-collagenous
UCR-3806	Left tibia, cnemial crest	2.3%	–10.3	Non-collagenous
		2000 Analyses		
UCR-3882	Left tibia, cnemial crest	18.2%	–13.8	Collagenous
UCR-3875	Left 3rd metacarpal	46.2%	–12.9	Collagenous
UCR-3878	Right 3rd metacarpal, proximal end	10.1%	–13.6	Non-collagenous
UCR-3879	Right 3rd (distal end) metacarpal	32.6%	–13.3	Collagenous
UCR-3876	Right 8th rib, vertebral end	30.9%	–13.2	Collagenous
UCR-3877	Right 8th rib, sternal end	20.9%	–13.4	Collagenous
UCR-3881	Left 2nd metatarsal, distal end	24.1%	–13.2	Collagenous
UCR-3880	Right 2nd metatarsal	5.7%	–15.1	Non-collagenous
		Averaged collagenous δ^{13}C values	**–13.5‰**	
		Averaged non-collagenous δ^{13}C values	**–12.4‰**	

[1]Taylor 2000

TABLE 3.3 Corps (WES) and Chatters' ^{14}C dates on shell and sediment from the Kennewick Man discovery site.

Lab No.	Sample No.	Sample	Profile	Core	Depth (BGL)	Elevation (MSL)[1]	Stratigraphy	Stratum	Fraction Dated	Method	^{14}C Age, RC yr ± 1 SD	$\delta^{13}C$‰ (VPDB)	Reference
Corps (WES) Shell and Sediment Radiocarbon Measurements													
Beta-113977	CPP-200-60-65	Shell	Immediately east of CPP-200	—	60-65 cm	345.0 ft (105.16 m)	Shell midden immediately above Mazama Ash[2,3] (Stratum III)	Between Units II and IV	Leached shell carbonate	Decay Counting	6090 ± 80	-8.3	Wakeley et al. 1998:I2
Beta-113838	CPP-005-60-80	Shell	CPP-005	—	60-70 cm	342.7 ft (104.45 m)	Immediately above Mazama Ash[2,3] (Stratum III)	Between Units II and IV	Leached shell carbonate	AMS	6510 ± 60	-8.2	Wakeley et al. 1998:I2
WW-1626	CPC-059.5 10-20cm	Sediment (Dense silt-loam)	—	CPC-059.5	10-20 cm	342.3 ft (104.33 m)	Base of carbonate concretion zone	Base of Stratum IV	HCl-digested, <63 μm sediment	AMS	9010 ± 50[4]	-25[5]	Wakeley et al. 1998:I1
WW-1737	CPC-059.5 130-138cm	Sediment (Dense silt-loam)	—	CPC-059.5	130-138 cm	338.6 ft (103.21 m)	—	Stratum V	HCl-digested, <63 μm sediment	AMS	12,460 ± 50[4]	-25[5]	Wakeley et al. 1998:I3
WW-1627	CPC-059.5 190-200cm	Sediment (Dense silt-loam)	—	CPC-059.5	190-200 cm	336.8 ft (102.66 m)	—	Stratum V	HCl-digested, <63 μm sediment	AMS	15,330 ± 60[4]	-25[5]	Wakeley et al. 1998:I1
WW-1738	CPC-059.5 220-229cm	Sediment (Dense silt-loam)	—	CPC-059.5	220-229 cm	335.7 ft (102.32 m)	—	Stratum V	HCl-digested, <63 μm sediment	AMS	14,560 ± 50[4]	-25[5]	Wakeley et al. 1998:I3
Chatters's Sediment Radiocarbon Measurement													
CAMS-67692	97KN-17	Sediment	CPP-044	—	90-92 cm	—	—	Stratum IV	Humic Acids	AMS	6280 ± 50	—	Pers. Comm.

Abbreviations: CAMS – Center for Accelerator Mass Spectrometry; Lawrence Livermore National Laboratory, Livermore, California; Conventional Decay Counting (Scintillation); WW – ^{14}C Laboratory of the U.S. Geological Survey, Reston, Virginia

[1] Elevations were originally given as feet above mean sea level (MSL) in Wakeley et al. 1998:41.
[2] Mazama Ash identified petrographically by Andrei Sarna-Wojcicki (Wakeley et al. 1998: Appendix H1–H3).
[3] The Mazama Ash bed has been radiocarbon dated as 6730 ± 40 RC yr (Hallett et al. 1997).
[4] CAMS measurement.
[5] Assumed $\delta^{13}C$ value according to Stuiver and Polach 1977.

as Mazama ash. The Mazama tephra has been dated to 6845 ± 50 RC yr BP (7787–7591 CAL BP) (Bacon 1983; Bacon and Lanphere 2006) and alternatively to 6730 ± 40 (7668–7513 CAL BP) (Hallet et al. 1997). Based on its presence in the GISP2 ice core, the age of Mazama ash is 7627 ± 150 CAL BP (Zdanowicz et al. 1999). Consequently, the two shells yielded ^{14}C ages consistent with their stratigraphic position immediately above the Mazama tephra.

Four ^{14}C dates were measured on four stratigraphic positions in Vibracore CPC–059.5. The resulting ages on HCl-digested, <63 μm sediment from depths of 10–20 cm, 130–138 cm, 190–200 cm, and 220–229 cm below ground level (BGL) were 9010 ± 50 (WW-1626), 12,460 ± 50 (WW-1737), 15,330 ± 60 (WW-1627) and 14,560 ± 50 (WW-1738) RC yr, respectively (Table 3.3). The stratigraphically highest date, 10–20 cm below ground level (core top), was approximately the elevation of the geologic profile's base, at 104.33 m elevation. The 9010 ± 50 RC yr BP age is approximately that expected for sediments stratigraphically beneath the horizon containing the Kennewick Man skeleton.

Chatters (2000) identified the Columbia Park terrace containing the Kennewick Man skeleton as early Holocene alluvium that postdates the Glacier Peak ash fall, which has been dated to 11,600 ± 50 RC yr (Kuehn et al. 2008). Unless the three older dates from core CPC–059.5 are from an older terrace, e.g., Qfg3 or Qfg4 alluvium (Huckleberry et al. 2003), their assumed early Holocene alluvium assignment precludes the sediments dating older than the 11,600 RC yr Glacier Peak ash. Consequently, the 12,460 to 15,330 RC yr ages in core CPC–059.5 are too old by approximately 860 to 3700 RC years. Because bulk organic matter within the silts and clays were dated, the sediments could have contained ancient carbon reworked from older alluvium upstream in the Columbia River basin. While the uppermost horizon of core CPC–059.5 dated at the expected age for this elevation, the apparently erroneous three underlying ages make the 9010 ± 50 RC yr age suspect (Chatters 2000).

One final sediment sample was dated by Chatters at the site's geologic profile, CPP–044 (90–92 cm BGL). The age on humic acids was 6280 ± 50 (CAMS-67692) (Table 3.3). This sample was taken at the interpreted stratigraphic position of the Kennewick Man skeleton and is approximately 2130 RC yr younger than the skeleton's 8410 RC yr measurement.

Conclusions from dating shells and sediments at the discovery site are that shell radiocarbon dates yielded similar ages expected for fossils lying immediately above the 6730 RC yr-old Mazama Ash and that dates on bulk organic matter were significantly too old, while humic acid dates were too young.

National Park Service Sponsored AMS ^{14}C Dating of the Skeleton

The NPS contracted with three radiocarbon laboratories (University of Arizona, UC-Riverside, and Beta Analytic) to date four samples extracted from the skeleton by the Department of the Interior (DOI) and the Corps in September 1999. Ages were reported in McManamon (2000). Two different bones, the right first metatarsal and portions of the left tibia's cnemial crest, were each dated by two different laboratories. Dates on the metatarsal were 8140 ± 40 RC yr (Beta-133993) and 8130 ± 40 RC yr (UCR-3807). Dates on the tibia were 5750 ± 100 RC yr (AA-34818) and 6840 ± 30 RC yr (UCR-3806). Both the Arizona and UC-Riverside laboratories expressed doubts about the measurements and termed their results a "minimum age" and "apparent radiocarbon age," respectively (Table 3.1). Only one laboratory, Beta Analytic, reported bone having sufficient carbon for dating and that the dated material was collagen.

PLAINTIFFS' RADIOCARBON DATING

The purposes of the radiocarbon, isotope, and biogeochemical analyses reported in this chapter were to establish an accurate and precise radiocarbon age for the Kennewick Man skeleton, discover why previous ^{14}C dates on the skeleton varied by 2660 RC years, make ^{14}C measurements on both bone collagen and carbonates precipitated on and within the bones, study how collagen preservation differed among skeletal elements, use δ^{13}C and δ^{15}N values to estimate the marine reservoir correction needed to calculate the calendar age, determine if amino acid analyses and other measurements could assist with aDNA sampling strategies, and apply the data to improving dating programs for future human skeleton discoveries.

Sample Selection and Methods

In 2006, Stafford collected fifteen samples from Kennewick Man bones archived at the Burke Museum. The samples were remnants returned by radiocarbon and aDNA laboratories and were materials remaining from their respective analyses during NPS-sponsored 1998–1999 studies. No new bone samples from the Kennewick Man skeleton were used or requested; all bone fragments were those previously returned to the Burke Museum by other laboratories. The bone fragments collected by the plaintiffs as potential samples for chemical analyses are listed in Table 3.4.

A subset of the fifteen bone samples collected was used for radiocarbon and geochemical analyses (Tables 3.5, 3.6, 3.7); the remainder were examined and returned to the Burke Museum without further testing. Samples selected were 2 cm in length or smaller and had characteristic pedogenic carbonate, significant amounts of sediment infilling, or physical traits such as hard, waxy bone that indicated good collagen preservation (Figures 3.1–3.6). For samples analyzed in this report, all attempts were made to archive a portion of that material for future confirmatory testing.

Before testing, all samples were photographed with a Canon EOS 60 digital camera to document physical properties and taphonomic characteristics. Results of chemical pretreatments were photographed to record how the collagen extractions proceeded, how chemical processes removed different contaminants, and how collagen preservation varied among bones. Photomacrographs and laboratory data are archived at the Smithsonian Institution, National Museum of Natural History, Washington, DC, the Burke Museum, Seattle, WA, and the Corps St. Louis District, MO.

Pretreatment of Bone Collagen for AMS ^{14}C Dating and Stable Isotope Analysis

Preparation of bones for AMS ^{14}C dating follow Stafford et al. 1991 and Waters and Stafford 2007. Bone samples were broken into approximately 4–5 mm fragments and decalcified in 4°C 0.5N HCl over 3 to 5 days; after washing in deionized water, the decalcified collagen was extracted with 0.1% KOH at 4°C for 24–30 hours and washed to neutrality with DI water (Table 3.8). The KOH-extracted, decalcified collagen's percent pseudomorph was recorded, and the collagen freeze-dried to determine percent yield of collagen relative to modern bone. Approximately 20–50 mg of decalcified, KOH-extracted collagen were heated at 90°C in 0.02N HCl to dissolve (gelatinize) the collagen. Heating continued only until the collagen dissolved, usually 5 to 30 minutes. After filtering the gelatin solution through a 0.45 μm Millex Durapore filter, the solution was freeze-dried and a percent gelatinization and percent weight yield were determined. Approximately 5–10 mg of gelatin were hydrolyzed 22 hours at 110°C in distilled 6N HCl. The hydrolyzate, containing free amino acids, fulvic acids, and insoluble inorganic and organic detritus, was passed through a 2 cm long × 5 mm diameter bed of XAD-2 resin in a solid phase extraction (SPE) column attached to a 0.45 μm Millex filter. The XAD column contained 100–200 μm diameter research grade XAD-2 from Serva Biochemicals (Cat. No. 42825). The bulk resin was initially wetted with acetone and washed voluminously with DI water and finally multiple washes with distilled 1N HCl. Individual SPE columns were packed with the XAD-2 as a slurry of resin and HCl. Each column was equilibrated with 50 ml of distilled 6N HCl and the washings discarded. The collagen hydrolyzate as approximately 1 to 2 ml of HCl was pipetted onto the SPE XAD column and eluted into a glass tube. Following the initial sample aliquot, the column was washed with 5 ml of 6N HCl that was added to the original eluate. The XAD-purified collagen hydrolyzate was dried by passing ultra high purity N_2 gas over the HCl solution, resulting in a viscous syrup. The dry hydrolyzate was diluted with DI water and approximately 2 to 3 mg of amino acids were transferred to 6 mm OD × 4 mm ID × 20 mm long quartz tubes and dried under vacuum. Approximately 50 mg of purified CuO wire and 5–10 mg Ag were added to each quartz tube. Stock CuO wire (Fisher Scientific, Cat. No. C474-500) was first combusted in crucibles at 900°C and stored in Pyrex tubes that were subsequently combusted at 570°C immediately before each use. Aesar 99.9995%, 30-60 mesh silver powder (Cat. No. 11408) was used without additional purification. All glass pipettes, beakers, and tubes were combusted at 550°C for 30 minutes before use. After evacuation to <20 millitorr (mtorr) by vacuum pumping through a liquid nitrogen trap, the tubes were sealed with a H_2/O_2 torch. The quartz sample tubes were

TABLE 3.4 Kennewick Man bone samples collected during the plaintiffs' studies at the Burke Museum for potential ^{14}C and stable isotope analyses.

SR-	NPS Study (2000) ID	Skeletal Element	Corps Number	Field Number	Lab Number	Previous Analysis ID	Mass	^{14}C Collagen	^{14}C CO_3^{-2}	$\delta^{13}C$ $\delta^{15}N$ $\delta^{18}O$	AAA	Comments
7082	AZ AMS Samples (A3-c)	Left Tibia	97.L.20b	Microsample No. 1	2004 KW-51	AA-34818c	1.364 g		X			Three, 7–10 mm long × 5 mm wide × 2–3 mm thick, bone fragments. White to extr pale yell brn. Outer 20–50 μm l. brn to l. yell brn; 5% dark reddish brn.
7083	AZ AMS Samples (A3-c)	Left Tibia	97.L.20b	Microsample No. 2	2004 KW-52	AA-34818d	1.760 g			X	X	One, 17 mm long × 15 mm wide × 11 mm thick, cortical fragment.
7084	AZ AMS Samples (A3-c)	Left Tibia	97.L.20b	Microsample No. 3	2004 KW-53	AA-34818e	0.492 g	X	X	X	X	90% white powder with 9 cortical fragments 2 mm × 4 mm diameter.
7085	UC-Riverside AMS μSamples (A3-d)	Left 2nd Metacarpal	97.L.16(MCb)	Microsample No. 4	2004 KW-57	UCR-3881	0.379 g					7–8 mm diameter fragment, disaggregated. Articular surface remains; well preserved.
7086	UC-Riverside AMS μSamples (A3-d)	Left Tibia	97.L.20b	Microsample No. 5	2004 KW-58	UCR-3882	2.401 g	X	X	X	X	Nine, 10–25 mm long × 3–4 mm square, elongate cortical fragments; saw cuts length-wise. Extr pale brn (~ white). Exterior patina 20–100 μm, yell brn to l. yell brn.
7087	UC-Riverside AMS μSamples (A3-d)	Rib	97.I.12d(13)	Microsample No. 6	2004 KW-54	UCR-3876	0.175 g					2–4 mm long × 0.2–0.3 mm thick cortical fragments; white to extr pale brn. Exterior mottled white & light reddish brn.

(continued)

TABLE 3.4 (continued) Kennewick Man bone samples collected during the plaintiffs' studies at the Burke Museum for potential ^{14}C and stable isotope analyses.

SR-	NPS Study (2000) ID	Skeletal Element	Corps Number	Field Number	Lab Number	Previous Analysis ID	Mass	^{14}C Collagen	^{14}C CO_3^{-2}	$\delta^{13}C$ $\delta^{15}N$	$\delta^{18}O$	AAA	Comments
7088	UC-Riverside AMS μSamples (A3-d)	Rib	97.I.12d(13)	Microsample No. 7	2004 KW-55	UCR-3877	0.193 g						7 mm long × 4–5 mm diameter fragment filled 100% with grey sediment. Cortical bone 0.2–0.3 mm thick, ~white. Exterior surface mottled 70% reddish brn & white.
7089	UC-Riverside AMS μSamples (A3-d)	Right 3rd Metacarpal	97.R.16(Mca)	Microsample No. 8	2004 KW-56	UCR-3878	0.0902 g						Twelve, 2–4 mm long × 1–2 mm wide, cortical fragments; white; no cortex.
7090	UC-Riverside AMS Sample (A3-b)	Left Tibia	97.L.20b	Microsample No. 9	2004 KW-59	UCR-3806				X		X	
7091	Yale University aDNA μSamples (A3-e)	Left 3rd Metacarpal	97.L.16(Mca)	Microsample No. 10	2004 KW-61		1.926 g					X	Interior bone white; cortex mottled 10% white, 40% pale yell, 50% l. red brn.
7092	Yale University aDNA μSamples (A3-e)	Rib	97.I.12d(13)	Microsample No. 11	2004 KW-60		4.245 g	X	X	X		X	42 mm long × 11 mm wide × 9 mm thick; 3.3 g (bone+sediment). Interior white, slightly waxy; cortex 45% white, 40% l. yell brn, 15% dk red brn. Sediment vf-f quartz sand, $CaCO_3$ drusy calcite in 2–5 mm voids.
7093	UC-Davis aDNA μSamples (A3-g)	Left 5th Metacarpal	99.I.16b and 99.I.16c	Microsample No. 12	2004 KW-62		0.128 g						Three, 2–12 mm long × 0.5 mm thick, cortical fragments; white. Cortex dk red brn. Negligible mottling; bone reportedly inside cranium. 100 μm diameter rootlet groove.

(continued)

TABLE 3.4 (continued) Kennewick Man bone samples collected during the plaintiffs' studies at the Burke Museum for potential ^{14}C and stable isotope analyses.

SR-	NPS Study (2000) ID	Skeletal Element	Corps Number	Field Number	Lab Number	Previous Analysis ID	Mass	^{14}C Collagen	^{14}C CO_3^{-2}	$\delta^{13}C$ $\delta^{15}N$	$\delta^{18}O$	AAA	Comments
7094	University of Michigan aDNA μSamples (A3-f)	Left 3rd Metacarpal	97.L.16(Mca)	Microsample No. 13	2004 KW-63		0.277 g					X	Five, 4–20 mm long × 0.5–0.7 mm thick, fragments.
7095	University of Michigan aDNA μSamples (A3-f)	Left 3rd Metacarpal	97.L.16(Mca)	Microsample No. 14	2004 KW-64		0.727 g			X		X	Cancellous & cortical bone fragments. Waxy, pale yell inner surfaces.
7096	Beta Analytic AMS	Right 1st Metatarsal	97.R.24(Mta)	Microsample No. 15	2004 KW-65	Beta-133993	3.184 g		X	X	X		White, cancellous fragments; $CaCO_3$ cemented sand.
7270	Yale University aDNA μSamples (A3-e)	Rib	97.L12d(13)	Microsample No. 11	SR-7092 (2004 KW-60)			X					Interior $CaCO_3$ cemented sand in cancellous zones. Clear, colorless glue.
7324	—	Bone & caliche	NONE		CENWW97.L.20B								
7325	—	Bone & caliche	NONE		CENWW97.R.24 (Mta) (3807)		2.685 g		X				
7326	—	Bone & caliche	NONE		CENWW97.R.16 (Mca) (3879)		0.220 g		Too small				
7327	—	Bone & caliche	NONE		CENWW97.A.I.25C (3880)		0.098 g		Too small				
7328	—	Bone & caliche	NONE		CENWW97.L.20b								
8125	University of Michigan aDNA μSamples (A3-f)	Left 3rd Metacarpal	97.L.16(Mca)	Microsample No. 13	SR-7094 2006 KW-63							X	Used for 1st aDNA analysis at GeoGenetics, Denmark
8330	Yale University aDNA μSamples (A3-e)	Left 3rd Metacarpal	97.L.16(Mca)	Microsample No. 10	SR-7091 2004 KW-61		1.926 g					X	Used for 2nd aDNA analysis at GeoGenetics, Denmark. See SR-7091

combusted at 820°C for 2 hours and cooled from 820°C to 150°C at 60°C per hour.

Preparation of Carbonates for AMS ^{14}C Dating

$CaCO_3$, carbonate cemented sands, and bone for apatite carbonate extraction were processed using the following procedures. Authigenic, euhedral $CaCO_3$ crystals from medullary cavities (Figures 3.5B and 3.6) were dissected from the voids and weighed; they received no further chemical or physical treatment. Carbonate-cemented sands were similarly removed as exterior cementations of sand on bone cortex or as carbonate-cemented sands occluding medullary cavities; these sands also received no additional physical or chemical processing. Apatite carbonate was extracted by first crushing bone to <500 μm, extracting the granules with 0.5M acetic acid, and finally washing in DI water to neutrality. Twenty to forty mg of each of the three carbonate types were weighed into Vacutainers (Santos et al. 2004), evacuated to <20 mtorr, reacted with 85% phosphoric acid, and the resulting CO_2 retained and purified for graphitization.

Graphitization of CO_2 and AMS ^{14}C Measurements

Following purification of the combustion and phosphoric acid hydrolysis products to remove water and non-condensable gases, 0.5 to 1 mg of carbon as CO_2 was converted into graphite by the Fe-H_2 method (Santos et al. 2004). Contemporary ^{14}C standards used for normalization are Oxalic Acid-I. Known-age bones used for calibrations included VIRI-I whale bone (consensus age = 8331 ± 6 RC yr) (Scott et al. 2010) and Dent Mammoth bone (10,950 ± RC yr) (Waters and Stafford 2007). Respective chemistry and combustion backgrounds and blanks were determined by using >70 ka collagen isolated from fossil *Eschrichtius* (Gray Whale) bone (Stafford et al. 1987; Waters and Stafford 2007) and Sigma Chemical Company alanine (Sigma A-7627). Graphite targets were prepared and analyzed at the Keck Carbon Cycle AMS Facility, Earth System Science Department, University of California-Irvine.

Measurement of Amino Acid Compositions for Bone and Collagen

Quantitative amino acid analyses were made by Margaret Condron at the Biopolymer Laboratory, David Geffen School of Medicine at University of California–Los Angeles. Approximately 1 mg of sample was hydrolyzed by using 6N vapor-phase HCl. The hydrolyzate was derivitized and subsequently analyzed by reverse phase (RP)-HPLC using a highly fluorescent amino-reactive probe on a Waters Alliance HPLC system. Detection limits were approximately 1 picomole. Analyses are reported as residues per thousand (R/1000) for each amino acid and as total nanomoles (nmol) of amino acids per mg (Table 3.6).

RESULTS
AMS ^{14}C Measurements

Seventeen radiocarbon measurements were made on carbonates and collagen fractions from five different samples representing a right third metatarsal (97.R.24(MTa)), a rib (97.I.12d(13)), and the cnemial crest of the left tibia (97.L.20b) (Table 3.5).

Carbonate fractions were from secondary geological carbonates and bone apatite. Geological carbonates comprised mineralizations that were unquestionably secondary, e.g., euhedral $CaCO_3$ crystals (Figures 3.5B, 3.5C, and 3.6) calcium carbonate that had cemented sand inside the bone or pedogenic, carbonate (caliche) concretions on the undersurfaces of bones. AMS ^{14}C measurements on three interior, secondary calcites ranged from 2170 ± 20 to 2340 ± 15 RC yr while the fourth, on external, pedogenic carbonate, measured 3900 ± 15 RC yr (Figure 3.8; Table 3.5); AMS ^{14}C measurements on the carbonate fraction of bone apatite before and after acetic acid leaching were 2285 ± 15 and 2660 ± 20 RC yr, respectively (Figure 3.8; Table 3.5).

Standard methods for extracting bone collagen for radiocarbon dating or stable isotope analysis yield several different chemical fractions as purification proceeds. In order of first-to-last, the chemical fractions were decalcified collagen, KOH-extracted decalcified collagen, gelatin, and finally XAD-purified gelatin hydrolyzate.

Nine AMS ^{14}C measurements made on these chemical fractions from a rib and the tibia ranged from 8095 ± 25 to 8360 ± 30 RC yr (Table 3.5). Dates on the most highly purified chemical fraction, XAD-hydrolyzate, were 8360 ± 30 RC yr (UCIAMS-116396) for the tibia and 8355 ± 30 (UCIAMS-116397) RC yr for the rib. The average of these measurements was 8358 ± 21 RC yr BP

TABLE 3.5 Stafford AMS ^{14}C dates on Kennewick Man bone collagen and carbonate samples.

	SR-	CHEM-	AMS Lab Number	Fraction Dated	Fm ± SD	^{14}C Age (RC yr ± SD)
				Bone Geological Carbonates		
Carbonate ^{14}C Measurements	7096	8543	UCIAMS-25734	Carbonate-cemented VF sand	0.7634 ± 0.0015	2170 ± 20
	7086	8590	UCIAMS-26951	Intermedullary carbonate-cemented sands	0.7572 ± 0.0013	2235 ± 15
	7082	8591	UCIAMS-26952	Calcite and vuggy calcite within bone	0.7475 ± 0.0013	2340 ± 15
	7086	8608	UCIAMS-29127	Pedogenic, carbonate-cemented sand adhering to underside of bone	0.6155 ± 0.0010	3900 ± 15
				Bone Apatite Carbonate		
	7084	8576	UCIAMS-26949	Total-bone carbonate	0.7524 ± 0.0013	2285 ± 15
	7084	8589	UCIAMS-26950	HAc-leached bone powder	0.7182 ± 0.0018	2660 ± 20
				Bone Collagen		
	Left Tibia, Cnemial Crest (97.L.20b)					
	7084	8573	UCIAMS-26937	KOH-extracted collagen	0.3612 ± 0.0009	8180 ± 25
	7086	8682	UCIAMS-116379	HCl-decalcified collagen	0.3651 ± 0.0010	8095 ± 25
	7086	8684	UCIAMS-116391	KOH-extracted collagen	0.3573 ± 0.0012	8265 ± 30
	7086	8686	UCIAMS-116385	Gelatin from KOH-extracted collagen	0.3537 ± 0.0010	8350 ± 25
Bone Collagen ^{14}C Measurements	7086	8688	UCIAMS-116396	** XAD-Gelatin from KOH-collagen	0.3532 ± 0.0012	8360 ± 30
	Rib (97.I.12d(13))					
	7092	8826	UCIAMS-116380	HCl-decalcified collagen	0.3574 ± 0.0010	8265 ± 25
	7092	8718	UCIAMS-116392	KOH-extracted collagen	0.3577 ± 0.0012	8260 ± 30
	7092	8719	UCIAMS-116386	Gelatin from KOH-extracted collagen	0.3547 ± 0.0010	8325 ± 25
	7092	8721	UCIAMS-116397	** XAD-Gelatin from KOH-collagen	0.3535 ± 0.0011	8355 ± 30
				Averaged XAD Measurements 8358 ± 21 RC yr BP		
				Marine Reservoir Corrected Age[2] 9075–8935 CAL yr BP		

[1]Calculated with OxCal 4.2, using Marine 09 (Ramsey 2013).
[2]Based on 90% marine diet[1], ΔR = 0 (400 RC YR Marine Offset), 2σ (95.4%) probability.

TABLE 3.6 Quantitative amino acid analyses on modern bone and collagen, Kennewick Man bone, and fossil comparisons.

									AMINO ACID RESIDUES PER THOUSAND R/1000												nmol AA's/ mg bone	Proportion of Modern (Modern = 100%)	Protein Description
SR-	CHEM-	Sample	HYP	ASP	THR	SER	GLU	PRO	GLY	ALA	VAL	MET	ILE	LEU	TYR	PHE	HIS	HYL	LYS	ARG			
									Modern Bone and Modern Collagen														
—	—	Modern bovid bone	90	53	19	35	71	124	322	89	27	7	14	32	6	16	5	5	28	56	2500	100%	Modern Bone (Collagen)
—	—	Type V collagen	90	47	17	31	73	114	338	113	26	7	12	26	4	20	4	16	21	39	—	100%	Modern Collagen
									Kennewick Man Bone														
7083	8644	1 mm from cortex	64	62	11	8	78	132	337	125	30	4	13	24	1	12	24	2	28	46	182	7.9%	Collagenous
7083	8645	2 mm from cortex	65	69	12	9	82	108	329	128	32	4	13	25	1	12	30	3	30	46	94	4.0%	Collagenous
7083	8646	3.5 mm from cortex	67	55	11	10	82	116	349	125	30	4	12	24	1	12	25	2	28	46	136	5.8%	Collagenous
7083	8647	5 mm from cortex	68	52	11	9	77	112	369	128	29	4	12	24	0	12	17	2	28	45	196	8.4%	Collagenous
7083	8695	KOH-Gelatin	71	50	11	12	80	123	336	125	27	4	10	22	0	11	40	3	28	46	—	—	Collagenous
7086	10083	Whole bone	74	47	10	17	78	106	372	127	29	2	12	24	1	10	17	0	32	41	439	17.6	Collagenous
7086	8682	Decalcified collagen	64	50	9	9	83	122	325	130	29	3	11	24	0	13	51	4	29	45	—	—	Collagenous

(continued)

TABLE 3.6 (*continued*) Quantitative amino acid analyses on modern bone and collagen, Kennewick Man bone, and fossil comparisons.

SR-	CHEM-	Sample	AMINO ACID RESIDUES PER THOUSAND R/1000																	nmol AA's/ mg bone	Proportion of Modern (Modern = 100%)	Protein Description
			HYP	ASP	THR	SER	GLU	PRO	GLY	ALA	VAL	MET	ILE	LEU	TYR	PHE	HIS	HYL	LYS	ARG			
7086	8683	HCL Soluble Fraction	67	59	10	9	86	85	341	130	31	5	10	25	0	10	59	4	36	33	—	—	Collagenous
7086	8686	KOH-Gelatin	66	50	10	8	78	112	364	133	28	4	10	23	0	11	30	0	30	43	—	—	Collagenous
7086	8690	Decalcified collagen	70	51	12	9	77	104	375	132	32	2	13	24	0	10	19	0	31	38	—	—	Collagenous
7091	8639	Whole bone	73	44	10	9	72	114	392	116	27	3	11	26	0	15	15	0	25	48	315	12.6%	Collagenous
7091	8725	Whole bone	73	47	14	11	68	114	379	124	27	3	10	23	0	15	19	0	23	48	394	15.8%	Collagenous
7094	8636	Cortex bone	75	51	11	13	79	124	328	124	28	4	11	23	0	12	41	3	26	47	188	8.1%	Collagenous
Fossil Bones with Varying Protein Preservation																							
5378	—	Bone	79	48	19	12	80	129	334	114	31	4	11	28	3	15	5	4	32	51	1606	72.7%	Collagenous-I
6361	—	Bone	55	74	16	14	96	98	354	127	28	1	11	28	4	12	4	2	36	40	338	15.3%	Collagenous-II
5172	—	Bone	47	95	27	18	109	89	265	118	40	17	21	40	6	19	8	6	38	35	113	5.1%	Collagen-derived
5379	—	Bone	56	104	22	20	94	82	318	120	36	0	20	41	8	17	7	0	35	19	20	0.9%	Collagen-derived
6009	—	Bone	20	147	19	16	118	92	267	117	45	0	24	45	8	17	9	0	23	33	3	0.1%	Collagen-derived
6309	—	Bone	0	198	26	20	164	44	203	101	55	0	32	52	11	24	12	0	27	31	2	0.1%	Non-collagenous

TABLE 3.7 Total isotope analyses measured on Kennewick Man bone collagen.

SR-	CHEM-	Fraction Analyzed	Skeletal Element	δ¹³C‰ (VPDB)	δ¹³C‰ SD	δ¹⁵N‰ (AIR)	δ¹⁵N‰ SD
7083	8692	HCl decalcified collagen	Left tibia	−14.55	0.02	21.47	0.03
7083	8694	KOH-extracted collagen	Left tibia	−14.29	0.04	21.38	0.1
7083	8695	Gelatin from KOH-extracted collagen	Left tibia	−13.68		21.71	
7090	8554	Powdered whole bone	Left tibia	−14.47	0.31	ND	
7095	8556	Powdered whole bone	Metacarpal	−16.42	0.17	19.56	1.257
7084	8552	Powdered whole bone	Left tibia	−14.82	0.16	ND	
7084	8640	HCl-soluble salts	Left tibia	−21.77	0.44	13.84	0.06
7086	8683	HCl-soluble salts	Left tibia	−20.57		17.20	
7086	8690	HCl decalcified "collagen"	Left tibia	−16.13		20.84	
7086	8682	HCl decalcified collagen	Left tibia	−13.81	0.04	21.63	0.030
7086	8684	KOH-extracted collagen	Left tibia	−13.76	0.03	21.31	0.021
7086	8686	Gelatin from KOH-extracted collagen	Left tibia	−13.78	0.04	21.26	0.037
7092	8718	KOH-extracted collagen	Rib	−13.81	0.01	21.09	0.050
7092	8719	Gelatin from KOH-extracted collagen	Rib	−13.84		21.15	
		Average of gelatin measurements (Schwarcz et al., Chapter 16)		−13.95‰		+21.26‰	
		Average of previous δ¹³C measurements (Taylor 2000) (Table 3.8)		−13.1‰		ND	

FIGURE 3.1 (A) Exterior showing secondary (pedogenic) calcium carbonate (CaCO₃) developed as nodules on the underside of the bone. (B) Exterior after removal of carbonate nodules for their AMS ¹⁴C measurement; exposed bone is reddish brown due to humates, and iron and manganese staining. (C) Cross section showing voids filled with both authigenic aragonite crystals and carbonate-cemented sand. Arrow indicates down-side of the bone.

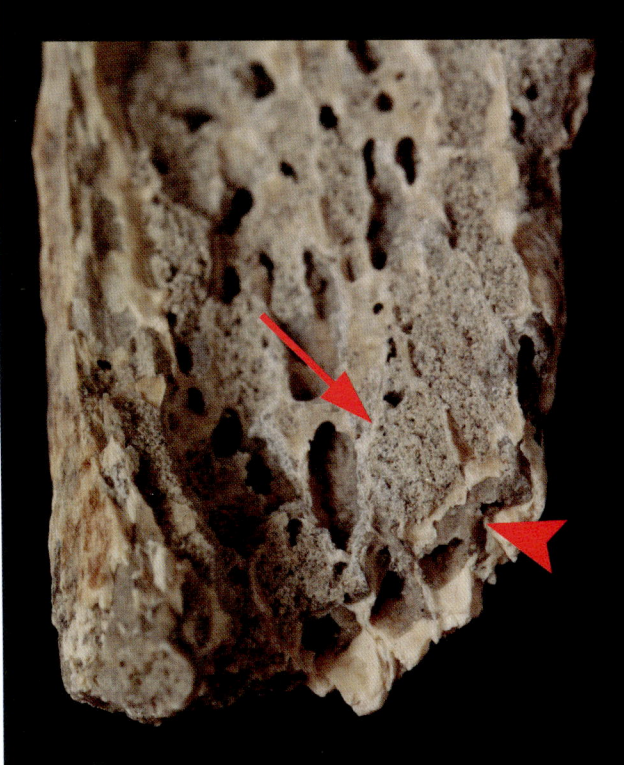

FIGURE 3.2 (A) Cancellous regions filled with carbonate-cemented, very fine-to-fine sand. (B) Interior, cancellous region showing two types (arrow and arrowhead) of secondary $CaCO_3$ precipitation.

FIGURE 3.3 (A) CaCO$_3$-cemented silty, very fine-to-fine sand overlays the thin, surficial patina. (B) A previous cut through the bone shows collagen polished during sawing.

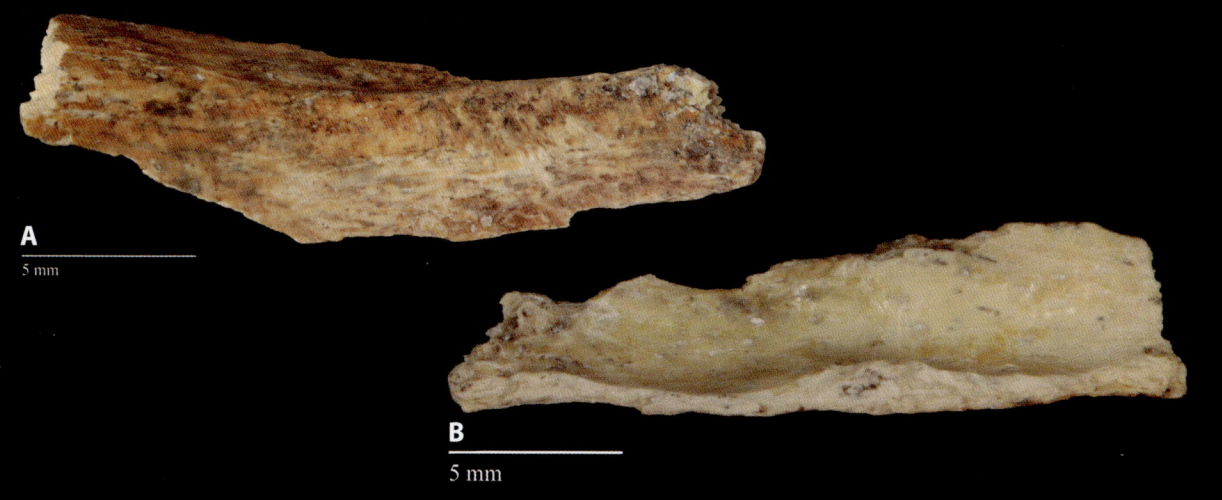

FIGURE 3.4 Bone fragments from the left third metacarpal (97.L.16(Mca)) used for aDNA analyses at the Centre for GeoGenetics, University of Copenhagen, Denmark. (A) and (B) are views of SR-8125 (SR-7094), which was previously tested for aDNA by the University of Michigan (Merriweather et al. 2000) and subsequently used as the first aDNA sample at GeoGenetics. (A) Overall view of SR-8125 (SR-7094) exterior with surficial patination. (B) Interior of SR-8125 (SR-7094) showing previous (University of Michigan) sampling and hard, waxy bone. (C) and (D) are views of sample SR-8330 (SR-7091), which had been previously analyzed for aDNA at Yale University (Kaestle et al. 2000) and was the second sample analyzed for aDNA at GeoGenetics. (C) Overall view of the proximal metacarpal. (D) Close-up view of medullary cavity after Yale's sampling, showing hard, waxy bone.

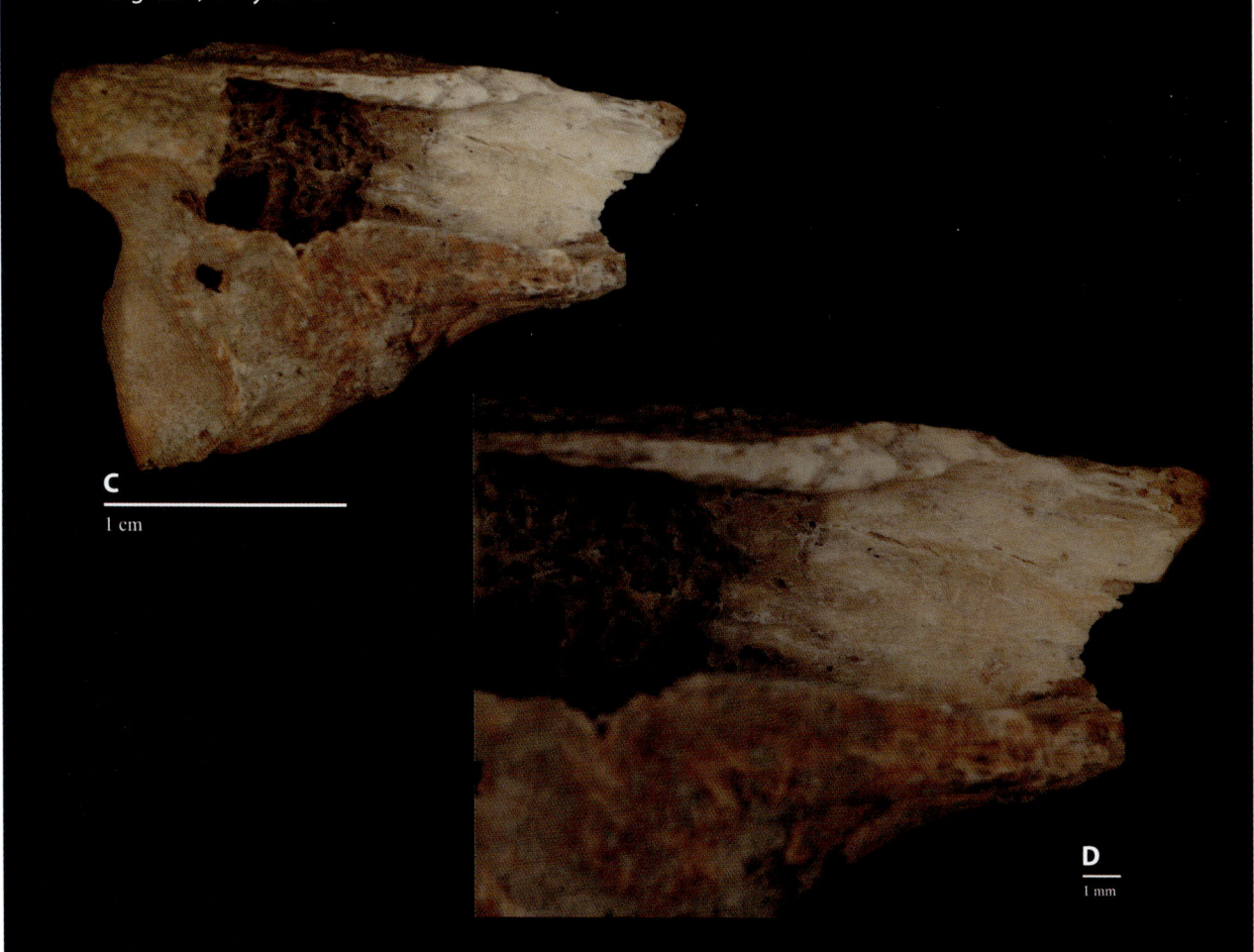

FIGURE 3.5 (A) Bone fragments typical of those used for this report's radiocarbon and stable isotope analyses. Arrow identifies bone fragment in (B) and (C). (B) Void filled with secondary aragonite crystals (arrow). (C) Close-up of secondary aragonite.

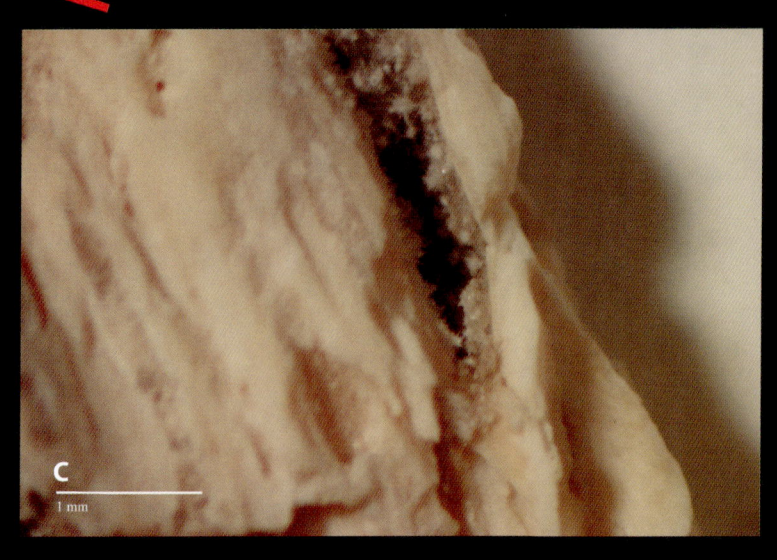

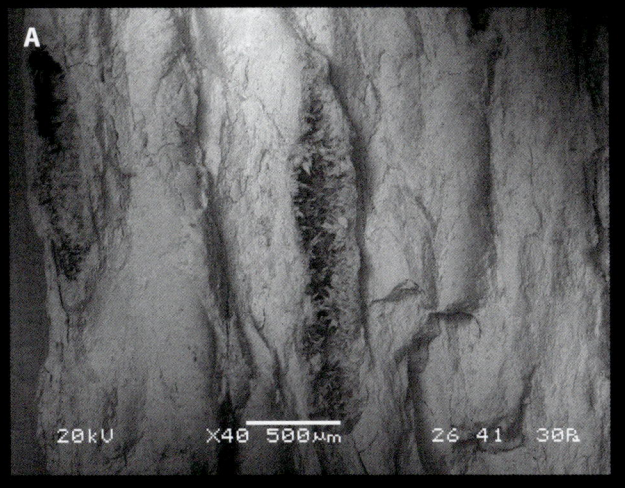

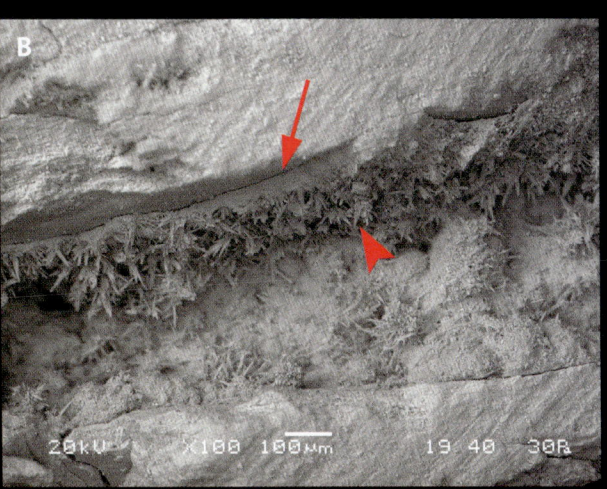

FIGURE 3.6 (A) Aragonite crystals at magnification X40 (SEM Image). (B) Two morphologies are present: equant $CaCO_3$ in a thin band along the bone cavity's wall (arrow) and aragonite blades (arrowhead). (C) Equant crystals have formed along a 25-to-50 μm wide band along the edge of bone at the void's wall.

FIGURE 3.7 Cross section of bone SR-7082 showing well-preserved Haversian canals that can be used to determine Kennewick Man's age at death.

TABLE 3.8 Yields of collagen from Kennewick Man bone samples as determined by wet chemistry and quantitative amino acid analyses.

SR-	Skeletal Element	Yield Decalcified Collagen (Modern = 20%)	Yield KOH-extracted Collagen (Modern = 20%)	Percent of collagen present in modern bone. Decalcified collagen (Modern = 100%)	Percent of collagen present in modern bone. KOH-extracted collagen (Modern = 100%)	Percent of collagen present in modern bone. By AAA (Modern = 100%)	Decalcified collagen pseudomorph (Modern = 100%)	KOH-extracted collagen pseudomorph (Modern = 100%)
7083	Left tibia, cnemial crest	1.7%	0.9%	8.5%	4.5%	3.8 to 7.3%	40%	30%
7084	Left tibia, cnemial crest	3.9%	1.5%	19.5%	7.5%	6.6%	20%	30%
7086	Left tibia, cnemial crest	9.2%	5.8%	46.5%	29.0%	17.6%	88%	88%
7090	Left tibia, cnemial crest	—	—	—	—	10.5%	—	—
7091	Left 3rd metacarpal	—	7.8%	—	39.0%	23.5%	88%	90%
7092	Rib	5.5%	4.9%	27.5%	24.5%	12.8%	88%	88%
7094	Left 3rd metacarpal	—	—	—	—	7.5%	—	—
7095	Left 3rd metacarpal	—	—	—	—	14.0 to 20.5%	—	—
7092	Rib	—	3.9%	—	19.5%	—	94%	85%

APATITE CARBONATE
Total CO$_3^{-2}$ 2285±15
HAc-leached 2660±20

DRUSY CALCITE
2340±15

CARBONATE CEMENTED SANDS
2170±20
2235±15

PROTEIN (XAD-GELATIN)
8355±25

CARBONATE CONCRETIONS (CALICHE)
3900±15

1 mm

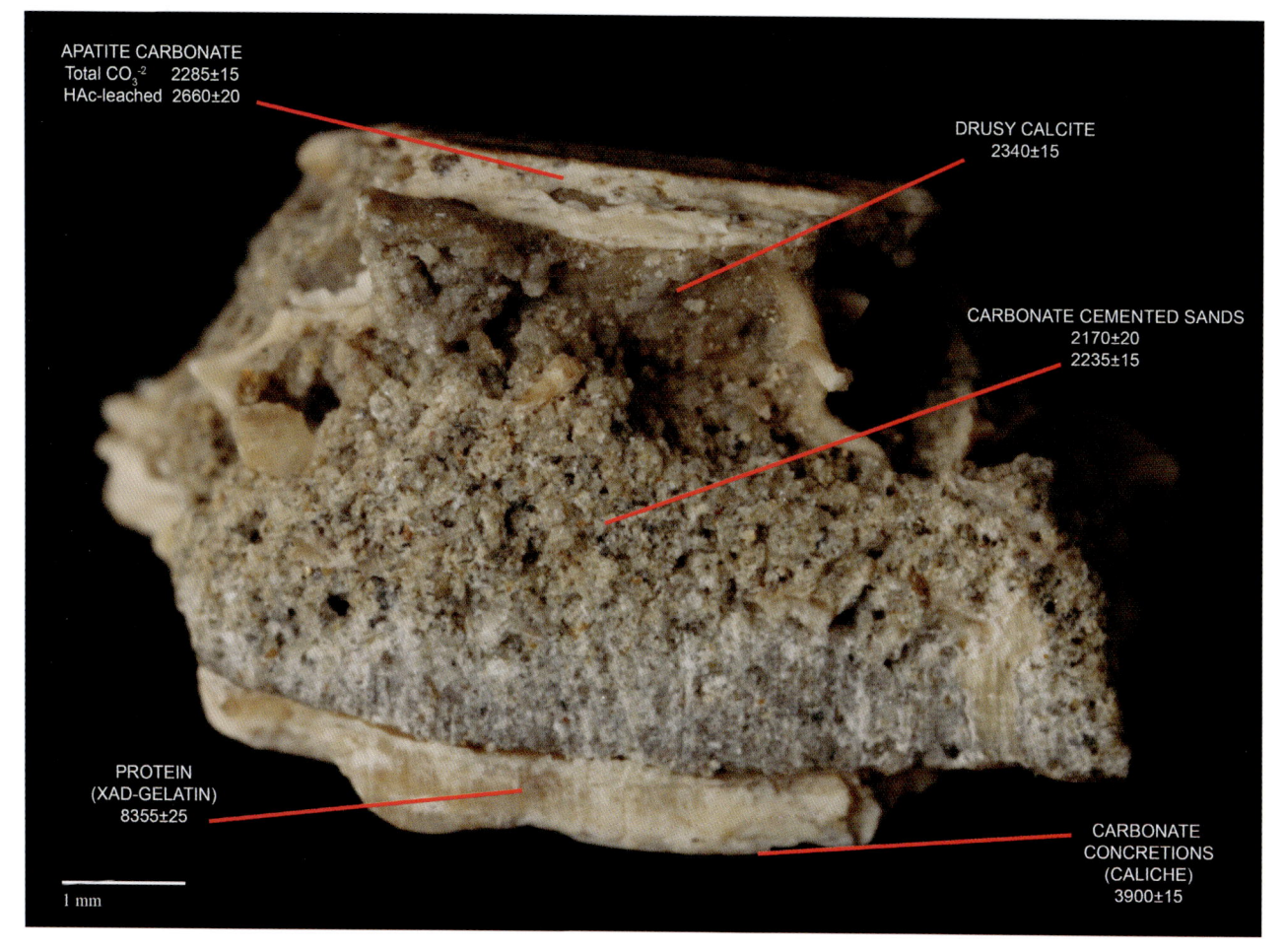

FIGURE 3.8   Fragment of SR-7096 showing all physical and chemical fractions dated on other Kennewick Man bones. The components were 1) bone protein (collagen), 2) apatite carbonate, 3) drusy calcite and euhedral aragonite, 4) carbonate cemented sands filling the intermedullary cavity and 5) pedogenic carbonate nodules (caliche) on the underside of many skeletal elements. These fractions were not dated on this specimen and their position is schematic only. Pedogenic carbonate is not present on this bone and its position is noted where it occurs on other skeletal elements. The radiocarbon age of XAD-purified collagen is in RC yr BP; carbonate ^{14}C dates are "apparent ^{14}C ages" and are not absolute age measurements due to uncertainty in the original Fraction Modern (Fm) value of secondary carbonate as it was being deposited.

(Table 3.5). Values on comparative chemical fractions from known-age bones and a >70 ka bone are listed in Table 3.9 and were used in calculating the ages for the unknown-age collagen samples.

Quantitative Amino Acid Analyses on Kennewick Bone and Bone Collagen

A quantitative amino acid analysis is the only assay that establishes conclusively whether or not a bone or tooth contains the protein collagen. These values, especially for the secondary (imino) amino acids hydroxyproline (HYP) and proline (PRO), enable vertebrate fossils to be ranked across a spectrum that ranges from collagenous to collagen-derived to non-collagenous (Table 3.6).

Modern bone contains approximately 2500 nmol of amino acids per mg of dry bone (Hare 1980). Five amino acids have characteristic abundances that are unique to collagen and include hydroxyproline (HYP) (9%), proline (PRO) (12%), aspartic acid (ASP) (5%), glutamic acid (GLU) (7%), and glycine (GLY) (33%). Of the 18 amino acids comprising collagen, these five provide the most definitive evidence for collagen's presence and its degree of degradation. Of these five, hydroxyproline is the most important for defining collagen.

TABLE 3.9 ^{14}C measurements on international standards, known-age, and background bones.

SR-	CHEM-	AMS Lab Number	Fraction Dated	Fm ± SD	^{14}C Age (RC yr ± SD)
			International Bone Dating Standard, VIRI-I (8331 ± 6 RC yr)[1]		
8134	10109	UCIAMS-127326	Whole, untreated bone	0.3639 ± 0.0011	8120 ± 25
8134	9772	UCIAMS-116381	HCl-decalcified collagen	0.3561 ± 0.0010	8295 ± 25
8134	9773	UCIAMS-116393	KOH-extracted collagen	0.359 ± 0.0012	8230 ± 30
8134	9775	UCIAMS-116387	Gelatin from KOH-extracted collagen	0.3544 ± 0.0010	8330 ± 25
8134	9776	UCIAMS-116402	XAD-gelatin (KOH-collagen)	0.3534 ± 0.0011	8355 ± 30
			Known-age Bone, Dent Mammoth, CO (10,980 RC yr)		
8135	9778	UCIAMS-116394	KOH-extracted collagen	0.2566 ± 0.0010	10,925 ± 35
8135	9780	UCIAMS-116388	Gelatin from KOH-extracted collagen	0.2539 ± 0.0009	11,015 ± 30
8135	9781	UCIAMS-116403	XAD-gelatin (KOH-collagen)	0.2555 ± 0.0012	10,960 ± 35
			^{14}C Background Bone, Beaufort Whale (Eschrichtius sp.) (>70 Ka)		
8136	9783	UCIAMS-116382	KOH-extracted collagen	0.0023 ± 0.0001	48,690 ± 240
8136	9785	UCIAMS-116389	Gelatin from KOH-extracted collagen	0.0023 ± 0.0001	48,730 ± 220
8136	9786	UCIAMS-116378	XAD-gelatin (KOH-collagen)	0.0034 ± 0.0001	45,580 ± 270

[1]Scott et al. 2010.

Stable Carbon and Nitrogen Isotope Measurements

Table 3.7 contains isotope measurements in addition to those given in Chapter 16. The additional measurements are on chemical fractions not suited for paleodiet research, but which provide information regarding mass balances of both isotopes and how diagenesis has affected Kennewick Man's bones. The established bone collagen isotopic values for Kennewick Man are $\delta^{13}C = -13.95‰$ and $\delta^{15}N = +21.26‰$ (Chapter 16) and are based on gelatin from KOH-extracted collagen. In contrast, decalcified collagen and KOH-extracted collagen have values differing by 1–2‰ of these values, while collagen-peptide-bearing decalcification salts were significantly depleted in both ^{13}C and ^{15}N and had values of $\delta^{13}C = -20.6$ to $-20.8‰$ and $\delta^{15}N = +13.8$ to $+17.2‰$) (Table 3.7).

Photomacroscopy Records Documenting Collagen's Chemical Purification

While significant information derives from quantitative measurements, considerable information results from recording how bones and collagen behave chemically. Increasingly older ^{14}C ages for collagen demonstrate that contaminants are being removed; however, visual proof of contaminants being removed is provided from photographs taken after each stage of purification chemistry.

Figures 3.9–3.11 document how the principal contaminants in bone—rootlets, sediment, and humic acids—are removed chemically and physically. Fossil bones and teeth are commonly filled and occluded with sediment and penetrated by rootlets. Humates that derive from enclosing sediments and leach downward from younger soil horizons are the primary organic contaminant.

Visual examination of bones or collagen is inadequate for identifying fine sediment and especially rootlets, particularly if the bone has been powdered before decalcification. This inorganic and organic detritus is readily evident in collagen from the bone samples from Kennewick Man's skeleton after decalcification and KOH extraction (Figures 3.9b, 3.9c; 3.10b–d). The best example of how much sedimentary and rootlet debris remains in decalcified and KOH-extracted collagen is Figure 3.11c, which shows detritus on the surface of a 0.45 μm filter after the gelatin solution was passed through the 0.45 μm membrane. Had decalcified collagen or KOH-extracted

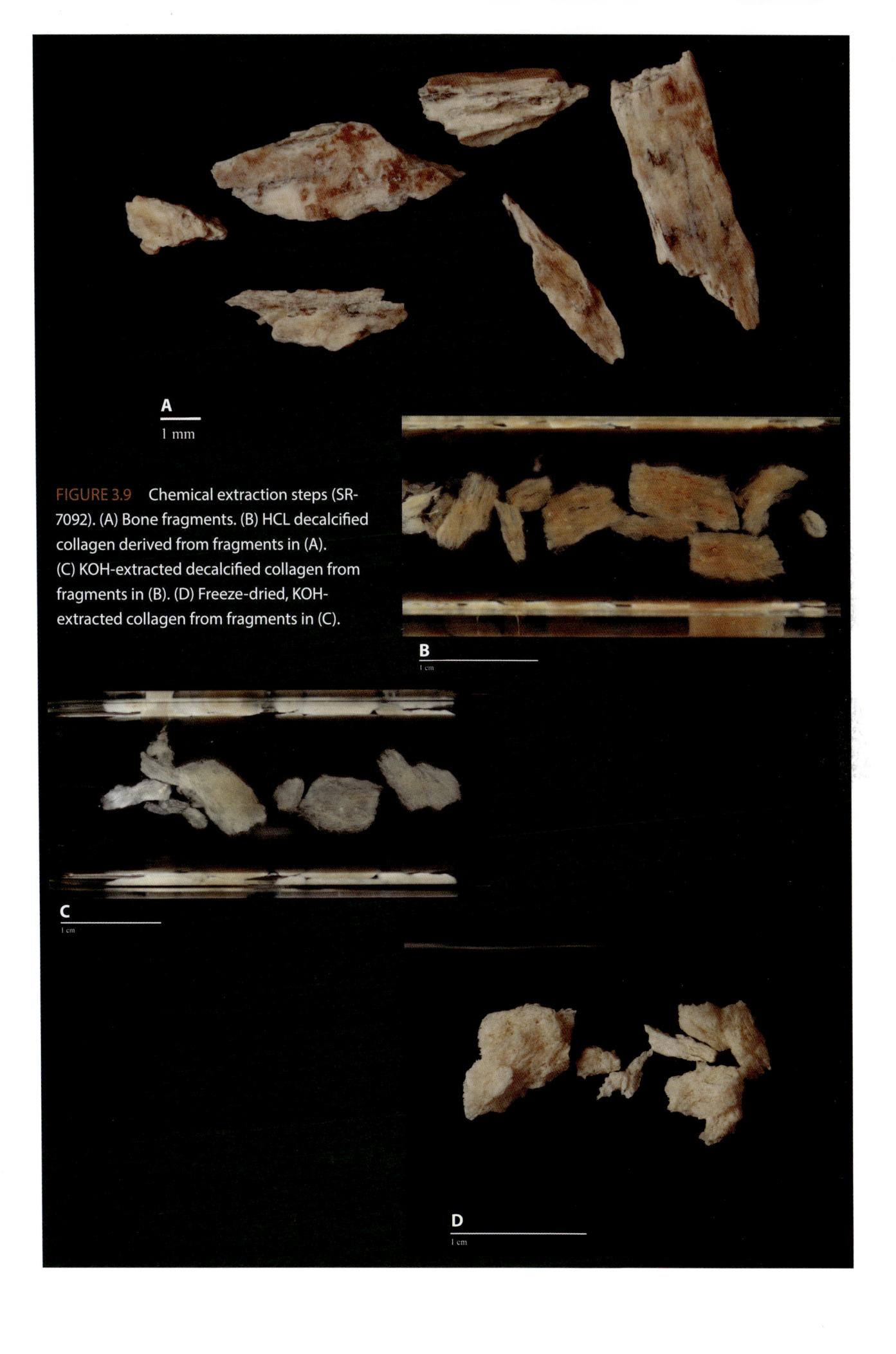

FIGURE 3.9 Chemical extraction steps (SR-7092). (A) Bone fragments. (B) HCL decalcified collagen derived from fragments in (A). (C) KOH-extracted decalcified collagen from fragments in (B). (D) Freeze-dried, KOH-extracted collagen from fragments in (C).

FIGURE 3.10 Chemical fractions from SR-7083. (A) HCl decalcified collagen from bone fragment. (B) HCl decalcified collagen that has been extracted with dilute KOH. (C) Close-up view of KOH-extracted decalcified collagen that contains rootlets. (D) Quartz sand, rootlets and other detritus insoluble during gelatinization of the collagen in 90°C, pH 2 water.

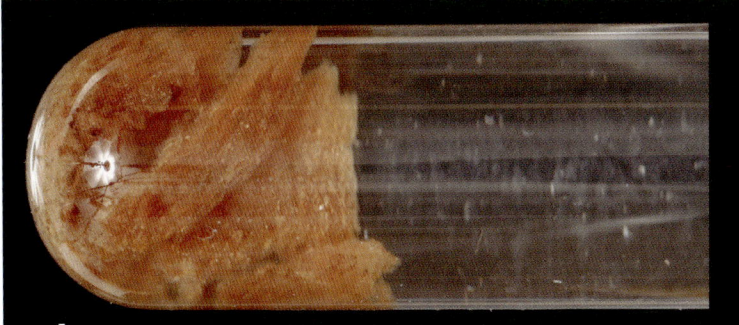

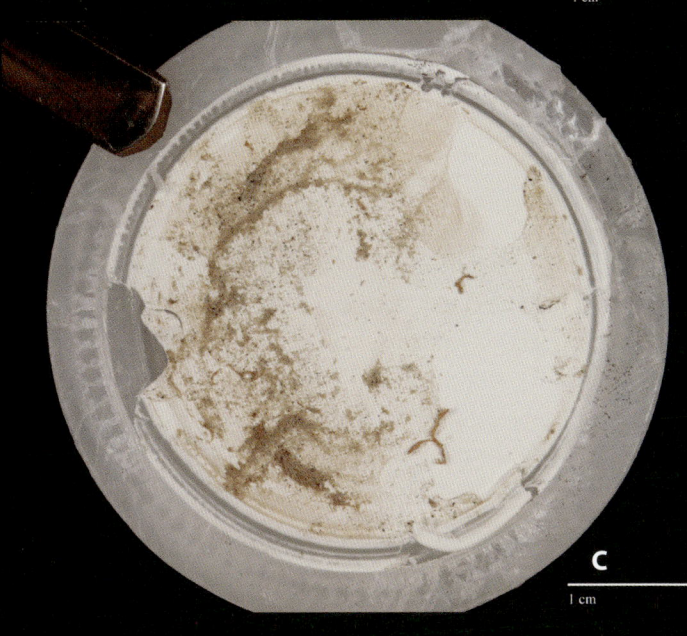

FIGURE 3.11 Chemical fractions from SR-7086. (A) Decalcified collagen. (B) Decalcified bone after extraction with dilute KOH. (C) Surface of a 25 mm diameter, Millipore 0.45 μm Durapore filter with silt and foreign organic matter. (D) Freeze-dried gelatin from KOH-extracted, decalcified collagen with contaminants removed.

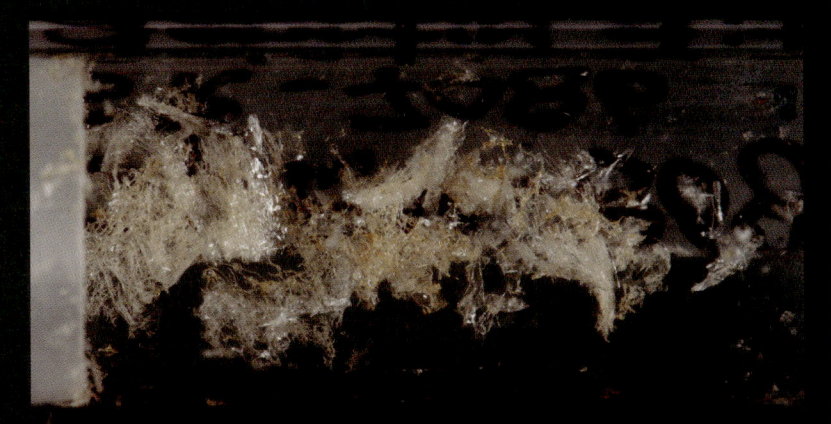

collagen been dated directly, carbon from these contaminants would have been added to the collagen's carbon.

Humate removal is demonstrated in photographs taken before and after dilute KOH was used on Kennewick collagens. Figures 3.9B and 3.9C are decalcified collagen and KOH-extracted collagen, respectively. Figures 3.10B, 3.10C, and 3.11A, further illustrate how decalcified collagen that is brown to light reddish brown becomes white after KOH removes a significant amount of humates (humic and fulvic acids).

After most humates are removed with dilute KOH, the next important purification step is dissolving collagen in hot, weakly acidic water. While collagen is soluble in acidic solutions, rootlets, clastic sediments and other detritus remain insoluble and are removed by filtering the solution through 0.45 μm membrane filters (Figure 3.11C). The filtered gelatin (Figure 3.11D) is hydrolyzed and purified by passing the solution through XAD resin.

DISCUSSION

The most important goal of this work was establishing an accurate ^{14}C age for the Kennewick Man skeleton. Based on that age, interpretations could be made about the site's stratigraphy, the skeleton's taphonomic history and diagenesis, his calendrical age, and eventually, how Kennewick Man compares to other skeletons from various time periods. For these discussions, calibrated ages in CAL BP are stated at 2 sigma (95.4% probability) based on OxCal 4.2 and use either IntCal 09 or Marine 09.

The established age for the Kennewick Man skeleton is 8358 ± 21 RC yr, which is the average of two measurements on XAD-purified gelatin: 8360 ± 30 (UCIAMS-116396) and 8355 ± 30 (UCIAMS-116397). The 8358 RC yr age is similar to the first ^{14}C age estimate for the skeleton, 8410 ± 60 (UCR-3476).

Doubts concerning the skeleton's geologic age were due to ^{14}C studies that produced values ranging from 5750 ± 100 (AA-34818) to 8410 ± 40 (Beta-133993) (Table 3.1; Figure 3.12) and comments by dating laboratories that the ages were "apparent radiocarbon ages" or were "too young" due to poor physical preservation, low collagen yields, or non-collagenous composition (Table 3.1).

Data from wet chemistry, visual inspection, and quantitative amino acid analyses indicate that all bones examined in this report have internal structures of modern

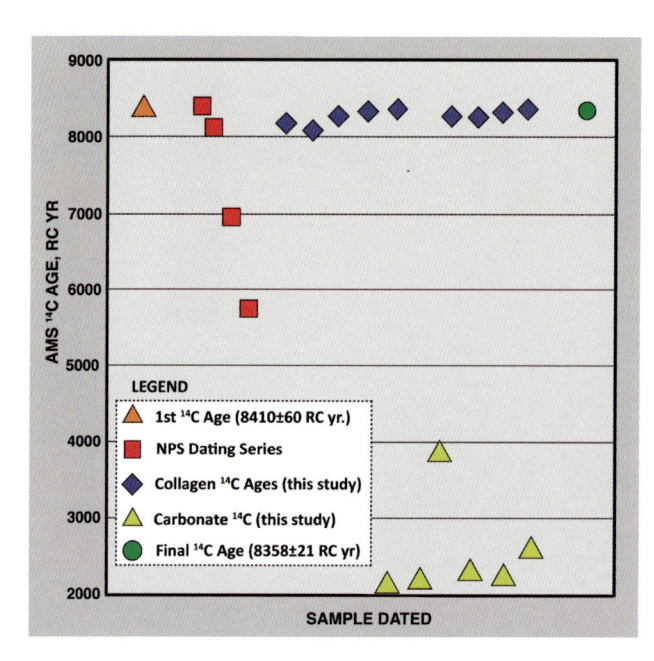

FIGURE 3.12 AMS ^{14}C measurements on carbonates and different collagen fractions from Kennewick Man bones.

bone, e.g., well-preserved Haversian canals (Figure 3.6), contain collagen and have collagenous amino acid compositions. The amino acid data (Table 3.6) are evidence collagen is preserved because the bones and collagen contain the expected values for hydroxyproline, proline, aspartic acid, glutamic acid and glycine. Although the bones contained from 4.5 to 39 percent of the collagen expected for a modern bone, the protein preserved is collagen.

The relative proportions of the five key amino acids (HYP, PRO, ASP, GLU, and GLY), determine if bone has a collagenous or non-collagenous composition. Collagenous compositions exist when hydroxyproline has 60–90 R/1000, GLY is ~333 R/1000, PRO is 120–90 R/1000, ASP is 45–55 R/1000, and GLU is 70–85 R/1000. The trends during collagen decomposition and mass loss are decreasing HYP, PRO, and GLY and elevated relative proportions of ASP and GLU. As these trends continue, hydroxyproline eventually is lost entirely (R/1000 = 0), aspartic and glutamic acids become elevated three to five times their normal values, and proline and glycine decrease to R/1000 values of 40 and 200–220, respectively (Table 3.6). As these amino acid values change with decreasing total protein content, bone changes from a collagen (modern bone) composition to collagenous, to collagen-derived, and finally to non-collagenous. By using quantitative amino acid analyses and monitoring the ratios of each of the five most important amino acids, protein contents of bones can be more accurately described (Table 3.6) than by using wet chemical yields and visual observations.

The ^{14}C dates on all collagen fractions measured for this report range from 8095 ± 25 (UCIAMS–116379) to 8360 ± 30 (UCIAMS–116396) (Table 3.5). The chemical fractions producing younger ^{14}C ages were decalcified collagen, KOH-collagen and gelatin. Their ages are 8 to 263 RC years less than the skeleton's 8358 RC age due to retention of humates in decalcified collagen, presence of rootlets, sediment and other detritus in KOH-collagen, and fulvic acids remaining in the gelatin.

The carbonate fractions from Kennewick Man bone have ^{14}C "ages" that are 4458 to 6228 RC yr less than the skeleton's 8358 RC yr age. The carbonate ^{14}C measurements are termed "apparent ^{14}C ages" because it is unknown if the CO_3^{-2} had the same fraction modern (Fm) as the atmosphere when the carbonate was being precipitated. Consequently, the ^{14}C measurements on carbonates do not provide accurate estimates for their time-of-formation.

AMS radiocarbon dates on sediments from the discovery site were general confirmation that the skeleton dated to the early Holocene; however, more accurate ages were not possible using alluvial carbon. Although the uppermost sediments in geological core CPC-059.5 ^{14}C dated as 9010 ± 50 RC yr (WW-1626), strata 120 to 190 cm deeper yielded ages that were 3450 to 6320 RC yr older and in reversed age-order for the two lower units. It is unknown if these ages are correct, if there is an unrecognized unconformity between 20 and 200 cm, or if the sediments contain geologically old, recycled alluvial carbon that is causing the sediments to date older than their time of deposition. The implication of these sedimentary ^{14}C dates is that using alluvial carbon or sedimentary humates for fluvial chronologies in this region of the Columbia River will have to be calibrated against shell or bones embedded within the strata.

The ultimate question—what is the absolute or calendar age for the Kennewick Man skeleton—cannot be determined as accurately as the skeleton's radiocarbon age. Assuming that the ^{14}C/^{12}C value for collagen is not affected by contaminating carbon, establishing an absolute age uses the ^{14}C age as the initial age value, incorporates any reservoir effects related to diet (marine or terrestrial), and uses this reservoir-corrected value for calculating a calendrical age in absolute years before present (1950 AD).

Based on a δ^{13}C value of –13.6‰, Taylor et al. (1998) assumed a 70 percent marine diet for Kennewick Man. They used 750 years as an oceanic offset versus the contemporary atmosphere, and after subtracting 70 percent of that value from the measured 8410 RC yr age, established the skeleton's reservoir-corrected ^{14}C age was 7780 ± 160 years and its calibrated age range was 8340–9200 CAL BP (Taylor et al. 1998).

Based on isotopic data showing that the Kennewick Man skeleton had a δ^{15}N value of +21.26‰ (Table 3.7; Chapter 16), the diet for Kennewick Man is estimated to have been at least 90 percent marine-derived foods. The actual reservoir effect for stream- and marine-derived foods consumed by Kennewick Man is poorly known. Estimates for marine reservoir corrections during the early Holocene range from 400 to 750 years; the latter value has been suggested as more appropriate (Dumond and Griffin 2002). For this report, the marine reservoir correction used is 750 years, which is the same value applied by Taylor et al. 1998. Using a 750 year marine reservoir correction and a 90 percent marine diet, Kennewick Man's 8358 ± 21 RC yr age yields a two sigma calendar age of 8690 to 8400 CAL BP (Table 3.5).

RECOMMENDATIONS FOR FUTURE RESEARCH

As technology advances, older methods become more accurate and precise as new techniques are developed. Since the discovery of the Kennewick Man skeleton in 1996, the precision of AMS ^{14}C dating has improved from ± 50 RC years to ± 20–25 RC years for late Pleistocene and younger samples, while sample size has decreased, from ~1 mg to 0.2 mg of carbon. With improved chemical purification methods (Waters and Stafford 2007), less fossil material is needed and ^{14}C measurements are more accurate.

Amino acid analyses readily determine the chemical preservation of bone and what bones contain the most collagen for radiocarbon, stable isotope, and biogeochemical analyses. Previous methods detected nanomole

(10^{-9}) levels of amino acids, whereas present techniques have picomole (10^{-12}) resolution, thereby enabling less than one mg of bone to be sampled with millimeter spatial resolution.

Stable isotope analyses (δ^{13}C, δ^{15}N) commonly use <0.5 mg of collagen to determine values on total collagen. The advent of compound-specific LC/IRMS (Barrie et al. 1984; Meier-Augenstein 1999) and GC/IRMS (Krummen et al. 2004) now enables δ^{13}C analysis of 16 or more specific amino acids and more accurate reconstruction of paleodiets on less than 0.01 mg (10 µg) of total collagen (Smith et al. 2009; Choy et al. 2010; Raghavan et al. 2010).

The above techniques will help clarify reconstructions and estimates of Kennewick Man's geographical origin, diet, and calibrated ^{14}C age. Calibrating a ^{14}C age requires accounting for reservoir effects—the difference in Fm value of dietary carbon versus the Fm of the animal's contemporaneous atmospheric CO_2. For Kennewick Man, the amount and type of marine foods establish his reservoir effect. Different Fm values will apply if his foods were primarily derived from the river or from the ocean, and each will change the calibrated age. Establishing Kennewick Man's geographic origin will consequently assist in identifying the proportion of time he resided in the continental interior versus time spent in coastal regions.

Kennewick Man's geographic origin could be better estimated by using $^{87/86}$Sr values from bone or enamel (Price et al. 1994a, 1994b) to identify the geological proveniences he inhabited. Another source of dietary information—marine foods versus riverine or terrestrial ones—is measuring carbon and nitrogen isotope values not only on total collagen in bone, but on individual amino acids in that protein (Choy et al. 2009, Raghavan et al. 2010). Finally, Fm and δ^{13}C and δ^{15}N values for major riverine and coastal foods must be measured to determine empirically what ocean and river reservoir effects are today and what the values were for these animals 8000 to 9000 RC yr ago.

Finally, successful aDNA analyses will be facilitated by more than thirteen years of advances since the first tests were made. When aDNA work was first proposed, it was assigned very low potential (Tuross and Kolman 2000), and when they were analyzed, the Kennewick Man bones failed to yield aDNA (Kaestle 2000; Merriweather et al. 2000; Smith et al. 2000). At the time of these reports, Tuross and Kolman (2000) correctly noted that modern DNA contamination, the age of the skeleton, and degradation of the organic constituents precluded successful aDNA analysis of the skeleton and that financial costs, laboratory time, and the mass of bone required would be very large. In the intervening thirteen years, aDNA analyses have been successful on single human hairs 4000 years old (Rasmussen et al. 2010), 38,000-year-old Neanderthal remains (Green et al. 2006) and a 700,000-year-old horse (Orlando et al. 2013). Advances will continue in this field and should lead to successful sequencing of aDNA from Kennewick Man's bones. As sample size further decreases for aDNA analyses, the recommendations to use only a tooth's pulp cavity (Smith 2000; Walker 2000) will be attainable.

CONCLUSIONS

The skeleton of Kennewick Man is physically and chemically well preserved, and due to these biogeochemical conditions, can be accurately ^{14}C dated. The established AMS ^{14}C age is 8358 ± 21 RC yr, which represents a two sigma calibrated date of 8690 to 8400 CAL BP using an estimated 90 percent marine reservoir effect. All bones examined and bone protein analyzed were collagenous. Consequently, stable isotopic and radiocarbon analyses have a high degree of accuracy.

REFERENCES

Bacon, Charles R. 1983. Eruptive history of Mount Mazama and Crater Lake Caldera, Cascade Range, U.S.A. *Journal of Volcanology and Geothermal Research* 18(1–4):57–115.

Bacon, Charles R., and Marvin A. Lanphere. 2006. Eruptive history and geochronology of Mount Mazama and the Crater Lake region, Oregon. *Geological Society of America Bulletin* 118(11–12):1331–1359.

Barrie, A., J. Bricout, and J. Koziet. 1984. Gas chromatography—stable isotope ratio analysis at natural abundance levels. *Biological Mass Spectrometry* 11(11):583–588.

Chatters, James C. 2000. The recovery and first analysis of an early Holocene human skeleton from Kennewick, Washington. *American Antiquity* 65(2):291–316.

———. 2001. *Ancient Encounters: Kennewick Man and the First Americans.* Simon & Schuster, New York.

Choy, Kyungcheol, Colin I. Smith, Benjamin T. Fuller, and Michael P. Richards. 2010. Investigation of amino acid δ¹³C signatures in bone collagen to reconstruct human palaeodiets using liquid chromatography-isotope ratio mass spectrometry. *Geochimica et Cosmochimica Acta* 74:6093–6111.

Donahue, Douglas. 2000a. Results of measurements on the Kennewick bone, equipment preparation, sampling procedure, and pretreatment procedure. Letter to Frank McManamon, 9 January 2000. National Park Service, http://www.cr.nps.gov/archeology/kennewick/donahue2.htm.

———. 2000b. Results of carbon-isotope measurements on the Kennewick bone sample. Letter to Francis P. McManamon, 10 January 2000. National Park Service, http://www.cr.nps.gov/archeology/kennewick/donahue1.htm.

Dumond, Don E., and Dennis G. Griffin. 2002. Measurement of the marine reservoir effect on radiocarbon ages in the eastern Bering Sea. *Arctic* 55(1):77–86.

Green, Richard E., Johannes Krause, Susan E. Ptak, Adrian W. Briggs, Michael T. Ronan, Jan F. Simons, Lei Du, Michael Egholm, Jonathan M. Rothberg, Maja Paunovic, and Svante Pääbo. 2006. Analysis of one million base pairs of Neanderthal DNA. *Nature* 444(16):330–336.

Hallett, D. J., L. V. Hills, and J. J. Clague. 1997. New accelerator mass spectrometry radiocarbon ages for the Mazama tephra layer from Kootenay National Park, British Columbia, Canada. *Canadian Journal of Earth Sciences* 34(9):1202–1209.

Hare, P. E. 1980. Organic geochemistry of bone and its relation to the survival of bone in the natural environment. In *Fossils in the Making: Vertebrate Taphonomy and Paleoecology*, edited by Anna K. Behrensmeyer and Andrew P. Hill, 208–219. University of Chicago Press.

Hood, Darden. 1999a. Report of radiocarbon dating result for bone sample "CENWW.97.R.24(MTa)/DOI1a." Letter to Dr. Francis P. McManamon, 17 October 1999. National Park Service, http://www.cr.nps.gov/archeology/kennewick/hood1.htm.

———. 1999b. Additional information regarding radiocarbon dating analysis of bone sample "CENWW.97.R.24 (Mta)/DOI1 a." Letter to Dr. Francis P. McManamon, 18 November 1999. National Park Service, http://www.cr.nps.gov/archeology/kennewick/hood2.htm.

Huckleberry, Gary, Thomas W. Stafford, Jr., and James C. Chatters. 1998. Preliminary Geoarchaeological Studies at Columbia Park, Kennewick, Washington, USA. Report submitted to U.S. Army Corps of Engineers, Walla Walla District, March 23, 1998.

Huckleberry, Gary, and Julie K. Stein. 1999. Chapter 3: Analysis of sediments associated with human remains found at Columbia Park, Kennewick, WA. In *Report on the Non-Destructive Examination, Description, and Analysis of the Human Remains from Columbia Park, Kennewick, Washington* [October 1999]. National Park Service, http://www.nps.gov/archeology/kennewick/huck_stein.htm.

Huckleberry, Gary, Julie K. Stein, and Paul Goldberg. 2003. Determining the provenience of Kennewick Man skeletal remains through sedimentological analyses. *Journal of Archaeological Science* 30(6):651–665.

Kaestle, Frederika A. 2000. Chapter 2: Report on DNA analysis of the remains of "Kennewick Man" from Columbia Park, Washington. In *Report on the DNA Testing Results of the Kennewick Human Remains from Columbia Park, Kennewick, Washington* [September 2000]. National Park Service, http://www.nps.gov/archeology/kennewick/Kaestle.

Krummen, Michael, Andreas W. Hilkert, Dieter Juchelka, Alexander Duhr, Hans-Jürgen Schlüter, and Reinhold Pesch. 2004. A new concept for isotope ratio monitoring liquid chromatography/mass spectrometry. *Rapid Communications in Mass Spectrometry*. 18(19):2260–2266.

Kuehn, Stephen C., Duane G. Froese, Paul E. Carrara, Franklin F. Foit, Jr., Nicholas J. G. Pearce, and Peter Rotheisler. 2008. Major-and trace-element characterization, expanded distribution, and a new chronology for the latest Pleistocene Glacier Peak tephras in western North America. *Quaternary Research* 71(2):201–216.

McManamon, Francis P. 2000. Determination that the Kennewick human skeletal remains are "Native American" for the purposes of the Native American Graves Protection and Repatriation Act (NAGPRA). Memorandum to Assistant Secretary, Fish and Wildlife and Parks, Donald J. Barry, concurred January 11, 2000. National Park Service, http://www.cr.nps.gov/archeology/kennewick/c14memo.htm.

Meier-Augenstein, W. 1999. Applied gas chromatography coupled to isotope mass spectrometry. *Journal of Chromatography A* 842(1–2):351–371.

Merriweather, D. Andrew, Graciela S. Cabana, and David M. Reed. 2000. Chapter 3: Kennewick Man ancient DNA analysis: Final report submitted to the Department of the Interior, National Park Service. In *Report on the DNA Testing Results of the Kennewick Human Remains from Columbia Park, Kennewick, Washington* [September 2000]. National Park Service, http://www.nps.gov/archeology/kennewick/Merriwether_Cabana.htm

Orlando, Ludovic, Aurélien Ginolhac, Guojie Zhang, Duane Froese, Anders Albrechtsen, Mathias Stiller, Mikkel Schubert, Enrico Cappellini, Bent Petersen, Ida Moltke, Philip L. F. Johnson, Matteo Fumagalli, Julia T. Vilstrup, Maanasa Raghavan, Thorfinn Korneliussen, Anna-Sapfo Malaspinas, Josef Vogt, Damian Szklarczyk, Christian D. Kelstrup, Jakob Vinther, Andrei Dolocan, Jesper Stenderup, Amhed M. V. Velazquez, James Cahill, Morten Rasmussen, Xiaoli Wang, Jiumeng Min, Grant D. Zazula, Andaine Seguin-Orlando, Cecilie Mortensen, Kim Magnussen, John F. Thompson, Jacobo Weinstock, Kristian Gregersen, Knut H. Røed, Véra Eisenmann, Carl J. Rubin, Donald C. Miller, Douglas F. Antczak, Mads F. Bertelsen, Søren Brunak, Khaled A. S. Al-Rasheid, Oliver Ryder, Leif Andersson, John Mundy, Anders Krogh, M. Thomas P. Gilbert, Kurt Kjær, Thomas Sicheritz-Ponten, Lars Juhl Jensen, Jesper V. Olsen, Michael Hofreiter, Rasmus Nielsen, Beth Shapiro, Jun Wang, and Eske Willerslev. 2013. Recalibrating *Equus* evolution using the genome sequence of an early Middle Pleistocene horse. *Nature* 499:74–81.

Price, T. Douglas, Clark M. Johnson, Joseph A. Ezzo, Jonathan Ericson, and James H. Burton. 1994a. Residential mobility in the prehistoric southwest United States: A preliminary study using strontium isotope analysis. *Journal of Archaeological Science* 21(3):315–330.

Price, T. Douglas, Gisela Grupe, and Peter Schröter. 1994b. Reconstruction of migration patterns in the Bell Beaker Period by stable strontium isotope analysis. *Applied Geochemistry* 9(4):413–417.

Raghavan, Maanasa, James S. O. McCullagh, Niels Lynnerup, and Robert E. M. Hedges. 2010. Amino acid $\delta^{13}C$ analysis of hair proteins and bone collagen using liquid chromatography/isotope ratio mass spectrometry: Paleodietary implications from intra-individual comparisons. *Rapid Communications in Mass Spectrometry* 24(5):541–548.

Ramsey, Christopher Bronk, Richard A. Staff, Charlotte L. Bryant, Fiona Bronk, Hiroyuki Kitagawa, Johannes van der Plicht, Gordon Schlolaut, Michael H. Marshall, Achim Brauer, Henry F. Lamb, Rebecca L. Payne, Pavel R. Tarasov, Tsuyoshi Haraguchi, Katsuya Gotand, Hitoshi Yonenobu, Yusuke Yokoyama, Ryuji Tada, and Takeshi Nakagawa. 2012. A complete terrestrial radiocarbon record for 11.2 to 52.8 kyr B.P. *Science* 338(6105):370–373.

Rasmussen, Morten, Yingrui Li, Stinus Lindgreen, Jakob Skou Pedersen, Anders Albrechtsen, Ida Moltke, Mait Metspalu, Ene Metspalu, Toomas Kivisild, Ramneek Gupta, Marcelo Bertalan, Kasper Nielsen, M. Thomas P. Gilbert, Yong Wang, Maanasa Raghavan, Paula F. Campos, Hanne Munkholm Kamp, Andrew S. Wilson, Andrew Gledhill, Silvana Tridico, Michael Bunce, Eline D. Lorenzen, Jonas Binladen, Xiaosen Guo, Jing Zhao, Xiuqing Zhang, Hao Zhang, Zhuo Li, Minfeng Chen, Ludovic Orlando, Karsten Kristiansen, Mads Bak, Niels Tommerup, Christian Bendixen, Tracey L. Pierre, Bjarne Grønnow, Morten Meldgaard, Claus Andreasen, Sardana A. Fedorova, Ludmila P. Osipova, Thomas F. G. Higham, Christopher Bronk Ramsey, Thomas v. O. Hansen, Finn C. Nielsen, Michael H. Crawford, Søren Brunak, Thomas Sicheritz-Pontén, Richard Villems, Rasmus Nielsen, Anders Krogh, Jun Wang, and Eske Willerslev. 2010. Ancient human genome sequence of an extinct Palaeo-Eskimo. *Nature* 463(11):757–762.

Santos, G. M., J. R. Southon, K. C. Druffel-Rodroguez, S. Griffin, and M. Mazon. 2004. Magnesium perchlorate as an alternative water trap in AMS graphite sample preparation; a report on sample preparation at KCCAMS at the University of California, Irvine. *Radiocarbon* 46(1):165–173.

Scott, Marian E., Gordon T. Cook, and Philip Naysmith. 2010. A report on phase 2 of the Fifth International Radiocarbon Intercomparison (VIRI). *Radiocarbon* 52(3):846–858.

Smith, Colin I., Benjamin T. Fuller, Kyungcheol Choy, and Michael P. Richards. 2009. A three-phase liquid chromatographic method for $\delta^{13}C$ analysis of amino acids from biological protein hydrolysates using liquid chromatography-isotope ratio mass spectrometry. *Analytical Biochemistry* 390:165–172.

Smith, David Glenn. 2000. Attachment A: A review of documents and evidence pertaining to the suitability of the skeletal remains from Kennewick, Washington, for DNA studies, to Chapter 1: Examination of the Kennewick Remains—Taphonomy, Micro-sampling, and DNA Analysis. In *Report on the DNA Testing Results of the Kennewick Human Remains from Columbia Park, Kennewick, Washington* [September 2000]. National Park Service, http://www.nps.gov/archeology/kennewick/GlennSmith.htm.

Smith, David Glenn, Ripan S. Malhi, Jason A. Esheman, and Frederika A. Kaestle. 2000. Chapter 4: Report on DNA analysis of the remains of "Kennewick Man" from Columbia Park, Washington." In *Report on the DNA Testing Results of the Kennewick Human Remains from Columbia Park, Kennewick, Washington* [September 2000]. National Park Service, http://www.nps.gov/archeology/kennewick/Smith.htm.

Stafford, Thomas W., Jr., A. J. T. Jull, Klaus Brendel, Raymond C. Duhamel, and Douglas Donahue. 1987. Study of bone radiocarbon dating accuracy at the University of Arizona NSF accelerator facility for radioisotope analysis. *Radiocarbon* 29(1):24–44.

Stafford, Thomas W., Jr., P. E. Hare, Lloyd Currie, A. J. T. Jull, and Douglas J. Donahue. 1991. Accelerator radiocarbon dating at the molecular level. *Journal of Archaeological Science* 18(1):35–72.

Stuiver, Minze, and Henry A. Polach. 1977. Discussion; Reporting of ^{14}C data. *Radiocarbon* 19(3):355–363.

Stuiver, Minze, G. W. Pearson, and Tom Braziunas. 1986. Radiocarbon age calibration of marine samples back to 9000 CAL YR BP. *Radiocarbon* 28(2B):980–1021.

Taylor, R. E. 1999. UCR Kennewick results and responses to inquiries of 12/7/99 and 12/17/99. Letter to Dr. Frank McManamon, 20 December 1999. National Park Service, http://www.cr.nps.gov/archeology/kennewick/taylor.htm.

Taylor, R. E. 2000. Attachment B: Amino acid composition and stable carbon isotope values on Kennewick skeleton bone to Chapter 1: Examination of the Kennewick Remains—Taphonomy, Micro-sampling, and DNA Analysis. In *Report on the DNA Testing Results of the Kennewick Human Remains from Columbia Park, Kennewick, Washington* [September 2000]. National Park Service, http://www.nps.gov/archeology/kennewick/taylor2.htm.

Taylor, R. E., Donna L. Kirner, John R. Southon, and James C. Chatters. 1998. Radiocarbon dates on Kennewick Man. *Science* 280(5367):1171–1172.

Taylor, R. E., David Glenn Smith, and John R. Southon. 2001. The Kennewick skeleton: Chronological and biomolecular contexts. *Radiocarbon* 43(2B):965–976.

Tuross, Noreen, and Connie J. Kolman. 2000. Potential for DNA testing of the human remains from Columbia Park, Kennewick, Washington [Feb. 3, 2000]. National Park Service, http://www.nps.gov/archeology/kennewick/tuross_kolman.htm.

Wakeley, Lillian D., William L. Murphy, Joseph B. Dunbar, Andrew G. Warne, Frederick L. Briuer, and Paul R. Nickens. 1998. *Geologic, Geoarchaeologic, and Historical Investigation of the Discovery Site of Ancient Remains in Columbia Park, Kennewick, Washington*. U.S. Army Corps of Engineers Technical Report GL-98-13. Prepared for U.S. Army Engineer District, Walla Walla. U.S. Army Corps of Engineers Waterways Experiment Station, Vicksburg, MS.

Walker, Phillip L., Clark Spencer Larsen, and Joseph F. Powell. 2000. Chapter 5: Final report on the physical examination and taphonomic assessment of the Kennewick human remains (CENWW.97.Kennewick) to assist with DNA sample selection. In *Report on the DNA Testing Results of the Kennewick Human Remains from Columbia Park, Kennewick, Washington* [September 2000]. National Park Service, http://www.nps.gov/archeology/kennewick/walker.htm.

Waters, Michael R., and Thomas W. Stafford, Jr. 2007. Redefining the age of Clovis: Implications for the peopling of the Americas. *Science* 315(5815):1122–1126.

Zdanowicz, C. M., G. A. Zielinski, and M. S. Germani. 1999. Mount Mazama eruption: Calendrical age verified and atmospheric impact assessed. *Geology* 27(7):621–624.

The Precedent-Setting Case of Kennewick Man   4

Alan L. Schneider and Paula A. Barran

James Chatters and Douglas Owsley had no inkling of what they were getting into when they made plans in August 1996 to investigate a skeleton that Chatters had recovered the preceding month at a site in eastern Washington State. The only issues that they could foresee at the time were scientific—what could be learned from this nearly complete, more than 9,000-year-old skeleton. Only one other existing North American skeleton, Spirit Cave Man from Nevada, was as old or older and as complete as this one. Like Spirit Cave Man and many other early New World skulls, the Kennewick Man skeleton (as it would soon be called) did not look Native American. It had a stone projectile point in its hip and other signs of an eventful, trauma-filled history.

Chatters' and Owsley's plans never had a chance. The federal government had its own plans for the skeleton, and they did not include study by scientists. Those government plans and decisions precipitated a precedent-setting lawsuit that dragged on for more than eight years. The safety of the skeleton was jeopardized, its discovery site ruined, and millions of dollars wasted before a series of court decisions forced the government to give scientists access to the skeleton.[1]

Those court decisions are the subject of this chapter. In terms of eventual impact, they may be as significant as the scientific study results described in this book. New precedents were set placing important limitations on what federal agencies must do, or not do, when exercising their regulatory powers over archaeological materials and sites. These precedents have the potential to influence how cultural resource laws in the United States will be interpreted and applied for decades to come.

SUMMARY OF CASE EVENTS
Discovery of the Skeleton

The skeleton was discovered on July 28, 1996, by two young men who were walking along the Columbia River near Kennewick, Washington, on their way to watch hydroplane races (Chatters 2001:19, 23–24). They found a human cranium in the shallow water near the riverbank, and after the races were over they reported their discovery to the police, who referred the matter to the county coroner. He called on Chatters to investigate the find. Since the skeleton had been found on federal land managed by the US Army Corps of Engineers (Corps), Chatters applied to the Corps on July 30, 1996, for a permit under the Archaeological Resources Protection Act to authorize collection of whatever could be found at the site (Chatters 2001:36). The permit was issued and made retroactive to July 28, 1996 (COE 9485, 9494, 9500). Over the next several weeks, Chatters gathered approximately 350 bones and bone fragments, representing at least 143 skeletal elements (Chatters 2000:301).

To resolve questions about the skeleton's age,[2] Chatters sent a metacarpal to the University of California, Riverside for radiocarbon dating. Tests there produced a radiocarbon age of 8,410 ± 60 years (equivalent to approximately 8,340 to 9,200 calendar years). Chatters contacted Owsley and arrangements were made for Chatters to fly the skeleton to the Smithsonian Institution in Washington, DC, where he and Owsley would study it with the help of other scientists (Benedict 2003:107–108). Chatters was scheduled to leave on September 8 (Chatters 2001:74).

Chatters informed the Corps of the radiocarbon dating results (Chatters 2001:57). On August 30, the Corps had the skeleton taken from Chatters' possession and transferred to an evidence locker at the sheriff's office (Benedict 2003:109–114; Chatters 2001:76, 83–84).[3] Chatters was also ordered to stop further testing of the metacarpal bone, which already had been sent to the University of California at Davis for DNA testing (Chatters 2001:48–50, 76).

Upon learning of the skeleton's age, local area American Indian tribes[4] immediately demanded that it be given to them for burial. The Corps was resolved to demonstrate that it was "a compassionate and supportive partner" of the tribes, and it did not want this matter to interfere with other problems involving the Tribal Claimants (Chatters 2001:97). Without making any real effort to investigate the facts of the matter and the Tribes' claims,[5] it assured them that "the skeleton would not be subjected to further desecration via scientific study" (ACOE 0759). On September 17, the Corps published a newspaper announcement that the skeleton would be given to a coalition of four tribes and one unrecognized band (the "Tribes" or "Tribal Claimants") (COE 9270). The date set for the transfer was October 24 (COE 9277). Owsley and other scientists requested permission to examine the skeleton before it was given to the Tribal Claimants for burial. The Corps refused because of Tribal religious objections.

When protests from scientists, members of Congress, and the public failed to persuade the Corps to change its mind,[6] eight scientists (the "plaintiffs") filed a lawsuit in federal district court in Portland, Oregon, to stop the transfer and obtain access to study the skeleton. An emergency hearing was held on October 23, 1996. The Corps agreed at the hearing to postpone the transfer indefinitely, and it was ordered not to release the skeleton without prior notice to the plaintiffs (Court Docket 29).

The Litigants

The plaintiffs in the lawsuit were: Robson Bonnichsen, Oregon State University, Corvallis; C. Loring Brace, University of Michigan, Ann Arbor; George W. Gill, University of Wyoming, Laramie; C. Vance Haynes, University of Arizona, Tucson; Richard Jantz, University of Tennessee, Knoxville; Douglas Owsley, Smithsonian Institution; Dennis Stanford, Smithsonian Institution; D. Gentry Steele, Texas A&M University, College Station.[7] Plaintiffs filed the lawsuit as individual persons; their employing institutions were not parties to the litigation.

The initial defendants in the case were the United States, the Department of Defense, the Corps, and various government officials. After the Secretary of the Interior Gale Norton's September 2002 determinations in favor of the Tribal Claimants, plaintiffs amended their compliant to add the Department of the Interior (Interior), the National Park Service, and some of their officials as defendants in the lawsuit.

The Tribal Claimants were a coalition consisting of the Confederated Tribes of the Colville Reservation, the Confederated Tribes of the Umatilla Indian Reservation, the Confederated Tribes of the Yakama Reservation, the Nez Perce Nation, and the Wanapum Band.[8] Although the Tribal Claimants were not parties to the lawsuit, they participated in the case as amici, which allowed them to take part in hearings and file briefs with the court.[9]

1997 Court Decisions

Various motions were filed in court and decided. Defendants started with a motion to dismiss plaintiffs' complaint, which was denied in February 1997 (District Court 1997a). Plaintiffs then filed a motion to obtain access to study the skeleton. Defendants countered with a motion for partial summary judgment. Another hearing was held, and on June 27, 1997, the court set aside all of the Corps' decisions relating to the skeleton (District Court 1997b).

The court found that the Corps had reached a "premature" decision without adequately considering all relevant factors or aspects of the matter. The Corps also prejudged the outcome, "assumed facts that proved to be erroneous," and "failed to articulate a satisfactory explanation for its actions" (District Court 1997b:642, 645). The court remanded the matter to the Corps for reconsideration. The Corps was directed "to fully reopen this matter, to gather additional evidence, to take a fresh look at the legal issues involved," and to reach new decisions based on all relevant evidence. It was also instructed to store the remains "in a manner that preserves their potential scientific value" (District Court 1997b:654). Plaintiffs' motion for access to study the remains was denied, without prejudice to the possibility of renewing their motion after the Corps concluded its new proceedings.

Efforts to Study the Discovery Site

Two months after the June 1997 court hearing, a team of scientists organized by plaintiffs applied to the Corps for a permit to investigate the skeleton's discovery site (COE 7176). The team consisted of Bonnichsen, Chatters, Haynes, Gary Huckleberry, and Thomas Stafford (Plaintiffs' Site Team). Their plan called for the excavation of a 50- to 100-foot-long trench near the riverbank where the skeleton had been found, and auger-hole or shovel-probe tests elsewhere throughout the site. The objective was to obtain detailed information about site geology, and to determine if any intact cultural deposits were present.

The Tribes objected to the permit application,[10] and the Corps decided to conduct its own investigations of the site (U.S. Army Corps of Engineers 1997). Plaintiffs' Site Team and Tribal representatives were allowed to participate as assistants to the Corps investigators. Fieldwork conducted on December 13–17, 1997, consisted primarily of a surface survey (COE 4250, 4896, 5815). No trench was excavated, and subsurface testing was limited to Vibracore samples taken at only five locations (none drilled in the terrace immediately above the location where the skeleton had been found) (COE 4250, 5815). Efforts to create a coherent stratigraphic profile of the sediments in the riverbank were frustrated by restrictions limiting the width of sediment exposures to not more than one meter and requiring that they be spaced at intervals of 10 to 40 meters along the bank. Moreover, many were only approximations because the Tribes objected to having the exposures cut back to produce clean vertical faces for observation.

Despite recommendations from Plaintiffs' Site Team and its own investigators, the Corps refused to authorize any further studies of the site. It later admitted that its investigations had failed to establish the site's characteristics or boundaries or to determine whether the site contained "any additional *in situ* and significant archaeological resources" (COE 5673).

Interior Gets Involved

The Corps requested advice from the Interior concerning some of the issues involved in the case. On December 23, 1997, Interior's chief consulting archaeologist, Francis McManamon, issued an opinion letter stating that all human remains that predate "the historically documented arrival of European explorers" would be deemed "Native American" and thus subject to disposition under the Native American Graves Protection and Repatriation Act (NAGPRA) (DOI 2128). This rule would apply even if the remains in question had no demonstrated connection to present-day American Indians. On March 24, 1998, the Corps and Interior entered into an agreement assigning to Interior responsibility for determining whether the Kennewick Man skeleton was Native American under NAGPRA and, if it was, to whom it should be given.

Mishandling of the Skeleton

Concerns quickly arose about the safety of the skeleton. Information surfaced in September 1997 revealing that the Corps had given Tribal representatives access to the skeleton (COE 6961). The Corps dismissed these incidents as posing no threat to the skeleton (COE 6730). However, they included religious ceremonies with the remains during which Tribal representatives were allowed to handle the skeleton, to place cedar boughs in the storage container, and to burn plants over the bones. These revelations were followed by news that parts of both femurs could not be found (Court Docket 132).[11] A few weeks later, it was learned that several bone fragments thought to belong to the skeleton were taken by Tribal representatives from the Corps repository cabinet at Pacific Northwest National Laboratory and secretly buried.

In response to these developments, the court ordered a hearing for May 28, 1998 (Court Docket 146). The Corps was ordered to move the skeleton to a safer repository. The Burke Museum of Natural History and Culture, Seattle, was later selected as the new repository (Court Docket 168).[12] The court ordered that plaintiffs be allowed to inspect the skeleton before it was moved. Chatters and Owsley examined it on October 28, 1998 (COE 4304).[13] They compiled a detailed inventory of the skeleton and identified and determined the body side of many of the fragments. Chatters observed that over the preceding two years new cracks had developed in the skull and other bones and that some existing cracks had increased in size.

Burial of the Discovery Site

The skeleton's discovery site was also mishandled. In September 1996 the Corps began working on plans to "stabilize" the site by covering it with a combination of sand, rock fill, and revegetation (COE S 409, S 937). Tribal representatives were concerned that the plan lacked sufficient "armoring" or "hardening" (COE S 907, S 939), and it was shelved. When Plaintiffs' Site Team filed their application to investigate the site, a new plan was quickly developed. This plan called for the placement of five parallel rows of coconut fiber logs in a stairstep pattern along a 237-foot span of the riverbank. The intervals between the log rows would be filled with rocks and small boulders and then covered with dirt and erosion control mats. The project would be capped with rows of willow trees (COE S 672).

The new plan was discussed with Tribal representatives (COE S 450). They wanted the site "covered as quickly as possible" (COE 5766), and so did the White House (COE S 821).[14] Plaintiffs were given only three days to comment on the project before it was sent out for bids (COE S 614, S 596).[15] They responded with numerous objections (COE S 534, S 596). A Corps expert on site preservation also expressed reservations about the plan.[16] These complaints were brushed aside,[17] and work on the project continued.

Bills were introduced in Congress to prohibit the Corps from carrying out the project without prior approval from the district court. The bills passed both houses of Congress and were waiting only for a conference committee to resolve differences in unrelated parts of the legislation.[18] The Corps announced that it would comply with the bills, but when Congress recessed for Easter with the legislation still in conference committee, the Corps reversed itself and ordered the project to go forward. Work began on April 6, 1998, and was substantially completed by April 14 (COE S 272, S 278). Approximately one million pounds of materials were transported to the site by helicopter and put in place (Figure 5.1).[19]

Interior's Investigations

After consulting with the Tribal Claimants, Interior began its investigation of the skeleton in February 1999.[20] Scientists were hired to analyze the skeleton's physical characteristics, the projectile point embedded in its hip, and the sediments adhering to the bones. They reported that the skeleton did not resemble modern American Indians. Craniometric analyses found that the skeleton's closest associations were with Polynesian and Ainu samples. However, those associations were remote, and probabilities were low of membership in any modern population. Analyses of cranial and dental discrete traits produced similar results. Odontometric data were more suggestive of a possible connection between the skeleton and modern American Indians, but the typicality probabilities were still very low. Interior's investigators did not rule out the possibility that the skeleton was ancestral to modern American Indians (DOI 10684). However, they also stated that "[o]nly a time series analysis of populations . . . extending from earliest occupation to the historic period, can provide a statistically valid means of assessing morphometric continuity of populations through time. Data for performing such an analysis are currently unavailable" (DOI 10685).[21]

Examination of the projectile point in the skeleton's hip indicated that it was "a possible or probable Cascade point." The investigator on this issue was unable to give a more definite opinion because of insufficient detail in the X-ray and CT scan images taken of the hip fragment containing the projectile point. Analyses of the sediments adhering to the bones produced few definitive results other than support for the hypothesis that the skeleton had been deposited at the site at least 7,600 years ago.

In fall 1999 additional radiocarbon dating tests were conducted.[22] Only one of the four samples tested had enough bone collagen to produce a date that was considered reliable. That date, 8410 ± 40 radiocarbon years ago, was virtually identical to the date obtained in 1996 ($8,410 \pm 60$ radiocarbon years ago). Attempts at DNA testing were unsuccessful.

Investigators were hired to gather and report on existing published information relating to regional archaeology, bioarchaeological and mortuary data, linguistics, and Tribal oral traditions. Interior's archaeological investigator concluded that "empirical gaps" in the archaeological record made it impossible to establish "cultural continuities or discontinuities" in the region over the past 9,500 years. The review of existing data on regional mortuary practices and bioarchaeological topics was equally inconclusive. The linguistics investigator was more positive. He asserted that the Tribal Claimants were descendants of a people who spoke a "proto-Penutian" language and who had resided in the region for at least the past 8,000 to 9,000 years (DOI 10310-10311, 10323). However, he also conceded that the existence of an actual proto-Penutian language phylum had yet to be established and that it was possible that Kennewick Man spoke a non-Penutian language (DOI 10326).[23] Interior's folklore investigator declared that the Tribal Claimants' oral traditions included accounts that related to events "that occurred in the distant past" and "are highly suggestive of a long-term establishment of the present-day tribes" in the Columbia Plateau for over 10,000 years. He admitted, however, that events described in "oral traditions" cannot be dated with precision and that attempting to use them to establish actual time sequences "does not meet with much success" (DOI 10298).[24]

New Agency Findings

On January 13, 2000, Interior announced that the skeleton would be deemed to be Native American under NAGPRA because it was pre-Columbian in age. On September 21, 2000, Secretary of the Interior Bruce Babbitt issued a determination letter awarding the skeleton to the Tribal Claimants. He said that the skeleton was culturally affiliated to them because "collected oral tradition evidence suggests a continuity between the cultural group represented by the Kennewick human remains and the modern-day claimant Indian tribes." As an alternative basis for awarding them the skeleton, he also determined that it had been found on land aboriginally occupied by the ancestors of one of the coalition members, the Umatilla. Based on this determination, the Corps reaffirmed its refusal to approve plaintiffs' study requests.

Plaintiffs then amended their complaint and filed a motion to set aside the new agency findings (Court Docket 416). A hearing was held on June 19–20, 2001 (Court Docket 456, 466); on August 30, 2002, the district court issued a 51-page opinion and order setting aside the agencies' decisions. The court found that the Secretary's award of the skeleton to the Tribal Claimants was contrary to both applicable law and the evidence in the case. In addition to many mistakes made under NAGPRA, the agencies were found to have violated a number of other federal substantive and procedural statutes. Rather than remanding the case to the agencies for new administrative proceedings as is customary when agency determinations are set aside, the court entered a final judgment in favor of plaintiffs allowing them to study the skeleton.

The agencies and three of the Tribal Claimants—Nez Perce, Umatilla, and Yakama—appealed to the Ninth Circuit Court of Appeals. Arguments presented to the appeals court included amicus briefs from a number of individuals and organizations.[25] Three supported the agencies and the Tribal Claimants,[26] and six supported the plaintiffs.[27] On February 4, 2004, the Ninth Circuit affirmed the district court's judgment and order (Ninth Circuit 2004).[28] The Tribes filed a petition with the Ninth Circuit for a rehearing en banc, and the petition was denied (Ninth Circuit 2004 at 868).[29]

Plaintiffs' Studies

Even with two court orders in plaintiffs' favor, gaining access to the skeleton was not routine or easy. The agencies raised numerous objections to plaintiffs' study plan[30] and until November 4, 2004, would not even agree to permit a visual inspection of the skeleton. An inspection visit was eventually conducted on December 14–15, 2004, to assess the skeleton's condition and to obtain information for refining details of plaintiffs' study plan.[31] The plan was approved on March 29, 2005,[32] and studies of the skeleton were conducted in July 2005 (taphonomic investigation) and February 2006 (general studies).

APPLICABLE STATUTES

The lawsuit required interpretation and application of a number of federal statutes. For archaeologists and other prehistorians the most significant issues in the case involved questions arising under: the Native American Graves Protection and Repatriation Act of 1990, the National Historic Preservation Act of 1966, the Archaeological Resources Protection Act of 1979, and the Administrative Procedure Act of 1946.

If they are followed by other courts, the rulings that were made in the lawsuit under these statutes will require major changes in the way archaeological materials and sites are administered in situations covered by federal law. Museums and federal agencies will have to change their criteria for making NAGPRA classifications of ancient human remains, and for how they evaluate repatriation claims based on cultural affiliation or aboriginal occupation of an object's discovery site. Federal agencies will have to take more care when undertaking activities that might adversely affect an archaeological site. They will also have to reassess their policies and procedures to ensure that they act as truly neutral and unbiased decision makers when confronted with challenges to repatriation claims.[33]

Rulings were also made in the lawsuit on issues arising under the Freedom of Information Act and the Equal Access to Justice Act. Those rulings are not reviewed here.

NAGPRA Rulings

Under NAGPRA, Native American and Native Hawaiian individuals, tribes, and organizations can claim ownership of human remains and other "cultural items" found on federal or tribal land or held in a federal or museum collection.[34] Such ownership claims can be allowed on the basis of where the object was found or the claimant's genealogical or cultural connection to the object. In the Kennewick Man case, the Tribal Claimants were said to be entitled to the skeleton on two separate grounds—cultural affiliation and aboriginal occupation of the discovery location.[35] Having reached this conclusion, the government then declared that the skeleton could not be studied without the Tribal Claimants' consent.[36] The Tribal Claimants opposed study of the skeleton for religious reasons.

Three issues were thus raised. Did NAGPRA apply to the Kennewick Man skeleton? If it did apply, did the Tribal Claimants have a valid claim to the skeleton? If NAGPRA applied, did it preclude study of the skeleton by plaintiffs?

Meaning of "Native American"

NAGPRA applies only to remains and objects that are "Native American." Consequently, the meaning of that term was one of the pivotal issues in the case. If the Kennewick skeleton was not Native American (or could not be shown to be Native American), it could not be given to the Tribal Claimants under NAGPRA and the statute could not be used as a justification for denying plaintiffs' study requests.

The statute defines Native American as meaning "of, or relating to, a tribe, people, or culture that is indigenous to the United States" [25 USC §3001(9)]. The government took the position that any item that predates European exploration automatically qualifies as Native American even if it lacks any verifiable relationship to present-day American Indians. Plaintiffs argued that this position was contrary to the statute. The district court agreed. It held that the government's interpretation failed to give appropriate meaning to the words of the statute. Congress' use of the word *is* in the phrase "that is indigenous," together with the statute's purpose and legislative history, indicate that NAGPRA was intended to apply only to those objects that are related to present-day American Indians.[37] To hold otherwise would be to assume that Congress intended to create an "odd or absurd" result since it would arbitrarily classify as Native American the remains of "possibly long-extinct immigrant peoples" who have no significant connection, genetically or culturally, to any surviving group (District Court 2002:1137–1138). Consequently, the Kennewick Man skeleton could not be considered Native American unless it was shown to have some cultural relationship to modern American Indians.

The court then looked to see if the evidence was sufficient to establish a connection between the Kennewick Man skeleton and modern American Indians. It was not. Evidence from linguistics and archaeology were not helpful since no one knew what language Kennewick Man spoke and since there were too many gaps and discontinuities in the archaeological record to draw any reliable conclusions. Bone samples from the skeleton had failed to produce any replicable DNA, and the characteristics of the skull were dissimilar to those of all modern populations—including American Indians. Nor could any reliance be placed on the Tribal Claimants' oral traditions that were said to show that the Tribes and their ancestors have always lived in the area. The relevance of such accounts was problematic since no one knew when they first arose (District Court 2002:1154). And, even if they were as old as they were claimed to be, it could not be assumed that they accurately reflected actual historical events. As the court noted, "we cannot know who first told a narrative, or the circumstances, or the identity of the intervening links in the chain, or whether the narrative has been altered, intentionally or otherwise, over time" (District Court 2002:1152).

The Ninth Circuit agreed with these conclusions. It found that the statute's language, objectives, and legislative history all clearly demonstrate that "Congress enacted NAGPRA to give American Indians control over the remains of their genetic and cultural forbearers, not over the remains of people bearing no special and significant genetic or cultural relationship to some presently existing indigenous tribe, people, or culture" (Ninth Circuit 2004:879). The Ninth Circuit agreed with the district court that such a relationship had not been established in this case:

> because Kennewick Man's remains are *so* old and the information about his era is *so* limited, the record does not permit the Secretary to conclude reasonably that Kennewick Man shares special and significant genetic or cultural features with presently existing indigenous tribes, people, or cultures. We thus hold that Kennewick Man's remains are not Native American human remains within the meaning the NAGPRA and that NAGPRA does not apply to them (Ninth Circuit 2004:880).

The potential consequences of this relationship test are significant. The assumption that there is a descendant-ancestor relationship between a modern tribe and every prehistoric skeleton found in the area that the tribe once occupied has been used to dispose of some of the oldest remains ever discovered in this country.[38] In addition, NAGPRA does not always require a demonstration of cultural affiliation in order for a tribe to claim human remains and other cultural items that qualify as Native American. For example, any cultural item deemed to be Native American can be claimed by a tribe if it was found on either the tribe's "tribal land"[39] or on land that the tribe aboriginally occupied. In such cases, the tribe does not have to demonstrate that it has any particular connection to the item being claimed. In addition, Regulations adopted by Interior since the Kennewick Man case was decided contemplate that virtually all culturally unidentifiable human remains[40] found in the United States will ultimately be transferred to Tribes or coalitions of Tribal claimants [43 CFR §10].[41] These provisions only apply, however, to remains that are determined to be "Native American" which can be problematic under the Kennewick Man relationship test. In many cases, particularly those involving older remains, the information available for determining whether they are, or are not, related to living Native Americans is meager at best and in most cases nonexistent.

No other reported case decision has had occasion to consider this issue. Consequently, the relationship test is the applicable rule only in those states and territories located in the Ninth Circuit.[42] However, in testimony before the Senate Committee on Indian Affairs in July 2005, the Department of the Interior announced that it concurred with the Ninth Circuit's relationship test (Hoffman 2005).[43]

On the political front, the Senate Committee on Indian Affairs has entertained several bills designed to overturn the Ninth Circuit's decision.[44] On the other hand, a bill was introduced in the House in 2006 to make the Ninth Circuit's decision binding nationwide.[45]

Cultural Affiliation

In order to have a complete record for possible appellate review, the district court considered the question of whether the Tribal Claimants would have had valid grounds for claiming the skeleton if it had been found to be Native American within the meaning of the statute. Addressing this issue helped increase the chances that a final judgment could be reached in the case without the need for any more administrative proceedings. Even if the appellate court were to disagree with the district court's ruling on the Native American issue, the court's judgment could still be affirmed if there were other meritorious grounds for setting aside the Secretary of the Interior's determination.

Cultural affiliation was one basis used by the Secretary for deciding that the Tribal Claimants were entitled to the Kennewick Man skeleton. The district court rejected this finding. Under the statute, claims based on cultural affiliation require proof of "a relationship of shared group identity which can be reasonably traced" between the claiming tribe and "an identifiable earlier group" to which the remains or object in question once belonged [25 USC §3001(2)]. The court found that the Secretary's determination was wrong for a number of reasons.

One reason was his failure to specify the "identifiable earlier group" to which Kennewick Man belonged. His passing reference to "the cultural group that existed in the Columbia Plateau region during the lifetime of the Kennewick Man" was inadequate since the record in the case indicated that as many as 20 different groups may have resided in the region around that time. The Secretary was unable to say which of these groups was Kennewick Man's. Equally unhelpful was the Secretary's reference to several names—"Windust Phase" and "Early Cascade Phase." The court noted that these terms were merely "broad labels" used by archaeologists to describe various projectile points and other artifacts that were used by ancient people over eras spanning thousands of years (District Court 2002:1144). Such labels do not necessarily equate to group identities, and in any event there was no evidence that Kennewick Man's group actually fabricated any of the items used to distinguish the archaeological phases in question. These defects were fatal. "[W]ithout proof of a link between Kennewick Man and an 'identifiable earlier group' there is no reasoned starting point from which to evaluate a shared group identity between the present-day Tribal Claimants and a particular earlier group" (District Court 2002:1147).

The Secretary's cultural affiliation determination was invalid for a second reason. Even if his vague references were considered sufficient to identify Kennewick Man's group, there was no evidence to show that the latter had a "shared group identity" with the Tribal Claimants. A shared group identity requires at a minimum "some common elements of language, religion, customs, traditions, morals, arts, cuisine, and other cultural features; a common perspective on the world and the group's role within it; and shared experiences that are part of the group's perception of its history" (District Court 2002:1147–48). Here there was no evidence to show that the Tribal Claimants and Kennewick Man's group shared such a common culture. This failure of proof could not be remedied by the Tribes' oral traditions concerning how long they and their ancestors have resided in the region. Even if such narratives were assumed to be true, they did not establish that the Tribal Claimants' ancestors had the same culture as Kennewick Man's group (District Court 2002:1155–1156).

The Secretary's cultural affiliation determination was also defective because he accepted a coalition of four tribes (and one unrecognized band) as a proper claimant under NAGPRA. The district court ruled that such an interpretation was inconsistent with the plain language of the statute, which refers to a claimant as a "tribe" in the singular tense. While there may be some exceptional circumstances in which joint claims would be proper, such circumstances were not presented in this case. Moreover, even when such circumstances are present, the claims of each coalition member must be separately analyzed to see if at least one of them meets all requirements of the cultural affiliation standard (District Court 2002:1142, 1143). Such an analysis had not been conducted by the Secretary. Contrary to arguments made by the government's attorneys, the Secretary had expressly rejected any need to evaluate each tribe's claim on its individual merits. Instead he had treated the Tribal Claimants as a single, collective entity.

Aboriginal Occupation
The Secretary's determination that the Tribal Claimants were entitled to the skeleton based on aboriginal occupation of its discovery location also failed to conform to statutory requirements. NAGPRA provides that under some circumstances a tribe can claim cultural items that are "discovered on Federal land that is reorganized by a final judgment of the Indian Claims Commission or the United States Court of Claims as the aboriginal land" of the claimant [25 USC §3002 (a)(2)(C)]. The Secretary had concluded that the statute's reference to a "final judgment" did not preclude reliance upon a case that had been settled before entry of a final judgment as long as the case included findings of aboriginal occupation. The District Court ruled that this interpretation was contrary to the statute's unambiguous language. Moreover, contrary to the Secretary's assumptions, the Indian Claims Commission case that was at issue did not make any findings of aboriginal occupation in favor of any of the Tribes that claimed the Kennewick Man skeleton. And, in any event, what findings it did make were vacated by the subsequent settlement (District Court 2002:1158–1160).

Although the government did not appeal any of these rulings on cultural affiliation and aboriginal occupation, they were appealed by three of the Tribal Claimants. The Ninth Circuit declined to rule on these matters since the Native American issue was dispositive of the case.

Standing
The agencies argued that the merits of their NAGPRA decisions could not be questioned in this lawsuit because

plaintiffs lacked the necessary legal standing to challenge those decisions. Standing is a judicial test used to ensure that the plaintiff in a lawsuit is someone who has a definite and personal interest in the matter being disputed as opposed to a mere meddler whose interests are abstract or insignificant.[46] Defendants contended that the only persons who have standing to challenge NAGPRA decisions by a museum or a federal agency are persons who have a right under the statute to claim the remains or objects in dispute. In other words, it was their position that NAGPRA created a private club that was limited exclusively to museums, federal agencies, and Native Americans, and whose decisions could not be reviewed or questioned by any outsider.

The district court rejected such a restrictive interpretation of the statue and the federal standing requirements. It found that plaintiffs' interests in the skeleton were not remote or speculative: "The plaintiffs have asserted a personal interest in this controversy. They propose to personally conduct tests on the remains, and to analyze the results of those tests. This data will then be used to further their ongoing research. That is a sufficient nexus to confer standing" (District Court 1997b:634).

The fact that plaintiffs could not assert any ownership rights to the skeleton under NAGPRA did not mean that they were outside the zone of interests regulated by that statute:

> A person potentially may be injured by either *under*-enforcement or *over*-enforcement of a law.... The language of NAGPRA, 25 U.S.C. §3013—the district courts have "jurisdiction over *any* action brought by *any person* alleging a *violation of this chapter*" [emphasis added]—is broad enough to encompass such an over-enforcement claim, provided the plaintiffs allege a direct injury to themselves as a result of the misapplication of the statute and not just a general grievance or remote injury" (District Court 1997b: 636–637).

The Court of Appeals agreed (Ninth Circuit 2004:367).

NAGPRA's Effect on Study Rights

The question of what effect, if any, NAGPRA has on scientists' study rights was not resolved in the Kennewick Man case. Neither the district court nor the Ninth Circuit expressed any opinion on this issue, since both concluded that NAGPRA did not apply to the skeleton.

The issue may be unavoidable in future cases. Many museums and federal agencies take the position that items subject to possible claims under NAGPRA are off-limits to study by researchers. Study is sometimes prohibited even in cases where the item in question has not been claimed or where it has been determined to be culturally unaffiliated. Such prohibitions can have significant adverse consequences on the ability of scientists to carry out their research programs. In the absence of relevant case precedents, and currently there are none, the legitimacy of these practices is unclear.[47]

ARCHAEOLOGICAL RESOURCES PROTECTION ACT RULINGS

The Archaeological Resources Protection Act protects sites and objects, including human skeletal remains, found on federal land that are more than 100 years old and are "capable of providing scientific or humanistic understanding of past human behavior, cultural adaptation, and related topics .." [16 USC §470bb; 43 CFR §7.3(1)(a) and (3)(vi)]. Since the Kennewick Man skeleton was found on federal land, was more than 100 years old, and came from a critical but poorly understood period in North American prehistory, its coverage by the statute would seem obvious. Government attorneys tried to dispute this point, but they could not overcome the fact that the Corps had issued a permit under this Act for Chatters' collection of the skeleton in 1996. As the district court observed, issuance of this permit constituted "an implicit determination" that the skeleton met all statutory requirements, including the requirement of archaeological significance (District Court 2002:1165–1166).

The regulations implementing the Archaeological Resources Protection Act mandate, among other things, that objects removed from a site must be preserved and protected for the benefit of the American public, including study by qualified researchers [36 CFR §§79.10(a), (b)]. Federal agencies customarily and routinely allow scientists to work with skeletal remains and other objects from their collections. All that is generally required is a written or telephone request describing the researcher's qualifications, his or her research objectives, and the activities that will be conducted with the objects to be studied.[48]

Applying these standards and practices to the circumstances of the Kennewick Man case, the district court found that no legitimate grounds existed for rejecting plaintiffs' study request. It stated:

> Plaintiffs are clearly the kind of 'qualified professionals' referenced in the regulations. . . . The record establishes that Plaintiffs are eminent scientists in the field of "First American Studies" who have written hundreds of scientific articles, papers, and monographs, and have examined thousands of human skeletal remains. The record also establishes that, but for Defendants' assumption that NAGPRA applies, Plaintiffs almost certainly would have been allowed access to study the remains (District Court 2002:1166).

It therefore ordered the Corps to allow plaintiffs to study the skeleton subject only to normal and reasonable terms and conditions for studies under the Archaeological Resources Protection Act (District Court 2002:1167). The Ninth Circuit rejected the agencies' objections and affirmed the study order (Ninth Circuit 2004:882).

NATIONAL HISTORIC PRESERVATION ACT VIOLATIONS

The National Historic Preservation Act was an issue in the Kennewick Man case because of the Corps' actions in burying the skeleton's discovery site. Under the NHPA, federal agencies must take special precautions before undertaking any activity that might adversely affect, either physically or visually, a property that is listed on or eligible for listing on the federal register of historic places. The Kennewick Man discovery site is so eligible. The district court found that the Corps violated the NHPA by failing to: adequately investigate the site before covering it, allow plaintiffs a fair opportunity to present their objections to the project,[49] consider the information provided by plaintiffs, properly evaluate all potential adverse effects of the project, and consider less damaging alternatives (District Court 2002:1163–1164).[50]

The Corps attempted to justify its burial of the discovery site by arguing that the project was undertaken to protect the site from erosion. The court rejected this argument. It noted that one Corps scientist had written that erosion at the site was "not as serious as that occurring" at many other reservoirs managed by the Corps (District Court 2002:1163). Rather than erosion control, the primary objective of the project was to prevent further investigation of the site.[51] Burial of the site "hindered efforts to verify the age of the Kennewick Man remains, and effectively ended efforts to determine whether other artifacts are present at the site which might shed light on the relationship between the remains and contemporary American Indians" (District Court 2002:1126).

The district court's findings relating to the National Historic Preservation Act were not appealed.

ADMINISTRATIVE PROCEDURE ACT ISSUES

The Administrative Procedure Act (APA) sets standards that must be met by federal agencies when they make decisions that affect the rights or interests of third parties. Agency decision-making processes that are arbitrary and capricious, contrary to law, or otherwise unfair violate the APA. Plaintiffs claimed that the methods used by the Corps and Interior in this case violated some of the APA's most important and fundamental standards,[52] and the district court agreed.

The court found that the agencies had failed to act as impartial and unbiased decision makers in a fair process as required by the APA. Their lack of neutrality and fairness was demonstrated by the following conduct: secretly furnishing the Tribal Claimants with advance copies of documents such as expert reports while withholding copies from the plaintiffs until after the administrative record was closed; allowing the Tribal Claimants to rebut the agencies' reports and submit reports of their own before the administrative record was closed; secretly meeting with the Tribal Claimants to discuss the views of Interior's decision makers and the potential weaknesses in the Tribes' claims; giving the Tribal Claimants secret opportunities to influence the agency's decision makers and to supplement the record in response to their concerns; secretly sending letters to the Tribal Claimants; secretly notifying the Tribal Claimants that the aboriginal occupation issue was once again under consideration despite earlier assurances to plaintiffs and the court that it was not an issue in the case; refusing plaintiffs' requests for clarification of the issues being considered by the agencies; and burying the skeleton discovery site to prevent any further investigation of it (District Court 2002:1133).

In response to defendants' argument that their conduct was justified because they had an obligation under NAGPRA to consult with the Tribal Claimants, the court observed that nothing in the statute requires that an agency "side with claimants in any dispute or litigation, or prevents an agency from furnishing the same information to tribal claimants and others interested in the agency's determination" (District Court 2002:1134).

When a court sets aside an agency's administrative decisions, the customary remedy is to remand the matter to the agency for new proceedings consistent with the court's findings. Here, however, the district court concluded that the usual remedy would be inappropriate since "[t]his is far from the usual case" (District Court 2002:1164). The court had already remanded the case once before (in 1997) and had given the agencies detailed directions on how they should conduct their new proceedings. Those directions had been ignored. Having observed the agencies' conduct over the preceding five years, the court was not convinced "that, if this action were remanded yet again, plaintiffs' request to study would be evaluated in a fair and appropriate manner" (District Court 2002:1165). Accordingly, it decided to end the case itself by entering a judgment for plaintiffs.

Although the Ninth Circuit did not comment on the district court's findings, it apparently agreed with them since it did not order a new remand but instead affirmed the judgment for plaintiffs.[53]

CONCLUSION

The Kennewick Man lawsuit was not the first time a federal agency has been challenged in court over its NAGPRA decisions.[54] However, it was the first time that such a challenge was undertaken by scientists or anyone seeking to assert the interests of science in the NAGPRA process. The Kennewick Man plaintiffs did so without assistance from their employing institutions or the support of any professional association. Few observers thought in 1996 that they could win, and many pressures were brought on them to quit.

It was fortunate for scientific research on Paleoamericans that they persisted. Their lawsuit changed the course of federal cultural resource management in ways scarcely imagined a decade before. Among other things, it resulted in the first reported court decisions to:

1. decide that scientists have standing to challenge agency over-interpretation of NAGPRA;
2. rule on the meaning of the term "Native American";
3. determine what aboriginal occupation means for purposes of NAGPRA;
4. analyze and discuss all the elements needed to establish a cultural affiliation claim;
5. rule on the probative value of tribal oral traditions that are offered as proof of a claim;
6. consider the validity of coalition claims;
7. find that agencies must consider the concerns of scientists when undertaking activities that affect important archaeological sites;
8. order a federal agency under the Archaeological Resources Protection Act to allow scientific study of one of its collections; and
9. rule that federal agencies must deal fairly and openly with scientists who dispute cultural resource decisions.

While all these precedents are important, two of them are particularly significant for archaeologists and other scholars who study American prehistory. One is the ruling that was made on the meaning of the term "Native American." As a result of that ruling, agencies and museums can no longer assume that every prehistoric skeleton found on federal or tribal land in this country is subject to NAGPRA. Instead, they must produce credible, reliable evidence that the remains have a special and significant genetic or cultural relationship to present-day American Indians (or Native Hawaiians or Native Alaskans). In the case of ancient skeletal remains such as Kennewick Man, such evidence is difficult to find. If it cannot be found, NAGPRA cannot be used as a pretext to block scientific study of the remains in question. Instead, the controlling federal statute is the Archaeological Resources Protection Act, which was enacted for the specific purpose of promoting scientific study of archaeological sites and objects for the benefit of all Americans.

Equally significant is the ruling that scientists can have standing to challenge wrongful NAGPRA decisions.

NAGPRA was intended as compromise legislation that balanced the interests of Native Americans, museums, scientists, and the general public (Ousley et al. 2005:2). In actual practice, the balance intended by Congress has often failed to materialize because agencies and museums have disregarded the interests of scientists when making decisions under NAGPRA. The Kennewick Man rulings have reinserted scientists into the NAGPRA process and given them the power to ensure that their interests be taken into account when decisions are made.

Prior to the Kennewick Man litigation, little thought or discussion was given to the specific meaning of many key terms used in NAGPRA and how that statute interacts with other federal cultural resource laws. As the first reported decisions to consider many of the critical issues involved in this area, the district court and Ninth Circuit opinions are an important source of guidance for other courts and decision makers in future controversies.[55]

NOTES

1. Estimated federal government expenditures on the lawsuit and other matters relating to the skeleton were $3 million (or more) through September 2000 (Schneider 2004). Total expenditures since then are unknown, but include more than $2 million in attorney fees and costs awarded to the plaintiffs (District Court 2004).

2. The skeleton was originally thought to be the remains of an early European pioneer because of its cranial features and the presence of historic era artifacts at the site (Benedict 2003:99; Chatters 2001:20).

3. Chatters was given only a few hours to pack the skeleton and prepare it for transport. It was later moved by the Corps to an artifact repository at the Pacific Northwest National Laboratory in Richland, Washington (Chatters 2001:76, 82).

4. The court opinions in the case used the term "American Indian" rather than "Native American" for references of this kind since one of the disputed issues was the meaning of the term Native American as defined by federal law (District Court 2002:1120). This chapter will do likewise.

5. There were obvious defects and inconsistencies in the claims. For example, one Tribe's claim declared that the skeleton's cultural affiliation could not be determined (COE 9316). This Tribe asserted that it was entitled to ownership of the skeleton because an Indian Court of Claims decision had determined that the Tribe's ancestors had aboriginally occupied the skeleton's discovery location. However, the case cited by the Tribe did not support its claim.

6. Protests received by the Corps included a joint letter from Sen. Slade Gorton and Reps. Doc Hastings, Jack Metcalf, and George R. Nethercutt, Jr., urging the Corps to "reconsider its decision" (COE 9129).

7. Chatters was not named as a plaintiff because of concerns that unnecessary issues might be created if someone were included in the lawsuit who had already examined the skeleton.

8. The Wanapum Band is not recognized by the United States as an Indian tribe. Federal law allows only recognized tribes to pursue repatriation claims for human remains and other cultural items held by museums and federal agencies (District Court 2002:1141).

9. Following its August 2002 decision, the district court allowed three of the Tribal Claimants to intervene as full parties for purposes of appealing the court's decision to the Ninth Circuit.

10. The Tribes argued that investigation of the site was unnecessary, but if any work were to be authorized it should be done by them (COE 6899).

11. The missing pieces were found several years later in a different box in the coroner's evidence locker. Their disappearance has never been explained.

12. The Museum's storage charges to the Corps for the first four years were more than $260,000. The Smithsonian Institution would have stored the skeleton for free (Schneider 2004:205).

13. Like most other aspects of the case, the inspection visit was a contentious matter. Owsley was not allowed to have the help of an assistant to record his observations, and he and Chatters were required to complete their assessment and inventory in a single session that began at 7:00 a.m. and did not end until approximately 4:00 a.m. the next morning (COE 4304).

14. A special task force was created to coordinate matters relating to the skeleton and its discovery site. Members included representatives from the Corps, Dept. of the Army, Interior, Dept. of Justice, and the White House (COE 2504; DOI 2002).

15. Plaintiffs had made repeated requests starting in November 1996 for information about any Corps plans for work on the site (COE S 596).

16. He stated that erosion at the discovery site was "not as

serious as that occurring at many other Corps of Engineers Reservoirs" and advised that "it would seem advisable to be cautious about long term deleterious effects of engineering site protection measures" (COE S 431).

17. The Corps declared that it would not respond to plaintiffs' objections "point by point" (COE S 450).

18. The bills were attached to House and Senate appropriations legislation (COE S 329, S 331, S 348).

19. The project cost at least $166,000 (Schneider 2004:206). For only $5,000 to $10,000, the site could have been protected from erosion by planting shallow-rooted vegetation and anchoring hay bales along the shoreline. This technique has been used at other Corps managed sites.

20. The Tribal Claimants had differing opinions about these investigations. Some opposed any study of the skeleton (Yakama and Wanapum); one was undecided (Nez Perce); one favored noninvasive examinations (Umatilla); another wanted DNA testing (Colville) (COE 5137; DOI 2719, 2962, 3381). Members of the Colville later reversed their position and opposed DNA testing (DOI 6712).

21. One of these investigators (Joseph Powell) has since declared that any assertion that Kennewick Man was "unrelated to contemporary Native American tribes . . . is absurd" (Edgar et al. 2007:107). He did not explain why a "time series analysis" was no longer needed to reach this conclusion.

22. Bone samples taken for these tests totaled approximately 30 grams (COE 3857). Two years earlier, defendants had told the district court that plaintiffs' proposed radiocarbon dating tests were excessively destructive because they would have required use of four to six grams of bone (COE 7713, 7733). Plaintiffs' actual request was for only two grams of bone (COE 7980).

23. For a detailed critique of this proto-Penutian hypothesis, see Goddard and Shipley 2003.

24. For an analysis of factors that can affect the historical accuracy of oral traditions, see Custred and Simic 2003.

25. Amicus briefs are briefs filed by persons who are not parties to the lawsuit but believe that they have arguments that may be helpful to the court. Such briefs often focus on one or a limited number of issues and discuss them in greater detail than the parties' briefs.

26. These briefs were from Sherry Hutt (standing and deference to agency findings), the Association of American Indian Affairs and the Morning Star Institute (interpretation of NAGPRA), and the Haudenosaunee Standing Committee on Burial Rules and Regulations (interpretation of NAGPRA).

27. These briefs were from the Ethnic Minority Council of America (general policy considerations), Ives Goddard and William Shipley (linguistics), Ohio Archaeological Council (cultural affiliation), Pacific Legal Foundation (First Amendment issues), Andrei Simic and Harry Glynn Custred (oral traditions), and the Texas Historical Commission (standing and importance of scientific study). Copies of these briefs can be found at www.friendsofpast.org. In addition, the Society for American Archaeology filed an amicus brief that challenged the Secretary's decision on cultural affiliation but did not address the core issue in the case: the meaning of the term Native American. Two years earlier, the Society filed an amicus brief with the district court arguing that the Secretary was correct in finding that the skeleton was Native American, but wrong for deciding to award it to the Tribal Claimants (Court Docket 443).

28. A new opinion was issued on April 19, 2004 that modified one footnote in the original opinion.

29. A petition for a rehearing en banc is a request that the appeal be reheard and decided by all or a larger panel of the appellate court's judges. The order denying the Tribes' petition stated that "no judge of the court has requested a vote on the Petition" (Ninth Circuit 2004:868).

30. Among other things, the agencies objected that plaintiffs' study plan did not "take sufficient account" of the investigations previously conducted, would require excessive handling of the skeleton, was "overly redundant," and did not call for reimbursement of all "associated costs" that might be incurred by the Burke Museum and the government (correspondence on file with the authors).

31. Plaintiffs' inspection team consisted of Bonnichsen, Owsley, Thomas Stafford (geochronologist), Hugh Berryman (anthropologist), Chip Clark (photographer), Cleone Hawkinson (recorder), and C. Wayne Smith (conservator).

32. Almost all of the agencies' objections to the study plan were eventually withdrawn or substantially modified. Plaintiffs were not required to pay any associated costs incurred by the Burke Museum or the government. A copy of plaintiffs' study plan can be found at www.friendsofpast.org.

33. For an analysis of some of the potential impacts of the case, see Seidemann 2003b.

34. NAGPRA lists four types of "cultural items" that can

be claimed under the statute: human remains, funerary objects, objects of cultural patrimony, and sacred objects [25 USC §3001(3)].

35. Both alternatives were originally articulated by the Corps in its September 1996 Notice of Intent to Repatriate which was vacated by the court in June 1997. The Secretary of the Interior took the same position in his September 2000 determination decision.

36. Government attorneys conceded that nothing in NAGPRA expressly prohibits scientific studies of skeletal remains for research purposes such as those proposed by plaintiffs. Nonetheless, they contended that such studies would be inconsistent with the potential ownership rights of the Tribal Claimants. However, scientists hired by the government were allowed to study the skeleton despite Tribal objections. Their examinations included many of the same kinds of studies that were proposed by plaintiffs and originally opposed by the government.

37. For a discussion of NAGPRA's legislative history indicating that the statute was not intended to apply to remains as ancient as the Kennewick Man skeleton, see Seidemann 2003a.

38. Included were Pelican Rapids Woman (7,800 years old), the Prospect skeleton (probably greater than 6,800 years old), and the Browns Valley remains (8,900 years old). All were declared to be Native American solely because of their age (Schneider and Bonnichsen 2005:299).

39. 25 USC §3002(a)(2)(A). "Tribal lands" are Indian reservations, dependent Indian communities, and certain Native Hawaiian trust lands [25 USC §3001(15)].

40. These are remains for which no kinship or cultural affiliation can be established. See 43 CFR §§10.9(e)(6) and 10.10(g).

41. Earlier, more restrictive Regulations were used to dispose of remains belonging to more than 1,000 culturally unidentifiable individuals (Ousley et al. 2005:11).

42. The states and territories included in the Ninth Circuit are: Alaska, Arizona, California, Idaho, Hawaii, Montana, Nevada, Oregon, Washington, Guam and the Northern Mariana Islands.

43. Interior was also the agency that originally formulated the pre-Columbian age test rejected by the courts in the Kennewick Man case.

44. Senate Bill 2843, sponsored by Colorado Sen. Ben Nighthorse Campbell in 2004, would have changed NAGPRA's definition of the term Native American to read as follows: "'Native American' means of, or relating to, a tribe, people, or culture that is *or was* indigenous to the United States" (italics indicates proposed addition). Arizona Sen. John McCain proposed a similar amendment in 2005 (S.B. 536).

45. H.R. 6043, sponsored by Washington Rep. Doc Hastings, would have changed the definition of Native American to read as follows: "'Native American' means cultural items that have a significant and substantial genetic or cultural relationship, based on factors other than geography alone, to a presently existing tribe, people, or culture that is now indigenous to the United States." Hastings sponsored a bill in 1999 that would have made even more sweeping changes to NAGPRA by expressly authorizing excavation and study of cultural items found on federal land regardless of whether they can, or cannot, be culturally affiliated to a specific tribe (H.R. 2643).

46. There are actually two separate standing tests. One is the "case" or "controversy" requirement of Article III of the U.S. Constitution that has been interpreted as prohibiting the federal courts from entertaining lawsuits that are speculative in nature. The other is a court-developed test requiring that persons who seek to challenge enforcement of a federal statute must be within the "zone of interests" protected or regulated by the statute in question. Plaintiffs passed both these tests.

47. For more information about NAGPRA and its impact on scientific research, see Ousley et al. 2005 (use of skeletal studies to obtain data for NAGPRA determinations), Schneider 1995 (general overview of NAGPRA), Seidemann 2003b (ambiguities in the statute that require clarification), and Springer 2006 (impacts of NAGPRA and other repatriation laws on scientific study of archaeological remains).

48. Interview on 8/23/03 with George W. Gill, then Curator of Human Osteology Collections, University of Wyoming. Gill managed skeletal collections held for federal agencies by the University of Wyoming, and studied many other federal collections held by other institutions.

49. Defendants argued that they had no obligation under the statute to solicit plaintiffs' views on the project. The court disagreed.

50. For examples of possible alternatives, see Court Docket 418 at 36.

51. The Corps' actions would not have been justified even if erosion control had been its primary (or sole) objective. Nothing in the Act exempts site protection projects from the need to comply with all statute requirements.

52. The Society for American Archaeology thought otherwise. It told the district court that Interior "generally has made a reasonable effort" to compile evidence in the case (Court Docket 443 at 1).

53. One of the issues raised in defendants' appeal to the Ninth Circuit was a claim that the district court had erred in not remanding the matter to them for more administrative proceedings.

54. See, e.g., *Idrogo* 1998 (non-Indians lacked standing to use NAGPRA to claim possession of Geronimo's remains); *NA IWI O NA KUPUNA O MOKAPU* 1995 (agency could conduct skeletal studies to obtain information for NAGPRA purposes despite objections from a Native Hawaiian organization).

55. See, e.g., *Fallon Paiute-Shoshone Tribe* 2006 (tribal challenge to agency determination of non-affiliation), where the court stated: "as the *Bonnichsen* line of cases represent the primary, and most complete, prior treatment of issues arising under NAGPRA, the court finds all aspects of that case which have not been overturned on appeal persuasive" (Order, fn. 6 at p. 21). None was overturned on appeal.

REFERENCES

References listed below under the headings "COE," "COE S" and "DOI" (with a number rather than a date) are documents from the administrative records filed with the district court by the US Army Corps of Engineers ("COE" and "COE S") and the Department of the Interior ("DOI"). References listed under the heading "ACOE" (with a number rather than a date) are documents that were produced by the Corps in response to plaintiffs' discovery motion, but were not included in the administrative records filed with the court. References listed under the heading "Court Docket" (with a number rather than a date) are other documents filed with the district court in the Kennewick Man lawsuit.

ACOE

0759 Army Corps Information Paper dated October 16, 1996 (summarizing events relating to the skeleton).

Administrative Procedure Act, 5 U.S.C. §701 et seq.

Archaeological Resources Protection Act of 1979, 16 U.S.C. §470aa et seq.

Benedict, Jeff. 2003. *No Bone Unturned: The Adventures of a Top Smithsonian Forensic Scientist and the Legal Battle for America's Oldest Skeleton*. HarperCollins, New York.

Chatters, James C. 2000. The recovery and first analysis of an early Holocene human skeleton from Kennewick, Washington. *American Antiquity* 65(2):291–316.

———. 2001. *Ancient Encounters: Kennewick Man and the First Americans*. Simon & Schuster, New York.

Code of Federal Regulations

36 CFR   §§79.10 et seq. (regulations implementing ARPA)

43 CFR   §§10 et seq. (regulations implementing NAGPRA)

COE

2504 Memorandum by Paul Rubenstein entitled "Report on White House Meeting Regarding the Ancient Human Remains found near Kennewick, Washington," 30 October 1997.

3857 Plaintiffs' October 1, 1999 Status Report.

4250 Stafford, Thomas W., *Analysis of Geoarchaeological Data and Research Objectives for the Kennewick Man Discovery Site, Columbia Park, Washington,* December 30, 1998.

4304 Plaintiffs' Supplemental Report on Transfer of the Skeleton.

4896 U.S. Army Corps of Engineers, Waterways Experiment Station, Technical Report GL-98-13, *Geologic, Geoarchaeological, and Historical Investigation of the Discovery Site of Ancient Remains in Columbia Park, Kennewick, Washington,* September 1998.

5128 Agency notes of meeting held with Tribal representatives on 14 July 1998.

5137 Agency notes of meeting held with Tribal representatives on 14 July 1998.

5673 U.S. Army Corps of Engineers letter, undated, from Lt. Colonel Donald R. Curtis, Jr. to Donald L. Klina, Director, Office of Planning and Review, Advisory Council on Historic Preservation.

5766 Information News Release dated April 2, 1998 from Sheila Whitelaw, Colville Communication Services, Confederated Tribes of the Colville Reservation.

5815 Huckleberry, Gary, and Thomas W. Stafford and James C. Chatters, *Preliminary Geoarchaeological Studies at Columbia Park, Kennewick Washington,*

	USA, Report Submitted to U.S. Army Corps of Engineers, March 23, 1998.
6730	Federal Defendants' Response to Bonnichsen Plaintiffs' Status Report.
6899	Letter dated October 7, 1997 from Donald G. Sampson, Chairman, Board of Trustees, Confederated Tribes of the Umatilla Indian Reservation, to the U.S. Army Corps of Engineers.
6961	Plaintiffs' Status Report for October 1, 1997.
7176	Application for a Federal Permit Under the Archaeological Resources Protection Act, dated August 26, 1997 from Gary A. Huckleberry.
7713	Federal Defendants' Response to Plaintiffs' Motion for Access to Study Remains.
7733	Declaration of Donald E. Tyler, attached to Federal Defendants' Response to Plaintiffs' Motion for Access to Study Remains.
7980	Plaintiffs' Motion for Order Granting Access to Study.
9129	Letter dated October 11, 1996 from Slade Gorton, Doc Hastings, Jack Metcalf and George R. Nethercutt, Jr. to Lt. Gen. Joe Ballard.
9270	Note to Advertising Department, 9/17/96, Notice of intent to repatriate human remains in the possession of the Walla Walla District, U.S. Army Corps of Engineers.
9277	Note to Advertising Department, 9/24/96, Notice of intent to repatriate human remains in the possession of the Walla Walla District, U.S. Army Corps of Engineers.
9316	Letter dated September 9, 1996 from the Confederated Tribes of the Umatilla Indian Reservation to Colonel Bartholomew B. Bohn, II.
9485	Letter dated August 5, 1996, with attachments, from Richard Carlton, Chief, Real Estate Division, U.S. Army Corps of Engineers, to James C. Chatters.
9494	Letter dated July 30, 1996, with attachments, from Richard Carlton, Chief, Real Estate Division, U.S. Army Corps of Engineers, to James C. Chatters.
9500	Letter dated July 30, 1996, from James C. Chatters to Richard Carlton, Chief, Real Estate Division, U.S. Army Corps of Engineers.

COE S

272	U.S. Army Corps of Engineers e-mail dated April 14, 1998 from Linda Kirts to Martin R. Cohen.
278	U.S. Army Corps of Engineers e-mail dated April 6, 1998 from Mark C. Charlton to David N. Keller and Brian Bryson.
329	Facsimile transmission dated 3/31 from the office of Congressman Doc Hastings to Lt. Col. Curtis.
331	Copy of portions of S.1768.
348	Second Supplement to Plaintiffs' Second Quarterly Status Report.
409	U.S. Army Corps of Engineers letter dated October 9, 1996, with attachments, to Rex Buck, Wanapum Band.
431	Preliminary Report of the Archaeological Component to the Geological Investigation of Site 45BN52, Columbia Park, Kennewick Washington, December 26, 1997, by Frederick L. Briuer, Director, Center for Cultural Site Preservation Technology, US Army Engineer Waterways Experiment Station.
450	U.S. Army Corps of Engineers letter dated March 6, 1998 to Don Klima, Director, Advisory Council, Historic Preservation [sic].
534	Letter dated January 20, 1997, with enclosures, from Alan L. Schneider to Linda Kirts, District Counsel, U.S. Army Corps of Engineers.
596	Letter dated December 29, 1997 from Alan L. Schneider to Linda Kirts, District Counsel, U.S. Army Corps of Engineers.
614	U.S. Army Corps of Engineers letter dated December 23, 1997 to Alan L. Schneider.
672	BCOE review plans and specs for the Equipment Rental Contract for the Site Preservation Work for Ancient Remains at Columbia Park in Kennewick, WA.
821	U.S. Army Corps of Engineers e-mail dated November 10, 1997 from Linda S. Carter to Col. J. Christianson.
907	Fax letter dated December 13, 1996 from the Confederated Tribes and Bands of the Yakama Indian Nation to Ray Tracy.
937	Fax letter dated October 4, 1996 from the Confederated Tribes and Bands of the Yakama Indian Nation to Ray Tracy.
939	Letter dated October 4, 1996 from the Confederated Tribes of the Colville Reservation to Lieutenant Colonel Donald L. Curtis, Jr.

Court Docket

29	Transcript of Proceedings held on 10/23/96 before Judge John Jelderks.
132	Federal Defendants' Supplement to Second Quarterly Status Report.
146	Court order dated May 29, 1998.
168	Court order dated August 31, 1998.

416 Plaintiffs' Motion to Vacate Second Administrative Action.

418 Plaintiffs' Memorandum In Support of Motion to Vacate Second Administrative Action.

443 Memorandum of Law in Support of the Society for American Archaeology's *Amicus Curiae* Submission, June 1, 2001.

456 Record of Proceedings (Day 1) before Judge John Jelderks.

466 Transcript of Proceedings held on 6/20/01 Volume 1.

Custred, Harry Glynn, Jr., and Andrei Simic. 2003. Brief of amicus curiae, submitted to the U.S. Court of Appeals for the Ninth Circuit in case no. 02-35996, Robson Bonnichsen et al. v. United States of America et al., May 2003. http://www.friendsofpast.org/kennewick-man/court/amici/amici.html

District Court

1997a *Bonnichsen et al. v. U.S.*, 969 F. Supp. 615 (D.Or.)

1997b *Bonnichsen et al. v. U.S.*, 969 F. Supp. 628 (D.Or.)

2002 *Bonnichsen et al. v. U.S.*, 217 F. Supp.2d 1116 (D.Or.)

2004 *Bonnichsen et al. v. U.S.*, 2004 WL 2901204 (D.Or.) (awarding attorney fees)

DOI

2002 Memorandum from Paul Rubenstein entitled "Interagency Meeting on the Treatment of Ancient Human Remains and the Site of Discovery of Such Remains Near Kennewick, Washington," 6 November 1997

2128 DOI letter dated December 23, 1997, from Francis P. McManamon to Donald Curtis, Jr., U. S. Army Corps of Engineers

2719 Letter dated April 3, 1998 from the Confederated Tribes of the Colville Reservation to Joe N. Ballard, Lieutenant General, U.S. Army

2962 Letter dated June 9, 1998 from the Confederated Tribes of the Colville Reservation to Bruce Babbitt, Secretary of the Interior

3381 Letter dated August 4, 1998 from the Confederated Tribes of the Umatilla Indian Reservation to Francis P. McManamon

6712 Letter dated January 25, 2000 from the Confederated Tribes of the Colville Reservation to Francis P. Mc-Manoman (sic)

10298 Boxberger, Daniel L., Review of Traditional Historical and Ethnographic Information

10310–10326 Hunn, Eugene S., Review of Linguistic Information

10684–10685 Powell, Joseph F. and Jerome C. Rose, Report on the Osteological Assessment of the Kennewick Man skeleton (CENWW.97.Kennewick)

Edgar, Heather J. H., Edward A. Jolie, Joseph F. Powell, and Joe E. Watkins. 2007. Contextual issues in Paleoindian repatriation: Spirit Cave Man as a case study. *Journal of Social Archaeology* 7(1):101–122.

Equal Access to Justice Act, 28 USC §2412.

Fallon Paiute-Shoshone Tribe v. U.S. Bureau of Land Management, U.S. District Court, District of Nevada, Case No. 03:04-CV-0466-LRH (RAM), order dated September 21, 2006.

Freedom of Information Act, 5 USC §552.

Goddard, Ives, and William Shipley. 2003. Brief of amicus curiae, submitted to the U.S. Court of Appeals for the Ninth Circuit in case no. 02-35996, Robson Bonnichsen et al. v. United States of America et al., May 2003. http://www.friendsofpast.org/kennewick-man/court/amici/amici.html.

Hoffman, Paul. 2005. Statement of Paul Hoffman, Deputy Assistant Secretary for Fish and Wildlife and Parks, Department of the Interior, before the Senate Committee on Indian Affairs concerning the oversight of the Native American Graves Protection and Repatriation Act, July 28, 2005. http://www.friendsofpast.org/nagpra/news/Hoffman.html.

H.R.

6043 A bill to amend the Native American Graves Protection and Repatriation Act, 109th Congress, 2nd Session.

2643 A bill to amend the Native American Graves Protection and Repatriation Act to provide for appropriate study and repatriation of remains for which a cultural affiliation is not readily ascertainable, 106th Congress, 1st Session.

Idrogo v. United States Army, 18 F. Supp.2d 25 (D.C. 1998).

National Historic Preservation Act, 16 USC §470 et seq.

Native American Graves Protection and Repatriation Act, 25 USC §3000 et seq.

NA IWI O NA KUPUNA O MOKAPU v. Dalton, 894 F. Supp. 1397 (D. Hawaii 1995).

Ninth Circuit. 2004. *Bonnichsen et al. v. U.S.*, 367 F.3d 864, rehearing en banc denied 367 F.3d 868 (9th Cir.).

Ousley, Stephen D., William T. Billeck and R. Eric Hollinger. 2005. Federal repatriation legislation and the role of physical anthropology in repatriation. *Yearbook of Physical Anthropology* 48:2–32.

Schneider, Alan L. 1995. NAGPRA and First American studies. *Current Research In the Pleistocene* 12:117–123.

———. 2004. Kennewick Man: The three-million-dollar skeleton. In *Legal Perspectives on Cultural Resources*, edited by Jennifer R. Richman, and Marion P. Forsyth, 202–215. Alta Mira Press, Walnut Creek, CA.

Schneider, Alan L., and Robson Bonnichsen. 2005. Where are we going? Public policy and science. In *Paleoamerican Origins: Beyond Clovis*, edited by Robson Bonnichsen, Bradley T. Lepper, Dennis Stanford, and Michael R. Waters, 297–312. Texas A&M Press, College Station.

Seidemann, Ryan M. 2003a. Congressional intent: What is the purpose of NAGPRA? *Mammoth Trumpet*, 18(3):1–2, 19–20.

———. 2003b. Time for a change? The Kennewick Man case and its implications for the future of Native American Graves Protection and Repatriation Act. *West Virginia Law Review* 106(1):149–175.

Senate Bill

536   Native American Omnibus Act of 2005, 109th Congress, 1st Session.

2843   Native American Technical Corrections Act of 2004, 108th Congress, 2nd Session.

Springer, James W. 2006. Scholarship vs. repatriationism. *Academic Questions* 19(1):6–36.

U.S. Army Corps of Engineers. 1997. U.S. Army Corps of Engineers letter dated October 31, 1997 to Dr. Gary Huckleberry, with attachments. Copy on file with the authors.

Reflections of a Former US Army Corps of Engineers Archaeologist

Larry D. Banks

Just four months earlier it was the site of Kennewick Man, one of the most important archaeological discoveries in North America. In August 1998 I saw tons of riprap covering this flat, well-drained, secondary terrace along the Columbia River system. Large stones, dumped by the US Army Corps of Engineers (Corps) hours before the signing of legislation that would have prevented it, obliterated any chance that archaeologists could ever effectively excavate what was an ideal site location (Figure 5.1). This was hard for me to accept. The Corps had been my employer for 27 years. For 23 of those years I served as senior archaeologist. In this role I helped the Corps develop a national program of archaeological excavation and mitigation and had been involved with every aspect of the Corps' program concerning cultural resources. On my recommendation, the Corps adopted the phrase *cultural resources* in 1971 because it was congressionally barred from funding archaeology under the Historic Sites Act of 1935. That authority was reserved solely for the National Park Service. To avoid this congressional constraint, the Corps' first archaeological contracts out of the Tulsa District used the term *cultural resources*, drawing from the National Environmental Policy Act. This terminology enabled the Corps to fund archaeology and assume responsibility for handling and holding historically significant items discovered while performing their primary mission. This effective Cultural Resources Management Program put in place regulations or directives that outlined Corps necessary actions when cultural resources were found.

How could adherence to these strict regulations result in the events of the Kennewick case as we now know them? It couldn't, but the erosion of these policies over time created an environment in which politics took precedence. I discovered to what extent this had occurred when I was asked to conduct a review and critique of the Walla Walla District's Cultural Resources Management Program, ordered by the Assistant Secretary of the Army Louis Caldera. While I was prohibited by the District's legal counsel from referring to the discovery of Kennewick Man in my 1998 final report, that issue was clearly the stimulus and basis for my review. The Kennewick case is undeniably the most critical lawsuit in the history of American archaeology, and I was dismayed to learn the conditions under which the Walla Walla District operated when the remains were found.

My extensive review of the District's procedures, along with lengthy interviews with key personnel, showed that critical decisions had decreased the effectiveness of the program. After 1993 the Corps rescinded traditional engineering regulations, allowing administrators in the field more individual initiative while also increasing the possibility of inaction or negligence. The replacement nonregulatory documents contained suggestions and advice, but not directives. For example, a North Pacific Division Native American policy statement that I read in 1998 was only for management *guidance* and it provided flexibility to work with and around legal or situational problems. This was clearly a weakening of the once-strong policies of the Corps.

In my view, transformation of the District's Cultural Resources Program happened largely as a response to the Native American Graves Protection and Repatriation Act of 1990 (NAGPRA). While the Corps normally operated in compliance with 12 public acts, the first dating as far back as 1906, NAGPRA became the preeminent law for the Walla Walla District. Its provisions substantially altered the relationship between the Corps of this District and Native Americans. Across the United States, tribes recognized as nations were now treated as sovereignties; federal agencies were instructed to work with them on a government-to-government basis. This laudable ap-

FIGURE 5.1 Helicopter dumping sand and rocks onto the discovery area in April 1998. View looking east toward the site.

proach unfortunately resulted in noncompetitive-bid contracts with Native American groups for site surveys on Walla Walla Corps projects and development of management plans. In some cases, the tribes won contracts by arguing that they were best qualified to work on archaeological projects simply because they were Native (an argument that I, as a rigorously trained Registered Professional Archaeologist, refute.)

The Corps' Native American coordinators and the tribes themselves effectively directed the cultural resources affairs within the District. The two archaeologists in the Walla Walla District had been removed from all responsibilities and decisions being made regarding Kennewick Man. They could not speak for the Walla Walla Corps on related archaeological matters and had no clear role.

From a larger perspective, seeming abdication of responsibility on the part of the Walla Walla District extended even further. Both the Office of the Chief of Engineers, Washington, DC Headquarters, and the Northwestern Division Office neglected to provide specific directives. The Walla Walla Corps apparently accepted direction from the Bonneville Power Administration (BPA), the electricity-generating company in the region. The BPA has had broad discretionary powers since its creation in 1937. Since 1992, the BPA has been directly funding activities associated with hydropower projects, including the management of cultural resources. As a result, the BPA also seemed to control spending on cultural resources by directing the Corps to disburse the program funds it provided to specific tribes for specific activities without complying with the Corps' contract law. These various conditions caused the District to defer to the tribes when there were no clear instructions to do otherwise.

If a well-organized Army Corps of Engineers Cultural Resources Program, built upon clearly assigned roles and responsibilities for all individuals, had been followed, it is likely that the Kennewick Man issue would have been treated quite differently, with a more positive public relations outcome for all involved, Native Americans, the District Corps, and the scientific community. What has turned out to be almost disastrous for all parties could have been a source of pride if everyone had cooperated for a better understanding of the scientific and cultural contributions that reasoned treatment of the Kennewick remains could have provided (Banks 2011). As important evidence of our Nation's past continues to be discovered, this case should cause all to reflect on the origins of cultural resource management policies—why they were created and why they need to be followed in the future.

REFERENCE

Banks, Larry D. 2011. "Memoirs of a Corps Archaeologist." Ms. on file at National Museum of Natural History, Washington, DC.

Curation History and Overview of the Plaintiffs' Studies

Cleone H. Hawkinson

Kennewick Man's skeleton is in remarkable condition, but a careful look reveals that the bones are fragmented. An intact and complete human skeleton can usually be arranged anatomically in 30 minutes or less, but the precise arrangement of this nearly complete skeleton took a full day. The opportunity to place the elements in full anatomical order was the first of many rewards for the nine years of effort spent to gain permission for scientists to study Kennewick Man. The two-week taphonomic study in July 2005 established a baseline of assessment for the specialized scientific studies in February 2006.

TAPHONOMIC STUDY: 2005

Taphonomic study typically includes the examination of the gravesite as soon after discovery as possible. However, in 1998 the US Army Corps of Engineers (Corps) buried the discovery site under tons of rock. The study team's first opportunity to examine the skeleton in depth began in a classroom more than 200 miles away and nine years after the remains were recovered.

The classroom in Seattle's University of Washington Burke Museum was set up with multiple work areas so the scientists could collect data independently but easily interact (Figure 6.1). This arrangement allowed the specialists to discuss observations while referring to the bones. Participants in the 2005 study included Douglas Owsley and his team from the National Museum of Natural History, Smithsonian Institution, David Hunt (physical anthropologist), Karin Bruwelheide (physical anthropologist), Chip Clark (scientific photographer), and Aleithea Williams (data management specialist). Other participants included Hugh Berryman (forensic anthropologist and bone trauma specialist), Thomas Stafford (geochemist), C. Wayne Smith (conservator), and Cleone Hawkinson (facilitator). Assisting the scientists were Burke Museum personnel and Corps curation staff and contract conservators. Three attorneys were present during the two weeks as their schedules allowed: Corps attorney Jennifer Richman and the scientists' attorneys Alan Schneider and Paula Barran.

Burke Museum personnel and the Corps established security and safety measures during this analysis. A security guard was posted outside the classroom to control access. The window in the door was covered and only persons pre-approved by the Corps were allowed inside. Each person was required to sign an access log when entering and leaving the classroom and to wear name badges while in the classroom. Corps staff monitored the temperature and relative humidity in the classroom during the day. After working hours, a data logger continued to record, but did not control, these two environmental conditions (Chapter 31). Although the Corps and the Burke Memorandum of Agreement stated that all activities involving handling the skeleton would be videotaped (DOI 2976), the scientists' request to bring a professional videographer to record the study was denied. In a letter to the plaintiffs' attorneys, the Corps stated that videotaping the studies might jeopardize the safety of the skeleton or disrupt the study. The conservators placed sheets of acid-free paper at each workspace. As the scientists examined the elements, the conservators or Corps staff regularly collected and bagged any fine bits of soil from the paper and added it to the collection.

Examination of the Skeleton

The classroom provided plenty of space for the scientists to work. Owsley led the team through a defined set of observations on the skeletal elements, analyzing each one in detail (Chapters 7 and 17). Williams or Hawkinson entered the data and verbal descriptions

FIGURE 6.1 Classroom at the University of Washington's Burke Museum used for the 2005 study of the Kennewick Man skeleton. Schneider (left) observes as Owsley and Berryman evaluate femur fragments. Williams (at right) records their descriptions and observations. In the background, Bruwelheide (right) records observations of rib fragments and Hunt (standing) prepares to reconstruct the photopolymer models of the cranium.

into electronic spreadsheets (Figure 6.2). Bruwelheide identified and positioned rib fragments and many of the previously unidentified small bone fragments (Figures 6.3–6.5. Berryman analyzed fracture patterns to establish anatomical positioning in situ and the process of the skeleton's erosion from the bank (Chapter 19) (Figure 6.6). Hunt compared photo polymer replicas produced by scanning and stereolithography to actual fragments of the cranium and right hip for measurement, fit, and accuracy of detail (Chapter 21). Stafford and Smith evaluated postmortem staining and the deposition of calcium carbonate concretions on bone surfaces. Throughout the study, elements of special interest to the scientists were taken to Clark's work area to be photographed (Figure 6.7).

Arranging the Skeleton

During the negotiations to finalize the study plan, the scientists learned they would not be allowed to use a non-permanent adhesive to reconstruct broken bones. Knowing they would need to assess the full skeleton in anatomical order, the scientists proposed a sand bed to support the fragments in position. The scientists designed a custom-built eight-foot long plywood frame to put on the classroom floor. They lined the structure with heavy fabric and filled it with 400 pounds of sand. Although Clark was prepared to cover the sand with special black photographic fabric, the conservators were concerned

FIGURE 6.2 Williams (bottom left) records the scientists' descriptions and observations, such as color variations on each bone piece using the Munsell color chart. Left to right: Williams, Stafford, C. W. Smith, Schneider, Owsley, and Berryman.

Cleone H. Hawkinson

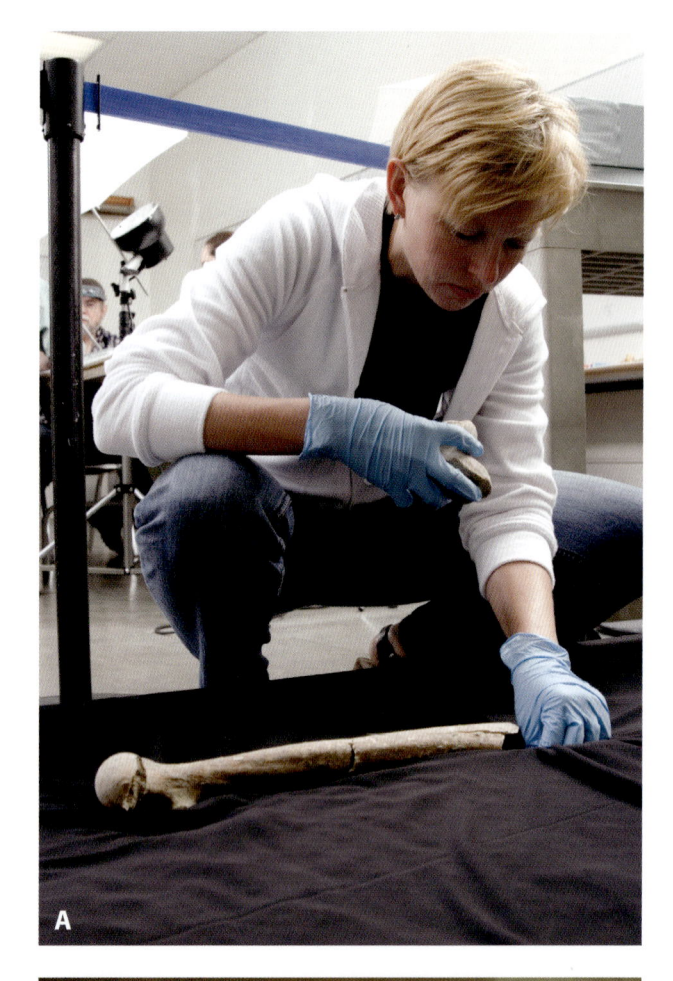

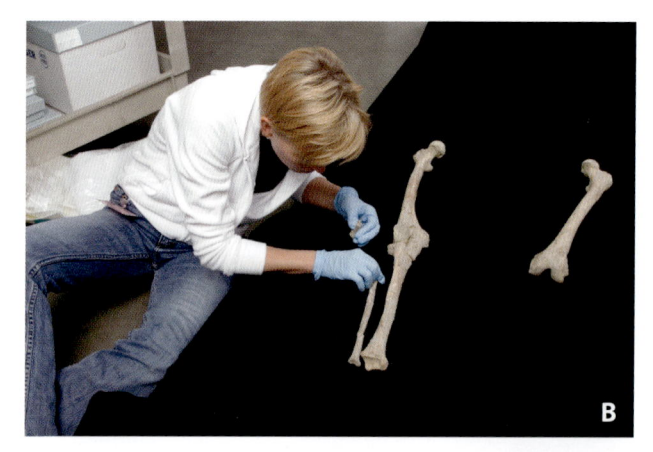

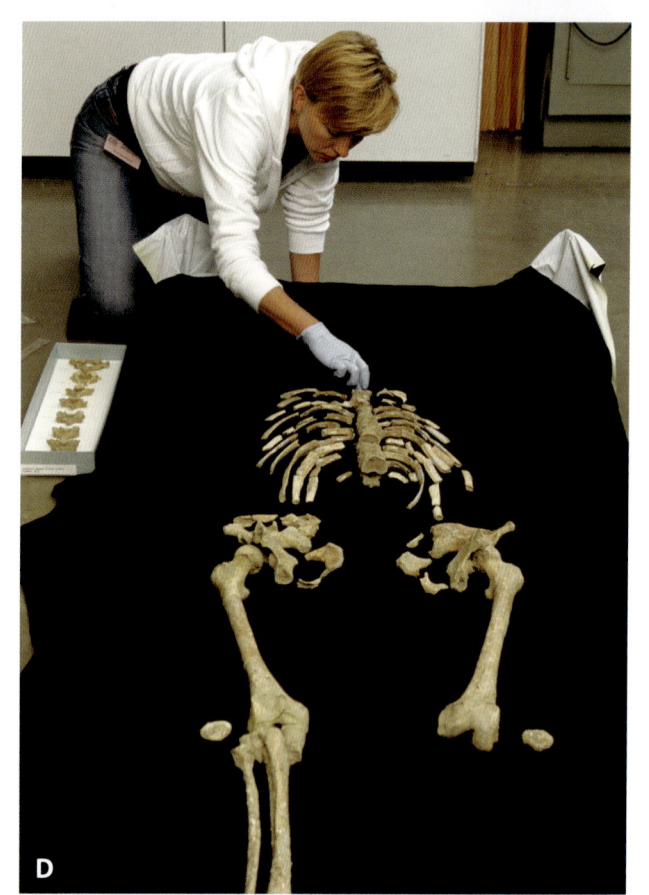

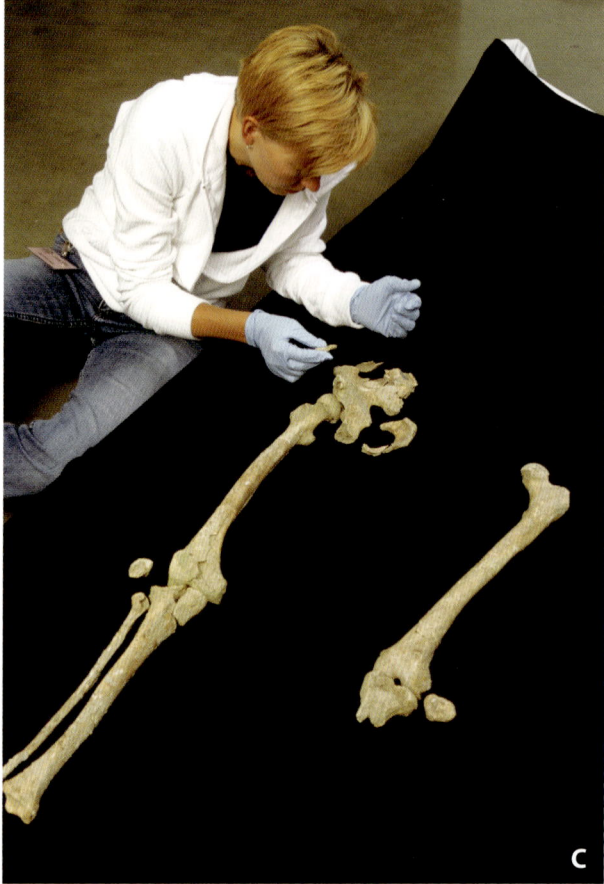

FIGURE 6.3 (A–F) The full skeleton, arranged for the first time in anatomical position, allowed the study team to check identification and placement of fragments, recognize patterns of breakage and staining across bones, and complete a comprehensive tophonomic analysis. Various team members positioned the fragments. Here, Bruwelheide arranges long bones, vertebral, and rib fragments in the soft cushion of fabric-covered sand devised in response to conservation concerns.

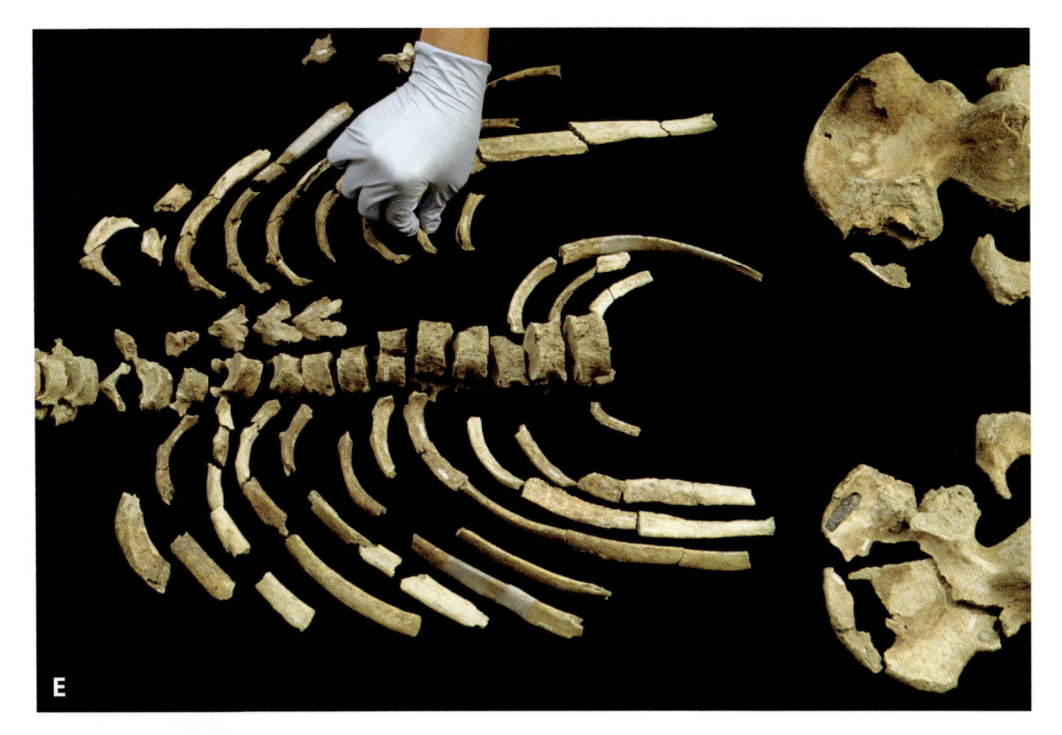

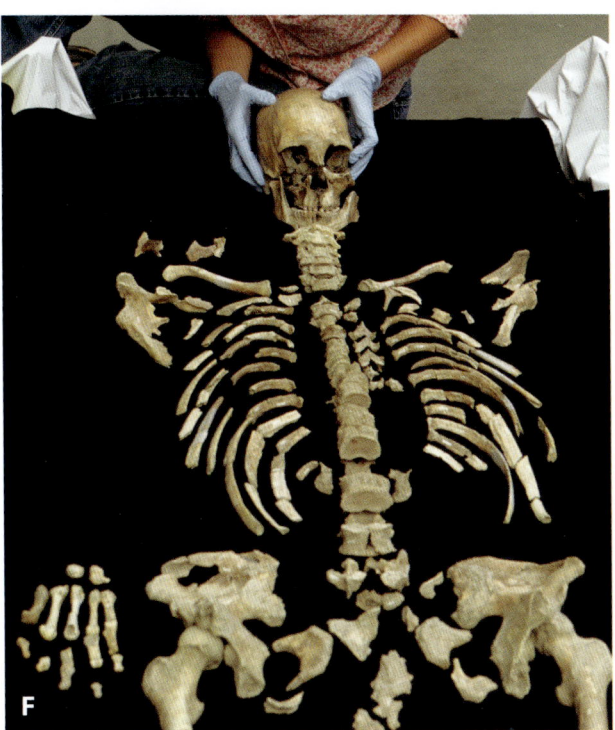

FIGURE 6.3 (A–F) (continued)

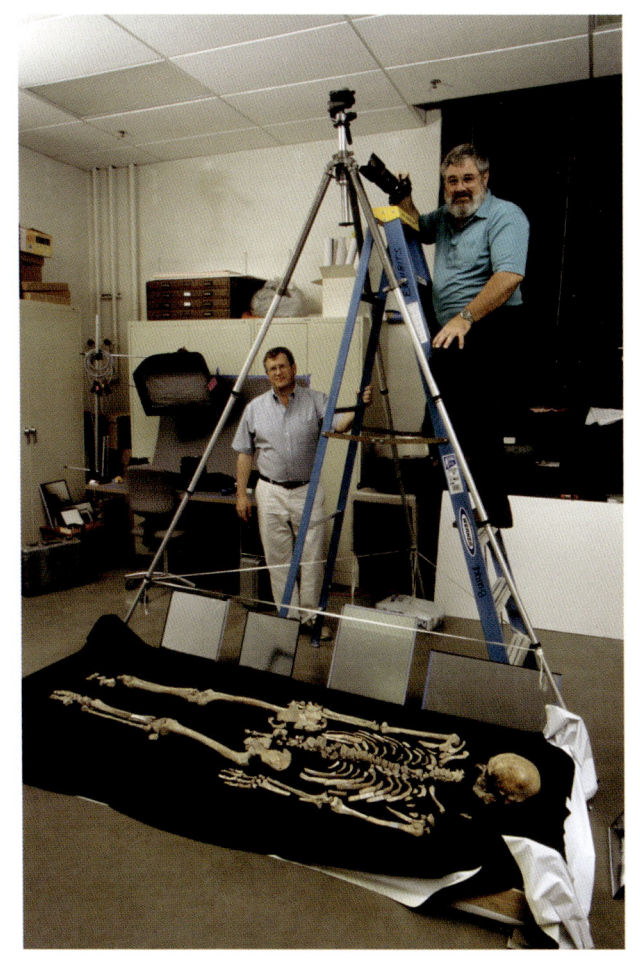

FIGURE 6.4 Clark on a ladder to photograph the layout of Kennewick Man's skeleton with a wide angle lens. Owsley steadies the ladder.

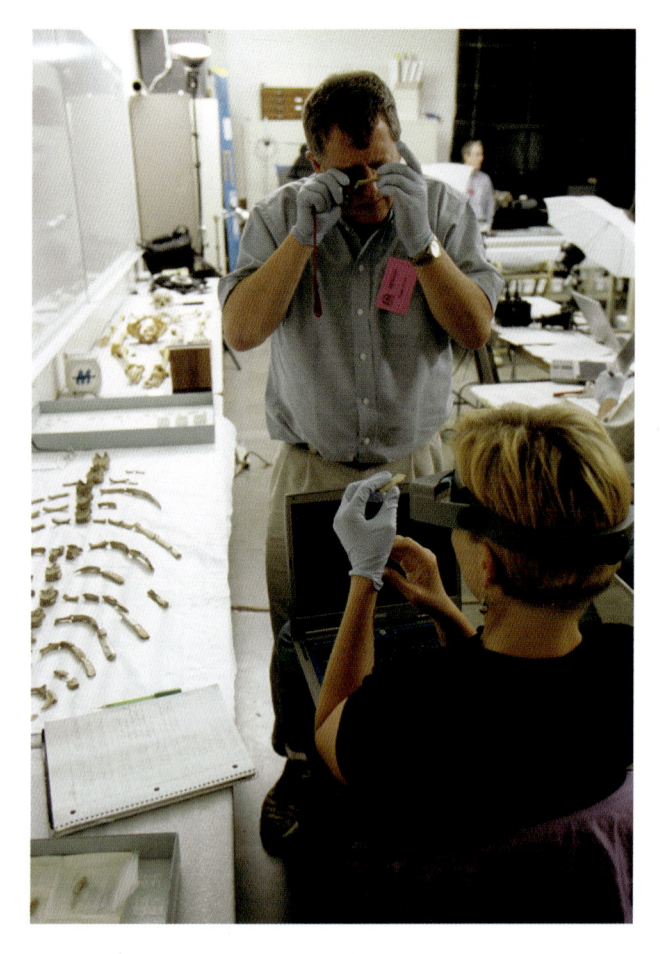

FIGURE 6.5 Bruwelheide (right) examines a rib fragment while Owsley (standing) looks for evidence of a healed fracture in a rib using a magnifier lens.

that the fabric might leave fibers on the bone fragments. However, the conservators tested a new, unwashed length of fabric and found it acceptable.

With the frame in place, Bruwelheide positioned each fragment in anatomical order by manipulating the sand under the fabric to support each piece in proper position. Clark took a series of photographs capturing the process and the final arrangement from an elevated view, using a wide-angle lens (Figures 6.3 and 6.4). The opportunity to observe the full skeleton allowed scientists to verify suspected patterning of the staining and sediments they had been documenting on individual elements. The conservators placed barricades around the sand table to protect the bones, monitored the scientists' interaction with the bones, and continued to collect minute bits of debris from the fabric.

The taphonomic study was the beginning of a carefully planned series of studies of Kennewick Man. An overview of events during the nine years leading to the taphonomic evaluation contributes to understanding the condition of the skeleton.

FROM EASTERN WASHINGTON TO SEATTLE: 1996–2005

Responsibility for the security, curation (administrative control), and conservation (issues related to preserving condition) of the Kennewick Man skeleton rests with

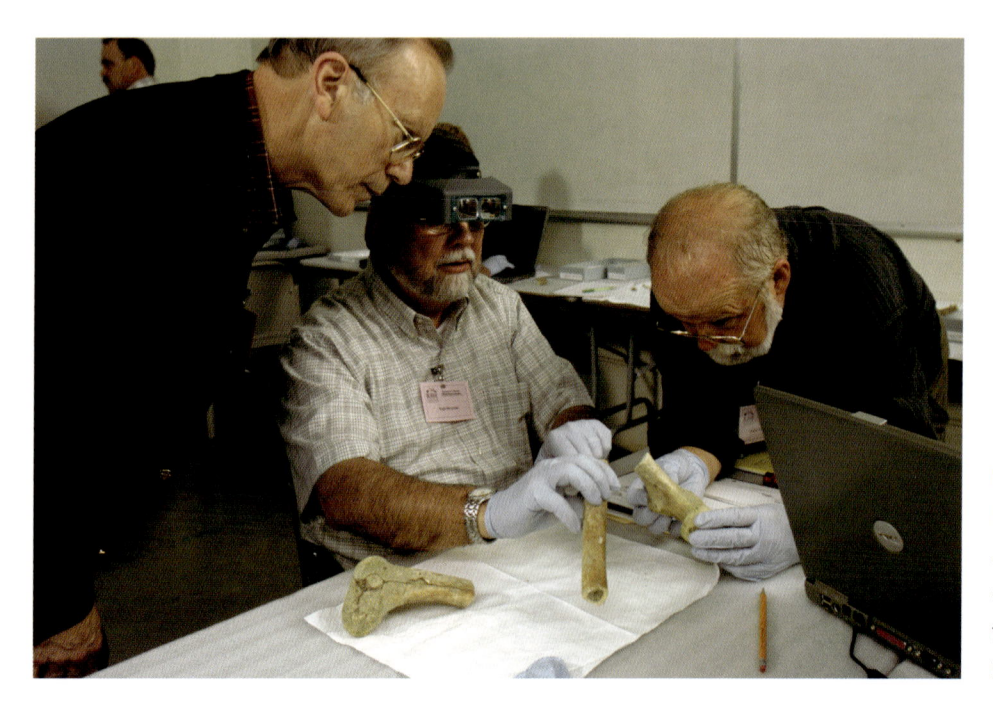

FIGURE 6.6 Schneider listens as Berryman and C. W. Smith discuss evidence of postmortem fracturing on sections of Kennewick Man's femur.

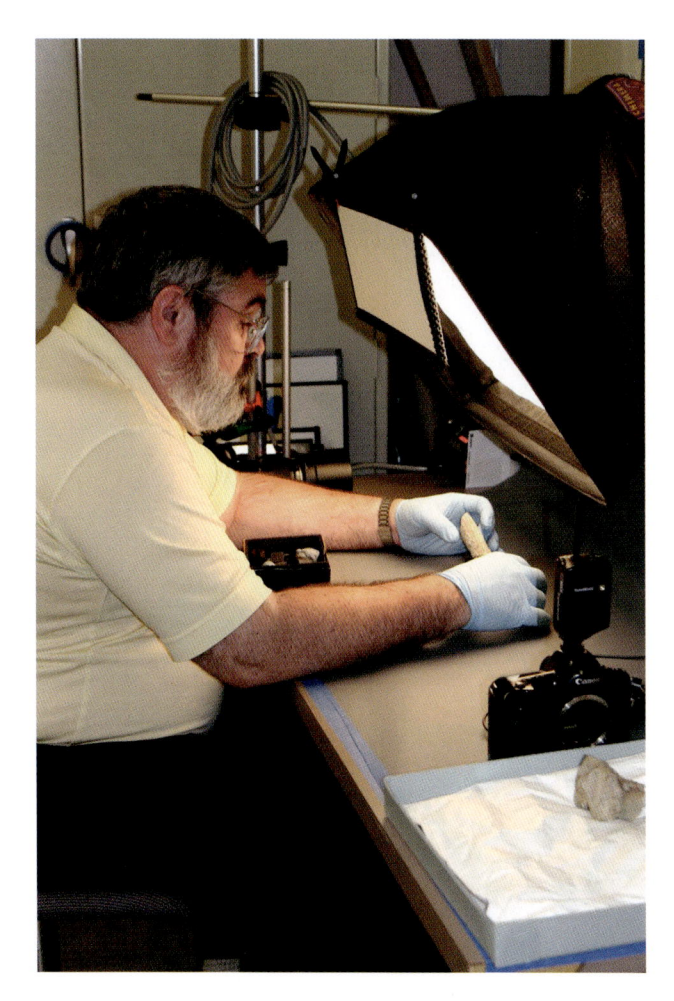

FIGURE 6.7 Clark at his portable photography setup. A Canon EOS® camera was used with a Chimera® softbox for sweep lighting, producing a graded background that enhanced depth.

the Corps. The bones have been exposed to the environments of a river terrace, a wet shoreline, the desert-dry climate of eastern Washington and the moderate Seattle climate. After recovery, the elements were allowed to begin to dry in James Chatters's laboratory in Richland, Washington. For more than a year, the bones were kept in plastic bags in a wooden box at the Pacific Northwest National Laboratory (PNNL). Since 1999 the collection has been housed in acid-free boxes in a Delta® cabinet in the basement of a museum (Chapter 31). During this time, documents show that events involving contact with the remains included dozens of people with a wide range of interests and expertise, and with varying expectations about the future of the bones and the care they would need.

Summer 1996 to Fall 1997
Discovery and Recovery of the Skeleton

When the bones were discovered on July 28, 1996, legal control was clear. The Benton County Coroner, Floyd Johnson, had jurisdiction under Washington State law. He asked archaeologist James Chatters to recover and document the remains. Both the coroner (COE 9365) and Chatters recorded the recovery process, which took place over several weeks (COE 9308; Chatters 2000, 2001). Because bones were recovered from the water and mud along the shoreline, conservation measures began at the time of recovery by allowing them to dry slowly to minimize cracking. Chatters (2000; 2001) describes collecting the bone fragments and subsequent activities in his laboratory in two publications.

On July 29, Chatters took the right ilium to Kennewick General Hospital to radiograph the embedded projectile point. The next day he took fragments of the skull, femora, and both innominates to Central Washington University in Ellensburg for analysis. Catherine MacMillian, a physical anthropologist, gave her opinion on age, ancestry, and sex. Steve Hackenberry, chair of the Anthropology Department, was also present. When Chatters returned to his laboratory in Richland, Ken Reid, an Idaho State Archaeologist, evaluated the projectile point in the right ilium fragment (Chatters 2001). That evening Chatters returned to the hospital for computed tomography (CT) scans of selected elements, including the piece of ilium with the projectile point, a humerus, and a femur. Radiographs were also taken of the mandible, maxillae, and the other femur (COE 9308; Chatters 2001).

The coroner's report on the chronology of events describes his approval on July 31 of Chatters's request for radiocarbon (^{14}C) and DNA testing (COE 9365). Chatters chose the left fifth metacarpal because it appeared intact and was found in sediment inside the cranium, confirming its direct association with the skeleton. On August 8, the University of California, Riverside, laboratory confirmed that the bone was well preserved and indicated that only part of the bone (500 mg) should be needed for testing (COE 630).

In the days that followed, Chatters reconstructed the cranium and made a polyurethane cast. A dentist radiographed and evaluated the teeth. From July through mid-August, Chatters discovered more bone fragments at the

Cleone H. Hawkinson

site and added them to the collection. He later provided a chronology of his involvement with the skeleton to the Corps (COE 9308).

The coroner received the results from the University of California, Riverside, laboratory on August 26, 1996 (COE 9474). The ^{14}C results revealed that the bones were approximately 9,200 calendar years old, immediately setting off a clash among legal jurisdictions. Although the coroner had not yet completed his investigation, the Corps asserted its jurisdiction and claimed that the ancient remains were subject to both the Archaeological Resources Protection Act (ARPA) and the Native American Graves Protection and Repatriation Act (NAGPRA). As early as July 29, the record shows communication between Corps staff and the Confederated Tribes of the Umatilla Indian Reservation (the Umatilla) and other local tribes (COE 9501; 9489; 9473). The Corps assumed that the bones were Native American and did not attempt to establish whether or not the skeleton met the criteria required for this determination as defined by NAGPRA (Chapter 4).

With the news of the skeleton's ^{14}C age, Chatters immediately sought more expertise in handling the skeleton. He contacted Owsley at the Smithsonian Institution's National Museum of Natural History to arrange for a study of the remains. On August 30, arrangements were confirmed for Chatters to fly with the skeleton to Washington, DC Also on that day, Grover Krantz, a retired physical anthropologist from Washington State University, Pullman, examined the bones in Chatters's laboratory. Krantz concluded the remains differed morphologically from modern American Indians and ought to be studied further (COE 9396).

At 5:00 p.m. on August 30, the Corps directed the Benton County coroner to remove the bones from Chatters's laboratory (COE 9308). Before turning over the remains, Chatters used a video camera to document the collection and its condition, creating a visual inventory that included the presence of two complete, though fragmented, femora. This documented inventory would become important in the years to come when segments of each femur were reported missing.

In the presence of a sheriff's deputy, Chatters packed the still-damp bone fragments, element by element, in unsealed plastic bags, placing the heaviest pieces (the femora and tibiae) in the bottom of a wooden box. Next he added the cranium and then placed the lighter bones in bags around and on top of the cranium. After sealing the wooden box with evidence tape, Chatters noticed a bag of ribs that he had overlooked. A cardboard box containing unrelated historic-era artifacts collected at the recovery site was still open. Because of the urgency to finish, Chatters placed the bag of ribs on top of the artifacts in the cardboard box (Chatters 2001). The deputy then sealed this box and took the two boxes to the evidence locker at the Benton County Sheriff's Office.

On August 31, Chatters joined a deputy in the evidence locker to transfer the bag of ribs to the wooden box containing the nearly complete skeleton. Chatters set the lightweight bag of ribs on top of the other bags (James Chatters, personal communication 2001). The deputy then resealed both boxes, which remained in the evidence locker for six days. Chatters's direct involvement with the skeleton was over for the time being. The plaintiffs' lawsuit to halt repatriation would not be filed for another two months.

Transfer of Custody to the Corps of Engineers

Throughout the early years of litigation, the Corps released little information about the skeleton and repeatedly assured the court that the bones were safe and well cared for. As a result, much of what is known of events is detailed in documents not available until well after the fact.

In the first week of September 1996, the Corps took custody of the remains from the Benton County coroner. The Corps had contracted with the PNNL in Richland, Washington, to act as a temporary curation facility (COE 9312). Since 1965, Battelle Memorial Institute, a nonprofit organization, has operated the PNNL for the US Department of Energy. The Corps had already made arrangements to repatriate the skeleton (Chapter 4) and therefore did not seek advice from experts on properly conserving the ancient remains in the interim.

The two boxes containing the skeleton were transferred to the PNNL. A coroner's report describes the event:

> On the 4th of September, 1996, transfer of custody and control of the remains to the Battelle Laboratories was

conducted at 1000 hours. The remains were secured in the "SIGMA FIVE" building at the laboratory. Paul Nickson accepted custody for the laboratory. The transfer was witnessed by Jeff Van Pelt of the Umatilla Tribe, two representatives of the Yakima (sic) Tribe, and Deputy Coroner Larry Duncan. A short meeting was held after the transfer. Jeff Van Pelt wanted assurance from Paul Nickson that no one would have access to the remains until jurisdiction over them was established (COE 9365).

Ray Tracy, staff archaeologist for the Corps, reported the transfer date as Thursday, September 5, 1996. Tracy stated:

> . . . Van Pelt and Mr. Johnson carried the two containers from the coroner's vehicle to the repository and placed them on a locked shelf. . . . I didn't ask to view the remains. I had decided that, in the absence of an inventory list, I could make no determination of their completeness (COE 9938).

The following morning Tracy conferred with Van Pelt about the possibility of reopening the containers to formally determine their contents and to establish a chain of custody. Van Pelt said he would put that question to his governing body (COE 9938).

Housing and Access to the Skeleton at Pacific Northwest National Laboratory

In 1998, the scientists learned from documents received through a Freedom of Information Act request that the first inventory of the bones at the PNNL was conducted on Tuesday, September 10, 1996, during a second visit by tribal representatives. Ten individuals from the Umatilla, the Confederated Tribes and Bands of the Yakama Nation, the Confederated Tribes of the Colville Reservation, the Wanapum Band, and the Nez Perce Tribe signed the log that documented individuals granted access to the skeleton on that day. This authorization stipulated that "any handling deemed necessary by the parties present will be minimized to avoid damage to the remains" (COE 7343; 9315). The remains were removed from their boxes and handled in order to complete the inventory. An archaeologist for the Umatilla Tribe, Julie Longenecker, was listed on the inventory as the data recorder. The names of other individuals, if any, who assisted with the inventory are not included on this document (COE 7344).

Longenecker's inventory lists 68 numbers, corresponding to numbers they wrote on the plastic bags containing the bones as they were removed from the wooden box. Information recorded about each fragment varies in detail. Longenecker recorded specifics about some fragments and for others gave only the name of the bone. The list includes only two femur fragments: "47. Distal Femur" and "56. Prox 1/2 of Femur (LT)." This is the first known indication that only two of the six femur fragments were in the collection. The tibiae were listed as "61. Distal ¾," "62. Prox Tibia = ½ – 2 pieces," and "63. Complete tibia 3 pieces." These numbers represent the heaviest bones, which Chatters states he placed in the bottom of the box. On the inventory, the cranium was listed last: "68. Cranium—was not taken out of box." Notes written on the inventory describe the condition of the bone:

> Bone is in fairly good condition—some pieces [strike out: are] still have dirt [strike out: mud] on them— Most of the bone has been cleaned—can't tell which pieces have been chemically treated. Black Basalt point lodged in innominate. Bone has tried to grow over the stone (COE 7344).

Neither the Corps nor PNNL reported receiving a copy of this inventory from the Umatilla. More details about Longenecker's inventory were given in an October 1997 report by Madeline Fang, a University of California, Berkeley, conservator and US Department of Justice consultant (COE 5334).

The day after the inventory, at the Corps' request, Chatters provided copies of his handwritten field and laboratory notes, images (photographs, radiographs, CT scans), reports, and other records to the Corps. These items eventually became part of the curation record for the collection (COE 9293).

Access logs and later reports show that unidentified bones were added to the collection after the transfer of custody to the Corps. Van Pelt and Longenecker, accompanied by Corps and PNNL representatives, added a brown lunch bag with unidentified contents to the collection on September 17. Corps Lieutenant Colonel Donald Curtis, Jr., drafted a letter of permission to access the collection in order to reunify the newly collected human bone and "for handling deemed necessary," (COE 7342). In October 1996, Van Pelt, Longenecker, and a Corps representative added unidentified bone fragments to the box of remains (COE 7341). Nearly a year later, on September 4, 1997, a Corps representative added two wet unidentified bone fragments (COE 7337), and one vertebra was added at an unknown date (COE 5332; 5334).

The record shows that the Corps approved requests for seven visits for religious reasons during 1996 and 1997. By August 1997, 17 individuals from five American Indian tribes and five members of the Asatru Folk Assembly had signed access logs. The access form for each of the seven visits included permission to handle the bones as deemed necessary, but the extent of handling by these visitors is unknown.

For a full year after taking possession of the skeleton, the Corps curation records did not include an inventory. Records reveal no evidence that any effort was made to improve the storage or to conserve the condition of the remains, which were still being stored in plastic bags.

Repatriation Claims

The Corps published in the *Tri-City Herald* of Kennewick, Washington, on September 17, 1996, and again on September 24, its intent to repatriate the bones under the provisions of NAGPRA (COE 9270, 9277). The date for transfer of the remains to the tribes was set for October 24, 1996. The Corps ordered a halt to the coroner-approved DNA scientific tests at the University of California, Davis (COE 8720), and stated they would allow no further tests (COE 9030). The extracted residue from the metacarpal intended for DNA testing was eventually returned to the collection in May 1998 (COE 2358). An emergency hearing resulting from the scientists' lawsuit was held in the Federal District Court in Portland, Oregon, on October 23, 1996 and, as a result, the Corps agreed to indefinitely postpone the transfer (Chapter 4).

Fall 1997 to Fall 1998

In the fall of 1997 Fang was hired by the Department of Justice, the agency representing the Corps in court, in response to the District Court's request for information about the skeleton. On September 17, 1997, just prior to Fang's visit, Corps and PNNL staff took photographs of the inside of the curation facility, the box holding the remains, and the box with the lid removed. Representatives of the Corps and the PNNL accompanied Fang during her site visit on October 20, 1997. Fang's report entitled "Site Visit to Pacific Northwest National Laboratory" describes the security and storage conditions:

> The box containing the Kennewick Man's remains are [sic] housed on the top shelf of the rolling unit, middle aisle.... While PNL [sic] staff do not access the Kennewick box, they access collections on both sides of [the] unit where it is stored. Hence the unit is rolled open routinely by staff working on the collection. Ray Tracy reported that he had unpacked the bones in September 1996 in the presence of Julie Longenecker, a member of the cultural resources staff of the Umatilla Nation, and Harvey Pete Reese, a Colville cultural resource staff member (both archaeologists and not tribal members). At that time they 1) labeled zip lock bags with numbers, e.g., "1," and 2) made a quick written inventory of the contents of each bag—number of specimens and tentative ID when they could. The written list was thought to be with Julie Longenecker.
>
> This inventory work coincided with the second visit by reps of the five tribes and the only time the remains were taken out of the box. In repacking the remains, long bones were placed at the bottom, skull bones on top, and bags with smaller fragments on top. At this time cedar twigs were burned (presumably over the bones) and other ones left on top of the remains. Twigs were present when we opened the box, as well as ashes from the smoking (COE 5332).

Fang's report left little doubt that improvements were needed despite the Corps' repeated assurances that the bones were secure and properly stored. She recommended new housing to protect the bones, that both an inventory and condition assessment be conducted, and that the cedar twigs placed in the box during the tribes' religious ceremonies be removed.

In response, the Corps called in Michael Trimble, Director, Mandatory Center of Expertise for the Curation and Management of Archaeological Collections,

St. Louis (Missouri) District of the Corps. Trimble and osteologist Teresa Militello joined six other Corps staff and two PNNL employees to inspect procedures in place to curate the skeleton. They had permission to remove some or all of the remains from the storage container and to take photographs. Militello's Trip Report and Trimble's Memorandum for the Record describe their activities and recommendations. Trimble commented:

> b. Security: Security for the remains is excellent . . . Only five staff members have key card access to the curation room. Security is equal to or exceeds many top museums in the United States (COE 6716a).

In early November, 11 individuals, including the St. Louis team, local Corps staff, two PNNL escorts, Department of Justice and Corps attorneys, and Fang returned to begin the curation of the remains (COE 6712). This work continued in mid-December when the St. Louis team traveled again to Richland (COE 6533).

Untangling the record of events of the spring of 1998 is challenging. In mid-March, the St. Louis team returned to continue preparing their records, to address inconsistencies in their notes, and to conduct an inventory. They paid particular attention to the identification of the femur fragments (COE 5373). A Corps press release confirmed that portions of both femurs were missing. The release stated that during the September 10, 1996 inventory

> . . . the Umatilla representative also noted fragments were not among the remains. Since the skeleton was nearly complete, the absence of the fragments of the femur bones was not considered significant then (COE 5996).

The release did not explain how the Umatilla representative recognized that femur fragments were missing at the first inventory or why being "nearly complete" excused the need for the Corps to investigate the matter.

In late March 1998, the access permission form was revised revealing new restrictions on the terms and conditions for access to the remains (COE 5314). Corps attorneys from the Walla Walla District accompanied Trimble's team during their second visit, which began March 24, to assess the skeleton. The Corps continued to include the tribes, but not the plaintiffs, in the approval process for access to and activities concerning the remains.

> Access is granted for the purpose of: Inspecting the ancient remains and updating/correcting accompanying records and files. Terms and conditions/purpose of access: No testing, measurement, photography, or other disturbance of the remains other than that agreed to by all parties will be performed. No portion of the remains will be removed from the curation facility. Any handling deemed necessary by the parties present will be minimized to avoid damage to the remains (COE 5314).

In late March, apparently more fragments from the discovery site were brought to the PNNL. The Corps did not document the identity of the fragments, who found them, or where they were stored. The index of the Corps' Administrative Record only lists "Fragments to collection March 27 1998," (COE 5809). According to Tracy, however, the fragments were not left at the facility at that time:

> We didn't gain access to the file shelves. We didn't leave any skeletal material in the curation facility (COE 5809).

A few days later, a document shows permission was granted to place new bone fragments in storage. Tracy and a PNNL escort signed in but no further details were given (COE 5808).

In April 1998 three events influenced the future curation and conservation decisions and the potential scientific understanding of the Kennewick Man skeleton. The Corps formally delegated responsibility to the US Department of the Interior to determine the disposition of the skeleton. Interior personnel from the National Park Service had been advising the Corps on NAGPRA issues since 1996. Now the Interior would determine what studies would be conducted, who would participate, and the extent of the sampling and tests allowed. They also controlled the content and release of the conclusions and scientific reports (DOI 5780). Then, on Monday, April 6, the Corps ordered tons of rock dropped from the air on the discovery site. This layer of rock was followed by gravel, then landscaping with coconut fiber logs, sand, and the planting of thousands of willow trees, thus precluding future opportunities to recover more contextual information about the skeleton (Chapter 4). The third event concerned the security of the skeleton. Early in the morning of April 27, and despite US Army Lieutenant General Joe N. Ballard's orders to the contrary, unauthorized tribal members entered the room where the remains were stored. Neither PNNL nor Corps employees already in the room asked them to return to the hall. The tribal members removed a box containing bone fragments and, in a ceremony later that morning, reportedly buried the contents with other more recent human remains in a common pit. The contents of the box they removed were described in part:

> . . . human bone contained in the box consisted of one human rib fragment found by James Chatters in mid-December of 1997 during the geomorphological investigation by W. E. S., one or two small fragments of possible human bone found by Leier and Wolf before the mobilization of the site protection . . . (COE 5651).

Trimble had made the decision not to incorporate the rib fragment into the main collection despite being told that Chatters had recovered it from the discovery site. The record does not reflect that Trimble or his staff made any attempt to match the bone with rib fragments in the collection before making this decision:

> Because it could not be determined that this bone was in any way connected with the Kennewick collection, I placed it in a small plastic bag, which was then placed in the same box as Collection 2 (COE 2349).

The unauthorized removal is described in detail in Corps records (COE 5652, COE 5656). A PNNL employee's account of the removal includes Walla Walla District Corps of Engineers attorney Linda Kirts' comments:

> . . . Battelle shared partial responsibility for allowing tribes in the room while the Corps was transferring human remains. She [Kirts] stated that the political issues associated with this case made it highly likely that the court would rule in favor of plaintiffs (scientists) due to poor security and sloppy handling (COE 5652).

The security measure that required a key-badge to access the storage room did not mean that access to the remains was restricted once in the room. Trimble's assessment (COE 6716a) that security was excellent because access was limited to five key badges was questionable. Between 1996 and May 1998, documents in the Corps record show that 44 individuals had signed access logs with specific permission to access the remains as they deemed necessary. More than a dozen names reflect repeat visits.

- 17 were from the various American Indian tribes
- 14 were Corps staff
- 4 were PNNL staff
- 5 were from the Asatru Folk Assembly
- 2 were Justice staff
- 2 were Interior staff

The scientists' security and curation concerns were aired during a Federal District Court hearing in Portland, Oregon, on May 28, 1998 (Court Docket 147). The Court ordered that the scientists be allowed to inspect the PNNL facility and that the skeleton be moved to a neutral curation facility (Chapter 4; Court Docket 146).

As ordered by the Court, mediation on these issues between the Justice attorneys and the scientists' attorneys was held June 17–19, 1998, in Portland, Oregon. Robson Bonnichsen, a plaintiff, Owsley, and Hawkinson were also present. Any agreement between the parties proved elusive. To begin to address the unresolved mediation issues, the Court further ordered the skeleton be inventoried before it was moved to a new facility.

The first task was to define the process to inspect the PNNL facilities (COE 2308). Because he was local to the area, Chatters conducted the inspection for the plaintiffs on June 26, 1998. He raised these security concerns in his report concerning access to the building and to the repository room (COE 2308):

- possible entry by unauthorized individuals (anyone with a valid access card can enter regardless of who the card was issued to)
- building and room exits are not monitored

- bags are not checked upon entry or exit
- storage cabinet and unit are easily accessible
- door to repository room left open when people are working in the room
- possible unauthorized entry with someone who has a pass or when someone is exiting the room
- the number of individuals with authorized access to the repository room included 10 cultural resource people and 10 to 15 security guards.

Chatters concluded that even with the access log it is impossible to know how many individuals may have taken the opportunity to access the Kennewick Man skeleton between September 4, 1996, and June 1998.

Through the summer of 1998, the Corps assessed museums to find a secure, neutral repository with adequate facilities and staff to support their special requirements. The Corps rejected the plaintiffs' recommendation to consider the Smithsonian Institution's Museum Support Center in Suitland, Maryland, which was designed especially for such use (COE 5115). The Corps instead selected the Burke Museum of Natural History and Culture, University of Washington, in Seattle.

Fall 1998 to Summer 1999

Plaintiffs' Inventory

In the fall of 1998 the Corps, with input from the plaintiffs, defined the process, conditions, and restrictions that would govern the plaintiffs' court-ordered inventory of the skeleton (COE 2273; COE 2178). The Corps approved the participation of Trimble's curation staff, various Corps, Interior, and PNNL staff, and allowed tribal representatives from the five tribes to be observers in the room. To clarify the earlier Order, the District Court stated that the plaintiffs could have two representatives. Owsley was the plaintiffs' choice to inventory the skeleton. They were given two options for the second representative. Chatters could evaluate condition changes in the skeleton. Alternatively, an experienced osteological assistant could record Owsley's observations on a laptop computer while he handled the bone fragments with gloved hands. This was Owsley's standard protocol when evaluating archaeological and forensic remains.

The need for Chatters to evaluate changes in the condition of the skeleton took priority. The Corps tried to block Chatters's participation and denied Owsley permission to bring in a computer stating that only a pencil could be used during the inventory. The District Court ruled otherwise. Chatters would be allowed to participate, and Owsley could bring a tape recorder in lieu of a computer and assistant (Court Docket 187).

Under Corps supervision Owsley and Chatters conducted their assessment on October 28, 1998. Because of the extensive fragmentation of Kennewick Man's bones and the limited time allotted, Owsley's inventory session began at 7:00 a.m. and ended at dawn the next day, 22 hours later. His recommendations for stabilizing the Kennewick Man skeleton are documented in his report, "Inventory Observations on the Kennewick Man Skeleton, October 28–29, 1998" (COE 2106) and in his "A Summary of Lost Inventory Observations," (COE 2599). Owsley created the second document from memory immediately following the examination because one of the audiotapes used to record his observations had been inadvertently taped over. Chatters assessed taphonomic changes, including expanding and lengthening cracks in the cranium (COE 1983).

The Corps defined the protocol for transferring the skeleton to the Burke Museum (COE 4788). On October 29, 1998, after the remains were packed for transport, the fabric-draped container was loaded into the back of a Corps van. The Asatru Folk Assembly conducted a sunrise religious ceremony not far from the PNNL with the van parked nearby on an access road. The van returned to the PNNL parking lot for the tribal ceremonies while security officers ensured that the press and other observers remained at a respectful distance. The Kennewick Man skeleton and the Corps curation team were then transported in a caravan with full security to the Burke Museum. The plaintiffs would not be permitted to see the remains again for six years.

Burke Museum Curation

An incoming inventory was conducted at the museum on October 29. Representatives from the various agencies, the museum, the tribes, and the plaintiffs observed the inventory. A plaintiffs' representative and Trimble for the Corps each recorded the events of the transfer (Court Docket 193; COE 1387). The Corps curation team, with the assistance of Burke staff, unpacked the remains onto trays to allow the fragments to acclimate to the Burke's laboratory conditions. Polaroid® photographs taken by the Corps during Owsley's inventory were used to track the identification of fragments.

By late January 1999, the curation team had housed the fragments, by element, in customized, acid-free boxes lined with DuPont™ Tyvek pads. The conservators created a cutout for each fragment to provide a snug fit. This system was designed to prevent fragments from touching each other or shifting in the box. The fragments and their box cutouts were labeled with Master Catalog numbers typed on nonpermanent adhesive labels. Contents of each box and its lid were similarly labeled.

Attachment "C" of the legal agreement, Memorandum of Understanding for Curatorial Services stated: "All access to, and handling of, the Remains will be recorded on video tape," (Stipulation 4k (Collection Management Protocol, Special Procedures, DOI 2976)). However, neither the incoming inspection nor the rehousing activities was recorded on videotape.

The boxes were stored in the drawers of a Delta® storage cabinet in a secure closet in the basement of the Burke Museum. Reports, notes, and a variety of other media were also stored inside the cabinet in the same environment as the bones. A Repository Summary described the environmental conditions at the Burke: "relative humidity levels are stable in the Archaeology section at 45%–50% and the average temperature range is 67–70 degrees Fahrenheit," (COE 5203). This document was incorporated into the Memorandum of Understanding for Curatorial Services. The Corps curation team placed a data logger inside the cabinet to record relative humidity and temperature on an hourly basis, seven days a week (Chapter 31).

Government Phase One Studies 1999

On February 25, 1999, the National Park Service began limited Phase One studies on the skeleton (all study reports are posted on the Department of the Interior's National Park Service website, http://www.nps.gov/archeology/kennewick). No sampling for "destructive tests," such as DNA or ^{14}C analysis, was allowed. The government's team included two physical anthropologists (Jerome Rose and Joseph Powell), two geoarchaeologists (Julie Stein and Gary Huckleberry), and a lithic specialist (John Fagan). Two observers, Hawkinson for the plaintiffs and a rotating representative from each of the five tribes, were allowed to watch the studies from inside the laboratory doorway. Trimble's curation team and Burke Museum staff assisted the scientists. Various Corps, Interior, and Justice staff also signed the access log over the five-day study. To maintain environmental controls and to ensure the safety of the skeleton in a congested room, Trimble limited the number of people in the laboratory at any one time. The studies were not recorded on videotape.

As part of the study, scientists were asked to identify the elements of interest for analysis and for imaging. Corps and Burke Museum staff moved the boxes holding the skeletal elements in and out of the laboratory as needed. Scientists were directed to keep handling to a minimum but were permitted to bring laptop computers, cameras, and standard osteological tools into the laboratory. Temperature and humidity data loggers continued to record conditions in the room while the studies were underway.

Some activities, such as removing the soil samples, the first of two temporary cranial reconstructions, and the photographic work area, were out of the line of sight of the observers who sat inside the doorway of the ell-shaped laboratory. Occasionally, a member of the Corps curation team gave updates on the activities to the observers.

Assisted by the conservators, Stein and Huckleberry removed soil from the surfaces and medullary cavities of the bones (Figure 6.8). They removed sediment from inside the cranial vault with a Dremel® tool inserted into the foramen magnum. The sediment from the medullary cavities and the cranium were later combined into one sample to compare with core samples taken from the discovery site in December 1997. The sediment samples removed from the bones to determine the age of the skeleton yielded inconclusive results (DOI 10739). This sediment sampling from the skeleton occurred before the Interior conducted a taphonomic assessment. To the extent possible, Corps staff took photographs before and after these activities. However, the staff had no means of taking close-up photographs to document the condition inside the vault before or after using the Dremel® drill.

Fagan examined radiographs taken of the stone point in the right innominate, while visually comparing them to the point embedded in the bone fragment. Remodeled bone concealed the view of both ends of the point. Rose

FIGURE 6.8 Medullary cavity of the distal end of the middle segment (97.L.13b) of Kennewick Man's left humerus. Sediment was removed from this internal region during the government's Phase One study using a Dremel® tool. The removed soil sample was tested to help determine the stratigraphic placement of the skeleton within the soil bank.

showed Fagan how to orient the fragment on a comparative skeleton to determine the direction from which the point entered the hip. Rose concluded that the point entered from the back. This differed from Chatters's assessment that the point entered at an angle from the front (Chatters 2000). Fagan was unable to positively identify the type of projectile point (DOI 10809).

The Corps allowed no permanent reconstruction of the bone fragments. The conservators advised Powell to use jeweler's wax and wooden dowels during his temporary reconstruction of the pieces of the cranium. After Powell took standard cranial measurements, the cranium and other bones were packed and transported to the University of Washington Hospital for imaging. At the hospital, Powell verified his measurements to ensure that the temporary reconstruction remained intact during transport. Upon his return to the laboratory, Powell continued his evaluation at a different table in full view of the observers because tribal members had earlier objected that they could not see the scientists' activities. As Powell checked his measurements, another reconstruction was necessary after heat from the nearby lamp softened the wax and the facial bones slumped onto the table.

While Powell worked with the cranium, Rose concentrated on the postcranial skeleton and took measurements. When feasible, Corps staff held pieces of postcranial bones together for the scientists to measure. Rose and Powell switched places to complete their assessments of discrete traits and to take photographs. Rose spent a brief time assessing the small bags of unidentified fragments, but records do not show that he made any new identifications (Powell and Rose 1999).

The Corps allowed the tribal representatives to conduct religious ceremonies in the laboratory after the observers were excused. Hawkinson was not told of the ceremony at the time.

On July 27, 1999, the National Park Service announced that further tests were needed to confirm the age of the skeleton (DOI 4244). Principal Investigator Francis P. McManamon from the National Park Service summed up the results of the Phase One studies:

> After a thorough assessment of the data and reports produced by the examining team, the DOI, which is assisting the COE in determining what disposition is appropriate for the remains, determined that radiocarbon testing was needed to adequately establish the chronological placement of the remains. Nondestructive tests alone were insufficient to answer the question of the age of the skeletal remains reasonably (DOI 10660).

Fall 1999 to Summer 2000
Government Phase Two Study

The Phase Two Study for ^{14}C sampling began on September 8, 1999. McManamon, Trimble, Vicki Cassman, a

Corps contract conservator, and Kathy Taylor, a University of Washington graduate student, participated in this session with assistance from Corps and Burke Museum staff. As in Phase One, observers were allowed to view the activity from inside the laboratory doorway.

As principal investigator, McManamon made the final decisions about which bones to sample and the placement of the cuts. He rejected the plaintiff's recommendation to first take microsamples to establish the sample's suitability for successful test results. The process was photographed using 35mm film and a Polaroid® backup. K. Taylor used a Dremel® tool and a jeweler's saw to cut samples. The Department of the Interior and Corps administrative records contain several documents related to the sampling process (e.g., DOI 4865).

Two bones were sampled, a foot bone and a tibia. The right first metatarsal (segment number CENW-W.97.R.24(MTa)), which had been complete and intact, was cut into two pieces. Prior to sectioning, this metatarsal weighed 18.5 grams. The other bone sampled was the proximal left tibial shaft (sample number CENW-W.97.L.20b), the only reasonably complete leg bone in the collection (Figures 6.9 and 6.10). The weight of this sample was 12 grams. K. Taylor made a cut along an existing longitudinal crack on the tibia, with the goal of removing the anterior crest, and used the jeweler's saw to continue where the circular blade could not reach. The entire crest was detached from the main body of the bone. She then used the Dremel® tool to cut the detached crest into two pieces, cutting at the midsection of the anterior tibial crest, a landmark area used in measuring this bone. Before the separation was complete, two small fragments of bone broke away. The detached sections for sampling weighed 6.5 grams and 5.3 grams. Cassman regularly brushed bone dust from K. Taylor's face shield and the acid-free paper covering the work surface. She then bagged the dust to store with the collection. When sectioning was complete, more than 30 grams of bone were taken from the two bones for ^{14}C testing. The session was not video recorded. The tribes held a religious ceremony in the laboratory after the bones had been put away. This time other observers were invited to remain nearby.

On April 26 and 27, 2000, the Phase Two study by government-selected scientists continued with the objective of completing a taphonomic assessment. A stated purpose of the taphonomic study was to identify potential bones for DNA sampling (DOI 317). Although Powell and Rose had (1999) made a cursory taphonomic assessment at the close of their studies, National Park Service personnel concluded that they needed more guidance. Several documents detailing this matter are available in the Interior administrative record and National Park Service reports (McManamon 2004).

The full Phase Two study team included McManamon as Principal Investigator; physical anthropologists Clark Larsen, Joseph Powell, and Philip Walker; David G. Smith and R. E. Taylor; Trimble's curation team; and Burke Museum support staff. The Corps invited repre-

FIGURE 6.9 Medial surface of Kennewick Man's left tibia after the removal of the proximal anterior crest during the government's Phase Two study. The bone was sectioned for ^{14}C testing.

FIGURE 6.10 Proximal section of Kennewick Man's left tibia (97.L.20b) showing removal of the anterior crest from this section of bone during the government's Phase Two study. A partial transverse section is visible as an indentation near the upper end of the bone, and the longitudinal cut extends the entire length of the bone piece. The newly exposed surface is light in color relative to the ground-stained, unmodified lateral surface.

sentatives of the plaintiffs, tribes, and the Asatru Folk Assembly to observe.

During the laboratory orientation, Trimble and the conservators expressed concerns about the fragile condition of the bones and asked the scientists to minimize handling them. No video recordings were taken during the Phase Two taphonomic sessions.

Larsen, Powell, and Walker devised a five-color coding scheme using a Munsell chart to assess variations, one box at a time, noting anomalies in bone color. They coded peri- and postmortem breaks and assessed the visible surfaces of the fragments in each box. Occasionally they removed a piece of bone to inspect surfaces not in view, following Trimble's request to handle the remains only when necessary. Larsen, Walker, and Powell did not lay out the skeleton in anatomical order. The scientists took impressions of various bone surfaces for later analysis and used a camera mounted on a stereo microscope to photograph areas of macroscopic abrasions, possible rodent gnawing identified during the Phase One studies, and other surface anomalies. Their conclusions are reported in DOI 8481.

D. Smith and R. E. Taylor conducted the sampling for DNA testing. The identity of all the bone samples and the weights removed for both types of tests are listed in Table 6.1 (COE 317; DOI 8481). The conservators retrieved and bagged bone dust and dirt loosened during sawing. The sampling process was not videotaped. The

TABLE 6.1 Samples taken by National Park Service teams for radiocarbon dating and DNA analysis.

	Samples for ^{14}C tests
Sample 1	Right first metatarsal (CENWW.97.R.24(MTa) complete 18g
Sample 2	Left proximal tibial shaft (CENWW.97.L.20b) 12 g
	Samples for DNA tests
Sample 1	Left 3rd metacarpal 97.L.16(MCa), distal end, 0.5g
Sample 2a	Right 8th rib 97.I.12d(13), vertebral end, 0.4g
Sample 2b	Right 8th rib 97.I.12d(13), vertebral end, 0.4g
Sample 3	Right 8th rib 97.I.12d(13), sternal end, 1.0g
Sample 4	Right 3rd metacarpal 97.R.16(MCa), proximal end, 0.4g
Sample 5	Right 3rd metacarpal 97.R. 16(MCa), distal end, 1.1g Note: During sawing, the right 3rd metacarpal bone snapped in half near midshaft from the pressure. Weight of bone following sampling: 5.3g (weight before sampling: 6.8g).
Sample 6	Right 2nd metatarsal 97A.I.25c, midshaft, 0.5g
Sample 7	Left 2nd left metacarpal 97.L.16(MCb), distal end, 1.1g
Sample 8	Left tibia 97.L.20b, proximal end, 5.8g Note: A piece from proximal end adjacent to area from which one of the 1999 ^{14}C sample was taken.

contracted laboratories were not able to replicate DNA from any of the samples, but their reports provide full descriptions of the tests (McManamon 2000; McManamon et al. 2000).

During the Phase One studies and the subsequent sampling for ^{14}C and DNA, Corps staff and the curation team took before and after pictures of the fragments. An image record of more than 1,600 pictures documented research activities in the lab. The curation team used the photos to note changes in the condition of fragments during inspections conducted twice a year. The early photos are kept in acid-free binders, and some have been digitized. These pictures provide a historical record of activities associated with these studies.

Fall 2000 to Summer 2001
Department of the Interior NAGPRA Determination
On September 21, 2000, Secretary of the Interior Bruce Babbitt determined that the remains were culturally affiliated to the claiming tribes and no further study was necessary (Babbitt 2000). A press release issued on September 25 announced this decision to the public (DOI 10815). The Secretary sent a report with the reasons for this decision to Secretary of the Army Louis Caldera (DOI 10012). Had the Federal District Court not ruled otherwise, this determination would have resulted in the prompt burial of Kennewick Man. The Court's decision meant the litigation could continue (Chapter 4).

In December 2000 Justice filed the federal defendants' Administrative Record with the Federal District Court in Portland, Oregon. The more than 20,000 pages of documents from the Corps and Interior did not include any photographs. The plaintiffs' attorneys requested copies of the photos as an important part of the Administrative Record. The tribes raised religious objections to Justice, who then told the District Court that the costs of making copies would be prohibitive. The Court directed that the plaintiffs could send representatives to review the image archive at the Burke Museum.

Image Record Evaluation
On April 2, 2001, Owsley, science photographer Clark, and Hawkinson arrived at the Burke Museum with copies of the Corps reports filed with the District Court. These documents described the contents of the image record. Rhonda Lueck (Corps) brought the specific items requested to the work area and kept a list of the requests for photocopies, while Trimble observed the assessment. Clark evaluated hundreds of Corps photographs and found none of the images to be useful for scientific purposes. Issues included poor lighting, lack of standard scientific orientation, inconsistent use of scales, faulty shutter speed, and other film and camera deficiencies (Court Docket 419).

Owsley focused on the rest of the image record, which included radiographs taken during the 1999 studies. He found they lacked a key to identify the fragments. Furthermore, the CT images available were copies, which had lost the detail of the original scans.

Hawkinson evaluated pictures of the fragments that conservators had reported as deteriorating from minor to severe cracking. She discovered discrepancies in fragment identification from one photo series to the next; that is, the same fragment had different labels. The adhesive labels, chosen because they were nonpermanent, had fallen off and someone had confused the identity of the fragments when replacing them. Although earlier pictures were available, the individual who reaffixed the labels did not properly identify the fragments. After these mix-ups were resolved, the "deteriorating" fragments actually had no identifiable changes in condition that could be seen in the photos.

Affidavits by Owsley and Hawkinson detailing the shortcomings in the Corps image archives were filed as addendums to the Plaintiffs' Reply to Motion on June 4, 2001 (Court Docket 447, 450, 451).

Oral Arguments and Missing Femur Fragments Found
Oral arguments were held in Federal District Court on June 19 and 20, 2001, in Portland, Oregon (Court Docket 466, 467). The missing femur fragments surfaced in Richland, Washington, the same week. The Benton County Sheriff notified the Federal Bureau of Investigation (FBI) that a box had been discovered in the coroner's evidence locker of the justice facility. The box contained four femur fragments that appeared to be those in Chatters's 1996 video of the Kennewick bones. Their identity was confirmed when the fragments clearly articulated with the pieces in the collection. The FBI turned over the femur fragments to Trimble, who reunited them with the

collection housed at the Burke Museum (Court Docket 469). The FBI investigation was brief and inconclusive. The details of their disappearance and storage conditions during the five years they were missing are not documented in the public record.

Fourteen months later, on August 30, 2002, Magistrate Judge John Jelderks, Federal District Court of Oregon, ruled in favor of the plaintiffs' request to study the bones of Kennewick Man (Court Docket 496) and other issues raised by the plaintiffs (Chapter 4). The Corps required the plaintiffs to file a detailed study plan prior to allowing access to the skeleton.

Fall 2002 to Winter 2004
Study Plan Developed and the Opinion Appealed
In September 2002 Owsley, Stafford, Dennis Stanford, Hawkinson, and attorney Schneider met for two days with plaintiff Bonnichsen and his staff at the Center for the Study of First Americans at Texas A&M University, College Station. Their purpose was to take a fresh look at their original study plan, then nearly five years old. Progress had been made in research in the various scientific fields, particularly with the aid of improved technology. The government-conducted studies had raised new issues. After deciding which studies to pursue and the key scientists to participate, the team determined the order of studies and identified the details they would need for each study (Figure 6.11).

Upon returning home, the team invited various specialists to participate in the upcoming study and to provide descriptions for the required study plan. The plaintiffs met the 45-day deadline for filing a detailed, updated study plan with the Corps (Court Docket 495). The study plan was placed on hold when the government and three tribes appealed to the Ninth Circuit Court of Appeals. Over a year later, on February 4, 2004, the three-judge panel from the Ninth Circuit Court of Appeals upheld the lower court's decision allowing the plaintiffs to study the remains. A tribal petition for the appeal to be heard from a broader panel of judges was denied (Chapter 4).

Spring 2004 Access Stalled
The scientists requested permission from the Corps to examine the skeleton briefly so that they could update the study plan prepared in 2002. The Corps refused to allow any access to the skeleton until a plan was approved. Further, the Corps required specific details from the scientific specialists before approving a study plan.

Only one plaintiff, Owsley, had seen the remains during his inventory six years before. Since then, the museum storage cabinet in the Burke Museum had been opened in 1999 for the Phase One studies, in 2000 for the Phase Two studies, in 2001 to add the recovered femur fragments to the collection, and twice yearly for Corps condition inspections. The only information available to the plaintiffs were biannual reports filed with the Court

FIGURE 6.11 Meeting in September 2002 at the Center for the Study of the First Americans at Texas A&M University, College Station, to discuss revisions to the original Kennewick skeleton study plan. Standing: Hawkinson, Stafford, Bonnichsen; seated: Schneider, Stanford, Waters, and Owsley.

(Chapter 31). Without access to the skeleton, the scientists were unable to provide the level of detail that the Corps required. Negotiations remained deadlocked for 10 more months.

December 2004 to Spring 2005
Plaintiffs Gain Access

The Corps granted brief access to the skeleton on December 14 and 15, 2004. This assessment was held at the Burke Museum in the same laboratory used for the National Park Service studies. The Corps curation team followed the same procedures for security, sign-in, and support as they had during previous studies.

Those present to review the remains were Owsley, Berryman, Bonnichsen, Stafford, and C. Wayne Smith. Also in the room were Clark, Hawkinson, Corps staff and contract conservators, and rotating Burke Museum staff (Figure 6.12). The scientists' attorney, Schneider, accompanied the team. A Department of Justice attorney was also present. The Corps limited the number of people in the laboratory at one time, so the attorneys waited in the hallway until someone stepped out so they could enter.

The data gathered were instrumental in helping the plaintiffs provide details to their proposed study plan. This was the first opportunity for Owsley to see the four femur fragments that had been missing from the collection during his 1998 inventory. Berryman, an expert in assessing fracture patterns, was optimistic about the potential for obtaining taphonomic information from the postmortem breaks. Bonnichsen's visual inspection of the projectile point embedded in the ilium fragment reinforced the plaintiffs' need for clearer images of the broken base and tip. He took photographs of the point from different angles. With the help of a light probe, Bonnichsen tried, albeit unsuccessfully, to see into the narrow gap between the bone and point. Stafford assessed stains and patterned concretions deposited on surfaces

FIGURE 6.12 Skeleton access in December 2004. The plaintiff study team collected details on the bones in order to refine their original study plan. Left to right: Bonnichsen, Schneider, Owsley, Stafford, and Berryman.

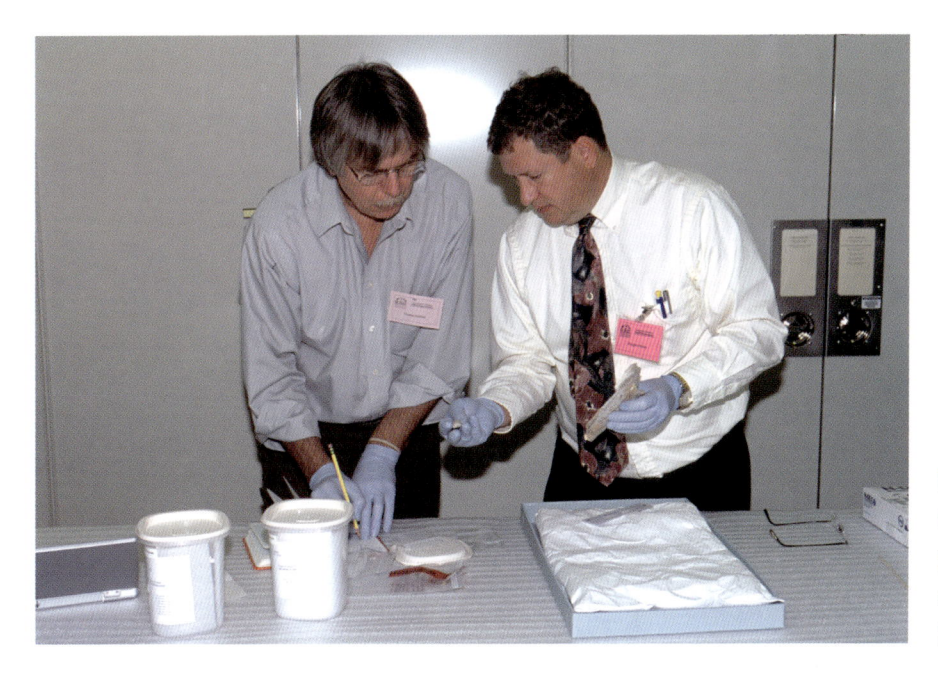

FIGURE 6.13 Stafford (left) and Owsley assessing cortical bone preservation, stain, and surface concretions during the December 2004 preliminary survey.

of the bones (Figure 6.13). He concluded that it should be possible to reconstruct the position of the remains when they were in the ground if the entire skeleton could be viewed in anatomical position. C. Wayne Smith was experienced in assessing bone exposed to water and was particularly interested in the condition of the cranium. He positioned a light probe to shine through the foramen magnum to view the inside of the cranial vault. The team was able to discern the ancient and recent soil deposits inside. The effects of sampling the sediment (in the form of gouges) during the government's Phase One 1999 studies were evident.

After the evaluation, the Corps curation team, scientists, and the attorneys met in a conference room in the museum to discuss the proposed components of the study plan and Corps requirements. The plaintiffs recommended reversible reconstruction of some elements using a conservator-approved adhesive. As they had during Powell's 1999 study, the Corps would allow only temporary reconstructions. They recommended using Parafilm® and jeweler's wax (Chapter 21). When the scientists pointed out the risks, the Corps conservators replied that the scientists should find a "creative" alternative.

The skull needed to be reconstructed for accurate measurement. In addition, the bone fragments needed reassembly to be viewed in full anatomical context. In a private session, the scientists discussed the limitations of using Parafilm® and the instability of jeweler's wax for this purpose. If the cranium with the face was reconstructed each time it was studied, the results would almost certainly vary as the delicate articulations would degrade with each refit. The group discussed Powell's experience of needing a second reconstruction when the jeweler's wax gave way under heat. The scientific team agreed that the conservators' method should be replaced with a safer solution. Two ideas emerged: to investigate industrial modeling technology and to build a sand table to hold bone fragments in anatomical position.

The scientific team researched new modeling techniques, identified a suitable facility, and obtained private funding to support a modeling project. They proposed a plan to the Corps to take high-resolution industrial CT scans of the separated fragments of the cranium and right innominate. They hoped to produce accurate copies of the original pieces through virtual and physical modeling (additive) technology.

The Corps approved this proposal, which was conducted at a facility near Chicago (Chapter 20). Rapid prototypes of the cranial pieces and fragments of the right hip were created to compare with actual bones during the taphonomic study in 2005. Hunt reconstructed the cranium and right innominate from these prototypes, which were then cast and made available to the study teams

(Chapter 21). A duplicate set of these casts was given to the Corps free of charge to be added to the archive for future scientists to use.

The plaintiffs' July 2005 taphonomic study was more comprehensive than previous examinations conducted by the government's study team (Powell and Rose 1999; Walker et al. 2000). This study was described and illustrated at the beginning of this chapter and their results are presented in Chapters 17, 18, and 19. In February 2006 specialized analyses were conducted on skeletal pathology, functional morphology, cranial morphometrics, and dentition by the plaintiffs' team of experienced scientists and doctoral students.

Scientific photographer Chip Clark documented the plaintiffs' studies. During the plaintiffs' condition assessment in December 2004, the CT scanning in 2005 (cranium and right innominate) and 2006 (long bones), and the scientific studies in July 2005 and February 2006, Clark took nearly 900 images including the full anatomical layout of the skeleton. Complete sets, some as 35 mm color slides but most in digital format, were provided to the Corps. Individual scientists from the study teams also gave copies of their images to the Corps.

FEBRUARY 2006 STUDIES

Following the taphonomic study session in 2005, further specialized studies on the skeleton were conducted in 2006. On February 16 and 17, Dennis Stanford (a plaintiff) evaluated the projectile point in the right hip fragment (Chapter 23). Richard Jantz (a plaintiff) arranged for CT and plain film imaging of selected bones at the University of Washington Hospital Department of Radiology, Seattle. He later took his standard suite of cranial measurements using a 3D digitizer (Chapter 25) (Figure 6.14).

The classroom at the Burke Museum was available to the scientific team for two days beginning February 21. The remaining studies were scheduled in the basement laboratory and ended on the afternoon of Friday, February 24. The Corps curation team and the Burke staff assisted. The curation team brought the boxes of elements to the scientists as requested, but the skeleton was not laid out in anatomical order as it was in 2005. The security and sign-in procedures that had been implemented in previous studies were followed and Corps staff regu-

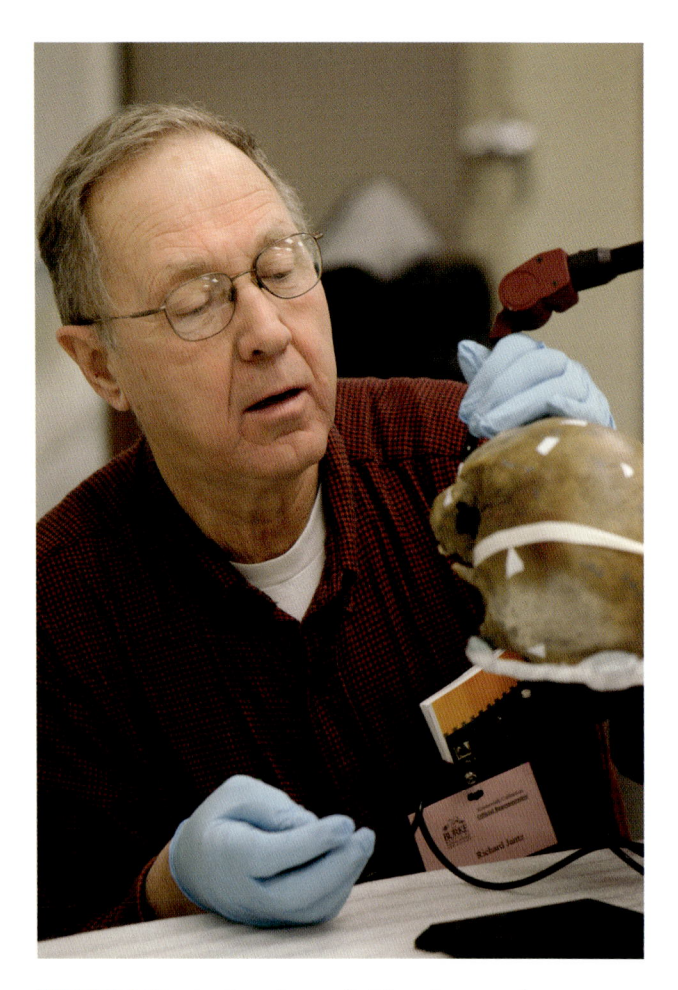

FIGURE 6.14 Jantz using a digitizer to record measurements of Kennewick Man's cranial vault at the Burke Museum, February 2006.

larly monitored temperature and humidity in both the classroom and the basement laboratory.

Three members of the taphonomy team (Owsley, Berryman, and Stafford) returned to address specific questions resulting from the 2005 study. Owsley and Berryman checked details of the fracture patterns. With this further study, Berryman was able to establish the order in which the sections of terrace collapsed onto the shore (Chapter 19).

Stafford arranged to conduct further dating and isotope analyses on fragments returned from previous analyses sponsored by the government. Stafford assessed the bag of debris and dust collected during sampling and study over the years but found nothing else he could use for geochemical analysis. The Corps denied requests to acquire new bone for testing.

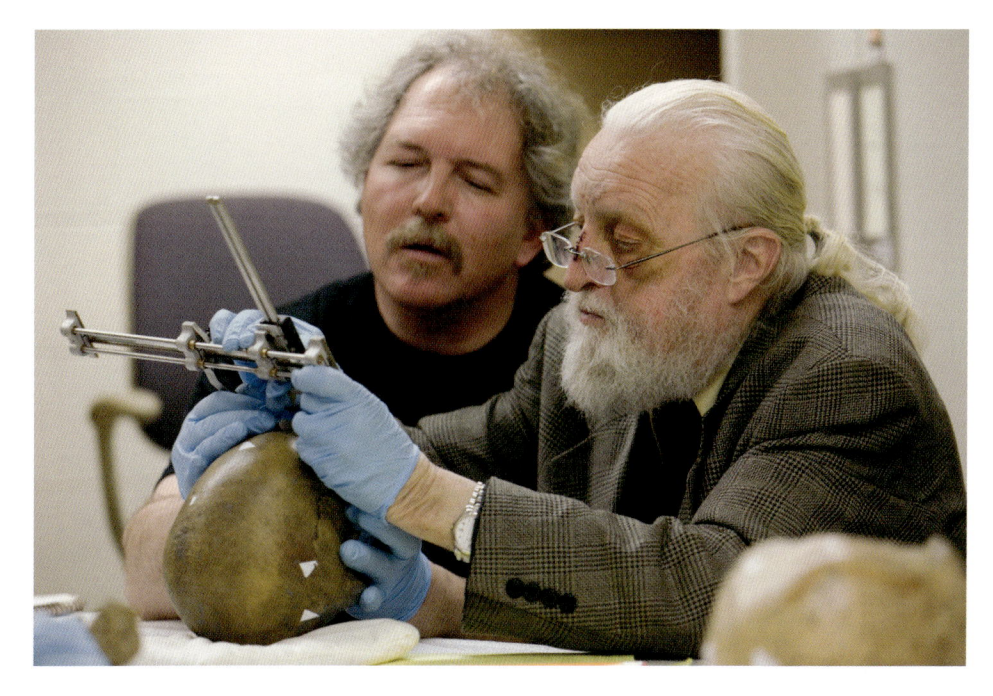

FIGURE 6.15 Nelson (left) assists Brace in recording measurements on a model of Kennewick Man's cranium.

Della Cook from Indiana University, Bloomington (Figure 6.16), evaluated bone pathology (Chapter 14) and George Gill (Figure 6.17) (a plaintiff), assessed cranial and postcranial discrete traits (Chapter 27). Christy Turner assessed dental traits and wear (Chapter 8) (Figure 6.18). Chatters joined the study team to confirm and expand his analysis, with emphasis on skeletal indicators of functional morphology and activity (Chapter 15) (Figure 6.19). When the scientists were finished working individually, they discussed their ideas with Owsley and other team members.

The final two days of study conducted in the basement laboratory provided fewer opportunities for the scientists to interact with each other in the presence of the bones. Team members were scheduled at specific times for their analyses to limit the number of people in the laboratory. Russell Nelson, Noriko Seguchi, and Daris Swindler helped C. Loring Brace (a plaintiff) collect cranial measurements and dental data consistent with his standard research protocol (Chapter 24) (Figure 6.15). Troy Case assessed morphological features of the hand and foot bones (Chapter 13) (Figure 6.20). Case used Corps' photographs to assess bones removed for destructive testing. Mark Teaford and Johns Hopkins University doctoral candidate Benjamin Auerbach were individually scheduled for specific blocks of time. Teaford took tooth enamel impressions for microwear analysis (Chapter 9). Auerbach measured the postcranial skeleton in order to determine stature, body proportions, and body mass (Chapter 11). As in previous studies, Clark provided photographic support.

Hawkinson remained in the hall outside the laboratory to facilitate the schedule and to ensure the scientists were aware of archived photographs and other information relevant to their studies. The sessions were not videotaped.

Summer 2006

On June 23, 2006, representatives from the Corps and Burke Museum met with Bart Cannon, a geologic analyst working on behalf of the scientists. They visited the Washington State Patrol Crime Laboratory in Marysville. The Corps took the cast of the hip fragment containing the projectile point to evaluate possible future instrumental analysis of the lithic point at the facility. This preliminary evaluation confirmed that there was no risk to Kennewick Man's ilium. However, to date, the Corps has not approved the plan proposed for this evaluation. On June 24, 2006, Chatters conducted a two-hour study session in the basement laboratory of the Burke Museum to finalize his documentation of skeletal indicators of functional morphology and activity. Since then, no other member of the plaintiffs' scientific team has been granted access by the Corps to study the skeleton.

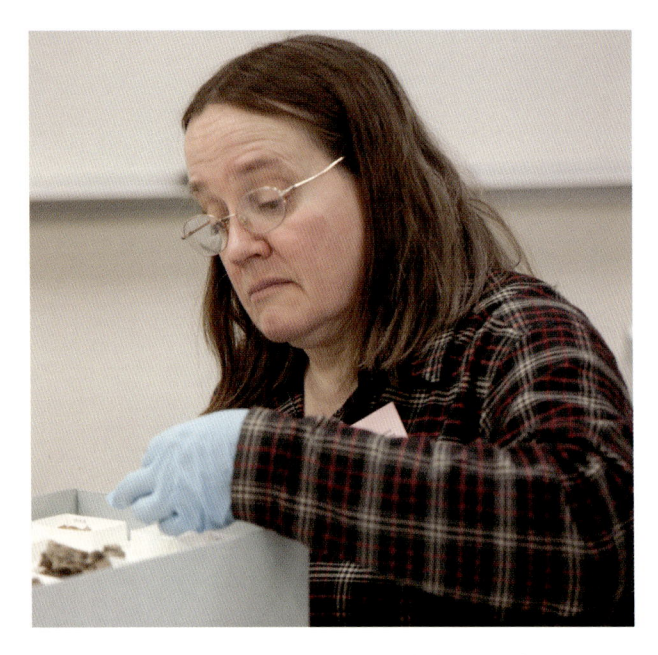

FIGURE 6.16   Cook, a specialist in skeletal pathology, evaluates pieces of Kennewick Man's skeleton during the February 2006 study.

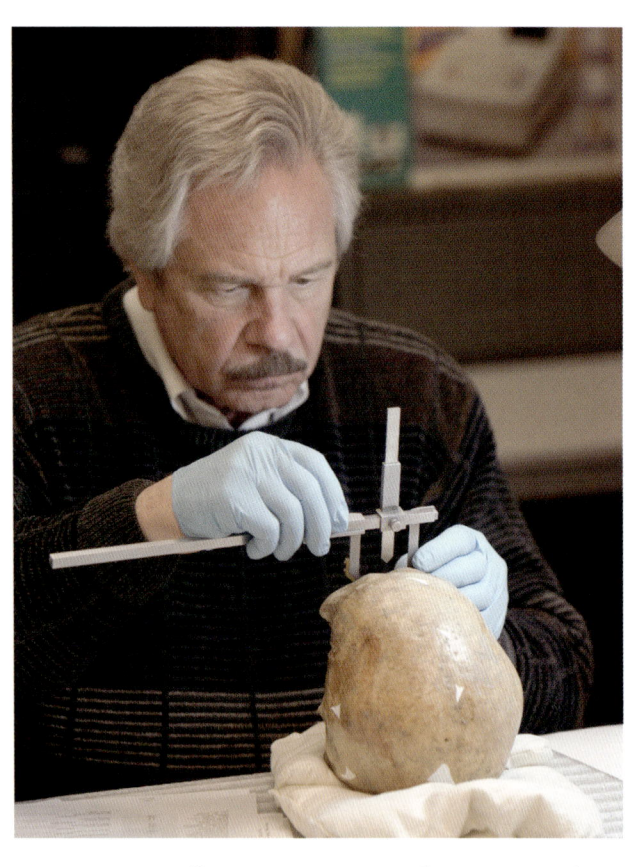

FIGURE 6.17   Gill measures Kennewick Man's cranial vault with a custom-designed radiometer during the February 2006 study.

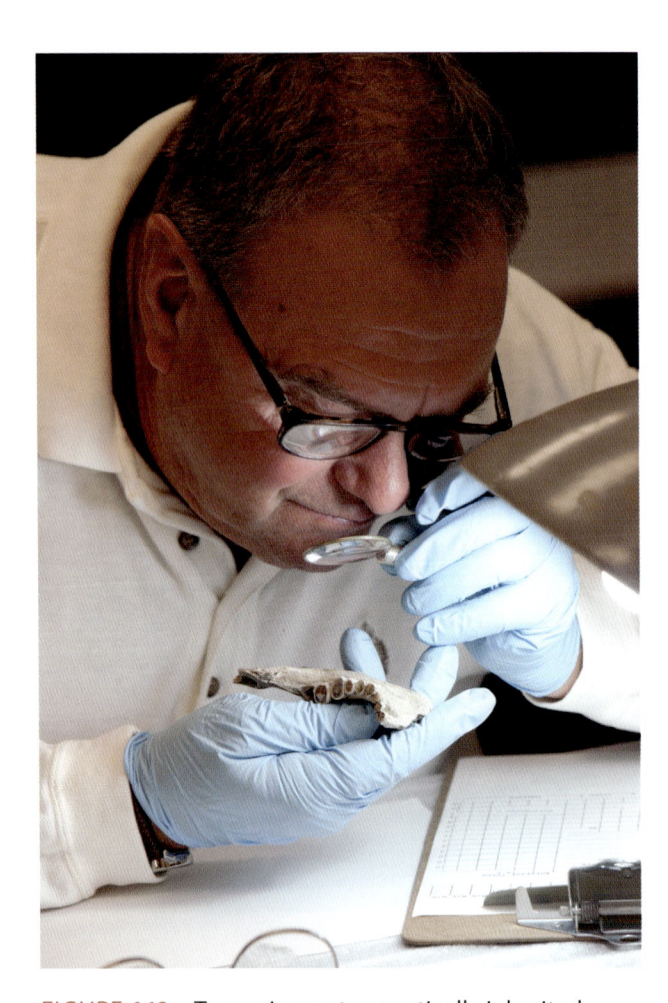

FIGURE 6.18   Turner inspects genetically inherited dental traits and the degree of tooth wear on a portion of Kennewick Man's left mandible, February 2006.

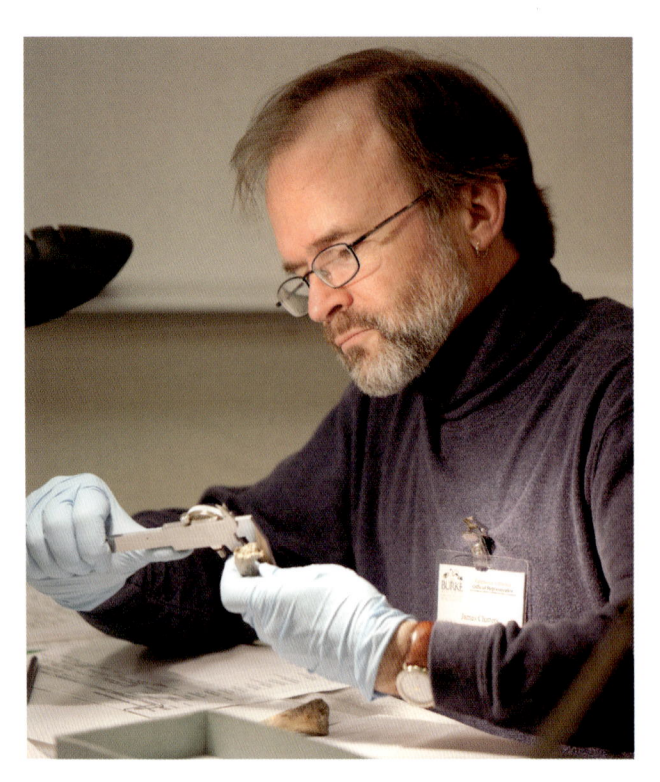

FIGURE 6.19   Chatters measures sections of bone as confirmation of previously recorded data and earlier observations, February 2006.

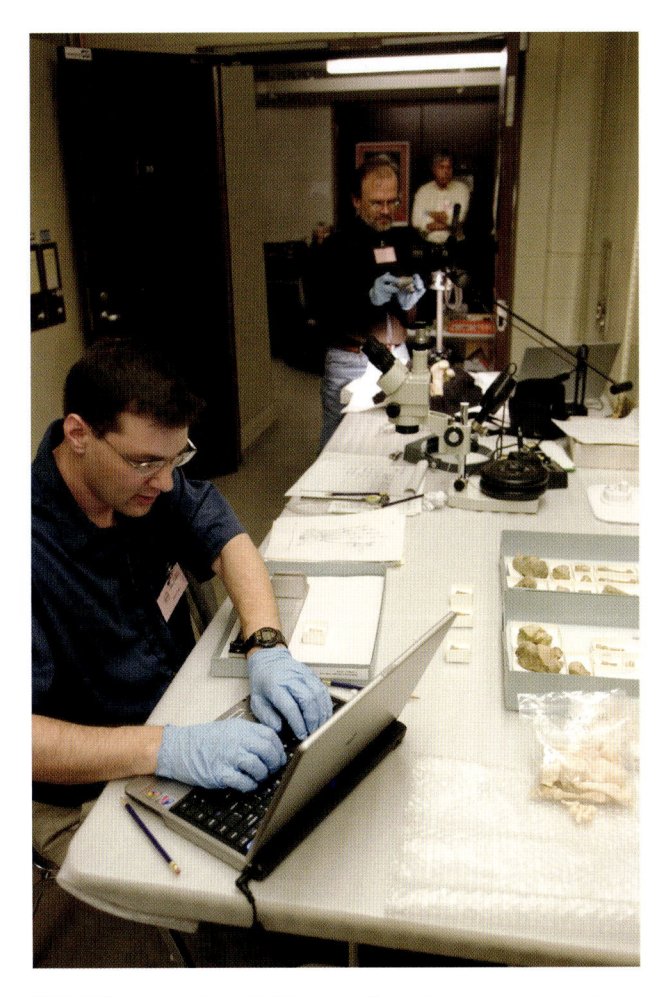

FIGURE 6.20 Case (left) recording observations and measurements on Kennewick Man's hand and foot bones, while Chatters conducts separate analyses, February 2006.

SUMMARY

This chapter describes the curation history of the Kennewick Man skeleton between 1996 and 2006 and presents an overview of the plaintiffs' study process. Gaining access to the skeleton took nearly nine years to accomplish.

Over two dozen scientists took part in the studies that contributed to the chapters in this volume. Five of the eight plaintiffs examined the skeleton and participated in the 2005 and 2006 studies. C. Vance Haynes, geoarchaeologist, was unable to investigate the discovery site following its burial under tons of rock. A plaintiff and major consultant to the litigation over the right to study the skeleton, Gentry Steele from Texas A&M University, was unable to travel to the Burke Museum due to declining health. Robson Bonnichsen participated in the December 2004 preliminary survey, but passed away prior to the 2005 taphonomic study. Chip Clark, whose exceptional photography is a major component of the scientific record, passed away in 2010. Clark's work presents clear documentation of the scientists' observations and their work as it progressed. The plaintiffs and their study team are committed to telling Kennewick Man's story, a legacy important to all Americans and of world-wide interest.

REFERENCES

References for this chapter are grouped by type: books and articles; administrative records for the US Army Corps of Engineers (COE numbers) and Department of the Interior (DOI numbers); the Court docket for *Bonnichsen et al.,* for the Federal District Court of Oregon; and documents available only on the Department of Interior website.

The information in the 20,000 pages of Administrative Records is incomplete. Key plaintiff documents appear only in their Excerpt of Record, which is listed in the Court Docket. Further complicating the task: not every document is listed individually in the indexes and a group of documents may have an index title that does not reflect the content. Frequently duplicate copies of the same document may appear multiple times in the Administrative Records of the COE or the DOI so a page may have more than one identification number.

Babbitt, Bruce. 2000. Letter to Louis Caldera, 21 September 2000. National Park Service, http://www.nps.gov/archeology/kennewick/babb_letter.htm.

Chatters, James C. 2000. The recovery and first analysis of an early Holocene human skeleton from Kennewick, Washington. *American Antiquity* 65(2):291–316.

———. 2001. *Ancient Encounters: Kennewick Man and the First Americans.* Simon & Schuster, New York.

McManamon, Francis. 2000. Memorandum to Assistant Secretary, Fish and Wildlife and Parks regarding "Determination That the Kennewick Human Skeletal Remains are 'Native American' for the Purposes of the Native American Graves Protection and Repatriation Act (NAGPRA)." [January 1, 2000]. http://www.nps.gov/archeology/kennewick/c14memo.htm.

McManamon, Francis P., Jason C. Roberts, and Brooke S. Blades. 2000. Chapter 1: Examination of the Kennewick

Remains—Taphonomy, Micro-sampling, and DNA Analysis. In *Report on the DNA Testing Results of the Kennewick Human Remains from Columbia Park, Kennewick, Washington* [September 2000]. National Park Service, http://www.nps.gov/archeology/kennewick/fpm_dna.htm.

McManamon, F. P. 2004. Kennewick Man [May 2004] National Park Service, www.nps.gov/history/archeology/kennewick/.

Powell, Joseph F., and Jerome C. Rose. 1999. Chapter 2. Report on the Osteological Assessment of the "Kennewick Man" Skeleton (CENWW.97. Kennewick). In *Report on the Non-Destructive Examination, Description, and Analysis of the Human Remains from Columbia Park, Kennewick, Washington* [October 1999]. National Park Service, http://www.nps.gov/archeology/kennewick/powell_rose.htm.

Walker, Phillip L., Clark Spencer Larsen, and Joseph F. Powell. 2000. Chapter 5. Final Report on the Physical Examination and Taphonomic Assessment of the Kennewick Human Remains (CENWW.97.Kennewick) to Assist with DNA Sample Selection. In *Report on the DNA Testing Results of the Kennewick Human Remains from Columbia Park, Kennewick, Washington* [September 2000]. National Park Service, http://www.nps.gov/archeology/kennewick/walker.htm.

COE numbers: US Army Corps of Engineers Administrative Record

317 Francis P. McManamon (DOI) notes: List of Ranking of candidate bone or teeth for DNA sampling and notes from microsampling on 4/26–4/27/00 (Draft).

630 James Chatters (Plaintiffs) description of the fifth metacarpal, December 20, 1999.

1387 Memorandum for Record: Trimble; Transfer of Walla Walla District Collection from Richland, Washington, to the Burke Museum, November 4, 1998.

1983 Photocopies of handwritten notes recorded by J. Chatters during osteological inventory, October 28–29, 1998.

2106 Inventory Observations on the Kennewick Man Collection, October 28–29, 1998. [Author Douglas W. Owsley]

2178 Douglas W. Owsley (Plaintiffs) to Michael K. Trimble (CEMVS) outlining pre-move inventory of remains, October 27, 1998. [Letter]

2273 Douglas W. Owsley (Plaintiffs) to Michael Trimble (CEMVS) and Robin Michael (DOJ) outlining pre-move procedures, September 17, 1998. [Letter]

2308 Cleone Hawkinson to Michael Trimble (CEMVS) regarding site inspection checklist used by James Chatters and comments on exit inventory proposed for transfer of collection, July 21, 1998. [Letter]

2349 Memorandum for Record: Review of Curation Procedures associated with human remains housed in Richland, WA (Status report). [Author Michael K. Trimble (CEMVS)]

2358 Memorandum for Record: Michael K. Trimble (CEMVS) regarding the transport of human skeletal remains from the University of California at Davis to PNNL, April 17, 1998.

2599 Summary of the Lost Inventory Observations, Douglas Owsley.

4788 Revised Joint Memorandum of Agreement Regarding the Transfer of the Remains to the Burke Museum, October 1, 1998.

5115 Letter from A. Schneider to R. Michael, USDOJ, re: Plaintiffs request that defendants review Smithsonian Institution as repository with enclosures, July 29, 1998.

5203 Federal Defendants Fourth Quarterly Status Report, July 1, 1998. Repository Summary Thomas Burke Memorial Washington State Museum, University of Washington, Seattle.

5314 Paper signed by LTC Curtis, Jr. listing persons granted access to remains, March 24, 1998.

5332 Dr. Madeline Fang Site Visit to Pacific Northwest National Laboratory, October 20, 1997.

5334 Letter from R. Michael USDOJ to P. Barran, enclosing report by Dr. Madeline Fang Site Visit, 9/10 Inventory, October 20, 1997, logs, and other documents, June 5, 1998.

5373 Memorandum for Record Synopsis of Site Visit to Battelle Laboratories 9 March 1998 to Review human remains from collection CENWW.97.KENNEWICK. [Author: Michael Trimble]

5651 Memorandum for Record: Status of "Additional" Human Remains from Columbia Park Site One as of 4/1/98. [Author: Ray Tracy]

5652 Statement by M. K. Wright, Battelle Laboratories, April 29, 1998.

5656 Memorandum for File Reburial of Multiple Remains

Recovered from Columbia Park, April 27, 1998. [Author: John Leier]

5808 Paper signed by LTC Curtis, Jr., listing persons granted access to remains, March 31, 1998.

5809 Paper signed by LTC Curtis, Jr., listing persons granted access to remains, March 27, 1998.

5996 News release, Army Corps notifies court of skeletal inventory discrepancy, March 10, 1998.

6533 Memo from D. Curtis, Jr. USACE approving access for December 15–19, 1997.

6712 Memo from D. Curtis, Jr. USACE approving access for November 3–7, 1997.

6716a Trip Report (Militello) and Memorandum for the Record, October 24–25, 1997.

7337 Memo from D. Curtis, Jr. USACE approving access to reunite additional fragments of the human remains of great antiquity collected at the discovery site, September 4, 1997.

7341 Memo from D. Curtis, Jr. USACE approving access to reunite recently collected skeletal elements with the Kennewick remains, October 18, 1996.

7342 Memo from D. Curtis, Jr. USACE approving access to reunite recently collected human remains with those presently held at PNNL, September 17, 1996.

7343 Memo from D. Curtis, Jr. USACE approving access for visual examination of the Columbia Park remains September 10, 1996.

7344 CTUIR inventory Richland Man Box, Recorded by J. Longenecker, PNNL, September 10, 1996.

8720 Memo for Record from R. Tracy, USACE, re: Columbia Park skeleton at U of CA, Davis, telephone conversation with Dr. Smith, October 25, 1996.

9030 Letter from L. Kirts, USACE to Dr. Kirner, University of California, re: DNA testing on human remains.

9270 Note to Advertising Department, September 17, 1996, Notice of intent to repatriate human remains in the possession of the Walla Walla District, US Army Corps of Engineers.

9277 Note to Advertising Department, September 24, 1996, Notice of intent to repatriate human remains in the possession of the Walla Walla District, US Army Corps of Engineers.

9293 Fax from J. Chatters, Applied Paleoscience, to R. Tracy, USACE, Enclosing information on skeletal remains. September 11, 1996.

9308 Chronology of Activities Associated with the 1996 Columbia Park Skeletal Finds, James C. Chatters, September 10, 1996.

9312 Memo of Understanding for Curatorial Services between USACE, WW, PNNL, September 9, 1996.

9315 Memo from LTC Curtis, Jr., USACE, re: Persons granted access to remains.

9365 Benton County Coroners (sic) Chronological Report of events relating to discoveries of human remains in Columbia Park, WA, September 6, 1996.

9396 Letter from A. Miller, Prosecuting Attorney for Benton County, WA to L. Kirts USACE with enclosed letters re: findings on examination of human remains, September 3, 1996.

9473 Memo from R. Tracy, USACE to A. Minthorn, J. Meninick, re: Columbia Park, Coroners news conference, August 27, 1996.

9474 Letter from Dr. Kirner, University of California, to Mr. Johnson, Coroner, Benton County, re: radiocarbon results, August 26, 1996.

9489 E-mail between C. Lohman, USACE, and R. Trace USSACE, re: human remains, August 5, 1996.

9501 E-mail from R. Tracy, USACE, to C. Lohman, USACE, re: finding of skeletal remains at Columbia Park, July 30, 1996.

9938 Memorandum for Record: Tracy Meeting at Battelle concerning human remains transfer Columbia Park discovery, undated (Sept 1996).

Court Docket

146 Order dated May 29, 1998, U. S. District Court, District of Oregon (Portland) Civil Docket for Case #: 3:96-cv-01481-JE.

147 Record of Proceedings before Judge John Jelderks: Curation hearing held May 28, 1998.

187 Record of Order by Judge John Jelderks, October 20, 1998.

193 Supplemental Report on *Transfer of the Skeleton*.

419 Affidavit of Douglas W. Owsley, filed by all plaintiffs (Entered April 17, 2001).

447 Reply to Motion related documents, June 4, 2001.

450 Affidavit of Cleone H. Hawkinson, filed by all plaintiffs (Entered June 5, 2001).

451 Affidavit of Douglas W. Owsley, filed by all plaintiffs (Entered June 5, 2001).

466 Transcript of Proceedings (Day 1) before Judge John

	Jelderks: Oral Arguments on the Briefings held June 19, 2001.
467	Transcript of Proceedings (Day 2) before Judge John Jelderks: Oral Arguments on the Briefings held June 20, 2001.
469	Notice Regarding Transfer of Femur Bones (Entered July 24, 2001).
495	Order Granted request to vacate decision to remand. Proposed study plan due in 45 days; defendants shall respond within 45 days; other topics. August 30, 2002.
496	Judgment: Pursuant to the Opinion and Order August 30, 2001.

DOI numbers: Department of the Interior Administrative Record

317	Francis P. McManamon (DOI) notes: Listing of Ranking of candidate bone or teeth for DNA sampling and notes from microsampling on April 26–27, 2000.
2976	Memorandum of Agreement for Cultural Services between the Army Corps of Engineers and the Thomas Burke Memorial Washington State Museum Attachment C: Collection Management Protocol included Special Procedures.
4244	Defendant 8th Quarterly Status Report with Declaration of FPM [McManamon] re: ^{14}C testing necessary to determine Native American status.
4865	Memorandum for Record Extracting bone samples from the Kennewick human remains 8 September 1999, Burke Museum, Seattle, Washington. September 23, 1999. [Author: Jason C. Roberts]
5780	Interagency Agreement Between DOA and DOI on the Delegation of Responsibilities for the Kennewick Human Remains (March 24, 1998).
8481	Email from FPM [McManamon] to Cleone Hawkinson / R. E. Taylor re: micro-samples extracted from K-remains. May 1, 2000.
10012	Letter from the Secretary to L. Caldera re: Disposition of the Kennewick Human Remains, September 25, 2000.
10660	The Initial Scientific Examination, Description, and Analysis of the Kennewick Man Human Remains. Author: F. P. McManamon. September 13, 2000.
10739	Analysis of Sediments Associated with Human Remains Found at Columbia Park, Kennewick, WA, National Park Service. Authors: Gary Huckleberry and July K. Stein. October 1999.
10809	Analysis of Lithic Artifact Embedded in the Columbia Park Remains. Author John L. Fagan. 1999.
10815	DOI determines "Kennewick Man" remains to go to five Indian tribes. September 25, 2000.

STUDYING SKELETAL EVIDENCE

The skeletal record is a source of information about demography, activities, diet, health, traumatic injuries, and even sociocultural interactions. Each examination begins with an inventory to determine the numbers of teeth and bones present and their completeness. At the most basic level, these inventories, as well as associated taphonomic observations, can be used to define the minimum number of individuals recovered from a given site when remains are commingled or the archaeological context is disturbed. More detailed analysis can lead to interpretations of pathology, burial context, postmortem processes, and diagenesis.

This second section provides detailed inventory documentation and interpretation of the Kennewick Man skeleton. Each bone piece is listed and described. This is done in an attempt to provide the reader, and future researchers, with enough background data to understand the composition of the skeletal remains, thus allowing for independent evaluations to be made. In this regard, multiple viewpoints are intentionally presented in the included chapters, some with differing opinions and interpretations of the observed skeletal features and chemical analyses. The chapters on taphonomy build upon the detailed bone descriptions and lead to interpretations of burial context and the postmortem processes that led to the recovery of the remains from the shoreline of Lake Wallula, an impound of the Columbia River. This level of interpretation of postmortem processes was developed in paleontology and forensic anthropology, but is relatively new within the field of Paleoamerican studies, and within bioarchaeology in general. Reading the skeleton for minute clues about in situ position and evidence for intentional burial demonstrates the level of information that can be recovered, even from ancient remains.

Skeletal Inventory, Morphology, and Pathology

7

Douglas W. Owsley, Aleithea A. Williams, and Karin S. Bruwelheide

The timely recovery of over 300 bones and fragments resulted in the remarkably complete skeleton of Kennewick Man. An inventory of those well-preserved skeletal elements, and their reassembly into articulated form, provides opportunity for close observation that leads to informed assessment of the skeleton. This overview presents the results of a comprehensive inventory of the remains, assesses morphological features that provide insight into Kennewick Man's activities in life, references previous and current interpretations of pathology, and explains the physical evidence for injuries, infection and arthritic changes. Chapters 8, 9, 10, and 13 present additional detailed information on topics not included here.

INVENTORY

A precise inventory of a set of human remains is the foundation for specific analyses, such as taphonomic assessments and evaluations of bone and dental pathology, as well as the reconstruction of fragmented remains that enables cranial and postcranial measurements (Buikstra and Ubelaker 1994; Owsley and Jantz 1989). James Chatters was the first to assess Kennewick Man's skeletal remains in 1996. Subsequent inventories were conducted by an archaeologist for the Umatilla Tribe in September 1996 and also by the US Army Corps of Engineers (Corps) in March 1998.

Owsley, Chatters, and Corps personnel and conservators completed a more detailed, court-ordered inventory on October 28–29, 1998, to establish an accurate record of all bones and incomplete fragments present, to determine whether any pieces were added or missing since the recovery by Chatters, and to verify that the assemblage represented only a single individual. Tasks included unpacking the remains from two Rubbermaid ActionPacker® storage containers, confirming the presence of all pieces labeled with previously assigned Corps catalog numbers, taking Polaroid® camera photographs of each piece, arranging the elements in anatomical order to determine whether duplicates were present, identifying and briefly describing each fragment, conservators reviewing their condition, and repacking. Restrictions imposed by the Corps limited Owsley and Chatters to only one day for the review and prohibited them from taking any measurements or computerized notes. The Corps originally stipulated that only pencil and paper could be used, but a favorable court ruling on a petition by the plaintiffs allowed them the use of an audio cassette recorder. Uncertain as to whether full access to the remains would be granted in the future, each fragment was methodically examined. The entire evaluation took about 21 hours.

The October 1998 inventory determined that the skeleton of Kennewick Man is comprised of about 265 bones and bone fragments, most of which were recovered in July and August 1996 (Table 7.1). Seventeen of 18 bones collected on September 17, 1996 (maintained separately as Collection A), and one of two bones collected on September 4, 1997 (Collection B) were integrated into the main collection as associated human bone based on their morphology, size, color, and preservation. Seven non-human bone fragments were reclassified as unassociated faunal specimens, and separated from the human remains for storage. Corps staff had tentatively identified an extra pubic bone (97.A.I.17.a) in the assemblage collected on September 17, 1996, suggesting the presence of a second individual. Upon inspection, this piece was identified as an ischium fragment that fit into an area of missing bone in Kennewick Man's left innominate (COE 4323). This inventory also confirmed that two-thirds of both femora (four pieces) recovered by Chatters were missing, as were bone

TABLE 7.1 Bone fragment identifications and catalog numbers assigned by the Corps. These catalog numbers are used as tracking guides in this and subsequent chapters to ensure clarity.

Specimen #	Bone	Description	Category
97.U.1a	Cranial vault	Frontal, parietals, temporals, occipital, partial nasals	Cranial and Dentition
97.U.1b	Left cranial vault fragment	N/A	Cranial and Dentition
97.U.1c	Left cranial vault fragment	N/A	Cranial and Dentition
97.R.1a	Right malar segment	Right malar and the articulating adjacent zygomatic process of the right maxilla	Cranial and Dentition
97.U.1c	Left malar	The frontal process and anterior malar are present	Cranial and Dentition
97.I.25v	Left maxilla fragment	Posterior maxillary sinus	Cranial and Dentition
97.R.2b	Right maxilla fragment	N/A	Cranial and Dentition
97.U.2a	Left and right maxillae	N/A	Cranial and Dentition
97.R.50a	Left maxillary M3	N/A	Cranial and Dentition
97.R.75a	Right mandibular M3	N/A	Cranial and Dentition
97.U.3a	Mandible	Left body	Cranial and Dentition
97.U.3b	Mandible	Right anterior body containing teeth I2 through M2	Cranial and Dentition
97.U.3c	Mandible	Right ascending ramus with the posterior M3 socket	Cranial and Dentition
97.U.6f	Sternum	N/A	Thorax: Sternum
97.I.25a	Sternum	N/A	Thorax: Sternum
97.L.11a	Left clavicle	Sternal end	Thorax: Clavicles
97.L.11b	Left clavicle	Middle fragment	Thorax: Clavicles
97.L.11c	Left clavicle	Lateral end	Thorax: Clavicles
97.R.11a	Right clavicle	Medial half	Thorax: Clavicles
97.R.11b	Right clavicle	Lateral half	Thorax: Clavicles
97.L.10a	Left scapula	Lateral spine and acromion process	Thorax: Scapulae
97.L.10b	Left scapula	Glenoid fossa	Thorax: Scapulae
97.L.10c	Left scapula	Fragment of coracoid process	Thorax: Scapulae
97.I.10c	Left scapula	Lateral, midaxillary border	Thorax: Scapulae
97.I.17e	Left scapula	Superior portion of the glenoid cavity	Thorax: Scapulae
97.I.10a	Left scapula	Lateral border	Thorax: Scapulae
97.I.25hh	Left scapula	Fragment of mid vertebral border	Thorax: Scapulae
97.I.25r	Left scapula	Fragment of body	Thorax: Scapulae
97.R.10a	Right scapula	The lateral spine and acromion process	Thorax: Scapulae
97.R.10b	Right scapula	Glenoid and coracoid process	Thorax: Scapulae
97.R.10c	Right scapula	Lateral, midaxillary border	Thorax: Scapulae
97.R.10d	Right scapula	Mid-lateral (root) of the spine	Thorax: Scapulae
97.I.25jj	Right scapula	Mid vertebral border	Thorax: Scapulae
97.U.4(C1a)	C1	Single, complete bone	Thorax: Vertebrae
97.U.4(C2a)	C2	Single, complete bone	Thorax: Vertebrae
97.U.4(C3a)	C3	Single, nearly complete bone	Thorax: Vertebrae
97.U.4(C4a)	C4	Single, nearly complete bone	Thorax: Vertebrae
97.A.U.4(C5a)	C5	Single, complete bone	Thorax: Vertebrae
97.U.4(C6a)	C6	Single, nearly complete bone	Thorax: Vertebrae
97.U.4(C7a)	C7	Single, nearly complete bone	Thorax: Vertebrae
97.U.5(T1a)	T1	N/A	Thorax: Vertebrae

(continued)

TABLE 7.1 (*continued*)   Bone fragment identifications and catalog numbers assigned by the Corps.

Specimen #	Bone	Description	Category
97.U.5d	T1	N/A	Thorax: Vertebrae
97A.U.5a	T1	N/A	Thorax: Vertebrae
97.U.5(T2a)	T2	N/A	Thorax: Vertebrae
97.U.5b	T2	N/A	Thorax: Vertebrae
97.U.5(T3a)	T3	N/A	Thorax: Vertebrae
97.U.5(T3b)	T3	N/A	Thorax: Vertebrae
97.U.5(T5a)	T4	N/A	Thorax: Vertebrae
97.U.5(T5b)	T4	N/A	Thorax: Vertebrae
97.U.5(T4a)	T5	N/A	Thorax: Vertebrae
97.U.5(T4b)	T5	N/A	Thorax: Vertebrae
97.U.5(T6a)	T6	N/A	Thorax: Vertebrae
97.U.5a	T6	N/A	Thorax: Vertebrae
97.U.5(T7a)	T7	N/A	Thorax: Vertebrae
97.U.5(T7b)	T7	N/A	Thorax: Vertebrae
97.U.5g	T7	N/A	Thorax: Vertebrae
97.U.5c	T7	N/A	Thorax: Vertebrae
97.U.5(T9a)	T8	Single, nearly complete bone	Thorax: Vertebrae
97.U.5(T8a)	T9	Single, mostly complete bone	Thorax: Vertebrae
97.U.5(T10a)	T10	Only the posterior third, both transverse processes, and the tip of the spinous process	Thorax: Vertebrae
97.U.5(T11a)	T11	Single, mostly complete bone	Thorax: Vertebrae
97.U.5(T12a)	T12	Single, mostly complete bone	Thorax: Vertebrae
97.U.6(L1b)	L1	Small portion of the inferior endplate	Thorax: Vertebrae
97.U.6c	L2	N/A	Thorax: Vertebrae
97.U.6(L4b)	L2	N/A	Thorax: Vertebrae
97.U.6(L1a)	L2	N/A	Thorax: Vertebrae
97.U.6(L4a)	L2	N/A	Thorax: Vertebrae
97.U.6(L3a)	L3	Single, mostly complete bone	Thorax: Vertebrae
97.U.6(L2b)	L4	N/A	Thorax: Vertebrae
97.U.6b	L4	N/A	Thorax: Vertebrae
97.U.6(L2a)	L4	N/A	Thorax: Vertebrae
97.U.6(L5b)	L5	N/A	Thorax: Vertebrae
97.U.6(L5a)	L5	N/A	Thorax: Vertebrae
97.U.6a	L5	N/A	Thorax: Vertebrae
97.I.25Z	L5	N/A	Thorax: Vertebrae
97.U.6(L5c)	L5	N/A	Thorax: Vertebrae
97.U.6(L2c)	Sacrum	Left articular facets of the first sacral unit (sacral promontory)	Thorax: Vertebrae
97.U.6(L5d)	Sacrum	Right articular facets of the first sacral unit (sacral promontory)	Thorax: Vertebrae
97.U.7a	Sacrum	Right margin of the sacral promontory and the superior-anterior aspect of the right ala	Thorax: Vertebrae
97.U.7b	Sacrum	Spinous processes of the sacral units	Thorax: Vertebrae
97.U.7c	Sacrum	Anterior aspect of the fifth sacral unit and a small portion of the fourth sacral unit	Thorax: Vertebrae
97.U.6d	Sacrum	Sacral fragment	Thorax: Vertebrae

(*continued*)

TABLE 7.1 (*continued*) Bone fragment identifications and catalog numbers assigned by the Corps.

Specimen #	Bone	Description	Category
97.I.12b(6)	Left rib 1	N/A	Thorax: Ribs
97.I.12b(1)	Left rib 1	N/A	Thorax: Ribs
97.I.12a(5)	Left rib 2	N/A	Thorax: Ribs
97.I.12f(3)	Left rib 2	N/A	Thorax: Ribs
97.I.12e(13)	Left rib 3	N/A	Thorax: Ribs
97.I.12e(10)	Left rib 3	N/A	Thorax: Ribs
97.I.12e(14)	Left rib 3	N/A	Thorax: Ribs
97.I.12a(2)	Left rib 3	N/A	Thorax: Ribs
97.I.12e(15)	Left rib 4	N/A	Thorax: Ribs
97.I.12e(9)	Left rib 4	N/A	Thorax: Ribs
97.I.12e(6)	Left rib 4	N/A	Thorax: Ribs
97.I.12f(1)	Left rib 4	N/A	Thorax: Ribs
97.I.12c(10)	Left rib 4	Two pieces held together by Parafilm®	Thorax: Ribs
97.I.12e(16)	Left rib 5	N/A	Thorax: Ribs
97.I.12e(17)	Left rib 5	N/A	Thorax: Ribs
97.I.12e(18)	Left rib 5	N/A	Thorax: Ribs
97.I.12e(8)	Left rib 5	N/A	Thorax: Ribs
97.I.12c(4)	Left rib 5	N/A	Thorax: Ribs
97.I.12c(6)	Left rib 5	N/A	Thorax: Ribs
97.I.12c(7)	Left rib 6	N/A	Thorax: Ribs
97.I.12f(11)	Left rib 6	N/A	Thorax: Ribs
97.I.12d(3)	Left rib 6	N/A	Thorax: Ribs
97.I.12e(2)	Left rib 7	N/A	Thorax: Ribs
97.I.12e(4)	Left rib 7	N/A	Thorax: Ribs
97.I.12f(13)	Left rib 7	N/A	Thorax: Ribs
97.I.12f(9)	Left rib 7	N/A	Thorax: Ribs
97.I.12e(12)	Left rib 8	N/A	Thorax: Ribs
97.I.12a(4)	Left rib 8	N/A	Thorax: Ribs
97.I.12c(2)	Left rib 8	N/A	Thorax: Ribs
97.I.12c(8)	Left rib 8	N/A	Thorax: Ribs
97.I.12c(5)	Left rib 8	N/A	Thorax: Ribs
97A.I.12b	Left rib 8	N/A	Thorax: Ribs
97.I.12e(1)	Left rib 9	Only the head and neck portion	Thorax: Ribs
97.I.12b(10)	Left rib 10	N/A	Thorax: Ribs
97.I.12f(10)	Left rib 10	N/A	Thorax: Ribs
97.I.12d(8)	Left rib 10	N/A	Thorax: Ribs
97.I.12c(3)	Left rib 10	N/A	Thorax: Ribs
97.I.12e(7)	Left rib 11	N/A	Thorax: Ribs
97.I.12c(1)	Left rib 11	N/A	Thorax: Ribs
97.I.12f(5)	Left rib 12	N/A	Thorax: Ribs
97.I.12f(6)	Left rib 12	N/A	Thorax: Ribs
97.I.12e(3)	Right rib 1	N/A	Thorax: Ribs
97.I.12b(7)	Right rib 2	N/A	Thorax: Ribs
97A.I.12c	Right rib 2	N/A	Thorax: Ribs

(*continued*)

TABLE 7.1 (*continued*)   Bone fragment identifications and catalog numbers assigned by the Corps.

Specimen #	Bone	Description	Category
97.I.12a(7)	Right rib 2	N/A	Thorax: Ribs
97.I.12e(5)	Right rib 3	N/A	Thorax: Ribs
97.I.12b(4)	Right rib 3	N/A	Thorax: Ribs
97.I.12b(3)	Right rib 3	N/A	Thorax: Ribs
97A.I.12d	Right rib 3	N/A	Thorax: Ribs
97.I.12a(9)	Right rib 3	N/A	Thorax: Ribs
97.I.12b(5)	Right rib 4	N/A	Thorax: Ribs
97.I.12b(15)	Right rib 4	N/A	Thorax: Ribs
97.I.12b(2)	Right rib 4	N/A	Thorax: Ribs
97.I.12a(1)	Right rib 4	N/A	Thorax: Ribs
97.I.12b(8)	Right rib 5	N/A	Thorax: Ribs
97A.I.12a	Right rib 5	N/A	Thorax: Ribs
97.I.12f(14)	Right rib 5	N/A	Thorax: Ribs
97.I.12f(8)	Right rib 5	N/A	Thorax: Ribs
97.I.12a(6)	Right rib 5	N/A	Thorax: Ribs
97.I.12a(8)	Right rib 5	Tentatively associated with this rib	Thorax: Ribs
97.I.12b(11)	Right rib 6	N/A	Thorax: Ribs
97.I.12d(10)	Right rib 6	N/A	Thorax: Ribs
97.I.12d(5)	Right rib 6	N/A	Thorax: Ribs
97.I.12f(12)	Right rib 6	N/A	Thorax: Ribs
97.I.12b(13)	Right rib 7	N/A	Thorax: Ribs
97.I.12d(14)	Right rib 7	N/A	Thorax: Ribs
97.I.12a(3)	Right rib 7	Tentatively associated with this rib	Thorax: Ribs
97.I.12b(14)	Right rib 8	N/A	Thorax: Ribs
97.I.12b(9)	Right rib 8	N/A	Thorax: Ribs
97.I.12d(2)	Right rib 8	N/A	Thorax: Ribs
97.I.12d(1)	Right rib 8	N/A	Thorax: Ribs
97.I.12d(7)	Right rib 8	N/A	Thorax: Ribs
97.I.12b(12)	Right rib 9	N/A	Thorax: Ribs
97.I.12d(12)	Right rib 9	N/A	Thorax: Ribs
97.I.12d(11)	Right rib 9	N/A	Thorax: Ribs
97.I.12d(9)	Right rib 10	N/A	Thorax: Ribs
97.I.12f(7)	Right rib 10	N/A	Thorax: Ribs
97.I.12d(6)	Right rib 10	N/A	Thorax: Ribs
97.I.12d(4)	Right rib 10	N/A	Thorax: Ribs
97.I.12e(11)	Right rib 12	N/A	Thorax: Ribs
97.L.17b	Left innominate	Nearly complete ilium and portions of the ischium and pubis	Pelvis
97A.I.17a	Left innominate	The ischiopubic ramus of ischium	Pelvis
97.L.17a	Left innominate	Left pubis including the symphyseal face	Pelvis
97A.I.25b	Left innominate	Small piece of superior posterior iliac crest	Pelvis
97.I.17a	Left innominate	May not belong with the innominate; may be calcaneus	Pelvis
97.I.25d	Left innominate	Most likely is a portion of the ischium	Pelvis
97.R.17a	Right innominate	Pubic bone	Pelvis
97.R.17b	Right innominate	Anterior ilium including the anterior superior and anterior inferior spines, the iliac fossa, and the anterior iliac crest	Pelvis

(*continued*)

TABLE 7.1 (*continued*)   Bone fragment identifications and catalog numbers assigned by the Corps.

Specimen #	Bone	Description	Category
97.R.17c	Right innominate	Posterior ilium in two segments; one contains the embedded projectile point	Pelvis
97.R.17d	Right innominate	Main portion containing the auricular surface, sciatic notch, complete acetabulum, and superior ischium	Pelvis
97.I.17d	Right innominate	Anterior middle aspect of the iliac crest	Pelvis
97.I.10b	Right innominate	Middle aspect of the iliac crest	Pelvis
97.I.17c	Right innominate	Ischial fragment	Pelvis
97.I.17b	Right innominate	Ischio-pubic ramus fragment	Pelvis
97.L.13a	Left humerus	Epiphysis and upper three-fifths of the diaphysis	Appendicular: Upper
97.L.13b	Left humerus	Lower portion of the diaphysis (about one-fourth)	Appendicular: Upper
97.L.13c	Left humerus	Most inferior one-fourth of the diaphysis and distal joint	Appendicular: Upper
97.R.13a	Right humerus	Proximal epiphysis and upper three-fourths of the diaphysis	Appendicular: Upper
97.R.13b	Right humerus	Lower one-fourth of the diaphysis and distal joint	Appendicular: Upper
97.L.14a	Left radius	Proximal one-third of the diaphysis including a nearly complete radial head	Appendicular: Upper
97.L.14b	Left radius	Distal one-half of the radius including the joint surface	Appendicular: Upper
97.R.14a	Right radius	Head and two-thirds of the diaphysis	Appendicular: Upper
97.R.14b	Right radius	Lower one-third of the diaphysis and the distal end	Appendicular: Upper
97.L.15a	Left ulna	Proximal one-third of the ulna including the joint surface	Appendicular: Upper
97.L.15b	Left ulna	Middle one-third of the diaphysis	Appendicular: Upper
97.L15c	Left ulna	Distal one-third of the diaphysis; the epiphysis is missing	Appendicular: Upper
97.R.15a	Right ulna	Proximal one-third of diaphysis including the intact joint	Appendicular: Upper
97.R.15b	Right ulna	Middle one-third of the diaphysis	Appendicular: Upper
97.R.15c	Right ulna	Distal one-third of the diaphysis including the distal joint	Appendicular: Upper
97.L.16(CRa)	Left Hand	Lunate	Appendicular: Hands
97.L.16(CRc)	Left Hand	Capitate	Appendicular: Hands
97.R.16(MCd)	Left Hand	Metacarpal 1	Appendicular: Hands
97.L.16(MCb)	Left Hand	Metacarpal 2	Appendicular: Hands
97.L.16(MCa)[1]	Left Hand	Metacarpal 3	Appendicular: Hands
97.L.16(MCc)	Left Hand	Metacarpal 4	Appendicular: Hands
97.L.16(MCd)	Left Hand	Metacarpal 4	Appendicular: Hands
97.A.I.24(Pb)	Left Hand	Proximal phalanx 1	Appendicular: Hands
99.I.16b	Left Hand	Metacarpal 5, residual sample from University of California Davis evaluated by Thomas W. Stafford (SR-7093); original Chatters sample	Appendicular: Hands
99.I.16c	Left Hand	Metacarpal 5, residual sample from University of California Davis evaluated by Thomas W. Stafford (SR-7093); original Chatters sample	Appendicular: Hands
97.L.16(Pf)	Left Hand	Proximal phalanx 2	Appendicular: Hands
97.L.16(Pa)	Left Hand	Proximal phalanx 4	Appendicular: Hands
97.A.I.16(Pa)	Left Hand	Proximal phalanx 5	Appendicular: Hands
97.R.16(Pe)	Left Hand	Intermediate phalanx 2	Appendicular: Hands
97.R.16(Pf)	Left Hand	Intermediate phalanx 3	Appendicular: Hands
97.L.16(Pd)	Left Hand	Intermediate phalanx 5	Appendicular: Hands
97.L.16(Pe)	Left Hand	Distal phalanx 1	Appendicular: Hands

(*continued*)

TABLE 7.1 (continued)   Bone fragment identifications and catalog numbers assigned by the Corps.

Specimen #	Bone	Description	Category
97.R.16(Pc)	Left Hand	Distal phalanx 4	Appendicular: Hands
97.L.16(CRb)	Right Hand	Scaphoid	Appendicular: Hands
97.R.16(CRb)	Right Hand	Lunate	Appendicular: Hands
97.R.16(CRa)	Right Hand	Hamate	Appendicular: Hands
97B.L.16(MCa)	Right Hand	Metacarpal 1	Appendicular: Hands
97.R.16(MCc)	Right Hand	Metacarpal 2	Appendicular: Hands
97.R.16(MCa)	Right Hand	Metacarpal 3	Appendicular: Hands
97.R.16(MCe)	Right Hand	Metacarpal 4	Appendicular: Hands
97.R.16(MCb)	Right Hand	Metacarpal 5	Appendicular: Hands
97.L.24(Pd)	Right Hand	Proximal phalanx 1	Appendicular: Hands
97.L.16(Pc)	Right Hand	Proximal phalanx 2	Appendicular: Hands
97.R.16(Pi)	Right Hand	Proximal phalanx 3	Appendicular: Hands
97.R.16(Pd)	Right Hand	Proximal phalanx 4	Appendicular: Hands
97.R.16(Pb)	Right Hand	Proximal phalanx 5	Appendicular: Hands
97.L.16(Pb)	Right Hand	Intermediate phalanx 4	Appendicular: Hands
97.R.16(Pg)	Right Hand	Intermediate phalanx 5	Appendicular: Hands
97.R.16(Ph)	Right Hand	Distal phalanx 1	Appendicular: Hands
97.R.16(Pa)	Right Hand	Distal phalanx 3	Appendicular: Hands
97.L.18a	Left femur	Proximal two-fifths	Appendicular: Lower
01.L.18b	Left femur	Middle and upper lower diaphysis	Appendicular: Lower
01.L.18a	Left femur	Portion of lower diaphysis and distal joint surface	Appendicular: Lower
01.R.18a	Right femur	Proximal one-fourths	Appendicular: Lower
01.R.18b	Right femur	Midshaft (middle one-third and lower proximal one-fourth)	Appendicular: Lower
97.R.18a	Right femur	Portion of lower diaphysis (one-third) and distal joint surface	Appendicular: Lower
97.L.20a	Left tibia	Proximal joint, epiphysis, and most superior diaphysis	Appendicular: Lower
97.L.20b	Left tibia	Two-fifths of the proximal diaphysis	Appendicular: Lower
97.L.20c	Left tibia	Lower three-fifths of diaphysis and distal joint	Appendicular: Lower
97.R.20a	Right tibia	Medial condyle and epiphysis; the anterior half of this bone segment is missing postmortem	Appendicular: Lower
97.R.20b	Right tibia	Lateral condyle and epiphysis; the anterior half is missing	Appendicular: Lower
97.R.20c	Right tibia	Proximal and middle one-third of the diaphysis including most of the tibial tuberosity	Appendicular: Lower
97R.20d	Right tibia	Distal one-third of the diaphysis including the joint	Appendicular: Lower
97.L.21d	Left fibula	Proximal diaphysis	Appendicular: Lower
97.L.21c	Left fibula	Middle diaphysis	Appendicular: Lower
97.R.21b	Right fibula	N/A	Appendicular: Lower
97.L.23a	Left talus	Incomplete, only the anterior head and neck	Appendicular: Lower
97.L.22a	Left calcaneus	Nearly complete	Appendicular: Lower
97.R.23a	Right talus	Complete	Appendicular: Lower
97.R.22a	Right calcaneus	Complete	Appendicular: Lower
97.L.24(TRb)	Left Foot	Navicular	Appendicular: Lower
97.L.24(Tra)	Left Foot	Cuboid	Appendicular: Lower
97.L.24(TRc)	Left Foot	Cuneiform 1	Appendicular: Lower
97.L.24(TRe)	Left Foot	Cuneiform 2	Appendicular: Lower
97.L.24(TRd)	Left Foot	Cuneiform 3	Appendicular: Lower

(continued)

TABLE 7.1 (continued) Bone fragment identifications and catalog numbers assigned by the Corps.

Specimen #	Bone	Description	Category
97.L.24(MTa)	Left Foot	Metatarsal 1	Appendicular: Lower
97.L.24(MTb)	Left Foot	Metatarsal 2	Appendicular: Lower
97.L.24(MTc)	Left Foot	Metatarsal 3	Appendicular: Lower
97.L.24(MTd)	Left Foot	Metatarsal 4	Appendicular: Lower
97.A.I.24(Pa)	Left Foot	Proximal phalanx 2	Appendicular: Lower
97.L.24(Pa)	Left Foot	Proximal phalanx 3	Appendicular: Lower
97.R.24(Pa)	Left Foot	Proximal phalanx 4	Appendicular: Lower
97.L.24(Pc)	Left Foot	Distal phalanx 1	Appendicular: Lower
97.R.24(TRa)	Right Foot	Navicular	Appendicular: Lower
97.R.24(Mta)[2]	Right Foot	Metatarsal 1	Appendicular: Lower
97.R.24(MTc)	Right Foot	Metatarsal 2	Appendicular: Lower
97.R.24(MTb)	Right Foot	Metatarsal 3	Appendicular: Lower
97.A.I.25c	Right Foot	Metatarsal 4	Appendicular: Lower
97.L.24(Pe)	Right Foot	Proximal phalanx 1	Appendicular: Lower
97.L.24(Pb)	Right Foot	Proximal phalanx 2	Appendicular: Lower
97.R.24(Pb)	Right Foot	Proximal phalanx 5	Appendicular: Lower
97.I.12d(13)	Unknown	Residual bone dust and small fragments from University of California Riverside evaluated by Thomas W. Stafford (SR-7087)	Unknown

[1]This bone was removed for DNA testing and is no longer in the collection. Identification made from photographs (Chapter 13).
[2]This bone was removed for AMS testing by Beta Analytic and is no longer in the collection. Identification made from photographs (Chapter 13).

fragments removed from the collection on April 27, 1998 (Chapter 6).

The plaintiffs secured another court-authorized permission to study the skeleton and conducted a preliminary inspection in December 2004 in order to refine their research design (Chapter 6). In July 2005 and February 2006, scientific teams comprised of the plaintiffs and other scientists, including doctoral students, were allowed to fully evaluate the skeleton at the Burke Museum with oversight by Corps personnel and their contract conservators.

The initial step in this examination was to validate and perfect the inventory as an essential baseline for future studies and comparative research. Additional study time helped to secure the identification and refitting of small bone pieces, many of which were less than an inch in length, including anatomical placement of more than 80 rib fragments (Figure 7.1). Simultaneously completed procedures involved systematic recording of taphonomic observations, identifying potential bone and sediment samples for specialized analyses (e.g., radiocarbon dating, stable isotopes, and osteon counting), temporarily reconstructing long bones to enable measurement (Figure 7.2; Table 7.2), identifying and describing pathological conditions, and estimating Kennewick Man's chronological age.

To aid the examination, x-ray and computed tomography scans of the limb bones were taken at the University of Washington Medical Center, Radiology Department using a GE Light Speed VCT/64 Slice Detector set at 0.625 mm slice intervals with a 0.984:1 pitch and GE DICOM viewing software. Digital radiographs were also taken of the mandible, maxillae, scapulae, clavicles, selected left and right ribs, limb bones, and the right ilium fragment with the embedded projectile point.

FIGURE 7.1　Reassembly of six fragments comprising the left eighth rib, confirmed by rearticulation of broken edges. Lighter-colored fragments of the sternal half (right side) lost surface staining due to wave-induced movement and tumbling along the shore.

A DESCRIPTIVE AND VISUAL GUIDE
Cranium and Mandible

The Kennewick skull consists of the cranial vault (97.U.1a), facial bones, and mandible. The frontal bone, right and left parietals, right and left temporals, and occipital bone are complete and exhibit normal articulation at the sutures (Figure 7.3). Partial nasal bones remain attached to the frontal bone and show slight antemortem deviation toward the left side. Their distal halves are missing due to postmortem breakage. Small portions of the frontal processes of the maxillae are also present, but with postmortem breakage inferior to the fronto-nasal suture. Features indicative of male sex include the large size and robusticity of the vault, development of the supraorbital ridges, a low, sloping frontal bone, large mastoid processes, and development of the nuchal ridge on the occipital bone. The nuchal ridge is slightly raised and the protuberance shows slight to moderate development.

The external surface of the greater wing of the left sphenoid, evaluated macroscopically and under stereozoom microscopy, shows loss of patination and exposed underlying cancellous bone. These alterations reflect postmortem damage caused by corrasion while the cranium was eroding from the soil bank, as dis-

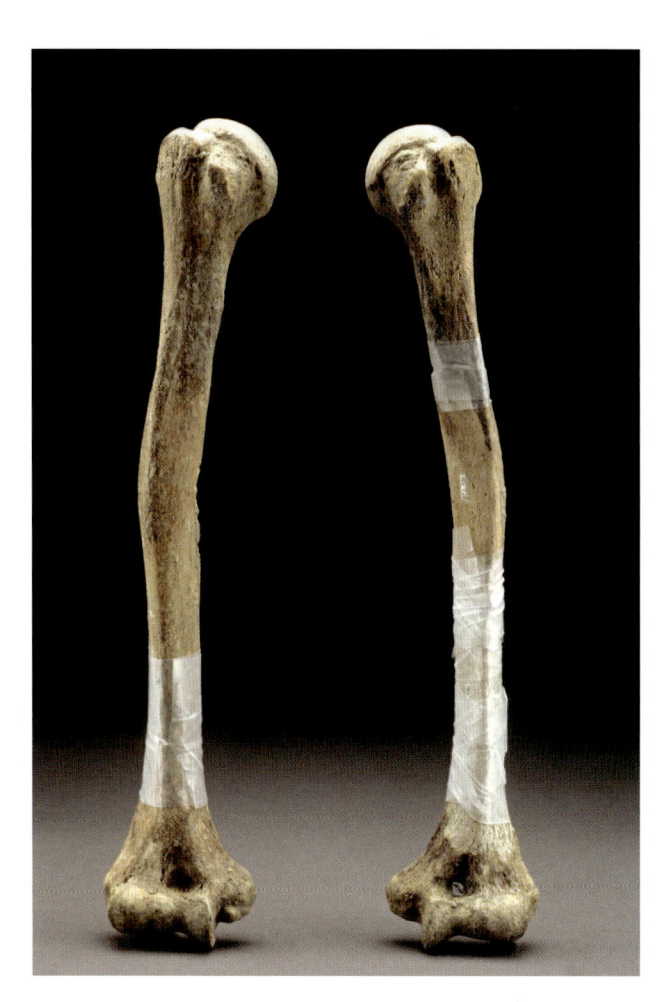

FIGURE 7.2　Reconstructed right and left humeri refit and temporarily secured for measurement, digital X-ray, and computed tomography using Parafilm®. This technique allowed temporary rearticulation of the separated bone segments without the use of adhesive.

TABLE 7.2 Postcranial measurements (millimeters), based on Zobeck (1983).

	Description	Left	Right		Description	Left	Right
1	Clavicle maximum length	—	153	37	Ulna least circumference of shaft	39	40
2	Clavicle diameter midshaft (A-P)	11	12	38	Sacrum anterior height	—	—
3	Clavicle diameter midshaft (S-I)	13	12	39	Sacrum anterior-superior breadth	—	—
4	Scapula maximum height	—	—	40	Sacrum maximum breadth of S1	—	—
5	Scapula maximum breadth	—	—	41	Innominate height	—	—
6	Scapula spine length	—	—	42	Iliac breadth	—	—
7	Scapula supraspinous length	—	—	43	Pubis length	—	—
8	Scapula infraspinous length	—	—	44	Ischium length	—	—
9	Scapula glenoid cavity breadth	28	28	45	Femur maximum length	473	466
10	Scapula glenoid cavity height	40	42	46	Femur bicondylar length	468	464
11	Scapula glenoid to inferior angle	—	—	47	Femur trochanteric length	452	450
12	Manubrium length	—	—	48	Femur subtrochanteric diameter (A-P)	28	28
13	Mesosternum length	—	—	49	Femur subtrochanteric diameter (M-L)	34	33
14	Stenebra 1 width	—	—	50	Femur diameter at midshaft (A-P)	36	34
15	Stenebra 3 width	—	—	51	Femur diameter at midshaft (M-L)	29	27
16	Humerus maximum length	341	340	52	Femur maximum vertical diameter of head	—	(48)
17	Humerus proximal epiphyseal breadth	48	51	53	Femur maximum horizontal diameter of head	48	48
18	Humerus maximum diameter midshaft	22	(24)	54	Femur diameter of lateral condyle (A-P)	65	67
19	Humerus minimum diameter midshaft	15	18	55	Femur diameter of medial condyle (A-P)	67	(67)
20	Humerus maximum diameter of head	44	46	56	Femur epicondylar breadth	—	89
21	Humerus epicondylar breadth	65	66	57	Femur bicondylar breadth	79	79
22	Humerus least circumference of shaft	60	69	58	Femur minimum vertical diameter of neck	32	32
23	Radius maximum length	—	261	59	Femur circumference at midshaft	104	100
24	Radius maximum diameter of head	24	24	60	Tibia condylo-malleolar length	398	—
25	Radius diameter midshaft (A-P)	—	11	61	Tibia maximum breadth of proximal epiphysis	82	—
26	Radius diameter of midshaft (M-L)	—	16	62	Tibia maximum breadth of distal epiphysis	53	55
27	Radius neck shaft circumference	43	41	63	Tibia diameter at nutrient foramen (A-P)	—	39
28	Ulna maximum length	—	287	64	Tibia diameter at nutrient foramen (M-L)	—	24
29	Ulna physiological length	—	252	65	Tibia position of nutrient foramen	—	—
30	Ulna maximum breadth of olecranon	26	27	66	Tibia circumference at nutrient foramen	—	100
31	Ulna minimum breadth of olecranon	20	21	67	Fibula maximum length	392	388
32	Ulna maximum width of olecranon	26	27	68	Fibula maximum diameter at midshaft	16	17
33	Ulna olecranon-radial notch	36	36	69	Calcaneus maximum length	89	86
34	Ulna olecranon-coronoid length	25	25	70	Calcaneus middle breadth	—	44
35	Ulna diameter at maximum crest (Dorso-volar)	18	17				
36	Ulna diameter at maximum crest (Transverse to dorso-volar)	15	16				

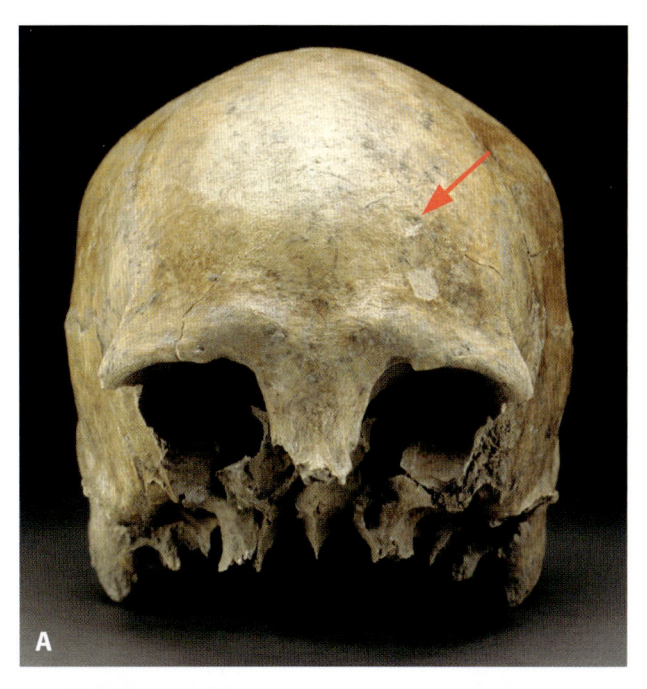

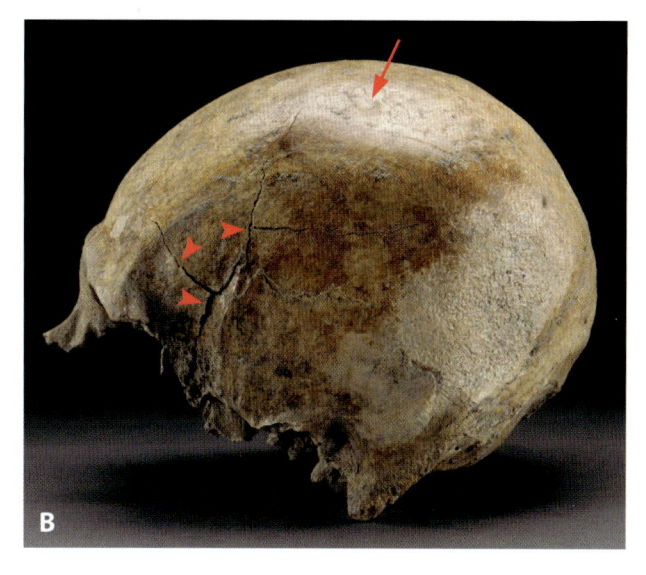

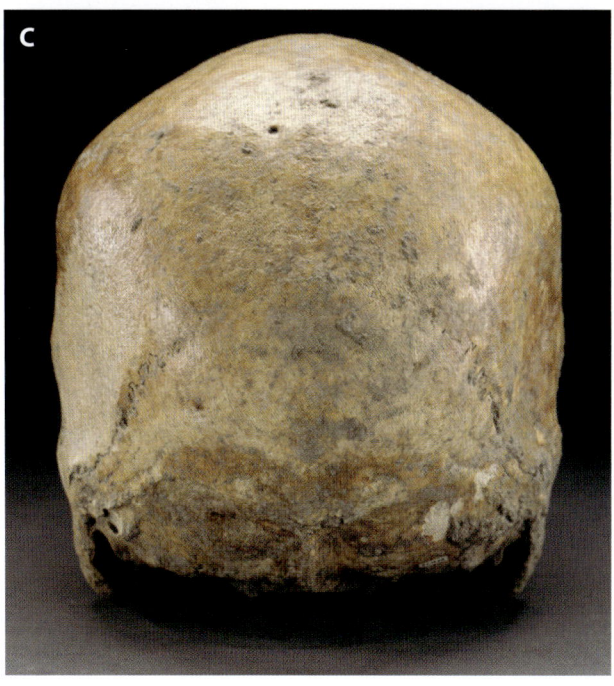

FIGURE 7.3 Frontal (A), left lateral (B), and posterior (C) views of the cranial vault. Healed, shallow depression injuries are present in the frontal and left parietal (arrows). Radiating fractures (arrowheads) in the left anterior vault and erosion of the greater wing of the sphenoid occurred postmortem. The right lateral view is shown in Figure 14.1.

cussed in Chapter 17. A postmortem radiating fracture with branching extensions emanates superiorly from this region. No reactive woven bone indicative of infection or remodeling during life as described by Chatters (2000:309, 2001:138–139) was identified. On the base of the cranium, the left occipital condyle has a small area of postmortem erosion, not arthritic eburnation (i.e., a polished appearance of the joint caused by contact with the atlas, cf. Chatters 2000:309, 2001:140).

A healed depression injury is evident in the left side outer table of the frontal bone. This small depression, measuring 7 mm superior-inferior by 6 mm transverse, is located approximately 36 mm above the medial third of the left orbital rim and 26 mm from the temporal line (Figure 7.3A). The shallow defect is depressed about 1 mm relative to the surrounding surface and has no associated radiating fracture lines. Powell and Rose (1999), Chatters (2000, 2001), and Walker et al. (2000) note the presence of this healed injury. While Powell and Rose (1999) did not identify any irregularities in the surrounding bone tissue, Chatters (2001:131) interpreted an irregular bone surface surrounding the indentation as evidence of a minor infection in the scalp. However, the distortion surrounding the depression is more consistent with postmortem deterioration and not antemortem bone formation in response to infection.

Another shallow, healed, depression fracture is present in the outer table of the mid-anterior left parietal immediately above the superior temporal line, 67 mm above the mid-left squamosal suture and 109 mm posterior to the mid-left orbit (Figure 7.3B). This subtle, irregular depression has an approximate anterior-posterior length of 13 mm, a maximum transverse width of 9 mm, and a depth of about 0.6 mm.

Both external auditory meatii (ear canals) have moderately large exostoses. The right exostosis is more visible than the left, with an approximate width of 4.5 mm, a raised height of 2.5 mm, and an estimated length of 10 mm (Figures 14.1 and 14.2). One study has suggested that these exostoses were large enough to impair hearing (Chatters 2000:309, 2001:140). However, the condition was likely not that severe (Chapter 14).

The facial bones are represented by five separate pieces. The right malar (97.R.1a) is nearly complete and articulates with the adjacent maxilla along a postmortem break. The zygomaxillary suture is fused. The left malar (97.U.1c) is partially represented by its frontal process and anterior portion. The temporal process of the left zygoma and the posterior portion of the maxillary sinus cavity are missing. The left and right maxillae are primarily represented by piece 97.U.2a (Figure 7.4). Also present are separate fragments of the right (97.R.2b) and left (97.I.25v) maxillae. Most of the left maxillary sinus and the anterior floor of the right sinus are visible. The sinus floors and left sinus walls are covered by carbonate cemented sediment (Figure 7.5). Where the sinus walls are visible, they show normal morphology with no apparent floor disruptions due to abscessing. The facial surfaces of the maxillae show normal morphology without evidence of age-related or pathology-induced involution. The right palatine bone is incomplete.

The mandible is nearly complete with postmortem breakage resulting in three articulating segments (97.U.3a, 97.U.3b, and 97.U.3c) (Figures 7.4 and 7.6). The superior half of the left ascending ramus, including the condyle, is missing (Figure 27.4). The represented right mandibular condyle and the left and right temporal fossae have normal morphology with no pathology. The anterior right fossa is covered with carbonate cemented sediment, but the exposed aspect shows normal anatomy.

Dentition

All teeth are present with the exception of the right maxillary third molar, which was lost in life, and the left mandibular third molar, which was lost postmortem (Figures 7.7 and 7.8). No carious lesions or active abscesses were present and calculus deposits are minimal. Secondary dentin from *in vivo* infilling of the pulp chambers is visible in teeth with pronounced wear. The maxil-

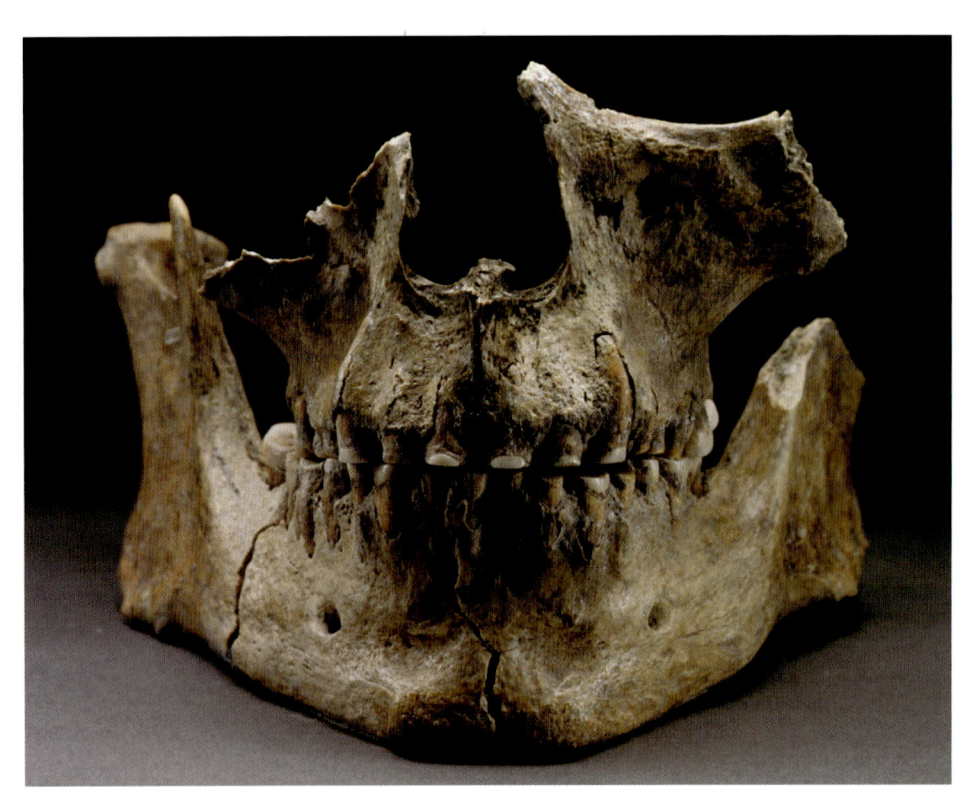

FIGURE 7.4 Frontal view of the midface and mandible. Breakage and separation from the cranial vault occurred during erosion from the terrace; note, for example, the light-colored (recent) transverse break across the left ascending ramus. A left oblique view is shown in Figure 27.4.

lary and mandibular alveolar ridges show minor damage from postmortem breakage, but age- or disease-related alveolar resorption is limited, providing no indication of periodontal disease (Figure 7.9). Radiographs of the dentition reveal the absence of root hypercementosis (Figures 7.5 and 7.6). Most pulp chambers and root canals show postmortem occlusion by radio-opaque sediment and calcium carbonate. The maxillary canines, left maxillary third molar, and the right mandibular second premolar and second molar show less deposition, and

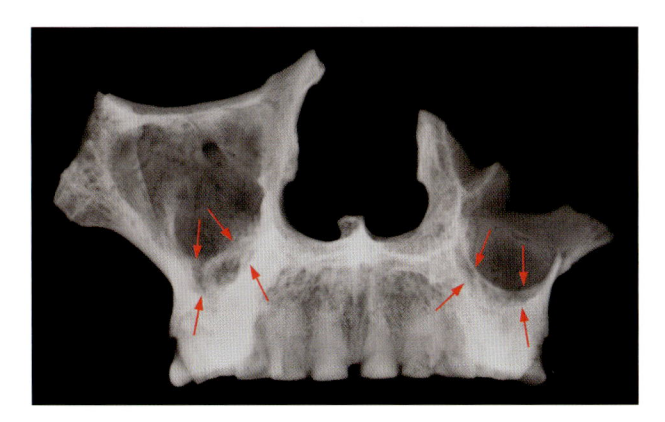

FIGURE 7.5 Posterior-anterior radiograph of the maxillae and anterior dentition showing carbonate cemented sediment accumulation on the floors of the maxillary sinuses. The symmetrical distribution of these deposits (arrows) has taphonomic significance (Chapter 17).

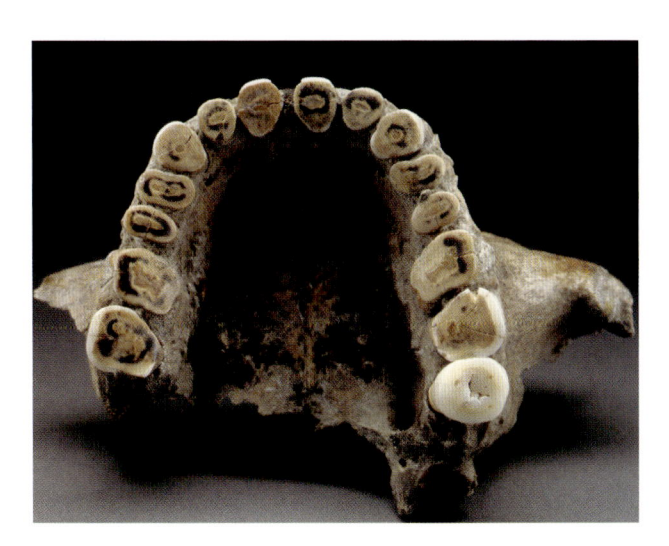

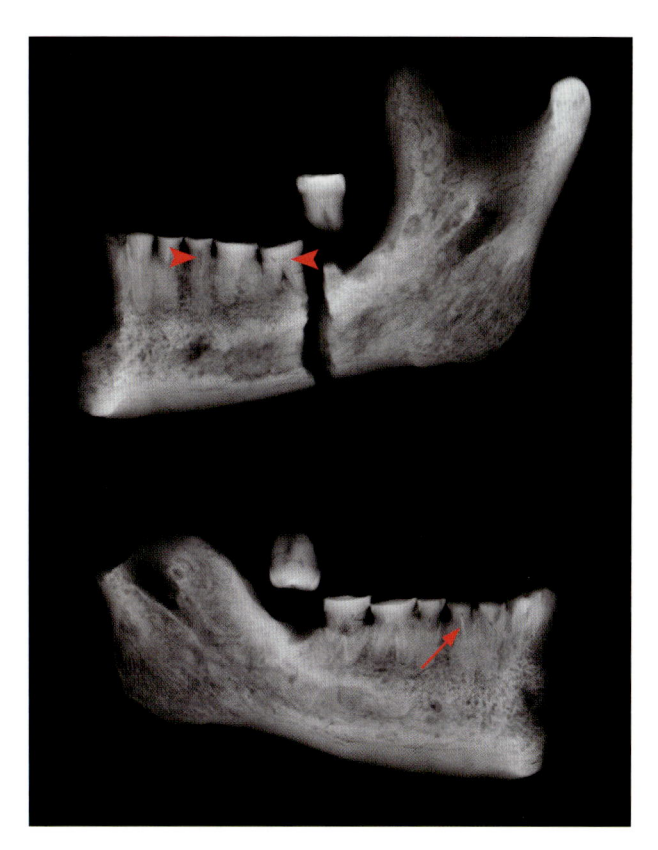

FIGURE 7.6 Radiograph of the three mandible pieces. Loose left maxillary and right mandibular third molars are placed near actual anatomical positions. Radio-opaque pulp chambers and root canals reflect infilling with extremely fine sediment and calcium carbonate, as seen in the left first premolar (arrow). Radiolucent pulp chambers of the right second premolar and second molar (arrowheads) and the left maxillary third molar indicate that they are partially open.

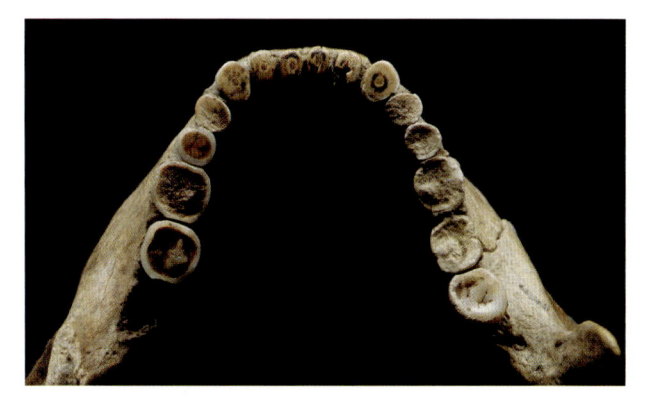

FIGURES 7.7 and 7.8 Occlusal views of the maxillary (left) and mandibular (right) dentitions. Postmortem pyrolusite staining (manganese dioxide) caused black discoloration in the primary dentin of the maxillary and anterior mandibular teeth. Pyrolusite stains show a circumferential distribution around in-filled pulp chambers (secondary dentin).

FIGURE 7.9 Lateral view of the left mandibular body. Postmortem breakage has resulted in a slightly irregular alveolar ridge, but in-life margin recession was minimal.

in the future may be possible sources for the recovery of mitochondrial DNA.

Tooth wear is pronounced and ranges in degree from stages three through eight according to the eight-stage Smith (1984:46) system of coding dental wear. Attrition scores and directionality (orientation) of the wear planes are listed in Table 7.3. The highest level of crown wear (stage 8) is noted for the maxillary first molars, first and second premolars, and lateral incisors, and the mandibular central and lateral incisors, sometimes with simultaneous buccal and lingual orientations (Figures 7.7, 7.8, and 7.10). The right mandibular third molar has the least amount of wear (stage 3) due to earlier loss of the occluding tooth in the maxilla. Microscopic evidence indicates that the extreme molar attrition was caused by mastication of foods contaminated by extremely fine abrasives with occasional coarser particles (Chapter 9). Facial margin rounding of the mandibular incisors and excessive anterior tooth wear of the first premolars reflects use of the teeth in task activities (Figures 7.10 and 7.11). However, no distinct incisive surface grooves are present.

TABLE 7.3 Degree and planes of dental wear.

	RM3	RM2	RM1	RP2	RP1	RC	RI2	RI1	LI1	LI2	LC	LP1	LP2	LM1	LM2	LM3
							Maxilla									
Stage of Wear[a]	—	7	8	8	8	7	8	7	7	7	7	8	8	8	7	3
Plane of Wear[b]	—	4	11	11	11	11	11	4	4	4	4	11	4	4	4	—
							Mandible									
Stage of Wear	3	7	7	7	7	7	8	8	8	8	7	7	6	7	6	—
Plane of Wear	—	9	10	4	4	1	11	11	11	11	11	1	1	13	13	—

[a]Based on Smith (1984).
[b]Plane of wear: 1. flat; 2. concave; 3. buccal slope (inclination toward the cheek; maxillary directed superiorly, mandibular directed inferiorly); 4. lingual slope (inclination toward the tongue; maxillary directed superiorly, mandibular directed inferiorly); 5. mesial slope (anterior inclination; maxillary directed superiorly, mandibular directed inferiorly); 6. distal slope (posterior inclination; maxillary directed superiorly, mandibular directed inferiorly); 7. concave-buccal; 8. concave-lingual; 9. concave-mesial; 10. concave-distal; 11. buccal-lingual; 12. buccal-mesial; 13. buccal-distal; 14. lingual-mesial; 15. lingual-distal; 16. distal-mesial

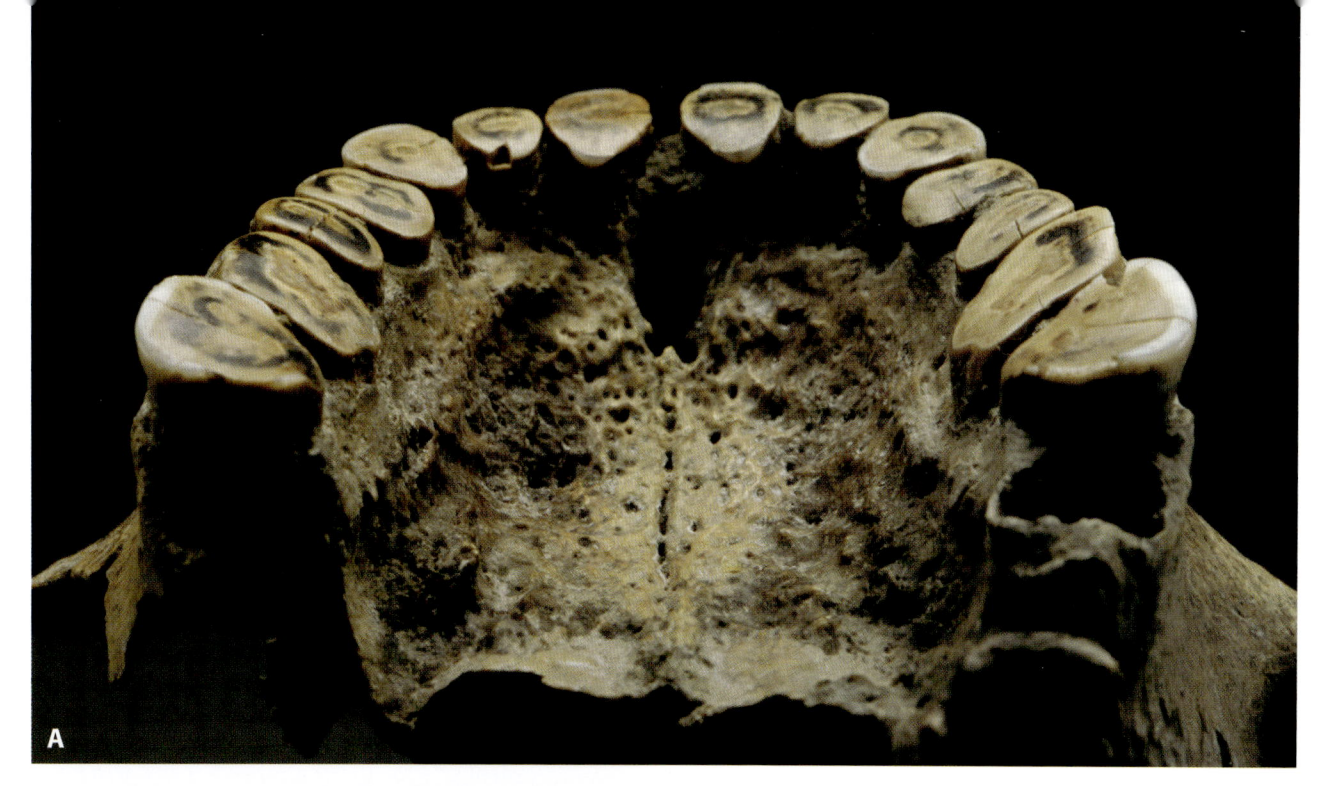

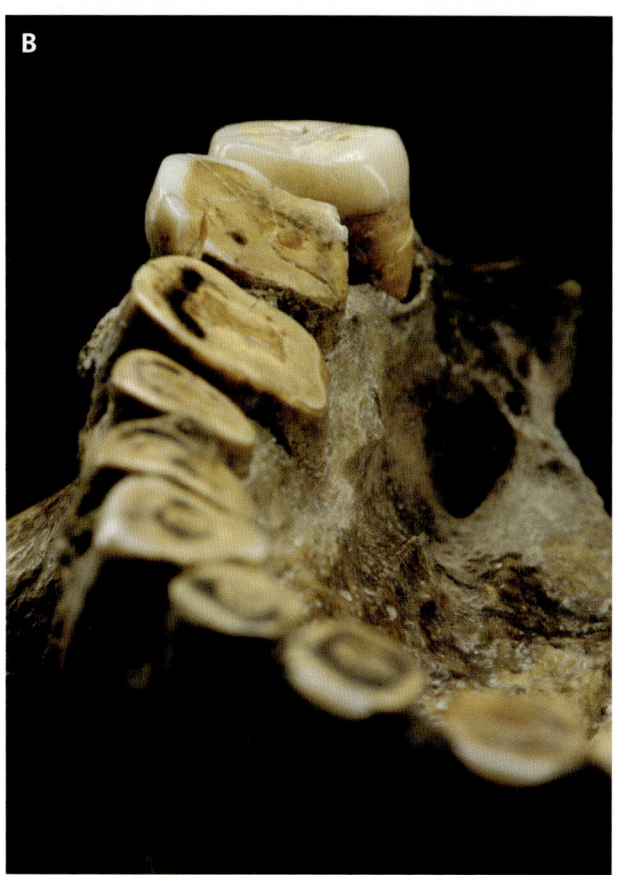

FIGURE 7.10 Posterior (A) and anterior (B) views showing pronounced, lingually-oriented wear in the maxillary premolars and first and second molars. The refit loose, left third molar has comparatively little occlusal wear (B).

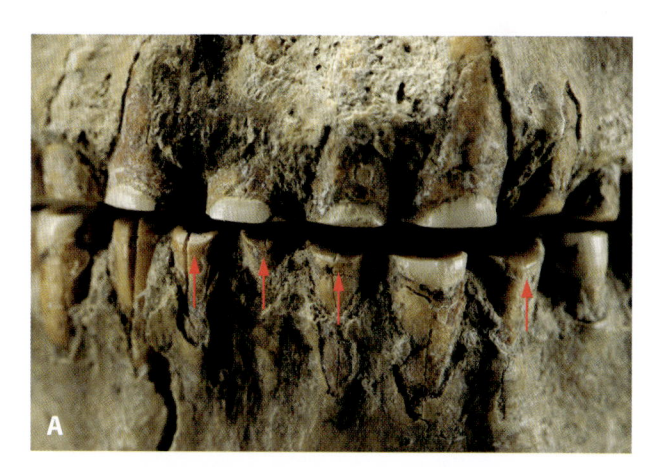

FIGURE 7.11 Edge-to-edge bite with pronounced anterior tooth wear (A) including facial margin rounding of the lower central and left lateral incisors (arrows) and left first premolar crowns (B) (arrows). Cracks are postmortem.

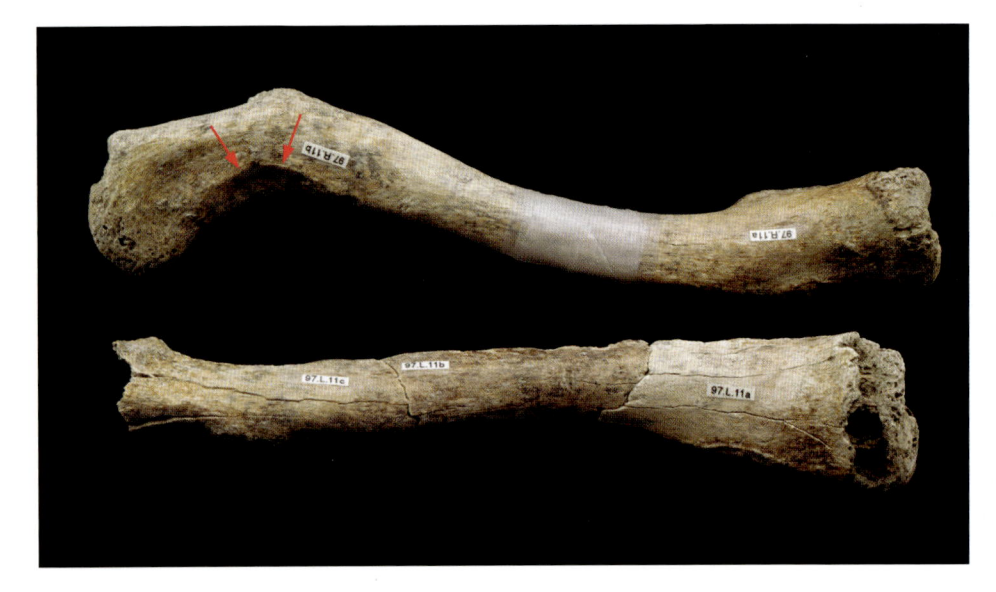

FIGURE 7.12 Right (top) and left clavicles. The two segments of the right clavicle are joined near the center with Parafilm®. On the lateral aspect of the right clavicle, the origin for the deltoid muscle is defined by a raised ridge and slight surface concavity (arrows).

Clavicles

Both clavicles are present (Figure 7.12). The left clavicle is represented by three articulating pieces. The most medial segment (97.L.11a) measures 56.4 mm in maximum length and includes a largely intact medial epiphysis. The middle fragment (97.L.11b) measures 49 mm in maximum length, and the most lateral piece (97.L.11c) has a maximum chord length of 55.5 mm. The lateral segment displays postmortem breakage of the acromial end with an estimated 8 to 15 mm of bone missing. This loss was estimated using the right clavicle for comparison.

The better-preserved right clavicle is represented by two articulating pieces that form a nearly complete bone. The sternal fragment (97.R.11a) measures 57.5 mm in length and the acromial fragment (97.R.11b) is 95 mm in length as measured along the anterior surface. Slight postmortem breakage is evident on the lateral end and trace erosion with cancellous bone exposure is present on the posterior and inferior borders of the medial epiphysis.

The sternal epiphysis of the left clavicle is fully united and its surface is concave. The left middle fragment shows flattening and ridge development related to the attachment (origin) of pectoralis major. The lateral end of the 97.L.11c segment has ridge development and surface concavity at the origin for the deltoid muscle.

As with the left side, the sternal epiphysis of the right clavicle has united and the surface is concave. The attachment for the sternocleidomastoid muscle on the superior surface of the medial end is identified by a raised ridge. The attachment site for the costo-clavicular ligament is roughened with pronounced ridge development. The ventral aspect of the medial body is flattened with a sharp inferior crest. This morphology reflects development and use of the pectoralis major muscle.

The right clavicle has slightly greater development and definition of bone at the muscle attachments and is larger than the left. The right conoid tubercle is more developed than on the left side and the origin for the deltoid muscle on the right clavicle is represented by a slightly deeper trough. The deltoid depressions measure 20 mm in length by 7 mm anterior-posterior on the right side and 5 mm anterior-posterior on the left.

Powell and Rose (1999) note asymmetry in the clavicles. They also describe the right clavicle as having a reactive area measuring 11 mm on the medial end "just lateral to the articular surface" and "the left clavicle has a similar, but smaller (width of 7.1 mm), lesion in the same location." Their assessment that the costo-clavicular ligament attachment areas show evidence of pathology is dismissed by Cook (Chapter 14). In fact, landmarks at the costo-clavicular attachment site on the clavicle, including tubercles, impressions, and all sizes and depths of grooves or fossae are usually not pathological, but reflect normal variation (Rogers et al. 2000). Deep costo-clavicular (rhomboid) fossae are more common in males between the ages of 20 and 30 years and are more often found on the right clavicle than the left. Features in Kennewick Man's clavicles are a product of normal variation, not pathology, and the bilateral asymmetry reflects function. Developmental differences, along with related indicators in the upper limb bones, suggest right handedness with strenuous use of the arms, especially the right arm.

Scapulae

The bones of the thorax include partial scapulae. The scapulae are represented by the glenoid fossae, lateral borders of the bodies, and acromial ends of the spines. The medial borders of the scapula bodies are comprised of small, unarticulated fragments. Eight segments represent the left scapula: (97.L.10a [lateral spine and acromion], 97.L.10b [glenoid fossa], 97.L.10c [fragment of the coracoid process], 97.I.10c [lateral, midaxillary border], 97.I.17e [superior portion of the glenoid cavity], 97.I.10a [lateral border], 97.I.25hh [mid-vertebral border], and 97.I.25r [fragment of the body]). Five pieces represent the right scapula (97.R.10a [lateral spine and acromion process], 97.R.10b [glenoid and coracoid process], 97.R.10c [lateral, midaxillary border], 97.R.10d [mid-lateral root of the spine], and 97.I.25jj [mid vertebral border]).

The left glenoid cavity has normal morphology (Figure 7.13). In contrast, the right glenoid cavity shows some degenerative "wear and tear" from use. Slight arthritic lipping of the supero-anterior and inferior rim with expansion and surface irregularity along the posterior joint margin are noted (Figure 7.14). The center of the cavity has a nearly-circular depressed surface disruption with an approximate diameter of 5 mm, possibly an indicator of glenoid cartilage erosion.

Right scapula damage also includes a posterior glenoid rim fracture. A flake of bone had spalled off the mid-posterior glenoid margin in life (Figure 7.15). This avulsion produced a pressure flake scar (i.e., a negative bulb of the flake) on the adjacent neck of the glenoid (Figure 7.16). The margin and adjacent posterior surface show subtle remodeling indicative of healing. Chatters (2000:309; 2001:139) identified this articular rim disruption, but classified it as perimortem or immediately postmortem, possibly from manipulation of the corpse.

Imaging studies of dominant-shoulder abnormalities in athletes provide a sports medicine perspective for understanding the causation of this injury. The motion of throwing places tremendous stress on the bones and

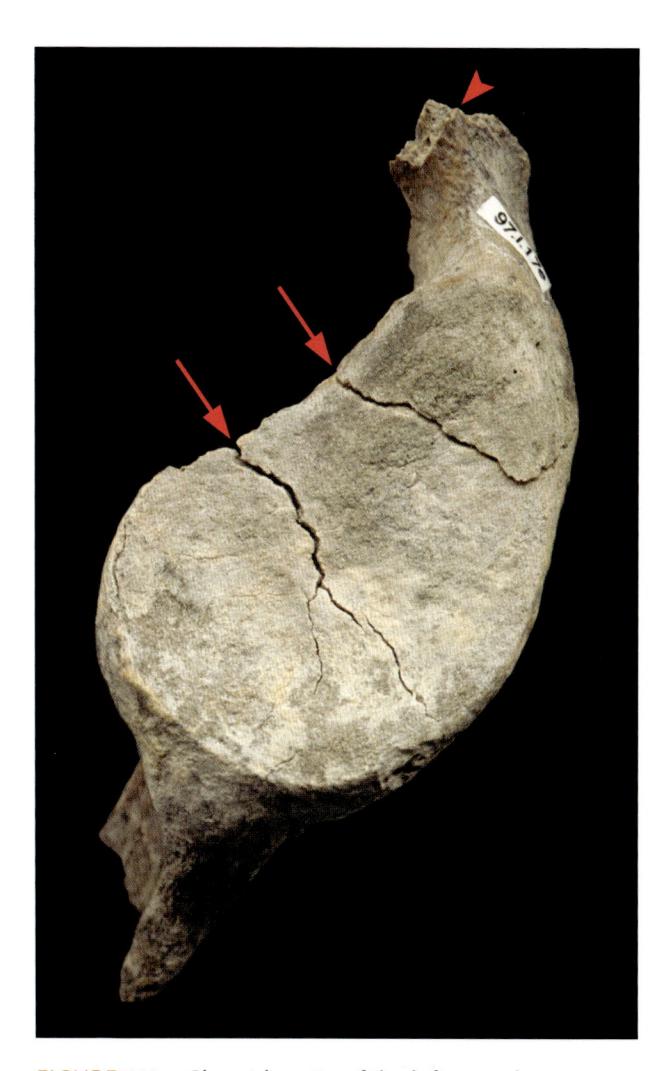

FIGURE 7.13 Glenoid cavity of the left scapula. Breakage of the coracoid process (arrowhead) and cracks in the glenoid surface (arrows) occurred when the skeleton eroded from the terrace.

FIGURE 7.14 Lateral view of the right glenoid cavity. Arrows point to slight arthritic lipping; arrowheads define fracture damage of the posterior rim.

FIGURE 7.15 The posterior margin of the glenoid shows antemortem loss of a rim segment that extends onto the posterior surface as a concavity (delineated by arrowheads). Calcium carbonate partially obscures the surface.

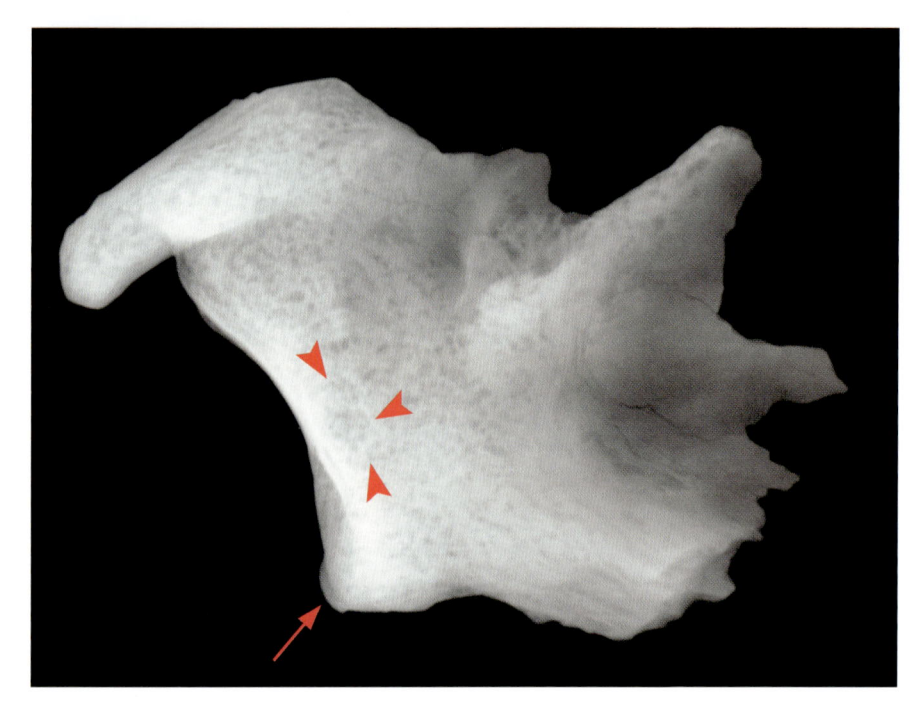

FIGURE 7.16 Anterior-posterior radiograph of the right scapula showing slight lipping (arrow) along the infraglenoid margin and the area of avulsed bone on the posterior surface (arrowheads).

soft tissues of the shoulder (Roger et al. 1999:1371). Stress fractures of the glenoid rim are relatively rare (McFarland et al. 2008), but have been observed in patients with years of throwing activity as a result of the "impaction of the humeral head against the glenoid rim" (Roger et al. 1999:1380). As such, posterior glenoid damage has a biomechanical etiology linked to overhead motion involving abduction and external rotation of the arm combined with overuse.

Kennewick Man's glenoid rim fracture can therefore be linked to activity involving rapid, vigorous movement of the arm and shoulder, such as the swift, snapping motion of throwing a dart with an atlatl (Figure 7.17). This movement could have produced an unsustainable force during the interaction between the raised, posteriorly-directed humerus and the posterior rim of the glenoid. This injury was undoubtedly a source of pain and discomfort that affected Kennewick Man's

FIGURE 7.17 Time-lapse photograph showing the range of motion while throwing a spear with an atlatl (Stanford 1979). The right hand swings back and then forward over the head with the wrist and elbow snapping straight at the time of release. The left leg steps forward to catch and support the body.

ability to hunt and fish since posterior shoulder pain is typically heightened by activities involving arm abduction and external rotation.

Humeri

The left and right humeri are complete (Figure 7.18). Four bone pieces represented by three catalogued segments form the left humerus: the proximal two-thirds (97.L.13a), lower middle diaphysis (97.L.13b), and distal fourth (97.L.13c). The right humerus has one postmortem fracture in the distal diaphysis that divides the proximal three-fourths (97.R.13a) from the distal one-fourth (97.R.13b) of the bone. The medial surface of the right diaphysis, beginning 14.5 cm below the head, has an area of damaged cortex that is illustrated and discussed under the topic of macroabrasions in Chapter 17, Figures 17.31 and 17.34. This localized area of postmortem damage has a superior-inferior length of 3.9 cm and a maximum transverse breadth of 1.5 cm.

The humeri are asymmetrical in size and shape. The right humerus has a slightly larger head and greater tubercle (Figure 7.19), as well as longer and larger shaft dimensions. The attachment site for the medial head of the triceps on the anteromedial proximal right diaphysis is represented by a slightly raised vertical ridge. The deltoid tuberosity and insertion for pectoralis major are more prominent on the right side (Figure 7.20). The deltoid tuberosity has a maximum width of 2.3 cm on the lateral surface with well-defined, angular margins. The deltoid tuberosity on the left humerus is not as distinct (width of 2.0 cm) with less dense internal cortical bone, as noted in the radiograph. (Measurement of the left deltoid tuberosity compensates for a recent, midline, vertically-aligned postmortem fracture that bisects the tuberosity's anterior and posterior surfaces.) The attachment ridge for pectoralis major on the right humerus is raised to a height of 2 mm above the flat anterior surface. The same measurement on the left humerus is 0.7 mm. (These measurements were taken on the anterior surface by orienting the caliper perpendicular to the bone such that the caliper extension is in contact with the anterior surface, near the latissimus dorsi attachment, with the pectoralis major insertion providing elevation on the lateral margin.) Likewise, the insertion for teres major is more prominently defined on the right humerus as a raised area. The left teres major insertion shows less

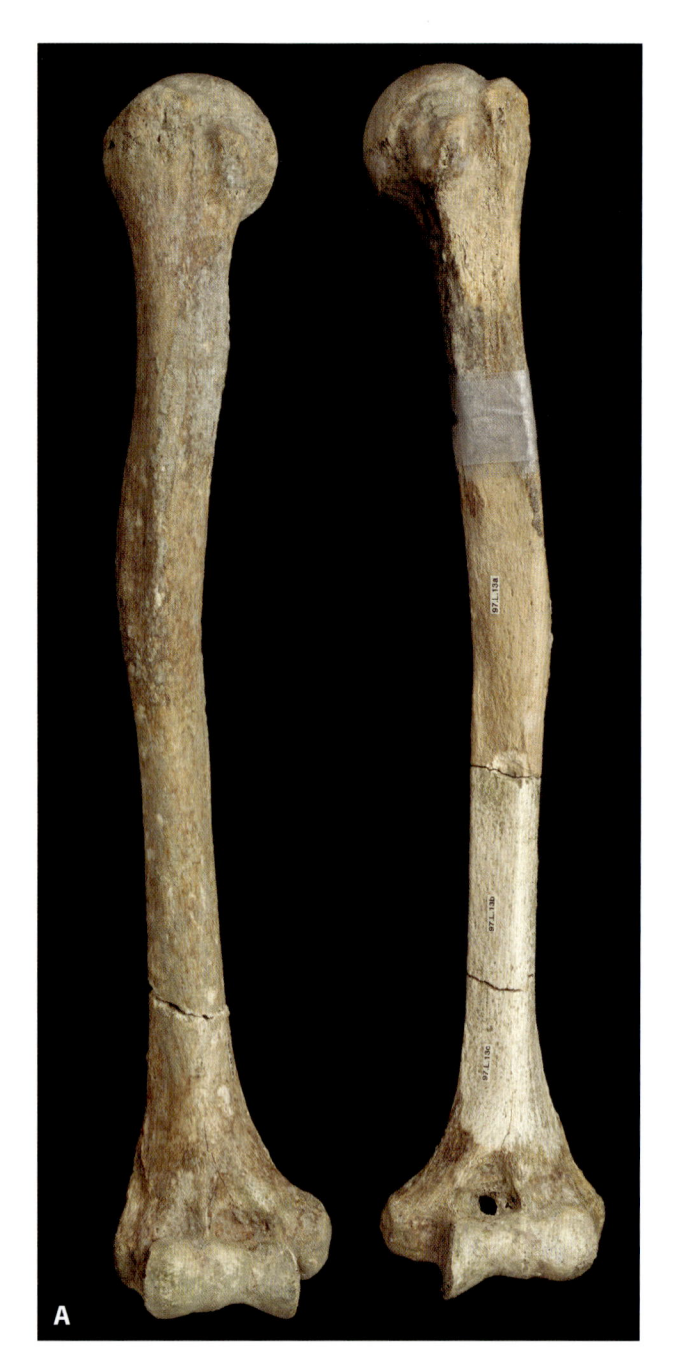

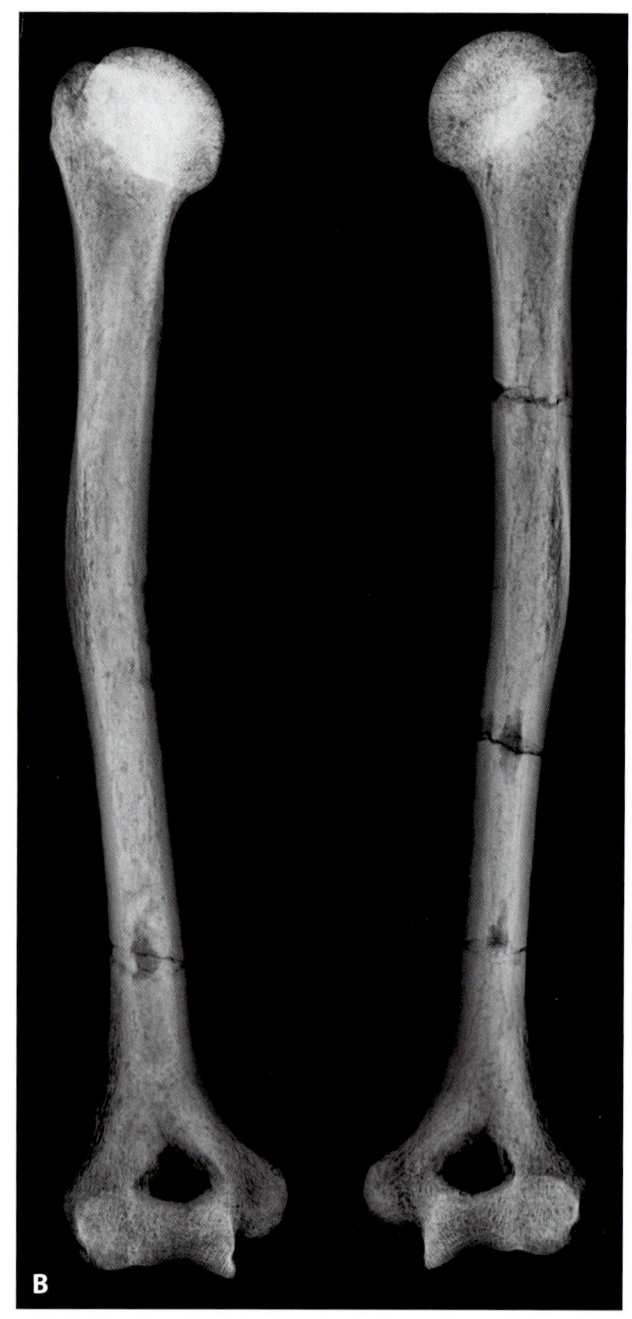

FIGURE 7.18 Right and left humeri showing bilateral differences in size, robusticity, and shape (A). The radiograph (B) reveals compact internal cancellous bone, a normal feature of healthy, young-to-middle-aged adults.

elevation from the surface, but is identified by dark staining from postmortem iron impregnation. The distal humeri are also slightly asymmetrical in size with greater development on the right side (Figure 7.21). In addition to differences in size, the diaphysis of the right humerus shows accentuated medial bending of its distal half, which is visually enhanced by the development of the right deltoid tuberosity. The left humerus shows bowing but to a lesser degree.

Atrophy of the left humerus has been offered as an explanation for the difference in size between the humeri (Chatters 2000:309; 2001:133–134). The size and musculoskeletal markings of the left arm, however, show development within the normal range (Chapter 12). Right

158 Chapter 7

FIGURE 7.19 A side by side comparison of the left and right humeral heads shows slightly greater development on the right side.

side bowing has been attributed to a "well-healed fracture from early life" and also to "hyper development from extensive usage" (Powell and Rose 1999). This study supports a functional/developmental basis linked to handedness and preferential, strenuous use of the right side, instead of pathology to account for the morphological differences in the humeri. Radiographically, the right cortex and medullary cavity lack internal disruptions indicative of a healed, antemortem fracture (Figure 7.18).

Extremely minor pathological changes are on the right humerus. The posterior-superior margin of the humeral head along the border with the greater tubercle has a small area of localized porosity consistent with a cystic lesion (see Roger et al. 1999). At the elbow, the anterior joint surface has a small focal area of osteophytic spicule formation. Articulating ulna and radius surfaces in the right elbow joint show more obvious degenerative changes (see description of the right radius, below).

The distal left humerus has a small supratrochlear foramen (septal aperture), a non-pathological anomaly, with a transverse diameter of 4 mm and a superior-inferior height of 5 mm (Figure 7.21). This perforation, closed in life by a membrane, varies in frequency among populations, and is more common on the left side than the right (Grant 1972). Although not in this study, this foramen has been linked to pathology in previous evaluations of Kennewick Man. Powell and Rose (1999) note this perforation in the left olecranon fossa along with a "resorptive reaction area" suspected to have been caused by hyper-

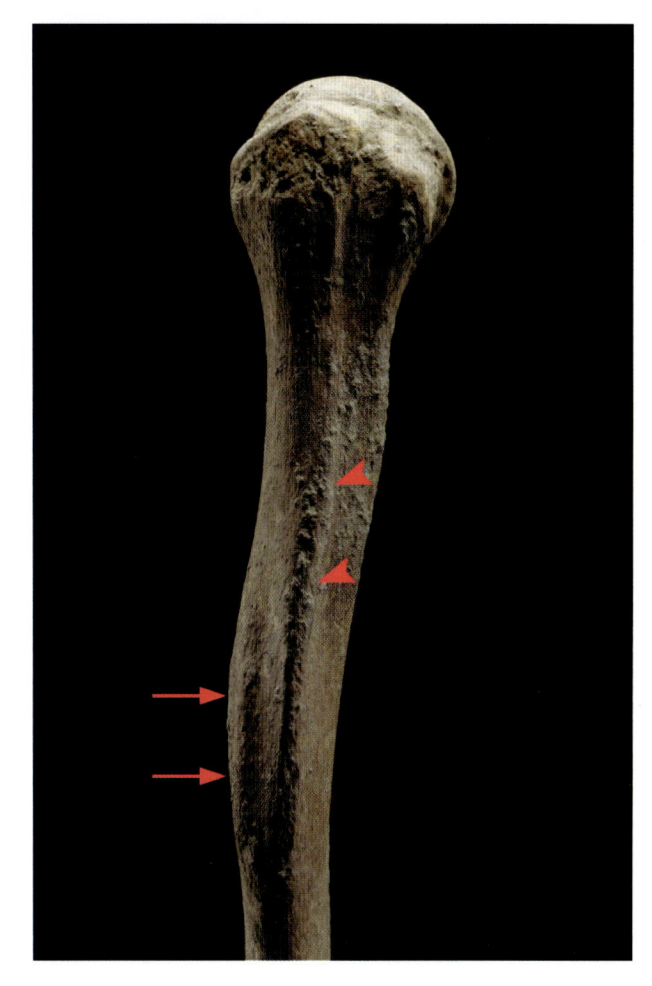

FIGURE 7.20 The proximal diaphysis of the right humerus has visible curvature along with a large, well-defined deltoid tuberosity (arrows) and insertion ridge for pectoralis major (arrowheads).

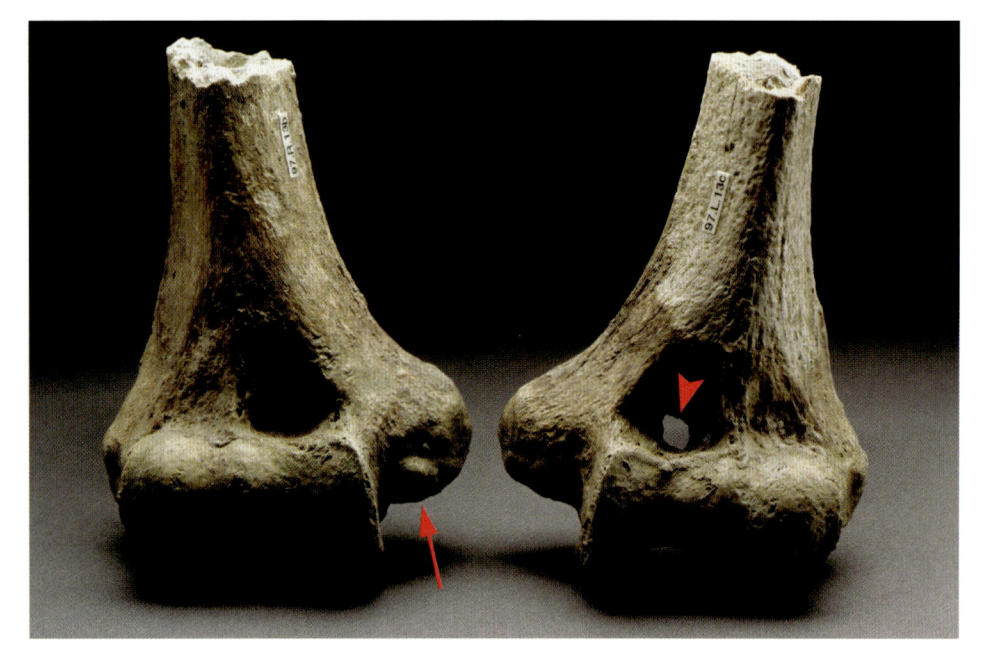

FIGURE 7.21 Side by side comparison of the distal humeri showing that the right medial epicondyle is slightly larger than the left. It has an inferiorly projecting tubercle (arrow) attachment for the ulnar collateral ligament. The left humerus has a small supratrochlear foramen (olecranon fossa) (arrowhead).

extension. Chatters (2000:308–309; 2001:132–134) and Walker et al. (2000) also describe an area of inflammation and reactive bone in the olecranon fossa. Chatters hypothesizes that the inflammation was associated with a suspected radial head chip fracture (see description of the left radius below). The distal joint surface of the left humerus has trace degenerative lipping on its margin along with minute surface porosity.

Radii and Ulnae

The left radius is incomplete and represented by proximal (97.L.14a) and distal (97.L.14b) segments with approximately 4.5 cm of the upper middle diaphysis missing. Estimation of the missing segment is based on comparison of the left radius with the right, which is complete and represented by proximal (97.R.14a) and distal (97.R.14b) segments (Figures 7.22 and 7.23).

The left radial tuberosity insertion for biceps brachii is moderately defined. A slight ridge has formed at this location and is visually emphasized by brown staining. The radial tuberosity on the right radius is prominent, but without the formation of a small enthesophyte.

An oval-shaped, porous area measuring 13 mm anterior-posterior by 6.5 mm superior-inferior is evident on the medial-superior joint surface of the left radial head. It is partially filled with carbonate cemented sediment. This postmortem erosion of the joint surface has been interpreted as an antemortem radial head fracture (Chapter 15; Chatters 2000:309, 2001:132–134; Walker et al. 2000). A radial head fracture was not confirmed during this assessment, although the medial margin does have slight degenerative (arthritic) lipping and porosity. The affected area represents that portion of the head in contact with the radial notch of the ulna when the elbow is in anatomical orientation (with the forearm supinated).

The medial and adjacent proximal surfaces of the right radial head show minor arthritic degeneration with surface spicule formation and porosity. The affected area has a chord length measurement along the radial head's medial surface of 13.5 mm with a superior-inferior chord measurement of 7 mm. Corresponding changes are present on the joint surface margin of the right ulna's radial notch and the articulating anterior joint surface of the right humerus at the transition between the capitulum and trochlea. The distal joint surface of the right radius is intact. It has slight degenerative lipping on the ulnar facet margin.

The left ulna is nearly complete with proximal (97.L.15a), midshaft (97.L.15b), and distal segments (97.L.15c) (Figure 7.23). Approximately 14 mm of the distal end and joint are missing, as estimated by comparison with the complete right ulna's three segments (proximal, 97.R.15a; midshaft, 97.R.15b; and distal, 97.R.15c).

FIGURE 7.22 Anterior and posterior views of the right radius.

The supinator crest on the lateral surface of the proximal left ulna is moderately developed. The origin for pronator quadratus on the anterior distal diaphysis is a well-defined ridge measuring 29 mm in length with a maximum thickness of 3.5 mm. The height of this ridge is difficult to measure due to carbonate cemented sediment on the anterior surface, but is greater than 1 mm. The supinator crest and tuberosity for brachialis on the proximal right ulna are moderately developed. The distal diaphysis (97.R.15c) has raised and roughened bone at the attachment for pronator quadratus. The medial ridge that defines this muscle's origin is prominent, having a length of 26 mm, a thickness of 3.5 mm, and a height of 1.2 mm.

The distal aspect of the left interosseous crest was scored for a small enthesopathy due to the presence of a prominently raised ridge. This ridge on the crest's anterior surface, lateral to the insertion for pronator quadratus, measures 22 mm in length with a maximum thickness of 2 mm and a height of 2 mm. This ridge is not as developed on the right ulna.

The proximal joint of the left ulna has slight degenerative lipping along a small portion of the inferior margin of the trochlear notch (the anterior joint margin of the coronoid process). The surfaces of the radial and trochlear notches are not affected. The right ulna has slight lipping along the inferior margin of its trochlear notch and the superior border of the radial notch.

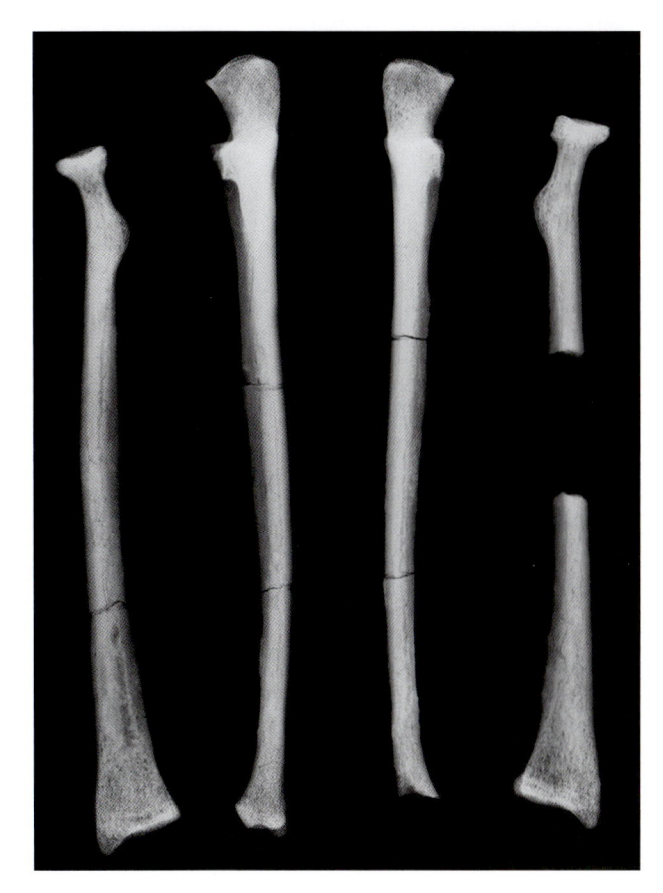

FIGURE 7.23 A radiograph of the radii and ulnae shows normal joint surfaces, moderately compact cancellous bone, and partial infilling of the medullary cavities with very fine silt. Surface irregularities are caused by adhering calcium carbonate concretions. The left radius is incomplete.

Owsley, Williams, and Bruwelheide

Sternum

Two small, non-articulating fragments (97.U.6f, 97.I.25a) represent the manubrium. Partial facets for the costal cartilage of the first ribs are visible on both pieces. Each fragment is comprised of cancellous bone and posterior surface cortical bone, as indicated by their slightly concave surfaces. The larger fragment, 97.I.25a, measures 35 mm by 31 mm. The sternal body and xiphoid process were not recovered.

Ribs

Rib fragment identification and side determinations were based on anatomical structure, including head and sternal end morphology, neck and body curvature, and size and thickness, as gauged by adjacent ribs within a side, as well as compared to the opposing rib. The length of each fragment was measured and Munsell colors and taphonomic features, such as loss of surface patina from abrasion and exposure, were documented.

The left first rib is represented by two articulating fragments forming the head, neck, and vertebral third of the body (97.I.12b(6), 97.I.12b(1)). The left second rib is represented by two body fragments (97.I.12a(5), 97.I.12f(3)), associated by color patterning, morphology, and flatness. The left third rib is represented by four articulating fragments that form the head, neck and vertebral half of the body (97.I.12e(13), 97.I.12e(10), 97.I.12e(14), 97.I.12a(2)). Six segments with joining articulations (97.I.12e(15), 97.I.12e(9), 97.I.12e(6), 97.I.12f(1), 97.I.12c(10)) comprise the left fourth rib. The head, neck and vertebral half of the rib body are present. The two most sternal pieces had been united with Parafilm® and assigned a single catalog number (97.I.12c(10)). The left fifth rib is represented by six fragments (97.I.12e(16), 97.I.12e(17), 97.I.12e(18), 97.I.12e(8), 97.I.12c(4), 97.I.12c 6)). The four vertebral-end fragments fit together. The two sternal fragments also articulate with one another, but do not articulate with the other section. The two sternal pieces (97.I.12c(4), 97.I.12c(6)) were placed with this rib on the basis of body morphology and similar preservation. The left sixth rib has three non-articulating fragments (97.I.12c(7), 97.I.12f(11), 97.I.12d(3)) matched on the basis of morphology relative to the right sixth rib. All three fragments show similar preservation. The left seventh rib is comprised of four fragments representing the head, neck, and vertebral half of the body (97.I.12e(2), 97.I.12e(4), 97.I.12f(13), 97.I.12f(9)). All four connect and two had been joined together using Parafilm®. The entire left eighth rib is formed by six articulating pieces (97.I.12e(12), 97.I.12a(4), 97.I.12c(2), 97.I.12c(8), 97.I.12c(5), 97A.I.12b)) (Figure 7.1). The ninth left rib is represented by one fragment (97.I.12e(1)). Four fragments form almost all of the left tenth rib with the exception of the sternal end (97.I.12b(10), 97.I.12f(10), 97.I.12d(8), 97.I.12c(3)). All four pieces can be refit and three had been joined with Parafilm®. The eleventh left rib is represented by two fragments forming two-thirds of the rib including the head (97.I.12e(7), 97.I.12c(1)). The two pieces do not articulate, but match in size, thickness, and body height. Two articulating fragments of the left twelfth rib are present (97.I.12f(5), 97.I.12f(6)).

An irregular depression in the sternal piece of the left second rib may indicate a well-healed fracture. An X-ray, however, revealed no osteological remodeling or anomalous formation. A well-healed fracture is present in the comparable section of the underlying left third rib (Figures 7.24 and 7.25). The diagonally oriented depression in the external surface measures about 7 mm across.

On the right side, most of the body of the first rib is present including the sternal end (97.I.12e(3)), but not the head. The right second rib is comprised of three fragments (97.I.12b(7), 97A.I.12c, 97.I.12a(7)). The two vertebral segments articulate to form the head and neck. The third fragment (97.I.12a(7)) is a portion of the sternal body including the end. The separated edge opposite the sternal end is pinched, as if compressed along the posterior surface of the margin.

Five right third rib fragments are present (97.I.12e(5), 97.I.12b(4), 97.I.12b(3), 97A.I.12d, 97.I.12a(9)); the four vertebral-end pieces articulate. The fifth (97.I.12a(9)) is a non-articulating body segment located adjacent to the sternal end, which is missing. Segment 97A.I.12d has a small, healed, resorptive lesion characterized by irregularity and thinning of the inferior margin. The affected area measures 10 mm in length by 5 mm superior-inferior. In addition, the non-articulating medial margin of the fragment 97.I.12a(9) is tapered on the internal and external surfaces from remodeling in life (Figure 7.26).

The fourth right rib is partially represented by four fragments (97.I.12b(5), 97.I.12b(15), 97.I.12b(2),

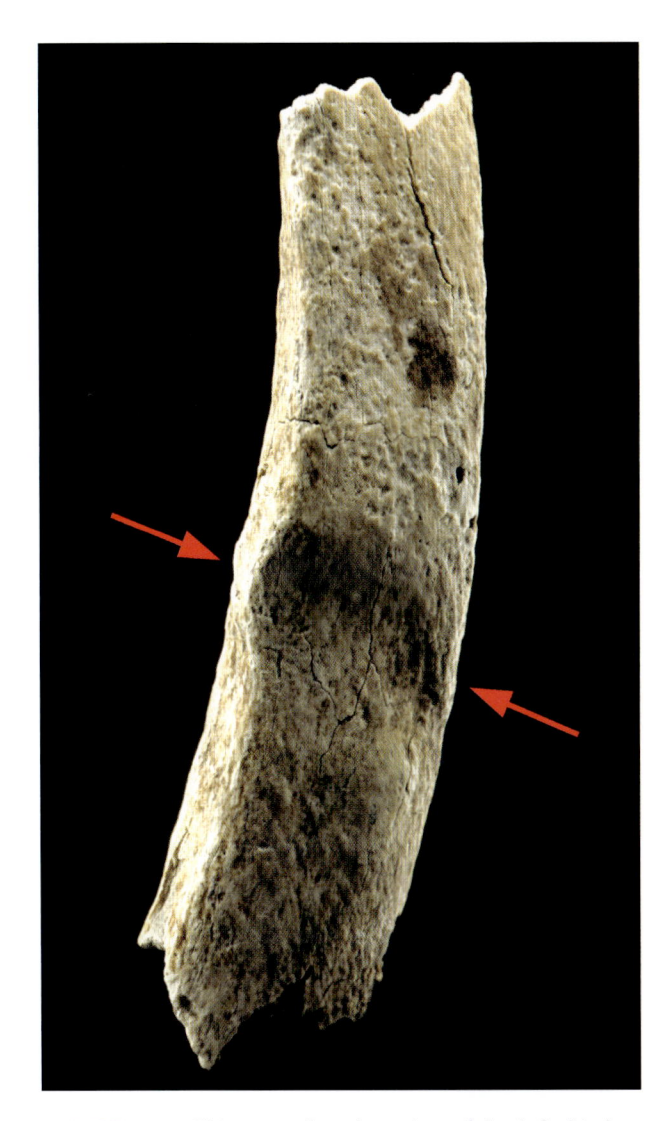

FIGURE 7.24 This sternal-end section of the left third rib (97.I.12a(2)) has a healed fracture evident as an external depression (arrows). This piece represents about one-third of the rib body.

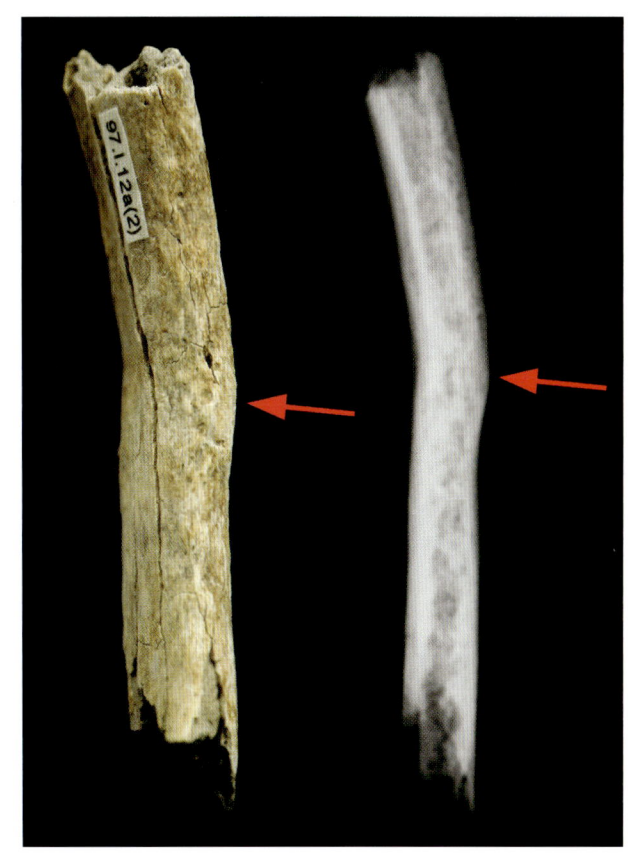

FIGURE 7.25 The internal surface photograph and corresponding radiograph of the left third rib (97.I.12a(2)) shows the fracture disconformity (arrow) and a slight increase in body thickness opposite the depression.

97.I.12a(1)). At least two pieces are missing: a body segment and the sternal end. The three vertebral end components articulate, but vary in color due to minor differences in the degree of weathering. The head and neck are lighter in color than the adjoining third piece. The more anteriorly positioned fourth piece (97.I.12a(1)) is polished and weathered, and cannot be refit. It also has a compressed margin along its medial edge with a discernible decrease in thickness (Figure 7.26).

Six pieces of the right fifth rib are present (97.I.12b(8), 97A.I.12a, 97.I.12f(14), 97.I.12f(8), 97.I.12a(6), and tentatively assigned 97.I.12a(8)). Two articulate. The most vertebral fragment represents the rib neck. The medial edge of the more sternal piece (97.I.12a(8)) is tapered and pinched off.

The right sixth rib is incompletely formed by four fragments (97.I.12b(11), 97.I.12d(10), 97.I.12d(5), and 97.I.12f(12)). Three sternal body pieces articulate and are held together with Parafilm®. A vertebral fragment represents the rib angle and includes the tubercle.

Three non-articulating pieces of the right seventh rib (97.I.12b(13), 97.I.12d(14), 97.I.12a(3)) are present and identified based on comparison with left rib seven. Fragment 97.I.12a(3) was tentatively matched to this rib due to similarity in color with the larger body fragment (97.I.12d(14)). Fragment 97.I.12a(3) is remodeled with a tapered sternal margin resembling that noted in fragments of the third, fourth, and fifth right ribs (Figure 7.27).

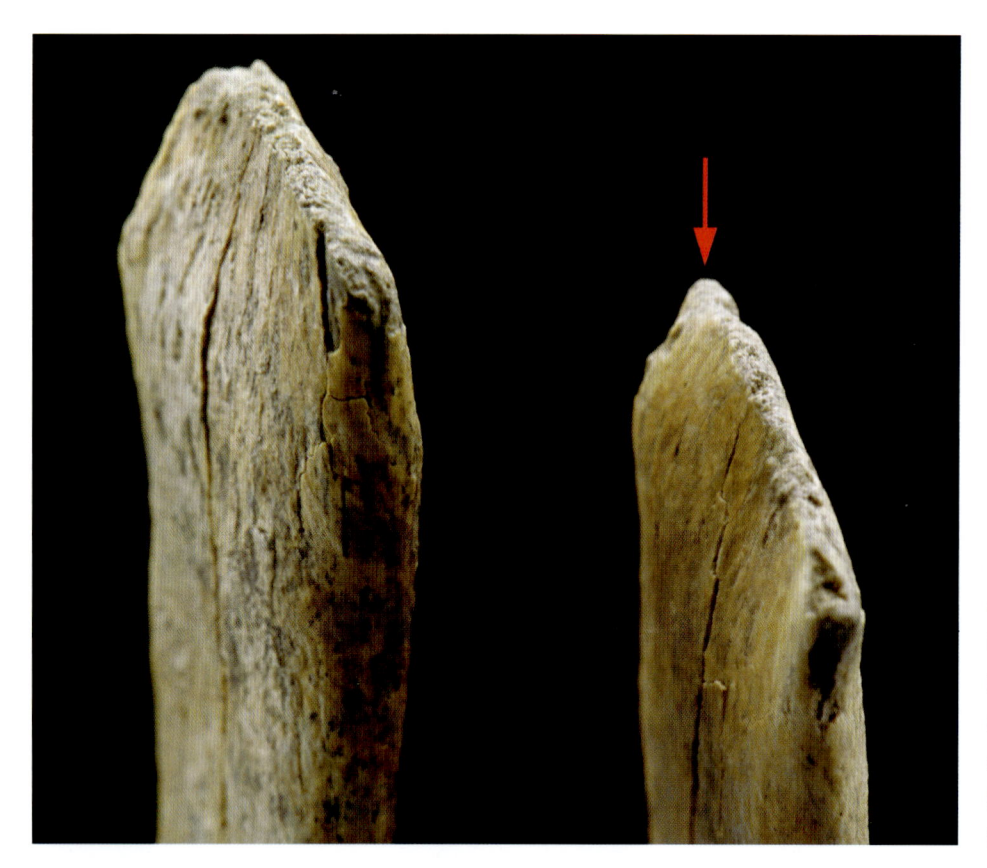

FIGURE 7.26 The tapered ends of healed fractures in the third (arrow) and fourth right ribs represent atypical remodeling and non-union of the broken ends due to fracture instability during the healing process.

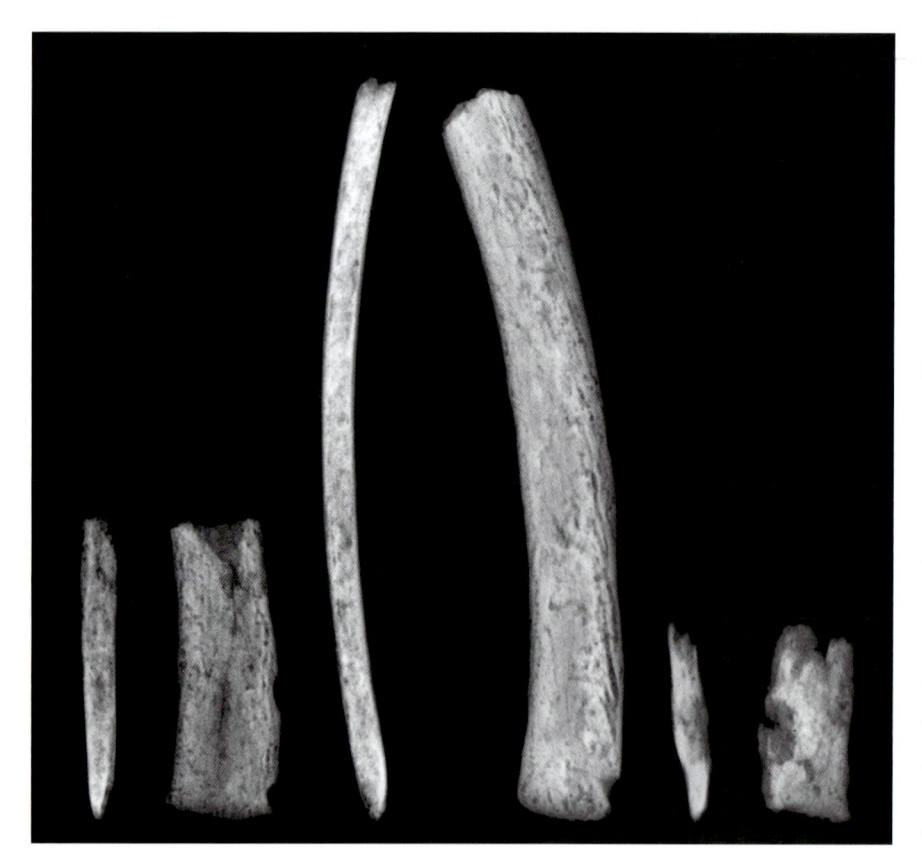

FIGURE 7.27 Adjacent superior-inferior and anterior-posterior radiographs of the right third (left), fourth (middle), and seventh (right) ribs show tapered, fractured edges without significant internal remodeling or callus formation. A slight increase in density at the tapered ends is distinguishable in the superior-inferior views. Rib body heights are not reduced at the fracture margins.

The entire eighth right rib is represented by five articulating fragments (97.I.12b(14), 97.I.12b(9), 97.I.12d(2), 97.I.12d(1), 97.I.12d(7)). The ninth rib is comprised of three articulating fragments (97.I.12b(12), 97.I.12d(12), 97.I.12d(11)) that, with the exception of the head, form most of the rib. The right tenth rib has four fragments (97.I.12d(9), 97.I.12f(7), 97.I.12d(6), 97.I.12d(4)). Three articulate to form the sternal half of the body. The eleventh rib is missing. The right twelfth rib is represented by one fragment (97.I.12e(11)) identified through size comparison with left rib twelve.

Anomalous "pinching" on the ends of some rib fragments has been noted, although with differing interpretations. Powell and Rose (1999) identify the left third rib (97.I.12a(2)) as having a "vertically oriented depression" indicative of a "normally healed fracture." In contrast, they consider the observed tapered ends atypical rib fractures, healed pseudarthroses caused by an antemortem injury years before death. A pseudarthrosis involves inadequate mineralization of callus allowing continued movement of the fractured ends, which can be bridged or held in position by fibrous tissue (Ortner 2003). Powell and Rose (1999) note four fragments with "gradual narrowing of the bone thickness" without loss of superior-inferior height ((97.I.12a(1), 97.I.12a(7), 97.I.12a(9), and 97.I.12a(3)). Unspecified right rib fragments 97.I.12a(1) and 97.I.12a(7) are thought by Powell and Rose to represent adjoining halves of a pseudarthrosis in rib 6, 7, or 8. Their report also matches fragment 97.I.12a(1) to fragment 97.I.12a(3), two sides of a pseudarthrosis, possibly of rib 6.

To the contrary, Walker et al. (2000) saw no evidence of new bone formation on the fragments with narrowed edges, and disagreed with the finding of pseudarthroses. They attributed the "pinched" ends to an undefined postmortem event that involved compression: " . . . the condition previously diagnosed as rib pseudarthroses in the Kennewick skeleton is a result of a post-mortem process perhaps resulting from ground pressure or some other mechanical process operating within the depositional environment." Walker et al. (2000:Table 1) were ultimately noncommittal, suggesting that the aforementioned breakage, as well as changes in another fragment (97.I.12a(8)), could have occurred perimortem or postmortem.

Chatters (2001:134–135) recognized seven fractures in at least six ribs on both sides of the chest. "Fractured ends of at least three and possibly four ribs, two right and one or two left, failed to heal together, forming pseudarthroses. This indicated the man suffered a condition known as flail chest, in which a segment of chest wall moves independently and counter to the chest wall as a whole (Beeson and Saegester 1983). It reduces breathing efficiency and coughing effectiveness (Kirsch and Sloan 1977), and would have affected his capability for strenuous exercise, at least until his body adapted" (Chatters 2000:309).

Flail chest syndrome is rejected by Cook (Chapter 14), since flail chest involves fractures in multiple adjacent (or opposing) ribs in two or more places each, as medically illustrated in Figure 7.28 (Bastos et al. 2008; Bjerke 2009; Hodgkinson et al. 1993; Qasim and Gwinnutt 2009). "[F]lail occurs when the integrity of the thoracic rigid cage is compromised and a segment, its size depending on the number of fractured ribs or costal cartilages, loses its continuity with the rest of the thoracic cage and moves paradoxically during spontaneous respiration" (Borman et al. 2006:903). The fractures in Kennewick Man's rib cage are not consistent with this description.

Incomplete recovery of the rib fragments hinders interpretation, and the degree of fragmentation makes placement of individual pieces challenging. However, this study finds Kennewick Man's fractures to be vertically aligned within the right antero-lateral aspect of the rib cage (Figure 7.29). Right ribs 2, 3, 4, 5, and 7 have fractures in the sternal thirds of their bodies. These changes are not the result of a perimortem injury, as there is evidence of healing, resulting in narrowed, remodeled edges from atrophy following nonunion. Further, it is difficult to imagine an in situ postmortem force that could produce a tapered appearance with collapse of the marrow space across multiple ribs without causing dry-bone crushing or irregular breakage.

That said, with the exception of the radiographically confirmed healed fracture in the left third rib, the appearance of these multiple segments with tapered edges is atypical relative to most healed rib fractures seen in anatomical and archaeological skeletal collections, as well as in contemporary forensic anthropology case work. Most fractures consolidate (rejoin) without medical

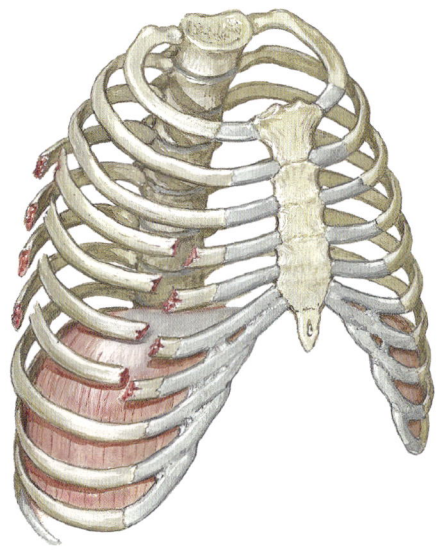

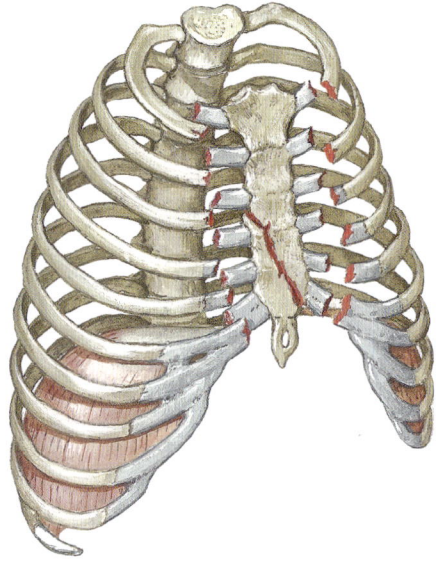

FIGURE 7.28 Medical diagrams illustrating fracture patterns associated with flail chest syndrome. Netter illustration from www.netterimages.com. © Elsevier Inc. All rights reserved.

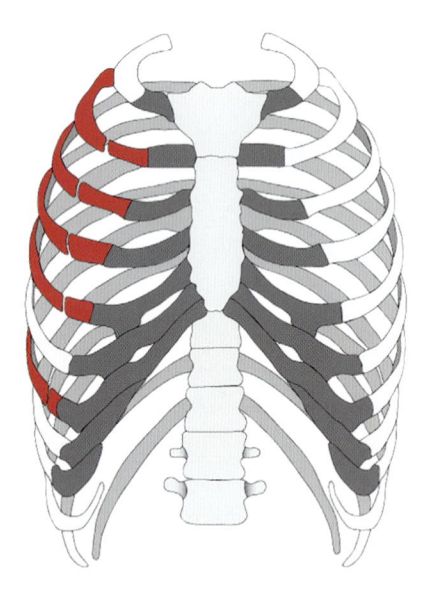

FIGURE 7.29 Approximate locations of right rib cage fractures in Kennewick Man with anomalous healing and non-union.

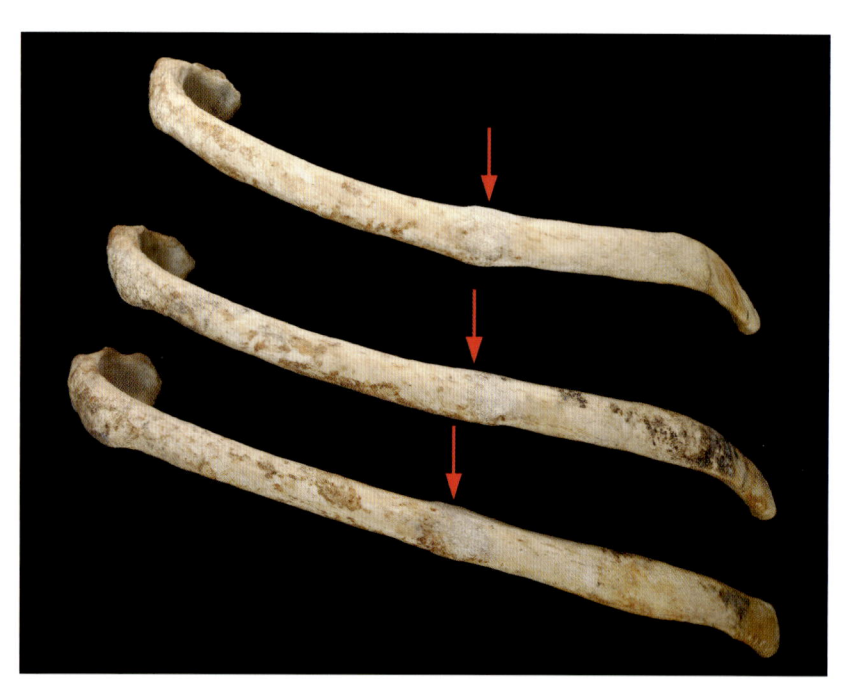

FIGURE 7.30 Anatomically aligned right ribs 4, 5 and 6 of an adult male with healed mid-body fractures (arrows) following a traumatic injury (Forensic case 44YorkRiver-VAOCME-52309).

intervention (Figure 7.30). Rib pseudarthroses have been reported in the clinical literature, but these are exceptions. They occur because cross-sectional surface areas of ribs are generally less than 60 square millimeters, and because fracture contact surfaces are sometimes insufficient for normal healing (Gardenbroek et al. 2009:1479). Rib atrophy, including fracture margin thinning following trauma is one expression (Figure 7.31). Mobile, hypertrophic pseudarthrosis with expanded fracture margins is another. Both occurrences are indicative of fracture instability. As Gardenbroek et al. (2009:1479) comment: "Given that ribs move an average of 25,000 times per day with respiration, it is surprising that pseudarthrosis of the ribs is not seen more frequently."

No indication of infection or significant interruption of the blood supply is noted in Kennewick Man's ribs,

166 *Chapter 7*

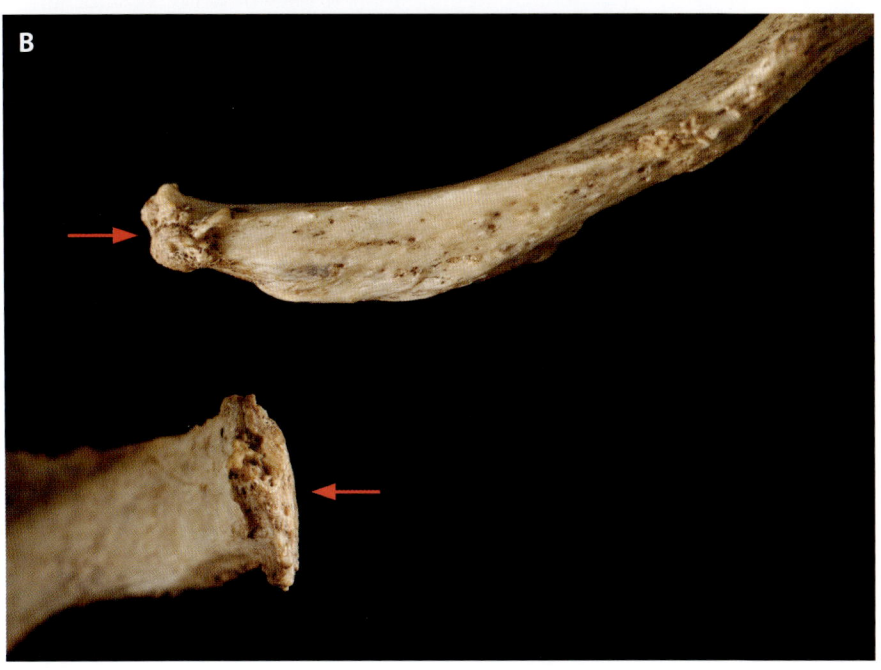

FIGURE 7.31 Superior and lateral views of a healed, atrophied, left eighth rib with multiple fractures (A) (Forensic case SI1997-02). The head and neck were less affected than the greatly diminished body, although the fracture margin joining the neck and body shows some atrophic pinching (B, arrows).

but the calcifying stimulus to the periosteum did not generate the callus needed for successful reunion of the right side fractures. Visual inspection and radiological examination of the fractured ends (Figure 7.27) fail to reveal callus formation typical of most healing or healed rib fractures. In chest X-rays, even rib pseudarthrosis is generally manifest by a vertical radiolucent line with dense, sclerotic borders (Guttentag and Salwen 1999). Nevertheless, the opposing ends of Kennewick Man's broken ribs healed while maintaining separation by the interposition of muscle tissue or loosely connected fibrous bridges. They also maintained normal dimensional development relative to the left side, suggesting that this injury occurred after growth had ceased, probably when Kennewick Man was a young adult.

In contemporary society, there are three main mechanisms for the type of chest trauma exhibited in the rib cage of Kennewick Man: low velocity impact, such as from a kick; high velocity impact, generally during traffic accidents; and crushing injuries (Hodgkinson et al. 1993). Kennewick Man sustained some type of direct blow to the right anterior third of the chest that fractured at least five ribs. It is possible that the left third rib was also broken at this time. Chest pain, especially during deep inhalation and coughing, and dyspnea (shortness of breath) are expected symptoms following this type of severe blunt trauma injury (Liman et al. 2003). In cases of nonunion, pain can persist for months during movement and/or deep respiration (Gardenbroek et al. 2009). Failed union of rib fractures is usually the result of inadequate immobilization during the healing phase (Ortner 2003:66). Kennewick Man's social and personal circumstances apparently did not allow sufficient time for recuperation and healing.

Cervical Vertebrae (C1–C7)

The first cervical vertebra (97.U.4(C1a)) is complete and in good condition with only trace erosion of the left transverse process. The second cervical (97.U.4(C2a)) is complete with minimal postmortem erosion of the spinous process. Old, postmortem breakage has damaged the right transverse process of the third cervical vertebra (97.U.4(C3a)) resulting in an open transverse foramen. A small surface osteophyte is present on the right inferior articular facet.

Cervical vertebra four is nearly complete (97.U.4(C4a)). An open left transverse foramen is the result of old, postmortem damage. Additional damage to the inferior endplate has destroyed most of the antero-inferior margin exposing the internal cancellous bone. Approximately one half of the inferior endplate remains. A small indentation in the superior right articular facet corresponds with an osteophyte on the matching right inferior facet of the third cervical vertebra. Degeneration of the right inferior articular facet has produced slight cupping of the surface.

Cervical vertebra five (97A.U.4(C5a)) is complete. The left and right transverse processes and the tip of the spinous process show minor breakage. Compression of the inferior facets of the fourth cervical vertebra into the superior facets of the fifth cervical vertebra is expressed by buttressing (build-up of a bony ridge) along the inferior margins of the fifth cervical's superior articular facets. Trace porosity is present along the anterior-inferior endplate margin, but no degenerative lipping is noted on the centrum.

Cervical vertebrae six (97.U.4(C6a)) and seven (97.U.4(C7a)) are nearly complete, although ancient postmortem breakage has damaged the transverse processes resulting in open foraminae. Trace degenerative lipping is present along the anterior-inferior endplate margin of the sixth cervical, but no degenerative changes are noted on the articular facets. The seventh cervical retains a remnant line of union for the anterior epiphyseal ring and no degenerative changes are noted on either the vertebra's body or facets.

Thoracic Vertebrae (T1–T12)

Fragments of all twelve thoracic vertebrae are present. Although incomplete in their representation, no degenerative lipping, joint surface porosity, or eburnation (bone-on-bone joint surface polish from loss of articular cartilage) are evident.

The first thoracic vertebra is comprised of three pieces (97.U.5(T1a), 97.U.5d, 97A.U.5a). The vertebral body is complete and in good condition. The superior articular facets were broken off postmortem and are missing. The inferior articular facets, which are part of the fragment representing the neural arch and anterior aspect of the spinous process, are present and nearly complete. The tip of the spinous process is missing. The smallest of the three first thoracic fragments represents the left transverse process.

Thoracic vertebra two is comprised of two pieces (97.U.5(T2a), 97.U.5b). The right side is represented by a partial superior articular facet and pedicle attached to the right half of the centrum. The left half of the body is missing. However, the left superior and inferior articular facets are still attached to a fragment of left lamina. The spinous process and right half of the neural arch are missing. The detached inferior edge of the lower left articular facet of the first thoracic is fused by carbonate cemented sediment to the left superior facet of the second thoracic vertebra.

The third thoracic vertebra is represented by two pieces (97.U.5(T3a), 97.U.5(T3b)). The centrum is complete and in good condition with attached pedicles and superior articular facets. The second piece is comprised of the lamina attached to complete inferior articular facets and the spinous process, which is missing the tip. These two pieces are separated through the lamina and can be rearticulated. The body shows no reduction in height, but the superior and inferior endplates show trace concavity. Component 97.U.5(T3b) has a well-defined epiphyseal ring and fine porous endplate texture. "Neural canal spurs" and "para-articular processes" (Nathan 1959) are evident on the anterior neural arch. These small projections of bone form at the insertion of the ligamentum flava near the articular facets of the thoracic vertebrae along the neural arch. They have been described as the result of ossification of the ligamentum flava (Mann and Murphy 1990), or "OLF" (Hukudu et al. 2000) and are generally considered a non-pathological condition (normal variation) unless the spurs are significant enough to impinge on the neural canal. Modern studies of the condition are linked to patients with severe expression of the condition.

The fourth thoracic vertebra is comprised of two pieces (97.U.5(T5a), 97.U.5(T5b)), the largest being the nearly complete centrum, although postmortem erosion has damaged the anterior body and inferior endplate margin. The second articulating fragment represents the spinous process *sans* tip, the right superior articular facet, and both inferior articular facets. Ossification of the ligamentum flava is noted in the form of bony spicules between the superior articular facets. Body height is normal, although there is trace endplate concavity.

Thoracic vertebra five is comprised of two pieces (97.U.5(T4a), 97.U.5(T4b)). The largest piece is a nearly complete body. Trace wedging of the left side is noted, but body height is essentially normal. This piece articulates with a second fragment at the base of the right pedicle. The second fragment is the neural arch with its right pedicle, a partial left pedicle, and the base of the spinous process. The tip of the spinous process is missing. Relatively complete superior and inferior articular facets are present, but the transverse processes have broken off postmortem. The ligamentum flava shows a moderate degree of ossification.

Two pieces form thoracic vertebra six (97.U.5(T6a), 97.U.5a). The largest piece is a nearly complete vertebral body. The second fragment is a portion of the right neural arch including the superior and inferior articular facets. The endplates are slightly concave and there is possible minor wedging of the centrum's left side. Slight ossification of the ligamentum flava is present.

Thoracic vertebra seven is comprised of four fragments (97.U.5(T7a), 97.U.5(T7b), 97.U.5g, 97.U.5c). The largest piece represents the left side of the vertebral body, neural arch, and a portion of the left side of the spinous process. Complete left superior and inferior articular facets are attached. This fragment articulates with a second piece representing the right side of the vertebral body. When joined at the point of contact in the posterior body, most of the anterior surface is missing. The third and fourth fragments represent the right side of the neural arch and right portion of the spinous process along with complete right superior and inferior articular facets. Body height is maintained, although trace concavity of the endplates may be present.

A single piece (97.U.5(T9a)) represents the eighth thoracic vertebra. Postmortem erosion has damaged the anterior aspect of the body with destruction of the superior and inferior endplate margins. The tip of the spinous process and the transverse processes are missing due to postmortem breakage. Complete articular facets are present. This vertebra has a well-defined remnant line of union for the epiphyseal ring and the endplates have a fine porous texture.

Thoracic vertebra nine is represented by a single piece of bone (97.U.5(T8a)). Although mostly complete, the transverse processes, tip of the spinous process, and anterior aspect of the vertebral body are missing from postmortem breakage.

The tenth thoracic vertebra is represented by a single bone piece (97.U.5(T10a)). This vertebra is missing most of its body (only the posterior third is present), both transverse processes, and the tip of the spinous process. The articular facets are present.

Thoracic vertebra eleven is nearly complete (97.U.5(T11a)). The anterior face of the body is eroded and the transverse processes and tip of the spinous process are missing from postmortem damage. Slight wedging of the centrum is characterized by reduced anterior body height and there is trace ossification of the ligamentum flava.

Thoracic vertebra twelve (97.U.5(T12a)) is nearly complete and in fair condition. A postmortem puncture is present in the inferior endplate. Both transverse processes and the tip of the spinous process are missing due to postmortem breakage. The height of the vertebral body seems normal and without depression (concavity) of the endplates.

Lumbar Vertebrae (L1–L5)

Fragments of all five lumbar vertebrae are present. The better preserved lumbar vertebrae show minor degenerative changes characterized by slight lipping of the endplate margins.

The first lumbar vertebra is represented by a small portion of the inferior endplate (97.U.6(L1b)). This fragment is identified as the first on the basis of elimination; inferior endplates of the second through the fifth lumbar vertebrae are partially or completely represented. The slightly larger size and the shape of the better-represented inferior endplates are consistent with identification of this piece as part of the first lumbar vertebra.

The second lumbar vertebra is comprised of four articulating pieces of bone (97.U.6c, 97.U.6(L4b), 97.U.6(L1a), 97.U.6(L4a)). The largest piece represents the spinous process, superior and inferior articular facets, and most of the neural arch. The other fragments represent portions of the vertebral body. No significant pathological changes are noted on the facets, although the centrum margin has slight degenerative lipping.

The third lumbar vertebra (97.U.6(L3a)) is the most complete. Only the transverse processes and tip of the spinous process are missing from postmortem breakage. No degenerative changes are evident on either the lumbar body or its facets.

The fourth lumbar vertebra is represented by three articulating pieces (97.U.6(L2b), 97.U.6b, 97.U.6(L2a)). The two largest pieces represent the left and right sides of the vertebral body, each with a superior articular facet attached. The third and smallest fragment is a portion of the inferior endplate. The transverse processes and spinous process are missing due to postmortem breakage. Slight degenerative lipping is present on the left inferior endplate margin. Powell and Rose (1999) also note the presence of lipping on the second and fourth lumbar vertebrae.

The fifth lumbar vertebra is represented by five separate pieces of bone (97.U.6(L5b), 97.U.6(L5a), 97.U.6a, 97.I.25Z, 97.U.6(L5c)). One piece includes a portion of the right vertebral body (superior endplate), right transverse process, and superior right articular facet. A second fragment represents a portion of the left vertebral body (superior endplate), left transverse process, and a partial left superior articular facet. A third piece is a partial neural arch, including an incomplete spinous process and complete left inferior articular facet. The two remaining fragments are portions of the vertebral body. None of these fragments articulate, but are identified as belonging to the fifth lumbar vertebra based on morphology.

Four bone fragments originally listed as lumbar vertebrae are not associated with L1 through L5. Two of these fragments (97.U.6(L2c) and 97.U.6(L5d)) represent the left and right articular facets of the first sacral unit. The right articular facet offers a point of contact with a large fragment of sacrum that includes the right ala and a portion of the first sacral segment. Neither articular facet of the sacrum displays degenerative changes. A small bag containing two fragments of vertebral endplate (labeled 97.U.6(L5c)) cannot be assigned to a specific vertebra.

Sacrum

The sacrum was originally identified as being represented by four fragments (97.U.7a, 97.U.7b, 97.U.7c, 97.U.6d). Two pieces previously misidentified as lumbar vertebrae, 97.U.6(L5d) and 97.U.6(L2c), represent additional segments of the sacrum. The right articular facet of the sacrum (97.U.6(L5d)) was identified based on morphology and its articulation with the spinous processes (97.U.7b), as well as with the fragment representing the right margin of the sacral promontory and the superior-anterior aspect of the right ala (97.U.7a). Fragment 97.U.7b articulates with sacral fragment 97.U.6d. Although the coccyx is missing, a piece of sacrum representing the anterior aspect of the fifth sacral unit and a small portion of the fourth sacral unit (97.U.7c) displays a well-preserved articulation surface with the coccyx.

The represented portion of sacral promontory has no degenerative changes. The segment of ala has an area of irregularity in its superior aspect. This region is occluded with sediment and its original surface is obscured. However, the location of the irregularity and its apparent morphology are consistent with a point of contact with the right transverse process of the fifth lumbar vertebra. This cannot be confirmed due to erosion of the fifth lumbar vertebra's transverse process.

Os Coxae (Innominates)

The left innominate is represented by three main pieces (97.L.17b, 97A.I.17a, 97.L.17a). A nearly complete ilium along with portions of the ischium and pubis comprise the largest piece (97.L.17b). The second largest segment (97.L.17a) represents the left pubis, including the symphyseal face. The inferior aspect of the ischium is represented by the third, and smallest, piece (97A.I.17a).

The largest piece of the left innominate is relatively well preserved. Postmortem breakage has resulted in damage to the superior-medial (internal) surface of the ilium (iliac fossa). The damaged area measures approximately 55 mm by 40 mm with the maximum measurement extending from the iliac crest toward the greater sciatic notch. This region is characterized by missing bone and exposed internal cancellous bone. The posterior portion of the ilium including the iliac tuberosity and posterior-superior iliac spine (retroauricular area) has broken off postmortem and is missing. The auricular surface is complete, although carbonate impregnated sediment obscures some areas from observation. A small piece of sacrum has broken off of the left ala and is fixed to the anterior margin of the sacroiliac joint by carbonate cemented sediment. Postmortem breakage along a relatively vertical plane has separated the pubic bone from the ilium through the superior ramus of the pubis, and the ischium from the ilium at the ischial tuberosity. The tip of the ischial spine broke off postmortem. The second largest piece, the pubic bone (97.L.17a), has fracturing across the superior and inferior rami. The symphyseal face is complete, with some sediment present that partially obscures details of the surface. A small section of the ischio-pubic ramus (97A.I.17a) is also present.

Three additional small pieces had been labeled as left innominate prior to the present study (97A.I.25b, 97.I.17a, and 97.I.25d). A small piece (97A.I.25b) represents a portion of superior-posterior iliac crest. A second fragment (97.I.25d) is tentatively identified as a segment of ischium. The third fragment (97.I.17a) may not belong with the innominate and instead appears more consistent with a thicker bone, such as a calcaneus.

The right innominate is represented by nine pieces identified by eight catalog numbers. The largest segment (97.R.17d) includes the auricular surface, sciatic notch, acetabulum, and superior ischium. This piece articulates with the next two largest fragments: the anterior ilium, including the anterior superior and inferior spines, the iliac fossa, and the anterior iliac crest (97.R.17b); and a dorsal segment of the ilium with an embedded projectile point (97.R.17c). Prior to this study, but after the segment numbers had been assigned, a small fragment of 97.R.17c had detached, but was not assigned another unique number. The fragment is part of the internal surface of the mid-posterior ilium. The middle aspect of the iliac crest is represented by two small pieces: segment 97.I.17d is more anterior, and 97.I.10b articulates with 97.R.17c. A sacral ala fragment is secured to the inferior auricular surface by heavy carbonate cemented sediment accumulation. Also present are the pubic bone (97.R.17a), an ischial fragment (97.I.17c), and a fragment of the ischio-pubic ramus (97.I.17b).

The acetabulum of the left innominate is complete and exhibits minor osteophytic lipping of its margin. On the right innominate, two small osteophytic bone spicules have formed along the anterior margin of the obturator foramen of the pubic fragment (97.R.17a). Otherwise, the anterior half of the foramen rim is smooth.

A projectile point is embedded in the right posterior ilium (97.R.17c) (Figure 7.32). A visual inspection of the point suggests it was made from an igneous rock, likely dacite or possibly basalt (Cannon personal communication 2013). The base, portions of both sides of the blade, and a serrated edge are visible through resorption windows in the internal (ventral) and external (dorsal) bone surfaces. The opening in the ventral surface measures 34 mm in length by 9 mm in height (Figure 7.33). A smaller opening in the opposing surface measures 17 mm in length by 16 mm in height (Figures 7.34 and 7.35). The faces of the point are aligned with the internal and external surfaces of the innominate. Specialized CT scans revealed the full dimensions of the embedded point to be 53 mm in length, a maximum width of 20 mm, and a maximum thickness of 6.7 mm. The tip of the projectile, if originally present, was not recovered. However, detached minute lithic fragments may be embedded in the ilium. The base of the point appears to be essentially intact (Chapter 20). As visualized in a digital radiograph

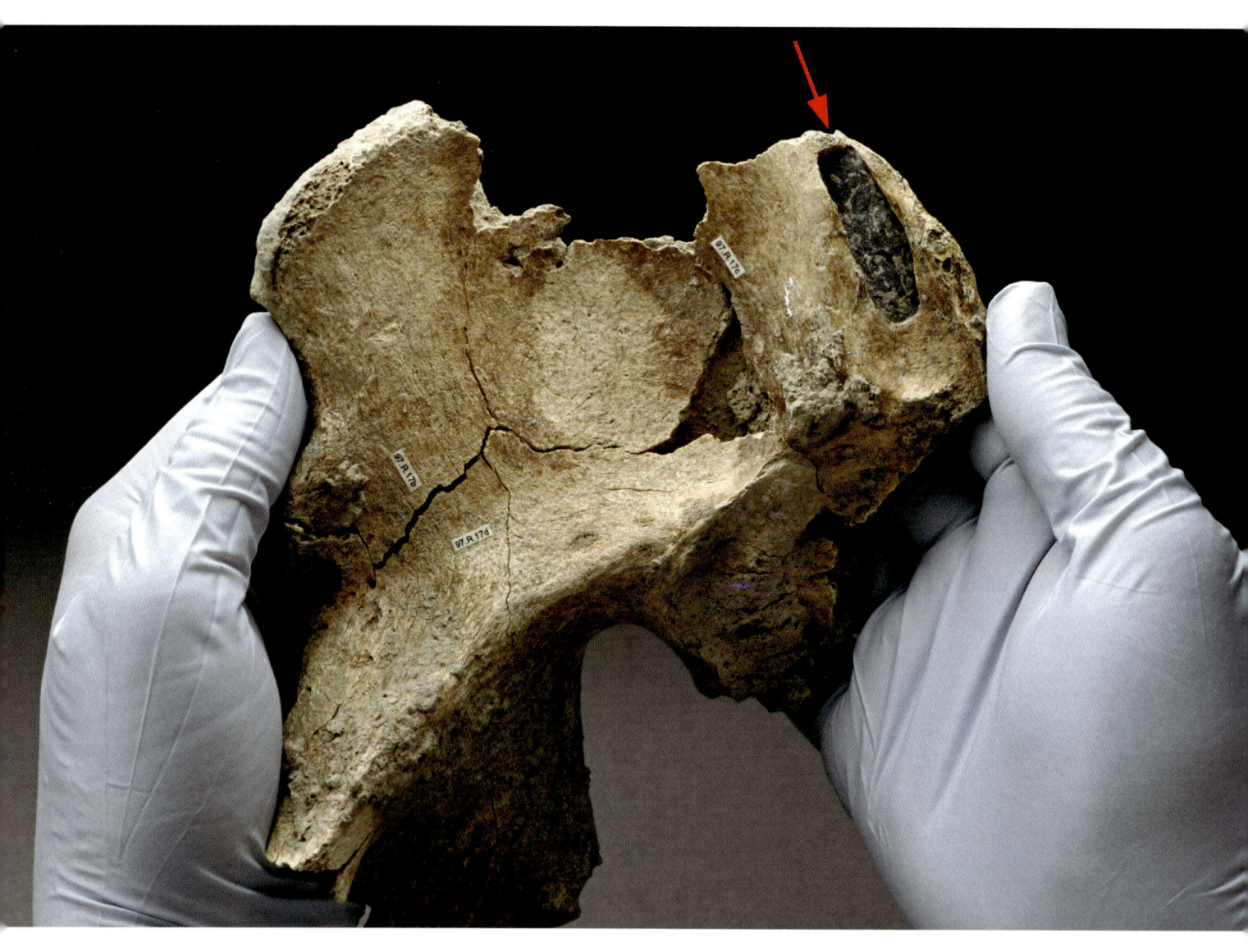

FIGURE 7.32 Internal view of the three main articulating pieces of the right ilium. The posterosuperior fragment (97.R17c) has an embedded stone projectile point, the base of which is identified by an arrow.

(Figure 7.36), the density of the projectile point and intruding carbonates are similar to bone, precluding clear definition of the outline of the point.

The orientation and direction of the projectile's entry has been assessed previously by others with conflicting interpretations. According to Powell and Rose (1999), the projectile entered the body posteriorly with an angle of entry that was "slightly below the horizontal." Chatters (2000:309) posited a right anterior entry from an atlatl propelled dart or spear thrown from "about 60 degrees to the right of front center and about 45 degrees above the horizontal" (2001:138).

Radiography and computed tomography imaging used to evaluate the placement and orientation of the embedded projectile within the pelvis indicate that the projectile forcefully struck the posterior right ilium traveling front to back from Kennewick Man's right side and slightly medially. The point entered his side with a downward descent of approximately 29 degrees above the horizontal (Figure 7.37) and 65 degrees to the right of the mid-sagittal plane (Figure 7.38). Whether the assailant was to his front or right side is undetermined. Kennewick Man, if aware of the dart hurtling toward him, may have pivoted to narrow his profile in an attempt to dodge the dart.

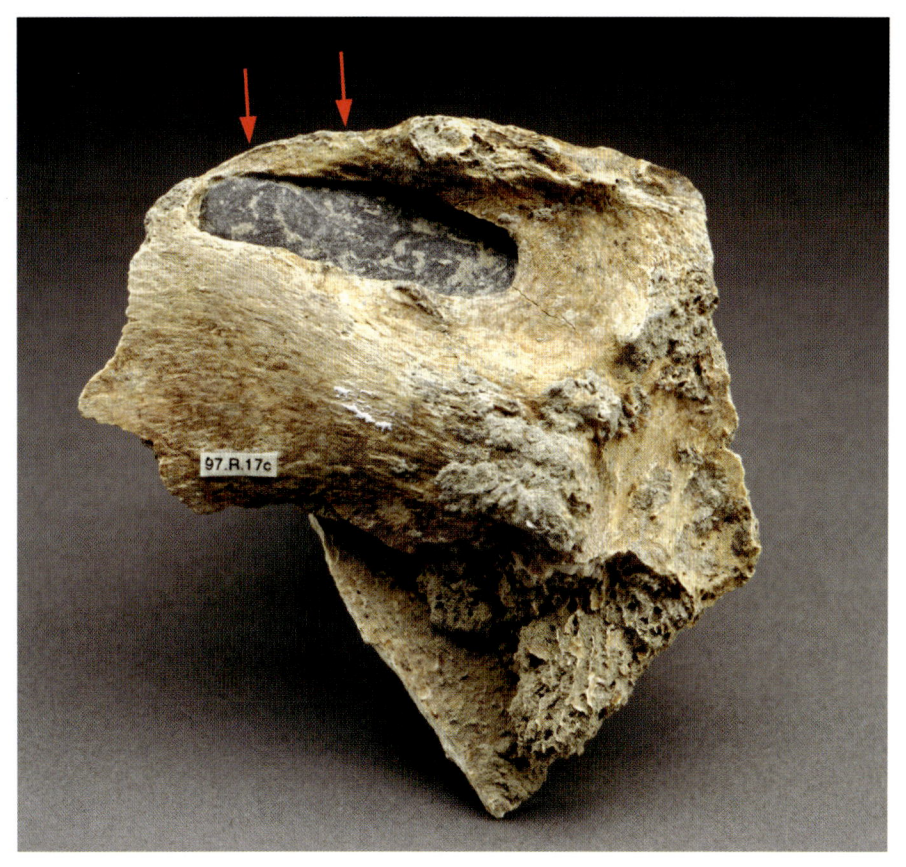

FIGURE 7.33   The projectile point is partially visible through an opening in the internal surface of the ilium (the anterosuperior margin of the fragment is on the left). The smooth, remodeled margin around the opening indicates advanced healing. A segment of the iliac crest was fractured off by the impact (arrows). New cortical bone has formed over the once-exposed cancellous bone.

FIGURE 7.34   A smaller window in the external surface of the posterior ilium reveals a portion of the reverse side of the point.

FIGURE 7.35 Slightly different views (A, B) of the external window showing pocketing from bone resorption around the serrated projectile point. Minor external iliac crest porosity above and anterior to the point is due to postmortem erosion.

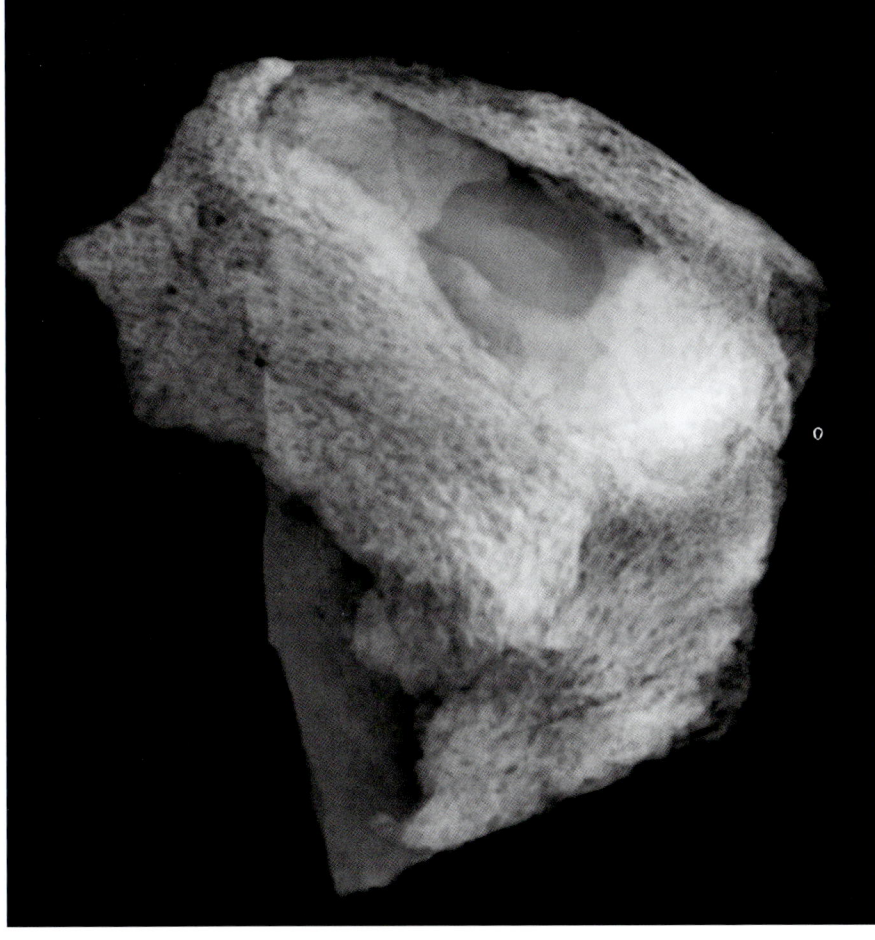

FIGURE 7.36   No abnormal, subperiosteal bone formation or draining sinuses from a chronic bacterial infection (osteomyelitis) are evident in this radiograph of the ilium fragment and embedded point. The embedded point is not well defined, although the superimposed internal and external areas of missing cortex are partially visible. Adhering carbonates and sediment obscure the posteroinferiorly encapsulated portion of the point.

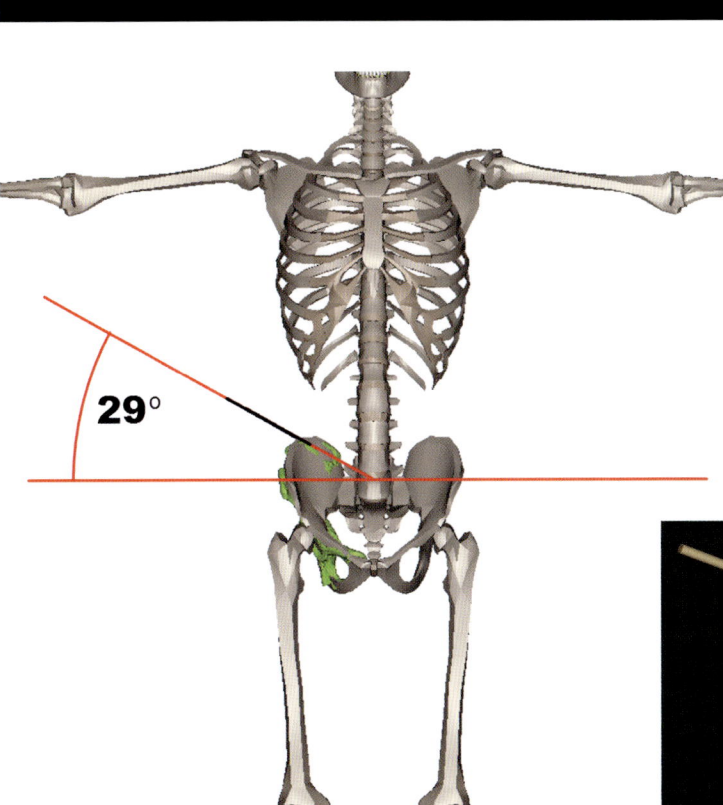

FIGURE 7.37   Kennewick Man's innominate (green) superimposed on a 3D skeletal model to measure the downward angle of impact (A). Frontal view (B) of Kennewick Man's right innominate (red) articulated with a modern sacrum (blue) for orientation. The embedded projectile point and reproduced foreshaft are illustrated as cylinders.

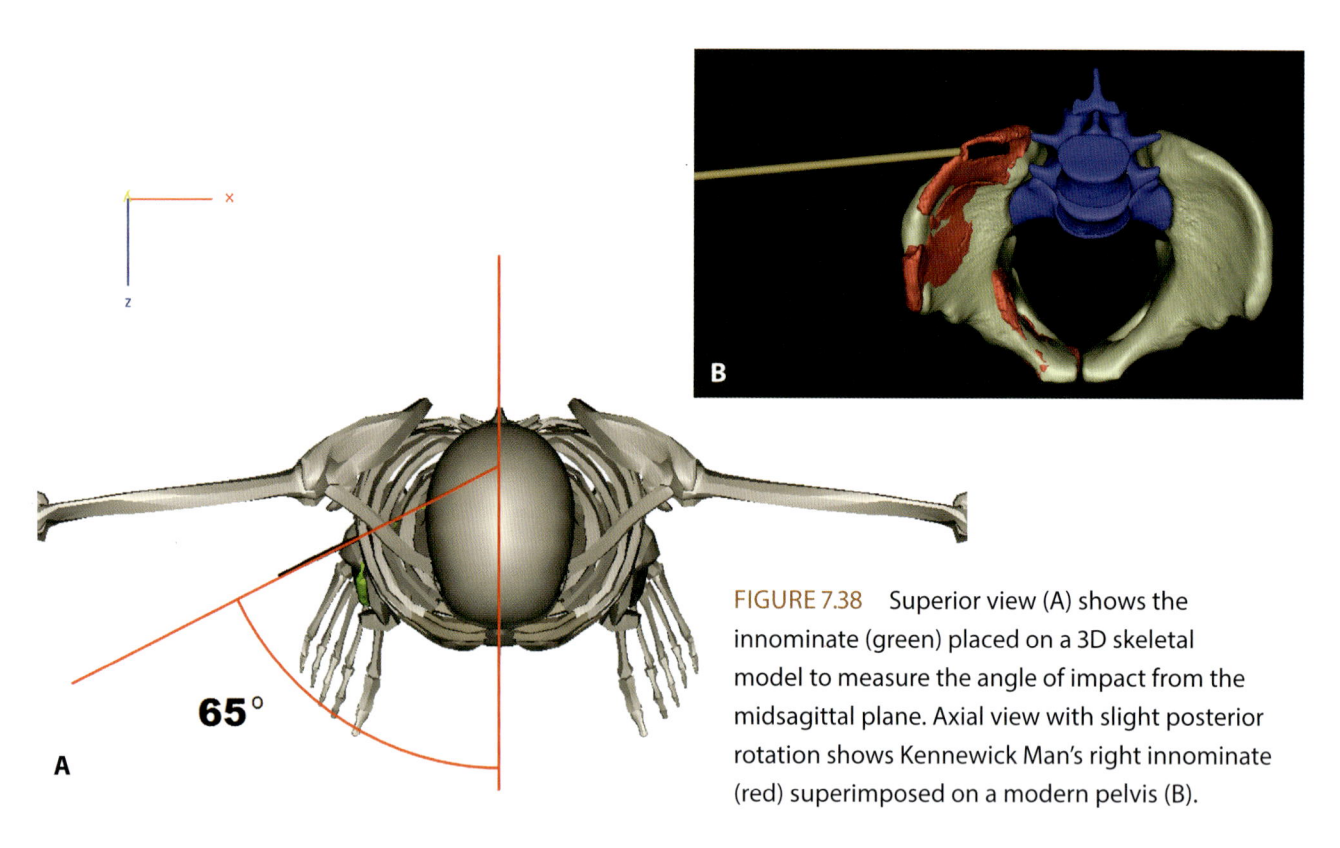

FIGURE 7.38 Superior view (A) shows the innominate (green) placed on a 3D skeletal model to measure the angle of impact from the midsagittal plane. Axial view with slight posterior rotation shows Kennewick Man's right innominate (red) superimposed on a modern pelvis (B).

Previous studies have discussed the pathological consequences of this injury. Chatters (Chapter 15, 2000:309, 2001:136–137) states that the wound shows signs of chronic infection from osteomyelitis. Walker et al. (2000) indicate that the projectile wound caused a "secondary infection," but do not elaborate. Powell and Rose (1999), on the other hand, state that radiographs of the injury show normal bone formation around the projectile indicating "complete healing and no sign of infection."

Radiographic and visual assessment of the right ilium indicate that the iliac fossa and impact fracture margins are smooth and remodeled, and lack vascular macroporosity, subperiosteal hypertrophic expansion, and draining sinuses. A minimum of 22 mm of the medial margin of the iliac crest fractured off at the time of impact and has remodeled smoothly (Figure 7.33). The missing fragment of the crest was not recovered and likely had resorbed. A new layer of cortical bone formed above the external basal half of the stone point during the healing process.

The injury healed with direct contact between remodeled bone and portions of the base and tip end of the point, along with osteoclastic resorption of bone away from visible edges and flat surfaces of the projectile (Figure 23.5). It is not clear why some bone remodeled to partially envelop the point while it resorbed elsewhere. The latter suggests fibrous capsule formation along the tissue/material interface (fibrosis). This systemic foreign-body reaction is a defense mechanism designed to eliminate exogenous materials by rejection, dissolution, resorption, or in this case demarcation (Donath et al. 1992). Given the size, fixed position, and immutability of the stone, the surrounding regenerative tissue likely produced a collagen-rich fibrous scar around the point (fibrous encapsulation) as a means of isolating the foreign body from surrounding normal tissue (Anderson 2001). Bone can repair without the formation of this intervening fibrous tissue capsule (Spector et al. 1989). Silica in wounds, however, is an irritant that can lead to nonhealing or excessive fibrosis (Silver 1984:61). The mineralogical composition of the projectile point needs to be analyzed using x-ray fluorescence and other tests (Chapters 23 and 32). Both dacite and basalt contain silica, 62 to 69 percent, and 45 to 53 percent, respectively (USGS Volcano Hazards Program).

Kennewick Man's right ilium healed without developing a chronic infection from bacterial contamination. The physical manifestation of infections in the pelvis can be dramatic, as illustrated in Civil War soldiers

FIGURE 7.39 Osteomyelitis of the right innominate and sacrum (A) of a Civil War soldier wounded by a lead, conoidal ball that penetrated the ilium near the sacro-iliac symphysis (gunshot wound, 1865, National Museum of Health and Medicine, AMM-1000137 [PS 3232]). The close-up (B) shows extensive new bone formation from a chronic infection that persisted for 82 weeks until death.

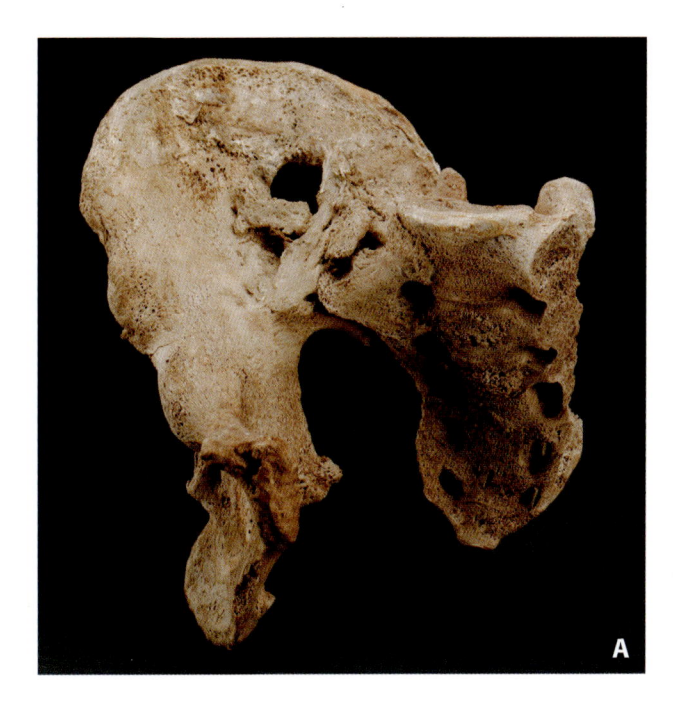

with gunshot wounds (Figures 7.39 and 7.40). In contrast, Kennewick Man's right ilium shows no indication of inflammatory reaction, necrosis, osteolytic (carious) destruction, draining sinuses, or deposition of reactive bone, all characteristics of osteomyelitis (see Rosenberg 1999; Ortner 2003).

Kennewick Man's pelvic injury can be more directly compared to healed wounds caused by embedded atlatl points and more commonly observed arrow wounds. In documented cases, iron points perforate large bones without producing comminuted fractures, causing at most only a slight fissure due to their shape, force, and velocity (Otis 1883; Wilson 1901). Stone points similarly cause little damage, but are liable to break upon impact with bones, and their bases often snap when being extracted. In nonfatal injuries, the healing process includes inflammatory reactions involving new bony growth, osteolysis and pocketing around points, remodeled smooth and rounded surrounding bony surfaces and openings, and sclerosis around cystic cavities containing points (Gregg et al. 1981; Hoppa et al. 2005; Polet et al. 1996; Ryan and Milner 2006; Vegas et al. 2012; Wilson 1901). Military surgeons dealing primarily, but not exclusively, with iron projectile points during the Plains and Southwest Indian Wars of the 1860s and 1870s reported that suppuration (pus) and abscess formation were typical, along with the migration of lithic debris toward the surface, and therefore worked diligently to extract arrowheads (Coues 1866; Milner 2005; Otis 1883). Coues (1866:322) explains:

> [I]t is exceedingly rare that these sharply triangular and jagged bits of stone excite so little disturbance that the encysting process can be accomplished. This can only take place under some peculiarly favorable circumstances, as to position, etc. The irritation is almost invariably sufficient to excite suppuration; consequent abscesses, (unless the wound is very open); and thus a gradual working of the head toward the surface; this 'pointing' being often at a considerable distance from the place where it might have been expected.

Estimates based on both prehistoric archaeological skeletons from eastern North America and nineteenth century military records indicate that around 11 percent of individuals with arrowheads embedded in bones survived for extended periods of time (Milner 2005).

Kennewick Man was lucky with regard to the location of the impact and the final outcome. The dart struck the posterior intermediate line of the iliac crest, having penetrated skin, subcutaneous tissue, the superficial layer

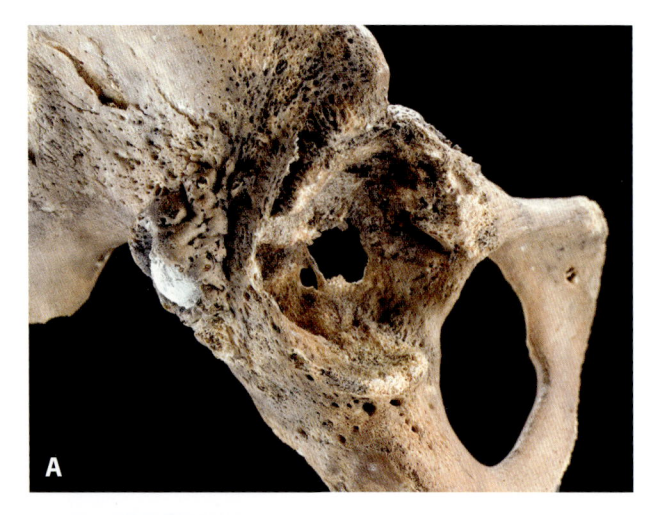

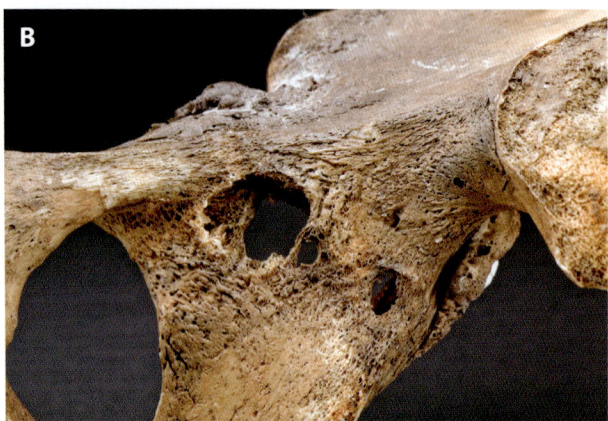

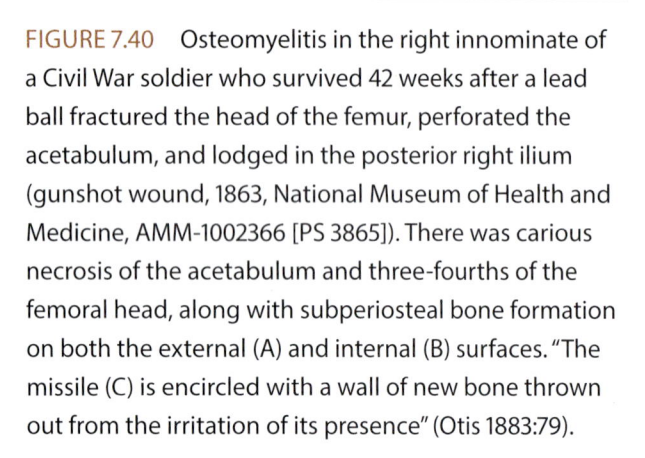

FIGURE 7.40 Osteomyelitis in the right innominate of a Civil War soldier who survived 42 weeks after a lead ball fractured the head of the femur, perforated the acetabulum, and lodged in the posterior right ilium (gunshot wound, 1863, National Museum of Health and Medicine, AMM-1002366 [PS 3865]). There was carious necrosis of the acetabulum and three-fourths of the femoral head, along with subperiosteal bone formation on both the external (A) and internal (B) surfaces. "The missile (C) is encircled with a wall of new bone thrown out from the irritation of its presence" (Otis 1883:79).

of the thoracolumbar facia, and fracturing off the inner lip of the crest. As the point of entry was above the origin of gluteus medius, superior and anterior to gluteus maximus, and behind the internal and external obliques, pelvic musculature was largely unaffected, with the exception of quadratus lumborum. This deep posterior muscle of the abdominal wall bends the trunk toward the same side, has its origin along the inner iliac crest in this area, and would have been injured by shearing off a segment of the crest. The near-surface injury in the iliac crest narrowly missed internal organs and healed by encapsulating the embedded point without severe infection. Perforation of the abdomen would have resulted in peritonitis which, without advanced medical intervention, is usually fatal. That is not to say, however, that the injury did not temporarily cause significant regional pain, as well as general discomfort. Distress experienced by soldiers with arrow wounds was described by Coues (1866:324):

> The constitutional disturbances following these arrow-wounds, even when the injury is confined to bone or muscle, are liable to be out of all proportion to the apparent amount of damage done. There is almost always considerable febrile action, more or less complete anorexia and sleeplessness, and derangement of all the secretions. To these may be added great irrita-

bility and intolerance of a moderate degree of pain, much dejection of spirits, haggard and anxious contenance, etc. The tendency to despondence becomes frequently a prominent symptom . . .

Kennewick Man's injury was long standing and likely occurred when he was a young adult. Powell and Rose (1999) interpret an asymmetry of the sciatic notches of the innominates as a possible effect of the disruption of the iliac crest during a period of growth. If true, this would indicate that the injury to the hip would have taken place when Kennewick Man was a teenager. However, the crest and ilium show no radiographic evidence of developmental malformation, only localized bone loss due to the dart injury with extensive remodeling of the fractured portion of the crest. Analysis of the leg bones (Chapter 12) identified no lasting or significant impact on his mobility or activity.

Femora

Three segments represent the nearly complete left femur (97.L.18a (proximal), 01.L.18b (midshaft), and 01.L.18a (distal)). The superior-posterior and inferior third of the femoral head, along with most of the medial aspect of the distal joint epiphysis, are missing due to postmortem breakage. The right femur is essentially complete and represented by proximal (01.R.18a), midshaft (01.R.18b), and distal (97.R.18a) segments. Minor breakage has damaged the superior aspect of the femoral head and external cortical bone along the posterior intertrochanteric crest and adjacent area.

A posterior-anterior radiograph of the femora reveals moderately compact cancellous bone in the heads, thick shaft cortices, and the locations of postmortem breaks (Figure 7.41). Carbonate cemented sediment is present in the medullary cavities. A small void in the distal right femur medullary cavity reflects sediment removal by the government team. Midshaft subperiosteal and medullary cavity diameters can be measured directly, along with other cortical dimensions through computed tomography (Chapter 12).

The fovea capitis of the left femur has a slightly raised rim with a porous interior. The posterior-superior aspect of the lateral condyle's distal joint surface of the lateral condyle is porous from joint cartilage deterioration in life, and has a single small projecting spicule (Figure 7.42). This localized area of degenerative erosion has a superior-inferior measurement of 15 mm, a transverse width of 15 mm, and a maximum depth of exposed cancellous bone of 1.4 mm. The adjacent medial margin of the lateral condyle has a slightly raised rim and moderate porosity. The anterior margin of the intercondylar fossa also has slight degenerative lipping.

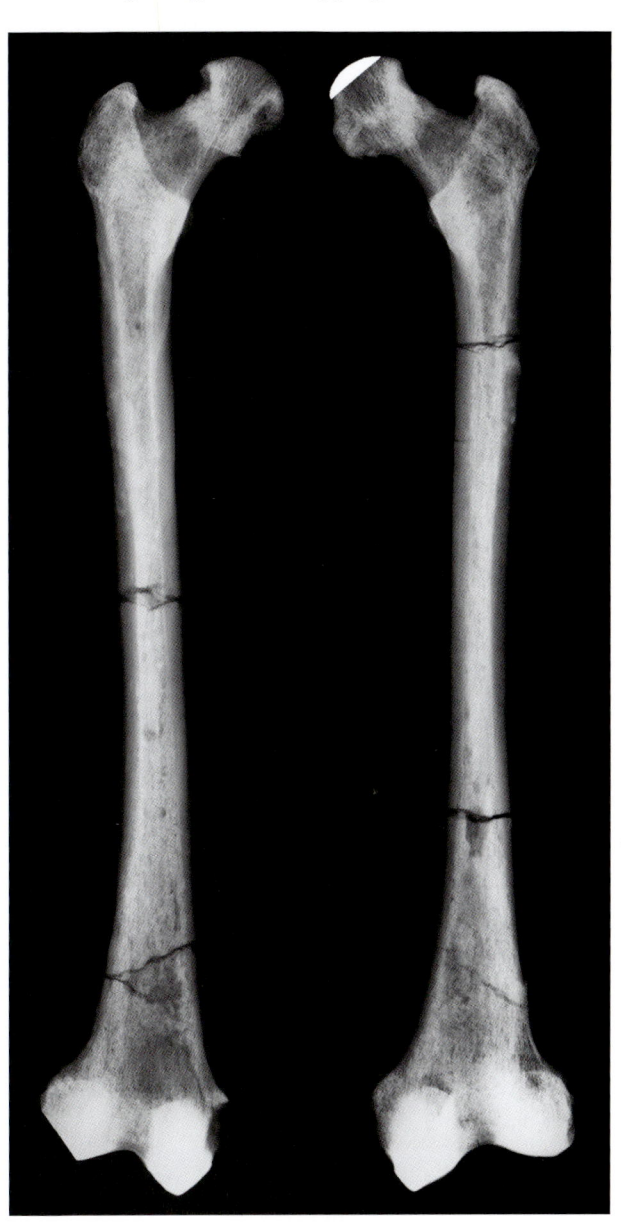

FIGURE 7.41 Posterior-anterior femur radiograph shows thick cortices, radio-opaque carbonate cemented silt and very fine sand in the medullary cavities, and complete and incomplete breaks that occurred when the skeleton eroded from the terrace.

FIGURE 7.42 Degenerative changes in the knee characterized by porosity and surface osteophytes on the left femur lateral condyle and proximal tibia.

On the right femur, the fovea capitis shows postmortem deterioration, although there is some antemortem porosity and slight rim development. The surface of the trochanteric fossa is visible and no intertrochanteric spicules are noted. The gluteal tuberosity on the posterior surface, the insertion for gluteus maximus, is slightly raised and roughened. The linea aspera, the attachment ridge for the adductor muscles ("hamstring" muscles) shows moderate definition in the midshaft along the middle two-thirds of the diaphysis (nearly the entire length of the middle segment). The linea aspera has an approximate midshaft width of 6.9 mm with a height of 4.0 mm.

The distal joint surface of the right femur exhibits slight degenerative lipping and porosity. Lipping is evident along the anterior margin of the intercondylar fossa and medial margin of the lateral condyle. As with the left side, the posterior-superior lateral condyle has an area of porosity and slight spicule formation. This area of cartilage erosion has a transverse width of 13 mm and a vertical height of 7 mm.

Patellae

Both patellae are present and complete. Neither bone is labeled. The patellae do not exhibit pathological changes with the exception of very minor degenerative lipping along the joint margins.

Tibiae and Fibulae

The left tibia is nearly complete and represented by proximal (97.L.20a), midshaft (97.L.20b), and distal (97.L.20c) segments. The anterior margin of the proximal joint is missing. Bone sampling in the upper third of the left tibia by the government team removed a large section of the anterior, proximal diaphysis including the lateral half of the tibial tuberosity. The area affected by this dissection is 146 mm in length by 18.7 mm in breadth with a maximum depth of 18 mm (Figure 6.10).

The right tibia has postmortem damage, but is fairly complete. Four pieces respresent the bone: the medial half of the proximal joint (97.R.20a), the lateral half of the proximal joint (97.R.20b), upper midshaft (97R.20c), and distal end (97R.20d). The anterior metaphysis is missing from the two pieces of proximal epiphysis.

The soleal line, the attachment for the superficial and deep foot flexor muscles on the posterior surface of the proximal left tibia, is moderately raised with a maximum width of 5 mm, a height of 2.5 mm, and a length of 69 mm (Figure 7.43). The shaft of the right tibia also has a well-defined line including a very slightly devel-

FIGURE 7.43 The soleal ridge (arrows), the origin of a superficial flexor muscle of the foot, and a line marking the origin of the deeper calf flexor muscle, tibialis posterior (arrowheads), are evident on the posterior surface of the left tibia.

oped enthesophyte. This raised ridge courses superiorly and diagonally across the proximal posterior shaft for 54 mm with a maximum width of 2.6 mm, and a height of 1.6 mm. The adjacent lateral attachment area for the tibialis posterior muscle is demarcated by a vertically oriented ridge that measures 33 mm in length, with a width of 3 mm and a height of 1.7 mm.

The proximal joint of the left tibia (97.L.20a) has very slight lipping along the margins of the tubercles of the intercondylar eminence. The posterior aspect of the lateral condyle has slight lipping and a localized area of macroporosity on the joint surface measuring 14.5 mm transverse by 8 mm anterior-posterior. A small squatting facet is located on the anterior rim of the distal tibio-talar joint. This notch has a width of 9 mm; only a small zone along the joint margin was affected.

The posterior surface of the lateral condyle of the right proximal tibia has localized spicule formation and microporosity from knee activity wear and tear. The affected area has a transverse width of approximately 8 mm and an anterior-posterior measurement of 6 mm. The lateral tubercle of the intercondylar eminence has extremely minor degenerative lipping. The mid-lateral third of the anterior margin of the tibiotalar joint has a small squatting facet characterized by a depressed, semi-lunar notch with a width of 8 mm and a superior-inferior height of 2.5 mm.

Posterior articular surface expansion and erosion on the proximal tibiae correspond with deterioration on the femoral condyles caused by tight flexion at the knees. Hyperflexion from habitual squatting likely caused these changes. The presence of squatting facets on the distal tibiae lends some support to this interpretation (Chapter 14).

The left fibula is complete with five segments present: the proximal joint (97.L.21a), the distal joint (97.L.21e), and the diaphysis formed by three fragments (97.L.21d, 97.L.21c, and 97.L.21b). The right fibula is also complete and represented by five pieces: the proximal joint (97.R.21a), the distal joint (97.R.21e), and the diaphysis formed by three fragments (97.R.21b, 97.R.21c, and 97.R.21d). No pathological changes are evident.

Foot Bones

The calcanei and tali are present. The medial surface of the left calcaneus (97.L.22a) exhibits recent postmortem loss of the sustentaculum tali including all of the middle articular facet and the medial surface below it. The right calcaneus and left and right tali are complete. The right calcaneus has slight degenerative lipping of the joint margins. Additional information on the tarsals, metatarsals, and phalanges are presented in Chapter 13.

AGE

Determining chronological age from the bones of middle-aged to older adults is one of the more challenging tasks faced by bioarchaeologists and forensic anthropologists. This process is especially difficult when working with prehistoric remains of people who experienced lifestyles and activities that differed greatly from contemporary groups upon which skeletal aging standards are based. In addition to features in bone, both dental attrition and pathology (e.g., numbers of abscesses, antemortem tooth loss, and alveolar margin resorption) can be informative. But again, aging standards do not exist for ancient populations and, of course, would need to be tailored for temporal and regional differences in diets, food processing techniques, and general levels of abrasive contaminants in foods. As such, dental age interpretations must be approached cautiously when estimating the age of a single individual.

The skeleton of Kennewick Man presents conflicting evidence with regard to his age, resulting in varying age estimates. Powell and Rose (1999) assigned an age of 45 to 50 years based on pubic symphysis and ilium auricular surface morphology and ectocranial suture closure. Chatters (2000, 2001) estimated 35 to 45 years using several criteria including suture closure, sternal end morphology of a rib, and the innominates.

Overall, Kennewick Man's skeleton exhibits complete growth indicative of an adult. The sternal epiphysis of the clavicles and iliac crests are fully united with no remnant lines of union. Few characteristics of advanced age are noted and most of the skeletal features are consistent with middle age.

Sternal rib end depth and shape, rim morphology, and the overall texture and quality of the bone can be reliable indicators of age (İşcan et al. 1984) despite the subjective nature of gross observation (Fanton et al. 2010). The sternal ends of Kennewick Man's five right ribs show only slight cupping, including rib 1 (97.I.12e(3)), rib 2 (97.I.12a(7)), rib 3 (97.I.12b(5)), rib 7 (97.I.12b(13)), and rib 8 (97.I.12b(14)) (Figure 7.44). No irregular bony projections are noted along the sternal end rims. These can occur with advancing age from the ossification of costal cartilage (İşcan et al. 1984; Kerley 1970; McCormick 1980; Semine and Damon 1975).

Age-related changes in the pubic symphysis and auricular surface of the ilium are considered among the more reliable criteria for estimating age-at-death in adults (Buikstra and Ubelaker 1994; Lovejoy et al. 1985; Meindl and Lovejoy 1989). The innominates of Kennewick Man exhibit characteristics indicative of early middle age. The ventral rims and upper extremities of the symphyseal surfaces appear to have been forming, rather than breaking down, with retention of ridges (billows) (Figure 7.45). Observed porosity is not age-related degeneration, but

FIGURE 7.44 Sternal rib cupping in the right second (A) and eighth (B) ribs. The degree of cupping and absence of bony projections from the rim suggest a younger middle age.

rather a postmortem alteration. Transverse organization and residual striae are still evident on the auricular surfaces with no apical margin activity (Figure 7.46). Again, surface porosity is due to postmortem deterioration.

Degenerative changes in the Kennewick Man skeleton are minor and not especially suggestive of age-advanced osteoarthritis. In the spine, degenerative changes include a small osteophyte on the third cervical vertebra and trace lipping of the anterior-inferior endplate of the sixth cervical vertebra. The seventh cervical vertebra has a remnant line of the anterior epiphyseal ring. The seventh thoracic vertebra displays slight concavity of the body, but otherwise body heights of the thoracic vertebrae are normal. Of the lumbar vertebrae, the endplate of the fourth has slight arthritic lipping. No evidence of pathological degeneration was observed on the represented portion of the sacral promontory.

Minor degenerative changes in the shoulder and long bone joints show patterning indicative of activity-induced wear and tear. The left glenoid of the scapula shows no degenerative changes, but the right glenoid fossa shows lipping and porosity of the joint margin and surface along with an associated rim fracture. The right arm bones show slight osteophytic spicule formation in the distal humeral joint, spicule formation and porosity of the adjacent radial head, and slight arthritic lipping of the radial notch. Slight lipping of the trochlear notch of the right ulna and distal joint of the right radius is also present. These degenerative changes are not repeated to the same degree in the left arm, which exhibits only trace arthritic lipping in the left radial head and the coracoid process of the ulna. The pelvic acetabulae are free of degenerative changes and neither of the iliac crests shows entheses formation. The femora display slight rim definition of the fovea capitis and the lateral condyles of the distal femora exhibit minor porosity. Corresponding porosity and spicular changes on the tibia joint surfaces link the degeneration to habitual behaviors involving the knee joints.

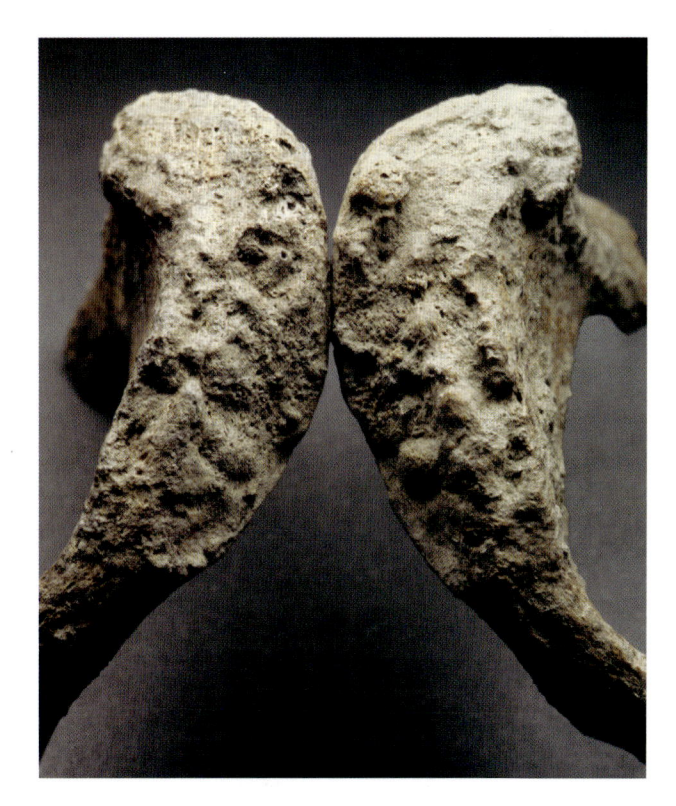

FIGURE 7.45 The pubic symphyseal surfaces of Kennewick Man have residual billowing and incomplete development of the ventral rims.

FIGURE 7.46 The auricular surfaces of the innominates show retention of residual striae with no significant apical rim development. Along with the pubic symphyseal surfaces, they indicate early middle age.

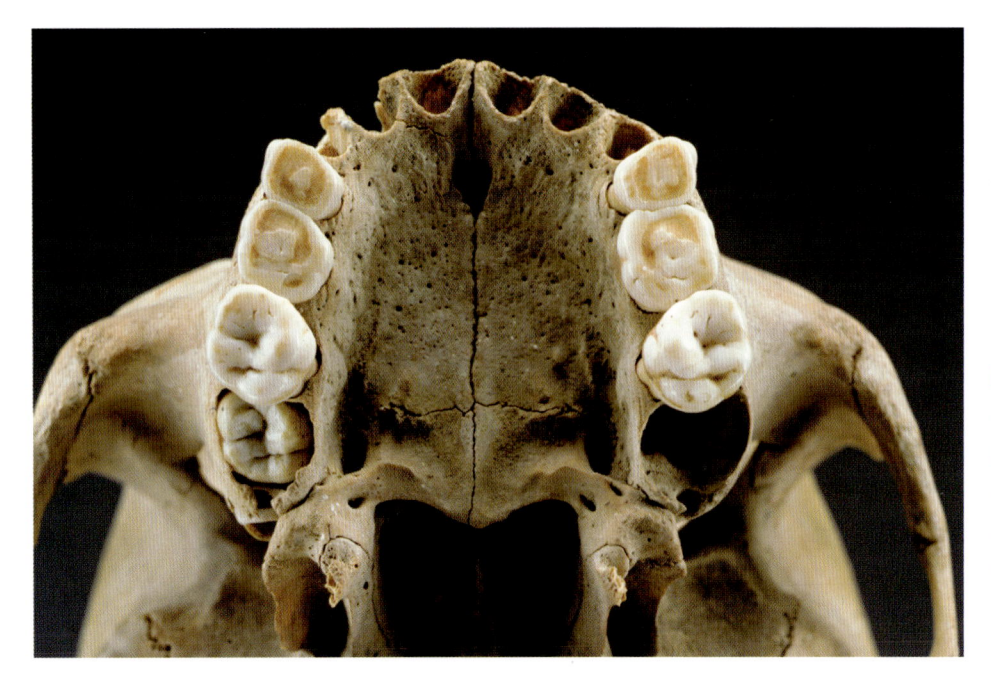

FIGURE 7.47 A rapid rate of tooth wear is indicated by dentin exposure in the deciduous first and second molars of a 9 year old child (AHUR 743) from the Grimes Shelter, Nevada, dated 9470 ± 60 RC yr. BP.

Advanced cranial suture closure and dental attrition are features in Kennewick Man suggesting an older age. There is almost complete obliteration of the cranial sutures; however, cranial suture closure has a high degree of variability. Kennewick Man's tooth wear is moderately heavy, but can be attributed in part to the use of teeth in task activities and to a diet containing abrasive contaminants. Paleoamerican children, for example, illustrate that tooth wear began early and progressed rapidly (Figure 7.47).

Examination of the tooth sockets may aid in assessing age. In general, the gingiva and alveolus tend to recede with advancing age, even in the absence of periodontal disease. In Kennewick Man, the level of alveolar ridge recession, as measured from the cemento-enamel junction to alveolar ridge, averages 3.3 mm for the four molars that can be measured. This modest amount, along with the loss of only one tooth in life, seems inconsistent with a somewhat advanced age.

The skeleton exhibits evidence of physical activity involving vigorous use of the right arm that resulted in damage to the right shoulder joint. Traumatic injuries resulting in an embedded stone point in the right hip bone and multiple fractures to the ribs, mostly on the right side, show advanced healing without chronic infection. Bone remodeling and development suggest a sustained level of physical activity subsequent to injury. Based on this assessment of dental and skeletal features, Kennewick Man's age was about 40 years, possibly 35–39 years, when he died.

Future studies, drawing from both inventory and interpretation, may refine these results as technology and methodology continue to develop.

REFERENCES

References listed below under the headings "COE" (with a number rather than a date) are documents from the administrative records filed with the district court by the Army Corps of Engineers.

Anderson, James M. 2001. Biological responses to materials. *Annual Review of Materials Research* 31:81–110.

Bastos, Renata, John H. Calhoon, and Clinton E. Baisden. 2008. Flail chest and pulmonary contusion. *Seminars in Thoracic and Cardiovascular Surgery* 20(1):39–45.

Beeson, Augustin, and Frederic Saegesser. 1983. *Color Atlas of Chest Trauma and Associated Injuries Volume 1*. Medical Economics Books, Oradell, NJ.

Bjerke, H. Scott. 2009. Flail Chest. eMedicine. http://emedicine.medscape.com/article/433779-overview.

Borman, J. B., L. Aharonson-Daniel, B. Savitsky, and K. Peleg. 2006. Unilateral flail chest is seldom a lethal injury. *Emergency Medicine Journal* 23(12):903–905.

Buikstra, Jane E., and Douglas H. Ubelaker, eds. 1994. *Stan-*

dards for Data Collection from Human Skeletal Remains. Research Series 44. Arkansas Archaeological Survey, Fayetteville.

COE 4323. Inventory Observations on the Kennewick Man Collection 10/28–29, 1998 [Author Douglas W. Owsley]

Chatters, James C. 2000. The recovery and first analysis of an early Holocene human skeleton from Kennewick, Washington. *American Antiquity* 65(2):291–316.

———. 2001. *Ancient Encounters: Kennewick Man and the First Americans.* Simon & Schuster, New York.

Coues, Elliott. 1866. Some notes on arrow-wounds. *The Medical and Surgical Reporter* 14:321–324.

Donath, Karl, Monika Laaß, and Hans-Joachim Günzl. 1992. The histopathology of different foreign-body reactions in oral soft tissue and bone tissue. *Virchows Archiv A Pathological Anatomy and Histopathology* 420:131–137.

Fanton, Lauren, Marie-Paule Gustin, Ulysse Paultre, Bettina Schrag, and Daniel Malicier. 2010. Critical study of observation of the sternal end of the right 4th rib. *Journal of Forensic Sciences* 55(2):467–472.

Gardenbroek, T. J., M. Bemelman, and L. P. H. Leenan. 2009. Pseudarthrosis of the ribs treated with a locking compression plate, a report of three cases. *Journal of Bone and Joint Surgery* 91(6):1477–1479.

Grant, J. C. Boileau. 1972. *An Atlas of Anatomy.* 6th edition. The Williams and Wilkins Co., Baltimore.

Gregg, John B., Larry J. Zimmerman, James P. Steele, Helen Ferwerda, and Pauline S. Gregg. 1981. Ante-mortem osteopathology at Crow Creek. *Plains Anthropologist* 26(94):287–300.

Guttentag, Adam R., and Julia J. Salwen. 1999. Keep your eyes on the ribs: The spectrum of normal variants and diseases that involve the ribs. *RadioGraphics* 19:1125–1142.

Hodgkinson, D. W., B. R. O'Driscoll, P. A. Driscoll, and D. A. Nicholson. 1993. ABC of emergency radiology. Chest radiographs: II. *British Medical Journal* 307(6914):1273–1277.

Hoppa, Robert D., Laura Allingham, Kevin Brownlee, Linda Larcombe, and Gregory Monks. 2005. An analysis of two late Archaic burials from Manitoba: The Eriksdale site (EfLl-1). *Canadian Journal of Archaeology* 29(2):234–266.

Hukuda, Sinsuke, Koji Inoue, Toshio Ushiyama, Yasuo Saruhashi, Atushi Iwasaki, Jie Huang, Akira Mayeda, Masashi Nakai, Fang Xiang Li, and Zhao Qing Yang. 2000. Spinal degenerative lesions and spinal ligamentous ossifications in ancient Chinese populations of the Yellow River civilization. *International Journal of Osteoarchaeology* 10(2):108–124.

İşcan, M. Yaşar, Susan R. Loth, and Ronald Wright. 1984. Age estimation from the rib by phase analysis: White males. *Journal of Forensic Sciences* 29(4):1094–1104.

Kerley, Ellis R. 1970. Estimation of skeletal age: After about age 30. In *Personal Identification in Mass Diasters*, edited by T. D. Stewart, 57–70. Smithsonian Institution, Washington, DC.

Kirsch, Marvin M., and Hebert Sloan. 1977. *Blunt Chest Trauma. General Principles of Management.* Little, Brown, Boston.

Liman, Serife Tuba, Akin Kuzucu, Abdullah Irfan Tastepe, Gulay Neslihan Ulasan, and Salih Topcu. 2003. Chest injury due to blunt trauma. *European Journal of Cardio-thoracic Surgery* 23(3):374–378.

Lovejoy, C. Owen, Richard S. Meindl, Thomas R. Pryzbeck, and Robert P. Mensforth. 1985. Chronological metamorphosis of the auricular surface of the ilium: A new method for the determination of adult skeletal age at death. *American Journal of Physical Anthropology* 68(1):15–28.

Mann, Robert W., and Sean P. Murphy. 1990. *Regional Atlas of Bone Disease: A Guide to Pathological and Normal Variation in the Human Skeleton.* Charles C. Thomas, Springfield, IL.

McCormick, W. F. 1980. Mineralization of the costal cartilages as an indicator of age: Preliminary observations. *Journal of Forensic Sciences* 25(4):736–741.

McFarland, Edward G., Miho J. Tanaka, and Derek F. Papp. 2008. Examination of the shoulder in the overhead and throwing athlete. *Clinics in Sports Medicine* 27(4):553–578.

Meindl, Richard S., and C. Owen Lovejoy. 1989. Age changes in the pelvis: Implications for paleodemography. In *Age Markers in the Human Skeleton*, edited by M. Yaşar İşcan, 137–168. Charles C. Thomas, Springfield, Illinois.

Milner, George R. 2005. Nineteenth-century arrow wounds and perceptions of prehistoric warfare. *American Antiquity* 70(1):144–156.

Nathan, Hilel. 1959. The para-articular process of the thoracic vertebrae. *Anatomy Records* 133(4):605–618.

Ortner, Donald J. 2003. *Identification of Pathological Conditions in Human Skeletal Remains.* 2nd edition. Academic Press, New York.

Otis, George A. 1883. *A Report of Surgical Cases Treated in the Army of the United States from 1865 to 1871.* Circular No. 3,

War Department, Surgeon General's Office. Government Printing Office, Washington, DC.

Owsley, Douglas W., and Richard L. Jantz. 1989. *A Systematic Approach to the Skeletal Biology of the Southern Plains, from Clovis to Comanchero: Archaeological Overview of the Southern Great Plains*. Arkansas Archaeological Survey Research Series 35. Arkansas Archaeological Survey, Fayetteville.

Polet, Caroline, Olivier Dutour, Rosine Orban, Ivan Jadin, and Stéphane Louryan. 1996. A healed wound caused by a flint arrowhead in a Neolithic human innominate from the Trou Rosette (Furfooz, Belgium). *International Journal of Osteoarchaeology* 6:414–420.

Powell, Joseph F., and Jerome C. Rose. 1999. Chapter 2. Report on the Osteological Assessment of the "Kennewick Man" Skeleton (CENWW97. Kennewick). In *Report on the Non-Destructive Examination, Description, and Analysis of the Human Remains from Columbia Park, Kennewick, Washington* [October 1999]. National Park Service, http://www/nps.gov/archeology/kennewick/powell_rose.htm.

Qasim, Zaffer, and Carl Gwinnutt. 2009. Flail chest: Pathophysiology and management. *Trauma* 11(1):63–70.

Roger, Bernard, Abdalla Skaf, Andrew W. Hooper, Nittaya Lektrakul, Lee RenYeh, and Donald Resnick. 1999. Imaging findings in the dominant shoulder of throwing athletes: Comparison of radiography, arthrography, CT arthrography, and MR arthrography with arthroscopic correlation. *American Journal of Roentgenology* 172(5):1371–1380.

Rogers, N. L., L. E., Flournoy, and W. F. McCormick. 2000. The rhomboid fossa of the clavicle as a sex and age estimator. *Journal of Forensic Sciences* 45(1):61–67.

Rosenberg, Andrew. 1999. Bones, joints, and soft tissue tumors. In *Pathological Basis of Disease*, edited by Ramzi S. Cotran, Vinay Kumar, and Tucker Collins, 1215–1268. W.B. Saunders Company, Philadelphia.

Ryan, Timothy M., and George R. Milner. 2006. Osteological applications of high-resolution computed tomography: A prehistoric arrow injury. *Journal of Archaeological Science* 33:871–879.

Semine, A. A., and A. Damon. 1975. Costochondral ossification and aging in five populations. *Human Biology* 47:101–106.

Silver, Ian A. 1984. Cellular microenvironment in healing and non-healing wounds. In *Soft and Hard Tissue Repair: Biological and Clinical Aspects*, edited by Thomas K. Hunt, R. Bruce Heppenstall, Eli Pines, and David Rovee, 50–66. Praeger, New York.

Smith, B. Holly. 1984. Patterns of molar wear in hunter-gatherers and agriculturalists. *American Journal of Physical Anthropology* 63(1):39–56.

Spector, Myron, Cynthia Cease, and Xia Tong-Li. 1989. The local tissue response to biomaterials. *Critical Reviews in Biocompatibility* 5:269–295.

Stanford, Dennis. 1979. Bison kill by ice age hunters. *National Geographic* 155(1):114–119.

USGS Volcano Hazards Program. 2013. "USGS: Volcano Hazards Program Glossary." Accessed July 24. http://volcanoes.usgs.gov/vsc/glossary.html.

Vegas, José Ignacio, Ángel Armendariz, Francisco Etxeberria, María Soledad Fernández, and Lourdes Herrasti. 2012. Prehistoric violence in northern Spain: San Juan ante Portam Latinam. In *Sticks, Stones, and Broken Bones: Neolithic Violence in a European Perspective*, edited by Rick Schulting and Linda Fibiger, 265–302. Oxford University Press, Oxford.

Walker, Phillip L., Clark Spencer Larsen, and Joseph F. Powell. 2000. Chapter 5: Final Report on the Physical Examination and Taphonomic Assessment of the Kennewick Human Remains (CENWW.97.Kennewick) to Assist with DNA Sample Selection. In Report on the DNA Testing Results of the Kennewick Human Remains from Columbia Park, Kennewick, Washington [September 2000]. National Park Service, http://www.nps.gov/archeology/kennewick/walker.htm.

Wilson, Thomas. 1901. Arrow wounds. *American Anthropologist* 3(3):513–531.

Zobeck, T. S. 1983. "Postcraniometric Variation among the Arikara." Ph.D. dissertation, University of Tennessee, Knoxville.

Dentition

Christy G. Turner II

This examination of Kennewick Man's dentition was based on procedures and observational methodology defined in the Arizona State University Dental Anthropology System, Tempe (ASU DAS) (Turner et al. 1991). This system is designed both to deal with commonly worn, broken, and usually incomplete archaeologically recovered dental remains and to provide physical standards to help minimize within- and between-observer error so that rigorous comparisons can be made for a series of nonmetric crown and root morphological traits at individual, local, regional, and global scales. The system is widely used by dental anthropologists and bioarchaeologists concerned with affinity assessment and microevolution of anatomically modern humans.

In this system, the morphological traits which show a strong genetic component in both occurrence and expression have been analyzed by classic and statistical procedures (Scott 1973; Harris 1977; Nichol 1990; Turner 1991). Their inherited quality makes the system valuable for highly comparable, inexpensive, biological affinity assessment through time as well as space, provided enough traits are utilized for sound statistical analysis. Most important, when sample sizes are sufficient, the findings almost always meet expectations based on independent archaeological, linguistic, and ethnographic information. Major works that illustrate the utility of the system include Irish (1993), Haeussler (1996), Weets (1996), Lipschultz (1996), Scott and Turner (1997), Hawkey (1998), Burnett (1998), Lee (1999), Lincoln-Babb (1999), Adler (1999, 2005), Reichart (2000), Ullinger (2002), and Martin (2003).

Ancient skeletal remains or slightly more recent finds in North America are, with a few exceptions, markedly incomplete. The exceptions studied with the Arizona State system include Gordon Creek Woman in Colorado (Breternitz et al. 1971), Horn Shelter No. 2 near Waco, Texas (Redder 1985), and Kennewick Man.

OBSERVATIONS

Kennewick Man had a normal set of permanent teeth, with no congenitally absent, dwarfed, malformed, maloccluded, impacted, retained deciduous, or supernumerary teeth—conditions that might suggest developmental problems in fetal life or childhood. The well-preserved dentition is consistent with an intentional, encapsulating, and protective burial as suggested by Owsley (Doughton 2006), rather than with a corpse abandoned on an exposed ground surface or a river-transported and beached carcass exposed to various destructive physical and biological taphonomic agencies.

The first examination of Kennewick Man's teeth and skeleton was made in 1996 by Chatters (2001). The next examination was carried out under the supervision of the National Park Service by Joseph F. Powell and Jerome C. Rose, both well known for their work in dental anthropology (Powell in morphology and Rose in pathology). Their findings were posted on the National Park Service web page (Powell and Rose 1999); however, there was no peer review of this report. Kennewick Man's teeth have been examined with differing techniques by these scholars. In general, the amount of dental morphological information obtainable is dependent to a large degree on the severity of tooth wear, oral pathology, and survivorship.

TOOTH WEAR

Most of Kennewick Man's tooth crowns are severely worn, making the roots near the crown-root junction the functional chewing surfaces of the majority of the remaining teeth (Figures 7.7 and 7.8). In other words, there is no crown morphology left to determine the expression of the Arizona State system key crown features.

With the extreme wear, even the crown-size measurements, often used in affinity assessment, could not be made. Powell and Rose (1999) did conduct a comparative odontometric analysis of Kennewick Man, but their analysis has no scientific value. The measurements have to be small and are not comparable with any published crown measurements that are based on unworn or minimally worn teeth.

Table 8.1 presents the wear scores for each tooth. The mean wear score for the observed 30 teeth is 3.6, a value indicating that nearly all the crowns have worn off. These findings can be compared to those from a study of tooth wear in 29 living Batak hunter-gatherers of the Philippines (Turner and Eder 2006). Their sex and ages are known and their diet and tooth-use behavior were observed by Eder to be only moderately affected by surrounding agricultural groups. The largest mean wear score for upper teeth was 2.4, which occurred in a 50-year-old female. For lower teeth, the largest Batak mean wear score was 2.1, found in a 46-year-old male.

Although dental wear varies with diet and tooth use, both within and between population groups, Kennewick Man's high wear score as compared with that of the Batak forcefully suggests that he was not a young adult (ca. 18–30 years) but more likely middle-aged (ca. 31–45 years) when he died and that his tooth-use behavior and dietary materials were more abrasive than those of the Batak. Schmucker (1983) found very pronounced wear and dentine exposure in the molars of middle-aged (31–45 years) hunter-gatherers of prehistoric California whose ages at death were estimated by osteological criteria. All Kennewick Man's observable pulp chambers had filled in with secondary dentine, another indication that he was past young adulthood at the time of death.

There is minor rounding of the otherwise flatly worn crowns on the lingual-occlusal borders of the lower-left and right-central incisors and the lower-left lateral incisor (Figure 7.11a). In addition, there is slight rounding of the buccal-occlusal border of the upper-right first molar. This rounding is presumably task-related. Tasks such as preparation of string or twine or repeatedly holding some material such as fishing line in the mouth come quickly to mind. Some manner of hide preparation or meat cutting is equally possible. The rounded right-molar wear hints at right-handedness.

There is very little heavy scratching of the occlusal surface as far as could be determined with a 20-power hand lens. Activity-induced chipping or fracturing—found frequently in prehistoric North American Aleuts and Eskimos, and to a lesser degree, northern Indians (Turner and Cadien 1969)—are nearly absent in Kennewick Man. Only one small, chipped area was found on the lower-left second premolar, a tooth in the postcanine dental arch that is commonly used as a vice or grip in combination with its opposing tooth in the upper jaw.

ORAL PATHOLOGY

There are no identifiable carious lesions of the tooth crowns, although the lower-left canine has what appears to be minor erosion on the buccal surface at the crown-root junction. Caries of the crowns occur infrequently in hunter-gatherers (Turner 1979; Schollmeyer and Turner 2004). Studies of prehistoric hunter-gatherers of northern California (Schmucker 1983; Adler and Turner 2000) found caries were more common in the acorn-using groups than in those groups whose food refuse suggests that they subsisted largely on marine and riverine resources. The oral health and tooth wear of

TABLE 8.1 Kennewick Man's dental wear.

I1R	I1L	I2R	I2L	CR	CL	P1R	P1L	P2R	P2L	M1R	M1L	M2R	M2L	M3R	M3L
								Upper Jaw							
4	4	4	4	4	4	4	4	4	4	4	4	3.3	3.5	A	1.5
								Lower Jaw							
4	4	4	4	3.5	3.5	3.5	3.5	3.5	3.5	3.5	3.5	3.5	3.5	1.5	P

Tooth ID: I, incisor; C, canine; P, premolar; M, molar. L, left; R, right. Wear scale: 0, no wear; 1, dentine exposed; 2, cusps worn off; 3, pulp exposed; 4, root stumps functional. A denotes antemortem loss; P denotes postmortem loss.

Kennewick Man indicate that his diet was noncariogenic, fairly abrasive, and likely comprised primarily of animal tissue including aquatic species.

Kennewick Man was free of other oral health considerations, such as generalized periodontal disease, periodontal pockets, and alveolar abscessing. Such is typically not the case for Aleuts and Eskimos, who subsisted almost exclusively on marine resources. Despite their lack of dental caries, they commonly suffered from abscessing and periodontal disease (Turner 1991).

CULTURAL TREATMENT

The Arizona State coding system indicates two kinds of cultural treatment: intentional modification of labial surfaces of anterior permanent teeth such as filing, incising, coloring, or ablation, and unintentional, mainly interproximal, toothpick grooves. The former could not be assessed, with the exception of ablation, because of wear. There were no identifiable examples of the latter as far as the adhering interproximal calcified earth matrix permitted study. Cases of toothpick grooves have been found in Archaic crania in North America and even earlier in the Old World (Turner and Cacciatore 1998).

CROWN AND ROOT MORPHOLOGICAL TRAITS

Little morphological information could be gleaned from the severely worn tooth crowns. For the maxillae, only the central incisors and the left third molar provided anatomical information. The incisors were apparently straight in their alignment, not in the V-shaped position called "winging." This interpretation is based on the relatively flat labial surface of the incisor roots that are parallel to the arch in the nonwinged expression. Third molars are generally less useful for morphological comparisons because of their lack of developmental stability (Dahlberg 1951). For the record, the upper-left third molar had a grade 3.5 metacone; apparent absence of the hypocone (grade 0); no Carabelli trait expression (grade 0); and no parastyle (grade 0). For the mandible, the lower-right third molar has six cusps; no deflecting wrinkle (grade 0); no protostylid (grade 0); and cusp 6 has grade 3 expression.

Linear extensions of crown enamel can project toward the root tip on the buccal surfaces of postcanine teeth. For Kennewick Man, enamel extensions could be scored for the upper-left and upper-right second molars (grade 1), and left third molar (grade 0). For the lower teeth, scores are: left and right first and second premolars (grade 0), right first molar (grade 3), left second molar (grade 2), and right third molar (grade 0).

The X-ray films were not informative about root number due to the presence of radio-opaque clay or silt that had infiltrated the spaces between the roots and alveolar sockets (University of Washington Hospital, Seattle). Table 8.2 interprets the available films. Detection of root number without X-rays was possible only for the maxillary left third molar (1 root, 5 radicals). For the mandibular teeth it appears that the left and right first and second molars had two roots each, and the right third molar possessed one root. The left third molar may have two roots.

Three osteological traits were absent (grade 0) in Kennewick Man: palatine torus, mandibular torus, and rocker jaw.

INTERPRETATION

Kennewick Man's dentition was examined for a series of standardized crown and root morphological traits according to the Arizona State system and then with those of thousands of prehistoric and historic North and South Americans, Arctic Eskimos and Aleuts, Siberians, Northeast and Southeast Asians, and the peoples of Oceania including Australia and New Guinea.

From these global surveys emerged a multiregional hypothesis that, in part, envisions two major dental divisions in eastern Asia and the Americas (Turner 1983) and elsewhere. The term *Sundadonty* designates the older set of dental trait frequencies that occurs all over prehistoric and recent Southeast Asia, except for recent Chinese emigrants. Sundadonty is also found in Polynesia and in the Jōmon-Ainu of Japan. *Sinodonty* refers to the trait frequency pattern present throughout prehistoric and recent Northeast Asia and *all* of the Americas. Sundadonty is characterized by a relatively simple and retained pattern, whereas Sinodonty has a more specialized and complex character. These differences can be seen, for example, in the values for three-rooted lower first molars and pronounced upper-incisor shoveling. Both are rela-

TABLE 8.2 Radiographic assessment of Kennewick Man's tooth root number.

Film No.	Jaw	Tooth	Root Number	Radical Number	Comment
SE1 M11	Upper	All	Indeterminate	Indeterminate	
SE1 M12	Upper	L&RI1	1	Indeterminate	
SE1 M13	Upper	All	Indeterminate	Indeterminate	
SE1 M14	Upper	All	Indeterminate	Indeterminate	
SE1 M15	Upper	All	Indeterminate	Indeterminate	
SE1 M16	Upper	All	Indeterminate	Indeterminate	
SE1 M17	Upper	All	Indeterminate	Indeterminate	
Visual	Upper	LM3	1	5	Loose tooth
SE1 M9	Lower	LC	1	Indeterminate	
		LM1	2	Indeterminate	
		LM2	2	Indeterminate	
SE1 M10	Lower	L&RC	1	Indeterminate	
		LP1	1	3 or 4*	
		RP1	1	Indeterminate	
		L&RP2	1	Indeterminate	
		L&RM1	2	Indeterminate	
		L&RM2	2	Indeterminate	
		LM3	2?	Indeterminate	
		RM3	1	3	

Note: Radical number based on direct visual observation.
*LP1 has a grade 3 or greater expression of Tome's root.

tively more common in Sinodonts and less frequent in Sundadonts. There are several other differences between Sundadonty and Sinodonty, but the eight traits listed in Table 8.3 are most significant, statistically speaking.

Comparing Table 8.3 with Table 8.4 shows that the teeth of North and South Americans are more like those of Northeast Asians than Southeast Asians. Even though the means of the frequencies of three-rooted lower first molars in North and South America are lower than those of Northeast Asia and the American Arctic, they are nevertheless identical (6.0%). While there are statistically significant differences in some of the key trait frequencies in North and South Americans, the differences are usually not great, that is, for Trait 8, $\chi^2 = 8.4$, 1. d.f., $P < 0.01$. Along with most other dental morphological traits, this suggests that North and South America were colonized by a single founding population. The slight dental frequency differences from Northeast Asian Sinodonts could have been caused by environmental restrictions on migrant group size in the Beringian transit

TABLE 8.3 The eight major features of Sundadonty and Sinodonty.

Trait	Sundadonty Mean %	Sinodonty Mean %
1. UI1 shovel	30.8	71.7
2. UI1 double-shovel	22.7	55.8
3. UP1 single root	70.6	78.8
4. UM1 enamel extension	26.4	50.1
5. UM3 P/R/CA*	16.3	32.4
6. LM1 deflecting wrinkle	25.5	44.1
7. LM1 three roots	8.8	24.7
8. LM2 four cusps	30.7	15.5

Source: Turner 1990. Includes many East Asian locations plus Arctic Eskimo and Aleut; omits American Indians.
*P/R/CA denotes peg-shaped, reduced in size, congenital absence complex of the upper third molar.

TABLE 8.4 Dental morphological traits of American Indians and Kennewick Man.

Trait	North American Indians* Mean %	South American Indians* Mean %	Kennewick Man
1. UI1 shovel	93.5	92.0	?
2. UI1 double-shovel	81.5	90.0	?
3. UP1 single root	85.5	87.0	?
4. UM1 enamel extension	40.7	49.0	?
5. UM3 P/R/CA*	18.7	25.0	0
6. LM1 deflecting wrinkle	38.2	38.0	?
7. LM1 three roots	6.0	6.0	0 (two roots)
8. LM2 four cusps	11.6	8.0	?

Source: More than 4,000 individuals are in the North and South American data sets, including Archaic assemblages. North America includes eastern U.S. and Canada, Southwest U.S., California, and Mesoamerica (Aleut-Eskimo are excluded). South America includes Ecuador, Peru, Chile, and Brazil.

*P/R/CA denotes peg-shaped, reduced in size, congenital absence complex of the upper third molar.

(founder's effect). Or there could have been slightly different gene pools in the Siberian river basins of the Lena, Ob, Yenisei, Angara, and Amur, as might be expected since late-Pleistocene alpine glaciers blocked mountain passes, except in the south near present-day Mongolia.

Repeated sampling of South American preceramic and preagricultural skeletal assemblages, which include Brazil (Lagoa Santa, Corondo, and Sambaqui Boguassi), Chile (Cuchipuy, Cerro Sota, Palli Aike, and Punta Teatinos), Ecuador (Santa Elena), and Peru (La Paloma, Huaca Prieta), turned up no indications of Sundadonty in recent Patagonian crania as claimed by Lahr (1995). In preceramic and preagricultural North American crania, all dental samples appear to be Sinodonts.

Without identifying which traits they used, Powell and Rose (1999) suggest that Kennewick Man's dental morphology strongly resembles that found in "Polynesia and southern Asia." Since Polynesians and Southeast Asians possess the Sundadont dental pattern, Powell and Rose are implying that Kennewick Man is a Sundadont although they explicitly state that "it is simply not possible to attribute Kennewick Man's dental discrete traits to either the Sinodont or Sundadont groups based on gross morphological observations" (Powell and Rose 1999). Why do these authors, along with Chatters (1996, 2001:182), imply that they have identified Sundadonty in North and South American crania? I believe the answer has two parts.

Kennewick Man provides an empirical example of faulty observations, because these authors have not given crown wear the consideration it deserves. Some of the crown traits, such as shoveling, appear simpler on worn than unworn teeth, which can mislead the observer into identifying Sundadonty rather than Sinodonty. Ignoring wear is most undeniably shown in the Powell and Rose (1999) tooth size analysis of essentially crownless teeth. To his credit, Chatters (2001:182) lists the traits that he used to identify Sundadonty in Kennewick Man: two-rooted upper first premolars, two-rooted lower first and second molars, full-sized third molars, and non-winged upper central incisors. Chatters is correct on these identifications, with the exception of the upper first premolar root number. From the X-ray films available, the number of roots could not be determined. More importantly, of the eight key traits that distinguish Sundadonty from Sinodonty, Chatters could work with only three (two if the premolar root number is ambiguous). This is an insufficient number from which to make a determination, especially for traits that tend to have intermediate frequencies rather than distinctively high or low occurrences, such as the Bushman canine or the Uto-Aztecan premolar. Moreover, Kennewick Man is a single individual so one must recognize that his trait expressions may not match the average trait frequencies of his contemporaries. Kennewick Man's teeth cannot be classified as being Sinodont, Sundadont, or any of the other major world dental patterns (Scott and Turner 1997).

Ever since the discovery of Clovis stone spear points in stratigraphic association with extinct mammoths, there has been much speculation and numerous claims of a pre-Clovis occupation of the New World (Haynes 2002; Fiedel 2004), both by archaeologists and physical anthropologists. Some physical anthropologists have claimed craniometric resemblances between Kennewick Man and Caucasians (Chatters 2001); Kennewick Man and Ainu and Polynesians (Powell and Rose 1999); and at least one claim for resemblances to Africans and Australian Aborigines (Neves et al. 1999). In a principal components analysis of the skull (mainly size), Chatters, Neves, and Blum (Chatters 2001:177–179) found Kennewick Man to be most like Polynesian crania of New Zealand and, to a lesser degree, Polynesian crania from Mokapu, Hawaii, Easter Island, and a series of Ainu from Japan. The similarity is, however, *less* than that calculated between Zulus and Tasmanians, and *less* than that between Peruvians and Europeans—hardly compelling findings for speculations about origins and affinity. For component II (mainly shape), Kennewick Man is clearly an outlier; that is, he resembles no one closely in the Chatters, Neves, and Blum worldwide data set. Since the shape component is generally held to be a better indicator of affinity, it is puzzling why these analysts went on to make a sweeping conclusion about the peopling of the New World.

Despite the inconclusiveness of their findings, they proceed to say that the three-migration hypothesis is not adequate and "a more complex model for the peopling of the Americas is needed" (Chatters et al. 1999:89). The evidence of one skull that fails to cluster closely with any group cannot overthrow the evidence based on hundreds of skeletons, which went into the formulation of the dental contribution to the three-migration hypothesis (Greenberg et al. 1986). Analysis of Kennewick Man's dentition is neither sufficient nor satisfactory to permit such a sweeping claim.

Kennewick Man's teeth provide no support for the claims that this ancient person was a Sundadont. Daris R. Swindler, who independently examined the teeth in February 2006, agrees that it is impossible to identify the teeth as fitting into the Sundadont pattern because of the extreme crown wear. This is a critical issue, because if Sundadonty could be demonstrated, then on dental grounds Kennewick Man would not be a good sub-fossil candidate for an early ancestor of modern Native Americans, all of whom are Sinodonts. He may well not have left any genetic or cultural (including language) descendants. However, there is no dental evidence to rule him out as a descendant of the Siberian Sinodonts who were direct ancestors of later Indians of North and South America.

At this point it is worth noting that all of the few late Pleistocene and early Holocene anatomically modern human teeth from north China and eastern Siberia were Sinodont (Turner 1992; Turner et al. 2000). Inasmuch as Siberia is regarded as the homeland for the ancestors of Native Americans, claiming that Kennewick Man's dentition, or any early New World dentition, has a non-Sinodont pattern has no support in what little is known about the late Pleistocene and early Holocene Siberians.

The claim that Kennewick Man was a Sundadont is not well founded. This is because not all the eight minimally defining dental traits could be scored due to their destruction by wear. Of those that were scored, their expression as a set could be found in many individuals belonging to any population in the world, including Sinodont Northeast Asians and all Native Americans.

REFERENCES

Adler, Alma Julia. 1999. "The Dentition of Contemporary Finns." Master's thesis, Arizona State University, Tempe.

———. 2005. "Dental Anthropology in Scotland: Morphological Comparisons of Whithorn, St. Andrews, and the Carmelite Friaries." Ph.D. dissertation, Arizona State University, Tempe.

Adler, Alma Julia, and Christy G. Turner II. 2000. Dental caries in prehistoric California Indians: A comparative test of the Jōmon agricultural hypothesis. *American Journal of Physical Anthropology* 111(S30):91–92.

Breternitz, David A., Alan C. Swedlund, and Duane C. Anderson. 1971. An early burial from Gordon Creek, Colorado. *American Antiquity* 36(2):170–182.

Burnett, Scott E. 1998. "Maxillary Premolar Accessory Ridges

(MxPAR): Worldwide Occurrence and Utility in Population Differentiation." Master's thesis, Arizona State University, Tempe.

Chatters, James C. 1996. "Kennewick Man." Smithsonian Institution, National Museum of Natural History, Arctic Studies Center. http://www.mnh.si.edu/arctic/html/kennewick_man.html.

———. 2001. *Ancient Encounters: Kennewick Man and the First Americans.* Simon & Schuster, New York.

Chatters, James C., Walter Neves, and Max Blum. 1999. The Kennewick Man: A first multivariate analysis. *Current Research in the Pleistocene* 16:87–90.

Dahlberg, Albert A. 1951. The dentition of the American Indians. In *Papers on the Physical Anthropology of the American Indian*, edited by W. S. Laughlin, 138–176. Viking Fund, New York.

Doughton, Sandi. 2006. "Kennewick Man Yields More Secrets." *The Seattle Times*, February 24. A8.

Fiedel, Stuart J. 2004. The Kennewick follies: "New" theories about the peopling of the Americas. *Journal of Anthropological Research* 60(1):75–110.

Greenberg, Joseph H., Christy G. Turner II, and Stephen L. Zegura. 1986. The settlement of the Americas: A comparison of the linguistic, dental, and genetic evidence. *Current Anthropology* 27:477–497.

Haeussler, Alice Marie Frances. 1996. "Dental Anthropology of Russia, Ukraine, Caucasus, Central Asia: The Evaluation of Five Hypotheses for Paleo-Indian Origins." Ph.D. dissertation, Arizona State University, Tempe.

Harris, Edward Frederick. 1977. "Anthropologic and Genetic Aspects of the Dental Morphology of Solomon Islanders, Melanesia." Ph.D. dissertation, Arizona State University, Tempe.

Hawkey, Diane Elizabeth. 1998. "Out of Asia: Dental Evidence for Microevolution and Affinities of Early Populations from India/Sri Lanka." Ph.D. dissertation, Arizona State University, Tempe.

Haynes, Gary. 2002. *The Early Settlement of North America: The Clovis Era.* Cambridge University Press, Cambridge.

Irish, Joel D. 1993. "The Dentition of Africa." Ph.D. dissertation, Arizona State University, Tempe.

Lahr, Marta Mirazón. 1995. Patterns of modern human diversification: Implications for Amerindian origins. *Yearbook of Physical Anthropology* 38:163–198.

Lee, Christine. 1999. "Origins and Interactions of the Caddo: A Study in Dental and Cranial Nonmetric Traits." Master's thesis, Arizona State University, Tempe.

Lincoln-Babb, Lorrie. 1999. "The Dental Morphology of the Yaqui Indians: An Affinity Assessment." Master's thesis, Arizona State University, Tempe.

Lipschultz, Joshua G. 1996. "Who Were the Natufians?: A Dental Assessment of their Population Affinities." Master's thesis, Arizona State University, Tempe.

Martin, Letresha. 2003. "On the Biological Affinity Between the Populations of Iron Age Roman Central Southern England, Using Dental Non-metrics as Evidence." Ph.D. dissertation, Australian National University, Canberra.

Neves, W. A., J. F. Powell, and E. G. Ozolins. 1999. Extracontinental morphological affinities of Lapa Vermelha IV, Hominid 1: A multivariate analysis with progressive numbers of variables. *Homo—Journal of Comparative Human Biology* 50(3):263–282.

Nichol, Christian R. 1990. "Dental Genetics and Biological Relationships of the Pima Indians of Arizona." Ph.D. dissertation, Arizona State University, Tempe.

Powell, Joseph F., and Jerome C. Rose. 1999. Chapter 2. Report on the Osteological Assessment of the "Kennewick Man" Skeleton (CENTWW.97.Kennewick). In *Report on the Non-Destructive Examination, Description, and Analysis of the Human Remains from Columbia Park, Kennewick, Washington* [October 1999]. National Park Service, http://www.nps.gov/archeology/kennewick/powell_rose.htm.

Redder, Albert J. 1985. Horn Shelter No. 2: The south end, a preliminary report. *Central Texas Archeologist* 10:37–65.

Reichart, Stephen C. 2000. "The Woodland Iroquoian People of Southern Ontario: A Dental Assessment of their Population Affinity." Master's thesis, Arizona State University, Tempe.

Schmucker, Betty J. 1983. "Dental Attrition: A Correlative Study of Dietary and Subsistence Patterns." Master's thesis, Arizona State University, Tempe.

Schollmeyer, Karen Gust, and Christy G. Turner II. 2004. Dental caries, prehistoric diet and the pithouse-to-Pueblo transition in southwestern Colorado. *American Antiquity* 69(3):569–582.

Scott, G. Richard. 1973. "Dental Morphology: A Genetic Study of American White Families and Variation in Living Southwest Indians." Ph.D. dissertation, Arizona State University, Tempe.

Scott, G. Richard, and Christy G. Turner II. 1997. *The Anthropology of Modern Human Teeth: Dental Morphology and its Variation in Recent Human Populations.* Cambridge University Press, Cambridge.

Turner, Christy G. II. 1979. Dental anthropological indications of agriculture among the Jōmon people of central Japan. X. Peopling of the Pacific. *American Journal of Physical Anthropology* 51(4):619–635.

———. 1983. Sinodonty and Sundadonty: A dental anthropological view of Mongoloid microevolution, origin, and dispersal into the Pacific Basin, Siberia, and the Americas. In *Late Pleistocene and Early Holocene Cultural Connections of Asia and Americas*, edited by R. S. Vasilievsky, 72–76. U.S.S.R. Academy of Sciences, Siberian Branch, Novosibirsk [in Russian].

———. 1990. The major features of Sundadonty and Sinodonty, including suggestions about East Asian microevolution, population history, and late Pleistocene relationships with Australian Aboriginals. *American Journal of Physical Anthropology* 82:295–317.

———. 1991. *The Dentition of Arctic Peoples.* Garland Publishing, New York.

———. 1992. New World origins: New research from the Americas and the Soviet Union. In *Ice Age Hunters of the Rockies*, edited by Dennis J. Stanford and Jane Stevenson Day, 7–50. Denver Museum of Natural History and University of Colorado Press, Niwot.

Turner, Christy G. II, Christian R. Nichol, and G. Richard Scott. 1991. Scoring procedures for key morphological traits of the permanent dentition: The Arizona State University dental anthropology system. In *Advances in Dental Anthropology*, edited by Marc A. Kelley and Clark S. Larsen, 13–31. Wiley-Liss, New York.

Turner, Christy G. II, and Erin Cacciatore. 1998. Interproximal tooth grooves in Pacific Basin, East Asian, and New World populations. *Anthropological Science* 106(supplement):85–94.

Turner, Christy G. II, and James D. Cadien. 1969. Dental chipping in Aleuts, Eskimos and Indians. *American Journal of Physical Anthropology* 31(3):303–310.

Turner, Christy G. II, and James F. Eder. 2006. Dental pathology, wear, and diet in a hunting and gathering forest-dwelling group: The Batak people of Palawan Island, The Philippines. *Dental Anthropology* 19(1):15–22.

Turner, Christy G. II, Yoshitaka Manabe, and Diane E. Hawkey. 2000. The Zhoukoudian Upper Cave dentition. *Acta Anthropologica Sinica* 19:253–268.

Ullinger, Jaime M. 2002. "A Dental Reconstruction of Biological Relationships in the Late Bronze-Early Iron Transition of the Southern Levant Using Dental Morphological Traits." Master's thesis, Arizona State University, Tempe.

Weets, Jaimin D. 1996. "The Dental Anthropology of Vanuatu, Eastern Melanesia." Master's thesis, Arizona State University, Tempe.

Dental Microwear

Mark F. Teaford and Sireen El Zaatari

This chapter is focused on the teeth of Kennewick Man, in particular, the question of whether dental microwear analyses can shed light on the diet and tooth use of this individual. Dental microwear analysis involves the study of microscopic wear on the enamel of tooth surfaces. Unlike gross morphological studies of teeth, there are no evolutionary assumptions required to interpret microwear data (Teaford 1994). Therefore, if postmortem effects can be ruled out, features on the enamel are assumed to be direct evidence of tooth use, at least in part reflective of diet (Teaford 1988, 1991; Ungar 2002). Since 1980, analyses of microscopic wear patterns on teeth have yielded insights into diet and tooth use in a variety of living and fossil animals (Daegling and Grine 1994; Danielson and Reinhard 1998; El Zaatari et al. 2005; Grine 1981, 1986; Grine et al. 2006; King et al. 1999a; Leakey et al. 2003; Merceron et al. 2005; Nystrom et al. 2004; Puech 1984; Puech and Prone 1979; Rensberger 1978, 1986; Rivals and Deniaux 2003; Rose et al. 1981; Ryan 1981; Ryan and Johanson 1989; Scott et al. 2005; Semprebon et al. 2004; Silcox and Teaford 2002; Solounias and Moelleken 1992; Solounias and Semprebon 2002; Strait 1993; Teaford 1988; Teaford and Glander 1991; Teaford and Lytle 1996; Teaford and Walker 1984; Teaford et al. 2008; Ungar 1990, 1994; Ungar et al. 2008; Walker 1976; Walker 1981; Walker et al. 1978). Studies of dental microwear are limited by the fact that most foods are not hard enough to scratch enamel (Lucas 1991; Peters 1982). Thus, many microwear patterns are actually caused by abrasives on foods rather than by the foods themselves.

The fact that abrasives rather than most foods are responsible for scratching dental enamel can be a curse or a blessing in the analysis of human populations. For example, the transition from hunting-gathering to agriculture has left a complex signal in the microwear record, because it seems to depend on which populations are examined and in which habitats (Bullington 1991; El Zaatari 2008, 2010; Organ et al. 2005; Pastor 1992; Pastor and Johnston 1992; Schmidt 2001; Teaford 1991, 2002; Teaford et al. 2001). In central North America, there seems to have been a transition in pre-agricultural populations to a slightly harder diet before the actual switch to agriculture (Schmidt 2001), with subsequent agricultural populations having a softer, less variable diet (Bullington 1991). Along the southeastern coast of the United States, microwear evidence for that dietary transition is complicated by local differences in microhabitat; most notably, variation occurs between coastal and inland sites, with the former evidently including significant amounts of large-grained abrasives in foods (Teaford 1991, 2002; Teaford et al. 2001). In other regions of the world (e.g., the Indian subcontinent), interpretations are further complicated by changes in food-processing techniques, as stone grinding tools introduced abrasives into what would have otherwise been a fairly non-abrasive diet (Molleson and Jones 1991; Pastor 1992; Pastor and Johnston 1992), while the cooking of foods, in other populations, led to a marked *decrease* in the amount of molar microwear (Ma and Teaford 2010; Molleson et al. 1993).

Of course, once the change to agriculture was made, human diets did not simply become stagnant. Some became more homogeneous, as evidenced by fairly uniform microwear patterns, while others became more variable (Molleson and Jones 1991). Some cereal diets left characteristic microwear patterns remarkably similar to those documented in laboratory studies (Gugel et al. 2001). Changes in food processing, most notably the cleaning and boiling of foods, certainly led to a marked decrease in the amount of microwear at some sites (Ma and Teaford 2010; Molleson et al. 1993). However, those changes also left microwear researchers with a dilemma: how to distinguish the microwear signatures of meat-eating versus the consumption of cooked food (Organ et al. 2005; Molleson et al. 1993). The effects of meat-eating on dental microwear depend on a number of factors. Meat is generally recognized as a tough substance because it requires a significant amount of work to cut or fracture it (Lucas and Peters 2000). However, it is *not* a hard substance. Thus meat, like so many other primate foods, cannot be expected to scratch enamel (Lucas 1991). As a result, the effects of meat on dental microwear patterns are likely to be indirect ones, such as the abrasive effects of grit adhering to it (Daegling and Grine 1994; El Zaatari 2010; Lalueza et al. 1996; Lalueza and Pérez-Pérez 1993; Pastor 1992), or the adhesive effects of tooth-on-tooth contacts occurring during its mastication (Organ et al. 2005; Puech et al. 1981, 1986; Teaford and Runestad 1992; Ungar et al. 2006). These could lead to very different microwear patterns, with the former yielding numerous scratches or pits of varying numbers and size (depending on the size, shape, and amount of abrasives), and the latter yielding many small pits (as seen on interstitial facets) (Radlanski and Jäger 1989) on sites of tooth-tooth contact like the canine-premolar complex in some primates (Walker 1984).

Kennewick Man had a normal set of teeth that were rather well preserved considering the roughly 9,000-year age of the skeleton. Although his teeth were in good condition from a preservation perspective, Kennewick Man's dentition showed extreme wear for a man of 35 to 45 years of age (Chapter 8). In fact, tooth wear was so severe that applying most standard dental metrics was impossible. Still, the degree and pattern of wear did provide some insights (Chapter 24); it was very reminiscent of certain Eskimo groups, and unlike other groups such as the Maori who often show severe wear associated with lingual dislocation of the teeth (Taylor 1963) as items are pushed against the buccal sides of the teeth during ingestion. Kennewick Man, by contrast, shows vertical wear with the maintenance of vertical roots. This particular wear pattern prompts two main questions: is there any enamel left for dental microwear analyses, and if so, what can it tell about tooth use in this individual?

MATERIALS AND METHODS

Before enamel impressions were taken for this project, tooth surfaces were closely examined, macroscopically and microscopically, to determine what might, and might not, be advisable, given the condition of the teeth and jaws.

Macroscopic examination

The mandible was in three pieces: the largest piece (97.U.3a) had the first right incisor through the second left molar plus the socket for left third molar; the remainder of the right tooth row (the second right incisor through the second right molar) was in another piece (97.U.3b); and the right ascending ramus and socket for the third molar formed the third piece (97.U.3c). To the naked eye, these pieces showed heavier tooth wear on the right side than on the left side, with the first and second molars on the right side reduced to thin enamel rims, while those on the left side had noticeably thicker rims and an enamel island on the left second molar. Most of the enamel was missing from the lingual sides of the left second incisor through the right first molar, and the same was true for the labial enamel on the incisors. The molars on the left side seemed cleaner than those on right, but all the post canine teeth on the right side were more weathered. The second right incisor had a large crack running labiolingually across most of the crown. The first left incisor had a similar but smaller crack, while the other mandibular incisors had hints of cracks. The second left premolar had a mesiodistal crack on the mesial portion of the crown.

The maxillae were in one piece and included all teeth except for the right third molar, which was lost in life. The left third molar was loose in a separate bag. Again, the wear was extremely heavy, with the premolars and first molars reduced to enamel rims. The canines and incisors had some enamel on the lingual and labial surfaces, and the second molars had the most enamel of any of the teeth. The right-left difference in the amount of tooth wear was not as evident in the maxillary teeth. There were cracks running mesiodistally on the right first incisor, canine, and second premolar as well as the left second incisor and second premolar. The left second molar had a Y-shaped crack across the crown, and the left first molar had a buccolingual crack on the buccal side. The loose third molars were easily in the best condition of any of the teeth, as the left third molar had only a tiny dentin exposure, while the right third molar had a larger one.

Microscopic examination

In preparation for cleaning and casting, each tooth was examined under a dissecting microscope at varying magnifications. In the left mandibular piece, the first right incisor looked too fragile for casting, while the left incisors were cracked and had preservative on them. The enamel of the canine, premolars, and first molar seemed to be coated with preservative, albeit with some visible scratches. The buccal side of the first molar was etched. Thus, the left second molar was in the best condition of any in this piece. For the right mandibular fragment, the buccal sides of most teeth were etched, with the only exceptions being the canine and first premolar. The right second incisor essentially had no enamel, and a crack in the crown left it too risky to cast. As for the occlusal surfaces, there were hints of microwear on the canine, first premolar, and second molars, of which the canine and second molar seemed the best. The lingual occlusal edge on the first premolar had odd gouges on it that could be either pre- or postmortem, and the first molar was too badly etched to be worth casting.

In the maxillae, many teeth had preservative on them, and the right second incisor was deemed too risky to cast due to the combination of cracks running through it and the preservative. The enamel surfaces on the second molars seemed the best, along with tiny portions on the canines and first incisors. The loose third molars had etching around most of their rims, and preservative in their basins, but the enamel in the basins looked good otherwise.

Based on these examinations, casting and analyses focused on the second and third molars, specifically, the left mandibular second molar, the right mandibular third molar, and left maxillary third molar.

COMPARATIVE SAMPLES

A total of 154 first and second molars (only one molar, either second or first, per individual) of one hunter-horticulturist and six hunter-gatherer groups with known but different diets were analyzed for comparative purposes. Combining data from both molar types is possible as it has been shown that there are no significant microwear differences between them among hunter-gatherer groups (El Zaatari 2010). These groups consisted of Andamanese ($n = 30$), Chumash ($n = 13$), Fuegians ($n = 6$), Tikigagmiut (Tigara) ($n = 25$), Aleut ($n = 21$), Arikara ($n = 16$), and the Khoe-San ($n = 43$).

Andamanese

The sample of Andamanese is comprised of protohistoric individuals who lived in the area of Port Blair, South Andaman Island, in the Bay of Bengal, immediately after the first permanent European settlement of the islands in 1858. These dental remains are housed at the British Museum of Natural History, London.

Historic records indicate that the Andamanese had a mixed diet. They relied on fish, dugong, and sea turtles (Myka 1993). Other sources of meat included pig (most frequently hunted in the rainy season), civets, iguana, turtle, and shellfish (Man 1883). They ate edible roots, fruits, and honey, all of which formed one-third of their daily food consumption. Andamanese food was almost always cooked, but eaten half roasted, such that it remained relatively tough (Pal 1984). The food was usually cooked uncovered, which enabled contamination by windborne sand (Pal 1984).

Chumash

The skeletons of this collection were recovered from site SCrI-3 on the Island of Santa Cruz, one of the four Northern Channel Islands 20 miles off the coast of southern California. This series, which dates from 4,000 to 5,000 years ago (Walker 1986), is at the British Museum of Natural History, London.

The Chumash diet relied heavily on marine protein coming from fish, marine mammals, and other marine resources such as shellfish (Walker 1996). They also ate small terrestrial animals available on the island (Fagan 1995). Ethnohistoric accounts for the more recent Chumash reported the consumption of several kinds of plants including hollyleaf cherry, manzanita, mangle berries, tarweed seeds, tubers, and sage (Timbrook 1993; Walker 1996).

Fuegians

The Fuegians included in this study date to 1880 (Manzi 1986). They were members of the Yamana tribe that occupied the Beagle Channel islands and the islands of the Chilean archipelago in the southwestern part of Tierra del Fuego (Manzi 1986; Yesner et al. 2003). This material is housed at the Museum of Anthropology at the University of Rome, Italy.

Stable isotope analysis of Yamana skeletons indicate that their diet was based on marine resources, but that terrestrial animals such as guanaco (*Lama guanicoe*) also formed an important part of their diet (Yesner et al. 2003). The ethnohistoric record of the late 1800s indicates that the Yamana tribe relied on fishing and seal hunting, with seal hunting of less importance (Bridges 1885; Snow 1861).

Tikigagmiut

The Tikigagmiut individuals were recovered during the 1939–1941 excavations at Point Hope, Alaska. These human remains date to 1200–1700 (Larsen and Rainey 1948) and are housed at the American Museum of Natural History, New York. They are the same individuals labeled "Tigara" in previous microwear work by El Zaatari on these samples (2008, 2010).

The similarity of the Tikigagmiut material culture to that of the Point Hope Eskimos who inhabited the area after European contact suggests similar dietary habits (Costa 1980). The diets of the twentieth-century Point Hope Eskimos consisted of 35 to 60 percent fat, 35 to 65 percent protein, and very little carbohydrate, derived from mostly from raw roots (Waugh 1930). The meat component of their diet consisted mainly of fish, whale, and seal. Meat was eaten fresh, dried on open racks, or kept frozen underground as a year-round staple (Draper 1978; Giddings 1967). Uncooked seal skin with attached subcutaneous fat was often chewed for prolonged periods of time (Balikci 1970).

Aleut

The Aleuts in this series were recovered by Hrdlička (1945) between 1936 and 1938 from the Aleutian Islands of Agattu, Amaknak, Kagamil, Umnak, and Unalaska. They are mostly protohistoric, dating to after 1700 (Ungar and Spencer 1999). These remains are housed at the Smithsonian Institution's National Museum of Natural History, Washington D.C.

The Aleutian Islanders are known from ethnohistorical reports to have relied almost exclusively on marine foods for their subsistence, including fresh and dried fish, mollusks, and sea mammals (Hrdlička 1945; Laughlin 1963). Land resources, such as edible tubers, rodents, and foxes, were occasionally eaten (Hrdlička 1945; Laughlin 1963). The Aleut are known for their extensive chewing of frozen and dried animal meat and skin.

Arikara

The sample of Arikara included here is from Mathew Stirling's exploration of the Mobridge site (39WWI), South Dakota, and dates to 1600–1700 (Jantz 1973; Wedel 1955). This series is part of the Smithsonian Institution's National Museum of Natural History collection.

Archaeological data indicate that the diet of the Arikara consisted of a mix of wild and cultivated plant and animal foods. The Arikara were semi-sedentary villagers who lived in compact, usually fortified villages. They were not technically hunter-gatherers, but the Arikara lifeway, with its emphasis on big-game hunting, makes them a useful reference series for this comparative analysis. They hunted bison as well as deer, antelope, and jackrabbit, whose meat was cut into strips and dried (Meyer 1977). The Arikara gathered black cherries, grapes, and goosefoot; they cultivated maize, beans, squash, peppers, and sunflowers (Blakeslee 1994; Hurt 1969; Meyer 1977; Tuross and Fogel 1994).

Khoe-San

These tooth samples derive from one historic (Riet River) and three prehistoric (Oakhurst Shelter, Matjes River, and Elands Bay) South African sites. The Oakhurst Rock Shelter lies 13 miles east of George. Excavations at this site were undertaken in 1932–1935 by A.J.H. Goodwin of the University of Cape Town. The site is dated to between 9,000 and 5,000 years ago (Morris 1992a). Matjes River Rock Shelter lies on the western side of the mouth of the Matjes River, east of Robberg. The skeletal remains from this site were excavated in the 1920s and in the 1950s (Sealy 2006). The individuals from Matjes River date to between 8,000 and 2,000 years ago (Sealy and Pfeiffer 2000). Elands Bay cave lies in the area of Cape Deseada at the southern end of Elands Bay. The human remains date to two periods, between 11,000–9,000 and 4,000–2,000 years ago. The Riet River skeletons come from the Orange River Valley, Northern Cape province, and date to between the 1100 and 2000 (Morris 1992b).

Stable isotope analyses of human bones from these

four sites indicate a mixed diet (Lee Thorp et al. 1993; Sealy 2006; Sealy and Pfeiffer 2000; Sealy and van der Merwe 1986, 1988; Sealy et al. 1992). Shellfish were found in these four sites, indicating fishing activity (Goodwin 1938; Sealy 2006). Ethnographic studies show that 80 percent or more of the diet of modern Kalahari hunter-gatherers consists of plant food (Silberbauer 1981). Previous dental microwear texture analyses of these samples demonstrated no significant differences among them, probably due to considerable variability in diet in each group (El Zaatari 2010). Thus, they were treated as one sample here.

DENTAL MICROWEAR TECHNIQUES
Dental Impressions

For the purposes of this study, molds of permanent tooth crowns of Kennewick Man were taken at the Burke Museum at the University of Washington, Seattle. Comparative samples were prepared in the American and European institutions where the collections are housed. All molds were prepared following the same procedure. Specimens were first cleaned with cotton swabs soaked in distilled water. Acetone and/or ethyl alcohol were used when the occlusal surface was covered with a preservative that could not be removed with water. Impressions were made using Coltene President Microsystems™ (polysiloxane vinyl) impression material. Positive casts were then made at Johns Hopkins University, Baltimore, or the State University of New York at Stony Brook with Epo-Tek 501 epoxy resin and hardener (Epoxy Technologies). This procedure allows for a reproduction of features with a fraction of a micron resolution (Beynon 1987; Teaford and Oyen 1989).

Dental Microwear Texture Analysis

Unlike many previous dental microwear analyses, which used scanning electron microscopy coupled with computerized digitization of individual features (e.g., Teaford 1988; Teaford et al. 2001; Ungar et al. 2006), this study used dental microwear texture analysis, a combination of confocal microscopy and scale-sensitive fractal analysis, because it allows for quicker, more objective analyses (Scott et al. 2005, 2006; Ungar et al. 2008). Specimens were examined using a Sensofar Plu white light scanning confocal microscope (Solarius Development Inc., Sunnyvale, California). The entire occlusal surface was first scanned at low magnification (10 x) to eliminate teeth that showed obvious postmortem damage and lacked good microwear preservation. Postmortem damage, which can occur during and after deposition/burial or during collection and preparation of the specimen, is generally easily identified at low magnification (Grine 1986; King et al. 1999b; Teaford 1988).

Following Scott et al. (2005, 2006), four adjoining scans of Phase II (crushing/grinding) facets (i.e., facets 9, 10n, or x) of the specimens with good wear surfaces were taken at 100 × magnification, with a lateral sampling interval of 0.18 μm and a vertical resolution of 0.005 μm, resulting in a total scanned area of 276 × 102 μm. Microscopy was conducted at the University of Arkansas at Fayetteville. The scans were leveled, modified (to erase defects), and 3D images were generated using the Solarmap Universal software (Solarius Development Inc., Sunnyvale, Calif.). The images present the 3D data in the form of a model of the surface being examined. Resulting data were then analyzed in Toothfrax and SFrax (Surfract). Three variables were considered: surface complexity (Asfc), scale of maximum complexity (Smc), and anisotropy (epLsar). Each has already been shown to yield insights into different aspects of food and abrasive properties (Scott et al. 2005, 2006; Ungar et al. 2007, 2008). Surface complexity (Asfc) is measured as the change in surface roughness at different scales, with more "complex" surfaces exhibiting a multiplicity of features (e.g., pits and scratches) of varying sizes overlying each other. The scale of maximum complexity (Smc), by contrast, is the scale at which the surface is most complex. So, a surface could be relatively uncomplex, yet have a high value for Smc, indicating less wear at very fine scales, and/or more wear at coarser scales. Conversely, a surface could be relatively complex, yet have a low value for Smc, indicating a varied assortment of features at relatively small scale. Anisotropy is a measure of the orientation concentration of surface roughness. So, if a surface has discernible directionality (e.g., high proportions of parallel scratches), it would be anisotropic and have a higher value for epLsar.

As Kennewick Man comprises a sample size of one, formal statistical tests were not conducted. However, as previous publications have documented differences (in rank-transformed values for microwear texture variables) among the comparative samples (e.g., El Zaatari 2010), means and standard deviations are presented for reference purposes in tables and figures.

RESULTS

The surfaces of Kennewick Man's left mandibular second molar and left maxillary and right mandibular third molars were relatively smooth. Since similar values for the microwear texture variables were obtained for both molars (Table 9.1), and since the comparative samples focused on first and second molars (with no significant differences between values for each tooth) (El Zaatari 2010), values for Kennewick Man's second molar are used in all figures.

Microwear texture values for Kennewick Man showed anisotropy values similar to those of most of the comparative samples except the Fuegians. Kennewick Man showed relatively uncomplex molar wear surfaces, most similar to those of the Arikara and Fuegians, but with a relatively large scale of maximum complexity as seen among the Aleut and Fuegians (Figures 9.1, 9.2, 9.3; Table 9.1).

DISCUSSION AND CONCLUSIONS

Molar microwear of Kennewick Man shows a combination of low surface complexity, but a relatively high scale of maximum complexity. This indicates that there were few overlapping features of different shapes and sizes and relatively little microscopic pitting, but those microwear features that were present tended to be relatively large. Most food items ingested by modern humans are not hard enough to scratch molar enamel. Thus, molar microwear differences among such groups often hinge upon differences in the cleanliness of food and food preparation techniques. The information here is probably a case in point.

The comparative populations can be roughly divided into three groups based on their complexity values, that is, those with high surface complexity (the Tikigagmiut and Andamanese), those with medium surface complexity (Khoe-San, Chumash, and Aleut), and those with low surface complexity (Fuegians and Arikara). Kennewick Man falls with the last. As noted by El Zaatari (2008, 2010), the high complexity values for the Tikigagmiut and Andamanese were probably due to large-grained abrasives adherent to food that was either dried or cooked outside in sandy environments. Interpretations of the intermediate group are more complex. The Aleut live in volcanic landscapes that lack sandy beaches, yet the relatively high scale of maximum complexity for the microwear on their molars indicates that, while the amount of abrasives was lower than in, for example, the Tikigagmiut, those abrasives that did strike the teeth were larger than in the Chumash, Khoe-San, Tikigagmiut, or Andamanese. The Chumash and Khoe-San may be linked by a greater proportion of plant material in their diets, but clearly, more work with a wider range of teeth would be valuable.

The groups with the lowest surface complexity are an intriguing mix, because the Fuegians have a much greater

TABLE 9.1 Molar microwear texture values for Kennewick Man and comparative samples.

	Asfc (mean ± sd)	Smc (mean ± sd)	epLsar (mean ± sd)
Tikigagmiut ($n = 25$)	6.569 ± 5.8	0.213 ± 0.07	0.0029 ± 0.0015
Andamanese ($n = 30$)	6.345 ± 4.2	0.213 ± 0.22	0.0026 ± 0.0012
Khoe-San ($n = 43$)	3.616 ± 2.4	0.194 ± 0.05	0.0026 ± 0.0013
Chumash ($n = 13$)	2.787 ± 2.3	0.190 ± 0.05	0.0023 ± 0.0007
Aleut ($n = 21$)	2.590 ± 1.6	0.454 ± 0.73	0.0028 ± 0.0011
Arikara ($n = 16$)	1.152 ± 1.1	0.820 ± 0.97	0.0027 ± 0.0015
Fuegians ($n = 6$)	0.948 ± 0.3	0.400 ± 0.13	0.0044 ± 0.0014
Kennewick Man–M$_2$	1.143	0.551	0.0025
Kennewick Man–M3	1.690	0.321	0.0019

Source: El Zaatari 2010

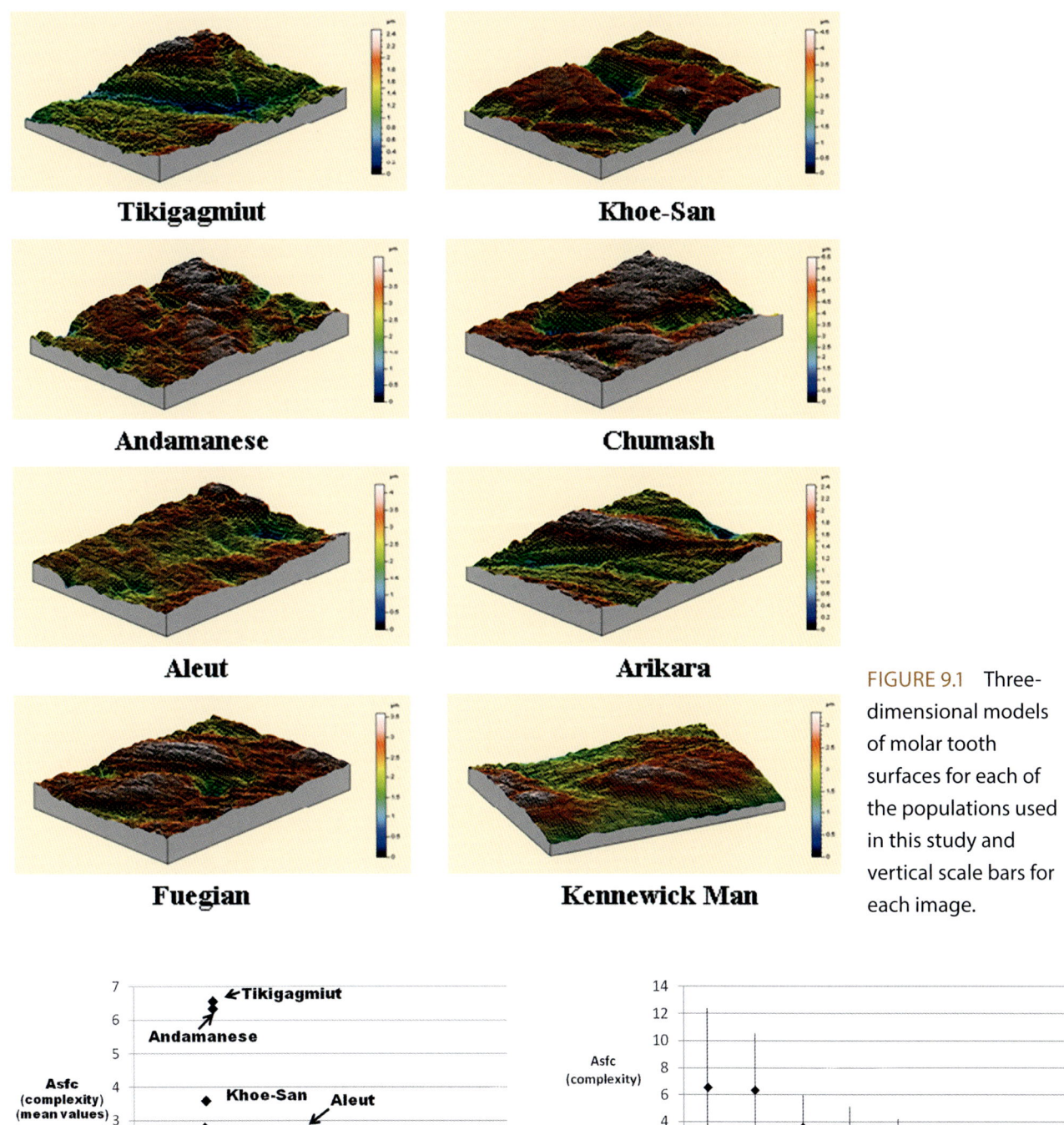

FIGURE 9.1 Three-dimensional models of molar tooth surfaces for each of the populations used in this study and vertical scale bars for each image.

FIGURE 9.2 Bivariate plot of mean values for microwear complexity vs. scale of maximum complexity for comparative samples and Kennewick Man.

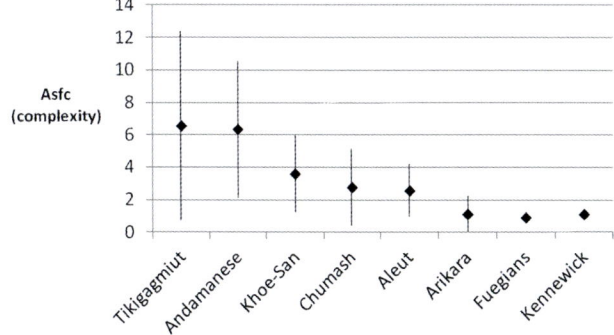

FIGURE 9.3 Plot of microwear complexity for comparative samples and Kennewick Man (mean ± sd).

emphasis on meat-eating than the Arikara. Not coincidentally, as noted by El Zaatari (2010), what separates the Fuegians is their distinct values for anisotropy (epLsar in Table 9.1). Unlike all the other samples in this study (including Kennewick Man), the Fuegians showed higher values for epLsar and, thus, a more distinct alignment of the scratches on their molars. This is probably a reflection of the fact that they habitually fed on tough foods that required more precise jaw movements than other populations. Other groups, like the Tikigagmiut, Andamanese, and Aleut, may have also eaten tough foods. However, the abrasive load of their diet probably obliterated some scratches and made their microwear textures more complex, effectively removing that signal from their teeth. The Arikara, by contrast, had a variable diet, with relatively few abrasives in it. If they had habitually eaten tough foods, their teeth would have shown that, but their values for anisotropy fall right with the other samples, indicating that was not the case. Instead, the relatively high values for scale of maximum complexity probably indicate that abrasives were only ingested occasionally, again probably adhering to the food rather than being part of the food itself.

Where does that leave Kennewick Man? Going into these analyses, one might have expected certain comparative populations to have shown similar microwear patterns to that of Kennewick Man. For instance, as noted in Chapter 24, based on gross molar wear alone, the Aleut or Tikigagmiut might be expected to be similar to Kennewick Man. Yet, Kennewick Man's low surface complexity indicates that either his foods generally did not include so many abrasives as those of the Tikigagmiut and Aleut, or he was very adept at cleaning those foods. Based on the habitat of Kennewick Man, which was dry and windy with exposed sandy and silty soils, and associated information suggesting a mixed diet from riverine and steppe sources, the Arikara or Khoe-San might be expected to be similar. As can be seen in Figures 9.2 and 9.3, microwear texture analyses show that Kennewick Man was most similar to the Arikara or Fuegian comparative samples. However, his diet was probably not exceptionally tough, as his anisotropy values are indistinguishable from those of the other samples except the Fuegians. Does this mean he habitually cooked his food to reduce its toughness? Evidence from the microwear does not allow us to say with certainty. However, of the samples analyzed, his most likely modern analogue was probably the Arikara.

REFERENCES

Balikci, Asen. 1970. *The Netsilik Eskimo*. Natural History Press, Prospect Heights, IL.

Beynon, A. D. 1987. Replication technique for studying microstructure in fossil enamel. *Scanning Microscopy* 1(2):663–669.

Blakeslee, Donald J. 1994. The archaeological context of human skeletons in the northern and central Plains. In *Skeletal Biology in the Great Plains: Migration, Warfare, Health and Subsistence,* edited by Douglas W. Owsley and Richard L. Jantz, 9–32. Smithsonian Institution Press, Washington, DC.

Bridges, Thomas. 1885. The Yahgans of Tierra del Fuego. *The Journal of the Anthropological Institute of Great Britain and Ireland* 14:288–289.

Bullington, Jill. 1991. Deciduous dental microwear of prehistoric juveniles from the lower Illinois River valley. *American Journal of Physical Anthropology* 84(1):59–73.

Costa, Raymond L., Jr. 1980. Incidence of caries and abscesses in archaeological Eskimo skeletal samples from Point Hope and Kodiak Island, Alaska. *American Journal of Physical Anthropology* 52(4):501–514.

Daegling, D. J., and F. E. Grine. 1994. Bamboo feeding, dental microwear, and diet of the Pleistocene ape *Gigantopithecus blacki*. *South African Journal of Science* 90(10):527–532.

Danielson, Dennis R., and Karl J. Reinhard. 1998. Human dental microwear caused by calcium oxalate phytoliths in prehistoric diet of the lower Pecos region, Texas. *American Journal of Physical Anthropology* 107(3):297–304.

Draper, H. H. 1978. Nutrition studies: The aboriginal Eskimo diet—A modern perspective. In *Eskimos of Northwestern Alaska: A Biological Perspective,* edited by Paul L. Jamison, Stephen L. Zegura, and Frederick A. Milan, 139–144. Dowden, Hutchinson and Ross Inc., Stroudsburg, PA.

El Zaatari, Sireen. 2008. Occlusal molar microwear and the diets of the Ipiutak and Tigara populations (Point Hope) with comparisons to the Aleut and Arikara. *Journal of Archaeological Science* 35(9):2517–2522.

———. 2010. Occlusal microwear texture analysis and the diets of historical/prehistoric hunter-gatherers. *International Journal of Osteoarchaeology* 20(1):67–87.

El Zaatari, Sireen, Frederick E. Grine, Mark F. Teaford, and Heather F. Smith. 2005. Molar microwear and dietary reconstructions of fossil Cercopithecoidea from the Plio-Pleistocene deposits of South Africa. *Journal of Human Evolution* 49(2):180–205.

Fagan, Brian. 1995. *Time Detectives*. Simon & Schuster, New York.

Giddings, J. Louis. 1967. *Ancient Men of the Arctic*. Secker and Warburg, London.

Goodwin, A. J. H. 1938. Archaeology of the Oakhurst Shelter, George. *Transactions of the Royal Society of South Africa* 25(3):229–257.

Grine, Frederick E. 1981. Trophic differences between "gracile" and "robust" australopithecines: A scanning electron microscope analysis of occlusal events. *South African Journal of Science* 77:203–230.

———. 1986. Dental evidence for dietary differences in *Australopithecus* and *Paranthropus*: A quantitative analysis of permanent molar microwear. *Journal of Human Evolution* 15(8):783–822.

Grine, Frederick E., Peter S. Ungar, Mark F. Teaford, and Sireen El Zaatari. 2006. Molar microwear in *Praeanthropus afarensis*: Evidence for dietary stasis through time and under diverse paleoecological conditions. *Journal of Human Evolution* 51(3):297–319.

Gugel, Irene L., Gisela Grupe, and Karl-Heinz Kunzelmann. 2001. Simulation of dental microwear: Characteristic traces by opal phytoliths give clues to ancient human dietary behavior. *American Journal of Physical Anthropology* 114(2):124–138.

Hrdlička, Aleš. 1945. *The Aleutian and Commander Islands and their Inhabitants*. Wistar Institute of Anatomy and Biology, Philadelphia.

Hurt, Wesley R. 1969. Seasonal economic and settlement patterns of the Arikara. *Plains Anthropologist* 14(43):32–37.

Jantz, Richard L. 1973. Microevolutionary change in Arikara crania: Multivariate analysis. *American Journal of Physical Anthropology* 38(1):15–26.

King, Tania, Leslie C. Aiello, and Peter Andrews. 1999a. Dental microwear of *Griphopithecus alpani*. *Journal of Human Evolution* 36(1):3–31.

King, Tania, Peter Andrews, and Basak Boz. 1999b. Effect of taphonomic processes on dental microwear. *American Journal of Physical Anthropology* 108(3):359–373.

Lalueza, Carles, and Alejandro Pérez-Pérez. 1993. The diet of the Neanderthal child Gibraltar 2 (Devil's Tower) through the study of the vestibular striation pattern. *Journal of Human Evolution* 24(1):29–41.

Lalueza, Carles, Alejandro Pérez-Pérez, and Daniel Turbón. 1996. Dietary inferences through buccal microwear analysis of middle and upper Pleistocene human fossils. *American Journal of Physical Anthropology* 100(3):367–387.

Larsen, Helge, and Froelich Rainey. 1948. *Ipiutak and the Arctic Whale Hunting Culture*. Anthropological Papers of the American Museum of Natural History 45. American Museum of Natural History, New York.

Laughlin, William S. 1963. Eskimos and Aleuts: Their origins and evolution. *Science* 142(3593):633–645.

Leakey, Meave G., Mark F. Teaford, and Carol V. Ward. 2003. Cercopithecidae from Lothagam. In *Lothagam: Dawn of Humanity in Eastern Africa*, edited by Meave G. Leakey and John M. Harris, 201–248. Columbia University Press, New York.

Lee Thorp, Julia A., Judith C. Sealy, and Alan G. Morris. 1993. Isotopic evidence for diets of prehistoric farmers in South Africa. In *Prehistoric Human Bone: Archaeology at the Molecular Level*, edited by Joseph Lambert and Gisela Grupe, 99–120. Springer Verlag, Berlin.

Lucas, P. W. 1991. Fundamental physical properties of fruits and seeds in the diet of Southeast Asian primates. In *Primatology Today*, edited by A. Ehara, T. Kimura, O. Takenaka, and M. Iwamoto, 152–158. Elsevier, Amsterdam.

Lucas, P. W., and C. R. Peters. 2000. Function of postcanine tooth shape in mammals. In *Development, Function and Evolution of Teeth*, edited by Mark F. Teaford, Moya M. Smith, and Mark W. J. Ferguson, 282–289. Cambridge University Press, New York.

Ma, Peter, and Mark F. Teaford. 2010. Dental microwear and diet in antebellum Baltimore. *American Journal of Physical Anthropology* 141(4):571–582.

Man, Edward H. 1883. On the aboriginal inhabitants of the Andaman Islands (Part III). *Journal of the Anthropological Institute of Great Britain and Ireland* 12:327–434.

Manzi, G. 1986. I Fuegini del Museo di Antropologia dell'Università di Roma "La Sapienza:" Significato di una collezione osteologica. *Museologia Scientifica* V (Supplement):43–54.

Merceron, Gildas, Cecile Blondel, Louis De Bonis, Georges D. Koufos, and Laurent Viriot. 2005. A new method of dental microwear analysis: Application to extant primates and

Ouranopithecus macedoniensis (Late Miocene of Greece). *Palaios* 20(6):551–561.

Meyer, Roy Willard. 1977. *The Village Indians of the Upper Missouri.* University of Nebraska Press, Lincoln.

Molleson, Theya, and Karen Jones. 1991. Dental evidence for dietary change at Abu Hureyra. *Journal of Archaeological Science* 18(5):525–539.

Molleson, Theya, Karen Jones, and Stephen Jones. 1993. Dietary change and the effects of food preparation on microwear patterns in the Late Neolithic of Abu Hureyra, northern Syria. *Journal of Human Evolution* 24(6):455–468.

Morris, Alan G. 1992a. *A Master Catalogue of Holocene Human Skeletons from South Africa.* Witwatersrand University Press, Johannesburg.

———. 1992b. *Skeletons of Contact: A Study of Protohistoric Burials from the Lower Orange River Valley, South Africa.* Witwatersrand University Press, Johannesburg.

Myka, Frank P. 1993. *Decline of Indigenous Populations: The Case of the Andaman Islanders.* Rawat Publications, Jaipur.

Nystrom, Pia, Jane E. Phillips-Conroy, and Clifford J. Jolly. 2004. Dental microwear in anubis and hybrid baboons (*Papio hamadryas*, sensu lato) living in Awash National Park, Ethiopia. *American Journal of Physical Anthropology* 125(3):279–291.

Organ, Jason M., Mark F. Teaford, and Clark Spencer Larsen. 2005. Dietary inferences from dental occlusal microwear at Mission San Luis de Apalachee. *American Journal of Physical Anthropology* 128(4):801–811.

Pal, Anadi. 1984. Molar tooth attrition among the Andaman Negritos. *Human Science, Journal of the Anthropological Survey of India* 33:313–321.

Pastor, Robert F. 1992. Dietary adaptations and dental microwear in Mesolithic and Chalcolithic South Asia. *Journal of Human Ecology* 2(special issue):21–228.

Pastor, Robert F., and Thomas L. Johnston. 1992. Dental microwear and attrition. In *Human Skeletal Remains from Mahadaha: A Gangetic Mesolithic Site*, edited by Kenneth A.R. Kennedy, 271–304. Cornell University Press, Ithaca.

Peters, Charles R. 1982. Electron-optical microscopic study of incipient dental microdamage from experimental seed and bone crushing. *American Journal of Physical Anthropology* 57(3):283–301.

Puech, Pierre-François. 1984. A la recherche du menu des premiers hommes. *Cahiers Ligures de Préhistoire et Protohistoire* 1:46–53.

Puech, P., F. Cianfarani, and H. Albertini. 1986. Dental microwear features as an indicator for plant food in early hominid: A preliminary study of enamel. *Human Biology* 1(6):507–515.

Puech, Pierre-François, and A. Prone. 1979. Reproduction expérimentale des processus d'usure dentaire par abrasion: Implications paléoécologique chex l'Homme fossile. *Comptes Redus de l'Academie des Sciences, Paris* 289:895–898.

Puech, Pierre-François, A. Prone, and H. Albertini. 1981. Reproduction expérimentale des processus d'altération de la surface dentaire par friction non abrasif et non adhésive: Application à l'étude de alimentation de L'Homme fossile. *Comptes Redus de l'Academie des Sciences, Paris* 293:729–734.

Radlanski, R. J., and A. Jäger. 1989. Zur mikromorphologie der approximalen Kontaktflächen und der okklusalen Schliffacetten menschlicher Zähne. *Deutsch Zahnärztl Zeitschrift* 44:196–197.

Rensberger, John M. 1978. Scanning electron microscopy of wear and occlusal events in some small herbivores. In *Development, Function and Evolution of Teeth*, edited by Percy M. Butler and Kenneth A. Joysey, 415–438. Academic Press, New York.

———. 1986. Early chewing mechanisms in mammalian herbivores. *Paleobiology* 12:474–494.

Rivals, Florent, and Brigitte Deniaux. 2003. Dental microwear analysis for investigating the diet of an argali population (*Ovis ammon antiqua*) of mid-Pleistocene age, Caune de l'Arago cave, eastern Pyrenees, France. *Palaeogeography, Palaeoclimatology, Palaeoecology* 193(3–4):443–455.

Rose, Kenneth D., Alan Walker, and Louis L. Jacobs. 1981. Function of the mandibular tooth comb in living and extinct mammals. *Nature* 289(5798):583–585.

Ryan, Alan S. 1981. Anterior dental microwear and its relationship to diet and feeding behavior in three African primates (*Pan troglodytes troglodytes*, *Gorilla gorilla gorilla*, and *Papio hamadryas*). *Primates* 22(4):533–550.

Ryan, Alan S., and Donald C. Johanson. 1989. Anterior dental microwear in *Australopithecus afarensis*: Comparisons with human and nonhuman primates. *Journal of Human Evolution* 18:235–268.

Schmidt, Christopher W. 2001. Dental microwear evidence for a dietary shift between two nonmaize-reliant prehistoric human populations from Indiana. *American Journal of Physical Anthropology* 114(2):139–145.

Scott, Robert S., Peter S. Ungar, Torbjorn S. Bergstrom, Christopher A. Brown, Frederick E. Grine, Mark F. Teaford, and Alan Walker. 2005. Dental microwear texture analysis shows within-species diet variability in fossil hominins. *Nature* 436:693–695.

Scott, Robert S., Peter S. Ungar, Torbjorn S. Bergstrom, Christopher A. Brown, Benjamin E. Childs, Mark F. Teaford, and Alan Walker. 2006. Dental microwear texture analysis: Technical considerations. *Journal of Human Evolution* 51(4):339–349.

Sealy, Judith C. 2006. Diet, mobility, and settlement pattern among Holocene hunter-gatherers in Southernmost Africa. *Current Anthropology* 47(4):569–595.

Sealy, J. C., M. K. Patrick, A. G. Morris, and D. Alder. 1992. Diet and dental caries among later Stone Age inhabitants of the Cape Province, South Africa. *American Journal of Physical Anthropology* 88(2):123–134.

Sealy, Judith C., and Susan Pfeiffer. 2000. Diet, body size, and landscape use among Holocene people in the southern Cape, South Africa. *Current Anthropology* 41(4):642–655.

Sealy, Judith C., and Nikolaas J. van der Merwe. 1986. Isotope assessment and the seasonal-mobility hypothesis in the southwestern Cape of South Africa. *Current Anthropology* 27(2):135–150.

———. 1988. Social, spatial and chronological patterning in marine food use as determined by $\delta^{13}C$ measurements of Holocene human skeletons from the south-western Cape, South Africa. *World Archaeology* 20(1):87–102.

Semprebon, Gina M., Laurie R. Godfrey, Nikos Solounias, Michael R. Sutherland, and William L. Jungers. 2004. Can low-magnification stereomicroscopy reveal diet? *Journal of Human Evolution* 47(3):115–144.

Silberbauer, George B. 1981. *Hunter and Habitat in the Central Kalahari Desert*. Cambridge University Press, Cambridge.

Silcox, Mary T., and Mark F. Teaford. 2002. The diet of worms: An analysis of mole dental microwear and its relevance to dietary inference in primates and other mammals. *Journal of Mammalogy* 83(3):804–814.

Snow, W. Parker. 1861. A few remarks on the wild tribes of Tierra del Fuego from personal observation. *Transactions of the Ethnological Society* 1:261–267.

Solounias, Nikos, and Sonja M. C. Moelleken. 1992. Tooth microwear analyses of *Eotragus sansaniensis* (Mammalia: Ruminantia), one of the oldest known bovids. *Journal of Vertebrate Paleontology* 12(1):113–121.

Solounias, Nikos, and Gina Semprebon. 2002. Advances in the reconstruction of ungulate ecomorphology with application to early fossil equids. *American Museum Novitates* 3366:1–49.

Strait, Suzanne G. 1993. Molar microwear in extant small-bodied faunivorous mammals: An analysis of feature density and pit frequency. *American Journal of Physical Anthropology* 92(1):63–79.

Taylor, R. M. S. 1963. Cause and effect of wear of teeth: further non-metrical studies of the teeth and palate in Moriori and Maori skulls. *Acta Anatomica* 53(1–2):97–157.

Teaford, Mark F. 1988. A review of dental microwear and diet in modern mammals. *Scanning Microscopy* 2(2):1149–1166.

———. 1991. Dental microwear: What can it tell us about diet and dental function? In *Advances in Dental Anthropology*, edited by Marc A. Kelley and Clark Spencer Larsen, 341–356. Alan R. Liss, New York.

———. 1994. Dental microwear and dental function. *Evolutionary Anthropology: Issues, News, and Reviews* 3(1):17–30.

———. 2002. Dental enamel microwear analysis. In *Foraging, Farming and Coastal Biocultural Adaptation in Late Prehistoric North Carolina*, edited by Dale L. Hutchinson, 169–177. University Press of Florida, Gainesville.

Teaford, Mark F., and Kenneth E. Glander. 1991. Dental microwear in live, wild-trapped *Alouatta palliata* from Costa Rica. *American Journal of Physical Anthropology* 85(3):313–319.

Teaford, Mark F., Clark Spencer Larsen, Robert F. Pastor, and Vivian E. Noble. 2001. Pits and scratches: Microscopic evidence of tooth use and masticatory behavior in La Florida. In *Bioarchaeology of Spanish Florida: The Impact of Colonialism*, edited by Clark Spencer Larsen, 82–112. University Press of Florida, Gainesville.

Teaford, Mark F., and James D. Lytle. 1996. Diet-induced changes in rates of human tooth microwear: A case study involving stone-ground maize. *American Journal of Physical Anthropology* 100(1):143–147.

Teaford, Mark F., and Ordean J. Oyen. 1989. Live primates and dental replication: new problems and new techniques. *American Journal of Physical Anthropology* 80(1):73–81.

Teaford, Mark F., and Jacqueline A. Runestad. 1992. Dental microwear and diet in Venezuelan primates. *American Journal of Physical Anthropology* 88(3):347–364.

Teaford, Mark F., Peter S. Ungar, and Richard F. Kay. 2008. Molar shape and molar microwear in the Koobi Fora monkeys: Ecomorphological implications. In *Koobi Fora Research Project*, vol. 6: *The Fossil Monkeys*, edited by Nina G. Jablonski and Meave G. Leakey, 337–358. Occasional Paper of the California Academy of Sciences, San Francisco.

Teaford, Mark F., and Alan Walker. 1984. Quantitative differences in dental microwear between primate species with different diets and a comment on the presumed diet of *Sivapithecus*. *American Journal of Physical Anthropology* 64(2):191–200.

Timbrook, Janice. 1993. Island Chumash ethnobotany. In *Archaeology on the Northern Channel Islands of California: Studies of Subsistence, Economics and Social Organization*, edited by Michael A. Glassow, 47–62. Coyote Press, Salinas.

Tuross, Noreen, and Marilyn L. Fogel. 1994. Stable isotope analysis and subsistence patterns at the Sully Site. In *Skeletal Biology of the Great Plain: Migration, Warfare, Health and Subsistence*, edited by Douglas W. Owsley and Richard L. Jantz, 283–289. Smithsonian Institution Press, Washington, DC.

Ungar, Peter S. 1990. Incisor microwear and feeding behavior in *Alouatta seniculus* and *Cebus olivaceus*. *American Journal of Primatology* 20(1):43–50.

———. 1994. Incisor microwear of Sumatran Anthropoid primates. *American Journal of Physical Anthropology* 94(3):339–363.

———. 2002. Reconstructing the diets of fossil primates. In *Reconstructing Behavior in the Primate Fossil Record*, edited by J. Michael Plavcan, Richard F. Kay, William L. Jungers, and Carel P. van Schaik, 261–296. Plenum, New York.

Ungar, Peter S., Frederick E. Grine, and Mark F. Teaford. 2008. Dental microwear and diet of the Plio-Pleistocene hominin *Paranthropus boisei*. *PLoS One* 3(4):e2044. doi:10.1371/journal.pone.0002044

Ungar, Peter S., Frederick E. Grine, Mark F. Teaford, and Sireen El Zaatari. 2006. Dental microwear and diet in African early *Homo*. *Journal of Human Evolution* 50(1):78–95.

Ungar, Peter S., Gildas Merceron, and Robert S. Scott. 2007. Dental microwear texture analysis of Varswater bovids and early Pliocene paleoenvironments of Langebaanweg, Western Cape Province, South Africa. *Journal of Mammalian Evolution* 14(3):163–181.

Ungar, Peter S., Peter S. Scott, Jessica R. Scott, and Mark F. Teaford. 2008. Dental microwear analysis: Historical perspectives and new approaches. In *Technique and Application in Dental Anthropology*, edited by Joel D. Irish and Greg C. Nelson, 389–425. Cambridge University Press, New York.

Ungar, Peter S., and Mark A. Spencer. 1999. Incisor microwear, diet, and tooth use in three Amerindian populations. *American Journal of Physical Anthropology* 109(3):387–396.

Walker, Alan. 1981. Diet and teeth, dietary hypotheses and human evolution. *Philosophical Transactions of the Royal Society of London* 292(B):57–64.

———. 1984. Mechanisms of honing in the male baboon canine. *American Journal of Physical Anthropology* 65(1):47–60.

Walker, Alan, Hendrick N. Hoeck, and Linda Perez. 1978. Microwear of mammalian teeth as an indicator of diet. *Science* 201(4359):908–910.

Walker, Phillip L. 1976. Wear striations on the incisors of ceropithecid monkeys as an index of diet and habitat preference. *American Journal of Physical Anthropology* 45(2):299–308.

———. 1986. Porotic hyperostosis in a marine-dependent California Indian population. *American Journal of Physical Anthropology* 69(3):345–354.

———. 1996. Integrative approaches to the study of ancient health: An example from the Santa Barbara channel area of Southern California. In *Notes on Population Significance of Paleopathological Conditions: Health, Illness and Death in the Past*, edited by Alejandro Pérez-Pérez, 97–105. Fundació Uriach, Barcelona.

Waugh, L. 1930. A study of the nutrition and teeth of the Eskimo of North Bering Sea and Arctic Alaska. *Journal of Dental Research* 10(3):387–393.

Wedel, Waldo R. 1955. Archaeological materials from the vicinity of Mobridge, South Dakota. *Smithsonian Institution American Ethnological Bulletin* 154:85–167.

Yesner, David R., Maria J. Figuerero Torres, Ricardo A. Guichon, and Luis A. Borrero. 2003. Stable isotope analysis of human bone and ethnohistoric subsistence patterns in Tierra del Fuego. *Journal of Anthropological Archaeology* 22(3):279–291.

Orthodontics

10

John L. Hayes

The recovery of Kennewick Man's intact dental arches presented a remarkable opportunity to study his dentition and reach some conclusions regarding his dental health. The teeth, dental arches, and facial-skeletal landmarks were measured and compared with two ancient American skulls of adult males from Horn Shelter No. 2 (Texas) and La Jolla W-12 (Chancellor site, California). To provide insight into modern health concerns, Kennewick Man was compared with three present-day North American population groups of European ancestry (hereafter, Caucasian).

TOOTH ATTRITION

Kennewick Man's teeth exhibit generalized chewing-surface wear far beyond what is seen in industrialized society. The excessive attrition was caused by a combination of abrasive substances in the diet and task-related activities. Little enamel remains on the occlusal surfaces; the relatively soft underlying dentin substructure is nearly completely exposed.

The maxillary and mandibular incisors lost 90 percent (8 to 9 mm) of their enamel crown height because of attrition. The maxillary first molars lost four to five millimeters of crown height on the buccal side and five to seven millimeters on the lingual side, resulting in a shortening of the length of the teeth and a canting of the occlusal surfaces. Most of the posterior teeth remain upright (without buccal or lingual tooth inclination) and are well centered on the alveolar ridge crests, the exception being the maxillary first molars.

Angular measurements of the occlusal surfaces indicate more wear of the maxillary molars than the corresponding mandibular molars. The maxillary first and second molars wore 28 to 30 degrees and 18 degrees, respectively, from the horizontal. The corresponding mandibular molars wore two to three degrees and three to six degrees respectively. Wear was essentially equal bilaterally.

Kennewick Man exhibits dental attrition well into the former pulp chambers of a number of teeth; however, the pulp chambers are filled with reparative dentin that can form in response to slow occlusal wear. Deep and rapid wear of the dentin layer can lead to pulp exposure—a potentially life-threatening event if an abscess occurs. The maxillary right third molar had been lost for some time. This troubled tooth likely became mobile during the process of attrition. In contrast, the Horn Shelter No. 2 skeleton features several pulp exposures on the mandibular teeth from attrition, with no exposures evident on the maxillary arch. The La Jolla W-12 skull features exposures on every mandibular tooth that was present, but no pulp exposures on the maxillary arch.

Kennewick Man's maxillary first molars are very slightly rotated lingually on the left and right sides. Prior to enamel deterioration, these crowns exited the bony arches vertically and would have been upright. Canted occlusal surfaces developed over time as the enamel attrition became more severe on the lingual side of these teeth. As the enamel wore down, occlusal forces were increasingly directed lingual to the center of rotation of the first maxillary molars. These forces caused those molars to rotate lingually, resulting in the development of a molar crossbite on the right side at the first molars. A molar crossbite occurs when a maxillary molar's buccal cusp tips have "crossed over" from their normal position (buccal to the buccal cusp tips of the mandibular molar) to a position where the maxillary cusp tips are not buccal to the mandibular molar.

The loss of crown height from attrition led to compensatory alveolar ridge bone growth of apparently equal magnitude and direction. This can be a physiological response to slow dental structure loss (Crothers 1992).

A cephalometric study of Kennewick Man determined that alveolar ridge growth offset the crown height loss, preserving facial proportions.

Although significant enamel attrition occurred over time, Kennewick Man would have had normally shaped and sized teeth, a determination based on visible root morphology and remaining crown structure. His teeth were well aligned throughout most of his life with a stability in alignment unique to ancient dentitions. Before the effects of attrition gained the upper hand, he would have had a pleasing smile.

With the combined effects of enamel attrition and alveolar bone growth, Kennewick Man would have had an excessive gingival display—a significant amount of gingiva visible while speaking or smiling. With age, the incisors became short in clinical crown height and featured only a one-millimeter narrow band of enamel remaining adjacent to the cemento-enamel junction.

OCCLUSION

A cephalometric study using Dolphin® software was conducted on Kennewick Man for this analysis. The digitized landmarks from this study are listed in Table 10.1. The original incisor position, prior to the effects of severe attrition, was approximated. After years of attrition, Kennewick Man's lower jaw may have worked into a more prognathic position, a condition that has been reported with severe dental attrition. It is also possible he may have had a slightly prognathic mandible from the start. His dental wear pattern suggests a "long centric"—a

TABLE 10.1 Skeletal and dental measurements of Kennewick Man and comparative populations.

Variable		Broadbent et al. 1975	Basyouni and Nanda 2000	58 male patients Hayes 2010	Kennewick Man	Horn Shelter No. 2	La Jolla W-12
				Skeletal values (mm)			
1	Mx-Mx	68	69 to 83	N/A	65	64.5	63
2	Go-Go	98.5	89 to 106	N/A	110	100	100
3	Ag-Ag	88	N/A	N/A	92	86	86
4	Differential (3 less 1)	20	N/A	N/A	27	21.5	23
5	Ln-Ln	N/A	27 to 34	N/A	25	27	23
				CAC-CAC (mm)			
6	Mx-Mx	N/A	N/A	40 average	51.5	52	54
7	Md-Md	N/A	N/A	45 average	47	47	49.5
8	Differential (6 less 7)	N/A	N/A	neg. 5 maxillary deficiency	4.5	5	4.5
				Dental values			
9	centroid (6-6) mm	N/A	N/A	N/A	47	N/A	N/A
10	lingual (6-6) mm	N/A	N/A	N/A	37	N/A	42
11	buccal (6-6) mm	N/A	N/A	N/A	57	N/A	N/A
12	SNA°	82			84[a]		
13	SNB°	80			81[a]		
14	ANB°	2			3.5		
15	FMA (MP-FH)°	23			11		
16	U1-L1°	130			139		
17	IMPA°	95			98		
18	MP-Op°	16			9		
19	Wits mm	-1			-0.4		

[a]Estimated sella position.

smooth forward and backward slide of two millimeters in which the jaw movement would have been unimpeded by cusp tips.

Maxillary first molars were positioned slightly forward of an ideal occlusion. By measuring the teeth, it was found that the maxillary posterior teeth (molars and premolars) lost a slight amount of mesial/distal crown structure as the occlusal surfaces wore down into the narrower root portion of the teeth. However, the mandibular posterior teeth lost slightly less interproximal tooth structure. As a result, the maxillary posterior teeth were able to migrate approximately one to 1.5 millimeters more than the mandibular teeth. Anterior migration of teeth and the closing of interdental spaces have been reported to be common in human dentition. The differential migration caused a gradual change in his occlusion.

Curve of Spee (COS) is a measure of the concavity of the occlusal plane of the mandible in an anteroposterior direction. Many skulls, from hominins to early *Homo sapiens*, as well as most present-day dental patients, display some degree of COS ranging from slight to marked concavity. Kennewick Man's COS is flat, meaning there is no concavity. No functional implications were drawn from this finding.

When dental casts of well-worn arches are held in proper apposition to each other, it is expected that opposing teeth will be in contact with one another. Kennewick Man's arches fit that expectation with the exception of the first premolars. Kennewick Man may have habitually held or manipulated something in his mouth, possibly cordage, with enough frequency to cause a bilateral intrusion on his lower first premolars, and forcing them out of contact with the maxillary dentition.

Dental casts were mounted on a dental articulator in an attempt to approximate mandibular excursive movements. During left and right side lateral excursions in the range of up to 10 millimeters, only the second molars were in contact (molar guidance occlusion). A more common expectation would have been lateral excursions guided either by the cuspids alone, or given severe attrition, by the cuspids, premolars, and molars (group guidance occlusion). During protrusive excursions, Kennewick Man's incisors acted in concert as evidenced by their wear facets.

DENTAL AND SKELETAL ANALYSIS

Table 10.1 provides dental and skeletal measurements for Kennewick Man, Horn Shelter No. 2, and La Jolla W-12. The table shows comparison values derived from the Bolton Standards study (Broadbent et al. 1975) of modern Caucasians from Cleveland, Ohio, as well as transverse facial measurements of contemporary males of northern and western European ancestry (Basyouni and Nanda 2000). Measurements of bimaxillary width (Mx-Mx), bigonial width (Go-Go), biantegonial width (Ag-Ag), and bilateronasal width (Ln-Ln) were taken (Table 10.1, variables 1, 2, 3, and 5; Figure 10.1). Kennewick Man's skeletal and dental values are remarkably consistent with the norms used today for Caucasians as revealed by the Bolton Standards study. Kennewick Man's facial skeletal morphology would be considered by most orthodontists to be well proportioned and, in several respects, ideally formed.

The Center of Alveolar Crest (CAC) is a new technique used to measure the skeletal width of both maxillary and mandibular arches at the alveolar crests (Hayes 2003) (Figure 10.2). Kennewick Man's arches were compared

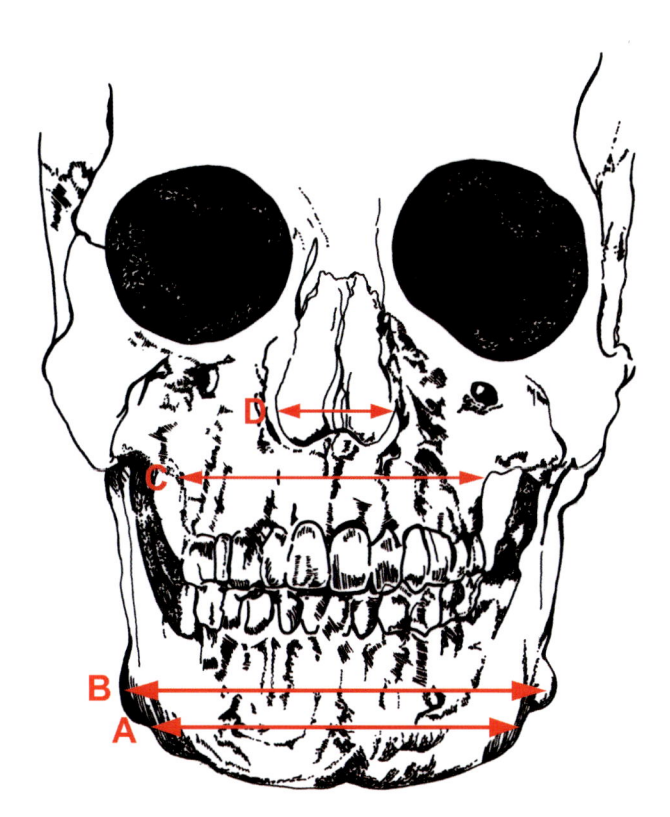

FIGURE 10.1   A: Biantegonial width (Ag-Ag); B: Bigonial width (Go-Go); C: Bimaxillary width (Mx-Mx); D: Bilateronasal width (Ln-Ln).

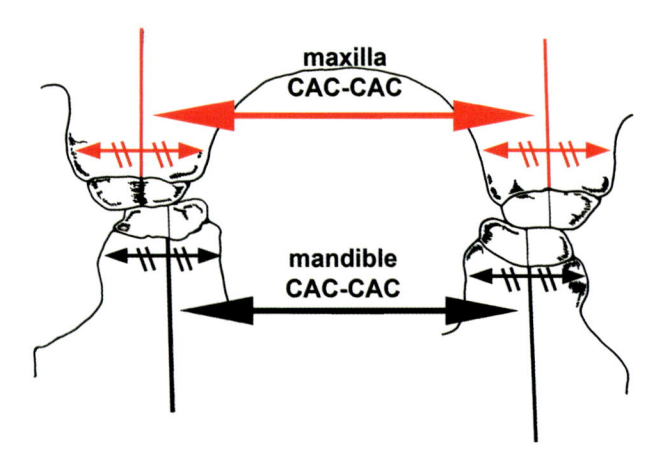

FIGURE 10.2 Cross-sectional view of CAC-CAC taken on 6-year molars.

with a sample of 58 modern Caucasian males from central Pennsylvania (Hayes 2010). Kennewick Man's CAC values were 51.5 mm (maxillary) and 47 mm (mandibular) resulting in a differential measurement (maxilla versus mandible) of 4.5 millimeters. This differential value is consistent with those of Horn Shelter No. 2 and La Jolla W-12 (Figure 10.3). Modern maxillary arches, on average, may be judged to be deficient based on the skeletal arch harmony seen with these ancient North Americans and with other skulls approximately 200 years old and older (Hayes 2003, 2008, 2010).

SUMMARY

In the prehistoric past, the condition of one's teeth was a prominent factor in determining everyday pain as well as longevity. Kennewick Man had to cope with dental problems, as did his contemporaries. During the early Holocene, those fortunate enough to reach what is now considered middle age were often unlucky with their teeth. Kennewick Man had numerous dental problems that probably caused varying levels of daily discomfort and even pain. Sensitive dentin was exposed on all teeth. Chewing would have become more difficult once the surface enamel was breeched and the softer underlying dentin wore flat.

By modern standards, his smile would have been pleasant into his twenties, but it began to degrade shortly thereafter because of abrasive contaminants in the diet and use of his teeth in task-related activities. By the last decade of his life, Kennewick Man would have had excessive gingival display from the effects of attrition and compensatory alveolar bone growth.

Facial proportions of Kennewick Man are remarkably consistent with modern norms. The Horn Shelter No. 2 and La Jolla W-12 skulls were not complete enough to draw conclusions regarding their orthodontic facial proportions.

Kennewick Man's teeth reveal unusual alignment stability, considering modern expectations, until affected

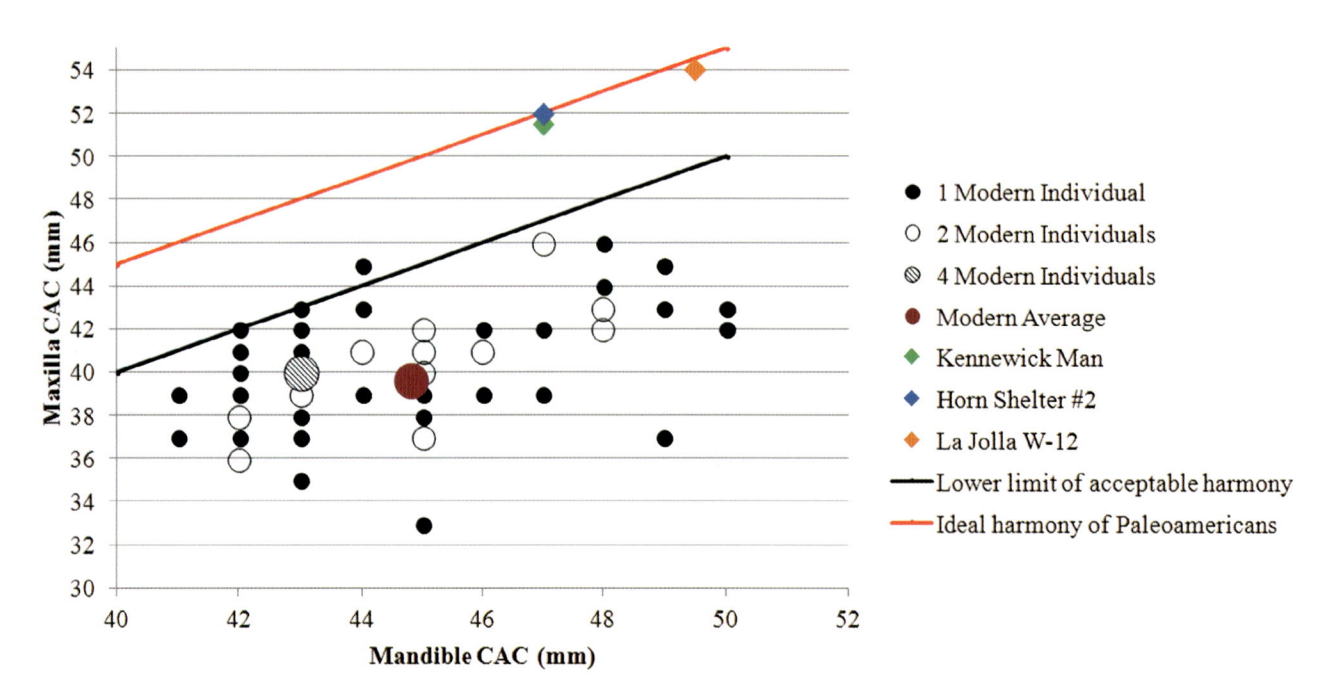

FIGURE 10.3 Caucasian male arches characterized by CAC ($n = 58$).

by severe dental attrition. Dental and skeletal analyses reveal that, prior to dental attrition, Kennewick Man's posterior teeth, and the teeth of Horn Shelter No. 2 and La Jolla W-12 were upright and well centered on their alveolar crests without either buccal dental inclination (maxillary arch) or lingual dental inclination (mandibular arch). This was likely possible due to skeletal transverse harmony resulting from their given maxillary and mandibular skeletal differential as measured by the CAC technique. It is a harmony not found with modern arches.

REFERENCES

Broadbent, B. Holly, Sr., B. Holly Broadbent, Jr., and William H. Golden. 1975. *Bolton Standards of Dentofacial Developmental Growth*. The C.V. Mosby Company, Saint Louis, MO.

Basyouni, Ahmed A., and Surender K. Nanda. 2000. *An Atlas of the Transverse Dimensions of the Face*. Center for Human Growth and Development, University of Michigan, Ann Arbor.

Crothers, A. J. R. 1992. Tooth wear and facial morphology. *Journal of Dentistry* 20(6):333–341.

Hayes, John L. 2003. "A Clinical Approach to Identify Transverse Discrepancies." Presentation to the Pennsylvania Association of Orthodontists, Philadelphia.

———. 2008. *The Williamsport Orthodontic Study*. Practice Based Research Network, University of Pennsylvania, School of Dental Medicine, Orthodontic Department.

———. 2010. In search of improved skeletal transverse diagnosis. Part 2: A new measurement technique used on 114 consecutive untreated patients. *Orthodontic Practice US* 1(4):34–39.

Body Mass, Stature, and Proportions of the Skeleton

Benjamin M. Auerbach

Body mass, stature, and the relative lengths of body sections (body proportions: distal limb elements to proximal elements, torso height relative to lower limb length) are important ecological indicators for various species (Allen 1877; Bergmann 1847; Peters 1983; Schmidt-Nielsen 1984), including humans (Coon 1955; Roberts 1978; Ruff 1991). Variations in morphology potentially indicate adaptation to environmental factors (such as climate, subsistence, mobility) and the affinities of groups to each other. In studies of the earliest inhabitants of the Americas, reconstructions of these morphological traits from the most ancient human remains could aid in the interpretation of origins and adaptations for past peoples. Such skeletons are rare, and the preservation of the required elements to estimate the morphological traits listed above is unusual.

Kennewick Man's skeletal remains are very complete. With the exception of the Spirit Cave skeleton recovered in Nevada, no other adult humans in North America dating to circa 9,000 years old or older are as complete as Kennewick Man's skeleton. A number of morphological traits can be estimated from Kennewick Man with greater accuracy than is possible for many skeletons from even half the antiquity of either Kennewick Man or Spirit Cave (Auerbach 2007). For this reason, Kennewick Man's skeleton provides a unique and valuable glimpse into the physical appearance of humans occupying the Americas during the Boreal period, following the last major glaciations of the Pleistocene.

This chapter discusses the methods for calculating and the resulting estimations of stature, body breadth, body mass, and body proportions of Kennewick Man's skeleton. These estimations are compared with other early Holocene humans from the Americas who are more often referred to as Paleoamericans or First Americans. The morphologies of the early Holocene humans are placed into context by comparing their variation with the diversity found in more recent precontact groups from the Americas.

OSTEOMETRY

Kennewick Man's skeleton was made available to the author for observation and measurement on February 24, 2006. All measurements taken were part of an established measurement protocol (Auerbach 2007) that consisted of standard osteometric measurements (Martin 1928; Raxter et al. 2006; Ruff 2002a) taken three times each and averaged. An extensive testing of these measurements on multiple skeletal samples has ensured low measurement errors (< 1 percent) for all of them. These measurements were obtained with a 500-millimeter field osteometric board, a pair of 300-millimeter spreading calipers (both manufactured by Paleo-Tech Concepts), and a pair of 150-millimeter Mitutoyo digital calipers, linked directly to a notebook computer. All data were directly entered into a Microsoft Excel spreadsheet.

Generally, all measurements could be obtained by direct measurement of skeletal elements. The long bones of Kennewick Man's skeleton had been broken taphonomically into contiguous pieces, which were carefully reassembled with the aid of conservators. Some measurements were estimated. Although carbonate concretions prevented the direct measurement of some skeletal dimensions, none of the osteometrics considered in this chapter was among those.

Stature

Stature is calculated in humans using mathematical or anatomical methods (Dwight 1894; Lundy 1985; Raxter et al. 2006). The mathematical method is based on the proportional correlation of element dimensions with stature (most commonly using regression equations). This technique requires knowledge of the ratio proportion of long bone lengths—namely femora—to stature, and benefits from knowledge of intralimb element proportions (brachial and, more apropos, crural indices), as these significantly vary among populations (Holliday and Ruff 1997). In some instances, the mathematical method has been used by researchers despite a lack of population-specific regression formulae. For example, Trotter and Gleser's (1952) stature estimation equations based on people of European descent have been used on Inuit skeletons, despite considerable differences in ancestry. This is justified by citing similar crural indices between the reference sample (from which the estimation equations were developed) and the test sample (Auerbach and Ruff 2004). It is noteworthy that the relative length of the torso to the lower limbs varies among human groups (Holliday 1997). Thus, the use of stature estimation formulae not specific for a given group, though based on another group with similar crural indices, is likely to result in indefinite estimation error (Auerbach and Ruff 2010).

The anatomical method involves the summation of the contributory heights of the skeletal elements composing stature, from cranium to calcaneus. This inherently takes into account the variation in limb and body proportions among human groups. Although other researchers developed anatomical stature estimation methods (Stewart 1979), Fully's (1956) method is increasingly used and regarded to be the best method for the approximation of stature from skeletal remains (Raxter et al. 2006). This method and its soft tissue correction are universally applicable to skeletons of all ancestries and both sexes. In a reexamination of the Fully (1956) method for anatomical stature estimation (Raxter et al. 2006, 2007), a revision of the Fully technique was demonstrated to produce more accurate estimations than mathematical method equations on a sample of skeletons with known statures. The disadvantage of the Fully method is its dependence on nearly complete skeletons; mathematical stature estimation methods are less time consuming (one or two long bone length measurements versus multiple elements) and do not rely on the preservation of multiple skeletal elements. Lundy (1985) and Auerbach (2011) discuss the relative merits of applying these techniques.

Only two mathematical method stature estimation equations based on Native New World reference samples are widely employed (Genovés 1967; Sciulli et al. 1990), in addition to equations more recently developed by Auerbach and Ruff (2010). These are inappropriate for use with Kennewick Man's skeleton, or any other early Holocene skeleton. The Hispanic sample used by Genovés, the Prairie archaeological sample used by Sciulli et al. (1990), and the various North American samples used by Auerbach and Ruff (2010) postdate Kennewick Man by millennia, during which time climatic, subsistence, and various other substantial environmental changes occurred (Stahl 1996) in addition to changes in genetic variation and patterns of gene flow (Schurr 2005; Kemp and Schurr 2010). These changes undoubtedly affected body proportions (Ruff 2002b). The equations also assumes genetic continuity between the humans of the early Holocene and those from more recent centuries. Thus, the use of the anatomical method for stature estimation is favored, as it makes no assumptions about body and limb proportions.

Kennewick Man's skeleton preserves a majority of the elements necessary for the anatomical stature estimation technique. Raxter et al.'s (2006) revised Fully method was chosen due to its demonstrated accuracy. This technique uses the following osteometric measurements in calculating skeletal stature, from which living stature is derived:

- basion-bregma height of the cranium,
- the maximum anterior superoinferior height of the second cervical vertebra (C2), including the odontoid process (dens),
- the maximum superoinferior height of the anterior margins of vertebral centra, anterior to the pedicles and rib facets, for all inferior cervical (C#), thoracic (T#), and lumbar (L#) vertebrae (C3 to L5),
- the maximum anterior superoinferior height of the first sacral element (S1),
- femoral bicondylar (physiological) length, averaged between left and right sides,
- tibial "Fully" length, from the most distal point of the medial malleolus to the most proximal point of the lateral condyle (generally at its most lateral aspect),
- talocalcaneal height, measured by rearticulating the talus and calcaneus and measuring them from

trochlea to tuber, orienting them so that the talar trochlea forms a 90-degree tangent with the osteometric board at its midpoint (Raxter et al. 2006 gives a diagram of this measurement).

In the case of Kennewick Man's skeleton, most of these standards were directly measurable at the time of observation. The individual measurements are provided in Table 11.1 (in addition to upper limb bone lengths used in other analyses). Basion-bregma height was taken on both the original cranium and on an exact rapid-prototype replica created from computed tomography (CT) scans of the original skull bones. Among the vertebrae, C4 and T10 were both eroded along their anterior midline margins, but maximum centra heights were still observable. T2 and T5 were cleaved sagittally through their centra, but the maximum observable heights were taken where they occurred. The superior margin of L1 was obliterated taphonomically, but, using the intact T12 to judge its maximum superior margin by rearticulating the elements, its height could be estimated. L2 was shattered and heavily eroded, prohibiting any measurements. L3 and L4 were largely intact and could be measured directly. The maximum height of L5 was estimated after a partial reconstruction of the shattered element using the large available fragments. A significant portion of S1 was preserved, still attached to the intact right ala, such that estimation of its height was feasible. All remaining vertebrae could be measured without estimation. Both femora were broken into three contiguous pieces and were reassembled with help for measurement, though the distal end of the right femur had experienced minor distortion

TABLE 11.1 Measurements of Kennewick Man's skeletal elements (mm).

Measurement[1]	Side (when applicable)		Measurement[1]	Side (when applicable)	
	Left	Right		Left	Right
Basion-bregma	142.00		T10	22.73	
C2 (including dens)	38.32		T11	22.88	
C3	12.71		T12	25.05	
C4	11.96		L1	24.55*	
C5	11.57		L2	*estimated*	
C6	12.68		L3	27.98	
C7	14.85		L4	28.88	
T1	17.16		L5	28.93*	
T2	17.14		S1	30.61*	
T3	17.56		Humeral maximum length	340.50*	341.00
T4	19.07		Radial maximum length	255.50[†]	262.00
T5	20.69		Bi-iliac breadth	281.00*	
T6	20.48		Femoral bicondylar length	465.00	463.50*
T7	22.23		Tibial maximum length	403.00	397.00*
T8	22.08		Tibial "Fully" length	395.50	393.00*
T9	23.01		Talocalcaneal height	*not available*	77.00

[1]See Raxter et al. (2006) for a description of all dimensions except the maximum lengths of the humeri, radii, and tibiae, which were measured using established methods (Martin, 1928).
*These dimensions involved some estimation in their measurements, due to taphonomic breakage and minor distortion. See the text for detailed descriptions of their estimations.
[†]Note that this measurement was estimated by aligning the proximal and distal fragments of the left radius with the intact left ulna, rearticulating the radial head with the radial notch and the ulnar notch with the radial articulation, and approximating maximum length of the radius. The completely intact right forearm bones were used to inform this reconstruction and estimation in the left forearm bones.

as a result of cracking and some postmortem breakage. Measurement of the bicondylar length of the right femur took this distortion into consideration, and an average of both sides' femoral bicondylar lengths is used for stature estimation. The left tibia had minimal erosion (except on the superior anterior aspect) and was broken into three contiguous pieces, which were reassembled for measurements. However, the right tibia—despite also consisting of contiguous pieces—had experienced more extensive erosion on its proximal end, which had broken in half parasagittally. An estimation of the right tibial Fully length was made at the time of measurement by attaching the lateral half of the proximal epiphysis to the intact diaphysis and distal epiphysis. An average of both tibial lengths is used in stature reconstruction. Finally, the talocalcaneal height could only be obtained for the right side, where both tarsals were completely intact; the left talar trochlea was missing, despite a completely intact left calcaneus. Three sets of measurements, taken in separate sessions, yielded very similar results for all of these estimations; the estimations are deemed reliable.

Thus, the only measurement that was not directly measured or estimated at the time of observations was the maximum height of the centrum of L2. However, protocols have been developed to estimate measurements of missing elements using the measurements of other elements taken from the same skeleton (Auerbach 2011). One method for estimating the heights of missing vertebrae is to average the known heights of adjacent vertebrae (Scuilli et al. 1990). This method is logical, as vertebral heights generally increase superoinferiorly, but is limited because it assumes a serial equal linearity in vertebral height increase. That is, it assumes that all vertebrae are of increasing and intermediate heights relative to the adjacent inferior and superior vertebrae. Though this is the trend for most vertebrae, there are important exceptions (Auerbach 2011). Table 11.2 reports the mean vertebral centra maximum anterior heights for all vertebrae, taken from a sample of 860 New World skeletons with complete vertebral columns. Note that most of the thoracic vertebrae and the lower lumbar vertebrae do increase at a nearly equal interval proceeding inferiorly. However, there are significant unequal intervals among the cervical vertebrae and transitional vertebrae in the thorax and lumbar regions. Table 11.2 shows the mean estimation errors for vertebral maximum heights estimated using adjacent vertebrae. Mean measurement errors with shaded cells in the table indicate estimations that are within measurement error (0.54 percent, on average, for all vertebrae; Auerbach 2007), and so are appropriate for the application of the averaging technique. Alternative methods to the averaging of adjacent vertebral heights are discussed in detail elsewhere (Auerbach 2011). L2 height estimations are highly accurate, and so the averaging of adjacent heights has been chosen for its estimation. This yields an estimated height of 26.27 millimeters for L2.

Adding together the measurements reported in Table 11.1 and the estimated height of L2 yield a "skeletal" stature of 159.691 centimeters for Kennewick Man's skeleton. This stature does not account for soft tissues, such as intervertebral discs, the heel pad of the sole of the foot, menisci on joints of the lower limb, or the scalp. In addition, there are gaps and overlaps between skeletal elements measured using this technique: namely, the space between the odontoid process and the foramen magnum, a gap between the inferior margin of S1 and the superior aspect of the femoral head, and an overlap between the talar trochlea and the malleolus (Raxter et al. 2006). Fully (1956) devised a discrete correction term to be added to skeletons between specific skeletal statures to account for these discrepancies. However, this correction term is inadequate and leads to underestimations of living statures. In addition, stature is known to decrease with age (Trotter and Gleser 1951) and therefore also needs to be accounted in the estimation of living stature from skeletal stature. Thus, the ordinary least square (OLS) regression equation that incorporates an age factor in its computation, developed by Raxter et al. (2006), is used instead to determine Kennewick Man's living stature:

$$\text{living stature} = 1.009 \times \text{skeletal stature [cm]} \\ - 0.0426 \times \text{age [years]} + 12.1.$$

Raxter et al. (2006) originally provided an alternative equation that left out the age factor, for use with archaeological samples in which age is estimated within a range and is not known precisely. This formula incorrectly assumes a mean age of 54 (the mean age for the sample from which the equation was derived) for the skeletons to which it is applied; this results in underestimates for the statures of many archaeological samples, wherein

TABLE 11.2 Mean maximum vertebral centra heights (mm) and mean estimation errors of estimations using the mean of adjacent vertebrae of Kennewick Man.

Vertebra[1]	Mean Maximum Centrum Height (mm)			Mean estimation error, all individuals[2]
	All (860)	Males (486)	Females (374)	
C1	10.67	11.12	10.11	N/A
C2	35.69	36.97	34.12	N/A
C3	12.48	12.97	11.88	N/A
C4	12.21	12.66	11.65	−.003 [0.024]
C5	11.94	12.32	11.46	−.257 [0.023]
C6	12.18	12.53	11.76	−.624 [0.022]
C7	13.67	14.03	13.23	−.285 [0.023]
T1	15.73	16.22	15.14	.193 [0.023]
T2	17.41	17.99	16.70	.804 [0.023]
T3	17.48	18.03	16.79	−.187 [0.022]
T4	17.92	18.49	17.21	−.059 [0.020]
T5	18.47	19.11	17.69	−.057 [0.021]
T6	19.15	19.87	18.26	.063 [0.023]
T7	19.69	20.44	18.78	.039 [0.022]
T8	20.16	20.90	19.27	−.064 [0.022]
T9	20.76	21.51	19.84	−.023 [0.021]
T10	21.41	22.13	20.52	.025 [0.025]
T11	22.00	22.65	21.20	−.597 [0.033]
T12	23.78	24.30	23.16	.191 [0.030]
L1	25.19	25.61	24.67	.405 [0.029]
L2	25.78	26.05	25.47	−.018 [0.029]
L3	26.42	26.71	26.06	−.053 [0.031]
L4	27.16	27.56	26.66	.194 [0.033]
L5	27.50	27.90	27.01	N/A

[1]Bolded vertebrae indicate vertebral heights that are able to be estimated reliably from the average of adjacent vertebral heights

[2]Shaded cells indicate vertebral heights with estimation errors within measurement error for the dimension.

mean individual age was likely less than 54. Even using approximate ages, as explained by Raxter et al. (2007), yields more accurate estimations than those of the equation without an age factor. Kennewick Man's age has been determined to range between 35 and 44 years, and so an approximate age of 35 was used for the calculation of his living stature. Using this approximate age, Kennewick Man's living stature is estimated to be 171.74 centimeters, which is equivalent to 67.61 inches.

Given the necessity of estimating the vertebral heights of L1, L5, and S1, and additional minor measurement error due to reassembly of the femora and tibiae, it is possible that the skeletal stature estimation is inaccurate by a few millimeters. However, it is unlikely that the early Holocene human found at Kennewick differed significantly in stature from the estimation derived above; the revised Fully method and its associated correction formula yielded stature estimates within 4.4 centimeters of known stature for 95 percent of the Terry Collection skeletons (National Museum of Natural History, Smithsonian Institution) on which it was tested, with a median absolute error of less than 1.5 centimeters. Thus, the estimate given *could* have ranged between 167.34 and 176.14 centimeters, though it is more likely that it differed by less than two centimeters, based on the results reported by Raxter et al. (2006). Therefore, the stature of Kennewick Man is estimated to be 171.74 ± 2 centimeters or between 66.83 and 68.40 inches.

Body Mass

Various methods are available for estimating human body mass (Auerbach and Ruff 2004). Although cranial measurements are used as a proxy for body mass, body mass estimates obtained using postcranial dimensions are substantially more accurate (Delson et al. 2000; Jungers 1988; Porter 2002; Ruff et al. 1999). Auerbach and Ruff (2004) discuss the relative merits of the two general techniques available for estimating body mass from postcranial dimensions: mechanical methods, which rely on dimensions functionally related to weight support (e.g., femoral head diameter, knee joint breadth, or talar trochlear diameter); and morphometric methods, which attempt to derive body shape and mass from direct reconstructions based on skeletal dimensions (Ruff 2002b; Ruff et al. 2005).

Mechanical methods for estimating body mass (Grine et al. 1995; McHenry 1992; Ruff et al. 1991) have some advantages over the morphometric methods. The measurements used in mechanical methods generally relate to the mechanical support of body mass and are articular dimensions or diaphyseal breadths or circumferences. Articular dimensions are unquestionably more favorable, namely because these are less sensitive to mechanical loads experienced after the cessation of primary growth (Auerbach and Ruff 2006; DeLeon and Auerbach 2007; Lieberman et al. 2001; Ruff 1988; Trinkaus et al. 1994). Furthermore, these estimates require far fewer observable skeletal dimensions (often only one measurement) than are necessary for morphometric methods (which require two or more dimensions, obtained from at least four elements). Femoral head diameter has become a commonly used estimator for body mass in humans (Sládek et al. 2006), as this articulation is subjected to the support of body weight and its external dimensions are subject to little change (other than pathological or traumatic causes) after the cessation of primary growth (Ruff et al. 1991).

Auerbach and Ruff (2004) demonstrated that, given the option, morphometric methods for estimating body mass, using stature and bi-iliac breadth (BIB) (Ruff et al. 1997; Ruff et al. 2005) produce less systematic bias. This is credited to their inclusion of more information about body shape in their calculation, their avoidance of complications created by the positive allometry associated with femoral head diameters, and their utilizing a more diverse human sample for the original equations to develop these estimations than those available for the mechanically based femoral head methods. A drawback for these equations, though, is the necessity of preserved os coxae and sacra, in addition to reliable methods for estimating stature. Therefore, morphometric methods are often not applicable to archaeological samples.

Kennewick Man's skeleton is exceptional. The man's stature is reliably estimated using the revised Fully anatomical stature estimation method. In addition, his pelvis is reasonably well preserved to allow for the estimation of his BIB. This measurement is the transverse (coronal) breadth of the articulated pelvis between the iliac (cristal) tubercles. The pelvis is articulated by matching the sacral auricular surfaces to the corresponding auricular surfaces of the os coxae. The pubic symphyses are articulated as well, and the sacral articulations are adjusted so that the sacrum fits as a keystone between the os coxae (Figure 11.1).

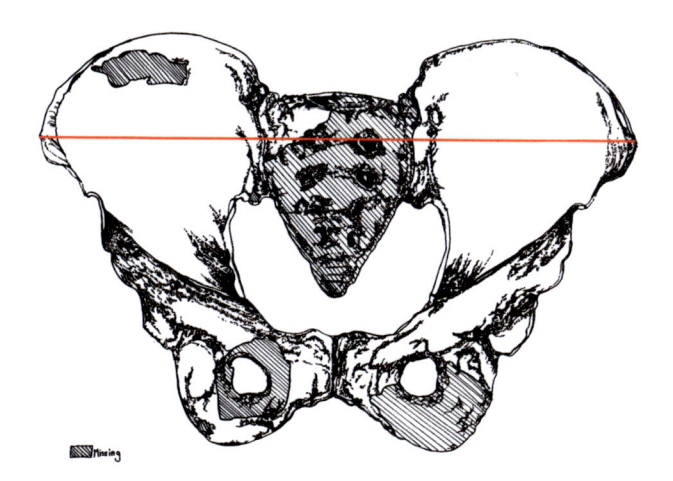

FIGURE 11.1 Bi-iliac breadth measurement on an articulated pelvis. The solid red line depicts the measurement. Unshaded regions represent portions of Kennewick Man's pelvis present at the time of observation.

Because of postmortem breakage, the measurement of BIB required more extensive reconstruction in Kennewick Man's skeleton at the time of measurement. His sacrum consisted of three large fragments, one of which included most of the right auricular surface and half of S1. The right os coxa, shattered into eight large fragments and found with a lithic point embedded in the iliac blade, had been virtually reconstructed and recreated by rapid prototyping as an intact model at the time of observation. Unshaded regions of the os coxae and sacrum in Figure 11.1 exhibit the portions of the pelvis intact at the time of measurement.

BIB was taken using three approximations. First, the width of the left ilium from the SI midpoint of the auricular surface to the lateral edge of the iliac crest was taken (47 mm) and added to the width of the right sacral fragment from the ala to the midpoint of S1 (90 mm). (The midpoint of S1 was still present on the sacral fragment.) This produces an estimate of a half of BIB, 137 millimeters. Second, the intact left os coxa was placed on the osteometric board, with the pubis reattached. The anteroposterior plane of the pubic symphysis was oriented to be perpendicular to the board, and the measurement from the pubic symphysis to the lateral edge of the iliac blade was taken (143 mm, again half of BIB). Third, the right os coxa rapid prototype model was placed in the same orientation on the osteometric board as the left had been placed in the second method, and the sacral fragment containing the right ala and S1 fragment was articulated with the os coxa. The midpoint of the S1 was aligned sagittally with the pubic symphysis and the measurement was taken (141 mm, also one-half BIB). The second and third measurements were nearly identical, and, when doubled, resulted in estimates that differed by only four millimeters. The summed measurements from the first estimation technique were similar to those of the second and third (8–12 mm different when doubled). A final check was performed at the time measurements were taken, prior to calculating BIB from the half-BIB estimates, in which the rapid prototype replica of the right os coxa, the left os coxa (with pubis reattached), and sacrum were all articulated in their relative anatomical positions (with the aid of two conservators). This reconstruction was measured to be 284 millimeters. Thus, the BIB taken has been judged to be a good estimation of the measurement were the pelves intact. For the purposes of this study, the average of the first three estimations (281 mm) will be used.

The morphometric method for body mass estimation is based, in principle, on a cylindrical model of human body shape, in which the diameter of the cylinder is equal to body breadth (BIB) and the length of the cylinder is equal to stature (Ruff 1991; Ruff and Walker 1993). Using this model, Ruff devised a relationship between these dimensions and body mass using a multiple OLS regression (Ruff et al. 1997), which he revised with colleagues to include a sample from high latitude (Ruff et al. 2005). The male formula devised in this revision is:

$$(0.422 \times \text{Living stature}) + (3.126 \times \text{Living BIB}) - 92.9.$$

Living BIB is skeletal BIB with a correction to account for soft tissue overlying the iliac tubercles (Ruff et al. 1997):

$$\text{Skeletal BIB [cm]} \times 1.17 - 3.0$$

Using these equations, the living stature determined in the preceding section, and the skeletal BIB explicated above, the body mass for Kennewick Man is determined to be 72.97 kilograms or approximately 160.9 pounds.

In order to assess the accuracy of this estimated body mass, the result of the morphometric method is compared with the body mass estimated from femoral head diameter, as these two methods are based on different augments (mechanical support versus body shape) (Auerbach and Ruff 2004; Ruff et al. 1997). Three femoral head estimation equations, using different samples, have been devised (Grine et al. 1995; McHenry 1992; Ruff et al. 1991). In Auerbach and Ruff's (2004) paper, reduced major axis regressions were generated to compare body masses estimated using these three femoral head equations and their average against body masses estimated using the morphometric method for body mass estimation; the results indicated that using an average of the three femoral head estimations yields body masses subject to the least systematic bias and closest in correspondence to the body masses calculated using the morphometric method. However, this was assessed using a global human sample that included skeletons outside the body mass range found in the Americas. Thus, using 835 New World skeletons with anatomically estimated statures and intact pelves, the same method was used (Auerbach 2007) to determine which femoral head equation was most appropriate for New World skeletons; this reassessment demonstrated that, in samples from the Americas, body mass is best approximated using the femoral head equation devised by Grine et al. (1995)[1], and not by the average of the three femoral head body mass estimation equations. The Grine et al. (1995) equation is:

Femoral head diameter [mm] × 2.268 − 36.5.

Using this equation on the average of the left and right femoral head diameters, Kennewick Man's skeleton has a body mass of 73.66 kilograms (approximately 162.4 pounds). This is less than one kilogram different from the body mass estimated using the morphometric method, suggesting that the male buried at Kennewick had a body mass around 73 kilograms. The close correspondence of results, then, is cited as validation of the body mass estimated using the stature and BIB morphometric method. Furthermore, it argues in favor for the accuracy of the BIB and stature estimations.

Limb and Body Proportions

Researchers in human morphological variation have documented the importance of variation in limb proportions (brachial and crural indices) and body proportions (relative torso length, relative body breadth) in models of human ecology (Danforth 1999; Holliday 1995; Jacobs 1985; Newman 1953; Ruff 1994; Ruff 2002b). Variation in limb proportions is thought to reflect morphological changes in total surface area, which in turn are hypothesized to be influenced—at least in part—by adaptation in response to climatic factors. Similarly, changes in body breadth influence the surface area/mass ratio of the body (Figure 11.2) and therefore are also hypothesized to vary in response to climatic factors (Ruff 1991). These proportions have been demonstrated to vary clinally between Europe and Africa, and through time in Europe (Holliday 1997, 1999). Holliday (1997) demonstrated that the relative length of the torso to the lower limbs is related to climatic factors, though other research has suggested that lower limb length is also sensitive to nutritional and related stresses (Bogin 1999; Tanner et al. 1982). Although the relative influence of environmental factors (e.g., climate, mobility and subsistence), genetic and epigenetic factors, and developmental factors on variation in these proportions is not fully understood, these and other studies indicate the utility of knowing morphological proportions in deciphering information about past climatic and subsistence adaptations among human groups.

The good preservation of Kennewick Man's skeleton allows for the estimation of many such proportions—namely, brachial and crural indices, relative torso to lower limb length, and relative body breadth. Humeral and radial maximum lengths (Martin 1928) on the right side were easily measured. Breakage and subsequent distortion in the proximal left humerus added some uncertainty (potentially .5 to 1 mm) to its maximum measurement, and the left radius was missing its diaphysis approximately from 66 to 45 percent of the maximum length, as measured from the distal point of the styloid process. Its maximum length was estimated by comparison with the right radius, and articulation of the proximal and distal fragments of the left radius with the left ulna. The estimated length of the left radius is reported in Table 11.1, along with the maximum length measure-

Lateral surface area:	πDL	$\pi 2DL$
Volume (mass):	$\pi/4 \, D^2 L$	$\pi/4 \, D^2 2L$
Surface area / mass :	$\dfrac{4}{D}$	$\dfrac{4}{D}$

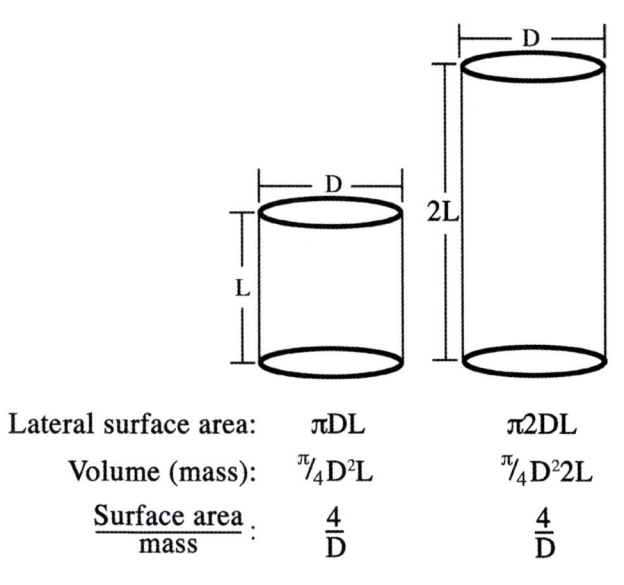

FIGURE 11.2 The cylindrical model proposed by Ruff (1991) for representing surface area-to-volume (mass) ratios among humans. Note that only variation in cylindrical diameter, and not height, results in changes in the ratio of surface area to volume. This Figure is redrawn after Ruff (1991) with permission.

ments taken for the other three upper limb bones. Only the bones of the right upper limb are used in determining brachial index (radial/humeral maximum lengths). The right femur exhibited minor distortion distally, and this was accounted for in the measurement of bicondylar length. Despite some erosion and the breakage in the proximal right tibia, its maximum length (which includes the intercondyloid eminence) could be judged based on its morphology and comparison with the left (intact) tibia. Averages of the left and right femora and of the left and right tibiae are used in calculating the crural index (tibial maximum/femoral bicondylar lengths). The average of the summed left lower limb elements and summed right lower limb elements provides a lower limb length (without tarsal height). This limb dimension is used in the examination of relative torso height, which is calculated by a modified version of Holliday's (1995) equation:[2]

$$\Sigma(T1 \text{ through } L5) + (FBL + TFL).$$

Finally, relative body breadth (Ruff 1991) is calculated by dividing the BIB by the living estimated stature. For Kennewick Man, this value is 0.1636.

COMPARISONS WITH OTHER EARLY HOLOCENE SKELETONS FROM NORTH AMERICA

Skeletons from the first two millennia of the Holocene in North America are exceedingly rare, and, as noted at the beginning of this chapter, are often incomplete. Comparative samples are restricted in this analysis to North America because of the uncertainty of the relationship of and likely significant difference in subsistence patterns of South American humans from the early Holocene with North American humans (Gruhn 2005; Munford et al. 1995). Only one other North American skeleton—found in Spirit Cave, Nevada—is as complete as Kennewick Man's skeleton.[3] Given the antiquity of these skeletons and their generally unplanned or unintentional discovery (Powell 2005), this is an unfortunate reality that limits the available sample of contemporaneous male skeletons, even in a broad temporal sense, to contrast with the morphological traits reported for Kennewick Man's skeleton.

Fortunately, four adult male skeletons (in addition to Kennewick Man's skeleton), dating from between 10,000 and 8,000 radiocarbon years before the present, have been excavated that preserve bones from which some of these morphological traits may be extrapolated. Kennewick Man's skeleton dates to circa 9,400 years ago. Central western Nevada's Spirit Cave skeleton, dated using preserved textiles, has an antiquity of circa 9,430 years ago (Tuohy and Dansie 1997). Near to the burial of Spirit Cave, and approximately dating to 100 years more recent than Spirit Cave, the remains from Wizards Beach were found close to Pyramid Lake, Nevada (Dansie 1997). Early Holocene remains dating to approximately 8,300 years ago were uncovered in south-central British Columbia along Gore Creek (Cybulski et al. 1981). Finally, the adult skeletal remains from Horn Shelter (in central Texas along the Brazos River) date to a time period very close to Spirit Cave and Wizards Beach, at 9,980 ± 370 radiocarbon years (Young et al. 1987).

Multiple other isolated skeletal elements and fragmentary individuals have been uncovered in North America that preserve some postcranial elements (Warm Mineral Springs, Little Salt Spring, Wilson-Leonard, the La Brea skeleton, Gordon Creek, etc.), yet few of these present any dimensions for useful comparison with Kennewick Man's skeleton's morphology. Furthermore, some of these other early Holocene skeletons were female (Wilson-Leonard, La Brea, and Gordon Creek), and using them in comparisons would be confounding variation with potential sex effects. For example, brachial indices are known to vary significantly between males and females within the same population; females, regardless of environment, have short forearms (relative to arms) compared with males (Ruff 1991; Trinkaus 1981).

Measurements for Spirit Cave and Horn Shelter skeletons were taken using the same methods as those used with Kennewick Man's skeleton. The Spirit Cave skeleton, which had been mummified (the right arm of which still is articulated), is in very good condition, and all measurements were taken without need for estimation.[4] The Horn Shelter skeleton, however, has fragmentary pelvic bones, few vertebrae, a distally broken and eroded left radius, a shattered left femoral neck, and some postmortem breakage on both tibiae; the tibiae lengths were estimated by comparing both sides. Although the remains from Wizards Beach are still under the receivership of the Nevada State Museum, Carson City (where Spirit

Cave is also held), permission was not obtained to measure these remains. Instead, measurements taken by Drs. Douglas Owsley and Richard Jantz were provided for this skeleton. Direct observations by the author on the Gore Creek skeleton were not possible, as these remains have been reburied; however, a number of measurements were taken from this skeleton (Jerome Cybulski, personal communication 2006).

Body masses, statures, and proportions for the five early Holocene male skeletons are given in Table 11.3. As no reliable mathematical stature estimation equations exist for skeletons of their antiquity, and as they do not preserve the necessary elements by which to employ the revised Fully technique, statures are not available for Horn Shelter, Wizards Beach, or Gore Creek. In addition, only Spirit Cave has a measured BIB. Because of these limitations, the Grine et al. (1995) femoral head body mass estimation formula has been chosen to calculate body masses for three of the skeletons. Gore Creek's only applicable measurements are tibial maximum length and femoral maximum length, which was converted to femoral bicondylar length using an OLS regression formula calculated using an equation provided in Auerbach (2011). (The Spirit Cave body mass estimations using the mechanical and morphometric methods differ by less than 5 kg.)

Comparing the morphological traits in Table 11.3 reveals that, even within this very small sample, the morphologies of these early Holocene males were not homogenous. Spirit Cave and Horn Shelter, despite a separation of nearly 2,000 kilometers between their burials, had similar brachial and crural indices. Likewise, Kennewick Man and Wizards Beach had similar limb proportions (though their geographic separation was only one-fourth of the distance between Horn Shelter and Spirit Cave, their temporal separation was much greater). These similarities are borne out in Figure 11.3, in which radius maximum length is plotted against humerus maximum length (brachial index), and tibia maximum length is plotted against femoral bicondylar length (crural index). It is noteworthy that Gore Creek's crural index is more similar to those of Spirit Cave and Horn Shelter even though he lived in a similar environment to Kennewick Man (but 1,000 years later and may not have been related.) Despite these differences, both Kennewick and Spirit Cave had wide bodies, and all but Spirit Cave

TABLE 11.3 Estimated morphological traits for American pre-contact skeletons.

Sample	n[1]	Temporal Period (RC yr. BP)	Brachial Index	Crural Index	Relative Torso Length	Estimated Living Stature (cm)	Estimated Body Mass (Femoral Head) (kg)	BIB (mm)	BIB / Stature
Early Holocene North American skeletons									
Horn Shelter	1	9710 ± 40	.7906	.8368			69.14		
Spirit Cave	1	9414 ± 25	.7933	.8292	0.4933	160.84	61.63	279.0	.173
Kennewick	1	8358 ± 30	.7683	.8616	0.4504	171.74	73.66	281.0	.164
Gore Creek	1	8340 ± 115		.8447					
Wizards Beach	1	9225 ± 60	.7699	.8622			76.90		
Early to Middle Archaic and recent groups									
Sadlermiut	30	500–0	.7197	.8049	.4609	158.74	73.20	273.1	.172
Point Hope	22	500–0	.7491	.8286	.4650	159.29	70.43	280.1	.176
Interior Salish	8	500–0	.7819	.8276	.4701	157.16	63.42	264.4	.167
Indian Knoll	31	6000–4000	.7714	.8451	.4599	161.35	60.05	258.5	.161
Ellis Landing	12	6000–4000	.7873	.8382			71.11	277.1	
Windover	32	8000–6000	.7947	.8548	.4474	166.32	65.00	265.4	.161
Ancón	28	1500–500	.7786	.8558	.4915	158.16	68.02	269.2	.169

[1]This is the maximum number of skeletons used in any group sample, though in the cases of relative torso length, BIB, and stature, not all the measured skeletons had the necessary preserved elements.

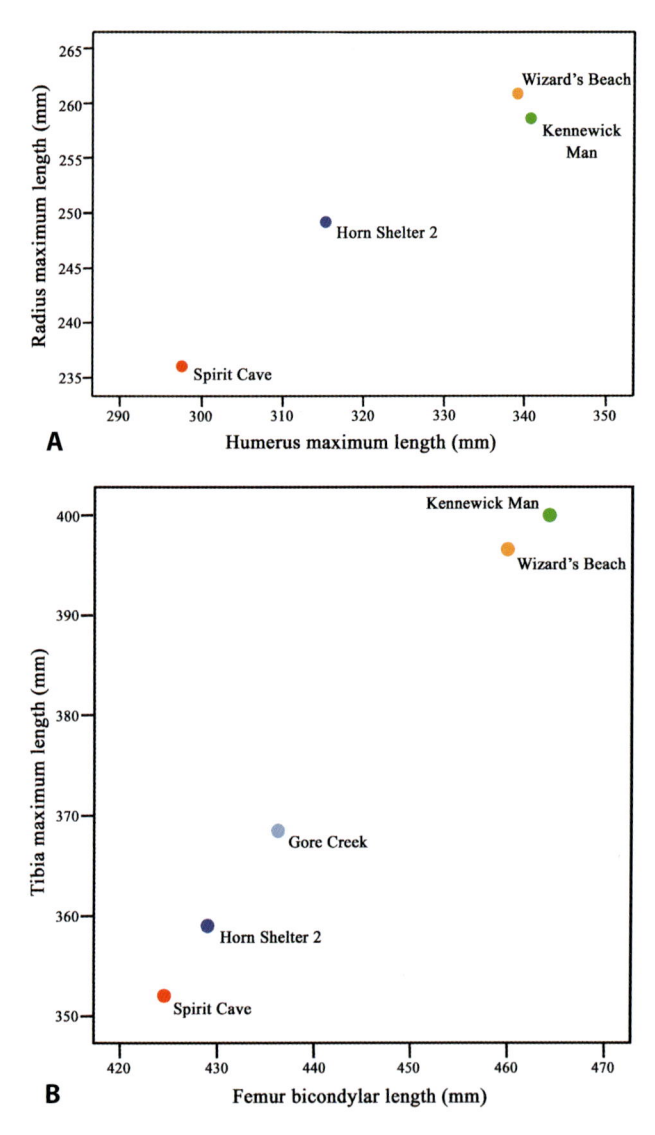

FIGURE 11.3 Brachial (A) and crural (B) indices of male North American early Holocene skeletons.

were massive individuals, compared to the mean male body masses for more recent groups (Auerbach and Ruff 2004; Auerbach 2007).

Without comparisons to patterns of variation in more recent humans, these conclusions lack context; it cannot be asserted that this variation represents as extensive a diversity as that found in more recent groups. Also, as these are individual samples, no statistical tests may reliably be used to assess their differences. It is interesting, though, that Kennewick Man and Wizards Beach seem to have similar morphologies (more than Horn Shelter and Spirit Cave to each other or to these other two skeletons), insofar that both were massive individuals with moderate brachial indices and high crural indices, as well as lower weight for their statures than Horn Shelter.

VARIATION IN NORTH AMERICAN EARLY HOLOCENE SKELETONS

In order to place the variation among the early Holocene males into context, their morphological traits are compared with those of males from seven more recent American groups (Table 11.3). The Windover site (8BR246) is the most ancient large cemetery excavated in North America, dating to circa 8,000 years ago, with over 140 burials (half of which remain unexcavated) (Doran 2002). Given the antiquity of this site and the large sample size it provides, it is useful to compare variation at Windover to variation in early Holocene skeletons. As the Windover site is located along the Atlantic coast of Florida, which was and remains a warm environment, samples from two other ancient sites from more temperate regions were chosen for analysis: the Indian Knoll cemetery (15OH2) from central western Kentucky (along the Green River) (Snow 1948; Webb 1946); and the Ellis Landing site (CA-CCo-295), which was located on the San Francisco Bay in California (Stewart and Praetzellis 2003). These two sites represent distinct Archaic cultures existing between at least 5,000 and 4,000 years ago. No cemeteries of similar antiquity have been discovered in higher latitudes. Rather, skeletons from three recent (within the last 500 years) higher-latitude samples are employed: the Arctic Sadlermiut excavated from Native Point, Southampton Island (Merbs 1974); skeletons from the Tigara culture of the Point Hope, Alaska, site (Larsen and Rainey 1948); and various Interior Salish individuals buried near Kamloops, British Columbia, on the Plateau (Fitzhugh and Krupnik 2006). The last individuals are of special relevance; they lived 9,000 to 8,000 years later in the same region as Kennewick Man's and Gore Creek skeletons, both of which were discovered on the Plateau. Finally, individuals from the Huari Empire necropolis at Ancón, Peru (ca. 1,300 to 700 years ago) are included as New World representatives of a more tropical morphology. The Sadlermiut, Point Hope, and Huari groups are chosen to provide examples of extreme morphologies found in the late Holocene of the Americas and therefore provide a broader context by which to judge earlier variation. With the exception of the skeletons from

Ancón, all the groups were hunter-gatherers, albeit with different dietary compositions and habitation patterns. The groups selected should minimize the effects of subsistence differences in the ensuing comparisons.

A couple of caveats should be noted before analyses using these samples ensue. Comparisons among these groups and with the early Holocene skeletons are merely informative of morphological variation. A number of factors, including specific climatic variables, temporal differences, variations in subsistence, cultural influences, and genetic differences, are not considered but undoubtedly had a significant effect on morphological variation. In addition, these samples do not likely represent biological populations; instead they are aggregations of multiple generations who buried their dead in a common locus (Larsen 1997). They do indicate the morphological trends for the biological populations from which they were drawn, which makes comparisons among them valid and useful.

Table 11.3 presents the means of the morphological traits of these more recent groups, estimated using the same methods described above for Kennewick Man. Examination indicates greater ranges among these samples than are present among the early Holocene males. The low brachial and crural indices in the two Arctic samples are notable, as are their high average body masses. A Kruskal-Wallis test with a Nemenyi post-hoc test for group differences indicates that the Sadlermiut have significantly lower brachial and crural indices than all other groups (including the Point Hope Eskimo) and significantly greater body masses than all other groups except the Point Hope people and individuals from Ellis Landing. Interestingly, only Windover males were significantly taller than the other groups, and both Windover and Indian Knoll males had significantly narrower bodies relative to stature. In fact, the individuals buried at Windover, based on their means, had morphological traits indicating an adaptation to warm climate compared to the other North American groups. This is further indicated by the relatively short torsos of Windover males. Unexpectedly, however, the Ancón skeletons have significantly longer torsos, relative to lower limb length, than the North American groups. As relative lower limb length is sensitive to nutrition and stress, and as the Huari were agriculturists, no environmental cause could be hypothesized as the primary contribution to this difference.

How does this variation among more recent groups compare with the differences observed among the five skeletons from the early Holocene? Figure 11.4 presents $\log_e$ bivariate plots of radius maximum length against humerus maximum length, and of tibia maximum length against femoral bicondylar length. (Natural logs are used to reduce heteroscedasticity.) The OLS regression lines are for two recent groups at the extremes for the ranges of these indices—the Sadlermiut and the Huari. The most ancient skeletons—the early Holocene and Windover males—tend to have high brachial indices, but more intermediate (though still high) crural indices. This is also

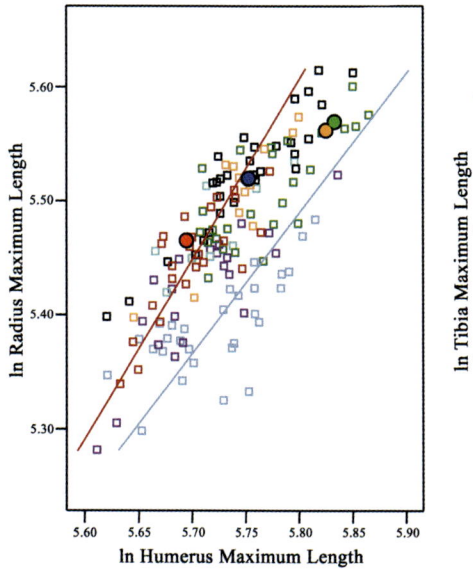

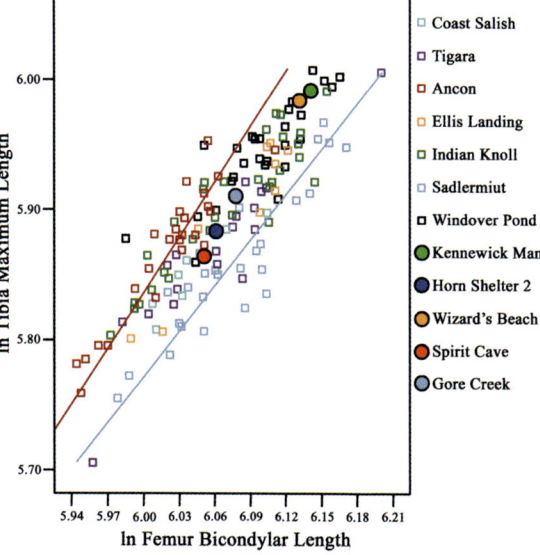

FIGURE 11.4 Brachial (left) and crural (right) indices for early Holocene, Archaic, and recent male skeletons. Two OLS regression lines are plotted: blue line, Sadlermiut; red line, Huari.

Benjamin M. Auerbach 223

the pattern for the Interior Salish. The early Holocene skeletons all plot close to the range of variation among the skeletons from Windover in both indices.

The intralimb patterns suggested previously among the early Holocene skeletons are upheld. Spirit Cave and Horn Shelter have brachial indices similar to those found among the Huari, while the Kennewick Man and Wizards Beach skeletons are intermediate (though outside the size range) between the Sadlermiut and the Peruvians. In contrast, the crural indices of all five early Holocene males have more intermediate proportions, though Kennewick Man and the Wizards Beach male have more similar proportions to the Huari than to Spirit Cave, Horn Shelter, or Gore Creek.

An examination of group means for estimated body mass in Table 11.3 reveals that the skeletons of Horn Shelter, Kennewick Man, and Wizards Beach are as massive as the Arctic groups, whereas Spirit Cave's body mass was among the lower end of the distribution, closest to the mean for males from Indian Knoll. Curiously, the body masses of males from Ancón were also among the highest in more recent groups. As demonstrated by Ruff (1994) and by Holliday (1995), in two groups with similar mean body masses, it is expected that those living in warmer environments would be taller, on average, than those living in colder climates. Table 11.3 reveals that the individuals from Windover and Indian Knoll were taller, relative to their body masses, than individuals in the Arctic examples and, surprisingly, Ancón. As these morphological characteristics are sensitive to a number of environmental factors, the reasons for the relatively high body mass, relative to stature, among the Huari remain elusive. It is evident that, although Spirit Cave is considerably shorter than Kennewick Man and Wizards Beach, all three have similar body masses relative to their statures, which in turn are similar to the individuals at Windover and Indian Knoll.

This pattern seems to contrast with the body breadths of Spirit Cave and Kennewick Man: Spirit Cave was more stocky—wider-bodied relative to his stature—compared with Kennewick Man. Assessing their relative body breadths against the more recent groups in Table 11.3, this assertion is maintained; Spirit Cave had a relative body breadth similar to those found in the Arctic, whereas Kennewick Man has a similar morphology to Windover and to Indian Knoll, as well as the Interior Salish. Figure 11.5 presents this proportion visually. Despite the body shape difference between these two early Holocene individuals, their purported climatic adaptation is similar. Ruff (1991, 1994) demonstrated that, within similar climatic conditions, body breadth remains constant, even as stature varies. However, in colder climates, body breadth increases, therefore decreasing the surface area relative to mass (Figure 11.2). Spirit Cave and Kennewick Man both have body breadths like those found among the Point Hope Eskimo—the widest-bodied more recent sample—and less like those of Windover or Indian Knoll. Thus, Kennewick Man may have had a similar appearance in relative body breadth to the individuals living in eastern Florida 1,000 years later, but his overall breadth, in context, implies an adaptation to colder climates, or a conserved morphology from his ancestors, who had been adapted to colder climates. Indeed, compared to the body breadths of humans in a global sample (Auerbach and Ruff 2004), all the skeletons, except those from Indian Knoll, fall within the range of higher latitude groups.

This is a curious assertion, as, in a greater context, the

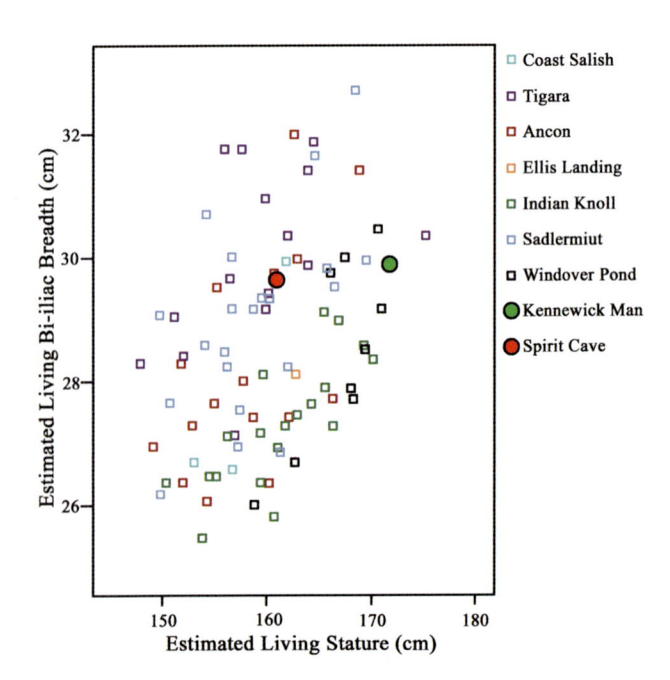

FIGURE 11.5 Relative body breadth of early Holocene, Archaic, and recent male skeletons.

relative body breadths of these New World skeletons—ancient and recent—are on the higher end of the relative body breadths reported for living Africans (.148–.174) and on the lower half of those reported for living Europeans (.160–.188) (Ruff and Walker 1993). Some individuals from the Arctic have high relative body breadths (.190 in one Point Hope male), and a few males from Indian Knoll and Windover range close to the lowest relative body breadths observed among Africans (e.g., .153 in a male from Indian Knoll). Despite these extremes, the American groups are as broad-bodied but not so short as higher-latitude groups from Eurasia.

Given the subtle but notable variation in specific morphological traits among the early Holocene individuals in these comparisons, how do they relate holistically? A discriminant function analysis (DFA) is one method to address this question. Ratio values (without arcsine transformation), unfortunately, invalidate the statistical assumptions of DFA, and so linear measurements have been chosen for the analysis. $\log_e$ conversions of humeral maximum length, radial maximum length, femoral bicondylar length, femoral head diameter, and tibial maximum length are entered into the DFA. BIB was not used to include a larger portion of the comparative sample, as well as Horn Shelter and Wizards Beach. BIB was also considered unnecessary because variation in body breadth only significantly differs between the Arctic groups and Indian Knoll. The early Holocene skeletons are not factored into determining the discriminant functions. The resulting analysis produces five functions, the first two of which explain 89.6% of the total variance and are chosen for further investigation. The correlation coefficients of variable loadings on these first two functions and the percent probabilities of assignment of early Holocene males to more recent groups are listed in Table 11.4; their discriminant scores are presented in a bivariate plot in Figure 11.6. Function 1's highest correlations are with radial length and femoral head diameter; individuals plotting higher on this axis have relatively longer forearms and are less massive (and therefore are, ostensibly, more warm climate-adapted). It is interesting that femoral length has little effect on this function, implying that variation in stature again does not discriminate strongly among

TABLE 11.4A Function loadings for discriminant function analysis.

Variable	Function 1	Function 2
$\log_e$ Humerus maximum length	.194	–.370
$\log_e$ Radius maximum length	.539*	–.066
$\log_e$ Femur bicondylar length	.040	–.391
$\log_e$ Tibia maximum length	.247	–.142
$\log_e$ Femoral head diameter	–.415	.305

*Largest absolute correlation between each variable and any discriminant function.

TABLE 11.4B Highest assigned probabilities of group membership of early Holocene males to more recent samples.

Early Holocene skeleton	Recent sample affiliation	Percent probability
Kennewick Man	Indian Knoll	66.80
Horn Shelter	Indian Knoll	42.65
Wizards Beach	Indian Knoll	68.53
Spirit Cave	Windover	22.27

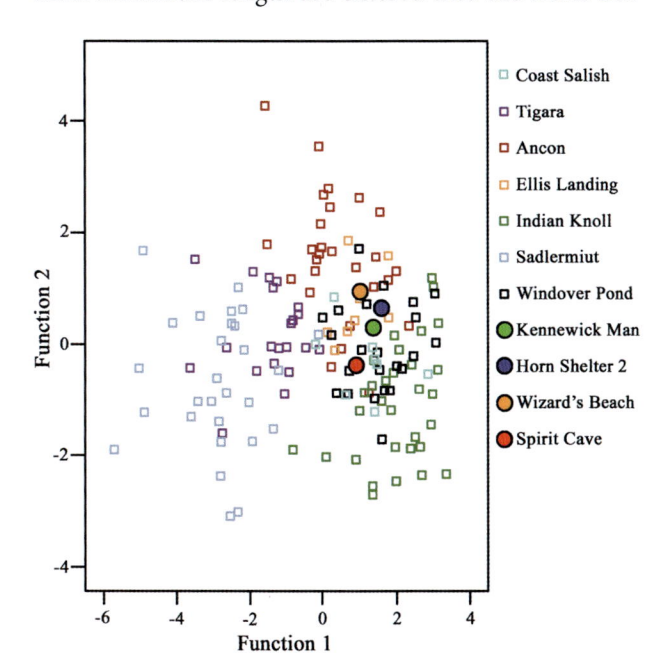

FIGURE 11.6 Function 1 ($\lambda = 3.47$, 71% of total variance) and Function 2 ($\lambda = .908$, 18.6% of total variance) bivariate plot of discriminant function analysis. See Table 11.4 for function loading correlation values.

these groups. The higher contribution of tibial length, relative to femoral length, may be an effect of greater variability in the length of the tibia (Auerbach 2010; Auerbach and Sylvester 2011; Holliday and Ruff 2001). Function 2 is more ambiguous, though it appears to relate to size: individuals plotting positively on this axis are more massive relative to their body lengths, though these are admittedly weak relationships.

The bivariate plot in Figure 11.6 demonstrates that the DFA is able to yield some group distinctions. The Arctic groups generally separate out from the lower-latitude groups on Function 1, while there is little separation in the second function. It is interesting that the early Holocene males clearly plot with the lower latitude groups. Moreover, the early Holocene males plot, again, within the range of males from the Windover site, as well as close to the range of those from Ellis Landing. Both of these, perhaps coincidentally, were Archaic coastal-dwelling groups. Note that Kennewick Man and Spirit Cave also plot within the range of the more recent Interior Salish of the Plateau.

CONCLUSION

Cautious reserve must be used in interpreting the morphological patterns described here. As with any analysis including isolated specimens, generalizations are forfeit to the effects of adding additional isolated specimens. Were another highly complete male skeleton from the early Holocene discovered and added to these analyses, the interpretation of overall morphological diversity could be drastically altered. After all, despite the great geographic distance among the five early Holocene skeletons' burials, none lived in the extreme climatic zones typically used to contrast the great differences in human morphological traits and proportions. Attributing the morphological appearances of these males to climatic adaptations of these ancient individuals, their populations, or their ancestors would be premature and potentially erroneous.

Nevertheless, some preliminary assertions about males living at the beginning of the Holocene may be made. Some morphological variation was already present in North America at the Pleistocene-Holocene transition. This agrees with the variation implied by studies of crania from the earliest American skeletons (Jantz and Owsley 2001, 2005; Powell and Neves 1999). Yet, diversity in body morphological traits—stature, body mass, and proportions—was not so extensive in the early Holocene as has been present in later time periods. Interestingly, the patterns for the earliest Holocene skeletons also do not match clinal expectations based on patterns documented for more recent humans (Jantz et al. 1992; Newman 1953; Roberts 1978; Ruff 1994, 2002b; Stinson 1990); higher brachial and crural indices would be expected in individuals living in putatively warmer climates, but only Horn Shelter exhibits a high brachial index. Kennewick Man, Gore Creek, and Wizards Beach, in contrast, have high crural indices. This is significant, as Ruff (1994) suggested that crural indices are more sensitive to climatic factors but may take longer to adapt than brachial indices. Therefore, one might interpret these results to suggest that Horn Shelter and Spirit Cave were from a population or populations different from Kennewick Man and Wizards Beach, especially as all four lived within a few centuries of each other. Undeniably, the Kennewick Man and Wizards Beach skeletons are consistently similar in their overall morphologies, more than any other paired comparisons among early Holocene skeletons.

Among other morphologies, a general lack of variation among the earliest individuals is similar to more recent patterns in the Americas. Ruff (1994) hypothesized that body breadth requires more generations to be affected by climatic factors, perhaps due to locomotor or obstetric constraints. Most groups in the Americas maintained generally wide bodies, despite living in varied environments. This supports Ruff's (1994) assertions concerning bi-iliac breadth in the Americas, and furthermore argues that the colonizing ancestors of all of the skeletons considered here ultimately passed through a cold filter, most likely in the Beringian Arctic. Stature also did not vary significantly among the groups considered (with the exception of the tall males from Windover), but this is at least in part a result of restricted variation in subsistence among the groups considered. In contrast, despite putatively similar subsistence strategies, the Spirit Cave and Kennewick Man skeletons have considerably different statures and relative torso lengths.

Also with the exception of Spirit Cave, body masses for early Holocene males are high when compared to more recent groups and to a global sample (Auerbach and Ruff 2004). More research into the differences in diet and habitation patterns among these individuals will help to clarify whether subsistence, climate, or genetics was more influential in these differences. As an aside, in contrast to these traits, males living near Lake Baikal in southern Siberia circa 7,000 years ago had considerably narrower body breadths than Spirit Cave and Kennewick Man, intermediate intralimb indices for both upper and lower limbs compared with the total sample presented in this chapter, and body masses intermediate between Spirit Cave and the three other early Holocene males (Auerbach 2007). How this might relate to ancestor-descendant relationships between contemporaneous groups in Siberia and the Americas cannot be addressed, especially because the specific ancestry of none of these groups is known.

In conclusion, what may be stated about the body mass, stature, and proportions of Kennewick Man? Compared to more recent groups in the Americas, he was tall, broad-bodied, and massive. He had a relatively short torso, though compared with his body breadth, he was not stocky. His legs were long compared to his thighs, but his forearms were not especially long relative to his arms' lengths. At least one other early Holocene male, Wizards Beach, had a similar morphology to his. Notably, his overall morphology was not much different from the appearance of males from the Windover site, though he was considerably broader and heavier than most of these individuals.

Early Holocene skeletons are uniquely important in providing an understanding of human diversity and origins in the Americas. As this chapter demonstrates, much may be gleaned about their variation from only a few postcranial measurements. The acquisition and analysis of more postcranial measurements from existing early Holocene skeletons is crucial to future investigations, especially given legal and cultural controversy concerning their patrimony. Moreover, the continued access and care of all human remains from the Americas is essential, not only to allow researchers to apply new techniques and reassess previous research, but also to provide a contextual framework for understanding variation and origins in the oldest skeletons. This is evident when contrasting the information previously available on early Holocene remains with the knowledge presented on Kennewick Man due to more data collection by multiple researchers. What may yet be uncovered from continued analysis of his remains and those of his contemporaries?

NOTES

1. This difference is likely a result of the large mean body mass of the reference sample employed by Grine and colleagues. Precontact humans from the Americas were generally large (mean femoral head estimated body mass of 66.5 kg, $n = 462$ males) compared with other groups (Auerbach and Ruff 2004). For example, the mean body mass for native Australian males was 61.7 kg, for Pre-Dynastic Egyptian males was 61.8 kg, and for Iron Age and Early Medieval European males was 65.2 kg.

2. Holliday (1995) included the ventral length of the sacrum in his calculation of torso length. In the calculation of torso length for this chapter, this measurement has been left out, as this measurement is not preserved in Kennewick Man's skeleton, and as the distal portion of the sacrum and proximal portion of the femora overlap. Sacral ventral length cannot be reliably predicted from any other measurements due to a great amount of variability in the sacral length.

3. The Buhl skeleton, found in southern Idaho in 1989 and reburied before extensive investigations could be conducted, may have been as complete as Kennewick Man's and Spirit Cave skeletons prior to excavation (Green et al. 1998). However, the majority of the Buhl skeleton's lower limbs were missing, in addition to a number of vertebrae and the os coxae, by the time researchers obtained it.

4. Spirit Cave is unusual, however, as his skeleton has an extra vertebral element. This is either a thoracized L1 or an additional thoracic vertebra. If it were a thoracized L1, he would have had six lumbar vertebrae in addition to a cranial shift in vertebrae. Otherwise, his vertebral column exhibits a caudal shift.

REFERENCES

Allen, Joel A. 1877. The influence of physical conditions on the genesis of species. *Radical Review* 1:108–140.

Auerbach, Benjamin M. 2007. "Human Skeletal Variation in the New World During the Holocene: Effects of Climate and Subsistence Across Geography and Time." Ph.D. dissertation, Johns Hopkins University School of Medicine, Baltimore.

———. 2010. Giants among us? Morphological variation and migration on the Great Plains. In *Human Variation in the Americas: The Integration of Archaeology and Biological Anthropology*, edited by Benjamin M. Auerbach, 72–214. Center for Archaeological Investigations, Carbondale, IL.

———. 2011. Methods for estimating missing human skeletal osteometric dimensions employed in the revised Fully technique for estimating stature. *American Journal of Physical Anthropology* 145(1):67–80.

Auerbach, Benjamin M., and Christopher B. Ruff. 2004. Human body mass estimation: A comparison of "morphometric" and "mechanical" methods. *American Journal of Physical Anthropology* 125(4):331–342.

———. 2006. Limb bone bilateral asymmetry: Variability and commonality among modern humans. *Journal of Human Evolution* 50(2):203–218.

———. 2010. Stature estimation formulae for indigenous North American populations. *American Journal of Physical Anthropology* 141(2):190–207.

Auerbach, Benjamin M., and Adam D. Sylvester. 2011. Allometry and apparent paradoxes in human limb proportions: Implications for scaling factors. *American Journal of Physical Anthropology* 144(3):382–391.

Bergmann, Carl. 1847. Über die Verhältnisse der wärmeökonomie der Thiere zu ihrer Größe. *Göttinger Studien* 3:595–708.

Bogin, Barry. 1999. *Patterns of Human Growth*. 2nd edition. Cambridge Studies in Biological and Evolutionary Anthropology 23. Cambridge University Press, Cambridge.

Chatters, James C. 2000. The recovery and first analysis of an early Holocene human skeleton from Kennewick, Washington. *American Antiquity* 65(2):291–316.

Coon, Carlton S. 1955. Some problems of human variability and natural selection in climate and culture. *American Naturalist* 89(848):257–279.

Cybulski, Jerome S., Donald E. Howes, James C. Haggarty, and Morley Eldridge. 1981. An early human skeleton from south-central British Columbia: Dating and bioarchaeological inference. *Canadian Journal of Archaeology* 5:49–59.

Danforth, Marie E. 1999. Coming up short: Stature and nutrition among the Ancient Maya of the southern lowlands. In *Reconstructing Ancient Maya Diet*, edited by Christine D. White, 103–117. University of Utah Press, Salt Lake City.

Dansie, Amy. 1997. Early Holocene burials in Nevada: Overview of localities, research and legal issues. *Nevada Historical Society Quarterly* 40(1):4–14.

DeLeon, Valerie B., and Benjamin M. Auerbach. 2007. Morphological variation in human long bones. *American Journal of Physical Anthropology* 132(S44):96.

Delson, Eric, Carl J. Terranova, William L. Jungers, Eric J. Sargis, Nina G. Jablonski, and Paul G. Dechow. 2000. Body mass in *Cercopithecidae* (Primates, Mammalia): Estimation and scaling in extinct and extant taxa. Anthropological Papers 83. American Museum of Natural History, New York.

Doran, Glen H. 2002. *Windover: Multidisciplinary Investigations of an Early Archaic Florida Cemetery*. University Press of Florida, Gainesville.

Dwight, Thomas. 1894. Methods for estimating height from parts of the skeleton. *Medical Record of New York* 46:293–296.

Fitzhugh, William W., and Igor Krupnik. 2006. *Gateways to Jesup II: Franz Boas and the Jesup North Pacific Expedition*. University of Washington Press, Seattle.

Fully, Georges. 1956. Une nouvelle méthode de détermination de la taille. *Annales des Médicine Légale* 35:266–273.

Genovés, Santiago. 1967. Proportionality of the long bones and their relation to stature among Mesoamericans. *American Journal of Physical Anthropology* 26(1):67–78.

Green, Thomas J., Bruce Cochran, Todd W. Fenton, James C. Woods, Gene L. Titmus, Larry Tieszen, Mary Anne Davis, and Suzanne J. Miller. 1998. The Buhl burial: A Paleoindian woman from southern Idaho. *American Antiquity* 63(3):437–456.

Grine, F. E., W. L. Jungers, P. V. Tobias, and O. M. Pearson. 1995. Fossil *Homo* femur from Berg Aukas, northern Namibia. *American Journal of Physical Anthropology* 97(2):151–185.

Gruhn, Ruth. 2005. The ignored continent: South America in models of earliest American prehistory. In *Paleoamerican Origins: Beyond Clovis*, edited by Robson Bonnichsen, Bradley T. Lepper, Dennis Stanford, and Michael R. Waters, 199–208. Texas A&M Press, College Station.

Holliday, Trenton W. 1995. "Body Size and Proportions in the Late Pleistocene Western Old World and the Origins of Modern Humans." Ph.D. dissertation, University of New Mexico. University Microfilms, Ann Arbor, MI.

———. 1997. Body proportions in Late Pleistocene Europe and modern human origins. *Journal of Human Evolution* 32(5):423–447.

———. 1999. Brachial and crural indices of European Late Upper Paleolithic and Mesolithic humans. *Journal of Human Evolution* 36(5):549–566.

Holliday, Trenton W., and Christopher B. Ruff. 1997. Ecogeographical patterning and stature prediction in fossil hominids: Comment on M. R. Feldesman and R. L. Fountain, *American Journal of Physical Anthropology* (1996) 100:207–224. *American Journal of Physical Anthropology* 103(1):137–140.

———. 2001. Relative variation in human proximal and distal limb segment lengths. *American Journal of Physical Anthropology* 116(1):26–33.

Jacobs, Kenneth H. 1985. Climate and the hominid postcranial skeleton in Würm and early Holocene Europe. *Current Anthropology* 26(4):512–514.

Jantz, Richard L., David R. Hunt, Anthony B. Falsetti, and P. J. Key. 1992. Variation among North Amerindians: Analysis of Boas's anthropometric data. *Human Biology* 64(3):435–461.

Jantz, Richard L., and Douglas W. Owsley. 2001. Variation among early North American crania. *American Journal of Physical Anthropology* 114(2):146–155.

———. 2005. Circumpacific populations and the peopling of the New World: Evidence from cranial morphometrics. In *Paleoamerican Origins: Beyond Clovis*, edited by Robson Bonnichsen, Bradley T. Lepper, Dennis Stanford, and Michael R. Waters, 267–275. Texas A&M Press, College Station.

Jungers, William L. 1988. New estimates of body size in australopithecines. In *Evolutionary History of the "Robust" Australopithecines*, edited by Fredrick E. Grine, 115–125. Aldine de Gruyter, New York.

Kemp, Brian M., and Theodore G. Schurr. 2010. Ancient and modern genetic variation in the Americas. In *Human Variation in the Americas: The Integration of Archaeology and Biological Anthropology*, edited by Benjamin M. Auerbach, 12–50. Center for Archaeological Investigations, Carbondale, IL.

Larsen, Clark S. 1997. *Bioarchaeology: Interpreting Behavior from the Human Skeleton*. Cambridge Studies in Biological Anthropology 21. Cambridge University Press, Cambridge.

Larsen, Helge, and Froelich Rainey. 1948. *Ipiutak and the Arctic Whale Hunting Culture*. Anthropological Papers of the American Museum of Natural History 42. American Museum of Natural History, New York.

Lieberman, Daniel E., Maureen E. Devlin, and Osbjorn M. Pearson. 2001. Articular area responses to mechanical loading: Effects of exercise, age, and skeletal location. *American Journal of Physical Anthropology* 116(4):266–277.

Lundy, John K. 1985. The mathematical versus anatomical methods of stature estimate from long bones. *American Journal of Forensic Medical Pathology* 6(1):73–75.

Martin, Rudolf. 1928. *Lehrbuch der Anthropologie in Systematischer Darstellung. Mit besonderer Berücksichtigung der anthropologischen Methoden für Studierende, Ärzte und Forschungsreisende*. Vol. 2. 2nd edition. Gustav Fischer, Jena.

McHenry, Henry M. 1992. Body size and proportions in early hominids. *American Journal of Physical Anthropology* 87(4):407–431.

Merbs, Charles F. 1974. The effects of cranial and caudal shift in the vertebral columns of northern populations. *Arctic Anthropology* 11(S):12–19.

Munford, Danusa, Maria do Carmo Zanini, and Walter Ales Neves. 1995. Human cranial variation in South America: Implications for the settlement of the New World. *Brazilian Journal of Genetics* 18:673–688.

Newman, Marshall T. 1953. The application of ecological rules to the racial anthropology of the Aboriginal New World. *American Anthropologist* 55(3):311–327.

Peters, Robert H. 1983. *The Ecological Implications of Body Size*. Cambridge University Press, Cambridge.

Porter, Alan. 2002. Estimation of body size and physique from hominin skeletal remains. *Homo-Journal of Comparative Human Biology* 53(1):17–38.

Powell, Joseph F. 2005. *The First Americans: Race, Evolution, and the Origin of Native Americans*. Cambridge University Press, Cambridge.

Powell, Joseph F., and Walter A. Neves. 1999. Craniofacial morphology of the first Americans: Pattern and process in the peopling of the New World. *American Journal of Physical Anthropology* 110(S29):153–188.

Raxter, Michelle H., Benjamin M. Auerbach, and Christopher B. Ruff. 2006. Revision of the Fully technique for estimating statures. *American Journal of Physical Anthropology* 130(3):374–384.

Raxter, Michelle H., Christopher B. Ruff, and Benjamin M. Auerbach. 2007. Technical Note: Revised Fully stature estimation technique. *American Journal of Physical Anthropology* 133(2):817–818.

Roberts, Derek F. 1978. *Climate and Human Variability*. Cummings, Menlo Park, CA.

Ruff, Christopher B. 1988. Hindlimb articular surface allometry in Hominoidea and *Macaca*, with comparisons to diaphyseal scaling. *Journal of Human Evolution* 17(7):687–714.

———. 1991. Climate and body shape in hominid evolution. *Journal of Human Evolution* 21(2):81–105.

———. 1994. Morphological adaptation to climate in modern and fossil hominids. *American Journal of Physical Anthropology* 37(S19):65–107.

———. 2002a. Long bone articular and diaphyseal structure in Old World monkeys and apes. I: Locomotor effects. *American Journal of Physical Anthropology* 119(4):305–342.

———. 2002b. Variation in human body size and shape. *Annual Review of Anthropology* 31:211–232.

Ruff, Christopher B., Brigitte M. Holt, Vladimír Sládek, Margit Berner, William A. Murphy Jr., Dieter zur Nedden, Horst Seidler, and Wolfgang Recheis. 2006. Body size, body proportions, and mobility in the Tyrolean "Iceman." *Journal of Human Evolution* 51(1):91–101

Ruff, Christopher B., Henry M. McHenry, and J. Francis Thackeray. 1999. The "robust" australopithecine hip: Cross-sectional morphology of the SK 82 and 97 proximal femora. *American Journal of Physical Anthropology* 109(4):509–521.

Ruff, Christopher B., Markku Niskanen, Juho-Antti Junno, and Paul Jamison. 2005. Body mass prediction from stature and bi-iliac breadth in two high latitude populations, with application to earlier higher latitude humans. *Journal of Human Evolution* 48(4):381–392.

Ruff, Christopher B., William W. Scott, and Allie Y.-C. Liu. 1991. Articular and diaphyseal remodeling of the proximal femur with changes in body mass in adults. *American Journal of Physical Anthropology* 86(3):397–413.

Ruff, Christopher B., Erik Trinkaus, and Trenton W. Holliday. 1997. Body mass and encephalization in Pleistocene *Homo*. *Nature* 387:173–176.

Ruff, Christopher B., and Alan Walker. 1993. Body size and body shape. In *The Nariokotome* Homo erectus *Skeleton*, edited by Alan Walker and Richard Leakey, 234–265. Harvard University Press, Cambridge, MA.

Schmidt-Nielsen, Knut. 1984. *Scaling: Why is Animal Size So Important?* Cambridge University Press, Cambridge.

Schurr, Theodore G. 2005. Tracking genes through time and space: Changing perspectives on New World origins. In *Paleoamerican Origins: Beyond Clovis*, edited by Robson Bonnichsen, Bradley T. Lepper, Dennis Stanford, and Michael R. Waters, 221–242. Texas A&M Press, College Station.

Sciulli, Paul W., Kim N. Schneider, and Michael C. Mahaney. 1990. Stature estimation in prehistoric Native Americans of Ohio. *American Journal of Physical Anthropology* 83(3):275–280.

Sládek, Vladimír, Margit Berner, and Robert Sailer. 2006. Mobility in central European Late Eneolithic and Early Bronze Age: Femoral cross-sectional geometry. *American Journal of Physical Anthropology* 130(3):320–332.

Snow, Charles E. 1948. Indian Knoll skeletons of site OH 2, Ohio County, Kentucky. *The University of Kentucky Reports in Anthropology and Archaeology* 4:371–532.

Stahl, Peter W. 1996. Holocene biodiversity: An archaeological perspective from the Americas. *Annual Review of Anthropology* 25:105–126.

Stewart, Suzanne, and Adrian Praetzellis. 2003. *Archaeological Research Issues for the Point Reyes National Seashore—Golden Gate National Recreation Area*. Anthropological Studies Center, Sonoma State University, CA

Stewart, T. Dale. 1979. *Essentials of Forensic Anthropology*. Charles C. Thomas, Springfield, IL.

Stinson, Sara. 1990. Variation in body size and shape among South American Indians. *American Journal of Human Biology* 2(1):37–51.

Tanner, James M., T. Hayashi, Michael A. Preece, and Noel Cameron. 1982. Increase in length of leg relative to trunk in Japanese children and adults from 1957 to 1977: Comparison with British and with Japanese Americans. *Annals of Human Biology* 9(5):411–423.

Trinkaus, Erik. 1981. Neanderthal limb proportions and cold adaptation. In *Aspects of Human Evolution*, edited by Christopher B. Stringer, 187–224. Taylor and Francis, London.

Trinkaus, Erik, Steven E. Churchill, and Christopher B. Ruff. 1994. Postcranial robusticity in *Homo*. II: Humeral bilateral asymmetry and bone plasticity. *American Journal of Physical Anthropology* 93(1):1–34.

Trotter, Mildred, and Goldine C. Gleser. 1951. The effect of aging on stature. *American Journal of Physical Anthropology* 9(3):311–324.

———. 1952. Estimation of stature from long bones of American Whites and Negroes. *American Journal of Physical Anthropology* 10(4):463–514.

Tuohy, Donald R., and Amy Dansie. 1997. New information regarding early Holocene manifestations in the western Great Basin. *Nevada Historical Society Quarterly* 40(1):24–53.

Webb, William S. 1946. Indian Knoll, Site OH 2, Ohio County, Kentucky. *The University of Kentucky Reports in Anthropology and Archaeology* 4:115–365.

Young, Diane E., Suzanne Patrick, and D. Gentry Steele. 1987. An analysis of the Paleoindian double burial from Horn Shelter No. 2, in Central Texas. *Plains Anthropologist* 32(117):275–298.

Reconstructing Habitual Activities by Biomechanical Analysis of Long Bones

12

Daniel J. Wescott

Patterns of mechanical loading greatly influence long-bone shaft strength and cross-sectional shape. As a result, human skeletal remains provide direct evidence for reconstructing activity patterns and intensity in past populations and individuals, especially when combined with archaeological and other biological information (Ruff et al. 2006a). No prior study has attempted to reconstruct the lifeways of early Holocene peoples in North America using long-bone cross-sectional rigidity and strength within a biomechanical framework. Paleoamerican skeletons are rare, and most studies have been descriptive or primarily focused on chronology, origins, and cranial morphology (Cybulski et al. 1981; Green et al. 1998; Howells 1938; Huckleberry et al. 2003; Jantz and Owsley 2001; Powell and Neves 1999; Powell and Rose 1999; Schurr and Sherry 2004; Waters et al. 2008).

MATERIALS AND METHODS
Computed Tomography Diaphyseal Cross-Sections

Kennewick Man's long bones were scanned in air using a General Electric® computed tomography (CT) scanner by D. Owsley and R. Jantz at the Department of Radiology, University of Washington, Seattle. Each bone was then digitally positioned with respect to a sagittal and a coronal plane using defined anatomical landmarks (Ruff 2002). Geometric cross-sectional properties were obtained from the CT scans at the midshaft and middistal (35 percent length from the distal end) humeri, subtrochanteric and midshaft for the femora, and midshaft for the tibiae. The right tibia is broken, and its midshaft was estimated by measuring half the left tibia length from the distal end. The cross-sectional oblique slices were imported into *ImageJ* (National Institute of Health 2008) and analyzed using a PC version of *MomentMacroJ* (Eschmann 1992) that calculates geometric properties from two-dimensional cross-sections.

The medullary cavities of the long bones are filled with sediment although the margins of the endosteal surfaces are distinguishable. By manually using the paintbrush tool in *ImageJ*, medullary cavity sediment was removed from the digital files. In the same way, small calcium carbonate deposits adhering to the external cortical bone that are apparent on the CT scans were also removed. This digital process was repeated three times for each cross-section, and each of the three images was then analyzed using *MomentMacroJ*. Maximum differences between attempts are approximately one percent, suggesting that the endosteal surfaces were sufficiently perceptible to define the medullary cavity manually with precision. The mean value of the three attempts was used in further analyses.

A biomechanical model assumes that the processes of modeling sculpt bones during life so that their architecture (cross-sectional diameter, shape, and cortical thickness) reflects the loading forces (tension, compression, shear, bending, and torsion) placed on each bone (Frost 2003; Pearson and Lieberman 2004; Ruff et al. 2006a). In life, bone adapts to the functional demand. As a consequence, cross-sectional geometric properties (Table 12.1) can be used to measure bone strength and rigidity (Ruff 2008) that, in turn, provide information for interpreting activity patterns and motion intensity. The cortical area (CA) of a bone provides information regarding strength in pure compression and tension loads. Bending rigidity can be estimated from second moments of area (SMA) calculated through the section. The bending SMA (I) can be calculated around any axis and, in this case, was calculated about the mediolateral (ML) and anteroposterior (AP) anatomical planes and the minimum and maximum bending rigidities (I_{min} and I_{max}, respectively).

TABLE 12.1 Structural properties calculated from bone cross-sections and external dimensions.

Measurement/Property[1]	Definition[2]
Total Area (TA)	Total subperiosteal area including cortical bone and medullary areas
Cortical Area (CA)	Area of cortical bone—indicator of compressive and tensile strength
Percent Cortical Area (%CA)	Percent cortical bone area relative to total subperiosteal area (CA/TA*100)
Medullary Area (MA)	Area within the medullary cavity of the diaphysis
Second Moments of Area, ML Axis (I_x)	AP bending rigidity
Second Moments of Area, AP Axis (I_y)	ML bending rigidity
Second Moments of Area, Maximum Axis (I_{max})	Maximum bending rigidity
Second Moments of Area, Minimum Axis (I_{min})	Minimum bending rigidity
Polar Moment of Area (J)	Torsional rigidity
Section Modulus, ML Axis (Z_x)	AP bending strength
Section Modulus, AP Axis (Z_y)	ML bending strength
Polar Section Modulus (Z_p)	Torsional strength; estimated as $J^{0.73}$
Diaphyseal Circularity Index, I_{max}/I_{min}	Ratio of the maximum and minimum bending strength
Diaphyseal Shape Index, I_x/I_y	Ratio of the AP to ML bending strength
Humeral Midshaft Shape (HMS)	Maximum divided by minimum diaphyseal midshaft diameter
Femur Subtrochanteric Shape (FSS)	External measure of femur subtrochanteric shape—AP/ML
Femur Midshaft Shape (FMS)	External measure of femur midshaft shape—AP/ML
Humerus Midshaft J Estimated (HMJ)[3]	$10^{-0.5227}$(maximum diameter)$^{2.4867}$(minimum diameter)$^{1.0775}$
Femur Midshaft J Estimated (FMJ)[4]	$10^{-1.0618}$(midshaft AP diameter)$^{1.6852}$(midshaft ML diameter)$^{2.3011}$
Femur Subtrochanteric J Estimated (FSJ)[5]	−124812+2925(AP diameter)+3360 (ML diameter)
Tibia Midshaft J Estimated (TMJ)[4]	$10^{-0.5184}$(AP diameter)$^{2.2915}$(ML diameter)$^{1.2546}$

[1]Areas in mm², Second Moments of Area in mm⁴, section moduli in mm.³
[2]AP = anteroposterior; ML = mediolateral.
[3]Osbjorn Pearson, personal communication 2008.
[4]Equation from Pearson et al. (2006).
[5]Equation from Wescott (2001).

The SMA about the ML axis (I_x) is proportional to the bone's bending rigidity in the AP plane, while the SMA about the AP axis (I_y) is proportional to the rigidity in the ML plane (Ruff 2008). Bending strength can be estimated using the section modulus in the appropriate axis. To calculate the section modulus (Z), the SMA is divided by the distance from the surface to the axis in the plane of interest. Torsional rigidity and strength are estimated from the polar moment of area (J), which is the SMA about the cross-section centroid. The polar section modulus (Z_p) estimates overall torsional and bending strength of the bone. The polar section modulus is calculated by dividing J by the average ML and AP radius of the section but can also be approximated by raising J to the power of 0.73 ($J^{0.73}$) (Ruff 2002, 2008).

Subperiosteal or total area (TA) and CA, second moments of area (I_x, I_y, I_{max}, and I_{min}), and section moduli (Z_x, Z_y) were calculated for each section using *MomentMacroJ*. In addition, the polar moments of area (J), the area of the medullary cavity (MA), the percent cortical bone (%CA), and the ratios of the SMA (I_x/I_y and I_{max}/I_{min}) were also derived (Table 12.1). Since bone breadths were not available for all comparative samples, the polar section modulus (Z_p) was estimated by taking J to the 0.73 power.

Due to the relationship between body size and bone shaft strength, cross-sectional properties must be scaled to some measure of body mass and shape (e.g., dividing the cross-sectional property by the product of body mass and bone length or by bone length raised to a power). Doing so allows comparison of bone strength between populations that differ significantly in body size and shape (Holliday 2002; Ruff 2000). In this study SMA values were divided by the product of body mass and squared bone length, section moduli were divided by the product of body mass and bone length, and cross-sectional areas (total, cortical, and medullary areas) were divided by body mass (Ruff 2000, 2008). In order to utilize as much comparative data as possible, body mass was calculated from the femoral head diameter using the male equation provided by Grine et al. (1995). When pelvic breadth is not available, this is the most appropriate method for estimating body mass from Native American skeletons (Chapter 11).

External Bone Dimensions and Derived Properties

Stature was estimated from femoral length using the Genovés (1967) male equation. Genovés' data were derived from a Hispanic sample and may not be appropriate for Holocene Americans, especially if they differ significantly in body proportions from the Hispanic sample. However, using the Genovés equations is likely the most accurate method available. For the Paleoamerican skeletons included in this study, the Genovés male equation provided an estimated stature close to the stature predicted using the anatomical method (Table 12.2).

Several variables were calculated that provide a proxy for long bone biomechanical properties (Table 12.1) (Wescott 2001). Studies have demonstrated that external dimensions can be used as surrogates or to estimate certain cross-sectional properties and, therefore, can be used to examine functional similarities or differences among groups (Jungers and Minns 1979; Pearson 2000; Pearson et al. 2006; Wescott 2001, 2006a). A bone shaft shape index was constructed for the femur, tibia, and humerus midshafts. Shape was calculated as the ratio of the AP and ML diaphyseal diameters for the femur, while the ratio of the maximum and minimum shaft diameters was used for the humerus and tibia. The shape index provides a proxy for the I_x/I_y and I_{max}/I_{min} ratios obtained from diaphyseal cross-sections (Pearson et al. 2006; Ruff 1987; Stock and Shaw 2007; Wescott 2001, 2006a). While the shape index generally underestimates the I_x/I_y ratio, it does provide a similar pattern when examining different populations using the femur and humerus (Wescott 2001, 2006a, 2008). Conversely, the shape index may not provide the same pattern as the I_{max}/I_{min} when examining oddly shaped bones such as the tibia. For the femur and humerus, a circular shaft will have a shape index of 1.0, while a shaft with greater bending force in the AP plane compared to the ML plane (or maximum versus minimum) will have a shape index greater than 1.0. Likewise, a shape index less than 1.0 indicates that greater ML than AP bending forces were placed on the shaft during life. An estimate of the polar second moment of area (J) was calculated at each diaphyseal level using the equations provided by Pearson and colleagues (Osbjorn Pearson, personal communication 2008; Pearson et al. 2006).

Comparative Samples

Comparative samples were obtained for adult males from North America, Asia, and Europe. Cross-sectional data (obtained from CT scans) for Plains and Aleutian Island populations, as well as the Horn Shelter No. 2, Texas, Paleoamerican skeleton were collected previously. The Plains sample is comprised of prehistoric, protohistoric, and historic Arikara Indian males from South Dakota. The Arikara practiced a mixed horticulture and

TABLE 12.2 Mean body size comparisons.

Sample	Stature (cm)[a]	Body Mass (kg)[b]	Bi-iliac Breadth (mm)[c]
Kennewick Man	172.7	73.6	281.0
Gore Creek	166.0	—	—
Spirit Cave	163.6	61.6	279.0
Wizards Beach	171.2	76.9	—
Horn Shelter	164.5	69.1	—
Prospect Man	162.4	61.0	—
Alaska	162.4	68.6	272.9
Aleutian Is.	160.7	66.6	265.3
British Columbia	162.8	—	276.7
Washington	162.5	65.9	263.8
Ky. Archaic	165.8	60.4	259.0
Great Basin BSHG[d]	165.8	68.7	—
Plains Village	167.7	67.9	277.5

[a]Stature calculated from femoral length using Genovés (1967) male equation; data for BC from Cybulski et al. (1981).
[b]Body mass calculated from male equations provided by Grine et al. (1995) for femoral head diameter.
[c]Bi-iliac breadth for Kennewick Man, Spirit Cave, and B.C. from Auerbach (2007).
[d]Broad-Spectrum Hunter Gatherer.

bison-hunting subsistence pattern, with females responsible for horticultural activities (Wescott 2001; Wescott and Cunningham 2006). The Washington sample, curated by the American Museum of Natural History, is comprised of prehistoric Native American skeletons from the Priest Rapids, Stanwood, Coupeville, Ellensburg and Gray's Harbor sites. The Aleutian Island sample represents historic and prehistoric skeletons of male coastal hunters from islands in the Aleutian archipelago (Hunt 2002). European Upper Paleolithic and Mesolithic cross-sectional data were obtained from Holt (1999), while cross-sectional data for the Tyrolean Iceman were acquired from Ruff et al. (2006b). Holt (1999) used a latex molding method, which provides an excellent estimate of true cross-sectional properties (O'Neill and Ruff 2004).

Long bone measurements of Native American males were used for comparisons. Measurements of Native American skeletons were obtained from the University of Tennessee/Smithsonian Institution (UT/SI) database (Wescott 2001). The UT/SI database includes measurements of several early Holocene skeletons including Horn Shelter No. 2, Texas; Wizards Beach, Nevada; the Spirit Cave mummy, Nevada; and Prospect Man, Oregon. Metric data for the Gore Creek skeleton, an early Holocene male from British Columbia, were obtained from Cybulski et al. (1981). The UT/SI database includes measurements for broad-spectrum hunter-gatherers from the Great Basin, Archaic hunter-gatherers from the Southeast (primarily Indian Knoll, Kentucky) and equestrian bison-hunters from the Plains (Wescott 2001). Comparative long bone measurement means for other populations from North America, Asia, and Europe were obtained from literature (Auerbach 2007; Cybulski et al. 1981; Marchi et al. 2007; Pearson 2000; Ruff 1999; Sakaue 1997; Schmitt et al. 2003; Stock and Pfeiffer 2004; Trinkaus et al. 1994; Wescott and Cunningham 2006).

RESULTS

Kennewick Man was relatively tall, broad bodied, and robust compared to most prehistoric populations along the Northwest Coast (Table 12.2). He was approximately 10 centimeters taller, his hips slightly wider, and more than 5 kilograms heavier than average for Native Americans from Alaska, Washington, and British Columbia. Kennewick Man was also taller and heavier than other early Holocene males from North America except Wizards Beach. The estimated stature based on femoral length is within the 169.7–173.7 centimeter range estimated using the anatomical method, and the estimated body mass of 72.4 kilograms based on femoral head diameter is nearly identical to the 72.9 kilogram predicted value using stature and hip breadth (Chapter 11).

Bone rigidity estimates suggest that Kennewick Man had very strong bones (Table 12.3). The percent cortical bone (%CA) ranges from 66.7 to 79.5 percent and averages 73 percent for all locations examined. The relatively high values of J and Z_p for Kennewick Man's femora and tibiae suggest that he participated in habitual activities that required powerful lower limbs. The I_x/I_y and I_{max}/I_{min} femoral ratios indicate that Kennewick Man's femora sustained greater AP than ML bending loads. His tibiae have high section moduli values, but the shape is more consistent with torsional loading instead of bending predominately in one direction.

Kennewick Man exhibits both upper and lower limb asymmetry. The asymmetry in the upper body is striking. The section modulus for the right humerus is 30 percent greater at midshaft and 22 percent greater at middistal than for the left humerus (Table 12.3). In the lower limbs, the left femur is 16.9 percent stronger at midshaft than the right. The right tibia was slightly stronger (3.6 percent) than the left at midshaft, but this may in part reflect error associated with the estimation of midshaft (Sládek et al. 2010).

INTERPRETATION

Long-bone diaphyseal morphology is determined by the interaction of numerous factors, including genetics, hormones, diet, disease, body size and shape, and mechanical loading history from routine activities. Bone modeling induced as a result of mechanical loading throughout life, but especially before adulthood, causes changes in the cross-sectional diameter, shape, and cortical thickness of long bones (Bridges 1995; Frost 2003; Pearson and Lieberman 2004; Ruff et al. 2006a). As a consequence, cross-sectional geometric properties of long bones reflect prior loading history and can be used to determine the type and intensity of habitual activity patterns in individuals or populations (Ruff et al. 2006a, 2006b).

Early Holocene inhabitants of North America are

TABLE 12.3 Kennewick Man standardized cross-sectional properties.

Section	TA[1]	MA[1]	%CA	J	Z_p[2]	Z_{est}[2,3]	Shape Index[4]	I_x/I_y	I_{max}/I_{min}
L Femur Subtrochanteric	1009.8	336.6	66.7	2275.0	108.9	106.1	0.80	1.16	1.61
L Femur Midshaft	919.7	188.4	79.5	2144.4	104.3	116.0	1.26	1.66	1.72
R Femur Subtrochanteric	921.1	241.7	73.8	2039.8	100.6	105.5	0.85	1.41	1.65
R Femur Midshaft	833.9	200.1	76.0	1700.1	88.0	95.7	1.24	1.52	1.61
Femoral Subtrochanteric Asymmetry[5]	−9.2	−32.8	10.0	−10.9	−7.9	−0.56	3.6	19.6	2.32
Femoral Midshaft Asymmetry[5]	−9.8	6.04	−14.3	−23.1	−16.9	−19.2	−1.6	−8.7	−6.55
L Tibia Midshaft	921.6	259.5	71.5	2361.8	117.0	121.2	—	1.34	1.62
R Tibia Midshaft	950.8	279.2	70.3	2481.5	121.3	125.6	1.63	1.29	1.62
Tibial Midshaft Asymmetry[5]	3.1	7.3	−1.7	4.9	3.6	3.6	—	−3.3	0.2
L Humerus Midshaft	373.4	101.8	72.7	464.9	37.3	42.0	1.53	0.87	2.02
L Humerus Middistal (35%)	346.3	81.1	76.6	397.6	33.2	—	—	0.93	1.42
R Humerus Midshaft	461.4	131.2	72.2	697.4	50.0	58.3	1.32	0.75	1.93
R Humerus Middistal (35%)	449.9	126.6	72.6	636.6	46.8	—	—	0.98	1.40
Humeral Midshaft Asymmetry[5]	21.1	21.8	0.3	40.8	30.0	32.5	−14.7	−15.3	−4.6
Humeral Middistal Asymmetry[5]	26.0	26.9	−6.1	47.0	21.9	—	—	3.0	−1.0

[1]Standardized by body mass estimated from femur head diameter.
[2]Standardized by the product of body mass and femur maximum length.
[3]Section modulus calculated from polar moment of area estimated from external dimensions raised to the 0.73 power.
[4]Shape index calculated as AP/ML for femur and as maximum/minimum for humerus and tibia.
[5]Asymmetry calculated as (right − left)/((right + left)/2)*100.

characterized as hunter-gatherers or broad-range foragers with high residential and logistical mobility (Kelly and Todd 1988). Paleoamericans are often characterized as having great terrestrial logistic mobility (distance covered by an individual or group daily from the residence and back) with dependence on megafauna. Kelly and Todd (1988) argued that Paleoamericans, especially Clovis, probably relied more on fauna than botanical resources although this specialized dependence on megafauna is likely overemphasized (Byers and Ugan 2005) and more appropriately applies to late Pleistocene populations. Like other hunter-gatherers, early Holocene populations in the Americas foraged for berries, seeds, roots, nuts, and other plant products, as well as hunted large, medium, and small animals.

Regional variation in subsistence is well documented historically and archaeologically and occurred during the early Holocene as well. Paleoamericans in the late Pleistocene and early Holocene likely exploited local environments and hunted a broad range of terrestrial and aquatic game. Spirit Cave and Wizards Beach, early Holocene males from the Great Basin, are both buried near ancient lakes. Archaeological evidence suggests that small animals and marsh flora were staples in the Great Basin diet during the early Holocene (Barker et al. 2000). Paleofecal material from the Spirit Cave mummy indicates consumption of small mass-captured fish and marsh flora (Eiselt 1997). In British Columbia, Cybulski et al. (1981) found that the male from Gore Creek differed in body build and skeletal robusticity from nearby coastal hunter-gatherers. They argued that the Gore Creek male had "an inland hunting adaptation where the emphasis for strength and agility is placed on the lower rather than the upper limbs" (Cybulski et al. 1981:54).

In the Columbia River Gorge, life was oriented around rivers and their vast resources (Ames 2000; Butler 1993; Butler and O'Connor 2004; Cressman 1960). Early Holocene sites on the Plateau are almost always near one of the major rivers (Chatters et al. 2012; Ames 2000). The frequency of faunal remains present in early Holocene Northwest Coast and Plateau sites is consistent with a pattern of broad-spectrum foraging (Butler and Campbell 2004). A wide variety of terrestrial prey (bison, elk, deer, antelope, and rabbit) was actively hunted and contributed significantly to the number of identified species in many Plateau sites (Ames 2000; Butler and Campbell 2004; Chatters et al. 2012). Chatters et al. (2012) contend that Kennewick Man and contemporary hunter-gatherers in the Northwest interior placed greater emphasis on terrestrial large ungulates than on fish and other prey. While there is less evidence that fish were a major source of protein for early Holocene foragers (Ames 2000; Chatters et al. 2012), fish do dominate the vertebrate assemblage at specific sites along the Columbia River (Butler and Campbell 2004). Butler and O'Conner (2004) demonstrate at The Dalles site on the Columbia River that salmon and other fish were a dietary resource as early as 9,300 years ago. Isotopic analysis of Kennewick Man's bone indicates that 70 percent or more of his diet may have come from marine sources (Chapter 3; Taylor et al. 1998). More recent evidence from carbon and nitrogen isotope values suggests an even greater reliance on fish and especially marine mammals as a source of protein (Chapters 3, 16, and 32). It is highly likely that spear fishing was common in clear rivers and streams of the Northwest Coast and Plateau during the early Holocene (Butler and O'Connor 2004; Stewart 1982), but there is no evidence of fish storage during the time Kennewick Man was alive (Ames 2000; Chatters et al. 2012).

The strong lower limbs of Kennewick Man are consistent with a lifeway involving habitual activities that bilaterally produced substantial AP bending stress in the femora and torsion in the tibiae. The asymmetry in his upper limbs indicates that he was involved in activities that placed greater mechanical loads, primarily torsion, on his right, and dominant, arm. To provide a more definitive reconstruction of the habitual activities of Kennewick Man, it is necessary to more closely examine the pattern of strength and asymmetry in his long bones and to compare him to populations with documented subsistence and activity patterns.

Lower Limb Strength, Shape, and Asymmetry

Kennewick Man had strong lower limbs (standardized for body size), especially compared to Archaic foragers (Indian Knoll) and most Plains bison hunters (Arikara) (Figure 12.1), indicating that he participated in intense habitual activities that placed considerable loads on his lower limbs, especially AP bending loads on his femora and torsion loads on the tibiae. The AP/ML ratio of Kennewick Man's femora stand out as distinct from most early North American Holocene males (Prospect Man, Gore Creek, Horn Shelter, Spirit Cave, and Wizards Beach) and from later hunters from Alaska, the Aleutian Islands, Washington, Plains (Arikara), and Southeast (Indian Knoll) (Figure 12.2). Kennewick Man had generally large femora, but the primary difference is due to their considerable AP diameter.

Long-bone midshaft shape correlates with directional changes in locomotor habits (Carlson and Judex 2007), and multiple researchers have demonstrated a link between AP buttressing of the femur and levels of terrestrial logistic mobility (Holt 2003; Marchi 2008; Ruff 1987; Stock 2006; Stock and Pfeiffer 2001, 2004). Bone

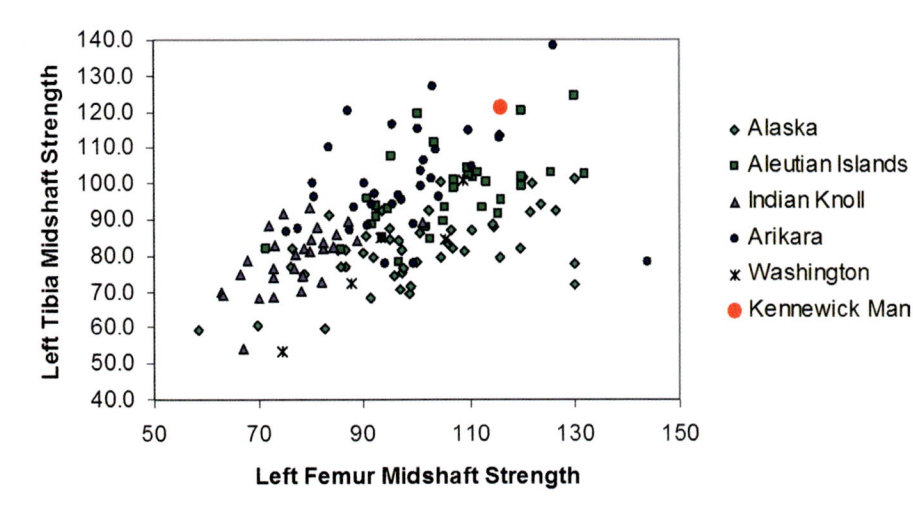

FIGURE 12.1 The relationship between tibial and femoral strength ($Z_p = J^{0.73}$) in Kennewick Man and Native American groups.

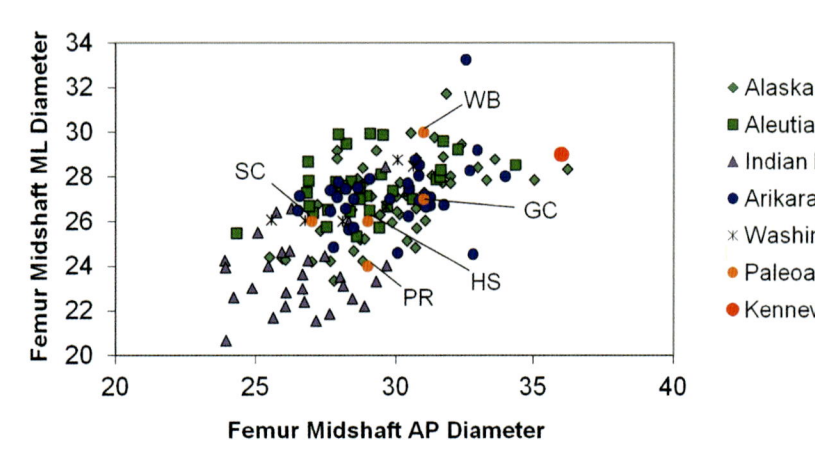

FIGURE 12.2 Bivariate plot of AP and ML midshaft femur external dimensions comparing Kennewick Man to Native American groups and other Paleoamericans (GC = Gore Creek, HS = Horn Shelter, PR = Prospect Man, SC = Spirit Cave, WB = Wizards Beach).

strength in both the femur and tibia correlate positively with terrestrial mobility levels among hunter-gatherers, and femur shape is a good indicator of mobility levels (Stock 2006). Activities such as long-distance running result in greater AP than ML bending loads on the femoral diaphyses. In response, modeling processes lay down more bone in the AP than ML plane of the diaphyseal cross section. Highly mobile groups and individuals have greater AP than ML bending rigidity in the lower limb bones than sedentary groups and individuals (Stock 2006; Shaw and Stock 2009a).

The extremely sturdy lower limbs and high femoral SMA ratios seen in Kennewick Man are consistent with a life that involved intense terrestrial mobility (Chatters et al. 2012). Cybulski et al. (1981) contended that the strong lower limbs and AP elongated cross-section of the femora, and especially the tibiae, of Gore Creek were associated with high mobility and dependence on foraging of terrestrial-littoral resources. However, while Kennewick Man's femora are very strong and AP expanded, his tibiae do not exhibit the AP buttressing expected for a highly mobile hunter-gatherer. In fact, Kennewick Man's tibiae exhibit nearly equal rigidity in the maximum (I_{max}) and minimum (I_{min}) planes. Figure 12.3 compares Kennewick Man's I_{max}/I_{min} values for the femur and tibia midshaft to European Upper Paleolithic (20,000–30,000 yr. BP), Late Upper Paleolithic (10,000–19,000 yr. BP), and Mesolithic (5,300–9,000 yr. BP) hunter-gatherers, which were highly mobile populations (Holt 2003; Marchi et al. 2007; Ruff et al. 2006b). Kennewick Man's femoral strength exceeds his European Mesolithic counterparts and most of the early Paleolithic hunters, but his tibial I_{max}/I_{min} ratio is well below the average for Paleolithic and Mesolithic males.

His tibial-to-femoral polar section modulus ratio (tibia Z_p/femur Z_p) is also high (1.25 average) compared to European hunters. In most individuals the tibial-to-femoral Z_p ratio is less than one, but Ruff et al. (2006b:95) found that greater size-standardized strength in the tibia than the femur is "unusual but not unprecedented" among European hunter-gatherers.

Shaw and Stock (2009a) compared tibia cross-sectional properties of cross-country runners, field hockey players, and sedentary males. They found that tibial rigidity was greater in runners and field hockey players compared to sedentary males. However, young adult runners exhibited more pronounced rigidity in the maximum plane (I_{max}) of the tibia compared to field hockey players. The tibiae of field hockey players were shaped like an equilateral triangle due to the more comparable rigidity in the minimum and maximum planes. Kennewick Man's tibiae exhibit more equal rigidity in both planes. Therefore, while overall strength of the lower limbs and femoral shape in Kennewick Man are consistent with high mobility, his tibial shape does not exhibit the AP buttressing expected for a person executing frequent long-distance terrestrial travel (Figure 12.3).

Several studies (Wescott 2001, 2006a; Sládek et al. 2006a) have found discrepancies between levels of mobility and femoral cross-sectional shape. Furthermore, it is likely that the structure of the femur is greatly influenced by body size and shape (Ruff et al. 2006b; Wescott 2006b). Tibial shape should be less affected by body shape than the femur (Ruff et al. 2006b; Stock 2006; Sládek et al. 2006b). Femoral morphology is affected by pelvic proportions since the long axes of the femora are at an angle to the forces of body mass, while the long axes of the tibiae are

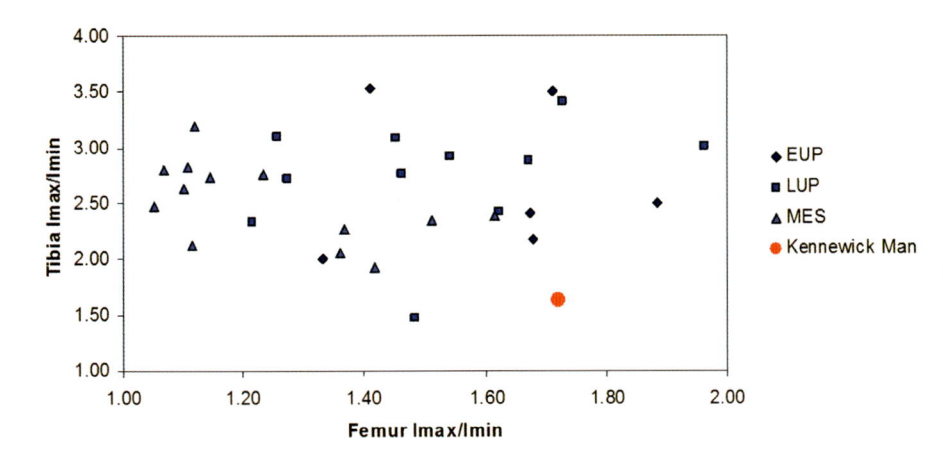

FIGURE 12.3 Kennewick Man's femoral and tibial midshaft I_{max}/I_{min} compared to prehistoric European Early Upper Paleolithic (EUP), Late Upper Paleolithic (LUP) and Mesolithic (MES).

nearly parallel and centered under the center of gravity during weight support. As a consequence, activity may be more clearly interpreted from distal rather than proximal limb bones (Stock 2006; Ruff et al. 2006b). The Neolithic Tyrolean Iceman, for example, has a relatively circular femoral middiaphyseal cross-section but an extremely AP buttressed tibial cross-section. Ruff et al. (2006b) argued that the AP buttressed tibia of the Tyrolean Iceman is consistent with a highly mobile lifeway such as seasonal migration from high to low elevations. The Iceman's relatively round femora, primarily associated with greater ML bending strength, reflect his short stature and stocky body build (Ruff et al. 2006b). If this association is valid, AP elongation of Kennewick Man's femora may reflect his relatively tall stature and greater body mass. The femoral AP/ML ratio has increased among recent Americans in the United States with escalation in stature and body mass (Rockhold 1998). Likewise, Gruss (2007) points out that the greater AP-oriented shape of the femoral midshaft in early *Homo sapiens* compared to Neanderthals may reflect greater AP forces applied to the femur of *H. sapiens* due to longer limbs. However, Shaw and Stock (2008) found no correlation between femoral AP strength and femur length.

The Tyrolean Iceman's extreme tibial AP buttressing may reflect high mobility over rugged mountainous terrain (Ruff et al. 2006b). Ruff (1999) and Marchi et al. (2007) have argued that groups living in rugged terrains, regardless of subsistence strategy, have stronger and more AP buttressed lower limb bones than do groups living in less rugged terrain. Similarly, clinical studies show that loads on the knee are considerably greater when climbing stairs than when walking (Taylor et al. 2004). However, Wescott (2001, 2008) found no correlation between local terrain relief and femoral cross-sectional shape. Likewise, Ruff (1999) observed that populations in highly rugged mountainous terrain have more robust but similarly shaped femora compared to subsistence-matched populations in flatter terrains. This suggests that femoral rigidity but not shape is affected by terrain type (Wescott 2006a). However, if tibial strength and shape, especially AP buttressing, reflect terrestrial mobility over rugged terrain, the cross-sectional morphology of Kennewick Man's lower limbs is not consistent with this behavior.

The terrain in the Columbia River Basin where Kennewick Man was discovered is not extremely rugged, and the strength and shape pattern in Kennewick Man's lower limbs may reflect high mobility but over less rugged terrain. There are steep canyon walls, but most of the landscape of the Plateau is relatively flat, hummocky terrain. The more symmetrical hypertrophy (similarity in I_{max} and I_{min}) of Kennewick Man's tibiae could reflect the relatively flat terrain. Shaw and Stock's (2009a) study of runners and field hockey players found that cross-country runners have extremely AP elliptical tibia diaphyses compared to field hockey players or controls. Nikander et al. (2010) examined cross-sectional geometry in the tibia of female athletes with different exercise loading types. Sports with high-impact activities (volleyball, hurdling, and triple jump) involving jumping and odd-impact sports (soccer and field hockey) where the participants make rapid turns and stops while running resulted in greater torsional strength than controls. The increase in robusticity was due to relatively equal periosteal expansion in all planes. Women in repetitive low-impact sports such as endurance running also had significantly greater tibia midshaft rigidity than controls, but the difference was primarily expansion of the tibial shaft in the AP plane (Nikander et al. 2010). While the terrain of the Plateau and Great Basin may not be so rugged as that experienced by the Tyrolean Iceman or other European hunter-gatherers, it was likely as rugged as the running paths used by modern cross-country runners. In both European prehistoric hunters and modern cross-country runners the tibiae are robust and AP elliptical in shape. Therefore, the relatively small I_{max}/I_{min} ratio of Kennewick Man's tibia is not consistent with long distance running or walking, even over relatively flat terrain.

Kennewick Man's pattern of femoral and tibial strength and shape is more consistent with terrestrial activities that involved high- or odd-impact loading (Nikander et al. 2010; Shaw and Stock 2009a). This may have involved explosive acceleration along with rapid turns and sudden stops. Birds and small mammals such as rabbit were being hunted by Paleoamericans in the Plateau. Faunal assemblages from central Oregon are often dominated by rabbits (Chatters et al. 2012). While the method used to hunt these animals is unknown, throwing sticks have been used in the Great Basin, California, and other locations in the west for thousands of years (Koerper 1998; Tuohy 2004) and may have been used by early Holocene hunters on the Plateau. This type of hunting would require rapid bursts over short distances so the hunter can dispatch the injured animal. High intensity running or sprinting with sudden turns and stops would require significant lower limb muscle strength and result in high- and odd-impact loads on the lower limb. Muscle forces on bone are a major stimulus for bone adaptation (Frost 2003). Distance running places more lower muscle strain intensity on the femora than sprinting, which may explain why Kennewick Man's femora are strong compared to European Mesolithic and Paleolithic hunters. Likewise, sprinting with rapid turns and possibly leaping over bunch grass would place high- and odd-impact loads on the tibia, which leads to strong tibiae with an equilateral triangular shape. Distance running and walking results in lower impact stress but with a greater number of strain cycles (Shaw and Stock 2009a).

Another possible explanation for the pattern of Kennewick Man's lower limb bone strength and shape is that he habitually walked in shallow, swift-moving water. This is consistent with isotopic evidence that suggests a reliance on marine resources. In the Plateau, probably even during the early Holocene, spearing of fish took place primarily in rivers and streams, and fishermen commonly walked up the stream or along its banks to find a suitable place to fish (Stewart 1982). The mechanical forces placed on lower limbs differ when walking in water than when walking on land. Barela, Stolf, and Duarte (2006) found that walking in shallow water results in drag force and reduced body weight. The greater drag force of walking in water likely causes greater AP bending force on the femora, especially in flowing water. However, the AP bending forces on the tibiae may be reduced because of the decreased body weight. Walking and standing in shallow water for long periods of time would be consistent with the very strong lower limb bones and the femoral AP buttressing seen in Kennewick Man. This is especially true if Kennewick Man began this activity at a young age. Mechanical loading associated with habitual activities before adulthood results in greater increases in cross-sectional properties than do activities started as an adult (Bradney et al. 1998; Haapasalo et al. 1996, 2000; Kannus et al. 1995; Pearson and Lieberman 2004; Trinkaus et al.1994). The size, rigidity, and shape of Kennewick Man's lower limb bones and his isotopic signature are consistent with shoreline hunting and fishing.

While Kennewick Man does not exhibit the lower limb morphology expected for a hunter who was actively involved in long-distance hunting of large ungulates, it does not rule out that possibility. However, his morphology does suggest that he was either involved in seasonal activities that placed different loads on his lower limbs than other large prey hunters from Europe or the Americas or he hunted large ungulates using a different technique. European Paleolithic hunters consistently exhibit AP buttressed tibia. Similarly, many of the Paleoamericans with tibial measurements (Horn Shelter, Gore Creek, and Wizards Beach) have a greater proximal shaft AP/ML ratio than Kennewick Man. The two exceptions are Spirit Cave and Prospect Man, both of whom were likely involved in fishing and hunting of smaller mammals and waterfowl. However, it might be seasonal activities that actually have the largest effect on the morphology of the skeleton (Frost 2003).

Kennewick Man has relatively high asymmetry in the cross-sectional properties of his lower limbs. His left femoral midshaft is 9.8 percent larger (TA) than the right. More striking is the 25.5 percent asymmetry in femoral midshaft I_{max} and the 23.1 percent asymmetry in J, and the 16.9 percent asymmetry in Z_p. Average femoral midshaft asymmetry in Arikara males is 3.0 percent for TA, 5.6 percent for I_{max}, and 7.1 percent, for J. Wescott and Cunningham (2006) found high levels of femoral asymmetry in historic Arikara females after they began to grow surplus crops for trade. They argued that intensifying field activities such as right-handed hoeing and raking placed greater mechanical loads on the left femur than the right. The relatively high asymmetry in Kennewick Man's femoral cross-sectional properties is unexpected, but consistent with spearing fish and throwing darts with an atlatl. The greater strength of his left femur compared to his right resulted from the ground force reaction caused by supporting his weight during the rapid deceleration following a throw. Another possibility is that Kennewick Man was throwing a spear or throwing stick while on the run. That is, he was running prior to his throw and planted his left leg causing sudden lower body deceleration just prior to throwing a spear or throwing stick.

The healed projectile wound in Kennewick Man's right ilium may have potentially contributed to the observed femoral asymmetry. Chatters (2000) asserted that the large projectile wound resulted in great discomfort while walking. Although it is possible that Kennewick Man temporarily favored his right leg, the injury does not appear to have affected the overall strength of his lower limbs. His right leg is very strong relative to other hunter-gatherers. The extremely strong lower limbs of Kennewick Man suggest a lifetime involved in vigorous activities that placed high mechanical loadings on the lower limbs.

Along with indications of patterns of activity, Kennewick Man's lower limb strength reflects the temperate climate in which he lived. Long bone robusticity correlates negatively with mean annual temperature (Pearson 2000; Stock 2006). Body mass, which is highly correlated with climate, explains more than 80 percent of the variance in femoral second moments of area during growth and development (Ruff 2003). Kennewick Man would be expected to exhibit relatively robust limb bones based on geographical and climatic location. Yet, climate cannot explain all of Kennewick Man's body size and bone strength. Although the climate may have been relatively warmer and dryer during the early Holocene than it is today (Whitlock 1992), Kennewick Man displays greater stature and body mass than other native groups in the region (Table 12.2). Weaver (2003) discovered that populations in cold climates have more circular femoral midshafts compared to populations in warm climates. The more circular femoral midshaft probably reflects climatically induced body shape differences that influence mediolateral mechanical loading on the bone.

Femoral to Humeral Diaphyseal Strength Proportions

The relative strength of Kennewick Man's upper and lower limbs is more comparable to inland hunter-gatherers and Plains bison hunters than to Northwest Coast hunter-gatherers. A comparison of Kennewick Man's size-standardized section moduli humeral-to-femoral midshaft ratios to Native Americans and other Paleoamericans shows that Kennewick Man, and most of the Paleoamericans, had relatively weak humeri in relation to their femora when compared to coastal hunters (Figure 12.4). This is even more dramatic when comparing the humeral 35 percent strength to the femoral midshaft strength. Kennewick Man's humeral-to-femoral ratios (0.32 left and 0.53 right) are considerably lower than average for hunter-gatherer males from Alaska (0.70), the Aleutian Islands (0.72), Washington (0.68), Indian Knoll (0.67), and Arikara horticulturalists (0.57). His humeri are within the normal range in midshaft strength for hunter-gatherers, though his left humerus is weak relative to his left femur (Figure 12.4). Stock and Pfeiffer (2001) compared the relative upper and lower limb strengths of primarily maritime mobile Andaman Islanders and terrestrially mobile Late Stone Age South Africans. The ratio of humerus to femur strength shows that the Andaman Islanders had much greater relative humeral strength, while femur strength was greatest among the South Africans. Stock (2006) later found that rowing is reflected in both the proximal and distal upper limb bones. Cybulski et al. (1981) also argued that well-developed and robust humeri are likely an adaptation to paddling canoes, which may have been an important mode of transportation for Kennewick Man. Weiss (2003) found that ocean-rowing

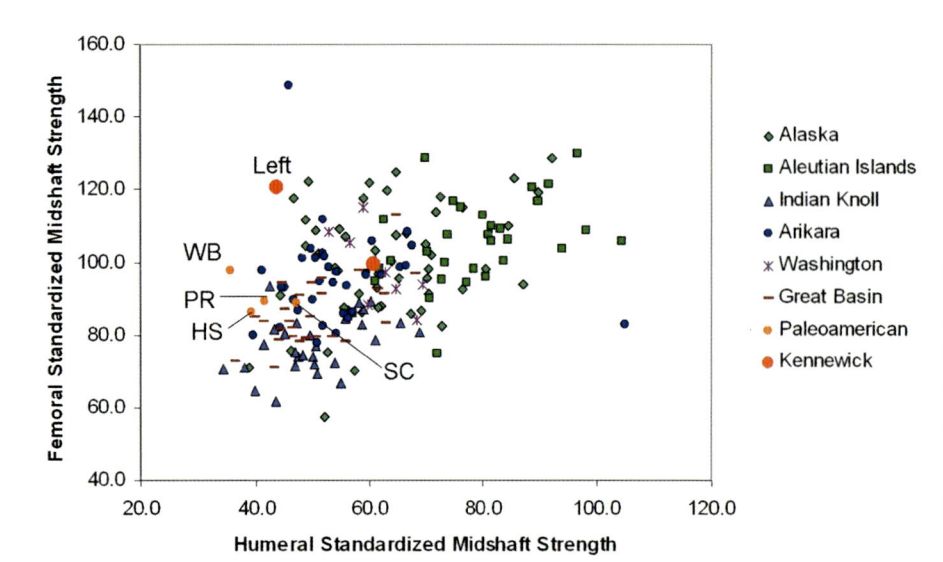

FIGURE 12.4 Relationship between humeral and femoral midshaft strength in Paleoamericans (HS = Horn Shelter, PR = Prospect Man, SC = Spirit Cave, WB = Wizards Beach; Left = Kennewick Man left side) and Native American groups.

populations have much greater humeral robusticity than river-rowing and nonrowing populations, though some of the greater robusticity was probably due to environmental and other biomechanical factors. Therefore, it is not surprising that Kennewick Man's humeral strength falls below that of Alaska and Aleutian Island ocean-rowing populations. The extreme humeral asymmetry seen in Kennewick Man (Table 12.3) further suggests that his humeral strength was not due to the use of watercraft since ocean-rowing populations such as the Jōmon from Japan, Inuits of Alaska, and Aleuts of the Aleutian Islands have strong but relatively symmetrical humeri (Table 12.4).

Humeral Strength, Shape, and Asymmetry

The analysis of bilateral asymmetry in upper limb diaphyseal cross-sectional geometric properties is often used in biomechanical studies to help interpret activity patterns (Rhodes and Knüsel 2005; Ruff and Jones 1981; Trinkaus et al. 1994). Kennewick Man had moderate to high levels of bilateral humeral strength asymmetry that indicate engagement in activities that required asymmetrical loading of his upper limbs. He had much greater strength in his right arm than in his left, indicating right-handedness. Asymmetry in J was 47 percent at middistal and 40.8 percent at midshaft (Table 12.3). In CA, his right humerus was 26.0 percent at middistal and 21.1 percent larger at midshaft than his left humerus. This level of asymmetry is greater than the mean for most hunter-gatherer males but within the range of inland hunter-gatherers. It is comparable to Upper Paleolithic males and forest foragers from South Africa, and much greater than the averages for maritime foragers and modern Japanese and Americans (Table 12.4).

The high level of asymmetry in hunter-gatherer males is generally attributed to unilateral loading due to habitual spear throwing (Shaw and Stock 2009b; Stock and Pfeiffer 2004; Trinkaus et al. 1994). Neanderthals have pronounced humeral asymmetry averaging between 24 and 57 percent as a result of thrusting their spear into large mammals (Schmitt et al. 2003; Trinkaus et al. 1994). Likewise, South African foragers and Late Upper Paleolithic Europeans who commonly hunted with a dart thrown with an atlatl also have high humeral asymmetry but not at the level seen in Neandertals and professional tennis players (Table 12.4). The humeral asymmetry seen in Kennewick Man is consistent with spear fishing and with using an atlatl or a throwing stick. Throwing a spear or stick regularly would place unilateral torsional loading on the dominant limb (Shaw and Stock 2009b; Robling et al. 2002). When spear fishing, the lightweight shaft must be thrust with enough power to impale the fish but not so much that it breaks the point or shaft if the fisherman misses (Stewart 1982). Once impaled, the fish was thrown to the shore, which requires strong bilateral use of the hands (Chapter 13). Propelling a spear with an atlatl requires significant strength in the dominant arm and places great torsional loading on the humerus.

Humeral shape may also provide vital information regarding activity patterns (Churchill et al. 1996; Schmitt et al. 2003; Weiss 2005). Among Neanderthals and Early

TABLE 12.4 Comparison of populations in humeral middistal (35%) diaphyseal strength and shape.

Region	Sample	N	% Asymmetry[1]				I_x/I_y[2]			
			CA		J		Right		Left	
South Africa[3]	Forest forager	5–6	9.5	6.4	33.9	12.0	1.11	0.17	1.05	0.16
South Africa[3]	Fynbos forager	7–9	10.8	4.2	15.2	3.8	1.12	0.11	1.17	0.23
Europe[4]	Neanderthals	6/5	16.5	11.5–47.4	53.1	21.9–98.5	1.27	0.15	1.30	0.16
Europe[4]	Early Upper Paleolithic	5–6	13.1	11.5–47.4	21.7	2.4–68.0	1.31	0.18	1.22	0.11
Europe[4]	Late Upper Paleolithic	7–9	31.7	18.9–50.7	50.9	8.5–101.9	1.16	0.13	1.12	0.19
Europe[5]	Neolithic (Italy)	4	8.03	—	15.8	—	—	—	—	—
Japan[4]	Jōmon foragers	10	6.7	0.4–15.0	6.4	1.6–16.7	0.90	0.13	0.90	0.13
Japan[6]	Japanese industrialists	46	4.4	—	8.1	—	1.10	0.14	1.13	0.14
North America[4]	Georgia Coast foragers	6	—	—	—	—	1.00	—	1.00	—
North America[4,7]	Georgia Coast agriculturalists	11–19	4.9	—	9.9	—	1.12	—	1.05	—
North America[4]	Aleuts maritime foragers	24	9.5	0.2–23.9	16.4	1.1–40.1	1.04	0.13	1.02	0.78
North America[7]	Great Basin hunter-gatherer	12	7.7	—	10.9	—	—	—	—	—
North America[7]	California hunter-gatherers	31	8.3	—	19.5	—	—	—	—	—
North America[8]	Arikara horticulturist-hunter	129	5.5	0.0–22.9	11.3	0.5–44.7	1.07	0.13	0.99	0.15
North America[4]	Euro-American industrialists	19	5.9	0.4–26.5	7.5	1.6–41.6	1.21	0.10	1.19	0.17
North America	Horn Shelter	1	14.4	—	30.4	—	1.15	—	1.18	—
North America	Kennewick Man	1	26.0	—	47.0	—	0.98	—	0.93	—
	International tennis players[9]	34	39.9	—	74.6	—	—	—	—	—

[1]The median and range are provided for % asymmetry in CA and J except for South African and modern Japanese populations where the mean and standard deviation are provided.
[2]Ratio of the second moment of inertia (I) in the AP plane to I in the ML plane, mean and standard deviation.
[3]Data from Stock and Pfeiffer (2004).
[4]Data from Schmitt and Churchhill (2003).
[5]Marchi et al (2006).
[6]Data from Sakaue (1997).
[7]Data from Ruff (1999).
[8]Wescott and Cunningham (2006).
[9]Trinkaus et al. (1994).

Upper Paleolithic European hunters, the AP bending strength is greater than the ML bending strength in the middistal humerus (indicated by high I_x/I_y values in Table 12.4), while the humeral shaft tends to be much rounder in Late Upper Paleolithic European hunters (Table 12.4) (Churchill et al. 1996). The more circular humeral cross-section of Late Upper Paleolithic hunters indicates they were performing tasks that required greater torsional loads than seen in earlier groups. Schmitt et al. (2003) showed experimentally that humeral asymmetry in Neanderthals was likely due to spear thrusting rather than throwing. They demonstrated that spear thrusting (common in close contact killing of prey) results primarily in AP bending forces on the dominant humerus. Spear throwing, on the other hand, results in torsion of the humerus. As a result, spear throwers will have more circular shafts (I_x/I_y values near one) than spear thrusters. Stock and Pfeiffer (2004) found significantly higher asymmetry in the humeral cross-sectional properties of males from the South African forest biome compared to males from the fynbos biome, a characteristic that they attributed to greater unilateral loading associated with spear use in forest males. In both groups, the I_x/I_y ratio was low, suggesting that the activities causing greater asymmetry in forest males result in torsional dynamic loads on the humerus rather than bending loads in one plane. There is clinical evidence that the twisting of the humeral shaft during throwing can lead to spiral fractures of the shaft indicating the extreme torsional strain placed on the bone (Soeda et al. 2002). In addition, Shaw and

Stock (2009b) found that the dominant humeri of cricket players are robust and relatively round compared to the nondominant arm due to throwing and batting the ball.

Kennewick Man appears to have been placing more ML bending force on his humeri than South African hunters, Late Upper Paleolithic Europeans, or the Horn Shelter male, all of whom were likely throwing spears with the aid of an atlatl (Churchill et al. 1996; Stock and Pfeiffer 2004). Hunters that partake in river- or ocean-rowing (Georgia Coast foragers, Jōmon foragers, and Aleuts) tend to have more circular and symmetrical humeri shafts than nonrowers. Though not directly tested, short-range spear throwing or thrusting in shallow rivers and streams may have placed greater ML bending stress on the humerus than did long-range spear throwing commonly used for terrestrial animals. Similarly, the use of a throwing stick would likely cause ML bending strain on the humerus of the dominant arm.

Kennewick Man's humeral strength asymmetry was not caused by disuse associated with a pathological or traumatic event. Poor diet, pathology, trauma, age, and sex all affect long bone strength and asymmetry (Lieverse et al. 2008; Ruff and Jones 1981; Weiss 2005). Chatters (2000) described a healed left radial head fracture and remodeling of the humeral septum and olecranon fossa, which he argued would have resulted in diminished use of the left arm. Walker et al. (2000) reported that the left radial head exhibits a defect caused by localized trauma and that reactive bone in the olecranon fossa suggest an inflammatory response. However, the bony changes in the elbow were unlikely severe enough to cause immobilization of the joint for any significant period (Chapter 7). Examination of the skeleton revealed no evidence of a healed radial fracture as described by Chatters (2000). Dimensions of the left radius and ulna do not indicate wasting as would be expected if the arm was not being used (Churchill and Formicola 1997; Lieverse et al. 2008). The moderate humeral asymmetry seen in Kennewick Man resulted from behaviors that caused differential loading of the arms from throwing a spear and using an atlatl.

CONCLUSION

Little attention has been paid to the information that can be revealed about early Holocene individuals in North America by examining long bone cross-sectional morphology. As a result, knowledge about the regional variation in subsistence technology among ancient Americans is inadequate. Based on the pattern of long bone strength and asymmetry in conjunction with stable isotope (Chapter 16; Taylor et al. 1998) and archaeological evidence (Butler 1993; Butler and Campbell 2004; Butler and O'Connor 2004; Cressman 1960; Chatters et al. 2012), it seems unlikely that Kennewick Man was a broad-spectrum forager. It seems more likely that he was engaged in habitual spearing or harpooning of fish and mammals in fast flowing rivers and streams and along the ocean coast. However, the relative upper-to-lower limb bone strength of Kennewick Man is similar to that of Gore Creek and other Paleoamericans, suggesting that he used his upper body more like these inland hunters than like coastal populations that use watercraft to hunt and fish. Tibial morphology suggests that daily long-distance terrestrial travel may have been less important to Kennewick Man than to some early Holocene hunters in North America and Europe. Kennewick Man's moderate to high humeral asymmetry is more similar to inland hunter-gatherer males than to coastal ocean-rowing hunters and indicates he was involved in activities that placed unilateral torsional loading on the right humerus. In many ways Kennewick Man's morphological pattern is unique and probably reflects regional variation in the way he exploited the local environment compared to early Americans in the Great Basin, Plains, and other regions. While the exact activities that Kennewick Man participated in are unknown, his skeletal morphology and isotopic signature are more consistent with exploiting marine sources rather than large terrestrial fauna for protein.

REFERENCES

Ames, Kenneth M. 2000. Chapter 2. Review of the Archaeological Data. In *Cultural Affiliation Report* [September 2000]. National Park Service, http://www.nps.gov/archeology/kennewick/ames.htm.

Auerbach, Benjamin M. 2007. "Human Skeletal Variation in the New World During the Holocene: Effects of Climate and Subsistence Across Geography and Time." Ph.D. dissertation, Johns Hopkins University School of Medicine, Baltimore.

Barela, Ana M. F., Sandro F. Stolf, and Marcos Duarte. 2006. Biomechanical characteristics of adults walking in shallow water and on land. *Journal of Electomyography and Kinesiology* 16(3):250–256.

Barker, Pat, Cynthia Ellis, and Stephanie Damadio. 2000. *Determination of Cultural Affiliation of Ancient Human Remains from Spirit Cave, Nevada.* Submitted to Bureau of Land Management Nevada State Office.

Bradney, M., G. Pearce, G. Naughton, C. Sullivan, S. Bass, T. Beck, J. Carlson, and E. Seeman. 1998. Moderate exercise during growth in prepubertal boys: Changes in bone mass, size, volumetric density, and bone strength: A controlled prospective study. *Journal of Bone and Mineral Research* 13(12):1814–1821.

Bridges, Patricia S. 1995. Skeletal biology and behavior in ancient humans. *Evolutionary Anthropology: Issues, News, and Reviews* 4(4):112–120.

Butler, Virginia L. 1993. Natural versus cultural salmonid remains: Origin of the Dalles Roadcut Bones, Columbia River, Oregon, U.S.A. *Journal of Archaeological Science* 20(1):1–24.

Butler, Virginia L., and Sarah K. Campbell. 2004. Resource intensification and resource depression in the Pacific Northwest of North America: A zooarchaeological review. *Journal of World Prehistory* 18(4):327–405.

Butler, Virginia L., and Jim E. O'Connor. 2004. 9000 years of salmon fishing on the Columbia River, North America. *Quaternary Research* 62(1):1–8.

Byers, David A., and Andrew Ugan. 2005. Should we expect large game specialization in the late Pleistocene? An optimal foraging perspective on early Paleoindian prey choice. *Journal of Archaeological Science* 32(1):1624–1640.

Carlson, Kristian J., and Stefan Judex. 2007. Increased non-linear locomotion alters diaphyseal bone shape. *Journal of Experimental Biology* 210:3117–3125.

Chatters, James C. 2000. The recovery and first analysis of an early Holocene human skeleton from Kennewick, Washington. *American Antiquity* 65(2):291–316.

Chatters, James C., Steven Hackenberger, Anna M. Prentiss, and Jayne-Leigh Thomas. 2012. The Paleoindian to Archaic transition in the Pacific Northwest: In situ development or ethnic replacement? In *From the Pleistocene to the Holocene: Human Organization and Cultural Transformations in Prehistoric North America,* edited by C. Britt Bousman and Bradley J. Vierra, 37–66. Texas A&M Press, College Station.

Churchill, Steven E., and Vincenzo Formicola. 1997. A case of marked bilateral asymmetry in the upper limbs of an upper Paleolithic male from Barma Grande (Liguria), Italy. *International Journal of Osteoarchaeology* 7(1):18–38.

Churchill, Steven E., Ann H. Weaver, and W. A. Niewoehner. 1996. Late Pleistocene human technological and subsistence behavior: Functional interpretations of upper limb morphology. *Quaternaria Nova* 6:413–447.

Cressman, L. S. 1960. *Cultural Sequence at the Dalles, Oregon: A Contribution to Pacific Northwest Prehistory.* Transactions of the American Philosophical Society 50, Part 10. American Philosophical Society, Philadelphia.

Cybulski, Jerome S., Donald E. Howes, James C. Haggarty, and Morley Eldridge. 1981. An early human skeleton from south-central British Columbia: Dating and bioarchaeological inference. *Canadian Journal of Archaeology* 5:49–59.

Edgar, Heather J. H. 1997. Paleopathology of the Wizard's Beach Man (AHUR 2023) and the Spirit Cave Mummy (AHUR 2064). *Nevada Historical Society Quarterly* 40(1):57–61.

Eiselt, B. Sunday. 1997. Fish remains from the Spirit Cave paleofecal materials: 9,400 year old evidence for Great Basin utilization of small fishes. *Nevada Historical Society Quarterly* 40(1):117–139.

Eschman P. 1992. SLCOMM. Eschman Archaeological Services, Albuquerque, NM.

Frost, Harold M. 2003. Bon's Mechanostat: A 2003 Update. *The Anatomical Record* 275A:1081–1101.

Genovés, Santiago. 1967. Proportionality of the long bones and their relation to stature among Mesoamericans. *American Journal of Physical Anthropology* 26(1):67–78.

Green, Thomas J., Bruce Cochran, Todd W. Fenton, James C. Woods, Gene L. Titmus, Larry Tieszen, Mary Anne Davis, Susanne J. Miller. 1998. The Buhl burial: A Paleoindian woman from southern Idaho. *American Antiquity* 63(3):437–456.

Grine, F. E., W. L. Jungers, P. V. Tobias, and O. M. Pearson. 1995. Fossil *Homo* femur from Berg Aukas, northern Namibia. *American Journal of Physical Anthropology* 97(2):151–185.

Gruss, Laura Tobias. 2007. Limb length and locomotor biomechanics in the genus *Homo*: An experimental study. *American Journal of Physical Anthropology* 134(1):106–116.

Haapasalo, H., S. Kontulainen, H. Sevänen, P. Kannus, M. Järvinen, and I. Vuori. 2000. Exercise-induced bone gain is due to enlargement in bone size without a change in volumetric bone density: A peripheral quantitative computed tomography study of the upper arms of male tennis players. *Bone* 27(3):351–357.

Haapasalo, Heidi, Harri Sievänen, Pekka Kannus, Ari Heinonen, Pekka Oja, and Ilkka Vuori. 1996. Dimensions and estimated mechanical characteristics of the humerus after long-term tennis loading. *Journal of Bone and Mineral Research* 11(6):864–872.

Holliday, Trenton W. 2002. Body size and postcranial robusticity of European Upper Paleolithic hominins. *Journal of Human Evolution* 43(4):513–528.

Holt, Brigitte M. 1999. "Biomechanical Evidence of Decreased Mobility in Upper Paleolithic and Mesolithic Europe." Ph.D. dissertation, University of Missouri, Columbia.

———. 2003. Mobility in upper Paleolithic and Mesolithic Europe: Evidence from the lower limb. *American Journal of Physical Anthropology* 122(3):200–215.

Howells, William W. 1938. Crania from Wyoming resembling "Minnesota Man." *American Antiquity* 3(4):318–326.

Huckleberry, Gary, Julie K. Stein, and Paul Goldberg. 2003. Determining the provenience of Kennewick Man skeletal remains through sedimentological analyses. *Journal of Archaeological Science* 30(6):651–665.

Hunt, David R. 2002. Aleutian remains at the Smithsonian Institution. In *To the Aleutians and Beyond: The Anthropology of William S. Laughlin*, edited by Bruno Frohlich, Albert B. Harper, and Rolf Gilberg, 137–153. Department of Ethnography, The National Museum of Denmark, Copenhagen.

Jantz, Richard L., and Douglas W. Owsley. 2001. Variation among early North American crania. *American Journal of Physical Anthropology* 114(2):146–155.

Jungers, William L., and Robert J. Minns. 1979. Computed tomography and biomechanical analysis of fossil long bones. *American Journal of Physical Anthropology* 50(2):285–290.

Kannus, Pekka, Heidi Haapasalo, Marja Aankelo, Harri Sievanen, Matti Pasanen, Ari Heinonen, Pekka Oja, and Ilkka Vuori. 1995. Effect of starting age of physical activity on bone mass in the dominant arm of tennis and squash players. *Annals of Internal Medicine* 123:27–31.

Kelly, Robert L., and Lawrence C. Todd. 1988. Coming into the country: Early Paleoindian hunting and mobility. *American Antiquity* 53(2):231–244.

Koerper, Henry C. 1998. A game string and rabbit stick cache from Borrego Valley, San Diego County. *Journal of California and Great Basin Anthropology* 20(2):252–270.

Lieverse, A. R., M. A. Metcalf, V. I. Bazaliiskii, and A. W. Weber. 2008. Pronounced bilateral asymmetry of the complete upper extremity: A case from the early Neolithic Baikal, Siberia. *International Journal of Osteoarchaeology* 18(3):219–239.

McHenry, Henry M. 1992. Body size and proportions in early Hominids. *American Journal of Physical Anthropology* 87(4):529–534.

Marchi, Damiano. 2008. Relationships between lower limb cross-sectional geometry and mobility: The case of a Neolithic sample from Italy. *American Journal of Physical Anthropology* 137(2):188–200.

Marchi, Damiano, Vitale S. Sparacello, Brigitte M. Holt, and Vincenzo Formicola. 2007. Biomechanical approach to the reconstruction of activity patterns in Neolithic western Liguria, Italy. *American Journal of Physical Anthropology* 131(4):447–455.

National Institute of Health. 2008. ImageJ. Public domain image processing software. https://rsb.info.nih.gov.

Nikander R., P. Kannus, T. Rantalainen, K. Uusi-Rasi, A. Heinonen, and H. Sievänen. 2010. Cross-sectional geometry of weight-bearing tibia in female athletes subjected to different exercise loadings. *Osteoporosis International* 21(10):1687–1694.

O'Neill, Matthew C., and Christopher B. Ruff. 2004. Estimating human long bone cross-sectional geometric properties: A comparison of noninvasive methods. *Journal of Human Evolution* 47(4):221–235.

Pearson, Osbjorn M. 2000. Activity, climate and postcranial robusticity: Implications for modern human origins and scenarios of adaptive change. *Current Anthropology* 41(4):569–607.

Pearson, O. M., R. M. Cordero, and A. M. Busby. 2006. How different were Neanderthals' habitual activities? A comparative analysis with diverse groups of recent humans. In *Neanderthals Revisited: New Approaches and Perspectives*, edited by Katerina Harvati and Terry Harrison, 135–156. Springer, New York.

Pearson, Osbjorn M., and Daniel E. Lieberman. 2004. The aging of Wolff's "Law": Ontogeny and responses to mechanical loading in cortical bone. *American Journal of Physical Anthropology* 125(S39):63–99.

Powell, Joseph F., and Walter A. Neves. 1999. Craniofacial morphology of the First Americans: Pattern and process in the peopling of the New World. *American Journal of Physical Anthropology* 110(S29):153–188.

Powell, Joseph F., and Jerome C. Rose. 1999. Chapter 2.

Report on the Osteological Assessment of the "Kennewick Man" Skeleton (CENWW.97.Kennewick). In *Report on the Non-Destructive Examination, Description, and Analysis of the Human Remains from Columbia Park, Kennewick, Washington* [October 1999]. National Park Service, http://www.nps.gov/archeology/kennewick/powell_rose.htm.

Robling, Alexander G., Felicia M. Hinant, David B. Burr, and Charles H. Turner. 2002. Improved bone structure and strength after long-term mechanical loading is greatest if loading is separated into short bouts. *Journal of Bone and Mineral Research* 17(8):1545–1554.

Rockhold, Lauren A. 1998. "Secular Change in External Femoral Measures from 1840 to 1970: A Biomechanical Interpretation." Master's thesis, University of Tennessee, Knoxville.

Rhodes, Jill A., and Christopher J. Knüsel. 2005. Activity-related skeletal change in medieval humeri: Cross-sectional and architectural alterations. *American Journal of Physical Anthropology* 128(3):536–546.

Ruff, Christopher. 1987. Sexual dimorphism in human lower limb bone structure: Relationship to subsistence strategy and sexual division of labor. *Journal of Human Evolution* 16(5):391–416.

———. 1999. Skeletal structure and behavioral patterns of prehistoric Great Basin populations. In *Prehistoric Lifeways in the Great Basin Wetlands: Bioarchaeological Reconstruction and Interpretation*, edited by Brian E. Hemphill and Clark Spenser Larsen, 290–320. University of Utah Press, Salt Lake City.

———. 2000. Body size, body shape, and long bone strength in modern humans. *Journal of Human Evolution* 38(2):269–290.

———. 2002. Long bone articular and diaphyseal structure in Old World monkeys and apes. I: Locomotor effects. *American Journal of Physical Anthropology* 119(4):305–342.

———. 2003. Growth in bone strength, body size, and muscle size in a juvenile longitudinal sample. *Bone* 33(3):317–329.

———. 2008. Biomechanical analyses of archaeological human skeletons. In *Biological Anthropology of the Human Skeleton*, 2nd edition, edited by M. Anne Katzenberg and Shelley R. Saunders, 183–206. Wiley, New York.

Ruff, Christopher, Brigitte M. Holt, and Erik Trinkaus. 2006a. Who's afraid of the big bad Wolff?: "Wolff's Law" and bone functional adaptation. *Journal of Physical Anthropology* 129(4):484–498.

Ruff, Christopher B., Brigitte M. Holt, Vladimír Sládek, Margit Berner, William A. Murphy, Dieter zur Nedden, Horst Seidler, and Wolfgang Recheis. 2006b. Body size, body proportions, and mobility in the Tyrolean "Iceman." *Journal of Human Evolution* 51(1):91–101.

Ruff, Christopher B., and Henry H. Jones. 1981. Bilateral asymmetry in cortical bone of the humerus and tibia—Sex and age factors. *Human Biology* 53(1):69–86.

Ruff, Christopher B., William W. Scott, and Allie Y.-C. Liu. 1991. Articular and diaphyseal remodeling of the proximal femur with changes in body mass in adults. *American Journal of Physical Anthropology* 86(3):397–413.

Sakaue, Kazuhiro. 1997. Bilateral asymmetry of the humerus in Jōmon people and modern Japanese. *Anthropological Science* 105(4):231–246.

Schmitt, Daniel, Steven E. Churchill, and William L. Hylander. 2003. Experimental evidence concerning spear use in Neanderthals and early modern humans. *Journal of Archaeological Science* 30(1):103–114.

Schurr, Theodore G., and Stephen T. Sherry. 2004. Mitochondrial DNA and Y chromosome diversity and the peopling of the Americas: Evolutionary and demographic evidence. *American Journal of Human Biology* 16(4):420–439.

Sládek, Vladimír, Margit Berner, Patrik Galeta, Lukáš Friedl, and Šárka Kudrnová. 2010. The effect of midshaft location on the error ranges of femoral and tibial cross-sectional parameters. *American Journal of Physical Anthropology* 141(2):325–332.

Sládek, Vladimír, Margit Berner, and Robert Sailer. 2006a. Mobility in central European late Eneolithic and early Bronze Age: Femoral cross-sectional geometry. *American Journal of Physical Anthropology* 130(3):320–332.

———. 2006b. Mobility in central European late Eneolithic and early Bronze Age: Tibial cross-sectional geometry. *Journal of Archaeological Science* 33(4):470–482.

Shaw, Collin N., and Jay T. Stock. 2008. The relationship between lower limb length and midshaft diaphyseal shape of the femur and tibia among modern humans. *American Journal of Physical Anthropology* 135(S46):192.

———. 2009a. Intensity, repetitiveness, and directionality of habitual adolescent mobility patterns influence the tibial diaphysis morphology of athletes. *American Journal of Physical Anthropology* 140(1):140–159.

———. 2009b. Habitual throwing and swimming correspond with upper limb diaphyseal strength and shape in modern human athletes. *American Journal of Physical Anthropology* 140(1):162–172.

Soeda, Tsunemitsu, Yasuaki Nakagawa, Takashi Suzuki, and Takashi Nakamura. 2002. Recurrent throwing fracture of the humerus in a baseball player. *American Journal of Sports Medicine* 30(6):900–902.

Stewart, Hilary. 1982. *Indian Fishing: Early Methods on the Northwest Coast.* University of Washington Press, Seattle.

Stock, Jay T. 2006. Hunter-gatherer postcranial robusticity relative to patterns of mobility, climatic adaptation, and selection for tissue economy. *American Journal of Physical Anthropology* 131(2):194–204.

Stock, Jay T., and Susan K. Pfeiffer. 2001. Linking structural variability in long bone diaphyses to habitual behaviors: Foragers from the southern African later Stone Age and the Andaman Islands. *American Journal of Physical Anthropology* 115(4):337–348.

———. 2004. Long bone robusticity and subsistence behaviour among later Stone Age foragers of the forest and Fynbos Biomes of South Africa. *Journal of Archaeological Science* 31(7):999–1013.

Stock, Jay T., and Colin N. Shaw. 2007. Which measures of diaphyseal robusticity are robust? A comparison of the external methods of quantifying the strength of long bone diaphyses of cross-sectional geometric properties. *American Journal of Physical Anthropology* 134(3):412–423.

Taylor, R. E., Donna. L. Kimer, John. R. Southon, and James C. Chatters. 1998. Radiocarbon dates of Kennewick Man. *Science* 280(5367):1171–1172.

Taylor, William R., Markus O. Heller, Georg Bergmann, and Georg N. Duda. 2004. Tibio-femoral loading during human gait and stair climbing. *Journal of Orthopaedic Research* 22:625–632.

Trinkaus, Erik, Steven E. Churchill, and Christopher B. Ruff. 1994. Postcranial robusticity in *Homo*. II: Humeral Bilateral Asymmetry and Bone Plasticity. *American Journal of Physical Anthropology* 93(1):1–34.

Tuohy, Donald R. 2004. Archaeological curved throwing sticks from fish cave, near Fallon, Nevada. *Journal of California and Great Basin Anthropology* 24(1):13–20.

Walker, Phillip L., Clark Spenser Larsen, and Joseph F. Powell. 2000. Chapter 5. Final Report on the Physical Examination and Taphonomic Assessment of the Kennewick Human Remains (CENWW.97.Kennewick) to Assist with DNA Sample Selection. In *Report on the DNA Testing Results of the Kennewick Human Remains from Columbia Park, Kennewick, Washington* [September 2000]. National Park Service, http://www.nps.gov/archeology/kennewick/walker.htm.

Waters, Michael R., Jason Wiersema, and Thomas W. Stafford, Jr. 2008. Brazoria Woman: A geoarchaeological evaluation of a reported late Pleistocene human burial. *Journal of Archaeological Science* 35(9):2425–2433.

Weaver, Timothy D. 2003. The shape of the Neanderthal femur is primarily the consequence of a hyperpolar body form. *Proceedings of the National Academy of Sciences* 100(12):6926–6929.

Weaver, Timothy D., and Karen Steudel-Numbers. 2005. Does climate and mobility explain the differences in body proportions between Neanderthals and their Upper Paleolithic successors? *Evolutionary Anthropology* 14(6):218–223.

Weiss, Elizabeth. 2003. Effects of rowing on humeral strength. *American Journal of Physical Anthropology* 121(4):293–302.

———. 2005. Humeral cross-sectional morphology from 18th century Quebec prisoners of war: Limits to activity reconstruction. *American Journal of Physical Anthropology* 126(3):311–317.

Wescott, Daniel J. 2001. "Structural Variation in the Humerus and Femur in the American Great Plains and Adjacent Regions: Differences in Subsistence Strategy and Physical Terrain." Ph.D. dissertation, University of Tennessee, Knoxville.

———. 2006a. Effect of mobility on femur midshaft shape and robusticity. *American Journal of Physical Anthropology* 130(2):201–213.

———. 2006b. Ontogeny of femur subtrochanteric shape in Native Americans and American blacks and whites. *Journal of Forensic Sciences* 51(6):1240–1245.

———. 2008. Biomechanical analysis of humeral and femoral structural variation in the Great Plains. *Plains Anthropologist* 53(207):333–355.

Wescott, Daniel J., and Deborah L. Cunningham. 2006. Temporal changes in Arikara humeral and femoral cross-sectional geometry associated with horticulture intensification. *Journal of Archaeological Science* 33(7):1022–1036.

Whitlock, Cathy. 1992. Vegetational and climatic history of the Pacific Northwest during the last 20,000 Years: Implications for understanding present-day biodiversity. *The Northwest Environmental Journal* 8:5–28.

Young, Diane E., Suzanne Patrick, and D. Gentry Steele. 1987. An analysis of the Paleoindian double burial from Horn Shelter No. 2, in Central Texas. *Plains Anthropologist* 32(117):275–298.

Bones of the Hands and Feet

13

D. Troy Case

It is rare to have the opportunity to study well-preserved hand and foot bones from an ancient skeleton. Even in more recent archaeological samples, these bones are frequently missing or poorly preserved. Good preservation of the carpals, tarsals, and many of the digital bones in the Kennewick Man skeleton, along with the surprising number of bones recovered, provides an unprecedented opportunity to learn about the morphology of the hands and feet of a Paleoamerican, and to gain an understanding of the way in which Kennewick Man used his appendages in everyday life. Included here is a survey of skeletal defects and pathology, an evaluation of asymmetry to assess handedness, and an analysis of taphonomic changes, such as calcium carbonate deposition, bone adhesions, loss of bone surface patina, and algal staining, to better understand positioning of the hands and feet in situ, and the extent to which these bones were exposed to the elements once the remains began eroding into the lake.

The data gathered from Kennewick Man's carpals, tarsals, and digital bones provide baseline information against which less well-preserved skeletons from the early Holocene period can be compared. One such skeleton is the adult male from Horn Shelter No. 2 in Texas, dating to about the same time period as Kennewick Man. Relatively good recovery of bones from the Horn Shelter skeleton allows detailed comparison of the metric dimensions of these two skeletons and indicates a high degree of similarity among these two individuals in terms of relative dimensions.

INVENTORY

The first objective was to position and side each hand and foot bone (Tables 13.1–13.2). Positioning entails identification of bone type (e.g., scaphoid or proximal foot phalanx) and, for the digital bones, the ray to which each bone belongs. The carpals, tarsals, metacarpals, and metatarsals were positioned and sided using standard techniques (Bass 2005). Since no standard techniques exist for positioning and siding the manual and pedal phalanges, these bones were assigned to the appropriate ray and side using a combination of techniques from several studies (Ricklan 1988; Case 1996, 2003; Case and Heilman 2006). The error component to positioning and siding phalanges is relatively minor in most cases, and it is further reduced when all or most of the bones of the same row are present, as is the case for Kennewick Man.

The positions and sides determined for the "complete" and "near complete" bones, with the few exceptions noted below, are accurate. Position and side determinations for the proximal, intermediate, and first distal phalanges of the hands are almost certainly correct. Distal phalanges of the second through fifth digits are more difficult to position with certainty, but enough of these bones are present that the determinations made here are likely correct, albeit with some possibility of error in both ray and side.

For the foot phalanges, experience suggests that ray and side are easily determined for "complete" or "near complete" first proximal and first distal phalanges, as well as fifth proximal phalanges. Determination of side is straightforward for the proximal phalanges in general. There is some possibility of error in terms of correctly identifying the third and fourth proximal phalanges, as their relative length and robusticity varies considerably among individuals. The probability of error is reduced because an example of each of the four lesser toe phalanges is present for comparison. Determination of side in these bones is most challenging for the second proximal phalanx. The intermediate and distal phalanges of the lesser toes were not recovered.

US Army Corps of Engineers (Corps) catalog numbers for each hand and foot bone indicate which side they

TABLE 13.1   Inventory of Kennewick Man's hand bones.

Bone Type	Left Side Catalog Number	Preservation	Right Side Catalog Number	Preservation
Scaphoid			97.L.16(CRb)	complete
Lunate	97.L.16(CRa)	complete	97.R.16(CRb)	complete
Capitate	97.L.16(CRc)	complete		
Hamate			97.R.16(CRa)	complete
Metacarpal 1	*97.R.16(MCd)*	complete	*97B.L.16(MCa)*	complete
Metacarpal 2	97.L.16(MCb)	2 pieces	97.R.16(MCc)	complete
Metacarpal 3	97.L.16(MCa)*	photo only	97.R.16(MCa)	2 pieces
Metacarpal 4	97.L.16(MCc)/(MCd)	2 pieces	97.R.16(MCe)	complete
Metacarpal 5			97.R.16(MCb)	complete
Proximal phalanx 1	*97A.I.24(Pb)*	complete	*97.L.24(Pd)*	distal damage
Proximal phalanx 2	97.L.16(Pf)	distal 1/3	*97.L.16(Pc)*	complete
Proximal phalanx 3			97.R.16(Pi)	distal 1/2
Proximal phalanx 4	97.L.16(Pa)	complete	97.R.16(Pd)	prox. damage
Proximal phalanx 5	*97A.I.16(Pa)*	some erosion	97.R.16(Pb)	complete
Intermediate phalanx 2	*97.R.16(Pe)*			
Intermediate phalanx 3	*97.R.16(Pf)*	complete		
Intermediate phalanx 4			*97.L.16(Pb)*	complete
Intermediate phalanx 5	97.L.16(Pd)	complete	97.R.16(Pg)	complete
Distal phalanx 1	97.L.16(Pe)	complete	97.R.16(Ph)	complete
Distal phalanx 3			97.R.16(Pa)	complete
Distal phalanx 4	97.R.16(Pc)	complete		
Distal phalanx 5				

*Removed for DNA analysis and no longer in collection. Observations made from photographs.
Bones belonging to positions differing from those indicated by the Corps labels are italicized.

were originally assigned to, as well as whether they were identified as belonging to the hand or foot. An L indicates left, an R indicates right, and an I indicates indeterminate. The number 16 indicates a hand bone, and the number 24 indicates bones belonging to the feet. Several misidentifications were corrected (Tables 13.1–13.2). Nine hand phalanges and three foot phalanges are assigned to digits different from the Corps inventory, and two phalanges originally placed with the feet have been moved to the hands. Five hand phalanges and two foot phalanges have been reassigned to the opposite side, and four phalanges marked indeterminate have now been placed on one side or the other. Proper positioning, particularly of those bones with better preservation, allows a more thorough analysis of Kennewick Man's hands and feet than would have been possible otherwise.

The hand and foot bones are generally well preserved. Recovery of bones was surprisingly good given the skeleton's antiquity and the circumstances of its recovery, suggesting a short duration between exposure and discovery. In the hands, 32 of 54 bones were recovered, including an example of nearly every hand bone except four carpals and the distal fifth phalanx. Most of these bones are complete or nearly so, with a nearly equal number of elements representing each side. Fifteen bones were recovered from the left side, and 17 from the right. The feet are less well represented, with 25 of 52 bones present. None of the intermediate or distal phalanges was recovered except the left distal first phalanx. An example of each of the other bones was recovered except for the fifth metatarsal. Preservation in the feet has favored the left side over the right, with 14 of 15 left side bones being

TABLE 13.2 Inventory of Kennewick Man's foot bones.

Bone Type	Left Side		Right Side	
	Catalog Number	Preservation	Catalog Number	Preservation
Talus	97.L.23a	head/neck	97.R.23a	complete
Calcaneus	97.L.22a	near complete	97.R.22a	complete
Navicular	97.L.24(TRb)	near complete	97.R.24(TRa)	complete
Cuboid	97.L.24(Tra)	complete		
Cuneiform 1	97.L.24(TRc)	complete		
Cuneiform 2	97.L.24(TRe)	complete		
Cuneiform 3	97.L.24(TRd)	complete		
Metatarsal 1	97.L.24(MTa)	complete	CENWW.97.R.24(Mta)*	photo only
Metatarsal 2	97.L.24(MTb)	complete	97.R.24(MTc)	head only
Metatarsal 3	97.L.24(MTc)	complete	97.R.24(MTb)	proximal 1/2
Metatarsal 4	97.L.24(MTd)	complete	CENNWW.97A.I.25c**	diaphysis only
Proximal phalanx 1			97.L.24(Pe)	plantar 1/2
Proximal phalanx 2	*97A.I.24(Pa)*	complete	97.L.24(Pb)	complete
Proximal phalanx 3	97.L.24(Pa)	complete		
Proximal phalanx 4	*97.R.24(Pa)*	complete		
Proximal phalanx 5			97.R.24(Pb)	near complete
Distal phalanx 1	97.L.24(Pc)	near complete		

*Removed for AMS dating and no longer in collection. Observations made from photographs.
**Diaphysis only. Could be MT5 as well, although shape does not seem right.
Bones belonging to positions that differ from those indicated by Corps labels are italicized.

complete or nearly so, while only six of ten bones are as well preserved on the right.

SKELETAL DEFECTS

The bones were next examined for presence of congenital or developmental defects that might impact the asymmetry and pattern profile studies discussed below. Congenital defects are of interest because some, such as the os intermetatarseum accessory bone and certain forms of tarsal coalition, may cause dysfunction that can impact skeletal and muscular symmetry. These defects were also of interest because some occur in higher frequencies among Native American groups than among individuals of European or African ancestry. These include thumb polydactyly, abnormal shortening of the metacarpals or metatarsals (known as brachydactyly E), presence of the os intermetatarseum accessory bone, presence of the calcaneus secundarius accessory bone, and coalition between the third metatarsal and third cuneiform (Cybulski 1988; Mann 1990; Tenney 1991; Case 1996; Case et al. 1998, 2006; Regan et al. 1999). Many of these traits show good evidence of heritability (Case 2003) so their presence in relatively high frequencies among Native Americans could be the result of founder effect, or perhaps long-term genetic drift either prior to or following migration to North America. Although the presence or absence of one of these traits in Kennewick Man tells little about his affiliation with later Native American groups, it is a step toward characterizing the types of defects present among the earliest inhabitants of North America as well as the frequency of these traits. If Kennewick Man were to exhibit multiple congenital defects that are fairly uncommon in populations from other parts of the world, but significantly more common in later Native American groups, it would support arguments of relatively close affinity between Kennewick Man and later Native Americans.

Examination of the hands and feet for congenital defects was carried out by visual inspection following specific scoring criteria (Case 2003:Appendix II). The de-

TABLE 13.3 Presence/absence of skeletal defects in the hands and feet of Kennewick Man.

Bone	Defect/Trait	Left	Right	Notes
Calcaneus	Calcaneonavicular coalition	P	P	probably fibrous
	Calcaneus secundarius	A	nd	
Talus	Talonavicular coalition	A	A	
	Talocalcaneal coalition	nd	A	
	Os trigonum	nd	P	trigonal process absent
Navicular	Os tibiale externum	A	A	
	Calcaneonavicular coalition	P	P	
Cuboid	Cubonavicular coalition	A	nd	
	Calcaneocuboid coalition	A	nd	
Cuneiform 1	Bipartition	A	nd	
	Os intermetatarseum	A	nd	
Cuneiform 2	CF1-CF2 coalition	A	nd	
Cuneiform 3	MT3-CF3 coalition	A	P	mild indentation and surface irregularity
Metatarsal 1	Brachydactyly	A	nd	
	Os intermetatarseum	A	A	
Metatarsal 2	Brachydactyly	A	nd	
Metatarsal 3	Os intermetatarseum	A	nd	
	MT3-CF3 coalition	A	nd	
Metatarsal 4	Brachydactyly	A	nd	
Distal Phalanx 1	Brachydactyly	A	nd	

P = present, A = absent, nd = not scorable

fects scored are listed in Table 13.3. Most of these defects occur in frequencies of less than 5 percent in typical populations, and it is not surprising that most were absent in Kennewick Man. However, two types of nonosseous coalition were identified. The most interesting of the two is nonosseous calcaneonavicular coalition, which is present bilaterally (Figures 13.1–13.5). The other is a probable case of third metatarsal-third cuneiform coalition on the right side. This defect is represented by a slight indentation of the plantar portion of the third metatarsal proximal facet with slight irregularity in the floor of the indentation. There are similar lesions in a few cases where the opposite side exhibits a more typical lesion with a pitted floor; however, in this case, the opposite foot is not affected, and the third cuneiform for the right side, which should also exhibit the lesion, was not recovered. Coalition of the third metatarsal and third cuneiform is one defect that tends to affect East Asians and later Native American groups in much higher frequencies than in other world populations, with frequencies reaching as high as 26 percent among some prehistoric California Indians (Regan et al. 1999). Since the defect is also found in people of European and African ancestry, however, its presence alone reveals little about population affinity.

Calcaneonavicular coalition is an uncommon condition that affects less than 3 percent of most populations (Pfitzner 1896; Cooperman et al. 2001; Case and Burnett 2011), although a frequency of nearly 10 percent has been reported for a small sample of Australians of European descent. The term "coalition" describes an abnormal union between two bones resulting from incomplete separation during the embryonic or early fetal period. These coalitions begin as a bridge of mesenchyme between the bones, often near the edge of a joint surface, but are converted to cartilage along with the rest of the primordial bones through the chondrification process by the early fetal period. A small percentage of coalitions become osseous bridges during adolescence, but the majority are nonosseous, with the bridge joining the two bones composed either of cartilage, fibrocartilage, or fibrous tissue (Pfitzner 1896; Kumai et al. 1998; Regan et al. 1999; Cooperman et al. 2001; Rühli et al. 2003; Burnett and Case 2005). Based on histological studies of nonosseous coalitions, Kumai et al. (1998) suggest that coalitions are often converted over the course of a lifetime from pure cartilage to fibrocartilage, and perhaps even to pure fibrous tissue, as a result of repeated microtrauma to the nonosseous bridge between the bones.

Coalitions typically occur at specific locations between two bones. Calcaneonavicular coalitions occur between the anterior beak of the calcaneus, and the lateroplantar surface of the navicular (Figure 13.1). The affected portion of the navicular takes on a characteristic squared off shape caused by the presence of additional bone along the margin of the coalition. The anterior portion of the calcaneus also tends to look distinctive, exhibiting an oblique outline at the distal end when viewed superiorly. This is in contrast to the rather straight mediolateral outline of this surface in unaffected individuals. The point of contact between the calcaneus and navicular can range from a large and deeply pitted lesion on both bones, probably indicative of a primarily cartilaginous bridge, to a somewhat irregular but unpitted lesion, probably indicative of a primarily or exclusively fibrous bridge (see Figure 13.1 for an example of each). A similar range of variation was presented by Regan et al. (1999) for third metatarsal-third cuneiform coalitions.

As the tarsals ossify and the joint surfaces develop, the cartilaginous or fibrous bridges between adjacent bones sometimes become painful. The pain appears to be associated with microtrauma to the nonosseous bridge caused by abnormal motion and subsequent mechanical stress in the tarsals. Since the coalitions are initially composed of cartilage, they are fed by a blood supply through small vessels that can rupture under stress (Kumai et al. 1998). Strenuous activity adds to the probability of microtrauma, and many coalitions become symptomatic in late childhood or adolescence as the bones take on adult forms (Stormont and Peterson 1983). For some affected individuals, coalitions remain asymptomatic. A small proportion of individuals with calcaneonavicular coalition develop a variety of flatfoot in which the peroneal muscles are shortened. For these individuals, repeated inversion of the foot causes the peroneal muscles to overstretch, leading to peroneal spasm.

Clinicians once thought that peroneal spastic flat-

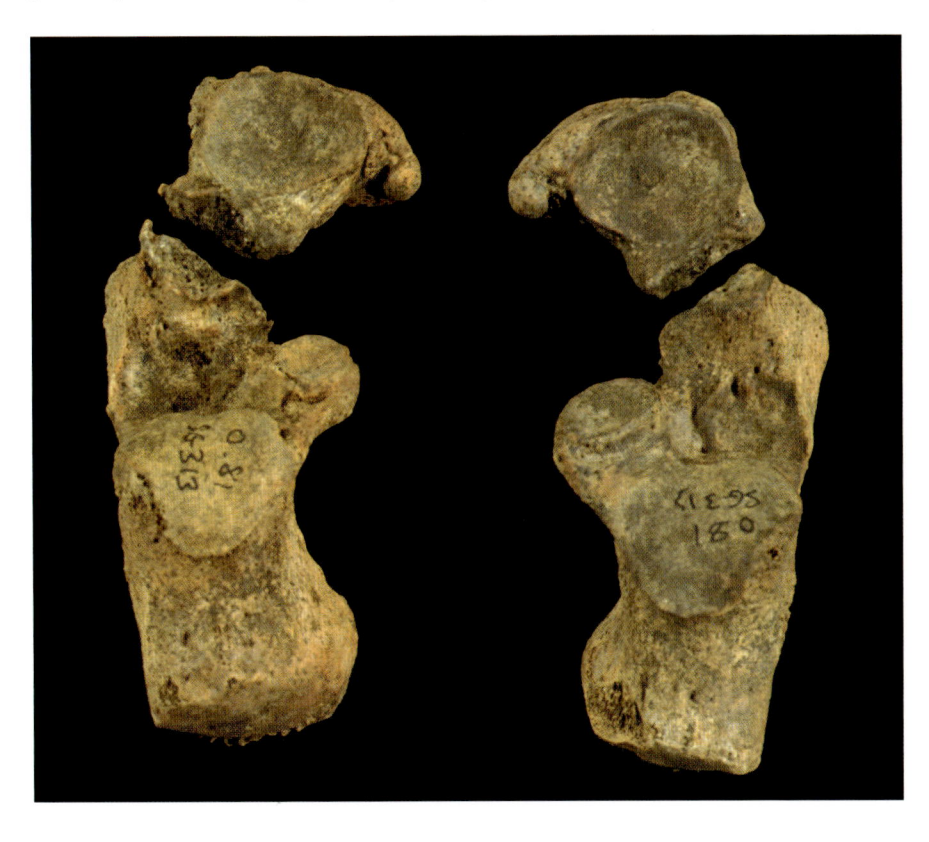

FIGURE 13.1 Bilateral example of calcaneonavicular coalition from a medieval Danish cemetery. Note how the shapes of the naviculars and anterior calcanei are altered. Also note the porosity of the lesion on the left compared to the right. Catalog number SBT081-313 from the Anthropological Data Base, Odense, Denmark. Reprinted from Case and Burnett (2011).

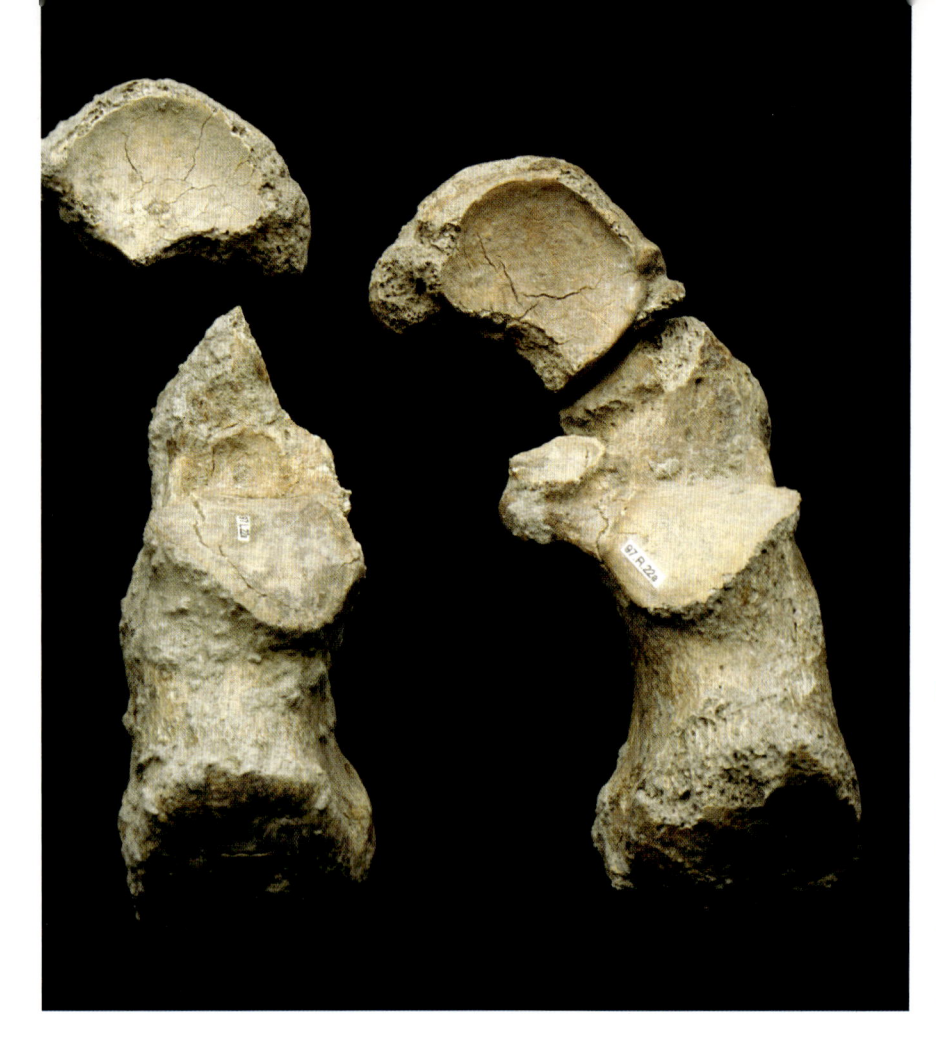

FIGURE 13.2 Bilateral calcaneonavicular coalition in Kennewick Man's feet. The left calcaneus is damaged, but the left navicular exhibits the characteristic "squared off" appearance. Catalog numbers 97.L.22a, 97.L.24(TRb), 97.R.22a, 97.L.24(TRa).

FIGURE 13.3 Left and right calcanei showing calcaneonavicular coalition lesion on Kennewick Man's right side, rim formation on the middle facet of the right side, and differences in deposition of carbonate cemented sediment. Catalog numbers 97.L.22a, 97.R.22a.

FIGURE 13.4 Kennewick Man's left and right naviculars showing a smooth lesion resulting from tarsal coalition, as well as bone adhesion from the calcaneus on the right side, and calcium carbonate buildup on the left. Catalog numbers 97.L.24(TRb), 97.R.24(TRa).

foot was a common symptom of tarsal coalitions, but later research suggests that its association with flatfoot and peroneal spasm is weak and more likely to occur in cases of osseous coalition rather than nonosseous ones (Stormont and Peterson 1983). Individuals who suffer from peroneal spasm often exhibit osteophytic lipping of the dorsolateral margin of the head of the talus, referred to as a "talar beak" (Webster and Roberts 1951; Conway and Cowell 1969). The lateral process of the talus may be broadened in these individuals (Conway and Cowell 1969). Both features result from abnormal subtalar motion associated with the coalition and peroneal spasm, and should be recognizable in skeletal remains. Arthritic changes on joint surfaces around the coalition site are sometimes seen (Berkey and Clark 1984). Such changes in the tarsals of Kennewick Man would suggest that he suffered from peroneal spasm in addition to calcaneonavicular coalition.

Kennewick Man appears to exhibit a bilateral case of nonosseous calcaneonavicular coalition, but likely without peroneal spasm (Figure 13.2). The left calcaneus is missing the sustentaculum tali and much of the anterior beak, making it difficult to say for certain whether it shows the appropriate morphology for a coalition. The only indicator in the left calcaneus is an abnormal extension of the distolateral corner of the calcaneus adjacent to the missing segment of bone (Figure 13.3). The navicular is more obviously affected on this side, exhibiting the characteristic squared-off look between the plantar-most and lateral-most points of the bone (Figure 13.4). The area of the coalition lesion is heavily encrusted with calcium carbonate, but enough of the surface is visible to indicate that there is a lesion on the left side that is similar to the lesion on the right.

The right side may have been more severely affected by calcaneonavicular coalition than the left (Figures 13.2, 13.4). There is a greater quantity of extra bone on the navicular adjacent to the coalition lesion, and the overall size of the lesion is large compared to other archaeological examples. As can be seen in Figure 13.2, a small piece of the calcaneus is still attached to the lateral edge of the lesion. This could be a small point of osseous coalition between the two bones, but a few similar adhesions of bone have been found in other bones of the hands and feet, suggesting the likelihood that these are the result of the cement-like effect of carbonate cemented sediment. A radiograph would help verify whether the calcaneus and talus are fused at this location, or whether the fairly

D. Troy Case

smooth lesion representing nonosseous coalition is present beneath the adhering bone.

These coalitions may have caused Kennewick Man discomfort at times, as is true for many clinical patients with similar nonosseous coalitions. It does not appear that he suffered from the more serious problem of peroneal spasm. There is no evidence of the talar beaking commonly associated with individuals suffering from peroneal spasms secondary to tarsal coalition, and the lateral process of the one intact talus appears normal. However, there is mild thickening of the articular rim in two tarsal bones—the right navicular and the middle facet of the calcaneus (Figures 13.3–13.4). The thickening of the rim on the middle facet of the calcaneus, though mild, is the most dramatic such change in the joint surfaces of the hands and feet. These changes, being found in the two bones joined by nonosseous coalition, may represent a mild reaction to abnormal motion between the talus and its neighbors caused by the presence of the coalition.

Calcaneonavicular coalition frequencies vary dramatically in different populations. Pfitzner (1896) found this type of coalition in 3.5 percent of 313 dissected cadavers from a German population; Rühli et al. (2003) found calcaneonavicular coalition in 9.7 percent of 62 dissected cadavers of European ancestry living in Australia. Case and Burnett (2011) found calcaneonavicular coalition in 2.2 percent of 502 medieval Danish skeletons, 2.1 percent of 341 Euro-Americans from the Terry Collection (National Museum of Natural History, Smithsonian Institution), and 0.2 percent of 531 South African Bantu skeletons from the Dart Collection (University of Witwatersrand, South Africa). No frequency data have been published for Native American samples. There is one unilateral example of calcaneonavicular coalition in an Arikara Indian from the Sully site in South Dakota, and another from the Vosberg site in Arizona (catalog number 70/1521). Cultural affiliation of the Vosberg people is unclear (Bueschgen and Case 1996). The Vosberg individual exhibited both calcaneonavicular and third metatarsal-third cuneiform coalition in the same foot, as appears to be the case with Kennewick Man. In the case of the Vosberg individual, the coalition between the third metatarsal and third cuneiform was osseous rather than nonosseous.

OTHER PATHOLOGY

Kennewick Man exhibits very little pathology in the hands and feet aside from the congenital defects described above. A few of his joint surfaces show mild rim development, but no evidence of pitting or eburnation is seen. The most extensive marginal joint changes are observed in the right calcaneus and navicular, and probably resulted from abnormal motion caused by the calcaneonavicular coalition (Figures 13.2 and 13.3). A pronounced rim is present around the margin of the middle facet of the right calcaneus, and mild rim formation is evident along the superior surface of the navicular's proximal facet. The only other marginal lipping in the right foot is found along the anterior base of the right second proximal phalanx. In the left foot, the navicular shows slight rim development along the plantar and medial portion of the distal facet. This, too, is likely a result of abnormal motion caused by calcaneonavicular coalition.

In the hands, both lunates exhibit similar mild rim development along the margins of the capitate and triquetral facets, though the right hand exhibits slightly more extensive lipping than the left. In the left hand, the capitate also exhibits some marginal lipping along the rim of the trapezoid facet. Unfortunately, the right capitate was not recovered. Slight rim formation is also evident on the bases of the first and second metacarpals of the left hand. The relatively minor expression of marginal lipping in Kennewick Man generally, and the fact that its greatest expression is found only in areas with congenital pathology, would seem to indicate a relatively young age at death for this individual.

ASYMMETRY AND HANDEDNESS

Examination of the humeri strongly suggests that Kennewick Man was right-handed. Evidence supporting this conclusion includes considerable asymmetry in total subperiosteal area at midshaft and middistal locations, favoring the right side (Chapter 12), as well as asymmetry in linear measurements such as humeral head diameter, and maximum and minimum diameter at midshaft (Chatters, 2000). The least circumference of the right ulna is also greater than that on the left.

Two types of data were gathered to assess side asymmetry. First, measurements of each bone were taken (Tables 13.4–13.6) and the sides compared. Measure-

TABLE 13.4 Carpal and tarsal measurements of Kennewick Man (in mm).

Bone Type	Catalog Number	Side	Length	Breadth	Height
Lunate	97.L.16(CRa)	L		17.5	18.5
		R		17.5	18.6
Capitate	97.L.16(CRc)	L	23.3		19.4
Hamate	97.R.16(CRa)	R	23.3		22.4
Talus	97.L.23a	L			
	97.R.23a	R	62.5	45.9	33.7
Calcaneus	97.L.22a	L		37.2	46.4
	97.R.22a	R		37.9	
Navicular	97.L.24(TRb)	L	21.1		32.2
Cuboid	97.L.24(TRa)	L	40.1	33.2	26.9
Cuneiform 1	97.L.24(TRc)	L		18.4	34.5
Cuneiform 2	97.L.24(TRe)	L	19.5	17.9	
Cuneiform 3	97.L.24(TRd)	L	28.4	18.8	25.3

TABLE 13.5 Bone measurements for the hands of Kennewick Man (in mm).

Bone Type	Catalog Number	Side	Length	Base Width	Base Height	Head Width	Head Height
Metacarpal 1	97.R.16(MCd)	L	50.0	16.4	16.3	17.2	14.5
	97.B.L.16(MCa)	R	49.4	16.4		17.8	14.5
Metacarpal 2	97.L.16(MCb)	L		18.3	18.0		
	97.R.16(MCc)	R	73.5	19.4	17.0	15.8	14.7
Metacarpal 4	97.L.16(MCc)/(MCd)	L		12.8	13.4	14.1	12.9
	97.R.16(MCe)	R	62.2	12.3		13.6	13.0
Metacarpal 5	97.R.16(MCb)	R	56.5		12.2	13.5	12.1
Proximal phalanx 1	97.L.24(Pd)	R	35.2	17.2	11.6	13.6	9.4
	97.A.I.24(Pb)	L		17.3	11.5		
Proximal phalanx 2	97.L.16(Pc)	R	43.1	18.1	11.9		
Proximal phalanx 4	97.L.16(Pa)	L	44.2			12.3	8.3
	97.R.16(Pd)	R	43.6		11.5	12.4	8.1
Proximal phalanx 5	97.A.I.16(Pa)	L				9.4	6.2
	97.R.16(Pb)	R	35.1	14.4		9.7	6.7
Intermediate phalanx 2	97.R.16(Pe)	L	25.7	14.1	9.3	10.6	5.9
Intermediate phalanx 3	97.R.16(Pf)	L	30.9	14.7	10.5		
Intermediate phalanx 4	97.L.16(Pb)	R	29.6	13.5	10.0	10.6	5.4
Intermediate phalanx 5	97.L.16(Pd)	L	20.4	11.3	7.9	9.4	4.8
	97.R.16(Pg)	R	20.3	11.1	7.8	8.4	4.6
Distal phalanx 1	97.L.16(Pe)	L	22.5	14.5		8.4	
	97.R.16(Ph)	R	24.0	15.2		8.7	
Distal phalanx 3	97.R.16(Pa)	R	18.3	11.1		7.4	
Distal phalanx 4	97.R.16(Pc)	L	17.9	10.6		6.6	

TABLE 13.6   Bone measurements for the feet of Kennewick Man (in mm).

Bone Type	Catalog Number	Side	Length	Base Width	Base Height	Head Width	Head Height
Metatarsal 3	97.L.24(MTc)	L	74.8	15.8	23.0		
	97.R.24(MTb)	R		14.9	22.8		
Metatarsal 4	97.L.24(MTd)	L	72.2	17.9			
Proximal phalanx 2	97.A.I.24(Pa)	L		13.3		10.9	
Proximal phalanx 3	97.L.24(Pa)	L		11.9	11.7	10.0	6.7
Proximal phalanx 4	97.R.24(Pa)	L	27.6			9.4	6.2
Distal phalanx 1	97.L.24(Pc)	L	24.7	19.8			

ments were made to the nearest 0.01 of a millimeter on a mini-osteometric board from Paleo-Tech Concepts, and rounded to the nearest 0.1 of a millimeter. Measurements for the carpals and tarsals were devised specifically for this project. Base width, base height, head width, and head height measures on the digital bones are similar dimensions to those commonly taken with calipers and often used for sex determination from hand and foot bones. In this case the mini-osteometric board was used in place of calipers. Definitions of the length measurements, identical to those described in Case and Ross (2007), were used primarily for the pattern profile analysis. Measurements of the heads and bases of the digital bones were used in the asymmetry analysis. They were chosen because measures of robusticity are influenced to some extent by lifetime activity (Case and Ross 2007). Although midshaft diameters are best for evaluating bilateral asymmetry as it relates to activity (Ruff et al. 1991; Lieberman et al. 2001), these measurements were abandoned because of Corps restrictions associated with handling the Kennewick Man skeleton. Length measurements were excluded from the asymmetry study because they tend to be affected primarily by genetics and nutritional factors rather than lifetime activity (Case and Ross 2007).

The second type of data used to assess hand asymmetry was visual examination and evaluation of specific muscle attachment sites (Table 13.7). Attachment sites for dorsal interosseous muscles along the shafts of the metacarpals and attachments for the medial and lateral flexors of the phalanges were examined. Left and right side attachment sites were checked for differences in size or rugosity. An ordinal scoring system denoting whether an attachment site on a particular bone was "smaller," "slightly smaller," "similar," "slightly larger," or "larger" than on the contralateral bone was used. Many bones could not be compared because one member of the pair was absent or damaged, and some better-preserved bones were excluded due to accumulation of calcium carbonate around attachment sites.

The hand and foot bones of Kennewick Man are generally symmetrical in size. To be classified as asymmetrical, measures have to differ by more than 0.5 of a millimeter (to avoid misinterpretation of simple measurement error), and the magnitude of the difference between the two sides has to equal or exceed 2.0 percent of total size. Few measurements could be compared within the feet. Among the tarsals, only the breadth of the calcaneus could be measured on both sides. These two bones differed by only 1.8 percent (0.7 mm), with the right side being larger. Even if this difference had exceeded 2.0 percent, it would be treated with caution with respect to assessing side dominance, as the calcanei exhibit tarsal coalition, which might have affected dimensions. As no other measures are available to assess asymmetry in the tarsals, no conclusions can be offered. Among the digital bones of the feet, only the third metatarsal bases could be compared. Their measurements are similar in terms of base height but differ by 5.7 percent in base width, the left side being larger. The base of the left third metatarsal was mildly affected by congenital coalition with the third cuneiform, which may have contributed to the somewhat large difference in base width.

Among the carpals, only the lunates could be compared (Table 13.4). Breadth and height measures were virtually identical for these bones, which suggests little

TABLE 13.7 Comparison of muscle and ligament attachment sites on Kennewick Man's bones.

Bone Type	Catalog Number	Side	Attachment 1	Attachment 1 Comparison	Attachment 2	Attachment 2 Comparison
Metacarpal 1	97.R.16(MCd)	L	1st dorsal interosseous	similar		
	97B.L.16(MCa)	R	1st dorsal interosseous	similar		
Metacarpal 2	97.L.16(MCb)	L	1st dorsal interosseous	similar	2nd dorsal interosseous	similar
	97.R.16(MCc)	R	1st dorsal interosseous	similar	2nd dorsal interosseous	similar
Metacarpal 4	97.L.16(MCc)/(MCd)	L	3rd dorsal interosseous	similar	4th dorsal interosseous	similar
	97.R.16(MCe)	R	3rd dorsal interosseous	similar	4th dorsal interosseous	similar
Proximal Phalanx 2	97.L.16(Pf)	L	collateral ligament (medial)	similar		
	97.L.16(Pc)	R	collateral ligament (medial)	similar		
Proximal Phalanx 4	97.L.16(Pa)	L	lateral flexor	smaller		
	97.R.16(Pd)	R	lateral flexor	larger		
Proximal Phalanx 5	97A.I.16(Pa)	L	collateral ligament (medial)	similar		
	97.R.16(Pb)	R	collateral ligament (medial)	similar		
Intermediate Phalanx 5	97.L.16(Pd)	L	medial flexor	slightly larger	lateral flexor	similar
	97.R.16(Pg)	R	medial flexor	slightly smaller	lateral flexor	similar

asymmetry in the wrist. Among the digital bones of the hand, the greatest variation between the sides is found in the base width and head width measurements. These two measures, at least among metacarpals, seem to show the greatest variability among populations (Case and Ross 2007), suggesting that they are the ones most likely influenced by differences in lifetime activity.

Base widths could only be compared for three metacarpals, and head widths for only two. Base width differences of 4.4 percent and 5.6 percent were seen between the second and fourth metacarpals respectively. The right second metacarpal and the left fourth metacarpal were larger. Head width differed between the two metacarpals that could be compared. The right side has a wider first metacarpal head by about 3 percent, while the left fourth metacarpal has a wider head by nearly 4 percent. These data seem to suggest random fluctuation in asymmetry between sides. However, they could reflect a pattern related to differential use of the dominant and nondominant hand.

Manipulative behavior in humans is often studied in terms of two major types of tool handling behavior: the precision grip and the power grip. The power grip is employed when the fingers of the hand encircle a handle or similarly shaped object to hold it firmly against the palm (Long et al. 1970). These activities include opening a jar lid, grasping a tool handle, or carrying a heavy object at one's side using a cylindrical handle. Two types of power grip are important in considering tool use in the distant past: the simple squeeze grip and the hammer squeeze grip. The simple squeeze grip is associated with tightly holding a spear, javelin, or any other cylindrical object in the hand. The hammer squeeze grip involves gripping a hammer, branch, or club and striking something with the object (Long et al. 1970; Marzke et al. 1992). Regardless of type, the power grip places stress primarily on specific bones and muscles, and asymmetry in the size of these bones or muscles could add support to the contention that Kennewick Man was right-handed.

Marzke et al.'s (1992) analysis of evolutionary adaptation to the hammer squeeze grip among hominids suggests that a firm hold on a tool such as a hammer or club comes from fourth and fifth finger flexion. The four medial metacarpals in humans are distinctively divided into two size groups, MC2-MC3 and MC4-MC5, distinguishing them from the other primates. This size difference is important in the squeeze grip, because it allows the fingers to flex around a tool handle at the metacarpophalangeal joints while maintaining a diagonal orientation of the tool across the palm. Most loading from the squeeze grip is concentrated on the fifth finger, and to a large degree on the fifth metacarpal, and some additional loading may also be distributed to the fourth metacarpal. Another area of the hand that generates relatively higher loads while engaged in a power grip is the base of the second metacarpal where it articulates with the first carpal row. This area of the wrist is also important in forceful precision grips (Marzke et al. 1992), so differences noted there could be indicative of either activity. The final area of increased loading with the power grip is along the axis of the first metacarpal. These forces are transmitted through the trapezium and received by the scaphoid. Unfortunately, none of the relevant carpals were observable bilaterally, and only the digital bones can provide clues to the manipulative behavior of Kennewick Man.

The stresses incurred by the power grip affect the fifth digit differentially but also indicate more broad muscle involvement affecting many of the digits. The main intrinsic muscles involved in activities related to the simple squeeze grip are the interosseous muscles (both dorsal and palmar), the flexor digitorum superficialis, which attaches to the anterior base of the middle phalanges, and the abductor minimi digiti of the fifth finger, which attaches to the medial side of the fifth proximal phalanx (Long et al. 1970; Marzke et al. 1992). For the specialized hammer squeeze grip, there is additional and prominent flexion activity by the intrinsic muscles of the index finger.

Taken together, the data on skeletal loading and muscle use during power grip activities suggest that bone size asymmetry in the fifth finger, and to a lesser extent, the fourth finger, could be indicative of differences in power grip activities between the hands. One could also expect to see differences in development of the dorsal interosseous muscle markings.

The precision or "pinch" grip is associated with tool making as well as tool use (Marzke et al. 1992; Marzke 1997). This special kind of grip typically involves the ability to firmly pinch large objects in one hand between the thumb and one or more fingers. Precision grips that are most effective in the creation and use of stone tools without handles involve pinching an object between the thumb and the side of the index finger, both the index and middle finger, or all four fingers (Marzke 1997:Figure 2). Individuals who habitually use such precision grips as they make or manipulate stone tools should have the greatest functional loading on the thumb of the dominant hand, substantial loading of the index and middle fingers, and somewhat less loading of the fourth and fifth fingers. Marzke et al. (1998) tentatively confirmed this expectation for the kind of precision grip used when a hammerstone strikes a stone core to produce a stone tool. Measuring muscle activity during tool production, they found that four muscles were most heavily recruited in the dominant hand during this activity. Two were muscles associated with the thumb (flexor pollicis brevis, first dorsal interosseous) that attach to the first proximal phalanx and the first metacarpal. A third is the flexor digitorum profundus of the index and fifth fingers. The tendon of this muscle attaches to the anterior base of the distal phalanx. During the same tool-making activity, only the flexor pollicis brevis of the thumb is heavily recruited in the nondominant hand. If muscle activity of this kind translates into differences in bone robusticity over time, then one might expect the metacarpal and proximal phalanx of the thumb, and perhaps the distal phalanges of the second and fifth digits, to show the greatest degree of increased robusticity on the dominant side. This differs somewhat from the power grip, which primarily involves the whole of the fifth digit, particularly the fifth metacarpal, and to a lesser extent, the fourth and first digits (Marzke et al. 1992). Therefore, evidence of greater loading in the thumb without additional loading in the fifth finger is more likely to indicate a forceful precision grip.

Although asymmetry in the hands of Kennewick Man was generally mild, robusticity measurements provide some support for the right hand being the one using the precision grip. The right first metacarpal showed greater robusticity in head width, while in the left hand, the fourth metacarpal, a bone that appears to be more important for power grip activities, was more robust. As already noted, measurements were not available for the most important bone in assessing a power grip: the fifth metacarpal. Among the hand phalanges, asymmetry was identified only between the two first distal phalanges and the two fifth intermediate phalanges. The first distal phalanx from the right side has a greater base width by 4.7 percent, which supports the contention of right-hand dominance. A more robust thumb could result from either precision- or power-grip behavior occurring primarily on one side. In contrast, the left fifth intermediate phalanx has a head width greater than the right by an impressive 10.8 percent, although there was no difference in base width between these two bones. Thus, the right hand seems to exhibit greater stresses on the thumb, while the left hand shows hints of greater stress in the fourth and fifth fingers, which might be expected in power grip type activities. Absence of the fifth metacarpal on the left side is unfortunate, but the data available indicate that Kennewick Man used his right hand for many precision activities, while the left hand may have been used more often with a power grip, perhaps to manipulate or hold objects in place. Given the results of the studies presented in Chapters 12 and 15, it seems unlikely that the left hand was used for activities involving spears or atlatls, forms of simple power grip activity.

Analysis of the muscle attachments (Table 13.7) provides some support for the interpretations about handedness and manipulative behaviors of Kennewick Man. Most of the muscle attachment sites in Kennewick Man's hands are similarly developed on the left and right sides, suggesting only mild differences in loading. Exceptions were in the fourth proximal phalanx, in which the medial flexor is larger on the right than the left side, and the fifth intermediate phalanx, in which the left medial flexor shows slightly greater development than the right. One of the muscle types heavily recruited in the power grip are the fifth digit flexors. Of note is an anomalous ridge of bone on the base of the right fifth metacarpal, just distal to and along the posterior half of the articulation for the fourth metacarpal. This ridge appears to have developed in response to abnormal motion between the right fourth and fifth metacarpal bases. It suggests heavy functional stress on the fourth and fifth metacarpals of the right side compared to the left, further indicating that the right hand was heavily engaged in precision grip activities. Although Kennewick Man used his left hand more than his right for certain activities, the evidence for probable use of a precision grip on the right side, the presence of stress-related pathology on the right, and the greater size of the humerus on the right are sufficient to conclude that Kennewick Man was right-hand dominant.

COMPARISON TO THE HORN SHELTER SKELETON

Another early Holocene skeleton with good preservation of the hands and feet comes from Horn Shelter No. 2 in Texas (Young et al. 1987). The skeleton, which dates to about 9700 RC yr. BP, is male with an estimated age of mid-30s to early 40s. This skeleton allows comparison to another male with the same chronological age from a similar time period.

The two individuals differed considerably in stature. Stature for the Horn Shelter skeleton has been estimated at 5 feet 4 inches, while Kennewick Man was relatively tall at 5 feet 7.5 inches (Chapter 11). This difference translates to a body size differential of approximately 5.0–6.0 percent, which is similar to differences in individual bone lengths of the arms. Kennewick Man's humerus is 8.4 percent longer and his ulna is 7.0 percent longer than that of the Horn Shelter man.

The Horn Shelter and Kennewick skeletons have similar numbers of bones preserved in the hands and feet. The number of foot bones is identical (Table 13.8). The Horn Shelter skeleton has a few more preserved hand bones than does Kennewick Man, mostly due to near complete recovery of a right hand (Table 13.9). Nevertheless, the Kennewick Man skeleton has a few more measurements available for analysis because his bones are somewhat less damaged (Tables 13.10–13.12).

The primary comparison made between the two skeletons was metric. The goal was to determine whether the relative sizes of the different bone rows in the hands and feet were similar. Relative lengths of the digital bones are controlled to a great extent by genes. These genes affect the bone rows differently, and individuals from distantly related populations tend to differ from one another in the pattern of these lengths (Poznanski et al. 1972; Takai 1978; Arias-Cazorla and Rodriguez-Larralde 1987; Mat-

TABLE 13.8 Kennewick Man and Horn Shelter foot bones present.

Bone Type	Left Side		Right Side	
	Kennewick Man	Horn Shelter	Kennewick Man	Horn Shelter
Talus	x	x	x	x
Calcaneus	x	x	x	x
Navicular	x	x	x	x
Cuboid	x	x		x
Cuneiform 1	x	x		x
Cuneiform 2	x	x		x
Cuneiform 3	x			x
Metatarsal 1	x		x	x
Metatarsal 2	x	x	x	x
Metatarsal 3	x	x	x	x
Metatarsal 4	x	x	x	x
Metatarsal 4		x		x
Proximal phalanx 1		x	x	
Proximal phalanx 2	x		x	
Proximal phalanx 3	x			x
Proximal phalanx 4	x	x		
Proximal phalanx 5			x	
Distal phalanx 1	x			

X denotes present.

TABLE 13.9 Kennewick Man and Horn Shelter hand bones present.

Bone Type	Left Side		Right Side	
	Kennewick Man	Horn Shelter	Kennewick Man	Horn Shelter
Scaphoid		x	x	x
Lunate	x	x	x	
Trapezium		x		x
Trapezoid				x
Capitate	x			x
Hamate		x	x	x
Metacarpal 1	x	x	x	x
Metacarpal 2	x	x	x	x
Metacarpal 3	x		x	x
Metacarpal 4	x	x	x	x
Metacarpal 5		x	x	x
Proximal phalanx 1	x		x	
Proximal phalanx 2	x	x	x	x
Proximal phalanx 3		x	x	x
Proximal phalanx 4	x	x	x	x
Proximal phalanx 5	x	x	x	x
Intermediate phalanx 2	x			x
Intermediate phalanx 3	x	x		x
Intermediate phalanx 4		x	x	
Intermediate phalanx 5	x		x	x
Distal phalanx 1	x		x	x
Distal phalanx 2				x
Distal phalanx 3			x	x
Distal phalanx 4	x			x
Distal phalanx 5				x

X denotes present.

suura and Kajii 1989). If Kennewick Man and the Horn Shelter male show similar proportionality among the various bone rows, it could suggest that they were from a closely related gene pool.

Table 13.13 presents comparative measurements for the two skeletons. When these percentages are combined into groups by bone row, some patterns emerge.

Kennewick Man exhibits greater stature and larger bones. If the two skeletons were fully proportional, one would expect the measurements of the hands and feet to differ by about the same amount. This is particularly true of the length measures, as they seem to be most strongly influenced by genes and nutrition, and less influenced by lifetime activity (Case and Ross 2007).

The width and height measures of the hand fit this expected difference. Taking the median of the base and head width percentage differences (Table 13.13), the hand bones of the Horn Shelter skeleton are about 8 percent smaller than those of Kennewick Man. For height measures, the Horn Shelter skeleton is about 7.5 percent smaller. Among length measures in the hand, the difference is somewhat greater. Taking the median of the length differences, the Horn Shelter skeleton's hand bones are nearly 11.0 percent shorter than those of Kennewick Man.

This difference suggests that the hands of the Horn Shelter male were somewhat more robust relative to length than the hands of Kennewick Man. Greater robusticity in the Horn Shelter skeleton is confirmed when

TABLE 13.10 Horn Shelter carpal and tarsal measurements (in mm).

Bone Type	Side	Length	Breadth	Height
Scaphoid	L		26.0	
Lunate	L		15.3	16.4
Hamate	L		18.6	22.0
	R			22.7
	R	54.4	40.4	
Cuboid	L	35.1		23.4
	R	34.6	27.5	
Cuneiform 1	L		17.3	32.3
	R		16.9	31.9
Cuneiform 2	L	18.8	15.9	
	R		16.1	
Cuneiform 3	L	25.0	17.3	25.3
	R	26.9	17.2	24.8

an index of length/width is calculated for three bones from both skeletons for which length and base width measures are available. The first and second metacarpals have about 5 percent greater relative base robusticity in the Horn Shelter skeleton, and the third metatarsal has 19.0 percent greater robusticity. The robusticity difference in the third metatarsal, which is surprisingly large, may be a result of the nonosseous tarsal coalition affecting the right foot of Kennewick Man skeleton, perhaps causing the width of the third metatarsal base to be abnormally small relative to that of the Horn Shelter skeleton.

It is unfortunate that metacarpal asymmetry could not be assessed in the Horn Shelter male. Asymmetry data would have allowed an assessment of the impact of this role on skeletal dimensions on the dominant side, relative to those described for Kennewick Man. The greater base robusticity relative to length in the first and second metacarpals from the right hand of the Horn Shelter

TABLE 13.11 Bone measurements for the Horn Shelter hands (in mm).

Bone Type	Side	Length	Base Width	Base Height	Head Width	Head Height
Metacarpal 1	L				15.5	
	R	43.0	15.2	14.9		
Metacarpal 2	R	67.4	18.9		14.6	13.6
Metacarpal 3	R			17.5		
Metacarpal 4	L		11.1	12.5	12.0	12.0
	R	55.5			12.5	
Metacarpal 5	L		14.0	12.2		
	R				11.7	11.2
Proximal phalanx 2	R	38.4				
Proximal phalanx 3	R	43.5				
Proximal phalanx 4	L	40.7				
	R	39.9				
Intermediate phalanx 2	R	22.4				
Intermediate phalanx 3	R	25.9				
Intermediate phalanx 4	L	24.3				
	R	24.4				
Intermediate phalanx 5	R	17.2				
Distal phalanx 1	R	21.3	13.6			
Distal phalanx 2	R	17.0				
Distal phalanx 3	R	17.3				
Distal phalanx 4	R	17.2				
Distal phalanx 4	R	15.4				

TABLE 13.12 Bone measurements for the Horn Shelter feet (in mm).

Bone Type	Side	Length	Base Width	Base Height	Head Width	Head Height
Metatarsal 1	R	60.8		28.5		
Metatarsal 2	L		16.3			
	R	73.4	16.7	21.1		
Metatarsal 3	L	66.3				16.0
	R	66.8	16.8	20.3	10.4	15.5
Metatarsal 4	L				10.6	
	R	66.7	13.1	17.7	10.6	14.0
Metatarsal 5	L	64.0			11.4	
	R	66.8	19.9	14.9	11.5	13.8
Proximal phalanx 1	R	30.3	18.5	16.4	16.5	10.2
Proximal phalanx 2	R	28.6				
Proximal phalanx 3	R	27.4				
Proximal phalanx 4	R	24.8				
Proximal phalanx 5	R	23.0				

TABLE 13.13 Percentage difference between Horn Shelter and Kennewick Man bone dimensions.

Bone	Length	Base Width	Head Width	Base Height	Head Height
			Hands		
Lunate		–12.6%		–11.6%	
Hamate				–1.3%	
MC1	–12.9%	–7.8%	–10.2%	–8.3%	
MC2	–8.3%	–2.8%	–7.4%		–7.6%
MC4	–10.8%	–13.3%	–8.2%	–7.3%	–6.5%
MC5			–13.0%	0.0%	–7.5%
PP2	–10.8%				
PP4	–8.7%				
IP2	*–12.9%*				
IP3	*–16.2%*				
IP4	*–17.3%*				
IP5	–15.4%				
DP1	–11.4%	–10.4%			
DP3	–5.6%				
DP4	*–4.1%*				
			Feet		
Talus	–13.1%	–12.1%			
Cuboid	–12.5%	*–17.1%*		–13.1%	
CF1		–5.7%		–6.4%	
CF2	–3.5%	–11.0%			
CF3	–12.0%	–7.8%		+2.1%	
MT3	–11.4%	+12.8%		–10.8%	
MT4	*–7.6%*				
PP4	–10.2%				

A negative value indicates a larger size for Kennewick Man. Italicized entries indicate comparison made using same bones but from opposite sides.

male seems consistent with repetitive use of a precision grip, as it is the first and second digits in particular that would have experienced increased loading (Marzke 997).

The other interesting trend in the bone dimension comparison is variation by row in the length data. The metacarpals and proximal phalanges show a length difference between the two skeletons of about 11 percent, which is nearly the same as that estimated for the entire hand. However, among the intermediate phalanges, for which there are four bones that can be compared, the difference rises to nearly 16 percent, then drops to about 6 percent in the distal phalanges. This length variation between bone rows indicates a difference in proportionality between the hands of the two skeletons.

The pattern in the feet is similar to that in the hands. Taking the median of the base widths, the foot bones of the Horn Shelter skeleton are about 8 percent smaller than those of Kennewick Man, while the height measures are about 8.6 percent smaller. Length measures show a greater difference between the skeletons. The Horn Shelter skeleton is about 11.4 percent smaller in overall bone length. Excluding the tarsals, the difference in the digital bone lengths alone is 10.2 percent. These results show that there are some differences in the proportions of bones between the two skeletons, with the digital bones of both the hands and feet of Kennewick Man being relatively longer compared to body size than in the Horn Shelter male.

Pattern Profile Analysis

One of the best ways to compare relative bone lengths among individuals is through metacarpophalangeal pattern profile analysis. Pattern profile analysis was originally developed by radiologists to allow simultaneous evaluation of gross and subtle variations of digital bone length in the entire hand (Poznanski et al. 1972). This technique is most often used by radiologists to assist diagnoses of congenital syndromes that affect relative bone lengths, but it can be applied to questions of similarity among species, populations, or individuals (Arias-Cazorla and Rodriguez-Larralde 1987; Smith 1995; Case 2003).

To perform pattern profile analysis, bone lengths are converted to standard deviation units (z-scores) using separately developed length standards for each individual bone (Poznanski et al. 1972). Typically, when individuals from the same population are compared, the length standards are drawn from the same population so that the resulting z-scores show how similar or different a particular pattern of relative lengths is in comparison to the mean pattern for the population. When individuals from different populations are compared, it makes no difference which set of length standards is used, so long as the same standards are used for each individual. Separate standards are needed for males and females, particularly if the sexes are to be compared to each other. In this case, length standards from the Euro-American subset of the Terry Anatomical Collection were used. Data for 163 to 170 individuals are available, depending on the bone (Case 2003).

Length standards are presented in Table 13.14, along with z-scores for Kennewick Man, the Horn Shelter male, and males from two Native American groups. One set is from the Zuni Pueblo of Hawikku in New Mexico, which has prehistoric and historic components. The other was drawn from the protohistoric Mobridge and Sully Arikara sites in South Dakota. These sites were included to gain a better understanding of how similar one might expect individuals from different, but fairly contemporaneous, populations to be in terms of patterns of relative bone lengths. The z-scores presented in Table 13.14 for the Zuni and Arikara samples are calculated from the mean length for each bone and represent an average for the total male sample. Depending on the bone, the sample size for the Hawikku measures ranged from 15 to 33 individuals. The sample size for the Arikara varies from 16 to 33 individuals.

After the bone lengths are converted to z-scores, they can be compared graphically by plotting the individual bones and their z-scores by row (Figure 13.6), and visually assessed based on the relative rise and fall of the pattern. It is the pattern of relative lengths that are of interest, not absolute lengths. If the plotted lines for two individuals have the same shape, regardless of how much distance separates them on the graph, their pattern profiles are similar. Another way to assess this similarity is by means of Pearson's correlations (Sokal and Rohlf 1995). A higher correlation indicates greater similarity in the relative lengths of the bones included in the analysis, while a lower correlation suggests less similarity. A negative correlation suggests dissimilarity.

TABLE 13.14 Length standards and associated z-scores for pattern profile analysis.

Bone	Terry Mean (mm)	Terry S.D. (mm)	Kennewick Man Z-score	Horn Shelter Z-score	Zuni Mean Z-score	Arikara Mean Z-score
RMC1	47.08	3.45	0.68	−1.17	−0.54	−0.22
RMC2	69.97	4.04	0.87	−0.64	−0.78	−0.18
RMC4	58.65	3.22	1.11	−0.98	−0.63	−0.37
RPP2	42.06	2.42	0.43	−1.50	−1.21	−0.02
RPP4	43.61	2.43	0.01	−1.55	−0.94	0.12
RIP4	29.44	2.14	0.05	−2.34	−1.42	−0.60
RIP5	21.04	2.21	−0.33	−1.74	−1.23	−0.54
RDP1	24.08	1.83	−0.04	−1.54	−1.32	0.43

Z-scores are standard deviation units.

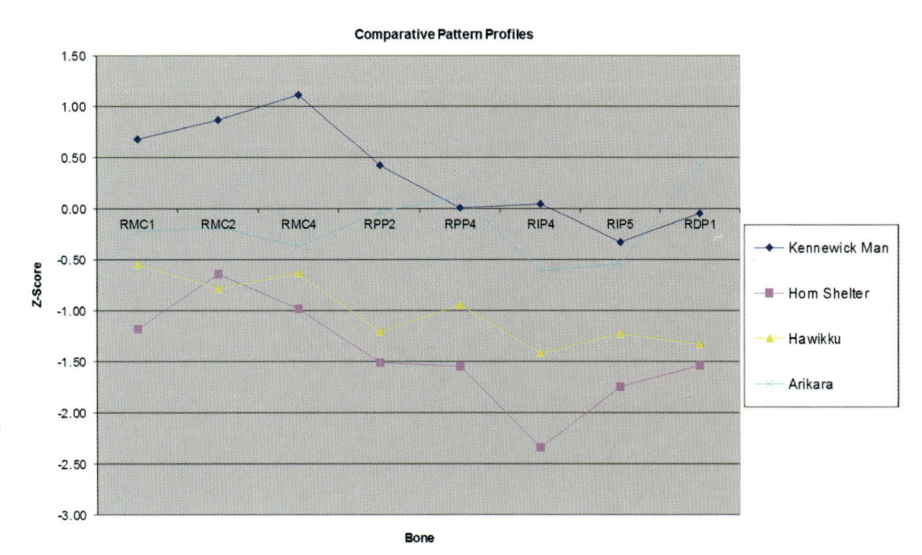

FIGURE 13.5 Metacarpophalangeal pattern profile for Kennewick Man, the Horn Shelter skeleton, and a mean pattern profile for the Arikara and Hawikku samples.

Eight hand bones from the Kennewick Man and the Horn Shelter skeletons were available for comparison of their relative lengths (three metacarpals, two proximal phalanges, two intermediate phalanges, and one distal phalanx). The graph of this comparison shown in Figure 13.5 indicates a somewhat similar pattern. The Pearson's correlation between the two skeletons is surprisingly high (r = 0.77). Both skeletons show a high correlation with the mean pattern profile for the Hawikku sample (0.78 for Kennwick Man, and nearly 0.80 for Horn Shelter). In contrast, the mean pattern profile for the Arikara sample shows a mildly negative correlation with Kennewick Man (r = −0.10) and only a slightly positive correlation with the Horn Shelter skeleton (r = 0.23).

To assess the significance of this degree of correlation between the two early Holocene skeletons, Kennewick Man and the Horn Shelter male were compared to each individual in the Terry, Zuni, and Arikara samples to determine whether such a high correlation would be common or rare in samples from different populations. If, as previously suggested, pattern profiles tend to be similar among individuals from similar populations, and if Kennewick Man and the Horn Shelter skeleton are from populations that diverged from each other more recently than did the Terry, Hawikku, and Arikara populations, then the early Holocene individuals should show higher correlations with each other than they would with individuals from these other samples. They should also show equal or greater similarity to one another than do the two Native American groups.

A total of 154 individuals from the Terry Collection have data available for all eight of the relevant bones needed for this analysis. By contrast, only six Arikara and four skeletons from Hawikku have every bone present. When Kennewick Man was compared to the individuals from the Terry Collection, 9/154 comparisons (5.8 percent) achieved a higher correlation than the value of 0.77 found between Kennewick Man and the Horn Shelter skeleton. The median correlation with the Terry sample was close to zero, suggesting only random similarity with the sample as a whole. Thus, there is some probability that an individual from a vastly different population and time period will show substantial similarity in pattern profiles. However, this relatively small probability of error suggests that the similarity indicated by the high correlation between Kennewick Man and the Horn Shelter male reflects somewhat close genetic ancestry

Of the four Hawikku skeletons that could be compared to Kennewick Man, one showed a higher correlation than between Kennewick Man and the Horn Shelter skeleton. The median correlation between the Hawikku sample and Kennewick Man was a relatively high 0.45. None of the six Arikara individuals showed a higher correlation with Kennewick Man than 0.77. The median correlation between Kennewick Man and the Arikara skeletons was negative (r = –0.21). By comparison, Hawikku individuals show a median correlation of 0.69 when compared to one another, and the Arikara have a median correlation of 0.49 when compared to others from the same population. When the Arikara and Hawikku individuals are compared to one another, the highest correlation is 0.69, but the median of all these pairwise comparisons is slightly negative (r = –0.06). Again, this seems to suggest that the correlation of 0.77 between Kennewick Man and the Horn Shelter skeleton is atypical.

The high correlation between the pattern profiles of Kennewick Man and the Horn Shelter male can be understood in terms of a fairly simple similarity in relative lengths compared to the Terry reference sample from which the length standards were calculated. In both early Holocene skeletons, the metacarpals are long relative to the phalanges. The difference in relative size between the second metacarpal and second proximal phalanx is close to 0.5 standard deviations compared to the Terry sample, while the difference between the fourth metacarpal and fourth proximal phalanx is greater than one standard deviation. There is an even larger difference between the fourth metacarpal and the fourth intermediate phalanx in both skeletons. Kennewick man has a fourth metacarpal that is over one standard deviation longer, and the Horn Shelter male's fourth metacarpal is about 1.5 standard deviations longer than in the Terry sample. Similar differences in relative lengths exist within the Zuni sample, except that the scale of the difference is only about half that seen in Kennewick Man.

Although pattern profile analysis is not a typical approach for assessing population affinity, results seem to indicate that Kennewick Man and the Horn Shelter male are derived from fairly closely related populations, in which the metacarpals tend to be long relative to the phalanges. The fact that these two ancient Americans, who lived in Washington State and Texas, exhibit greater similarity in their pattern profiles than any of the fairly contemporaneous individuals from South Dakota or New Mexico, suggests a considerable degree of similarity.

Another interesting feature of Kennewick Man's hands is their relatively large absolute size compared to either of the two more recent Native American groups or the Terry Collection individuals. The height of the Kennewick Man pattern profile in Figure 13.5 indicates that his metacarpals were substantially longer than those of the Zuni or Arikara skeletons. The three metacarpals are all close to one standard deviation (or more) longer than the Arikara mean, and 1.5 to 2 standard deviations longer when compared to the Zuni mean and the Horn Shelter skeleton. The phalanges of Kennewick man are more similar to the Arikara mean in absolute size, but are still 1.5 to 2 standard deviations longer than the phalanges of the Horn Shelter male and the means of the Zuni, despite the much greater similarity in the overall pattern profile shape between Kennewick, Horn Shelter, and the Zuni group.

Because the pattern profile was created using length standards from the Terry Collection, any bone length above the zero line on the graph represents a length greater than the mean for the Terry sample. In this case, all the Kennewick Man metacarpals are longer than the Terry Collection mean, while those of the more distal elements are either similar in size, or slightly shorter than the Terry mean. The Horn Shelter skeleton shows a somewhat different pattern, in which the metacarpals are 1 to 1.5 standard deviations shorter than the Terry mean, while the phalanges are all more than 1.5 standard deviations shorter. Considering these results in terms of growth gradients, Kennewick Man's metacarpals are longer relative to those of the Terry Collection toward the medial side of the hand, and shorter on the lateral side. This pattern is reversed in the proximal and intermediate phalanges, where the bones are relatively longer on the lateral side and shorter on the medial side.

TAPHONOMY

Four primary types of taphonomic data were gathered from the hand and foot bones of Kennewick Man. (Table 17.2) The distribution and relative size/thickness of calcium carbonate deposits were examined to help clarify positioning of the hands and feet after deposition and to determine whether or not they were articulated or disarticulated prior to eroding from the soil. The presence of carbonate cemented sediment in a few cases caused adjacent bones to stick together. Small bits of adjacent bones broke off and remained attached via the sediment to a neighbor during the erosion process. Such adhesions were noted because they indicate that bones were in close proximity to one another after the body had decayed and as the calcium carbonate deposits began to form. Green algal stains were also noted, as was bone surface patina and color loss.

Calcium Carbonate Deposition

In the left hand, calcium carbonate deposits are generally light and often found in small, isolated globs rather than widely distributed across segments of bones. Among the three metacarpals, calcium carbonate deposits are light and mostly found around the bases of the second and fourth metacarpals. On the fourth metacarpal, these deposits are mostly found in the hollows at the proximal and distal ends of the shaft near, but not on, the articular surfaces. Among the phalanges, the pattern changes to light to moderate deposition of calcium carbonate, which tends to be heaviest on the palmar sides of the bones. This pattern holds for the proximal, intermediate, and distal phalanges.

In the right hand, the first and fifth metacarpals respectively exhibit heavy and moderate calcium carbonate deposits on their palmar surfaces, while the other three metacarpals are affected only moderately on the medial and/or lateral sides of the proximal and distal ends, as was true of the metacarpals from the left side. Like the left side, the proximal and distal phalanges exhibit the greatest amount of calcium carbonate deposition on their palmar surfaces, only in this case, that deposition is typically moderate to heavy (Figure 13.6).

Two conclusions can be drawn from the calcium carbonate deposition on the hand bones. First, the right hand was more heavily affected by calcium carbonate than the left hand, based on the fact that deposition tended to be moderate to heavy on the right and light to moderate on the left. This difference might reflect slight

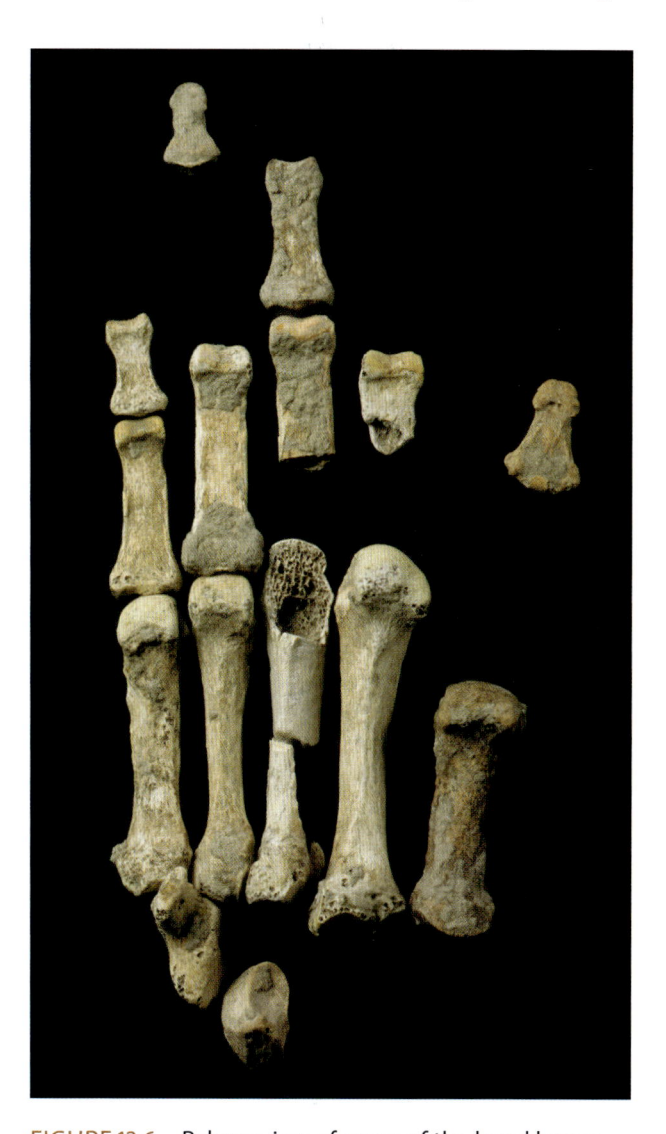

FIGURE 13.6 Palmar view of some of the hand bones of Kennewick Man showing location of calcium carbonate buildup.

differences in the depths at which these two hands were buried. Second, both the left and right hands were almost certainly positioned palm down, allowing calcium carbonate to accumulate on the undersides of these bones.

In the left foot, the calcaneus shows moderate to heavy deposition of carbonate cemented sediment concentrated along the lateral side and extending in some places onto the superior part of the body (Figure 13.7). In contrast, there is little calcium carbonate on the plantar portion of the body (Figure 13.8). This suggests that the calcaneus, and presumably the entire foot, came to rest on its side after the body decayed. Evidence from the other tarsals support this contention. The most laterally positioned tarsals, including the cuboid, second cuneiform, and third cuneiform, all show heaviest calcium carbonate concentrations on their superolateral surfaces. The first cuneiform shows heaviest accumulation on the lateral side, while the medially positioned navicular shows greatest accumulation of calcium carbonate along its plantar surface. From this pattern, the foot was situated somewhat on its side when the longitudinal arch collapsed from decay, causing the lateral bones to rest on their lateral and superior surfaces, and the most medial bones to rest on their plantar surfaces. The metatarsals show a pattern reminiscent of the metacarpals. The first metatarsal shows greatest calcium carbonate accumulation on the plantar surface (Figures 13.9 and 13.10), but the three remaining metatarsals show greatest deposition along the medial and lateral sides of the proximal and distal ends, with limited or no accumulation along the plantar portion. It seems likely in both the hands and the feet that the close association of the proximal and distal ends of these bones may have acted to channel water in such a way that it left deposits in the spaces between them.

As seen in the two calcanei (Figure 13.3), there is much less calcium carbonate deposited on the lateral and superior portions of the right foot than the left. In the right foot, the heaviest concentration is on the plantar part of the calcaneal body, suggesting that this foot was not lying over on its side after skeletonization. The right first metatarsal shows considerable calcium carbonate on the plantar surface, although the heaviest concentration is on the lateral side of the shaft. The third metatarsal shows the heaviest concentrations on the lateral side, although in this case it is restricted to the proximal end. Pedal phalanges from the first and second digits exhibit fairly heavy concentrations of calcium carbonate on their plantar surfaces. Unfortunately, data on the right foot are limited somewhat to the more medial bones of the foot, but the pattern suggests that these bones came to rest more on their plantar surfaces.

Adhesions

The bones of the hands and feet remained in close articulation after Kennewick Man became skeletonized. Calcium carbonate deposits are rarely found on the joint surfaces of the bones, suggesting close contact. Exceptions are the hamate facet of the left capitate, the capitate facet of the right lunate, and the trapezium facet of the left second metacarpal base. Adhesions between neighboring bones also confirm sustained association. In the right hand, the second and third metacarpals were joined at their proximal ends by carbonate cemented sediment, as evidenced by a remnant of the third metacarpal base attached to the second metacarpal at the location expected if the two bones had been in proper articulation. There are three examples of adhesions in the feet. On the left side, the first and second metatarsals are joined by carbonate cemented sediment at both the proximal and distal ends (Figures 13.9 and 13.10). In addition, a small piece of the third metatarsal is attached to the base of the second metatarsal. These three bones were in close articulation after the flesh had decayed and as water began to invade the grave. In the right foot, a piece of the calcaneus is attached to the navicular, indicating that these bones also remained in close association after burial and decay (Figure 13.11).

Loss of Patina and Color

Loss of surface texture and color has affected many more bones in the hands than the feet, and the right hand has been affected to a greater extent than the left. This loss of texture and color appears to have been caused by tumbling of these smaller bones against the shallow river bottom as a result of currents and wave action (Chapter 17). In the feet, none of the bones of the left side shows evidence of color loss, and only three bones from the right (second metatarsal, third metatarsal, and fifth proximal phalanx) have the characteristic stark white appearance.

FIGURE 13.7 Lateral view of Kennewick Man's left calcaneus showing moderate to heavy accumulation of calcium carbonate. Catalog number 97.R.22a.

FIGURE 13.8 Plantar view of left calcaneus showing much lighter calcium carbonate accumulation. Catalog number 97.R.22a.

In the hands, the right side shows color loss of at least parts of four metacarpals, one proximal phalanx and the fourth and fifth intermediate phalanges. Two other bones, the right hamate and fifth proximal phalanx, appear to exhibit mild color loss, though not so obviously as in the other bones. By contrast, none of the metacarpals shows evidence of color loss on the left side, and only four phalanges (three proximal and one intermediate) are affected. The distal first phalanx appears to show mild color loss in this hand. The right hand and foot are both more affected by color loss than the left sides, indicating that these bones were exposed to greater tumbling and polishing than the left hand and foot.

Algal Staining

Green algal stains were identified on the left cuboid (superomedial portion of proximal facet), the right talus (on trochlea and neck), and three right metacarpals. The second metacarpal showed light algal staining on the proximal facet only, the third metacarpal showed staining on the medial side of the proximal base, and the fifth metacarpal showed probable algal staining on the head. The location of the algal stains on the second and third metacarpal bases suggest that they were exposed but still in articulation when the algae were growing.

CONCLUSION

The hand and foot bones of Kennewick Man are surprisingly well represented and well preserved. In particular, the digital bones of both hands, and the tarsals and proximal phalanges of the left foot, are mostly complete and well enough preserved that it was possible to assign

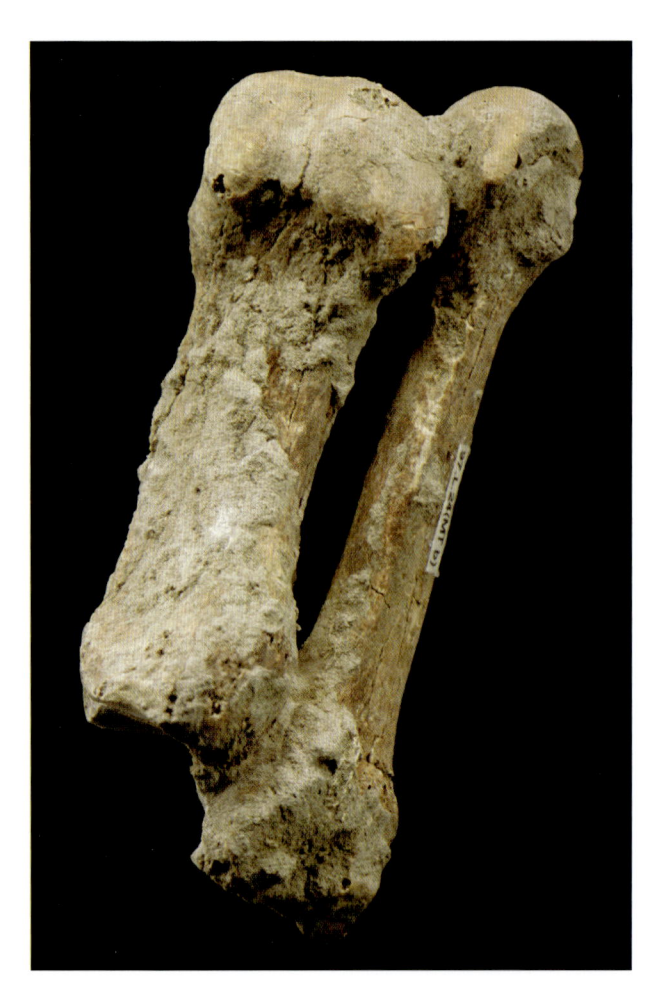

FIGURE 13.9 Plantar view of Kennewick Man's left first and second metatarsals showing heavy accumulation of calcium carbonate on the first metatarsal, and lighter accumulation on the second metatarsal. The two bones are fused by means of carbonate cemented sediment. Catalog numbers 97.L.24(MTa) and 97.L.24(MTb).

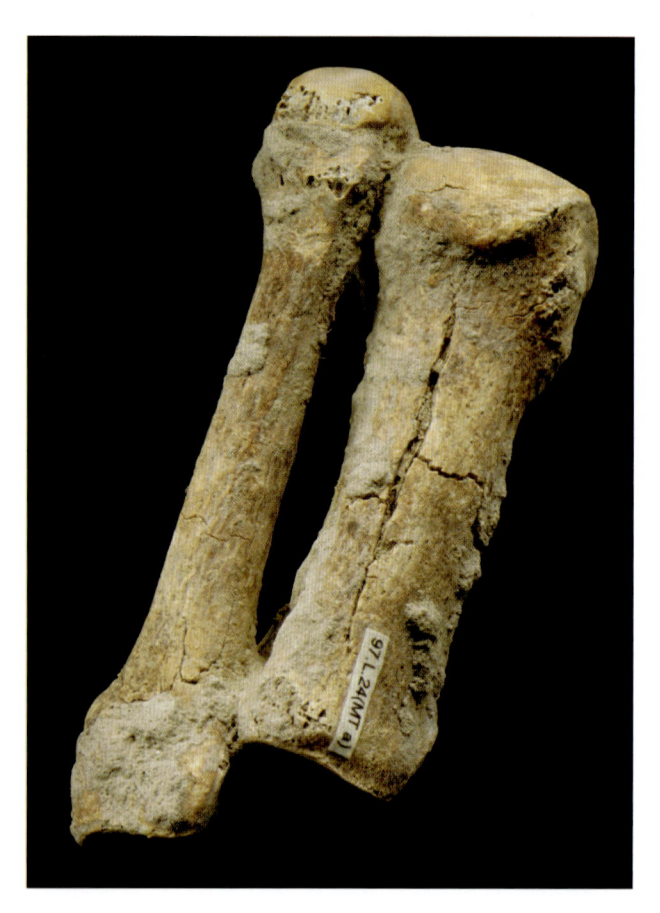

FIGURE 13.10 Dorsal view of left first and second metatarsals, showing lighter accumulation of calcium carbonate.

nearly all of them to the correct side and digit with high confidence. Sufficient hand bones were recovered from both sides to allow assessments of asymmetry in length, robusticity, and relative size of certain muscle attachments. It was also possible to compare many hand and foot dimensions of Kennewick Man to the Paleoamerican male from the Horn Shelter No. 2 site. The incredible recovery of so many small bones, despite exposure of the grave and erosion into the lake margin, suggests a short duration between exposure and recovery.

Kennewick Man exhibits two relatively uncommon congenital defects in the feet. Both are examples of developmental errors in which two bones fail to separate properly during embryonic development. The more common of the two defects is nonosseous coalition between the third metatarsal and third cuneiform, which appears to have been present in the right foot but not the left. This coalition results in a bridge of cartilage or fibrous tissue between the plantar joint spaces of the third metatarsal and third cuneiform. The form of the lesion was more subtle than usual in Kennewick Man but is similar to examples seen by the author in other more obvious contexts. Despite its subtlety, the defect may have been responsible for surprising asymmetry in the dimensions of the third metatarsal base. Frequencies of third metatarsal–third cuneiform coalition vary quite significantly in different populations. It is present in low frequencies in all samples of reasonable size, but tends to be present in particularly high frequencies among Native Americans and Japanese. Its frequency in other East Asian groups is unknown.

The less common congenital defect exhibited by Kennewick Man is nonosseous calcaneonavicular coalition, in which the calcaneus and navicular remain connected by a bridge of cartilaginous or fibrous tissue. This condition was evident in both feet, though more obvious on the right because of postmortem damage to the left calcaneus. In modern clinical studies, calcaneonavicular coalition can sometimes lead to pain or dysfunction. The only evidence of pathology associated with the coalition in Kennewick Man was lipping of the joint margins of both naviculars and the middle facet of the right calcaneus, which probably resulted from abnormal motion caused by the restrictive nature of the nonosseous bridge. The only other joint margin pathology in the foot was mild rim development or lipping around the base of the right second proximal phalanx.

FIGURE 13.11 Distal view of Kennewick Man's right navicular showing the small piece of the calcaneus still attached to the plantar portion of the bone by means of carbonate cemented sediment. Note the abnormal flattened surface to which this bone segment is attached. This flattened surface is the result of nonosseous calcaneonavicular coalition. Catalog number 97.R.24(TRa).

Mild rim development was seen around some of the joint surfaces in the hands. These include both lunates, the left capitate, and the left first and second metacarpal bases. None was so extensive as even the moderate lipping associated with the tarsal coalition. The left hand exhibited a greater number of joints with mild lipping, but the lipping of the right hand, at least in the lunate, was somewhat more pronounced. These results indicate significant use of both hands for activities that placed abnormal stresses on the joints, though the nature of the activities may have been different, resulting in the different patterns of affected joints.

An attempt to assess handedness through bone size and muscle attachment asymmetry suggests more similarity than difference in the hands. No asymmetry was evident in the wrist, although only one pair of carpals, the lunates, was available for comparison. Greatest differences between the sides were found in the first, second, and fourth metacarpals. Unfortunately, the fifth metacarpal on the left side was not available for analysis. The right hand exhibited greater robusticity in the first and second metacarpals (base width), while the left exhibited greater robusticity in the fourth metacarpal (head width and base width). Although these differences could represent random fluctuations between sides, they can also be interpreted as representing greater engagement in precision grip activities on the right, and perhaps power grip activities on the left. A greater emphasis on precision grip activities on the right side is consistent with evidence for right-hand dominance found in other studies in this volume.

Comparison of Kennewick Man's hand bones to those of the similarly ancient Horn Shelter male suggests great similarity in terms of patterns of relative bone length, but considerable differences in relative bone robusticity. Kennewick Man's estimated stature and his long bone lengths are approximately 5–8 percent greater than those of the Horn Shelter male. Head and base width measures in the hand tend to follow this trend in size difference, while the length measures are relatively longer in Kennewick Man, exceeding Horn Shelter by about 11 percent. Kennewick Man's hands were somewhat less robust for their size compared with Horn Shelter. Comparative data in the feet are limited to the third metatarsal base, which was pathological and therefore should not be used to compare relative size.

Although Kennewick Man's digital bones were substantially longer than those of the Horn Shelter male, the two skeletons were surprisingly similar in terms of relative bone lengths. This similarity is of particular interest because bone length appears to be much more influenced by genes than by lifetime activity. Pattern profile analysis was used to compare Kennewick Man to Horn Shelter, and to two small samples of protohistoric Native American skeletons of Zuni and Arikara ancestry. They were also compared to a modern sample of Euro-Americans. Kennewick Man's pattern profile was highly correlated with that of the Horn Shelter male (0.77) and was fairly highly correlated with the mean pattern for the Zuni sample from New Mexico. Conversely, Kennewick Man's pattern profile correlation with the Arikara sample was slightly negative. The great similarity between Kennewick Man and Horn Shelter can be attributed to a pattern of relatively long metacarpals compared to proximal phalanges. A similar pattern of relative lengths, though less dramatic, is seen among the Zuni. It was also interesting to discover that Kennewick Man's metacarpals were considerably longer than the Euro-American mean in the Terry Collection, while his phalangeal lengths were quite similar.

Taphonomic analysis of the hands and feet provides good evidence of Kennewick Man's burial position. Deposition of calcium carbonate was located primarily on the palmar surfaces of the digital bones of the hands, suggesting that Kennewick Man was buried with his hands at his sides, his palms downward. There also tends to be heavier accumulation of calcium carbonate on the right side than the left, suggesting slight differences in the depths of the two hands. Calcium carbonate deposition on the foot bones indicates that the left foot was positioned on its lateral side after the body decayed, while the right foot bones came to rest mostly on their plantar surfaces.

There is little doubt that Kennewick Man's burial was a primary interment. Calcium carbonate deposition was mostly nonexistent on the joint surfaces of most of the hand and foot bones, and several examples of bone adhesions, in which two adjacent bones are cemented together by soil mixed with calcium carbonate, were identified. These include adhesions between the second and third metacarpals of the right hand, the first, second, and third metatarsals of the left foot, and the calcaneus and navicular of the right foot. Some of the bones of the hands and feet spent time exposed to sun and damp conditions. Algal staining was found on a left tarsal, a right tarsal, and three right metacarpals. Several bones, particularly of the right hand, exhibit significant loss of color and surface patina probably caused by tumbling and polishing due to wave action.

APPENDIX A: DESCRIPTIONS OF HAND AND FOOT BONE MEASUREMENTS

Measurements from the Feet: Taken with a Mini-Osteometric Board

1. Calcaneus:

Tuberosity maximum breadth: Measured from the most laterally projecting point of the tuberosity to the most medial point. The long axis of the calcaneus should be parallel to the stationary upright of the board, and the calcaneus should be resting on the medial and lateral tuberosities.

Body Height (liberally modified from Steele): Place the bone on its medial side with the sustentaculum tali hanging over the edge of the board, and the calcaneus resting against the stationary upright at medial tubercle. Measure the maximum dimension between the medial tubercle of the tuberosity and the most superior point on the tuberosity.

2. Talus:

Maximum length (modified from Steele): Maximum length between the most posterior point of the trigonal process, to the most anterior point on the navicular facet.

Width of Talus (modified from Steele): Place the talus with the inferior face lying on the base of the board and the medial edge resting against the stationary upright. Measure to the most lateral point on the lateral side.

Height of Talus: Place the talus on the board resting on its lateral side. The bone should contact the stationary upright at a medial and lateral point on the posterior, inferior facet, and an anterior point beneath the head. Measure to the most superior point on the trochlea.

3. Navicular:

Length (proximodistal): Place the proximal face of the bone against the stationary upright, with the medial tuberosity contacting the board as much as possible. The proximal face should contact the stationary upright at two or more points along the superior margin of the proximal facet. Hold the bone in place as the movable upright is brought into position.

Height: Place the proximal face of the bone so that it is lying on the base of the board. If the bone has a tendency to lie flat in two different planes, then select the plane where the superior portion of the proximal face is lying flat. The superior edge from the medial side should be flat against the stationary upright. Measure to the inferior edge.

4. Cuboid:

Length (proximodistal): The bone should be lying on its medial side (third cuneiform facet down). Distal end against the stationary upright of the osteometric board. Take maximum length measurement between plantar edge of distal facet and the proximal beak.

Breadth (mediolateral): The bone should be lying on its superior surface with the medial side (third cuneiform facet side) resting against the stationary upright of the osteometric board. The bone should be held in place while sliding the movable upright into position against the lateral edge of the inferior tuberosity.

Height: The bone should be lying on its medial side (third cuneiform facet down). Position the superior surface against the stationary upright such that, at minimum, the superior portion of the distal articular margin touches the upright. Hold the bone in place as the movable upright is brought into position.

Cuneiform I:

Length (proximodistal): Place the lateral side so that it is lying on the base of the board. The distal face should rest against the stationary upright along most of its length. Measure to the proximal edge.

Breadth (mediolateral): Place the distal end so that it is lying on the base of the board. The lateral side should rest against the stationary upright along most of its length. Hold the bone against the stationary upright. Measure to the most medial side.

Height: Place the lateral side so that it is lying on the base of the board. The plantar portion of the bone should rest on two points toward the proximal and distal edges of the bone. Measure to the superior end.

Cuneiform II:

Length (proximodistal): Place the lateral side so that it is lying on the base of the board. The proximal end should rest against the stationary upright. Try to make three points (two superior and one plantar) come into contact with the stationary upright. Apply pressure to the bone with a finger to make sure it remains flat against the stationary upright and the base of the board. Slide the moveable upright against the distalmost point.

Breadth (mediolateral): Place the proximal end so that it is lying on the base of the board. The lateral edge (curved side of facet) should be lying against the stationary upright. Measure to the most medial point.

Cuneiform III:

Length (proximodistal): Place the medial side so that it is lying on the base of the board. The distal end should rest against the stationary upright, with the plantar edge also resting near or at the base of the board. Measure to the proximalmost point.

Breadth (mediolateral): Place the distal end so that it is lying on the base of the board. The superior third of the medial side should rest flat against the stationary upright. Measure to the lateralmost point.

Height: Place the medial side so that it is lying on the base of the board. The superomedial edge should rest against the stationary upright. Measure to the plantarmost point.

Metatarsals and Phalanges:

Base width: For this measurement, the base of the bone is gently clamped between the stationary and movable uprights of the board while the shaft and head of the bone project from the far side of the board away from the person taking the measurement. The bone should be held horizontally such that the superior portion of the shaft faces upward and the long axis of the shaft is perpendicular to the long axis of the Mini-Osteometric board. The medial and lateral surfaces of the base are then gently clamped between the two uprights and the measurement recorded.

Base height: For this measurement, the base of the bone is gently clamped between the stationary and movable uprights of the board while the shaft and head of the bone project from the far side of the board away from the person taking the measurement. The bone should be held horizontally such that one side of the shaft faces upward and the long axis of the shaft is perpendicular to the long axis of the Mini-Osteometric board. The superior and inferior surfaces of the base are then gently clamped between the two uprights and the measurement recorded.

Head width: For this measurement, the head of the bone is gently clamped between the stationary and movable uprights of the board while the shaft and base of the bone project from the far side of the board away from the person taking the measurement. The bone should be held horizontally such that the superior portion of the shaft faces upward and the long axis of the shaft is perpendicular to the long axis of the Mini-Osteometric board. The medial and lateral surfaces of the head are then gently clamped between the two uprights and the measurement recorded.

Head height: For this measurement, the head of the bone is gently clamped between the stationary and movable uprights of the board while the shaft and base of the bone project from the far side of the board away from the person taking the measurement. The bone should be held horizontally such that one side of the shaft faces upward and the long axis of the shaft is perpendicular to the long axis of the Mini-Osteometric board. The superior and inferior surfaces of the head are then gently clamped between the two uprights and the measurement recorded.

Measurements from the Hands: Taken with a Mini-Osteometric Board

1. Scaphoid:

Breadth (mediolateral): Place the distal side so that it is lying on the base of the board. The medial (rounded edge) should rest against the stationary upright. Measure maximum width to the lateral tubercle.

2. Lunate:

Breadth (mediolateral): Place the proximal side so that it is lying on the base of the board (This is the side with the best view of the siding articulation). The two edges of the lunate facet should rest against the stationary upright. The articulation for the radius should point upward. Measure to the apex of the lunate.

Height (posteroanterior): Place the proximal side so that it is lying on the base of the board. The flattest portion of the superior margin should rest against the stationary upright. In this position, a line drawn through the inferior edge of the lunate facet (from the stationary upright) will come close to bisecting the bone. Measure to the inferior edge of the lunate facet.

3. Capitate:

Length: Place the bone such that the hamate facet is lying on the base of the board, and the distal surface is resting against the stationary upright at two points. Measure to the head of the bone.

Height: Place the bone such that the hamate facet is lying on the base of the board, and the posterior edge is resting against the stationary upright. Measure to the posterodistal edge (where the MC3 styloid sometimes fuses).

4. Hamate:

Length: Place the bone such that the lunate and triquetral facets are lying on the base of the board, and the edges of the metacarpal facets, or one edge of the metacarpal facet and the edge of the hamulus, are resting against the stationary upright. Measure to the wedge-shaped edge of the bone.

Height: Place the bone such that the lunate and triquetral facets are lying on the base of the board, and the posterior portion of the bone rests against the stationary upright at its flattest point. Measure to the tip of the hamulus.

Metacarpals and Phalanges:

Base width: For this measurement, the base of the bone is gently clamped between the stationary and movable uprights of the board while the shaft and head of the bone project from the far side of the board away from the person taking the measurement. The bone should be held horizontally such that the superior portion of the shaft faces upward and the long axis of the shaft is perpendicular to the long axis of the Mini-Osteometric board. The medial and lateral surfaces of the base are then gently clamped between the two uprights and the measurement recorded.

Base height: For this measurement, the base of the bone is gently clamped between the stationary and movable uprights of the board while the shaft and head of the bone project from the far side of the board away from the person taking the measurement. The bone should be held horizontally such that one side of the shaft faces upward and the long axis of the shaft is perpendicular to the long axis of the Mini-Osteometric board. The superior and inferior surfaces of the base are then gently clamped between the two uprights and the measurement recorded.

Head width: For this measurement, the head of the bone is gently clamped between the stationary and movable uprights of the board while the shaft and base of the bone project from the far side of the board away from the person taking the measurement. The bone should be held horizontally such that the superior portion of the shaft faces upward and the long axis of the shaft is perpendicular to the long axis of the Mini-Osteometric board. The medial and lateral surfaces of the head are then gently clamped between the two uprights and the measurement recorded.

Head height: For this measurement, the head of the bone is gently clamped between the stationary and movable uprights of the board while the shaft and base of the bone project from the far side of the board away from the person taking the measurement. The bone should be held horizontally such that one side of the shaft faces upward and the long axis of the shaft is perpendicular to the long axis of the Mini-Osteometric board. The superior and inferior surfaces of the head are then gently clamped between the two uprights and the measurement recorded.

REFERENCES

Arias-Cazorla, Sergio, and Alvaro Rodriguez-Larralde. 1987. Metacarpophalangeal pattern profiles in Venezuelan and northern Caucasoid samples compared. *American Journal of Physical Anthropology* 73(1):71–80.

Bass, William M. 2005. *Human Osteology: A Laboratory and Field Manual.* Fifth edition. Special Publication 2. Missouri Archaeological Society, Columbia.

Berkey, S. F., and B. Clark. 1984. Tarsal coalition. Case report and review of literature. *Journal of the American Podiatry Association* 74(1):31–37.

Bueschgen, Wolf D., and D. Troy Case. 1996. Evidence of prehistoric scalping at Vosberg, Central Arizona. *International Journal of Osteoarchaeology* 6(3):230–248.

Burnett, S. E., and D. T. Case. 2005. Naviculo-cuneiform I coalition: Evidence of significant differences in tarsal coalition frequency. *The Foot* 15(2):80–85.

Case, D. Troy. 1996. "Developmental Defects of the Hands and Feet in Paleopathology." Master's thesis, Arizona State University, Tempe.

———. 2003. "Who's Related to Whom? Skeletal Kinship Analysis in Medieval Danish Cemeteries." Ph.D. dissertation, Arizona State University, Tempe.

Case, D. T., and S. E. Burnett. 2011. Identification of tarsal coalition and frequency estimates from skeletal samples. *International Journal of Osteoarchaeology*. doi: 10.1002/oa.1228.

Case, D. T., and J. Heilman. 2006. New siding techniques for the manual phalanges: A blind test. *International Journal of Osteoarchaeology* 16(4):338–346.

Case, D. T., R. J. Hill, C. F. Merbs, and M. Fong. 2006. Polydactyly in the prehistoric American Southwest. *International Journal of Osteoarchaeology* 16(3):221–235.

Case, D. Troy, Nancy S. Ossenberg, and Scott E. Burnett. 1998. Os intermetatarseum: A heritable accessory bone of the human foot. *American Journal of Physical Anthropology* 107(2):199–209.

Case, D. Troy, and Ann H. Ross. 2007. Sex determination from hand and foot bone lengths. *Journal of Forensic Sciences* 52(2):264–270.

Chatters, James C. 2000. The recovery and first analysis of an early Holocene human skeleton from Kennewick, Washington. *American Antiquity* 65(2):291–316.

Conway, James J., and Henry R. Cowell. 1969. Tarsal coalition: Clinical significance and roentgenographic demonstration. *Radiology* 92:799–811.

Cooperman, Daniel R., Bruce E. Janke, Allison Gilmore, Bruce M. Latimer, Mark R. Brinker, and George H. Thompson. 2001. A three-dimensional study of calcaneonavicular tarsal coalitions. *Journal of Pediatric Orthopaedics* 21(5):648–651.

Cybulski, Jerome S. 1988. Brachydactyly, a possible inherited anomaly at prehistoric Prince Rupert Harbour. *American Journal of Physical Anthropology* 76(3):363–376.

Kumai, T., Y. Takakura, K. Akiyama, I. Higashiyama, and S. Tamai. 1998. Histopathological study of nonosseous tarsal coalition. *Foot and Ankle International* 19(8):525–531.

Lieberman, Daniel E., Maureen J. Devlin, and Osbjorn M. Pearson. 2001. Articular area responses to mechanical loading: Effects of exercise, age, and skeletal location. *American Journal of Physical Anthropology* 116(4):266–277.

Long, Charles II, P. W. Conrad, E. A. Hall, and S. L. Furler. 1970. Intrinsic-extrinsic muscle control of the hand in power grip and precision handling. *Journal of Bone and Joint Surgery* 52(5):853–867.

Mann, Robert W. 1990. Calcaneus secundarius: Description and frequency in six skeletal samples. *American Journal of Physical Anthropology* 81(1):17–25.

Marzke, M. W., Wullstein, K. L., and Viegas, S. F. 1992. Evolution of the power ("squeeze") grip and its morphological correlates in hominids. *American Journal of Physical Anthropology* 89(3):283–298.

Marzke, Mary W. 1997. Precision grips, hand morphology, and tools. *American Journal of Physical Anthropology* 102(1):91–110.

Marzke, Mary W., N. Toth, K. Schick, S. Reece, B. Steinberg, K. Hunt, R. L. Linscheid, and K.-N. An. 1998. EMG study of hand muscle recruitment during hard hammer percussion manufacture of Oldowan tools. *American Journal of Physical Anthropology* 105(3):315–332.

Matsuura, Shinya, and Tadashi Kajii. 1989. Radiographic measurements of metacarpophalangeal lengths in Japanese children. *Japanese Journal of Human Genetics* 34(2):159–168.

Pfitzner, W. 1896. Beiträge zur Kenntnis des Menschlichen Extremitätenskelets: VII. Die Variationen in Aufbau des Fussskelets. In *Morphologische Arbeiten*, edited by G. Schwalbe, 245–527. Gustav Fischer, Jena.

Poznanski, Andrew K., Stanley M. Garn, Jerrold M. Nagy, and John C. Gall, Jr. 1972. Metacarpophalangeal pattern profiles in the evaluation of skeletal malformations. *Radiology* 104:1–11.

Regan, Marcia H., D. Troy Case, and Juliet C. Brundige. 1999. Articular surface defects in the third metatarsal and third cuneiform: Nonosseous tarsal coalition. *American Journal of Physical Anthropology* 109(1):53–66.

Ricklan, D. E. 1988. "A Functional and Morphological Study of the Hand Bones of Early and Recent South African Hominids." Ph.D. dissertation, University of Witwatersrand, Johannesburg, South Africa.

Ruff, Christopher B., William W. Scott, and Allie Y.-C. Liu. 1991. Articular and diaphyseal remodeling of the proximal femur with changes in body mass in adults. *American Journal of Physical Anthropology* 86(3):397–413.

Rühli, F. J., L. B. Solomon, and M. Henneberg. 2003. High prevalence of tarsal coalitions and tarsal joint variants in a recent cadaver sample and its possible significance. *Clinical Anatomy* 16(5):411–415.

Smith, Shelley L. 1995. Pattern profile analysis of hominid and chimpanzee hand bones. *American Journal of Physical Anthropology* 96(3):283–300.

Sokal, Robert R., and F. James Rohlf. 1995. *Biometry*. 3rd edition. W. H. Freeman and Company, New York.

Stormont, Daniel M., and Hamlet A. Peterson. 1983. The relative incidence of tarsal coalition. *Clinical Orthopaedics and Related Research* 181:28–36.

Takai, S. 1978. Metacarpal and phalangeal lengths are influenced by row-related factor. *Human Biology* 50(1):51–56.

Tenney, James M. 1991. Comparison of third metatarsal and third cuneiform defects among various populations. *International Journal of Osteoarchaeology* 1(3-4):169–172.

Webster, Frederick S., and William M. Roberts. 1951. Tarsal anomalies and peroneal spastic flatfoot. *Journal of the American Medical Association* 146(12):1099–1104.

Young, Diane E., Suzanne Patrick, and D. Gentry Steele. 1987. An analysis of the Paleoindian double burial from Horn Shelter No. 2, in central Texas. *Plains Anthropologist* 32(117):275–298.

14

The Natural Shocks That Flesh Is Heir To

Della Collins Cook

Pathology in the Kennewick skeleton was evaluated three times prior to the plaintiff's team investigation in 2006, with results that differ in a surprising number of details. Chatters (2000, 2001) describes and stages ten injuries, incurred from adolescence to the perimortem interval, and calls longstanding injuries to the elbow debilitating. Powell and Rose (1999) add several subtle lesions to that list. Walker et al. (2000) document the lesions thoroughly but are more circumspect in their interpretations. This chapter presents an independent examination of the Kennewick skeleton and focuses on evidence for disability.

CHRONOLOGICAL AGE

Preservation, conservation, and postmortem damage are critical to the observation of bone lesions, but these conditions are problematic in Kennewick Man. Some of the surface of the skeleton is covered with carbonate-cemented sediment, a gray, calcareous deposit that fills most foramina and exposed trabecular space and obscures periosteal texture in many areas. The cranial vault is partially filled with this deposit, and the endocranial surfaces cannot be seen. Where the gray calcareous deposit is broken away, the red-brown bone surface has exfoliated in some areas, exposing yellow to ivory cortical bone. This freshly broken tissue has angular fracture surfaces when viewed under a 20x lens, and it is highly crystalline, which suggests taphonomic alteration.

Age at death is vital for interpreting bone lesions. In Kennewick Man there is a discrepancy between reportedly young age and advanced joint margin changes that together suggested an arthritis syndrome. Judged from textual evidence alone, the various opinions concerning his age may transform normal age change into evidence for unusual disease patterns. Previous accounts of early humans from the New World have generated similar controversies. For example, Tepexpan Man, which dates to the Archaic period, was initially aged as 55 to 65 years (de Terra et al. 1949). His age was revised down to 25–30 years by observers who discounted cranial suture closure and attributed dental wear and tooth loss to periodontal disease (Genovés 1960; Gonzales et al. 2003; Moss 1960). Other experienced bioanthropologists who reviewed this skeleton inclined toward the original opinion regarding age (see Brooks 1962).

Kennewick Man's age was initially reported as 35 to 45 years, with "incipient osteoporosis" suggesting an older age (Chatters 2000:305). Teams of later observers found no evidence for loss of bone density. Cortical bone cross-sections of the left femur, which could be observed directly, are quite thick. Powell and Rose (1999) were impressed with the extensive cranial suture closure, leading them to suggest an age of 45 to 50 years. Recent skepticism about suture closure as a source of age information discounts several quite reliable studies (e.g., Acsádi and Nemeskéri 1970). Competing reconstructions of Kennewick Man's facial features present quite different concepts of his aging process, which may complicate the picture at a subliminal level (Chatters 2001; Lemonick and Dorfman 2006).

Evaluation of the pubic symphysis is complicated by calcareous deposits on all surfaces, but there was clearly a well-formed ventral rampart not quite complete at its superior limit, indicating an age of at least 35 years. Ring epiphyses of the vertebral bodies are distinct but well fused. Carbonate-cemented sediment obscures much of the surface of the auricular processes, but the visible areas show fine porosity rather than the irregular large pitting characteristic of advanced age. The medial epiphyses of the clavicles are well fused. Epiphyseal lines on the femoral head are still visible in exposed trabecular bone, but are fully fused and obliterated where the external surface is preserved. A conservative interpretation of this evidence in light of the extensive suture closure would indicate an age range of 35 to 50 years. A plausible narrow estimate is 35 to 40 years, but the wider range must be kept in mind.

CRANIAL INJURIES

Chatters (2000), Powell and Rose (1999), and Walker et al. (2000) all discuss a healed, depressed lesion on the left frontal. This lesion is small, confined to the outer table, and well healed. Because it does not appear to penetrate to the inner table, it is unlikely to have posed any long-term problem. Small depressed fractures of the cranial vault are a common finding. In his model study of vault injuries in California Channel Island crania, Walker (1989) explores the patterning by age, sex, and location that might be expected in accidental injuries, various kinds of culturally constrained interpersonal violence, and self-injury. This approach is not feasible when examining a single individual. Chatters marshaled evidence for vault injury in other Paleoamerican remains to make a case for interpersonal violence. This approach is interesting, but any pattern generated from individuals widely separated in time and space must be viewed with caution.

The active lesion that Chatters and the Walker team describe on the left inferior aspect of the greater wing of the sphenoid is postmortem damage. Examination with a 20x lens shows that the outer table is broken away and the apparently coarsened trabeculae are thickened with mineral deposits. If there is an abnormality here, it is limited to coarsened trabeculation, and it does not indicate "acute osteomyelitis from infection originating in adjacent soft tissue" (Chatters 2000:308). It should be noted that this part of the skull is well protected by the overlying temporalis muscle and that neither the zygomatic arch nor the boundaries of the temporalis show any alteration that would support Chatters's opinion.

AUDITORY EXOSTOSES

Kennewick Man has auditory exostoses, although it is excessive to characterize these as "severe" (Chatters 2000:309). They occlude about two thirds of the external auditory meatus and probably did not impact hearing or subsistence activity (Taylor 2000) (Figures 14.1 and 14.2). The exclusive attribution of auditory exostosis to diving in cold water has been challenged (Hutchinson et al. 1997; Okumura et al. 2007), but the association remains a strong one (Capasso et al. 1999). The presence of auditory exostoses in Kennewick Man augments the isotopic

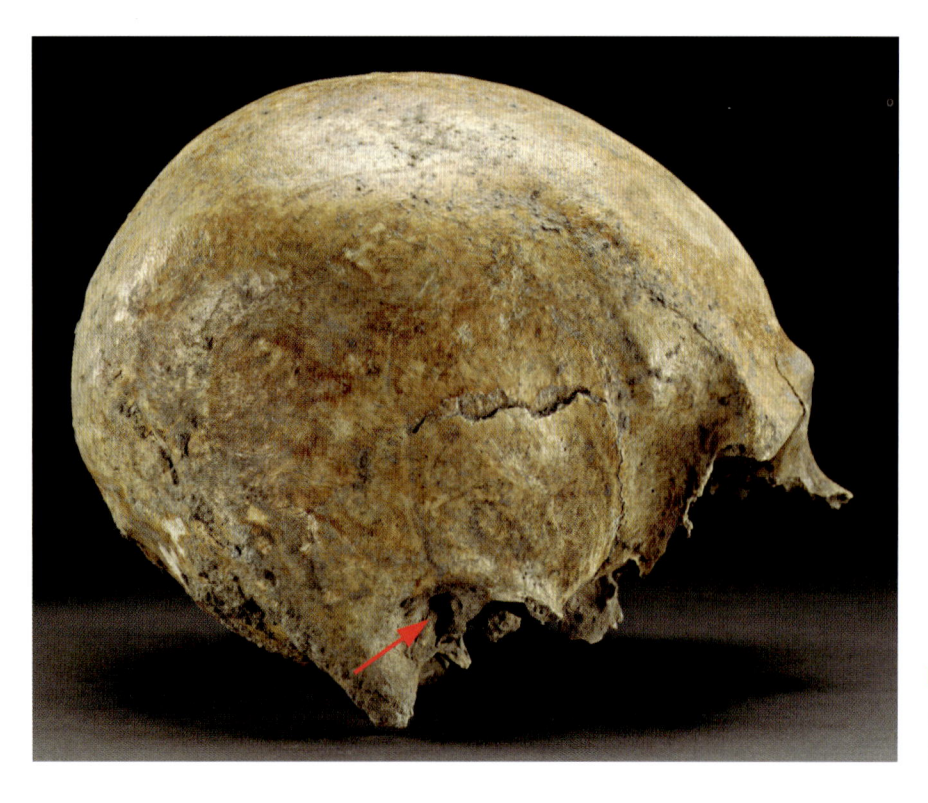

FIGURE 14.1 Right view of vault showing auditory exostosis (arrow).

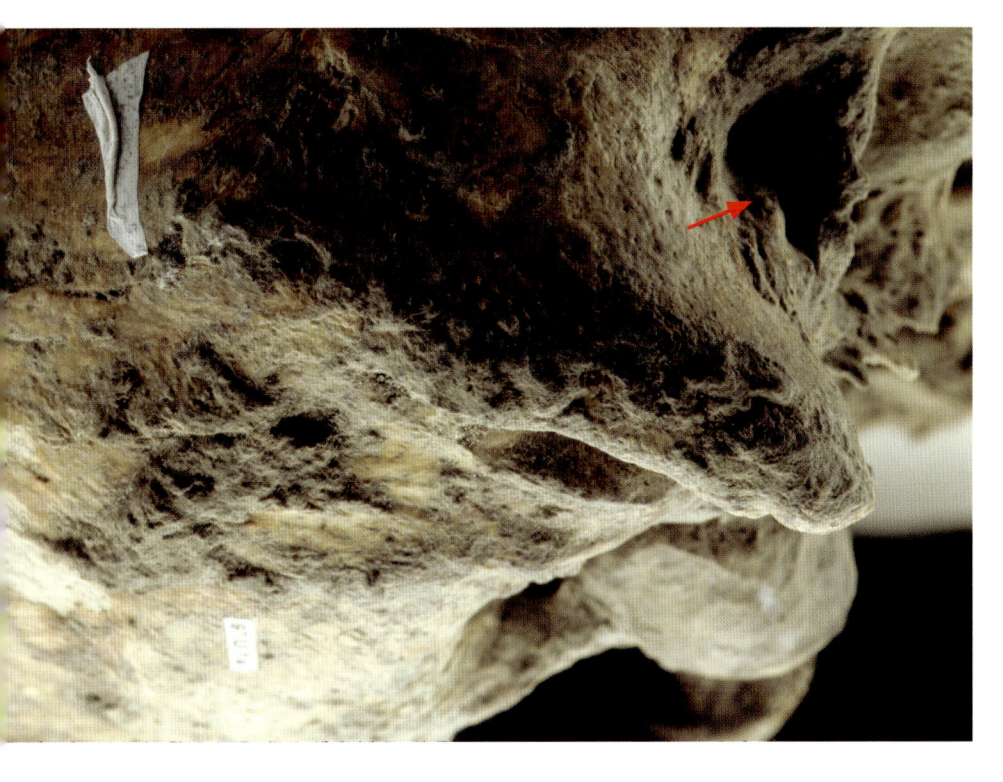

FIGURE 14.2 Detail of right mastoid region showing auditory exostosis (arrow), gray calcareous deposits, red-brown bone surface, and ivory underlying cortical bone.

evidence for hunting and fishing in wet environments as an important part of his way of life.

TEETH

Oral infection was as potentially life-threatening as the lesions that have attracted greater interest in the Kennewick case. He experienced heavy dental wear with considerable hypereruption of all teeth (Levers and Darling 1983; Danneberg et al. 1991). There are no pulp exposures because secondary dentin filled in the pulp cavities as the crowns were lost to wear. The right maxillary third molar was lost in life, but the socket had begun to fill in. There is slight evidence of periodontal regression around the remaining molars, particularly in the interdental septum between the right mandibular second and third molars. There are no pulp cysts or abscesses, and the incipient apical lesions that Chatters (2000:309) reports on radiographs are not evident. The dense maxillary sinus walls may be somewhat thickened as part of the mechanism of hypereruption. New bone formation visible in the sinus walls can also be seen as a marker of sinus infection, whether secondary to dental disease or respiratory infection, and it is more common in hunter-gatherers than in recent groups (Roberts 2007). Kennewick Man illustrates the view that heavy wear with consequent hypereruption is the normal adaptive pattern of the human species (Kaifu et al. 2003). He coped successfully with the demands on his teeth.

Wear on the anterior teeth is particularly heavy, and the enamel on the lower incisors has been lost, a pattern that has been interpreted reflecting use of the teeth as tools, as well as abrasive content of foods, particularly dried fish (Hinton 1981; Molnar 1972). There is isotopic evidence for high consumption of salmon in Kennewick Man (Taylor 2000), and fish and other meats dried in open-air circumstances are notoriously abrasive.

Powell and Rose (1999) point out the absence of calculus and suggest that it may have been removed when Chatters made a cast of the skull. Remnants of calcareous matrix in the dental interspaces suggest that this was not the case, and that Kennewick's teeth were remarkably free of calculus. In this regard he resembles another ancient hunter of the Columbia River (Pennefather-O'Brien and Strezewsky 2002) and calls into question generalizations linking calculus with meat consumption that have appeared in the paleopathology literature (Lieverse 1999).

Was the striking longitudinal fracture in the right mandibular lateral incisor antemortem (Figure 7.11)? At the occlusal surface the edges are rounded, and the wedge-shaped defect suggests a fracture while the collagen was still intact. If this fracture was antemortem, it may well have been quite painful and could have been a portal for systemic infection. Careful cleaning, surface replication, and scanning electron microscopy of microwear could shed light on whether the fracture occurred during life.

HIP WOUND

The projectile point embedded in the right iliac crest is by far the most controversial aspect of Kennewick Man's pathology (Figure 7.32). Chatters (2000:309) argued that the projectile entered from the side "into or anterior to the iliac crest" along the blade of the ilium and that it "drained pus from the man's hip throughout his life." Powell and Rose (1999) interpreted the point as having entered from the dorsal end of the iliac crest and as completely healed with "no sign of past infection." Walker et al. (2000) did not comment on the direction of the wound but did find "secondary infection." They did not remark on whether the infection was active at the time of death.

The wound was from the side. The weapon entered the iliac crest just anteriolateral to the flexure that marks the lateral margin of the erector spinae muscles and passed into the trabecular space in the dorsal third of the ilium blade, angled slightly below the transverse plane and toward the auricular process. The aperture of the surface wound was lateral and opened where the iliac crest is superficial. Only the thin margins of the gluteus medius and maximus muscles could have been impacted, and the wound as positioned allowed ready drainage. A small portion of the epiphysis of the iliac crest is missing at the point of impact.

Staging the wound is important to evaluating its impact on Kennewick Man's life as well as the circumstances under which he may have been wounded. The iliac blade is not greatly deformed, as one would expect if the wound had occurred before most of the upward growth of the iliac blade was complete. The iliac crest is interrupted and very slightly displaced by the wound, which suggests that the epiphysis had ossified but had not fused when the wound was sustained. A wound that was largely in cartilage might be at some increased risk of developing a persistent fibrous capsule or cyst. Appearance and closure of the iliac crest epiphysis spans the ages of 14 to 22 in males (Scheuer and Black 2000), and the region in which the projectile point is embedded closes late relative to the remainder of the epiphysis (McKern and Stewart 1957), suggesting that this wound occurred in late adolescence.

The literature of paleopathology is replete with survived projectile wounds that have left foreign bodies in bone with little or no evidence of resultant handicap (Capasso et al. 1994; Hoppa et al. 2005; Jurmain 1999, 2001; Pales 1930; Ryan and Milner 2006). Even open abdominal wounds are occasionally survived in the absence of effective medical care (Rutkow 1998).

In Kennewick Man's case there is no evidence that the abdominal cavity was breached. The lack of new or remodeled periosteal reactive bone on the internal surface of the ilium and sacrum suggests that the wound drained externally and healed well. There is an opening through the internal cortical bone of the ilium that is smooth walled, and a gap of 1–4 millimeters separates the margins of this aperture from the projectile point (Figure 7.23). No newly formed fiber bone is present on the inner side aspect of the wound cavity. There is no increased porosity around the wound margins at the opening through the external cortex in the dorsal aspect of the ilium. There is coarse, remodeled reactive bone within the cavity surrounding the embedded projectile point, but there is no recently formed fiber bone. This lesion does not appear to have been active at the time of death, which suggests that it had closed externally and healed, and it resulted in no persisting periosteal elevation, bleeding, or infection on either the internal or external surface of the ilium. There are no alterations to the insertions of the gluteal muscles that would point to functional consequences of the injury.

The opening around the projectile point embedded in Kennewick Man's iliac crest may have been occupied by fibrous tissue, with or without a fluid-filled cyst. A bone cyst formed in reaction to a foreign body might be quiescent for many years, or it might become inflamed, expansive, and painful (see Geschickter and Copeland 1936:263 for an example of a foreign body cyst quiescent for 18 months, and Chikkamuniyappa et al. 2005 for a shrapnel cyst becoming symptomatic after 60 years, both wounds predating antibiotic therapy). The absence of recent new bone formation suggests that this lesion was no longer infected. Its location in the iliac crest would make formation of a cyst relatively easy in that this area consists of intersecting layers of fibrous tissue inserting on the bone. Drainage during the healing process would have been relatively easy due to the superficial location. The iliac crest is a common location for bone marrow biopsy and bone graft harvesting for this reason. Avulsion fractures of the iliac crest generally heal spontaneously without displacement, and young athletes who suffer this injury can walk without pain after a few weeks (Steerman et al. 2008; Valdés et al. 2000). Kennewick Man may have experienced a longer period of disability because his wound was open, and it would have been a nuisance with respect to clothing until it closed. In two French Neolithic examples that closely resemble Kennewick Man, in the lower limb the ilium is often struck, but its wounds are not serious; they have cured with the inclusion of arrow points in reactive tissue (Pales 1930:91).

Because injuries are one of the few windows to past human behavior, there is considerable temptation to overinterpret them. Chatters (2001) develops a romantic scenario of intergroup conflict over Kennewick Man's wife and supposes that he was nearly successful in dodging the weapon. Kennewick Man's healed hip wound has been listed as evidence for violence in an edited volume on the origins of war (Ferguson 2006). The presumption that this wound resulted from interpersonal violence rather than from a hunting accident or any other scenario is an overinterpretation of the evidence. The angle and location of Kennewick Man's wound is particularly interesting in the context of ethnographic films that show young men and boys avoiding injury by turning the back obliquely to oncoming arrows while they observe the arrow's flight and attempt to dodge it. *Dead Birds* (1963), *Arrow Game* (1975), and *A Man Called "Bee": Studying the Yanomamo* (1974) are films that depict successful avoidance as well as less successful feints that result in superficial wounds in nonvital dorsal locations. These oblique wounds correspond closely to the one seen in Kennewick Man. Dodging a bullet is a bad joke; dodging an arrow or an atlatl dart is a skill that takes practice. New Guinea highlanders and Amazonian Indians, farmers engaged in continuous war with their neighbors, are inappropriate models for Paleoamerican hunters, but these films demonstrate that arrows and spears can be dodged effectively and that a part of learning this technology was practice in avoiding or minimizing wounds.

Frison (2004) has made a strong case for the pervasiveness of group hunting strategies in western North America from Paleoamerican times on, and he is one of the few archaeologists to remark on either the necessity for practicing hunting skills or the risk of accidental wounds during cooperative hunts. Ethnographic literature abounds in observations, such as this from the 1600s: "The usual and daily practice of the boys was to shoot their bows and arrows . . . they learned to throw the prong with which fish are speared and practiced other sports and exercises" (Tooker 1991:124). There are quantitative data on resulting injuries in ethnographic sources. For example, among Shiwiar foragers accidental knife, ax, lance, harpoon, and blowgun wounds outnumber wounds resulting from interpersonal violence (Sugiyama 2004), and deaths resulting from hunting and gathering accidents outnumber violent deaths among Aka pygmies (Hewlett et al. 1986). High risk for arrow wounds resulting from hunting accidents is noted for the Dobe !Kung, but these are often fatal because the arrows are poisoned (Howell 1979). It is possible to attribute Kennewick Man's wound to an accident rather than to interpersonal violence.

OSTEOCHONDRITIS AND LONGSTANDING ELBOW INJURY

Chatters (2000:308) calls injury to the left humerus "crippling." Rose and Powell (1999) see healed fractures of both distal humeri. The left humerus is smaller than the right, not only in length but also in head and shaft diameters (Figures 7.18–7.19). This asymmetry could be activity-related. Although there is no evidence of fracture of the distal shaft, such fractures may heal with little distortion. The left medial epicondyle is smaller than the right but normal in surface morphology, as are the other distal elements; thus, there is no evidence for a full-fledged avulsion injury. There are no osteophytes or remnants of the epiphyseal line. It seems plausible that there was a low-grade disruption of the distal epiphyses on the left side that limited growth, perhaps, as Chatters suggests, in association with the injury to the head of the radius. However, another examination found no evidence of antemortem injury to the left radial head (Chapter 7) and concluded the observed changes to be postmortem, along with slight arthritic lipping. It is more plausible that the asymmetry of the humeri is related to biomechanical stress.

Epiphyseal injuries with varying degrees of vascular necrosis are associated with a wide variety of strenuous activities in children and adolescents. There is an interesting literature on differences in distribution of growth plate disorders across modern human populations, but their natural history has received little attention in paleopathology.

The injury to the left radial head has been called a chip fracture, which, if real, raises the issue of the fate of the fragment. Chatters (2001) suggests that the fracture resulted from a "fall on the outstretched hand." An alternative diagnosis is osteochondritis that affected the palmar or medial edge of the radial head and the rim of the proximal surface. Trabecular tissue is exposed and coarsened, and there is a dense, elevated margin. This assessment indicates chronic overuse rather than acute trauma. There are no corresponding lesions on either the capitulum of the left humerus or the radial notch of the ulna. Muscle attachments are well developed on both arms, suggesting that this putative injury was not debilitating.

Chatters (2000:308) lists "extensive remodeling in humeral septum, olecranon fossa" as a pathological change and relates it to the left radial head fracture. This is puzzling because there is no direct relationship between the radial head and the humeral septum. Walker et al. (2000) agree that there is a pathological change: "the olecranon fossa is partially filled with reactive bone indicative of an inflammatory process." The humeral septum develops in adolescence from the bone separating the coracoid fossa from the olecranon fossa. Children's bone in this region is always thick. As the distal humerus remodels and the joint margins increase in rugosity, the area thins, which sometimes results in a septal aperture, a septum that consists of the fused ventral and dorsal periosteum with no intervening bone. The fossae that contribute to the aperture accommodate the ulna's coracoid process in flexion and the olecranon process in extension. Kennewick Man has a septal aperture on the left side, and there is porous bone indicating resorption surrounding it (Walker et al. 2000). This is a normal phenomenon and not indicative of inflammation. Previous researchers may have considered this feature a pathology because septal apertures are an uncommon anomaly in males in recent human populations. Higher frequencies and little sexual dimorphism has been reported in ancient North African foragers (Dutour 1989) and in Neanderthals (Trinkhaus 1983), among others. It is noteworthy that Kennewick Man's left humerus is more gracile, a usual association of septal aperture (Benfer and McKern 1966).

The consequence of identifying the radial injury as osteochrondritis—or osteonchondrosis, to use the British term that avoids the mistaken understanding that infection is implied—is to substantially downgrade the level of disability one expects Kennewick Man to have experienced. This diagnosis is consistent with the absence of extensive ossification of soft tissue surrounding the joint and with the lack of evidence for loss of bone density or muscle marking in the right arm. Rothschild (1997) dismissed subchondral pitting as nonpathological. While his opinion is perhaps overstated, it does suggest how minor Kennewick Man's problems arising from the radial injury may have been. The conclusions in Chapter 7 are more conservative, acknowledging only the presence of minor degenerative changes.

SHOULDER GIRDLE

Modifications of the costoclavicular ligament attachments at the medial ends of the clavicles were "inflammatory" according to Powell and Rose (1999). This word choice carries with it the inference that there was pain, heat, and swelling in these joints. Stirland (1998) has pointed out the variability in this enthesis in the *Mary Rose* seamen, and she cautions against a narrow interpretation. She and others see deeply excavated and roughened costoclavicular ligament attachments as musculoskeletal stress markers rather than as pathological changes and describe a range of associated activities from kayak paddling to plowing (Capasso et al. 1999). Clinical correlation studies that would support labeling them as pathological or painful are lacking.

The scapulae are fragmentary. Both glenoid surfaces are robust with dense, lipped margins and central rugosities. There is carbonate-cemented sediment on the dorsal inferior margin of the right glenoid fossa where Chatters identified a chip fracture, and it is unclear whether this is in fact an antemortem fracture rather than an irregularity in formation of the lipped margin.

OSTEOCHONDRITIS OF THE KNEE

The dorsal aspects of the lateral condyles of both femora show osteochondritis (Figure 14.3). There is erosion at the point of greatest convexity and the surface surrounding the eroded area is dense and elevated, more so on the left than on the right (Figure 14.4). There are also facets extending the margins of the articular surfaces into the popliteal space.

Osteochondritis at the knee is most commonly a consequence of damage to the meniscus, which then erodes the cartilage and eventually the underlying bone. This process is painful. High frequency of osteochondritis dessicans has been reported for young seamen and archers recovered from the wreck of the *Mary Rose* (Stirland 1984), where the location is the common one in modern populations, the distal aspect of the medial condyle.

The dorsal aspect of the femoral condyle is an unusual location for osteochondritis in modern, clinical literature, but examples are not difficult to find in prehistoric North American remains, often in association with other markers of flexion of the knee in the form of Charles facets proximal to the condyles. Lesions similar to those in Kennewick Man have been reported in series ranging from Nigerian farmers to Neanderthals, although there has been controversy about what specific activities result in these changes (Capasso et al. 1999). The meniscus moves with the tibia as the knee flexes and corresponds to the dorsal surface of the condyle in tight flexion of the knee. Presumably osteochondritis at this location reflects damage to the meniscus related to habitual loading of the tightly flexed knee. The fact that Kennewick Man's lesions are limited to the lateral condyles suggests that outward rotation at the knees was a feature of the causal activities as well.

RIB FRACTURES

Unusual rib lesions in Kennewick Man have prompted disagreement among students of his pathological conditions. Nontraumatic discontinuities in ribs are reported as a rarity in the radiography literature while stress fractures of ribs are relatively common. They occur gradually and may not result in periosteal elevation or extensive hematomas that then ossify as fracture callus. Stress fractures of the ribs are reported in golfers and backpackers as well as in cases following strenuous coughing (Resnick and Niwayama 1988).

Kennewick Man exhibits several healed and healing rib fractures. The most puzzling ones have no unremodeled fracture callus, but the inner and outer cortices are pinched together like a metal pipe that has been bent to the point of crimping. These may represent stress fractures and/or false joints or pseudarthroses occurring so early in life that all fracture calluses were remodeled and broken edges were overgrown with fibrous tissue.

It was not possible during the 2006 study to arrange

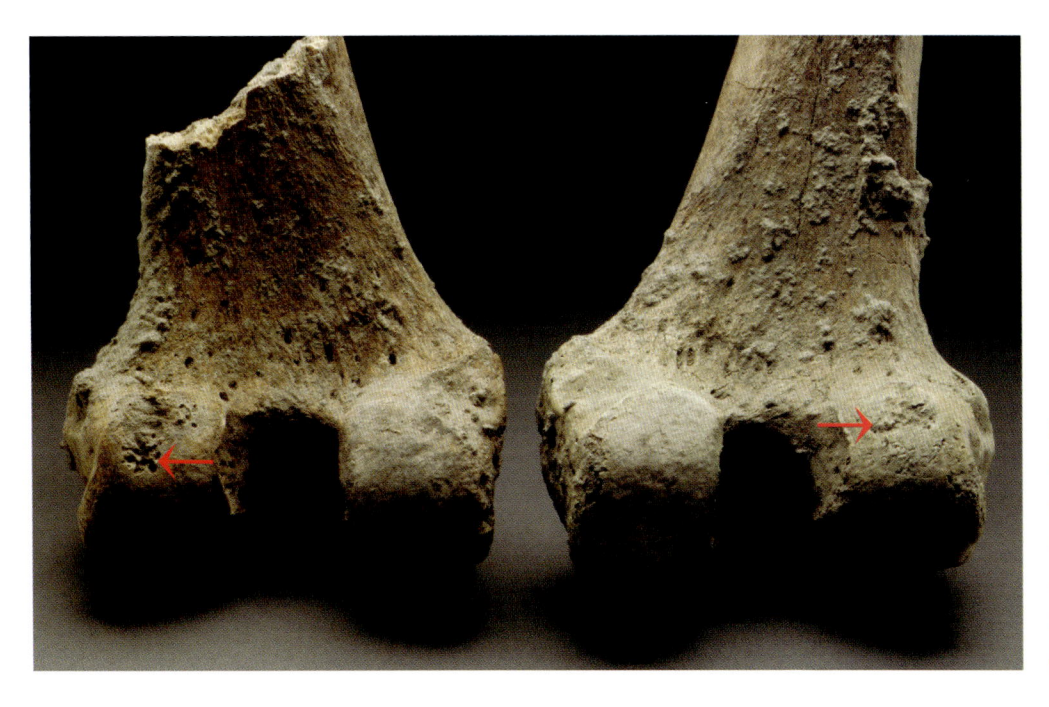

FIGURE 14.3 Distal femora showing degenerative changes from osteochondritis of the dorsal aspect of both lateral condyles (arrows).

FIGURE 14.4 Detail of the left femur and tibia showing the erosive lesion of the lateral femoral condyle and corresponding nodular deposit on the lateral condyle of the tibia.

the rib fragments by rib and side, and several of the interesting fragments are so short that determining side is problematic. One of the pinched or crimped ribs is identified as right rib 5 (97.1.12A(8)) (Chapter 7). The interpretation that there were two episodes of chest injury, an early one resulting in pseudarthroses and a later one resulting in healing fractures with abundant callus, is attractive but difficult to prove. The argument that the rib injuries are so extensive that Kennewick Man suffered from an unstable "flail chest" (Chatters 2001:135) is not supported by the evidence. One would expect to see poorly consolidated new callus if this were the case.

SPINE

Chatters (2000:308) reports "minor to moderate lipping and pitting of the upper neck, upper thoracic vertebrae, shoulders, elbows, and knees" and diagnoses this as osteoarthritis. Extensive lipping of joint margin changes in an adult of less than 50 years of age might be seen as more suggestive of an arthritis syndrome than of wear-and-tear joint damage. There are several genetically mediated joint diseases that are relatively common in modern American Indian groups from the Northwest (Gofton et al. 1972), and several of these are quite variable in their distribution across human populations.

Marginal lesions of vertebral and appendicular joints in Kennewick Man are quite minor and do not support a diagnosis of any of the early arthritis syndromes. There are small osteophytes of the ligamenta flava at several levels. The only potentially important spinal lesions are override facets of the left dorsal intervertebral joint between cervical vertebrae C4 and C5 and the left and right dorsal intervertebral joints between C5 and C6. These lesions reflect excessive dorsiflexion of the neck and could point to pressure on the dorsal nerve roots at this level. This would affect sensory function in the hand.

UNDISCOVER'D COUNTRY FROM WHOSE BOURN NO TRAVELER RETURNS

The Kennewick skeleton provides no information relevant to cause and manner of death. This is not an unusual circumstance, even in modern forensic investigations. However, this fascinating skeleton is remarkably informative about his life experience. Kennewick Man is an exceptionally eloquent example of osteobiography. His injuries tell a tale of a strenuous life, of the "natural shocks that flesh is heir to," as Shakespeare put it in *Hamlet*. As such, this person is a poignant focus for demonstrating the importance of study and restudy; various opinions about his life story are tenable among similarly trained anthropologists in the decade following his discovery, and this variety will necessarily increase with the passage of time (Buikstra and Gordon 1981). The expectation that limited access to a handful of scholars will adequately document ancient skeletons is demonstrably wrong. New technologies emerged during that decade, and older ones have become more refined, and they will, of course, continue to do so. Reburial poses incalculable losses to our future knowledge of the past, the "undiscover'd country from whose bourn" this exceptional traveler returned to tell his tale.

REFERENCES

Acsádi, György, and János Nemeskéri. 1970. *History of Human Lifespan and Mortality*. Akadémiai Kiadó, Budapest.

Anderson, John Y., and Erik Trinkaus. 1998. Patterns of sexual, bilateral and interpopulational variation in human femoral neck-shaft angles. *Journal of Anatomy* 192:279–285.

Benfer, Robert A., and Thomas W. McKern. 1966. The correlation of bone robusticity with the perforation of the coronoid-olecranon septum in the humerus of man. *American Journal of Physical Anthropology* 24(2):247–252.

Brooks, Sheilagh T. 1962. Review of *The Human Skeleton in Forensic Medicine*, by Wilton Marion Krogman. *American Anthropologist* 64(6):1356–1358.

Buikstra, Jane E., and Claire C. Gordon. 1981. The study and restudy of human skeletal series: The importance of long term curation. *Annals of the New York Academy of Sciences* 376:449–465.

Capasso, Luigi, Luisa Di Domenicantoni, G. di Tota, and Caterina Scarsini. 1994. A bronze javelin point in a femur from the necropolis of Pontecagnano (Salerno, Southern Italy, IV century BC). *Homo: Journal of Comparative Human Biology* 45(S):530.

Capasso, Luigi, Kenneth A. R. Kennedy, and Cynthia A. Wilczak. 1999. *Atlas of Occupational Markers on Human Remains*. Edigrafital S.p.A., Teramo, Italy.

Chatters, James C. 2000. The recovery and initial analysis of an early Holocene human skeleton from Kennewick, Washington. *American Antiquity* 65(2):291–316.

———. 2001. *Ancient Encounters: Kennewick Man and the First Americans*. Simon & Schuster, New York.

Chikkamuniyappa, Chandrasekar, Rami Hussein, and Shirzad Houshian. 2005. Shrapnel causing a cyst in a tendon: Presenting 60 years later in a World War II veteran. *Journal of Orthopaedic Science* 10(5):537–539.

Danenberg, P. J., R. S. Hirsch, N. G. Clarke, P. I. Leppard, and L. C. Richards. 1991. Continuous tooth eruption in Australian aboriginal skulls. *American Journal of Physical Anthropology* 85(3):305–312.

Danforth, Marie E., Della C. Cook, and Stanley G. Knick III. 1994. The human remains from Carter Ranch Pueblo, Arizona: Health in isolation. *American Antiquity* 59(1):88–101.

de Terra, Helmut, Javier Romero, and T. D. Stewart. 1949. *Tepexpan Man*. Publication in Anthropology 11. Viking Fund, New York.

Dutour, Olivier. 1989. *Hommes Fossiles du Sahara: Peuplements Holocènes du Mali Septentrional*. Éditions du CNRS, Paris.

Ferguson, R. Brian. 2006. Archaeology, cultural anthropology, and the origins and intensification of war. In *The Archaeology of Warfare: Prehistories of Raiding and Conquest*, edited by Elizabeth N. Arkush and Mark W. Allen, 469–523. University Press of Florida, Gainesville.

Flecker, H. 1942. Time of appearance and fusion of ossification centers as observed by roentgenographic methods. *American Journal of Roentgenology and Radium Therapy* 47(1):97–159.

Frison, George C. 2004. *Survival by Hunting: Prehistoric Human Predators and Animal Prey*. University of Illinois Press, Urbana.

Genovés, Santiago T. 1960. Revaluation of age, stature and sex of the Tepexpan remains, Mexico. *American Journal of Physical Anthropology* 18(3):205–217.

Geschickter, Charles F., and Murray M. Copeland. 1936.

Tumors of Bone (Including the Jaws and Joints). Revised Edition. American Journal of Cancer, New York.

Ghibely, A. 1990. La maladie luxante chez les Indiens du Quebec. *Acta Orthopaedica Belgica* 56(1):37–42.

Gofton, J. P., H. A. Bennett, H. A. Smythe, and J. L. Decker. 1972. Sacroiliitis and ankylosing spondylitis in North American Indians. *Annals of Rheumatic Diseases* 31(6):474–481.

Gonzalez, Silvia, José Concepción Jiménez-López, Robert Hedges, David Huddart, James C. Ohman, Alan Turner, and José Antonio Pompa y Padilla. 2003. Earliest humans in the Americas: New evidence from México. *Journal of Human Evolution* 44(3):379–387.

Hewlett, Barry S., Jan M. H. van de Koppel, and Maria van de Koppel. 1986. Causes of death among Aka Pygmies of the Central African Republic. In *African Pygmies*, edited by Luigi Luca Cavalli-Sforza, 45–63. Academic Press, San Diego, CA.

Hinton, Robert J. 1981. Form and patterning of anterior tooth wear among aboriginal human groups. *American Journal of Physical Anthropology* 54(4):555–564.

Hoppa, Robert D., Laura Allingham, Kevin Brownlee, Linda Larcombe, and Gregory Monks. 2005. An analysis of two Late Archaic burials from Manitoba: The Eriksdale site (EfLl-1). *Canadian Journal of Archaeology* 29(2):234–266.

Howell, Nancy. 1979. *Demography of the Dobe !Kung.* Academic Press, New York.

Hutchinson, Dale L., Cristopher B. Denise, Hal J. Daniel, and Gerhard W. Kalmus. 1997. A reevaluation of the cold water etiology of external auditory exostoses. *American Journal of Physical Anthropology* 103(3):417–422.

Johnsen, Knut, Rasmus Goll, and Olav Reikerås. 2008. Acetabular dysplasia in the Sami population: A population study among Sami in north Norway. *International Journal of Circumpolar Health* 67(1):147–153.

Jurmain, Robert. 1999. *Stories from the Skeleton: Behavioral Reconstruction in Human Osteology.* Gordon and Breach, Amsterdam.

———. 2001. Paleoepidemiolgical patterns of trauma in a prehistoric population from central California. *American Journal of Physical Anthropology* 115(1):13–23.

Kaifu, Yousuke, Kazutaka Kasai, Grant C. Townsend, and Lindsay C. Richards. 2003. Tooth wear and the "design" of the human dentition: A perspective from evolutionary medicine. *Yearbook of Physical Anthropology* 46:47–61.

Lafferty, Christopher M., David J. Sartoris, Rose Tyson, Donald Resnick, Sevil Kursunoglu, Deborah Pate, and David Sutherland. 1986. Acetabular alterations in untreated congenital dysplasia of the hip: Computed tomography with multiplanar re-formation and three-dimensional analysis. *Journal of Computer Assisted Tomography* 10(1):84–91.

Lazenby, Richard A., and Peter McCormack. 1985. Salmon and malnutrition on the Northwest Coast. *Current Anthropology* 26(3):379–384.

Lemonick, Michael D., and Andrea Dorfman. 2006. Who were the first Americans? *Time* 167(11):44–52.

Levers, B. G. H., and A. I. Darling. 1983. Continuous eruption of some adult human teeth of ancient populations. *Archives of Oral Biology* 28(5):401–408.

Lieverse, Angela R. 1999. Diet and the aetiology of dental calculus. *International Journal of Osteoarchaeology* 9(4):219–232.

McKern, Thomas W., and T. D. Stewart. 1957. *Skeletal Age Changes in Young American Males Analyzed From the Standpoint of Age Identification.* Quartermaster Research & Development Command, Environmental Protection Research Division, Technical Report EP-45. Headquarters, Quartermaster Research & Development Center, U.S. Army, Natick, MA.

Molnar, Stephan. 1972. Tooth wear and culture: A survey of tooth functions among some prehistoric populations. *Current Anthropology* 13(5):511–526.

Moss, Melvin L. 1960. A reevaluation of the dental status and chronological age of the Tepexpan remains. *American Journal of Physical Anthropology* 18(1):71–72.

Okumura, Maria M., Célia C. H. Boyadjian, and Sabine Eggers. 2007. Auditory exostoses as an aquatic activity marker: A comparison of coastal and inland skeletal remains from tropical and subtropical regions of Brazil. *American Journal of Physical Anthropology* 132(4):558–567.

Pales, Léon. 1930. *Paleopathologie et Pathologie Comparative.* Masson, Paris.

Pennefather-O'Brien, Elizabeth E., and Michael Strezewski. 2002. An initial description of the archaeology and morphology of the Clark's Fork skeletal material, Bonner County, Idaho. *North American Archaeologist* 23(2):101–115.

Powell, Joseph F., and Jerome C. Rose. 1999. Chapter 2. Report on the Osteological Assessment of the "Kennewick Man" Skeleton (CENWW.97. Kennewick). In *Report on* the *Non-Destructive Examination, Description, and Analysis of the Human Remains from Columbia Park, Kennewick, Washington* [October 1999]. National Park Service, http://www.nps.gov/archeology/kennewick/powell_rose.htm.

Rabin, David L., Clifford R. Barnett, William D. Arnold, Robert H. Freiberger, and Gyla Brooks. 1965. Untreated congenital hip disease. *American Journal of Public Health* 55(S):1–44.

Resnick, Donald, and Gen Niwayama. 1988. *Diagnosis of Bone and Joint Disorders*. 2nd edition. W. B. Saunders, Philadelphia.

Roberts, Charlotte A. 2007. A bioarcheological study of maxillary sinusitis. *American Journal of Physical Anthropology* 133(2):792–807.

Rothschild, Bruce M. 1997. Porosity: A curiosity without diagnostic significance. *American Journal of Physical Anthropology* 104(4):529–533.

Rutkow, Ira M. 1998. Beaumont and St. Martin: A blast from the past. *Archives of Surgery* 133:1259.

Ryan, Timothy M., and George R. Milner. 2006. Osteological applications of high-resolution computed tomography: A prehistoric arrow injury. *Journal of Archaeological Science* 33(6):871–879.

Scheuer, Louise, and Sue Black. 2000. *Developmental Juvenile Osteology*. Academic Press, New York.

Steerman, James G., Michael T. Reeder, Brian E. Udermann, Robert W. Pettitt, and Steven R. Murray. 2008. Avulsion fracture of the iliac crest apophysis in a collegiate wrestler. *Clinical Journal of Sport Medicine* 18(1):102–103.

Stirland, A. 1984. The burials from King Henry VIII's ship, the *Mary Rose*: An interim statement. *Paleopathology Newsletter* 47:7–10.

———. 1998. Musculoskeletal evidence for activity: Problems of evaluation. *International Journal of Osteoarchaeology* 8:354–362.

Sugiyama, Lawrence S. 2004. Illness, injury, and disability among Shiwiar forager-horticulturalists: Implications of health-risk buffering for the evolution of human life history. *American Journal of Physical Anthropology* 123(4):371–389.

Taylor, R. E. 2000. Chapter 1: Examination of the Kennewick Remains—Taphonomy, Micro-sampling, and DNA Analysis, Attachment B: Amino Acid Composition and Stable Carbon Isotope Values on Kennewick Skeleton Bone. *Report on the DNA Testing Results of the Kennewick Human Remains from Columbia Park, Kennewick, Washington* [September 2000]. National Park Service, http://www.nps.gov/archeology/kennewick/taylor2.htm.

Tooker, Elisabeth. 1991. *An Ethnography of the Huron Indians, 1615–1649*. Syracuse University Press, Syracuse, NY.

Trinkhaus, Erik. 1983. *The Shanidar Neandertals*. Academic Press, New York.

Valdés, M., J. Molins, and O. Acebes. 2000. Avulsion fracture of the iliac crest in a football player. *Scandinavian Journal of Medicine & Science in Sports* 10(3):178–180.

Walker, Phillip L. 1989. Cranial injuries as evidence of violence in prehistoric southern California. *American Journal of Physical Anthropology* 80(3):313–323.

Walker, Phillip L., Clark Spencer Larsen, and Joseph F. Powell. 2000. Chapter 5. Final Report on the Physical Examination and Taphonomic Assessment of the Kennewick Human Remains (CENWW.97.Kennewick) to Assist with DNA Sample Selection. In *Report on the DNA Testing Results of the Kennewick Human Remains from Columbia Park, Kennewick, Washington* [September 2000]. National Park Service, http://www.nps.gov/archeology/kennewick/walker.htm.

Occupational Stress Markers and Patterns of Injury

15

James C. Chatters

One of the greatest scientific and humanistic values of a skeleton as ancient and as complete as Kennewick Man's is the window it provides into a life lived in early Holocene North America. Even though excavation and analysis of stone and bone tools, cooking and dwelling features, and food remains offer a general understanding of how people made a living, sheltered themselves, and moved across the landscape through the seasons, they never truly give a sense of what it was like to *be* one of those people. The skeleton, on the other hand, is a very personal representation of a past life. Far from being merely a genetically dictated structure for supporting muscles and organs, the skeleton is a plastic medium that physically maintains a journal of an individual existence (İşcan and Kennedy 1989; Jurmain 1999). Episodes of hunger or illness during childhood and youth leave growth interruptions in the form of Harris lines in long bones and hypoplasias in teeth. Habitual activities leave patterns in muscle-attachment surfaces, cortical bone thickness, and joint surfaces. Injuries bespeak occupational hazards. Wounds testify to interpersonal conflict.

This chapter documents and interprets what are collectively known as skeletal markers of occupational stress (Kennedy 1983; see Jurmain 1999). Three sets of skeletal characteristics are of interest: musculoskeletal stress markings (MSM), changes in joint surfaces, and traumatic injuries and their patterns of healing.

Two questions are central to this analysis of occupational stress. First, in what subsistence activity or activities did Kennewick Man most often engage? Given the skeleton's 8,400-year radiocarbon age, this man was probably a participant in the Western Stemmed Tradition (Chapter 2). The evidence for Western Stemmed Tradition subsistence in the southern Columbia Plateau is somewhat contradictory. Faunal assemblages are consistently dominated by large game, specifically bison, elk, and deer, which points to hunting as an occupational emphasis. This would indicate spear or atlatl use as a primary occupational behavior. Bolas may also have been used. In apparent contradiction, bone collagen from both Kennewick Man (Taylor et al. 1998) and his near-contemporary, Stick Man of eastern Washington (Chatters et al. 2000), contains high levels of $\delta^{13}C$, which may identify anadromous fish, such as salmon, as the two men's primary source of protein. Thus, it is possible that Kennewick Man engaged in fishing-oriented activities, such as use of a fish spear or nets. However, high $\delta^{13}C$ ratios may also indicate consumption of bison and elk that subsisted mainly on warm season grasses with the C4 photosynthetic pathway (Ames 2000). This would be consistent with paleoecological evidence that a grassland ecosystem occupied the Columbia Plateau in the markedly seasonal early Holocene (Chapter 2; Chatters 1998). The most often stressed muscle attachments and joints and their degree of left/right asymmetry should differ depending on whether hunting or fishing activities were emphasized. Net fishing, a strongly bilateral activity, would result in symmetrical muscle development; spear fishing and hunting with a spear (javelin), atlatl, or bola would result in left/right asymmetry. Overhand throwing, as with a javelin, atlatl, or bola, would use different muscles than would downward thrusting with a fish spear (Hawkey and Merbs 1995). Patterns of injury and healing could also result from these patterns of movement.

Second, did the projectile wound in Kennewick Man's right hip affect his gait? The wound in the ilium resulted in significant bone necrosis and remodeling on the visceral surface of the upper iliac fossa and iliac crest, the surface from which *iliacus*, the major thigh flexor, originates. Whether or not the wound and subsequent infection damaged the periosteum (Chapter 14), the effectiveness of *iliacus* is highly likely to have been diminished because of pain, loss of origin surface, and/or anterior and downward movement of a part of the origin surface as *iliacus* contracted. Hence, the left leg would have been involved to some degree to pull the right forward, placing additional stress on muscle attachments and joint surfaces on the left side. If the injury occurred early in life, such differences are likely to be manifest; if late in life, after habitual behaviors had left their mark, the injury should leave little or no trace.

The skeleton of Kennewick Man represents the remains of a large male, estimated by various investigators to have died between the ages of 35 and 55 years (Chapter 7; Chatters 2000; Powell and Rose 1999; Walker et al. 2000). Differences in bone size and rugosity between right and left arms indicate he was right-handed. The skeleton is largely complete and in good to excellent condition, although some elements, such as the scapulae, are too fragmentary for an analysis of muscle attachment sites. Other elements, most notably the tibiae and femora, have muscle attachment sites that suffered postmortem damage, were sampled for radiocarbon analysis by the US Department of Interior, or are obscured by calcium carbonate and silica cement that investigators did not have permission to remove. Study of lower limb flexors was particularly limited by the extensive damage and cementation of the proximal tibiae. All joint surfaces of the appendicular skeleton, excluding the hands and feet, were more or less intact and accessible. The skeleton exhibits numerous traumatic injuries (Chatters 2000, 2004; Powell and Rose 1999; Walker et al. 2000).

MUSCULOSKELETAL STRESS MARKERS

Inference of activity patterns from musculoskeletal markers is based on Wolff's (1892) bone remodeling theory, known as Wolff's Law, which states: when a muscle insertion site is under repeated or prolonged stress, the site becomes increasingly vascularized. This elevated blood flow stimulates osteoblastic activity, which leads to hypertrophy of the bone at the attachment site (Chamany and Tschantz 1972; Ruff et al. 2006; Wolff 1892; Woo et al. 1981). Habitually stressed sites thus become more elevated and rugose than attachment sites for less heavily used muscles. As stress continues or increases, individual muscle fibers may tear away from the periosteum. Subsequent reattachment of the muscle fibers interrupts local blood flow, triggering necrosis and resorption of the bone, and producing lytic-like lesions on the muscle attachment surface. If continued overuse does not allow the stress to be relieved, resorption can cause deep, sharp-edged pits and grooves to form. The size and depth of such lesions is a measure of the severity and duration of muscle stress. Highly developed attachment surfaces, manifested as elevated, roughened ridges or deep, sharp-edged troughs, are considered to be the result of a lifetime of intensive activity. More severe traumas can produce a more extreme osteoblastic response, causing bone spurs (enthesophytes) to extend into the affected muscle or tendon. However, these are generally attributed to individual injuries, such as a muscle pull or tendon strain rather than to habitual muscle reuse (Hawkey and Merbs 1995), although it is reasonable to assume that such injuries occur in proportion to the frequency or intensity of muscle stress (Dutour 1986).

The extent of MSM development is not strictly proportional to activity-related muscle use. Most studies of large skeletal collections have shown a correlation between MSM prominence and age (al-Oumaoui et al. 2004; Lieverse et al. 2009). Some researchers have found that MSMs also tend to be more prominent in males than in females and to vary in proportion to body size (Molnar 2006; Nagy 1998; Weiss 2004, 2007; Wilzack 1998). These tendencies should not affect interpretation of MSM in the Kennewick skeleton, except to the extent that as a large man, he tends to have well-developed markers.

MSM analysis has been used extensively as a means to infer activity patterns from skeletal remains of ancient populations. Some studies have focused on general levels of activity or evidence for a sexual division of labor (al-Oumaoui et al 2004; Churchill and Morris 1998; Eshed et al. 2004; Lieverse et al. 2009; Weiss 2007). Others have striven to identify individual behaviors, such as watercraft propulsion and hunting practices (Dutour 1986; Hawkey and Merbs 1995; Kennedy 1983; Nagy 1999; Peterson 1998). These studies have all directly addressed or identified evidence for the overhand throwing action of atlatls or spears in foraging populations.

The majority of MSM research has been directed at the upper limbs, although a few studies have included the lower limbs. For example, Weiss (2004) studied lower limb markings in precontact British Columbians and 18th-century Euro-Canadian prisoners to understand the effect of age, sex, and body size on MSM expression. Al-Oumaoui et al. (2004) included eight attachment sites in their analysis of activity patterns in Copper Age through Medieval populations of the Iberian Peninsula, where it proved useful for distinguishing farmers from pastoralists.

MSM analyses have typically been conducted at the population level, which, by enabling researchers to statistically identify central tendencies, is a more reliable approach to understanding collective activities of extinct cultural groups. This raises questions about the reliability of an MSM analysis of just one skeleton. MSM has, however, been used effectively at the individual level, primarily as an identification tool in forensic investigations (Kennedy 1989). The best-known such case is the use of MSM in phalanges in an Egyptian mummy to identify it as the remains of a royal scribe (Kennedy et al. 1986). Hawkey (1998) was able to infer the limited movements of a disabled man from Gran Quivira Pueblo, New Mexico, including a comparison between the subject and other males in the same skeletal series. No such comparison is possible in the present case.

METHODS

The method used to measure MSM in the Kennewick skeleton is based on that of Hawkey and Merbs (1995). They have shown their approach to have a high degree of both inter- and intra-investigator replicability, minimizing the subjectivity for which MSM studies have been criticized (Stirland 1998). This approach addresses three dimensions of musculoskeletal stress: robusticity, stress lesions, and ossifications (exostoses).

Robusticity (R) and stress lesions (S) are each scored from 0 to 3, as follows.

- R0 – No marking is visible.
- R1 – Cortex is only slightly raised and palpable, but no distinct crests or ridges are present. There is no external roughening. Lines, if present, are discontinuous.
- R2 – Cortical surface is roughened. A mounded elevation, if characteristic of the site, is easily visible, but there are no sharp ridges or crests. Lines, if present, are continuous.
- R3 – Cortical surface is elevated, with sharp crests and/or ridges.
- S1 – Cortex exhibits lytic-like furrowing or pitting below the surface of the cortex less than 1 millimeter deep.
- S2 – Lytic-like pitting or furrowing is deeper (>1 mm but < 3 mm) and more extensive (up to 5 mm long).
- S3 – Pitting or furrowing is marked (3+ mm deep and > 5 mm long).

The sequence is considered continuous, with increasing expression from R0 through S3. If expression of robusticity far exceeded the minimum criteria for a given score but did not reach the minimum criteria for the next higher score, it was scored a half-step higher. For example, if a site exhibited an elevated surface and crests or ridges, but the ridges could not be called sharp, it was scored at R2.5, rather than R2. This allowed greater exactness in comparisons among attachment sites and between left and right expressions. For computation purposes, MSM scores were converted to a seven-level scoring system, beginning with 0 for R0 and ending with 6 for S3. Exostoses were described in detail.

Each muscle attachment was measured, under the assumption that attachment size is proportional to the size of the muscle itself (Zumwalt et al. 2000). The size of the muscle attachment sites allowed consideration of the degree of laterality in individual muscles that had similar MSM scores on left and right sides. To document muscle attachment size, the length and width were measured in millimeters of surface roughness and/or elevation for each attachment surface, then attachment surface area was computed simply as length multiplied by width (understanding that this is not a true measure of the area of an irregular surface). Other means, such as those employed by Zumwalt et al. (2000), could not be used in the time and setting available for this study.

Scoring and measuring of 22 upper limb and 16 lower limb sites were accomplished on at least one side (Table 15.1). Insertion sites were given preference over origins because the maximum amount of muscular pull is usually exerted at its insertion (Hawkey and Merbs

TABLE 15.1 Muscle and ligament attachment sites addressed in this analysis.

Element	Origin/insertion/ligament	Function[1]
Scapula	pectoralis minor	pulls shoulder down and forward
Clavicle	subclavius	draws shoulder down and forward
	costoclavicular lig.	resists elevation, rotation of scapula
Humerus	supraspinatus	abducts, laterally rotates humerus
	subscapularis	medially rotates humerus; draws it forward and down when arm is raised
	infraspinatus	laterally rotates head of humerus
	teres minor	adducts, laterally rotates humeral head
	teres major	adducts, medially rotates, and draws back humerus
	latissimus dorsi	adducts, extends, medially rotates humerus
	pectoralis major	adducts, draws forward, medially rotates arm
	detoideus	abducts, flexes, extends humerus
	common flexor origin	flexes wrist, fingers
Radius	biceps brachii	flexes forearm when supinated, supinates forearm when flexed
	pronator teres	pronates forearm, assists in flexion
	pronator quadratus	pronates forearm
	brachioradialis	flexes forearm when semipronated or semisupinated; supinates forearm
Ulna	triceps brachii	extends forearm
	anconeus	extends forearm
	supinator	supinates forearm
	pronator teres/flexor digitorum superficialis (origin)	pronate, assist in flexing forearm/flexes fingers
	pronator quadratus (origin)	pronates forearm
	brachialis	flexes pronated forearm
Ilium	rectus femoris (origin)	flexes thigh, extends lower leg
	sartorius (origin)	adducts and flexes leg
Pubis	adductor magnus	adducts thigh; adducts and flexes leg
Femur	gluteus maximus	extends thigh
	gluteus medius	abducts, medially rotates thigh
	gluteus minimus	abducts, medially rotates thigh
	iliopsoas	flexes thigh
	iliofemoral ligament	holds femoral head in acetabulum
	vastus medialis	extends leg
	linea aspera	origin of tibia extensors *vastus medialis* and *vastus lateralis*
	adductor magnus	adducts thigh
Patella	quadriceps tendon	extends leg
Tibia	soleus (origin)	extends foot
	patellar ligament	leg extensors (quadriceps group)
Calcaneus	gastrocnemius, soleus	extend foot
	extensor digitorum brevis	extends toes

[1] Marieb 1995; Warfel 1974.

1995). Where either the insertion site of a muscle could not be scored because it was missing or encrusted by carbonate-cemented sediment, or the origin site was anatomically the more focused osseous contact, the origin site was scored. When the identification of origin versus insertion was essentially arbitrary, as in *pronator quadratus* and *supinator*, both were scored.

The results were analyzed to highlight the most intensively used muscles and the muscles that were most strongly lateralized. To identify the most heavily used muscles on one side's upper and lower extremities, the mean of all attachment site scores for a particular limb and side was computed, then each attachment was compared with that mean. Muscles associated with attachment sites that have scores greater than the mean are inferred to have been the most heavily used in life. To factor out lateralization and identify specific muscle groups potentially associated with habitual movements of the dominant arm, the left-side score was subtracted from the right-side score to determine the degree of difference (δ). Those muscles with a δ greater than the mean of the absolute value of the individual δ scores indicate heavier utilization of the right side; those with a negative δ above the mean indicate an emphasis on left side musculature.

RESULTS
Upper Extremities

Twenty attachment sites could be scored on the left arm and shoulder, but only 17 on the right (Table 15.2; Figure 15.1). Postmortem damage to the scapula prevented scoring of the origin for *pectoralis minor* and the radial insertion of *pronator teres* on the left; damage and coatings of carbonate-cemented sediment made it impossible to score insertions for the scapular muscles, *teres minor*,

TABLE 15.2 Results of MSM analysis for Kennewick Man's upper extremities. Numbers in boldface exceed arm and δ means.

Element	Origin/insertion	Right Score	Right Area[1]	Left Score	Left Area	δ MSM R-L	δ Area R-L	% δ area
Scapula	pectoralis minor	1	84	...	...	...	...	...
Clavicle	*subclavius*	1.5	354	1	348	0.5	6	2
	costoclavicular ligament	**3**	448	1.5	300	**1.5**	148	49
Humerus	*supraspinatus*	...	...	1	108	...	...	...
	subscapularis	...	...	1	276	...	...	...
	infraspinatus	...	...	1	276	...	...	...
	teres minor	...	...	1	273	...	...	...
	teres major	2	424	1	468	1	−44	11
	Latissimus dorsi	...	...	1	438	...	...	...
	pectoralis major	2.5	544	1	504	**1.5**	40	8
	detoideus	**3**	1204	1.5	504	**1.5**	700	**139**
	common flexor	**3**[2]	320	1.5	280	**1.5**	40	14
Radius	*biceps brachii*	1	387.8	1	336	0	51.8	15
	pronator quadratus	1.5	...	1	...	0.5	...	...
	pronator teres	2	72	...	...	...	...	...
	brachioradialis	2	...	2	45	0	...	...
Ulna	*triceps brachii*	2.5	572[3]	2	572[3]	0.5	0	0
	anconeus	1	840	1	840	0	0	0
	supinator	**3**	315	1	82	**2**	233	**284**
	pronator teres	**6**	117	**4**	84	**2**	33	39
	pronator quadratus	**4**	...	1.5	...	**2.5**	...	...
	brachialis	**4**	540	**2**	432	**2**	108	25
MEAN		2.5	...	1.4	...	1.1	...	49

[1]Area in square millimeters.
[2]Ossification is also present on this origin site of the common flexor tendon.
[3]The area marked by distinct A-P ridges is 140 mm² on the left, 264 mm² on the right.

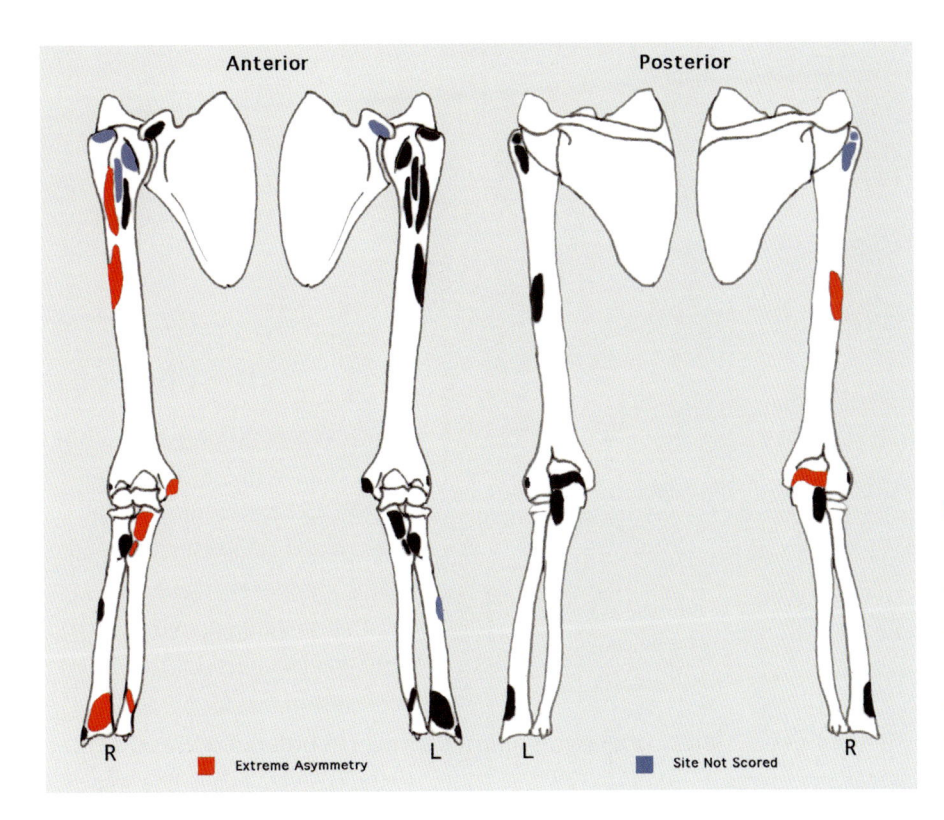

FIGURE 15.1 The distribution of scored origin and insertion sites in the upper extremities showing a pattern of asymmetry favoring muscles on the right side. The clavicles, with the sternoclavicular ligament and *subclavius* MSM sites are not included.

and *latissimus dorsi* on the right. Thus, only 15 sites could be scored bilaterally. Areas of attachment surfaces were more difficult to measure because they were obscured by coatings of carbonate-cemented sediment, resulting in only 12 bilateral measurements.

Right side scores are generally, but not always, greater than those of the left side. *Pronator teres* (ulnar head origin) scores the highest on both arms, with a score of 6 in the right arm and 4 in the left. The costoclavicular ligament, *deltoideus*, common flexor origin, *pronator quadratus*, and *brachialis* score above the mean for both arms (Figure 15.2), although none of these sites on the left arm scores higher than 2. Only *brachioradialis* and *triceps brachii* exceed the mean on the left but not the right, although the scores for both muscles are similar on the left and right.

The muscles that score highest in both arms are those that adduct the shoulder, pronate the forearm, flex the forearm when pronated, and flex the fingers and wrists. The costoclavicular ligament resists the rotation and elevation of the scapula. The triceps, the primary extensors of the forearms, are also well developed on both sides, although not excessively on the right.

This pattern of muscle development indicates an action that involved holding a vertically oriented object in front of the body with the palms facing one another, while repeatedly forcibly lifting and depressing that object. Poling a watercraft or using a long-handled dipnet could produce such a pattern if the netting were practiced in relatively still water. In moving water, a net would be forced laterally or away from the holder, necessitating engagement of *latissimus dorsi* and *teres major* to keep the shaft vertical. Attachments for those muscles do not appear to have been stressed. The pattern is not what would be expected from retrieving a seine or other large net because that action would heavily involve the *biceps brachii*, *teres major*, and *latissimus dorsi*, none of which exhibits intensive bilateral use. Since these three muscles, along with the costoclavicular ligament, *brachialis*, *triceps*, *brachialis*, and the pronators, would also have been engaged in paddling a watercraft (Hawkey and Merbs 1995; Steen and Lane 1998), that activity is also not indicated.

Dipnetting, or a similar movement, is not the only possible explanation for the bilateral MSM pattern in the upper limbs. There is at least one plausible alternative for the bilateral expression of the pronators. The pressure flaking method of stone tool production includes holding

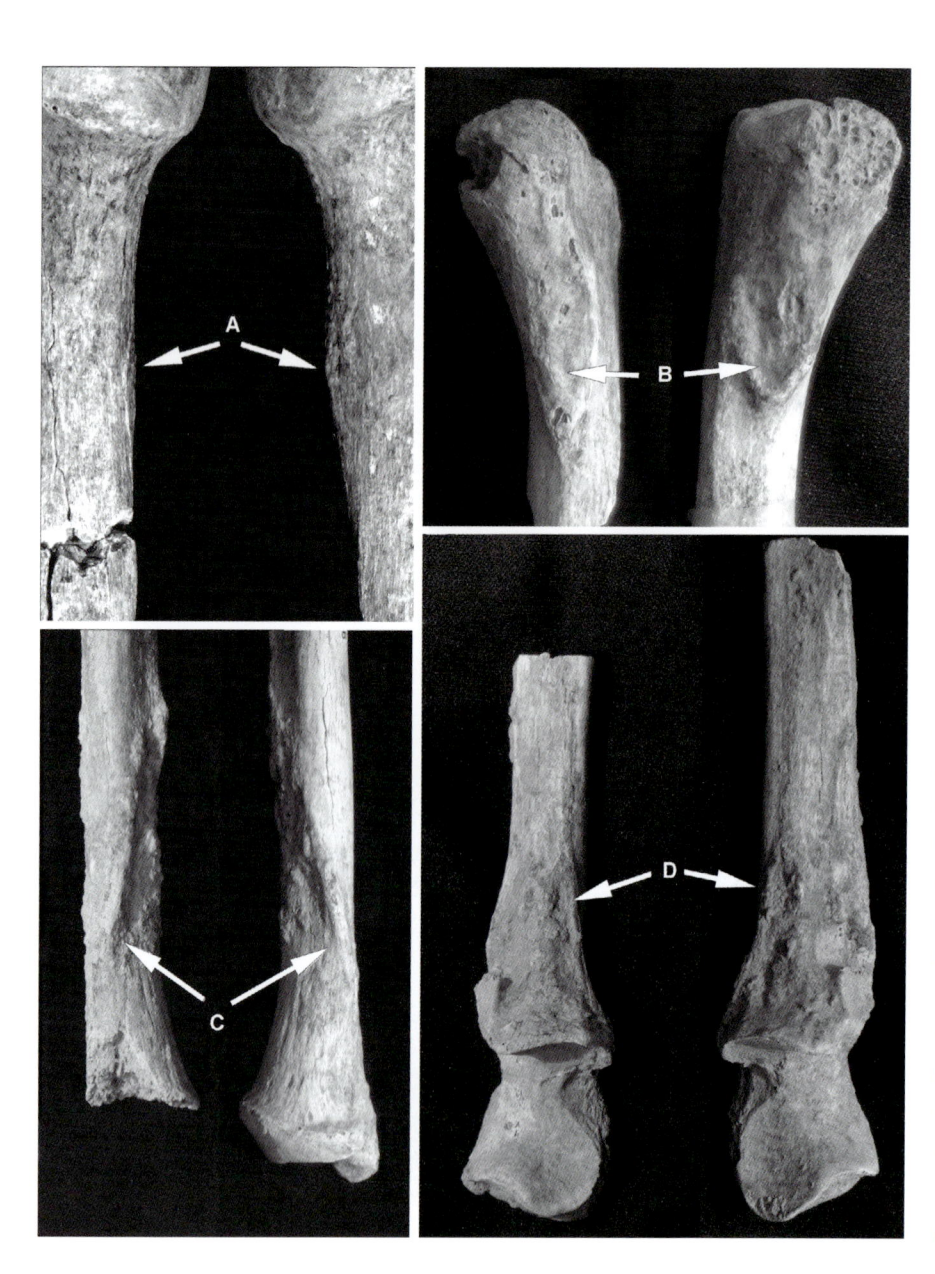

FIGURE 15.2 Musculoskeletal stress markers in Kennewick Man's upper extremities, illustrating differences between right and left sides. Included are attachments for A) *teres major* insertions on the humeri; B) the costoclavicular ligament; C) *pronator quadratus* on the distal ulna; and D) *brachialis* insertion on the proximal ulna. Paired elements shown in this and other figures are compiled from separate photos so the features are illuminated on each side from the same anatomical orientation and angle. Left side bones are on the left in all photographs.

the tool being worked against the nondominant hand, rotating both hands medially by forcefully pronating both forearms, and flexing the wrists to press flakes from the tool edge. Any fabrication activity that uses the dominant hand as the manipulator of a tool and the nondominant hand to hold the subject firmly requires similar movements. It would also account for the high level of supinator development in the left hand but not the right. The pronounced development of *brachialis, brachioradialis,* and *deltoideus* could be produced by repeatedly lifting burdens, although such an activity would also engage *biceps brachii,* for which Kennewick Man has a low MSM score in both arms.

Differences of MSM expression between the right and left arms present a pattern that is strongly indicative of another subsistence activity. In order of their δ scores, *pronator quadratus, pronator teres, supinator, brachialis, deltoideus, pectoralis major*, the common flexors, and the costoclavicular ligament show above average right side lateralization. Two of these, *deltoideus* and *supinator*, also have above average attachment areas. The movements represented by this pattern of development are forceful supination of the forearm; elevation, abduction, adduction, and forceful medial rotation of the humerus; pronation and flexion of the forearm; and flexion of the wrist. This group of actions is much like that described in the sports medicine literature for the overhand motion used by baseball pitchers and javelin throwers. Similar muscle expression, although not in this precise constellation, has been interpreted (Hawkey and Merbs 1995; Kennedy 1983; Nagy 1999; Peterson 1998) as evidence of spear, atlatl, or harpoon throwing.

The overhand throw is a complex action that includes a cocking phase, an acceleration phase, and a deceleration phase (Andrews et al. 1995; McGinnis 2005; Pappas and Walzner 1995). In the acceleration phase, the muscles and ligaments are under the greatest strain, since they are moving under load. Therefore, it is that phase that is most likely to promote MSM development. The thrower begins with torso turned at right angles to the target, with the dominant side away from the target. With the dominant arm cocked in a supinated, extended, posteriorly elevated pose, he pushes laterally off the dominant side foot and strides forward onto the nondominant foot, simultaneously beginning to rotate the torso, and thus the dominant shoulder, toward the target. Placement of the nondominant foot halts the body's forward motion, transferring its momentum, in sequence, into rotation of the shoulder, humerus, forearm, and hand. The forearm, which begins this sequence forcibly supinated, is pronated and flexed as it is brought forward, assisted by the medial rotation of the humerus produced by *pectoralis*. Finally, the forearm is extended and the wrist forcibly flexed as the propelled object is released. Throughout this action, *deltoideus* is keeping the arm elevated and abducted at the shoulder. Forearm extensors, although contracted, do not play a significant part in this action, since the forward motion of the forearm and hand is largely supplied by the leg, torso, and shoulder musculature.

Only one exostosis was found in the upper extremity, at the medial epicondyle of the right humerus—the common origin for the wrist and hand flexors. It consists of a 3-millimeter wide (mediolateral), 7-millimeter long (anteroposterior), 3-millimeter high projection from the distal edge, adjacent to the trochlear surface (Figure 7.21). This defect is consistent with the asymmetrical expression of MSMs. This defect has been reported in a Neolithic Sub-Saharan African skeleton by Dutour (1986), who notes that this enthesopathy is observed in modern javelin throwers and golfers.

Lower Extremities

Twelve of 15 attachment sites could be scored on both lower extremities (Table 15.3; Figure 15.3), although areas could rarely be measured due to the presence of carbonate-cemented sediment. Only the insertion site for *flexor digitorum brevis* and origin sites for *sartorius* and *adductor magnus* could not be observed bilaterally. MSM scores are uniformly high, averaging over 2.0 for both sides, and show a strong tendency for bilateral symmetry (Figure 15.4). In this case, it is the differences between the two sides that are most informative. The left leg exhibits higher MSM development in more sites than the right.

In the right leg, only *gluteus minimus* has an MSM δ above the mean for the two sides. The left leg exceeds the right for the *gluteus maximus*, the linea aspera, and the iliofemoral ligament. All differences are 1 point or less. The linea aspera is primarily the origin site for the *vastus medialis* and *vastus lateralis*, two major muscles of the quadriceps group. There is no evident difference in the *iliopsoas* insertion, which should be less prominent if the projectile wound in the ilium had a prolonged effect on that muscle.

Differences between right and left legs form a coherent pattern. *Gluteus minimus* abducts the thigh; *gluteus maximus* and the *quadriceps* are the primary extensors of the thigh and lower leg, respectively. Use of these muscles is again consistent with the overhand throwing motion. In the beginning of the throw's acceleration phase, the person drives laterally off the dominant side foot by engaging the large abductors of the thigh, producing forward motion. The body's forward motion is then stopped by the nondominant side leg, which must continue to absorb much of the forward momentum of the body throughout the remainder of the throw. The extensors of the leg must be fully engaged to keep the leg from collapsing under such force. This action puts the nondominant iliofemoral ligament under strain as it maintains the position of the femoral head in the olecranon fossa. Lower limb MSM asymmetries, expressed in a right side abductor and left side flexors, are thus entirely consistent with a repeated overhand throwing motion.

CHANGES TO JOINT SURFACES

Patterns of osteoarthritic pathologies frequently are used as indicators of occupational stress (Kennedy 1989; Nagy

TABLE 15.3 Results of MSM analysis for Kennewick Man's lower limbs. Numbers in boldface exceed leg and δ means.

Element	Origin/insertion	L. score	L. Area[1]	R. score	R. Area	δ MSM R-L	δ Area R-L	% δ area
Illium	rectus femoris	2	...	2	...	0	...	...
	sartorius	...	...	2	...	...	...	...
Ischium	adductor magnus	3	...	...	...	...	...	...
Femur	gluteus medius	2	...	2	144	0	...	...
	gluteus minimus	2	630	2.5	828	**0.5**	198	**31**
	iliopsoas	2.5	322	2.5	375	0	53	16
	iliofemoral ligament	2.5	182	2	200	-0.5	18	10
	gluteus maximus	3	...	2	...	-1	...	...
	linea aspera (vastus medialis and vastus lateralis)	3	...	2	...	-1	...	...
	adductor magnus	1	...	1	...	0	...	...
Patella	quadriceps tendon	1.5	656	1.5	640	0	0	0
Tibia	soleus	3	625	3	...	0	...	...
	patellar ligament	3[2]	...	3	220	0	...	...
Calcaneus	achilles tendon (gastrocnemius, solis)	2	...	2	...	0	...	...
	flexor digitorum brev.	...	...	2	...	...	...	...
MEAN		2.34		2.11		0.25		19

[1]Area in square millimeters.
[2]Most of this site was damaged by sample removal, making this score uncertain.

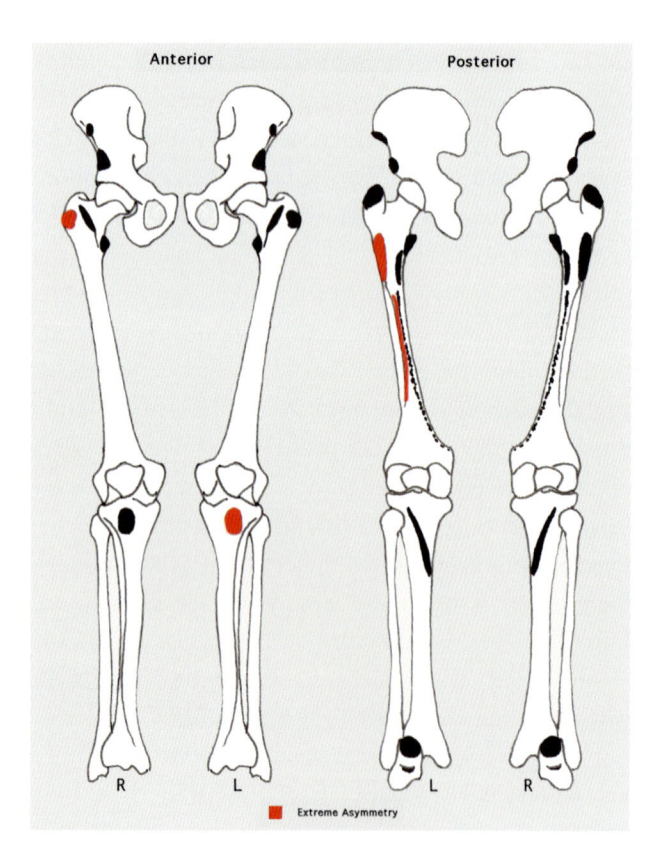

FIGURE 15.3 The distribution of scored MSM sites in the lower extremities showing a pattern of asymmetry favoring abductors of the right leg and thigh and lower leg extensors on the left side.

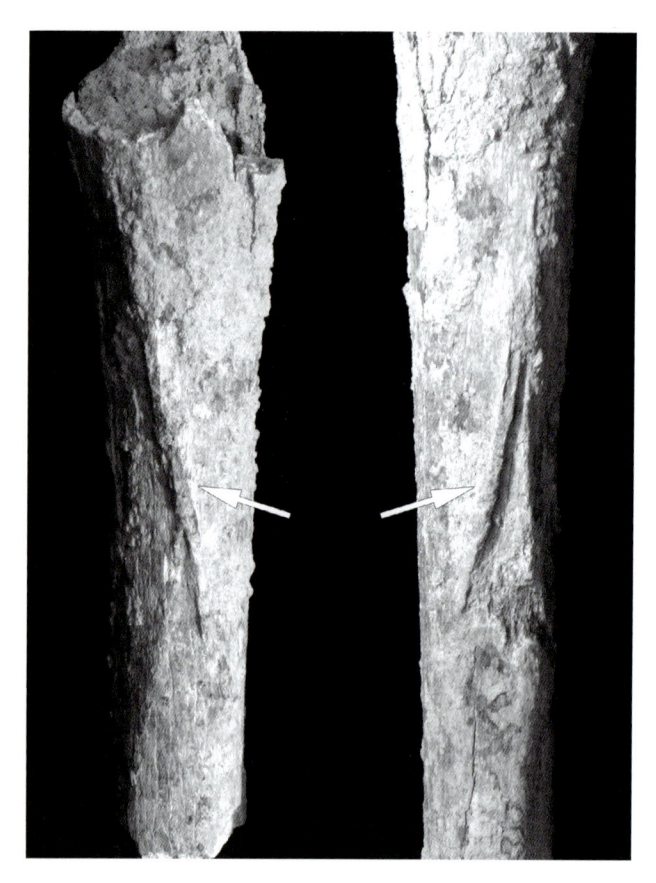

FIGURE 15.4 *Soleus* origins on Kennewick Man's proximal tibiae, illustrating the bilateral symmetry of MSM in the lower extremities.

1999) because repeated movements of and impacts to joints produce inflammation and cartilage damage that leads to lipping, pitting, and eburnation of the contact surfaces. For this analysis, all accessible elements were inspected and scored for kind and degree of defects of articular surfaces. (Elements of the hands, feet, and skull were not available because they were the principal foci of other investigators.)

Osteoarthritic defects in the Kennewick skeleton have been scored by multiple researchers (Chatters 2000; Powell and Rose 1999; Walker et al. 2000) and were documented separately during the 2006 study opportunity (Chapter 6). There is consensus that most of the osteoarthritic defects in the extremities are minor. There is, for example, no indication of "atlatl elbow" (Kennedy 1989). Moderate to severe defects occur in only two joint locations: the right shoulder and both knees.

Shoulder

The surfaces of both glenoid fossae are granular, but no lipping, pitting, or new bone formation is present on the left. In the precise center of the right glenoid fossa is a 6-millimeter diameter raised area of new bone. Inside this raised area is a pitted circle that is 4 millimeters in diameter. A broad, smooth exostosis occupies the posterior rim, separated from the articular surface of the glenoid by a narrow sulcus (Figure 15.5). Extending from the axillary border to the base of the coracoid process and parallel with the glenoid surface, this curved bar of dense bone is 5 millimeters wide mediolaterally and has an anteroposterior thickness of 4 millimeters. It is this exostosis that is of primary interest here.

Extensive extracapsular deposition of new bone along the posterior rim of the glenoid is known as thrower's exostosis, or thrower's shoulder. The defect is common in professional baseball players, who exhibit it most often in the inferior portion of the posterior rim (Andrews et al. 1995). During the cocking and acceleration phases of the throw, the humeral head rotates against the posterior glenoid rim (Pappas and Walzer 1995), which, following Wolff's Law, responds with an addition of bone mass.

Knees

Both knees exhibit almost precisely matching osteoarthritic defects, which occur with a posterior-distal extension of the articular surfaces of the tibiae. Each femur has an irregular ovoid area of remodeling, minor pitting, and sclerosis of exposed trabeculae on the posterior articular surface of the lateral condyle. The left defect measures 15 by 16 millimeters and the right, 14 by 17 millimeters. The right-side defect has a 1.5 millimeter-long osteophyte projecting from its medial edge. The tibiae are marked by similar, smaller lesions on the posterior medial aspects of their lateral condyles. The articular surface of the condyle extends down the posterior aspect of the epiphysis, distal to these defects.

These knee joint defects are unlikely to be associated with locomotion or other movements of the legs. Retroversion of the tibial head and defects of the posterior superior surface of the lateral condyle of the femur and the posterior surface of the lateral condyle of the tibia are well documented as outcomes of a lifetime of squatting (Chapter 14; Kennedy 1989:Table 1).

TRAUMA

Trauma encompasses defects in the bone that are presumed to result from injury due to extrinsic influences. A result of the interaction between an individual and the environment, trauma is closely associated with behavior and thus tends to vary with lifeways, sex, and the age of the individual at the time of death (Ortner and Putschar 1981). During the 2006 study, traumatic conditions in those portions of the skeleton that were then available were described. That inspection included observations at 10X and 45X magnification, inspection of radiographs available in the summer of 2005, and high-resolution digital color photography. During a second study in June 2006, the right ilium was inspected, and details of the projectile wound were recorded at the courtesy of the US Army Corps of Engineers and Burke Museum staff. During the July and August 1996 examination, the trauma to the right ilium was documented in both color and black-and-white photography and through computed tomography (CT). The skull was documented using high-resolution casting as well as color and black-and-white photography. For each defect attributable to trauma, its location, surface characteristics, and dimensions were recorded.

Skeletal trauma in the Kennewick remains (Chatters 2000, 2001, 2004; Powell and Rose 1999; Walker et al.

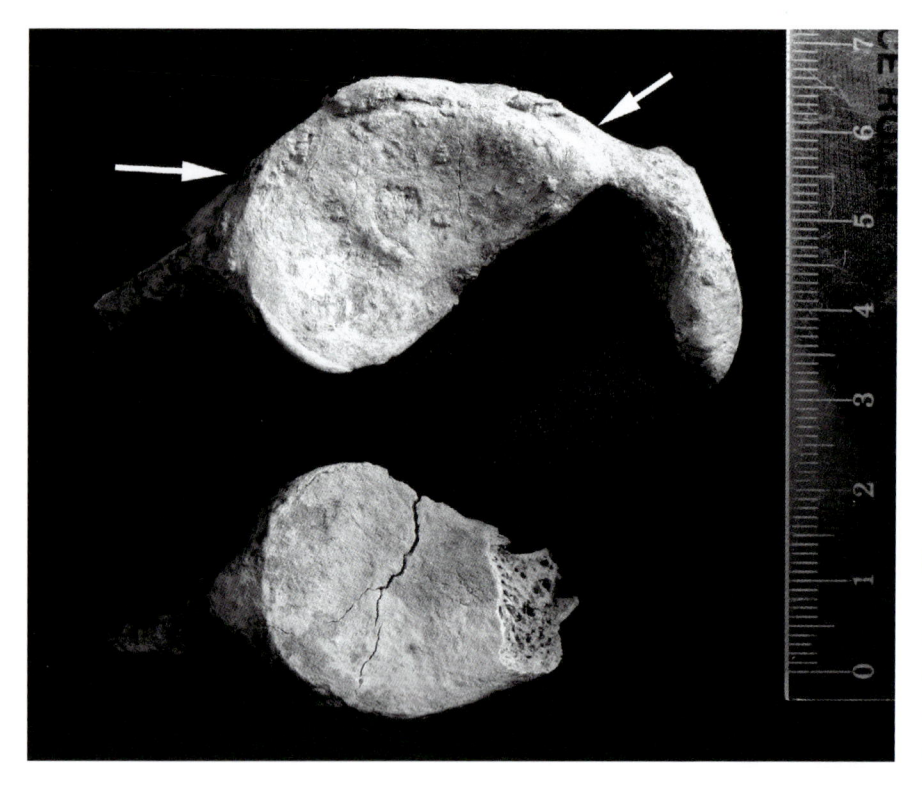

FIGURE 15.5 Left and right glenoid fossae of Kennewick Man's scapulae, showing new bone formation in the center and exostosis (between arrows) along the posterior rim of the right.

2000) is presented in Chapters 7 and 14. Traumatic defects are visible in the right ilium, cranium, right shoulder, left radius, and in multiple ribs. Previous authors have noted other defects, which, upon reexamination, do not appear to exist. Damage to the olecranon fossa of the left humerus that was previously interpreted, based on a review of photographs, to be antemortem remodeling (Walker et al. 2000) is better explained as postmortem erosion. A thickening in the shaft of the right humerus that Powell and Rose (1999) reported as a healed fracture is a profoundly hypertrophied deltoid tuberosity. Defects in the right ilium and ribs have engendered considerable interest and difference of opinion among researchers regarding their diagnosis.

Ribs

Seven rib fragments have characteristics indicative of fracture and subsequent healing. Two of these are from the left, three from the right; the rest are side-indeterminate. The left ribs are distal body fragments assigned to ribs 2 and 3; 97.I.12a(2), attributed to rib 3, exhibits a fully healed, displaced complete fracture; 97.I.12a(5), assigned to rib 2, has an irregular, thinned cross-section that may indicate past trauma, although no evidence of callus is present. It is the right and indeterminate ribs that exhibit an unusual pattern of healing.

Five fragments—97.I.12a(1), (3), (7), (8), and (9)—each have one tapered, pinched-looking edge (Figure 15.6A). The pieces most clearly fractured are numbers (1) and (9), which have been assigned to the right ribs 4 and 3, respectively. Each is a distal body fragment with a postmortem fracture at the vertebral end and a sclerosed sternal fracture end that is medially rounded in the cranio-caudal plane and tapered, or hemi-elliptical in the horizontal plane, giving it the "pinched" appearance referred to by Walker et al. (2000). As shown in Figure 15.6B and 15.6C, the tapered edge consists of continuous cortical bone; there is no seam or evidence of trabeculae that must exist if the cortical surfaces were simply crushed together. The bone surface is uniformly dense and the fragments appear uniformly opaque in radiographs, although the presence of carbonate-cemented sediment in the trabecular openings probably contributes to this apparent normalcy. The inferior corner of each of these healed edges is marked by a spherical opening, which may be a cloaca, but some postmortem damage obscures its original form. The three other defects occur in fragments from

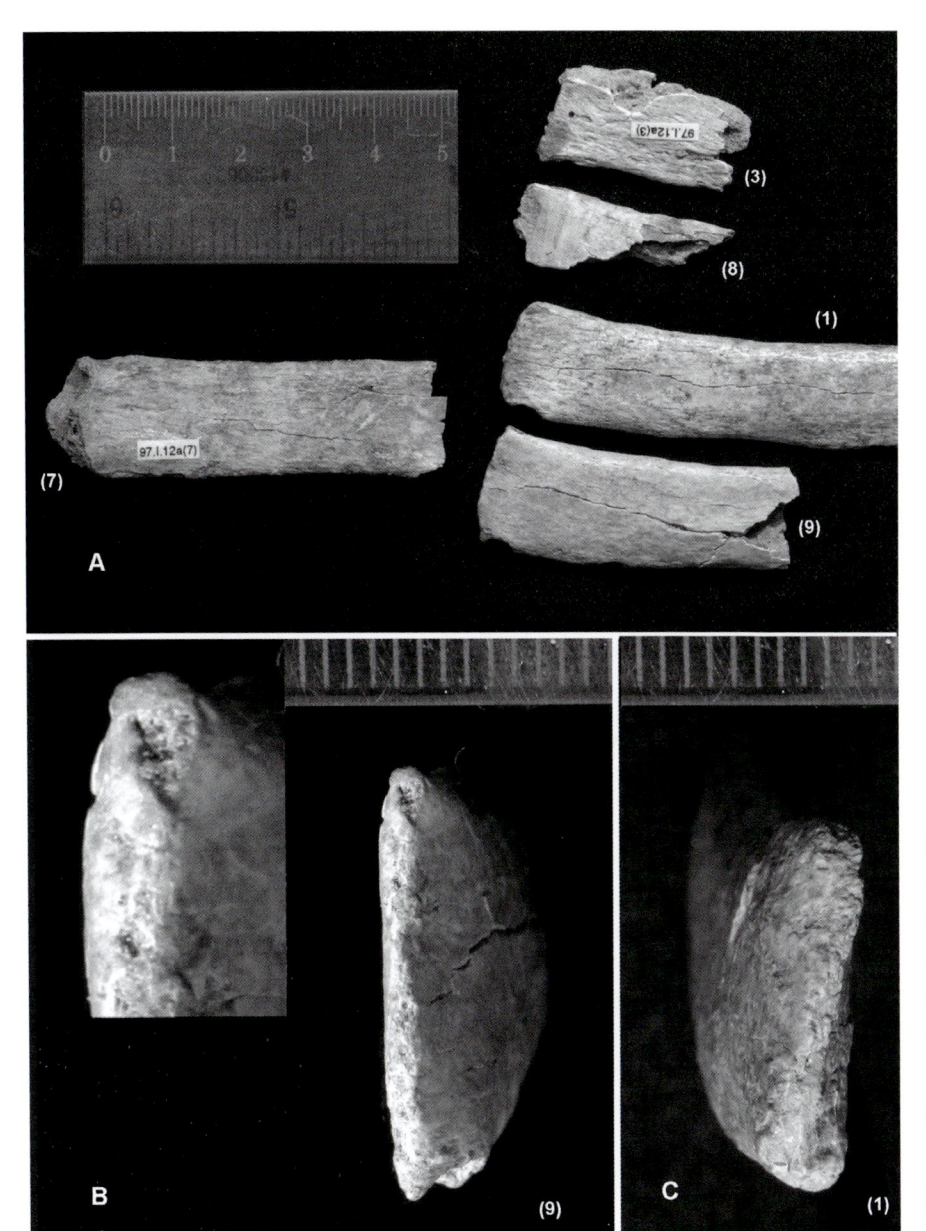

FIGURE 15.6 Kennewick Man's right rib fragments showing fracture defects in visceral (A) and edge-on views (B, C). The image at the left in 6B is an enlargement showing continuous dense bone along the healed fracture edge. Fragments are from catalog group 97.1.12a; numbers in parentheses are the individual inventory designations.

the upper right hemithorax, but they have been only tentatively assigned to specific ribs. Fragment (8), likely from rib 5, is remodeled the same way as fragments (1) and (9), differing only in that its visceral surface is beveled ventrally rather than having a symmetrical taper. Fragment (3), tentatively attributed to rib 7, is tapered, like (1) and (9), but its thinned edge has been broken postmortem. Finally, fragment (7) is the sternal end of a right rib thought to be rib 2. The vertebral end of this fragment has been beveled ventrally, like (8), but its thinned edge has also been broken postmortem. Thus, there is evidence of at least six fractured ribs, all but one in the upper hemithorax (five right, one left), but the fracture surfaces of at least three on the right side never rejoined.

The defects in the ribs have been interpreted differently by each observer. Chatters (2000) noted up to 7 fractures, with evidence of pseudoarthrosis in at least two ribs in the series 3 through 6. Powell and Rose (1999) saw only six fractured fragments but did see four of the tapered right-rib segments (1, 3, 7, 9) as long-healed pseudoarthrotic fractures to ribs 6 and 7. Walker et al. (2000), failing to see exuberant new bone or any sign of callus formation, concluded that no fractures or pseudo-

James C. Chatters

arthroses existed in the right ribs and sought to explain the phenomenon as a postmortem "pinching."

The defects in the right ribs can be explained as a form of nonunion fracture. Fracture edges fail to unite when motion, a large interfragment gap, or loss of blood supply from vascular damage or infection prevent fragment apposition and/or osteogenesis. Nonunions take three forms, any of which may be surrounded by a fluid-filled cavity with a synovial membrane, known as a pseudo-arthrosis (Rodriguez-Mechan and Forriol 2004). Hypertrophic nonunions occur in well-vascularized bone surrounded by tissues with osteogenic potential. They exhibit exuberant callus formation that takes a characteristic "elephant foot" or "horse hoof" form on the fracture edges (Canale and Beaty 2008; Mears and Kitchen 2007). Oligotropic nonunions are also vascular, occurring in cases where major displacement or large movements prevent active osteogenesis (Chen et al. 2002). Callus is absent, so fragments largely retain their original form (Canale and Beaty 2008; Mears and Kitchen 2007). Such fractures may become atrophic. Atrophic nonunion occurs when there is a large gap between fragments, poor blood flow due to associated soft tissue damage, or necrosis of an intervening fragment. The intervening space is filled by tissue without osteogenic potential. In this case, fragment ends become osteopenic or osteoporotic and take on a tapered appearance (Mears and Kitchen 2007). Ultimately all three types heal (Rodriguez-Mechan and Forriol 2004), and the surfaces of even the atrophic nonunions become sclerosed (Wheeless 2009). The tapered appearance and well-sclerosed surfaces of the fractured ends of Kennewick Man's right ribs are consistent with a well-healed atrophic nonunion. Atrophic nonunion is more common in fractures caused by high-energy impact, when there is considerable damage to the soft tissues and blood vessels, and comminuted free fragments occur (Rodriguez-Mechan and Forriol 2004).

Nonunion of multiple rib fractures, as seen in Kennewick Man, is rare, but a handful of cases are described in the medical literature (Cacchione et al. 2000; Leavitt 1942; Ng et al. 2001; Slater et al. 2001; Zuidema 2007). In most cases, the fractured ends exhibit the characteristic "elephant foot" or "horse hoof" masses (Ng et al. 2001: Figures 1, 2; Zuidema 2007: Figure 1). This was the form Walker et al. (2000) expected to see, along with a frayed, hinged fracture edge, rather than the nearly straight-edged break the Kennewick Man's ribs exhibit.

While multiple rib nonunions are rare, multiple atrophic nonunions are even rarer. Leavitt (1942) reports the case of a man who had suffered a fall, fracturing multiple ribs. One year after the injury, the 8th through 10th ribs had not reunited. A radiograph of the injury (Leavitt 1942: Figure 1) shows three ribs with right angle fractures having smooth, rather than frayed, hinged-looking edges. They appear virtually identical with those observed in the Kennewick skeleton. The bright radiographic images, and Leavitt's description of having to remove periosteum and split the fractured ends with an osteotome in order to insert grafts, indicate the fractured ends had sclerosed. The cause of this nonunion was continued separation of the fractured bone ends. During surgery, Leavitt observed a .5 to .75 inch excursion between the nonunited ends. He attributes this excursion and the resulting nonunion to an adhesive mass between the posterior diaphragm and the thoracic wall. This adhesion caused the posterior rib segments to be pulled away from their mates with each breath. A similar process, both in the initial injury and the mechanics of nonunion may account for the characteristics of the rib fractures in the Kennewick skeleton.

Right Ilium

The 54-millimeter long, 20-millimeter wide basal portion of a willow-leaf-shaped, straight-based, serrated-edged, andesite projectile point is embedded between the cortical tables of the iliac fossa, near the sacroiliac joint (Figure 7.32). The point is oriented parallel to the cortical surfaces, with the blade end medial, and the basal end lateral and slightly superior to the horizontal plane. Contrary to the conclusions of Fagan (1999) and Powell and Rose (1999), it did not enter from the rear, but rather from the right front. Bone around the entry wound and the enclosed stone projectile shows evidence of an initial phase of necrosis and resorption, with subsequent redeposition of cortical bone over what would have been exposed trabeculae.

Radiographic CT cross-sections indicate that, at the time of death, a void remained around the stone, leaving a gap of up to 3.4 millimeters between the projectile point and the new bone, and creating windows in both surfaces of the ilium through which the stone can be seen. The lateral window measures 17 millimeters anterior-posterior by 16 millimeters superior-inferior and has a 2-millimeter wide band of new cortical bone along its medial rim. New bone also appears to have been deposited around the projectile point on this surface, as indicated by the presence of a scalloped edge along the superior rim of the window that closely matches the adjacent serrations in the edge of the projectile blade. The visceral window is an oval measuring 37 millimeters in length by 12 millimeters superior-inferior. Medial to this opening is a concave, roughly crescent-shaped depression that extends 14 millimeters medial, 8 millimeters superior, and 12 millimeters inferior to the window's medial end. This area is composed of dense cortical bone marked by low, concentric ridges. CT radiographic cross-sections of the wound show small, roughly circular voids surrounded by reactive bone near the superior and inferior edges of the blade as well as sclerosis of the walls of the larger lesion.

Additional defects occur around the margins of the wound on the visceral surface of the ilium. Irregular exostoses occur medial and inferior to the lesion, adjacent to the sacroiliac joint. They may represent a partially remodeled area of involucrum. A broad concavity with a steep medial wall extends from the superior rim of the visceral window to the iliac crest, which thins to approximately one-third its normal thickness and is marked by dense callus. This appears to be the track of a cloaca (fistula) that at one time drained the sinus surrounding the projectile point.

The defects in the ilium had fully healed. The presence of a walled-off sinus between stone and bone and a large cloaca that drained the sinus along the rim of the iliac crest does not contradict this conclusion. The overall pattern seen in this defect is indicative of traumatically induced chronic suppurative osteomyelitis, in which the suppurative process is walled off by new bone and fibrous tissue; it is characterized by a chronically, or at least episodically draining sinus. Before antibiotics were developed, this condition was inevitable in such a wound (Waldvogel et al. 1971; Wilensky 1934).

Cranium

The frontal bone exhibits a 6-millimeter diameter, and 0.6 millimeter deep, circular depression in its left anterior aspect centered 40 millimeters above the superior rim of the orbit and 22 millimeters from the midsagittal plane. Although no fracture lines can be seen, the smooth appearance of the depression indicates a well-healed depressed fracture (Figure 7.3). The depression is in the center of a 2-millimeter-wide circle of surface irregularities, suggesting that the injury may have been associated with a secondary infection of the scalp. Injuries similar to this one have been reported for three of the five other males who are associated with the Western Stemmed Tradition in the Plateau and Great Basin. Stick Man (Chatters et al. 2000), the Spirit Cave mummy (Jantz and Owsley 1997), and Marmes III (Krantz 1979) also have minor fractures to the frontal. Stick Man's injury is on the right side, the others all on the left. Although these injuries may be due to interpersonal conflict, an occupational explanation is also possible.

Right Shoulder

A small conchoidal-shaped defect is present in the thickened posterior margin of the glenoid (Figure 15.7). This lesion is 18 millimeters wide in a cranio-caudal direction, 9 millimeters long in a mediolateral direction, and 1 to 2 millimeters deep. In lithic terminology, this lesion is a negative flake scar (Crabtree 1982). The negative bulb of percussion, the point at which force was applied, is oriented toward the glenoid surface, indicating that force came from anteriorly or anterolaterally. This defect was previously interpreted as a perimortem injury because the black and white photographs then available were not detailed enough to display bone remodeling (Chatters 2000, 2004). Microscopic examination at 45X revealed a sclerosed surface lacking any sharp edges (Figure 15.7, inset). The surface of the defect is smooth and appears polished. The exposed trabeculae are thickened and their broken edges rounded. The fractured edges of the scapular fragment on which this defect occurs are clear and sharp, indicating that postexposure erosion cannot account for the smoothing of the defect surface. The trauma thus appears to have occurred antemortem, but the chip that was broken away from the glenoid exostosis did not reunite with it.

Fracture of the posterior glenoid rim is uncommon (Rowe 1988), as it results from an at least partial dislocation of the gleno-humeral joint (Moseley 1969). Such a fracture has one of three causes: a fall on an outstretched

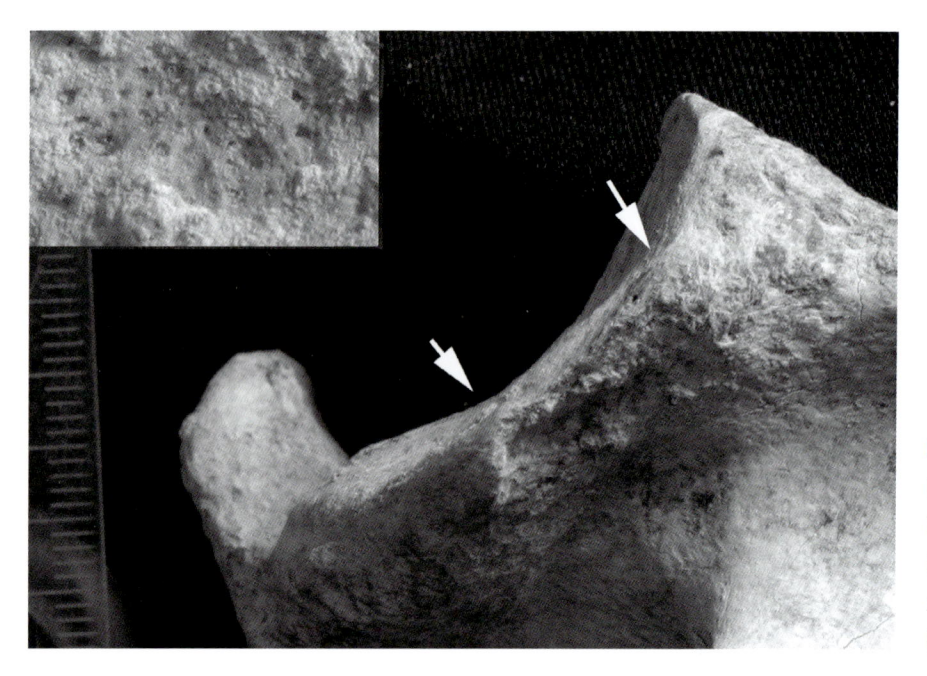

FIGURE 15.7 The fracture defect in the exostosis of Kennewick Man's right posterior glenoid (between arrows), showing sclerosis of the fracture surface (enlargement at upper left).

arm in forward flexion and internal rotation, adduction and forceful internal rotation of the humerus, or a direct blow to the anterior aspect of the shoulder. The failure of the chip to reunite with the glenoid is another example of oligotrophic or atrophic nonunion. As in any fracture, reattachment of the chip would have required immobilization, contact between the fracture surfaces, and adequate blood flow for proper healing. Blood flow to the chip of new bone from the glenoid exostosis would have been severed with the fracture and remained that way if contact between the fragments were not maintained, leading to necrosis.

Left Radius

A healed fracture is present in the head of the left radius. A chip of bone 12 millimeters wide and 5 millimeters thick had broken away from the posterior aspect of the articular circumference. The injury underwent considerable remodeling, and irregular new bone now occupies the area. A narrow crack extends laterally from the distal margin of the radial fracture along the epiphyseal contact.

Like the injury to the shoulder, radial head fractures of this kind usually result from a fall on an outstretched hand with the elbow partially flexed and pronated, which brings the posterolateral aspect of the radial head into contact with the capitulum of the humerus (Morey 2000). The crack along what appears to be the epiphyseal contact might be taken to mean that the epiphyses had only recently been closed when the injury occurred. However, radial head fracture often results in crushing of the proximal edge of the break, so the pattern of the fracture is not necessarily an indicator of the age when the injury occurred (Dr. David Mohler, personal communication 2000). Such crushing could explain the irregular pattern of healing seen in this defect.

INTERPRETATION

MSM analysis of the Kennewick skeleton found unilateral right-side hypertrophy in the attachment surfaces for *pronator quadratus, pronator teres, supinator, brachialis, deltoideus, pectoralis major*, the common flexors, and the costoclavicular ligament. In the legs, *gluteus minimus* was more developed on the right, while *gluteus maximus*, and *vastus medialis* and *vastus lateralis* of the quadriceps group were more developed on the left. These patterns are closely consistent with the forceful overhand motion of a javelin thrower or baseball pitcher; both drive laterally off their dominant leg and stop their torsos' forward motion with the extensors of the opposite leg as forceful rotation of the shoulder whips the arm and hand forward. Kennewick Man, it appears, spent a large proportion of his time throwing either a spear or an atlatl in the manner depicted in Figure 7.17. Much of that activity was probably practice to maintain the accuracy needed for success in the hunt. Other activities, including stone tool production and perhaps dip netting or poling a boat resulted in less well-developed bilateral musculature in the arms.

Many of the pathological defects in the skeleton are related in some way to the occupational emphasis on throwing. The exostosis on the posterior margin of the glenoid of the right scapula results from the rotation of the humeral head against the posterior surface of the fossa, a condition known as thrower's shoulder. When a blow to the shoulder joint drove a large chip from this exostosis, it failed to heal, probably in part because of continued practice of the throwing action. The manner in which the ribs failed to heal can also be attributed to this process.

Multiple ribs were fractured on both sides of the upper hemithorax, including at least the left second and perhaps third and the right third and fourth. A single sternal-end fragment from the right side broke approximately 55 millimeters from the costochondral junction. A review of other cases of rib nonunion indicates that ribs typically fracture in alignment, so all of the right fractures were probably in a similar location. This puts the breaks medial to the origin of *serratus anterior*, the large muscle that inserts in the vertebral border of the scapula and abducts it (Marieb 1995). *Serratus anterior* is in contraction during all phases of the overhand throw, helping to elevate the shoulder during the cocking phase, helping to stabilize the scapula and allow its distal tip to freely rotate during acceleration, and working with other scapular muscles to limit both rotation and lateral migration of the scapula during the deceleration phase (Perry and Glousman 1990). Thus, *serratus anterior* may have served the same purpose in Kennewick Man as the adhesive mass in the posterior diaphragm did in Leavitt's (1942) patient. A well-toned *serratus anterior* would have contracted when the fractures occurred, retracting the rib ends on the vertebral side of the break and fostering the conditions necessary for atrophic nonunion. Continued, repeated contraction of *serratus anterior* as the man hunted for subsistence would have maintained this condition.

The projectile wound in the ilium is also intimately related to the overhand throw. The 54-millimeter-long, 20-millimeter-wide, 5-millimeter-thick base and midsection of a serrated-edged projectile point is all but completely enclosed, having penetrated at or near the iliac crest *after* its tip had sheared away. This deep impact is indicative of high velocity and strongly indicates use of an atlatl. Therefore, although scientists are unable to distinguish atlatl use from use of a javelin-like spear, the depth of the pelvic wound suggests Kennewick Man was at least familiar with the atlatl concept.

The cranial injury, if independent of the other fractures, could also be occupationally related. Similar small depressed fractures occur in the frontal bones of most of the males associated with the Western Stemmed Tradition and are the only cranial injuries evident in that group. Western Stemmed Tradition people of the Pacific Northwest (where all but one of the injured men lived) used small, longitudinally girdled plumb-bob sized implements identified as bola stones. If these implements were indeed parts of bolas and not fishing weights, their use could account for the pattern of injury. A bola consists of two or three stone weights tied together by equal lengths of cord. It is thrown by holding one stone in the hand while whirling the other(s) above the head to gain kinetic energy before releasing. Shoulder and elbow structure allow this whirling action to go efficiently in only one direction—clockwise with the right hand and counterclockwise with the left. An error in the acceleration of the bola can give the user a sharp blow to the head. The direction of rotation means that a left-handed user would strike his right frontal, a right-handed user his left. Of the four known fractures, three occur in the left frontal, one in the right, which is close to the proportion of right- and left-handedness in humans. Despite this intriguing relationship between archaeological and skeletal data, interpersonal conflict remains an equally plausible explanation for the regional pattern of cranial injury.

The manner in which the nonprojectile injuries occurred, their locations, and the patterns of healing are indicative of how and when during the man's life they may have taken place. The fractures to the left radius and right scapula result from falls forward. The left radius probably fractured when Kennewick Man attempted to stop a fall with his outstretched hand. The damage to the right scapula apparently occurred when he fell on his inwardly bent right forearm. In both cases, Kennewick Man appears to have been attempting to break a fall with outstretched arms. It is also possible that the shoulder fracture resulted from a direct impact to the shoulder itself. The rib fractures, distributed as they are to either side of the sternum, resulted from a massive force to the upper midchest. Fractures of the upper three ribs are uncommon and usually occur in the context of severe trauma (Groskin 1991); most patients reported in the medical literature with multiple rib-fracture nonunions received their injuries in automobile accidents, farming accidents, or falls (Landercasper et al. 1984). The cranial injury is also frontal. It is entirely possible that this suite of injuries occurred in a single accident, such as a fall from a cliff. If that were the case, it must have taken place late in the man's life, not in his youth, as Powell and Rose (1999) suggest. The scapula injury could have occurred only after the glenoid exostosis had fully formed; the exostosis shows no sign of developing further after injury. Likewise, nonunion of the right ribs, which may have been caused in part by contraction of a muscle central to the overhand throwing action that was essential to Kennewick Man's subsistence strategy, would also have limited the effectiveness of *serratus anterior* to that very action. It is unlikely the high degree of MSM that is evident in the right arm and shoulder could have developed under such limitations.

This same reasoning can be applied to the projectile wound, which limited the effectiveness of *iliacus*, an important muscle for locomotion. MSM results and the distribution of osteoarthritis in the legs show no evidence that Kennewick Man's gait was affected. Thus, it appears more likely that this injury also occurred late in life, after MSMs and joint damage were well advanced.

CONCLUSION

Analysis of occupational stress markers sought to discern traces of Kennewick Man's subsistence behaviors and whether the leg bones bore evidence that the projectile wound in his pelvis affected his gait. Results show that the principal activity in his life was throwing a projectile, likely with the aid of an atlatl. Although there is secondary evidence of an activity similar to dipnetting, Kennewick Man was first and foremost a big game hunter. The throwing action required by that occupation resulted in bony changes to his right shoulder and contributed to nonunion of fractures in the shoulder and right upper thorax. There was no discernible effect of the projectile wound on the muscle use or development of osteoarthritis in the legs.

Of the traumatic defects in the skeleton, all but the projectile wound can be attributed to one or more forward falls, at least one of which led to high energy impact. If a single event accounts for the shoulder and rib injuries, both of which resulted from violent trauma and would have affected performance of his principal occupation, then the event occurred later in life. Lack of evidence that the pelvic wound affected gait indicates it, too, was more likely received in the man's later years.

REFERENCES

al-Omaoui, I., S. Jimenez-Brobeil, and P. du Souich. 2004. Markers of activity patterns in some populations of the Iberian Peninsula. *International Journal of Osteoarchaeology* 14(5):343–359.

Ames, Kenneth M. 2000. Chapter 2. Review of the Archaeological Data. In *Cultural Affiliation Report* [September 2000]. National Park Service, http://www.nps.gov/archeology/kennewick/ames.htm.

Andrews, James R., Laura A. Timmerman, and Kevin E. Wilk. 1995. Baseball. In *Athletic Injuries of the Shoulder*, edited by Frank A. Pettrone, 323–341. McGraw-Hill, New York.

Cacchione, Robert N., J. David Richardson, and David Seligson. 2000. Painful nonunion of multiple rib fractures managed by operative stabilization. *Journal of Trauma* 48(2):319–321.

Canale, S. Terry, and James H. Beaty, eds. 2008. *Campbell's Operative Orthopaedics*. 11th edition. Mosby Elsevier, Philadelphia.

Chamay, A., and P. Tschantz. 1972. Mechanical influences in bone remodeling. Experimental research on Wolff's law. *Journal of Biomechanics* 5(2):173–180.

Chatters, James C. 1998. Environment. In *Handbook of North American Indians,* vol. 12: *Plateau*, edited by Deward E. Walker, 29–48. Smithsonian Institution Press, Washington DC.

———. 2000. The recovery and initial analysis of an early Holocene human skeleton from Kennewick, Washington. *American Antiquity* 65(2):291–316.

———. 2001. *Ancient Encounters: Kennewick Man and the First Americans*. Simon & Schuster, New York.

———. 2004. Kennewick Man: A Paleoamerican skeleton from the Northwestern U.S. In *New Perspectives on the First Americans*, edited by Bradley T. Lepper and Robson Bonnichsen, 129–137. Texas A&M Press, College Station.

———. 2010. Patterns of death and the peopling of the Americas. In *Symposio International III: El Hombre Temprano en America*, edited by José C. Jiménez López, Carlos Serrano Sánchez, Arturo González González, and Felisa J. Aguilar Arellano, 53–74. Instituto Nacional de Antropología y Historia, Mexico City.

Chatters, James C., Steven Hackenberger, Alan Busacca, Linda S. Cummings, Richard L. Jantz, Thomas W. Stafford, and R. Ervin Taylor. 2000. A possible second early Holocene skull from eastern Washington, USA. *Current Research in the Pleistocene* 17:93–95.

Chen, Xinquan, Lewis S. Kidder, and William D. Lew. 2002. Osteogenic protein-1 induced bone formation in infected segmental rat femur. *Journal of Orthopaedic Research* 20:142–150.

Churchill, Steven E., and Alan G. Morris. 1998. Muscle marking morphology and labour intensity in prehistoric Khoisan foragers. *International Journal of Osteoarchaeology* 8:390–411.

Crabtree, Don E. 1982. *An Introduction to Flint Working*. 2nd edition. Idaho State University Museum of Natural History Occasional Paper 28. Idaho State University Museum of Natural History, Pocatello.

Dutour, O. 1986. Enthesopathies (lesions of muscular insertions) as indicators of the activities of Neolithic Saharan populations. *American Journal of Physical Anthropology* 71(2):221–224.

Eshed, Vered, Avi Gopher, Ehud Galili, and Israel Hershkovitz. 2004. Musculoskeletal stress markers in Natufian hunter-gatherers and Neolithic farmers in the Levant: The upper limb. *American Journal of Physical Anthropology* 123(4):303–315.

Fagan, John L. 1999. Analysis of the Lithic Artifact Embedded in the Columbia Park Remains. In *Report on the Non-Destructive Examination, Description, and Analysis of the Human Remains from Columbia Park, Kennewick, Washington* [October 1999]. National Park Service, http://www.nps.gov/archeology/kennewick/fagan.htm.

Groskin, Stewart A. 1991. *Radiological, Clinical, and Biomechanical Aspects of Chest Trauma*. Springer-Verlag, New York.

Harten, Louise B. 1980. The osteology of the human skeletal material from the Braden Site, 10-WN-117, in Western Idaho. In *Anthropological Papers in Memory of Earl H. Swanson, Jr.*, edited by Lucille B. Harten, Claude N. Warren, and Donald R. Tuohy, 130–148. Idaho Museum of Natural History, Pocatello.

Hawkey, Diane E. 1998. Disability, compassion, and the skeletal record: Using musculoskeletal stress markers (MSM) to construct an osteobiography from early New Mexico. *International Journal of Osteoarchaeology* 8(5):326–340.

Hawkey, Diane E., and Charles F. Merbs 1995. Activity-induced musculoskeletal stress markers (MSM) and subsistence strategy changes among ancient Hudson's Bay Eskimos. *International Journal of Osteoarchaeology* 5(4):324–338.

İşcan, Mehmet Yaşar, and Kenneth A. R. Kennedy, eds. 1989. *Reconstruction of Life from the Skeleton*. Alan R. Liss, New York.

Jantz, Richard L., and Douglas W. Owsley. 1997. Pathology, taphonomy, and cranial morphometrics of the Spirit Cave Mummy. *Nevada Historical Society Quarterly* 40(1):62–84.

Jurmain, Robert 1999. *Stories from the Skeleton: Behavioral Reconstruction in Human Osteology*. Gordon and Breach, Amsterdam.

Kennedy, Kenneth A. R. 1983. Morphological variations in ulnar supinator crests and fossae as identifying markers of occupational stress. *Journal of Forensic Sciences* 28(4):871–876.

———. 1989. Skeletal markers of occupational stress. In *Reconstruction of Life from the Skeleton*, edited by Mehmet Yaşar İşcan and Kenneth A. R. Kennedy, 129–160. Alan R. Liss, New York.

Kennedy, Kenneth A. R., Thomas Plumber, and John Chiment. 1986. Identification of the eminent dead: Penpi, a scribe of Ancient Egypt. In *Forensic Osteology: Advances in the Identification of Human Remains*, edited by Kathleen J. Reichs, 290–307. Charles C. Thomas, Springfield, IL.

Krantz, Grover S. 1979. Oldest human remains from the Marmes Site. *Northwest Anthropological Research Notes* 13(2):159-174.

Landercasper, J., T. Cogbill, and L. Lindesmith. 1984. Long-term disability after flail chest injury. *Journal of Trauma* 24(5):401–414.

Leavitt, Darrell G. 1942. Non-union of three ribs. *The Journal of Bone & Joint Surgery* 24(4):932–936.

Lieverse, Angela R., Vladimir I. Bazaliiskii, Olga Ivanova Goriunova, and Andrej W. Weber. 2009. Upper limb musculoskeletal stress markers among middle Holocene foragers of Siberia's Cis-Baikal region. *American Journal of Physical Anthropology* 138(4):458–472.

Marieb, Elaine N. 1995. *Human Anatomy and Physiology.* 3rd edition. Benjamin Cummings, Red City, CA.

McGinnis, Peter M. 2005. *Biomechanics of Sport and Exercise.* 2nd edition. Human Kinetics, Champaign, IL.

Mears, Simon C., and Melanie Kitchen. 2007. Nonunion of fractures. In *The 5-Minute Orthopaedic Consult*, edited by Frank J. Frassica, Paul D. Sponseller, and John H. Wilkens, 276–277. Lippincott Williams & Wilkins, Philadelphia.

Molnar, Petra. 2006. Tracing prehistoric activities: Musculoskeletal stress marker analysis of a stone age population on the Island of Gotland in the Baltic Sea. *American Journal of Physical Anthropology* 129(1):12–23.

Morey, Bernard F. 2000. Radial head fracture. In *The Elbow and Its Disorders*. 3rd edition. Edited by Bernard F. Morey, 383–401. W. B. Saunders Co., Philadelphia.

Mosley, Herbert. F. 1969. *Shoulder Lesions.* The Williams and Wilkins Company, Baltimore.

Nagy, B. L. B. 1999. Bioarchaeological evidence for atl-atl use in prehistory. *American Journal of Physical Anthropology* 108(S28):208.

———. 1998. Age, Activity, and Musculoskeletal Stress Markers. *American Journal of Physical Anthropology* 105(S26):168–169.

Ng, A. B. Y., P. V. Giannoudis, Q. Bismal, A. F. Hinsche, and R. M. Smith. 2001. Operative stabilisation of painful nonunited multiple rib fractures. *Injury* 32(8):637–639.

Ortner, Donald J., and Walter G. J. Putschar 1981. *Identification of Pathological Conditions in Human Skeletal Remains.* Smithsonian Institution Press, Washington, DC.

Pappas, Arthur M., and Janet Walzer. 1995. *Upper Extremity Injuries in the Athlete.* Churchill Livingston, Edinborough.

Perry, Jacquelin, and Ronald Glousman. 1990. Biomechanics of throwing. In *The Upper Extremity in Sports Medicine*, edited by James A. Nicholas, Elliott B. Hershman, and Martin A. Posner, 727–751. C. V. Mosby, St Louis, MO.

Peterson, Jane. 1998. The Natufian hunting conundrum: Spears, atlatls, or bows? Musculoskeletal and armature evidence. *International Journal of Osteoarchaeology* 8(5):378–389.

Powell, Joseph F., and Jerome C. Rose. 1999. Chapter 2. Report on the Osteological Assessment of the "Kennewick Man" Skeleton (CENWW.97. Kennewick). In *Report on the Non-Destructive Examination, Description, and Analysis of the Human Remains from Columbia Park, Kennewick, Washington* [October 1999]. National Park Service, http://www.nps.gov/archeology/kennewick/powell_rose.htm.

Rodriguez-Merchan, E. Carlos, and Francisco Forriol. 2004. Nonunion: General principles and experimental data. *Clinical Orthopaedics and Related Research* 419:4–12.

Rowe, Carter R. 1988. Dislocations of the shoulder. In *The Shoulder*, edited by Carter R. Rowe, 165–291. Churchill Livingstone, New York.

Ruff, Christopher, Brigitte Holt, and Erik Trinkhaus. 2006. Who's afraid of the big bad Wolff: "Wolff's law" and bone functional adaptation. *American Journal of Physical Anthropology* 129(4):484–498.

Slater, Mathew S., John C. Mayberry, and Donbald D. Trunkey. 2001. Operative stabilization of a flail chest six years after injury. *Annals of Thoracic Surgery* 72(2):600–601.

Steen, Susan L., and Robert W. Lane. 1998. Evaluation of habitual activities among two Alaskan Eskimo populations based on musculoskeletal stress markers. *International Journal of Osteoarchaeology* 8(5):341–353.

Stirland, A. J. 1998. Musculoskeletal evidence for activity: Problems of evaluation. *International Journal of Osteoarchaeology* 8(5):354–362.

Taylor, R. E., Donna L. Kirner, John R. Southon, and James C. Chatters. 1998. Radiocarbon dates of Kennewick Man. *Science* 280(5367):1171–1172.

Waldvogel, Francis A., Gerald Medoff, and Morton N. Schwartz. 1971. *Osteomyelitis: Clinical Features, Therapeutic Considerations, and Unusual Aspects.* Charles Thomas, Springfield, IL.

Walker, Phillip L., Clark Spencer Larsen, and Joseph F. Powell. 2000. Chapter 5. Final Report on the Physical Examination and Taphonomic Assessment of the Kennewick Human Remains. (CENW.97.Kennewick) to Assist with DNA Sample Selection. In *Report on the DNA Testing Results of the Kennewick Human Remains from Columbia Park, Kennewick, Washington* [September 2000]. National Park Service, http://www.nps.gov/archeology/kennewick/walker.htm.

Warfel, John H. 1974. *The Extremities.* 4th edition. Lea and Febiger, Philadelphia.

Weiss, Elizabeth. 2004. Understanding muscle markers: Lower limbs. *American Journal of Physical Anthropology* 125(3):232–238.

———. 2007. Muscle markers revisited: Activity pattern reconstruction with controls in central California Amerind Population. *American Journal of Physical Anthropology* 133(3):931–940.

Wheeless, Clifford R. 2009 *Wheeless' Textbook of Orthopaedics*. http://www.wheelessonline.com/ortho.

Wilensky, Abraham Owen. 1934. *Osteomyelitis, Its Pathogenesis, Symptomology, and Treatment*. The MacMillan Company, New York.

Wilzack, Cynthia A. 1998. Consideration of sexual dimorphism, age, and asymmetry in quantitative measurements of muscle insertion sites. *International Journal of Osteoarchaeology* 8(5):311–325.

Wolff, Julius. 1892. *Das Gesetz der Transformation der Knocken*. A. Hirchwild, Berlin.

Woo, S. L., S. C. Kuei, D. Amiel, M. A. Gomez, W. C. Hayes, F. C. White, and W.H. Akeson. 1981. The effect of prolonged physical training on the properties of long bone: A study of Wolff's Law. *Journal of Bone and Joint Surgery* 63(5):780–787.

Zuidema, W. P. 2007. Angle stable plate osteosynthesis for nonunion of rib fractures. *The Internet Journal of Orthopedic Surgery* 6(2). doi: 10.5580/9f9.

Zumwalt, A. C., C. B. Ruff, and C. A. Witczak. 2000. Primate muscle insertions: What does size tell you? *American Journal of Physical Anthropology*, 111(S30):331.

Stable Isotopic Evidence for Diet and Origin

Henry P. Schwarcz, Thomas W. Stafford Jr., Martin Knyf, Brian Chisholm, Fred J. Longstaffe, James C. Chatters, and Douglas W. Owsley

Multiple types of evidence are needed in order to understand Kennewick Man's history and origin. Stable isotopic composition of the skeleton provides information about his diet and geographical residence. In general, it is known that the ratios of the stable isotopes of carbon ($^{13}C/^{12}C$) and nitrogen ($^{15}N/^{14}N$) in the collagen of bone, as well as the $^{13}C/^{12}C$ of the carbonate in bone mineral (hydroxyapatite), reflect the corresponding isotope ratios in the diet (DeNiro and Epstein 1978, 1980; Katzenberg 2000; Schwarcz and Schoeninger 1991) weighted according to the amounts consumed. Reconstruction of paleodiet is based on the significant variation in isotope ratios among the foods consumed by humans.

Some initial information about Kennewick Man's diet was gleaned from the $^{13}C/^{12}C$ ratio measured while determining his radiocarbon age (Taylor et al. 1998). Collagen extracted from a bone sample gave a value of $\delta^{13}C = -14.9$ per mil (‰). Taylor et al. inferred that Kennewick Man's diet was approximately 70 percent marine foods, which resulted in correcting his radiocarbon age to account for the reservoir effect (nonzero age of marine carbon).

Analysis of the mineral component (hydroxyapatite) of bone can also provide important information about Kennewick Man's life. The $^{13}C/^{12}C$ ratio of the carbonate ions (CO_3^{2-}) in the crystal structure reflects the corresponding isotope ratio in the total diet (Schwarcz 2000). Also the $^{18}O/^{16}O$ ratio of hydroxyapatite, determined by analysis of either its constituent phosphate (PO_4^{3-}) or carbonate ions, would reflect the $^{18}O/^{16}O$ of the drinking water consumed through the last part of his life (Longinelli 1984). These data can be used to learn something about the geographic location he inhabited.

METHODS

Carbon and nitrogen isotope analyses were carried out on collagen that had been previously extracted for radiocarbon analysis. Samples of gelatin extracted from the collagen were analyzed, all coming from the cnemial crest of the left tibia. (Collagen was extracted using procedures described in Chapter 3) The freeze-dried samples were analyzed using a Costech ECS 4010 elemental analyzer connected to a ThermoFinnigan DeltaPlus XP stable isotope ratio mass spectrometer. The separated CO_2 and N_2 were carried in a helium stream to the mass spectrometer via a Conflo III coupling. Carbon and nitrogen were analyzed in the same sample. The results are reported in δ notation: $\delta(‰) = [(Rsample/Rstandard) - 1]$. For carbon, R is $^{13}C/^{12}C$ and the standard is VPDB, while for nitrogen, R is $^{15}N/^{14}N$ and the standard is atmospheric nitrogen (AIR). Precision of both δ values was ± 0.2‰. The analyses were compared to international standards, including NBS 21 (graphite) for carbon and USGS 40 (glutamic acid) for nitrogen and carbon.

The $^{13}C/^{12}C$ and $^{18}O/^{16}O$ ratios of the CO_3 component of the bone mineral were determined from a sample of metatarsal bone. A common problem encountered in carbonate analyses of buried bone is that the samples may contain calcium carbonate (calcite) that has grown on the surface and in the pores of the bone while buried. This contaminates the bone with an extraneous CO_3 component that cannot be distinguished from that in the bone mineral structure. In the case of Kennewick Man's samples, it was visually apparent that calcium carbonate concretions were attached to the bone and that calcite had precipitated in medullary cavities and within cancellous tissues. Most of these carbonates were removed physically and by treating the bone with 0.1 M acetic acid buffered with sodium acetate for four hours at room temperature (Koch et al. 1997).

The acid-treated bone was crushed to a fine powder and aliquots of two milligrams were reacted with phosphoric acid at 95°C in an Autocarb device attached to a

VG SIRA mass spectrometer. Both the $\delta^{13}C$ and $\delta^{18}O$ data are reported with respect to the VPDB standard. The precision of analyses is ± 0.1‰. The oxygen isotopic fractionation between bone CO_3 and the liberated CO_2 are assumed to be the same as for calcite reacted with phosphoric acid at the same temperature (Koch et al. 1997).

Carbon dioxide from the acid reaction of another portion of the bone was used to determine the $^{14}C/^{12}C$ ratio and the corresponding ^{14}C age. The ^{14}C analyses were carried out at the radiocarbon analysis laboratory at the University of California, Irvine.

Phosphate oxygen was isolated from a sample of 35.5 milligrams of trabecular bone from a metatarsal. There was only sufficient material for one analysis. The bone was dissolved in 15 mL of 3 M acetic acid at 22°C. Silver phosphate (Ag_3PO_4) was prepared following the procedure of Crowson et al. (1991). Approximately 24.2 milligrams of Ag_3PO_4 (a 68 percent yield from the original sample weight) were loaded into a nickel reaction vessel under a flow of dry nitrogen gas. The sample (and standards) were then heated under vacuum for two hours at 300°C, followed by reaction with bromine pentafluoride (BrF_5) at 600°C for 18 hours using an extraction line and procedures based on work by Tudge (1960), Clayton and Mayeda (1963), Crowson (1991), and Stuart-Williams and Schwarcz (1995). The released oxygen gas from the sample and standards was then quantitatively converted to CO_2 by reaction with incandescent carbon and cryogenically transferred to an evacuated gas tube for isotopic analysis using a Micromass Prism II dual-inlet, triple-collecting, stable isotope ratio mass spectrometer.

The CO_2 yield of 4.62 µmol/mg obtained for the sample is very close to the theoretical silver phosphate yield of 4.78 µmol/mg. Duplicate aliquots of the laboratory silver phosphate standard analyzed at the same time had an average CO_2 yield of 4.82 ± 0.05 µmol/mg. Likewise, concurrent duplicate analyses of the laboratory silver phosphate sample produced a $\delta^{18}O$ value of +11.2 ± 0.2‰ (VSMOW), which compares well with its accepted value of +11.2‰. A $\delta^{18}O$ value of +10.42‰ was obtained for the laboratory standard CO_2 gas during this same batch of mass spectrometric analyses, which also compares well with its accepted (VSMOW) value of +10.5‰.

For comparison of the oxygen isotopic analysis of bone, freshwater mussel shells that were found in situ at a site approximately 200 meters upstream of the Kennewick Man discovery site were analyzed. Additional samples of mussels from other localities along the Columbia River were also analyzed (Table 16.1), in order to track possible long-term changes in the $^{18}O/^{16}O$ ratio of the river through the mid-Holocene. Each shell was sectioned perpendicular to the growth layers. A powder sample was filed from the cut edge representing an average of the growth history of each shell. These were reacted with phosphoric acid at 95°C and analyzed as described above.

RESULTS

Each sample of collagen was analyzed in duplicate or triplicate for each isotope. The data are shown in Table 16.2. The average error of analysis (CV) in the replicates was 2.1 percent and 0.70 percent for $\delta^{13}C$ and $\delta^{15}N$, respectively. The C/N ratios of the samples were all close to the expected value of 3.2 (van Klinken 1999), indicating that the collagen was well preserved. The three pairs of $\delta^{13}C$ and $\delta^{15}N$ analyses agree closely but differ by more than the precision of analysis.

The sample of acid-treated metatarsal bone was analyzed in duplicate to determine $\delta^{13}C$ and $\delta^{18}O$ of the bone carbonate; those data are also presented in Table 16.2. For reference, analyses of a sample of pedogenic calcium carbonate that was associated with the bone are presented. This is likely the same composition as any carbonate precipitated inside the bone. An aliquot of the same bone was analyzed for $\delta^{18}O$ of its phosphate at the University of Western Ontario. That result is given in Table 16.2.

DISCUSSION
Diet

In general, $\delta^{13}C$ and $\delta^{15}N$ values of collagen are determined by the proportions and isotopic compositions of foods in the diet (Schwarcz 1991). In addition, different foods contain varying concentrations of protein (providing both C and N), which must be taken into account when calculating the effects of food on the isotopic composition of collagen (Phillips and Gregg 2003; Semmens and Moore 2008).

Two other factors also must be considered. First, there is a tendency for the amino acids from dietary protein to be used directly in the biosynthesis of collagen in

TABLE 16.1 Identification and ages of shells for $\delta^{18}O$ analysis.

Sample ID	Site	Sample Type	14C date (RC yr. BP)	Species	Reference
SHELL 1, 2	Kennewick Man Site	shell	6090 ± 80	Gonidea angulata	Wakeley et al. 1998
SHELL 3	450K420	bone collagen	5150 ± 40	Margaritinopsis falcata	Chatters 2003
SHELL 4	45D0373	charcoal	6580 ± 40	Margaritinopsis falcata	Chatters 2003
SHELL 5	Columbia River at Methow Rapids	shell	Modern	Anodonta kennerlyi	

TABLE 16.2 Results of isotopic analyses of Kennewick Man's bone collagen, carbonate, and phosphate and pedogenic calcium carbonate adhering to the bones.

	Bone Collagen Data						
Sample ID	$\delta^{13}C$ (‰, VPDB)	SD	$\delta^{15}N$ (‰, AIR)	SD	wt% C	wt% N	C/N
SIRA 27-07	−13.77		21.30		37.6	13.5	3.24
SIRA 27-07	−13.72		21.30		38.5	13.8	3.24
SIRA 27-07	−13.78		21.34		38.3	13.7	3.25
average	−13.76	0.03	21.31	0.02			
SIRA 27-10	−14.38		21.48		37.2	12.7	3.41
SIRA 27-10	−14.21		21.29		36.8	12.7	3.37
average	−14.29	0.04	21.38	0.1			
SIRA 27-18	−13.82		21.14		43.4	15.4	3.28
SIRA 27-18	−13.81		21.05		43.6	15.5	3.29
average	−13.81	0.01	21.09	0.05			
Global average	−13.95	0.29	21.26	0.15			

	Bone and Pedogenic Calcium Carbonate Data				
Sample ID	$\delta^{13}C$ (‰, VPDB)	$\delta^{18}O$ (‰, VPDB)	SR[1]	Catalog Number	Material
CHEM 9711	−5.92	−9.80	SR-7325	97.R.24 (MTa)	Trabecular Bone (metatarsal)
CHEM 9711	−6.11	−9.74			
average	−6.01	−9.77			
CHEM 8723	−7.16	−14.24	SR-7092	97.I.12d(13)	Pedogenic calcium carbonate
CHEM 8723	−7.14	−14.26			
CHEM 8724	−7.66	−13.99	SR-7092	97.I.12d(13)	Pedogenic calcium carbonate
CHEM 8724	−7.73	−14.10			
CHEM 9710	−7.96	−14.09	SR-7092	97.I.12d(13)	Pedogenic calcium carbonate
CHEM 9710	−7.94	−14.08			

	Bone Phosphate Data		
Sample ID	$\delta^{18}O$ (‰, VSMOW)	Yield (μmol/mg)	Accepted Value
Laboratory standard phosphate	11.00	4.87	11.2
Kennewick Man 97.R.24	5.31	4.62	
Laboratory standard phosphate	11.37	4.77	11.2
Laboratory standard carbon dioxide	10.18		10.10

[1]Stafford Research identification number.

humans. This is the so-called "routing" effect (Krueger and Sullivan 1984), which was demonstrated experimentally by Ambrose and Norr (1993). While this effect is moderated in humans with low-protein diets (Schwarcz 2000), it is probably relevant to the present study. In contrast, the $\delta^{13}C$ value of the CO_3 in bone mineral should be representative of the entire diet (Ambrose and Norr 1993; Schwarcz 2000).

Second, both the $\delta^{13}C$ and $\delta^{15}N$ values of collagen and carbonate are offset relative to those of the total diet. For collagen, the $\delta^{13}C$ offset is estimated to be 5‰ (that is, $\delta^{13}C$ [collagen] = $\delta^{13}C$ [diet] + 5‰) and arises principally because of the anomalously high proportion of ^{13}C-rich glycine in collagen. The offset for $\delta^{15}N$ is ~ 3‰ (Schwarcz and Schoeninger 1991) although higher values have been claimed (Hedges and Reynard 2007). This represents the trophic level effect, which enriches any consumer in ^{15}N relative to its diet. The $\delta^{13}C$ offset between bone-CO_3 and diet is not known precisely but is presumed to be close to 13‰ (Prowse et al. 2003).

The isotopic composition of bone collagen represents an average of the dietary intake over the 10 to 15 years prior to death (Hedges et al. 2007; Manolagas 2000). Assuming that Kennewick Man had not arrived at this site just before his death, his diet would have included some of the foods that were then abundantly available in the vicinity of the site where he was found. Local subsistence items include plant and animal foods from the terrain on either bank of the Columbia River, as well as fish and freshwater mollusks inhabiting the river. The isotopic compositions of some of these potential foods are shown on Figure 16.1. Specifically, they include: terrestrial game, terrestrial plants, salmon, white sturgeon, nonanadromous fish, mussels, and birds of prey.

A wide variety of herbivorous mammals, such as elk and deer, would have lived on the lands adjacent to the Columbia River valley. Two omnivorous mammals, black bear (*Ursus americanus*) and grizzly bear (*U. horribilis*) would also have been present. Part of the bears' diet is salmon, causing them to have anomalously high $\delta^{13}C$ values as well as being one trophic level (~ 3‰) higher in $\delta^{15}N$ than salmon. The flesh of any individual bear could range widely in $\delta^{15}N$ and $\delta^{13}C$ because their diets could have included both terrestrial and marine resources; therefore they are not shown on Figure 16.1.

Berries, roots, seeds, and leafy plants would have been available in the region.

Salmon were likely plentiful. The modern isotopic ranges of five varieties of *Oncorhynchus* are shown, taken from a synthesis of data sources compiled by Johnson and Schindler (2009). Increasing δ values along the sequence presumably correspond to increasing trophic level of the organism at maturity. However, these data are mainly for fish caught in the open ocean and may not be completely representative of the same species caught in rivers at the time of spawning. Analysis of salmon otoliths from midden deposits in this region dating to 7000 RC yr. BP shows only chinook salmon (Reid and Chatters 1997).

White sturgeon were present as anadromous fish before dams landlocked those caught inland. They were by far the biggest fish in the system and a top consumer.

Many species of freshwater fish inhabit the Columbia River. Excluding introduced species, the native fish

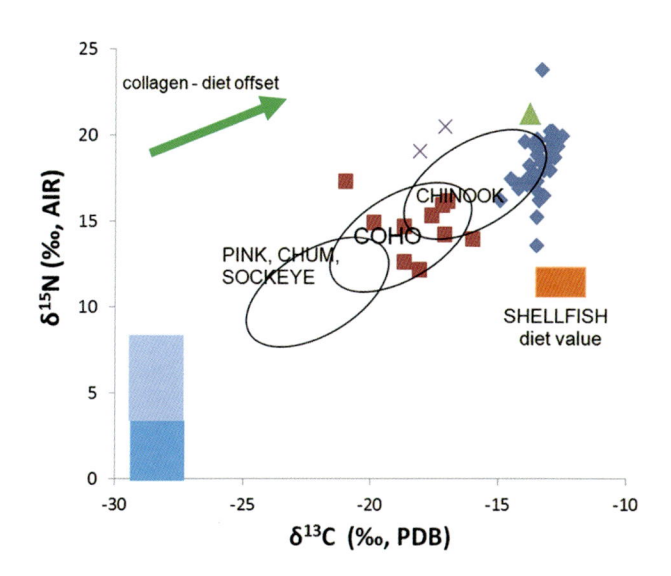

FIGURE 16.1 Isotopic compositions. Green triangle: Kennewick Man. Blue diamonds: humans from the northwest coast of B.C. (Chisholm, Schwarcz, and Knyf forthcoming). Blue rectangles: terrestrial foods (plants, dark blue; animal flesh, light blue). Red squares: marine fish. Ellipses: salmon ranges (from Johnson and Schindler 2009). X: seal, sea lion. Orange rectangle: mollusk range (mussel, clam) from B.C. coast, shown offset by fractionation between collagen and diet. Green arrow: offset between collagen (head of arrow) and diet.

included bull and cutthroat trout, suckers, northern pikeminnow (squawfish), chiselmouth, peamouth, and burbot. These fish obtained all of their nutrients in the freshwater environment and would thus be expected to have been significantly depleted of ^{13}C relative to the anadromous salmon, whose tissue mass is acquired while at sea. However, dead salmon may have contributed significant C and N to their food chain (Chapter 2).

Edible freshwater mussels also live in the waters of the Columbia River, and shell middens are common along its banks. As with the freshwater fish, significant absorption of C and N from the flesh of dead and decaying anadromous fish would have contributed a marine-like isotopic signal. However, mussels would also have acquired nutrients from local primary production by aquatic flora and from terrestrial detritus.

Bald eagles, gulls, condors, and cormorants were present in the immediate vicinity as shown by the presence of their bones in contemporaneous archaeological sites (Chapter 2). No data are included for birds in Figure 16.1 because there are no representative values for $\delta^{15}N$ or $\delta^{13}C$ of birds feeding in a fluvial/terrestrial environment. To the extent that birds were feeding on marine foods, they would have isotopic compositions resembling those of humans from the northwest coast of British Columbia. Humans subsisting entirely on birds of prey would be one trophic level higher.

Figure 16.1 also includes representative data for two groups of potential food sources that were unlikely to have been present in the vicinity of where Kennewick Man was found: marine mollusks and pinnipeds, such as seals and sea lions. Seals and sea lions were included in the diet of some prehistoric coastal peoples (Glassow and Wilcoxon 1988). Evidence from the archaeological record suggests that only harbor seals moved inland along the Columbia River; these predators were known to swim up the Columbia River at least as far as Celilo Falls (at The Dalles), more than 270 kilometers along the river from its mouth (Lyman et al. 2002).

Large shell middens along the coast of the Pacific Ocean from Santa Barbara, California (Erlandson et al. 1999) to British Columbia (Cannon and Burchell 2009) testify to the importance of clams and mussels as a part of prehistoric peoples' diet in this region, extending at least as far back as 9000 CAL yr. BP.

C_4 grasses are not included in Figure 16.1. Such foods would have even higher $\delta^{13}C$ values (~ −11‰) but lower $\delta^{15}N$ values (typically 2–5‰) than most marine foods. The main edible C_4 plant in North America is maize, but its domestication and consumption postdated Kennewick Man by at least 8,000 years (Blake 2006).

Figure 16.1 also shows the assumed offset between diet and collagen. Comparing the collagen isotopic data for Kennewick Man to possible foods using this offset indicates that the protein in his diet could be explained by a diet rich in salmon. Although the graph tends to point to coho as the most likely species to have been consumed, remains of this species are not found in older middens. Also, the isotopic compositions of salmon (including coho) that had swum 530 kilometers from the mouth of the Columbia to Kennewick is not known.

Besides being a likely component of Kennewick Man's diet, it is important to appreciate that salmon may have indirectly contributed to the isotopic composition of his diet in two ways—directly as prey or through the decaying flesh of spawning fish. In the first category are bears and birds of prey, such as gulls, bald eagles, cormorants, and condors, which are all part of a high trophic level and are known to consume large amounts of river fish, including salmon. They may have had high $\delta^{15}N$ values, comparable to or higher than that of salmon, while the $\delta^{13}C$ values of their flesh (especially that of bears, eagles, and cormorants) may have been shifted to somewhat lower values because their diet could have also included terrestrial mammals and birds. Chapter 2 notes that some sites representative of the Western Stemmed Tradition, (therefore potentially coeval with Kennewick Man) contain large numbers of bones of birds of prey (Cressman et al. 1960). Kennewick Man may have also hunted black bear, although bear bones are not reported as a part of any coeval midden assemblage from this region.

Kennewick Man could also have acquired marine-derived isotopic compositions through nutrients from dead salmon. These would enter the food web of resident, nonanadromous fish, which would have been plentiful. Gresh et al. (2000) estimate that spawning fish brought between 77 and 103 million kilograms of biomass into the Columbia River basin annually (present levels are only 6 to 7 percent of that amount). Dying and decomposing there, they were the foundation of the nutrient cycle in an otherwise nutrient-poor system. The carcasses of spawning salmon and other anadromous fish constitute a major source of nitrogen and organic carbon to the food web (Claeson et al al. 2006; Kline et al. 1993; Mathisen et al. 1988). As a result, the δ^{15}N and δ^{13}C of organisms living in that web will be increased, possibly to levels approaching those found in the salmon themselves (Kline et al. 1990, 1993; Gregory-Eaves et al. 2007).

Gregory-Eaves et al. (2007:Figure 1) found that in lakes in western North America in which high numbers of spawning salmon died, resident trout displayed δ^{15}N values comparable to or higher than that of the salmon. Their δ^{13}C values ranged up to –19‰, comparable to values observed in salmon (Figure 16.1). These data were recorded in fish living in lakes with 20,000 sockeye spawners per square kilometer. Likewise, Kline et al. (1990) studied resident trout in Sashin Creek, Alaska, where about 30,000 pink salmon (*O. gorbuscha*) return to spawn each year. They found δ^{15}N values for trout ranging from 10.9 to 12.9‰ (average = 12.6 ± 0.1‰), which is comparable to values for marine salmon. The average δ^{13}C value of these fish (–22.6 ± 0.9‰) is within the range of some marine salmon (Figure 16.1).

The δ^{13}C data observed in these two studies appear to be lower than Kennewick Man's inferred diet, but the δ^{15}N values correspond to his predicted diet. However, these are data for lakes and a small river. In both of these environments, the concentration of dead salmon would have been quite high relative to background sources of nutrients. It is not clear if similar conditions would have prevailed in the Columbia River when Kennewick Man was living there. We consider it possible that Kennewick Man's diet included, in addition to marine fish, some amounts of resident river fish whose isotopic signature had shifted upward as a result of deposition of decaying salmon.

Recycling of marine-derived nutrients may also influence the δ^{15}N (although not the δ^{13}C) of surrounding terrestrial vegetation, thereby affecting the fauna living in and on it, as well as the fauna which feed directly on the salmon, such as bears and eagles. However, the magnitude of the shift in δ^{15}N in the vegetation is comparatively small (a few ‰ at most; Drake et al. 2011) and shifts of comparable magnitude are observed in the fauna feeding on this vegetation (Ben-David et al. 1998). This would not have a noticeable effect on the δ^{15}N of humans consuming significant amounts of such fauna. An interesting comparison is provided by δ^{15}N and δ^{13}C data from bone collagen of ancient native people from the northwest coast of British Columbia (Figure 16.2). These data (from an unpublished study by Chisholm and Schwarcz 2012) show that Kennewick Man has higher δ^{15}N but almost identical δ^{13}C to this population. However, the δ^{15}N value of Kennewick Man is significantly higher than that of any of the people at the British Columbia sites (Figure 16.2). His δ^{13}C falls within the same range as those samples. The lower δ^{15}N observed for some of the people on the British Columbia coast is attributed to their consumption of lower trophic level shellfish (clams and mussels). This is indicated archaeologically by the presence of large shell middens (Cannon and Burchell 2009). Near the Kennewick discovery, however, only freshwater mollusks would have been available. Consumption of significant quantities of mollusks would have resulted in a lower δ^{15}N in Kennewick Man's bone collagen. A possible explanation for the higher δ^{15}N for Kennewick Man compared to people from British Columbia is that he also consumed significant amounts of the flesh of birds of prey, which were themselves feasting on salmon.

It is interesting to compare these data with Lovell et al.'s (1986) study of the δ^{13}C of human bone collagen from people living along the length of the salmon-bearing Thompson River in interior British Columbia (Fig-

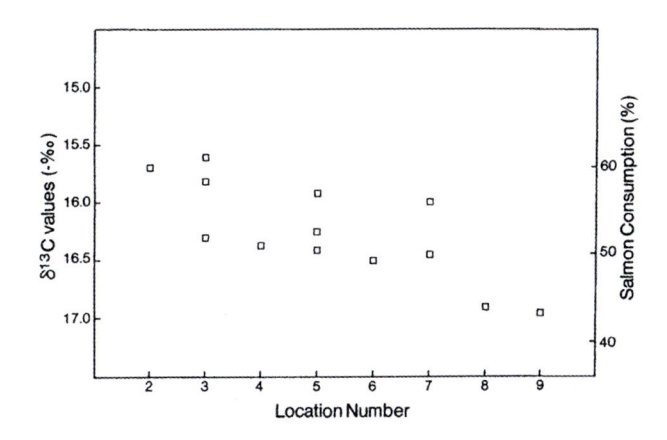

FIGURE 16.2 Gradient of δ^{13}C of collagen from humans living along the Thompson River, B.C., as it relates to salmon consumption. The sample sites are labeled with a location number, 2 being closest to the ocean and 9 being furthest upstream (Lovell et al. 1986).

ure 16.2). Unfortunately, no $\delta^{15}N$ data were collected. The Thompson is a tributary of the Fraser River. Salmon migrate seasonally up the Fraser and Thompson. People living near the mouth of the Thompson (at the Fraser River) exhibited $\delta^{13}C$ values suggestive of a diet including about 60 percent salmon, which was substantially less than that consumed by people on the coast (Lovell et al. 1986). At the headwaters of the river, about 450 kilometers from the mouth of the Fraser River, the average diet included about 45 percent salmon. Lovell et al. (1986) infer that the protein content in the diets of these people consisted of a mixture of terrestrial herbivore flesh, freshwater fish, and salmon, with the proportion of salmon decreasing with distance upstream.

The $\delta^{13}C$ of collagen should be largely representative of Kennewick Man's protein intake. This is especially true when the diet is rich in protein. This would have been true of Kennewick Man, given that his isotopic compositions indicate a diet composed almost entirely of marine food or the flesh of fish and other animals whose nutrient base was dominated by the decaying flesh of marine fish. In principle, it should also be possible to learn something about the $\delta^{13}C$ of the total diet consumed (including carbohydrates and fats), through analysis of structural carbonate in the bone mineral. In this case, however, there is evidence that the $\delta^{13}C$ of bone carbonate from Kennewick Man was significantly altered postmortem. This is indicated by the fact that the bone carbonate gives a radiocarbon age of 2285 ± 15 RC yr. BP, which is discrepant with the ^{14}C age of the bone collagen (8410 RC yr. BP). The younger age is similar to that obtained for calcite from the medullary cavity of the same bone (2235 ± 15 RC yr. BP). This result suggests that secondary calcite filling open spaces in the bone (e.g., Haversian canals) dominates the carbonate fraction of this sample. Even if an unaltered carbonate component remained in the bone mineral, it would be impossible to isolate it from secondary contamination; therefore, it is impossible to determine accurately the $\delta^{13}C$ of the total human diet. This evidence for postmortem alteration also indicates that the $\delta^{18}O$ of CO_2 liberated from bone carbonate should not be used to deduce geographic origin.

The isotopic composition of Kennewick Man's bone collagen does not show any evidence for consumption of terrestrial herbivore flesh even though he lived 530 kilometers upstream from the sea in a terrain with some terrestrial animals. He may have consumed plant foods growing along the banks of the river or on the surrounding plain.

Native people from coastal British Columbia also appeared to lack traces of terrestrial fauna in their diet, although the bones of terrestrial animals were relatively common in their middens (Ames and Maschner 1999). Fish would not have been as abundant in that environment, but they were also not a significant component of the diet (Chisholm et al. 1982).

Comparing Kennewick Man's isotopic data with available reference sources leads to the conclusion that the protein content of his diet was dominated by a mixture of marine foods plus the flesh of fish and birds whose food web was based on decaying salmon flesh. It seems probable that he also consumed some terrestrial berries, seeds, and greens. The greens contributed little protein to his diet, thereby having a negligible effect on the $\delta^{13}C$ of his bone collagen. His high $\delta^{15}N$ value implies that that his total protein intake was high enough that endogenous amino acid synthesis using carbon atoms from nonprotein foods was suppressed (see Schwarcz 2000). Therefore the $\delta^{13}C$ value of his collagen does not reflect any terrestrial sources of carbohydrates that he may have eaten.

The $\delta^{15}N$ and $\delta^{13}C$ data imply that an adequate supply of salmon as well as salmon-based (in the sense defined above) foods were available throughout the year. Currently, salmon are present in the river for half of the year. Chinook salmon pass through in three pulses between April and October, the main pulses being in April and August–September. Steelhead salmon are present over the same time interval. Sockeye salmon are present from May to September but mostly in the early part of that interval. Coho are found in a few tributaries of the Columbia between August and November. At present, no salmon are known to be in the river between November and April, so it is unclear what foods Kennewick Man would have eaten during those months in order to maintain his marine-like isotope signature. Some of the nutrient from dead salmon would have persisted in the stream for weeks to months after the end of each salmon run. It is conceivable that Kennewick Man and his brethren would already have possessed the technology for drying, smoking, and storing salmon flesh to tide them over periods of scarcity. That said, there is currently no archaeological evidence for food storage.

GEOGRAPHIC PROVENANCE

In principle, something about the geographic provenance of a human can be learned from isotopic analysis of oxygen in their skeleton. Oxygen is present in the crystal structure of hydroxyapatite in bone as a component of both phosphate (PO_4^{3-}) and carbonate (CO_3^{2-}) ions. Both ions carry an isotopic signal reflecting the $\delta^{18}O$ of body water during approximately the last 10 years of life. This signal is believed to largely represent the $\delta^{18}O$ value of local precipitation and has been used to trace the movements of humans in both archaeological and forensic contexts (Daux et al. 2008; Ehleringer et al. 2010; Prowse et al. 2007). This is possible because the $\delta^{18}O$ of precipitation varies regionally, generally decreasing with increasing latitude and distance from the coast (Gat 1996). However, large rivers can carry water over great distances so that a human living beside and consuming water from such a river may look isotopically quite different from someone living nearby who drinks local rainfall.

The easiest approach to determining body water history from the $^{18}O/^{16}O$ of bone is through the analysis of CO_2 from the CO_3^{-2} component of bone mineral. However, the radiocarbon analysis of such CO_2 from Kennewick Man strongly indicates that a large fraction of the bone mineral's carbonate was supplied by secondary carbonate that crystallized after his death. In general, studies suggest that the phosphate component is less susceptible to alteration than bone carbonate (Zazzo et al. 2004).

A value of $\delta^{18}O = 5.3‰$ (VSMOW) (Table 16.2) was obtained from the phosphate. This value is used to estimate the $\delta^{18}O$ of Kennewick Man's body water (and hence, drinking water). Daux et al. (2008) show that the $\delta^{18}O$ of human bone phosphate is related to that of local tap water, $\delta^{18}O(w)$, by the equation:

$$\delta^{18}O(w) = 1.54\, \delta^{18}O\,(PO_4) - 33.72$$

where both $\delta^{18}O$ values are given with respect to VSMOW. From this relationship, we calculate that Kennewick Man's $\delta^{18}O(w) = -25.5‰$. This is a remarkably low value by comparison with local rainfall. Figure 16.3 shows a compilation of modern $\delta^{18}O(w)$ values for the western United States (Kendall et al. 2010). The local precipitation averages around $-15‰$ (VSMOW). It is possible that this discrepancy could arise because Kennewick Man was mainly drinking water from the Columbia River whose $\delta^{18}O$ might be substantially different from that of local rain. An analysis of $\delta^{18}O$ of the Columbia River at Richland, Washington, a few kilometers from Kennewick, gives $\delta^{18}O = -17.2‰$ (Kendall and Coplen 2001), only slightly lower than local precipitation, and presumably reflecting the large contribution of water from the Columbia River drainage basin lying to the north. We can estimate changes in the $\delta^{18}O(w)$ of Columbia River water from $\delta^{18}O$ analyses of mollusk shells that grew in the river in past times. Table 16.3 summarizes analyses of one modern shell and a series of shells dating back to 6580 RC yr. BP and shows estimates of $\delta^{18}O(w)$ calculated assuming a growth temperature of 12 °C, which is the average annual temperature at the site today. The average value is $-15.4 \pm 0.6‰$. If the temperature had been lower (e.g., by 2–4 °C), then $\delta^{18}O(w)$ could have been lower but only by about 0.25‰ / °C. The turnover time for bone remodeling is estimated to be between 10 and 20 years (Hedges et al. 2007; Manolagas 2000). Therefore,

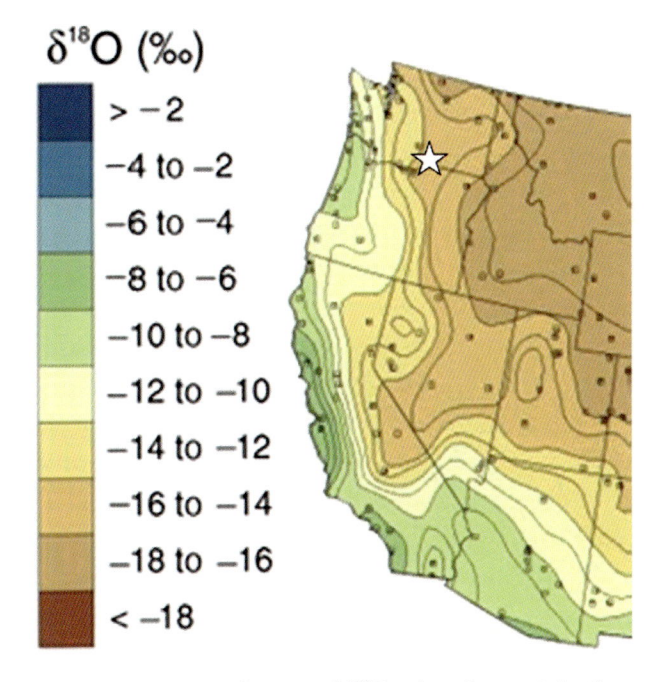

FIGURE 16.3 Distribution of $\delta^{18}O$ values in precipitation over the western United States (Kendall and Copland 2001). Star marks the position of Kennewick, Washington.

TABLE 16.3 Oxygen isotopic analyses of modern and ancient shells from the Columbia River.

Sample	$\delta^{13}C$ (‰, VPDB)	$\delta^{18}O$ (‰, VPDB)	$\delta^{18}O(w)$ (‰, VSMOW) (average T =12°C)*	Age
SHELL 1a	−8.0	−16.0	−15.0	6090 ± 80
SHELL 1b	−8.2	−16.0	−15.0	
SHELL 2a	−9.1	−15.7	−14.7	6090 ± 80
SHELL 2b	−9.2	−15.6	−14.6	
SHELL 3a	−9.0	−16.7	−15.7	5150 ± 40
SHELL 3b	−9.1	−16.9	−15.9	
SHELL 4a	−9.2	−17.1	−16.1	6580 ± 40
SHELL 4b	−9.2	−17.1	−16.1	
			Average −15.4	
			SD 0.6	
SHELL 5a	−7.9	−15.8	−14.8	Modern
SHELL 5b	−8.0	−15.8	−14.8	

*$\delta^{18}O$ of water calculated using equation in Kim and O'Neil (1997).

Kennewick Man lived at least his last 10 years in a region where he had access to drinking water substantially more depleted of ^{18}O than water flowing in the Columbia River over the past 6,600 years.

The headwaters of the Columbia River include the Columbia Ice Field, which straddles the Alberta-British Columbia border in Canada. One explanation for the lower $\delta^{18}O$ of the Columbia River at the time of Kennewick Man is that during that period the river was being fed by larger amounts of glacial meltwater than today. Present-day ice in the Columbia Ice Field has an average $\delta^{18}O$ of about −22‰ based on analyses of seepage water in Castleguard Cave (Yonge and Krouse 1987). Modern meltwater from the Athabasca Glacier (part of the Columbia Ice Field) is known to range in $\delta^{18}O$ from −23.1‰ in the spring to −21.3‰ in the fall (F. Longstaffe, personal communication 2012). Such compositions approach those of the water inferred from the $\delta^{18}O$ of bone phosphate from Kennewick Man. This estimation neglects any evaporative increase in $\delta^{18}O(w)$ that might have occurred as glacier-derived meltwater traveled downstream from the Columbia Ice Field to Kennewick. This meltwater scenario is a possible, but tentative, explanation for Kennewick Man's extremely low $\delta^{18}O(PO_4)$ value. Further analyses of additional material are needed to test the reproducibility of the single result obtained and to investigate postmortem alteration of the phosphate oxygen. In particular, it would be desirable to analyze the enamel of one of Kennewick Man's teeth, as this tissue may have preserved an unaltered $\delta^{18}O$ signal from his youth. Also, the $\delta^{13}C$ of tooth enamel would help resolve the question of whether Kennewick Man's diet included terrestrial sources of carbohydrate, such as berries and leafy plants.

The simplest explanation of the result is a drinking-water supply for Kennewick Man overwhelmingly dominated by glacial meltwater of the sort that would be expected from the Columbia Ice Field. However, given the paucity of evidence for a supply of this low-^{18}O water (Cordilleran glaciers had completely melted away and most alpine glaciers were near their postglacial minimum; see Chatters 1998), the isotope data cannot be strictly used to rule out the possibility that Kennewick Man actually migrated from some considerable distance away while remaining close to a supply of marine or isotopically marine-like foods. Thus we cannot rule out possible immigration from the Pacific coast.

CONCLUSION

What is interesting in this circumstance is not only what Kennewick Man ate but what, apparently, he did not eat. The high $\delta^{13}C$ and exceptionally high $\delta^{15}N$ value obtained for his bone collagen strongly suggests that the protein content of his diet was composed of salmon plus some unknown fraction of resident fish and/or other animals that fed on the anadromous fish-based food chain, such as birds of prey or perhaps even an occasional black bear. Significantly, his diet did not contain large amounts of the flesh of deer, hare, or other herbivorous terrestrial species available in the surrounding landscape to offset the effects of the marine and marine-based prey (Chapter 2; Butler 2000). A diet consisting of fish alone, supplemented by an approximately equal amount of low-protein berries or seeds, would have been nutritionally adequate.

The use of stable isotope data to reconstruct the past diet and geographic provenance of humans is normally based on analyses of large numbers of individuals. Typically such a study (Prowse et al. 2007; Walker and DeNiro 1986) involves the analysis of several tens of individuals, a process that generates a range of isotopic compositions for the population and thereby infers the range of foods consumed. This is not possible in the present instance. The stable isotopic data for the bone of Kennewick Man allows cautious reconstruction of his diet and geographic provenance.

The results suggest an unexpected degree of dietary selectivity by Kennewick Man. He was strictly dependent on fish and possibly birds, but not on terrestrial sources of protein. Kennewick Man seems to have been peculiarly selective in his diet compared with people living on a major inland river in British Columbia some thousands of years later (Lovell et al. 1986). Later inhabitants' diet consisted in large part of terrestrial foods as well as salmon and other fish.

One possible explanation for the results and his especially focused dietary behavior is that, in the years prior to his death, Kennewick Man had lived on the Pacific coast where high trophic level marine organisms (including pinnipeds such as seal and sea lion) were abundant. He could have travelled the 500+ kilometers upstream to the site where he was found before his bones remodeled to reflect his new environment. Pinnipeds are known to migrate up the Columbia River, but not so far as Kennewick.

The oxygen isotopic evidence does not rule out the possibility of a coastal origin, but suggests that Kennewick Man did not live near the coast within the last 10–15 years of his life. These data are still equivocal. The length of time needed to acquire an oxygen isotopic signal indicative of provenance is similar to the time needed to acquire a dietary signal, since both are coupled to the remodeling rate of bone. Further oxygen isotopic analyses, especially of tooth enamel, might be helpful to resolve this important issue.

REFERENCES

Ambrose, Stanley H., and Lynette Norr. 1993. Experimental evidence for the relation of the carbon isotope ratios of whole diet and dietary protein to those of collagen and carbonate. In *Prehistoric Human Bone: Archaeology at the Molecular Level,* edited by Joseph B. Lambert and Gisela Grupe, 1–37. Springer, Berlin.

Ames, Kenneth M., and Herbert D. G. Maschner. 1999. *Peoples of the Northwest Coast: Their Archaeology and Prehistory*. Thames and Hudson, London.

Ben-David, M., T. A. Hanley, and D. M. Schell. 1998. Fertilization of terrestrial vegetation by spawning Pacific salmon: The role of flooding and predator activity. *Oikos* 83(1):47–55.

Blake, Michael. 2006. Dating the initial spread of *Zea mays*. In *Histories of Maize*, edited by John E. Staller, Robert H. Tykot, and Bruce F. Benz, 55–72. Academic Press, Amsterdam.

Bowen, Gabriel J., and Bruce Wilkinson. 2002. Spatial distribution of $\delta^{18}O$ in meteoric precipitation. *Geology* 30(1):315–318.

Butler, Virginia. 2000. Resource depression on the Northwest Coast of North America. *Antiquity* 74:649–61

Cannon, Aubrey, and Meghan Burchell. 2009. Clam growth-stage profiles as a measure of harvest intensity and resource management on the central coast of British Columbia. *Journal of Archaeological Science* 36(4):1050–1060

Chatters, James C. 1998. Environment. In *Handbook of North American Indians,* vol. 12: *Plateau,* edited by Deward E. Walker, 29–48. Smithsonian Institution Press, Washington DC.

Chatters, James C. 2000. The recovery and initial analysis of an Early Holocene human skeleton from Kennewick, Washington. *American Antiquity* 65(2):291–316

Chatters, James C. 2003. "Archaeological Investigations at Okanogan Phase Sites 45OK373 and 45OK420, Okanogan County, Washington." Report prepared for Douglas County PUD by Applied Paleoscience, Bothell, Washington.

Chatters, James C., Steven Hackenberger, Alan Busacca, Linda S. Cummings, Richard L. Jantz, Thomas W. Stafford, and R. Ervin Taylor. 2000. A possible second early Holocene skull from eastern Washington, USA. *Current Research in the Pleistocene* 17:93–95.

Chisholm, Brian S., D. Erle Nelson, and Henry P. Schwarcz. 1982. Stable carbon isotope ratios as a measure of marine versus terrestrial protein in ancient diets. *Science* 216(4550):1131–1132.

Chisholm, Brian S., Henry P. Schwarcz, and Martin J. Knyf. Forthcoming. "Intense consumption of marine resources by coastal populations, British Columbia, Canada."

Claeson, Shannon M., Judith L. Li, Jana E. Compton, and Peter A. Bisson. 2006. Response of nutrients, biofilm and benthic insects to salmon carcass addition. *Canadian Journal of Fisheries and Aquatic Sciences* 63(6):1230–1241.

Clayton, Robert N., and Toshiko Mayeda. 1963. The use of bromine pentafluoride in the extraction of oxygen from oxides and silicates for isotopic analysis. *Geochimica et Cosmochimica Acta* 27(1):43–52.

Cressman, L. S., David L. Cole, Wilbur A. Davis, Thomas M. Newman, and Daniel J. Scheans. 1960. Cultural sequences at The Dalles, Oregon: A contribution to Pacific Northwest prehistory. *Transactions of the American Philosophical Society* 50(10):1–108.

Crowson, Ronald A., William J. Showers, Ellen K. Wright, and Thomas C. Hoering. 1991. Preparation of phosphate samples for oxygen isotope analysis. *Analytical Chemistry* 63(20):2397–2400.

Daux, Valérie, Christophe Lécuyer, Marie-Anne Héran, Roman Amiot, Laurent Simon, François Fourel, François Martineau, Niels Lynnerup, Hervé Reychler, and Gilles Escarguel. 2008. Oxygen isotope fractionation between human phosphate and water revisited. *Journal of Human Evolution* 55(6):1138–1147.

DeNiro, Michael J., and Samuel Epstein. 1978. Influence of diet on the distribution of carbon isotopes in animals. In *Geochimica et Cosmoschimica Acta* 42(5):495–506.

Drake, D. C., Paul J. Sheppard, and Robert J. Naiman. 2011. Relationships between salmon abundance and tree-ring $\delta^{15}N$: Three objective tests. *Canadian Journal of Forest Research* 41(12):2423–2432.

Dutton, Andrea, Bruce Wilkinson, Jeff Welker, Gabriel J. Bowen, and Kenneth Lohmann. 2005. Spatial distribution and seasonal variation in $^{18}O/^{16}O$ of modern precipitation and river water across the conterminous USA. *Hydrological Processes* 19(20):4121–4146.

Ehleringer, James, Alexandra H. Thompson, David W. Podlesak, Gabriel J. Bowen, Leslie A. Chesson, Thure E. Cerling, Todd Park, Paul Dostie, and Henry P. Schwarcz. 2010. A framework for the incorporation of isotopes and isoscapes into geospatial forensic investigations. In *Isoscapes: Understanding Movement, Pattern, and Process on Earth through Isotope Mapping*, edited by Jason B. West, Gabriel J. Bowen, Todd E. Dawson, and Kevin P. Tu, 357–388. Springer-Verlag, Heidelberg.

Erlandson, Jon M., Torben C. Rick, Rene L. Vellanoweth, and Douglas J. Kennett. 1999. Maritime subsistence at a 9,300 year old shell midden on Santa Rosa Island, California. *Journal of Field Archaeology* 26(3):255–265.

Gat, J. R. 1996. Oxygen and hydrogen isotopes in the hydrologic cycle. *Annual Review of Earth and Planetary Sciences* 24:225–262.

Glassow, Michael A., and Larry R. Wilcoxon. 1988. Coastal adaptations near Point Conception, California, with particular regard to shellfish exploitation. *American Antiquity* 53(1):36–44.

Gregory-Eaves, Irene, Marc J. Demers, Lynda Kimpe, Eva M. Krümmel, Robie W. Macdonald, Bruce P. Finney, and Jules M. Blais. 2007. Tracing salmon-derived nutrients and contaminants in freshwater food webs across a pronounced spawner density gradient. *Environmental Toxicology and Chemistry* 26(6):1100–1108.

Gresh, Ted, Jim Lichatowich, and Peter Schoonmaker. 2000. An estimate of historic and current levels of salmon production in the northeast Pacific ecosystem: Evidence of nutrient deficiency in the freshwater systems of the Pacific Northwest. *Fisheries* 25(1):15–21.

Hedges, Robert E. M., John G. Clement, C. David Thomas, and Tamsin C. O'Connell. 2007. Collagen turnover in the adult femoral midshaft: Modeled from anthropogenic radiocarbon tracer measurements. *American Journal of Physical Anthropology* 133(2):808–816.

Hedges, Robert E. M., and Linda M. Reynard. 2007. Nitrogen isotopes and the trophic level of humans in archaeology. *Journal of Archaeological Science* 34(8):1240–1251.

Iacumin, P., H. Bocherens, A. Mariotti, and A. Longinelli. 1996. Oxygen isotope analyses of co-existing carbonate and phosphate in biogenic apatite: A way to monitor diagenetic alteration of bone phosphate? *Earth and Planetary Science Letters* 142(1–2):1–6.

Johnson, Susan P., and Daniel E. Schindler. 2009. Trophic ecology of Pacific salmon (*Oncorhynchus spp.*) in the ocean: A synthesis of stable isotope research. *Ecological Research* 24(4):855–863.

Katzenberg, M. Anne. 2000. Stable isotope analysis: A tool for studying past diet, demography, and life history. In *Biological Anthropology of the Human Skeleton*, edited by M. Anne Katzenberg, and Shelley R. Saunders, 305–327. Wiley-Liss, New York.

Kendall, Carol, and Tyler B. Coplen. 2001. Distribution of oxygen-18 and deuterium in river waters across the United States. *Hydrological Processes* 15(7):1363–1393.

Kendall, Carol, Megan Young, and Steven Silva. 2010. Applications of stable isotopes for regional to national-scale water quality and environmental monitoring programs. In *Isoscapes: Understanding Movement, Pattern and Process on Earth Through Isotope Mapping*, edited by Jason B. West, Gabriel J. Bowen, Todd E. Dawson, and Kevin P. Tu, 89–111. Springer-Verlag, Heidelberg.

Kim, Sang-Tae, and James R. O'Neil. 1997. Equilibrium and nonequilibrium oxygen isotope effects in synthetic carbonates. *Geochimica et Cosmochimica Acta* 61(16):3461–3475.

Kline, Thomas C., Jr., John J. Goering, Ole A. Mathisen, Patrick H. Poe, and Patrick L Parker. 1990. Recycling of elements transported upstream by runs of Pacific salmon: I. $\delta^{15}N$ and $\delta^{13}C$ evidence in Sashin Creek, southeastern Alaska. *Canadian Journal of Fisheries and Aquatic Sciences* 47(1):136–144.

Kline, Thomas C., Jr., John J. Goering, Ole A. Mathisen, and Patrick H. Poe. 1993. Recycling of elements transported upstream by runs of Pacific salmon: II. $\delta^{15}N$ and $\delta^{13}C$ evidence in the Kvichak River, Bristol Bay, southwestern Alaska. *Canadian Journal of Fisheries and Aquatic Sciences* 50(11):2350–2365.

Koch, Paul L., Noreen Tuross, and Marilyn Fogel. 1997. The effects of sample treatment and diagenesis on the isotopic integrity of carbonate in biogenic hydroxylapatite. *Journal of Archaeological Science* 24(5):417–429.

Krueger, Harold W., and Charles H. Sullivan. 1984. Models for carbon isotope fractionation between diet and bone. In *Stable Isotopes in Nutrition*, edited by Judith R. Turnlund, and Phyllis E. Johnson, 205–222. American Chemical Society Symposium Series 258. American Chemical Society, Washington DC.

Longinelli, Antonio. 1984. Oxygen isotopes in mammal bone phosphate: A new tool for paleohydrological and paleoclimatological research? *Geochimica et Cosmochima Acta* 48(2):385–390.

Lovell, Nancy C., Brian S. Chisholm, D. Erle Nelson, and Henry P. Schwarcz. 1986. Prehistoric salmon consumption in interior British Columbia. *Canadian Journal of Archaeology* 10:99–106.

Lyman, R. Lee, Judith L. Harpole, Christyann Darwent, and Robert Church. 2002. Prehistoric occurrence of pinnipeds in the lower Columbia River. *Northwest Naturalist* 83(1):1–6.

Manolagas, Stavros C. 2000. Birth and death of bone cells: Basic regulatory mechanisms and implications for the pathogenesis and treatment of osteoporosis. *Endocrine Reviews* 21(2):115–137.

Mathisen, Ole A., Patrick L. Parker, John J. Goering, Thomas C. Kline, Patrick H. Poe, and Richard S. Scanlan. 1988. Recycling of marine elements transported into freshwater by anadromous salmon. *Verhandlungen der Internationalen Vereinigung für Theoretische und Angewandte Limnologie* 23:2249–2258.

Phillips, Donald L., and Jillian W. Gregg. 2003. Source partitioning using stable isotopes: Coping with too many sources. *Oecologia* 136(2):261–269.

Prowse, Tracy, Henry P. Schwarcz, Shelly Saunders, Roberto Macchiarelli, and Luca Bondioli. 2003. Isotopic paleodiet studies of skeletons from the Imperial Roman-age cemetery of Isola Sacra, Rome, Italy. *Journal of Archaeological Science* 31(3):259–272.

Prowse, Tracy L., Henry P. Schwarcz, Peter Garnsey, Martin Knyf, Roberto Macchiarelli, and Luca Bondioli. 2007. Isotopic evidence for large scale immigration to imperial Rome. *American Journal of Physical Anthropology* 132(4):510–519.

Reid, Kenneth C., and James C. Chatters 1997. "Kirkwood Bar: Passports in Time Excavations at 10IH699 in the Hells Canyon National Recreation Area, Wallowa-Whitman National Forest." Rainshadow Research Project Report 28 and Applied Paleoscience Project Report F-6, Pullman, WA.

Schwarcz, Henry P. 1991. Some theoretical aspects of isotope paleodiet studies. *Journal of Archaeological Science* 18(3):261–275.

———. 2000. Some biochemical aspects of carbon isotopic paleodiet studies. In *Biogeochemical Approaches to Paleodietary Analysis,* edited by Stanley H. Ambrose and M. Anne Katzenberg, 189–210. Kluwer Academic, New York.

Schwarcz, Henry P., and Margaret J. Schoeninger, 1991. Stable isotope analyses in human nutritional ecology. *American Journal of Physical Anthropology* 34(S13):283–321.

Semmens, Brice X., and Jonathan W. Moore. 2008. MixSIR: A Bayesian Stable Isotope Mixing Model, Version 1.04. http://www.ecologybox.org.

Stuart-Williams, Hilary Le Q., and Henry P. Schwarcz. 1995. Oxygen isotopic analysis of silver orthophosphate using a reaction with bromine. *Geochimica et Cosmochimica Acta* 59(18):3837–3841.

Taylor, R. E., Donna L. Kirner, John R. Southon, and James C. Chatters. 1998. Radiocarbon dates of Kennewick Man. *Science* 280(5367):1171–1172.

Tudge, A. P. 1960. A method of analysis of oxygen isotopes in orthophosphate—its use in the measurement of paleotemperatures. *Geochimica et Cosmochimica Acta* 18(1–2):81–93.

van Klinken, G. J. 1999. Bone collagen quality indicators for palaeodietary and radiocarbon measurements. *Journal of Archaeological Science* 26(6):687–695.

Wakeley, Lillian D., William L. Murphy, Joseph B. Dunbar, Andrew G. Warne, Frederick L. Briuer, and Paul R. Nickens. 1998. *Geologic, Geoarchaeologic, and Historical Investigation of the Discovery Site of Ancient Remains in Columbia Park, Kennewick, Washington.* U.S. Army Corps of Engineers Technical Report GL-98-13. Prepared for U.S. Army Engineer District, Walla Walla. U.S. Army Corps of Engineers Waterways Experiment Station, Vicksburg, MS.

Walker, Phillip L., and Michael J. DeNiro. 1986. Stable nitrogen and carbon isotope ratios in bone collagen as indices of prehistoric dietary dependence on marine and terrestrial resources in southern California. *American Journal of Physical Anthropology* 71(1):51–61.

Yonge, Charles J., and H. Roy Krouse. 1987. The origin of sulphates in Castleguard Cave, Columbia Icefields, Canada. Chemical Geology (Isotope Geoscience Section) 65:427–433.

Zazzo, Antoine, Christophe Lécuyer, Simon M. F. Sheppard, Patricia Grandjean, and André Mariotti. 2004. Diagenesis and the reconstruction of paleoenvironments: A method to restore original $\delta^{18}O$ values of carbonate and phosphate from fossil tooth enamel. *Geochimica et Cosmochimica Acta* 68(10):2245–2258.

Taphonomic Indicators of Burial Context

17

Douglas W. Owsley, Aleithea A. Williams, and Thomas W. Stafford Jr.

Taphonomic analysis is a well-established approach for understanding the processes that affect an organism's remains after death (Behrensmeyer and Hill 1980; Bonnichsen and Sorg 1989; Haglund and Sorg 1997). This chapter will present the effects of various actions—geologic, pedologic (soil formation processes), biologic (plant and animal), and human—on Kennewick Man's remains from the time of his death until the time of recovery.

Kennewick Man's bones were found after they eroded from the bank of Lake Wallula. Many questions arose because no one saw the skeleton in its original geologic context. Was Kennewick Man's body intentionally buried by other humans or by natural processes (Chatters 2001)? Perhaps Kennewick Man was a flood victim whose body floated down the Columbia River, ultimately coming to rest away from the main channel where it was subsequently buried by flood deposits. Can the stratigraphic placement of the remains within the terrace bank or the original positioning of the body be discerned? Is it possible to comment on the in situ orientation of the remains relative to the river (the contemporary lake shoreline)? Furthermore, what can be said about the process by which the remains eroded from the bank onto the shoreline (i.e., the mechanics of the erosion process or the timing required for this event to occur)?

To answer these questions, an interdisciplinary team conducted a detailed taphonomic analysis in 2005 and 2006 (Chapter 6). This evaluation was coordinated with the inventory, reconstruction, and measurement of individual bones, and the reassembly of the entire skeleton in anatomical position. First, the approximately 265 bones and fragments comprising the skeleton were reassembled into complete, or as complete as possible, individual elements. A standardized coding system was then used to record taphonomic observations, with emphasis on the long bones and cranium. The presence of sediments adhering to the surfaces or within the medullary cavities of the bones and the endocranium was documented, especially the density and surface distribution of calcium carbonate-cemented sediment. This survey also searched for evidence of skeletal disintegration (biological diagenesis) from such causes as root etching, animal scavenging, and weathering (cracking, surface deterioration, or sun bleaching).

PREVIOUS TAPHONOMIC ASSESSMENTS

To some extent, previous studies of Kennewick Man (Huckleberry et al. 2003:663; McManamon et al. 2000; Powell and Rose 1999; Walker et al. 2000) have examined aspects of the skeleton that relate to taphonomic processes. Multiple reports (Huckleberry et al. 2003:663; Powell and Rose 1999) have commented on the completeness of the remains. Colors of the bones have also been considered. According to Walker et al. (2000), a few of Kennewick Man's bones have a white or light-colored hue that can be interpreted as sun-bleaching. Powell and Rose (1999) identified red staining on the left third metacarpal (97.L.16.Mca), two right rib fragments (97.I.12d.11, 97.I.12d.5), 27 unidentified fragments of left ribs, an articular facet (97.U.6.(L2c)), the anterior and posterior surfaces of the shaft of the right tibia, two phalanges of the left foot (97.L.24.Pb, 97.L.24Pa), and nine other unidentified fragments. "The red staining on some bones *may* be cultural in origin, suggesting the application of red ochre pigment to the skin of the individual prior to interment. This determination will require confirmation of iron oxide levels in the matrix adhering to the bone and possibly chemical analysis of the bone itself" (Powell and Rose 1999, italics in original). Walker et al. (2000) also observed reddish discolorations on several bones: a complete right first metacarpal (97.R.16(MCd)); two complete right hand phalanges (97.R.16(Pe) and 97.R.16(Pf)); the right pubis (97.R.17a); and the right ilium fragment containing the embedded projectile point (97.R.17c). Unlike Powell and Rose (1999), Walker et al. (2002) reported that the stains likely did not have a cultural origin, but rather were "the result of natural processes operating after this person's burial," specifically decomposition of roots in contact with the skeleton. Chatters (2001), on the other hand, found no evidence for red pigment on the skeleton.

Both Powell and Rose (1999) and Walker et al. (2000) noted animal damage on the skeleton. Powell and Rose (1999) identified multiple examples: 1) gnaw marks made by a medium-sized animal on the proximal midshaft quarter of the right humerus; 2) marks made by a small animal on the proximal right ulna; 3) small rodent gnaw marks on the fourth metacarpal of the right hand; 4) small rodent gnaw marks on rib segment 97.I.12d.11 (identified by Owsley and Karin Bruwelheide as the right ninth rib); 5) a patch of large gnaw marks on the anterior crest on the proximal middle quarter of the left tibia; 6) animal activity on the right and left fibulae; 7) animal activity on the left fourth metatarsal (on three locations); and 8) animal activity on the left foot phalanges. Powell and Rose (1999) also indicated that the gnaw marks on the proximal shaft of the right ulna lacked staining and soil accumulation, suggesting that the gnawing occurred while the skeleton was exposed on the lake shore immediately before discovery. Walker et al. (2000: Figure 8) identified rodent gnawing on the medial surface of the proximal end of the right ulna (97.R.15a) and on a radius (side unknown, this segment was misidentified as 97.R.15a, which is the right ulna). Contrary to Powell and Rose (1999) and Walker et al. (2000), Huckleberry et al. (2003) found no evidence of carnivore scavenging.

Based on these aspects—near-completeness of the remains, bone color, and the evidence, or lack thereof, for rodent and carnivore gnawing—these studies have commented on whether Kennewick Man's burial was natural or intentional and on how the body was oriented. One Department of the Interior study team noted that while "both human burial and rapid alluvial sedimentation are viable hypotheses," the context of the skeleton likely represents an intentional burial because the remains are relatively complete and there is no evidence of carnivore scavenging (Huckleberry et al. 2003:663). No inferences were made about the in situ position of the body within the proposed grave. Powell and Rose (1999) agreed that the remains most likely derived from an intentional burial due to their completeness and the presence of what they believed to be red ochre staining on some bones.

Chatters (2001) tested the possibility that Kennewick Man was a natural (geologic) burial by commissioning a sediment analysis from an expert from Washington State University, Pullman. The analyst found that the carbonate-cemented sediment on the skeleton was more likely derived from the "uppermost layer of the river-laid stratum," rather than the stratigraphically higher, eolian strata (Chatters 2001:154). Chatters (2001:157) argued that an intentional burial should have a greater range of sediment types adhering to the skeleton, whereas the sediment analysis was only able to match sediment from the skeleton to one stratum at the site. However, he noted that there was evidence against accidental interment, including excellent preservation and the absence of damage caused by carnivore scavenging.

Walker et al. (1999) reported that Kennewick Man had been buried intentionally on his side in a flexed position, and that the skeleton came out of the bank during two erosion episodes. In their opinion, the presence of sun-bleaching, algae, and abrasions on the long bones indicated that the remains were in the fetal position with the limb bones facing the river. They suggested the initial phase of erosion exposed most of the appendicular skeleton followed "several weeks or months" later by a second collapse that deposited the axial skeleton on the shoreline.

Kennewick Man was buried intentionally, but this chapter will present evidence for a different body arrangement and orientation, as well as for a different order of bone erosion from the bank and a shorter time span for this process.

RECORDING TAPHONOMIC OBSERVATIONS

In this analysis of the Kennewick Man skeleton, general patterns of taphonomic change were noted with the goal of providing information about the circumstances under which the skeleton was buried (i.e., natural or intentional), the in situ positioning of the body, and its orientation relative to the lake.

The bones and fragments of the skeleton were inspected individually. Although the bones of the limbs were the primary foci of this study, special attention was also given to rib fragments. The sternal thirds of many ribs are missing, but most vertebral halves are present. Accurate identification and description of these fragments were necessary not only for the evaluation of complicated antemortem rib pathology but also for identifying postmortem rib fractures and other taphonomic indicators.

Using a specially designed coding system, taphonomic observations were recorded in spreadsheets—organized anatomically from head to foot—that note each segment of each long bone, its previously defined catalog number, and a description of the segment. One spreadsheet records each segment's Munsell color(s), including sun bleaching and stains (green from algae, red from ochre) and any surface damage such as cortical exfoliation, loss of patina, root etching, or animal activity. It also notes sediment intrusion into the medullary cavities and carbonate-cemented sediment on the surfaces.

A similar spreadsheet was developed for recording complete and incomplete postmortem breaks in the skeleton (Chapter 19). Observations included tension and compression surfaces along with their morphology, the location of the breakaway spur and fixed end, the break angle and shape, the presence and number of secondary breaks, and the direction of break propagation. Each fracture face (exposed internal cortical bone) was also assessed for color and staining using Munsell color charts (Figure 17.1). Color variations among fracture surfaces provide a means of recognizing fractures that occurred at the same time or experienced differences in postfracture microenvironments (Figure 17.2). As such, the postmortem fracture analysis provides insight into the order in

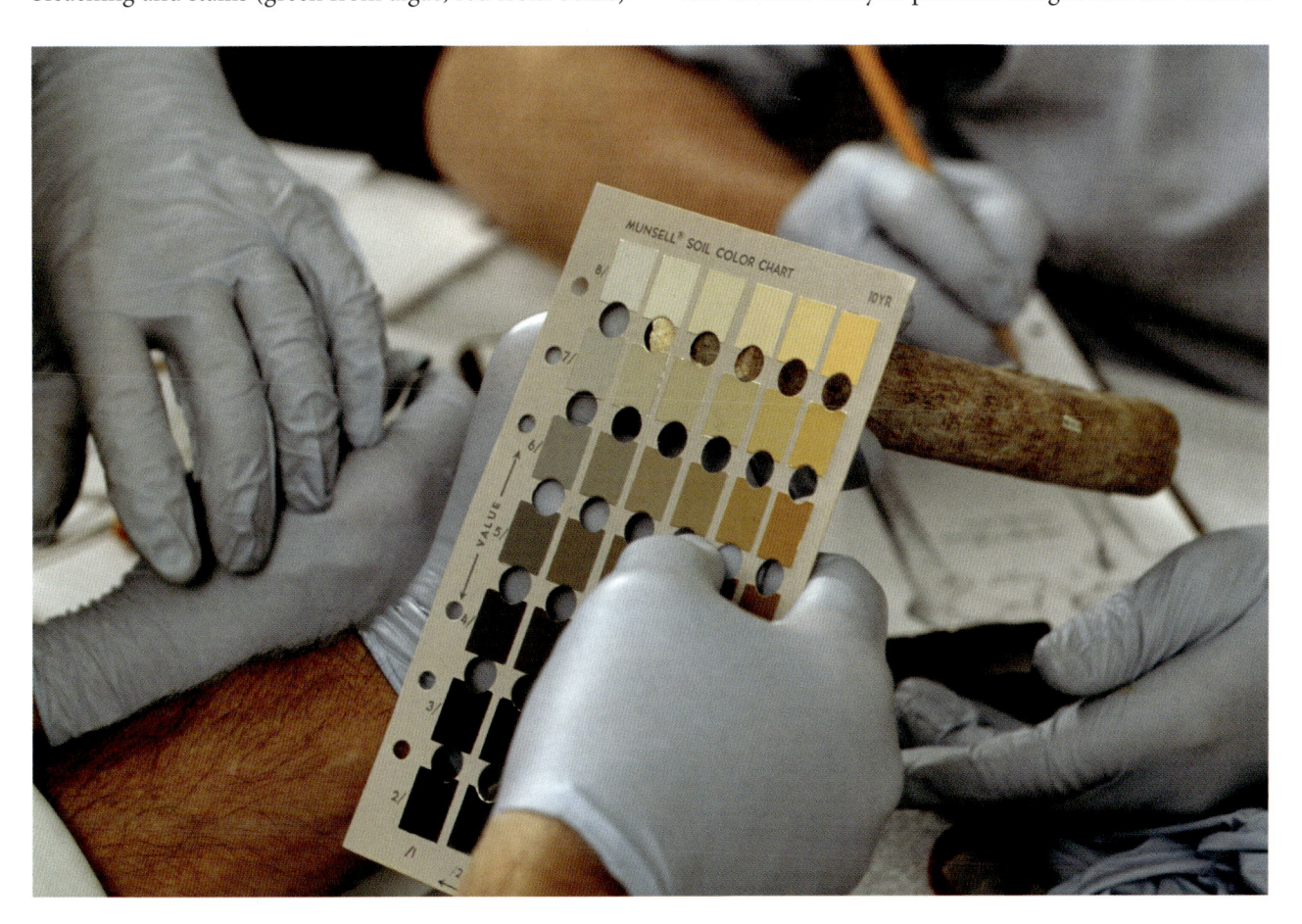

FIGURE 17.1 Evaluating the color of exposed femur cortex using a Munsell color chart. Compared to the bone's sediment-stained external surface, the color of the break visible through the chart's portal is much lighter, almost white, indicating recent postmortem breakage that occurred when the skeleton eroded from the bank.

FIGURE 17.2 Adjoining light-colored fracture margins in the right radius represent damage that occurred at the time Kennewick Man's skeleton eroded from the terrace. The medullary cavity is filled with fine sediment containing secondary calcite crystals.

which the remains eroded from the bank and contributes to an understanding of the body's in situ positioning.

Many of these observations, especially break characteristics and color gradients, were also recorded on bone diagrams showing the anterior, medial, lateral, and posterior surfaces of each long bone (Figure 17.3). Taphonomic observations for the scapulae, clavicles, ribs, and innominates were similarly illustrated on bone diagrams. The cranium and vertebrae were individually described, and the interior of the cranial vault was assessed using computed tomography (CT).

Munsell colors for the clavicles, scapulae, long bones, and fracture surfaces were recorded under indoor fluorescent lighting. The colors of the cranial bones and mandible were judged to be less true as a result of previous, standard surface treatments related to the preparation of the skull for casting. Chatters (2000:300) used a diluted water-based polymer to stabilize the cranium and bones of the face, reconstructed the mandible using Elmer's® glue to facilitate initial measurement and photography, and applied a release agent to the skull prior to the construction of a polyurethane blanket mold.

After assessing each bone or fragment, reassembled bones were arranged in anatomical position on a table padded with natural sand covered with a lint-free black fabric (Chapter 6). Rearticulating the skeleton enabled taphonomic indicators to be seen both in the context of individual bones as well as across the entire skeleton. Some bones were temporarily rejoined using Parafilm® to refit major fragments and stabilize certain elements to facilitate taphonomy and pathology analyses and measurements. No bones were glued together by this taphonomy study team.

NATURAL VERSUS DELIBERATE INTERMENT
Completeness and Lack of Animal Damage

Anatomical completeness of the skeleton, including many small hand and foot bones, and lack of animal damage are strong indicators for the intentional burial of Kennewick Man by humans. Many paleontological

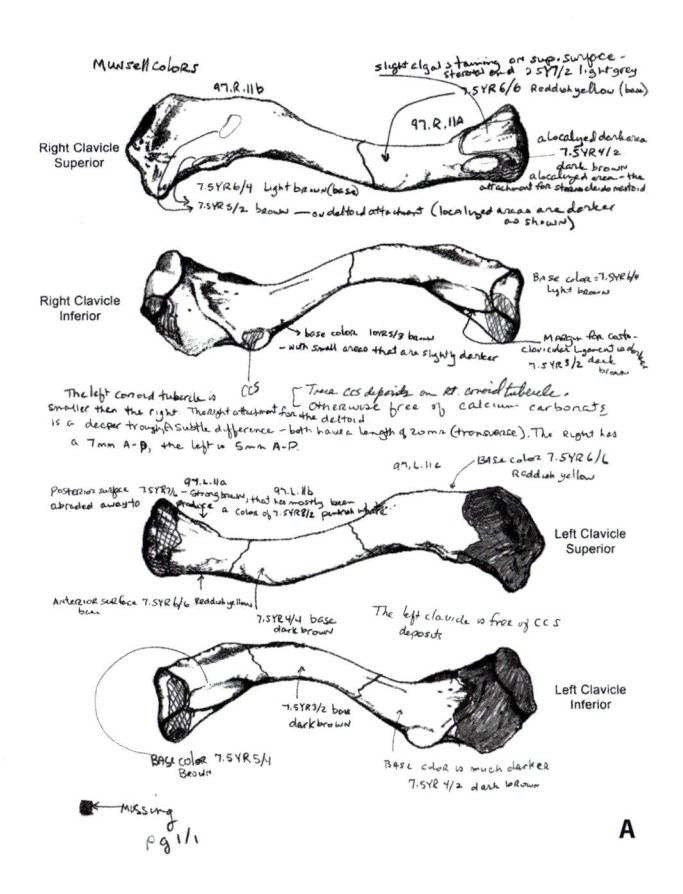

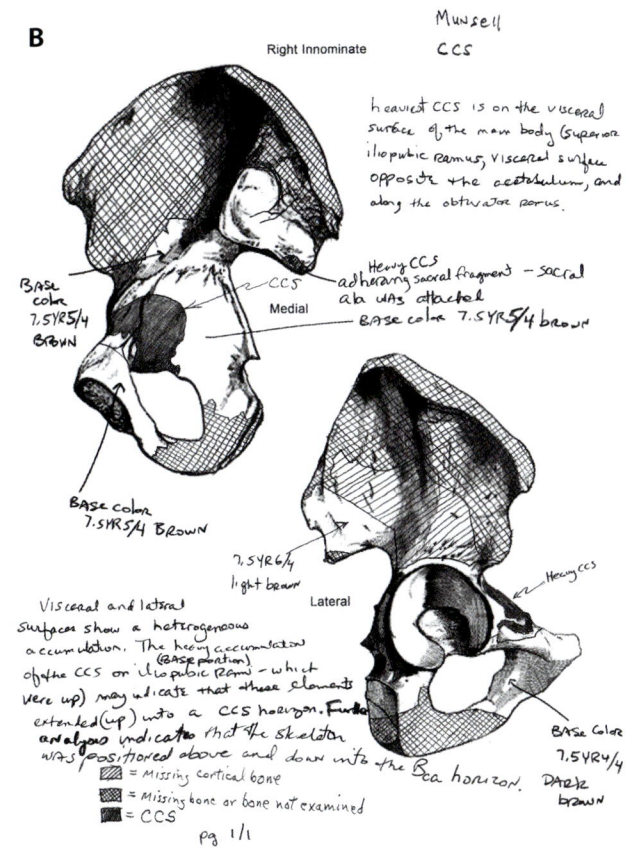

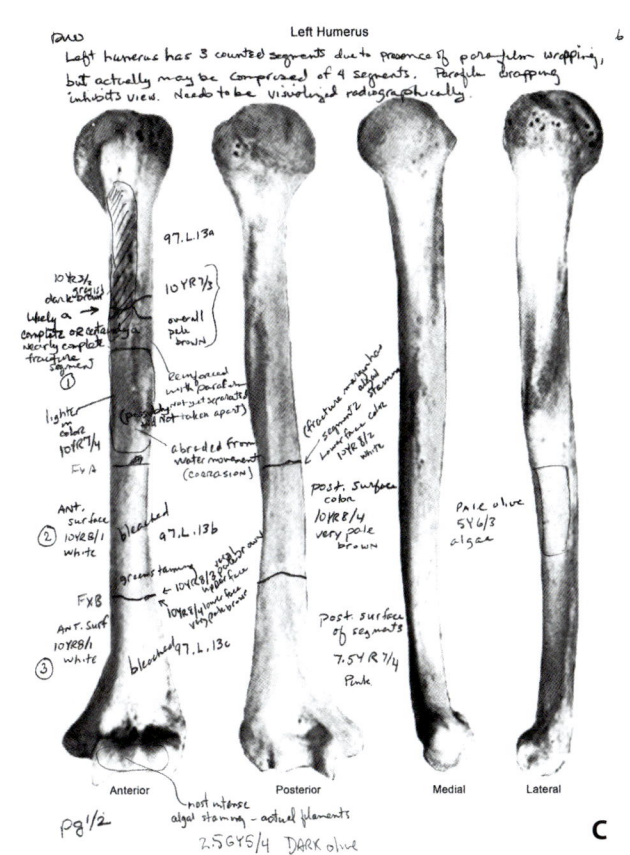

FIGURE 17.3 Diagrams of Kennewick Man's right and left clavicles (A, showing both superior and inferior surfaces), left humerus (B, showing all surfaces), and right innominate (C, showing medial and lateral surfaces) completed during the 2005 and 2006 studies. Attributes noted on the diagrams include: Munsell colors, carbonate-cemented sediment, staining, remarks on functional morphology, postmortem breaks, missing bone, and other taphonomic observations. These diagrams and corresponding written descriptions were critical in establishing taphonomic patterning throughout the skeleton.

sites contain mostly complete naturally buried fossil skeletons without evidence for animal damage; however, carnivores commonly find unburied remains soon after death. The incomplete skeleton of a male from On Your Knees Cave in Alaska, for example, shows damage from large carnivore scavenging, including gnawing and tooth punctures (Figure 17.4). Another example comes from Midland, Texas, where fragmentary human remains that may have been scattered and dismembered by animals

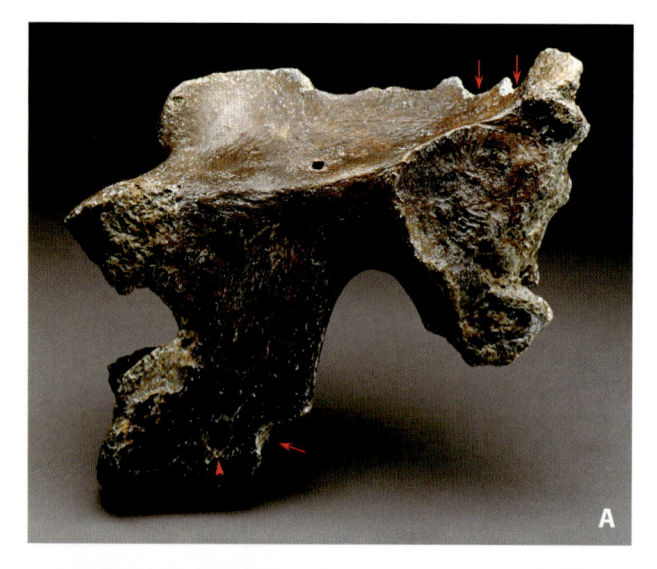

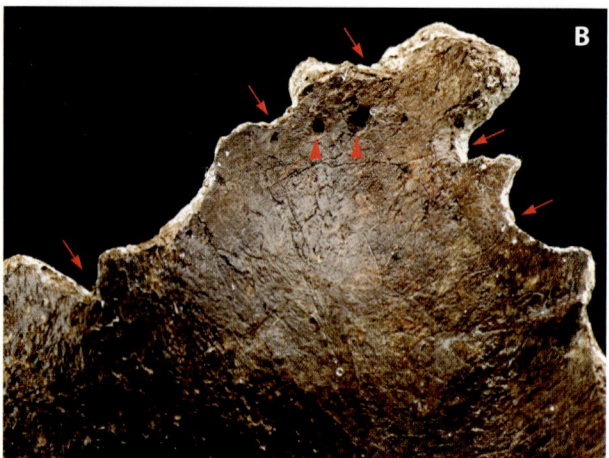

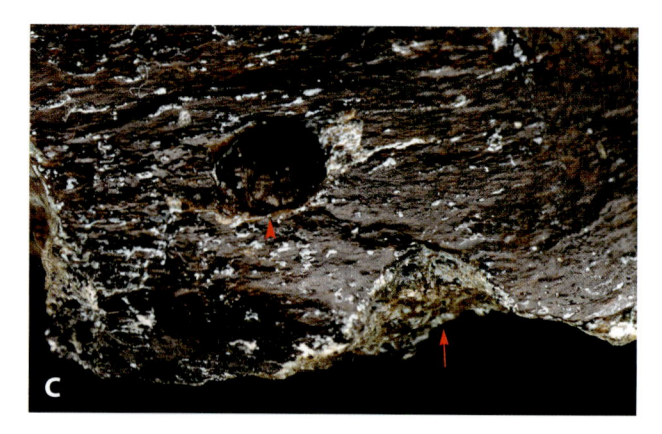

FIGURE 17.4 Internal and external views of a partial right innominate of a male, aged 20 to 22 years, from On Your Knees Cave showing solid, compact cancellous bone. Loss of the pubis and breakage of the ilium and ishium are due to large carnivore scavenging of these human remains. The remnant margin of the ilium has multiple, semicircular bear-sized tooth gouges (A and B, arrows) with a row of interior puncture marks on the external surface (A and B, arrowheads). The internal surface of the superior ischium (C) has a bear-sized canine tooth puncture (arrowhead) measuring 7 by 5 millimeters with a depth of more than 3 millimeters (the lumen is partially occluded). An adjacent puncture, located along the posterior margin of the ischium, shows as a semi-circular gouge with a superior-inferior diameter of about 8 millimeters (arrow).

FIGURE 17.5 In this modern forensic example, the complete skull, partial thorax, and left femur diaphysis were the only recovered elements of a male. The degree of bone breakage, though leaving a complete skull, reflects scavenging by a bear. The femur has irregular, jagged breakage of proximal and distal cortical bone with loss of the joint ends. Remnants of ribs are attached to the vertebral column, which also shows loss of the anterior centra.

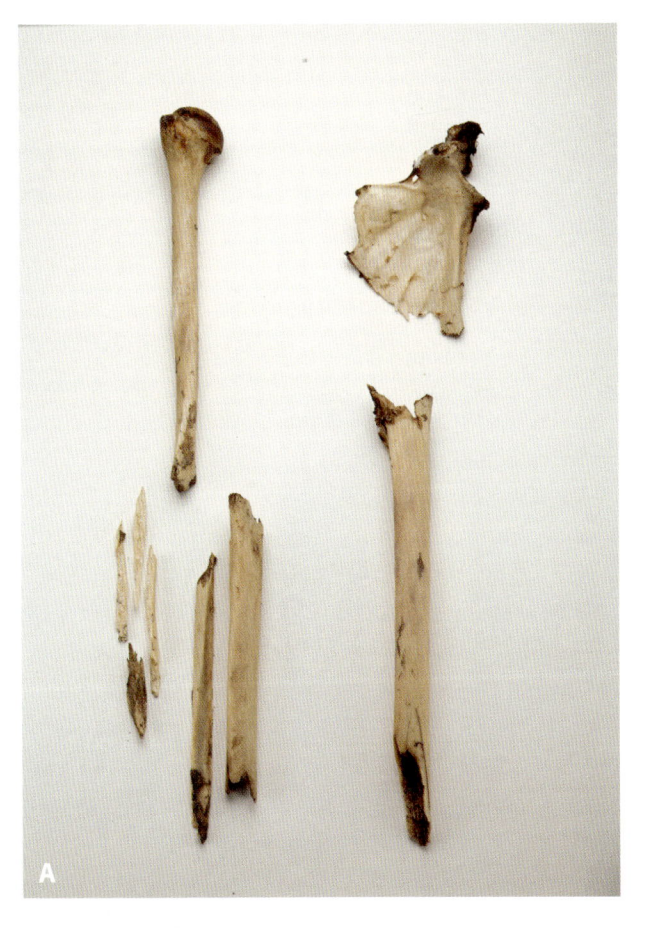

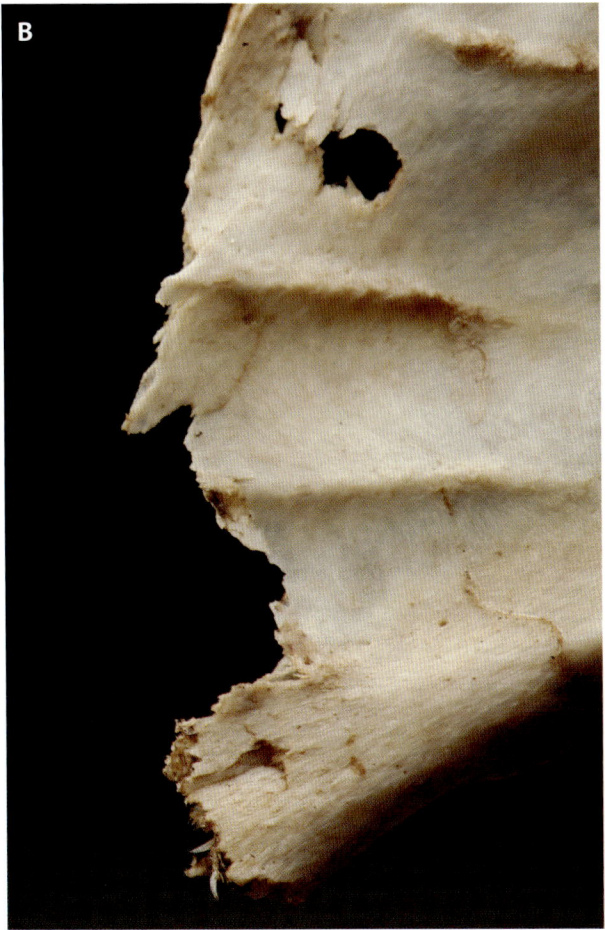

FIGURE 17.6 In this modern forensic example, chewed and splintered skeletal remains of a female were damaged by bear scavenging. The distal end of the left humerus was gnawed off (A) and the proximal and distal ends of the left tibia, right femur, and fibula are missing. The left scapula (B) has vertebral border damage with a puncture from a canine tooth measuring six millimeters in diameter (close-up of the internal surface). The articulated pelvis (C) has breakage of the pubic bones and iliac crests with multiple semi-circular gouges and tooth punctures, as illustrated in a close-up view of the external surface of the right ilium (D).

FIGURE 17.7 Stafford (center) and C. Wayne Smith (right) record Munsell colors of a bone segment; remarks were recorded by Aleithea Williams (left). The box containing bone segments shows that each piece is fitted into Styrofoam® specifically cut to its profile and separately labeled.

were discovered in 1953 (Wendorf et al. 1955:96). There, a fragmented, partial calvarium, teeth, two rib fragments, and three metacarpals were all that was found embedded in a late Pleistocene sand deposit.

These ancient examples come from different environmental contexts from those observed in contemporary forensic anthropology casework. Nevertheless, modern forensic experience from a variety of outdoor situations notes the frequent occurrence of animal scavenging when remains are exposed for any length of time (Haglund and Sorg 1997). Examples from forensic casework further illustrate the degree and type of damage that can occur when human remains are scavenged by large carnivores (Figures 17.5 and 17.6). Destruction of this type was not observed on Kennewick Man's bones, suggesting that his body was buried. The skeleton has suffered some physical damage in the form of macroabrasions that, as will be discussed in more detail later in this chapter, are clearly distinct from carnivore or rodent damage.

Bone Colors and Staining

The presence of red ochre staining would provide supporting evidence that Kennewick Man was intentionally interred. Red ochre (powdered hematite, Fe_2O_3), either covering remains or left as a grave offering, has been observed in early Holocene burials in the New World. Primary interments with red ochre dating from 9,500 to 10,680 RC yr. BP include adults from Arch Lake (New Mexico), Horn Shelter No. 2 (Texas), Gordon Creek (Colorado), and a 1.5-year-old child from the Anzick site (Montana) (Breternitz et al. 1971; Owsley and Hunt 2001; Owsley et al. 2010). There is also evidence for the use of red ochre in late Paleolithic burials from sites in Eurasia, including Sunghir, Dolní Věstonice, Paviland, and Abrigo do Lagar Velho (Trinkaus and Buzhilova 2010; Formicola et al. 2001; Aldhouse-Green and Pettitt 1998; Duarte et al. 1999).

The study team inspected each segment of the recovered bones for discolorations and staining (Figure 17.7). In no instances could verifiable red ochre or reddish stains be explained by human behavior. Although a few bones and bone fragments have a reddish hue, the stains observed are so subtle and occur so randomly that they are attributed to natural chemical processes, unlike the example shown in Figure 17.8.

In Kennewick Man's bones, the base color of the medial surface of the right tibia is a reddish yellow (7.5YR6/6) with localized dark red areas (7.5YR3/0) corresponding with areas of recent exfoliation of overlying carbonates. The lateral surface of the proximal and middle diaphysis of this tibia is a light brown color (7.5YR6/4). Reddish tones are also evident on the left clavicle. This hue appears to be a weathering effect involving loss of surface patina from light scouring and fine abrasion of surfaces briefly exposed to weathering and corrasion. The anterior aspect of the sternal end of the left clavicle is reddish yellow (7.5YR6/6), and the posterior aspect of this segment has a strong brown base color (7.5YR7/6) that

FIGURE 17.8 The bones of a possible Clovis-age male infant from the Anzick site in Montana were heavily stained by red ochre.

has abraded away due to exposure. Base colors of the superior and inferior surfaces of the sternal and middle segments of the left clavicle are brown (10YR5/4) and dark brown (7.5YR4/4), but they grade toward a reddish yellow (7.5YR6/6) on the superior surface of the acromial end, while remaining dark brown (7.5YR4/2) on the inferior surface. Although the reddish yellow tones are somewhat atypical, they appear natural in origin.

With few exceptions, the base colors of the majority of the Kennewick bones fall within brown tones ranging from pale and light to very dark (Table 17.1). The darkest colors observed were from manganese dioxide staining on the teeth (Figure 7.7). Raised and roughened ligament and muscle attachments are dark brown (7.5YR3/2, 7.5YR4/2, 7.5YR4/4), primarily due to greater surface irregularity and microporosity, which facilitated the uptake of pigments from the surrounding sediment. The overall color of the right humerus (97.R.13a and 97.R.13b), for example, is very pale brown (10YR7/3), while the deltoid tuberosity and the muscle attachments for pectoralis major, latissimus dorsi, and teres major are dark brown (7.5YR4/2). Decomposing organic matter is one source of brownish coloration commonly affecting buried bones; however, in the case of Kennewick Man, the presence of iron oxides in the soil is a more significant factor.

Subtle bone surface deterioration (categorized as weathering in Table 17.2) caused primarily by water abrasion and brief exposure while in situ during the period of erosion from the terrace bank, is partially reflected by lighter surface tones. Color differences between exposed and protected surfaces that remained buried for a short while longer can sometimes be used to infer in situ positioning. The inferior surface of the left clavicle, for example, has a darker color than the superior surface, suggesting the superior side was up and consequently more affected by weathering. This pattern is seen in other bones. The superior surface of the lateral half of the right clavicle is lighter in color than the inferior surface on which even darker staining appears on the rhomboid fossa, the attachment for the costoclavicular ligament. Both scapulae exhibit a loss of patination of their ventral surfaces, which contrasts with darker posterior surfaces.

Deposits of carbonate-cemented sediment create another color variant. A representative color of such an accumulation is light gray (10YR7/1), as observed on the medial surface of the left ulna. On carbonate-covered surfaces, underlying original colors were evident in locations where adhering sediments have flaked off without removing the superficial layer of bone. This pattern is evident on the left radius (10YR4/4, dark yellowish brown) and the lower midshaft of the right tibia (varying from 7.5YR3/0, very dark red, to 7.5YR3/2, dark brown), where

TABLE 17.1 Munsell colors of selected Kennewick Man bones.

Bone	Side	Specimen Number	Surface	Segment Number	Segment Description	Munsell Color
Clavicle	L	97.L.11a	Superior surface		Sternal end	Anterior surface is 7.5YR6/6 (reddish yellow); posterior surface is 7.5YR7/6 (strong brown) and has been mostly abraded away to produce a color of 7.5YR8/2 (pinkish white)
Clavicle	L	97.L.11b	Superior surface		Middle section	Base is 7.5YR4/4 (dark brown)
Clavicle	L	97.R.11c	Superior surface		Acromial end (missing epiphysis)	Base is 7.5YR6/6 (reddish yellow)
Clavicle	L	97.L.11a	Inferior surface		Sternal end	Base color is 10YR5/4 (brown)
Clavicle	L	97.L.11b	Inferior surface		Middle section	Base is 7.5YR3/2 (dark brown)
Clavicle	L	97.R.11c	Inferior surface		Acromial end (missing epiphysis)	Base is 7.5YR4/2 (dark brown)
Clavicle	R	97.R.11a	Superior surface		Sternal end	Base is 7.5YR6/6 (reddish yellow) with a localized area of 7.5YR4/2 (dark brown) at the attachment site for the sternocleidomastoid
Clavicle	R	97.R.11b	Superior surface		Acromial end	Base is 7.5YR6/4 (light brown) with a localized area of 7.5YR5/2 (brown) on the deltoid attachment
Clavicle	R	97.R.11a	Inferior surface		Sternal end	Base color is 7.5YR6/4 (light brown); margin for costoclavicular ligament is 7.5YR3/2 (dark brown)
Clavicle	R	97.R.11b	Inferior surface		Acromial end	Base color is 10YR5/3 (brown) with small areas slightly darker
Scapula	L		Anterior			Primarily 10YR7.3 (very pale brown) with about 30% coverage of 7.5YR4/2 (dark brown) below the glenoid in the attachment area of the long head of the triceps
Scapula	L		Posterior			Primarily 10YR8.3 (very pale brown) with 10YR5/2 (grayish brown) below the glenoid in the attachment area of the long head of the triceps
Scapula	R	97.R.10b	Ventral		Glenoid portion (including adjacent surfaces)	7.5YR5/4 (brown)
Scapula	R	97.R.10b	Dorsal		Glenoid portion (including adjacent surfaces)	7.5YR7/4 (pink)
Humerus	L	97.L.13a		1	Epiphysis and upper three-fifths of the diaphysis	Overall color is 10YR7/3 (very pale brown). The interior cortex and upper face of fracture A is 10YR8/3 (very pale brown). Dark staining of the anterior surface of the prominent muscle attachment areas is 10YR3/2 (very dark grayish-brown) from iron with the possible addition of manganese staining. The interior of the incomplete fracture located toward the proximal end of the humerus is 10YR8/2 (white).

(continued)

TABLE 17.1 (*continued*) Munsell colors of selected Kennewick Man bones.

Bone	Side	Specimen Number	Surface	Segment Number	Segment Description	Munsell Color
Humerus	L	97.L.13b		2	Lower portion of the diaphysis (about one-fourth)	Anterior surface is 10YR8/1 (white); posterior surface is 10YR8/4 (very pale brown)
Humerus	L	97.L.13c		3	Most inferior one-fourth of the diaphysis and distal joint	Anterior surface is 10YR8/1 (white); posterior surface is 7.5YR7/4 (pink)
Humerus	R	97.R.13a		1	Proximal epiphysis and upper three-fourths of the diaphysis	Overall color is 10YR7/3 (very pale brown) from iron oxide staining. Other color ranges from 7.5YR4/2 (dark brown) on the deltoid tuberosity in the attachment locations for pectoralis major, latissmus dorsi, and teres major (muscle attachment areas that have stained more darkly) to 7.5YR7/2 (pinkish-gray). The most interior cortical bone color is 10YR8/2 (white).
Humerus	R	97.R.13b		2	Lower one-fourth of the diaphysis and distal joint	Overall color is 10YR7/3 (very pale brown) from iron oxide staining. Other color ranges from 7.5YR4/2 (dark brown) on the posterior prominent ridge of the deltoid tuberosity (a muscle attachment area that has stained more darkly) to 7.5YR7/2 (pinkish-gray). Deeper cortical bone color is 10YR8/2 (white).
Radius	L	97.L.14a		1	Proximal one-third of the diaphysis including a nearly complete radial head	Overall color is 7.5YR7/4 (pink). Darkest areas are 7.5YR4/2 (brown). The darker areas are primarily located on the radial tuberosity and adjacent medial surface in the region surrounding the interosseous crest.
Radius	L	Not recovered		2	Not recovered	Not recovered
Radius	L	97.L.14b		3	Distal one-half of the radius including the joint surface	Overall color is 10YR8/4 (very pale brown). Small localized areas on the anterior surface of the distal diaphysis, which likely represent areas of recent loss of carbonate-cemented sediment, can be as dark as 10YR4/4 (dark yellowish brown).
Radius	R	97.R.14a		1	Head and two-thirds of the diaphysis	Overall color is 7.5YR7/4 (pink). Darker areas, colored 7.5YR3/2 (dark brown), are evident on the radial tuberosity and in a linear band on the proximal one-third of the posterior surface
Radius	R	97.R.14b		2	Lower one-third of the diaphysis and the distal end	Overall color is 7.5YR7/4 (pink). The darkest area is 7.5YR4/4 (dark brown) located on the lateral surface of the distal one-fourth.

(*continued*)

TABLE 17.1 (continued) Munsell colors of selected Kennewick Man bones.

Bone	Side	Specimen Number	Surface	Segment Number	Segment Description	Munsell Color
Ulna	L	97.L.15a		1	Proximal one-third of the ulna including the joint surface	Overall color is 10YR8/4 (very pale brown). The darkest color is 10YR3/2 (very dark grayish brown). The darker color is localized on the anterior surface of the coronoid process and on the tuberosity, the attachment site for brachialis. The color of the upper face of fracture A is 10YR8/3 (very pale brown); the color of the lower face of fracture A is 10YR8/2 (white).
Ulna	L	97.L.15b		2	Middle one-third of the diaphysis	The base color of the anterior surface is 7.5YR7/4 (pink). The base color of the posterior surface is 7.5YR6/4 (light brown). The posterior surface is slightly darker than the anterior surface. The medial surface is 7.5YR8/2 (pinkish-white).
Ulna	L	97.L15c		3	Distal one-third of the diaphysis; the epiphysis is missing	Anterior surface base color is 7.5YR5/4 (brown). That color applies to all surfaces of the segment (representing the underlying bone color). However, about 20% of the segment is covered with carbonate-cemented sediment that has a base color of 10YR7/1 (light gray). The distal diaphysis of segment 3 is darker in color than segments 1 and 2. This color, along with lack of algal staining and sun-bleaching, indicates less aerial exposure than segments 1 and 2.
Ulna	L	Not recovered		4	Distal epiphysis and joint surface	Not recovered
Ulna	R	97.R.15a		1	Proximal one-third of diaphysis including the intact joint	The base color of the anterior and lateral surfaces is 7.5YR6/4 (light brown). The muscle attachment areas on the anterior surface (the tuberosity) are 7.5YR4/4 (dark brown). The posterior and medial surfaces are a base color of 10YR7/6 (yellow).
Ulna	R	97.R.15b		2	Middle one-third of the diaphysis	The medial surface, at 7.5YR8/4 (pink), is the lightest of the three sides. The base color of the anterior surface is 7.5YR6/4 (light brown). The posterior surface base color is 7.5YR 5/4 (brown).
Ulna	R	97.R.15c		3	Distal one-third of the diaphysis including the distal joint	The anterior surface has a localized darker colored area, 7.5YR5/2 (brown), in the attachment area for the pronator quadratus. The base color for the anterior surface is 7.5YR7/4 (pink).
Femur	L	97.L.18a		1	Proximal two-fifths	Overall color ranges from 7.5YR6/4 (light brown) to 7.5YR5/4 (brown). Most prominent is the 7.5YR6/6 (yellowish-brown), which is probably caused by iron oxide staining.

(continued)

TABLE 17.1 (continued) Munsell colors of selected Kennewick Man bones.

Bone	Side	Specimen Number	Surface	Segment Number	Segment Description	Munsell Color
Femur	L	01.L.18b		2	Middle and upper lower diaphysis	Overall color of the segment is 10YR7/4 (very pale brown). Color ranges from 10YR4/4 (dark yellowish brown), which is directly under the sediment that has recently exfoliated (not sun-bleached) without taking outer bone surface, to 10YR8/3 (very pale brown), where localized areas of carbonate-cemented sediment have exfoliated with loss of superficial bone.
Femur	L	01.L.18a		3	Portion of lower diaphysis and distal joint surface	The sediment is 10YR7/1 (light gray). Bone making up the outer rim of the breakage surface is 10YR8/2 (white). The overall segment ranges from 10YR7/2 (light gray), on the metaphyseal surface, to 10YR6/4 (slight yellow brown) on the posterior surface.
Femur	R	01.R.18a		1	Proximal one-fourths	There are traces of green staining, colored 2.5Y8/2 (white), on the entire segment.
Femur	R	01.R.18b		2	Midshaft (middle one-third and lower proximal one-fourth)	The overall color of the segment is 7.5YR6/4 (light brown). The color ranges from 7.5YR4/4 (dark brown) to 7.5YR8/2 (white). The lightest color present on the segment is 7.5YR8/2 (white), where the cortical surface as been broken through and the inner cortical bone is visible.
Femur	R	97.R.18a		3	Portion of lower diaphysis (one-third) and distal joint surface	The overall color of the segment is 7.5YR5/4 (brown). The color ranges from 7.5YR8/2 (pinkish white) to 7.5YR4/4 (dark brown).
Tibia	L	97.L.20a		1	Proximal joint, epiphysis, and most superior diaphysis	The overall color of the segment is 10YR7/3 (very pale brown).
Tibia	L	97.L.20b		2	Two-fifths of the proximal diaphysis	The overall color of the segment is 7.5YR5/4 (brown). The lightest color present is 10YR7/2 (light gray).
Tibia	L	97.L.20c		3	Lower three-fifths of diaphysis and distal joint	The base color is 7.5YR6/4 (light brown). A significant amount of the bone is 7.5YR5/4 (brown). Dark spots (7.5YR4/4 (brown) to 7.5YR3/4 (dark brown) are present in a few locations.
Tibia	R	97.R.20a		1	Medial condyle and epiphysis; the anterior half of this bone segment is missing postmortem	Not recorded
Tibia	R	97.R.20b		2	Lateral condyle and epiphysis; the anterior half is missing	Not recorded

(continued)

TABLE 17.1 (continued) Munsell colors of selected Kennewick Man bones.

Bone	Side	Specimen Number	Surface	Segment Number	Segment Description	Munsell Color
Tibia	R	97.R.20c		3	Proximal and middle one-third of the diaphysis including most of the tibial tuberosity	The base color of the medial surface is 7.5YR6/6 (reddish yellow). A few localized areas of darker staining, color 7.5YR3/0 (very dark red), represent recent areas of carbonate-cemented sediment exfoliation. The surface also has areas of 7.5YR4/2 (dark brown) representing areas of surface carbonate-cemented sediment exfoliation that occurred earlier (areas of older loss) and probably are more oxidized. The lateral surface base color is 7.5YR6/4 (light brown). The darkest color, under recently exposed carbonate-cemented sediment, is 7.5YR3/2 (dark brown). The posterior surface base color is 7.5YR7/4 (pink). The posterior surface has localized areas as dark as 7.5YR3/2 (dark brown).
Tibia	R	97R.20d		4	Distal one-third of the diaphysis including the joint	The base color on the anterior surface is 7.5YR6/6 (reddish yellow) with light colored areas representing recent loss of the outer surface bone. There are also darker areas of carbonate-cemented sediment deposit, color 7.5YR3/2 (dark brown), where the underlying subperiosteal surface was left intact. The color of the posterior surface is 7.5YR6/4 (light brown). The posterior surface has very small, localized dark colored areas from recent exfoliation.

Munsell colors were determined under fluorescent interior lighting in the Burke Museum, Seattle. The bones were dry at the time of observation.

recent loss of carbonate-cemented sediment reveals dark cortical bone. The interior color is white (10YR8/2) or very pale brown (10YR8/3) (unless discolored by algae or slightly stained by submersion in the mudflat) in areas where exfoliation of carbonate-cemented sediment has removed surface bone or where breakage as a result of erosion has exposed interior cortical bone (Figure 17.2).

The absence of ochre staining does not negate the possibility of burial by humans. Other kinds of evidence, including color variations, help resolve the question of natural versus deliberate interment.

TAPHONOMIC INDICATORS FOR ANATOMICAL PLACEMENT OF THE BODY

Carbonate-Cemented Sediment Concretions

The Formation of Carbonate-Cemented Sediment

Microscopic analysis showed that clayey silt and very fine sandy-clayey silt were cemented to the majority of Kennewick Man's bones with secondary (pedogenic) calcium carbonate. The depth-to-carbonate horizon and the presence, form, and depth of pedogenic carbonate are important tools for studying paleoclimates (Tanner 2010). In this case, pedogenic carbonate, or carbonate-cemented sediment, is a major taphonomic indicator not only of burial position, but also of stratigraphic position.

Pedogenic carbonate develops in the soil below the A horizon, the uppermost horizon where organic matter accumulates. Pedogenic carbonate is deposited during soil formation processes that precipitate calcium carbonate in the B horizon, the soil zone where sediments are most altered chemically and where combinations of calcium carbonate, clays, or iron oxides accumulate (Deocampo 2010). The relative amounts of clay, carbonate, or iron oxide depend primarily upon climate and the mineralogy of local rock and sediments. These pedogenic carbonate horizons are historically termed Bca, although the revised term is now Bk. In this study, Bca is used because it is more familiar to the scientific community.

The genesis of pedogenic carbonate occurs when rain and atmospheric carbon dioxide (CO_2) combine to form dilute carbonic acid. This weakly acidic water dissolves minerals, yielding water soluble calcium carbonate, bicarbonates, and other salts capable of precipitation if ground water conditions are suitable (Deocampo 2010).

Low rainfall is the single most important condition for the development of pedogenic carbonate, which typically occurs in arid to semi-arid environments receiving less than approximately 25 inches of rain per year. As a result of low rainfall, pedogenic carbonate is deposited closer to the ground surface. "The precipitation of calcium carbonate at depth in a soil profile occurs as carbonate solubility is lost in meteoric waters, which are naturally acidic due to dissolution of atmospheric and soil-respired CO_2. In part, the depth at which conditions of calcium-carbonate solubility change to insolubility is controlled by the flow of water through the vadose zone, that is, meteoric precipitation" (Tanner 2010:185). At higher rainfalls, soluble salts are washed into the groundwater table and are removed from the sediments. At lower rainfalls, water penetrates a rainfall-dependent depth in the soil and evaporates, thereby precipitating $CaCO_3$. Due to a rainshadow effect caused by the Cascade Mountain Range, there is much lower precipitation on the eastern side of the range (Whitney and Sandelin 2003:10) where Kennewick Man was found.

Globally, soil carbonate horizons vary in thickness from a few millimeters to tens of centimeters depending on how many hundreds or thousands of years they were developing and under what rainfall conditions. Pedogenic carbonate horizons begin forming when thin (<50μm) films of $CaCO_3$ coat root hairs and the insides of root hair channels. As filamentous $CaCO_3$ accumulates, the thin threads coalesce and form irregular aggregates that are a few millimeters to one centimeter in diameter, a condition reached after several hundred to a few thousand years depending upon environmental conditions. Over thousands of years carbonate horizons can become massive, measuring up to a meter or more, and contain greater than 30 to 40 percent $CaCO_3$ by volume.

Pedogenic calcium carbonate accumulates on and between sediment grains, occluding and cementing the sediment as a result. On grains larger than a few millimeters, carbonate is initially deposited on the underside. With increasing time, sedimentary clasts larger than approximately five millimeters become enveloped on all sides with $CaCO_3$. Before pedogenic carbonate envelops an object on all sides, the initial location of the $CaCO_3$ is an excellent indicator of the vertical position of the clast within the sediments. Geologic indicators of the original

Z-orientation (up or down) of a fossil, called geopetal structures, enable the original position to be discerned regardless of how the fossil may have moved spatially at a later date. Because pedogenic carbonate deposits form first on the underside of objects, the presence of calcium carbonate and carbonate-cemented sediments on an artifact, or in this case, the bones of Kennewick Man, indicates which surface was down (Figure 17.9).

Overall Distribution of Carbonate-Cemented Sediment on the Skeleton

The in situ orientation of Kennewick Man's skeleton is firmly established by the location and distribution of carbonate-cemented sediment on the bones, including the degree of coverage and thickness (Table 17.2). Carbonate cementation fused bones in their original articulated positions. This is a particularly definitive indicator of the anatomical integrity of the skeleton prior to erosion. Figure 17.10 and Table 17.3 identify bone-to-bone adhesions. These adhesions and the distribution of cemented sediment accumulations serve as geopetal indicators, which reveal that Kennewick Man was buried in a position different from that postulated by Walker et al. (2000).

FIGURE 17.9 Two surfaces of a rhyolite core from the Mountaineer site, a Folsom-age occupation (10,600 RC yr. BP) in Gunnison, CO. Pedogenic carbonate forms a geopetal structure that accumulates first on the lowest part of a buried artifact and, over time, coats more-elevated areas. Regardless of the position in which the artifact was found, carbonate presence establishes the original up (top) and down surfaces (bottom).

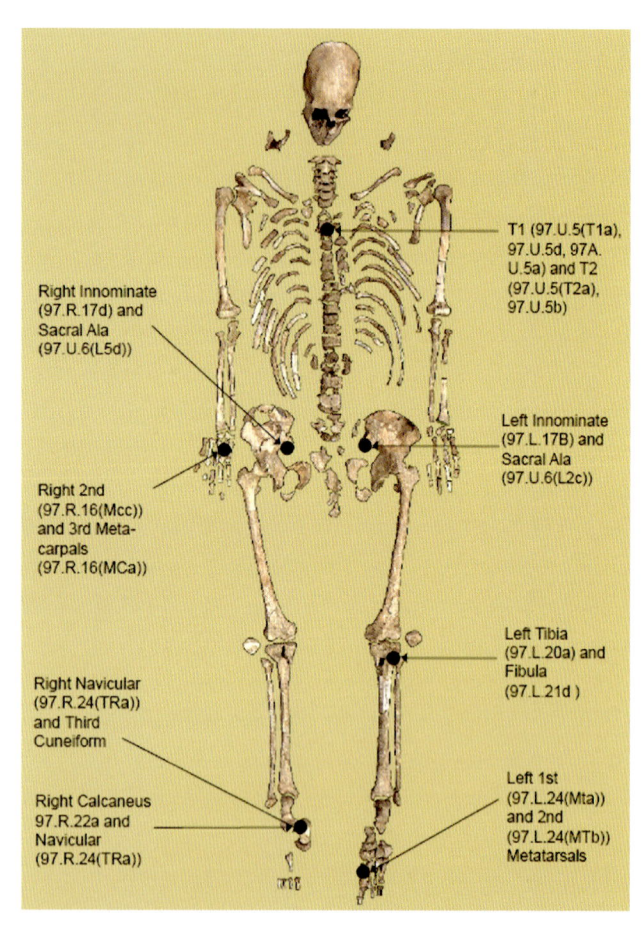

FIGURE 17.10 Diagram of bone-to-bone adhesions in Kennewick Man's skeleton. Adjoining or articulated skeletal elements were cemented together by calcium carbonate. As the skeleton disarticulated and fragmented during erosion, weaker bone components broke and the strong carbonate-impregnated sediment remained intact. As a result, small fragments were left adhering to the neighboring bone. Bone adhesions provide convincing evidence that the skeleton was articulated.

TABLE 17.2 In situ anatomical positioning of skeletal elements within the sediment.

Bone	Side	Bone Number	Specimen Number	Segment Number	Segment Description	Taphonomic Indicator	Taphonomy Description	"Down" Surface
Cranium			97.U.1a			CCS	CCS deposits were removed from the cranium in preparation for casting. Trace amounts, reflecting heavier accumulations, are still in place in the nuchal region, and heaviest below the inferior nuchal line, including around the external occipital protuberance and especially on the right lateral inferior nuchal region and posterolateral aspect of the right mastoid process.	Posterior/inferior
Maxilla	L		97.U.2a			CCS	Internal sinus walls are covered by CCS accumulations.	Inferior
Maxillae			97.U.2a			SED	Radiographs show thickened sediment accumulation on the floors of the maxillary sinuses without similar build-up apposition on the more superior medial and lateral walls.	Posterior
Sternum			97.I.25a			CCS	Slight formation on the posterior surface; anterior surface cannot be evaluated.	Posterior?
Clavicle	L		97.L.11b, 97.L.11c			CCS	Slight CCS accumulation with minor nodule formation on the inferior surface, especially on the lateral aspect.	Inferior
Clavicle	L		97.L.11b, 97.L.11c			WEA	Superior surfaces of the middle and lateral segments are lighter in color than their inferior surfaces.	Posterior?
Clavicle	R		97.R.11b (Lateral half)			CCS	Slight accumulation on the inferior margin of the lateral surface, specifically on the conoid tubercle.	Posterior
Clavicle	R		97.R.11b (Lateral half)			WEA	The superior surface middle third is lighter in color than the inferior surface. The rhomboid fossa, on the inferior sternal end, is dark brown in color.	Posterior
Scapula	L		97.L.10c			CPFX	Complete fracture of the superior portion of the joint, resulting in separation of the coracoid process from the scapula. Load direction: anterior to posterior.	Posterior
Scapula	L		97.L.10b, 97.L.10c, 97.I.17e, 97.I.10c			IPFX	Ventral surface IPFX courses anterior to posterior across the glenoid (bifurcating on the midglenoid face) with the most superior fracture continuing posteriorly and ending just short of the posterior rim with a single interruption. Load direction: anterior to posterior.	Posterior
Scapula	L		97.L.10b, 97.L.10c, 97.I.17e, 97.I.10c			WEA	The ventral surface of 97.I.10c has lost some iron oxide patination and is lighter in color. The posterior surface is slightly darker. The adjacent ventral and lateral surfaces of inferior fragment 97.I.10a are lighter in color, and its more inferior ventral aspect is a medium brown.	Posterior

(continued)

CCS: Carbonate-Cemented Sediment; SED: Sediment Accumulation; ALG: Algal Staining; CPFX: Complete Postmortem Fracture; IPFX: Incomplete Postmortem Fracture; WEA: Weathering

TABLE 17.2 (continued) In situ anatomical positioning of skeletal elements within the sediment.

Bone	Side	Bone Number	Specimen Number	Segment Number	Segment Description	Taphonomic Indicator	Taphonomy Description	"Down" Surface
Scapula	R		97.R.10a, 97.R.10d			CCS	The posterior superior margin of the spine has small CCS nodules; the posterior surface of the spine has slight, widespread CCS formation while the small, attached ventral surface of the body has only miniscule CCS accumulation.	Posterior
Scapula	R		97.R.10a: lateral spine and acromion process; 97R.10d: midlateral (root) of the spine			WEA	Superior surface of 97.R.10a and the immediately adjacent surface of 97R.10d are light in color. The inferior and posterior spine surfaces of these two pieces are darker in color (tan to medium brown).	Posterior
Cervical vertebra		C1	97.U.4(C1a)			ALG	Trace algae present on anterior surface of centrum.	Posterior
Cervical vertebra		C1	97.U.4(C1a)			CCS	Trace CCS is present on both the anterior and posterior external surfaces of the vertebra.	
Cervical vertebra		C2	97.U.4(C2a)			ALG	Slight to moderate algae is present, especially on sides of the dens process.	
Cervical vertebra		C3	97.U.4(C3a)			CPFX	Postmortem breakage of the right transverse process has resulted in an open transverse foramen.	
Cervical vertebra		C4	97.U.4(C4a)			CPFX	Postmortem damage of the left transverse process has resulted in an open left transverse foramen.	
Cervical vertebra		C5	97A.U.4(C5a)			CCS	Posterior surface of neural arch lamina has three-dimensional formation, measuring 2.3 mm thick.	Posterior
Cervical vertebra		C6	97.U.4(C6a)			CPFX	Postmortem breakage of the left and right transverse processes has resulted in open transverse foramina.	
Cervical vertebra		C7	97.U.4(C7a)			CPFX	Postmortem breakage of the left and right transverse processes has resulted in open transverse foramina.	
Thoracic vertebra		T2	97.U.5(T2a) and 97.U.5b			CCS	Present in the exposed cancellous bone adjacent to the superior facet (on the posterior surface of the vertebra).	
Thoracic vertebra		T4	97.U.5(T5a) and 97.U.5(T5b)			CCS	Present on the posterior surfaces of the neural tube and the neural arch.	Posterior
Thoracic vertebra		T4	97.U.5(T5a) and 97.U.5(T5b)			WEA	Erosion has damaged the anterior body and inferior endplate margin.	Posterior

(continued)

CCS: Carbonate-Cemented Sediment; SED: Sediment Accumulation; ALG: Algal Staining; CPFX: Complete Postmortem Fracture; IPFX: Incomplete Postmortem Fracture; WEA: Weathering

TABLE 17.2 (continued) In situ anatomical positioning of skeletal elements within the sediment.

Bone	Side	Bone Number	Specimen Number	Segment Number	Segment Description	Taphonomic Indicator	Taphonomy Description	"Down" Surface
Thoracic vertebra		T5	97.U.5(T4a) and 97.U.5(T4b)			CCS	Heavy on the posterior surface of the neural tube, the adjacent lamina, and inferior facets. Slight on the posterior surface of the neural arch	Posterior
Thoracic vertebra		T5	97.U.5(T4a) and 97.U.5(T4b)			WEA	Slight erosion on the anterior aspect of the body	Posterior
Thoracic vertebra		T6	97.U.5(T6a) and 97.U.5a			WEA	Erosion along the body's anterior face which has destroyed most of the anterior endplate margins	Posterior
Thoracic vertebra		T8	97.U.5(T9a)			WEA	Erosion on the anterior aspect of the body resulting in the destruction of the superior and inferior endplate margins	Posterior
Thoracic vertebra		T11	97.U.5(T11a)			WEA	The margin of the centrum has eroded away.	Posterior
Thoracic vertebra		T12	97.U.5(T12a)			WEA	Some erosion on the anterior aspect of the body	Posterior
Thoracic vertebra		T12	97.U.5(T12a)			CCS	Posterior surface	Posterior
Lumbar vertebra		L3	97.U.6(L3a)			WEA	Erosion is evident on the anterior aspect of the lumbar body	Posterior
Rib	R	Rib 3	97A.I.12d			CCS	Along the inferior edge and on the external surface along the inferior margin of a postmortem depression feature	
Rib	R	Rib 5				WEA		
Rib	R	Rib 7	97.I.12(3)			WEA		
Rib	R	Rib 8	97.I.12b(14), 97.I.12b(9), 97.I.12d(2), 97.I.12d(1), 97.I.12d(7)			CCS	Small, three-dimensional deposits on lateral (external) surface, particularly on the inferior middle third of the body	
Hand	L					CCS	Proximal, middle, and distal phalanges have light to moderate CCS on the palmar surfaces	Palmar
Hand	R					CCS	MC1, MC5, and proximal and distal phalanges have moderate accumulation on the palmar surfaces	Palmar
Humerus	L		97.L.13a	1	Epiphysis, and upper three-fifths of the diaphysis	CCS	One nodule is located on the superior margin of the lesser tubercle	Posterior
Humerus	L		97.L.13b	2	Lower one-fourth of the diaphysis	ALG	Slight on anterior and lateral surfaces and the lower face of the fracture between this segment and the proximal portion of the bone (97.L.13a)	Posterior

CCS: Carbonate-Cemented Sediment; SED: Sediment Accumulation; ALG: Algal Staining; CPFX: Complete Postmortem Fracture; IPFX: Incomplete Postmortem Fracture; WEA: Weathering

(continued)

TABLE 17.2 (continued)   In situ anatomical positioning of skeletal elements within the sediment.

Bone	Side	Bone Number	Specimen Number	Segment Number	Segment Description	Taphonomic Indicator	Taphonomy Description	"Down" Surface
Humerus	L		97.L.13c	3	Distal joint and most inferior one-fourth of the diaphysis	ALG	Slight on medial, anterior, and lateral surfaces, and moderate on metaphysis and joint surface. Green on the joint surface includes algal filaments located anterior-posterior along the circumference of the trochlea	Posterior
Humerus	R		97.R.13a	1		CCS	Light on anterior surface of the proximal ends at muscle attachments; single nodules appear on the posterior aspect of the superior surface of the greater tubercle and the posterior surface of the metaphysis; light on anterior surface of the deltoid tuberosity	
Humerus	R		97.R.13b	2	Lower one-fourth of the diaphysis and distal joint	ALG	Moderate on distal one-fourth of the segment, especially on the anterior surface and anterior metaphysis and joint	Posterior
Radius	L		97.L.14a	1	Proximal one-third of the diaphysis including a nearly complete radial head	CCS (single nodule)	Small, single localized nodule located on the medial surface of the head where it articulates with the radial notch of the ulna	
Radius	L		97.L.14b	3		CCS	Moderate, concentrated along the interosseous crest of the medial surface, in a small localized area on the distal anteromedial surface, and lateral surface of the distal metaphysis	
Radius	L		97.L.14b	3		IPFX	Incomplete fracture on the medial surface	Lateral
Radius	R		97.R.14a	1	Head and two-thirds of the diaphysis	CCS	Moderate on the medial aspect of the anterior surface along the interosseous crest; localized deposits on the posterior surface (some of which were previously sampled by the government study team)	Anterior
Radius	R		97.R.14b	2		CCS	Moderate deposits on the medial halves of both the anterior and posterior surfaces and along the interosseous crest	
Ulna	L		97.L.15a	1	Proximal one-third, including the joint surface	ALG	Slight to moderate on the olecranon and trochlear notch. Also on the face of the fracture between this segment and the more distal (midshaft) segment of the left ulna (97.L.15b)	Posterior
Ulna	L		97.L.15a	1		CCS	Trace deposits on the proximal end of the olecranon; moderate on anterior surface of the diaphysis in the area of the tuberosity and the lateral aspect along the interosseous crest	
Ulna	L		97.L.15a	1		IPFX	Incomplete fracture opening on the anterior surface then propagating posteriorly	Posterior

(continued)

CCS: Carbonate-Cemented Sediment; SED: Sediment Accumulation; ALG: Algal Staining; CPFX: Complete Postmortem Fracture; IPFX: Incomplete Postmortem Fracture; WEA: Weathering

TABLE 17.2 (continued) In situ anatomical positioning of skeletal elements within the sediment.

Bone	Side	Bone Number	Specimen Number	Segment Number	Segment Description	Taphonomic Indicator	Taphonomy Description	"Down" Surface
Ulna	L		97.L.15b	2		CCS	Moderate nodular deposits present on the medial, lateral, and posterior surfaces (medial surface previously sampled by the government study team)	
Ulna	L		97.L.15c	3	Distal one-third of the diaphysis; the epiphysis is missing	CCS	Moderate on all surfaces, heavier on the anterior surface in the attachment area of pronator quadratus	
Ulna	R		97.R.15a	1	Proximal one-third of diaphysis, including the intact joint	ALG	Moderate on trochlear notch, lateral and superior aspect of the olecranon	Posterior
Ulna	R		97.R.15a	1		CCS	Small concretions present on the lateral surface posterior to the supinator crest; slight to moderate on the anterior surface in the area of the tuberosity	
Ulna	R		97.R.15b	2	Middle one-third of the diaphysis	CCS	Slight on the anterior surface and on the posterior surfaces	
Ulna	R		97.R.15c	3		CCS	Localized, moderate formation in the attachment area for pronator quadratus on the anterior surface; slight on the medial rim of the joint surface (epiphysis); one localized deposit on the lateral surface	
Innominate	L		97.L.17b, 97.L.17a			CCS	Moderately heavy accumulations on the superior ilial component of the ilio-pubic ramus (above the anterior acetabulum), the anterior third of the external surface of the ilium (above the superior acetabulum), the posterior surface of the ischium, the posterior external surface of the ilium adjacent to the posterior fracture margin, and the external (anterior) surface of the pubic bone	Posterior
Innominate	R		97.R.17d			CCS	Widespread and rugose accumulations are present on the superior ilial portion of the ilial pubic ramus (above the anterior acetabulum), the visceral component opposite the acetabulum, the auricular surface and superior auricular surface rim, the visceral surface of the anterior inferior iliac spine, and the lateral surface adjacent to the anterior superior iliac spine	Posterior

(continued)

CCS: Carbonate-Cemented Sediment; SED: Sediment Accumulation; ALG: Algal Staining; CPFX: Complete Postmortem Fracture; IPFX: Incomplete Postmortem Fracture; WEA: Weathering

TABLE 17.2 (continued) In situ anatomical positioning of skeletal elements within the sediment.

Bone	Side	Bone Number Specimen Number	Segment Number	Segment Description	Taphonomic Indicator	Taphonomy Description	"Down" Surface
Femur	L	97.L.18a	1	Proximal two-fifths	CCS	Deposits in the trabecular bone, especially on the superior margin that was exposed when the superior and posterior margins and the inferior third of the femoral head broke off postmortem; present in the intertrochanteric fossa; slight on the anterior and lateral surfaces of the greater trochanter; heavy, three-dimensional formation on the posterior surface primarily along the linea aspera	Posterior
Femur	L	01.L.18b	2	Middle and upper lower diaphysis	CCS	Heavy, three-dimensional formation on the posterior surface primarily along the linea aspera; also on the distal end where the linea aspera becomes less pronounced and CCS formation begins to fan out	Posterior
Femur	L	01.L.18a	3	Lower diaphysis and distal joint surface	CCS	Heavy, three-dimensional formation on the lateral surface; trace on the posterior surface; slight, longitudinally-oriented CCS stria on the surface of a shallow depressed area of the medial condyle	Posterior
Femur	L	97.L.18a, 01.L.18b	1, 2	See descriptions above	CPFX	Proximal segment fixed and the middle segment is displaced anterolateral to posterolateral	Posterior
Femur	L	01.L.18b, 01.L.18a	2, 3	See descriptions above	CPFX	Breakage progressed from anterolateral to posterolateral	Posterior
Femur	R	01.R.18a	1	Proximal one-fourth	CCS	Moderate on the posterior surface of the proximal diaphysis in the region of the gluteal tuberosity	Posterior
Femur	R	01.R.18b	2	Midshaft (middle one-third and lower proximal one-fourth)	CCS	Moderate on the anterior surface of the proximal end with some small deposits on the distal end; small deposits on the medial and lateral surfaces; heavy, three-dimensional formation (which lessens as it progresses inferiorly) localized on the linea aspera and the portion of the lower gluteal tuberosity	Posterior
Femur	R	97.R.18a	3	Distal joint surface and portion of lower diaphysis (one-third)	CCS	Small, localized deposits on the anterior, medial, and lateral surfaces of the proximal diaphysis; slight to moderate formation on the medial, anterior, and lateral surfaces of the distal portion of the diaphysis; moderate formation on the posterior surface concentrated from the distal joint through the entire length of the segment (heaviest on the lateral aspect of the posterior surface)	Posterior
Femur	R	01.R.18a, 01.R.18b	1, 2	See descriptions above	CPFX	Fracture progressed from the medial side to the posterolateral side with the proximal segment fixed	Posterior

CCS: Carbonate-Cemented Sediment; SED: Sediment Accumulation; ALG: Algal Staining; CPFX: Complete Postmortem Fracture; IPFX: Incomplete Postmortem Fracture; WEA: Weathering

(continued)

TABLE 17.2 (continued) In situ anatomical positioning of skeletal elements within the sediment.

Bone	Side	Bone Number	Specimen Number	Segment Number	Segment Description	Taphonomic Indicator	Taphonomy Description	"Down" Surface
Femur	R		01.R.18b, 97.R.18a	2, 3	See descriptions above	CPFX	Fracture progressed anterolateral to posterior with the midshaft segment fixed relative to the distal segment	Posterior
Femur	R		97.R.18a	3	Distal joint surface and portion of lower diaphysis (one-third)	IPFX	On the anterior side of the bone, a horizontally-oriented break occurred that was bounded by longitudinal (aligned with the long axis of the bone) drying cracks that was produced by an anterior to posterior force	Posterior
Tibia	L		97.L.20a	1		CCS	Moderate formation on the medial surface inferior to the metaphysis	
Tibia	L		97.L.20a	1		WEA		
Tibia	L		97.L.20b	2	Two-fifths of the proximal diaphysis	CCS	Heavy, three-dimensional formation on the lateral, medial, and posterior surfaces (anterior surface was previously removed and cannot be evaluated)	Posterior
Tibia	L		97.L.20c	3	Lower three-fifths of diaphysis and distal joint	CCS	Moderate to heavy on the medial and lateral surfaces; moderately heavy, three-dimensional formation on the distal posterior surface	Posterior
Tibia	L		97.L.20b, 97.L.20c	2, 3	See descriptions above	CPFX	The distal half of the shaft (97.L.20c) was fixed while the middle segment (97.L.20b) was levered from anterior to posterior to produce a complete break	
Tibia	R		97.R.20a	1	Medial condyle and epiphysis. The anterior half of this bone segment is missing post-mortem	CCS	Only the medial external surface and the medial condyle are represented; moderate on the medial surface	
Tibia	R		97.R.20b	2	Lateral condyle and epiphysis. The anterior half is missing	CCS	Thin, widespread formation on all surfaces	
Tibia	R		97.R.20c	3	Proximal and middle one-third of the diaphysis including most of the tibial tuberosity	CCS	Moderate and widespread on the medial, lateral and posterior surfaces (anterior crest is free of CCS); heavy on the posterior one-third of the diaphysis; considerable amount of CCS has flaked off	

(continued)

CCS: Carbonate-Cemented Sediment; SED: Sediment Accumulation; ALG: Algal Staining; CPFX: Complete Postmortem Fracture; IPFX: Incomplete Postmortem Fracture; WEA: Weathering

TABLE 17.2 (continued) In situ anatomical positioning of skeletal elements within the sediment.

Bone	Side	Bone Number	Specimen Number	Segment Number	Segment Description	Taphonomic Indicator	Taphonomy Description	"Down" Surface
Tibia	R	97R.20d		4	Distal third of the diaphysis, including the joint.	CCS	Moderate to heavy on all surfaces	Posterior
Tibia	R		97.R.20c, 97.R.20d	3, 4	See descriptions above	CPFX	The more proximal segment (97.R.20c) was levered anteriorly to posteriorly. The break opened first on the anterior crest and propagated distally and posteriorly leaving a breakaway spur on the posterior fracture face of the fixed distal end (97.R.20d).	
First metatarsal	L		97.L.24(MTa)			CCS	MT1 shows heaviest accumulation on the plantar surface	Plantar
Navicular	L		97.L.24(TRb)			CCS	Navicular shows heaviest accumulation on the plantar surface	Plantar
Second through fourth metatarsals	L		97.L.24(MTb) 97.L.24(MTc) 97.L.24(MTd)			CCS	These metatarsals show the greatest accumulation on the medial and lateral sides of the proximal and distal ends	Medio-lateral
Lateral tarsals and calcaneus	L		97.L.22a (Calcaneus)			CCS	Moderate to heavy on the lateral side of the calcaneus; less on the plantar portion. The cuboid and cuneiforms 2 and 3 have heavy accumulations on the superolateral surfaces. First cuneiform shows heaviest accumulation on the lateral side	Lateral
First metatarsal	L		97.L.24(Mta)			IPFX	Recent transverse fracture initiated on the medial dorsal (superior) side at midshaft, extending into a preexisting longitudinal drying fracture located in the medial surface; the distal end is depressed in a plantolateral direction. Load direction is on the dorsal-medial surface (relaxed plantar flexion of the foot with approximately 45° lateral rotation).	Plantar/lateral
Calcaneus	R		97.R.22a			CCS	Heavy on the plantar surface	Plantar
Metatarsals	R					CCS	Heaviest on the plantar and lateral sides of the shaft of MT1	Plantar/lateral
Foot phalanges	R					CCS	Heavy on the phalanges of the first and second toes on the plantar surface	Plantar

CCS: Carbonate-Cemented Sediment; SED: Sediment Accumulation; ALG: Algal Staining; CPFX: Complete Postmortem Fracture; IPFX: Incomplete Postmortem Fracture; WEA: Weathering

Sediments adhering to the skeleton are hard, gray-to-white aggregates of clayey silt cemented with calcium carbonate. This pedogenic carbonate varies from white films 10 to 50 μm thick to rugose (nodular), angular, and equant accumulations one to three millimeters in diameter and up to three to five millimeters thick. In specific locations, these deposits envelop and encircle several square centimeters of the bones' exterior surfaces. Accumulation of pedogenic calcium carbonate was affected by each bone's position relative to the depth where pedogenic carbonate formed. Consequently, there are varying amounts of carbonate encrustations adhering to the bone surfaces.

Both visual inspection and computed tomography images of the cranium show that the heaviest deposits of carbonate-cemented sediment are in the inferior nuchal region, both ecto- and endocranially, and also on the right temporomandibular fossa. The deposits indicate the cranium was resting on the inferior nuchal region with the face positioned up and toward the feet (i.e., the chin resting on the throat). The frontal, maxillary, and zygomatic bones are free of such sediment, confirming that the anterior cranium (face) would have been higher than the posterior vault. The presence of sediment deposits on the inferior maxillary sinus walls, combined with the lack of build-up on the medial and lateral walls, is evidence that gravity caused this accumulation while the cranium rested face up. The posterior maxillary and mandibular dentition have carbonate-cemented sediment, but none has accumulated on the anterior teeth.

Both left and right clavicles have slight carbonate-cemented sediment on the inferior aspects, while the anterior, superior, and posterior surfaces are virtually free of concretions, indicating that the inferior surfaces were positioned down in situ. The scapulae are nearly free of carbonate-cemented sediment with the exception of slight accumulations on the superior border of the posterior surface of the right scapular spine, the adjacent posterior rim of the glenoid, and the medial anterior surface of the coracoid process. These occurrences

TABLE 17.3 Carbonate-cemented sediment causing adhesion of anatomically adjacent bones in situ.

Bone	Side	Bone Number	Specimen Number	Segment Number	Taphonomy Description
Hand	R				The proximal second and third metacarpals were together as evidenced by a remnant of the third metacarpal base attached to the second metacarpal.
Thoracic vertebra		T1, T2			Bottom edge of the left inferior articular facet of T1 is attached to the left superior facet of T2.
Innominate	L		97.L.17B		Small fragments of the left sacral ala are fused to the innominate; one fragment adheres to the superoanterior margin of the auricular surface, the other to posterior inferior auricular surface.
Innominate	R		97.R.17d		A small sacral ala fragment is fused to the inferior auricular surface.
Tibia	L		97.L.20a	1	The proximal left tibia retains tiny fragments of the left fibula in a small accumulation on the posterior surface.
Navicular/calcaneus	R		97.R.24(TRa) and 97.R.22a		A piece of calcaneus is attached to the navicular.
Navicular/third cuneiform	R		97.R.24(TRa)		A piece of the third cuneiform adheres to the navicular on the lateral, superior surface.
Metatarsals	L		97.L.24(Mta) and 97.L.24(MTb)		The first and second metatarsals are joined at their proximal and distal ends. In addition, a small piece of the third metatarsal is attached to the base of the second metatarsal.

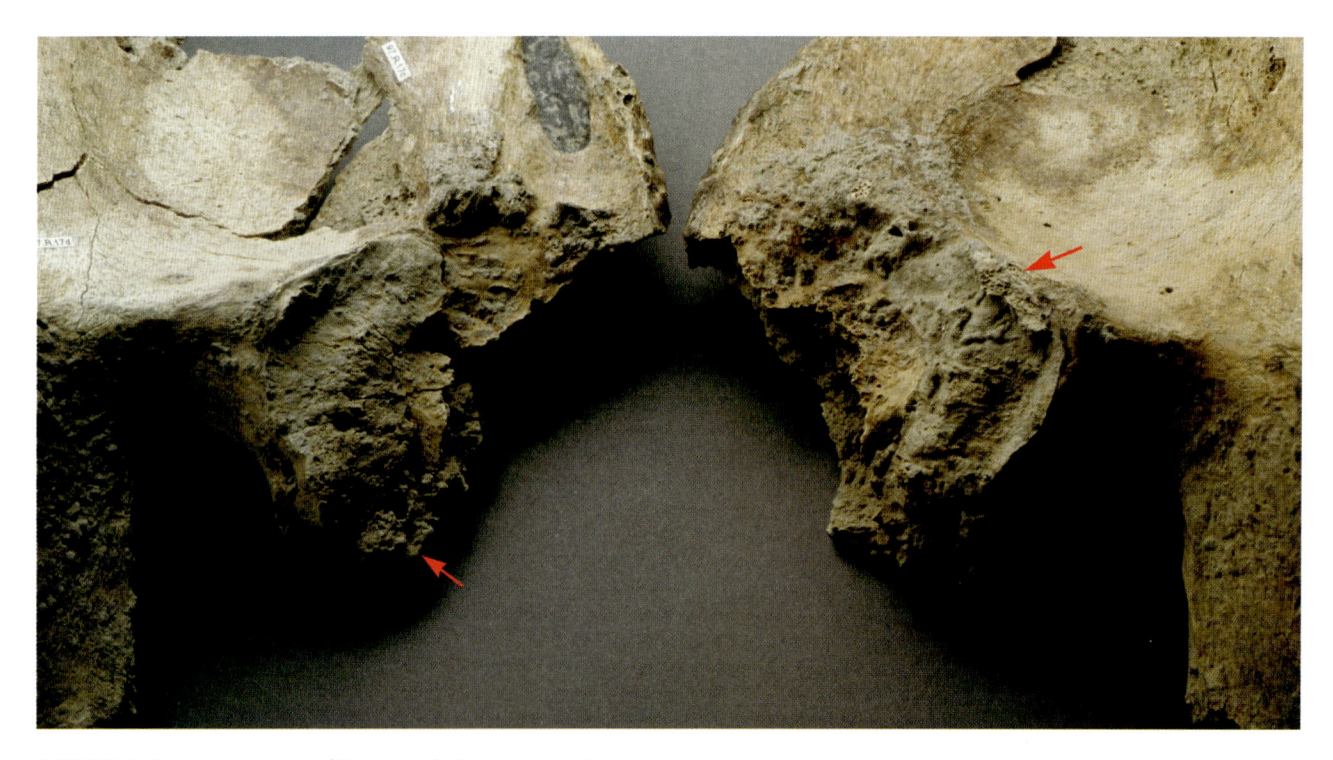

FIGURE 17.11 Fragments of the sacral ala cemented to the auricular surfaces of the innominates by carbonates (arrows). These bone-to-bone adhesions confirm anatomical articulation of Kennewick Man's pelvis.

suggest the posterior surface was resting down, but that the scapulae were essentially above the zone containing carbonate-cemented sediment. Carbonate on the posterior surfaces, but not on the anterior surfaces, of six vertebrae indicate that the spine was positioned with the posterior surface down, as in a supine, extended burial.

On the left innominate, accumulations are greatest on the exterior and visceral surfaces of the ischium and the exterior surface of the pubis. Less dense accumulations occur on the exterior surface of the ilium, while the visceral surface of this segment is virtually free of carbonate- cemented sediment. The visceral and lateral surfaces of the iliopubic ramus of the right innominate have heavy accumulation, while the visceral surface of the ilium (of the sections that could be observed) is free of carbonate-cemented sediment. Both sacroiliac articulations have fragments of the left and right sacral ala cemented to them, demonstrating that the pelvis was articulated during the period of carbonate precipitation (Figure 17.11).

The left and right humeri were located in an area that was essentially free of carbonate-cemented sediment, although trace accumulations are present on the anterior surface of the proximal right humerus.

Carbonate-cemented sediments on the lower right arm bone indicate in situ positioning of the forearm and hand in the prone position with the anterior (palm) surface down. On the distal right radius, carbonate-cemented sediment formed on the anteromedial and posteromedial surfaces (along the interosseous crest) and anterior surface of the distal metaphysis, but not on the lateral surface nor the anterolateral and posterolateral surfaces. The proximal segment of the right radius exhibits moderate carbonate-cemented sediment on the posterior surface near the attachment site for pronator teres.

Deposits of carbonate-cemented sediment on the right ulna correspond with carbonate-cemented sediment on the anteromedial and posteromedial surfaces of the right radius (Figure 17.12). The articulation of the pronated, right distal radius and ulna shows deposition continuity across the two bones. The juxtaposition provides further evidence for pronation of the right arm. The posterior surface of the right ulna is free of carbonate-cemented sediment.

The right hand shows marked carbonate-cemented sediment accumulation on the palmar surface, espe-

FIGURE 17.12 The distal right radius and ulna rotated to complete the pattern of knobby, rugose carbonate-cemented sediment show Kennewick Man's right forearm was pronated with the palm down.

cially on the phalanges, indicating that this surface was down. Additionally, there are no bone-to-bone adhesions or impressions of the right hand on the right or left innominates, specifically the ilio-pubic ramus. If the right hand had been draped over the innominates, imprints in the carbonate-cemented sediment due to contact with the phalanges would likely be present. Similarly, the right hand was not in contact with the thoracic region. These observations indicate that the right hand was resting alongside, not on, the body.

The carbonate-cemented sediment pattern on Kennewick Man's legs provides further evidence of a supine burial. The raised linea aspera (posterior surface) of the left femur shows heavy accumulation of carbonate-cemented sediment until reaching the more distal aspect where the linea aspera is less pronounced. At the distal end, carbonate-cemented sediment fans out and covers more of the posterior surface with pronounced formation. The anterior, lateral, and medial surfaces of the left femur show negligible carbonate-cemented sediment accumulation. A similar pattern is evident on the right femur. Nearly the full length of the posterior surface has thick, rugose, carbonate-cemented sediment localized on the linea aspera and a portion of the lower gluteal tuberosity. Carbonate-cemented sediment accumulation on the two proximal segments indicates that the posterior surfaces were down.

While the entire left tibia cannot be fully evaluated for carbonate-cemented sediment because of prior sectioning by the government study team, heavy deposits are present on the medial, lateral, and posterior surfaces. The lack of carbonate-cemented sediment on the anterior crest indicates supine positioning of this bone. If the anterior crest had been down, it would have functioned as a focal ridge similar to that of the lineae aspera of the femora. The right tibia also shows widespread, heavy accumulation of carbonate-cemented sediment on the medial, lateral, and posterior surfaces. Recent loss of carbonate-cemented sediment is reflected in areas free of accumulation. Color variation on this bone indicates that additional carbonate-cemented sediment was present at one time.

The cranium and bones of the upper body have less carbonate-cemented sediment adhering to them than the bones of the pelvis and legs. This gradual superior-to-inferior gradient of increasing carbonate precipitate along the length of the skeleton helps locate the interment in sediments at the top of the horizontal Bca horizon with the lower half of the body dipping into this zone. As such, this contrast in carbonate precipitation across a length of less than six feet (i.e., the horizontal burial length of Kennewick Man) implies a slight inclination in the bottom of the burial pit. The floor of the grave had a gradual slope with the end containing the upper body and head higher than the feet. Radiocarbon dating of the carbonates indicates this early Holocene Bca horizon formed 2,000–2,500 RC yr. BP, thousands of years after the date of interment (Chapter 3).

Bone-Specific Distribution of Carbonate-Cemented Sediment

Cranium   Some carbonate cementation was removed from the exterior of the cranium in 1996 in preparation for casting. Small amounts remain on the occipital bone below the protuberance, especially on the right lateral inferior nuchal region and posterolateral aspect of the right mastoid process. The right inferior nuchal region has three-dimensional carbonate cementation up to three milllimeters thick (Figure 14.2). In contrast, the frontal and zygomatic bones appear to have been free of carbonate-cemented sediment accumulation.

Loose to weakly consolidated sediment inside the cranial vault was removed following initial recovery. During the February 1999 government-sponsored study (Huckleberry and Stein 1999), additional sediment was extracted (Figure 17.13). At the time of this study, the vault was still partially filled with consolidated sediment along the endocranial margin, which is evident in computed tomography (CT) imaging.

Endocranially, the frontal bone contained vesicular (porous) sediment with voids 0.5 to two millimeters in diameter. These small, air-filled spaces are seen in sagittal-section CT images (Figures 17.13 and 17.14). In contrast, sediment located in the inferior nuchal region of the interior cranial vault is compact and lacks macroscopic porosity. Overall, the interface between the inner table (the endocranial surface) and the carbonate-cemented sediment is evident. Sediment accumulation within the semilunar cerebellar fossa of the occipital bone reflects an entrapment effect. Percolating water carrying silt, clay, and carbonates that entered the cranium through the optic foramina and superior orbital fissures was blocked by the relatively impermeable barrier of the occipital bone, which inhibited downward movement of water and facilitated consolidation of dense sediment.

The left and right maxillae are primarily represented by segment 97.U.2a. Gross observation of the maxillae reveals that the walls of the left sinus are lightly covered by carbonate-cemented sediment. Radiographic eval-

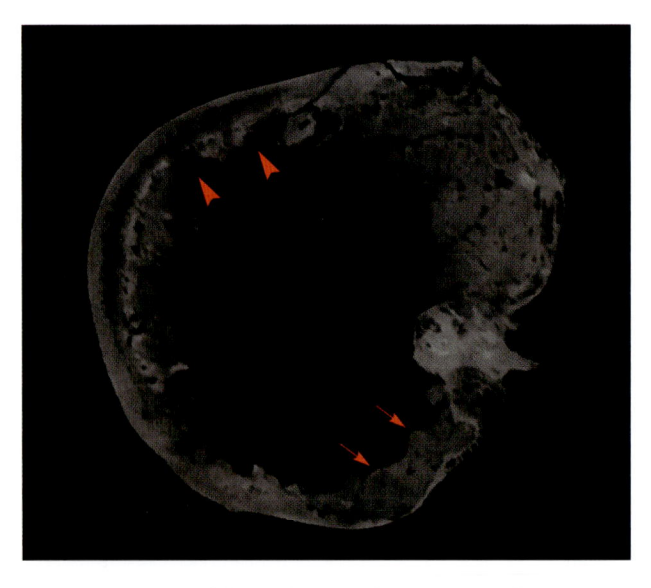

FIGURE 17.13   Left lateral sagittal section CT slice (318 of 840) of Kennewick Man's cranium shows sediment remaining inside the vault. The density of the calcium carbonate-impregnated sediment is similar to bone, but the inner table and diploe are identifiable. Sediment sampling by the government team is evident as gouges in the top of the image (arrowheads). This extraction was done using a Dremel® tool to loosen the sediment, then tilting the cranium to allow sediment to fall out (Huckleberry and Stein 1999). Small vesicles within the sediment are visible in the frontal and superolateral regions. Dense, more compact sediment lacking porosity had accumulated in the cerebellar region (arrows). Figures 17.13, 17.14, 17.28, and 17.29 are displayed in the approximate in situ position.

uation shows thickened accumulation on the floors of the maxillary sinuses. This is particularly evident in the lingual-facial radiograph of the anterior teeth, which shows accumulation on the inferior floor of the left sinus measuring two to three millimeters in depth (Figure 7.5). The floor of the right sinus shows slightly greater thickening, with accumulations measuring four to six millimeters. The upper surface of this sediment is smooth and concave. The medial and lateral walls of the sinuses lack this degree of accumulation and the posterior walls of the sinuses are missing. This relatively symmetrical pattern of nasal floor build-up reflects gravitationally driven sediment accumulation and provides information on the in situ positioning of the cranium.

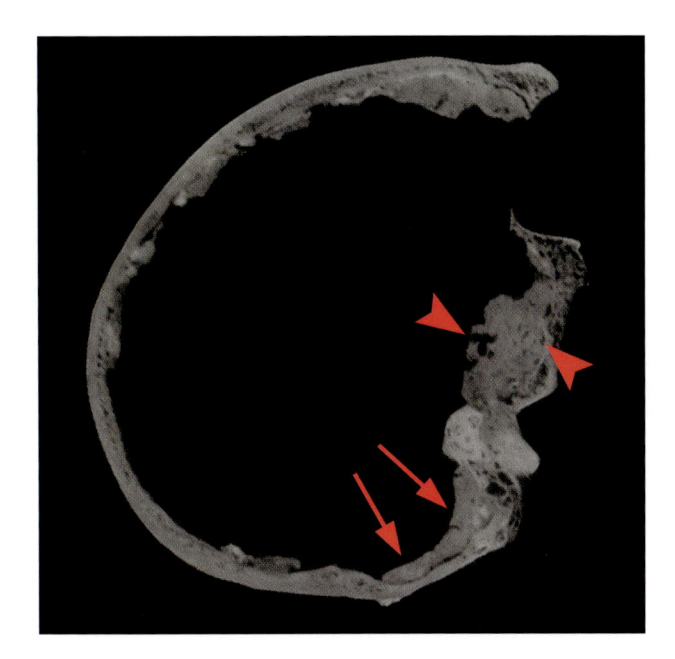

FIGURE 17.14 Right lateral sagittal section CT slice (661 of 840) shows dense sediment lacking porosity in the inferior nuchal region (arrows). Small vesicles are present inside sediment within Kennewick Man's cranium in the inferior temporal region (arrowheads). Voids are also present within the mastoid process and the sphenoid.

Dentition Three-dimensional carbonate deposits are present on both the buccal surface of the left maxillary first molar root and the partial socket of the right third molar. The left mandibular third molar is missing and its socket is also partially filled with carbonate cementation. Radiographs reveal that many anterior tooth chambers are free of sediment. The left maxillary canine and the right maxillary second incisor, canine, and second premolar pulp chambers are still partially open, as are the left mandibular canine and second premolar pulp chambers.

Sternum Two small fragments of bone that do not articulate represent the sternum. Both consist of cancellous and posterior surface cortical bone, as indicated by their slightly concave surface. The larger fragment (97.I.25a) is covered with some carbonate nodules. The largest nodule on this surface measures 4.3 millimeters across by 5.8 millimeters with a height of 1.1 millimeter. Without the presence of the anterior surface of the bone, anterior versus posterior coverage cannot be evaluated.

Clavicles The most lateral fragment of the left clavicle (97.L.11c) exhibits slight carbonate cementation with nodules on its inferior surface. There are no significant carbonate accumulations on the anterior, superior, or posterior surfaces. On the right clavicle, the lateral end (97.R.11b) has slight carbonate accumulation on its inferior surface, specifically on the conoid tubercle.

Scapulae No carbonate accumulations were noted on the left scapula. Carbonate-cemented sediment accumulations on the right scapula are relatively minor. The anterior surface of the right glenoid and the medial anterior surface of the coracoid process (97.R.10b) have small nodules and very light area coverage. The posterior rim of the glenoid has slightly larger carbonate nodule formations. The largest is on the superior posterior rim, measuring 6.7 millimeters in length (superior-inferior) and 2.9 millimeters in width (medial-lateral) with a maximum thickness of 1.6 millimeter. The posterior margin and adjacent surface of the glenoid has slight but widespread nodular carbonate-cemented sediment accumulation that partially obscures an antemortem chip fracture. Also on the posterior scapula, the superior margin of the lateral spine (97.R.10a) has small carbonate-cemented sediment nodules. The posterior surface of the midspine (97.R.10d) has minor but extensive carbonate-cemented sediment accumulation. The ventral surface of the body attached to this segment has miniscule accumulation.

Vertebrae Carbonate-cemented sediment is present on six vertebrae. The first cervical vertebra (97.U.4(C1a)) has trace accumulations on both the anterior and posterior external surfaces. A deposit is attached to the lower facet support (i.e., between the facet face and the neural arch of the first cervical vertebra). The fifth cervical vertebra (represented by 97A.U.4(C5a)), exhibits three-dimensional carbonate-cemented sediment on the posterior surface of the right lamina, measuring 2.3 millimeters in thickness (Figure 17.15). The bottom edge of the left inferior articular facet of the first thoracic vertebra (97A.U.5a) has broken off and is fused by carbonate cementation to the left superior facet of the second thoracic vertebra (97.U.5(T2a)) (Figure 17.16). Carbonates have also accumulated in the exposed cancellous bone of the posterior lamina adjacent to the superior facet

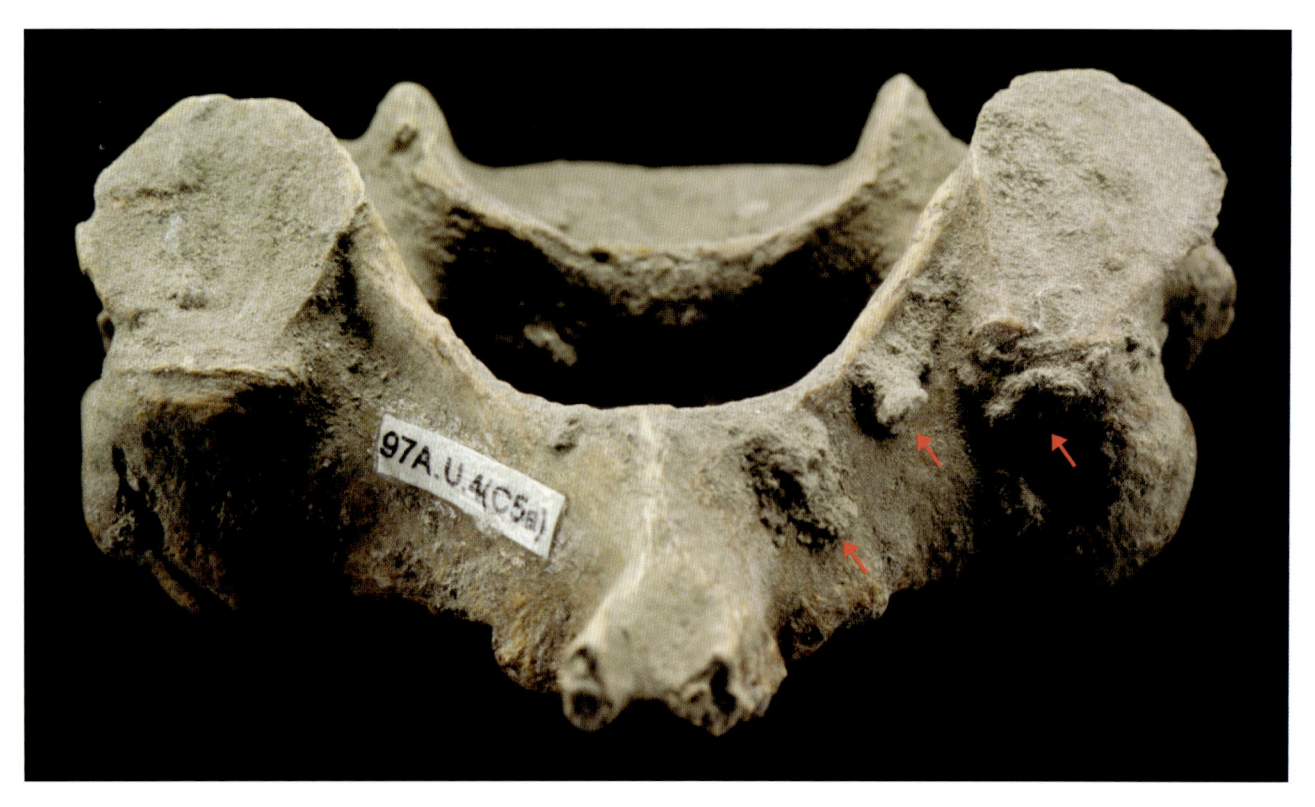

FIGURE 17.15 Carbonate precipitation formed nodules (arrows) on the right lamina of Kennewick Man's fifth cervical vertebra. This vertebra was positioned with the anterior surface (centrum) up for this accumulation to take place.

FIGURE 17.16 A remnant of an inferior facet (highlighted) of Kennewick Man's first thoracic vertebra (right) bonded by calcium carbonate to a superior facet of the second thoracic vertebra (left). Internal cancellous bone was exposed when this adhesion broke apart.

on the second thoracic vertebra. On the fourth thoracic vertebra, carbonate-cemented sediment is present on the posterior surface of the neural tube and on the posterior surface of the neural arch. Its centrum is free of carbonates. Thoracic vertebra five (97.U.5(T4a) and 97.U.5(T4b)) has been stabilized by carbonate-cemented sediment accumulation. The posterior surfaces of the neural tube, the adjacent lamina, and inferior facets show aggregated, heavy carbonate-cemented sediment, which is also present on the posterior surface of the neural arch to a lesser degree. The anterior surface of the fifth thoracic vertebra is free of carbonate-cemented sediment and the posterior surface of thoracic vertebra 12 exhibits some accumulation.

Ribs Only three ribs exhibit significant sediment accumulations. The surface of the right first rib (97.I.12e(3)) is partially covered with carbonate-cemented sediment. The right third rib is free of carbonates, except for nodule development on the inferior, lateral surface adjacent to the healed depression fracture (Chapter 7). Right rib eight has no carbonate accumulation on the visceral surface, but the lateral (external) surface has small, three-dimensional deposits, particularly on the inferior middle third of the body.

Innominates Fragments of the left sacral ala are cemented to the auricular surface of the left innominate (Table 17.3; Figure 17.11). One piece of ala adheres to the superoanterior margin of the auricular surface and measures 20 (superior-inferior) by 6.2 millimeters (anterior-posterior). The exposed fracture surface is light colored. A second, smaller fragment of ala is fused to the posteroinferior auricular surface. These small fragments of sacral ala broke off when the articulated pelvis eroded out of the bank.

The left innominate also has a large carbonate-cemented sediment nodule measuring 16 (anterior-posterior) by 7.6 millimeters (superior-inferior), with a height of 2.4 millimeters above the superior margin of the auricular surface. The visceral surface of the pubic bone (97.L.17a) is free of carbonate-cemented sediment except for one small nodule. There is three-dimensional, moderately heavy carbonate-cemented sediment on the external (anterior) surface of this segment. Other areas with moderately heavy carbonate-cemented sediment accumulation are the superior ilial component of the ilio-pubic ramus above the anterior acetabulum, the anterior external surface of the ilium above the superior acetabulum, the posterior surface of the ischium, and along the posterior external surface of the ilium adjacent to the posterior postmortem fracture margin. The main portion of the visceral surface of the ilium is free of carbonate-cemented sediment and the central ilium basin (fossa) is light in color from loss of patination. The lateral surface of the recovered ilium is darker in color with minor carbonate accumulation.

On the right innominate, carbonates accumulated on the visceral surface of the main body, including on the superior iliopubic ramus, the visceral surface opposite the acetabulum, and above the obturator foramen. Three-dimensional localized accumulations are present on 97.R.17d (made up of the auricular surface, sciatic notch, complete acetabulum, and superior ischium), on the superior ilial portion of the iliopubic ramus above the anterior acetabulum, and on the visceral surface opposite the acetabulum. The accumulation is widespread and rugose with nodular accumulations as thick as two millimeters. The auricular surface also has carbonate accumulation that partially obscures fine surface detail. Three-dimensional carbonate-cemented sediment accumulations are present above the superior auricular surface rim. Internally, a right sacral ala fragment is still fused to the inferior auricular surface (Figure 17.11). Other nodular, heavy carbonate-cemented sediment accumulations are present on the visceral surface of the anterior inferior iliac spine and the lateral surface adjacent to the anterior superior iliac spine. The latter has carbonate-cemented sediment as thick as 2.2 millimeters. The anterior and posterior (visceral) surfaces of the pubic bone are free of carbonate-cemented sediment. The visceral surface of the ilium in the ilial fossa (97.R.17b and 97.R.17d) is free of carbonate cementation. A stone point is embedded in a portion of the iliac crest; the ilium was not analyzed for carbonate concretions due to prior cleaning with a mild acid after recovery.

Humeri The humeri are relatively free of carbonate deposits (Figures 17.17 and 17.18). A nodule on the superior margin of the lesser tubercle of the left humerus

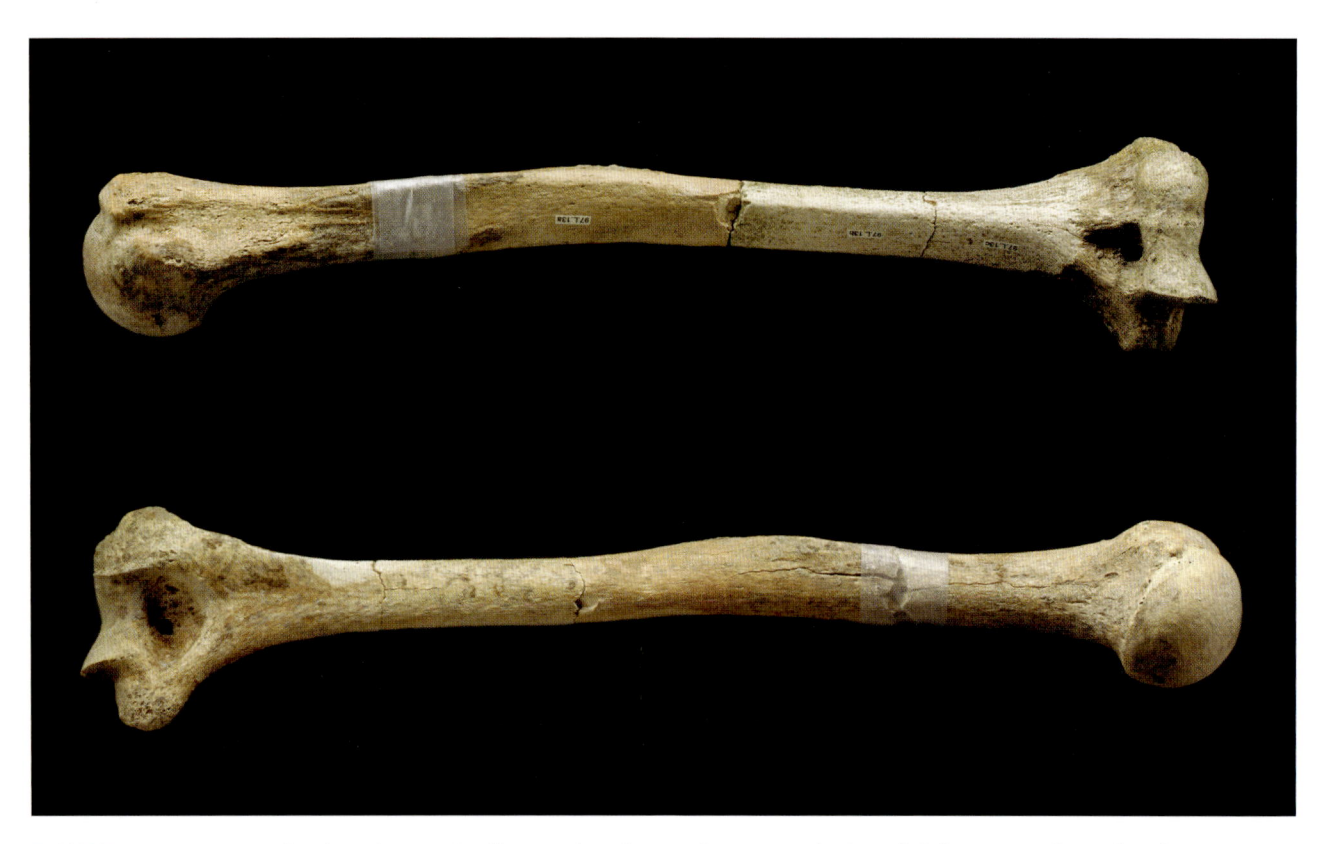

FIGURE 17.17 Anterior (top) and posterior (bottom) surfaces of Kennewick Man's left humerus show the absence of calcium carbonate precipitation. A break at the proximal end is lightly held together by Parafilm® applied by government scientists.

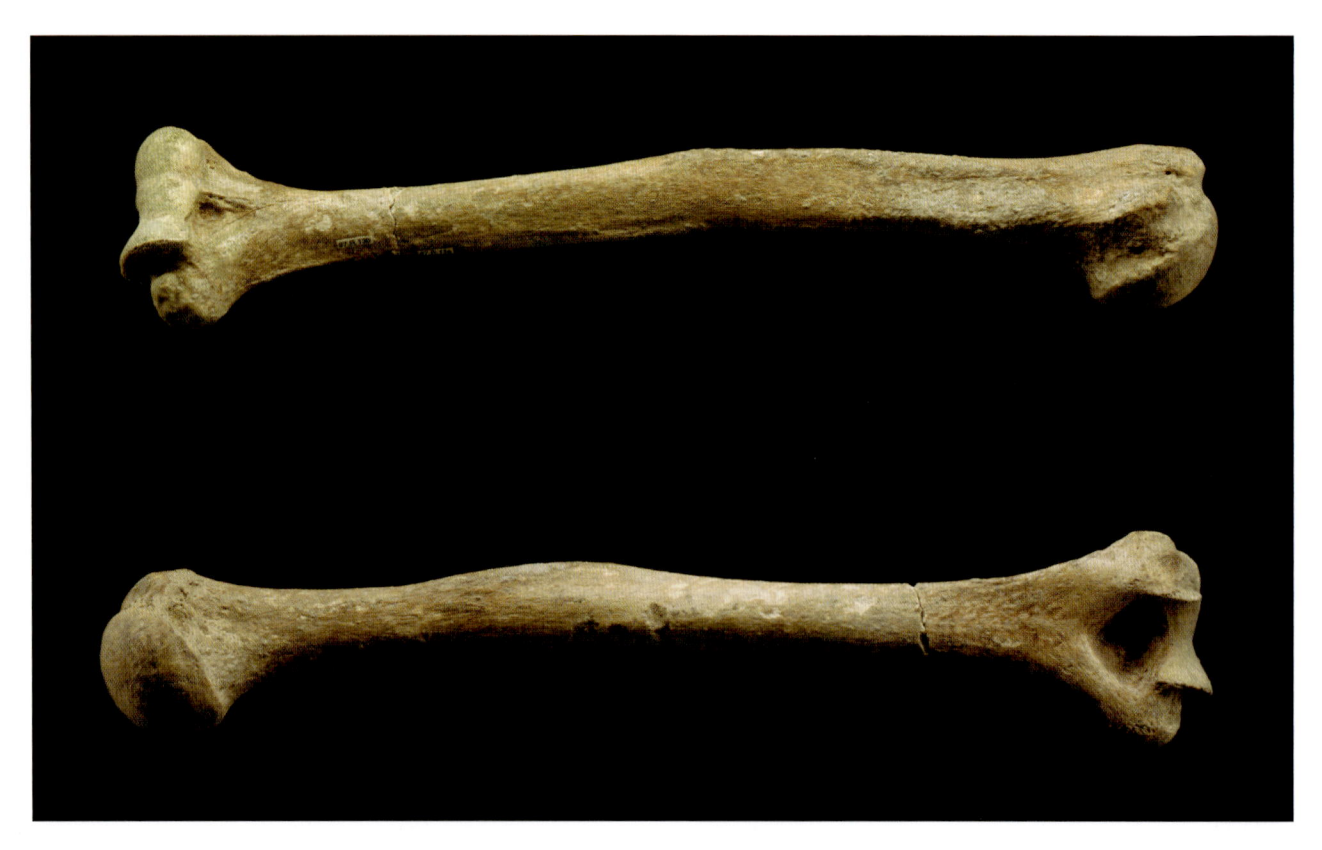

FIGURE 17.18 The anterio-medial (top) and posterior (bottom) surfaces of Kennewick Man's right humerus exhibit little to no calcium carbonate precipitation with the exception of tiny nodules on the head and thin accumulations on proximal muscle attachments.

has a maximum diameter of 9.7 millimeters with a height of 1.6 millimeters. On the right humerus, light carbonate-cemented sediment accumulations are evident on the anterior surface of the proximal end, specifically along the muscle attachment sites for pectoralis major, latissmus dorsi, and teres major, as well as on the anterior surface of the deltoid tuberosity. The posterior-superior surface of the greater tubercle and the posterior surface of the metaphysis have small, nodular, single, focal accumulations.

Radii A localized nodule measuring two millimeters across with a height of one millimeter is present on the medial surface of the left radial head where it articulates with the radial notch of the ulna. On the distal segment of this bone (a large section of the proximal diaphysis is missing postmortem), deposits of carbonate-cemented sediment are present on the lower interosseous crest and anterior surface. These accumulations are moderate, but concentrated in localized areas. No carbonate-cemented sediment is present on the lateral diaphysis or posterior surface. Accumulations are concentrated along the interosseous crest of the medial surface and in a small localized area on the distal anteromedial surface. This accumulation on the distal fourth of the left radius corresponds with three-dimensional, localized accumulation on the lateral surface of the distal ulna. A moderate amount of accumulation is also present on the lateral surface of the distal metaphysis.

On the right radius, moderate, isolated deposits of cemented sediment have accumulated on the medial aspect of the anterior surface along the interosseous crest. The posterior surface of the most proximal segment has localized sediment deposits, some of which were previously sampled by the government team, as evidenced by loss of characteristic external surface appearance. Distally, carbonates have accumulated on the medial halves of both the anterior and posterior surfaces along the interosseous crest. One localized deposit on the posterior surface, at the lower boundary of the lower one-third of the diaphysis, measures 15 (superior-inferior) by 7.5 millimeters transverse. The surface of this accumulation has been removed, leaving a finely textured, gray, shiny surface. The section remaining is approximately 0.8 millimeter thick. Adjacent and inferior to this is an area of lighter colored bone that indicates some outer cortex exfoliation. This exfoliation may be attributed to carbonate sampling.

On the lateral surfaces of the right radius, the majority of both the most proximal segment and the distal one-fourth are free of carbonate cementation.

Ulnae The left ulna has localized deposits of carbonate-cemented sediment. One ovoid concretion measuring 6.2 by 4.5 millimeters with a thickness of two millimeters is present on the proximal end of the olecranon. The anterior surface of the diaphysis has focal areas of moderate accumulation near the tuberosity and lateral aspect of the proximal interosseous crest. There are three-dimensional, moderate-level carbonate-cemented sediment deposits on the medial, lateral, and anterior surfaces of the midshaft. These deposits have approximate thicknesses of one millimeter and a nodular texture. The accumulation on the medial surface has been sampled previously, as indicated by scraping, apparent loss of thickness, and a lighter gray color. There is moderate carbonate-cemented sediment in several locations on the anterior, medial, and lateral surfaces of the tubular shaped distal ulna. These deposits are heavier on the anterior surface at the attachment area of pronator quadratus. Various locations along the posterior surface of the diaphysis also show localized areas of accumulation. These deposits are at least 0.9 millimeter thick.

The proximal third of the right ulna exhibits little carbonate-cemented sediment accumulation. Small concretions are present on the lateral surface posterior to the supinator crest. The anterior surface has slight to moderate coverage in the area of the tuberosity. The remainder of this segment of the ulna is free of carbonate-cemented sediment. On the middle one-third of the diaphysis, there is slight accumulation on the anterior surface, covering about 10 percent of the overall area. The medial surface is free of sediment. There is a small amount of carbonate-cemented sediment in the transition (i.e., the ridge) that divides the posterior and lateral surfaces. The anterior surface of the distal third of the diaphysis has localized, moderate carbonate adhesions at the attachment area for pronator quadratus. Concretions on the anterior surface are about one millimeter thick, slightly rugose, and confined to the muscle attachment area. The distal third is otherwise free of carbonate-cemented sediment except for the articular end (epiphysis) on the medial rim of the joint surface and a localized 10 by 13 millimeters deposit on the lateral surface. The posterior surface of the right ulna is generally free of carbonate-cemented sediment accumulation. Distal surface accumulations on the right ulna correspond with carbonate deposition on the anteromedial and posteromedial surfaces of the right radius (Figure 17.12).

Hands Moderately thick carbonate-cemented sediment is present on the palmar surface of the left hand bones. There are accumulations on the proximal phalanx of the thumb on the distal and proximal ends of the anterior surface and the third middle and fourth proximal phalanges. On the right hand, carbonate accumulation is also localized to the palmar surfaces. The corresponding dorsal surfaces are free of carbonate-cemented sediment. Concretions of slight to moderate thickness are present on the proximal ends of the first, second (one nodule, abraded), and fourth metacarpals; the diaphyses of these metacarpals are largely free of sediment. The distal phalanx of the thumb also exhibits a moderate amount of carbonate-cemented sediment on the anterior surface. Heavy, three-dimensional carbonate-cemented sediment accumulations are present on the anterior surfaces of proximal phalanges three and four, as well as the distal end of phalanx four.

Femora The left femur, which was broken into three segments, shows similar carbonate-cemented sediment patterning on the two proximal segments (representing about three-fourths of the bone), while the distal segment displays a slightly different pattern (Figure 17.19). The anterior and lateral surfaces of the greater trochanter have carbonate accumulations. The posterior surface of the neck is free of carbonate-cemented sediment, although there is accumulation in the trochanteric fossa, which continues into the damaged area on the superior-posterior margin of the femur head. There are 10 μm to 50 μm diameter rootlets adhering close to the carbonate-cemented sediment on the posterior surface of the proximal aspect. Overall, minimal to negligible carbonate-cemented sediment appears on the anterior surface of the proximal diaphysis, forming irregularly in areas of 100 μm to 200 μm in thickness.

This pattern on the proximal end was likely produced by the elevation of the neck and head of the femur as a result of anatomical articulation with the acetabulum, with the greater trochanter in a lower position. Carbonate cementation on the midshaft is strongly concentrated on the linea aspera, has a rugose appearance, and is one to two millimeters thick. The raised linea aspera shows the greatest concentration until reaching the more distal aspect of this segment, where the linea aspera is less pronounced. At this point, the carbonate-cemented sediment spread laterally and covers more of the posterior surface. It is still heavy, measuring approximately 0.7 millimeter thick. The bone has a light gray color where the carbonate has exfoliated along the linea aspera. The medial, anterior, and lateral surfaces of the midshaft segment are nearly completely free of carbonate-cemented sediment. Where present, they are represented by two by three millimeter ovoids.

On the distal segment of the left femur, including the lower portion of the diaphysis and the distal joint, heavy build-up (up to 1 millimeter thick) of carbonate-cemented sediment is present on the lateral surface. The relatively light accumulation on the posterior surface differs from pronounced accumulation on the linea aspera of the proximal segments. A groove in the border of the medial condyle with a maximum width of approximately 18.5 millimeters and maximum depth of approximately 2.6 millimeters has slight adhering carbonate-cemented sediment on the surface and appears to represent a contact depression (compression). The carbonate-cemented sediment striae at this location are longitudinally oriented.

All three segments of the right femur similarly show a moderate to heavy presence of carbonate-cemented sediment on the posterior surface (Figure 17.20). The proximal segment is represented by the ball, neck, and a small section of the diaphysis. Moderate carbonate-cemented sediment with a maximum thickness of 200 μm to 300 μm is present on the posterior surface in the region of the gluteal tuberosity, while the ball, anterior, lateral, and medial surfaces are free of concretions.

The upper 79 millimeters of the next more distal segment of the right femur, represented by a majority of the diaphysis, has moderate carbonate-cemented sediment on the anterior surface. A rugose pattern was beginning to

FIGURE 17.19 Kennewick Man's left femur. The anterior surface (A) has minimal accumulation of carbonate-cemented sediment compared to the posterior surface (B), which has dense accumulation, especially along the linea aspera. Middle anterior (C) and middle posterior (D) have the same pattern.

coalesce, covering approximately 30 percent of this area. The lower 114 millimeters of the anterior surface is essentially free of carbonate-cemented sediment, except for a few small focal areas with diameters of 1.5 to 3 millimeters. Nearly the full length of the posterior surface has rugose carbonate-cemented sediment localized on the linea aspera and the portion of the lower gluteal tuberosity present on this segment. It is thickest in the proximal aspect of the posterior surface and slightly less so along the raised linea aspera as it progresses inferiorly. Carbonate-cemented sediment accumulation on the proximal end of the linea aspera measures approximately three to four millimeters at its thickest. Progressing distally along the linea aspera, the average thickness is one to 1.3 millimeter. The medial and lateral surfaces of the segment are essentially free of accumulation except for small focal areas.

On the most distal segment, represented by a small section of diaphysis and the distal joint, a moderate amount of carbonate-cemented sediment is present near the metaphysis. On the posterior diaphysis, concretions are concentrated from the joint through the entire length of the segment. The thickest carbonate-cemented sediment deposits measure approximately two millimeters, many of which are located on the lateral aspect of the posterior surface. There is no carbonate-cemented sediment on the distal joint surface. The anterior surface of the upper distal diaphysis is relatively free of carbonate-cemented sediment with two to three millimeter raised ovoids covering approximately only 2 to 3 percent of this region. The lateral and medial surfaces show focal areas of carbonate-cemented sediment similar to the upper distal diaphysis.

Tibiae The left tibia has heavy cemented sediment accumulation on the medial, lateral, and posterior surfaces. The anterior surface of the proximal segment, represented primarily by the epiphysis, is difficult to evaluate because the metaphysis is missing postmortem. There is an eight by ten millimeter concretion on the posterior surface of the proximal segment that is approximately 2.1 millimeters thick. Particles of the anterior head of the fibula are in the matrix of carbonate-cemented sediment adhering to the surface of the tibia. This reflects in situ articulation of the tibia with the adjacent head of the fibula (Figures 17.21 and 17.22). About 80 to 90 percent of the medial surface

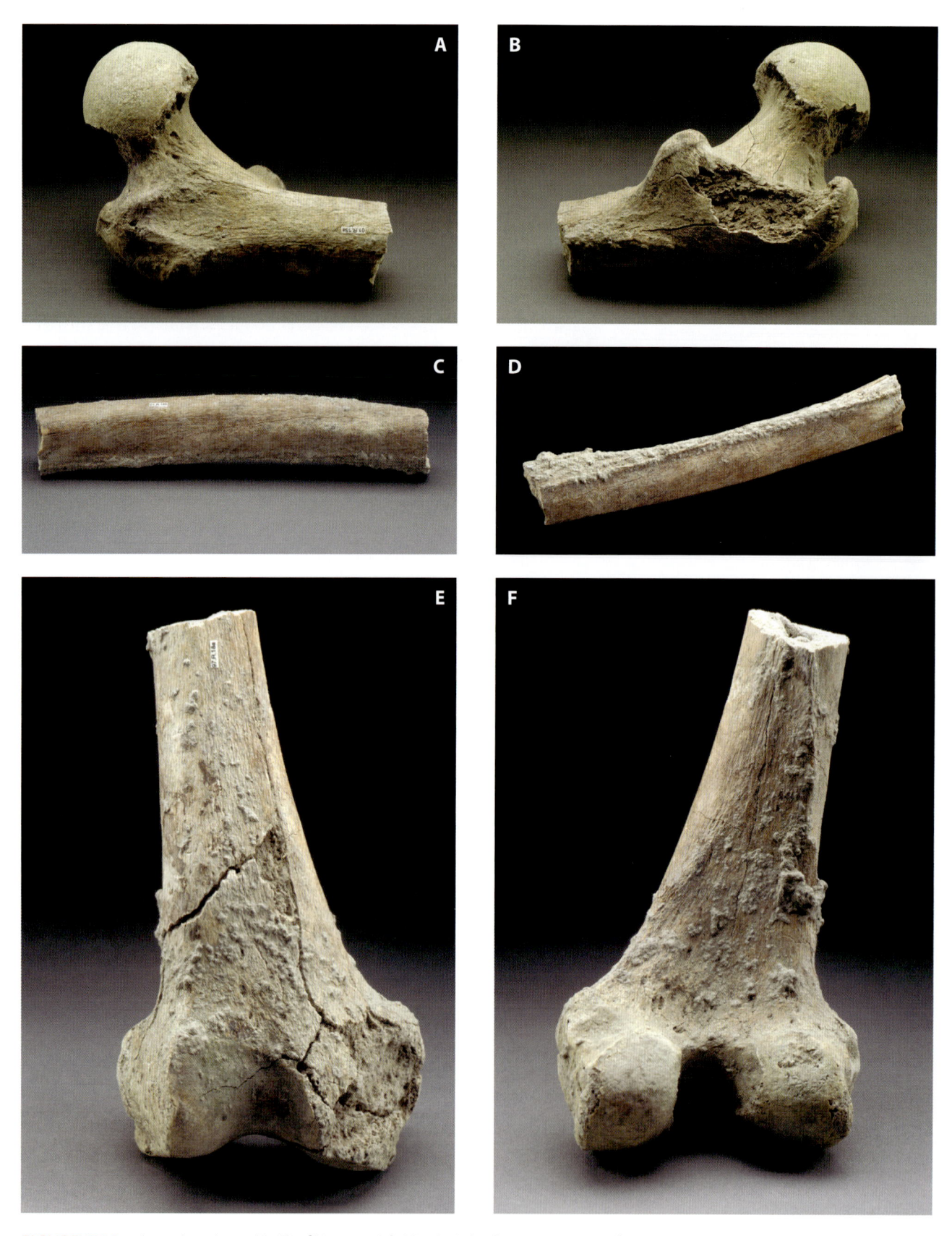

FIGURE 17.20 Anterior views (A, E) of Kennewick Man's right femur compared to posterior views (B, D, F). The anterior surface of the upper three-fourths of the femur has minimal carbonate-cemented sediment accumulation relative to the posterior surface (proximal and middle segments). The distal diaphysis and epiphysis have accumulations on both the anterior and posterior surfaces. The distal end was deeper in the Bca horizon where more carbonates precipitated. Midshaft (C) shows carbonate-cemented sediment at the bottom.

FIGURE 17.21 Posterior view of Kennewick Man's proximal left tibia epiphysis with nodular, carbonate-impregnated sediment adhering to the medial and lateral halves of the bone. Tiny, brown-colored fragments of the fibula are cemented onto the tibia (arrow; close-up shown in Figure 17.22).

FIGURE 17.22 Close-up of fibula fragments adhering to Kennewick Man's left tibia by carbonate-cemented sediment.

below the metaphysis is covered with cemented sediment that is 0.5 millimeter thick and slightly rugose.

The presence of carbonate-cemented sediment cannot be fully evaluated on the anterior aspect of the proximal diaphysis of the left tibia because of prior sectioning by government scientists (Figure 6.10). The more superior portion, which has the medial half of the tibial tuberosity, is present and it is free of carbonate-cemented sediment accumulation. The medial surface of the proximal end of the segment shows 80 to 90 percent coverage of carbonate-cemented sediment, while the lower two-thirds of the medial surface exhibit less (approximately 20 percent), although very rugose, coverage. Nearly the full length of the lateral surface shows rugose carbonate-cemented sediment accumulation with approximately 50 percent coverage. Similarly, there are rugose accumulations on the posterior surface.

Carbonate-cemented sediment is heaviest on the medial and lateral surfaces of the distal segment, representing approximately half of the diaphysis and an intact epiphysis (Figure 17.23). At midshaft, segment 97.L.20c has extensive carbonate coverage with thicknesses as great as

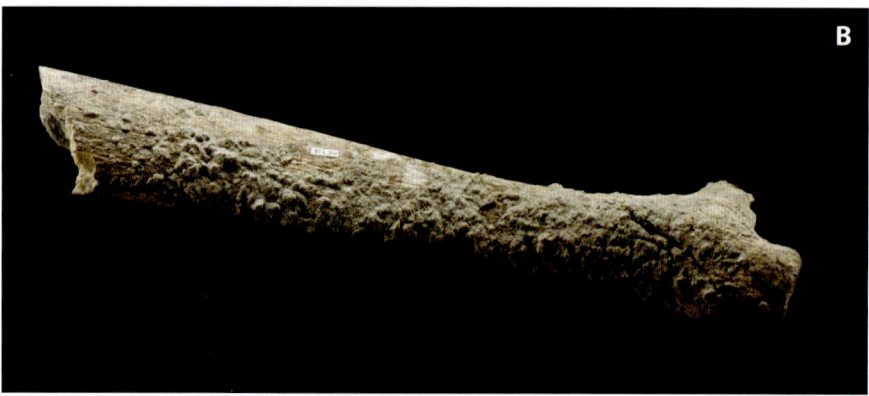

FIGURE 17.23 Rugose carbonate-cemented sediment formations on the anterior (A), medial (B), and posterior (C) surfaces of Kennewick Man's distal left tibia, including the medial malleolus.

2.5 millimeters. Area coverage and thickness is greater on the lateral and medial surfaces than on the posterior surface. There is a 70 by 15 millimeter area of heavy accumulation on the lateral surface. The medial surface, which has more rugose carbonate-cemented sediment than the lateral, exhibits 90 percent surface coverage with heavy deposits, some greater than one millimeter in thickness. The anterior crest of the middle to the upper-lower third of the diaphysis is relatively free of carbonate-cemented sediment. The posterior aspect of this segment has heavy deposits along the medial and lateral margins of most of the diaphysis. A majority of the intervening posterior surface is relatively free of carbonate-cemented sediment. Moderately rugose carbonate-cemented sediment covers approximately 50 to 60 percent of the surface of the distal one-fourth of the posterior surface.

On the right tibia, the proximal epiphysis is broken into medial and lateral segments. The surface of the medial half is covered by widespread, moderately rugose carbonate-cemented sediment with nodules up to one millimeter in thickness. The surfaces of the lateral segment are nearly entirely covered with the exception of the lateral joint surface, which is nearly free of carbonates. The external cortex is represented by a posterior lateral condyle. Cemented sediment is widespread on the posterior surface, with one localized area measuring 1.9 millimeter in thickness.

A more distal segment of the right tibia, representing the upper half of the diaphysis, has widespread carbonate-cemented sediment accumulations on the medial, lateral, and posterior surfaces, while the anterior tibial crest is relatively free of nodules. Nodules up to 2.4 millimeters thick are present on the lateral and medial surfaces. The proximal third of the posterior diaphysis has widespread and thick carbonate-cemented sediment. While it would have enveloped much of the bone surface, a considerable amount of carbonate-cemented sediment has flaked off. The presence of dark brown patches interspersed with light-colored areas indicates where the sediment and the outer cortex have deteriorated. The distal segment of the right tibia, comprising the lower one-fourth of the diaphysis and joint, has sediment nodules as thick as 1.7 millimeters on the medial malleolus and 2.7 millimeters on the medial surface near the postmortem fracture margin.

Feet The first and second metatarsals of the left foot are held together by carbonate-cemented sediment (Figures 13.10 and 13.11). In general, the metatarsals have carbonate accumulations on the medial and lateral surfaces. Carbonate-cemented sediment is especially heavy on the calcaneus, where coverage is three-dimensional and rugose on the lateral surface (Figure 13.8). Nearly half of the medial surface (which cannot be fully evaluated due to postmortem loss) is covered by three-dimensional accumulation, which is also heavy on the plantar surface of the bone. With the exception of the inferior aspect, the posterior talar articular surface is free of accumulation, as is most of the posterior surface. The range of accumulation on the medial, lateral, and plantar surfaces is between 0.6 and 1.5 millimeter thick. In general, accumulation is heavy on the lateral surface of the left foot, indicating lateral rolling at interment.

Carbonate-cemented sediment accumulation on the right foot occurs primarily on the plantar surfaces, indicating flexion of the foot in situ. A fragment of the third cuneiform is fused to the navicular on the lateral, superior surface by carbonates, as is a small piece of calcaneus (Table 17.3; Figures 13.5 and 13.12), demonstrating articulation of these bones within the burial.

Additional Indicators of In Situ Positioning

In addition to the presence of carbonate-cemented sediment, two other taphonomic assessments help determine the orientation and positioning of Kennewick Man's skeleton in situ and provide information about the erosion process. Bone fractures are described in Chapter 19. Algal stains are discussed in Chapter 18. These two chapters are briefly introduced here to complete and aid in interpreting the extensive data presented in Table 17.2.

Postmortem Fractures

Breakage characteristics and patterning were assessed to obtain information about the mechanisms that produced postmortem fractures in the long bones. Chapter 19 explains fracture production in living tubular bone as a step toward interpreting fracture production in dry (ancient) bones. The process is similar, but in old bones, fracture features are subtle and require experience to interpret. The key is determining which bone surfaces experienced either tension (i.e., pulling forces from bending) or compression stress (squeezing) from the weight of the soil and the collapse of the terrace bank. Modern bone has the biomechanical strength of mild steel and can withstand considerable compressive force. It is, however, significantly weaker under tensional force. Bone always fails first under tension.

Each fracture surface was examined to identify the fixed, or cantilevered, end versus the free end of the bone and also to determine the anatomical direction that the free end moved to produce the fracture. This distinction applies a model that recognizes that portions of a given bone (the fixed end) were secured by intact sediment as other sections of the bank gave way. The process of bank collapse applied a force (the bank overburden) to the end that was able to move because sediment was eroding out from under the skeleton.

The angle and contour of the fracture surface and the location of a breakaway spur help determine the fixed end of the bone. The morphology of the fracture surface, orientation of secondary fractures, and location of the breakaway spur identify the tension and compression sides of each fracture and thus the anatomical direction that the free end moved to produce the fracture. Data were analyzed to identify and determine the relationship between the tension versus compression sides of each fracture surface relative to anatomical orientation. Anatomical maps created by Berryman depict the way each bone was angled (bent) to produce each specific fracture (Figures 17.24 and 17.25).

Bone breakage that shows no evidence of healing or remodeling could represent either fractures that occurred at the time of death (perimortem) or fractures that occurred after death (postmortem). The characteristics of a fracture, such as the shape of the broken edges and color of the exposed margins, usually make it possible to distinguish perimortem from postmortem damage. No perimortem fractures are evident in Kennewick Man's skeleton. In contrast, bone damage was extensive during the postmortem interval, which spanned millennia.

Three categories of postmortem fractures that differ in the time of occurrence were observed in the skeleton. First, recent breaks, recognized by light-colored fracture surfaces (Figure 17.2), are extensive and widespread across the skeleton. Most of this breakage occurred when the bones eroded from the alluvial terrace. Second, some breakage represents damage that occurred in antiquity as the weight of overlying sediment continued to compress and slightly displace bones that had lost their strength. In such cases, fracture edges are stained the same brownish color as the soil-stained external surfaces of the bones. Flat bones, such as ribs, and irregular bones with thin cortices, such as vertebrae, were most susceptible to this type of damage. Observed examples include breakage of the pedicles of the thoracic vertebrae, separating the vertebral bodies from the neural arches, and breakage of some neural spines. Third, other fractures in antiquity are the longitudinally oriented drying cracks in some diaphyses, examples of which include the distal femora and the left first metatarpal (Figures 19.9 and 13.11).

Interpretations based on postmortem fractures cor-

FIGURE 17.24 Berryman annotating characteristics of postmortem breaks in Kennewick Man's left femur.

roborate the in situ positioning of the Kennewick Man skeleton as suggested by the patterning of carbonate-cemented sediment. Such fractures provide independent evidence that the body was supine with the forearms pronated and the legs extended, with slight lateral rotation of the left foot. Incomplete bone fractures leave bone sections intact; complete fractures involve breakage and separation into two or more pieces. There are more complete than incomplete fractures in the Kennewick Man skeleton. Particularly diagnostic incomplete (IPFX) and complete (CPFX) postmortem fractures are identified in Table 17.2. The presence of incomplete and complete fractures indicating anterior to posterior load direction provide compelling physical evidence that the posterior surface of the left scapula was down. Fracture data also confirm that the posterior surfaces of the left (Figure 17.26) and right femora and tibiae were down in situ.

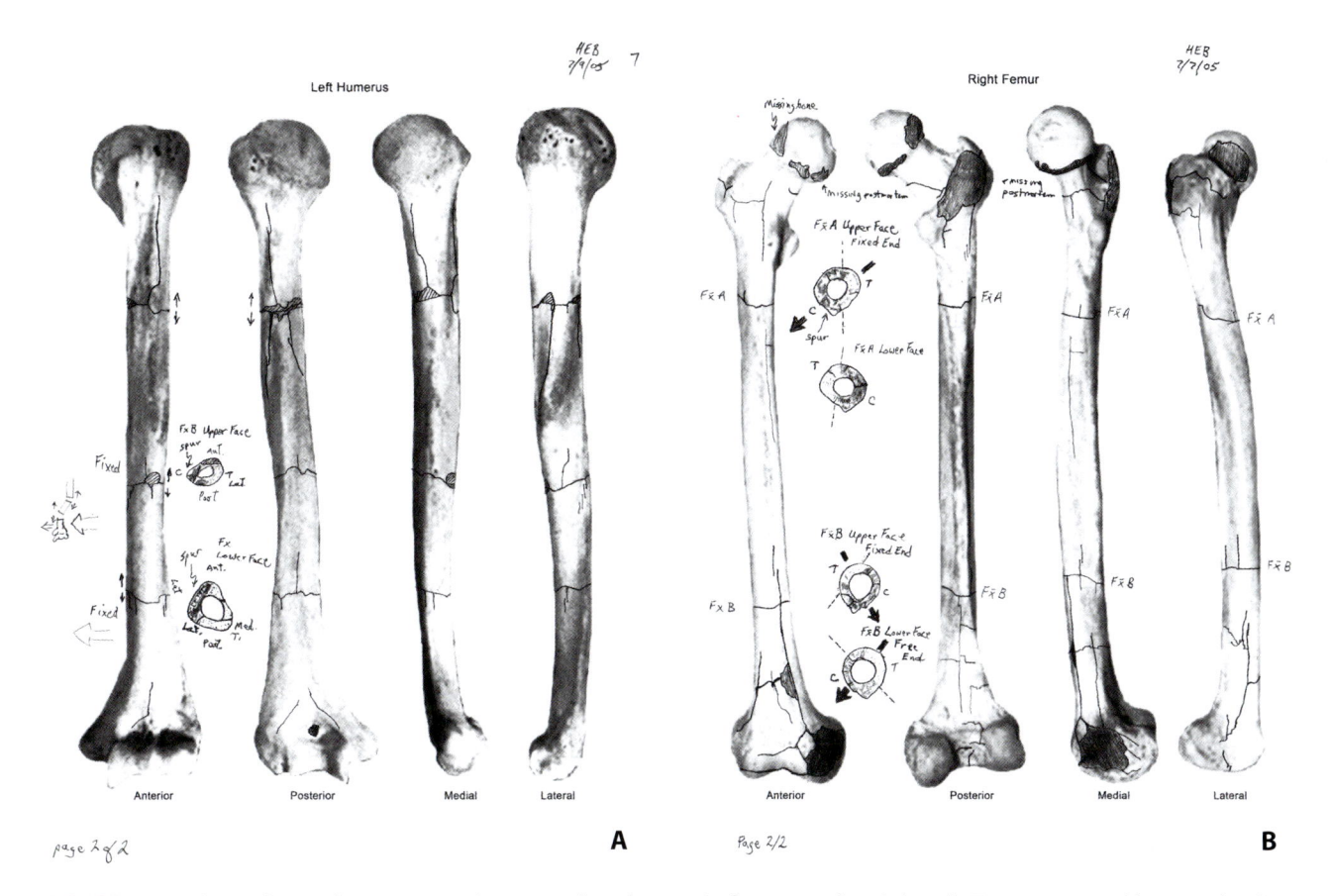

FIGURE 17.25 Long-bone diagrams used to record taphonomic features of each break. Transverse and longitudinal fractures were illustrated, as were breakaway spurs, compression and tension surfaces, and fixed ends. Diagrams of Kennewick Man's left humerus (A) and right femur (B) are shown.

FIGURE 17.26 Bank erosion produced an overburden force that pushed downward on the distal end of Kennewick Man's left femur. The three incomplete bending fractures that transect the anterior diaphysis and joint face show that this surface was up in situ.

Algae

Chlorophyll stains on bones are another source of information that can be used to determine the burial position of Kennewick Man. Identification of the algal species responsible for the staining can provide biological and ecological information about its habitat, leading to a more complete understanding of the bones' recent taphonomic history.

In the case of Kennewick Man's bones, subtle green stains were caused by benthic aquatic algae. Since all algae require sunlight, water, nutrients, and a substrate, such as bone, on which to grow, a bone's position can partially be gauged by noting which anatomical surfaces were colonized by algae. Evidence of algal growth, recorded as green staining on skeletal diagrams, helped determine which anatomical surfaces were exposed to light while the remains were in situ. Observations indicating that a specific surface was facing up are listed in Table 17.2.

Because Kennewick Man's limb bones are broken into fairly large pieces, the presence of green staining on larger and heavier segments was useful for orienting the limbs relative to the ground surface. A similar pattern of green staining on contiguous pieces of the same bone, in all probability, occurred during the initial phase of exposure while that portion was in situ and intact. The two distal segments of the left humerus (97.L.13b and 97.L.13c) provide the best example (Table 17.2). Both show slight green staining on the anterior and lateral surfaces, indicating that the anterior surface was up (Figure 17.17). Algal colonization of out-of-context segments would be less likely to show anatomical integrity (i.e., segments of the same bone scattered on the mudflat would be unlikely to show a matching algal pattern).

Staining of the same anatomical surface of adjacent bones also likely reflects the algae's growth on the skeleton while in situ. For example, the anterior surfaces of both arms show adjacent patterning at the humeral-ulnar joints; while the elbow joints were still articulated, the posterior surfaces of the humeri and ulnae were down and free of algae (Table 17.2).

ORIENTATION OF THE SKELETON RELATIVE TO THE WATER

Left side surface damage caused by corrasion, along with the patterned presence of macroabrasions, provides evidence of body alignment relative to the lake. Collectively, they indicate which side and which bone surfaces were closest to the lake and most subject to direct wave action (i.e., which skeletal surfaces experienced wave damage). The evidence suggests that the remains were exposed from left to right as the skeleton lay in a supine position, on its back, with the left side toward the Columbia River (Figure 17.27).

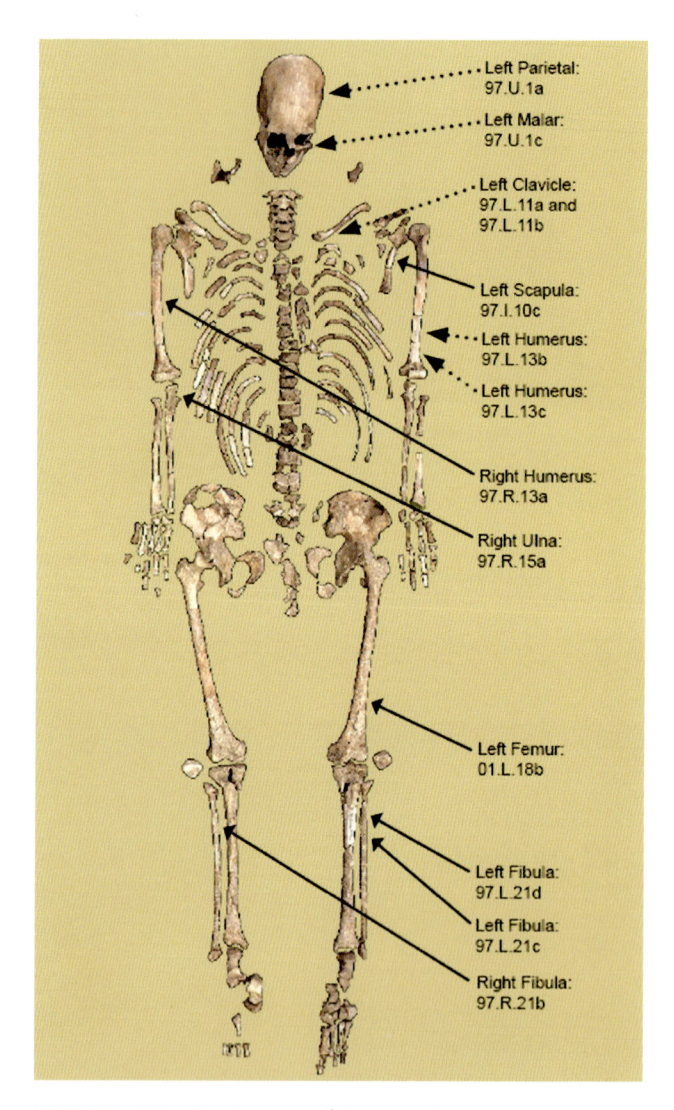

FIGURE 17.27 Taphonomic damage from macroabrasions (solid arrows) and corrasion (dotted arrows). Lateral bone surfaces on the left side of Kennewick Man's skeleton and medial surfaces on the right side were affected. This pattern reflects the progression of damage from left to right as the in situ skeleton was exposed to wave damage and debris in the water.

Corrasion

Early stages of bone weathering from exposure, including bleaching, mosaic cracking, and flaking or exfoliation, can eventually progress to advanced stages where the bone disintegrates (Behrensmeyer 1978). This process is informative when determining the length of time that a bone has been exposed. In forensic cases found in unprotected desert conditions, for example, bone bleaching can appear as early as the second month after death, but it is more common by the sixth month (Galloway 1997). However, Kennewick Man's bone surfaces do not show the characteristic appearance of sun-bleaching, mosaic cracking, or flaking. The localized, light-colored, burnished areas that were observed were produced by another type of environmental weathering: corrasion. Corrasion is the mechanical erosion (scouring and pitting) of bone surfaces by sand and silt moved by water or wind. In this case, waves carrying particulates lapped against partially exposed bones while still in situ, prior to the collapse of the bank. The abrasive effect of silt and moving water can rapidly scour and polish bone surfaces, as observed in modern forensic cases (Brooks and Brooks 1997).

The left side of the cranium (the malar, greater wing of the sphenoid, and posterior temporal and parietal), left clavicle, and left humerus were damaged by wave action (Table 17.4). The outer tables of the temporal and parietal bones show loss of patination and superficial erosion and their color is lighter than the right side of the cranium (Figure 7.3b compared to Figure 14.1). The color of the left malar has darkened due to casting in 1996. A photograph taken at that time shows that it was originally lighter in color.

The presence of dense sediment in the cerebellar fossae of the cranial vault has been presented as evidence that Kennewick Man's face was directed upward and toward the feet in the burial (Figures 17.28 and 17.29). Figure 17.28 shows rotation of the cranial vault to approximate this position. The external presence of carbonate nodules on the inferior nuchal region and mastoid processes corroborates this interpretation, as does the occurrence of demarcated corrasive damage on the posterior left temporal squamous and parietal. The left mastoid process and occipital squamous were still buried and protected by intact sediment as waves began to lap against the left side of the cranium, producing a line of demarcation between exposed (and altered) bone versus unexposed bone. Surface damage and postmortem fracturing of the left lateral wing of the sphenoid were also caused by this process.

The sternal end of the left clavicle (97.L.11a) has corrasion damage on its lateral side with decreasing effect toward the medial end. The middle segment of the left clavicle also exhibits subtle erosion due to corrasion that is less pronounced than on the sternal end (Figure 7.18).

Corrasion damaged the distal two-thirds of the left humerus but did not affect the right humerus (Figures 17.17 and 17.30). The left humerus has four segments, the top two of which are held together by Parafilm®. The proximal segment (97.L.13a) has a base color of very pale brown (10YR7/3) on all surfaces except for muscle attachment sites, which are darker. The overall color of 97.L.13a is markedly different from the anterior and anterolateral surfaces of the middle segment (97.L.13b) and the medial, anterior, and lateral surfaces of the distal segment (97.L.13c), all of which are a very pale white (10YR8/1). The corresponding posterior surfaces are darker (another shade of very pale brown, 10YR8/4) and pink (7.5YR7/4). The distal two-thirds of the left humerus experienced corrasive action during initial exposure in situ (Table 17.4; Figure 17.30).

Macroabrasions

Macroabrasions on bone surfaces were also observed (Figure 17.31). Each abrasion was described and measured, and its location recorded (Table 17.4). These macroabrasions are not typical of animal damage, such as rodent or carnivore activity. Rather, the appearance of the macroabrasions on the skeleton indicates that they were caused by the reciprocating movements of hard objects, likely plant detritus (i.e., brush and small branches drifting in the lake water), rubbing against the bone.

Abrasions are found on the medial surfaces of right-side bones and the lateral surfaces of left-side bones (Table 17.4). A small area of recent abrasion is located on the medial surface of the right ulna (97.R.15a) inferior to the semilunar notch. The right fibula (97.R.21b) exhibits abrasion on the posteromedial surface of the proximal end of the segment (Figure 17.32). There are two areas with grooves: one located approximately 18 millimeters from the proximal break and the other approximately

TABLE 17.4 Corrasion and macroabrasions as taphonomic indicators of in situ orientation.

Bone	Side	Specimen Number	Taphonomic Indicator	Segment Number	Taphonomy Description	Orientation
Cranium	L	97.U.1a	COR		The posterior-inferior left parietal shows outer table erosion and loss of iron/manganese oxide patination.	Left side toward river
Sphenoid	L		COR		External surface, loss of patination and exposed underlying cancellous bone.	Left side toward river
Malar	L	97.U.1c	COR		The external surface was lighter in color based on a 1996 photograph; the bone has subsequently darkened due to casting.	Left side toward river
Clavicle	L	97.L.I1a, 97.L.I1b: Sternal and middle segments	COR		On the lateral end fracture face of 97.L.I1a, adjacent superior surface, decreasing laterally to medially along the length of the segment (more apparent on lateral end); segment 97.L.I1b has less microabrasion than 97.L.I1a, appears on the superior surface.	Left side toward river
Humerus	L	97.L.I3b	COR	2	Anterior and antero-lateral surfaces.	Left side toward river; left humerus, but not the right, was water abraded
Humerus	L	97.L.I3c	COR	3	Anterior, medial, and lateral surfaces of the distal diaphysis.	Left side toward river; left humerus, but not the right, was water abraded
Scapula	L	97.L.I10c: Lateral, midaxillary border	ABR		Three areas affected: transverse striations (directed medial-lateral) on the lateral-ventral aspect and two others inferior to the first on the lateral aspect.	Left side toward river
Humerus	R	97.R.R.13a: Proximal epiphysis and upper three-fourths of the diaphysis	ABR	1	Localized area of microabrasion on the medial surface of the diaphysis beginning 145 mm below the head of the humerus, superior-inferior length of 39 mm with a maximum transverse breadth of 15 mm.	Left side toward river
Ulna	R	97.R.15a	ABR	1	Photograph in Walker et al. (2000) shows macroabrasion of the medial aspect of the proximal end of the ulna.	Left side toward river
Femur	L	01.L.18b: Middle and upper lower diaphysis	ABR	2	Distal lateral surface: grooves consisting of two lateral troughs with a slightly elevated ridge dividing the troughs; superior-inferior length of 18 mm, anterior-posterior length of 23 mm; a small, single gouge is located 18 mm above the superior margin of the larger abraded defect.	Left side toward river
Fibula	L	97.L.21d: Proximal diaphysis	ABR		Posterolateral surface (on carbonate-cemented sediment only).	
Fibula	L	97.L.21c: Middle diaphysis	ABR		Posterolateral surface: deeply abraded.	Left side toward river
Fibula	R	97.R.21b: Proximal diaphysis	ABR		Two areas affected on the posteromedial surface of the proximal end of segment.	Left side toward river

COR: Corrasion, ABR: Macroabrasions

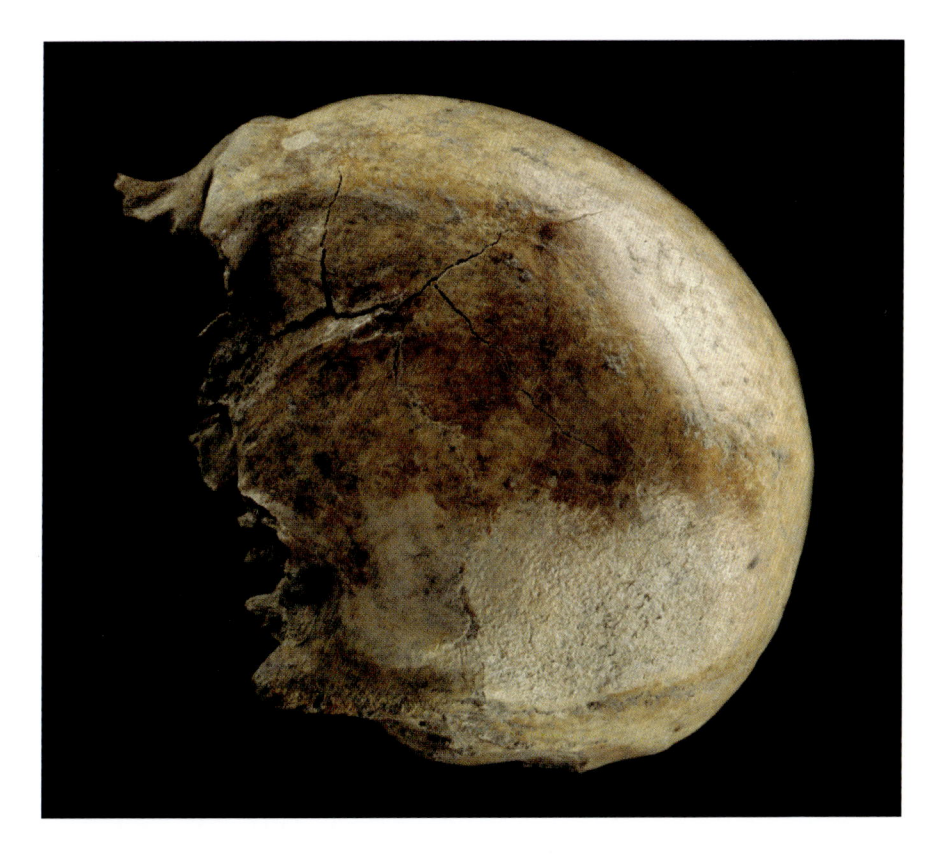

FIGURE 17.28　Left lateral aspect of Kennewick Man's cranial vault showing corrasion damage. The vault was photographed in the Frankfort horizontal and the resulting image was rotated approximately 55 degrees. In situ positioning is defined by a waterline that shows the occipital remained embedded and protected in sediment for slightly longer than adjacent bones. This brief exposure of the posterior parietal and temporal resulted in pitting and loss of surface color.

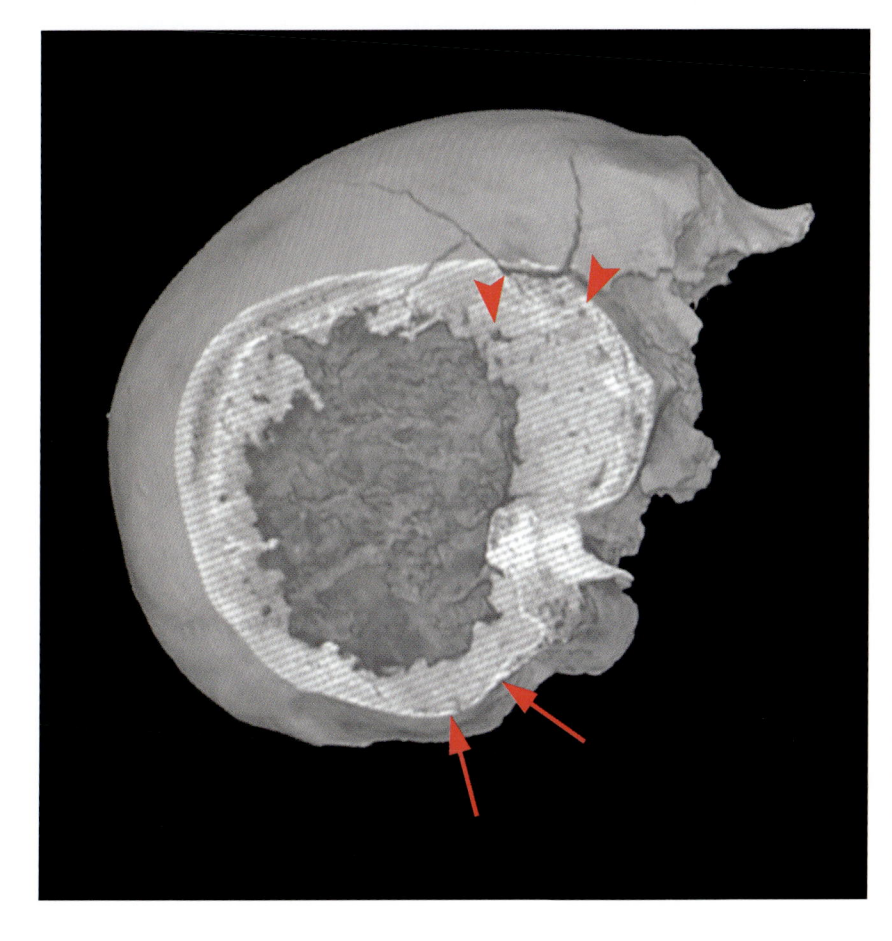

FIGURE 17.29　This CT 3D virtual reconstruction exposes the right lateral third of the cranial vault. In situ, Kennewick Man's face was directed upward with his chin toward his chest without significant tilting toward either side. Water carrying silt, clay, and carbonates entered the cranial vault through the optic foramina and orbital fissures. Dense sediment accumulated internally along the posterior base of the cranium (arrows), while carbonate concretions formed externally in the nuchal and mastoid regions. Lower density sediment in the frontal lobe and along the superior endocranial surface contains voids (arrowheads).

FIGURE 17.30 Kennewick Man's left humeral diaphysis, exhibiting both slight green algae staining and white, abraded bone caused by the forces of corrasion while in situ. (A) shows the distal two-thirds of the anterior surface; (B) is an anteromedial view. The largely unaffected posterior surface is shown in Figure 17.17.

FIGURE 17.32 A proximal segment of the diaphysis of Kennewick Man's right fibula, exhibiting postmortem grooves on the posteromedial surface.

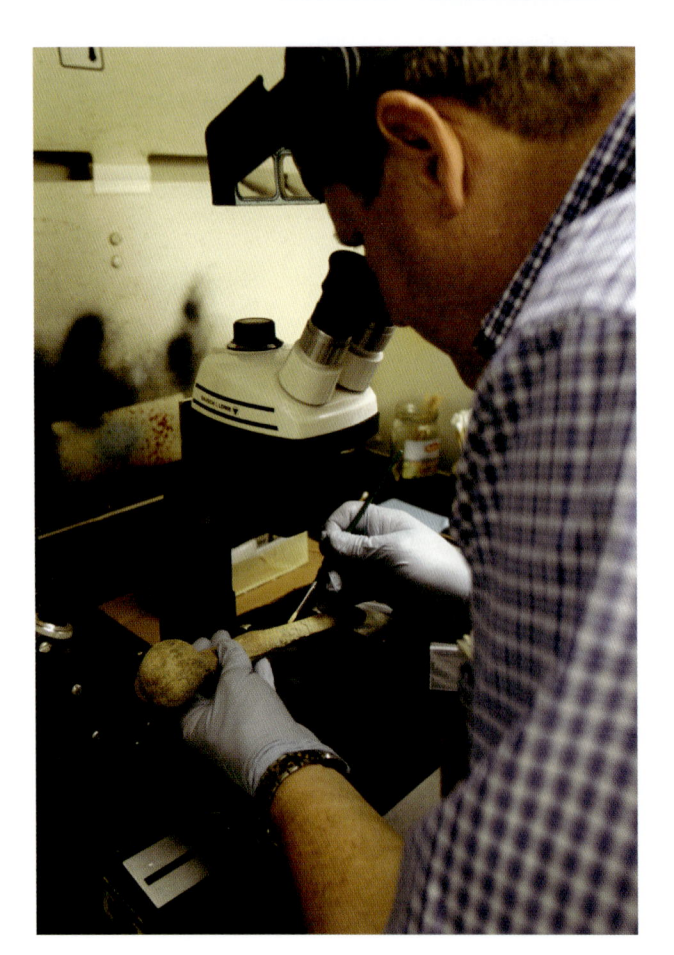

FIGURE 17.31 Using a fine brush and low-power stereozoom magnification, Owsley cleans and evaluates features of an abraded area on the medial surface of the right humerus.

FIGURE 17.33 The color difference between the unmarred bone cortex and the abraded areas of the lower lateral surface of the middle diaphysis of Kennewick Man's left femur resulted from recent damage.

four millimeters from the same break. The left scapula has three areas of abrasion confined to the lateral aspect of fragment 97.I.10c (the lateral, midaxillary border). The lower lateral surface of the middle diaphysis of the left femur (01.L.18b) has two small abrasions (Figure 17.33). The light pale brown color (10YR8/3) is similar to the transverse breakage surfaces of the femoral shaft. This area was only recently abraded. On the left fibula, macroabrasions are present on the posterolateral surfaces of adjoining pieces 97.L.21d (proximal diaphysis) and 97.L.21c (middle diaphysis). The bone surface of segment 97.L.21c was deeply abraded. On segment 97.L.21d, only the carbonate-cemented sediment is gouged, indicating that the bone was abraded after the formation of the carbonate deposits. In fact, all these abrasions occurred at the time the skeleton eroded out of the bank.

Consider, for example, the abrasion on the medial surface of the right humerus diaphysis (97.R.13a), which begins 14.5 centimeters below the head (Figure 17.34). This damaged area measures 3.9 centimeters (superior-inferior) by 1.5 centimeters in breadth. The exposed underlying cortical bone shows very slight staining. Smaller areas of more recent localized surface abrasions with lighter-colored surfaces are also present. The abraded area has shallow troughs with relatively smooth surfaces that are somewhat similar to medium-sized animal gnaw marks. However, this area lacks specific characteristics associated with carnivore scavenging: destruction of the less-dense articular ends, progressive reduction of shaft length, indentation pits caused by tooth cusps, and linear or parallel scoring marks that follow the contours of the bone (Haglund 1997a). Rather, the abraded area is characterized by thin, shallow, transverse, and somewhat parallel marks that gradually become larger, deeper, and more chaotic. This pattern suggests repetitive and disordered contact between the bone surface and a moving object, such as plant detritus in the slosh zone at the water's edge.

Haglund (1997b) characterizes rodent gnaw marks as "parallel marks, fan-shaped patterns, or totally disorganized striae overlying each other" depending on the character of the bone and how many times it has been gnawed. Typical examples of rodent damage observed in forensic casework are shown in Figure 17.35. Shipman and Rose (1983) delve into the morphology of rodent tooth activity and describe gnawing action either as the use of the upper jaw as an anchor or as the use of both jaws to draw across the bone towards each other to create the "chaotic" pattern of marks. Walker et al. (2000: Figure 8) illustrated a possible example of rodent damage, abrasions that appear "chaotic" with multiple intersecting grooves at various angles to one another along with parallel striae. However, the accompanying close-up photograph shows details inconsistent with rodent gnaw marks. The grooves are wide at one end and appear to converge to a point at the other, are curvilinear, and have a "wavy" appearance. Also missing are chattermarks, or striae that frequently appear parallel to the

FIGURE 17.34 Irregular, but relatively parallel, grooves of varying depths in the right humerus are characteristic of contact with water-borne detritus, not carnivore gnawing.

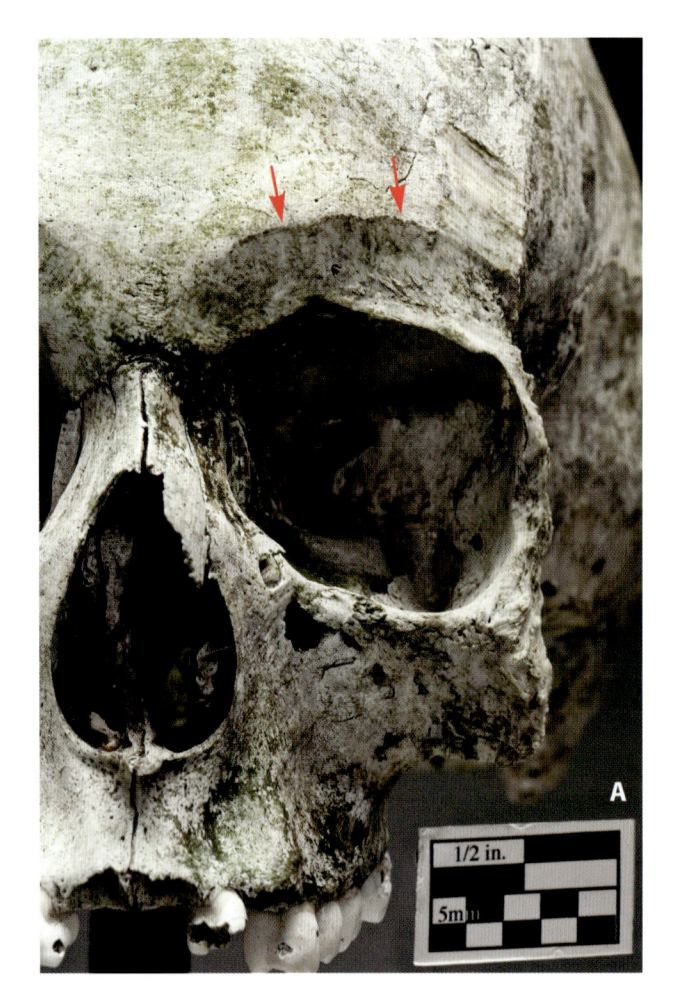

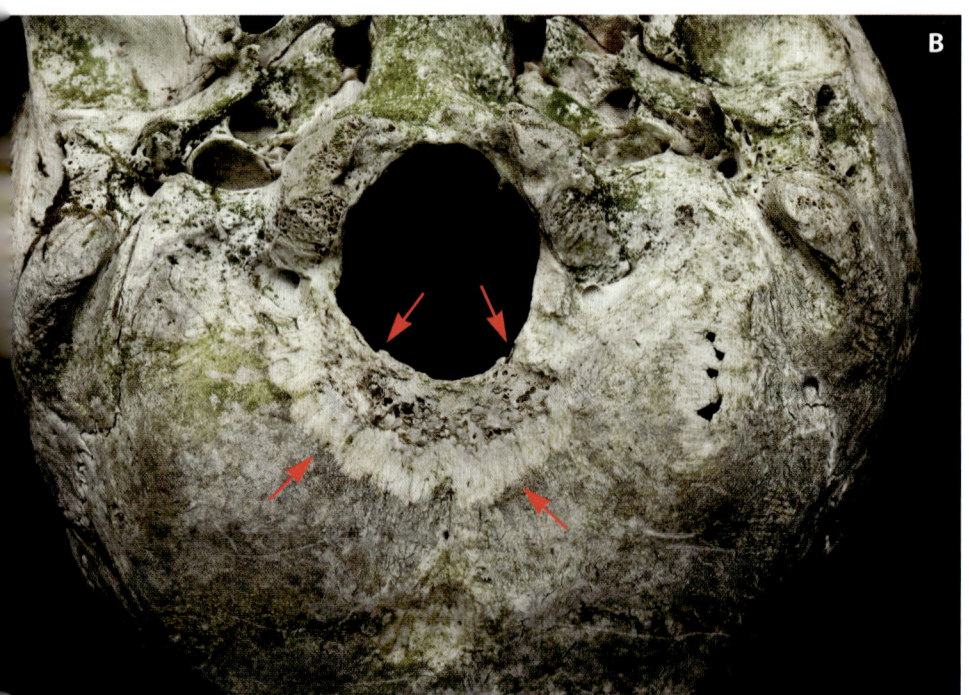

FIGURE 17.35 The frontal bone (A) of this cranium from a forensic case exhibits rodent gnawing characterized by parallel striae in the left supraorbital margin (arrows); the left zygomatic was almost completely destroyed. The posterior margin of the foramen magnum (B) and nearly the entire length of the right tibia (C) were damaged by gnawing.

bite direction in rodent gnawing (Miller 1969; Shipman and Rose 1983).

The macroabrasions are not consistent with rodent or carnivore activity. Furthermore, these postmortem abrasions are limited to the lateral surfaces of the left side of the body and the medial surfaces of right side bones. No abrasions were detected on the anterior or posterior surfaces of any skeletal elements. These abrasions resulted from contact erosion between the skeleton and wave-induced debris movement while the remains were exposed to lake waters. Initially the lateral surfaces of left side bones were impacted, followed by the medial surfaces of right-side bones.

Postmortem Breakage

The left side of the skeleton shows more postmortem damage than the right, as indicated by the number of primary and secondary fractures in major bones (Table 17.5; Figures 7.18, 7.19, and 7.23). This difference is further evidence that the left side experienced slightly longer exposure to wave damage.

STRATIGRAPHIC PLACEMENT OF THE BURIAL

In 1998, the US Army Corps of Engineers (Corps) buried the Kennewick Man site under tons of rocks. Knowledge of the stratigraphy, therefore, is based on information collected during previous studies. The first was by Chatters who took sediment samples in November 1996 and drew diagrams of stratigraphic sequences (Chatters 2001:154). The second study of site stratigraphy and sediment deposition was led by the Corps in October and December 1997 (Wakely et al. 1998:xiii). The field report by Wakely et al. was released in 1998, as was a report submitted to the Corps by Huckleberry, Stafford, and Chatters (1998) who accompanied Corps investigators. Huckleberry and Stein's (1999) report to the National Park Service and Huckleberry et al. (2003) were based on the data collected in 1997.

Huckleberry et al. (1998) divided the terrace into two layers, Lithostratigraphic Units I and II, a stratigraphy outline that was maintained in Huckleberry and Stein (1999). Wakely et al. (1998) divided the stratigraphic layers into five units: Units I, II, and III are the same as Huckleberry et al.'s (1998) Unit I. Wakely's units IV and V are identical to Huckleberry et al.'s Unit II (Huckleberry and Stein 2003:Table 1).

According to Huckleberry and Stein (1999), the uppermost unit, Lithostratigraphic Unit I, was between 25 and 80 centimeters thick. These measurements were made at two study areas, CPP-054 and CPP-059.5, along a 345-meter stretch of the lake bank. The Kennewick Man skeleton was discovered between test areas CPP-057 and CPP-093. Toward the base of Lithostratigraphic Unit I is a discontinuous bed of volcanic ash that the United States Geological Survey identified as the Mount Mazama tephra dating to a catastrophic eruption ca. 7600 CAL yr. BP (Chapter 3; Bacon 1983). Lithostratigraphic Unit II continues from 70 to 130 centimeters below the modern surface.

Huckleberry and Stein (1999) used several methods to match the sediment adhering to or inside the cranium and medullary cavities of Kennewick Man's long bones to profile samples taken from the two defined strata in the riverbank. These tests included granulometry, micromorphology, thermogravimetry, X-ray diffraction, and trace-element analysis. Granulometry and micromorphology tests did not conclusively match sediment from the skeleton to the site. The thermogravimetric analysis revealed that the skeleton most likely came from between the lower eight centimeters of Unit I to the upper 40 centimeters of Unit II (62–110 centimeters

TABLE 17.5 Counts of primary and secondary fractures in the clavicles and limb bones of Kennewick Man.

Bone	Number of Bone Segments		Secondary Fractures	
	Left	Right	Left	Right
Clavicle	4	2		
Humerus	4	2	10	5
Radius	3	2	5+	5
Ulna	4	3	11+	9
Femur	3	3	22	12
Tibia	3	3 (long axis segments)	13+	14+

Note: The number of primary fractures is one less than the number of segments comprising the whole bone.

below the surface). X-ray diffraction comparisons indicated that the skeleton could have derived from either 70 to 80 centimeters, or from a second best match of 95 to 135 centimeters below the surface. No sediment from the intervening layers matched sediment on the skeleton. The trace-element analysis only demonstrated that the dark soil adhering to the skeleton was most likely from the beach and not the terrace; the analysis gave no clues as to the stratigraphic positioning of the skeleton in the bank. Huckleberry and Stein (1999) concluded that the skeleton likely eroded from sediments located 70 to 120 centimeters below modern ground level, which includes a partial section of the concretion-bearing stratum at a depth of 70 to 140 centimeters.

Huckleberry et al. (2003) tested sediment from both the Kennewick Man skeleton and the discovery site. X-ray diffraction and thermogravimetry analyses affirmed the results of the earlier study by Huckleberry and Stein (1999). Their analysis indicated that the skeleton likely derived from an 80-centimeter thick section of the concretion-bearing zone in the upper part of Unit II (approximately 60 to 140 centimeters below modern ground level). These investigators also concluded that Kennewick Man was probably buried rapidly based on the well-preserved and intact remains. Sedimentary evidence from the site indicates "low-energy deposition with relatively small increments of vertical accretion," a process inconsistent with a rapid fluvial event as the means of interment (Huckleberry et al. 2003:663).

Chatters (2000) also investigated the question of the stratigraphic positioning of Kennewick Man. Soil samples from the skeleton and the discovery site were submitted to the Washington State University Department of Crop and Soil Sciences for textural analysis. This study concluded that sediment from the skeleton most closely resembled that from either of the two concretion-rich zones at 70 to 100 centimeters or at 130 to 160 centimeters. The presence of concretions adhering to the skeleton precludes association with the upper level stratum, Unit I, because of the lack of concretions in this stratum, and with the sediment layers below 160 centimeters in Unit II, due to the fine texture of this layer. Chatters (2000) concluded that the skeleton must have eroded from a level between 70 and 160 centimeters below modern ground level.

Following interment, eolian sediment accumulated over the site and included tephra from the eruption of Mount Mazama. The aforementioned studies have conservatively interpreted the overall range for the stratigraphic placement of the skeleton as somewhere between 60 to 160 centimeters below ground level. The results of this taphonomic study, along with the careful work of these preceding studies, offer further resolution of the question of placement.

In 1997, Stafford described two distinct Bca horizons, an upper and a lower. At CPP-044 (Figure 17.36), the depths below ground level of these two carbonate units are 75 to 100 centimeters and 124 to 154 centimeters, respectively. The upper Bca horizon dates to the late Holocene, approximately 2,000–2,500 RC yr. BP, and is the one that overprinted the Kennewick Man skeleton with pedogenic carbonate. The lower carbonate horizon dates to the early Holocene, approximately 9,500–10,000 RC yr. BP. Based on the pattern of carbonate accumulations, Kennewick Man's upper skeleton was at the top of the upper Bca horizon, while his feet were completely enclosed within the carbonate, indicating a slightly greater depth. The burial depth for the skeleton, therefore, can be interpreted to have been approximately 70 centimeters below ground level at the highest portion of the cranium (i.e., the frontal bone) and at most 85 to 90 centimeters below ground level at his feet.

The carbonate horizon in which Kennewick Man was buried is 25 centimeters thick (75 to 100 centimeters BGL at CPP-044, the approximate original location of his burial) and his head protruded roughly five centimeters or more above the top of the carbonate. The angle at which the skeleton was inclined in the upper layer of Bca can be estimated using the patterned accumulation of carbonate sediment, skeletal measurements, and the principles of geometry. Kennewick Man was 5'7" tall (170 centimeters). If his skeleton was inclined at an angle of 3 degrees from horizontal, the vertical distance between his feet to five centimeters above the upper carbonate layer would be ten centimeters. If Kennewick Man was inclined at an angle of 5 degrees, the vertical distance would be 15 centimeters $[X = 170 (\sin A)]$. A gradual inclination of only 3 to 5 degrees from horizontal could easily accommodate the observed distribution of carbonates adhering to the skeleton.

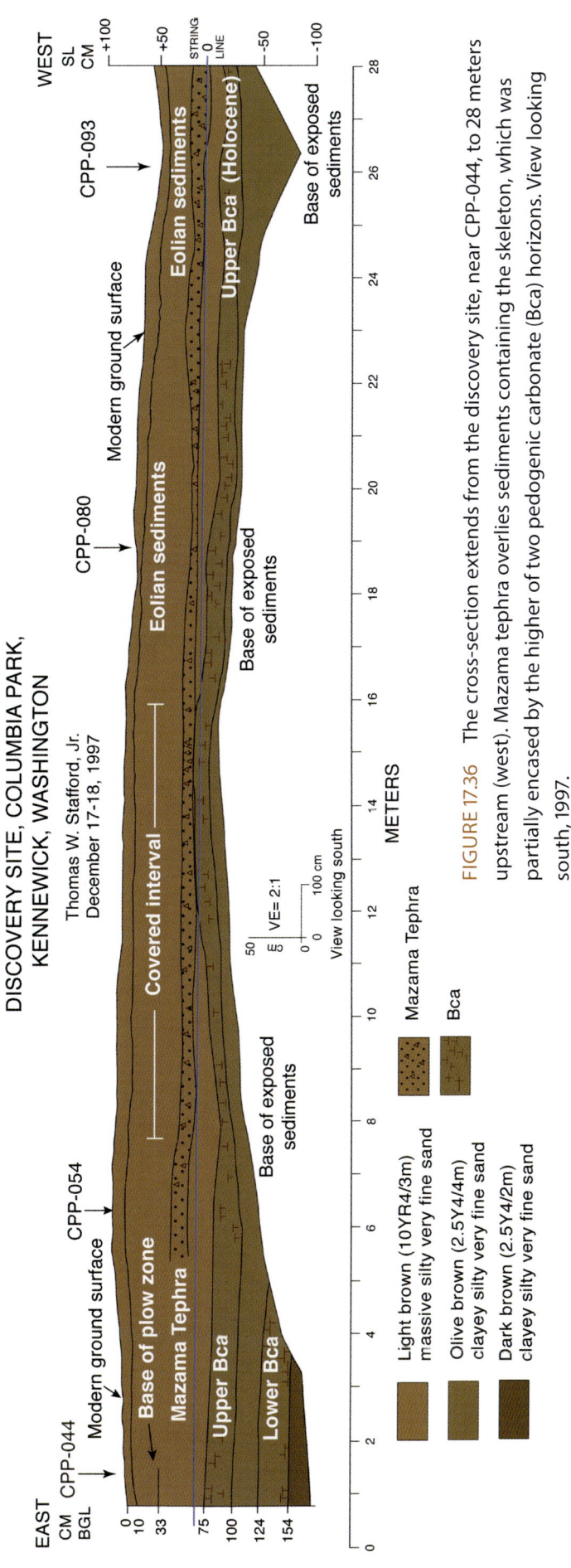

FIGURE 17.36 The cross-section extends from the discovery site, near CPP-044, to 28 meters upstream (west). Mazama tephra overlies sediments containing the skeleton, which was partially encased by the higher of two pedogenic carbonate (Bca) horizons. View looking south, 1997.

FIGURE 17.37 Fracture margin on the proximal end of Kennewick Man's right femur at left; adjoining segment at right. The proximal fracture face is white and slightly stained with green algae, while the distal face is slightly darker. The distal femur had broken away from the proximal end, which was fixed in the sediment and exposed to water and sun before it also calved out of the bank.

TIMING OF THE EROSION PROCESS
Evidence from Algae

Algal colonization of the bones, the presence of which has been used to determine the anatomical placement of the Kennewick Man skeleton, was possible during two consecutive wet intervals prior to recovery: first, the in situ period of progressing exposure prior to displacement of skeletal elements out of the bank and second, after the remains had calved out of the bank. The first interval covers the period of active erosion while exposed bones were still held or somewhat cemented in place by solidified sediments containing carbonates. The right femur provides an example. Opposing fracture surfaces separating the proximal (01.R.18a) and middle (01.R.18b) thirds of the right femur differ in the presence or absence of chlorophyll staining. The upper (proximal) surface has a greenish cast; the adjoining lower surface is light in color (Figure 17.37). Proximal segment staining indicates algal formation on the exposed fracture surface while the proximal femur was still in situ and embedded in the bank. The portion of the terrace containing the distal two-thirds of the femur had collapsed. Algae did not form on the broken end of the displaced middle segment because it was either submerged or partially buried in the mudflat.

The second interval of potential algal colonization began after the bones were dislodged from the bank, with some continuing movement along the shoreline mudflat, a process facilitated by their small size and low mass. Smaller bones and fragments, such as pieces of ribs, metacarpals, and phalanges were more easily tumbled and dispersed by wave action (Figure 17.38; Table 17.6). This second period was brief. The low density of the algal growth observed on the skeleton indicates that Kenne-

FIGURE 17.38 A sternal fragment of Kennewick Man's left fifth rib has been corraded on both the visceral (top) and external (bottom) surfaces. This pattern is consistent with tumbling along the lake margin.

TABLE 17.6 Microabraded fragments and small bones showing loss of surface color.

Bone	Side	Bone Number	Specimen Number	Taphonomy Description	Mechanism
Clavicle	L		97.L.11a	The lateral end fracture face, the adjacent superior surface, and the posterior face are light in color and microabraded; the polished appearance is more obvious on the lateral aspect and progressively decreases toward the medial aspect.	Due to in situ corrasion or post-erosion tumbling
Rib	L	Rib 1	97.I.12b(6) and 97.I.12b(1)	Body fragment exhibits loss of patina and surface color.	
Rib	L	Rib 3		Sternal half of both surfaces.	Tumbling action
Rib	L	Rib 5	Most lateral fragment	Fourth and most lateral fragment.	
Rib	L	Rib 8	Sternal end	The more anterior (sternal end) rib fragments are lighter in color and abraded.	
Rib	R	Rib 5	97.I.12a(6)	Both surfaces.	Tumbling action
Rib	R	Rib 7	97.I.12(3)	Both surfaces highly eroded.	Tumbling action
Hand	L			Three proximal and one intermediate phalanges.	Tumbling action
Hand	R			Four metacarpals, one proximal phalanx, and the fourth and fifth intermediate phalanges.	Tumbling action
Foot	R			Second and third metatarsals and fifth proximal phalanx	Tumbling action

wick Man's bones were exposed for only a short period, estimated at one to four weeks (Chapter 18).

Lake Levels

Kennewick Man's skeleton was discovered on the banks of the Lake Wallula reservoir on July 28, 1996, during a period of high water linked to hydroplane races. An alternating pattern of higher and lower lake levels during the preceding six months, in conjunction with the large amount of wake created by barge traffic and water events, such as the July races, actively eroded the skeleton from the lake terrace.

Figure 17.39 graphically illustrates Lake Walulla water levels in 1995 and 1996. These elevations were recorded by a Corps tracking station (Number 12514500) on Clover Island, just east of Columbia Park in Lake Wallula. Lake levels were measured against the National Geodetic Vertical Datum (NGVD29) standard (Kimbrough et al. 2002). This vertical control datum, established in 1929 by the United States Coast and Geodetic Survey, is the surface against which certain elevation data in the United States are referenced. Washington State experienced above average precipitation and stream flow from October 1995 through September 1996 (United States Geological Survey 1996 Water-Data Report). The average (arithmetic mean) water level for the calendar year 1995 was 339.41 feet and for 1996 was 339.99 feet. Flooding in November 1995 resulted from above-average stream flow, while in 1996, January, February, April, and June saw the highest levels above the average. Heavy rain, flood, and snow events during these months caused water levels to reach their peak at 341.50 feet during the second week of June.

Figures 17.40 and 17.41 illustrate how the bank was undercut at the approximate area where Kennewick Man's bones were discovered. Sloping units of the bank eventually calved onto the mudflat below it. The cranium, mandible, some cervical vertebrae, and small rib fragments were located approximately three meters downstream of the long bones, foot bones, and larger rib fragments (Chapter 2). This location suggests that the skeleton was positioned with his head downstream of the feet (i.e., the right side of his body was closest to the water). However, taphonomic evidence, including corrasion and recent postmortem breakage, indicates that the body was positioned at a specific level with the left side closest to the lake with the head upstream (Figures 17.42 and 17.43). As the body eroded from the terrace in units, some bones, including the skull, moved slightly downstream relative to the long bones.

Most of the bones were found in the immediate vicinity of each other, confirming that they had recently eroded from the bank. Active erosion exposed the skeleton in July 1996. Had the bones eroded out even two months earlier—given the highest water level on record that year was in June, followed by a steep decline, and then another peak in late July for the hydroplane races—the bones would have scattered with significantly less recovery. The skeleton is over 90 percent complete, including bones that are the most mobile in a fluvial environment—the bones of the foot (Boaz and Behrensmeyer 1976)—many of which were recovered at the discovery site.

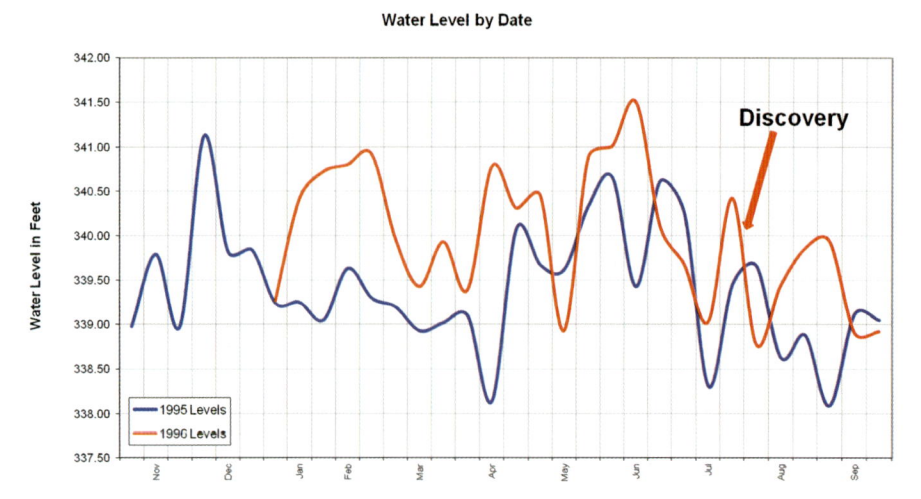

FIGURE 17.39 Water levels of Lake Wallula plotted using readings taken from the US Army Corps of Engineers elevation tracking station number 12514500 on Clover Island. Lake levels in 1996 were higher than the preceding year due to flooding in February and heavy rains and flooding in late May and early June.

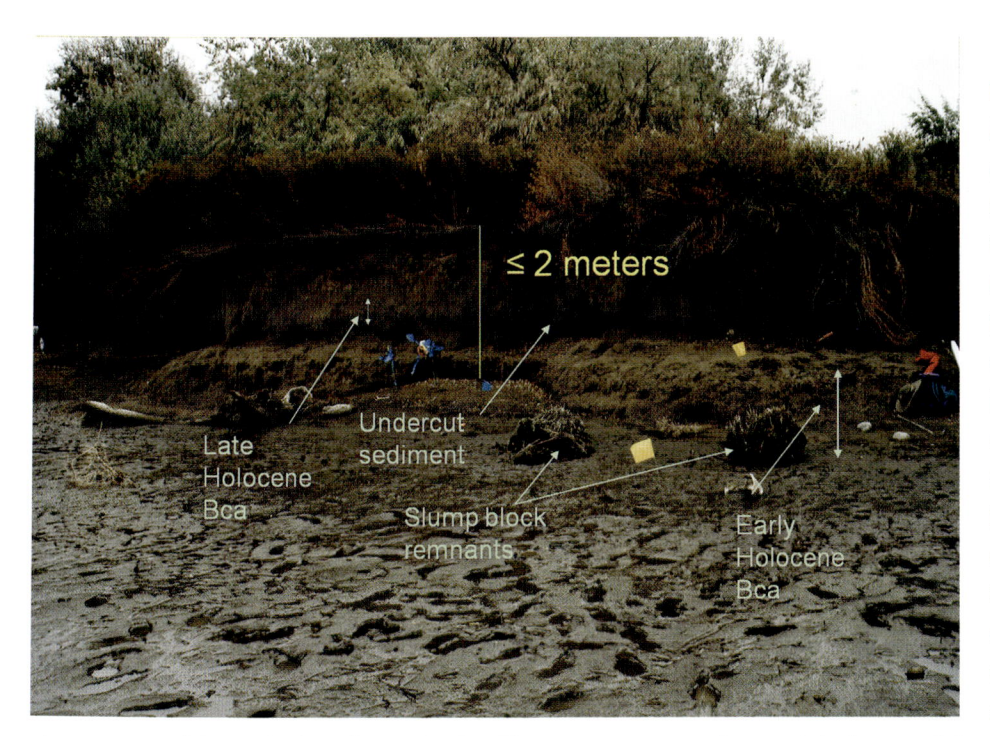

FIGURE 17.40 Early and later Holocene alluvium exposed at the Kennewick Man discovery site, 1997. View looking south shows locations of the late Holocene pedogenic Bca carbonate horizon and early Holocene sediments. Wave erosion of softer sediments (unconsolidated sand) immediately below the upper Bca horizon undercut the harder, more indurated calcium carbonate-cemented sediments. In addition, piping—erosion along root molds, vertical cracks, and animal burrows—occurred several decimeters inland from the outcrop face. As vertical piping planes intersected horizontal planes-of-erosion at the base of the late-Holocene Bca, 1 to 2 m thick (vertical height) by 1 to 2^+ m square blocks formed. These large blocks with hard carbonates at the base and dense mats of grass and woody plant roots at the top eventually collapsed by gravity onto the lakeshore. As wave action disaggregated the blocks of alluvium, the bones were released from enclosing sediments. The final remnants of the alluvial blocks are dense masses of plant roots that form tussocks along the shoreline.

FIGURE 17.41 View of the Kennewick Man discovery site looking upstream, approximately west-northwest, 1997. Numerous blocks of alluvium (A) have collapsed onto the lakeshore and are analogous to the blocks that contained the skeleton of Kennewick Man. The meters-long vertical cracks (B) parallel to the shoreline are formed by piping. These vertical cracks intersect with horizontal planes of erosion at the base of the Bca and form blocks that collapse onto the shoreline.

FIGURE 17.42 The stratigraphic profile of the Kennewick Man site, 1997. The outcrop is 165 centimeters high (modern ground level to the base of the vertical exposure). For comparison, the dimensions of the field book are 19 centimeters high by 12 centimeters wide. The upper carbonate horizon (Bca) is located approximately 75 to 100 centimeters below ground level. Placed in this topography, Kennewick Man's frontal was the highest bone in the strata with at least 5 centimeters of the anterior cranium above (outside) the carbonate horizon. The skeleton's highest-to-lowest elevations spanned 20 centimeters or less vertically, locating the skeleton approximately 70 to 90 centimeters below present ground level.

FIGURE 17.43 Reconstructed placement of the skeleton in the terrace bank based on taphonomic evidence. Kennewick Man's left side was closest to the lake with his head upstream.

CONCLUSION

This taphonomic study provides a detailed picture of the postmortem fate of Kennewick Man's skeleton, thereby resolving several unsettled issues. Results are based on a comprehensive evaluation of multiple taphonomic indicators. The findings are more definitive or differ from those reported by the government study team.

From the data, the authors conclude that Kennewick Man was intentionally buried by other humans. The physical preservation of the bones is exceptional with good anatomical completeness and no indication of animal gnawing or scavenging. No grave offerings or artifacts were discovered, and this study found no evidence for the use of red ochre. His body was buried close to the Columbia River (Chapter 2). The construction of McNary Dam created Lake Walulla, which eventually exposed the burial. Undercutting and vertical piping of the terrace bank calved off blocks of grass-covered masses, depositing the skeleton on the shoreline.

Physical preservation of individual bones is excellent, although there is extensive postmortem mechanical damage. This breakage occurred at different times, as indicated by either light (recent) or dark (more ancient) fracture margins. Postmortem fractures that occurred in 1996 when the bones calved out of the bank are whitish in color, and they are best observed in long bones. A few breaks with dark fracture surfaces are older. This type of damage, evident in some ribs and pedicles and spines of vertebrae, was caused by overlying sediments compressing the skeleton.

Multiple lines of evidence indicate that Kennewick Man was buried in an extended, supine position, with his left side toward the river and his head upstream. Cemented adhesions between some adjacent bones confirm that the body was articulated throughout most of its interment. The accumulation of carbonate-cemented sediment on the skeleton shows that the dorsal aspect (back) was down. There was lateral rotation of the left foot and his hands were by his sides, palm surfaces down (i.e., pronated). Algae and bone-breakage patterns support supine positioning. The two distal segments of the left humerus have algal staining on the anterior and lateral surfaces, but not the posterior. These surfaces were higher up in the strata and simultaneously exposed to both the sun and water so that algal growth could occur. The fracture pattern analysis (Chapter 19) determined that most of the long bones broke in an anterior to posterior direction, which could have only occurred if the anterior surfaces were up. Collectively, the lower legs fit a fracture pattern consistent with the bank below the knees giving way first, followed by areas under the feet and pelvis, and ultimately the upper body.

The remains were recovered in a relatively small scatter area, reflecting a brief period between erosion and discovery. Given the extreme fluctuation in the water level in July 1996 and even higher levels in June 1996, a longer period of exposure would have scattered the bones over a wider area with greatly reduced recovery. Minimal algal growth on the bones suggests a short interval between the erosion event and discovery.

Light bone colors were determined to be the result of corrasion, abrasion from particulates in waves lapping against the bones. The left side of the cranium and the lower half of the left humerus best illustrate this through loss of soil stain patina, along with polishing and superficial pockmarking. Abrasions caused by debris in the water rubbing against the bones damaged left-side lateral bone surfaces first, and then progressed to right bone medial surfaces. Such a pattern suggests that the left side of the skeleton was closer to the shoreline. As such, the head was positioned upstream relative to the feet. Sedimentary matrix kept the cranium, mandible, and some cervical vertebrae in close proximity until water action began to break them apart.

The patterned accumulation of carbonate-cemented sediment (i.e., more accumulation on posterior surfaces versus anterior) is very strong evidence that the skeleton came specifically from the Holocene stratigraphic layer in which this carbonate precipitated. Differential distribution of carbonates across the length of the skeleton, with more cemented sediment adhering to the lower legs and progressively less on the upper torso and head, allows a subtle distinction. The head, with the chin pressed onto the chest, and upper torso were slightly elevated relative to the lower limbs, an inclination of approximately five degrees. This reconstruction of the burial indicates that the remains were located about 70 to 90 centimeters below ground level (i.e., the lower centimeters of Unit I and top of Unit II).

REFERENCES

Aldhouse-Green, Stephen, and Paul Pettitt. 1998. Paviland Cave: Contextualizing the "Red Lady." *Antiquity* 72(278):756–772.

Bacon, Charles R. 1983. Eruptive history of Mount Mazama and Crater Lake Caldera, Cascade Range, U.S.A. *Journal of Volcanology and Geothermal Research* 18(1–4):57–115.

Behrensmeyer, Anna K. 1978. Taphonomic and ecological information from bone weathering. *Paleobiology* 4(2):150–162.

Behrensmeyer, Anna K., and Andrew P. Hill, eds. 1980. *Fossils in the Making: Vertebrate Taphonomy and Paleoecology*. The University of Chicago Press, IL.

Boaz, N. T., and K. Behrensmeyer. 1976. Hominid taphonomy: Transport of human skeletal parts in an artificial fluviatile environment. *American Journal of Physical Anthropology* 45(1):53–60.

Bonnichsen, Robson, and Marcella H. Sorg. 1989. *Bone Modification*. Center for the Study of the First Americans, Institute for Quaternary Studies, University of Maine, Orono.

Breternitz, David A., Alan C. Swedlund, and Duane C. Anderson. 1971. An early burial from Gordon Creek, Colorado. *American Antiquity* 36(2):170–182.

Brooks, Sheilagh, and Richard H. Brooks. 1997. The taphonomic effects of flood waters on bone. In *Forensic Taphonomy: The Postmortem Fate of Human Remains*, edited by William D. Haglund and Marcella H. Sorg, 553–558. CRC Press, Boca Raton, FL.

Chatters, James C. 2000. The recovery and first analysis of an early Holocene human skeleton from Kennewick, Washington. *American Antiquity* 65(2):291–316.

———. 2001. *Ancient Encounters: Kennewick Man and the First Americans*. Simon & Schuster, New York.

Deocampo, Daniel M. 2010. The geochemistry of continental carbonates. In *Carbonates in Continental Settings: Geochemistry, Diagenesis and Applications*, edited by A. M. Alonso-Zarza and L.H. Tanner, 1–45. Elsevier, Amsterdam.

Duarte, Cidália, João Maurício, Paul B. Pettitt, Pedro Souto, Erik Trinkaus, Hans van der Plicht, and João Zilhão. 1999. The early Upper Paleolithic human skeleton from the Abrigo do Lagar Velho (Portugal) and modern human emergence in Iberia. *Proceedings of the National Academy of Sciences* 96(13):7604–7609.

Formicola, Vincenzo, Antonella Pontrandolfi, and Jiři Svoboda. 2001. The upper Paleolithic triple burial of Dolní Věstonice: Pathology and funerary behavior. *American Journal of Physical Anthropology* 115(4):372–379.

Galloway, Alison. 1997. The process of decomposition: A model from the Arizona-Sonoran desert. In *Forensic Taphonomy: The Postmortem Fate of Human Remains*, edited by William D. Haglund and Marcella H. Sorg, 139–149. CRC Press, Boca Raton, FL.

Haglund, William D. 1997a. Dogs and coyotes: Postmortem involvement with human remains. In *Forensic Taphonomy: The Postmortem Fate of Human Remains*, edited by William D. Haglund and Marcella H. Sorg, 367–381. CRC Press, Boca Raton, FL.

———. 1997b. Rodents and human remains. In *Forensic Taphonomy: The Postmortem Fate of Human Remains*, edited by William D. Haglund and Marcella H. Sorg, 367–381. CRC Press, Boca Raton, FL.

Haglund, William D., and Marcella H. Sorg, eds. 1997. *Forensic Taphonomy: The Postmortem Fate of Human Remains*. CRC Press, Boca Raton, FL.

Huckleberry, Gary, and Julie K. Stein. 1999. Chapter 3. Analysis of Sediments Associated with Human Remains Found at Columbia Park, Kennewick, WA. In *Report on the Non-Destructive Examination, Description, and Analysis of the Human Remains from Columbia Park, Kennewick, Washington* [October 1999]. National Park Service, http://www.nps.gov/archeology/kennewick/huck_stein.htm.

Huckleberry, Gary, Julie K. Stein, and Paul Goldberg. 2003. Determining the provenience of Kennewick Man skeletal remains through sedimentological analyses. *Journal of Archaeological Science* 30(6):651–665.

Huckleberry, Gary, Thomas W. Stafford, Jr., and James C. Chatters. 1998. Preliminary Geoarchaeological Studies at Columbia Park, Kennewick, Washington, USA. Report submitted to U.S. Army Corps of Engineers, Walla Walla District, March 23, 1998.

Kimbrough, R. A., W. D. Wiggins, R. R. Smith, G. P. Ruppert, S. M. Knowles, and V. F. Renslow. 2002. Water Resources Data, Washington, Water Year 2002, Water-Data Report WA-02-1. U. S. Geological Survey, U.S. Department of the Interior, http://pubs.usgs.gov/wdr/WDR-WA-02-1/pdf/wdr-wa-02-1.pdf.

McManamon, Francis P., Jason C. Roberts, and Brooke S. Blades. 2000. Chapter 1. Examination of the Kennewick Remains—Taphonomy, Micro-sampling, and DNA Analysis. In *Report on the DNA Testing Results of the Kennewick*

Human Remains from Columbia Park, Kennewick, Washington [September 2000]. National Park Service, http://www.nps.gov/archeology/kennewick/fpm_dna.htm.

Miller, George J. 1969. A study of cuts, grooves, and other marks on recent and fossil bone: I. Animal tooth marks. *Tebiwa* 12(1):20–26.

Owsley, Douglas W., and David R. Hunt. 2001. Clovis and early Archaic crania from the Anzick site (24PA506), Park County, Montana. *Plains Anthropologist* 46(176):115–124.

Owsley, Douglas W., Margaret A. Jodry, Thomas W. Stafford, Jr., C. Vance Haynes, Jr., and Dennis J. Stanford. 2010. *Arch Lake Woman: Physical Anthropology and Geoarchaeology.* Texas A&M University Press, College Station.

Powell, Joseph F., and Jerome C. Rose. 1999. Chapter 2. Report on the Osteological Assessment of the "Kennewick Man" Skeleton (CENWW.97.Kennewick). In *Report on the Non-Destructive Examination, Description, and Analysis of the Human Remains from Columbia Park, Kennewick, Washington* [October 1999]. National Park Service, http://www.nps.gov/archeology/kennewick/powell_rose.htm.

Shipman, Pat, and Jennie Rose. 1983. Early hominid hunting, butchering, and carcass-processing behaviors: Approaches to the fossil record. *Journal of Anthropological Archaeology* 2(1):57–98.

Tanner, Lawrence H. 2010. Continental carbonates as indicators of paleoclimate. In *Carbonates in Continental Settings: Geochemistry, Diagenesis and Applications,* edited by A. M. Alonso-Zarza and L. H. Tanner, 179–206. Elsevier, Amsterdam.

Trinkhaus, E., and A. P. Buzhilova. 2010. The death and burial of Sunghir 1. *International Journal of Osteoarchaeology.* doi:10.1002/oa.1227.

United States Geological Survey. 1996. 1996 Washington State Hydrologic Summary. In *Water Resources Data—Washington, Water Year 1996: USGS Water-Data Report WA-96-1.* Prepared in cooperation with the State of Washington and with other agencies. http://vulcan.wr.usgs.gov/Volcanoes/Washington/Hydrology/Summaries/hydro_summary_1996.html.

Wakeley, Lillian D., William L. Murphy, Joseph B. Dunbar, Andrew G. Warne, Frederick L. Briuer, and Paul R. Nickens. 1998. *Geologic, Geoarchaeologic, and Historical Investigation of the Discovery Site of Ancient Remains in Columbia Park, Kennewick, Washington.* U.S. Army Corps of Engineers Technical Report GL-98-13. Prepared for U.S. Army Engineer District, Walla Walla. U.S. Army Corps of Engineers Waterways Experiment Station, Vicksburg, MS.

Walker, Phillip L., Clark Spencer Larsen, and Joseph F. Powell. 2000. Chapter 5. Final Report on the Physical Examination and Taphonomic Assessment of the Kennewick Human Remains (CENWW.97.Kennewick) to Assist with DNA Sample Selection. In *Report on the DNA Testing Results of the Kennewick Human Remains from Columbia Park, Kennewick, Washington* [September 2000]. National Park Service, http://www.nps.gov/archeology/kennewick/walker.htm.

Wendorf, Fred, Alex D. Krieger, and Claude C. Albritton. 1955. *The Midland Discovery: A Report on the Pleistocene Human Remains from Midland, Texas.* University of Texas Press, Austin.

Whitney, Stephen R., and Rob Sandelin. 2003. *Field Guide to the Cascades & Olympics.* 2nd edition. The Mountaineers Books, Seattle.

Benthic Aquatic Algae: Indicators of Recent Taphonomic History

James N. Norris and Douglas W. Owsley

Green stains attributable to algae were observed on some bones of Kennewick Man. This staining was studied with three objectives. The first was to identify the algal species present. The second was to derive taphonomic information about the in situ anatomical positioning of the skeleton from the patterning of the green staining. Among other requirements, algae need sunlight and water in order to grow. Based on this principle of algal ecology, areas of green staining indicate bone surfaces that had simultaneous exposure to sunlight and moisture. Mapping the distribution of algal growth helps to orient the bones and identify which surfaces were up and exposed directly or indirectly to sunlight. The third objective was to estimate the timeline from initial exposure to the discovery and collection of skeletal elements. The extent of algal colonization on the bones is compared to that at the discovery location. This comparison helps to approximate the length of time from the initial exposure to the recovery of the skeleton.

ASSESSMENTS OF GREEN STAINING AND ALGAL TAXONOMY

Green stains were initially observed on specific bones of the skeleton by the government-contract physical anthropologists (Powell and Rose 1999; Walker et al. 2000). Their reports tabularized their findings, which are listed in Table 1, and noted that the green stains on recovered bones indicated that they were from a wet environment and exposed to sunlight. "The algae stains indicate that after the grave was uncovered by natural processes it spent some time (at least several weeks) immersed in shallow water" (Powell and Rose 1999). Walker et al. (2000) supported this interpretation "by the fact that many of the bones with algae adhering to them also have bleached areas indicating several weeks or more of exposure to the sun." Powell and Rose (1999) made a contradictory remark in their report suggesting that the time interval for formation was shorter: "Algal staining on some elements is probably due to exposure of the remains in shallow water just prior to their recovery in along [sic] the Columbia River." The report continues, "The algae stain is the product of recent exposure within a damp/wet environment and occurred after the grave was disturbed and prior to discovery of the skeleton. Such staining is common on forensic cases where the body/skeleton has been exposed to the environment and becomes covered with water in a puddle or other body of water" (Powell and Rose 1999). These reports leave the time interval of skeleton exposure prior to recovery uncertain.

In December 2004, while completing a preliminary survey of the skeleton in order to develop their research design and study plan (Chapter 7), forensic anthropologists on the plaintiff's study team also noted this green staining. A comprehensive assessment of this staining was determined to be an essential component of their taphonomic study.

All surfaces of the skull, clavicles, vertebrae, ribs, and long bones were inspected in order to determine which elements showed visible discolorations. Observed green areas were mapped and the location and relative degrees of green staining were estimated and described (Figure 18.1; Table 18.2). Along with a few cervical and thoracic vertebrae and rib fragments, most of the long bones of the arms and legs show some green coloration. Examples of larger and heavier bones with green staining include the medial, anterior, and lateral surfaces of the distal left humerus (97.L.13b, 97.L.13c; Figure 18.2); the anterior surface of the distal right humerus (97.R.13b; Figure 18.3); the posterior surface of the distal half of the left radius (97.L.14b); the anterior olecranon and trochlear notch of the proximal left and right ulnae

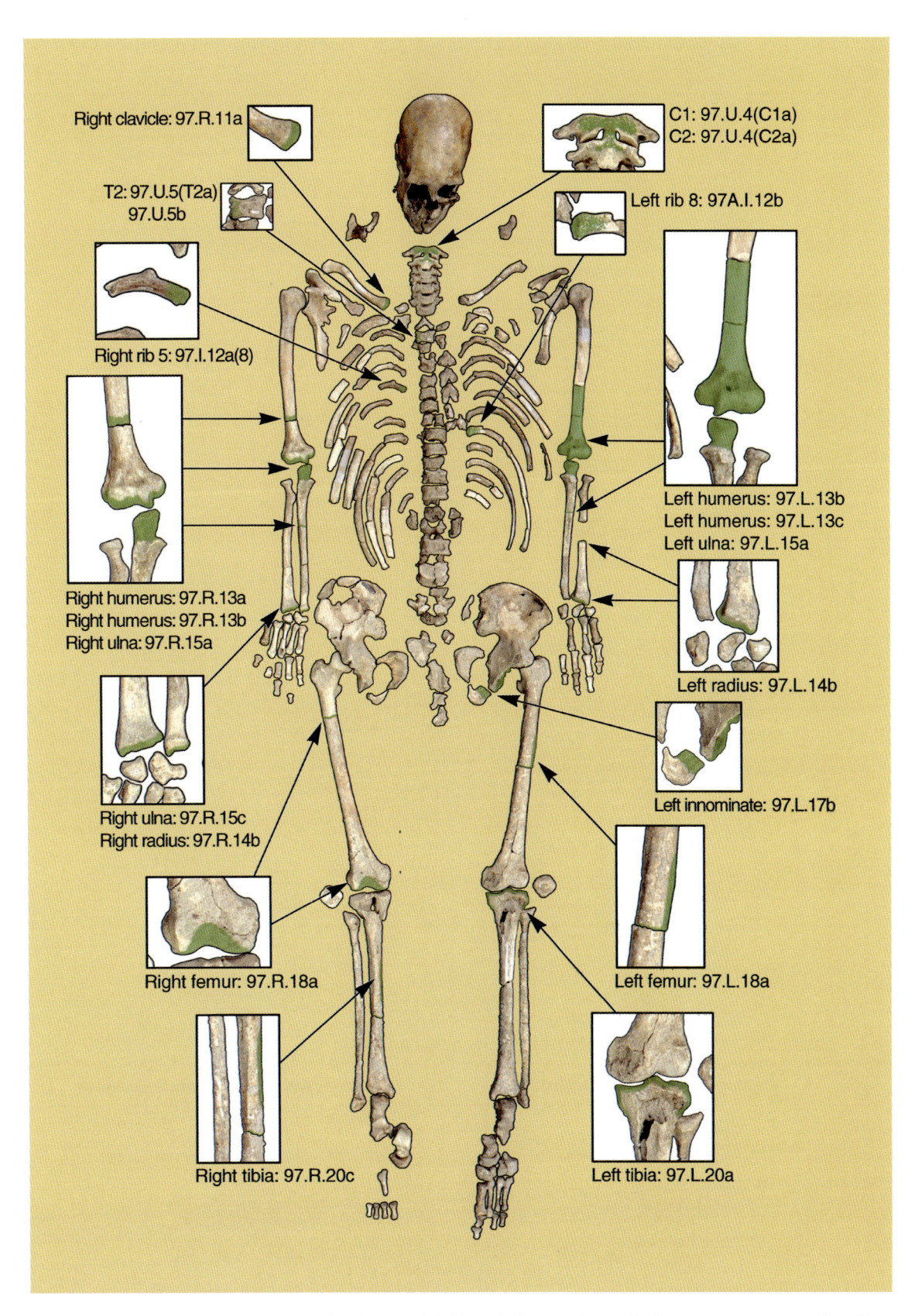

FIGURE 18.1 Faint green stains on the Kennewick Man skeleton where algal areas were mapped and the degree of coverage estimated; see Table 18.2.

TABLE 18.1 Bones of Kennewick Man identified in other studies as showing green stains from algae.

Source	Description	Segment Number
Powell and Rose 1999	Atlas, axis, 3rd cervical vertebrae	97.U.4 C1a–C3a
	Two thoracic vertebrae	Not identified
	Left ribs: two unidentified fragments	Not identified
	Right 10th rib: fragment	97.I.12d.4
	Left humerus: distal half	Not identified
	Left radius: proximal end	Not identified
	Left ulna	Not identified
	Left metacarpals	Not identified
	Left tibia: "proximal portions"	Not identified
	Right humerus: distal shaft and end	Not identified
	Right radius: distal segment	Not identified
	Right ulna: proximal and distal ends	Not identified
	Right hand: 2, 3, 4 metacarpals and 1st row 3rd phalanx	Not identified
	Right femur: anterior articular surface	Not identified
	Right talus	Not identified
	Three unidentified faunal fragments	Not identified
Walker et al. 2000	One complete axis	97 U.4(C2 a)
	One rib fragment	97 I.12a(8)
	One left seventh rib fragment	97A I.12b
	One left humerus fragment (shaft)	97 L.13b
	One left humerus fragment (distal)	97 L.13c
	One left ulna fragment (proximal)	97 L.15a
	One right humerus fragment (distal)	97 R.13b
	One right ulna fragment (proximal)	97 R.15a
	One right ulna fragment (distal)	97 R.15c
	One right femur fragment (distal)	97 R.18a
	One complete right talus	97 R.23a

TABLE 18.2 Kennewick Man bones with algal filaments or green staining.

Bone	Catalog Number and Description	Location and Degree of Algal Coverage
Cervical Vertebra 1	97.U.4(C1a)	Trace on anterior surface of centrum
Cervical Vertebra 2	97.U.4(C2a)	Slight to moderate, especially on sides of dens process
Cervical Vertebra 7	97.U.4(C7a)	Algae on sediment adhering to bone
Thoracic Vertebra 2	97.U.5(T2a), 97.U.5b	Slight on the anterior right half of the body
Right clavicle	97.R.11a: Sternal end	Slight on superior surface, and adjacent anterior and posterior margins of sternal end
Left rib 8	97.A.I.12b: Sternal end	Slight on visceral surface of the sternal end
Right rib 5	97.I.12a(8): Sternal end	Slight on external surface of the sternal end

(continued)

TABLE 18.2 (*continued*)   Kennewick Man bones with algal filaments or green staining.

Bone	Catalog Number and Description	Location and Degree of Algal Coverage
Left humerus	97.L.13b: Majority of the lower one-fourth of the diaphysis	Slight on anterior and lateral surfaces and the lower face of the fracture between this segment and the proximal portion of the bone (97.L.13a)
Left humerus	97.L.13c: Most inferior portion of the diaphysis and distal joint	Slight on medial, anterior, and lateral surfaces, and moderate on metaphysis and joint surface. Green on the joint surface includes algal filaments located anterior-posterior along the circumference of the trochlea
Left radius	97.L.14b: Distal one-half of the radius including the joint surface	Faint on distal joint surface and posterior surface; also on face of the fracture between this segment and the midshaft of the radius (segment not recovered)
Left ulna	97.L.15a: Proximal one-third of the ulna including the joint surface	Slight to moderate on the olecranon and trochlear notch. Also on the face of the fracture between this segment and the more distal (midshaft) segment of the left ulna (97.L.15b)
Left ulna	97.L.15b: Middle one-third of the diaphysis	Slight on face of fracture between this segment and the proximal segment of the left ulna (97.L.15a)
Left innominate	97.L.17b: Near complete ilium and portions of the ischium and pubis	Slight on the visceral and lateral surfaces inferior to the acetabular rim and posterior external surface of the acetabular rim
Left femur	97.L.18a: Proximal two-fifths of the entire bone	Slight on 4 distinct areas: distal anterolateral aspect (this area is less acutely green than others, may not be algae); posterior surface of greater trochanter; posterior surface of ball of the femur; and inferior surface of face of the fracture between this segment and next more distal segment (01.L.18b)
Left tibia	97.L.20a: Proximal joint, epiphysis, and most superior diaphysis	Slight on anterior third of joint surface, and on anterior aspects of the lateral and medial surfaces (anterior surface missing due to postmortem breakage)
Right humerus	97.R.13a: Proximal epiphysis and upper three-fourths of the diaphysis	Trace on distal anterior circumference near the fracture between this segment and the more distal segment (97.R.13b)
Right humerus	97.R.13b: Lower one-fourth of the diaphysis and distal joint	Moderate on distal one-fourth of the segment, especially on the anterior surface and anterior metaphysis and joint
Right radius	97.R.14b: Lower one-third of the diaphysis and the distal end	Slight on anterior surface of the distal metaphysis and joint surface
Right ulna	97.R.15a: Proximal one-third of diaphysis including the intact joint	Moderate on trochlear notch, lateral and superior aspect of the olecranon, and face of fracture between this segment and the more distal segment (97.R.15b)
Right ulna	97.R.15c: Distal one-third of the diaphysis including the distal joint	Slight on the joint surface
Right femur	01.R.18a: Proximal one-fourth of the entire bone	Slight on anterior and posterior ball and neck, slight on greater trochanter, trace on anterior surface of the diaphysis, and slight on upper face of fracture between this segment and the next more distal segment (01.R.18b)
Right femur	97.R.18a: Distal one-third of diaphysis and distal joint surface	Moderate on distal metaphysis and joint surface, and is strongest on the anterior aspect of the joint surface
Right tibia	97.R.20c: Proximal and middle one-third of the diaphysis, incl. most of the tibial tuberosity	Extremely faint on the medial anterior diaphysis and on the posterior surface. Also on the upper face of the fracture between this segment and the next more distal segment of the right tibia (97R.20d)

FIGURE 18.2 Distal left humerus (97.L.13b, 97.L.13c) of Kennewick Man showing trace evidence of green algae on the medial, anterior, and lateral surfaces but not the posterior surface. 97.L.13c was one of the bones from which algal specimens were removed for identification.

(97.L.15a, 97.R.15a); and anterior areas of the proximal left tibia (97.L.20a).

A few of the bones examined at the Burke Museum, Seattle, revealed the presence of very small filamentous green tufts (<1 mm in length) under low power magnification. Owsley selected and sampled the greens from the surfaces of two bones in preparation for microscopic analysis to identify the species responsible for the coloration. Two minute portions (<1 mm) of air-dried green areas were sampled by gently scraping the surface of the second cervical vertebra (97.U.4(C2a)) and the distal joint surface of the left humerus (97.L.13c; Figure 18.2) in February 2006 at the Burke Museum.

Microscope slides were made with portions of algae rehydrated with distilled water and mounted in a 10 percent clear Karo® syrup/90 percent distilled water solution, with phenol added as a preservative, and placed on microscope slides under 22 mm cover slips. Specimens (#US Alg. Coll.-209254, microscope slide numbers: −8466, −8467, and −8468) deposited in the Algal Collection of the U.S. National Herbarium, National Museum of Natural History, Smithsonian Institution (herbarium abbreviation follows Holmgren et al. 1990) were studied in the Algae Laboratory using a Zeiss Universal light microscope and identified using Cox and Bold (1966), Dillard (1989), John (2002, 2003), Prescott (1962, 1978), Printz (1964), Shubert (2003), Smith (1950), and Wehr and Sheath (2003), with additional information from

FIGURE 18.3 Distal right humerus (97.R.13b) of Kennewick Man showing evidence of green staining on anterior surface.

Guiry and Guiry (2007). Photomicrographic images were made with a digital Olympus Q-color 5 on a Zeiss Universal microscope, transferred to a Mac G-5 computer, opened with BRIDGE on Adobe Photoshop CS3, and printed with an Epson Stylus Photo 2200 printer.

Greenish areas on this skeleton could have resulted from contact with, or growth of, aquatic flowering plants (Spermatophyta), mosses (Bryophyta), liverworts (Hepatophyta), algae (Chlorophyta), or blue-greens (Cyanobacteria); all of which have chlorophyll. Minute portions (< 0.5 mm^2) removed from the second cervical vertebra and left humerus, however, were identified to be the benthic green alga *Stigeoclonium tenue* (C. Agardh) (Kützing 1843) (Chlorophyta, Chaetophorales, Chaetophoraceae). Specimens of *S. tenue* were uniseriate, terete filaments with slightly obtuse apices and were mostly unbranched (Figure 18.4), but a few were sparsely, oppositely, or irregularly branched (Figure 18.5). Lower attachment structures were not seen in the sampled specimens. A minute encrusting green alga in one sample resembled *Protoderma viride* (Kützing 1843), which is probably the young growth form of *S. tenue* (Prescott 1962; Simons et al. 1986; Simons and van Beem 1987). Cells of the narrower filaments and branches were 9–18 (–22) microns (μm) long by (4–) 5–6 μm in diameter, and those of the larger filaments were 40–60 μm long by 12–15 μm in diameter. Some of the filament cells were inflated in appearance (Figure 18.4), possibly reproductive sporangia. Cells of the filaments contained a single laminate chloroplast (Figure 18.5); pyrenoids were not evident.

Widespread in geographical distribution and ecologically tolerant, *Stigeoclonium tenue* is known from an extensive variety of aquatic freshwater habitats (Sarma 1986; John 2002; Wehr and Sheath 2003; Guiry and Guiry 2007). It is often found in running or still waters of creeks, rivers, and lake shallows growing attached to rocks, wood, or other substrata. It can be epiphytic or entangled on other aquatic plants. The erect filamentous portions of *S. tenue* are often eaten by herbivores (e.g., amphibians, fish, mollusks). However, the prostrate, grazer-resistant, basal portions of *S. tenue* remain and can regenerate new erect filamentous uprights (Rose-

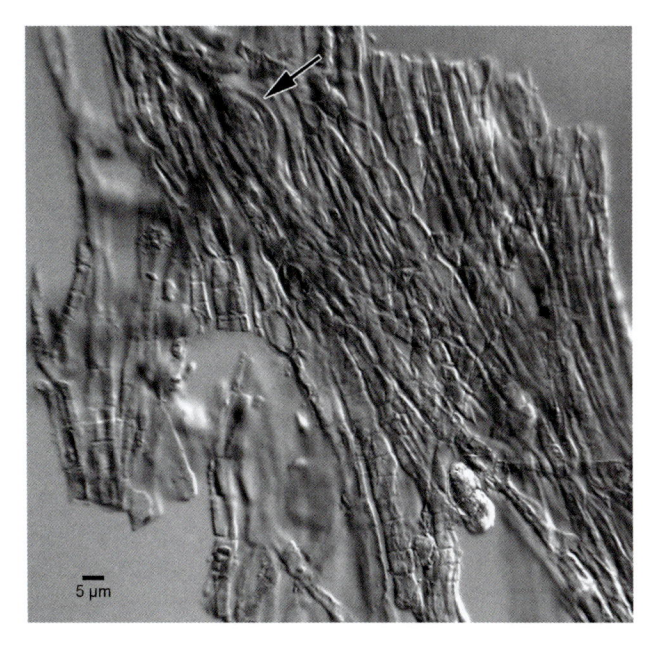

FIGURE 18.4 Filamentous green alga (*Stigeoclonium tenue* (C. Agardh) Kützing) collected from bone surface, showing portion of air-dried, matted uniseriate filaments, and enlarged cell, probably a reproductive structure (arrow).

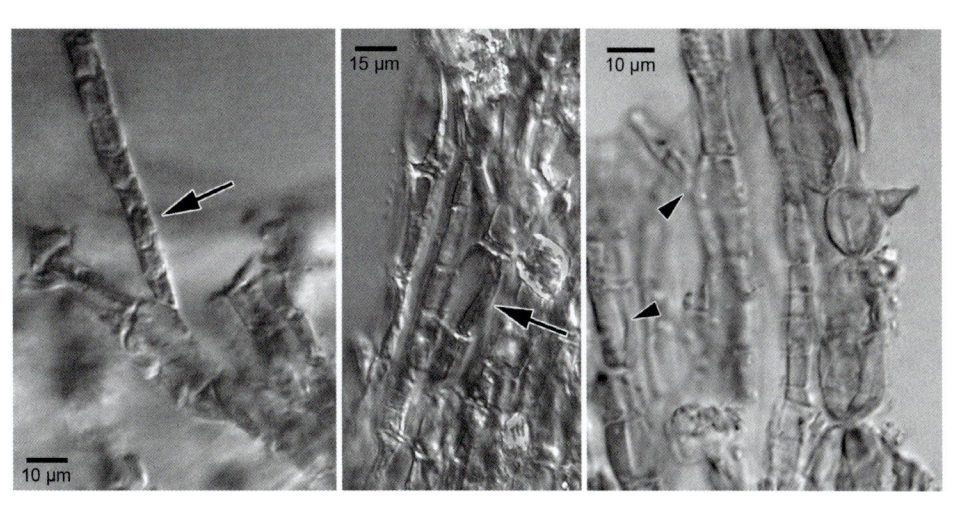

FIGURE 18.5 *Stigeoclonium tenue*. Left, branched filament (arrow); center, elongated cells of filaments with a single laminate chloroplast within a cell (arrow); right, branched filament (upper right arrowhead) and laminate chloroplast within cell of a filament (lower left arrowhead).

mond and Brawley 1996). This species grows especially well in environs with more sunlight and waters rich in organics or nutrients (Rosemond and Brawley 1996).

RECONSTRUCTING RECENT TAPHONOMIC HISTORY

The distribution of green staining on Kennewick Man's skeleton was mapped and described in order to provide information about the in situ anatomical positioning of the remains (Chapter 17). *Stigeoclonium tenue*, a freshwater green alga, requires sunlight, water, nutrients, and available substrata to grow. Algal colonization and growth are not possible on buried bone surfaces. Based on principles of algal ecology, the algae's comparative size and distribution on the skeleton, and photographs of the habitat and bones at the discovery site, it is possible to determine which bone surfaces were exposed to sunlight. The patterned presence of algal growth on specific bone surfaces indicates those that were exposed to light while the skeleton was exposed in situ, which helps to orient the bone's position relative to the sun.

A conservative approach was adopted in interpreting whether observed green staining on specific bones provided information about original subsurface positioning (Chapter 17). Contiguous bone segments of the elbow region of the left arm (Figures 18.1 and 18.2; Table 18.2), that is, the distal left humerus (97.L.13b, 97.L.13c), radius (97.L.14b), and ulna (97.L.15a) show similar green staining, reflecting algal colonization while the remains were in situ. Ventral and dorsal green staining on smaller bones, such as rib fragments, potentially includes algal colonization before, as well as after displacement, reflecting their tumbling or rolling around as a result of water motion. There was no evidence of dead or sun-bleached algae, which would appear whitish in color.

Species identification and information on the ecology of the source of the staining, *Stigeoclonium tenue*, a filamentous green alga, was essential, as it helped to determine the length of time that passed between initial exposure and the recovery of the remains. Laboratory identification of the species that caused these discolorations provides ecological information relevant to understanding the timing and dynamics of the erosion process and changing habitat.

Bank erosion due to high water levels, wave action, boat wakes, and variation in water heights caused a portion of the Columbia River terrace that contained the Kennewick Man skeleton to topple onto the shoreline (Wakeley et al. 1998; Chatters 2000). The drawdown of Lake Wallula revealed the near-shore environment in which Kennewick Man was found. The riparian habitat is dominated by grasses and reeds along the bank edge, and shrubs and trees farther back. The bones were dispersed along the mudflat about 10 feet offshore over an area of about 300 square feet (Chatters 2000). The bones were in relatively close proximity to each other and evidenced extensive postmortem breakage, but otherwise appeared in good condition. Most of the larger segments were picked up in shallow water in mud on the shoreline. Displaced smaller pieces, such as phalanges and fragments of ribs, were found along windrows near the shoreline. Bones of the postcranial skeleton were rinsed at the time of collection but otherwise were not brushed or cleaned (James Chatters, personal communication 2009).

Algal colonization could have occurred during two time-sequential intervals: from first exposure of individual bones to sunlight and wet conditions in situ, and subsequently by growing on exposed surfaces after the bones washed out of the bank. During the second event, new and previously buried bone surfaces became potential substrates for colonization. During this phase, displaced pieces of bones were either partially submerged or underwater along the shoreline, and they were affected by currents and wave action. Smaller and flatter fragments with little mass (e.g., hand bones and rib segments) were more prone to movement by water motion, while heavier and longer pieces became buried in the mudflat.

The time from exposure of the bones to sunlight and water prior to their removal can be estimated by comparing sizes and aerial (surface) coverage of the algae on the bones to that of algal populations in the surrounding habitat. Photographs of the habitat at the discovery site, taken when the broad mudflat abutting the vertical portion of riverbank was exposed in July 1996 and November 1996, show filamentous greens growing on the mudflat surface (Figures 18.6–18.8). These similar-appearing filamentous algae (presumed to be the same species) on the adjacent mudflat and a log visible in a later photograph are larger in size with more extensive coverage than that observed on the skeleton. Limited coverage and

FIGURE 18.6 Discovery site on shore of Lake Wallula showing exposed habitat where the bones of Kennewick Man were found. Photographed looking east, June 29, 1996.

FIGURE 18.7 Filamentous green algae growing on the Lake Wallula mudflat above a windrow at the time Kennewick Man's bones were recovered on June 29, 1996. Modern debris and bone fragments are commingled with calcareous concretions.

FIGURE 18.8 Green algae on the Lake Wallula mudflat in November 1996. Algae on the bones are small relative to those of the surrounding habitat. The smaller size of the algae on the bones indicates that they had less time to develop during their growing season in contrast to the larger, more extensive algal growth observed on surrounding substrata.

very short filaments on the bones indicate that they were exposed to the conditions required for algae colonization for only a short period of time prior to their discovery. The short lengths of the filaments and limited patch surface coverage indicate the algae had colonized and grown on the exposed surfaces of the bones approximately one to four weeks prior to the skeleton's recovery.

DISCUSSION

Postmortem pigmentation and staining of bones are occasionally observed on both surface-exposed and previously buried remains. Green stains are occasionally observed on bones recovered from archaeological contexts as well as in remains from contemporary forensic investigations. Depending upon the situation, postmortem green staining can be attributed to copper oxide staining from contact with metallic objects such as native copper artifacts, or during the historic period from shroud pins or other artifacts such as brass buttons. In forensic casework, greenish discolorations can potentially signal bone contact with copper-jacketed bullets or jewelry, or in the dentition, the presence of amalgam fillings containing elemental copper.

Under certain environmental conditions, and regardless of antiquity, green stains on bones can be caused by plant pigments, or the green color can indicate growth of microscopic or minute algae. Such is the case on some of the bones of the Kennewick skeleton. Botanical materials have been shown to be powerful tools for the interpretation of modern, historic, and prehistoric events, sites, and habitats. Terrestrial botanical information can contribute to the resolution of criminal cases (Lane et al. 1990; Bock and D. Norris 1997; Miller Coyle et al. 2001) and elucidate past environments and paleoclimates of archaeological and paleoarchaeological sites.

The use of biota from aquatic environments has been less common in forensic studies (Aumen and Gurtz, in Keiper and Casamatta 2001). However, algae have been used as biological indicators in water quality studies, drowning investigations, and ecological (spatial and temporal) habitat analyses. Microalgae, such as diatoms and other phytoplankton, have a long history of use in drowning investigations (Kobayashi et al. 1993; Matsumoto and Fukui 1993), especially in estimating time since death by drowning (Casamatta and Verb 2000). Benthic diatoms and other microalgae or macroalgae, including Cyanobacteria, can also help link individuals to the aquatic freshwater or marine habitat of a specific locale or crime scene (Siver et al. 1994; Miller Coyle et al. 2001; Norris 1981, 2007). Profiles of algal species composition and ecological studies of seasonal variation in abundance and diversity can be generated and then used to compare specimens collected from a victim's lungs, body tissues (Ludes et al. 1996), or clothing against specimens collected from suspected crime scenes (Norris 1984).

Taphonomic analysis of Kennewick Man's skeleton demonstrates an entirely different use of botanical evidence. Identification and ecological considerations of algae growing on objects recovered at archaeological sites can assist in determination of the position and orientation of objects of paleoarchaeological interest. Green stains on the bones of Kennewick Man were a benthic filamentous species of green algae *Stigeoclonium tenue* (Chlorophyta). Examination of photographs taken at the discovery site, the distribution of algae growing on the bones (side exposed to sunlight, and a knowledge of the basic ecological and physiological requirements of the alga [i.e., they are photosynthetic, requiring light, water, nutrients, specific temperature range, and substrate for growth]) helped determine the original position of some bones in situ prior to their erosion from the bank and removal from the site. Further, the extent of growth on the individual bones facilitated an estimate of the time the bones had been exposed to conditions suitable for algal growth prior to their discovery.

The minute sizes of the filamentous green tuffs (<1 mm in length) of *Stigeoclonium tenue* on the bones sampled, along with limited surface coverage resulting in subtle staining indicate that these green algae had most likely settled on the bones within one to four weeks prior to the time of their discovery. This is particularly apparent when the minute algae are compared to what appear to be much larger algae (i.e., in aerial coverage and apparent filament lengths of several cm) observed growing on the substratum in field photographs in the vicinity of the recovery site. Unfortunately, the Corps destroyed the site before this study was permitted, eliminating any possibility of extracting more precise quantitative ecological or algal data.

REFERENCES

Bock, Jane H., and David O. Norris. 1997. Forensic botany: An under-utilized resource. *Journal of Forensic Sciences* 42(3):364–367.

Casamatta, Dale A., and Robert G. Verb. 2000. Algal colonization of submerged carcasses in a mid-order woodland stream. *Journal of Forensic Sciences* 45(6):1280–1285.

Chatters, James C. 2000. The recovery and first analysis of an early Holocene human skeleton from Kennewick, Washington. *American Antiquity* 65(2):291–316.

Cox, Elenor R., and Harold C. Bold. 1966. *Phycological Studies VII. Taxonomic Investigations of* Stigeoclonium. University of Texas Publications 6618:1–167.

Dillard, Gary E. 1989. *Freshwater Algae of the Southeastern United States. Part 2. Chlorophyceae: Ulothrichales, Microsporales, Cylindrocapsales, Sphaeropleales, Chaetophorales, Cladophorales, Schzogoniales, Siphonales and Oedogoniales.* Bibliotheca Phycologica. Band 83. J. Cramer, Stuttgart.

Guiry, M. D., and G. M. Guiry. 2007. *AlgaeBase version 4.2.* National University of Ireland, Galway. http://www.algaebase.org.

Holmgren, Patricia K., Noel H. Holmgren, and Lisa C. Barnett. 1990. *The Herbaria of the World. Index Herbariorum, Part 1.* 8th edition. New York Botanical Gardens, New York.

John, David M. 2002. Orders Chaetophorales, Klebshormidiales, Microsporales, Ulotrichales. In *The Freshwater Algal Flora of the British Isles. An Identification Guide to Freshwater and Terrestrial Algae*, edited by David M. John, Brian A. Whitton, and Alan J. Brook, 433–468. Cambridge University Press, Cambridge.

———. 2003. Filamentous and plantlike green algae. In *Freshwater Algae of North America: Ecology and Classification*, edited by John D. Wehr, and Robert G. Sheath, 311–352. Academic Press, San Diego.

Keiper, Joe B., and Dale A. Casamatta. 2001. Benthic organism as forensic indicators. *Journal of the North American Benthological Society* 20(2):311–324.

Kobayashi, Masahiko, Yoshihiro Yamada, Wei-Dong Zhang, Yoshiyuki Itakura, Masataka Nagao, and Takehiko Takatori. 1993. Novel detection of plankton from lung tissue by enzymatic digestion method. *Forensic Science International* 60(1–2):81–90.

Kützing, Friedrich T. 1843. *Phycologia generalis oder Anatomie, Physiologie und Systemkunde der Tange.* F. A. Brockhaus, Leipzig.

Lane, Meredith A., Loran C. Anderson, Theodore M. Barkley, Jane H. Bock, Ernest M. Gifford, David W. Hall, David O. Norris, Thomas L. Rost, and William Louis Stern. 1990. Forensic botany: Plants, perpetrators, pests, poisons, and pot. *Bioscience* 40(1):34–39.

Ludes, B., M. Coste, A. Tracqui, and P. Mangin. 1996. Continuous river monitoring of the diatoms in the diagnosis of drowning. *Journal of Forensic Sciences* 41(3):425–428.

Matsumoto, Hiroshi, and Yuko Fukui. 1993. A simple method for diatom detection in drowning. *Forensic Science International* 60(1–2):91–95.

Miller Coyle, Heather, Carll Ladd, Timothy Palmbach, and Henry C. Lee. 2001. The green revolution; Botanical contributions to forensics and drug enforcement. *Croatian Medical Journal* 42(3):340–345.

Norris, James N. 1981. Unpublished forensic case report, Honolulu, Oahu. Ms. on file at National Museum of Natural History, Washington, DC.

———. 1984. Unpublished forensic case report, Arlington County, Virginia. Ms. on file at National Museum of Natural History, Washington, DC.

———. 2007. Unpublished forensic case report, Cleveland County, North Carolina. Ms. on file at National Museum of Natural History, Washington, DC.

Powell, Joseph F., and Jerome C. Rose. 1999. Chapter 2. Report on the Osteological Assessment of the "Kennewick Man" Skeleton (CENWW.97.Kennewick). In *Report on the Non-Destructive Examination, Description, and Analysis of the Human Remains from Columbia Park, Kennewick, Washington* [October 1999]. National Park Service, http://www.nps.gov/archeology/kennewick/powell_rose.htm.

Prescott, Gerald Webber. 1962. *Algae of the Western Great Lakes Area, with an Illustrated Key to the Genera of Desmids and Freshwater Diatoms*. Revised edition. William C. Brown, Dubuque, IA.

———. 1978. *How to Know the Freshwater Algae*. 3rd edition. William C. Brown, Dubuque, IA.

Printz, Henrik. 1964. Die Chaetphoralen der Binnengewasser (eine systematische Ubersicht). *Hydrobiologia* 24:1–376.

Rosemond, Amy D., and Susan H. Brawley. 1996. Species-specific characteristics explain the persistence of *Stigeoclonium tenue* (Chlorophyta) in a woodland stream. *Journal of Phycology* 32(1):54–63.

Sarma, Pranjit. 1986. The freshwater Chaetophorales of New Zealand. *Beihefte Nova Hedwigia* 58:1–169.

Shubert, L. Elliot. 2003. Nonmotile coccoid and colonial green algae. In *Freshwater Algae of North America: Ecology and Classification*, edited by John D. Wehr, and Robert G. Sheath, 253–310. Academic Press, San Diego.

Simons, J., and A. P. van Beem. 1987. Observations on asexual and sexual reproduction in *Stigeoclonium helveticum* Vischer (Chlorophyta) with implications for the life history. *Phycologia* 26(3):356–362.

Simons, J., A. P. van Beem, and P. J. R. de Vries. 1986. Morphology of the prostrate thallus of *Stigeoclonium* (Chlorophyceae, Chaetophorales) and its taxonomic implications. *Phycologia* 25(2):210–220.

Siver, Peter A., Wayne D. Lord, and Donald J. McCarthy. 1994. Forensic limnology: The use of freshwater algal community ecology to link suspects to an aquatic crime scene in southern New England. *Journal of Forensic Sciences* 39(3):847–853.

Smith, Gilbert M. 1950. *The Freshwater Algae of the United States*. 2nd edition. McGraw-Hill, New York.

Wakeley, Lillian D., William L. Murphy, Joseph B. Dunbar, Andrew G. Warne, Frederick L. Briuer, and Paul R. Nickens. 1998. *Geologic, Geoarchaeologic, and Historical Investigation of the Discovery Site of Ancient Remains in Columbia Park, Kennewick, Washington*. U.S. Army Corps of Engineers Technical Report GL-98-13. Prepared for U.S. Army Engineer District, Walla Walla. U.S. Army Corps of Engineers Waterways Experiment Station, Vicksburg, MS.

Walker, Phillip L., Clark Spencer Larsen, and Joseph F. Powell. 2000. Chapter 5. Final Report on the Physical Examination and Taphonomic Assessment of the Kennewick Human Remains (CENWW.97.Kennewick) to Assist with DNA Sample Selection. In *Report on the DNA Testing Results of the Kennewick Human Remains from Columbia Park, Kennewick, Washington* [September 2000]. National Park Service, http://www.nps.gov/archeology/kennewick/walker.htm.

Wehr, John D., and Robert G. Sheath, eds. 2003. *Freshwater Algae of North America: Ecology and Classification*. Academic Press, San Diego.

Postmortem Breakage as a Taphonomic Tool for Determining Burial Position

Hugh E. Berryman

The characteristics of bone, especially its appearance when it breaks, can be used to suggest Kennewick Man's burial position. Since all bone will fracture under sufficient force, the first determination must be whether the break occurred around the time of death or after. Two papers (Berryman et al. 1991; Berryman and Symes 1999) identify morphological characteristics that differentiate fresh (often referred to as perimortem) fractures from dry (postmortem) breaks. Fracture characteristics identified during these earlier research endeavors are used in this chapter to evaluate the breaks associated with Kennewick Man.

The lack of discoloration on a fractured bone surface, relative to the overall stained appearance of the bone, is commonly used to distinguish postmortem breakage from perimortem trauma (Ubelaker and Adams 1995). However, postmortem breaks, even those occurring hundreds of years after burial, will eventually become discolored if the bone remains in the staining environment. Although the length of time required for staining to occur is unknown, once stained, a postmortem break could be mistaken as perimortem when color alone is the criterion. Obviously, additional considerations need to be made to substantiate the perimortem or postmortem timing of a break.

Identifying a bone fracture as occurring perimortem or postmortem requires an understanding of the basic physical principles involved with fracture production. Bone resists fracturing through both its mineral and collagen components. The mineral component allows bone to withstand compressive forces while the collagen content gives visco-elastic qualities needed to allow bending. The viscous properties permit bone to deform and transmit the forces in many directions while the elastic properties allow bone to return to its original shape once the load is removed. If the load exceeds the elastic capacity, the bone will move into the plastic phase and will become permanently deformed. Bone can withstand far greater compressive forces than tensile forces. When the material strength is exceeded, fracture dynamics dictate that bone will always fail first on the surface under tension, not compression.

An understanding of the dynamics of fracture production in fresh, tubular bone, where the features are better defined, will facilitate their recognition in dry bone (i.e., bone with a diminished collagen level) where their expression is much more subtle. The following description of fracture dynamics is based upon the author's experience in the examination of bone trauma taken from forensic cases at autopsy, the experimental fracturing of bone, and research directed toward identifying the mechanics responsible for fracture formation. Table 19.1 lists the observations made and defines how the data are coded. Table 19.2 provides the results of the examination of the fracture patterns in the long bones of the arms, the femora, and the tibiae of Kennewick Man.

Figure 19.1 illustrates fracture production in fresh bone when a load is applied to the free end, displacing it downward (Figure 19.1A). The mineral component of fresh bone is capable of withstanding compressive forces while the collagen allows it to absorb energy through elastic and plastic deformation. As deformation occurs, the convex side of the bone experiences tensile stress and the concave side experiences compressive stress. The bone can better withstand compressive stress than tensile stress (Figure 19.1B). As the load increases and the bone bends, microscopic irregularities in the bone surface act to elevate tensile stress locally. Eventually, the tensile stress exceeds the strength of the bone at one of these microscopic features, and the primary crack or fracture initiates (Figure 19.1B). The fracture advances toward the compression side if the load pressure continues.

TABLE 19.1 Fracture characteristics coding format.

Name	Description
Location of fracture on bone	The general anatomic location of the fracture on the bone (e.g., distal end of proximal one-third of diaphysis, near midshaft, or distal metaphysis).
Color of fracture, upper face (Munsell)	Each fracture results in two faces—the upper face is closest to the proximal end of the bone. Munsell Color Chart designation for color on the proximal face of the fracture.
Color of fracture, lower face (Munsell)	Each fracture results in two faces—the lower face is closest to the distal end of the bone. Munsell Color Chart designation for color on the distal face of the fracture.
Primary fracture angle	The angle of the primary fracture is classified as straight (i.e., perpendicular to the long axis of the bone); diagonal (i.e., at an angle to the long axis of the bone regardless of how slight the angle might be); or indeterminate (i.e., obscured to the point that an accurate determination cannot be made).
Primary fracture shape	The overall morphology of the primary fracture surface is classified as straight (i.e., an overall flat surface); dentate (i.e., characterized as a surface with angular steps); curved (i.e., reminiscent of the sigmoid curve); and indeterminate.
Number of secondary fractures	Secondary fractures are problematic in dry bone since some may actually be preexisting longitudinal drying fractures. Fractures occurring perpendicular to the primary fracture are considered to be secondary fractures unless they can be identified as preexisting longitudinal drying fractures and eliminated from the count. The count is made for upper (i.e., those on the segment of bone proximal to the primary fracture) and lower (i.e., those on the segment of bone distal to the primary fracture).
Tension location	Anatomic and numeric location is indicated. Number is assigned by superimposing a clock face over the bone shaft as if viewed through the bone shaft from proximal to distal with 12 anterior, 6 posterior, 3 to right side, and 9 to the left side. This "clock face" approach facilitates the evaluation of fracture propagation when comparing left and right elements to determine the affects of external forces.
Tension surface morphology	The tension surface is classified as flat; irregular; or indeterminate.
Compression location	Anatomic and numeric location is indicated. Number is assigned by superimposing a clock face over the bone shaft as if viewed through the bone shaft from proximal to distal with 12 anterior, 6 posterior, 3 to right side, and 9 to the left side.
Compression surface morphology	The compression surface is classified as flat; irregular; or indeterminate.
Breakaway spur indicated	Breakaway spurs are sometime subtle in expression or may be fractured and missing. Obvious breakaway spurs are indicated with a yes; the lack of an obvious breakaway spur is indicated with a no; possible is entered when a breakaway spur is more subtle, broken away, or questionable; and indeterminate indicates that no decision can be made.
Breakaway spur face location	A primary fracture is composed of two fracture surfaces or faces formed where the segment proximal to the fracture site and the segment distal to the fracture side join. The breakaway spur is identified as being on the upper face (i.e., the face closest to the proximal end of the bone); the lower face (i.e., the face is closest to the distal end of the bone); or indeterminate.
Breakaway spur anatomic location	The anatomic location of the breakaway spur is identified as anterior; anterolateral; anteromedial; lateral; medial; posterior; posterolateral; and posteromedial. In addition, numbers are assigned that reflect the area of the surface covered by the breakaway spur.
Fixed segment	Each fracture separates two segments of bone—one proximal to the fracture site and one distal to the fracture site. Features such as tension location, compression location, and breakaway spur location are used to classify the fixed segment as proximal to the fracture; distal to the fracture; or indeterminate.
Fracture propagation direction	Fracture propagation direction is identified from the site of initiation on the bone to the site of completion at the breakaway spur, and is expressed in anatomic terms (e.g., anteromedial to posterolateral and distal to proximal).

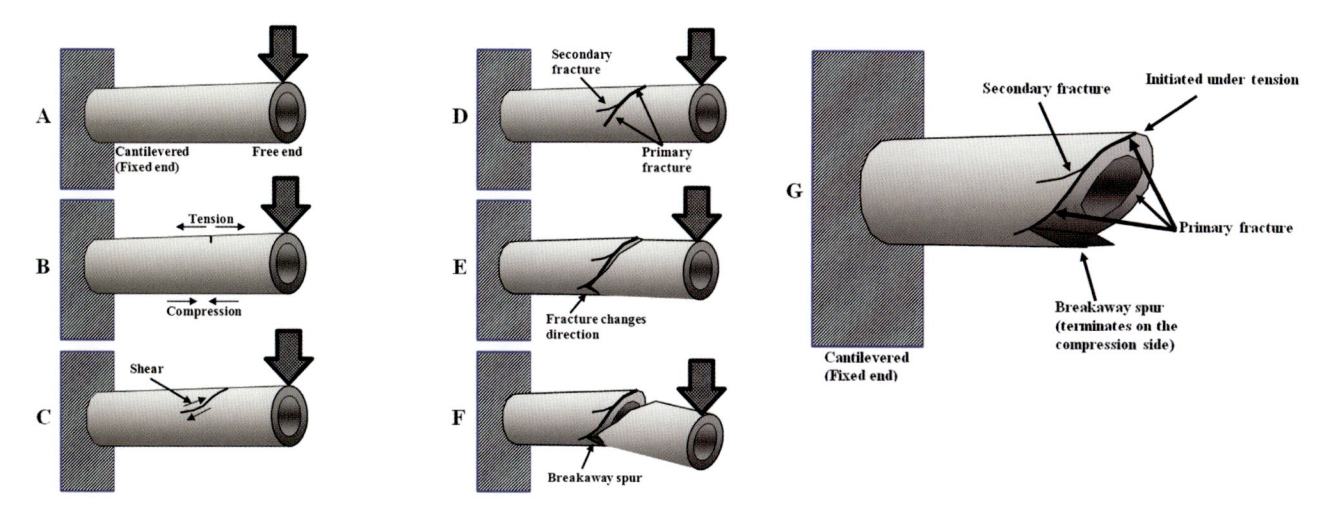

FIGURE 19.1 Tubular bone fracture. A through F illustrate successive stages in the fracturing of a tubular bone. G illustrates characteristics present on the cantilevered, or fixed, end of the bone.

As the fracture progresses (Figure 19.1C), shear forces direct it toward the cantilevered or fixed end. Fracture advance terminates when the load is eliminated and/or the fracture encounters an area of bone microstructure, morphology, or composition capable of absorbing or diverting the energy. When intrinsic factors halt the fracture but the load continues to displace the free end, a new fracture will open along the primary fracture surface and continue to advance at an angle toward the cantilevered end. These incomplete breaks, termed secondary fractures (Figure 19.1D), will develop at points of intrinsic resistance. The angle created by the intersection of the primary and secondary fracture always points toward the side of the bone with tension. A primary fracture may also advance across the bone shaft without the formation of secondary fractures. As the primary fracture nears the opposite side of the bone, it abruptly changes direction and completes the fracture (Figure 19.1E). The wedge of bone produced by this sudden change of direction is the breakaway spur (Figure 19.1F).

To determine the fracture dynamics in a situation involving perimortem trauma, the fracture site can be examined for the location of the breakaway spur, the angle of the primary fracture, and the orientation of the angle produced by the primary and secondary fractures (Figure 19.1G). The angle created by the primary and secondary fractures points to the tension side of the bone, while the primary fracture is directed toward the fixed end, and the location of the breakaway spur can be used to identify both the compression side of the bone and the segment of bone that is fixed or less mobile during fracture formation.

There is a significant difference in morphology between the fractures in fresh bone and the breaks in dry bone, pointing directly to the depletion of the elastic component with moisture loss over time as the major contributor (Hughes and White 2009; Wheatley 2008; Wieberg and Wescott 2008). The interplay of intrinsic and extrinsic factors dictates fracture morphology and, more specifically, the physical expression of associated features. The morphology in fresh and dry bones (Berryman et al. 1991, 1992; Berryman and Symes 1999) is directly related to the loss of the elastic/plastic components (Figure 19.2). With fractures in fresh bone, the diagonal angle is produced by the shear forces that drive the direction of fracture toward the fixed end of the bone. Fracture to a fresh bone is characterized by a smooth surface, often with a sigmoid shape. Although breaks across dry bones can be sigmoid in shape, most are straight (relatively flat) or dentate (angular), while the break angle is most frequently described as diagonal. The break angle in dry bone, important in confirming the direction the fracture propagated across the bone, is frequently a very subtle diagonal with only a slight variant from perpendicular.

Differentiation of the tension portion from the compression portion of the fracture is the key to determining the way a fracture propagates across a bone. A

TABLE 19.2 Descriptions and characteristics of long bone fractures (breakage) of Kennewick Man.

Bone	Proximal Segment No.	Distal Segment No.	Location of Fracture	Color of Fracture, Upper Face	Color of Fracture, Lower Face	Primary Fracture Angle	Primary Fracture Shape	Number of Secondary Fractures
Left Humerus	97.L.13a(1)	97.L.13b (2)	Near midshaft	10YR8/3 very pale brown	10YR8/2 white	Diagonal (slight)	Straight	Upper 3 Lower 2
Left Humerus	97.L.13b (2)	97.L.13c (3)	Middle distal one-third of diaphysis	10YR8/3 very pale brown	10YR8/4 very pale brown	Diagonal (slight)	Straight	Upper 1 Lower 4
Right Humerus	97.R.13a (1)	97.R.13b (2)	Distal one-fourth of diaphysis	10YR8/2 white	10YR8/2 white	Diagonal (slight)	Straight	Upper 3 Lower 2
Left Radius	97.L.14a (1)	Missing (2)	Proximal one-third of diaphysis	10YR8/3 very pale brown	Missing	Diagonal	Curved	Upper 3 Lower missing
Left Radius	Missing (2)	97.L.14b (3)	Midshaft	Missing	10YR8/2 white, with algal staining	Diagonal	Curved	Upper missing Lower 2
Right Radius	97.R.14a (1)	97.R.14b (2)	Distal one-fourth of diaphysis	10YR8/3 very pale brown	10YR8/3 very pale brown	Diagonal (slight)	Dentate	Upper 2 Lower 3
Left Ulna	97.L.15a (1)	97.L.15b (2)	Distal end of proximal one-third of diaphysis	10YR8/3 very pale brown	10YR8/2 white	Straight	Straight	Upper 3 Lower 3
Left Ulna	97.L.15b (2)	97.L.15c (3)	Proximal end of distal one-third of diaphysis	10YR8/2 white, largely obscured by sediment	10YR8/2 white	Straight	Dentate	Upper 3 Lower 2
Left Ulna	97.L.15c (3)	Missing (4)	Distal metaphysis	10YR6/1 grey	Missing	Diagonal	Dentate	Upper 0 Lower missing
Right Ulna	97.R.15a (1)	97.R.15b (2)	Just superior to midshaft	10YR8/3 very pale brown, partly obscured by sediment/algae	10YR8/3 very pale brown	Diagonal (slight)	Straight	Upper 2 Lower 3

Ind. = Indeterminate

Tension Location	Tension Surface Morphology	Compression Location	Compression Surface Morphology	Breakaway Spur Indicated	Breakaway Spur Location	Fixed Segment	Fracture Propagation Direction
Lateral (7 to 11)	Flat	Medial (11 to 7)	Irregular	Yes	Upper, Postero medial (2 to 7)	Proximal (segment 1)	Lateral to medial and distal to proximal
Medial (11 to 8)	Flat	Lateral (8 to 11)	Irregular	Yes	Lower, Lateral (8 to 11)	Distal (segment 3)	Medial to lateral and proximal to distal
Postero medial (6 to 10)	Flat	Antero lateral (10 to 5)	Irregular	Yes	Upper, Lateral (1 to 4)	Proximal (segment 1)	Postero medial to anterolateral and distal to proximal
Antero medial (11 to 4)	Irregular	Postero lateral (4 to 11)	Irregular	Yes	Upper, Posterior (4 to 8)	Proximal (segment 1)	Anteromedial to posterolateral and distal to proximal
Antero medial (12 to 4)	Irregular	Postero lateral (4 to 12)	Irregular	Ind.	Ind., upper face missing	Proximal (segment 2)	Anteromedial to posterolateral and distal to proximal
Antero medial (9 to 11)	Flat	Postero lateral (11 to 9)	Irregular	Yes	Upper, Lateral (1 to 6)	Proximal (segment 1)	Anteromedial to posterolateral and distal to proximal
Antero medial (12 to 3)	Flat	Postero lateral (3 to 12)	Slight irregular relief	No	None	Possibly proximal (segment 1)	Anteromedial to posterolateral
Posterior (5 to 8)	Flat	Anterior (8 to 5)	Irregular	Yes, but fractured and missing	Upper, but missing, antero-medial (12 to 2)	Possibly proximal (segment 2)	Posterior to anteromedial
Ind.	Ind.	Ind.	Ind.	Ind.	Ind.	Ind. due to thin cortical bone and irregular fracture surface	Not interpreted
Antero medial (8 to 12)	Flat	Postero lateral (12 to 8)	Irregular	Yes, but fractured and missing	Upper, Lateral (1 to 5)	Proximal (segment 1)	Anteromedial to posterolateral and distal to proximal

(continued)

TABLE 19.2 (*continued*) Descriptions and characteristics of long bone fractures (breakage) of Kennewick Man.

Bone	Proximal Segment No.	Distal Segment No.	Location of Fracture	Color of Fracture, Upper Face	Color of Fracture, Lower Face	Primary Fracture Angle	Primary Fracture Shape	Number of Secondary Fractures
Right Ulna	97.R.15b (2)	97.R.15c (3)	Proximal end of distal one-third of diaphysis	10YR8/2 white, largely obscured by sediment	10YR8/2 white	Diagonal (slight)	Dentate	Upper 3 Lower 1
Left Femur	97.L.18a (1)	01.L.18b (2)	Midshaft	5YR8/3 to 7/3 pale yellow	10YR8/4 very pale brown	Diagonal (slight)	Dentate	Upper 6 Lower 5
Left Femur	01.L.18b (2)	01.L.18a (3)	Distal end	10YR8/3 very pale brown	10YR7/4 very pale brown	Diagonal	Dentate	Upper 6 Lower 5
Right Femur	01.R.18a (1)	01.R.18b (2)	Proximal diaphysis	5Y7/3 pale yellow	10YR7/3 very pale brown	Diagonal (slight)	Dentate	Upper 3 Lower 2
Right Femur	01.R.18b (2)	97.R.18a (3)	Distal one-third of diaphysis	7.5YR8/4 very pale brown	10YR8/2 to 7/2 white to light gray	Straight	Straight	Upper 4 Lower 3
Left Tibia	97.L.20a (1)	97.L.20b (2)	Proximal diaphysis	10YR8/4 pink	7.5YR8/4 pink	Ind., obscured by prior sectioning by gov. for sampling	Ind.	Upper 4 Lower ind.
Left Tibia,	97.L.20b (2)	97.L.20c (3)	Slightly proximal midshaft	Anterior 10YR8/2 white; posterior 10YR7/2 light gray	Posterior 10YR8/3 very pale brown; anterior 10YR7/2 light gray	Diagonal	Dentate	Upper 2 (anterior portion missing) Lower 7
Right Tibia	97.R.20a (medial) and 97.R.20b (lateral) (1)	97.R.20c (3)	Transects the tibial tuberosity separating medial and lateral condyles	Omitted	Omitted	Ind.	Dentate	Upper ind. Lower 6
Right Tibia	97.R.20c (3)	97.R.20d (4)	Proximal end of the distal one-third of diaphysis	2.5Y8/2 white	10YR8/3 very pale brown	Diagonal	Curved	Upper 3 Lower 5

Ind. = Indeterminate

Tension Location	Tension Surface Morphology	Compression Location	Compression Surface Morphology	Breakaway Spur Indicated	Breakaway Spur Location	Fixed Segment	Fracture Propagation Direction
Antero lateral (1 to 3)	Flat	Postero medial (3 to 1)	Irregular	Yes, but fractured and missing	Upper, Postero medial (6 to 9)	Proximal (segment 2)	Anterolateral to postero medial and distal to proximal
Antero lateral (7 to 12)	Flat (fairly low relief)	Postero medial (12 to 7)	Irregular	Possibly	Upper, Posterior (6)	Proximal (segment 1)	Anterolateral to postero medial and distal to proximal
Antero lateral (7 to 12)	Flat	Postero medial (12 to 7)	Irregular	No	None	Proximal (segment 2)	Anterolateral to postero medial and distal to proximal
Medial (7 to 11)	Flat	Lateral (12 to 6)	Irregular	Yes	Upper, Postero-lateral (3 to 5)	Proximal (segment 1)	Medial to posterolateral and distal to proximal
Antero lateral (12 to 3)	Flat	Posterior (4 to 8)	Slightly irregular	Ind.	Probably upper, Position on bone ind.	Probably proximal (segment 2)	Anterolateral to posterior
Possibly postero lateral (7 to 9)	Irregular	Possibly medial (2 to 4)	Possibly irregular	Possibly	Upper, Possibly medial (2 to 4)	Possibly proximal (segment 1)	Possibly postero lateral to medial
Anterior (9 to 2)	Flat	Posterior (2 to 9)	Irregular	Yes	Lower, Posterior (5 to 7)	Distal (segment 3)	Anterior to posterior and proximal to distal
Ind.	Ind.	Ind.	Ind.	Ind.	Ind.	Ind.	Ind.
Anterior (10 to 2)	Slightly irregular	Posterior (2 to 10)	Flat	Yes	Lower, Posterior (4 to 7)	Distal (segment 3)	Anterior to posterior and proximal to distal

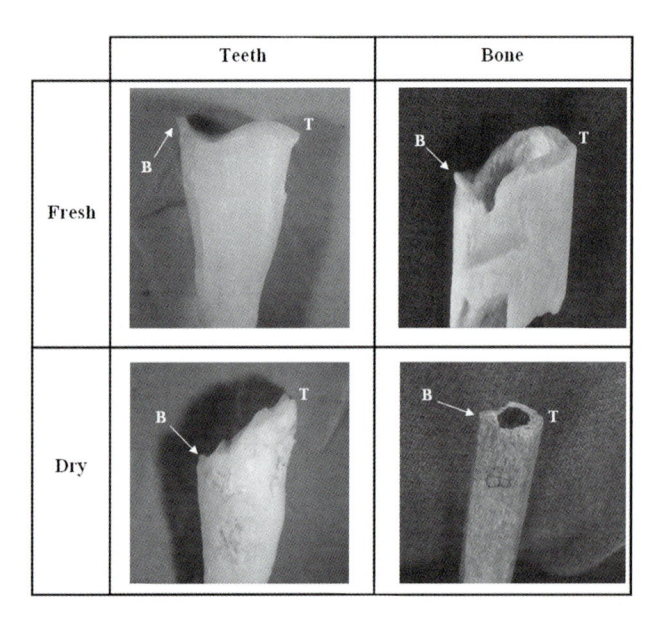

FIGURE 19.2 Fracture morphology characteristic of fresh teeth and bones and dry teeth and bones. Breakaway spurs are found on the compression side in both the fresh and dry stage. B indicates a breakaway spur, and T indicates tension.

well-defined breakaway spur provides the best indicator of the compression side, but, in its absence, the differences in surface morphology between the tension and compression bone portions can be used. The tension surface morphology tends to be flat (least irregular), while the compression surface morphology tends to be more irregular or angular.

This chapter uses observable fracture features to identify the way bone segments were displaced to produce breaks and to identify the most probable mechanics responsible for an individual break or group of breaks. In an attempt to be more objective and to standardize the data collection, the various characteristics associated with fracture initiation and fracture completion (i.e., the tension side of the bone and the compression side of the bone, respectively) are examined individually.

When comparing a fracture in fresh bone from a contemporary autopsy and a break in dry bone from the tibia of Kennewick Man (Figure 19.3), the dry bone exhibits the overall sigmoid curve of a fresh bone fracture, but it also possesses the more angular or steplike surface characteristic of a dry bone break. In Figure 19.3, the presence of a breakaway spur on a segment of tibia shaft confirms

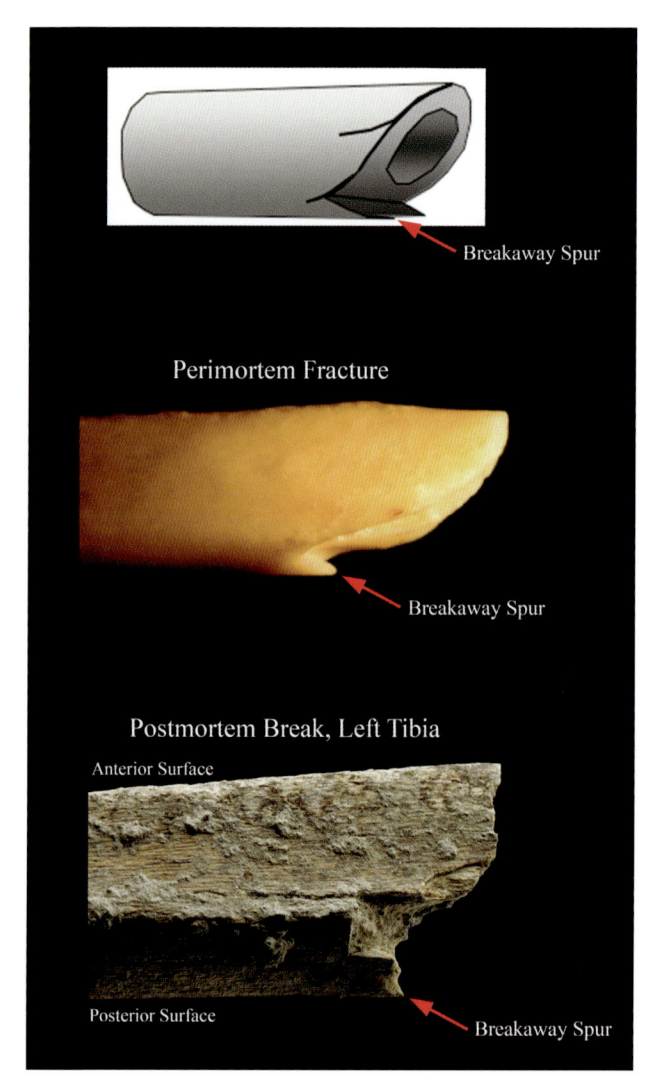

FIGURE 19.3 Comparison of perimortem fracture and postmortem break. Note that the postmortem break (dry bone) exhibits the same features as the perimortem fracture (fresh bone). Specifically, it is angled toward the fixed end and terminates with a breakaway spur. The surface of the break in dry bone is rough or steplike when compared to the smoother contour of a fracture in fresh bone.

exactly how the fracture occurred. This segment of bone was fixed relative to the more proximal segment (not shown) that moved from an anterior to posterior direction to produce the break.

TAPHONOMIC INTERPRETATION OF POSTMORTEM LONG BONE BREAKAGE

Due to time and access, this study focused on the long bones of the arms and legs (the humeri, radii, ulnae, femora, and tibiae), although a cursory examination was given to other bones. Of the bones examined, none exhibit fractures with features indicative of perimortem origin. All the fractures examined, comprising the long bones, metacarpals, metatarsals, and the scapulae, were clearly broken postmortem.

Postmortem movement or displacement of Kennewick Man's remains may have resulted from three natural forces: (1) subtle movement of the soil through great expanses of time due to periodic, micro-environmental events, (2) acute movement produced by burrowing animals that coincidentally displaced and broke bones, or (3) the catastrophic failure of the alluvial Holocene river terrace as shoreline erosion undercut the burial site. Periodic breaks in buried bone would include drying cracks—usually superficial cracks aligned parallel to the long axis of the long bones—and the cracks caused by long-term pressure and displacement from overburden. The morphology and more superficial location of drying cracks make them easily distinguishable, as are the cracks or crushing in the thinner cortex of the joint surfaces produced by the pressure of overburden. In some cases it is difficult to determine the cause of acute breaks, but they are usually easy to recognize. It can be anticipated that breaks occurring days to weeks before recovery would be lighter in color than the cortical surface; however, the length of time required for a bone surface to become soil-stained is unknown. Acute breaks produced by animal activity may have occurred hundreds or thousands of years prior to discovery and will be soil stained. Since no evidence of animal gnawing is present on Kennewick Man's bones and most of the broken surfaces are less stained than the cortex, the observed acute breaks were produced by the catastrophic event of the soil bank slumping vertically onto the shoreline with some possibility of damage after displacement. Some acute breaks may have occurred days to weeks prior to discovery due to subtle movement of earth as the undercut bank began to yield to its inevitable collapse.

BODY POSITION

In general, intentional burials are positioned either extended or flexed. Extended burials tend to be supine and flexed burials tend to be positioned on one side, although both types can vary. In either burial position, the upper arms could be extended, adducted, abducted, and rotated medially or laterally with the lower arms flexed or extended, and either pronate or supinate. The positioning of one arm can be independent of the other unlike, in general, the positioning of the legs. Regardless of burial type, the bones of the legs tend to be roughly parallel to each other (i.e., femur to femur and tibia to tibia), and, whether broken by ground pressure or by bank collapse, these bones should experience similar environmental pressures and should express similar fracture patterns. Since fracture variance is anticipated with the position of the arms, breaks to the bones of the legs, with their more limited range of movement relative to the body, provide the key for determining an extended or flexed body position. Extended burials exhibit breaks that indicate an anterior-to-posterior movement of shaft segments while flexed burials more commonly display a lateral, or side-to-side, movement.

LOWER BODY BONE BREAKS
Feet

The left first metatarsal exhibits a transverse break that initiated on the dorsomedial side of the shaft (Figure 19.4). The laterally directed end of the break crosses a preexisting longitudinal drying crack on the superior surface of the bone, and the medially directed end of the break extends inferiorly where it terminates into a preexisting longitudinal crack on the medial side of the bone. Careful examination of the bone indicates that the distal end is slightly depressed in a plantolateral direction. This is consistent with the overburden pressing down on the left foot while it rested in a plantar flexed position with slight lateral rotation (Table 17.2). This is a common positioning of the feet in extended, supine burials.

Tibias

The left tibia has two complete breaks, one at the proximal end and the other at the midshaft (Figure 19.5). The proximal end of the tibia is incomplete (i.e., missing a section from the anterior crest that was previously removed by the government study team for testing). The proximal break is in an area of relatively thin cortical bone and exposed cancellous bone with longitudinal drying fractures, making interpretation tentative and

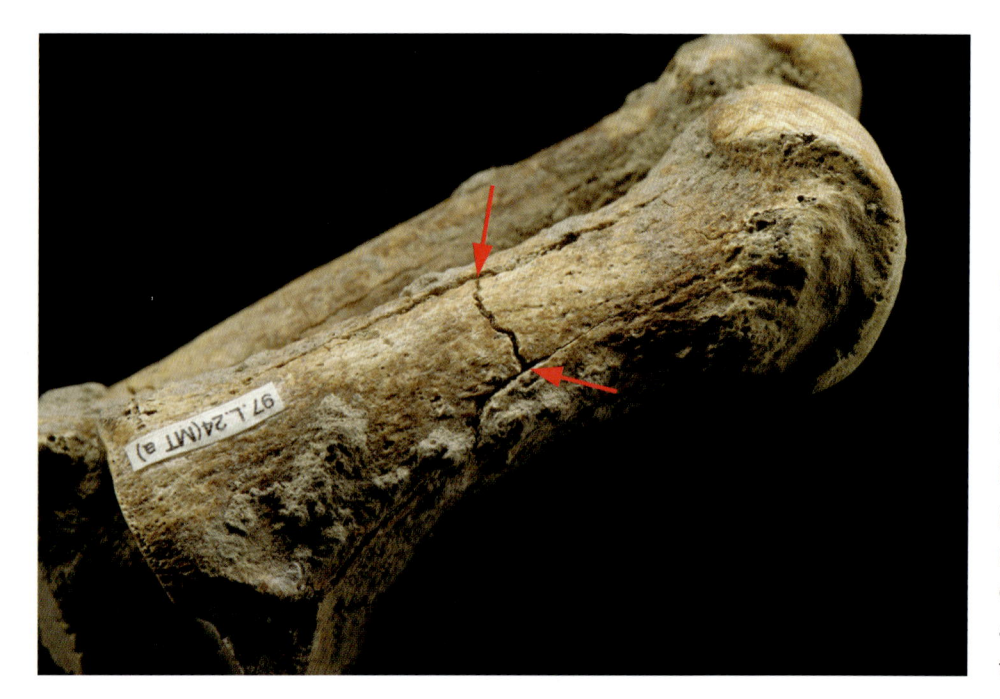

FIGURE 19.4 Left first and second metatarsal bones of Kennewick Man. Red arrows indicate the sites where the transverse break, produced by overlying ground pressure, terminates in the preexisting longitudinal drying cracks on the medial and superior surfaces of the bone.

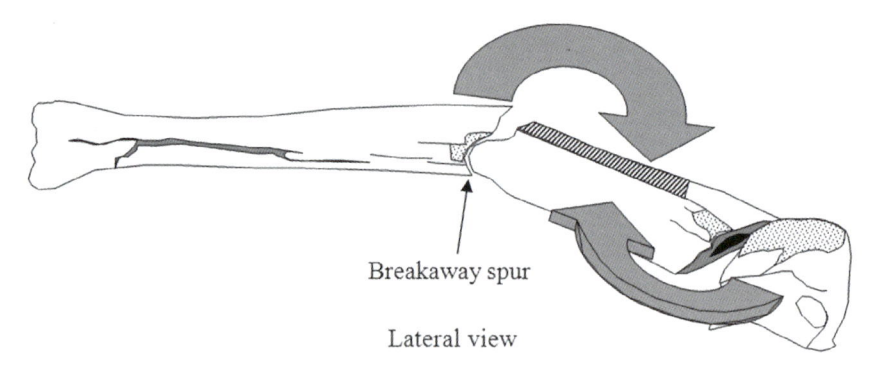

FIGURE 19.5 Breaks to Kennewick Man's left tibia, lateral view. The fractures indicate that the midshaft of the tibia broke as the proximal half of the shaft was displaced posteriorly. The location of the well-defined breakaway spur means that the distal end of the shaft was fixed while the proximal end was moving posteriorly. This is consistent with the body being in a supine position and the bank slumping under the knees. The most proximal break is difficult to interpret due to the thinner cortex, drying fractures and bone removed during sampling. It appears to have resulted from the distal end of the segment of shaft produced by the two breaks moving posterolateral to medial with the proximal end fixed. This could have resulted from the most proximal end moving laterally during the collapse of the riverbank.

Breakaway spur
Lateral view

Curved arrows indicate the fixed end of the shaft (blunt end of arrow) and the direction of movement of the free end of the shaft (point of arrow).

questionable at best. Considering these limitations, a possible breakaway spur may be present on the medial side of the proximal segment (97.L.20a). If this is accurate, it is speculated that the proximal end was fixed and the second segment (97.L.20b) may have moved in a posterolateral to medial direction.

The complete break of the midshaft has a breakaway spur on the posterior aspect of the most distal segment (97.L.20c) (Figure 19.6). It is clear that the distal half of the shaft was fixed while the middle segment was levered from anterior to posterior to produce the break. This break is one of the strongest confirmations that the tibia was resting on its posterior surface with the terrace bank giving away under the knees of Kennewick Man. An anterior to posterior movement to produce a break to the tibia is a strong validation of an extended and supine burial position.

The right tibia exhibits two complete breaks, one to

FIGURE 19.6 Kennewick Man's left tibia shaft, lateral view. This fracture clearly demonstrates the characteristics of a fixed segment.

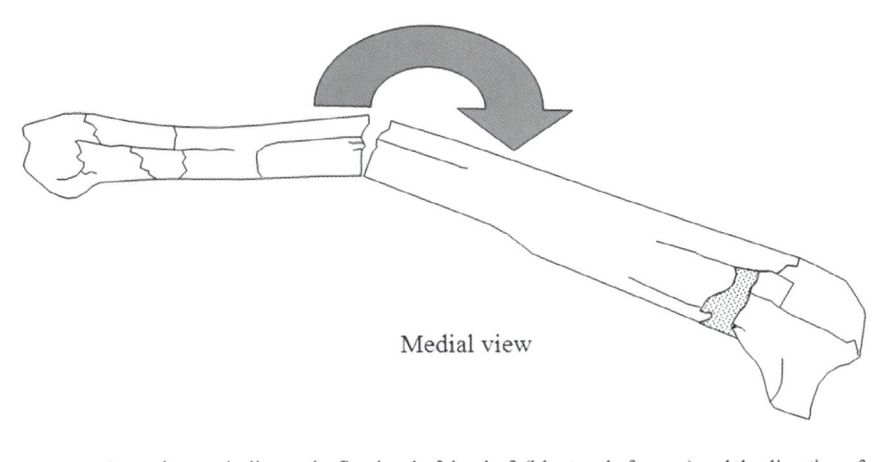

FIGURE 19.7 Breaks to Kennewick Man's right tibia, medial view. The break to the midshaft indicates that the distal end was fixed as the proximal half of the shaft was being displaced posteriorly. This is consistent with the body being in a supine position and the riverbank slumping under the knees. The most proximal fracture is difficult to interpret due to the thinner cortex.

Curved arrow indicates the fixed end of the shaft (blunt end of arrow) and the direction of movement of the free end of the shaft (point of arrow).

the proximal end at the metaphysis and the other located just distal to the midshaft (Figure 19.7). The proximal end of the bone (97.R.20a, medial, and 97.R.20b, lateral) is incomplete and the cortical bone is too thin for reliable interpretation of the dynamics involved in the production of the most proximal break.

The break to the tibial shaft was produced with the distal end of the bone fixed (97.R.20d) and the proximal shaft (97.R.20c) being levered from anterior to posterior. The break opened first on the anterior crest and propagated distally and posteriorly, leaving a breakaway spur on the posterior face of the distal segment (97.R.20d). The breaks to the right tibia mimic those to the left and further confirm an extended and supine burial position.

Femurs

The left femur exhibits two complete breaks—one located at the midshaft and the other at the distal shaft—dividing the shaft into three segments (Figure 19.8). The break to the midshaft opened first on the anterolateral side, and the most proximal segment (97.L.18a) has a breakaway spur on the posterior side. The area of compression appears to have been on the posteromedial side of the bone. This indicates that the fracture was produced with the proximal segment (97.L.18a) fixed and the middle segment (01.L.18b) displaced in an anterolateral to posteromedial direction.

The most distal break also advanced from an anterolateral to posteromedial direction, and its path was interrupted by a preexisting longitudinal fracture. As the most distal of the three segments (01.L.18a) was being displaced posteromedially, the break opened on the anterolateral surface, moving medial and posterior until it reached the preexisting drying fracture; that is, a fracture formed by the interplay of wet and dry phases that split the bone. The break traveled proximally along the drying fracture until it opened again and continued posteromedially transecting the shaft (Figure 19.9). The break traveled in a Z-shaped path across the anterior and medial surface of the distal shaft.

The right femur exhibits two complete breaks, one located at the proximal end of the shaft just distal to the lesser trochanter and the other located approximately three quarters down the length of the shaft (Figure 19.10). The most proximal break opened first on the medial side and completed with a breakaway spur on the posterolateral aspect of the proximal segment (01.R.18a). The bone broke when the proximal segment was fixed and the midshaft segment (01.R.18b) was displaced laterally and posteriorly.

The more distal break lacks a well-defined breakaway spur; however, the general morphology of the fracture surface exhibits a flatter area along the anterolateral side, which suggests tension, as well as a more irregular area on the posterior side of the shaft, indicating compression. This break appears to have progressed from anterolateral

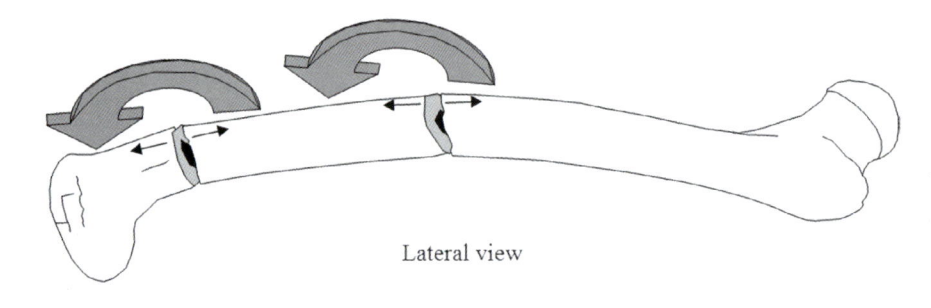

Lateral view

← → Double arrows indicate the location where the bone initially failed under tension.

↻ Curved arrows indicate the fixed end of the shaft (blunt end of arrow) and the direction of movement of the free end of the shaft (point of arrow).

FIGURE 19.8 Breaks to Kennewick Man's left femur, lateral view. The fractures indicate that the distal end of the femur and the lower middle segment between the two breaks were being displaced anterolateral to posteromedial. This is consistent with the body being in a supine position and the riverbank slumping under the knees.

FIGURE 19.9 Kennewick Man's left distal femur, anteriomedial view. The arrow points to the site where the transverse fracture on the anterior surface, produced by the downward force of ground pressure, enters the preexisting longitudinal drying fracture. This transverse fracture also enters a second preexisting longitudinal drying fracture (not pictured) on the lateral side of the bone.

to posterior with the midshaft segment (01.R.18b) being fixed relative to the distal segment (97.R.18a).

Anterior to posterior force directed to the distal femur is evident in a separate fracture configuration located on the anterior side of the bone, just proximal to the distal articular surface. This consists of a horizontally-oriented break bounded laterally and medially by drying cracks aligned with the long axis of the bone (Figure 19.11). This horizontal break likely opened from the pressure of the overburden pressing downward while the bone was positioned with its posterior surface down.

In a similar vein, transverse breaks in the anterior surface of the distal femora (Figures 19.9 and 19.11) indicate that overlying ground pressure or perhaps the collapse of the bank displaced the distal ends posteriorly. This indicates that both femora were lying on their posterior surfaces (i.e., in a supine position).

Analysis

With the posterior side of the femora down and the dorsomedial side of the left foot up, the position of the body is clearly supine and extended. The way the long bones fractured is consistent with the body being in a supine position and the riverbank slumping under the knees.

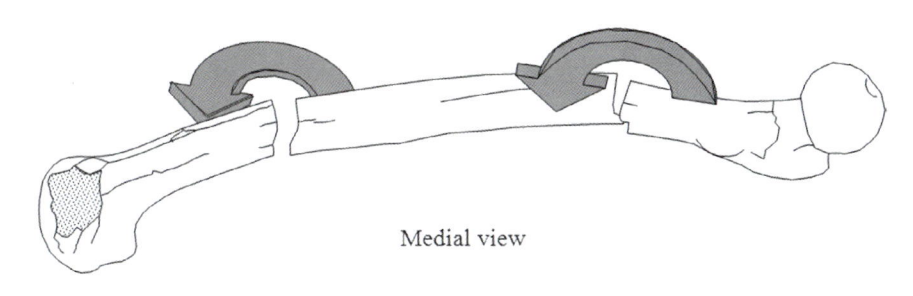

FIGURE 19.10 Breaks to Kennewick Man's right femur, medial view. The fractures indicate that the distal end of the femur was being displaced anterolateral to posterior. The midshaft segment was being displaced from medial to posterolateral with the proximal end fixed. This is consistent with the body being in a supine position and the riverbank slumping under the knees.

Curved arrows indicate the fixed end of the shaft (blunt end of arrow) and the direction of movement of the free end of the shaft (point of arrow).

Medial view

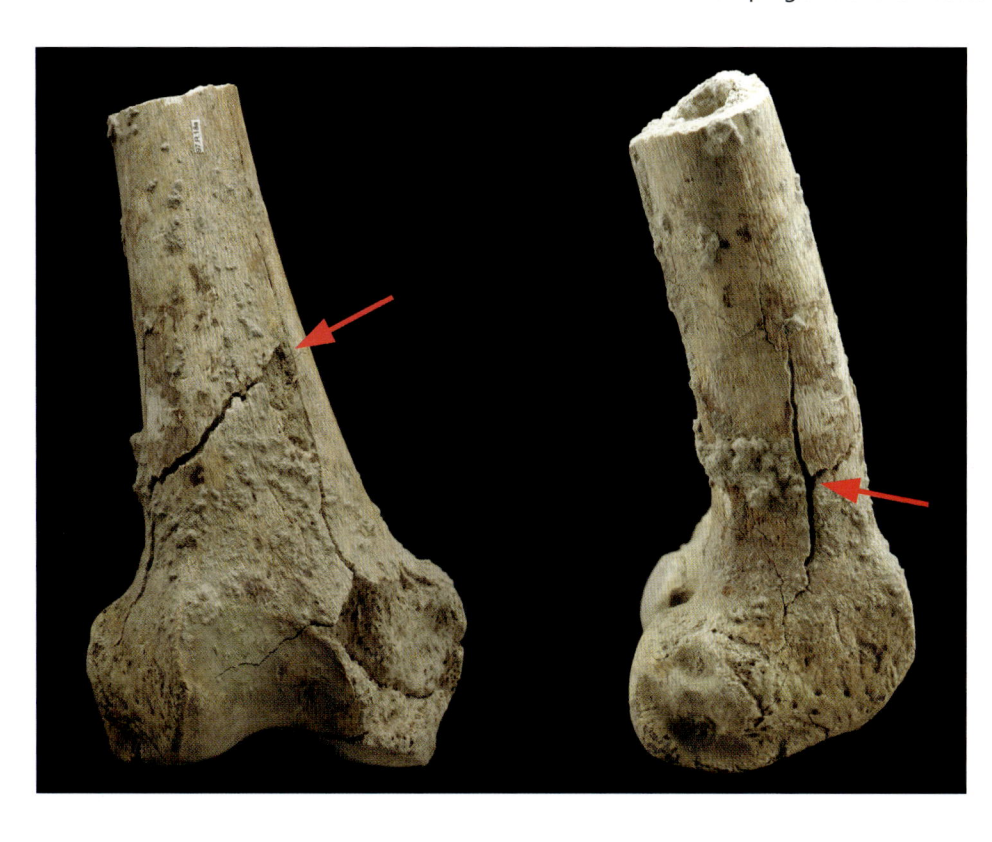

FIGURE 19.11 Kennewick Man's right distal femur, anterior and lateral views. The red arrows point to the sites where the transverse break on the anterior surface, produced by the downward force of ground pressure, terminates in the preexisting longitudinal drying cracks.

Hugh E. Berryman

bones of the lower limbs were displaced to produce the breaks (i.e., identification of the tension side of each break and the relationship of the fixed end to the free end) can be explained if the body was extended and supine. The breaks to the left tibia provide the strongest confirmation that the tibia was resting on its posterior surface with the bank giving way under the knees of Kennewick Man. Breaks to the bones of the lower extremities are consistent with the bank below the knees eroding away, followed by the areas under the feet and the pelvis.

LEFT SCAPULA

The glenoid cavity of the left scapula exhibits two breaks: an incomplete break extending from the anterior margin of the articular surface terminating just short of the posterior border and a complete break to the superior portion of the joint separating the coracoid process (Figure 19.12) (Table 17.1). Both of these breaks initiated under tension on the anterior border; the fracture to the middle of the joint surface is incomplete and a breakaway spur is present on the posterior joint surface. These two breaks are consistent with ground pressure produced by the overburden or from the slump of the bank as the scapula rested on its posterior surface and also consistent with an extended supine burial.

LEFT LOWER ARM BONE BREAKS
Ulna

The left ulna exhibits damage to the distal end and two complete breaks that separate the shaft into three approximately equal segments (Figure 19.13). The most proximal break opened first on the anteromedial surface, and, although there is no defined breakaway spur, the compression side of the bone seems to have been on the posterolateral side. This break occurred with the proximal end fixed (97.L.15a) while the midshaft segment (97.L.15b) moved in an anteromedial to posterolateral direction. The presence and configuration of an incomplete break on the anterior surface just proximal to this complete break (on segment 97.L.15a) provides additional proof that the proximal end of the bone was fixed and the distal end was displaced posteriorly (Figure 19.13). This incomplete break opened first on the anterior side of the bone then traveled posteriorly as shear forces directed it toward the proximal end.

The most distal break opened first on the posterior surface, propagating posterior to anterior across the shaft, and it appears to have been produced with the midshaft segment (97.L.15b) fixed relative to the distal end of the ulna (97.L.15c). The midshaft may have pivoted relative to the proximal and distal segments as slumping earth took the distal end downward toward the water, producing both complete breaks (Figure 19.14).

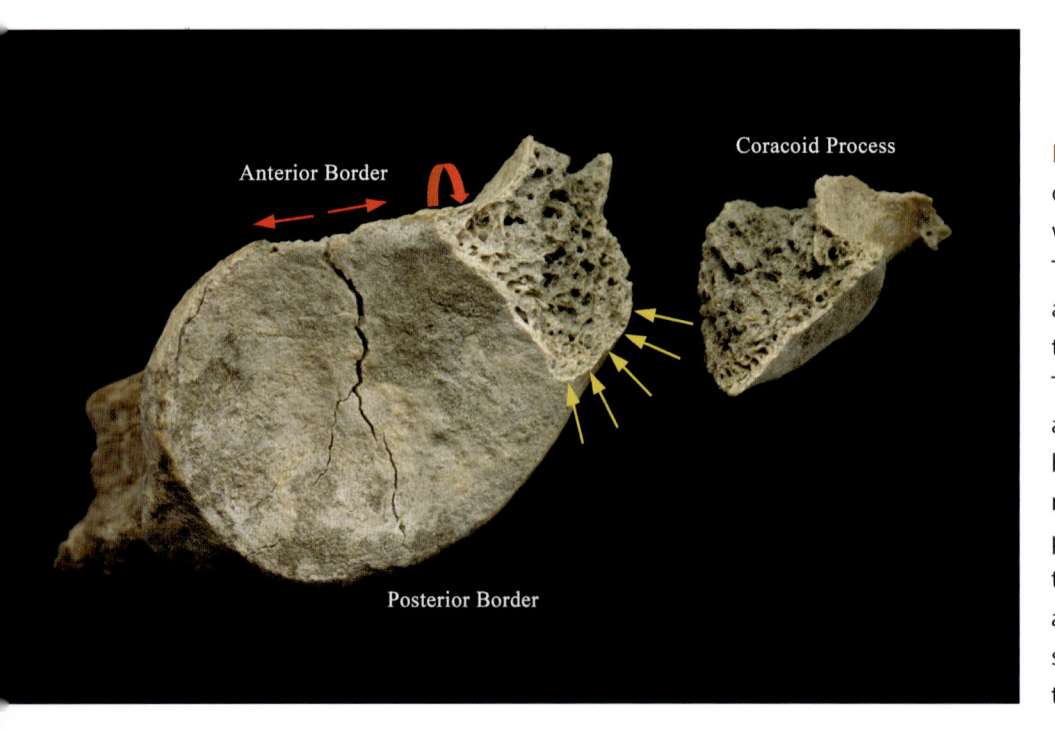

FIGURE 19.12 Glenoid cavity of Kennewick Man's left scapula with separated coracoid process. The straight red arrows indicate a fracture that opened under tension on the anterior border. The curved red arrow indicates an area of tension on the anterior border and the direction of movement of the coracoid process as it fractured free from the glenoid cavity. The yellow arrows point to the breakaway spur on the compression side of the bone.

Lateral view   Anterior view

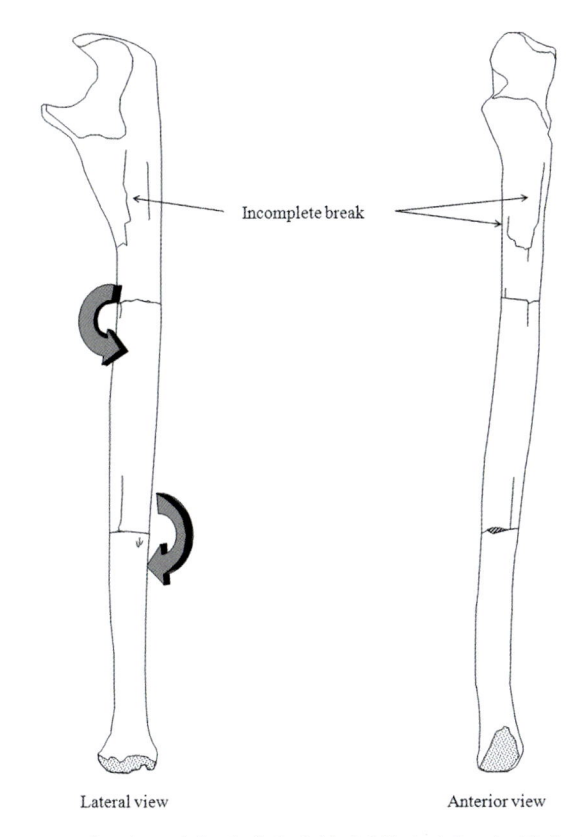

Curved arrows indicate the fixed end of the shaft (blunt end of arrow) and the direction of movement of the free end of the shaft (point of arrow).

FIGURE 19.13 Postmortem breaks to Kennewick Man's left ulna. Of primary importance is the presence of the incomplete break, which clearly illustrates that the distal end of the bone was being moved posteriorly, resulting in tension on the anterior surface.

Radius

The left radius exhibits two complete breaks and one incomplete break (Figure 19.15). The two complete breaks separated the shaft into three segments; unfortunately, the second segment—the midshaft component—was not recovered. The most proximal break was located approximately one-third of the way down the length of the shaft from the proximal end. This break opened first on the interosseous crest (the anteromedial side of the shaft) and propagated diagonally at a relatively steep angle in a distal to proximal direction. The break, when viewed in profile, displays a curved surface reminiscent of a fracture in fresh bone with a relatively indistinct breakaway spur. This break was produced with segment 97.L.14a (the proximal end of the bone) being fixed and segment two (the

← → Double arrows indicate the location of where the bone initially failed under tension.
▼▼ Block arrows indicate the direction of movement of the riverbank that produced the breaks.

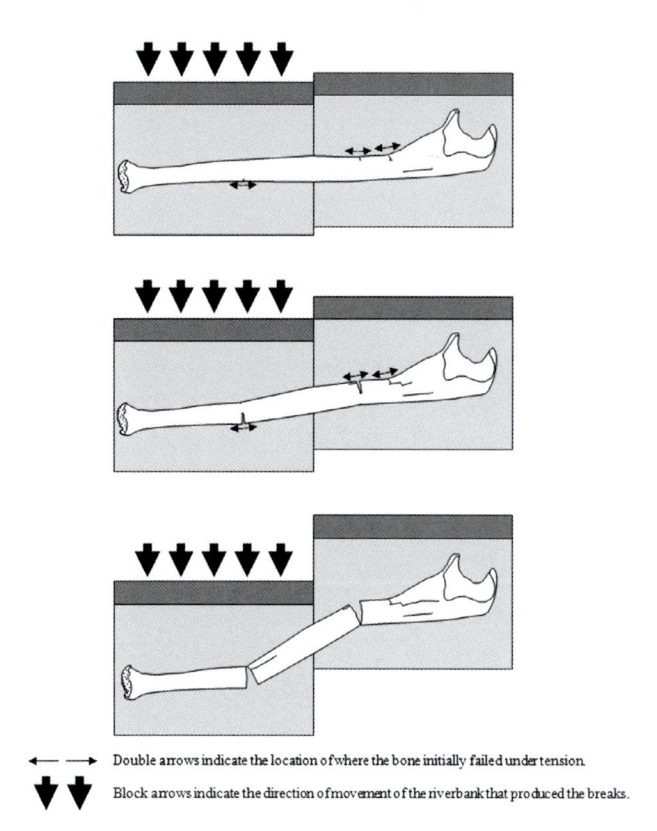

FIGURE 19.14 Possible mechanism for breaks to Kennewick Mans' left ulna. As the bank slumped vertically, displacing the distal end posteriorly, the bone failed under tension in two locations on the anterior surface (proximal end) and in one position on the posterior surface (distal end). This can be explained if the middle segment pivoted during the collapse of the bank.

missing segment) being displaced from anteromedial to posterolateral.

The more distal complete break is located near the midshaft. The characteristics of the second break and the mechanism that produced it are similar to the first with the break traveling at a sharp angle from distal to proximal and anteromedial to posterolateral. The break was produced by the second segment (the missing segment) being fixed and the third segment (97.L.14b), or distal end of the shaft, being displaced from anteromedial to posterolateral.

An incomplete break is located 22 millimeters distal to the second break (on segment 97.L.14b), as measured along the interosseous crest. This break initiated on the interosseous crest prior to or at the same time that the second break was forming. It propagated in a proximal

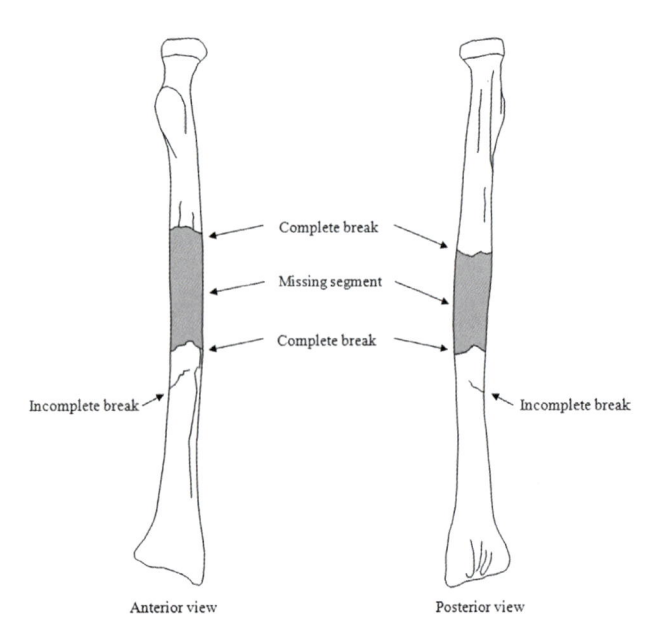

FIGURE 19.15 Breaks to Kennewick Man's left radius. Two complete breaks and one incomplete break are present. The missing segment of bone is shaded.

direction from the medial border, appearing on both the anterior and posterior surfaces of the shaft. Evidence of its preceding the second break is based upon the close proximity of the two breaks. Had the second break preexisted, the 22-millimeter segment would have been a part of the free (i.e., moving) distal end and would have been too short to become fixed to the extent that the incomplete break could initiate. The location and characteristics of this incomplete break confirm that the distal end of the bone experienced the initial displacement and was displaced in a medial to lateral direction while the mid and proximal shaft was fixed (Figure 19.15).

The fracture complex involving the left radius is remarkably uniform and consistent. Both complete breaks and the incomplete break indicate that the proximal end of the bone was fixed while the distal end was being displaced in a largely medial to lateral direction. These breaks are so similar that they were likely produced by the same mechanism—the soil as it slumped into the lake. With the proximal end fixed, the bank began to collapse, thereby increasing the tension along the interosseous crest (Figure 19.16). The incomplete break near the distal end probably formed first or at the same time as the second break near the midshaft, perhaps followed

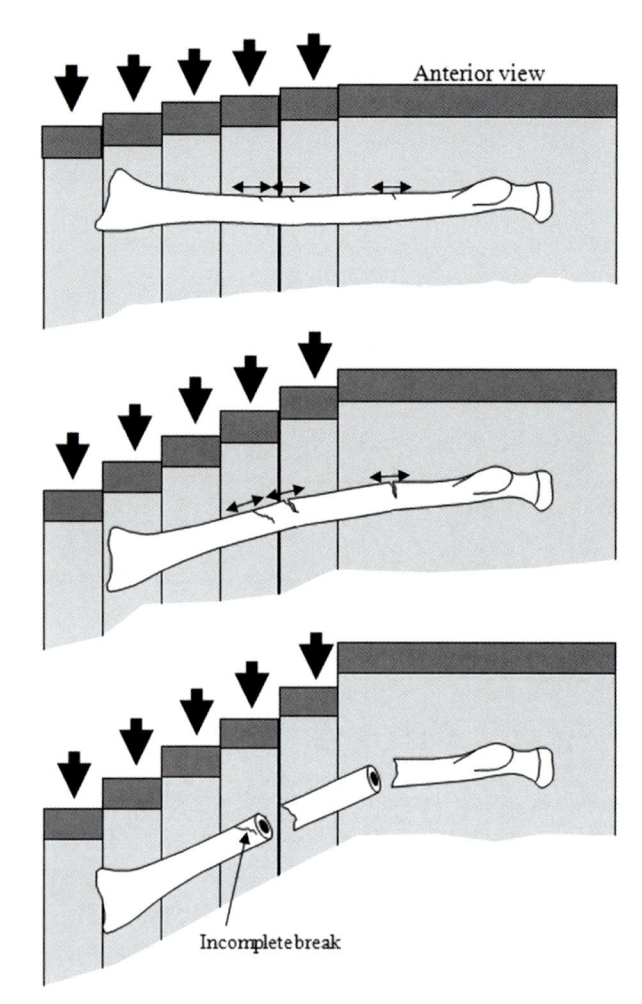

FIGURE 19.16 Mechanism for breaks to Kennewick Man's left radius, anterior view. The most distal incomplete break provides the clearest evidence of the medial to lateral direction of force responsible for the complex of fractures on this bone. With the body in the supine position, these fractures were produced with the left lower arm hyperpronated and the proximal end of the radius fixed as the bank collapsed vertically, forcing the distal end downward.

by the break near the proximal third of the shaft. The incomplete break encountered intrinsic resistance, allowing the midshaft break and ultimately the proximal break to complete.

Analysis

If Kennewick Man's body lay in a supine position, two possible scenarios may account for the fracture complex associated with the left radius: first, if his left forearm was

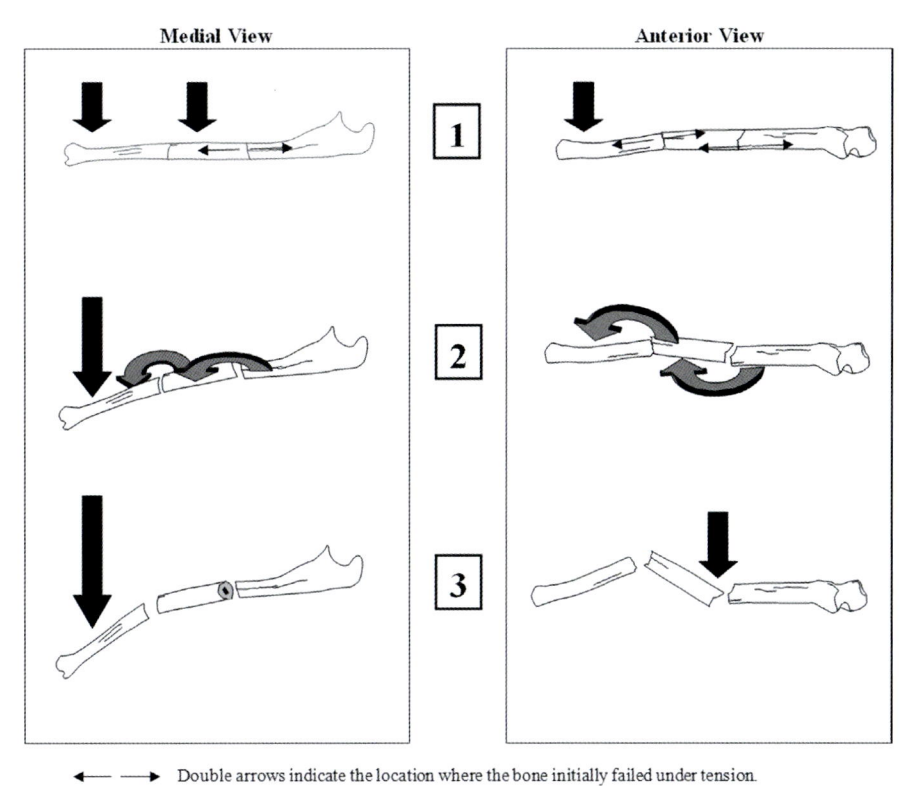

FIGURE 19.17 Breaks to Kennewick Man's right ulna. Numbers 1 through 3 represent vertical displacement (medial view) of the bone as it slumped into the river, and horizontal displacement (anterior view) as portions of the bank pulled away and fell into the river.

↔ Double arrows indicate the location where the bone initially failed under tension.

↓ Block arrows indicate the direction of movement of the riverbank that produced the breaks.

↻ Curved arrows indicate the fixed end of the shaft (blunt end of arrow) and the direction of movement of the free end of the shaft (point of arrow).

supinated, these breaks could have resulted as blocks of the bank pulled away horizontally and fell onto the lake shoreline; or second, if his left forearm was hyperpronated, the breaks could have formed as the bank slumped vertically. The incomplete break and most proximal complete break of the left ulna suggest that the proximal end of the ulna was fixed and the distal end was displaced posteriorly. This development would be inconsistent with the first scenario, but it fits the second. In archaeological settings, extended burials do not tend to be positioned with the forearms supine (in a typical anatomy textbook position) but are more commonly pronate, making the second scenario more likely.

RIGHT LOWER ARM BONE BREAKS
Ulna

The right ulna exhibits two complete breaks that divide the bone into three approximately equal segments (Figure 19.17). The most proximal break opened on the anteromedial surface with a breakaway spur scar (the spur having been fractured off and missing) located on and just posterior to the interosseous crest of the proximal segment. This indicates that the proximal break formed when the proximal end (97.R.15a) of the ulna was fixed and the midshaft segment (97.R.15b) moved from an anteromedial to a posterolateral direction (Figure 19.17).

The more distal break separates the distal end of the bone from the midshaft. It initiated on the lateral side and propagated proximally at a slight angle from an anterolateral to a posteromedial direction with a breakaway spur being located on the posteromedial side of the midshaft segment. The fracture formed with the midshaft (97.R.15b) fixed and the distal end (97.R.15c) being displaced from an anterolateral to posteromedial direction (Figure 19.17).

Radius

The right radius has one complete break to the distal third of the shaft (Figure 19.18). The fracture initiated on

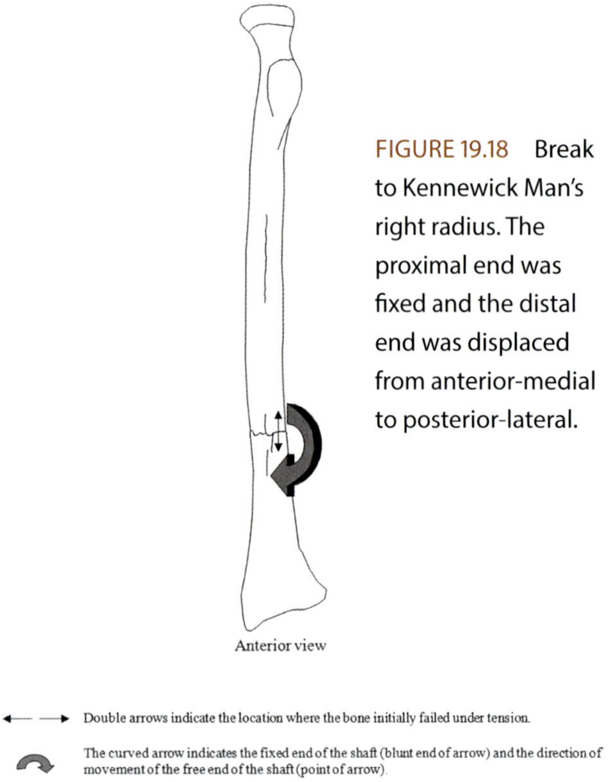

FIGURE 19.18 Break to Kennewick Man's right radius. The proximal end was fixed and the distal end was displaced from anterior-medial to posterior-lateral.

Anterior view

← → Double arrows indicate the location where the bone initially failed under tension.

↪ The curved arrow indicates the fixed end of the shaft (blunt end of arrow) and the direction of movement of the free end of the shaft (point of arrow).

the anteromedial surface and the proximal segment has a breakaway spur on the posterolateral side. The break was produced with the proximal end (97.R.14a) of the shaft fixed and the distal end (97.R.14b) displaced from an anteromedial to a posterolateral direction.

Analysis

Breaks to the distal ends of the right radius and ulna can be reconciled if produced by a single event, such as a vertical, right-to-left slump of the bank, with the right lower arm hyperpronated (Figure 19.19). In this position, the breaks could have formed if the proximal ends of radius and ulna were fixed while the distal ends were displaced in a downward and right-to-left direction.

LEFT AND RIGHT HUMERUS

The left humerus had three breaks that divide the bone into four approximately equal segments (Figure 19.20). The most proximal break is not presented in Table 19.2

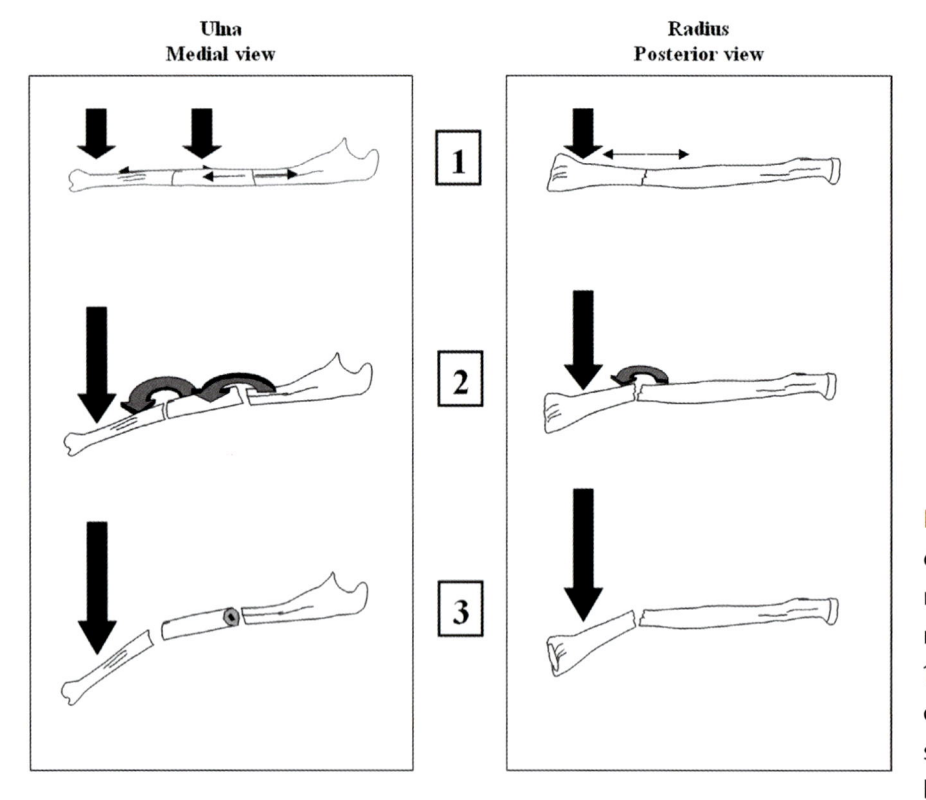

Ulna
Medial view

Radius
Posterior view

← → Double arrows indicate the location of where the bone initially failed under tension.

↓ Block arrows indicate the direction of movement of the riverbank that produced the breaks.

↪ Curved arrows indicate the fixed end of the shaft (blunt end of arrow) and the direction of movement of the free end of the shaft (point of arrow).

FIGURE 19.19 Fracture mechanics of the bones of Kennewick Man's right lower arm with the right radius hyperpronated. Numbers 1 through 3 represent vertical displacement in succession as seen from the river (assuming the head was oriented upstream). From this view and with the lower arm hyperpronate, the medial side of the right ulna and the posterior right radius would be exposed.

410 Chapter 19

but is represented by a crack that circumscribes the entire shaft of the humerus at a site approximately one-quarter down the shaft length from the proximal end (across the deltoid tuberosity). The distal end containing the inferior portion of the deltoid tuberosity was the fixed end of the bone. The free end of the bone was the proximal end with the humeral head (i.e., segment 97.L.13a contains the break). Location of the tension side of this fracture is difficult to interpret, and was initially thought to be on the anterolateral side. Further scrutiny suggests that the initial crack formed on the anteromedial surface when the head was displaced from an anteromedial to a posterolateral direction. The presence of numerous secondary fractures aligned with the long axis of the shaft that contain "bridging" flakes suggests that axial loading may have been a factor in the production of this break.

The second break is located at the midshaft and appears to have opened first on the lateral surface. This break formed with the proximal segment (97.L.13a) of the shaft fixed and the distal segment (97.L.13b) moving lateral to medial. The third and most distal break opened first on the medial surface. The distal segment (97.L.13c) was fixed and the segment (97.L.13b) proximal to the fracture moved laterally (Figure 19.20).

Assuming that the humerus was positioned with the posterior surface down, a simple vertical movement of earth (Figure 19.20, lateral view 1–3) cannot explain the fracture complex. Fracture dynamics indicate that both horizontal and vertical movements were likely involved. Perhaps the block of earth containing the humerus slumped into the lake by pulling away from the bank as it slid downward (Figure 19.20, anterior view 1–3). However, if the humerus had been rotated medially, the second and third fractures from the proximal end would be consistent with a vertical movement of earth.

The right humerus has a single complete break at the

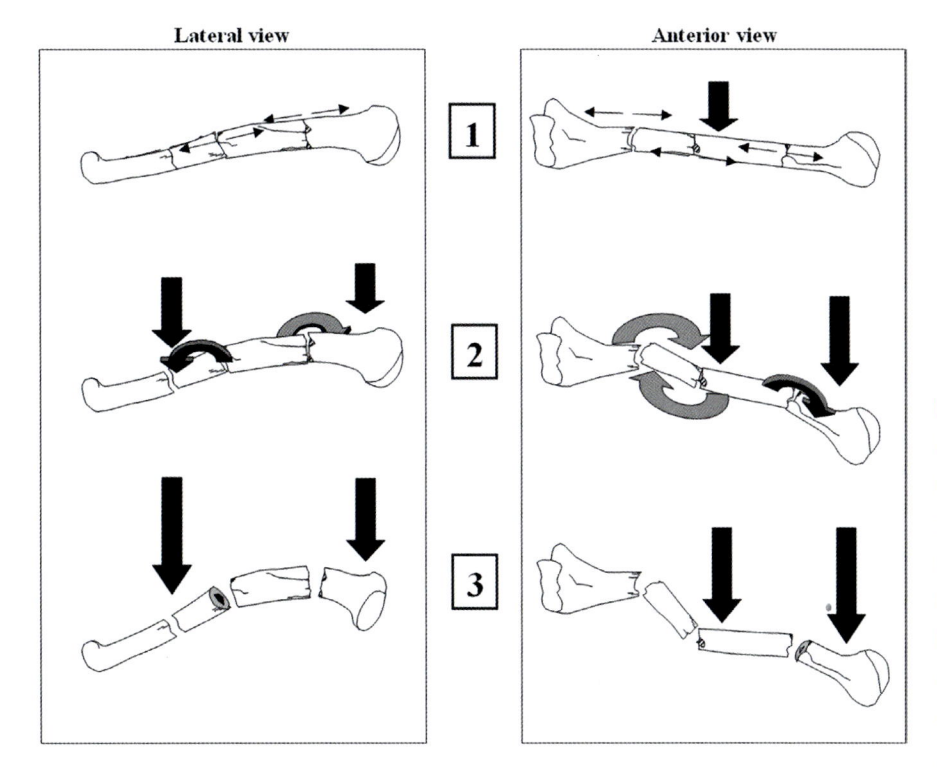

FIGURE 19.20 Postmortem breaks to Kennewick Man's left humerus. Numbers 1 through 3 represent a lateral view (as viewed river to bank with the body supine and oriented with head upstream, feet downstream) of the bone showing vertical displacement as the terrace slumped into the lake, and anterior view (as viewed vertically from the surface) showing horizontal displacement as portions of the bank pulled away and fell into the river.

Double arrows indicate the location where the bone initially failed under tension.
Block arrows indicate the direction of movement of the riverbank that produced the breaks.
Curved arrows indicate the fixed end of the shaft (blunt end of arrow) and the direction of movement of the free end of the shaft (point of arrow).

Hugh E. Berryman

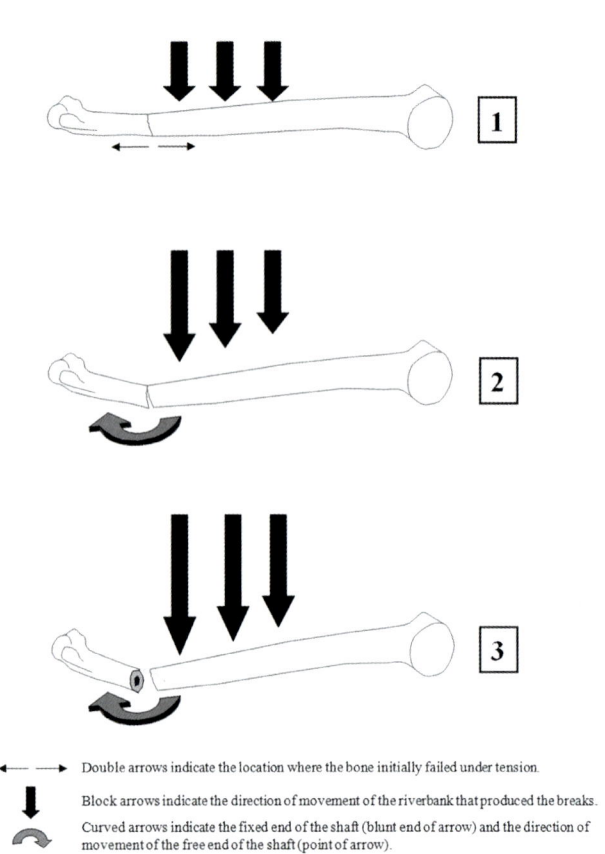

FIGURE 19.21 Break to Kennewick Man's right humerus, medial view. The proximal end of the shaft was fixed and the distal end was displaced in a posteromedial to anterolateral direction.

distal end of the shaft producing a proximal segment (97.R.13a) that comprises most of the shaft, and a much shorter distal segment (97.R.13b). A scar where a breakaway spur had been was present on the anterolateral aspect of the fracture face of the proximal segment. This indicates that the proximal end of the shaft was fixed and the distal end was displaced in a posteromedial to anterolateral direction (Figure 19.21). The forces necessary to produce this break are difficult to reconcile with the supine position proposed for Kennewick Man. If vertical slumping of the bank had produced this break, the bone should have been positioned with its anterolateral surface down. The color of the fractured surface is white, indicating minimal exposure to a staining environment and implying that the bone may have been broken closer to the time of recovery and in a separate event from the bank collapse.

CONCLUSION

Most of the breaks examined are consistent with Kennewick Man having been placed in an extended, supine position with his feet plantar flexed and the left foot laterally rotated, arms to his side, and lower arms hyperpronated. The location of breaks to the distal femora, glenoid cavity of the left scapula, and left first metatarsal support this conclusion. The downward orientation of the posterior surfaces of both femora, both tibiae, and the left scapula is impossible to accomplish in the typical flexed burial, but feasible for an extended burial that was supine. The breaks in the left first metatarsal are consistent with the foot being plantar flexed and laterally rotated, a position also consistent with an extended supine burial.

An extended, supine position with the lower arms hyperpronated and the body aligned, more or less, parallel to the river is evidenced by taphonomic indicators other than postmortem bone breakage. Although body position relative to the river's shoreline is difficult to prove with breaks alone, the number and location of secondary fractures may provide collaborative support. Most of the secondary fractures are probably a result of drying, and tend to occur with greater frequency on the left side of the body than the right. This may suggest that the left side was subjected to a slightly longer weathering period than the right due to earlier exposure. With the primary breaks, anatomic location of the tension side of the breaks along with the fixed and free ends of the bones suggest that the bank collapse began below Kennewick Man's knees, progressing distally and proximally. The bones of the arms were difficult to interpret. If the arms were extended as suggested by the left ulna (the incomplete break on the anterior surface indicates that the posterior surface was down), the lower arms would have been hyperpronated as evidenced by the left radius (the incomplete break on the medial surface indicates that the lateral surface was down). The right ulna and radius show the same general positioning, although they lack the incomplete breaks that would provide irrefutable evidence of the position of the tension side of the bone. The right humerus does not fit this scenario and may have been broken by other taphonomic forces after the alluvial terrace collapsed. Figure 19.22 provides an illustrative summary of bone breakage by the progressive erosion of the terrace bank as it calved onto the shoreline.

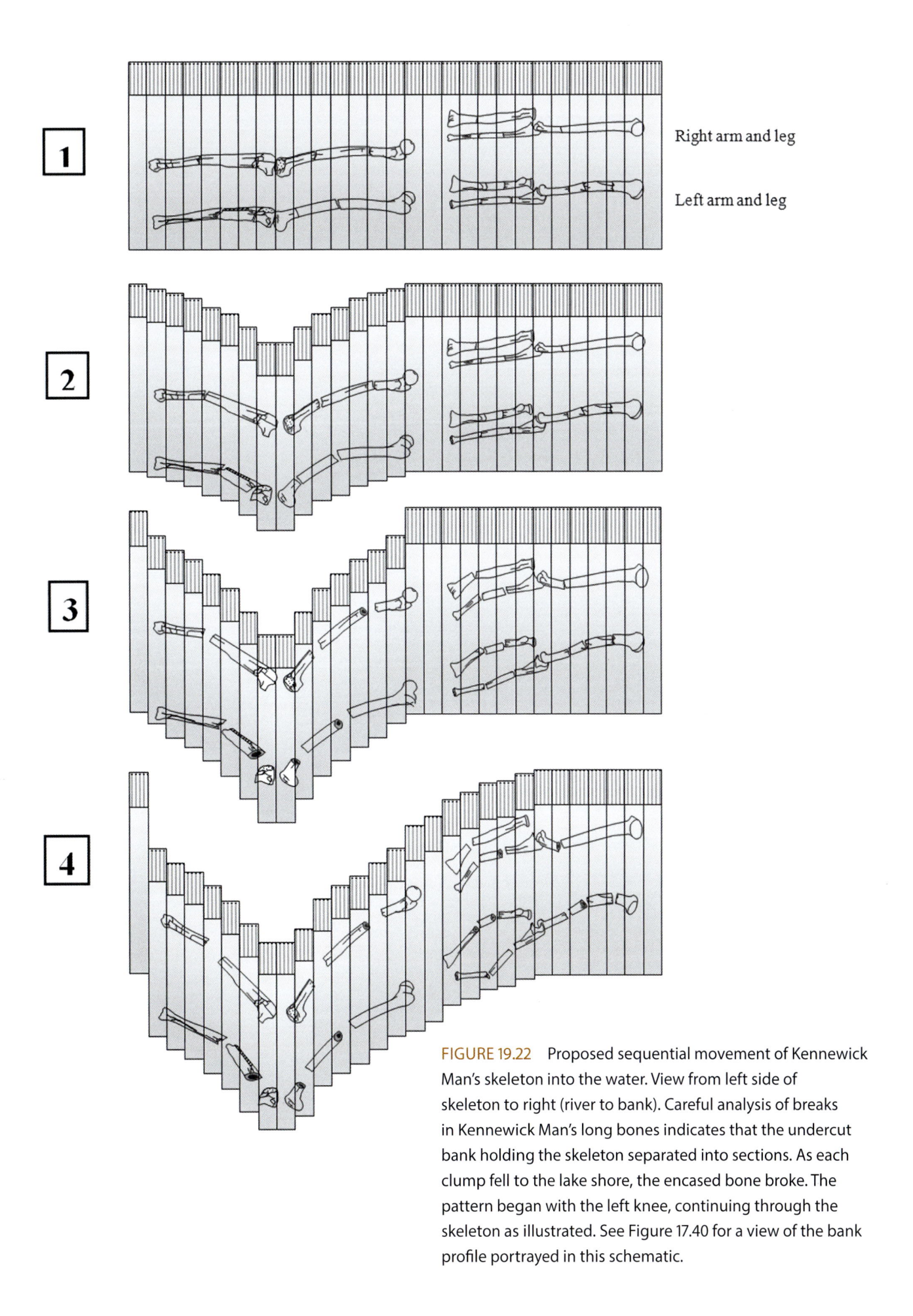

|1|

Right arm and leg

Left arm and leg

|2|

|3|

|4|

FIGURE 19.22 Proposed sequential movement of Kennewick Man's skeleton into the water. View from left side of skeleton to right (river to bank). Careful analysis of breaks in Kennewick Man's long bones indicates that the undercut bank holding the skeleton separated into sections. As each clump fell to the lake shore, the encased bone broke. The pattern began with the left knee, continuing through the skeleton as illustrated. See Figure 17.40 for a view of the bank profile portrayed in this schematic.

REFERENCES

Berryman, Hugh E., Harry H. Mincer, and S. J. Moore. 1991. "Forensic characteristics of antemortem, perimortem and postmortem dental fractures." Presented at the 43rd annual meeting of the American Academy of Forensic Sciences, Anaheim, CA.

Berryman, Hugh E., Harry Mincer, S. A. Symes, O. C. Smith, and S. J. Moore. 1992. "The length of the perimortem interval in dental fracture." Presented at the 44th annual meeting of the American Academy of Forensic Sciences, New Orleans, LA.

Berryman, Hugh E., and Steven A. Symes. 1999. "Differentiating perimortem from postmortem fractures in bone." Personal Identification in Mass Disasters Conference, 15–19 November, Honolulu, HI.

Hughes, Cris E., and Crystal A. White. 2009. Crack propagation in teeth: A comparison of perimortem and postmortem behavior of dental materials and cracks. *Journal of Forensic Sciences* 54(2):263–269.

Ubelaker, Douglas H., and Bradley J. Adams. 1995. Differentiation of perimortem and postmortem trauma using taphonomic indicators. *Journal of Forensic Sciences* 40(3):509–512.

Wheatley, Bruce P. 2008. Perimortem or postmortem bone fractures? An experimental study of fracture patterns in deer femora. *Journal of Forensic Sciences* 53(1):69–72.

Wieberg, Danielle A. M., and Daniel J. Wescott. 2008. Estimating the timing of long bone fractures: Correlation between the postmortem interval, bone moisture content, and blunt force trauma fracture characteristics. *Journal of Forensic Sciences* 53(5):1028–1034.

APPLYING TECHNOLOGY TO INTERPRETATION

Technological advancements provide an ever-expanding platform for skeletal studies. Over the past two decades, gross visual examinations of bones and human remains have benefited from the use of computed tomography, modeling, and rapid prototyping software. These systems capture both the internal and external morphology of skeletal elements, allowing for long-term access and even replication through additive manufacturing. These chapters detail the innovative technologies that supplemented the Kennewick Man study. In 2005, elements of the skeleton were scanned by an industrial computed tomography system, instrumentation that represented cutting-edge technology for the time. These scans were then used to create stereolithographic reproductions of select bones, as well as the projectile point embedded in Kennewick Man's hip. Scanning, virtual modeling, and 3D printing have enhanced various aspects of the research and offer an alternative method of visualization, and thus, interpretation. Replicas do not, however, provide an exact replacement for the real objects and should be used with appropriate caution. These rapidly developing fields will undoubtedly be critical to answering future questions about Kennewick Man.

Computed Tomography, Visualization, and 3D Modeling

20

Rebecca Snyder

This chapter presents technical information on the computed tomography (CT) scanning and three-dimensional (3D) visualization work performed on the remains of Kennewick Man so that future researchers can place both the raw and processed data and the digital reconstructions within their proper context and perform accurate analyses. For readers not fully versed in how CT and voxel-based data work, this chapter also includes: detailed explanations of how the chosen scanner collected and processed the data, how the data are structured, and resources for finding further information.

Computed tomography, often referred to as CAT scan or CT, is used to visualize both surface and internal details of a person or object in a noninvasive manner through the use of X-rays. Among many other applications, medical professionals use CT to see internal structures of patients and to create patient-specific surgical models and prosthetics. Industrial CT is based on the same principles but uses stronger X-ray sources for longer durations to yield more detailed visualizations. Though originally designed for industrial applications like materials strength testing and verifying the integrity of toxic waste containers, industrial CT machines are used by geologists, sedimentologists, physical anthropologists, and forensic scientists. The increased strength of the X-ray beam not only provides greater penetration of dense objects like sediments, bones, and teeth but also can yield finer details. CT has been a popular tool to capture, nondestructively, volumetric data that can be viewed, studied, and analyzed in ways not possible with the original.

In the case of the Kennewick Man, CT provided opportunities for enriched study, including the chance to view and digitally extract the stone projectile point embedded within part of the right ilium, the provision of a clear view of the soil deposition pattern inside the cranium, and the generation of physical 3D models that, unlike the original remains, could be used to reconstruct the cranium and right innominate.

Although a powerful research tool, CT data have some limitations. The detector, picking up the X-rays as they pass through a specimen, records changes in density and represents them on the computer screen in grayscale. All internal and external color information is lost. For example: surface paint or tattoos would appear in the same grayscale value as the surrounding material and therefore be indistinguishable in CT data, unless there is a density difference between two pigmentation sources. Biological and inorganic material of the same density will be indistinguishable from each other in CT data, so dense bone and teeth can appear similar to mineral and rock materials since their density values fall within the same range. The more powerful the X-ray source and the more sensitive the detector, the greater the graduation between density values can be made and small differences can be more easily detected.

Before detailing the parameters of the Kennewick Man CT data and the extraction process, it is important to understand how these data are generated and structured. The detector is comprised of a grid of photodiodes that independently record the strength of the X-ray beam as it strikes the surface of the detector. Each of these photodiodes represents a single pixel (Figure 20.1, step 1). Detectors used in CT are typically comprised of a grid of either 512x512 or 1024x1024 photodiodes. The physical dimensions these pixels represent depend on the selected field of view (FOV). Smaller FOVs provide more detailed scans, though the FOV is constrained on the lower end by the overall strength of the X-ray beam and the scanner's ability to focus the beam across small areas. The software intercepting the signals from the detector uses the selected scan thickness setting to apply a third dimension to these pixels, providing depth. With the

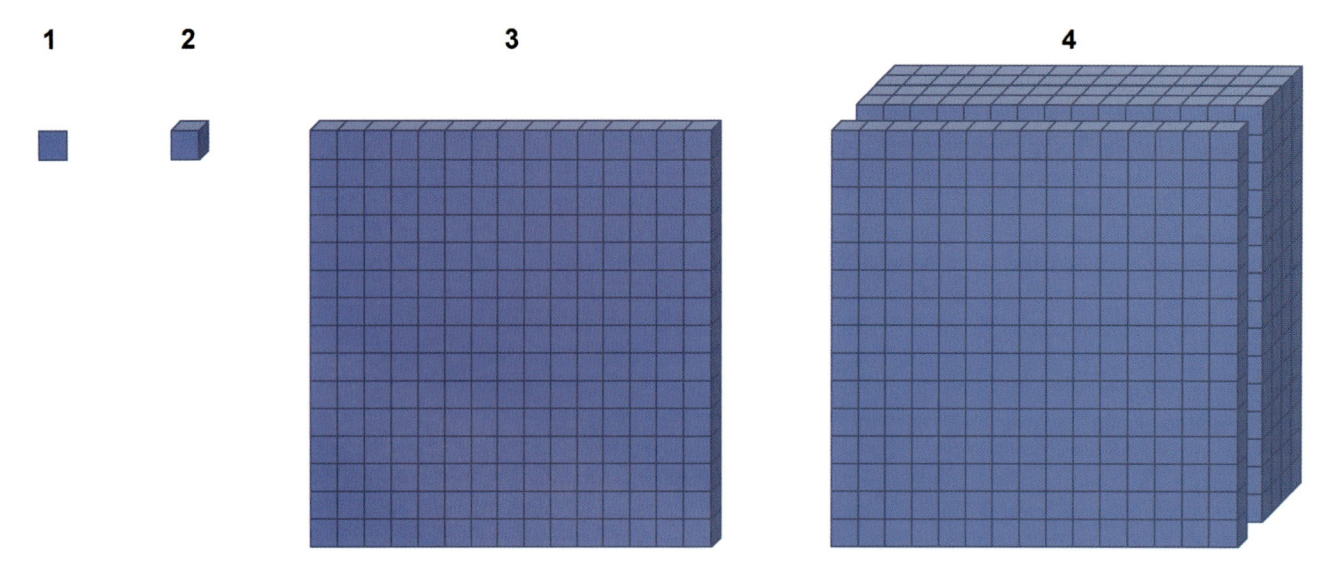

FIGURE 20.1 Voxel Primer. 1, single pixel; 2, pixel with depth applied resulting in a voxel; 3, all voxels from detector assembled into a single slice; 4, slices combined to reconstruct original scan volume.

thickness data applied, the 2D pixels become 3D voxels (Figure 20.1, step 2). The individual voxels are assembled into a single "slice" representing the entire surface of the detector (Figure 20.1, step 3). CT visualization programs can take successive slices of data, stack them back to back like a virtual deck of cards, and recreate the original scan volume (Figure 20.1, step 4).

In 1999, the US Army Corps of Engineers (Corps) had the partly reconstructed cranium and the fragment of the right ilium with the embedded point from the Kennewick remains scanned at the University of Washington Medical Center, Seattle. The GE® Genesis Hispeed RP medical CT machine generated slices in an alternating pattern of 1-millimeter and 5-millimeter slices. Although these scans provided a full view of the embedded stone projectile point, the data generated were not sufficiently detailed for comprehensive investigation into the point's origin and flint-knapping technique.

Digital reconstructions using the 1999 data can be identified by the overall coarseness of the mesh as well as by the banding, seen as stripes, which is caused by the visualization software performing the necessary extrapolations of data between slices that are far apart (Figure 20.2). The 1999 raw data are in DICOM format and consequently have additional information embedded in

FIGURE 20.2 Stone projectile point generated from the 1999 CT data from the Corps. The lithic fragment is shown in situ as a shaded polygonal mesh with the partial ilium viewed as a point cloud. Models generated from these data show the banding effect caused by the variable slice thickness. The ilium is not in anatomical position.

the header including scan date, scanner settings, and slice thickness (Figure 20.3).

Dr. Douglas Owsley received permission to scan, and in some cases rescan, 21 skull and right innominate fragments of the Kennewick Man's skeleton. In June 2005, Owsley's team met a Corps representative and a conservator for Kennewick Man at Bio-Imaging Research, Inc. (BIR) in Lincolnshire, Illinois, to perform the CT scanning. As detailed in the Corps-approved industrial CT Imaging proposal, Owsley and his team had intended several uses for the CT data, including but not limited to: digitally removing the glue used to piece together the cranium, separating the individual cranial fragments, and reconstructing them without damage or alteration to the original remains; digitally extracting the projectile point embedded in the ilium; visualizing the interior of the cranium to more readily examine the sediment deposition patterns; and creating a set of replicas to work from when the original remains are not accessible.

Digital reconstructions using the 2005 data provided much more detail of both internal and external features of the bone when viewing individual slices, without noticeable procedure-derived visual artifacts in 3D surface reconstructions. The 2005 data are available in the raw BIR proprietary format that can be read by some CT visualization software packages, TIFF format (which, in actuality, is an extensionless variant from the TIFF 6.0 standard using Private Tags), and Bitmap format, though detail is lost in the conversion to this format (16-bit to 8-bit). A limited amount of data are additionally available in DICOM and JPEG2000 formats generated by BIR at the time of scanning but do not represent the entire dataset and do not contain the standard embedded headers. The raw BIR, DICOM, and JPEG2000 formatted data are unprocessed detector dumps and thus represent the cylindrical data spiral. Unless the CT visualization program used by subsequent researchers can interpret BIR data, the researcher will be unable to convert the data to standard voxel-based slices. Since the TIFF data are not in a fully TIFF 6.0 compliant format, most image programs (e.g., Photoshop®) will not be able to open the files, nor will most TIFF Tag readers be able to view the embedded metadata (Private Tags are not required to be registered to the standards body). While CT data visualization programs can access this information, researchers can also use standard binary/hex readers to read the embedded information (Figure 20.4). The 3D work described later in this chapter used the TIFF data to generate the reconstructions. These data are unsigned 16-bit proprietary BIR TIFF format files with little-endian byte order (magic numbers: 4949 2A00). In the time since the scans were performed in 2005, BIR has been acquired by Varian Medical Systems, Inc., under their Security and Inspection unit. If future researchers desire to reprocess the raw BIR formatted data and are unable to identify a CT data visualization program that can both view and convert the data to voxel based slices, inquiries will need to be made to Varian to determine if their software offerings are compatible.

An industrial micro CT machine with ultra-high resolution was selected to perform the scans. The configuration of the CT machine results in cylindrical scan volumes so each scan set-up was created on a circular base. The scanning platform on which the base sits rotates in the path of the X-ray beam, so more accurate data will result when the overall density the beam encounters is

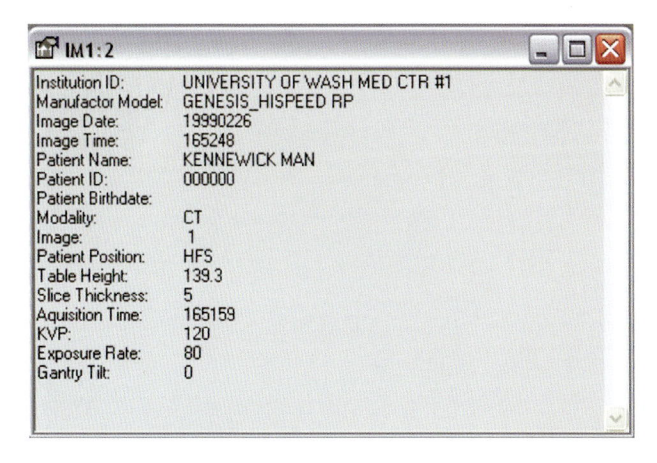

FIGURE 20.3 DICOM file header data from the 1999 CT scan of Kennewick Man.

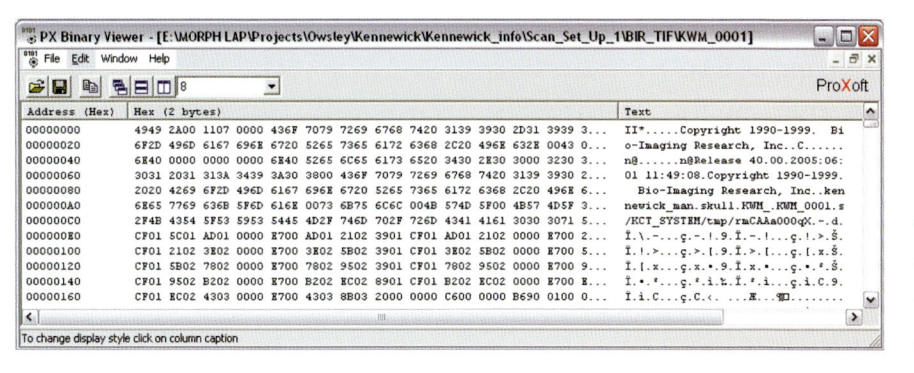

FIGURE 20.4 Binary/Hex view of BIR TIFF data. Note the magic numbers indicating TIFF format, and the embedded tag information in the right column.

Rebecca Snyder

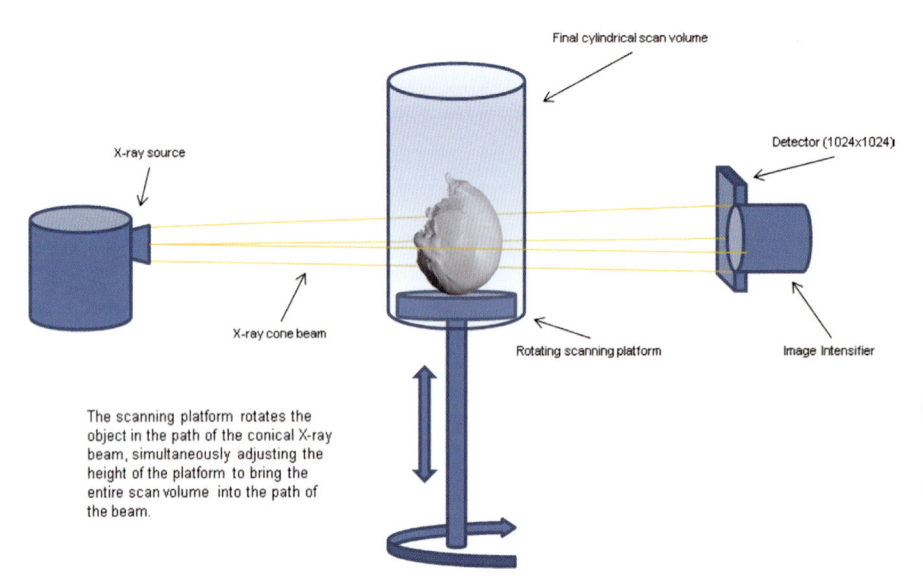

FIGURE 20.5 Schematic of how X-rays are projected through materials, intercepted by the detector and interpreted as a final volumetric reconstruction.

balanced. If balanced, the beam will be of equal strength as it passes through the remains from all angles. This is important since the more energy and focus the beam has when it hits the detector, the more detail and sharpness the resulting data achieves. If the path is too dense, too much of the beam will be absorbed or deflected by the objects and will result in a weak and/or diffused beam when it hits the detector. Conceptually, it is like balancing a tire on a rim (Figure 20.5).

The 21 individual fragments to be scanned were divided into two groups based on available scanning space. In order to protect the bones, they were arranged on a foam disk modified to accommodate each piece (Figure 20.6). Foam does not interfere with the scanning process because it has a density similar to air; digitally removing it is a simple process. Given the sensitive nature of the remains, they were returned to their specialized storage containers at the end of each scanning day. The first set-up included the cranium and 11 smaller fragments. The second set up contained the other nine fragments including the right innominate with embedded stone projectile point, detached maxillae, and mandibular elements. In the case of the second set-up, the first attempts to scan encountered problems, as detailed in the Error Log (Table 20.2), and a second day of scanning was needed, resulting in a slightly altered set-up. The aborted attempts for the second set-up are referred to in the documentation as Set-up 2a, and the final version labeled Set-up 2b (Table 20.1). Photographs were taken

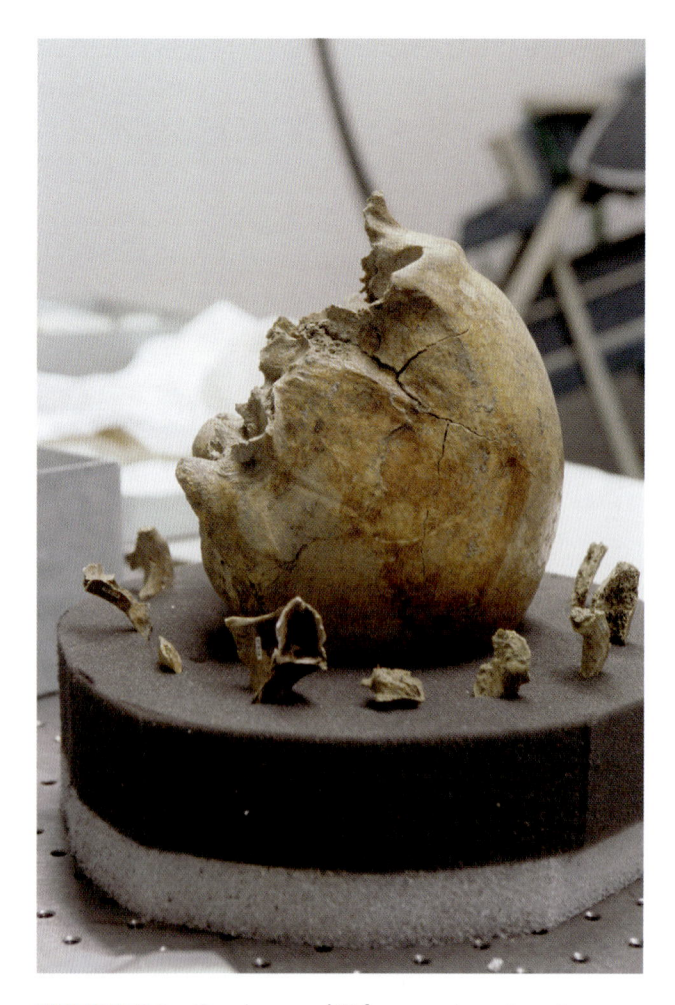

FIGURE 20.6 Cranium and 11 fragments, set-up 1, on foam base being prepared for scanning.

420 Chapter 20

TABLE 20.1 Scan set-ups.

Scan Set Up 1		Scan Set Up 2 (a and b)	
97I17B	Right ischium fragment	97R17C 'large'	Right ilium fragment with embedded point
97I17C	Os coxa fragment	97R75A	Mandibular right third molar
97I25V	Left maxilla fragment	97U3C	Right mandibular ramus
97R1A	Right malar	97U2A	Maxillae
97U1B	Cranial fragment	97R17B	Right ilium fragment
97R2B	Right maxillary fragment	97U3B	Right mandibular fragment
97U1C	Left cranial fragment	97R50A	Maxillary left third molar
97R17C 'small'	Right ilium fragment	97R17D	Right acetabulum
97R17A	Right pubis	97U3A	Left mandibular fragment
97I10B	Right iliac crest fragment		
97I17D	Iliac crest fragment		
97U1A	Cranial vault		

TABLE 20.2 Description of procedure.

	Location		
Bio-Imaging Research, Inc. (BIR), Lincolnshire, IL			
	Participants in 2005 Industrial CT Scanning		
Douglas Owsley	National Museum of Natural History	Forensic Anthropologist, Plaintiff	
Chip Clark	National Museum of Natural History	Photographer	
Rebecca Snyder	National Museum of Natural History	Technical Consultant	
Chris Pulliam	Corps of Engineers	Archaeologist, Corps Representative	
Vicki Cassman	University of Las Vegas	Corps Contract Conservator	
Art Andersen	Virtual Surfaces, Inc.	Contractor	
Barry Smith	Bio-Imaging Research, Inc.	Marketing Manager	
Chuck Smith	Bio-Imaging Research, Inc.	Vice President of Engineering	
Forrest Martin	Bio-Imaging Research, Inc.	Scanning technician and programmer	
	Equipment		
Scanner	MicroCT UHR/MCT Fast Scan		
Source	FienFocus 225kV		
Detector	Varian PaxScan 2520 flat panel		
	Software		
CT Interface and Data Processing Software	ACTIS r40.00	UNIX platform capture and analysis program designed by BIR for their scanners	
	Scanner Settings		
kV	170		
mA	100		
voxel	1024x1024x16		
scan mode	"3rd Gen, Volume Detector"	This value will be seen in the file headers of the BIR raw and BIR TIFF format data	
protocol	ken_testCRS	This value will be seen in the file headers of the BIR raw and BIR TIFF format data	
FOV	240.00mm	BIR uses the term 'FOR' for this value in their documentation	

(continued)

TABLE 20.2 (continued) Description of procedure.

		UNIX File Structure
File Cabinet	kennewick_man	
Folder	skull	
	assortment	

		Study
This creates the file name structure, i.e., each image is named KWM_0001, KWM_0002 . . . etc.		
6/1/2005	KWM_	First scan of Set Up 1, completed but banding issues
6/1/2005	KWM1_	First scan attempt of Set Up 2, aborted due to constrained FOV
6/2/2005	KWM2_	Second scan attempt of Set Up 2, completed but banding issues
6/2/2005	KWM3_	Scan of clear plastic cone for value comparison
6/2/2005	KWM4_	Test scan of Set Up 2, not full scan
6/2/2005	KWM5_	Third scan attempt of Set Up 2, aborted due to offset rotate UNIX memory issue
6/2/2005	KWM6_	Fourth scan attempt of Set Up 2, aborted due to 'softness'. System calibrations performed, it was off by approximately '1 ray'
6/2/2005	KWM7_	Fifth scan attempt of Set Up 2, completed
6/2/2005	KWM8_	Second scan attempt of Set Up 1

		Error Log
Times are approximate, and EST.		
6/1/2005	12:15PM	Scan aborted due to network communication error.
6/1/2005	12:40PM	Offset rotation UNIX memory issue at boundary between slices 0240/0241.
6/1/2005	1:45PM	Offset rotation UNIX memory issue at boundary between slices 0345/0346. Major visual artifacts at the boundary of slices 0345/0346. According to BIR staff it was due to a corrupted RAW file caused by a known UNIX/ACTIS memory issue. Scanning aborted. Problem resolved at restart of x-ray ("self-correcting").
6/2/2005	2:15PM	Program error. "System Status: Caught Signal 11" (Study KWM7_).
6/2/2005	2:50PM	X-ray source shut down (Study KWM7_).
6/2/2005	3:00PM	Offset rotation UNIX memory issue at boundary between slices 0315/0316 (Study KWM7_).
6/2/2005	4:15PM	Offset rotation UNIX memory issue at boundary between slices 0180/0181 (Study KWM8_).
6/2/2005	4:25PM	Offset rotation UNIX memory issue at boundary between slices 0210/0211 (Study KWM8_).
6/2/2005	5:15PM	Offset rotation UNIX memory issue at boundary between slices 0435/0436 (Study KWM8_).
6/2/2005	5:30PM	Offset rotation UNIX memory issue at boundary between slices 0510/0511 (Study KWM8_).

by National Museum of Natural History photographer Chip Clark and others present on all scanning days, so published photographs of the second set up may have come from either arrangement (see Figures 20.7–20.9 for set-up diagrams).

The machine used was a BIR Custom MicroCT UHR/MCT Fast Scanner with a Varian PaxScan 2520 medical flat bed detector (Figure 20.10). Data were processed by BIR-designed ACTIS r40.00, a UNIX platform capture and analysis program.

All scans were done at 170kV, 100 mA, with field of view set to 240.00 millimeter in custom software "scan mode" setting: "3rd Gen, Volume Detector." The scan technician created a custom protocol, saved as "ken_test-CRS." The detector captured the data as 1024x1024 pixels with 16 bit depth. See Table 20.2 for more complete technical details. The raw data were first reconstructed in ACTIS as data spiral volume reflecting the capture method (Figure 20.11). The volume data were then processed and divided into individual slices reflecting the

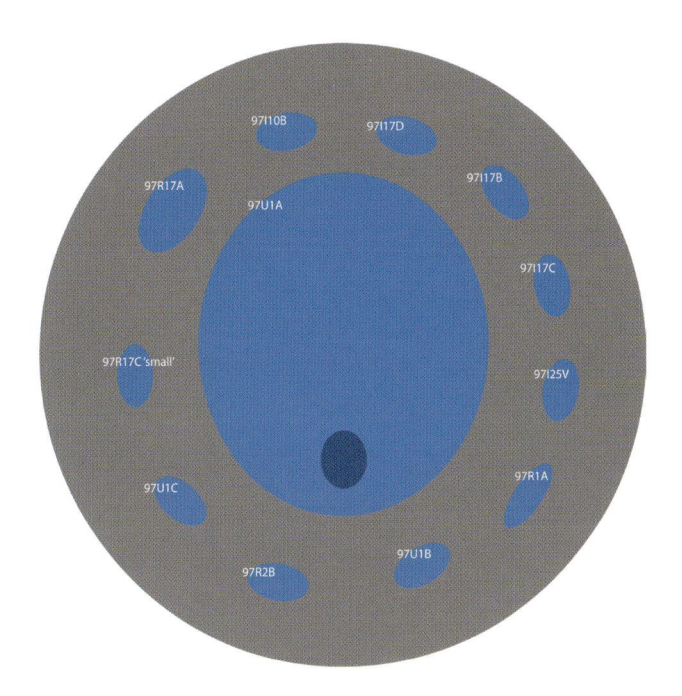

FIGURE 20.7 Scan Set-up 1.

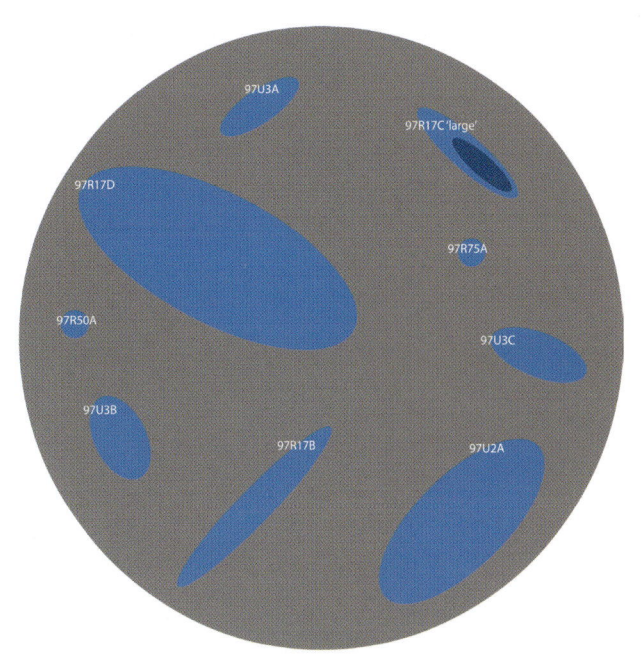

FIGURE 20.8 Scan Set-up 2a. Not used.

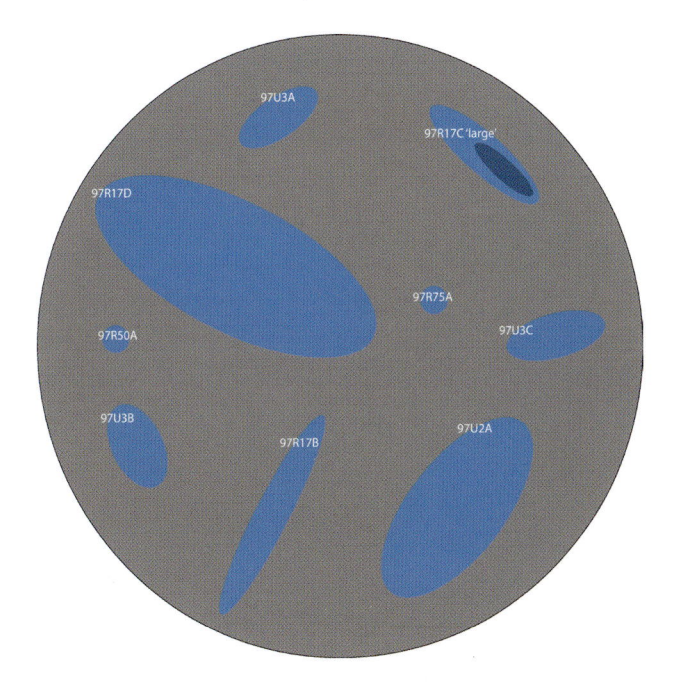

FIGURE 20.9 Scan Set-up 2b.

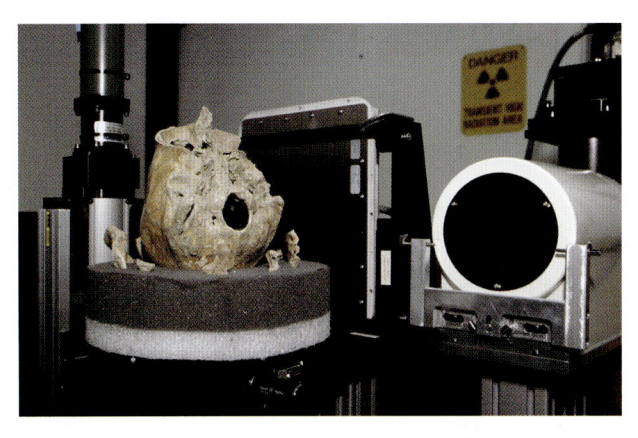

FIGURE 20.10 BIR MicroCT/UHR industrial scanner with Kennewick Man's remains.

calibrated voxel settings. For the Kennewick Man scans this resulted in the following voxel dimensions:

X: 0.234 mm/pix
Y: 0.234 mm/pix
Z: 0.39 mm thick

These dimensions are calculated in the following manner:

X and Y axes:
FOV = 240.0 mm, represented by 1024 pixels, which means each pixel represents 0.234375 mm, rounded to 0.234 mm.

Z Axis:
According to the documentation, the scanner was operated in line mode to achieve the slice thickness of 0.39 mm. 1 line = 0.195 mm with scan form set to 2 lines (2 × 0.195 = 0.39).

Rebecca Snyder

These axes are relative to the scanner and reflect the position of the remains at the time of scanning. When the raw data are processed into voxel data the axes are automatically assigned by the visualization software program. These assignments typically assume a standard reclining patient orientation and cannot be changed even though the labels may not match the anatomical position of the specimen scanned. In the case of Kennewick Man, balancing density and maintaining the stability and security of the remains were deemed more vital than aligning to arbitrary axis labels (Figures 20.12 and 20.13).

Using the converted slice data, a preliminary digital extraction of the embedded stone projectile point was generated by Art Anderson in Mimics using automated segmentation to distinguish between the stone of the projectile point and the surrounding bone. The resulting isolated stone projectile point data were then converted into a hollow wireframe model representing just the surface (Figure 20.14). This wireframe was exported from Mimics in a stereolithography compatible format, in this case the commonly used *.stl format. When the resulting 3D stereolithography model was sent to the National Museum of Natural History for approval, it was determined that finer detail would be more useful for possible identification of the point's production technique (Figure 20.15). The version of the Mimics software used was designed to process data from medical scanners, so its limited range of Hounsfield Units (HU), interpreted and displayed as grayscale, made sensitive determinations between objects of similar density more difficult. The author used VG Studio MAX, which has a greater HU range, to create a second model. After the innominate was isolated from the rest

FIGURE 20.11 Raw data from MicroCT/UHR. This data spiral is then processed within ARCTOS and converted into standard CT slice data.

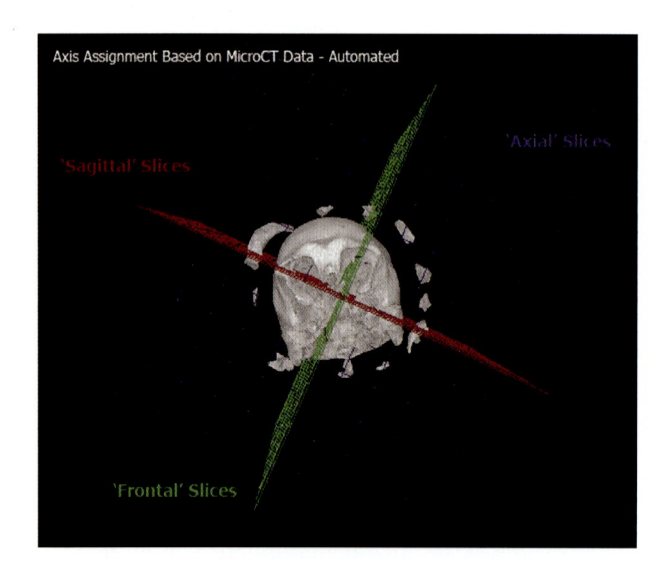

FIGURE 20.12 Automated axis assignment based on the 2005 CT data. These orientation assignments cannot be changed. As a consequence, transverse sections are labeled in the CT data visualization software as axial slices.

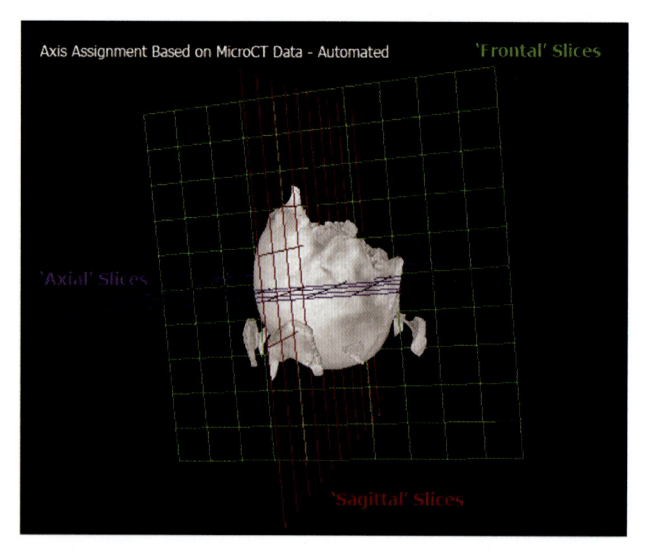

FIGURE 20.13 Automated axis assignment based on the 2005 CT data. These orientation assignments cannot be changed. As a consequence, transverse sections are labeled in the CT data visualization software as axial slices.

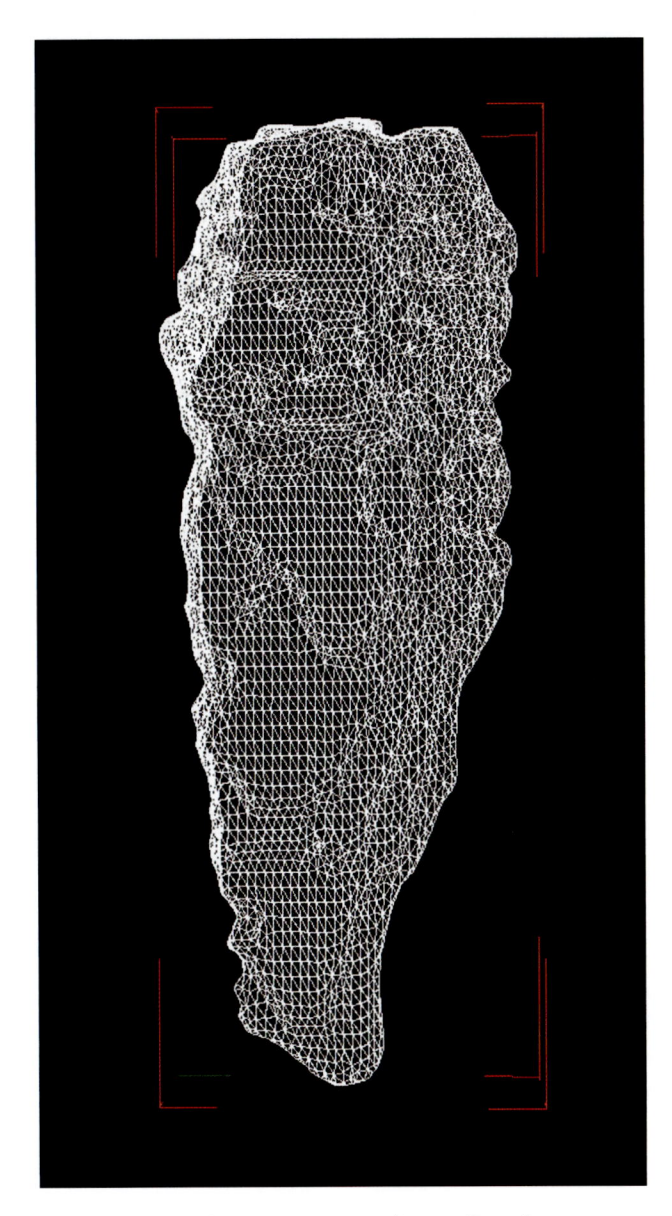

FIGURE 20.14 Stone point as polygonal mesh representing only surface data.

FIGURE 20.15 Initial digital extraction of the embedded stone projectile point.

of the study, the program's "classification" tool was used to restrict the range of data selected, leaving only bone, stone, and silt. Viewing the stone projectile point in transverse slices of the illium (Figure 20.16), an initial selection of the stone projectile point was created from a single slice and the program's segmentation propagation tool was used to generate the first base attempt at isolating the stone projectile point data from the surrounding bone and silt. Given the similarity between the density of the stone projectile point and the calcium carbonate of the silt in and around the wound, using the automated tools exclusively would not be sufficient to obtain a model unobstructed by silt and sand. Using the propagated selection as a baseline, the isolated data were refined by visually removing or adding voxels to the selection, one slice at a time. The resulting model was likewise converted into a wireframe model, exported in *.stl format, and a stereolithography 3D model was generated by Designcraft of Lake Zurich, Illinois. This 3D model was approved by Owsley and Dennis Stanford, also of National Museum of Natural History (Figure 20.17).

Stereolithography is a type of three-dimensional printing. A polygonal mesh surface model, imported into the printer software, is subsequently optimized and prepared for printing. Often this means ensuring there are no holes along the surface of the model (often referred to as being "watertight"). Small errors are corrected and the printing stages are calculated. Stereolithography divides the digital model into a series of slices, each one

FIGURE 20.16 A single transverse slice of the ilium fragment with the embedded stone projectile point. For the second digital extraction, the preliminary segmentation generated by the density differential was manually corrected for increased precision and detail. The corrections were done individually for all transverse slices and then exported as an isolated model.

sent in turn to the printer so the physical model builds layer by layer like a contour line map. Within the stereolithography printer, a vat of a photopolymer remains liquid until exposed to laser light. The printer directs the laser to draw the first layer of the digital model across the surface of the vat. Once it hardens, the completed layer is mechanically moved downward and a fresh layer of photopolymer is scraped across its surface so the next layer of the digital model may be drawn. By the time the last layer has been deposited, the entire physical model is immersed in the liquid. The model is then removed from the printer and is prepared for curing. First, it is washed to remove any remaining liquid, then support structures needed to keep the model from moving during printing are manually cut away, and finally the surface is smoothed (see Chapter 21 for further discussion on surface treatment of the models). The model is then cured in a sealed chamber of ultraviolet light to finish the catalytic reaction and fully harden the model.

Each layer of the stereolithography model was to be 0.5 millimeters thick; the final model was to be accurate within 0.1 millimeter of the original. For this project, all the stereolithography models but one were generated by

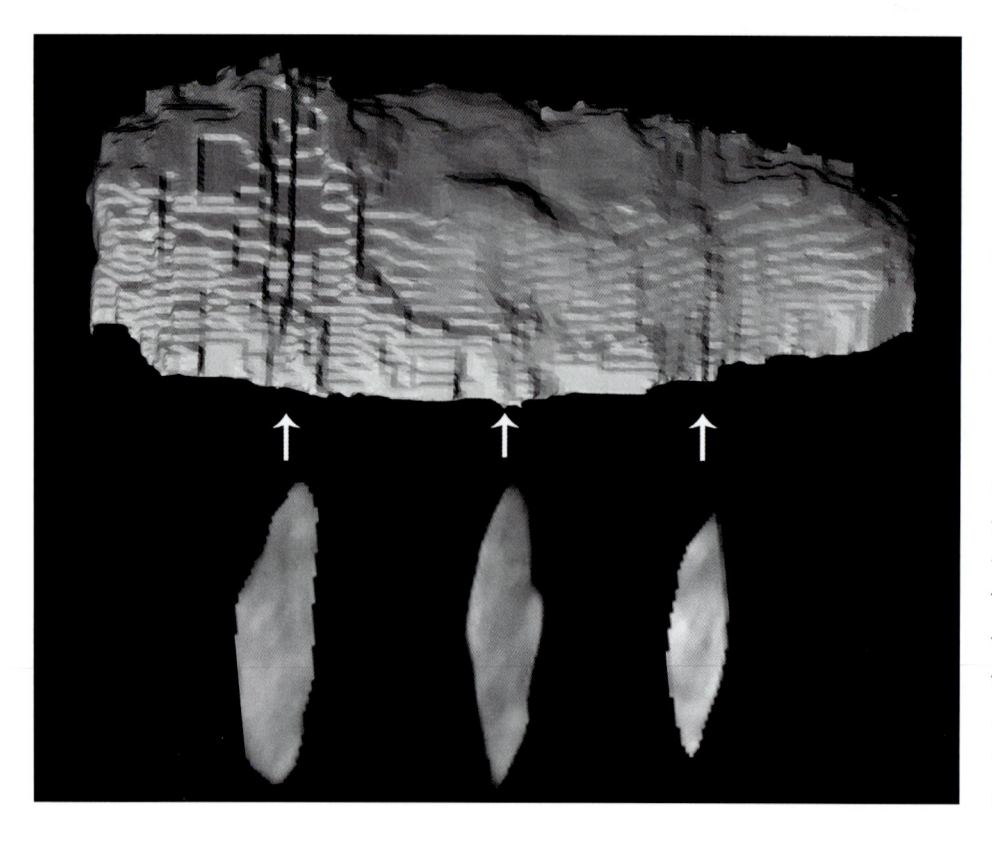

FIGURE 20.17 Isolated stone projectile point shown as a shaded polygonal mesh representing only surface data. The lines that appear similar to topography are from the slice thickness on the z-axis and are normal. The cross-sections are from transverse sections of the voxel data and placed in their corresponding locations along the projectile point length.

FIGURE 20.18 Stereolithography models of Kennewick Man's cranium, mandible, and right innominate fragments.

Brian Wilcox of Point Data Marketing, Inc. in Sparks, Nevada, using polygonal meshes generated by Art Andersen of Virtual Surfaces, Inc. in Mt. Prospect, Illinois (Figure 20.18). The second virtual extraction of the embedded stone projectile point was generated by Rebecca Snyder of NMNH and prepared for printing by Art Andersen (Figure 20.19).

During the course of studying the projectile point, both in situ and the extracted model, a question was raised: could fragments of the point that likely splintered off during entry into the ilium be found embedded within the interior of the bone? After detailed examination of the CT data, no definitive stone fragments could be identified given the close similarity in density between the calcium carbonate and the stone point. It is possible that some of the conglomerations of what was interpreted as calcium carbonate or dense bone could actually be small embedded lithic fragments of the shattered projectile point. However, alternative analysis methods will need to be performed, as the software used is unable to make a definitive determination of this possibility (Figures 20.20

FIGURE 20.19 Stereolithography model of the extracted projectile point.

Rebecca Snyder

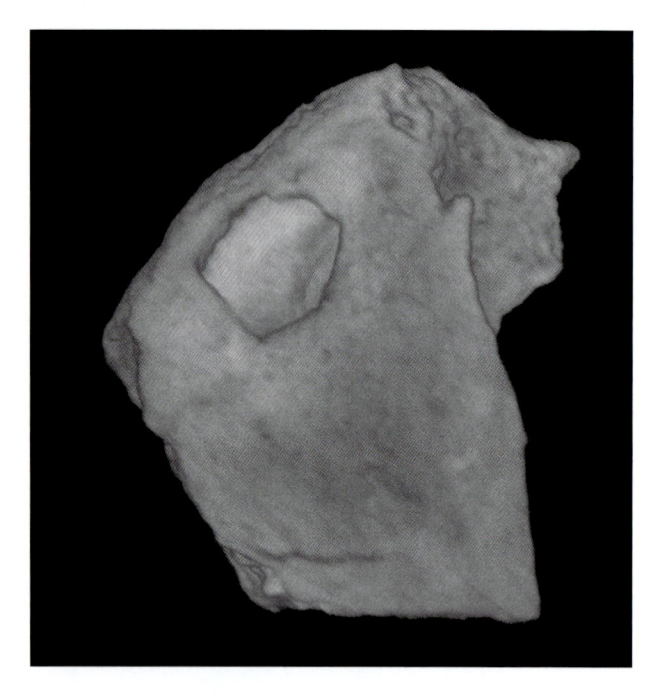

FIGURE 20.20 Exterior view of the right ilium fragment in anatomical position as viewed in VG Studio MAX.

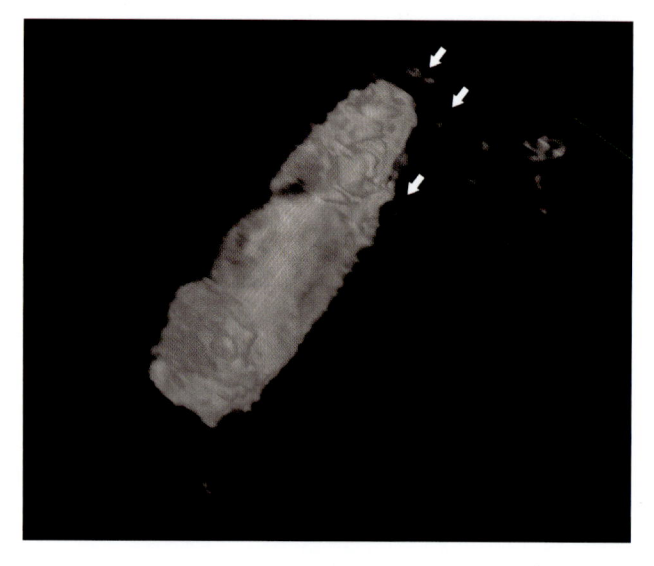

FIGURE 20.21 Right ilium fragment in anatomical position as viewed in VG Studio MAX with classification settings altered to remove most of the bone and silt to reveal the stone projectile point position in relation to anatomical position. While not completely isolated, the stone projectile point is sufficiently revealed to yield orientation and angle of insertion. Candidates for shatter particles from the tip are indicated by white arrows. Classification settings: 0-28325 suppressed, 28326 set to 0 with a linear progression to 42917 set to 100%.

and 20.21). This is at least one area where revisiting the original data in the future would continue to enhance the body of knowledge generated about Kennewick Man.

REFERENCES

Adobe Developers Association. 1992. *TIFF*. Revision 6.0. http://partners.adobe.com/public/developer/en/tiff/TIFF6.pdf.

Designcraft. 2011. "Designcraft, Inc." Accessed March 31. http://www.designcraft.com/.

Digital Imaging and Communications in Medicine. 2011. "DICOM Standards Group". Accessed March 31. http://medical.nema.org/.

Harvest Technologies. 2011. "Harvest Technologies." Accessed March 31. http://www.harvest-tech.com/.

ICT Imaging Proposal, Skull and Innominate, Kennewick Man Skeleton. Court document dated May 10th, 2005. On file at National Museum of Natural History, Smithsonian Institution, Washington, DC.

Kessler, Gary. 2011. "File Signatures Table, Magic Number Explanation and Key." Accessed Mar 31. http://www.garykessler.net/library/file_sigs.html.

Library of Congress. 2011. "Sustainability of Digital Formats, Planning for Library of Congress Collections, Image Format Descriptions". Accessed March 31. http://www.digitalpreservation.gov/formats/fdd/descriptions.shtml.

Materialise. 2011. "Materialise." Accessed March 31. http://www.materialise.com/.

Varian Medical Systems. 2011. "Bio-Imaging Research, Inc. (BIR)." Accessed March 31. http://www.varian.com/us.

Virtual Surfaces, Inc. 2011. "VSI." Accessed March 31. http://www.virtualsurfaces.com/.

Volume Graphics. 2011. "VG Studio MAX." Accessed March 31. http://www.volumegraphics.com/products/vgstudiomax/.

Prototype Accuracy and Reassembly

David R. Hunt

Kennewick Man's damaged skull needed to be reconstructed so that a full set of measurements could be taken for morphometric comparisons and for use in producing a facial reconstruction. The bones of the midface needed to be reassembled and then reattached to the cranial vault producing a dimensionally accurate reconstruction. The mandible was in three pieces, and its reconstruction was essential to the process of reattaching the midface by matching the occlusal surfaces of the upper and lower teeth relative to condyle articulations in the temporomandibular joints.

Virtual modeling through industrial computed tomography (CT) and the process of stereolithography—the hardening of minute layers of liquid resin—produced remarkable stand-ins for the original bones. This rapid prototyping system takes the polygonized point cloud developed during data processing and virtually slices it into layers around 0.1 millimeters thick. These slices are fed into a machine (a 3D printer) that uses a photopolymer liquid plastic to lay down consecutive layers until the complete data unit is fabricated (Chapter 20). The resulting model is extremely accurate and, in normal practice, this process is used to ensure form, fit, and function of manufactured parts. Applying this process to the Kennewick Man investigation enabled accurate reproductions of the bone pieces to be glued together and measured.

Resin prototypes of each fragment of Kennewick Man's skull and right innominate were prepared by Harvest Technologies, Belton, Texas, and delivered to the Burke Museum of Natural History and Culture, University of Washington, in Seattle, for assembly in the presence of the actual bone during the inventory and taphonomy studies in 2006. The prototypes were inventoried to confirm delivery of two copies of each of the scanned fragments. This paper compares measurements of the detached pieces of the actual skull and innominate to the resin prototypes in order to assess their dimensional accuracy for the purpose of reconstruction.

In normal industrial prototype production, the translucent resin pieces are sandblasted as part of the standard finishing process to increase surface contrast and improve surface texture. For this particular project, there was concern that too much resin might be removed, so no pieces were subjected to sandblasting. As a result, there is a subtle stepped—rather than smooth—feel to the pieces because the resin is deposited in incremental layers resembling a 3D topographic map. It was acknowledged that trace amounts of extra resin would potentially remain on the surfaces, which could affect the morphological characteristics and measurements.

QUALITY ASSURANCE AND THE RECONSTRUCTION PROCESS

To assess the effectiveness of prototyping this method for reconstructing the skull and right innominate, the prototypes and the original bone fragments were compared to ensure that the reproductions were morphologically and dimensionally accurate. This step was necessary to confirm that the models were accurate copies of the original fragments, thus providing a reliable basis for accurate reconstruction. This evaluation was achieved by measuring and comparing dimensions on the actual bones to their corresponding replicas. Determining the degree of accuracy was somewhat difficult, as the different fragments generally lacked standard morphological points for measurement. Specific dimensions of lengths, breadths, and thicknesses were identified on each original piece. Finding these same points of measurement on the clear resin prototypes was not always easy because of their translucency. Maximum lengths and specific shape measurements were taken on the original bone fragments and prototypes with a needle-nosed Mitutoyo© digital caliper accurate to 0.01 millimeter.

The measuring points, borders, and angles of these measurements were drawn and recorded on photographs of the original bone fragments. Examples are shown in Figures 21.1, 21.2, and 21.3. The value in parentheses is the corresponding measurement obtained from the prototype. The hope was that the prototype's measurements would closely replicate the values of the original bone. Table 21.1 compares measurements taken on the midface, mandible, and right innominate fragments and the corresponding prototypes. The intact cranial vault (97.U.1a) is not evaluated here; it is considered later as part of the finished reconstruction.

Overall, the prototypes are accurate size and shape reproductions of the original pieces. Of the 67 measurements taken in this set, most differences are extremely small. Only two innominate fragment measurements and four cranial and mandibular measurements differ by more than 1.0 millimeter. Prototype differences include a slightly wider maxillary alveolar breadth and higher mandibular body heights. The heights likely resulted from the replication of interstitial soil sediment that is not recognizable in the resin model. The pelvic bone prototype (97.R.17a) that is 1.65 millimeters smaller than the original reflects the complexity of reproducing a low-density material that was difficult to detect during the tomographic scan.

Assessed visually, the surfaces of the prototypes are not exact copies of the originals. Rather, the undulations and layered gradient of the surface are the result of the 0.39 millimeter slice thickness of CT data that are mathematically assembled and then converted into 0.1 millimeter thick resin layers. CT data are based on the measurement of density within the entire scanning volume, including the air surrounding the object, and are based on Hounsfield units (Kak and Slaney 2001:49–175). A decision has to be made by the CT technician during scanning, or by the manufacturing technician during production, as to what range of density values will be selected in order to best capture the most information without overproducing highly dense (opaque) structures or underproducing ("burning-out") more lucent (thin) structures. Setting the range too wide results in surrounding materials or air appearing solid and being reproduced in the model; setting the range too narrow renders less dense features invisible to the software and the model-making machinery. Thus, there are a few instances where the thin edge or a small projection may be underproduced in the prototype due to the restrictions of how density data are digitally represented and translated into a single density material.

Only gross morphology can be reliably detected within the confines of the 0.39 millimeters per slice

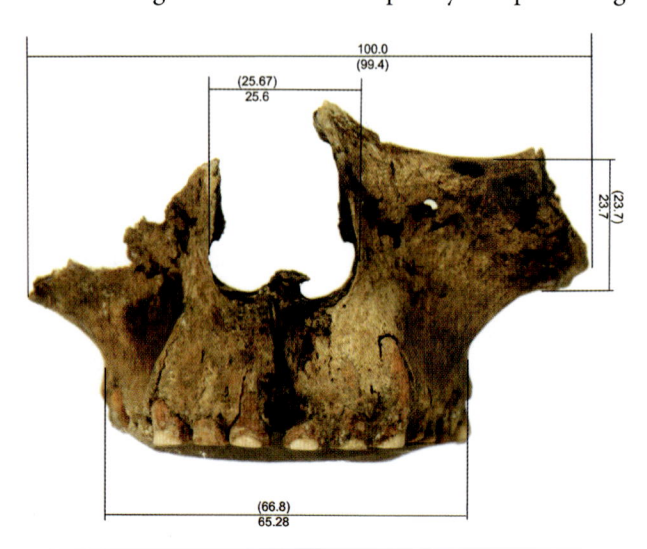

FIGURE 21.1 Photograph of maxilla fragment 97.U.2a, with maximum and specific metric chord measurements (millimeters) of Kennewick Man's original bone and corresponding measurements of the resin prototype (in parentheses).

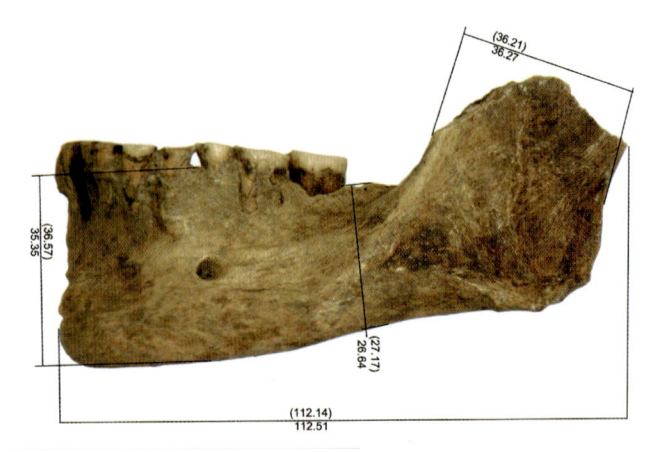

FIGURE 21.2 Photograph of mandible fragment 97.U.3a, with maximum and specific metric chord measurements (millimeters) of Kennewick Man's original bone and corresponding measurements of the resin prototype (in parentheses).

FIGURE 21.3 Photograph of innominate fragment 97.R.17d, with maximum and specific metric chord measurements (millimeters) of Kennewick Man's original bone and corresponding measurements of the resin prototype (in parentheses).

thickness yielded by the CT scanner. Subtle morphological features, such as those used for age determination from the pubic symphysis and auricular surfaces of the innominates or minor arthritic changes in the articular rims of the joints, cannot be reproduced in the CT prototypes. (This level of detail would more likely be captured by silicone molding and casting. This is also true for subtle features of the dentition and the fine, thin bones of the nose and palate or other radiolucent structures.)

The loss of microlevel structures became readily apparent when fitting together the prototype fragments of the cranium, mandible, and right pelvis during reconstruction. In reassembling original bone pieces, the subtle fingerlike projections of fractured edges on actual bone allow for a confirmed relocation of the pieces into their original alignment at the break (i.e., a precise refit). Subtle projections and fracture margin irregularities were not captured by the imaging, thus reducing the tactile and visual "sure fit" in the prototypes. The reattachment joint was smoothed over, or blurred, by the averaging requirements of the prototyping process. As a consequence, it was necessary to remove some of the resin material from the prototype fracture areas to make a fit that would be within the measurement tolerances of the reconstruction when compared to the original. Prototype refittings required constant and extensive comparison to the original bone pieces. This involved numerous rearticulations of the fracture edges to compare, gauge, and finely modify articulating surfaces to achieve accurate measurements and correct fitting of the prototype pieces.

An unexpected difficulty in the prototype reconstruction was found in working with the translucent resin fragments; the external surface, in essence, visually disappeared. The usual opaque nature of real bone, with the

TABLE 21.1 Size comparison of each prototype to Kennewick Man's actual bone.

Bone segment number	Segment description	Metric	Bone measurement (mm)	Prototype measurement (mm)	Measurement differences over 1.0 mm
		Cranium			
97.R.1a	Right malar segment (right malar and the articulating adjacent zygomatic process of the right maxilla)	Max. height of fragment	49.67	49.61	
97.R.1a	Right malar segment (right malar and the articulating adjacent zygomatic process of the right maxilla)	Max. breadth of fragment	41.25	41.17	
97.U.1b	Zygomatic arch fragment	Max. breadth of fragment	18.76	18.83	
97.U.1b	Zygomatic arch fragment	Max. height of fragment	23.59	23.01	
97.U.1c	Left malar (the frontal process and anterior malar are present)	Max. length of fragment	47.59	47.08	
97.U.1c	Left malar (the frontal process and anterior malar are present)	Max. breadth of fragment	28.98	29.91	
97.U.1c	Left malar (the frontal process and anterior malar are present)	Max. breadth from the lateral orbital rim to the lateral edge of the malar	15.36	16.64	1.28 mm
97.I.25v	Left maxilla fragment (posterior maxillary sinus)	Max. length of fragment	23.25	23.50	
97.I.25v	Left maxilla fragment (posterior maxillary sinus)	Max. breadth of fragment	23.52	23.52	
97.R.2b	Right maxilla fragment	Max. breadth of fragment	10.01	9.84	
97.R.2b	Right maxilla fragment	Max. length of fragment	42.69	42.32	
97.R.2b	Right maxilla fragment	Max. transverse breadth of fragment	13.94	13.89	
97.U.2a	Left and right maxillae (nasals glued to maxilla)	Max. breadth from fracture edge of the right and left lateral maxillae	100.00	99.40	
97.U.2a	Left and right maxillae (nasals glued to maxilla)	Max. nasal breadth	25.60	25.67	
97.U.2a	Left and right maxillae (nasals glued to maxilla)	Max. alveolar breadth	65.28	66.80	1.52 mm, attributable to material on the prototype surface from the CT scan and manufacture
97.U.2a	Left and right maxillae (nasals glued to maxilla)	Max. alveolar length	58.34	58.38	

(*continued*)

TABLE 21.1 (continued) Size comparison of each prototype to Kennewick Man's actual bone.

Bone segment number	Segment description	Metric	Bone measurement (mm)	Prototype measurement (mm)	Measurement differences over 1.0 mm
97.U.2a	Left and right maxillae (nasals glued to maxilla)	Max. cheek height, left side	23.70	23.70	
97.U.2a	Left and right maxillae (nasals glued to maxilla)	Max. breadth from buccal edges of right and left 2nd molar	65.73	66.63	
97.U.2a	Left and right maxillae (nasals glued to maxilla)	Max. breadth from lingual edges of right and left 2nd molar	42.16	41.54	
97.U.2a	Left and right maxillae (nasals glued to maxilla)	Max. cord from nasal spine to fracture edge of left maxilla	65.60	65.20	
97.U.2a	Left and right maxillae (nasals glued to maxilla)	Max. chord from nasal spine to fracture edge right maxilla	60.90	60.19	
97.U.2a	Left and right maxillae (nasals glued to maxilla)	Max. vertical height from 1st right central incisor to superior fracture edge of right maxillary rim of nasal aperture	47.90	47.70	
97.U.2a	Left and right maxillae (nasals glued to maxilla)	Max. breadth of left maxillary region from the medial fracture edge of the nasal aperture to the lateral fracture edge of the left inferior orbit rim	44.33	44.65	
		Mandible			
97.U.3a	Mandible (left body)	Corpus height at mental foramen (between P1 and P2)	35.35	36.57	1.22 mm, likely due to the interstitial sediment between the replicated teeth that was not distinguishable in the resin mold
97.U.3a	Mandible (left body)	Corpus height from the alveolus posterior to M2 to inferior corpus edge	26.64	27.47	
97.U.3a	Mandible (left body)	Max. length from mental eminence to most dorsal fracture edge of ascending ramus	112.51	112.14	
97.U.3a	Mandible (left body)	Minimum ascending ramus breadth	26.27	36.21	
97.U.3b	Mandible (right anterior body containing teeth I2 through M2)	Corpus height at mental foramen between P1 and P2	35.13	36.48	1.35 mm, likely due to the interstitial sediment between the replicated teeth that was not distinguishable in the resin mold

(continued)

TABLE 21.1 (continued) Size comparison of each prototype to Kennewick Man's actual bone.

Bone segment number	Segment description	Metric	Bone measurement (mm)	Prototype measurement (mm)	Measurement differences over 1.0 mm
97.U.3b	Mandible (right anterior body containing teeth I2 through M2)	Corpus height at mental foramen between P3 and P4	35.13	36.48	
97.U.3b	Mandible (right anterior body containing teeth I2 through M2)	Corpus breadth at mental foramen	10.42	10.90	
97.U.3b	Mandible (right anterior body containing teeth I2 through M2)	Max. corpus length from mental eminence fracture edge to dorsal corpus fracture edge, caliper flat on the dorsal fracture	56.58	56.39	
97.U.3b	Mandible (right anterior body containing teeth I2 through M2)	Max. length from labial edge of I2 to distal edge of M2	47.29	47.26	
97.U.3b	Mandible (right anterior body containing teeth I2 through M2)	Max. corpus length from mental eminence fracture edge to dorsal corpus fracture edge, caliper flat on the dorsal fracture	56.58	56.39	
97.U.3b	Mandible (right anterior body containing teeth I2 through M2)	Max. length from root apex of I2 to root apex of M2	43.00	43.29	
97.U.3c	Mandible (right ascending ramus with the posterior M3 socket)	Length from anterior corpus fracture, with caliper flat to anterior fracture to the dorsal edge of the gonal region	49.77	49.75	
97.U.3c	Mandible (right ascending ramus with the posterior M3 socket)	Max. ramus breadth from anterior edge of coronoid process to dorsal edge of condyle	50.30	50.48	
97.U.3c	Mandible (right ascending ramus with the posterior M3 socket)	Minimum ascending ramus breadth	35.50	35.83	
97.U.3c	Mandible (right ascending ramus with the posterior M3 socket)	Sagittal breadth of the condyle	10.70	11.36	
97.U.3c	Mandible (right ascending ramus with the posterior M3 socket)	Length from most anterior-inferior edge of corpus fracture to the superior-dorsal edge of the condyle	88.64	88.41	
		Innominate			
97.I.I7d	Iliac crest fragment	Max. breadth of fragment	18.92	19.02	
97.I.I7d	Iliac crest fragment	Max. length of fragment	40.31	40.24	
97.I.I7c	Unarticulated ischial fragment	Max. breadth of fragment	21.76	22.32	
97.I.I7c	Unarticulated ischial fragment	Max. length of fragment	49.89	49.23	

(continued)

TABLE 21.1 (continued) Size comparison of each prototype to Kennewick Man's actual bone.

Bone segment number	Segment description	Metric	Bone measurement (mm)	Prototype measurement (mm)	Measurement differences over 1.0 mm
97.R.17c	Visceral surface of mid-posterior ilium (small detached fragment from 97.R.17c)	Max. breadth of fragment	27.07	27.12	
97.R.17c	Visceral surface of mid-posterior ilium (small detached fragment from 97.R.17c)	Max. height of fragment	18.22	18.59	
97.I.10b	Fragment of iliac crest	Max. length of fragment	46.04	45.38	
97.I.10b	Fragment of iliac crest	Max. breadth of fragment	19.74	20.68	
97.I.17b	Ischio-pubic ramus fragment	Max. length of fragment	44.05	44.07	
97.I.17b	Ischio-pubic ramus fragment	Max. breadth from inner obturator edge to fragment edge	17.83	17.91	
97.R.17a	Pubic symphysis fragment	Max. length from superior pubis articular edge to fracture edge	63.12	62.99	
97.R.17a	Pubic symphysis fragment	Max. height of fragment	76.84	75.19	1.65 mm, attributable to difficulty reproducing the small finger projection at the mid-ascending ramus due to its low density
97.R.17a	Pubic symphysis fragment	Length from pubis articular face to lateral edge of obturator	32.26	31.78	
97.R.17b	Ilium anterior body segment	Max. height from superior iliac crest to inferior- anterior fracture edge	106.87	106.50	
97.R.17b	Ilium anterior body segment	Height superior iliac crest to inferior-dorsal fracture edge	91.41	90.82	
97.R.17b	Ilium anterior body segment	Anteroposterior length from anterior rim of the iliac edge between the superior and inferior spines to the dorsal fracture edge	51.21	50.69	
97.R.17c	Ilium dorsal segment with projectile point	Max. breadth of fragment	74.40	74.19	
97.R.17c	Ilium dorsal segment with projectile point	Max. height of fragment	75.28	74.82	
97.R.17c	Ilium dorsal segment with projectile point	Max. height of opening at projectile point	18.72	17.53	1.19 mm, attributable to differential densities that influence the production of the bone edge in relation to the lithic material

(continued)

TABLE 21.1 (*continued*) Size comparison of each prototype to Kennewick Man's actual bone.

Bone segment number	Segment description	Metric	Bone measurement (mm)	Prototype measurement (mm)	Measurement differences over 1.0 mm
97.R.17d	Innominate portion with auricular surface, sciatic notch, acetabulum, and superior ischium	Sciatic notch breadth from anterior-inferior to posterior-inferior spine	38.55	38.40	
97.R.17d	Innominate portion with auricular surface, sciatic notch, acetabulum, and superior ischium	Sciatic notch breadth across the notch from posterior-inferior spine to anterior edge	23.87	23.89	
97.R.17d	Innominate portion with auricular surface, sciatic notch, acetabulum, and superior ischium	Ischial portion from most posterior edge of obturator rim to anterior-inferior spine	50.67	49.55	
97.R.17d	Innominate portion with auricular surface, sciatic notch, acetabulum, and superior ischium	Max. height of acetabulum	56.49	56.29	
97.R.17d	Innominate portion with auricular surface, sciatic notch, acetabulum, and superior ischium	Max. breadth of acetabulum	51.11	50.19	
97.R.17d	Innominate portion with auricular surface, sciatic notch, acetabulum, and superior ischium	Max. height from anterior inferior iliac spine to most inferior edge of ischial fracture line	121.71	122.13	
97.R.17d	Innominate portion with auricular surface, sciatic notch, acetabulum, and superior ischium	Max. height from most superior fracture edge of the ilium body to the most inferior edge of ischial fracture line	137.19	136.79	
97.R.17d	Innominate portion with auricular surface, sciatic notch, acetabulum, and superior ischium	Max. breadth from inferior acetabular rim to edge if ishium inferior to the sciatic notch.	36.66	37.05	

information it holds, was not there (Figures 21.4, 21.5, and 21.6). Due to the translucency of the material, the surface morphology on the prototypes had to be identified by feel because the external surface was impossible to visually discern. This was compounded by the graduated characteristics of the surface as a result of the 3D layered reconstruction effect (Figure 21.7).

LOCATIONS OF POSTMORTEM CRANIAL DISTORTION

Three areas of postmortem distortion that could affect morphometric analyses of the reconstructed cranium were identified in the original bone. A fracture is evident along the anterior rim of the foramen magnum. It begins to the right of basion and extends to the left along the anterior rim of the left occipital condyle and down the lateral side of the condyle (Figures 21.8 and 21.9). Approximately three-quarters down the condyle, the fracture passes through the condyle and ends at the foramen magnum (Figure 21.10).

There is a 2-millimeter gap along the left side of the sphenoid body (Figure 21.11). This separation is associated with postmortem breakage in the left sphenoid wing and left frontal/temporal area. The gap, 1 millimeter wide in the fronto-temporal suture area, results in minor forward displacement of the face (Figure 21.12).

FIGURE 21.4 Right innominate fragments of Kennewick Man's original bone and the corresponding resin prototypes.

FIGURE 21.5 Kennewick Man's right pubic bone, beside the corresponding resin prototype, illustrating the loss of symphyseal surface recognition.

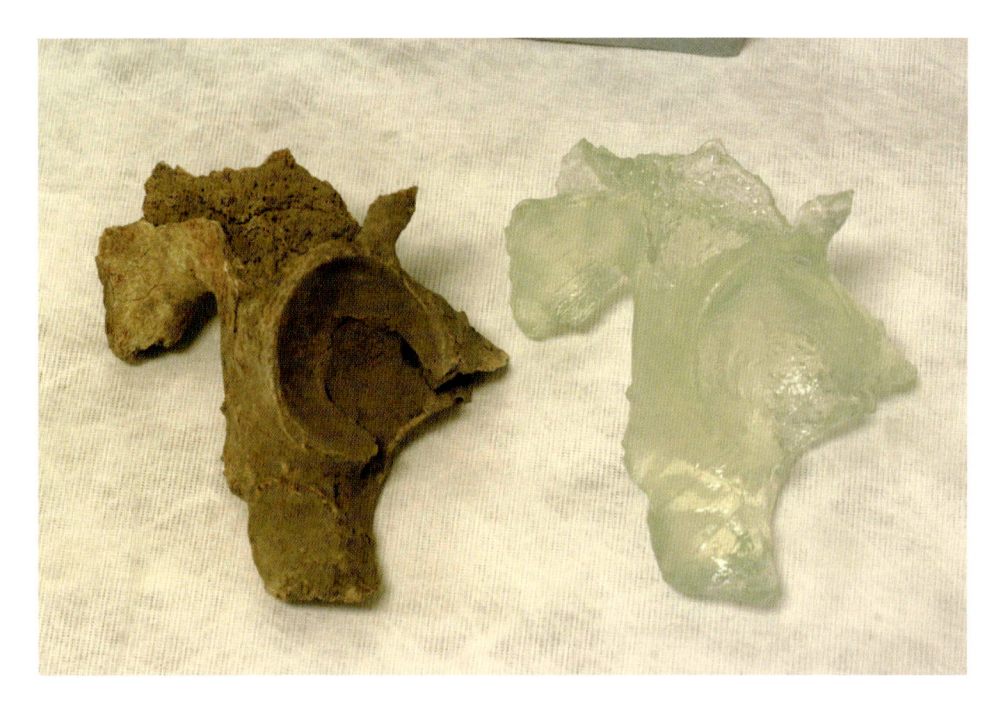

FIGURE 21.6 Lateral side of Kennewick Man's right innominate and the corresponding resin prototype illustrating the difficulty in identifying fracture areas and surface features.

FIGURE 21.7 Fracture edge of the inferior aspect of Kennewick Man's ischiopubic ramus compared to the corresponding resin prototype illustrating the layering effect of the computed tomography reconstruction and the loss of definition in the fractured area of the resin prototype.

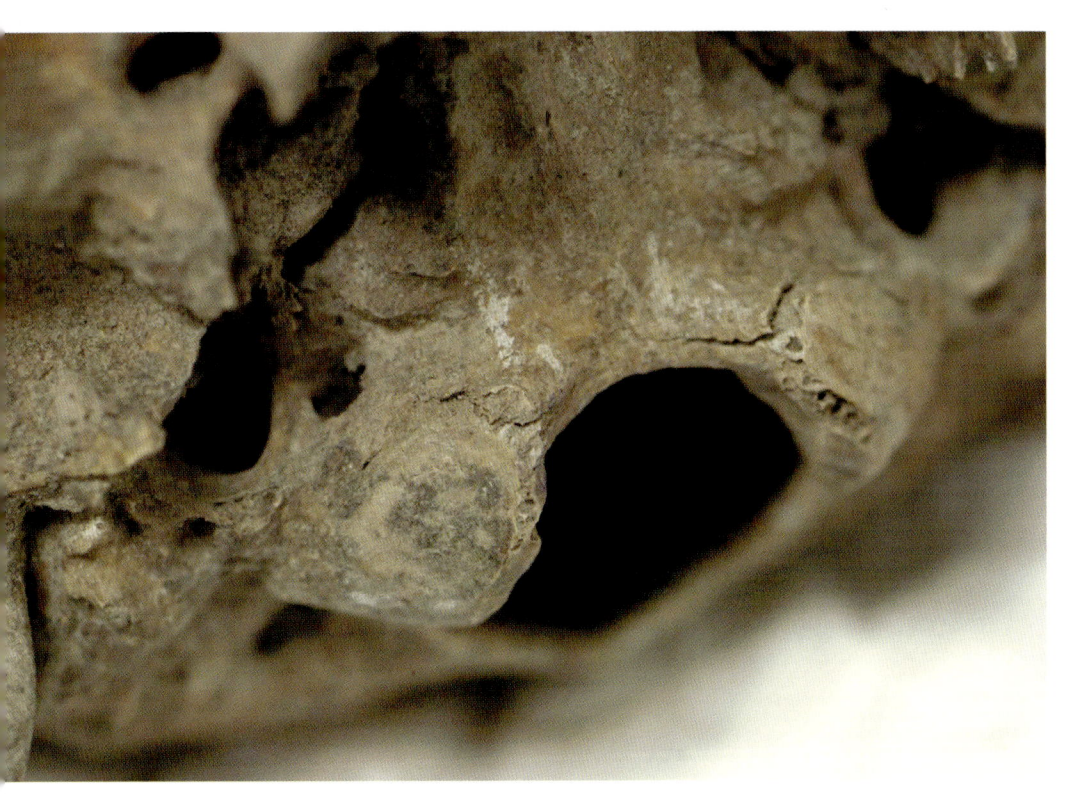

FIGURE 21.8 Crack along the anterior rim of Kennewick Man's foramen magnum, starting at the lateral anterior aspect of the right condyle and passing along the anterior region of basion to the left condyle.

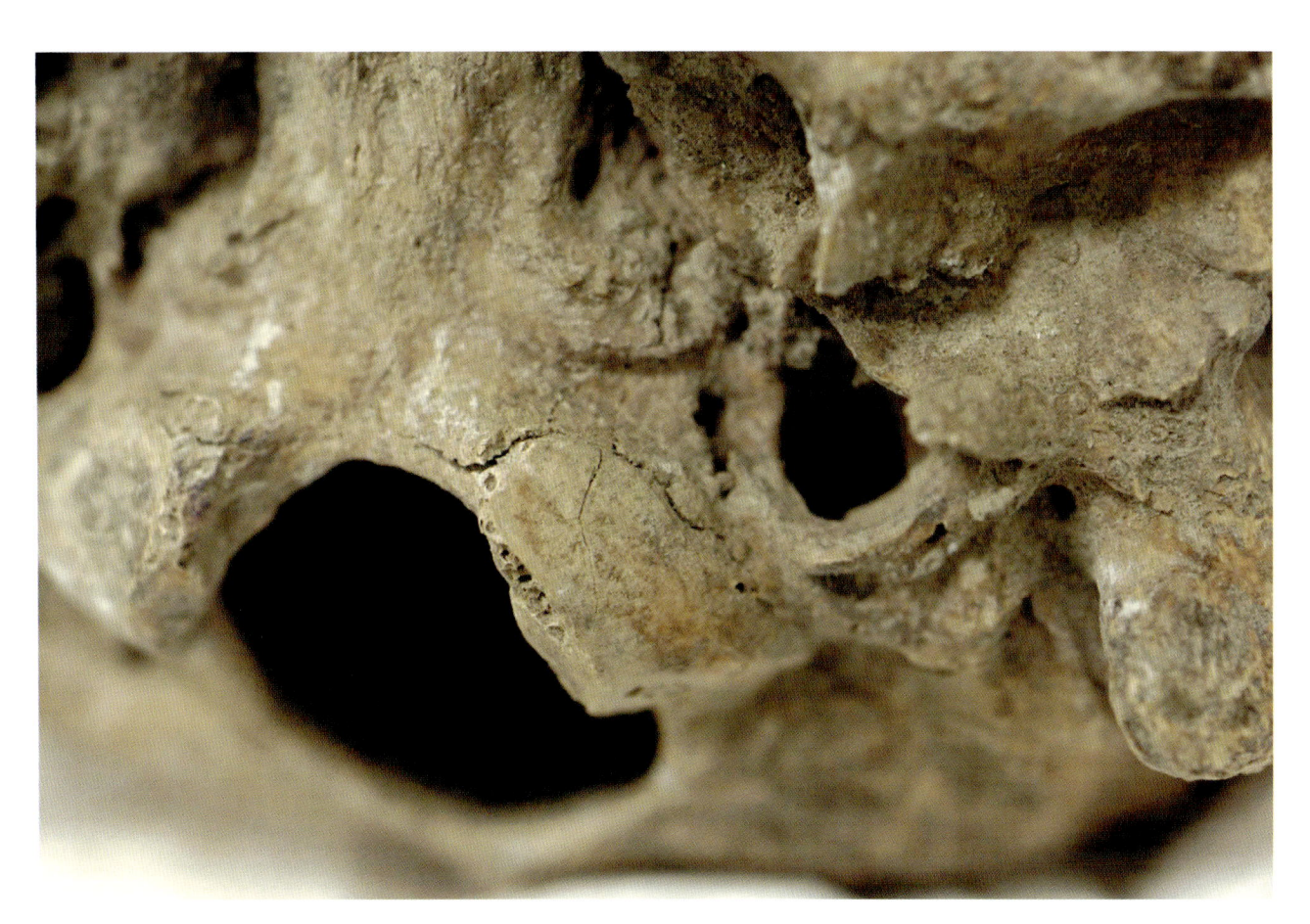

FIGURE 21.9 Crack along the anterior region of Kennewick Man's left condyle, passing dorsally along the lateral border of the condyle.

FIGURE 21.10 Crack in Kennewick Man's postero-lateral left occipital condyle.

A right frontal fracture is located mid-orbit and extends approximately 35 millimeters medial-superior into the forehead. This crack is 0.9 millimeters in breadth, contributing to the slight forward positioning of the mid-frontal, and affecting the placement of the lower region of the face when reattached. No other cracking or distortion is observed on the dorsal portion (occipital or dorsal parietals) of this vault.

RECONSTRUCTION
Cranium

Proper placement of the prototype face onto the prototype vault was based on measurements taken on the original cranium with and without the face balanced on the vault. The most anterior rim of the maxillary alveolar arch is 100 millimeters from basion on the reconstructed original. This is the length set to locate the face position in the prototype reconstruction. The basion-prosthion measurement from the reconstructed original is 113 millimeters. This measurement was used as another reference for positioning the face on the prototype of the vault. Orbit height (measured as the vertical superior-inferior chord) on the reconstructed original was 37.5 millimeters for the left and 38 millimeters for the right side. These measurements were used to position the face on the vault.

Twelve measurements of the reconstructed replica of the cranium were compared to corresponding data for the vault (97.U.1a) and temporarily reassembled cranium in order to gauge the resulting accuracy of the completed reproduction (Table 21.2). These measurements were taken separately by Jantz and Spradley using a digitizer (Chapter 25). The difference in techniques introduces minor variation. For example, the maximum length of the cranium normally measures about a millimeter shorter with a digitizer than when taken using calipers. Differences between the model and the actual cranium are not great. However, basion-nasion length, maximum cranial breath, and minimum frontal breadth are one to two millimeters larger in the model. The observed differ-

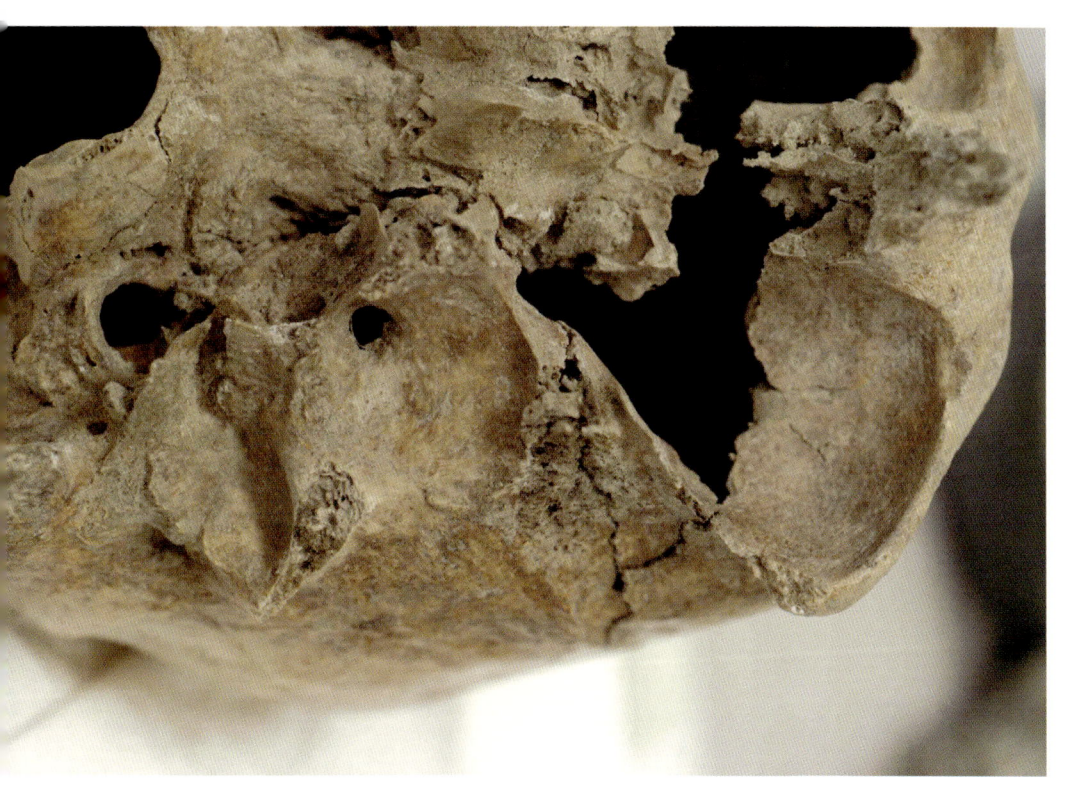

FIGURE 21.11 Crack in Kennewick Man's anterior body of the sphenoid, radiating to the left sphenoid wing and to the left frontal.

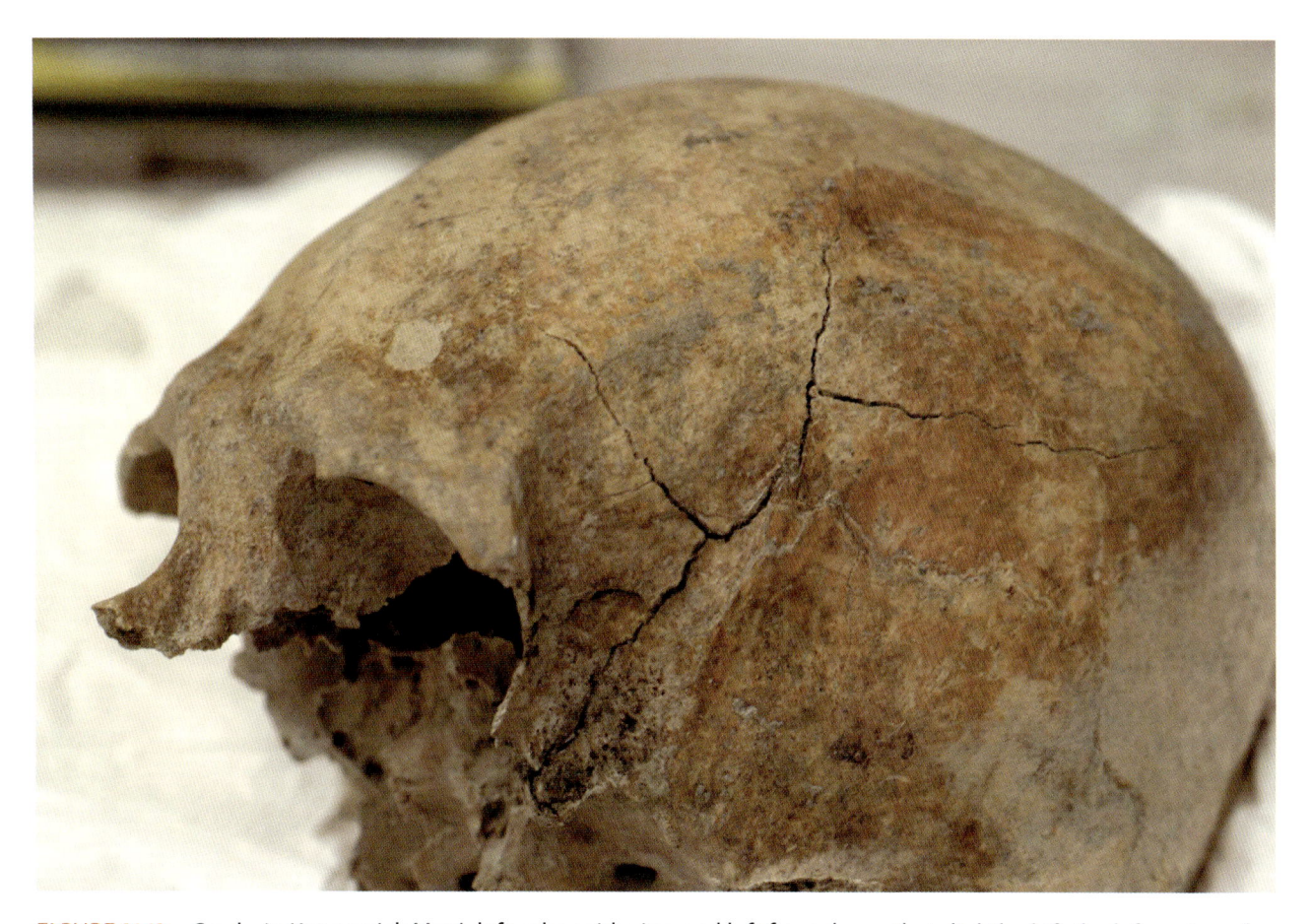

FIGURE 21.12 Cracks in Kennewick Man's left sphenoid wing and left frontal area that slightly shift the left region of the face anteriorly.

TABLE 21.2 Measurement comparison of dimensions of the reconstructed bone to the prototype reconstruction.

Measurement	Bone measurement (mm)		Prototype measurement (mm)		Measurement differences
	Hunt	Spradley/Jantz	Hunt	Spradley/Jantz	
Cranium					
Glabello-occipital length	190	188	190	189	
Maximum cranial breadth	141	141	143	142	~1 mm
Minimum frontal breadth	96	95	97	98	~1 mm
Basion-nasion length	111	112	113	115	~2 mm, in part due to difficulty in refitting the face to the vault
Basion-bregma height	142	141	142	141	
Foramen Magnum length	40	40	39	39	1 mm
Foramen Magnum breadth	32	33	32	32	
Bimaxillary breadth	99		99		
Orbit height (vertical) right	38		38		
Orbit height (vertical) left	37.5		37		
Biorbital breadth	105		104		1 mm, attributable to slight undersizing of prototype due to fitting of the fracture edges on the resin fragments
Nasion-prosthion height	76.2		76		
Innominate					
Iliac breadth	146		148		2 mm, attributable to slight oversizing of prototype due to the need for material removal for fitting the fracture edges of the resin fragments
Pubic length (from medial rim of acetabulum to the edge of the symphysis)	70		71		1 mm, attributable to slight oversizing of prototype due to the need for material removal for fitting the fracture edges of the resin fragments
Inlet length (from anterior-inferior edge of auricular surface to most posterior of the pubic symphysis)	119		119		
Obturator foramen diameter (from superior rim on ischium to inferior-superior edge of pubis at the rim edge)	56.5		55.5		1 mm, attributable to slight undersizing of prototype due to fitting of the fracture edges on the resin fragments

ences are patterned in the same direction across observers and measurement techniques, suggesting real, but unexplained, differences between the actual bone and the recreated model.

Mandible

The middorsal fracture of the right mandibular corpus and ascending ramus have an unfixed join, making a consistent repositioning of these fragments problematic. The "play" in the join caused the mandibular angle to vary from 119 to 122 degrees. Using the morphological characteristics of the mandible for refitting produced the greater end of the angle range (121 to 122 degrees) but also made this range the most appropriate measurement. This fluctuation affected the ramus height measurement from the mandibulometer. This difference was between 62.5 and 62.8 millimeters. All measurements of the mandible were made with sliding calipers and a mandibulometer made by the GPM company.

Right Innominate

The two locations in the innominate that required the most resin material removal from the prototypes were the large fracture joins of the ilium. One was just dorsal to the acetabulum and the other was in the central region of the iliac plate. The material was removed using dental scraping tools and a Dremel® tool. Numerous temporary refittings of the original bone pieces were required in order to visualize a proper fit of the prototype pieces as well as to confirm metric accuracy in the resulting prototype reconstruction. Metric comparison of the prototype reconstruction to the reconstructed bones is found in Table 21.2.

DISCUSSION

The US Army Corps of Engineers (Corps) prohibited reconstruction with adhesives, recommending that the desired cranial measurements be taken by temporarily fitting the fragments together by gravity balancing with support of the fracture joins by jeweler's wax. The author tried this approach during the reassembly process by repositioning the various fragments of the face while conservators Nancy Odegaard and Vicki Cassman applied jeweler's wax strips over each fracture join. The end result was a precariously balanced and fragile structure. These wax supports fatigued quickly, requiring rapid measurement of the various metric chords. The balanced fragments collapsed twice, requiring the author to catch the fragments before they fell to the table. Two repositionings of the fragments with wax applications were required to take a partial set of cranial measurements using traditional calipers (Figures 21.13 and 21.14). Dismantling the reconstruction back into individual fragments revealed that the jeweler's wax had adhered to the bone cortex and left wax residue. The conservators used bamboo picks in several attempts to remove as much of the adhering wax from the bone as possible. The end result was not any more damaging than would have occurred with the use of an appropriate reversible adhesive. The lack of permanence represented by this approach was an inadequate solution, as other scientists needed to measure and digitize the skull. Full evaluation of the projectile injury to the pelvis would similarly benefit from reassembly of the right innominate.

No reconstruction of a skull or single bone is likely to have precisely the same measurements as the unbroken original because small pieces of bone will flake off and/or the use of adhesives creates unintended deviation. Nonetheless, the most accurate results are obtained when pieces of the original bone are reassembled. When that is not possible, the next most precise reconstruction would be derived from high-resolution silicone molds of the original bone, producing casts to reassemble the skeletal part. Neither process was allowed during the study of Kennewick Man. Corps conservators were concerned about potential damage that could be caused by casting or by original bone reconstruction with a reversible adhesive. Acrylic resins, like B72 mixed with acetone and ethanol, are used by many federal agencies including the National Park Service (NPS) (Segal 1998:89). Thus, as an alternative, 21 critical bone pieces were digitally reproduced and stereolithographically regenerated in resin. This method provided the next best means for obtaining accurate replicas under the circumstances.

In the case of an important skeleton like Kennewick Man, this was the best way to proceed, resulting in accurate reproductions of the cranium, mandible, and right innominate. However, there are qualifications. Prototype morphology and surface features are

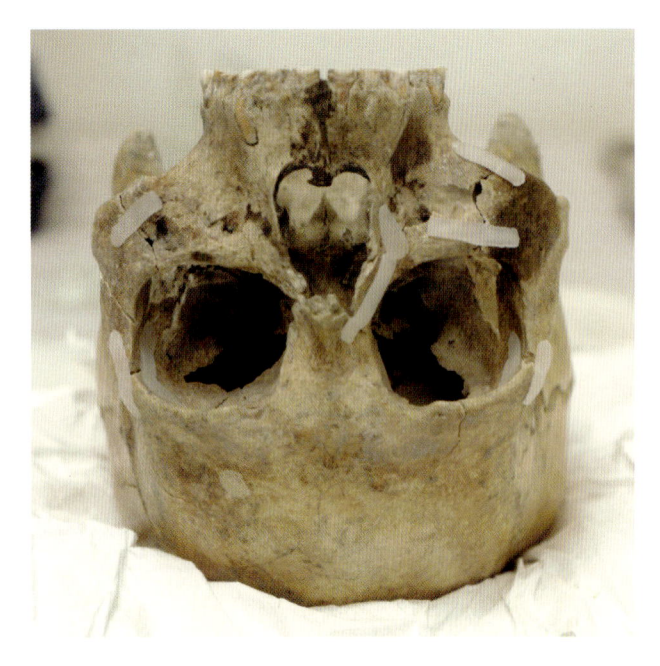

FIGURE 21.13 Kennewick Man's reconstructed midface held in place by jeweler's wax strips.

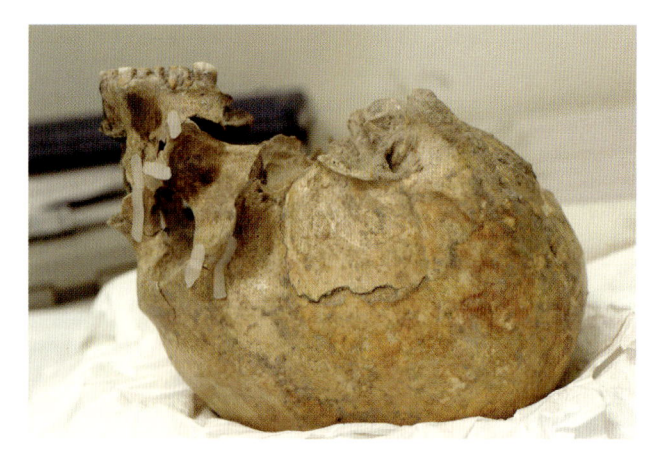

FIGURE 21.14 Kennewick Man's cranium reconstruction held in place by jeweler's wax strips.

the product of computerized reconstruction of density values. Even with highly sophisticated CT scanning, the resulting prototypes are not perfect replicas with exact representations of original surfaces. This became clear when comparing the fracture edges of the original bone to the prototypes; the subtle splinters of bone in the fractures were not reproduced. This made refitting the fracture joints less exact. Also, the lucency of the clear resin used to produce the prototypes made locating and identifying some reference points and certain surface morphology quite difficult. It may be advisable to use opaque resin material for better surface illustration in future prototyping projects. It should also be pointed out that it was necessary to remove slight amounts of the surface material at the fracture joints for a more precise fit. In this particular reconstruction effort, all fittings were constantly compared to the original bones and their refits to accurately replicate these alignments in the prototypes. These frequent comparisons were essential. This time-consuming, and sometimes precariously repetitive, refitting would not have been necessary if the original fragments could have been molded and cast, or the complete original bone reconstruction could have been silicone molded using well-established procedures.

Final metric comparisons of the completed models show that stereolithography enabled the production of very high quality, but not perfect, replicas of the original bones that can be used as proxies in metric analyses (Figure 21.15). However, as finished pieces, prototype reproductions lack sufficient surface detail for evaluating subtle skeletal and dental features, such as dental morphology, pathological conditions such as the shallow depression fracture on Kennewick Man's frontal bone, and the kinds of postmortem taphonomic observations described elsewhere in this volume. Other scanning techniques and high resolution photography can help in this regard.

Finally, it should be noted that industrial CT and stereolithography are cost-prohibitive for most academic use. Only because of extreme necessity in this highly significant investigation, and with the backing of private donors, could a $30,000 scanning and reproduction process be employed for the study of these remains.

REFERENCES

Kak, A. C., and Malcolm Slaney. 2001. *Principles of Computerized Tomographic Imaging.* Society of Industrial and Applied Mathematics.

Segal, Terry. 1998. Collections Management—Marking. In *The New Museum Registration Methods*, edited by Rebecca A. Buck and Jean Allman Gilmore, 65–94. American Association of Museums, Washington, DC.

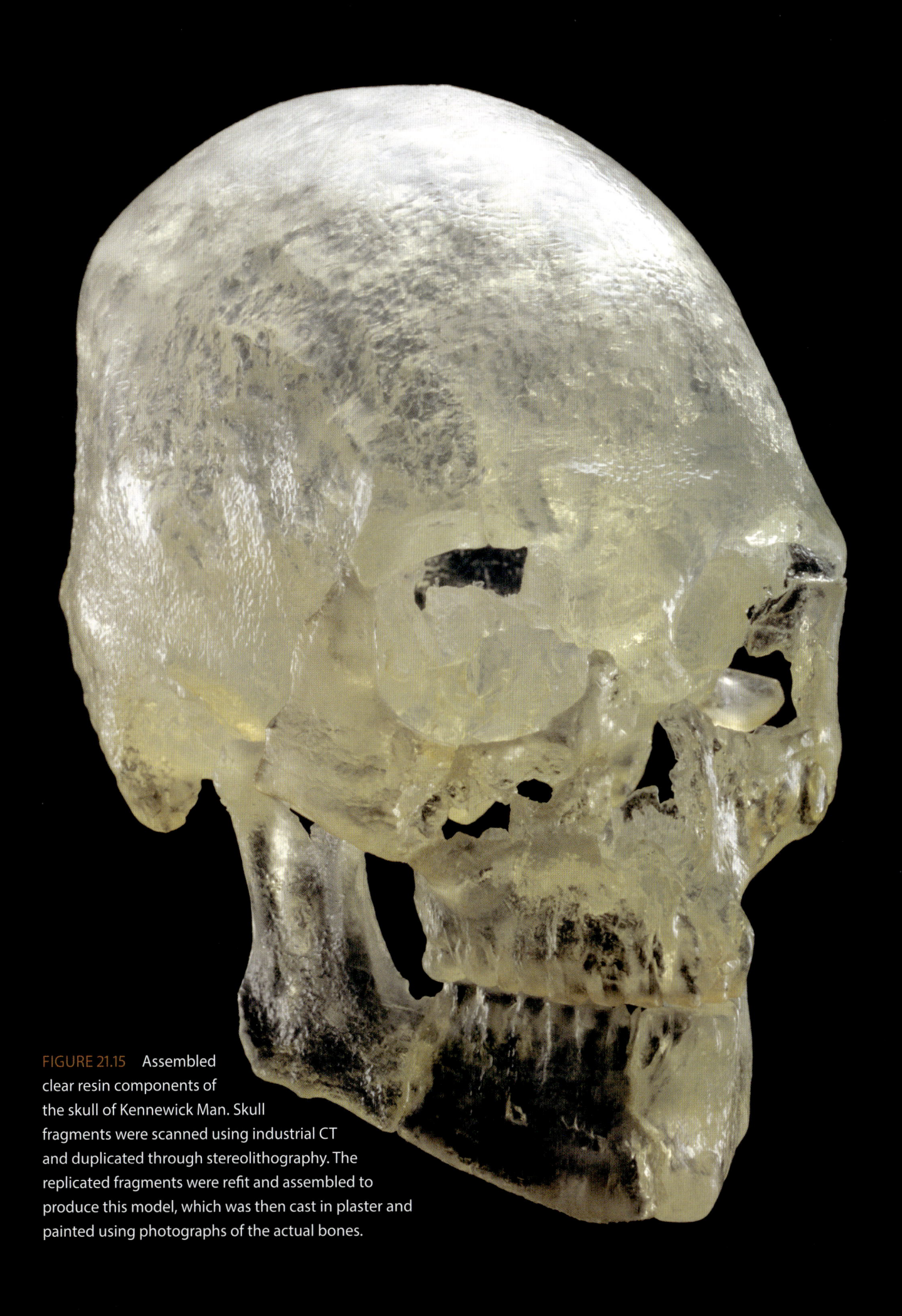

FIGURE 21.15 Assembled clear resin components of the skull of Kennewick Man. Skull fragments were scanned using industrial CT and duplicated through stereolithography. The replicated fragments were refit and assembled to produce this model, which was then cast in plaster and painted using photographs of the actual bones.

FIGURE 22.1 The painted and matte-sprayed cast of the cranium and mandible of Kennewick Man.

Molding and Casting Methods

22

Steven J. Jabo

When rare skeletal remains are found—such as those of Kennewick Man—it is often desirable to create precise reproductions for scientific study (Figure 22.1). Techniques for creating such replicas involve several steps. First, a computed tomography scan produces a digitized image of each bone or fragment. Next, a stereolithographic device generates a resin model, or prototype, of each skeletal element. The resin pieces can be fused together to reconstruct anatomical structures.

In the Smithsonian Institution's National Museum of Natural History, Vertebrate Preparation Lab experience with stereolithographic prototypes has shown that the resin is limited in terms of its fidelity to the measurements of the original artifact over time. For instance, a resin model of a *Triceratops horridus* femur with an overall length of 187.5 millimeters had shrunk by 4.5 millimeters, or 2.4%, over the course of a year. As with all epoxies, long-term exposure to ultraviolet light may degrade the resin, causing it to become brittle and friable. Extreme temperatures (above 48 degrees C) can also warp a prototype.

Thus, the standard protocol in the lab is to mold and cast the individual and reconstructed prototyped elements in a more stable medium, such as hydrocal cement, as soon as possible after production. The casts are then painted to match the original material. Specific procedures used to reproduce the Kennewick Man remains are discussed below.

MOLDING

Two-Part, Thin-Walled Molds

The cranium, mandible, innominate, and cranial vault were reconstructed using thin-walled molds. The process of fusing the prototyped elements resulted in gaps and open spaces on each skeletal element (Figure 22.2). These, along with foramina and other natural openings, were blocked out with clay to prevent penetration by the liquid molding compound. Care was taken to maintain the outline of each opening as it was being sealed to let the observer know it existed.

A parting seam line for the mold was constructed by affixing a clay flange around the element. The flange included indentations in the clay that would become registration keys in the mold (Figure 22.3).

The resin prototype was then coated with a paraffin-based, aerosol-release agent to prevent the mold rubber from sticking to the epoxy. A quantity of thin silicone molding compound was evacuated in a vacuum chamber and applied with a disposable acid brush to the areas

FIGURE 22.2 The reconstructed cranium of Kennewick Man made of fused prototyped bone fragments illustrating gaps and openings.

FIGURE 22.3 The reconstructed mandible set in clay and awaiting the first coat of silicone rubber. Circular indentations are registration keys to align mold sections.

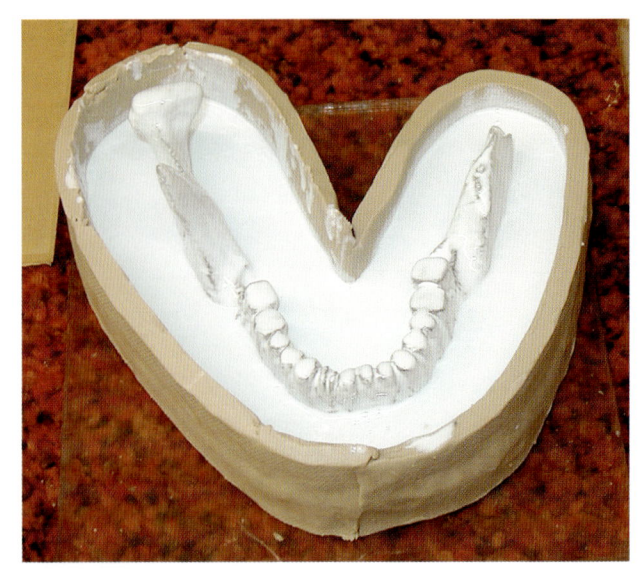

FIGURE 22.4 The mandible after the application of the detail coat of silicone.

above the seam line to capture the fine details of the prototype (Figure 22.4). This layer of silicone was allowed to cure at least 12 hours (Figure 22.5).

Batches of viscous silicone were mixed and applied over the detail coat to thicken the layer of mold rubber. In some areas, the mold rubber surface was smoothed with a finger coated in dish soap.

After the mold wall was built up to a thickness of at least 10 millimeters, any cavities or depressions, such as the orbital openings in the cranium, were filled with molding plaster to create support keys that maintained the shape of the cavities in the mold (Figure 22.6). A plaster mother mold was then constructed to cover all the silicone and support keys. The rigid mother mold maintained the shape of the thin, flexible rubber mold after it was pulled from the prototype (Figure 22.7).

The element was inverted and the clay flange that created the seam line was completely removed, exposing the cured silicone of the first half of the mold. A clay wall was constructed around the outer edge of this bottom surface in preparation to receive more liquid silicone that would make the second half of the mold (Figure 22.8).

The whole surface was then sprayed with the release agent to keep the rubber from sticking to the epoxy and, more importantly, from sticking to the first half of the

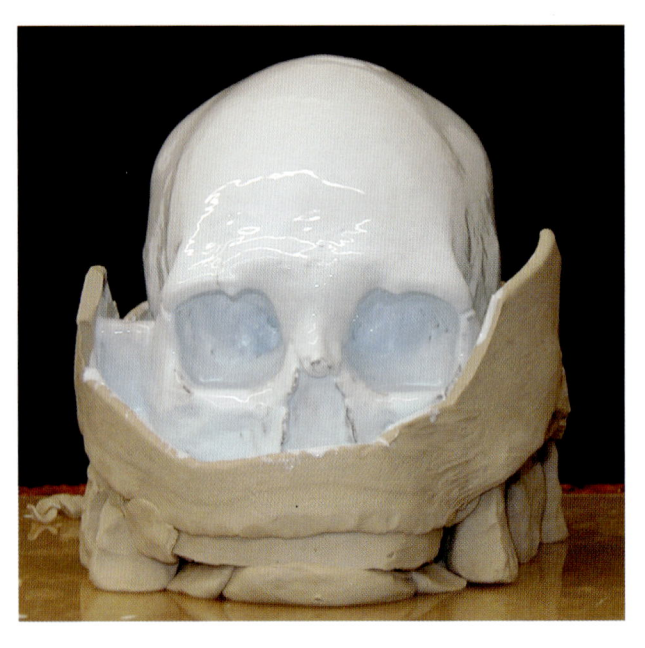

FIGURE 22.5 The cranium covered with cured silicone rubber.

mold. The same procedures were repeated to create the second half of the mold.

The mother mold was removed and the seam line between the two halves of the rubber mold was partially separated. The rubber was then carefully pulled off the prototypes and placed back into its mother mold. All excess clay was then cleaned off the seam lines and openings of each prototype.

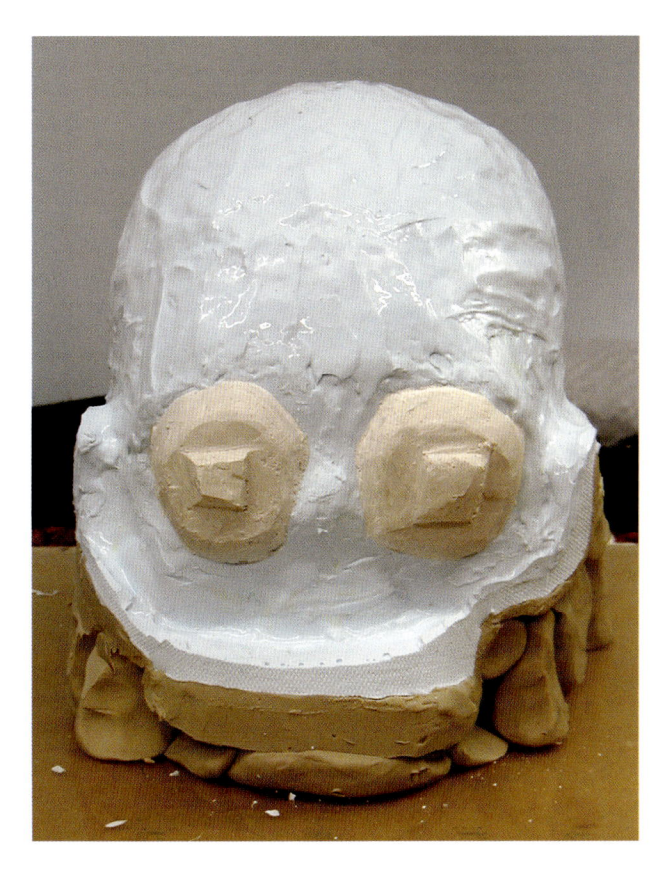

FIGURE 22.6 The cranium showing the thickened mold rubber and orbital support keys.

FIGURE 22.7 Left lateral view of the cranium and mother mold over the first half of the mold.

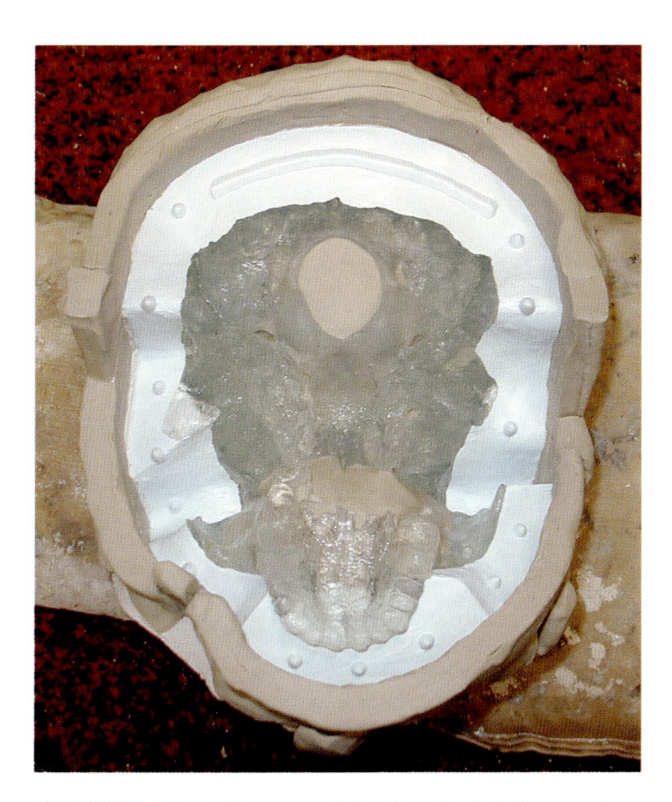

FIGURE 22.8 The bottom of the first half of the cranium mold after removal of the clay flange and construction of the clay wall.

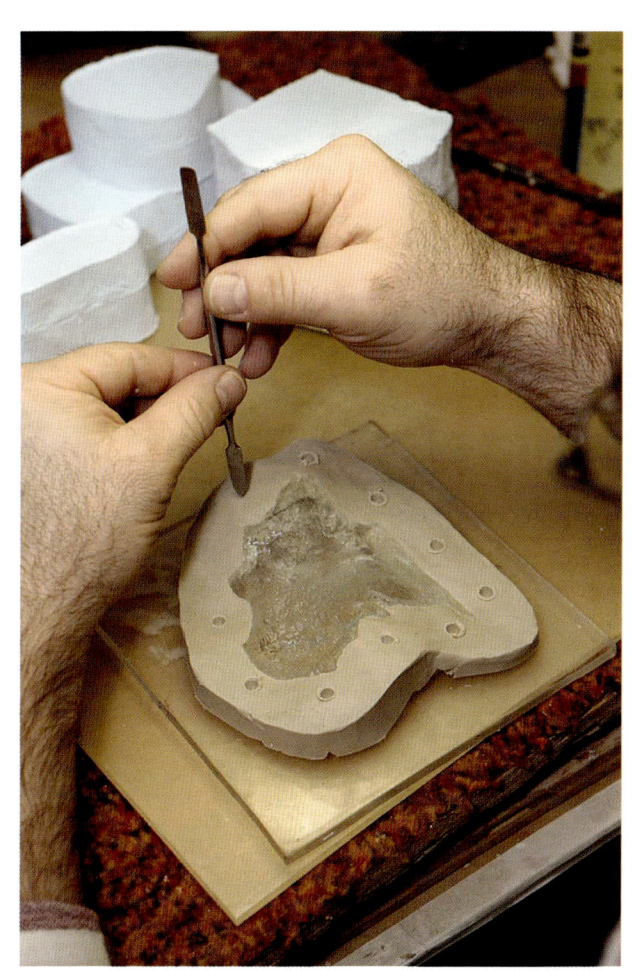

FIGURE 22.9 Setting a prototyped piece of innominate into clay in preparation for creating the first half of a two-part block mold.

Two-Part Block Molds

The fabricated, clear pieces of cranium, mandible, innominate, and the extracted projectile point were then reproduced using block molds. As in making the thin-walled molds, all cracks, gaps, foramina, and open spaces on each element were first blocked out with clay to prevent penetration by the liquid molding compound. Care was taken to maintain the outline of each opening as it was being sealed to let the observer know it existed.

The prototyped element was set into a clay base, the surface of which created the seam line of the mold. Indentations were carved into the clay surface; these would become registration keys in the mold (Figure 22.9).

The base was surrounded with a clay wall taller than the specimen and the release agent was sprayed onto the prototype to keep the mold rubber from sticking to it. A batch of liquid silicone molding compound was evacuated in a vacuum chamber and poured over the element until it was covered by at least 6 millimeters of silicone. The silicone was allowed to cure for at least 12 hours, creating a block that made up the first half of the mold.

The element was inverted and the clay base that created the seam line was completely removed. A dike wall was constructed around the first half of the mold, higher than the element, and the whole surface was sprayed with the release agent to keep the rubber from sticking to the epoxy and, more importantly, from sticking to the first half of the mold (Figure 22.10).

Evacuated silicone was poured over this side of the element until it was, again, completely covered (Figure 22.11). The silicone was allowed to cure for at least 12 hours, creating the block that made up the second half of the mold.

FIGURE 22.10   Jabo applies the release agent to the bottom of the first half of the block mold of a fragment of innominate after the dike wall was constructed.

FIGURE 22.11   Pouring the evacuated silicone to create the second half of a block mold.

The two halves of the mold were pulled apart, and the prototyped element was removed. All excess clay was then cleaned off the seam lines and openings of each prototype.

CASTING

All elements were cast in very hard, superfine-grained hydrocal gypsum cement reinforced with microfibers of fiberglass. Generally, the hydrocal was poured and brushed into each half of the mold. The liquid plaster was vibrated in order to work trapped air bubbles to the surface where they were removed with a fine paintbrush. The filled halves of the mold were then quickly and carefully placed together and laid flat on the table with a light weight placed on top of the mold to gently squeeze any excess plaster out through the seams. The plaster was allowed to set at least four hours. Then the molds were pulled apart and the casts removed (Figure 22.12). The flashing along the seam line was removed, and any tiny bubbles on the surface of the cast were filled with liquid plaster.

PAINTING

Acrylic and latex paints were used to color the plaster casts using photographs of the actual specimen as a guide. The tonal palate was designed to represent the brownish-tan color of the bone, adhering gray calcium carbonate deposits, and localized loss of the left-side bone surface patination, depicted as a lighter, bleached appearance (Figure 7.3B). A beige/almond base coat was applied; various secondary colors were thinned and dabbed over that. It was all washed with dilute black paint, and when that dried some of the secondary colors were lightly dry-brushed over the black wash to bring out the highlights and give the elements some three-dimensionality. Matte finish sealer was sprayed over all when the painting was completed.

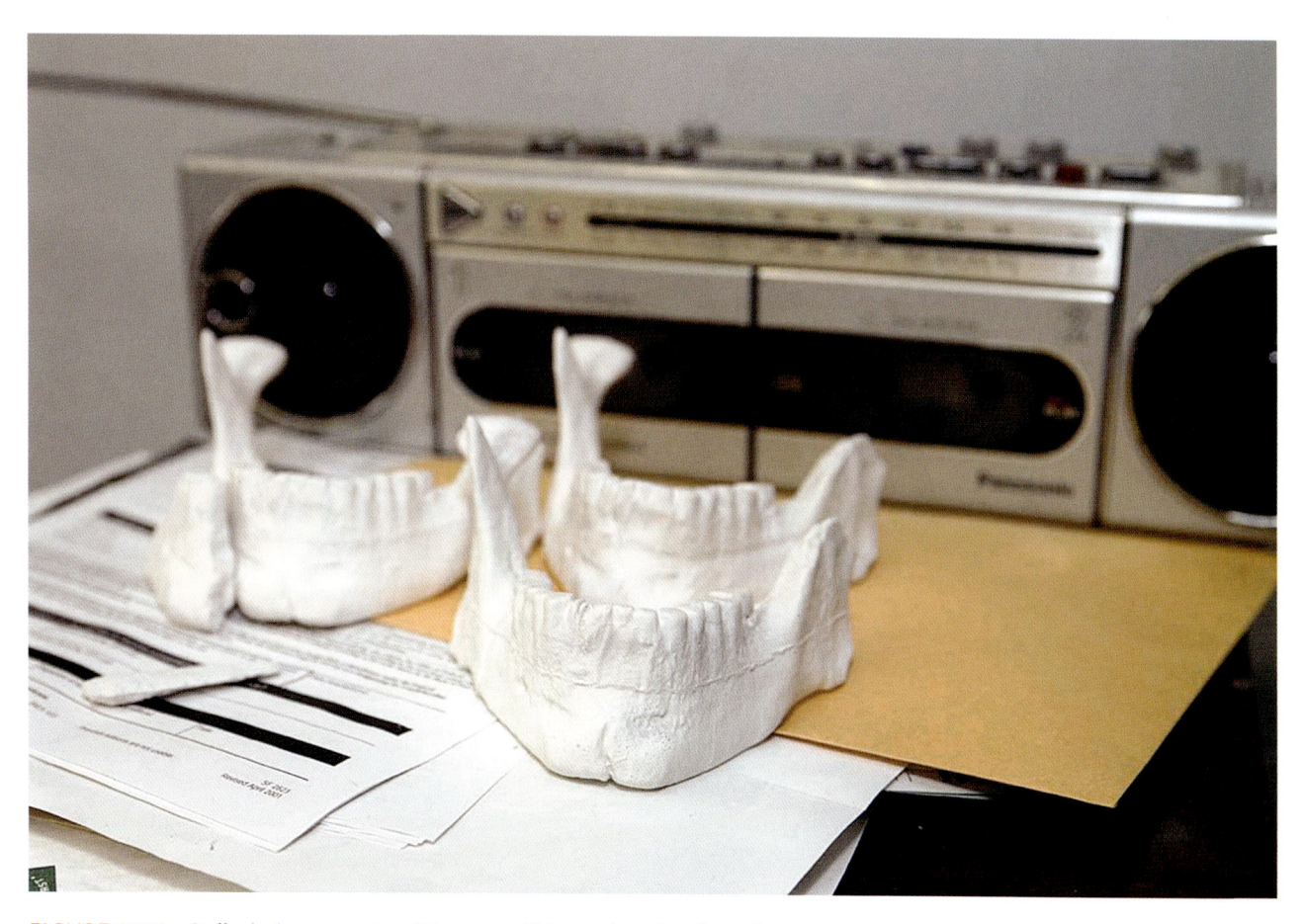

FIGURE 22.12 Pulled plaster casts of the mandible and projectile point.

Steven J. Jabo

23
The Point of the Story

Dennis J. Stanford

The human bones inadvertently found by Will Thomas and Dave Deacy along the shore of the Columbia River triggered the normal events that happen after such a discovery. The police were notified and the coroner was called to the scene. Based on historic artifacts found at the discovery site, and because of the skull's condition and physical characteristics, the individual was initially thought to be a white explorer or settler buried near a historic farmstead and recently exposed by river erosion. Lacking medical or legal significance, the case would have been closed, and the person now known as Kennewick Man probably would have been reburied in an unmarked grave in a local cemetery.

James Chatters, asked to recover and examine the remains, discovered a stone object embedded in the man's right hip (Figure 23.1). Chatters could see only a portion of the object, so it could not be readily identified (Chatters 2001). The stone appeared to be of volcanic origin, material commonly used by local prehistoric people for the manufacture of weapon tips, so the probability that it was a stone projectile point was high.

Shortly after the discovery in 1996, Chatters took the bone to the Kennewick General Hospital for computed tomography (CT) and x-ray scanning. The scans confirmed that the embedded object was a stone projectile point, but it appeared to be a type of weapon tip that was thousands of years old. Clearly, there was much more to the story. Kennewick Man had come to tell us about early Holocene life in America.

The limited views provided by the scans gave a general estimate of the projectile's shape and size, and showed that the point had partially serrated edges (Figure 23.2). These serrations, along with the general shape of the point, led to the conclusion that it was an early type known locally as Cascade (Butler 1961; Leonhardy and Rice 1970). Cascade points were used by groups, including members of the Old Cordilleran Tradition (OCT), from roughly 10,000–5,000 years ago who lived throughout the Northwest Coast region and Cascade Mountain Range from Alaska southward into the northern Great Basin and California (see Chapter 2 for a discussion of the OCT). There are three sub-varieties of Cascade projectile points termed Cascade A (dated to 8,000–4,000 RC yr. BP), Cascade B (8,500–6,500 RC yr. BP), and Cascade C (8,000–4,000 RC yr. BP) (Lohse 1995) (Figure 23.3).

The antiquity of the burial, as suggested by the projectile point, was verified through radiocarbon dating of bone samples from the skeleton, the results of which ranged between $8,130 \pm 40$ RC yr. BP and $5,750 \pm 100$ RC yr. BP, thus falling into the Cascade time period (Ames 2000). Stafford has dated the skeleton to $8,358 \pm 21$ RC yr. BP (Chapter 3). Since Kennewick Man died as many as 80 centuries before Europeans settled in the area, where did he come from and who threw a dart at him? Identifying the type and origin of the projectile point would help to answer these questions.

The present study reevaluates the classification of the projectile point embedded in Kennewick Man's right hip and also considers the weapon used to inflict this injury. Kennewick Man's right ilium was rescanned in 2006 using the latest industrial CT equipment available, which provided improved imagery (Figure 20.21). A relatively accurate plastic model of the point was created from the CT data, allowing for measurement and description based on the scans and the model (Figure 23.4). Any classification made without direct examination of the entire actual point, however, is equivocal.

The projectile point is nearly complete, with the exception of the tip, and has a length of 53.0 mm, a maximum width of 20.3 mm, and a maximum thickness of 6.7 mm. It is lanceolate in shape with a biconvex cross-section, a long, tapering stem, and a bulbous tip (Figure 23.5). It

FIGURE 23.1 Stone tool embedded in the internal surface of the right ilium.

FIGURE 23.2 External view of the exposed portion of the projectile point showing serration.

is also clear that the point has partially serrated edges. Serrations occur on points of various types, including broken points that were reworked to create new stems or tips. This rejuvenation process changes the width/thickness ratio of the artifact and, as a result, dulls the tool. Serrating the margins of a reworked point, however, produces a sharper cutting edge. During a comprehensive study of projectile points from Washington State's Puget Sound, serration was not used as a distinguishing trait since it occurs to some degree on points throughout most of the cultural sequence of the area (Croes et al. 2008).

Based on the characteristics of the point, it may be classified as something other than a Cascade. Cascade points, unlike the projectile embedded in Kennewick Man's right ilium, do not have long, tapering stems or bulbous tips. Because of its overall morphology, and re-sharpening events that resulted in a bulbous shape near the tip, I believe that the projectile point is more likely to

Dennis J. Stanford

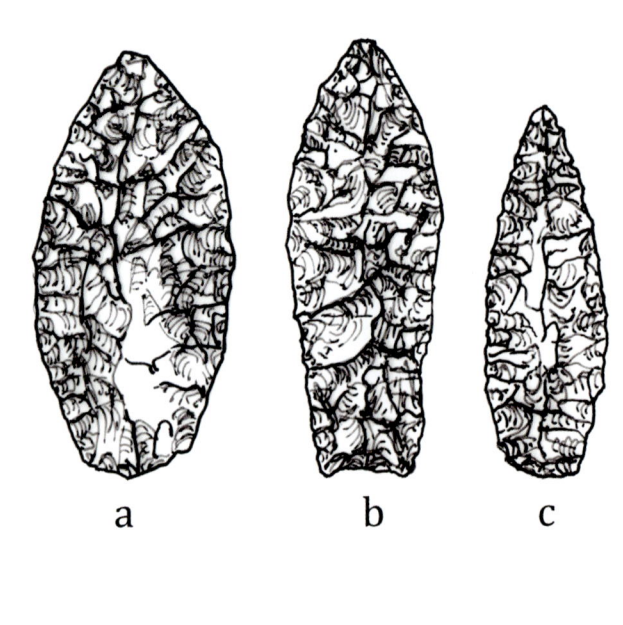

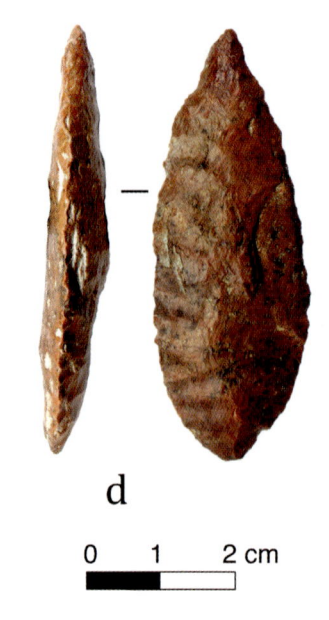

FIGURE 23.3 Three styles of Cascade projectile points: type a, type b, and type c (redrawn from Lohse and Schou 2008); d) Cascade point (side and face views) from a site near Vantage, Washington.

be a heavily reworked Haskett (Figure 23.6), rather than a Cascade point. As such, this provides an opportunity to consider multiple interpretations concerning the significance of the point and of Kennewick Man himself.

Haskett points were first made very long, so that they could be retipped or rebased when broken. They have relatively widely spaced flake scars that generally meet along the central long axis of each face of the point creating a diamond-shaped cross section. Broken Haskett points were refurbished by applying pressure flakes that followed the ridges between the prior flake scars. Each time the points were reworked their flake scars became narrower, thinner, and their cross-sections became more lenticular. When repointed they tended to develop foreshortened tips that were relatively bulbous, but were thinned to a needle-sharp point. The point lodged in Kennewick Man's hip was likely broken and reworked several times before it was eventually hafted to the dart shaft that was thrown at him.

Haskett projectile points are late Paleoindian weapon tips. They date slightly earlier than the Cascade and are thought to be associated with the final phase of the Western Stemmed Tradition (WST) (Bryan 1980) (see Chapter 2 for a discussion of the WST). If a WST occupation of Paisley Cave in Oregon is verified to date to 14,000 CAL yr. BP (Jenkins et al. 2012), then these people are the earliest group known to live in the region and their tradition lasted for more than 4000 years. Haskett

FIGURE 23.4 Both surfaces of the projectile point cast produced by CT scanning and additive manufacturing. Right angle arrows at the base indicate breakage that likely occurred during the attempt to remove the dart from the pelvis. Outlines show cross sections at marked locations. Schematic shows long profile illustrating the bulbous tip end.

sites dated between ca. 12,000 and 9,000 CAL yr. BP are found throughout the Great Basin and adjacent areas (Figure 23.7). One Haskett occupation site, with uncalibrated ^{14}C dates of ca. 10,500, was excavated at the Sentinel Gap near Yakima, just northeast of Kennewick, Washington (Huckleberry et al. 2003). It is possible, then,

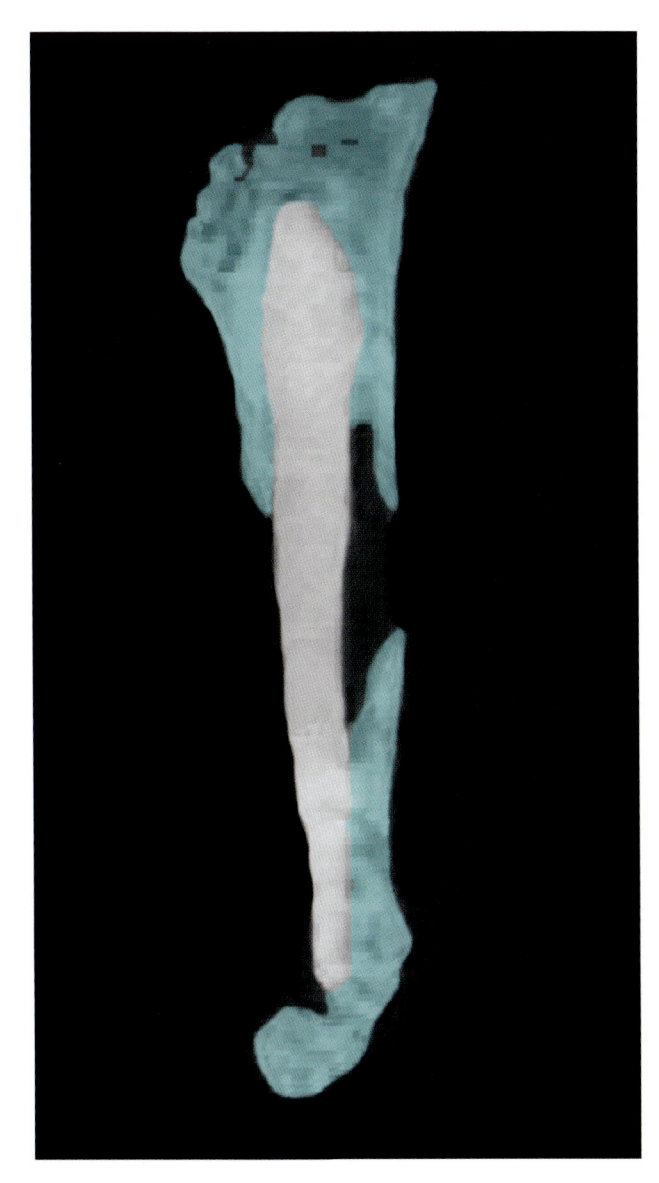

FIGURE 23.5 Digital CT image (not to scale) of the longitudinal cross section of the projectile point (gray) showing its bulbous tip end at the top. The bone (blue) has grown around the tip end and base leaving fenestrations in the external (right) and internal surfaces of the ilium.

that Kennewick Man lived in the area near the end of the Haskett phase, or perhaps he lived during the transition between the Western Stemmed and Old Cordilleran traditions. Was Kennewick Man a member of either of these traditions, or was he from some other place entirely? What sort of struggles might have been waged between members of these traditions during transitional periods?

One method of scientific analysis that would help with the identification of the projectile point is X-ray fluorescence (XRF). This non-destructive analytical technique measures the presence and percentages of elements that make up various kinds of materials, including stone. The point is made from a dark, fine grained, almost glassy igneous rock, possibly dacite or basalt. The ejecta from volcanic events have unique variations of elements, making it possible to identify with confidence the geographic locations of sources of volcanic stones. XRF analyses have been successfully used to identify these unique combinations of elements from specific volcanos. Over the years, large data banks characterizing the various flows have been developed by cooperating research labs. Volcanic sources are common throughout the entire Pacific Rim, and an XRF study of the projectile point embedded in Kennewick Man would tell where the source material originated. This information may not only shed light on the type of point in Kennewick Man's hip, but may also help to determine Kennewick Man's origin, where his travels may have taken him, and where he was when he received this injury. The Kennewick Man study team submitted a proposal to conduct XRF analysis of the projectile point to the Corps (Owsley 2006). XRF analysis was not permitted because of the possibility of damage to the ilium, despite the fact that the proposal identified four different non-destructive instruments that were available for this work. The study team's preference was the Edax Eagle II X-ray fluorescence spectrometer with element detection limits ranging down to 10 ppm, offered by the Washington State Crime Lab in nearby Marysville, Washington. The resulting spectrum would be highly diagnostic of rock types from specific formations, producing a "fingerprint" for provenance and source attribution purposes.

At the time, we saw little potential risk to the ilium, and now techniques utilized for XRF and mineralogical studies are even more refined, further minimizing any possibility of damage. This work should be done, as it could answer questions surrounding where and with whom Kennewick Man may have had his altercation (Chapter 32).

While there are still unresolved questions about the projectile point, this artifact's position in Kennewick Man's hip bone can tell us about the weapon used to inflict this injury and how the point entered his body (see

FIGURE 23.6 Reduction sequence of Haskett points: a) complete Haskett point from the type site in Idaho, b) resharpened tip on Haskett point from Fairfield, Idaho, c) Haskett point from McMinnville, Oregon, d) reconstruction of the point embedded in Kennewick Man. The arrow box identifies the tip digitally transferred from the type point (a) to illustrate how it might have looked before impact.

Chapter 7 for a detailed discussion of how Kennewick Man's body reacted to this injury). Stone points were carefully calibrated to the size and strength of materials used as foreshafts or shafts. A point was joined, or hafted, to its shaft or to a shorter detachable foreshaft with sinew, twine, or hide strips secured with mastic such as pine pitch. The haft width of the artifact's foreshaft or shaft would have been roughly 13 mm in diameter (Figures 7.37A and 7.38A). The momentum necessary to propel the point through skin and subcutaneous tissue while also taking out a piece of the iliac crest was likely accomplished with the aid of a spear thrower, also known as an atlatl (see Hrdlicka 2002, 2003 for discussions of atlatl ballistics). An atlatl is an extension of the user's arm that increases the thrust of a dart or spear (Figure 7.17). Although an atlatl made with primitive materials attained the world's record distance throw of 177.9 m (Clubb 1984), compilations of ethnographic records show that the normal distance for spear throwers was 10 to 30 m (Cunday 1989; Hutchings and Brüchet 1997). These shorter distances correspond to my own experiments in which accuracy and survival of weaponry decreased significantly at longer distances. At shorter distances, the dart impact angle was nearly parallel to the ground. Based on the orientation of the point in Kennewick Man's hip, it has been determined that the dart hit him from the right side at a downward arc-angle of 29 degrees (Figure 7.37A). This suggests that the dart was either thrown from a great distance and had begun to lose momentum, or that the dart was deliberately lobbed to hit a concealed Kennewick Man. An account relayed by George MacDonald of the Canadian Museum of Man suggests that this is a possibility. While working in Australia during the 1970s, MacDonald was in the bush with several aborigines armed with woomeras (atlatls). The relative effectiveness of his pistol and their woomeras was

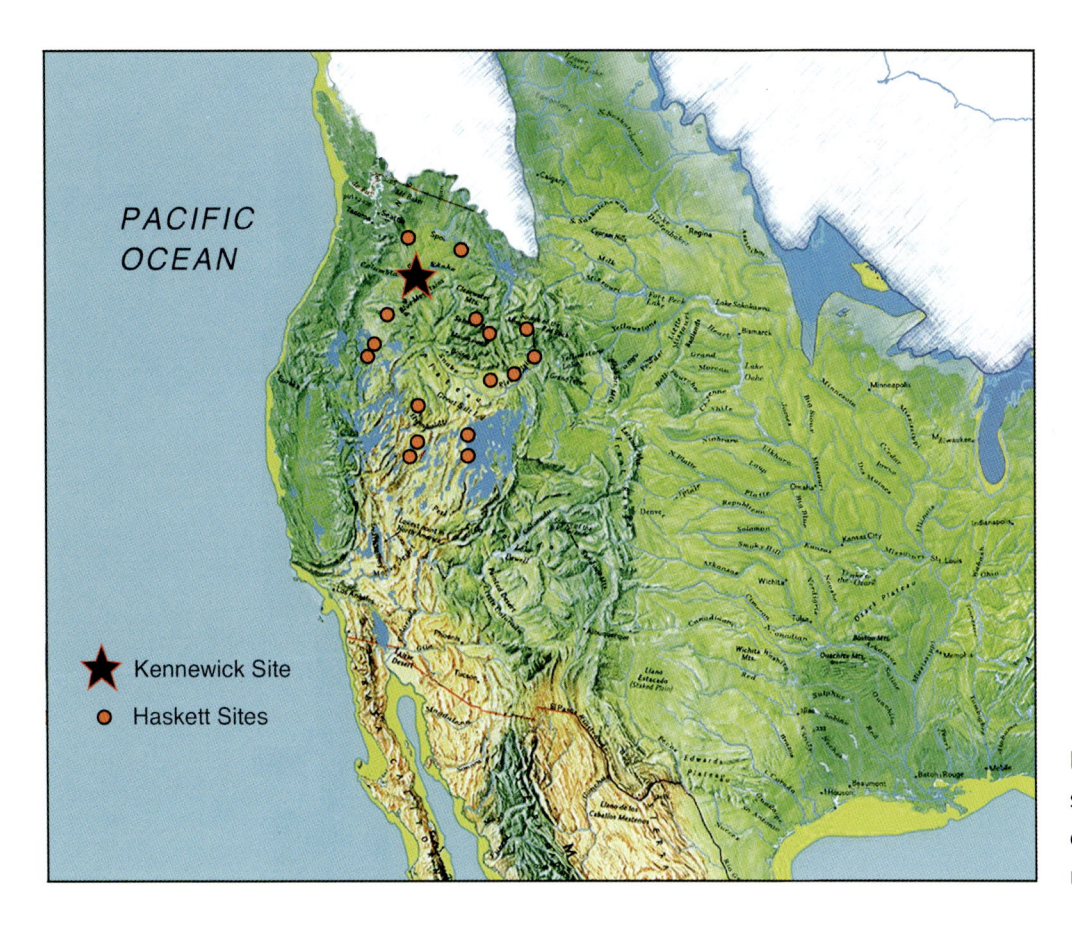

FIGURE 23.7 Map showing the distribution of reported Haskett sites.

considered; a contest was soon underway. In round one, a Styrofoam® cup was set up on a platform some 20 m away for each contestant and every contestant hit his cup. For round two, a large heavy rock was placed directly in front of each cup. The woomera darts arched over the rocks and hit their targets. Kennewick Man, therefore, could have been wounded even if he were crouched under cover.

When Kennewick Man was hit, about 2.9 mm of the projectile's tip broke off and is now missing. No lithic fragments are evident in the CT scan of the ilium, indicating fracture upon impact. Missing pieces were potentially embedded in the missing portion of the iliac crest, the location of initial impact, or deflected into adjacent soft tissue. Mobile tip fragments were likely excreted naturally by the body, or if not, were lost during the skeleton's erosion from the terrace. The base of the point was also damaged; a small rectangular flake was dislodged along its edge, perhaps when the dart shaft and its hafting material were removed from Kennewick Man's body (Figure 23.4). The point was so firmly embedded in the bone, however, that it remained lodged in Kennewick Man's hip throughout the rest of his life. The amount of bone growth around the artifact suggests that Kennewick man was wounded years before his death (Chapter 7).

While we can draw some conclusions about this injury, without an unobstructed view of the projectile point and the ability to conduct XRF analysis, variables necessary to learning about Kennewick Man's life, as well as the social and historic significance of the circumstances of this injury, are still unknown. Is the projectile point a Cascade point, as initially thought, a Haskett point, as this analysis suggests, or is it an unidentified type from a far off region of the Pacific Rim? Was he possibly wounded by friendly fire? Since he was a younger man with high testosterone levels, he may have been the victim of a local disagreement. On the other hand, he may have been a Cascade person, who during his early years was a long-distance ranger exploring new territory, and was wounded during an altercation with someone of the WST, with a Haskett point causing the

Dennis J. Stanford

wound. Or, if it is a Cascade point, and if he did spend the majority of his life near the coast as isotopic studies of his diet suggest (Chapter 32), why was his final resting place located far up the Columbia River? As he travelled along the Pacific coast of the Northwest, Kennewick Man may have been a trader and his wound inflicted by someone from a far-off region of the Pacific world such as California, Alaska, the Kamchatka Peninsula, or even Japan. An Asian origin of the stone is possible; thin, bifacial, lanceolate projectile points with tapering tangs were being used in Japan at about this same time period (Oda and Keally 1975). If he was a trader, he may have been a native from Asia, hence the importance of XRF, as well as DNA, testing in the future. Especially with the probability of continuing advances and the development of more sophisticated techniques for DNA recovery and analyses in the future, Kennewick Man's remains need to be preserved so that a more complete version of his story can eventually be told and lead to a to a deeper understanding of the complexity of the peopling of the west coast of the Americas. Many scenarios can be constructed, which only emphasizes the importance of future studies to obtain additional information about Kennewick Man and his injury.

REFERENCES

Ames, Kenneth M. 2000. Chapter 2. Review of the Archaeological Data. In *Cultural Affiliation Report* [September 2000]. National Park Service, http://www.nps.gov/archeology/kennewick/ames.htm.

Bryan, Alan L. 1980. The stemmed point tradition: An early technological tradition in western North America. In *Anthropological Papers in Memory of Earl H. Swanson, Jr.*, edited by Lucille B. Harten, Claude N. Warren, and Donald R. Tuohy, 77–107. Special Publication of the Idaho State University Museum, Pocatello.

Butler, B. Robert. 1961. *The Old Cordilleran Culture in the Pacific Northwest*. Idaho State College Museum Occasional Paper 5. Pocatello, Idaho.

Chatters, James C. 2001. *Ancient Encounters: Kennewick Man and the First Americans*. Simon & Schuster, New York.

Clubb, L. 1994. Untitled. *The Atlatl* 7(3):14.

Croes, D. R., S. Williams, L. Ross, M. Collard, C. Dennler, and B. Vargo. 2008. The projectile point sequences in the Puget Sound Region. In *Projectile Point Sequences in Northwestern North America*, edited by Roy L. Carlson and Martin Paul Robert Magne, 187–208. Archaeology Press, Simon Fraser University, Burnaby, BC.

Cunday, B. J. 1989. *Formal Variation in Australian Spear and Spearthrower Technology*. BAR International Series 546. British Archaeological Reports, Oxford.

Hrdlicka, D. 2002. How hard does it hit? *The Atlatl* 15(4):16–18.

———. 2003. How hard does it hit? A revised study of atlatl and dart ballistics. *The Atlatl* 16(2):15–18.

Huckleberry, Gary, Brett Lentz, Jerry Galm, and Stan Gough. 2003. Recent geoarchaeological discoveries in central Washington. In *Field Guide 4: Western Cordillera and Adjacent Areas*, edited by Terry W. Swanson, 237–250. The Geological Society of America, Boulder, CO.

Hutchings, W. Karl, and Lorenz Brüchet. 1997. Spearthrower performance: Ethnographic and experimental research. *Antiquity* 71(274):890–897.

Jenkins, Dennis L., Loren G. Davis, Thomas W. Stafford, Jr., Paula F. Campos, Bryan Hockett, George T. Jones, Linda Scott Cummings, Chad Yost, Thomas J. Connolly, Robert M. Yohe II, Summer C. Gibbons, Maanasa Raghavan, Morten Rasmussen, Johanna L. A. Paijmans, Michael Hofreiter, Brian M. Kemp, Jodi Lynn Barta, Cara Monroe, M. Thomas P. Gilbert, and Eske Willerslev. 2012. Clovis age Western Stemmed projectile points and human coprolites at the Paisley Caves. *Science* 337(6091):223–228.

Leonhardy, Frank C., and David G. Rice. 1970. A proposed culture typology for the lower Snake River, Southeastern Washington. *Northwest Anthropological Research Notes* 4(1):161–168.

Lohse, E. S. 1995. Northern intermountain west projectile point chronology. *Tebiwa* 25(1):3–51.

Lohse, E. S., and C. Schou. 2008. The Southern Columbia Plateau projectile point sequence: An informatics-based approach. In *Projectile Point Sequences in Northwestern North America,* edited by Roy L. Carlson and Martin P. R. Magne, 187–208. Archaeology Press, Simon Fraser University, Burnaby, BC.

Oda, S., and C. T. Keally. 1975. *Japanese Preceramic Cultural Chronology*. International Christian University, ICU Archaeology Research Center, Tokyo, Japan.

Owsley, Douglas W. 2006. Correspondence to Christopher B. Pulliam regarding projectile point analysis, dated February 10.

INCORPORATING POPULATION DATA

The following chapters investigate Kennewick Man's relationships to other populations, both ancient and modern, using primarily metric data from the cranium but also, to a lesser degree, nonmetric data from other skeletal elements. Fully appreciating the value of morphometric approaches requires some background, particularly with regard to the concept of cranial plasticity and the forces that do and do not affect cranial morphology.

Accumulating evidence supports the concept that cranial morphology provides as much insight into population structure and affinity as genetic data. Rogers and Harpending (1983), Relethford and Harpending (1994), and Relethford (2004) have specifically noted that, despite the possibility of developmental plasticity, signals of population structure and affinity are not significantly obscured. Roseman (2004), Roseman and Weaver (2007), and Betti et al. (2010) have observed that cranial morphology behaves according to neutral expectations. Finally, specific quantitative trait loci controlling facial, basi- and neurocranial morphology have recently been located (Sherwood et al. 2011). Discovery of specific genetic influences on craniofacial morphology continue (Liu et al. 2012), resulting in the expectation that we will soon have an even better understanding of genetic variation in cranial morphology.

Recent discoveries notwithstanding, it has been clear for some time that cranial variation tracks genetic variation better than it does environmental variation (Howells 1989). Hence arguments to the effect that " . . . we have very little research and data verifying the magnitude and the genetic mechanisms involved," and that "By contrast, we have volumes of data on how the cranium responds to nutritional, dietary . . . , and environmental forces within the life span, particularly during growth and development" (Swedlund and Anderson 2003:163) are inaccurate and misleading because they ignore the effects of evolutionary processes on populations.

Perceived cranial responses to environmental circumstances have frequently been used to argue against the idea that variation in cranial morphology could reflect genetic variation (Thomas 2000; Swedland and Anderson 2003; Armelagos and Van Gerven 2003). Franz Boas' (1911) study of immigrants to the United States is frequently, and mistakenly, considered sufficient evidence to make the case that the cranium is so plastic that it reflects little more than environmental variation. Thomas (2000:105) states "Boas showed that a broad headed population could become long headed in one generation." Armelagos and Van Gerven (2003:55) credit Boas with demonstrating that " . . . traits such as cephalic index (the most sacred of racial traits) changed by the magnitude of a race in one generation...." (The cranial index, or the cephalic index for the living, is the ratio of the breadth of the skull [head] to its length as a percentage.) Boas' case for change rested primarily on two of his study groups, Hebrews and Sicilians, the former becoming longer headed, the latter shorter headed after migrating to the United States. For Boas, this meant that they were converging toward an American type, at least as far as cranial index was concerned.

After closer assessments, it was clear that Boas' conclusions were not as clear as commonly thought. Sparks and Jantz (2002) demonstrate that the vast majority of variation among Boas' groups was accounted for by the ethnic group and only a small part to immigration status. Gravlee et al. (2003) also reanalyzed Boas' immigrant data, drawing conclusions apparently at odds with those of Sparks and Jantz (2002), but close examination of their results shows that the ethnic component dominates. Like Boas, Gravlee et al. (2003) showed that there were some differences between immigrants and their American born children, and, like Boas, chose to exaggerate the importance of the approximately 2.5 percent of among group variation in cephalic index accounted for by birthplace.

More recent evidence indicates that Boas' attribution of change to the American environment was questionable. What is puzzling about the long history of citing Boas in support of plasticity is that no one has thought to ask what caused the changes, beyond the "American environment." Boas even acknowledged that he had no specific causes in mind. He did consider, at some length, the possibility that changes in infant care, i.e., abandoning the practice of binding infants in cradles, were responsible for the lower cephalic index seen in American-born Hebrews, but ultimately rejected it. He also called attention to impaired growth of Sicilian children in New York, observing that it was the only group where American-born children were shorter than foreign-born.

The question of Hebrew and Sicilian changes has recently been reexamined by Jantz and Logan (2010). They conclude that the change in Hebrew cephalic index was in fact likely the result of discontinuing the practice of cradling infants in America. They further conclude that change in Sicilians resulted from poorer nutrition and impaired growth in America, where they were subject to discrimination and lower wages. Rather than the American environment being the responsible agent, changes in Hebrews and Sicilians resulted from specific behavioral and economic conditions unique to each group.

Critics of tracing genetic relationships from cranial morphology also argue that the transition to agriculture (and therefore a softer diet) modified crania through changes in masticatory stresses. The most commonly cited paper in support of this notion is Carlson and Van Gerven (1977), which purportedly shows that Neolithic Nubians have shorter, higher crania than Mesolithic Nubians, a change attributed to reduced functional demands on the masticatory process. Their work, however, relies on the assumption of evolutionary continuity between Mesolithic and Neolithic Nubians, despite a time difference of 10,000 years. Based on dental traits, Turner and Markowitz (1990) and Irish and Turner (1990) argue against population continuity. This is in line with Diamond and Bellwood's (2003) finding that the adoption of agriculture was often accompanied by migration. Along this line, Pinhasi and von Cramon-Taubadel (2009) conclude that cranial changes in Europe accompanying agriculture are due to demic diffusion. Attributing cranial change to the transition to agriculture involves an anti-migration bias that is extreme unless there is good evidence for continuity. In North America, population movements and replacement were the rule. For example, in the Northern Plains, Caddoan-speaking people moved into the Middle Missouri region, displacing and assimilating resident Siouan speakers, resulting in changes in cranial morphology (Jantz 1973; Key 1983). Subsequent migration of equestrian nomads into the Plains introduced new populations with identifiable cranial features. In the Great Basin of North America, the intrusion of Numic speakers is well-documented (Bettinger and Baumhoff 1982; Kaestle and Smith 2001).

Brachycephalization in Europe is another cranial modification often attributed to agriculture, acting through reduced masticatory demands (see Larsen 1997 for review). Yet it is now clear that brachycephalization did not really emerge in Europe until well after Medieval times (Boldsen 2000; Necrasov 1974). Twentieth-century Americans, surely placing lower demands on their masticatory apparatus than at any time in history, have lower cranial indices than nineteenth-century Americans (Jantz and Logan 2010), indicating a trend away from brachycephalization. A decrease in cranial index has also been observed in 20th-century Germans (Zellner et al. 1998).

Evidence that agriculture directly influences cranial morphology in any significant way is actually quite weak (Gonzalez-Jose et al. 2005; Baab et al. 2010). It has yet to be demonstrated that cranial form can be drastically altered by subsistence changes. The possible exception is the remarkable change in American cranio-facial morphology over the past 150 years (Jantz and Meadows Jantz 2000). However, two points need to be made in this regard. First, it has not yet been established that the observed changes are purely plastic in origin. Rapid evolutionary change has been documented in various organisms (Drake and Klingenberg 2008; Pergams and Lawler 2009; Herrel et al. 2008), and selection appears to be operating on modern Americans (Milot et al. 2011; Stearns et al. 2010). Second, to the extent that these changes in Americans are plastic, they can only be described as responses to extreme conditions. The modern American experience is unique in human history with respect to nutrition, disease, activity, infant mortality, and many other variables.

All this is to say that in the debate over cranial plasticity, it is not acceptable to merely cite Boas and others without serious consideration of the data on which their conclusions are based. The argument is not that cranial plasticity does not exist—but that there are definite limits to plasticity, and major morphological transformations continue to be primarily dictated by genetic influences.

REFERENCES

Armelagos, George J., and Dennis P. Van Gerven. 2003. A century of skeletal biology and paleopathology: Contrasts, contradictions and conflicts. *American Anthropologist* 105(1):53–64.

Baab, Karen L., Sarah E. Freidline, Steven L. Wang, and Timothy Hanson. 2010. Relationship of cranial robusticity to cranial form, geography and climate in *Homo sapiens*. *American Journal of Physical Anthropology* 141(1):97–115.

Betti, Lia, François Balloux, Tsunehiko Hanihara, and Andrea Manica. 2010. The relative role of drift and selection in shaping the human skull. *American Journal of Physical Anthropology* 141(1):76–82.

Bettinger Robert L., and Martin A. Baumhoff. 1982. The Numic spread: Great Basin cultures in competition. *American Antiquity* 47:485–503.

Boas, Franz. 1911. *Changes in Bodily Form of Descendents of Immigrants*. Final Report 38. Government Printing Office, Washington, DC.

Boldsen, Jesper L. 2000. Human demographic evolution: Is it possible to forget about Darwin? Research Report 14. Danish Center for Demographic Research, Odense University.

Carlson, David S., and Dennis P. Van Gerven. 1977. Masticatory function and post-Pleistocene evolution in Nubia. *American Journal of Physical Anthropology* 46(3):495–506.

Diamond, Jared, and Peter Bellwood. 2003. Farmers and their languages: The first expansions. *Science* 300(5619):597–603.

Drake, Abby G., and Christian P. Klingenberg. 2008. The pace of morphological change: Historical transformation of skull shape in St. Bernard dogs. *Proceedings of the Royal Society B* 275:71–76.

González-José, Rolando, Fernando Ramírez-Rozzi, Marina Sardi, Neus Martínez-Abadías, Miquel Hernández, and Hector M. Pucciarelli. 2005. Functional-cranial approach to the influence of economic strategy on skull morphology. *American Journal of Physical Anthropology* 128(4):757–771.

Gravlee, Clarence C., H. Russel Bernard, and William R. Leonard. 2003. Heredity, environment, and cranial form: A re-analysis of Boas's immigrant data. *American Anthropologist* 105(1):125–138.

Herrel, Anthony, Katleen Huyghe, Bieke Vanhooydonck, Thierry Backeljau, Karin Breugelmans, Irena Grbac, Raoul Van Damme, and Duncan J. Irschick. 2008. Rapid large-scale evolutionary divergence in morphology and performance associated with exploitation of a different dietary resource. *Proceedings of the National Academy of Science U.S.A.* 105(12):4792–4795.

Howells, William W. 1989. *Skull Shapes and the Map*. Peabody Museum of Archaeology and Ethnology, Harvard University, Cambridge, MA.

Irish, Joel D., and Christy G. Turner. 1990. West African dental affinity of Late Pleistocene Nubians—Peopling of the Eurafrican-South Asian Triangle II. *Homo* 41(1):42–53.

Jantz, Richard L. 1973. Microevolutionary change in Arikara crania: A multivariate analysis. *American Journal of Physical Anthropology* 38(1):15–26.

Jantz, Richard L., and Lee Meadows Jantz. 2000. Secular change in craniofacial morphology. *American Journal of Human Biology* 12(3):327–338.

Jantz, Richard L., and Michael H. Logan. 2010. Why does head form change in children of immigrants? A reappraisal. *American Journal of Human Biology* 22(5):702–707.

Kaestle, Frederika A., and David G. Smith. 2001. Ancient mitochondrial DNA evidence for prehistoric population movement: The Numic expansion. *American Journal of Physical Anthropology* 115(1):1–12.

Key, Patrick J. 1983. *Craniometric Relationships among Plains Indians: Cultural-Historical and Evolutionary Implications*. Reports of Investigation 34. Department of Anthropology, University of Tennessee, Knoxville.

Larsen, Clark S. 1997. *Bioarchaeology: Interpreting Behavior from the Human Skeleton*. Cambridge University Press, Cambridge.

Liu, Fan, Fedde van der Lijn, Claudia Schurmann, Gu Zhu, M. Mallar Chakravarty, Pirro G. Hysi, Andreas Wollstein, Oscar Lao, Marleen de Bruijne, M. Arfan Ikram, Aad van der Lugt, Fernando Rivadeneira, André G. Uitterlinden, Albert Hofman, Wiro J. Niessen, Georg Homuth, Greig de Zubicaray, Katie L. McMahon, Paul M. Thompson, Amro Daboul, Ralf Puls, Katrin Hegenscheid, Liisa Bevan, Zdenka Pausova, Sarah E. Medland, Grant W. Montgomery, Margaret J. Wright, Carol Wicking, Stefan Boehringer, Timothy D. Spector, Tomáš Paus, Nicholas G. Martin, Reiner Biffar, Manfred Kayser. 2012. A genome-wide association study identifies five loci influencing facial morphology in Europeans. *PLoS Genetics* 8(9):e1002932. doi:10.1371/journal.pgen.1002932.

Milot, Emmanuel, Francine M. Mayer, Daniel H. Nussey, Mireille Boisvert, Fannie Pelletier, and Denise Réale. 2011. Evidence for evolution in response to natural selection in a contemporary human population. *Proceedings of the National Academy of Science* 108(41):17040–17045.

Necrasov, Olga. 1974. The process of brachecephalisation in Rumanian populations from Neolithic to recent times. In *Biology of Human Populations*, edited by Bernhard Wolfram and Anneliese Kandler, 512–524. Gustav Fischer, Stuttgart.

Pergams, Oliver R. W., and Joshua J. Lawler. 2009. Recent and widespread rapid morphological change in rodents. *PLoS ONE* 4(7): e6452. doi:6410.1371/journal.pone.0006452.

Pinhasi, Ron, and Noreen von Cramon-Taubadel. 2009. Craniometric data supports demic diffusion model for the spread of agriculture into Europe. *PLoS ONE* 4(8):e6747. doi:10.1371/journal.pone.0006747.

Relethford, John H., and Henry C. Harpending. 1994. Craniometric variation, genetic theory, and modern human origins. *American Journal of Physical Anthropology* 95(3):249–270.

Relethford, John H. 2004. Boas and beyond: Migration and craniometric variation. *American Journal of Human Biology* 16(4):379–386.

Rogers, Alan R., and Henry C. Harpending. 1983. Population structure and quantitative characters. *Genetics* 105(4):985–1002.

Roseman, Charles C. 2004. Detecting interregionally diversifying natural selection on modern human cranial form by using matched molecular and morphometric data. *Proceedings of the National Academy of Sciences, U.S.A.* 101(35):12824–12829.

Roseman, Charles C., and Timothy D. Weaver. 2007. Molecules versus morphology? Not for the human cranium. *Bioessays* 29(12):1185–1188.

Sherwood, Richard J., Dana L. Duren, Michael C. Mahaney, John Blangero, Thomas D. Dyer, Shelley A. Cole, Stefan A. Czerwinski, William Cameron Chumlea, Roger M. Siervogel, Audren C. Choh, Ramzi W. Nahhas, Miryoung Lee, and Bradford Towne. 2011. A genome-wide linkage scan for quantitative trait loci influencing the craniofacial complex in humans (*Homo sapiens sapiens*). *The Anatomical Record: Advances in Integrative Anatomy and Evolutionary Biology* 294(4):664–675.

Sparks, Corey S., and Richard L. Jantz. 2002. A reassessment of human cranial plasticity: Boas revisited. *Proceedings of the National Academy of Science* 99(23):14636–14639.

Stearns, Stephen C., Sean G. Byars, Diddahally R. Govindaraju, and Douglas Ewbank. 2010. Measuring selection in contemporary human populations. *Nature Reviews Genetics* 11(9):611–622. doi:10.1038/nrg2831.

Swedlund, Alan, and Duane Anderson. 2003. Gordon Creek woman meets Spirit Cave man: response to comment by Owsley and Jantz. *American Antiquity* 68(1):161–167.

Thomas, David H. 2000. *Skull Wars: Kennewick Man, Archaeology, and the Battle for Native American Identity*. Basic Books, New York.

Turner, Christy G., and Michael A. Markowitz. 1990. Dental discontinuity between late Pleistocene and recent Nubians. I. Peopling of the Eurafrican-South Asian triangle. *Homo* 41(1):32–41.

Zellner, K., U. Jaeger, and K. Kromeyer-Hauschild. 1998. Das Phänomen der Debrachykephalisation bei Jener Schulkindern. *Anthropologischer Anzeiger* 56:301–312.

The Ainu and Jōmon Connection

24

C. Loring Brace, Noriko Seguchi, A. Russell Nelson, Pan Qifeng, Hideyuki Umeda, Margaret Wilson, and Mary L. Brace

When a photograph of Kennewick Man's skull taken by James Chatters appeared in a widely distributed public source (Egan 1996), many physical anthropologists familiar with worldwide human skeletal collections noticed a striking similarity to the indigenous Ainu peoples of northern Japan. Another Chatters photograph of Kennewick Man's cranium in the June 16, 1997, issue of *The New Yorker* magazine (Preston 1997) reinforced the initial impression. Kennewick Man's remains are 9,000 years older and clearly more robust, but a comparable reduction in robustness has occurred in all parts of the world where in situ continuity can be documented (Brace 2005; Brace and Tracer 1992; Brace et al. 1993).

The modern Ainu (Figure 24.1) cannot have given rise to Kennewick Man, but they may have shared a common ancestor. It has long been realized that the ancestors of the Ainu were the prehistoric Jōmon of the Japanese archipelago and the coastal area to the north and northeast (Koganei 1903, 1927, 1937; Hudson 1999). The early Jōmon could be the ancestors of Kennewick Man and possibly other early inhabitants of the western hemisphere. To test this hypothesis in a quantitative fashion, it is necessary to compare the dentition and craniofacial measurements of Kennewick Man to those of other populations.

DENTITION

The initial realization that Kennewick Man bore little resemblance to Native American groups led to the suggestion that the remains had Caucasoid features and may have been a European settler (Chatters 1997). Like those of Kennewick Man, the teeth of pre-nineteenth- century European settlers in the western hemisphere often reveal edge-to-edge bites (Brace 1977), but virtually none shows the degree of wear visible in Kennewick Man's dentition. His teeth are so heavily worn that few of them yield reliable measurements of the mesial-distal or buccal-lingual dimensions. The upper-right and lower-left third molars have only modest occlusal wear, but two third molars do not provide a firm basis for inferring anything about tooth size (Brace 1980).

In the eight-point scale (Smith 1983) to evaluate tooth wear, a score of 8 means the level of wear extends below the lowest reach of crown enamel. In Kennewick's mouth, 23 teeth are rated at the two most-extreme degrees of wear (Table 24.1). Among European remains, one must look well back into the Pleistocene to find comparable degrees of dental wear, and these have never been found in the edge-to-edge occlusion of the first European settlers in the western hemisphere.

Although the degree of wear is comparable to that of the "fern-root plane" in the Maori dentition (Taylor 1963), the form is very different. The beginning stage of his lingual wear on the second premolar and the first and second molars is shown in Figures 7.7 and 7.8. However, instead of pushing the tooth over on its side—so that the buccal surfaces of the lower molar crowns and roots and the comparable lingual surfaces of the upper molars function as the occlusal plane—the teeth are largely worn straight down with the roots remaining vertical. The result is highly reminiscent of the kind of wear seen in Eskimo dentitions prior to the introduction of canned foods (Waugh 1937; Poncins 1941). Mandibular tooth wear in Kennewick Man looks very much like that caused by hide-working on the Eskimo dentition. The extraordinarily "Eskimoid" maxillary-mandibular occlusal wear pattern is shown in Figures 7.10 and 7.11. Due to this severe wear, Kennewick Man's dentition provides little usable evidence of population affinity.

CRANIOFACIAL MEASUREMENTS

The next step in testing the relationship between Kennewick Man and various population groups was to com-

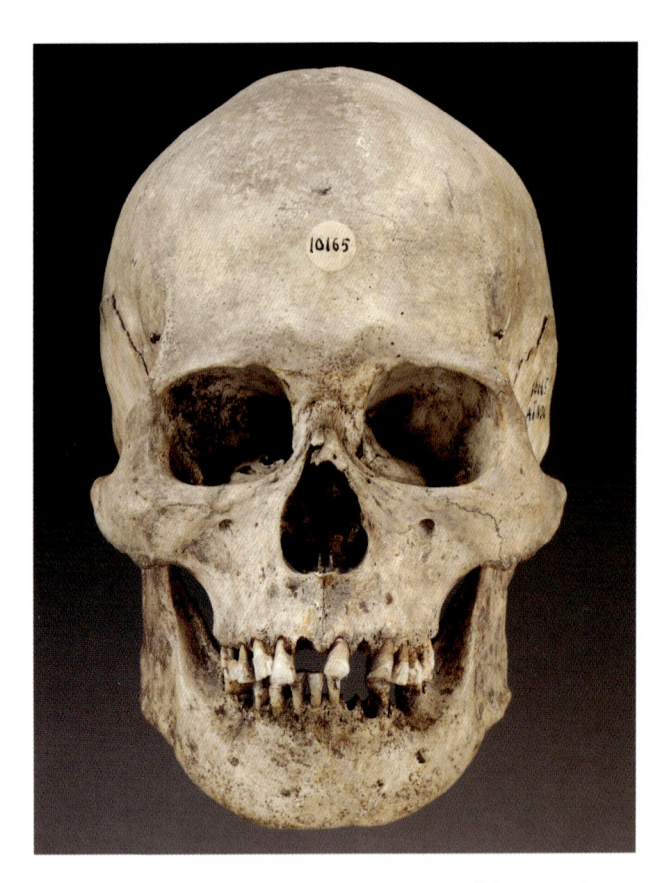

FIGURE 24.1 Frontal view of an Ainu skull (Musée de l'Homme, Paris, 10165).

TABLE 24.1 Kennewick Man's tooth wear.

	Upper		Lower	
	L	R	L	R
I^1	7	7	I_1 8	8
I^2	7	8	I_2 8	8
C	6	7	C 7	6
P^1	8	8	P_1 7	7
P^2	8	8	P_2 6	6
M^1	8	8	M_1 7	7
M^2	7	7	M_2 7	6
M^3	3	—	M_3 —	4

Note: Based on eight-point tooth-wear scale (Smith 1983).

TABLE 24.2 Craniofacial measurements of Kennewick Man in millimeters.

Nasal Height	(Martin No. 55)	55
Nasal Bone Height	(Martin No. 56 [2])	23
Piriform Aperture Height	(Martin No. 55 [1])	37
Nasion Prosthion Length	(Martin No. 48)	76
Nasion Basion	(Martin No. 5)	110
Basion Prosthion	(Martin No. 40)	113
Superior Nasal Bone Width	(Martin No. 57 [2])	14
Simotic Width	(Howells 1973)	9
Inferior Nasal Bone Width	(Martin No. 57 [3])	15
Nasal Breadth	(Martin No. 54)	25
Simotic Subtense	(Howells 1973)	3
Inferior Simotic Subtense	(Brace and Hunt 1990)	7
Fronto Orbital Width Subtense at Nasion	(Woo and Morant 1934)	17
Mid Orbital Width Subtense at Rhinion	(Woo and Morant 1934)	20
Bizygomatic Breadth	(Martin No. 45)	140
Basion Bregma	(Martin No. 17)	140
Basion Rhinion	(Brace and Hunt 1990)	116
Width at 13 (Fronto Malar Temporalis)	(Brace and Hunt 1990)	107
Width at 14 (Mid Orbital Width)	(Woo and Morant 1934)	60

Note: All Martin measurements are from Martin (1928).

pare craniofacial measurements. Converting the battery of measurements from all examples into Z-scores (Brace et al. 1989) makes it possible to combine male and female data without concern about sex-related differences in body size (Brace and Hunt 1990). Nineteen measurements were taken (Table 24.2) and compared to 48 groups and three individuals from the Americas, Europe, Africa, and Asia (Table 24.3).

The R- or Relationship-Matrix technique, originally applied to testing genetic elements (Harpending and Jenkins 1973), has been adapted for use on quantitative traits (Relethford and Blangero 1990). The genetic distances generated from R-matrices for samples from the different populations can be compared by the neighbor-joining procedure (Saitou and Nei 1987). This produces weblike trees (Hudson and Bryant 2006) or dendrograms in which the distances between twigs are proportional to the degree of relationship between the groups being compared.

The results of applying this technique to Kennewick Man and population reference groups are displayed in Figures 24.2–24.4. In each figure, the twig is long for Kennewick Man compared to those from the other sam-

TABLE 24.3 Population samples.

Population	No. of Females	No. of Males	Total	Population	No. of Females	No. of Males	Total
Ainu	23	33	56	Maryland	12	14	26
Aleut	15	17	32	Mexico	34	42	76
Atayal	14	22	36	Mississippi	30	25	55
Athabascan	23	19	42	Mongol	21	29	50
Blackfoot	17	15	32	Mongol Bronze	25	29	54
Borneo	5	7	12	Neolithic China	9	21	30
Buhl	1	0	1	North Australia	6	12	18
Buriat	14	12	26	North China	24	38	62
Burma	15	21	36	Patagonia	4	9	13
Bushman/Hottentot	3	9	12	Peru	29	26	55
China Bronze	16	45	61	Polynesia	87	74	161
Chukchi	7	12	19	Port au Choix	10	8	18
Chumash	35	36	71	Sepik River New Guinea	4	6	10
Contra Costa (SF Bay)	10	15	25	Snake River	8	6	14
Delaware	7	6	13	South Asia	20	31	51
Eskimo	63	77	140	South Australia	13	16	29
Great Lakes	7	8	15	South China	43	84	127
Haida	24	25	49	Spirit Cave	0	1	1
Heilongjiang	8	10	18	Sudan	5	10	15
Hopi	5	7	12	Tennessee Archaic	10	19	29
Indian Knoll	19	22	41	Thai	27	37	64
Japan	100	217	317	Tierra del Fuego	7	15	22
Java	1	4	5	West Africa	51	54	105
Jōmon	6	3	9	West Europe	111	152	263
Kennewick	0	1	1	Windover	17	15	32
Lagoa Santa	2	3	5	Wizards Beach	0	1	1

ples, because he is a unique individual whereas most of the twigs are for actual population samples and there are elements of common variance in the calculation of each, which cannot be the case when a single individual is compared.

Kennewick Man was tested against a worldwide set of human craniofacial samples as is shown in Figure 24.2. Kennewick Man lies on a twig with the Ainu and Polynesians. In this plot, the Jōmon sample is in another cluster with a 300- to 1,700-year-old sample from Contra Costa County in San Francisco Bay (Cook and Heizer 1962) and the 8,200 to 9,500-year-old Lagoa Santa sample from southeast Brazil (Neves et al. 2003). Other twigs in the Jōmon clusters are the Haida from the North American Northwest Coast as well as other North American samples, and the people of Tierra del Fuego and Patagonia at the southern end of South America.

Like Kennewick Man's remains, Archaic examples such as Buhl in Idaho (Green et. al. 1998) and Spirit Cave and Wizards Beach in Nevada (Jantz and Owsley 2001; Edgar et. al. 2007) appear at the end of long twigs. Although all these Paleoamericans lived about 9,000 to 10,000 radiocarbon years before the present, the Idaho and Nevada remains are more closely related to other Archaic and Mississippian groups than to the Jōmon and Kennewick Man.

Despite more than a century of claims to the contrary (Quatrefages 1881; ten Kate 1885; Verneau 1903; Rivet

FIGURE 24.2 Neighbor-joining dendrogram derived from cranial measurements of population samples from Europe, Africa, Australia, Polynesia, Asia, North and South American Archaic and Mississippian groups, and Kennewick Man's cranium.

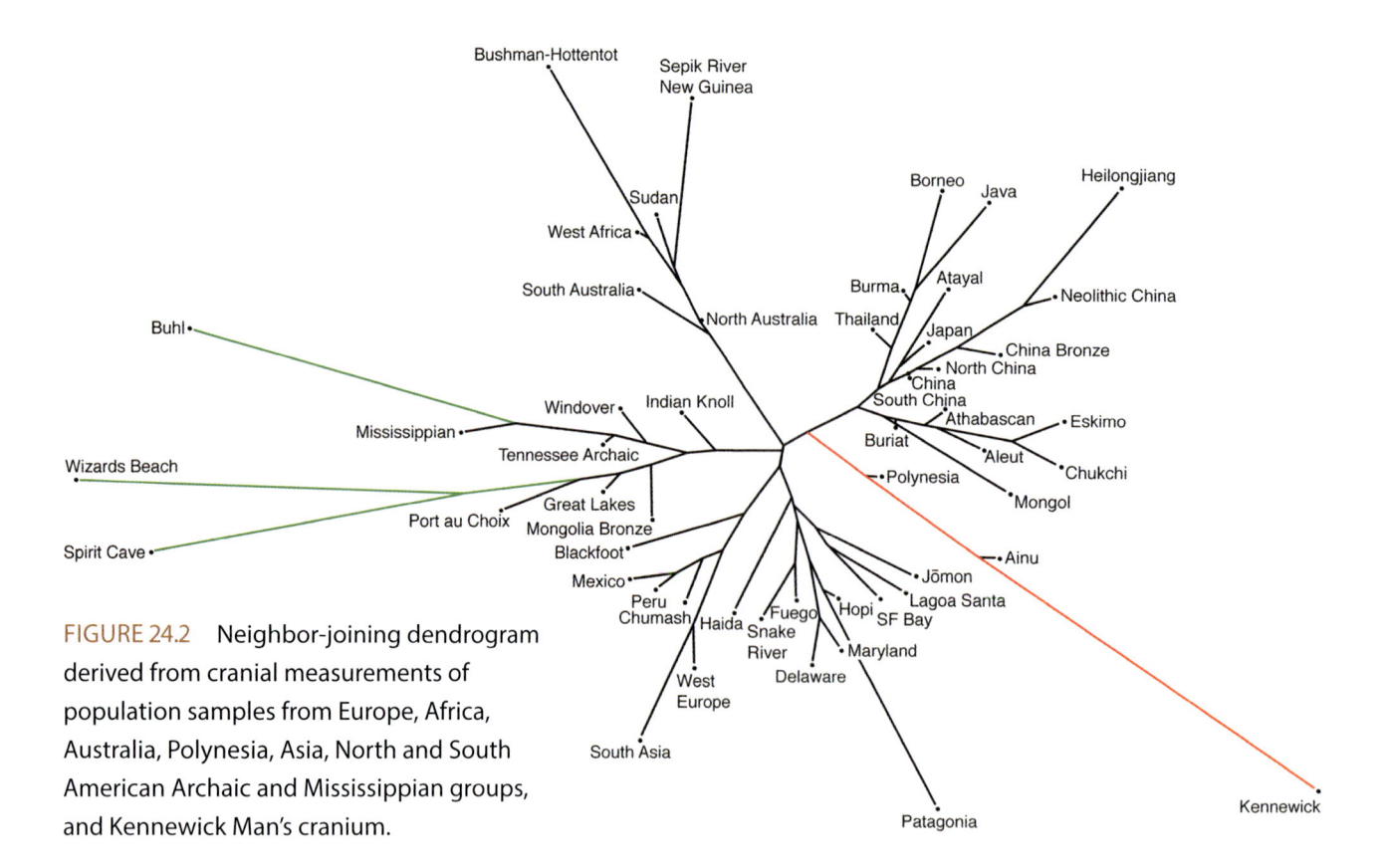

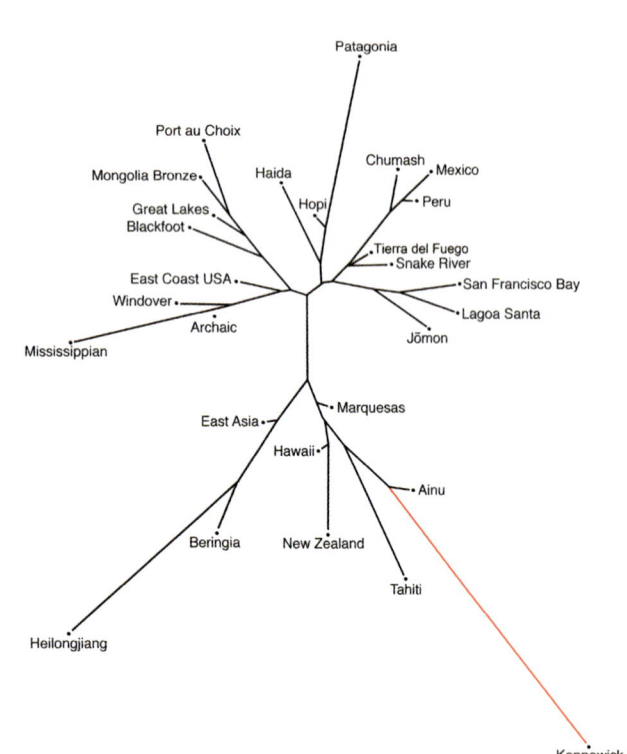

FIGURE 24.3 Neighbor-joining dendrogram with Asian and New World samples plus Kennewick Man, but omitting African, Australian, and European groups.

1908, 1925; Neves and Pucciarelli 1991; Dewar 2002; Neves et al. 2003; Neves and Hubbe 2005; Neves, Hubbe, and Correal 2007; Neves, Hubbe, and Piló 2007), it appears that neither Africans nor Australians contributed to the form of the first Americans. Figure 24.2 also reveals that the longstanding claims of a probable Melanesian or Australian source for the Lagoa Santa remains in Brazil are not supported by a comparison of relevant samples (Seguchi et al. 2006; Seguchi et al. 2011). At the time the first northeast Asians were moving along the coasts or across the Bering land bridge into the New World, tooth size of the Murray Valley Australians was the same as that of *Homo erectus* (Brace 1980). By the time that Europeans colonized Australia, it had reduced slightly to the size of the Neanderthals (Brace 2005), but nowhere in the entire western hemisphere was there ever a human population that had teeth with dimensions comparable to those of recent or prehistoric Australian aborigines.

Removing European, African, South Asian, Melanesian, and Australian samples results in Figure 24.3. Again, Kennewick is on the same twig with the Ainu and Polynesian samples. The East Asian samples branch off before the twig that ends in Kennewick. The Jōmon, as before,

466 *Chapter 24*

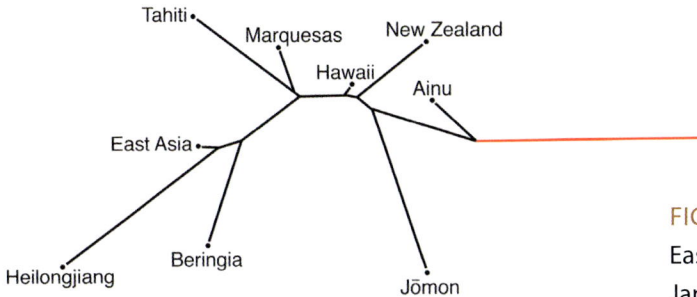

FIGURE 24.4 Neighbor-joining dendrogram for living East Asian and Oceanic groups, the prehistoric Jōmon of Japan, and Kennewick Man.

cluster tightly with the New World populations all the way from the Northwest Coast and the San Francisco Bay down to Brazil and the southern tip of South America.

In Figure 24.4, the various Chinese samples were similar enough to combine them into a single East Asia twig. The Beringia twig is comprised of Eskimos and Aleuts, and the twig labeled Heilongjian would be called Manchus by many observers. Heilongjiang (Black Dragon River) is the Chinese name for the Amur River, which separates northeast China from Siberia.

Kennewick Man is on the same twig as the Ainu and Polynesians. The Ainu are clearly the descendants of the prehistoric Jōmon (Brace and Nagai 1982; Fitzhugh 1999; Hudson 1999), although they have been historically identifiable for nearly 2,000 years. The case can therefore be made that the Jōmon of the northeast coast of Asia were an important source for the first known inhabitants of the western hemisphere.

ORIGINS OF THE JŌMON

Because of the limited archaeological research at the northeastern edge of Asia, the earliest traces of the Jōmon are difficult to identify. It is generally accepted that they descended from Late Pleistocene inhabitants of the eastern coast of Asia and farther west (Zhushchikovskāia 2005), but whether they had come north from Southeast Asia (Hanihara 1991) or south from Siberia (Kozintsev 1993) has been a subject of continuing debate (Hudson 1999; Fitzhugh 1999). As of 1850, the Jōmon's Ainu descendants were found from the Japanese island of Hokkaido to Sakhalin Island off the coast of Manchuria, and the Kurile island chain north to the southern part of the Siberian peninsula of Kamchatka (Fitzhugh 1999).

The Initial Jōmon of Japan has a calibrated date of 16,500 years ago and is associated with one of the oldest pottery-making traditions in the world (Habu 2004; Lapteff 2006). That was preceded, as was the case at the western edge of the Old World, by a prolonged reliance on earth-oven cookery during the last glaciation, to heat food that would otherwise remain frozen (Brace 1976). The result was a comparable amount of dental reduction among both the eastern and western populations of the northernmost inhabitants, a concept known as the "neutral theory" (Brace 2005). This reduction—rather than Southeast Asian origins and the Sundadont dental pattern as defined by Turner (1989)—may be why incisor shoveling is markedly less evident in the Jōmon and Ainu dentition than in most of the living inhabitants of East Asia.

Among the Ainu, average tooth size, tooth by tooth, is significantly less than that of any other Asian group (Brace and Nagai 1982; Brace et al. 1989). This suggests that the Ainu descended from people such as the Jōmon, who lived in the north temperate zone during the last glaciation.

Other characteristics of the Ainu also provide clues to their Jōmon ancestors' area of origin. The Ainu have the lightest average skin color of any Asians, which argues against tropical ancestry (Harvey and Lord 1978). Likewise, beards and body hair occur in abundance among the Ainu (Hutchinson et al. 1902) but are less common among tropical peoples.

THE JŌMON IN THE NEW WORLD

Watercraft capable of formidable oceanic voyaging were present in prehistoric Japan (Aikens et al. 1986). It is reasonable to assume that the Ainu canoe, built of planks sewn to a dugout bottom piece (Iwasaki-Goodman and Nomoto 1999), reflects the style of their Jōmon forebears. Canoes manufactured by essentially the same technique were used by Paleoamericans of the Pacific Northwest (Ohtsuka 1999), the southern California coast (Hudson

TABLE 24.4 Mahalanobis D^2 values between populations and Kennewick Man.

	Ainu	Beringia	Heilong	Jōmon	Kennewick Man	Tahiti	Hawaii	New Zealand	Marquesas	E. Asia
Ainu	0.00									
Beringia	14.88	0.00								
Heilong	19.29	13.94	0.00							
Jōmon	8.24	15.78	26.90	0.00						
Kennewick Man	16.70	38.05	45.47	34.68	0.00					
Tahiti	11.78	16.04	18.29	19.41	26.76	0.00				
Hawaii	5.75	10.75	16.74	9.00	26.73	8.44	0.00			
New Zealand	6.41	17.95	18.90	13.98	26.54	14.30	4.62	0.00		
Marquesas	10.00	10.36	15.36	16.40	26.76	7.69	5.32	7.32	0.00	
E. Asia	12.70	8.80	9.89	13.80	36.36	12.23	7.56	11.55	7.43	0.00

1976), Chile (Lothrop 1932), and past the Strait of Magellan to the west coast of Patagonia (Cooper 1917).

THE JŌMON IN OCEANIA

It has been suggested that the Jōmon gave rise to the peoples who moved out into the Pacific and settled the islands of Oceania (Brace and Nagai 1982; Brace et al. 1989; Brace and Hunt 1990; Brace and Tracer 1992). This is consistent with the picture presented in Figure 24.4 and Mahalnobis D^2 values (Table 24.4). However, such a possibility has also been vigorously denied (Pietrusewky 2005, 2006a, 2006b). If this connection is true, then the Jōmon spoke an Austronesian language. Although there is no direct evidence concerning their language, there are some hints in the language spoken by the Ainu.

Some analysts have treated Ainu as a "linguistic isolate" (Laufer 1917). However, the complete lack of voiced consonants or surds, the presence of a vigesimal number system (Laufer 1917; Sternberg 1929), and more than 70 word parallels between Ainu and "Malayo-Polynesian" (Gjerdman 1926) suggest that the Ainu language was derived from proto-Austronesian (Murayama 1992).

Japanese itself, while basically an Altaic language (Murayama 1972), has been called a mixed-speech because a very large portion of its basic vocabulary is of "Malayo-Polynesian" origin. This suggests long-time contact with an Austronesian-speaking people (Murayama 1976). And indeed, the Japanese came from the mainland into an island archipelago that had long been inhabited by a non-Japanese-speaking people who could be one of the main sources from which the Austronesian-speaking peoples arose. The Ainu language itself could be called a mixed-speech because of the overwhelming long-term influence of a more powerful Altaic-speaking people (Ohno 1970).

It has been suggested that the sewn-plank canoe was brought to the western hemisphere by Polynesians (Graebner 1913; MacLeod 1929). Given that the inhabitants of the northeast coast of Asia were in place long before humans arrived in Polynesia and that the Jōmon used their boat-making skills to travel to Beringia and down the full extent of the west coast of the New World long before the Polynesians used the same kinds of boats to arrive in the Pacific Islands (Fladmark 1983), the Jōmon may represent the population from which the western hemisphere was initially settled, and the seagoing capabilities of these people may have made their explorations possible.

Kennewick Man cannot be seen as the first Old World immigrant to the western hemisphere, nor can he be implicated as the bearer of the sewn-plank canoe. However, he can certainly be recognized as the direct descendant and morphological representative of the people who were responsible for these accomplishments and more.

REFERENCES

Aikens, C. Melvin, Kenneth M. Ames, and David Sanger. 1986. Affluent collectors at the edges of Eurasia and North America: Some comparisons and observations on the evolution of society among north-temperate coastal hunter-gatherers. In *Prehistoric Hunter Gatherers in Japan: New Research Methods*, edited by Takeru Akazawa and C. Melvin Aikens, 3–26. Bulletin 27. The University Museum, The University of Tokyo, Tokyo.

Brace, C. Loring. 1976. Tooth reduction in the Orient. *Asian Perspectives* 19(2):203–219.

———. 1977. Occlusal to the anthropological eye. In *The Biology of Occlusal Development*, edited by James A. McNamara Jr., 179–209. Monograph 7, Craniofacial Growth Series. Center for Human Growth and Development, University of Michigan, Ann Arbor.

———. 1980. Australian tooth size clines and the death of a stereotype. *Current Anthropology* 21(2):141–164.

———. 2005. "Neutral theory" and the dynamics of the evolution of "modern" human morphology. *Human Evolution* 19(1):19–38.

Brace, C. Loring, Mary L. Brace, and William R. Leonard. 1989. Reflections on the face of Japan: A multivariate craniofacial and odontometric perspective. *American Journal of Physical Anthropology* 78(1):93–113.

Brace, C. Loring, and Kevin D. Hunt. 1990. A non-racial perspective on human variation: A(ustralia) to Z(uni). *American Journal of Physical Anthropology* 82(3):341–360.

Brace, C. Loring, and Masafumi Nagai. 1982. Japanese tooth size, past and present. *American Journal of Physical Anthropology* 59(4):399–411.

Brace, C. Loring, and David P. Tracer. 1992. Craniofacial continuity and change: A comparison of Late Pleistocene and Recent Europe and Asia. In *The Evolution and Dispersal of Modern Humans in Asia*, edited by Takeru Akazawa, Kenichi Aoki, and Tasuku Kimura, 439–471. Hokusen-Sha, Tokyo.

Brace, C. Loring, David P. Tracer, Lucia Allen Yaroch, John Robb, Kari Brandt, and A. Russell Nelson. 1993. Clines and clusters versus "Race." A test in ancient Egypt and the case of a death on the Nile. *American Journal of Physical Anthropology* 36(S17):1–31.

Chatters, James C. 1997. Encounter with an ancestor. *Anthropology Newsletter* 38(1):9–10.

———. 2001. *Ancient Encounters: Kennewick Man and the First Americans*. Simon & Schuster, New York.

Cook, Sherburne F., and Robert F. Heizer. 1962. *Chemical Analysis of the Hotchkiss Site (CCo-138)*. University of California Archaeological Survey Report 57(1):1–25.

Cooper, John M. 1917. Analytical and critical bibliography of the tribes of Tierra del Fuego adjacent territory. *Bureau of American Ethnology Bulletin* 63.

Dewar, Elaine. 2002. *Bones—Discovering the First Americans*. Random House, Canada.

Edgar, Heather J. H., Edward A. Jolie, Joseph F. Powell, and Joe E. Watkins. 2007. Contextual issues in Paleoindian repatriation: Spirit Cave Man as a case study. *Journal of Social Archaeology* 17(1):101–122.

Egan, Timothy. 1996. "Tribe stops study of bones that challenge history." *The New York Times*, September 30, A12.

Fitzhugh, William W. 1999. Introduction. In *Ainu: Spirit of a Northern People*, edited by William W. Fitzhugh and Chisato O. Dubreuil, 8–26. Arctic Studies Center, National Museum of Natural History, Smithsonian Institution, University of Washington/Perpetua Press, Los Angeles.

Fladmark, Knut R. 1983. Times and places: Environmental correlates of Mid-to-Late Wisconsin human population expansion in North America. In *Early Man in the New World*, edited by Richard Shutler, Jr., 13–41. Sage Publications, Beverly Hills.

Gjerdman, Olof. 1926. Word parallels between Ainu and other languages. *Le Monde Oriental* 20:29–84.

Graebner, Fritz. 1913. Amerika und die Südseekultures. In *Ethnologica im Auftrage des Vereins für Főrderung des städtischen Rauten-straugh*, edited by W. Foy, 43–66. Joest-Museums für Vőlkerkunde un Cőln, Heft 1.

Green, Thomas J., Bruce Cochran, Todd W. Fenton, James C. Woods, Gene L. Titmus, Larry Tieszen, Mary Ann Davis, and Suzanne J. Miller. 1998. The Buhl burial: A Paleoindian woman from southern Idaho. *American Antiquity* 63(3):437–456.

Habu, Junko. 2004. *Ancient Jōmon of Japan*. Cambridge University Press, Cambridge.

Hanihara, Kazuro. 1991. Dual structure model for the population structure of the Japanese. *Japan Review* 2:1–33.

Harpending, Henry, and Trefor Jenkins. 1973. Genetic distance among southern African populations. In *Methods and Theories of Anthropological Genetics*, edited by Michael H. Crawford and Peter L. Workman, 177–198. New Mexico Press, Albuquerque.

Harvey, Robin G., and Jayne M. Lord. 1978. Skin colour of the Ainu of Hidaka, northern Japan. *Annals of Human Biology* 5(5):459–467.

Howells, William W. 1973. *Cranial Variation in Man: A Study by Multivariate Analysis of Patterns of Difference Among*

Hudson, Dee Travis. 1976. Chumash canoes of Mission Santa Barbara: The revolt of 1824. *The Journal of California Anthropology* 3(2):5–15.

Hudson, Mark J. 1999. Ainu ethnogenesis and the northern Fujiwara. *Arctic Anthropology* 36(1):73–83.

Hudson, Daniel H., and David Bryant. 2006. Application of phylogenetic networks in evolutionary studies. *Molecular Biology and Evolution* 23(2):254–267.

Hutchinson, Henry N., J. W. Gregory, and Richard Lydekker. 1902. *The Living Races of Mankind*. Hutchinson & Company, London.

Iwasaki-Goodman, Masami, and Masahiro Nomoto. 1999. The Ainu on whales and whaling. In *Ainu: Spirit of a Northern People*, edited by William W. Fitzhugh and Chisato O. Dubreuil, 222–226. Arctic Studies Center, National Museum of Natural History, Smithsonian Institution, University of Washington/Perpetua Press, Los Angeles.

Jantz, Richard L., and Douglas W. Owsley. 2001. Variation among early North American crania. *American Journal of Physical Anthropology* 114(2):146–155.

Koganei, Yoshikiyo. 1903. Über die Uberwohner von Japan. *Globus* 84(7):101–106.

———. 1927. Zur Frage der Abstammung der Aino und ihre Verwandtschaft mit anderen völkern. *Anthropologischer Anzeiger* 4(3):201–207.

———. 1937. Zur Frage des "sudlichen Elementes" in Japanischen Volke. *Zettschrift für Rassenkunde* 5(2):123–130.

Kozintsev, Alexander G. 1993. Ainu origins in the light of modern physical anthropology. *Homo—Journal of Comparative Human Biology* 44(2):105–127.

Lapteff, Sergey. 2006. Relationships between Jōmon culture and the cultures of the Yangtze, South China, and continental Southeast Asia. *Journal of the International Research Center for Japanese Studies* 18:249–286.

Laufer, Berthold. 1917. Vegesimal and decimal system in the Ainu numerals. *Journal of the American Oriental Society* 37(3):192–208.

Lothrop, Samuel K. 1932. Aboriginal navigation of the west coast of South America. *Journal of the Royal Anthropological Institute of Great Britain and Ireland* 62:229–256.

MacLeod, W. C. 1929. On the Southeast Asian origins of American culture. *American Anthropologist* 31(3):554–560.

Martin, Rudolf. 1928. *Lehrbuch der Anthropologie in Systematischer Darstellung. Mit besondereer Berücksichtigung der anthropologischen Methoden für Studierende, Ärzte und Forschungsreisende*. Vol. 2. 2nd edition. Gustav Fischer, Jena.

Murayama, Shichiro. 1972. Review of *Japanese and Other Altaic Languages* by Roy Andrew Miller. *Monumenta Nipponica* 27(4):463–467.

———. 1976. The Malayo-Polynesian component in the Japanese language. *Journal of Japanese Studies* 2(2):413–431.

———. 1992. *Ainugo no Kigen (The Origin of the Ainu Language)*. Sanichi Shobo, Tokyo.

Neves, Walter A., and Mark Hubbe. 2005. Cranial morphology of early Americans from Lagoa Santa, Brazil: Implications for the settlement of the New World. *Proceedings of the National Academy of Sciences U.S.A.* 102(51):18309–18314.

Neves, Walter A., Mark Hubbe, and Gonzalo Correal. 2007. Human skeleton remains from Sabana de Bogotá, Colombia: A case of Paleoamerican morphology survival in South America? *American Journal of Physical Anthropology* 133(4):1098–2007.

Neves, Walter A., Mark Hubbe, and Luís Beethoven Piló. 2007. Early Holocene human skeletal remains from Sumidouro Cave, Lagoa Santa, Brazil: History of discoveries, geological and chronological context and comparative cranial morphology. *Journal of Human Evolution* 52(1):16–30.

Neves, Walter A., André Prous, Roland González-José, Renato Kipnis, and Joseph Powell. 2003. Early Holocene human skeletal remains from Santana do Riacho, Brazil: Implications for the settlement of the New World. *Journal of Human Evolution* 45(1):19–42.

Neves, Walter A., and Hector M. Pucciarelli. 1991. Morphological affinities of the first Americas: An exploratory analysis based on early South American human remains. *Journal of Human Evolution* 21(4):261–273.

Ohno, Susumu. 1970. *The Origin of the Japanese Language*. Kokusai Bunka Shinkokai Okada, Tokyo.

Ohtsuka, Kazuyoshi. 1999. *Itaomachip*: Reviving a boat-building and trading tradition. In *Ainu: Spirit of a Northern People*, edited by William W. Fitzhugh and Chisato O. Dubreuil, 374–376. Arctic Studies Center, National Museum of Natural History, Smithsonian Institution, University of Washington/Perpetua Press, Los Angeles.

Pietrusewsky, Michael. 2005. The physical anthropology of the Pacific, East Asia, and Southeast Asia: A multivariate craniometric analysis. In *The Peopling of East Asia: Putting Together Archaeology, Linguistics, and Genetics*, edited by Larent Sagart, R. Blench, and Alicia Sanchez-Mazas, 201–229. Routledge-Curzon, London.

———. 2006a. The initial settlement of remote Oceania: The evidence from physical anthropology. In *Austronesian Diaspora and the Ethnogenesis of People in Indonesia Archipelago,* edited by T. Simunjuntak, I. H. E. Poojoh, and M. Hisyam, 320–347. Indonesian Institute of Sciences, LIPI Press, Jakarta, Indonesia.

———. 2006b. A multivariate craniometric study of prehistoric and modern inhabitants of Southeast Asia, East Asia and surrounding regions: A human kaleidoscope? In *Bioarchaeology of Southeast Asia,* edited by Marc R. Oxenham and Nancy Tales, 59–90. Cambridge University Press, Cambridge.

Poncins, Gontran de. 1941. *Kabloona.* Reynal and Hitchcock, Inc., New York.

Preston, Douglas. 1997. The lost man. *The New Yorker* 72(16):70–81.

Quatrefages, Armand de. 1881. L'homme fossile de Lagoa-Santa (Brésil) et ses descendants actuels. *Comptes Rendus de Sé ances de l'Academie des Sciences* 93(22):882–884.

Relethford, John H., and John Blangero. 1990. Detection of differential gene flow from patterns in quantitative variation. *Human Biology* 62(1):5–25.

Rivet, Paul. 1908. La Race de Lâgoa Santa chez les populations précolombiennes de l'Équateur. *Bulletins et Mémoires de la Société d'Anthropologie de Paris* Ve série Tome 9:209–274.

———. 1925. Les origins de l'homme américain. *L'Anthropologie* 35:293–319.

Saitou, Naruya, and Masatoshi Nei. 1987. The neighbor-joining method: A new method for reconstructing phylogenetic trees. *Molecular Biology and Evolution* 4(4):416–425.

Seguchi, Noriko, Hideyuki Umeda, A. Russell Nelson, and C. Loring Brace. 2006. Do early South Americans show biological similarity to Australians? Lagoa Santa in odontometric and craniometric perspective. *American Journal of Physical Anthropology* 129(S42):162.

Seguchi, Noriko, Ashley McKeown, Ryan Schmidt, Hideyuki Umeda, and C. Loring Brace. 2011. An alternative view of the peopling of South America: Lagoa Santa in craniometric perspective. *Anthropological Science* 119(1):21–38.

Smith, B. Holly. 1983. "Dental Attrition in Hunter-Gatherers and Agriculturalists." Ph.D. dissertation, University of Michigan, Ann Arbor.

Sternberg, Leo. 1929. The Ainu problem. *Anthropos* 24(5,6):755–799.

Taylor, R. M. S. 1963. Cause and effect of wear of teeth: Further non-metrical studies of the teeth and palate in Moriori and Maori skulls. *Acta Anatomica* 53(1–2):97–157.

ten Kate, Herman. 1885. Sur les crânes de Lagoa-Santa. *Bulletin de la Société d'Anthropologie de Paris* 3e série, Tome 8:240–244.

Turner, Christy G., II. 1989. Teeth and prehistory in Asia. *Scientific American* 26(2):88–96.

Verneau, René. 1903. *Les anciens Patagons, Contributions à l'Étude des Races Précolombiennes de l'Amérique du Sud.* Imprimerie de Monaco.

Waugh, Leuman M. 1937. Influence of diet on jaws and face of the American Eskimo. *Journal of the American Dental Association* 24(10):1640–1647.

Woo, Ting-Liang, and Geoffrey M. Morant. 1933. A biometric study of the "flatness" of the facial skeleton in man. *Biometrica* 26:196–250.

Zhushchikovskäia, Inna S. 2005. *Prehistoric Pottery-Making of the Russian Far East.* Translated and edited by Richard L. Bland and C. Melvin Aikens. Archaeopress, Oxford.

Cranial Morphometric Evidence for Early Holocene Relationships and Population Structure

Richard L. Jantz and M. Katherine Spradley

Although scientists have been aware of the existence of early American crania since the early twentieth century, little progress has been made in understanding their meaning until recently, primarily because of early anthropologist Aleš Hrdlička's insistence that they were all essentially the same as modern Native Americans (Hrdlička 1903, 1937). The American Homotype, as Hrdlička's ideas came to be called, held sway until well past the midpoint of the twentieth century. The first attempt to develop a synthesis on early crania did not appear until the early 1990s (Steele and Powell 1992), but it demonstrated that early crania differ systematically and significantly from modern Native Americans. In the meantime, additional American crania have come to light, either through redating existing remains, such as Nevada's Spirit Cave mummy (Tuohy and Dansie 1997), the cranium from Central Washington known as Stick Man (Chatters et al. 2000), and the Minnesota remains (Myster and O'Connell 1997), or discovery of new remains, such as Kennewick Man.

It is now broadly accepted that variation in cranial morphology is mainly a reflection of genetic variation, thus providing a basis for investigating human population structure and reconstructing population history (Roseman and Weaver 2007; Weaver et al. 2008; Smith 2009). Morphology of early crania can contribute to the increasingly complex issue of the colonization of the Americas and subsequent evolution of Native American populations. Evidence from cranial morphology has been used to argue in favor of the Pacific coastal movement of early populations into the New World (Jantz and Owsley 2005).

This chapter, focusing on morphometric features of the Kennewick Man cranium, addresses three somewhat independent issues regarding its assessment: measurement variation, mainly interobserver variation; the relationship of Kennewick Man to modern populations; and a comparison of Kennewick Man to other early crania with assessment of variation among the early populations they represent.

MEASUREMENTS AND INTEROBSERVER VARIATION

Howells' (1973) measurement protocol, used for this study of Kennewick Man, expanded the number of cranial measurements taken and modified some definitions. It established a more internally consistent measurement set designed for multivariate analysis. These measurements quantified overall morphology of the cranium as well as specific aspects of the vault and face, which enabled more specific analysis of variation. Howells' protocol has been widely adopted; it is the basis of the University of Tennessee database (Key 1983) and the Smithsonian National Museum of Natural History Repatriation Office's database. As originally designed, it utilized several special measuring instruments that were not widely available. The method of obtaining measurements has changed from manipulating several different calipers to applying a digitizer to record the x, y, and z coordinates of the cranial landmarks that are used as end points of traditional measurements. From these coordinates, interlandmark distances corresponding to Howells' measurements can be computed using the 3Skull software package (Ousley and McKeown 2001).

Since the discovery of Kennewick Man's skull in 1996, three investigators have studied and measured either the original or casts made from it. James Chatters was the only scientist to systematically observe and measure the original remains before they were seized by the US Army Corps of Engineers (Corps) at the end of August 1996. Chatters took an abbreviated set of measurements on the original, followed by a full set of Howells' measurements from a first generation cast (Chatters 2000). The next opportunity to measure the original cranium was afforded to Joseph Powell in February 1999 (Powell and Rose 1999), followed by members of the plaintiffs' group in February 2006. Alternative measurement sets were obtained by different scientists (Chapters 24 and 27). For this chapter another set of Howells's measurements was obtained by the authors using a digitizer.

Because of constraints imposed by Corps conservators, reconstruction of the original skull was not possible. Therefore, the original cranial vault, without the face attached, was digitized. The complete cranium was digitized using a cast made from a high-resolution computed tomography (CT) scan. David Hunt performed reconstruction of cast elements (Chapters 20, 21, and 22). Corps conservators also did not allow landmarks on the original skull to be lightly marked with a pencil in the usual fashion. Therefore, landmarks were marked with small triangular pieces of Teflon®, pointing toward each landmark (Figure 6.15).

The Kennewick cranium is a challenge to measure. Landmarks are difficult to locate on the original because of obscured sutures and because of color imparted by taphonomic processes. Measuring casts only increases the difficulty in locating landmarks, while constraints imposed by Corps conservators added to the challenge. Some investigators also had the added adversity of having to work under time constraints.

Table 25.1 presents measurements obtained by Spradley and Jantz on the original (cranial vault without the face attached) and on the reconstructed cast, by Chatters on a first-generation cast produced using a traditional molding process, and by Powell and Rose (1999) on the original. There are potentially obvious sources of variation among these three sets of measurements, relating to different reconstructions, different types of casts, and casts versus the original. Chatters's cast and the cast made from high-resolution CT involve different processes. The reconstructions were also different. Chatters's cast was his own reconstruction while Hunt reconstructed the CT-derived cast as described in this volume. The only direct comparisons are vault measurements by Spradley and Jantz versus those made by Powell on the original. Dimensions with excessive variation are identified in bold in Table 25.1.

Looking at measurements on the original, only the discrepancies in interorbital breadth (Dkb) and nasodacryal subtense (Nds) can be definitely attributed to observers and require special comment. Powell's interorbital measurement is four millimeters less than that of Spradley and Jantz; his nasodacryal subtense is two to four millimeters smaller. Chatters obtained similar measurements to Powell. This indicates that both Powell and Chatters were locating dacryon more medially and anteriorly than Spradley and Jantz. It seems likely that Powell and Chatters were not locating dacryon as defined by Howells, but rather the landmark maxillofrontale. Another striking difference can be seen in biasterionic breadth (Asb), where Powell's value is six to nine millimeters smaller than Spradley and Jantz. This difference is likely due to the difficulty of locating asterion on the Kennewick cranium. The location used in this analysis was chosen after considerable inspection and consultation with other team members, resulting in the most reliable estimate possible.

Casting apparently affected basion-nasion length (Bnl). The value from the cast exceeds the original by three millimeters, but data from the original and Powell's are similar. Other differences are probably attributable to difficulties in landmark location. On the whole, measurements from different casts and observers are in broad agreement, and characterization of the skull based on them would be similar. In the analysis that follows, Spradley and Jantz measurements on the original are used when the analysis is limited to the vault. When analysis involves the entire skull, Spradley and Jantz measurements of the cast made from the CT scan are used.

MORPHOMETRIC RELATIONSHIPS OF KENNEWICK MAN TO MODERN POPULATIONS

Assessing the relationships of early crania to modern populations can provide insight into whether early populations bear any similarity to modern people of the same region. This comparison in turn addresses questions of population history, such as whether there is evidence for continuity or replacement. Howells (1995) used this approach successfully in assessing crania from different parts of the world, but it may be worthwhile to briefly consider his results on American crania. Of 14 crania he

TABLE 25.1 Measurements of Kennewick Man's cranium. Dimensions with excessive variation are in bold.

Measurement	Spradley & Jantz cast	Spradley & Jantz original	James Chatters cast	Powell & Rose original	Measurement (cont.)	Spradley & Jantz cast	Spradley & Jantz original	James Chatters cast	Powell & Rose original
GOL	189	188	189	189	FRC	116	116	112	112
NOL	190	189	190	191	FRS	18	18	17	18
BNL	**115**	**112**	**110**	**113**	FRF	56	55	50	52
BBH	141	141	138	143	PAC	112	112	112	114
XCB	142	141	139	140	PAS	21	20	28	22
WFB	98	95	97	-	PAF	49	54	58	58
AUB	127	127	126	128	**OCC**	**102**	**105**	**107**	**109**
ASB	**125**	**122**	-	**116**	OCS	24	25	26	26
BPL	114		110	114	OCF	37	43	43	44
NPH	78		75	76	FOL	39	40	42	41
NLH	**60**		**55**	**55**	FOB	32	33		
JUB	122		122	123	NAR	106	103	103	103
NLB	26	26	26	25	SSR	114		111	112
MAB	68	67	66	66	PRR	118		115	116
MAL	55	56	58	56	DKR	92	91	94	96
MDH	34		34	33	ZOR	93		93	92
OBH	38		37	37	FMR	90	90		88
OBB	43		45	45	EKR	83		81	84
DKB	**21**	**21**	**18**	**17**	ZMR	86		84	87
NDS	**14**	**12**	**9**	**10**	AVR	96		90	
WNB	8	7.5	7.5	7.5	BRR	122	122	122	123
SIS	3	2.9	4.2	4	VRR	126	126	126	124
ZMB	**97**		**102**	**102**	LAR	113	115	113	117
SSS	29		28	29	OSR	42	45	46	45
FMB	104	104	104	107	BAR	19	20	18	18
NAS	17	15	16	17					
EKB	103		104	105					
DKS	10		12	13					
WMH	24	23	23	24					
GLS	5	3	3	3					

tested, nine aligned with populations outside America, leading him to the conclusion that the results were "essentially worthless" (1995:37). Six were from the eastern United States, including three Archaic period crania from Indian Knoll, Kentucky. The remainder came from Pecos Pueblo in New Mexico. These results are largely attributed to the paucity of North American reference samples used by Howells (limited to the Arikara from South Dakota and the Chumash from Santa Cruz Island, California). Such a limited comparative set could in no way adequately reflect the morphometric diversity of North America. Pecos Pueblo crania were the most vexing to Howells because he obtained a mix of Native American, circumpacific, and European classifications from a single late prehistoric-historic site. Pecos, however, has a historic component with Spanish remains, and Howells's analysis correctly identified a skull of European origin in his Pecos sample (see Bruwelheide et al. 2010). Beyond that, Pecos is exceedingly heterogeneous (Weisensee and Jantz 2010), which helps explain Howells's perception of odd results.

Stating that a fossil is similar to a particular recent group is not the same as saying the fossil belongs to that group. It is simply a statement of similarity. Related to the question of similarity is whether the fossil falls within the range of variation of any modern groups. A criticism often directed at such an analysis is that it "forces classification." The goal, however, is not classification, but ascertaining patterns of similarity and difference to modern groups. Today, biostatisticians have a better understanding of modern morphometric variation, which provides a framework for assessing earlier variation about which much less is known. At the same time, it is clear that modern human populations do not represent all variation that existed in past millennia. Ancient crania that differ from modern populations can provide information about differences between early and present-day variation.

Several morphometric analyses of Kennewick Man's measurements have been conducted using various measurement sets and different comparative and theoretical frameworks (Chapter 20) often emphasizing Kennewick Man's affinity with the Ainu of Japan. Powell and Rose (1999) concluded that Kennewick is most similar to crania from the South Pacific, Polynesia, and Ainu. Chatters et al. (1999) identified Kennewick as most similar to their Polynesian-Ainu cluster using size and shape, but as a distant outlier when using shape alone.

Statistical Procedure

The distance analyses presented, unless otherwise noted, are computed as follows:

1. Center reference samples on a grand mean of zero, separately by sex. This allows pooling males and females while eliminating sex differences that could potentially confound the analysis, particularly when sample sizes are unbalanced.
2. Form variance-covariance matrices for each sample, and pool, weighting by sample size.
3. Extract eigenvalues and eigenvectors from the pooled within-groups covariance matrix. Use these to obtain the mean principal component (PC) scores for each group.
4. Since the PC scores are orthogonal, the Mahalanobis distance (D^2) between each pair of groups can be computed as the sum of the squared distance between PC scores, divided by the variance, in this case the eigenvalue.
5. One or more fossil crania are added to the analysis by getting their sex-specific deviations from the grand mean, calculating PC scores and obtaining the Mahalanobis distance from the reference samples, and from each other, thereby increasing the size of the distance matrix.
6. The principal coordinates are obtained from the D^2 matrix using the procedure described by Gower (1972). Principal coordinates extracted from a Mahalanobis D^2 matrix are equivalent to canonical variates.

This procedure is only one of several approaches to obtaining Mahalanobis distances, but it does allow some options as to how they are obtained. When many variables are available, it is sometimes desirable to reduce them to avoid singular or near-singular covariance matrices, or over-fitting. Howells (1995) used the first 14 discriminant functions, out of a possible 27, to reduce dimensionality. The approach used here is more methodologically aligned with Key (1983), who used truncated principal components. Rather than using the correlation matrix as Key did, the covariance matrix was used and the principal components were truncated to include 95 percent of the within-group variance. This is an arbitrary truncation but has the advantage of retaining most of the variation while excluding components with very small eigenvalues. These typically draw contrasts between highly correlated dimensions and represent residual variation that is not biologically relevant and potentially a reflection of measurement error. In the case of fossil evaluations, particularly when some reconstruction has been necessary, it may also exclude reconstruction errors.

Analysis

The analysis begins with a comparison of Kennewick Man to populations from the major geographic regions of the world. Measurements from major geographic units, as shown in Table 25.2, were pooled. Howells' data form the core of these samples, but American and European

TABLE 25.2 World cranial samples used in this chapter.

	N males	N females	Measured by
Africa			
Dogon	47	52	WW Howells
Teita	33	50	WW Howells
Zulu	55	46	WW Howells
America			
Arikara	42	27	WW Howells
Santa Cruz Chumash	51	51	WW Howells
Peru	55	55	WW Howells
Blackfeet	25	45	RL Jantz
Numic (Paiute, Ute, and Shoshone)	37	24	RL Jantz
Sioux	23	9	RL Jantz, SD Ousley
Crow	11	10	RL Jantz
Grand Pawnee	18	12	RL Jantz
Ponca/Omaha	18	17	RL Jantz
Zuni	27	27	RL Jantz, MK Spradley
Asia			
N Japan	52	42	WW Howells
S Japan	50	41	WW Howells
Hainan	45	38	WW Howells
Atayal	29	18	WW Howells
Philippines	50		WW Howells
SW Pacific			
Australia	52	49	WW Howells
Tasmania	45	42	WW Howells
Tolai	56	54	WW Howells
Europe			
Norse	55	55	WW Howells
Zalavar	53	45	WW Howells
Berg	56	53	WW Howells
Croatians	31	3	RL Jantz, MK Spradley
Portuguese	69	64	KE Weisensee
Czechs	47	34	KE Weisensee, MK Spradley
English-17th Century	31	39	MK Spradley
Polynesia			
Moriori	57	51	WW Howells
Mokapu	51	49	WW Howells
Easter Island	49	37	WW Howells
Maori	26	3	WW Howells, RL Jantz

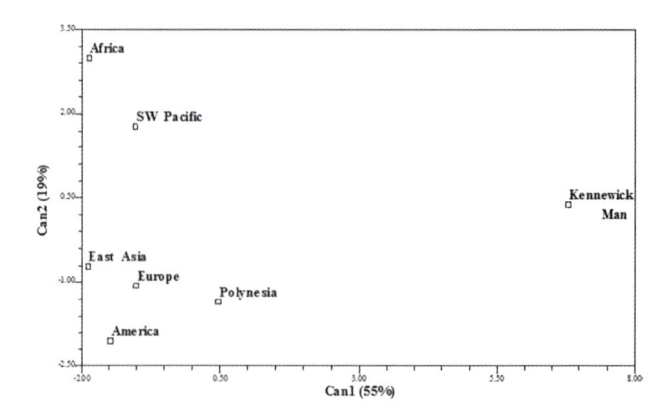

FIGURE 25.1 Canonical variates showing relationships of Kennewick Man to populations representing major regions of the world. Kennewick Man, while atypical of all modern reference groups, is most similar to Polynesians.

representation have been significantly increased; and Howells's Maori data have been expanded somewhat.

The two-dimensional canonical plot showing Kennewick Man's relationships to these major groups is shown in Figure 25.1. The two canonical vectors account for 70 percent of the variation among the groups. Kennewick Man dominates the first axis, being about equidistant from most major populations. The second axis arranges the major modern groups in about the same way as most previous analyses of world populations, grouping the Southwest Pacific and Africa on one end with Europe, East Asia, Polynesia, and America loosely grouped on the other. This corresponds to the first bifurcation in Howells's (1989) cluster analysis of major human groups. Kennewick Man's uniqueness on the first canonical axis is based on extreme facial forwardness, frontal and parietal flatness, a wide posterior vault, high face, and large eye orbits.

The only indication of greater affinity to a particular group is seen in the displacement of Polynesians toward Kennewick Man on the first axis. This is evident in the Mahalanobis distances, presented in Table 25.3. Kennewick Man's distance is lowest with Polynesians, indicating greater similarity, and his posterior probability high, although he would be atypical of all groups. In the broadest sense, Kennewick Man's morphology is slightly more similar to Europeans than it is to Native Americans.

Another approach is nearest-neighbor discriminant analysis, which finds individual crania with the lowest Mahalanobis distances to Kennewick man. Nearest-

TABLE 25.3 Mahalanobis distances of Kennewick Man from major geographically defined groups.

Group	Mahalanobis D²ᵃ	Posterior Probability	Typicality Probability
Polynesia	49.93	0.9999	0.0009
Europe	69.36	0.0001	0.0000
Southwest Pacific	69.64	0.0000	0.0000
America	77.05	0.0000	0.0000
East Asia	80.95	0.0000	0.0000
Africa	83.54	0.0000	0.0000

ᵃBased on 23 principal components extracted from 42 variables.

TABLE 25.4 Linear discriminant classification of Kennewick Man using Cranid and 36 reference samples.

Order	Sample	Probability
1	Moriori, Chatham Island, males	0.90088
2	Mokapu, Hawaii, males	0.06621
3	Moriori, Chatham Island, females	0.03046
4	N. Japan, Hokkaido, males	0.00127
5	Maori, New Zealand, males	0.00056
6	Mokapu, Hawaii, females	0.00031
7	S. Japan, Kyushu, males	0.00018
8	Santa Cruz Island, California, males	0.00005
9	Arikara, S. Dakota, males	0.00004
10	Philippines, males	0.00002
11	Ainu, Hokkaido, males	0.00001
12	Easter Island, males	0.00001

neighbor avoids a frequently criticized assumption of linear discriminant analysis, that groups must be *a priori* defined, by assessing affinity on the basis of the pattern of nearest neighbors. Cranid, a program written and distributed by Richard Wright (2007), was used to obtain Kennewick Man's nearest neighbors among modern populations. Cranid contains Howells' database plus 10 additional population samples, yielding 3,163 crania. It provides a more complete picture of world variation than Howells' data. However, it uses only 29 variables, a subset of Howells's measurements, omitting mainly radii and measurements that are difficult to take. Cranid also finds the closest population using linear discriminant function, which is also presented because of the wider choice of populations. Both Cranid's linear discriminant and nearest-neighbor analysis use local population samples, rather than pooling them into major geographical units.

Table 25.4 gives the linear discriminant results from Cranid, presenting all groups with posterior probabilities of 0.00001 or higher. The results replicate what was observed in the geographic analysis by strongly identifying Polynesians, and specifically the Moriori, as being the most similar to Kennewick Man (with a posterior probability of 0.9). Polynesians account for six of the 12 most similar population samples, the remainder being mostly other circumpacific populations, all with very low posterior probabilities. Only two Native American samples appear among the top 12, the Chumash from Santa Cruz and Arikara, in positions 8 and 9, respectively. Both have extremely low posterior probabilities.

Table 25.5 identifies the 56 individuals that constitute Kennewick Man's nearest neighbors. The number 56 is arbitrary, obtained as the square root of the 3,163 crania in Cranid's database (Wright 2007). From a global perspective, Kennewick Man's nearest neighbors are strongly patterned. Polynesians account for 38 of his 56 nearest neighbors, and all but four (one Norse, one Zulu, and two Arikara) are circumpacific populations. Only four of Kennewick Man's nearest neighbors are Native Americans: two Arikara, one Chumash from Santa Cruz Island, and an Eskimo. Kennewick Man's nearest neighbors are the Chatham Islands Moriori, both males and females. The only non-Polynesian groups with a significant number of "hits" are North and South Japanese. The single nearest neighbor is a Moriori male, with a Mahalanobis distance (D) of 6.661 from Kennewick Man. This value is 2.15 standard deviations above the average nearest neighbor distance in Cranid's database (Wright 2007). Kennewick Man's overall distance from the grand human centroid is 9.053, about 2.7 standard deviations from the mean distance (Wright 2007). These statistics make it clear that Kennewick Man is atypical among modern human crania. Under normal curve assumptions, Kennewick Man is more atypical than all but about ten of the 3,163 crania in Cranid's database. However, the nearest-neighbor analysis is in agreement with the linear discriminant analysis that Kennewick Man is most similar to the Moriori. Despite being on the extreme outer limits of modern human variation, Kennewick Man would be a less extreme member of

TABLE 25.5 Nearest neighbor classification of Kennewick Man's skull using Cranid and 36 reference samples. Kennewick Man is most closely aligned with Polynesians, particularly the Moriori.

Sample	Hits	Weighted Score
Moriori, Chatham Island, males	16	888
Moriori, Chatham Island, females	13	806
Mokapu, Hawaii, males	7	434
N. Japan, Hokkaido, males	5	288
S. Japan, Kyushu, males	3	190
Arikara, S. Dakota, males	2	151
N. Japan, Hokkaido, females	1	99
Hainan, China, males	1	70
Ainu, Hokkaido, males	1	66
Easter Island, males	1	65
Mokapu, Hawaii, females	1	65
Santa Cruz Island, California, males	1	62
S. Australia, males	1	61
Eskimo, Greenland, males	1	60
Norse, Norway, males	1	58
Zulu, S. Africa, males	1	58

the Polynesian Moriori than any other human group in Cranid's database.

These results make clear the circumpacific character of Kennewick Man's nearest neighbors, whether assessed from individual crania or sample means. Results from both discriminant analysis and nearest neighbor analysis emphasize two features of Kennewick Man's morphometric relationships to modern populations. First, Kennewick Man falls on the outer margins of the range of variation of modern groups. Second, allowing that Kennewick Man is on the outer margins of modern human variation, he is clearly more closely aligned with Pacific Rim populations, especially Polynesians, and specifically the Moriori.

As a third examination of Kennewick Man's affinities, a test was constructed that compares his cranial measurements to an extensive set of Native American populations, as well as with Polynesians. The Ainu were also included in this analysis, since his affinities have been linked to the Ainu. Results shown in Table 25.6 emphasize even more strongly Kennewick Man's Polynesian affinity. Again, the Moriori is the population to which

TABLE 25.6 Kennewick Man assessed against Polynesians and Native Americans. Kennewick Man's morphology is aligned with Polynesians and is atypical of all Native American groups in this analysis.

Group	Mahalanobis (D^2)	Posterior Probability	Typicality Probability
Moriori	52.163	0.771	0.145
Mokapu	54.616	0.226	0.100
Maori	63.787	0.002	0.026
Easter I.	67.403	0.000	0.009
Arikara	68.531	0.000	0.007
Ainu	70.001	0.000	0.005
Blackfeet	72.089	0.000	0.003
Sioux	72.815	0.000	0.004
Santa Cruz Chumash	76.475	0.000	0.001
Numic peoples	77.502	0.000	0.001
Pawnee	79.015	0.000	0.001
Crow	84.055	0.000	0.000
Ponca/Omaha	88.285	0.000	0.000
Zuni	103.487	0.000	0.000

Variables: Gol, Nol, Xcb, Aub, Asb, Nph, Nlh, Jub, Nlb, Mab, Obh, Obb, Dkb, Nds, Wnb, Sis, Zmb, Sss, Fmb, Nas, Ekb, Dks, Wmh, Frc, Frs, Pac, Pas, Occ, Ocs, Fol, Nar, Sssr, Prr, Dkr, Zor, Fmr, Ekr, Zmr, Brr, Lar, Osr, Bar. Distances calculated from 24 principal components.

he is most similar, followed closely by Mokapu Hawaiians. Together they account for 99.8 percent of the posterior probability. Moriori, Hawaiians, and the Maori of New Zealand are the only groups whose variation would accommodate Kennewick Man, if only marginally. All Polynesian groups have lower distances with Kennewick Man than do any of the Native American groups. The Ainu are not quite so similar, classifying after the Arikara sample. Typicality probabilities show that he would be an extremely atypical member of any Native American group used in this analysis.

MORPHOMETRIC RELATIONSHIPS OF KENNEWICK MAN TO OTHER EARLY HOLOCENE FOSSILS
Kennewick Man's Relationships Assessed by Distance

Ideally it should be possible to include Kennewick Man in an analysis with other ancient crania (Table 1.1; Jantz and Owsley 2001). Performing such an analysis requires a common set of variables, but the pattern of missing measurements among available crania is such that the resulting common dataset is small. In order to include as many fossil crania from North and Central America as possible, Kennewick Man was compared to each one individually, based on shared measurements. Kennewick Man is nearly complete, which allows comparisons using maximum variable sets. This analysis also identifies the population most similar to each fossil cranium tested. As in the previous analysis, comparative groups were limited to Polynesians, the Ainu, and Native Americans. Table 25.7 shows the Mahalanobis distance (D^2) of 14 early Holocene fossils to Native American and Polynesian groups. The D^2 and Population columns identify the group most similar to each fossil based on measurements in common with Kennewick Man. The Pacific affinity of the fossils is unmistakable. Eleven of 14 fossils have a Polynesian sample or the Ainu as the most similar group, while only three are more similar to Native American groups. Most would be atypical members of their most similar group, but San Miguel would be a typical Moriori while both Browns Valley and Horn Shelter No. 2 would fit comfortably within the Maori. The Moriori of the Chatham Islands and the Maori of New Zealand have similar cranial features and shared a common ancestor (Diamond 1999; Poll and Dennison 1992). Only Wizards Beach would be a typical Native American.

TABLE 25.7 Comparison of fossil crania to Polynesian and Native American samples and to Kennewick Man. Kennewick Man is most similar to San Miguel and Buhl.

Fossil	Variables	Lowest D^2	Population	D to Kennewick Man	D/ Variables
San Miguel, CA	17	11.32	Moriori	6.18	0.36
Buhl, ID	22	35.48[a]	Santa Cruz	7.89	0.36
Horn Shelter No. 2, TX	19	14.59	Maori	7.43	0.39
Wizards Beach, NV	24	25.09	Numic	10.44[b]	0.44
Chimalhuacan, MEX	19	30.057[a]	Maori	8.4[b]	0.44
Stick Man, WA	13	27.74[a]	Mokapu	5.92	0.46
Browns Valley, MN	18	14.97	Maori	8.25[b]	0.46
Gordon Creek, CO	20	35.02[a]	Maori	9.39[b]	0.47
Spirit Cave, NV	23	40.31[a]	Ainu	10.92[b]	0.47
Metro Balderas, MEX	16	30.26[a]	Ainu	7.45[b]	0.47
Peñón III, MEX	19	33.8[a]	Santa Cruz	8.9[b]	0.47
Arch Lake, NM	14	30.11[a]	Ainu	8.67[b]	0.62
Cueva del Tecolote, MEX	16	52.9[a]	Maori	11.02[b]	0.69
Tlapacoya I, MEX	11	35.88	Ainu	8.63[b]	0.78

[a] Atypical compared to the most similar population, at 0.05 level or below.
[b] Distance from Kennewick Man greater than expected for 2 crania drawn from the same population.

The last two columns of Table 25.7 compare Kennewick Man to each fossil to ascertain his nearest neighbors among early Holocene crania. Four crania—San Miguel, Stick Man, Horn Shelter No. 2, and Buhl—have distances no larger than the random distances expected between two crania drawn from a single population. Table entries are sorted by the last column, which is the distance from Kennewick Man divided by the number of variables to adjust for differing numbers of variables. By this criterion, San Miguel, Buhl, and Horn Shelter No. 2 are Kennewick Man's nearest neighbors among Paleoamerican crania. All others except Stick Man exceed the distance to be expected from crania drawn from a single population. Arch Lake and the two Mexican fossils, Cueva del Tecolote and Tlapacoya, are especially distant from Kennewick Man.

Cranial Size

Early Holocene Americans have large crania compared to later Native Americans, and indeed relative to other modern populations except for Polynesians (Jantz and Owsley 2005). Size is often considered a nuisance variable to be removed from analyses seeking to assess population structure, but size is worthy of investigation in its own right. A pattern of decreasing cranial capacity has been identified in Europe and Africa from Upper Paleolithic to modern times (Henneberg 1988; Henneberg and Steyn 1993).

Following Henneberg (1988) the cranial module was used as a measure of vault size as follows:

$$\text{Cranial module} = \frac{\text{Gol} + \text{Xcb} + \text{Bbh}}{3}$$

Cranial capacity (Henneberg's CCII) was estimated from the cranial module after adjusting for an estimate of vault thickness. Since early crania are frequently missing the cranial base, an upper cranial module was calculated using vertex radius in place of basion bregma height as follows:

$$\text{Upper cranial module} = \frac{\text{Gol} + \text{Xcb} + \text{Vrr}}{3}$$

Table 25.8 presents the cranial module, upper cranial module, and cranial capacity for all early Holocene crania from which the required measurements could be obtained. The ten crania from the United States were measured by the Spradley and Jantz team. Measurements from the five Mexican crania (Metro Balderas, Tlapacoya I, Chimalhuacan, Cueva del Tecolote, Peñón III)

TABLE 25.8   Measures of cranial size of early Holocene North Americans.

Fossil	Sex	Cranial Module	Upper Cranial Module	Cranial Capacity
Kennewick Man	M	156.67	151.67	1544
Spirit Cave	M	156.67	153.67	1536
Wizards Beach	M	152.33	148.33	1443
San Miguel	M	152.34[a]	148.67	1447
Horn Shelter No. 2	M	156.74[a]	153	1544
Browns Valley	M	156.06[a]	152.33	1523
Stick Man	M	161.48[a]	157.67	1642
Metro Balderas	M	160.80[a]	157.00	1634
Tlapacoya I	M	156.40[a]	152.67	1523
Chimalhuacan	M	153.69[a]	150.00	1469
Cueva del Tecolote	M	158.32[a]	154.67	1560
Gordon Creek	F	151.33	145.33	1435
Buhl	F	152.33	147.67	1453
Arch Lake	F	151.00		1442
Peñón III	F	149.33		1365

[a]Estimated from upper cranial module using regression: cranial module = 1.0147 x upper cranial module + 1.4885. R^2 = 0.974.

were obtained from González-José et al. (2005). Eight of the 11 male crania had missing bases, preventing calculation of the cranial module and hence the cranial capacity as defined by Henneberg (1988). In those cases the cranial module was estimated from the upper cranial module using a regression calculated from a world sample of modern human crania.

Table 25.9 compares the early Holocene Americans listed in Table 25.8 with both Paleolithic and Mesolithic Europeans, and with recent human populations. Early Holocene Americans have cranial capacities approximately equal to those of Upper Paleolithic and Mesolithic Europeans but well above those of recent Native North Americans. The individual values for several early Holocene crania (see Table 25.8) are completely outside the range of modern Native Americans. Paleoamericans are also homogeneous as judged by the standard deviation despite coming from locations ranging from the state of Washington to Mexico. The only modern groups with comparable cranial capacities are those from Polynesia, which exceed other modern groups by 85 to 145 cubic centimeters.

Cranial Size in Relation to Body Size

Paleoamerican skull sizes and cranial capacities exceed those of modern Native Americans and most modern populations from other regions of the world. These results raise the question as to whether body size contributes to cranial capacity. Auerbach (Chapter 11) estimated weight for four early skeletons (Kennewick, Horn Shelter No. 2, Spirit Cave, and Wizards Beach) and height for two (Kennewick and Spirit Cave) using the anatomical method. Body mass for three additional early skeletons from which a femur head diameter is available (Browns Valley, Gordon Creek, and Buhl) have been estimated, using the regression presented by Grine et al. (1995). Height was estimated for Wizards Beach, Horn Shelter No. 2, and Buhl using the femur bicondylar length in a formula presented by Auerbach and Ruff (2010). Gordon Creek has a tibia which lacks the malleolus. Therefore 12 millimeters were added, an estimate of average malleolus length (Jantz et al. 1995). The resulting estimated condylar-malleolar length was converted to maximum length using the regression formula presented by Auerbach (2011); the tibia formula from Auerbach and Ruff (2010) was used to estimate Gordon Creek's stature. Formulae for Plains groups were used because the relatively long-legged Plains groups are more likely to reflect the proportionality of Paleoamericans.

Table 25.10 presents height and body mass estimates for Paleoamericans. Estimates of height for Horn Shelter

TABLE 25.9 Mean cranial capacities of early Holocene Americans compared to Upper Paleolithic and Mesolithic Europeans, and to modern humans. Polynesians are the only group with cranial capacities comparable to early Holocene Americans.

Group	N	Males					Females				
		n	Cranial capacity	SD	Min	Max	n	Cranial capacity	SD	Min	Max
Early Holocene		11	1533.2	65.5	1443	1642	4	1424.0	39.9	1365	1453
Upper Paleolithic[a]		15	1569.0				9	1416.0			
Mesolithic[a]		35	1593				29	1502			
North America	7	334	1385.3	83.8	1139	1634	204	1242.4	82.4	1030	1424
Europe	6	341	1461.8	93.4	1000	1792	293	1295.4	96.4	969	1542
East Asia	5	229	1437.6	89.0	1144	1648	129	1278.4	91.9	1102	1489
Polynesia	4	182	1545.7	85.5	1313	1782	140	1384.9	100.2	1127	1646
West Africa	6	209	1390.5	103.2	1178	1921	178	1239.9	104.2	977	1500
Southwest Pacific	3	153	1404.9	84.6	1187	1629	145	1242.9	103.5	1022	1523

[a]from Henneberg 1988
N = number of samples, n = number of individuals

TABLE 25.10 Body size estimates of Paleoamericans.

Fossil	Sex	Height (cm)	Weight (Kg)
Kennewick Man	M	171.7[a]	73.7[a]
Wizards Beach	M	170.5	76.9[a]
Browns Valley	M		63.3
Spirit Cave	M	160.8[a]	61.6[a]
Horn Shelter No. 2	M	161.5	69.1[a]
Buhl	F	156.1	56.5
Gordon Creek	F	150.3	61.0

[a]From Chapter 11.

No. 2, Gordon Creek, and Buhl were presented in their original descriptions. Young (1988) gave a stature estimate for the Horn Shelter male of 162.4 centimeters using Genovés's (1967) formulae, and 166.3 centimeters using Trotter and Gleser's (1958) formulae for Mongoloids. Both are greater than the Spradley and Jantz estimate, although Genoves's by less than one centimeter. Breternitz et al. (1971) estimated Gordon Creek at 149.5 centimeters using Genovés's formulae, considering the ulna, humerus, and fibula jointly. This falls slightly below the estimate in Table 25.10. Green et al. (1998) present an estimate of 165 centimeters for Buhl, as opposed to the Spradley and Jantz estimate of 156.1 centimeters. The Green et al. (1998) estimate is almost certainly incorrect and even conflicts with their conversion to the English system of "about 5 feet 2 inches." This equals 157.5 centimeters, which is closer to the Spradley and Jantz estimate.

Kennewick Man and Wizards Beach are taller and heavier than Horn Shelter No. 2 or Spirit Cave. Browns Valley unfortunately lacks a height estimate but is similar to Spirit Cave in weight. The two females are a contrast in body build. The diminutive stature and robust build of Gordon Creek was observed by Breternitz et al. (1971). Buhl is taller and lighter, so the skeleton might be characterized as relatively gracile. Gordon Creek is about 10 centimeters shorter than the Spirit Cave male, but it weighs almost as much. This variation in body build is not patterned in any interpretable way, so it is difficult to know how to assess its meaning.

Relationships between brain size and body size in a recent Native American sample are shown in Figure 25.2. The sample is highly heterogeneous, consisting of Southwest horticulturists, Plains villagers, Plains equestrian

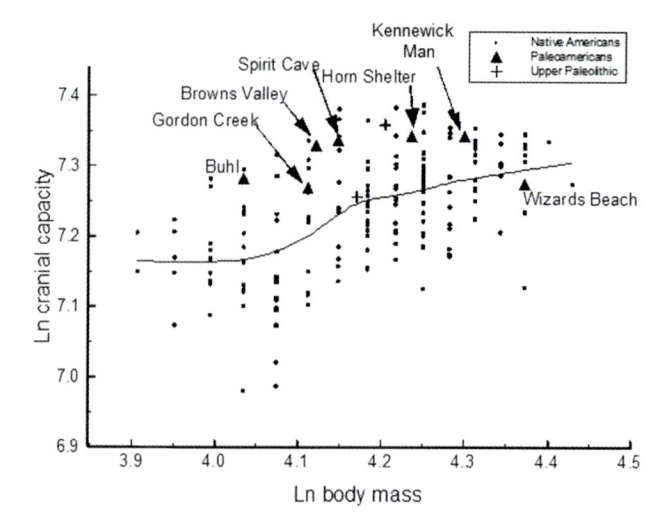

FIGURE 25.2 Natural log relationship of body mass (weight) to cranial capacity in recent Native American and Paleoamerican males and females, and means for Upper Paleolithic European males and females. Upper Paleolithic males and females are both designated by +, males are the uppermost. Paleoamericans have larger cranial capacities relative to body size.

nomads, and Great Basin hunters and gatherers. A loess regression line graphs the relationship between Native American brain size and body size. Paleoamericans and means for Upper Paleolithic males and females are superimposed on the plot but did not contribute to the regression. Upper Paleolithic cranial capacity was obtained from Henneberg (1988; cf. Table 25.9), while body mass estimates were obtained from Holliday (2002). Paleoamericans, with one exception, fall well above the regression line on the vertical axis within the range of variation of recent Native Americans, but mostly on the outer margin of the distribution. This means that the large cranial sizes of the Paleoamericans were not simply a function of larger body sizes, but rather a reflection of larger absolute cranial sizes.

Paleoamerican Population Structure

Population structure refers to the subdivision of populations of a region into smaller breeding groups. Such groups are able to differentiate, mainly through genetic drift. The magnitude of differentiation depends on the size of the local populations and the extent to which they are connected by gene flow. Interpretations of variation among groups must ultimately involve population structure.

The structure of early Holocene populations has not been addressed in any quantitative manner, presumably because small samples limit application of population structure models designed for modern populations. Historically, this issue relates to views of Native American diversity. Hrdlička (1914), author of the "American Homotype" idea, saw ancient crania as not differing from modern Native Americans. On the other extreme, Hooton (1930) saw extreme variability in the inhabitants of just one site, Pecos Pueblo, and attributed that representation to the diversity of the original founders. The idea of Native American diversity was carried forward in Neumann's (1952) typological synthesis of Native American crania. These arguments continue to the present. Fiedel (2004) has resurrected Hrdlička's American Homotype, and the single migration hypothesis is favored by some using mtDNA evidence (Merriwether et al. 1996), but not by others (Schurr 2004). Classical genetic markers (e.g., blood groups and serum proteins) show that modern Native American populations differ more widely than populations from any other region of the world (Cavalli-Sforza et al. 1994). This diversity is paralleled by anthropometric data (Jantz 2006). Native Americans contain greater variability than populations from the entire world in their trunk length-leg length proportions (Jantz et al. 2010). Such high levels of variability do not support derivation of the current population from a single migration unless there was a mechanism allowing Native Americans to generate more variability in less time than populations in other regions of the world.

Variability of early populations is obviously relevant to variability of later populations. The hypothesis that modern Native Americans are derived from early Holocene populations depends to some extent on high diversity among early Holocene peoples. Hence Powell (2005:228) has characterized the Paleoamerican sample as extremely diverse relative to recent Native Americans (see also Edgar et al. 2007). Powell (2005) apparently arrived at this conclusion by observing that the Paleoamerican sample was more variable than the Relethford-Blangero (1990) model would predict. But this model is designed for subdivided populations connected by gene flow, which cannot be the case for temporally distant populations.

This chapter addresses variability in a manner that avoids assuming that modern and early Holocene populations are part of the same system. It questions whether one can draw random samples from modern groups that are less variable than early crania, which would be expected if early variability exceeds modern. The procedure is as follows. First, obtain a within-group variance covariance matrix from nine modern Native American samples: the Arikara, Santa Cruz Chumash, Blackfeet, Numic peoples, Zuni, Sioux, Pawnee, and Ponca/Omaha. Their geographical distribution includes the Northern and Northwestern Plains, Central Plains, Great Basin, California coast, and Southwest. This distribution approximates that of early Holocene skeletons. In analyses incorporating Mesoamericans, a sample of Guatemalan Maya was included. The subsistence pattern was primarily horticultural, although the Blackfeet, Sioux, and some of the Numic speakers were equestrian nomads.

Second, calculate a co-phenetic matrix describing variation among early Holocene fossils as yy', where y is the centered principal component score matrix obtained using eigenvectors and eigenvalues from the within-covariance matrix computed from the modern samples. The diagonals of the co-phenetic matrix are then converted to a fixation index (F_{st}) as described by Relethford and Blangero (1990). Since the F_{st} is based on individual examples, it will be inflated due to sampling. There is no intent to use the F_{st} to estimate the actual divergence among the populations these fossils represent. It is a convenient index bounded by 0 and 1 that can be used to compare to values derived from the random samples from the modern groups. Random samples were obtained by drawing one example from each modern group and forming 499 F_{st} values.

Because the fossil crania have missing data, maximizing the number of fossil crania included in the analysis means using a smaller number of measurements. Analyses were therefore conducted in a stepwise manner, beginning first with all United States fossils and then removing those with the most missing data. Next Mesoamerican fossils from González-José et al. (2005) were added. Six runs were made with various combinations of fossils and variables. Results are presented in Table 25.11.

In all trials the fossil's F_{st} exceeds the mean F_{st} from 499 randomly drawn samples from modern groups

TABLE 25.11 Comparison of early Holocene cranial diversity with resampled modern diversity.

Analysis	No. variables	No. Principal Components	Fossils	Fossil F_{st}	Resampled F_{st}	Resampled S.D.	P fossil > random	Modern F_{st}
1	15	9	9	0.342	0.304	0.040	0.15	0.127
2	27	16	8	0.351	0.306	0.034	0.10	0.110
3	34	19	7	0.350	0.304	0.032	0.06	0.129
4	13	8	11	0.365	0.304	0.043	0.08	0.128
5	17	11	10	0.400	0.304	0.043	0.02	0.119
6	27	15	9	0.398	0.301	0.054	0.04	0.109

1. Kennewick Man, Gordon Creek, Sprit Cave, Wizards Beach, Horn Shelter No. 2, Browns Valley, Buhl, San Miguel, Stick Man.
2. Stick Man removed.
3. San Miguel removed.
4. Mexican Peñón III, Metro Balderas, Cueva del Tecolote, Chimalhuacan added.
5. Peñón III removed.
6. Cueva del Tecolote removed.

(Table 25.11). A test of the hypothesis that the fossil F_{st} exceeds the modern F_{st} was carried out by counting the number of random F_{st} values greater than the fossil values and dividing by 500. The first three runs with United States fossils are not significant, although the fossil F_{st} exceeds the resampled. Adding the Mesoamericans increases the diversity among fossil examples to the point where trials five and six are significant. Worth noting is that the average of random samples varies little by changing the number of variables, and the F_{st} for the modern groups is also minimally affected by different variables. As far as the American crania are concerned, the F_{st} among the populations represented by these fossils was not very different from that seen among recent Native Americans from the same region. Including Mesoamerican fossils is perhaps not a fair comparison because there is only one recent sample, modern Maya, so among-group variation is still mainly a reflection of variation among United States groups. The notion of extremely high variability of early crania, lacking common features, is not supported by this analysis.

There are some caveats required of this analysis. Early Holocene diversity is subject to an unknown amount of sampling error. Also, modern populations have very different population structures than early Holocene populations. Horticulturist populations, which normally have village endogamy, may be subject to greater among-group variation, although considerable intertribal gene flow occurred (Jantz 1973). Equestrian nomads had wider ranges and presumably a more open population structure. The population structure of early Americans is unknown; it can only be inferred from modern hunters and gatherers. Meiklejohn (1977) has argued that the boreal forest Athapaskan population functioned as a single biological unit with a constantly changing internal structure. Deme formation and local isolation were short-lived, and genes were extensively dispersed. Such systems have been found to characterize hunters and gatherers more generally (Hill et al. 2011). If such patterns were typical of early Americans, then one might expect to find limited local variation and regional homogeneity, since long distance gene flow would have prevailed.

In an attempt to search for spatial structure, the distances among eight United States fossils that allowed a reasonable set of measurements were calculated. Mahalanobis distances (D) are presented in Table 25.12. Pairs of crania more different than would be expected by drawing random pairs from a single population are considered significant. Most differences are not statistically significant. Those that are significant involve Gordon Creek, which differs from Spirit Cave, Browns Valley, and Kennewick Man, as well as Kennewick Man differing from Spirit Cave.

Figure 25.3 shows the interskull distances displayed on principal coordinates. Accounting for 29 percent of in-

TABLE 25.12   Mahalanobis distances (D) between early Holocene crania.

	Gordon Creek	Spirit Cave	Wizards Beach	Horn Shelter No. 2	Browns Valley	Buhl	Kennewick Man	San Miguel
Gordon Creek	0							
Spirit Cave	6.857	0						
Wizards Beach	6.888	4.075	0					
Horn Shelter No. 2	5.587	5.611	5.585	0				
Browns Valley	7.255[a]	6.135	6.568	4.902	0			
Buhl	7.009	6.202	5.939	4.463	5.472	0		
Kennewick Man	7.942[a]	8.199[a]	7.187	5.670	6.474	6.277	0	
San Miguel	7.296[a]	6.101	5.300	6.515	6.263	6.117	6.422	0

Expected distance (D) 5.5678.

Mean = 6.2253771

SD = 0.9478.

Distances >7.22 can be considered statistically significant.

[a] Distances greater than expected from the same population.

terskull variation, axis 1 separates the Great Basin crania (Spirit Cave and Wizards Beach) from Kennewick Man. The Kennewick cranium has marked facial forwardness, frontal flatness, and a broad posterior vault breadth, features that are least pronounced in Wizards Beach and Spirit Cave. Browns Valley, Buhl, and Horn Shelter No. 2 exhibit Kennewick Man's features to a lesser degree. The second axis, accounting for 23 percent of variation, separates Gordon Creek from the other crania, putting San Miguel Man on the opposite end of the axis. Gordon Creek's differentiation from San Miguel and the others results from a wide vault, greater upper facial flatness, a low face, and wide nose. Gordon Creek is unique among this group of fossils, but Arch Lake, which is too incomplete to include in this analysis, is also characterized by a broad and short vault (Owsley et al. 2010).

Spatial patterning is evident to some extent. The two Great Basin crania (Spirit Cave and Wizards Beach) are clearly a geographic cluster as reflected by their low distance in Table 25.11. This cluster might also include San Miguel, which would involve only a minimal geographic stretch. Buhl and Browns Valley might be construed as a Northern Plains/Plateau cluster, and Horn Shelter No. 2 could be a Southern Plains expression of this cluster. Horn Shelter No. 2 and Gordon Creek would form a more natural geographic cluster, and Horn Shelter is displaced slightly toward Gordon Creek.

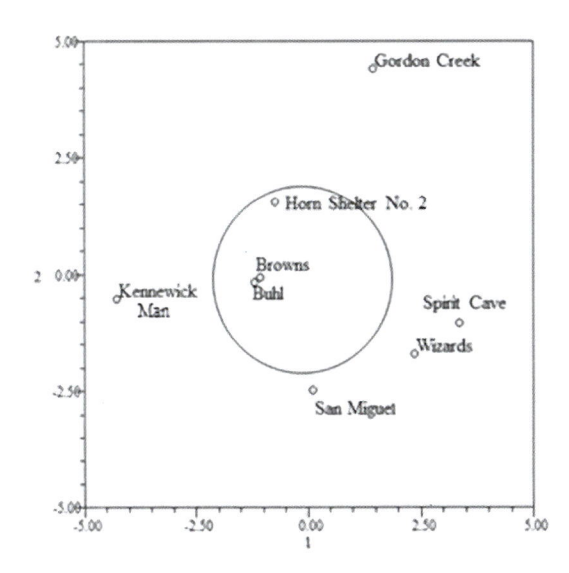

FIGURE 25.3   Principal coordinates plot of distances among early Holocene crania from North America. The circle has a radius of two standard deviations, which corresponds to average within-group variation.

The circle in Figure 25.3 has a radius of two standard units, which is a reflection of the range of within-group variation of modern tribes. It serves to illustrate how much the variation among early Holocene crania exceeds what one would expect from a single population. The circle would be expected to contain 95 percent of the members of a single population, but five of the eight early Holocene examples fall outside. This is not a manifestation of extreme variability among early Holocene crania, but rather an expression of differentiation, probably geographically patterned, among these early populations.

DISCUSSION

This analysis provides evidence that early Holocene Americans differ from recent Native Americans in a number of ways. Fossils of early Holocene Americans have larger cranial vaults, both absolutely and relative to body size, and their metric relationships to the Old World favor circumpacific populations. They are more or less equivalent to modern Native Americans in morphometric diversity.

The large cranial capacities in early Americans are consistent with what is seen in the Upper Paleolithic and Mesolithic of Europe and Africa. Reduction in cranial capacity is evident in modern Native Americans, as is also the case in Europe and Africa. While these changes are widely seen in both the Old and New Worlds, they were not universal. There is no evidence of reduction in Polynesians, who have maintained large cranial capacities, as have the Ainu and the Buriat from the southern end of Lake Baikal in Siberia (Jantz and Owsley 2005). One might argue that the large cranial capacity of the Buriat is a consequence of cold adaptation (Beals et al. 1984), but that would not account for Polynesian cranial size. Diet quality has been proposed as a determinant of brain size (Taylor and van Schaik 2007), so one might argue that Polynesian cranial size results from a high quality marine diet. If that is the case, then why has cranial size decreased dramatically in coastal California from Archaic times to the present (Jantz and Owsley 2005), where marine resources have been used continuously? A general model explaining variation in cranial capacity in modern humans does not yet exist, so it is not possible to say more about the American situation beyond noting that it agrees with the pattern seen in much of the Old World.

Early Holocene diversity is more or less comparable to that seen in modern tribes in the American West. This conclusion is in contrast to the suggestion that early Holocene crania are much more variable than late Holocene groups (Powell 2005:228) and that they display little internal consistency (Waguespack 2007). The pendulum has swung from Hrdlička's view of phenotypic uniformity to one of exceedingly high diversity, presumably driven by small, isolated groups subject to high levels of founder effect and genetic drift (Powell and Neves 1999).

As is often the case, the actual situation appears to lie between these extreme views. Nonetheless, it is remarkable that early Holocene people were as variable as modern Native Americans, considering that they had far less time to generate this variability. Interpreting that variability is constrained by unknowns. Did the crania composing the sample descend from some of the earliest colonists or from more recent arrivals or might some be migrants themselves? If the population structure was open and long-distance travel common (Boldurian 1991; Hill et al. 2011; Mahli et al. 2004; Meiklejohn 1977), it would have worked against drift and founder effect as processes for generating high levels of local variation. Early Holocene variability argues against Paleoamericans originating from a small founding group (Hey 2005) in favor of variability being rooted in ongoing immigration.

Results obtained from this analysis support and strengthen the earlier conclusion concerning coastal migrations which postulated that early Holocene crania would be more similar to Polynesians than to other recent human populations (Jantz and Owsley 2005). The current analysis provides additional clues about the sources of the migrants. Kennewick Man is the clearest example to date of the similarities between Polynesians and early Holocene crania observed by Jantz and Owsley (2005). This finding obviously does not suggest a direct connection with Polynesians. Rather, it suggests that early Americans and Polynesians have roots in the same Asian populations, probably those inhabiting coastal areas and using watercraft to exploit marine resources. It further supports the idea that the ancestors of these early Americans left Asia bearing a more generalized morphology, prior to the development of features now typical of East Asians (Lahr 1995; Steele and Powell 1999). This older morphology is preserved in Polynesians.

Morphology associated with recent Native Americans resulted from subsequent movement of people from Asia bearing morphology more typical of recent East Asians. These later migrants may have displaced the earlier ones, or more likely assimilated some, which would help explain the high variability among recent Native Americans. Evidence for a coastal route does not preclude the traditional route over the Bering landbridge, which may well have been the route of people more typically

Asian in morphology. Such a scenario was deftly stated by Haynes (2005:128): "Imagine the excitement if Clovis folk or their successors met other people, perhaps in the Great Basin, people who arrived along the West Coast via a coastal route." This excitement would presumably have been no less if the Clovis people were descendants of first arrivals from a westward expansion, having come to America via an Atlantic route (Stanford and Bradley 2012). Some of this interaction could well have included aggression, as indicated by traumatic injuries in at least three of the early skeletons (Owsley and Jantz 1999).

Ongoing immigration from Asia also likely contributed to variation of both early and recent Americans. Brace et al. (2004) report high diversity among prehistoric inhabitants of the Americas, going back to Archaic times. They also found that Archaic, Woodland, and even some Late Prehistoric samples display affinities to Old World populations, such as the Ainu-Jōmon and Polynesians. This demonstrates that the Polynesian affinities of early Holocene crania, especially Kennewick Man, are not anomalous findings, but rather part of a widespread pattern that persisted for some time and, in some areas, until rather recently (Gill 2005; González-José et al. 2003). The idea that people have been moving back and forth across the Bering Strait for millennia is not new (Fitzhugh and Krupnik 2001). Watercraft must have contributed to movements (Erlandson 2002), although some maritime migrations were likely accidental (Quimby 1985). A model of continuous interchange, based both on microsatellite data (Ray et al. 2010) and cranial morphology (Azvedo et al. 2011) provides a much better fit than discrete migration hypotheses.

This analysis of Kennewick Man has shown that he fits into the pattern of other early Holocene crania. Like most others, he has similarities to circumpacific populations, especially Polynesians. Kennewick Man increases the range of variation of early Holocene crania somewhat, but overall early Holocene populations do not appear to be significantly more variable than modern populations. Cranial morphology suggests that the early Pacific colonizers of America left Asia prior to the development of classic Asian cranial features, as summarized by Steele and Powell (1999). Kennewick Man fits well within this basic paradigm.

REFERENCES

Auerbach, Benjamin M. 2011. Methods for estimating missing human skeletal osteometric dimensions employed in the revised Fully technique for estimating stature. *American Journal of Physical Anthropology* 145(1):67–80.

Auerbach, Benjamin M., and Christopher B. Ruff. 2010. Stature estimation formulae for indigenous North American populations. *American Journal of Physical Anthropology* 141(2):190–207.

Azevedo, Soledad de, Ariadna Nocera, Carolina Paschetta, Lucía Castillo, Marina González, and Rolando González-José. 2011. Evaluating microevolutionary models for the early settlement of the New World: The importance of recurrent gene flow with Asia. *American Journal of Physical Anthropology* 146(4):539–552.

Beals, Kenneth L., Courtland L. Smith, and Stephen M. Dodd. 1984. Brain size, cranial morphology, climate and time machines. *Current Anthropology* 25(3):301–330.

Boldurian, Anthony T. 1991. Folsom mobility and organization of lithic technology: A view from Blackwater Draw, New Mexico. *Plains Anthropologist* 36(137):281–295.

Brace, C. Loring, A. Russell Nelson, and Pan Qifeng. 2004. Peopling of the New World: A comparative craniofacial view. In *The Settlement of the American Continents: A Multidisciplinary Approach to Human Biogeography*, edited by C. Michael Barton, Geoffrey A. Clark, David Yesner, and Georges A. Pearson, 28–38. University of Arizona Press, Tucson.

Breternitz, David A., Alan C. Swedlund, and Duane C. Anderson. 1971. An early burial from Gordon Creek, Colorado. *American Antiquity* 36(2):170–182.

Bruwelheide, Karin S., Douglas W. Owsley, and Richard L. Jantz. 2010. Burials from the fourth mission church at Pecos. In *Pecos Pueblo Revisited: The Biological and Social Context*, edited by Michele E. Morgan, 129–160. Peabody Museum of Archeology and Ethnology Paper 85. Peabody Museum of Archeology and Ethnology, Harvard University, Cambridge, MA.

Cavalli-Sforza, Luigi L., Paolo Menozzi, and Alberto Piazza. 1994. *The History and Geography of Human Genes*. Princeton University Press, Princeton.

Chatters, James C. 2000. The recovery and first analysis of an early Holocene human skeleton from Kennewick, Washington. *American Antiquity* 65(2):291–316.

Chatters, James C., Walter A. Neves, and Max Blum. 1999. The Kennewick Man: A first multivariate analysis. *Current Research in the Pleistocene* 16:87–90.

Chatters, James C., Steven Hackenberger, Alan J. Busacca, Linda Scott Cummings, Richard L. Jantz, Thomas W. Stafford, Jr., and R. Ervin Taylor. 2000. A possible second Early-Holocene skull from Central Washington. *Current Research in the Pleistocene* 17:93–95.

Diamond, Jared. 1999. *Guns, Germs, and Steel: The Fates of Human Societies.* W.W. Norton & Company, New York.

Edgar, Heather J. H., Edward A. Jolie, Joseph F. Powell, and Joe E. Watkins. 2007. Contextual issues in Paleoindian repatriation: Spirit Cave Man as a case study. *Journal of Social Archaeology* 7(1):101–122.

Erlandson, Jon. M. 2002. Anatomically modern humans, maritime voyaging, and Pleistocene colonization of the Americas. In *The First Americans: The Pleistocene Colonization of the New World*, edited by Nina G. Jablonski, 59–92. California Academy of Sciences Memoir 27. California Academy of Sciences, San Francisco.

Fiedel, Stuart. 2004. The Kennewick follies: "New" theories about the peopling of the Americas. *Journal of Anthropological Research* 60(1):75–110.

Fitzhugh, William W., and Igor Krupnik. 2001. Introduction. In *Gateways: Exploring the Legacy of the Jesup North Pacific Expedition, 1897–1902*, edited by William W. Fitzhugh and Igor Krupnik, 1–16. Contributions to Circumpolar Anthropology 1. Arctic Studies Center, National Museum of Natural History, Smithsonian Institution, Washington, DC.

Genovés, Santiago. 1967. Proportionality of the long bones and their relation to stature among Mesoamericans. *American Journal of Physical Anthropology* 26(1):67–78.

Gill, George W. 2005. Appearance of the "Mongoloid Skeletal Trait Complex" in the northwestern Great Plains: Migration, selection, or both? In *Paleoamerican Origins: Beyond Clovis*, edited by Robson Bonnichsen, Bradley T. Lepper, Dennis Stanford, and Michael R. Waters, 257–266. Texas A&M Press, College Station.

González-José, Rolando, Antonio González-Martín, Miquel Hernández, Hector M. Pucciarelli, Marina Sardi, Alfonso Rosales, and Silvina Van der Molen. 2003. Craniometric evidence for Paleoamerican survival in Baja California. *Nature* 425(6953):62–65.

González-José, Rolando, Walter Neves, Marta Mirazón Lahr, Silvia González, Héctor Pucciarelli, Miguel Hernández Martínez, and Gonzalo Correal. 2005. Late Pleistocene/Holocene craniofacial morphology in Mesoamerican Paleoindians: Implications for the peopling of the New World. *American Journal of Physical Anthropology* 128(4):772–780.

Gower, J. C. 1972. Measures of taxonomic distance and their analysis. In *The Assessment of Population Affinities in Man*, edited by J. S. Weiner, 1–24. Clarendon Press, Oxford.

Green, Thomas J., Bruce Cochran, Todd W. Fenton, James C. Woods, Gene L. Titmus, Larry Tieszen, Mary Anne Davis, and Susanne J. Miller. 1998. The Buhl burial: A Paleoindian woman from southern Idaho. *American Antiquity* 63(3):437–456.

Grine, F. E., W. L. Jungers, P. V. Tobias, and O. M. Pearson. 1995. Fossil *Homo* femur from Berg Aukas, northern Namibia. *American Journal of Physical Anthropology* 97(2):151–185.

Haynes, C. Vance Jr. 2005. Clovis, Pre-Clovis, climate change and extinction. In *Paleoamerican Origins: Beyond Clovis*, edited by Robson Bonnichsen, Bradley T. Lepper, Dennis Stanford, and Michael R. Waters, 113–132. Texas A&M University, College Station.

Henneberg, M. 1988. Decrease of human skull size in the Holocene. *Human Biology* 60(3):395–405.

Henneberg, M., and M. Steyn. 1993. Trends in cranial capacity and cranial index in Subsaharan Africa during the Holocene. *American Journal of Human Biology* 5(4):473–479.

———. 1995. Diachronic variation of cranial size and shape in the Holocene: A manifestation of hormonal evolution? *Rivista di Antropologia* 73:159–164.

Hey, Jody. 2005. On the number of New World founders: A population genetic portrait of the peopling of the Americas. *PLoS Biology* 3(6): e193. doi:10.1371/journal.pbio.0030193.

Hill, Kim R., Robert S. Walker, Miran Božičević, James Eder, Thomas Headland, Barry Hewlett, A. Magdalena Hurtado, Frank W. Marlowe, Polly Wiessner, and Brian Wood. 2011. Co-residence patterns in hunter-gatherer societies show unique human social structure. *Science* 331(6022):1286–1289.

Holliday, Trenton W. 2002. Body size and postcranial robusticity of European upper Paleolithic hominins. *Journal of Human Evolution* 43(4):513–528.

Hooton, Earnest A. 1930. *The Indians of Pecos Pueblo, A Study of Their Skeletal Remains.* Yale University Press, New Haven.

Howells, William W. 1973. *Cranial Variation in Man: A Study by Multivariate Analysis of Patterns of Difference Among Human Populations.* Papers of the Peabody Museum of Archaeology and Ethnology 67. Harvard University, Cambridge, MA.

———. 1989. *Skull Shapes and the Map: Craniometric Analyses in the Dispersion of Modern* Homo. Papers of the Peabody Museum of Archaeology and Ethnology 79. Harvard University, Cambridge, MA.

———. 1995. *Who's Who in Skulls: Ethnic Identification of Crania from Measurement.* Papers of the Peabody Museum of Archaeology and Ethnology 82. Harvard University, Cambridge, MA.

Hrdlička, Aleš. 1903. The Lansing skeleton. *American Anthropologist* 5(2):323–330.

———. 1914. Physical anthropology in America: An historical sketch. *American Anthropologist* 16(4):508–554.

———. 1937. The Minnesota 'Man.' *American Journal of Physical Anthropology* 22(2):175–199.

Jantz, Richard L. 1973. Microevolutionary change in Arikara crania: A multivariate analysis. *American Journal of Physical Anthropology* 38(1):15–26.

———. 2006. Anthropometry. In *Handbook of North American Indians,* vol 3: *Environment, Origins, and Population,* edited by Douglas H. Ubelaker, 777–788. Smithsonian Institution Press, Washington, DC.

Jantz, Richard L., David R. Hunt, and Lee Meadows. 1995. The measure and mismeasure of the tibia: Implications for stature estimation. *Journal of Forensic Sciences* 40(5):758–761.

Jantz, Richard L., Paul Marr, and Claire A. Jantz. 2010. Body proportions in recent Native Americans: Colonization history vs. ecogeographic patterns. In *Human Variation in the Americas: The Integration of Archaeology and Biological Anthropology*, edited by Benjamin M. Auerbach, 292–310. Occasional Paper No. 38. Center for Archaeological Investigations, Southern Illinois University, Carbondale.

Jantz, Richard L., and Douglas W. Owsley. 2001. Variation among early North American crania. *American Journal of Physical Anthropology* 114(2):146–155.

———. 2005. Circumpacific populations and the peopling of the New World: Evidence from cranial morphometrics. In *Paleoamerican Origins: Beyond Clovis*, edited by Robson Bonnichsen, Bradley T. Lepper, Dennis Stanford, and Michael R. Waters, 267–275. Texas A&M Press, College Station.

Key, Patrick J. 1983. "Craniometric Relationships among Plains Indians: Cultural-Historical and Evolutionary Implications." Ph.D. dissertation, University of Tennessee, Knoxville.

Lahr, Marta Mirazón. 1995. Patterns of modern human diversification: Implications for Amerindian origins. *American Journal of Physical Anthropology* 38(S2):163–198.

Malhi, Ripan S., Katherine E. Breece, Beth A. Schultz Shook, Frederika A. Kaestle, James C. Chatters, Steven Hackenberger, and David Glenn Smith. 2004. Patterns of mtDNA diversity in northwestern North America. *Human Biology* 76(1):33–54.

Meiklejohn, Christopher. 1977. Genetic differentiation and deme structure: Considerations for an understanding of the Athapascan/Algonkian continuum. In *Prehistory of the North American Sub-Arctic: The Athapaskan Question,* edited by J. Helmer, S. Van Dyke, and F. Mense, 106–110. The University of Calgary Archaeological Association, Calgary.

Merriwether, D. A., W. W. Hall, A. Vahlne, and R. E. Ferrell. 1996. mtDNA variation indicates Mongolia may have been the source for the founding population for the New World. *American Journal of Human Genetics* 59(1):204–212.

Myster, Susan M. Thurston, and Barbara O'Connell. 1997. Bioarchaeology of Iowa, Wisconsin, and Minnesota. In *Bioarchaeology of the North Central United States*, edited by Douglas W. Owsley and Jerome C. Rose, 147–239. Arkansas Archaeological Survey Research Series 49. Arkansas Archaeological Survey, Fayetteville.

Neumann, Georg K. 1952. Archeology and race in the American Indian. In *Archeology of Eastern United States*, edited by James B. Griffin, 13–34. University of Chicago Press, Chicago.

Ousley, Stephen D., and Ashley H. McKeown. 2001. Three dimensional digitizing of human skulls as an archival procedure. In *Human Remains: Conservation, Retrieval and Analysis*, edited by Emily Williams, 173–184. British Archaeological Reports International Series 934. Archaeopress, Oxford.

Owsley, Douglas W., and Richard L. Jantz. 1999. Databases for Paleo-American skeletal biology research. In *Who Were the First Americans? Proceedings of the 58th Annual Biology Colloquium, Oregon State University*, edited by Robson Bonnichsen, 79–96. Texas A&M University Press, College Station.

Owsley, Douglas W., Margaret A. Jodry, Thomas W. Stafford, Jr., C. Vance Haynes, Jr., and Dennis J. Stanford. 2010. *Arch Lake Woman: Physical Anthropology and Geoarchaeology*. Texas A&M Press, College Station.

Poll, Heinrich, and K. J. Dennison. 1992. *On Skulls and Skeletons of the Inhabitants of the Chatham Islands: Results of a Journey to the Pacific, Schauinsland, 1896–1897*. K. J. Dennison, Dunedin, New Zealand.

Powell, Joseph F., and Jerome C. Rose. 1999. Chapter 2. Report on the Osteological Assessment of the "Kennewick Man" Skeleton (CENWW.97. Kennewick). In *Report on the Non-Destructive Examination, Description, and Analysis of the Human Remains from Columbia Park, Kennewick, Washington* [October 1999]. National Park Service, http://www.nps.gov/archeology/kennewick/powell_rose.htm.

Powell, Joseph F. 2005. *The First Americans: Race, Evolution, and the Origin of Native Americans*. Cambridge University Press, Cambridge.

Powell, Joseph F., and Walter A. Neves. 1999. Craniofacial morphology of the First Americans: Pattern and process in the peopling of the New World. *American Journal of Physical Anthropology* 110(S29):153–188.

Quimby, George I. 1985. Japanese wrecks, iron tools, and prehistoric Indians of the Northwest Coast. *Arctic Anthropology* 22(2):7–15.

Ray, N., D. Wegmann, N. J. R. Fagundes, S. Wang, A. Ruiz-Linares, and L. Excoffier. 2010. A statistical evaluation of models for the initial settlement of the American continent emphasizes the importance of gene flow with Asia. *Molecular Biology and Evolution* 27(2):337–345.

Relethford, John H., and John Blangero. 1990. Detection of differential gene flow from patterns of quantitative variation. *Human Biology* 62(1):5–25.

Roseman, Charles C., and Timothy D. Weaver. 2007. Molecules versus morphology? Not for the human cranium. *Bioessays* 29(12):1185–1188.

Schurr, Theodore G. 2004. An anthropological genetic view of the peopling of the New World. In *The Settlement of the American Continents: A Multidisciplinary Approach to Human Biogeography*, edited by C. Michael Barton, Geoffrey Clark, David Yesner, and Georges A. Pearson, 11–27. University of Arizona Press, Tucson.

Smith, Heather F. 2009. Which cranial regions reflect molecular distances reliably in humans? Evidence from three-dimensional morphology. *American Journal of Human Biology* 21(1):36–47.

Stanford, Dennis J., and Bruce A. Bradley. 2012. *Across Atlantic Ice: The Origin of America's Clovis Culture*. University of California Press, Berkeley.

Steele, D. Gentry, and Joseph F. Powell. 1992. Peopling of the Americas: Paleobiological evidence. *Human Biology* 64(3):303–336.

———. 1999. Peopling of the Americas: A historical and comparative perspective. In *Who Were the First Americans? Proceedings of the 58th Annual Biology Colloquium, Oregon State University*, edited by Robson Bonnichsen, 97–127. Oregon State University, Corvallis.

Stewart, T. Dale. 1973. *The People of America*. Charles Scribner's Sons, New York.

Swedlund, Alan, and Duane Anderson. 1999. Gordon Creek Woman meets Kennewick Man: New interpretations and protocols regarding the peopling of the Americas. *American Antiquity* 64(4):569–576.

Taylor, Andrea B., and Carel P. van Schaik. 2007. Variation in brain size and ecology in Pongo. *Journal of Human Evolution* 52(1):59–71.

Trotter, Mildred, and Goldine C. Gleser. 1958. A reevaluation of estimation of stature taken during life and of long bones after death. *American Journal of Physical Anthropology* 16(1):79–123.

Tuohy, Donald R., and Amy J. Dansie. 1997. New information regarding early Holocene manifestations in the western Great Basin. *Nevada Historical Society Quarterly* 40(1):24–53.

Waguespack, Nicole M. 2007. Why we're still arguing about the Pleistocene occupation of the Americas. *Evolutionary Anthropology* 16(2):63–74.

Walker, Robert S., Kim R. Hill, Mark V. Flinn, and Ryan M. Ellsworth. 2011. Evolutionary history of hunter-gatherer marriage practices. *PLoS ONE* 6(4):e19066. doi:19010.11371/journal.pone.0019066.

Weaver, Timothy D., Charles C. Roseman, and Christopher B. Stringer. 2008. Close correspondence between quantitative- and molecular-genetic divergence times for Neandertals and modern humans. *Proceedings of the National Academy of Sciences* 105(12):4645–4649.

Weisensee, Katherine E., and Richard L. Jantz. 2010. Rethinking Hooton: A reexamination of the Pecos cranial and postcranial data using recent methods. In *Pecos Pueblo Revisited: The Biological and Social Context*, edited by Michele E. Morgan, 43–55. Papers of the Peabody Museum No. 85. Peabody Museum of Archeology and Ethnology, Harvard University, Cambridge, MA.

Wright, Richard K. 2007. Cranid. http://www.box.net/shared/h0674knjzl.

Young, Diane. 1988. An osteological analysis of the Paleoindian double burial from Horn Shelter No. 2. *Central Texas Archeologist* 11:11–115.

Two-Dimensional Geometric Morphometrics

M. Katherine Spradley, Katherine E. Weisensee, and Richard L. Jantz

In spite of Hrdlička's (1937) original assessment of no morphological difference, evidence now indicates that Paleoamericans have a craniofacial morphology different from modern Native Americans (Chapter 25; Neves and Blum 2000; González-José et al. 2001; Jantz and Owsley 2001; Neves et al. 2005). Paleoamericans tend to have long cranial vaults with shorter, narrower faces, while modern Native Americans typically exhibit shorter cranial vaults with taller, wider faces. The morphology of Kennewick Man's cranium needs to be integrated into this generalization. With a few inaccuracies, Kennewick Man's craniometric data were reported by government-appointed scientists in 1999. Analyses of these data (Powell and Rose 1999; Chatters 2000; Steele and Powell 2002; Herrmann et al. 2006) have used traditional interlandmark distances to assess Kennewick Man's biological relationships to other groups, particularly to recent Native Americans.

Interlandmark distances provide metric descriptions of overall characteristics of cranial size and shape, but they do so in only one dimension without providing information about relative positions of landmarks or the direction or magnitude of change. Geometric morphometric analyses, on the other hand, can use landmark data to explore variations in shape and size independent of one another or combined using two- or three-dimensional data. Kennewick Man's measurements have been included in only one study that used geometric morphometric methods (González-José et al. 2008). That study was not focused on Kennewick Man in particular, but rather on the process of peopling the New World with emphasis on South America. Their analysis focused on shape variation without incorporating size.

Geometric morphometric methods, which utilize landmark data, are more robust than techniques analyzing traditional interlandmark distances (Slice 2007; Mitteroecker and Gunz 2009). This approach captures the overall size, as well as the shape, that is, geometric properties "invariant to position, orientation and isometric size differences" (Slice 2007:262). Traditional interlandmark distances do not provide this information. Most geometric morphometric analyses in biological anthropology use standard landmarks, often the same used to measure traditional interlandmark distances. Geometric morphometric methods can incorporate two- or three-dimensional landmark data (i.e., Cartesian coordinates), but until the 1990s, such coordinates were not easily obtained. Portable digitizers, such as the Microscribe® 3Dx, facilitate landmark data collection, and these data can also be extracted from photographs using software. Landmark coordinates can be triangulated using linear measurements when at least three measurements originate from the same landmark.

Geometric morphometric methods are beneficial because shape and size can be discriminated and analyzed separately. When examining changes in populations over time, an investigator may want to determine whether observed differences in morphology are due to changes in size alone or to changes in size and shape. It is also possible to visualize changes in shape that are due to changes in size, that is, allometric. However, geometric morphometric methods are the only way to analyze and visualize shape changes and components of shape change, in other words, how relative positions of landmarks differ among groups.

In order to isolate the variables of size and shape, geometric morphometrics scale all specimens to a common centroid size. Analysis of shape differences takes place in a curvilinear space termed Kendall's shape space (Rohlf 1999; Slice 2001; Mitteroecker and Gunz 2009) in which only shapes can be compared. To address questions beyond mean shape differences through the application of statistical methods, the data must be projected onto a tangent Euclidean plane (Slice 2001). Instead of interpreting the statistical output for two-dimensional plots, shape differences associated with axes from canonical variates analysis can be viewed as wireframe models.

The assertion that Paleoamericans have a craniofacial morphology different from recent Native Americans is based on traditional interlandmark distances. This chapter applies geometric morphometric methods to evaluate shape variation in Paleoamerican and Archaic period crania relative to Historic North American groups, with emphasis on Kennewick Man. Centroid size is incorporated into the analysis. This evaluation also addresses size variation as well as the allometry of Paleoamerican, Archaic, and recent Native American crania.

COMPARATIVE SAMPLES AND METHODOLOGY
Cranial Samples

Groups used to assess biological relationships among Paleoamericans, Archaic period remains, and recent Native Americans are presented in Table 26.1. Four Paleoamerican crania are used in these analyses. Although additional crania are available, sufficient data for this type of analysis can only be collected from the selected four. Five Archaic period groups and seven recent Native American groups are included as comparative samples for a total sample of 483 crania. Two of the Paleoamericans were measured by Jantz, Spradley measured one, and the dataset for one was extracted from photographs (Hermann et al. 2006). The Arikara and Santa Cruz historic samples come from the Howells's craniometric database (http://konig.la.utk.edu/howells.htm). The remaining subsets were measured by Jantz, D. Owsley, and Jantz's students over the past 30 years. The population samples presented in Table 26.1 were pooled into Paleoamerican, Archaic, and Historic groups in this search for patterning or variation in size and shape within a diachronic framework.

Landmarks

Following advances in technical capability, more recently collected cranial data are available in landmark format. A potential limitation to geometric morphometric analysis is that previously collected reference data, such as

TABLE 26.1   Individual crania, groups, and sample sizes used in analyses.

Group	Temporal Group	N	Geographic Location	Time Period	Source
Kennewick Man	Paleoamerican	1	Kennewick, Washington	8410 RC yr. BP	Chatters 2000
Buhl	Paleoamerican	1	Buhl, Idaho	10,625 RC yr. BP	Green et al. 1998
Spirit Cave	Paleoamerican	1	Carson Sink, Nevada	9415 RC yr. BP	Dansie 1997
Wizards Beach	Paleoamerican	1	Pyramid Lake, Nevada	9225 RC yr. BP	Dansie 1997
Pelican Rapids	Archaic	1	Minnesota	7850 RC yr. BP	Myster and O'Connell 1997
La Jolla	Archaic	2	La Jolla, California	5460 and 7370 RC yr. BP	Shumway et al. 1961
Windover	Archaic	21	Titusville, Florida	8120 RC yr. BP	Dickel 2002
Eva	Archaic	24	Benton Co., Tennessee	7150 RC yr. BP	Herrmann et al. 2006
Indian Knoll	Archaic	27	Paradise, Kentucky	4180 RC yr. BP	Herrmann et al. 2006
Arikara	Historic	69	Fort Sully, South Dakota	A.D. 1600–1750	Herrmann et al. 2006
Zuni	Historic	47	Hawikuh, New Mexico	A.D. 1680	Herrmann et al. 2006, Hodge 1937
Numic	Historic	35	Great Basin (various locations)	A.D. 1600–1900	
Fremont	Historic	30	Utah (various locations)	A.D. 500–1250	Jennings 1978:155
Blackfeet	Historic	62	Montana (various locations)	A.D. 1500–1900	
Sioux	Historic	59	Plains (majority South Dakota)	A.D. 1750–1900	
Santa Cruz	Historic	102	Santa Cruz Island, California	A.D. 1700–1900	Howells 1973

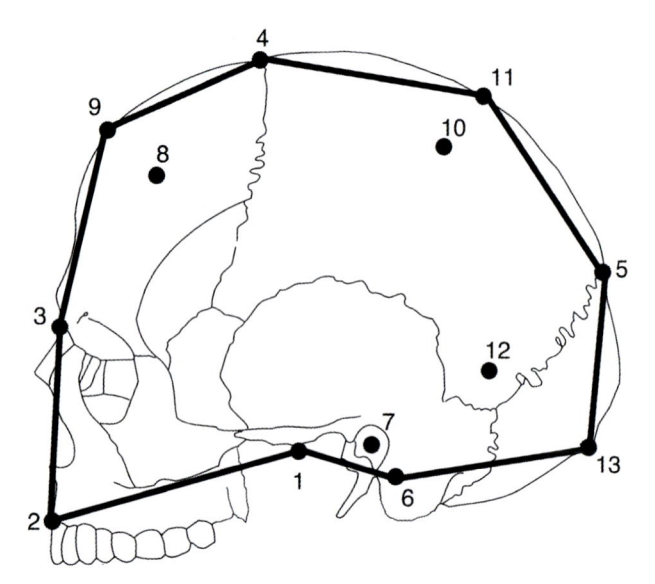

FIGURE 26.1 Anatomical location of 2-dimensional landmarks reconstructed from the Truss function: 1. Basion; 2. Prosthion; 3. Nasion; 4. Bregma; 5. Lambda; 6. Opisthion; 7. Radiometer point; 8. Nasion subtense fraction; 9. Metopion; 10. Bregma subtense fraction; 11. Parietal subtense fraction; 12. Lambda subtense fraction; 13. Occipital subtense fraction.

Howells' (1973) craniometric standards, are only available in traditional interlandmark distances. Traditional interlandmark distances can be calculated from two- or three-dimensional data using the Pythagorean Theorem. Nevertheless, it is more difficult and not always possible to calculate landmark data from interlandmark distances unless there are a number of redundant measurements. Fortunately, some of Howells's interlandmark distances contain measurements that share common origin points. This redundancy allows for the reconstruction of 13 two-dimensional landmarks (Figure 26.1) from the complete Howells's dataset using an iterative algorithm of the truss feature (Carpenter et al. 1996; Wescott and Jantz 2005) in Morpheus (Slice 2001, 2007).

Most of the landmarks shown in Figure 26.1 are commonly used in traditional craniometric data collection and analysis. The frontal, parietal, and occipital subtense fractions lie on the chord where the perpendicular to the bone surface is maximized, identifying the metopion, parietal, and occipital points. These internal points are calculated, not measured, because they are identified in space rather than on the actual bone surface.

Shape

The first step involves alignment of landmarks to a common coordinate system using a Generalized Procrustes analysis. This analysis arbitrarily fixes landmarks related to a specimen's size, location, and orientation in order to minimize the total sum of squares, which eliminates nonshape-related variation (size) (Bookstein 1991; Slice 2007; Mitteroecker and Gunz 2009). Following the Generalized Procrustes analysis, the landmarks are projected from Kendall's shape space into a Euclidean tangent plane, allowing standard multivariate statistics to be carried out (e.g., principal components, canonical variate analysis, and discriminant function analysis). Within Kendall's shape space it is possible to determine whether shape differences exist between groups, but to determine the statistical significance of the variation, the coordinates must be projected onto tangent space (Slice 2001).

Because of the small sample size of the Paleoamerican group, and because the Generalized Procrustes analysis removes size-related variation, males and females were pooled in order to examine shape variation. A canonical variate analysis was then performed to evaluate the relationship of each group to the others and to determine whether shape features distinguish the three temporal groups (Klingenberg 2011). Mahalanobis distances with associated p-values were used to assess the significance of group differences. Wireframe graphs depicting sagittal outlines of cranial shapes enable visualization of the associated shape differences.

Size and Allometry

To test for size differences among the Paleoamerican, Archaic, and Historic groups, the centroid size was obtained via Generalized Procrustes analysis in MorphoJ (Klingenberg 2011) for each individual within each of the three temporal groups. Centroid size is "the square root of the summed of squared deviations of the coordinates from their centroid" (Mitteroecker and Gunz 2009:238); it is not correlated with shape variables provided there is no allometry. Centroid size and log centroid size are included in the output from the Generalized Procrustes analysis and are used to explore size variation separately. An analysis of variance was used to determine whether significant differences in size exist and a Tukey multiple comparison test determined which groups differ significantly from one another. Because there is only one female Paleoamerican cranium, the pooled within-group standard deviation was used as the standard deviation. The analysis of variance was performed in SAS 9.1.3 (SAS 9.1.3, SAS Institute Inc. 2002–2004).

Multivariate regression of the Generalized Procrustes coordinates onto the centroid size was performed in order to explore whether changes in size resulted in changes in shape. A linear function of the shape variables was chosen to maximize regression on size. A significant regression means that change in size causes change in shape. Analyses were run separately for males and females because of differences in size between the sexes. Wireframe graphs allow visualization of the shape changes associated with size.

RESULTS
Shape

The Mahalanobis distances generated from the Generalized Procrustes coordinates indicate significant differences among groups at the 0.05 level (Table 26.2). The smallest distance is between the Historic and Archaic groups; the largest distance is between the Paleoamerican and Historic groups. The canonical variate analysis plot (Figure 26.2) indicates separation among all three groups. The first axis (92.45 percent variation) separates the Archaic and Historic groups; the second axis (7.55 percent variation) separates Paleoamerican individuals from both groups. Kennewick Man, Wizards Beach, Spirit Cave, and Buhl are most differentiated on the second axis and are located on the farthest end of the positive axis. The Paleoamerican group is outside the range of the Archaic and Historic groups, emphasizing their unique morphology.

The shape profiles are visualized in the wireframe graphs in units of Mahalanobis distances (Figure 26.3). The shaded area is the average shape of all groups and the wireframe represents a change in 10 Mahalanobis distance units from the geometric mean. When considering the mean shape of crania representing a particular group, it is important to remember that the mean shape does not exist in reality. Shape variation in the positive direction along the first canonical axis includes a longer and more forwardly projecting face (nasion to prosthion); this reflects morphological differences between Archaic (negative axis) and Historic (positive axis) groups (Figure 26.3a). Changes in the frontal bone include a more anterior and superiorly placed metopion (frontal subtense) and an inferior and posterior placement of bregma and parietal subtense. Lambda is also positioned anteriorly and the occipital subtense posteriorly. Basion moves from an inferior position to a superior position reflecting a flatter cranial base. However, the position of opisthion changes little, resulting in a steeper inclination of the basion-opisthion plane in the Historic group. There is also a flexion point at basion between basion-opisthion and basion-prosthion, prosthion being inferior to basion.

Shape differences associated with the second canonical variate (Figure 26.2) separate the Paleoamerican group from both the Archaic and the Historic. The Paleoamerican group is characterized by longer and lower

TABLE 26.2 Mahalanobis distances from Procrustes Coordinates and *P*-values.

Group	Archaic	Historic	Paleoamerican
Archaic		< .0001	0.0305
Historic	2.5830		0.0278
Paleoamerican	3.1892	3.316	

Note: Distances below diagonal, *p*-values above.

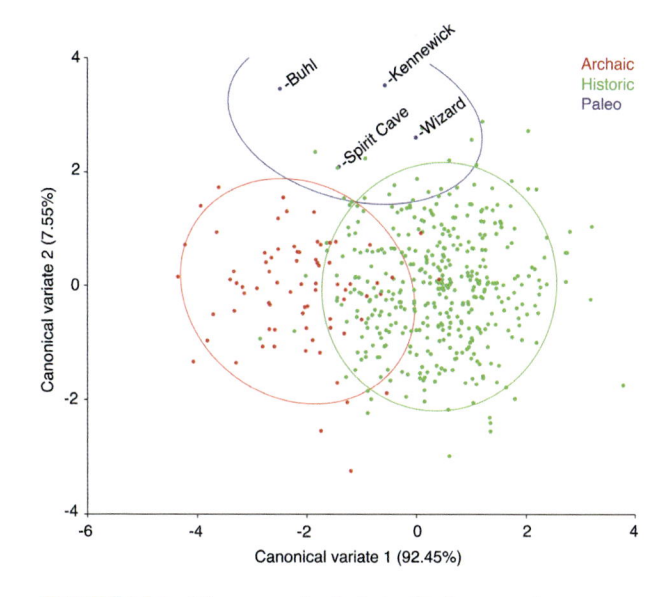

FIGURE 26.2 The canonical plot of Paleoamerican, Archaic, and Historic groups showing the separation among groups.

vaults compared to the average configuration with basion located more anteriorly and inferiorly (Figure 26.3b). The Paleoamerican face is more prognathic and the frontal is sloping with less anterior projection. The posteriorly sloped frontal, elongated parietals, and anterior locations of both basion and prosthion result in a prognathic lower face. In contrast to the Historic group, and to a lesser extent Archaic populations, the Paleoamerican group has virtually no flexion at basion. Prosthion, basion, and opisthion can be connected by a nearly straight line with prosthion superior to basion.

Size

Based on an analysis of variance test, significant differences in size are present in both males ($p < 0.0001$) and females ($p < 0.0001$). Means and standard deviations of the centroid sizes are presented in Tables 26.3 and 26.4 along with the Tukey pair-wise comparisons. For females (Table 26.3), Buhl has the largest centroid size at 281.43, followed by the Archaic group at 278.75, and the Historic at 271.08. With a Paleoamerican sample of one, the Tukey multiple comparison indicates that the only statistically significant difference is between the Archaic and Historic groups. Male centroid sizes show the same pattern as females: the Paleoamerican group has the largest crania, 297.05, followed by the Archaic, 287.56, and Historic groups, 280.28 (Table 26.4). For the male sample, Tukey multiple comparisons indicate that the Paleoamerican group is significantly different in centroid size from both the Archaic and Historic groups and that the Archaic group is different from the Historic samples. Male Paleoamerican and Archaic groups do not differ significantly in their centroid sizes.

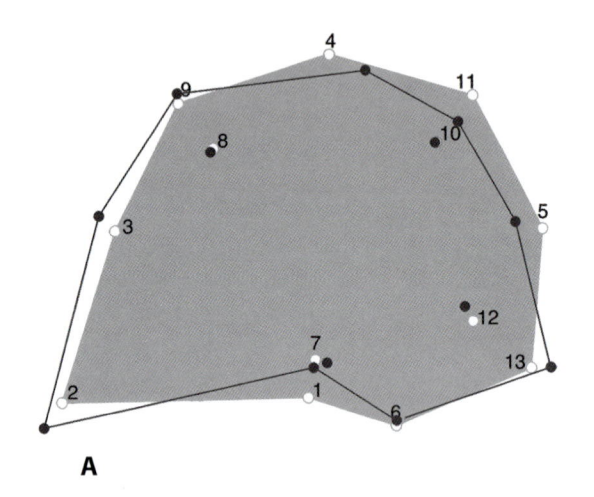

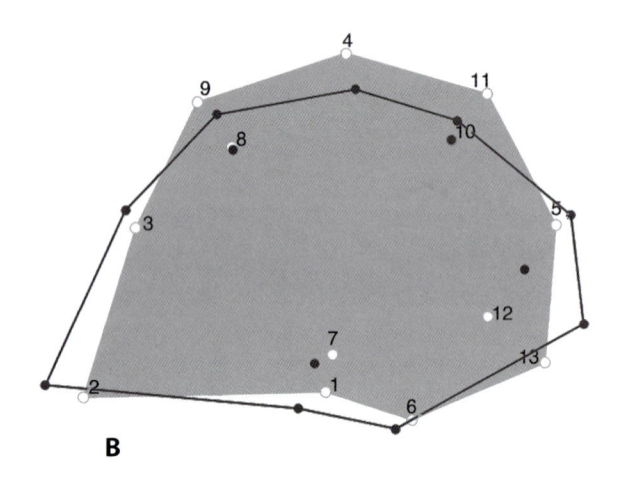

FIGURE 26.3 Shape changes associated with Canonical Variate 1 (Historic group) and 2 (Paleoamerican group) shown in Figure 26.2. The shaded areas represent the mean configuration of all groups and the wireframe outline represents shape changes in 10 Mahalanobis distance units in the positive direction on each axis. The wireframe outline in (A) represents the shape of the Historic group compared to the average shape of all groups. The wireframe outline in (B) represents the shape of the Paleoamerican group compared to the average shape.

TABLE 26.3 Female centroid size with Tukey results.

Group	N	Mean	SD	Tukey comparison
Paleoamerican	1	281.43	7.78[a]	Paleoamerican to Archaic
				Paleoamerican to Historic
Archaic	41	278.75	8.19	Archaic–Historic[b]
Historic	78	271.08	7.38	

[a]Obtained from the pooled within-group standard deviation.
[b]Indicates significant at the 0.05 level.

TABLE 26.4 Male Centroid Size with Tukey Results

Group	N	Mean	SD	Tukey comparison
Paleoamerican	3	297.05	3.96	Paleoamerican to Archaic
				Paleoamerican to Historic[a]
Archaic	34	287.56	8.48	Archaic–Historic[a]
Historic	202	280.28	7.53	

[a]Indicates significant at the 0.05 level.

Allometry

The regression of shape on centroid size is statistically significant in both males ($p < 0.0002$) and females ($p < 0.0001$). Centroid size accounts for 2.13 percent of shape variation in males and 1.76 percent in females. The relationships between cranial shape and centroid size represents changes associated with static allometry (Drake and Klingenberg 2008; Klingenberg 2011). Although statistically significant, the small percentage of variation accounted for and the large degree of scatter seen in the plots of centroid size with shape indicate that this static allometry is weak (Figures 26.4–26.5). Changes in cranial shape are not strongly correlated with changes in size in either males or females.

In order to visualize detected allometric changes, a magnification of 80 millimeters was necessary. Male changes in shape associated with size involve the cranial vault (Figure 26.5). There is little change in the facial area at prosthion and nasion. There is slight inferior placement of metopion and bregma resulting in a sloping appearance of the frontal bone and a shorter vault height. Changes in the posterior vault include posterior projection of lambda and superior placement of opisthion. In the females, allometric shape changes are similar but include a small posterior change in prosthion. Metopion and bregma are both slightly inferior. As in males, the most marked change is posterior projection of the vault involving lambda and the occipital subtense.

DISCUSSION

In concert with earlier morphometric studies, this set of Paleoamericans clearly has a different cranial morphol-

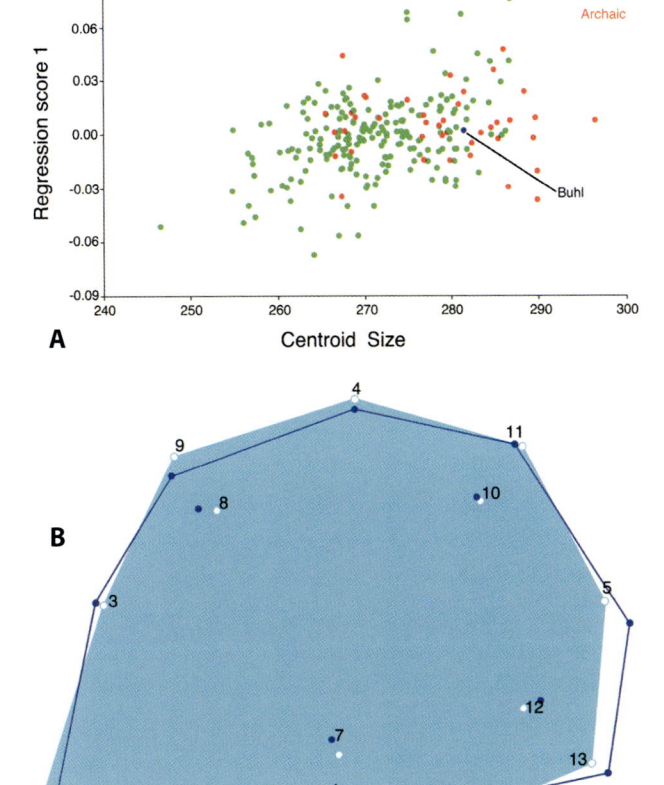

FIGURE 26.4 Regression of shape on size for females (A) and shape changes associated with changes in size magnified by 80 mm (B). The dark line represents the shape changes as compared to the average form shaded in blue.

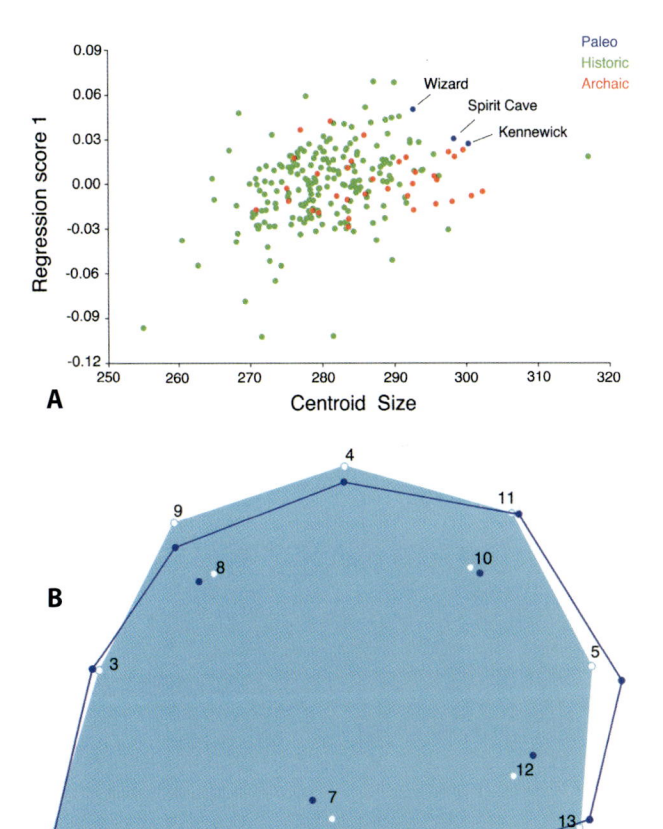

FIGURE 26.5 Regression of shape on size for males (A) and shape changes associated with changes in size magnified by 80 mm (B). The dark line represents the shape changes as compared to the average form shaded in blue.

ogy (long and narrow) from Historic Native Americans (short and broad). Kennewick Man's cranial form falls outside the range of variation of Historic period Native Americans (Powell and Neves 1999; Powell and Rose 1999; Herrmann et al. 2006). The results presented here diverge from previous analyses by allowing this shape variation to be visualized two-dimensionally in the sagittal plane. This refinement provides information regarding positioning of landmarks relative to one another. Paleoamericans differ not only in overall length and facial height. Rather, the entire craniofacial complex differs from the Archaic and Historic groups due to posteriorly sloping frontal bones, elongated and curved parietals, and anterior placement of both basion and prosthion resulting in a prognathic lower face.

In the shape analyses, Kennewick Man and the other Paleoamerican crania included fall outside of the range of variation of the Archaic and Historic groups. These shape variations are significantly different among all groups. The Paleoamericans also have larger crania. Although the regression of shape on centroid size is significant, the percent of variation explained by this regression is small. The morphological shape differences observed among the three groups, especially the differentiated craniofacial morphology of the Paleoamericans, is not primarily due to allometry. It is the case, especially in males, that their large centroid size is in part consistent with the observed Paleo/Historic difference, including a flatter frontal and longer cranial vault.

The presence of marked Paleo/Historic differences in cranial morphology is not in question, but it is subject to differing interpretations. Most previous analyses of Paleoamerican crania suggest that the differing craniofacial morphology supports a multiple migration model for the peopling of the New World (Lahr 1995; Steele and Powell 1999; Jantz and Owsley 2001; Steele and Powell 2002; Herrmann et al. 2006). Genetic evidence most often supports a single migration suggesting that morphological changes can be viewed as a continuum among Paleoamericans, Archaics, and Historic Native Americans (Zegura et al. 2004; Fagundes et al. 2008; Goebel et al. 2008). However, genetic evidence from Canada and South America suggests a minimum of three different migrations (Reich et al. 2012). Others cite 10,000 years of in situ change due to selection, genetic drift, gene flow, and nongenetic factors as the basis of these changes in cranial morphology (Carlson and Van Gerven 1977; Carlson and Van Gerven 1979; Larsen 1997; Armelagos and Van Gerven 2003; Powell 2005; Perez et al. 2009).

Swedlund and Anderson (2003:164) have suggested that "nutrition, and diet, and mechanical forces on the face related to diet and food processing can account for significant amounts of temporal variation in craniofacial morphology." Larsen (1997:233) emphasized the influence of changing masticatory function linked to dietary changes from the Holocene hunter-gatherer subsistence to late prehistoric agriculturists: "changes observed in Holocene craniofacial morphology and structure are related to biocultural factors," although processes such as gene flow can also shape the craniofacial complex.

Powell and Neves (1999) attribute some of the craniofacial variation present in the Americas to nongenetic or environmental factors such as developmental constraints and functional morphology of the facial region. Carlsen and Van Gerven (1977) were among the first to advance a masticatory functional hypothesis based on a study of Nubian populations spanning over 10,000 years. This study is often cited when invoking dietary explanations for the changes in craniofacial morphology over time (Powell and Neves 1999; Larsen 1997; Larsen 2002; Armelagos and Van Gerven 2003; Swedlund and Anderson 2003; Perez et al. 2009). Carlsen and Van Gerven, however, naively assume population continuity with little to no gene flow over 10,000 years, citing selection for reduced dentition size and dietary changes (the masticatory function hypothesis) as the primary factor for significant craniofacial change through time.

Research has determined that craniofacial morphology is moderately to highly heritable, with a neutral model of quantitative variation (Donnan 1935; Howells 1973; Harris and Hayes 1980; Devor 1987; Relethford 2004; Johannsdottir et al. 2005; Carson 2006; Martinez-Abadias et al. 2009; Holló et al. 2010). Craniometrics have proven useful for studying population relationships on a worldwide scale (Howells 1973; Relethford 2001; Roseman 2004), as well as within geographical regions (Sutter 2000; Schillaci and Stojanowski 2005; Perez et al. 2007; Ousley et al. 2009). Selective pressures and other nonrandom factors thought to influence phenotypic plasticity (e.g., dietary shifts) do not seem to obscure the underlying genetic structure, suggesting that population structure and history can be inferred from craniofacial morphology (Relethford 2004; Roseman 2004; Pucciarelli et al. 2006; Perez et al. 2007; Pinhasi and von Cramon-Taubadel 2009; Hubbe et al. 2009; Hubbe et al. 2010; Roseman et al. 2010). Although environmental changes may cause alterations in the craniofacial complex, it does not obscure phenotypic variability (Roseman 2010) or negate the underlying genetic structure (Relethford 2004).

The transition in morphology from the Paleoamericans to the Archaic series examined in this chapter certainly cannot be explained on the basis of a dietary shift. Both groups were hunter-gatherers. Because differences are observed in the positioning of all landmarks for Paleoamericans compared to the Archaic and Historic groups, not just those associated with differences in diet, it is unlikely that the observed changes were due to changes in subsistence strategies.

Perez et al. (2007), using landmark data and geometric morphometrics on prehistoric populations in South America, found that the facial region associated with masticatory function did not change significantly over time, even with a dietary change. In terms of experimental research, Roseman et al. (2010), using geometric morphometric methods, found that dietary differences between wild versus captive baboons (hard versus soft diets, respectively) did not cause phenotypic variability in the observed craniofacial traits.

A possible explanation for growth differences between Paleoamerican and Historic cranial differences may lie in the cranial base. There are insufficient landmarks to investigate the different components of the cranial base, but variation in the orientation of foramen magnum is evident. A relationship between orientation of the foramen magnum and flexion of the cranial base has been identified (Simpson 2005). Simpson also showed that Polynesians have higher cranial base angles (less flexion) than other human groups. Given the morphometric similarity of Paleoamericans to Polynesians (Chapter 25), one might infer that Paleoamericans are also characterized by high cranial base angles. This remains to be demonstrated, but given relationships between cranial base and craniofacial features generally, more attention to the cranial base is warranted. Access to a larger number of Paleoamerican crania is essential for this work.

CONCLUSION

Geometric morphometric methodology provides further insight into the morphological distinctiveness of Paleoamericans, as compared to Archaic and Historic Native Americans. The morphology changes presented in this chapter go beyond the traditional approach using interlandmark distances. The differences between Paleoamerican and Native American morphology are more than longer cranial vaults and shorter faces. The relative positions of all cranial landmarks used in the present analyses differ in orientation and provide initial unique insight into Paleoamerican morphology. The craniofacial shape differences observed among the Paleoamerican, Archaic, and Historic groups were undoubtedly influenced by a variety of complex processes including gene flow, genetic drift, and selection. Polygenic traits, such as craniofacial morphology, were also likely shaped by unknown environmental processes. However, invoking dietary explanations as the major force driving these shape changes is an oversimplification.

REFERENCES

Armelagos, George J., and Dennis P. Van Gerven. 2003. A century of skeletal biology and paleopathology: Contrasts, contradictions, and conflicts. *American Anthropologist* 105(1):53–64.

Bookstein, Fred L. 1991. *Morphometric Tools for Landmark Data: Geometry and Biology*. Cambridge University Press, Cambridge.

Carlson, David S., and Dennis P. Van Gerven. 1977. Masticatory function and Post-Pleistocene evolution in Nubia. *American Journal of Physical Anthropology* 46(3):495–506.

———. 1979. Diffusion, biological determinism and biocultural adaptation in the Nubian Corridor. *American Anthropologist* 81(3):561–580.

Carpenter, Kent E., H. Joseph Sommer III, and Leslie F. Marcus. 1996. Converting Truss interlandmark distances to Cartesian coordinates. In *Advances in Morphometrics*, edited by Leslie F. Marcus, Marco Corti, Anna Loy, Gavin J. P. Naylor, and Dennis E. Slice, 103–111. Plenum Press, New York.

Carson, E. Ann. 2006. Maximum likelihood estimation of

human craniometric heritabilities. *American Journal of Physical Anthropology* 131(2):169–180.

Chatters, James C. 2000. The recovery and initial analysis of an Early Holocene human skeleton from Kennewick, Washington. *American Antiquity* 65(2):291–316.

Dansie, Amy. 1997. Early Holocene burials in Nevada: Overview of localities, research, and legal issues. *Nevada Historical Society Quarterly* 40(1):4–14.

Devor, E. J. 1987. Transmission of human craniofacial dimensions. *Journal of Craniofacial Genetics and Developmental Biology* 7(2):95–106.

Dickel, David N. 2002. Analysis of mortuary patterns. In *Windover: Multidisciplinary Investigations of an Early Archaic Florida Cemetery*, edited by Glen H. Doran, 73–96. University Press of Florida, Gainesville.

Donnan, Elizabeth. 1935. *Documents Illustrative of the History of the Slave Trade to America*. Vol. 4. Publication 409. Carnegie Institute of Washington, DC.

Drake, Abby Grace, and Christian Peter Klingenberg. 2008. The pace of morphological change: Historical transformation of skull shape in St. Bernard dogs. *Proceedings of the Royal Society B: Biological Sciences* 275(1630):71–76.

Fagundes, Nelson J. R., Ricardo Kanitz, Roberta Eckert, Ana C. S. Valls, Mauricio R. Bogo, Francisco M. Salzano, David Glenn Smith, Wilson A. Silva Jr., Marco A. Zago, Andrea K. Ribeiro-Dos-Santos, Sidney E. B. Santos, Maria Luiza Petzl-Erler, and Sandro L. Bonatto. 2008. Mitochondrial population genomics supports a single pre-Clovis origin with a coastal route for the peopling of the Americas. *American Journal of Human Genetics* 82(3):583–592.

Goebel, Ted, Michael R. Waters, and Dennis H. O'Rourke. 2008. The late Pleistocene dispersal of modern humans in the Americas. *Science* 319(5869):1497–1502.

González-José, Rolando, Maria Cátira Bortolini, Fabrício R. Santos, and Sandro L. Bonatto. 2008. The peopling of America: Craniofacial shape variation on a continental scale and its interpretation from an interdisciplinary view. *American Journal of Physical Anthropology* 137(2):175–187.

González-José, Rolando, Silvia L. Dahinten, María A. Luis, Miquel Hernández, and Hector M. Pucciarelli. 2001. Craniometric variation and the settlement of the Americas: Testing hypotheses by means of R-matrix and matrix correlation analyses. *American Journal of Physical Anthropology* 116(2):154–165.

Harris, J. E., and M. D. Hayes. 1980. Heritability of the shape of the craniofacial complex. *Journal of Dental Research* 59(A):442.

Herrmann, Nicholas P., Richard L. Jantz, and Douglas W. Owsley. 2006. Buhl revisited: Three-dimensional photographic reconstruction and morphometric reevaluation. In *El Hombre Temprano en América y sus Implicaciones en el Poblamiento de la Cuenca de México*, edited by José Concepción Jiménez López, Silvia González, José Antonio Pompa y Padilla, Francisco Ortiz Pedraza, 211–220. Instituto Nacional de Antropologia e Historia, Mexico City.

Hodge, Frederick Webb. 1937. *History of Hawikuh, New Mexico: One of the So-called Cities of Cíbola*. The Southwest Museum, Los Angeles.

Holló, Gábor, László Szathmáry, Antónia Marcsik, and Zoltán Barta. 2010. Linear measurements of the neurocranium are better indicators of population differences than those of the facial skeleton: Comparative study of 1,961 skulls. *Human Biology* 82(1):29–46.

Howells, William W. 1973. *Cranial Variation in Man: A Study by Multivariate Analysis of Patterns of Difference Among Human Populations*. Papers of the Peabody Museum of Archaeology and Ethnology 67. Harvard University, Cambridge, MA.

Hrdlička, Aleš. 1937. Early man in America: What have the bones to say? In *Early Man: As Depicted by Leading Authorities at the International Symposium at the Academy of Natural Sciences*, edited by George Grant MacCurdy, 93–104. Lippencott Co., London.

Hubbe, Mark, Tsunehiko Hanihara, and Katerina Harvati. 2009. Climate signatures in the morphological differentiation of worldwide modern human populations. *The Anatomical Record: Advances in Integrative Anatomy & Evolutionary Biology* 292(11):1720–1733.

Hubbe, Mark, Walter A. Neves, and Katerina Harvati. 2010. Testing evolutionary and dispersion scenarios for the settlement of the New World. *Plos One* 5(6):e11105. doi:10.1371/journal.pone.0011105.

Jantz, Richard L., and Douglas W. Owsley. 2001. Variation among early North American crania. *American Journal of Physical Anthropology* 114(2):146–155.

Johannsdottir, Berglind, Freyr Thorarinsson, Arni Thordarson, and Thordur Eydal Magnusson. 2005. Heritability of craniofacial characteristics between parents and offspring estimated from lateral cephalograms. *American Journal of Orthodontics and Dentofacial Orthopedics* 127(2):200–207.

Jungers, William L., Anthony B. Falsetti, and Christina E. Wall. 1995. Shape, relative size, and size-adjustments in morphometrics. *American Journal of Physical Anthropology* 38(S2):137–161.

Klingenberg, Christian Peter. 2011. MorphoJ: An integrated software package for geometric morphometrics. *Molecular Ecology Resources* 11(2):353–357.

Lahr, Marta Mirazón. 1995. Patterns of modern human diversification: Implications for Amerindian origins. *Yearbook of Physical Anthropology* 38:163–198.

Larsen, Clark Spencer. 1997. *Bioarchaeology: Interpreting Behavior from the Human Skeleton*. Cambridge Studies in Biological Anthropology 21. Cambridge University Press, Cambridge.

———. 2002. Bioarchaeology: The lives and lifestyles of past people. *Journal of Archaeological Research* 10(2):119–166.

Martinez-Abadias, Neus, Mircia Esparza, Torstein Sjovold, Rolando González-José, Mauro Santos, and Miquel Hernandez. 2009. Heritability of human cranial dimensions: Comparing the evolvability of different cranial regions. *Journal of Anatomy* 214(1):19–35.

Mitteroecker, Philipp, and Philipp Gunz. 2009. Advances in Geometric Morphometrics. *Evolutionary Biology* 36(2):235–247.

Morpheus. Institute for Anthropology, University of Vienna, Vienna.

Myster, Susan M. Thurston, and Barbara O'Connell. 1997. Bioarchaeology of Iowa, Wisconsin, and Minnesota. In *Bioarchaeology of the North Central United States*, edited by Douglas W. Owsley and Jerome C. Rose, 147–239. Arkansas Archaeological Survey Research Series 49. Arkansas Archaeological Survey, Fayetteville.

Neves, Walter A., and Max Blum. 2000. The Buhl burial: A comment on Green et al. *American Antiquity* 65(1):191–193.

Neves, Walter A., Mark Hubbe, Maria Mercedes M. Okumura, Rolando González-José, Levy Figuti, Sabina Eggers, and Paulo Antonio Dantas De Blasis. 2005. A new early Holocene human skeleton from Brazil: Implications for the settlement of the New World. *Journal of Human Evolution* 48(4):403–414.

Ousley, Stephen, Richard Jantz, and Donna Freid. 2009. Understanding race and human variation: Why forensic anthropologists are good at identifying race. *American Journal of Physical Anthropology* 139(1):68–76.

Perez, S. Ivan, Valeria Bernal, and Paula N. Gonzalez. 2007. Evolutionary relationships among prehistoric human populations: An evaluation of relatedness patterns based on facial morphometric data using molecular data. *Human Biology* 79(1):25–50.

Perez, S. Ivan, Valeria Bernal, Paula N. Gonzalez, Marina Sardi, and Gustavo G. Politis. 2009. Discrepancy between cranial and DNA data of early Americans: Implications for American peopling. *Plos One* 4(5):e5746. doi: 10.1371/journal.pone.005746.

Pinhasi, Ron, and Noreen von Cramon-Taubadel. 2009. Craniometric data supports demic diffusion model for the spread of agriculture into Europe. *Plos One* 4(8):e6747. doi: 10.1371/journal.pone.0006747.

Powell, Joseph F. 2005. *The First Americans: Race, Evolution, and the Origin of Native Americans*. Cambridge University Press, Cambridge.

Powell, Joseph F., and Walter A. Neves. 1999. Craniofacial morphology of the first Americans: Pattern and process in the peopling of the New World. *American Journal of Physical Anthropology* 110(S29):158–188.

Powell, Joseph F., and Jerome C. Rose. 1999. Chapter 2. Report on the Osteological Assessment of the "Kennewick Man" Skeleton. In *Report on the Non-Destructive Examination, Description, and Analysis of the Human Remains from Columbia Park, Kennewick, Washington*. [October 1999]. National Park Service, http://www.nps.gov/archeology/kennewick/powell_rose.htm.

Pucciarelli, Héctor M., Walter A. Neves, Rolando González-José, Marina L. Sardi, Fernando Ramírez Rozzi, Adelaida Struck, and Mary Y. Bonilla. 2006. East-west cranial differentiation in pre-Columbian human populations of South America. *Homo—Journal of Comparative Human Biology* 57(2):133–150.

Reich, David, Nick Patterson, Desmond Campbell, Arti Tandon, Stéphane Mazieres, Nicolas Ray, Maria V. Parra, Winston Rojas, Constanza Duque, Natalia Mesa, Luis F. García, Omar Triana, Silvia Blair, Amanda Maestre, Juan C. Dib, Claudio M. Bravi, Graciela Bailliet, Daniel Corach, Tábita Hünemeier, Maria Cátira Bortolini, Francisco M. Salzano, María Luiza Petzl-Erler, Victor Acuña-Alonzo, Carlos Aguilar-Salinas, Samuel Canizales-Quinteros, Teresa Tusié-Luna, Laura Riba, Maricela Rodríguez-Cruz, Mardia Lopez-Alarcón, Ramón Coral-Vazquez, Thelma Canto-Cetina, Irma Silva-Zolezzi, Juan Carlos Fernandez-Lopez, Alejandra V.

Contreras, Gerardo Jimenez-Sanchez, Maria José Gómez-Vázquez, Julio Molina, Ángel Carracedo, Antonio Salas, Carla Gallo, Giovanni Poletti, David B. Witonsky, Gorka Alkorta-Aranburu, Rem I. Sukernik, Ludmila Osipova, Sardana A. Fedorova, René Vasquez, Mercedes Villena, Claudia Moreau, Ramiro Barrantes, David Pauls, Laurent Excoffier, Gabriel Bedoya, Francisco Rothhammer, Jean-Michel Dugoujon, Georges Larrouy, William Klitz, Damian Labuda, Judith Kidd, Kenneth Kidd, Anna Di Rienzo, Nelson B. Freimer, Alkes L. Price, and Andrés Ruiz-Linares. 2012. Reconstructing Native American population history. *Nature* 488:370-374. doi:10.1038/nature11258.

Relethford, John H. 2001. Global analysis of regional differences in craniometric diversity and population substructure. *Human Biology* 73(5):629–636.

———. 2004. Boas and beyond: Migration and craniometric variation. *American Journal of Human Biology* 16(4):379–386.

Rohlf, F. James. 1999. Shape statistics: Procrustes superimpositions and tangent spaces. *Journal of Classification* 16(2):197–223.

Roseman, Charles C. 2004. Detecting interregionally diversifying natural selection on modern human cranial form by using matched molecular and morphometric data. *Proceedings of the National Academy of Sciences* 101(35):12824–12829.

Roseman, Charles C., Katherine E. Willmore, Jeffrey Rogers, Charles Hildebolt, Brooke E. Sadler, Joan T. Richtsmeier, and James M. Cheverud. 2010. Genetic and environmental contributions to variation in baboon cranial morphology. *American Journal of Physical Anthropology* 143(1):1–12.

SAS 9.1.3 8.SAS Institute Inc., Cary, NC. 2002-2004.

Schillacil, Michael A., and Christopher M. Stojanowski. 2005. Craniometric variation and population history of the prehistoric Tewa. *American Journal of Physical Anthropology* 126(4):404–412.

Shumway, George, Carl L. Hubbs, and James R. Moriarty. 1961. Scripps estates site, San Diego, California: A La Jolla site dated 5460 to 7370 years before the present. *Annals of the New York Academy of Sciences* 93:41–131.

Simpson, Ellie Kristina. 2005. "Variation in the Cranial Base Flexion and Craniofacial Morphology in Modern Humans." Ph.D. dissertation, University of Adelaide, Australia.

Slice, Dennis E. 2001. Landmark coordinates aligned by Procrustes analysis do not lie in Kendall's shape space. *Systematic Biology* 50(1):141–149.

———. 2007. Geometric morphometrics. *Annual Review of Anthropology* 36:261–281.

Steele, D. Gentry, and Joseph F. Powell. 1999. Peopling of the Americas: A historical and comparative perspective. In *Who Were the First Americans? Proceedings of the 58th Annual Biology Colloquium, Oregon State University*, edited by Robson Bonnichsen, 97–126. Oregon State University, Corvallis.

———. 2002. Facing the past: A view of the North American human fossil record. In *The First Americans: The Pleistocene Colonization of the New World*, edited by Nina G. Jablonski, 94–122. Memoirs of the California Academy of Sciences No. 27. California Academy of Sciences, San Francisco.

Sutter, Richard C. 2000. Prehistoric genetic and culture change: A bioarchaeological search for Pre-Inka Altiplano colonies in the coastal valleys of Moquegua, Peru, and Azapa, Chile. *Latin American Antiquity* 11(1):43–70.

Swedlund, Alan, and Duane Anderson. 2003. Gordon Creek Woman meets Spirit Cave Man: A response to comment by Owsley and Jantz. *American Antiquity* 68(1):161–167.

Wescott, Daniel J., and Richard L. Jantz. 2005. Assessing craniofacial secular change in American blacks and whites using geometric morphometry. In *Modern Morphometrics in Physical Anthropology*, edited by Dennis E. Slice, 231–245. Kluwer Academic/Plenum Publishers, New York.

Zegura, Stephen L., Tatiana M. Karafet, Lev A. Zhivotovsky, and Michael F. Hammer. 2004. High-resolution SNPs and microsatellite haplotypes point to a single, recent entry of Native American Y chromosomes into the Americas. *Molecular Biology and Evolution* 21(1):164–175.

Morphological Features That Reflect Population Affinities

George W. Gill

Multivariate analysis of craniofacial metrics provides a basis for establishing population affinities among skeletal series of modern human populations, as well as with their late prehistoric antecedents (Howells 1973; Jantz and Owsley 2001). An assessment of certain characteristics of the skull (Brues 1990) that are difficult to evaluate metrically—for example, palate shape, nasal sill development, and shape of the nasal aperture—is also an effective method for establishing population affinity. Some of these morphological traits differ markedly among human populations. That is why Powell and Rose (1999) utilized both approaches in their initial analysis of Kennewick Man's skeleton, and that is why the team of scholars restudying the remains likewise incorporated both metric and nonmetric approaches. R. L. Jantz, C. L. Brace, and others utilize multivariate craniometric methods. This chapter focuses primarily on nonmetric, anthroposcopic approaches, along with selected measurements that are not included in the broader metric studies, as another means of evaluating Kennewick Man's similarity to or difference from Native Americans, Polynesians, and American Whites and Blacks.

The specialized approaches selected for this study have been found useful by forensic anthropologists in assessing ancestry of unidentified skeletons from contemporary populations. Particularly in cases where it is desirable to ascribe the biological relationship of a single skeleton to a major population, these specialized approaches (often nonmetric) are preferred. Some practicing forensic anthropologists prefer nonmetric approaches to traditional craniometrics because they are quicker and easier to apply and are perceived to be more accurate. Whether the latter is true or not remains to be proven through controlled testing and comparison. Clearly, both approaches are important in tracking biological relatedness. Characteristics of the femur have also been shown to be useful in reflecting population affinity (Baker et al. 1990; Gill 2001a; Wescott 2006).

This chapter presents results of the nonmetric trait assessment of Kennewick Man's skeleton made by the author in February 2006 along with specific metric tests conducted on the skull and femur. Comparisons are made to results from similar studies by the author and others on Paleoamerican and Archaic period skeletons and appropriate reference population samples. A secondary purpose of this study is to evaluate the reliability and accuracy of the stereolithographic model of Kennewick Man's skull for representing actual characteristics of the face and cranial vault.

METHODS

Interorbital projection measurements were taken according to the methods outlined by Gill (Gill et al. 1988; Gill and Gilbert 1990) utilizing a modified coordinate caliper known as the simometer (illustrations in Bass 2005:91; Byers 2011:144). The 57 standard craniofacial measurements of the University of Wyoming Human Osteology Laboratory checklist were collected when possible from both the original and the model of the reconstructed specimen. All measurements were taken with standard anthropometric sliding and spreading calipers, a palatometer, mandibulometer, coordinate caliper (simometer), and head spanner. All measurements in this protocol are defined in the literature by Bass (2005), Howells (1973), Gill et al. (1988), or Olivier (1969). Most of these data are not presented here since they duplicate measurements in the craniometric studies (Chapters 24 and 25).

The nonmetric trait observations are likewise from the University of Wyoming Human Osteology protocol. These consist of 32 discrete trait observations on the cranium and mandible (e.g., multiple or accessory foramina, epactal and other sutural bones, extra sutures) as well as 16 continuous anthroposcopic characteristics (e.g., palate form, chin projection, and nasal sill development). Most of the University of Wyoming discrete trait list is originally from the list prepared by Jantz (1970) for his study of the Northern Plains Arikara, and adapted by Gill (1971) for studies on Toltec Period skeletons from western Mexico. This discrete trait protocol has been applied to prehistoric skeletal samples from Peru (Baker 1981), Easter Island (Furgeson 2003; Furgeson and Gill 2005), and the Northwestern Plains of North America (Rogers 2008). Definitions and illustrations of these traits are found in most of these publications. Rogers (2008) provides the most current, complete, and accessible presentation of the traits as they have been defined and utilized by University of Wyoming physical anthropologists, and as they have been applied here.

The continuous, nonmetric traits of the cranium and mandible used in this study are also from the University of Wyoming checklist. These are defined and illustrated by Gill (1971) along with the discrete trait definitions. More comprehensive definitions of the categories of nasal sill development with photographic illustrations of each category are given by Gill and Deeds (2008). Rawlings (2002) also provides excellent illustrations of palate shapes, as well as commonly found patterns of palatine suture form within a number of human populations, and these are the approaches used here.

Femur metrics were taken according to Bass (2005) for maximum length, physiological length, midshaft perimeter, head diameter, and the two proximal diaphysis measurements of anterior-posterior diameter and medial-lateral diameter. Height of the intercondylar notch was taken according to the technique established by Baker (see Baker et al. 1990:91–95, and Bass 2005:237) and torsion height according to R. L. Weathermon's direct measurement method (see Gill 2001a). The only three instruments needed for this femur study were an osteometric board, flexible tape, and a sliding caliper.

Comparative samples of American Whites and North American Indians are drawn from the University of Wyoming database and represent groups largely from the Plains and southwest. For a description of the University of Wyoming Plains skeletal collections and database ($n =$ 600+), consult Weathermon and Miller (2008). Samples also included in the University of Wyoming database come from Mesoamerica (Postclassic West Mexico, $n =$ 245) and South America (late Prehistoric Peru, $n = 100$). Polynesians in the database are largely from Easter Island (1981 Easter Island Anthropological Expedition, $n = 426$; National Museum of Natural History, Santiago, Chile, $n = 47$) and the Marquesas (American Museum of Natural History, New York, $n = 25$).

RESULTS

This analysis focused upon seven primary areas: 1) interorbital projection metrics, 2) features of the nasal skeleton, 3) characteristics of the palate, 4) mandibular morphology, 5) discrete traits of the cranium, 6) traits of the femur, and 7) evaluation of the accuracy and utility of the stereolithographic cranial model.

Interorbital Projection

Table 27.1 shows the interorbital projection metrics obtained from the midface of Kennewick Man's cranium. The model loses accuracy with some of the refined features (such as the degree of nasal sill sharpness) and some of the smaller metric dimensions. The maxillofrontal subtense obtained from the model was thus ignored in Table 27.1 in favor of the much more accurate reading from the original. All measurements obtained

TABLE 27.1 Interorbital projection metrics of Kennewick Man.

	Original (mm)	Model (mm)
Measurements		
Maxillofrontal subtense	(8)	(5)*
Maxillofrontal breadth	(21)	(21)
Zygoorbital subtense	—	(22)
Zygoorbital breadth	—	(59)
Alpha subtense	—	(18)
Alpha cord	—	(29)
Simotic cord	(9)	(8)
Indices		
Maxillofrontal	(38.10)	(23.81)*
Zygoorbital	—	(36.07)
Alpha	—	(62.07)

*Rejected values due to imperfections in the model.

from landmarks defined by sutures are, of necessity, listed as estimates in Table 27.1 (signified by parentheses) since all of Kennewick Man's craniofacial suture lines are almost entirely obliterated. This unusual condition happened during the lifetime of the individual, and to a degree rarely seen in other human skulls, even those of quite elderly individuals. Because of this highly unusual situation, all of the interorbital projection metrics, and the indices derived from them, are estimates. These estimates, however, are considered to be accurate to within 1–2 millimeters of the actual value, or else no estimate was made. Figure 27.1 shows the placement of all three interorbital indices in relation to those of a sample of 300 modern individuals (125 historic and modern US Whites and 175 archaic and late prehistoric American Indians).

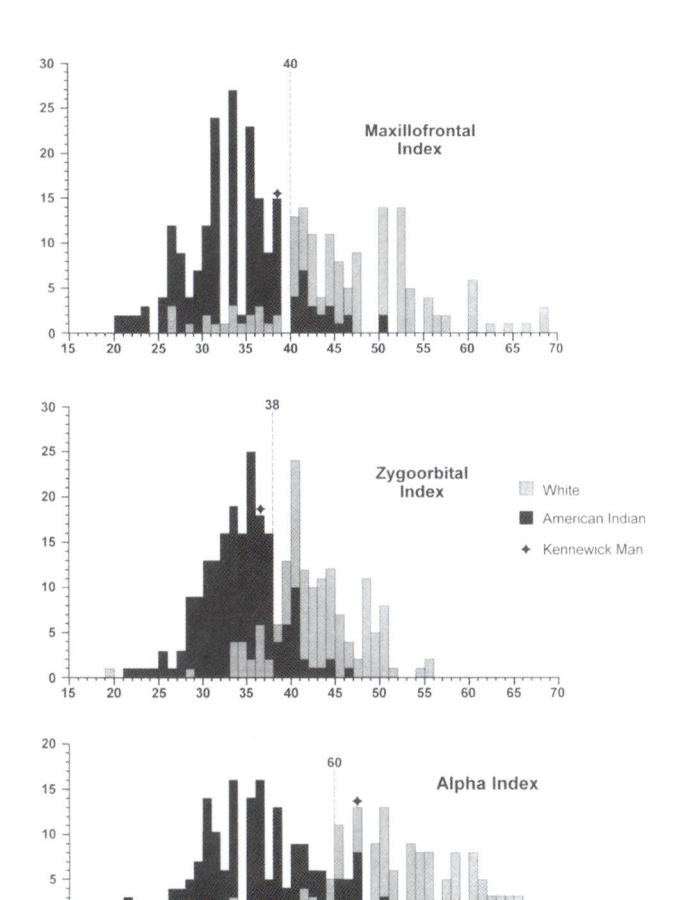

FIGURE 27.1 Placement of Kennewick Man's three interorbital index values in relation to those of 175 Archaic and late prehistoric American Indians and 125 modern US Whites. Graph adapted from Gill (1998).

Nasal Skeleton

A very large nasal spine is present and found in combination with a dull nasal sill. A dull nasal sill is the most common condition among Paleoamericans, but a large nasal spine is not. This combination of features is most unusual within modern populations except in Polynesians. Of interest too is that the lateral aspect of the sill is not dull at all, but sharp. Likewise, in its most medial aspect (along the borders of the nasal spine) the sill is also sharp. In other words, this "dull sill" is dull, or "blurred," in its mid aspect only, and not in its most lateral or most medial aspects.

Figure 27.2 shows the nasal profile, which is gently concave. The proportions of the nasal skeleton (breadth of 26 mm; height of 58 mm) support the general visual impression that the nose was long and narrow (leptorrhine nasal index of 44.83). The model yields a breadth measurement one millimeter greater (27 mm) but still produces a leptorrhine index value (46.55).

The angle into the orbit of the frontal processes of the maxillae, that is, the processes of bone lateral to the nasal bones, is intermediate between the steep angle seen on

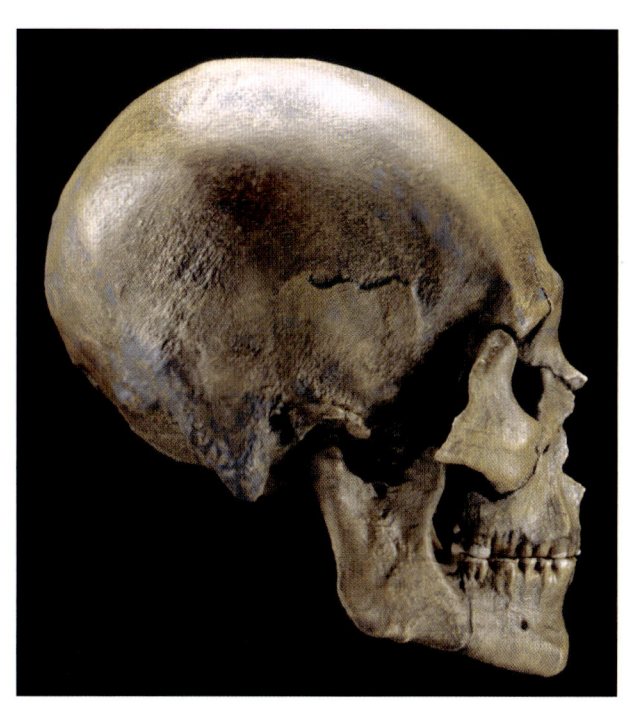

FIGURE 27.2 Right lateral view of Kennewick Man's skull and mandible showing the general sagittal outline, lack of alveolar prognathism, prominent chin, and nasal skeleton, along with the gently concave nasal profile.

nearly all crania of Europeans and their relatives, and the low angle seen on nearly all other human crania. This is fully consistent with the interorbital projection values for Kennewick Man that are borderline between those for American Whites and those for American Indians (and also African Americans). A study of 105 African Americans with regard to interorbital projection (Gill and Gilbert 1990) reveals index means nearly identical to those for the American Indians shown in Figure 27.1. Since East Asians have even lower values than the African Americans and American Indians (and also less angle to the frontal processes), this region of the face serves forensically to isolate quite well those of European ancestry from all other groups.

Palate

The slightly elliptic palate form of Kennewick Man is unusual in that it reveals the morphology of a broad parabolic palate except for the most posterior aspect of the alveolus, which does curve inward. The breadth measurement across the palate at the position of M3 is two millimeters less than at the position of M2, thus confirming that the palate is in fact elliptic. However, the overall impression when visually assessed is of a broad parabolic form as seen in most other Paleoamerican examples (Figure 27.3).

The palatine suture is likewise unusual in that it shows a jagged expression on the left side, as seen among Europeans and American Whites, and a nearly straight form on the right side, similar to that seen on crania of late prehistoric and modern American Indians. Such mixed patterns are not uncommon among Polynesians.

Table 27.2 presents the three measurements taken on the palate of Kennewick Man. The palatal index of 119.64 confirms the visual impression of a broad (brachyuranic) palate form. The model yields a slightly longer external alveolar length measurement, and therefore a slightly narrower index reading of 117.24, but this slight difference does not alter the classificatory category of this wide palate. Any index above 115 is classified within the "broad" range. Relatively short, broad palates, like Kennewick Man's, are a common condition among people of Eurasian origin (Europeans, both East and West Asians, and the various American Indian populations). The rather great palatal depth of 15 millimeters is also common to male individuals from these same populations.

Mandible

The results of the metric assessment of Kennewick Man's mandible are presented in Table 27.3. The high mandibular symphysis of 42 millimeters is consistent with the long upper facial height as revealed by both the nasion-alveolare measurement (80 mm) and the nasion-prosthion measurement (77 mm). The other measurements of this remarkably European-like mandible are not particularly noteworthy for a long-faced and fairly large-faced male individual.

Much more remarkable than any of the mandibular metrics are some of the morphological features of this jawbone. It reveals a very square, prominent, bilateral chin, indistinguishable from the chin form of most European male mandibles (Figure 27.4). This chin form is in stark contrast to the rounded, nonprominent, median chin of the vast majority of American Indian mandibles, especially those from late prehistoric contexts. Clear cupping is also evident below the incisors, and also above the corners of the strongly bilateral chin. In fact, actual fossae occur in that location as is sometimes encountered

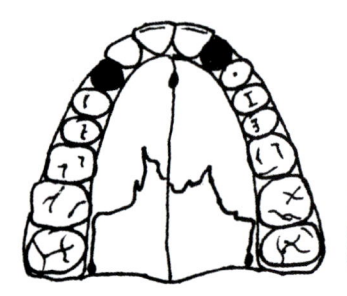

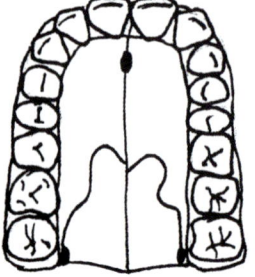

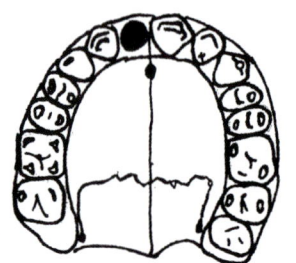

FIGURE 27.3 Human palate shapes: parabolic (left), hyperbolic (center), elliptic (right). Each is shown with the palatine suture shape commonly found associated with it jagged (left), arched (center), and straight (right). Adapted from Olivier (1969), Gill (1998), and Rawlings (2002).

TABLE 27.2 Palate metrics of Kennewick Man.

	Original (mm)	Model (mm)
Measurements		
External alveolar breadth	67	68
External alveolar length	(56)	(58)
Palatal depth	15	15
Index		
Palatal	(119.64)	(117.24)

TABLE 27.3 Mandibular metrics of Kennewick Man.

Measurements	Original (mm)	Model (mm)
Symphyseal height	42	41
Bigonial diameter	—	113
Bicodylar diameter	—	—
Breadth of ascending ramus (right side)	36	36
Height of ascending ramus	—	63
Corpal length	—	84
Gonial angle	—	35°

among the more extreme European mandibles, and also those of some Archaic period examples from western Nebraska and southeastern Wyoming, which lack entirely the "Mongoloid craniofacial trait complex" (Gill 2005, 2008).

Another remarkable feature is the very thin inferior margin of the horizontal ramus of the jaw. This too places the jawbone within the range of most European mandibles and stands in complete contrast to the condition found among the vast majority of American Indian mandibles, which reveal almost uniformly thick, heavy jaw rami.

Also observable on Kennewick Man's mandible is pronounced eversion of both gonial angles. This, coupled with the excessive dental attrition, indicates not only heavy use of the teeth as tools but also heavy masticatory activity.

The genial tubercles are observable and are small in size. An assessment of the form of the inferior margin of the horizontal jaw ramus shows it to be straight with a slight antegonial notch, therefore totally free of the

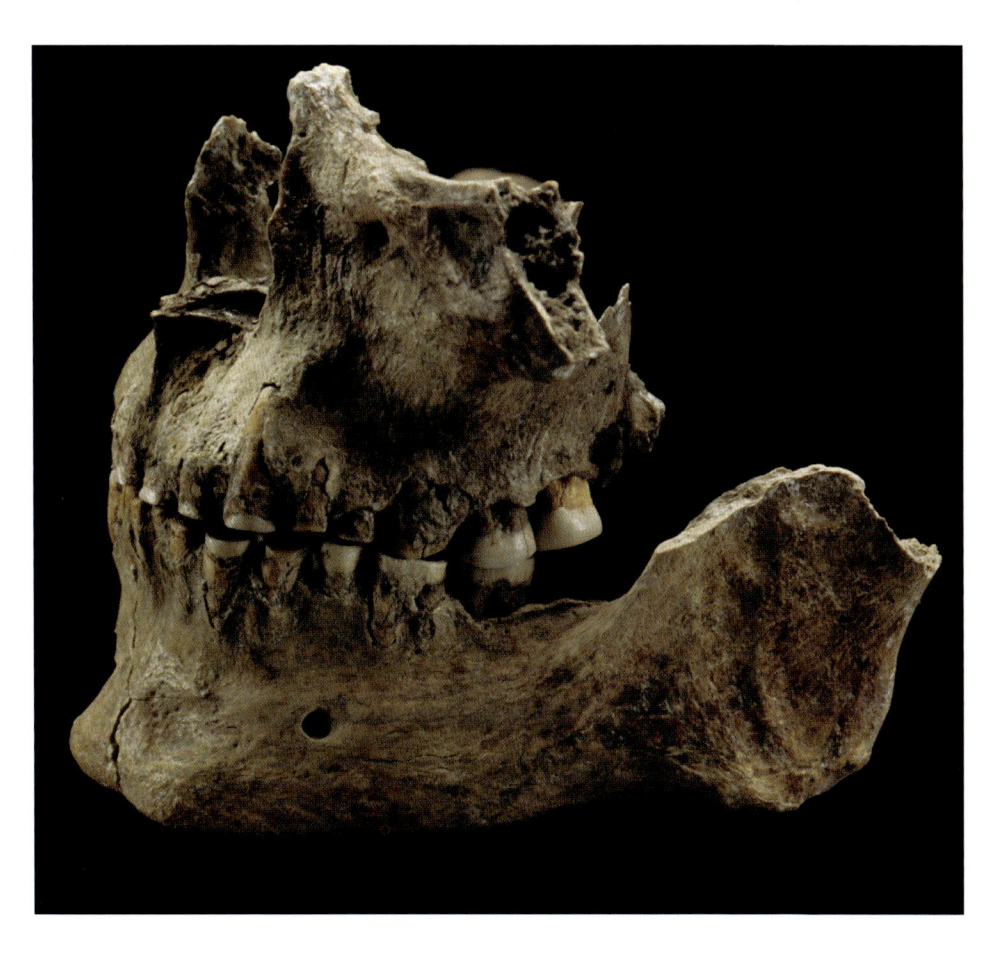

FIGURE 27.4 Maxilla and mandible of Kennewick Man showing a very square, prominent, bilateral chin and a large prominent nasal spine.

rocker jaw condition found on most Polynesian mandibles, and among some early East Asians.

An evaluation of some discrete nonmetric traits of the mandible (Table 27.4) reveals no mandibular tori, no accessory mental foramina, and a single, wide mylo-hyoid bridge on the left ascending mandibular ramus (none on the right side).

Cranium

In addition to those few discrete traits that were assessed from the mandible, 16 traits of the cranium were successfully scored. As indicated at the bottom of Table 27.4, a large number of other useful discrete traits could not be assessed due to almost complete obliteration of most cranial suture lines on Kennewick Man's skull. In a few instances, missing bone also precluded assessment.

Kennewick Man's skull is typified by a lack of many of the commonly occurring discrete cranial traits. For instance, no supraorbital, frontal, or malar foraminae occur. Likewise, both anterior condylar canals are single, and only one parietal foramen is present. Evidence suggests a lack of accessory infraorbital foramina as well.

A suture into the infraorbital foramen is detectable on the left side, but could not be assessed on the right. A slight pharyngeal fossa is likewise visible. No indications were found to suggest the presence of either an Inca bone

TABLE 27.4 Discrete nonmetric trait occurrences of Kennewick Man's skull.

Single Traits	Present	Absent
Inca bone		X
Epactal bone		X
Metopic suture		X
Pharyngeal fossa	X	
Mandibular torus		X
Palatal torus		X
Rocker mandible		X

	Present		Absent	
Paired Traits	Left	Right	Left	Right
Tympanic dehiscences			X	X
Double anterior condylar canal			X	X
Multiple lesser palatine foramina	X	unknown		
Supraorbital foramen			X	X
Frontal foramen			X	X
Malar foramina			X	X
Parietal foramen	X			X
Frontal grooves			X	X
Accessory mental foramina			X	X
Mylo-hyoid bridge	X			X
Suture into infraorbital foramen	X	unknown		
Accessory infraorbital foramina			X	unknown
Open coronal 3 suture	unknown	X		

Note: Discrete craniofacial traits that could not be assessed due to destruction or suture obliteration are the presence or absence of lambdoidal wormian bones, parietal notch bones, asterion ossicles, mastoid suture ossicles, epiteric bone, coronal ossicles, sagittal ossicles, bregmatic bone, a left turning superior sagittal sinus, os japonicum, curved zygomaticomaxillary suture, and fronto-temporal articulation.

or epactal bones, or of a metopic suture. No palatal torus exists nor tympanic dehiscences. Also there is a total absence of frontal grooves.

It is unfortunate that the shape of the zygomaticomaxillary suture could not be assessed. This trait has shown itself to be quite useful in forensic contexts, since the curved suture form occurs frequently among Polynesians and Whites, but is less frequently found within populations of late prehistoric American Indians and African Americans (Holborow 2002). Some Paleoamerican and Archaic period populations, on the other hand, do reveal curved zygomaticomaxillary sutures.

Femur

Reliable metrics were obtained from the fragmented femora through careful reconstruction of the broken elements. These large fragments fortunately have clean fracture lines, which makes rearticulation of the broken pieces easy and results in an accurate restoration. Table 27.5 lists the femoral metrics obtained. A somewhat unusual asymmetry in femoral length may be seen from the table with the left femur yielding an 8 millimeter greater length than the right. Physiological lengths are somewhat more comparable, but the left is still 6 millimeters greater in length.

The robusticity of the right femur, which can be calculated by the Olivier method (Olivier 1969) as 20.69, is relatively robust. Diameter of the femoral head, 49 millimeters (Table 27.5), is only one millimeter above average for American White and Black males (Stewart 1979) but three millimeters above average for East Polynesian males and male individuals within many populations of both Mesoamerican and North American Indians (Gill 2004).

Regarding the biological affinities and relationships of Kennewick Man, a much more revealing area of the femur than either robusticity of the head or diaphysis is the shape of the proximal femoral diaphysis. This is measured at a point below the lesser trochanter (Bass 2005:224). The degree of platymeria in this region of the femur can be helpful in assessing ancestry forensically, especially in many western regions of the United States where the American Indian populations tend to be quite platymeric. Whites, largely of western European origin, are almost the opposite (eurymeric). Table 27.5 shows the subtrochanteric measurements obtained on Kennewick Man's femora. Accurate assessments of both anterior-posterior diameter and medial-lateral diameter could be obtained on both femora. Figure 27.5 shows plot points obtained for both femora within the context of values for two modern populations: US Whites and western American Indians. Kennewick Man's femora align more closely with a majority of other Paleoamericans (see Gill and Jimenez 2004) and people of European origin than with those of the later American Indian populations.

Torsion height measurements of 63 and 61 millimeters reveal a basic symmetry for the two femora. Values this

TABLE 27.5 Femur metrics of Kennewick Man.

Measurements	Left (mm)	Right (mm)
Maximum length	472	464
Physiological length	468	462
Midshaft perimeter	—	96
Head diameter	49	49
Subtrochanteric anterior-posterior diameter	32	31
Subtrochanteric medial-lateral diameter	33	33
Torsion height	63	61
Intercondylar notch height	33	33

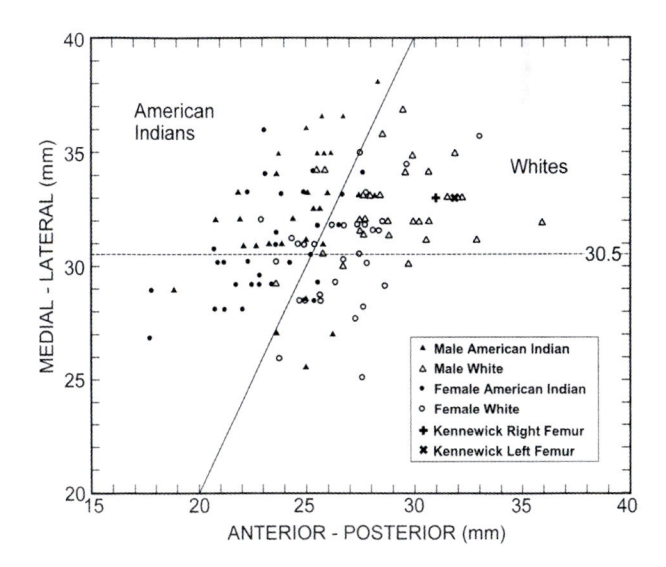

FIGURE 27.5 Plot points of Kennewick Man's subtrochanteric femur metrics shown in relation to those of modern Whites and American Indians. Scatter plot adapted from Gill and Rhine (1990).

high reveal a fairly high degree of femoral torsion. The work of Royer et. al (1997) with a sample of 56 Plains Indians and 27 Whites, utilizing this same approach, produced torsion height means of 62.15 for the Plains Indians and only 57.63 for the Whites.

The intercondylar notch reading of 33 millimeters is a very normal notch height for *Homo sapiens* males. As Baker et al. (1990) point out, that value would be slightly low for Black males and slightly high for White males, but only a millimeter from the modal value for each (34 for Black males and 31–32 mm for White males). Unfortunately, published data regarding notch height are limited and unavailable entirely for American Indian populations.

Stereolithographic Cranial Model

Clearly the cranial model provides many advantages for study, as would any carefully produced cast of the original. It allows some measurements to be taken that would be difficult or impossible to obtain from the disassociated fragments of the original. Also, within this study it allowed for assessment of some continuous nonmetric traits, such as orbital form, that could not be assessed accurately from the fragmented original bones. On the other hand, some small metric dimensions (e.g., maxillofrontal subtense) and some refined visual assessments (e.g., nasal sill development) are not reflected accurately in the model. That is true for some discrete trait evaluations as well. Table 27.6 lists all the discrete traits evaluated on Kennewick Man's skull for which discrepancies were found between the results obtained from the original and those from the model. In most instances the expression of the trait simply cannot be discerned from the model and therefore was assessed as "unknown." In two instances, this lack of refinement of the model actually led to a misreading of the physical trait (e.g., lack of a pharyngeal fossa when one exists, and lack of a left parietal foramen when one is present). Understandably a trait like the small suture into the left infraorbital foramen (scarcely observable on the original bone) is simply not to be found on the model. Perhaps more disappointing is the fact that no sign of the pharyngeal fossa was to be found on the model and no evaluation at all could be made of the double or single nature of the anterior condylar canals. Such traits, important in tracking genetic connections between individuals, lineages, and populations are disappointingly manifested, or not exhibited at all, on the model.

Regarding metrics, Table 27.7 lists all measurements within the University of Wyoming protocol that were taken on both the original and the model and that deviated 2 millimeters or more between the model and the

TABLE 27.6 Discrete trait evaluation comparison between Kennewick Man's skull and the model.

Single Traits	Original		Model	
	Present	Absent	Present	Absent
Pharyngeal fossa	X			X*

Paired Traits	Present		Absent		Present		Absent	
	Left	Right	Left	Right	Left	Right	Left	Right
Double anterior condylar canal			X	X			unknown	unknown
Multiple lesser palatine foramina	X	unknown			unknown	unknown		
Parietal foramen	X			X			X*	X
Accessory mental foramina			X	X			unknown	X
Mylo-hyoid bridge	X			X	unknown			X
Suture into infraorbital foramen	X	unknown					unknown	unknown
Open coronal 3 suture	unknown	X					unknown	unknown

*Wrong trait assessment yielded by the model.

TABLE 27.7 Craniometric discrepancies between Kennewick Man's skull and the model.

Measurements	Original (mm)	Model (mm)
Min. frontal breadth	96	98
Biasterionic breadth	(123)	(126)
Bistephanic breadth	(95)	(100)
Auricular height	121	124
Porion-bregma height	120	123
Maxillofrontal subtense	(8)	(5)
Mastoid length	33	30
Mastoid width	11	13
External alveolar length	(56)	(58)

Note: Figures in parentheses are estimates.

original. On the fragmented original only 29 measurements were able to be collected of the 57 measurements listed on the University of Wyoming checklist. Of these 29 variables 20 matched perfectly (exact same result or within 1 mm) the corresponding measurements collected from the model. Also the model allowed for the collection of 53 of the 57 measurements of the cranium and mandible, rather than only 29. Yet it is troubling that nine measurements deviate as far as they do from those taken from the original. Disregarding for the moment minimum frontal breadth and external alveolar length (which were only inflated by 2 millimeters on the model), and bistephanic breadth, which is sometimes difficult to ascertain even on a perfectly preserved skull, four problem areas on the model remain. The inflated value for biasterionic breadth could possibly be traceable to the difficulty in establishing precise suture locations on both the original and on the model. The 3-millimeter inflated values yielded by the model for auricular height and porion-bregma height, however, clearly reflect a more basic problem with the accuracy of the model. This is also true for the disappointingly low value for maxillofrontal subtense produced by the model. A 3-millimeter deviation in a measurement that rarely exceeds 10 millimeters within *Homo sapiens* is an intolerable error, especially considering the importance of maxillofrontal subtense in establishing human biological relationships (Gill et al. 1988). Also quite troubling is the very different morphology of the mastoid process suggested by the model when compared to the actual Kennewick Man mastoid process. A process 3 millimeters shorter and 2 millimeters wider than the original Kennewick Man mastoid process suggests a cranial feature quite different from the one observed in and measured on the original skull.

DISCUSSION

Comparison of the results of the interorbital projection metrics shown in Table 27.1 with those obtained by Powell and Rose (1999) in their study of Kennewick Man reflects a large disparity. They obtained a maxillofrontal index of 46.9 rather than 38.1, a zygoorbital index of 30.9 rather than 36.1, and an alpha index of 72.7 instead of 62.1. Their results indicate two things not suggested by this study: first, that two of three indices fall *above* the non-White/White cutoff, thus placing Kennewick Man with US Whites and Europeans on nasal projection; and second, a wide disparity between those two indices and the zygoorbital index well *below* the non-White/White cutoff. Even though this level of disparity is possible, it is quite unlikely to actually occur on a single individual cranium. The results arrived at in this study show not only much greater internal consistency between the three indices, but also show two of the three indices to be slightly below and not above the cutoff values between Whites and non-Whites. In short, Powell and Rose's results place Kennewick Man with Europeans and American Whites. The results here place him slightly in the non-White sector on both maxillofrontal and zygoorbital indices and slightly into the White sector only on the alpha index. This is a pattern reflected occasionally by Easter Islanders and other East Polynesians, that is, a non-White overall result but a high degree of projection of the distal aspect (tip) of the nasal bridge reflected by a fairly large alpha subtense and index.

One should hesitate to simply call Powell and Rose wrong in their interorbital projection metrics for the reason stated above, that is, the lack of visibility of craniofacial suture lines on Kennewick Man's cranium. Consequently, any effort to apply this method to the cranium must rely upon estimates of the precise locations of landmarks for measurements and not totally verifiable points of measurement. Suffice it to say only that the results from this study appear much more likely in terms of internal consistency than the results of the Powell and Rose (1999) study.

To substantiate the assumption that Kennewick Man reveals an interorbital projection pattern similar to that of the Polynesians, a sample of 91 well-preserved adult crania of late prehistoric-protohistoric Easter Islanders was examined with regard to these three indices. Of these 91 Polynesians only one individual placed above 40 on maxillofrontal index, the cutoff value for Whites, while two were above 38 on zygoorbital index, the sectioning point for that index. Yet twice as many ($n = 4$) fell, like Kennewick Man, with the Whites on alpha index (i.e., above 60). A more extensive survey of these same midfacial metrics among additional Easter Islanders ($n = 262$) and other Polynesian Island groups (Marquesas Islands, Society Islands, Tuamotus) has been conducted by Stefan (2000) and all groups show measurement means very close to those for the 91 Easter Islanders examined here. Table 27.8 shows the population means for these three interorbital indices for four major geographic populations of modern humans. For this study the Easter Islanders mentioned above are combined with additional ones from Stefan's larger sample ($n = 325$).

Other features of the nasal skeleton are consistent with the fairly high projection values. For instance, the leptorrhine nasal index and intermediate angle of the frontal processes (i.e., slope into the orbits) seem to fit this overall morphological pattern.

The dull nasal sill, or "blurred" sill, as it is referred to by Powell and Rose (1999), is a condition seen commonly on other Paleoamericans and Archaic North American crania (Gill 2005, 2008). It is, of course, common as well among modern Blacks and other populations from the equatorial latitudes, such as native Australians and Melanesians (Olivier 1969), but in these populations the dull sill is generally not combined with a prominent, leptorrhine nose. However, Polynesians do show this combination of nasal features (Gill 2001b), thus showing a strong similarity between Kennewick Man (and some other Paleoamericans) and Polynesians, at least Easter Islanders and those from the Marquesas Islands.

The combination of a long nasal spine with a dull nasal sill is not a common one within most modern populations, yet 35.8 percent of late period Easter Island males show similarly large nasal spines ($n = 53$) and 89.6 percent have dull (or absent) nasal sills. So, a rather large number of these Polynesians show the same unusual combination of features revealed by Kennewick Man.

The overall appearance of Kennewick Man's broad palate is likewise similar to those of the majority of Easter Islanders and other Polynesians. It is technically an elliptic one (exhibiting a narrowing at the position of M3), a condition not quite so common among the Polynesians as the truly parabolic palates. However, in the form of the palatine suture, with the left side showing a jagged pattern and a straight form on the right, it is fully reminiscent of the Polynesian condition. Chapman (1991) found these mixed patterns frequently in samples of prehistoric and protohistoric Easter Islanders and Marquesas Islanders.

The departure of Kennewick Man's palate form from that of most late prehistoric American Indians is fairly clear. Not only do most American Indians show straighter palatine sutures than Kennewick Man's (and those of the Polynesians), but also the elliptic palates of American Indians are more often highly curved (horseshoe-shaped) like the one illustrated in Figure 27.3. Yet it should be pointed out that the basic percentages of elliptic pal-

TABLE 27.8 Interorbital index values for four population samples.

Population	Maxillofrontal Index		Zygoorbital Index		Alpha Index	
	Mean	Standard deviation	Mean	Standard deviation	Mean	Standard deviation
US Whites ($n = 125$)	46.6	8.97	42.9	5.49	68.2	8.67
SECTIONING VALUES	**40.0**		**38.0**		**60.0**	
N. American Indians ($n = 173$)	33.6	5.74	34.0	4.33	51.3	7.46
US Blacks ($n = 100$)	34.0	8.00	35.0	9.00	49.0	9.00
Easter I. Polynesians ($n = 325$)	28.3	5.71	32.0	4.15	49.9	7.71

ates to parabolic ones are not that dissimilar between the various American Indian groups and the Polynesians. Peruvians ($n = 95$) show a 60:40 ratio of elliptic to broad parabolic palates (none is hyperbolic) with the males showing a slightly higher frequency of elliptic ones (65.9 percent) than the females (58.3 percent). Mesoamerican Indians from West Mexico ($n = 76$) show no sex differences, but the frequencies are rather like the Peruvian males (68 percent elliptic and 32 percent parabolic) (Gill and Case 2011). Rawlings (2002) reports 42 percent elliptic palates for North American Indians, and Chapman and Gill (1993) found 48 percent elliptic and 50 percent parabolic in a similar sample of mostly late prehistoric Plains Indians. Available data from Easter Island ($n = 106$) (Baker 1981) suggest 52 percent of males have parabolic palates, 47 percent elliptic ones, and one percent hyperbolic.

The mandible of Kennewick Man, with its very square and prominent bilateral chin and slender horizontal ramus, provides a most surprisingly European-like morphological pattern. An examination of this well-preserved jawbone in 2006 in the Burke Museum laboratory demonstrated how Grover Krantz and other anthropologists who examined Kennewick Man's skeleton shortly after its discovery could conclude that the skeleton revealed a "Caucasoid pattern" of characteristics (COE 9396). Their assessments created a considerable amount of confusion and criticism at the time, but in their defense they were interpreting much of the skeletal evidence quite correctly. No part of Kennewick Man's skeleton supports this interpretation better than the mandible. Not only do Europeans, American Whites, and other Caucasoids demonstrate these mandibular features, but so do certain Paleoamerican and Archaic Period North American skeletons. For instance, the male and female mandibles from the Late Plains Archaic Huntley site of southeastern Wyoming (Gill 2008) show these same features. So does the Harlan County Lake skeleton, an Archaic Period interment from central Nebraska (Baker 1989). To an only slightly lesser extent so do the Paleoamerican mandibles of Spirit Cave and Wizards Beach, Nevada (Gill 2001b, 2005; Gill and Jimenez 2004; Jantz and Owsley 1997). The Willson cranium, another Late Plains Archaic skull from southeastern Wyoming, has many of these mandibular features (Gill 2008).

Nearly all of these descriptions have been published since 1996; therefore, Krantz and the other researchers cannot be faulted for not knowing of these parallels in mandibular morphology (and some other craniofacial traits) between Europeans and certain Paleoamericans (and their Archaic period descendants).

In these mandibular features Kennewick Man's skeleton departs markedly not only from the late prehistoric American Indians but also to some extent from Polynesians. Most late prehistoric and protohistoric Easter Islanders, at least, demonstrate a heavy mandible, and although only 48.5 percent show a curved rocker jaw condition to the inferior margin of the horizontal jaw ramus (Gill 1990), many others show an anterior rocker condition. This creates a different profile from the jawbones of the Europeans and Kennewick Man's mandible. Also, very few Polynesian mandibles demonstrate the chin prominence (and related features) that cause Kennewick Man's mandible to mirror European morphology.

Other trait assessments within this study that reflect a somewhat European-like condition are the long nasal spine and simple cranial sutures. A leptorrhine nasal form too is quite common within European populations but is likewise found frequently within Polynesian groups and some American Indian populations. So, the narrow nose shape is not a very revealing characteristic in this particular instance.

More striking parallels to the Polynesian condition are revealed by nearly all aspects of the nasal skeleton. Visual inspection suggests, and the simometer metrics confirm, a strong nasal projection, but non-European in character. That is, the distal aspect of the nasal bones is prominent while the rest of the nasal bridge is much less projecting. This is the classic Polynesian condition with regard to nasal projection. The narrow, well-formed nasal bridge in combination with a dull nasal sill is likewise a common Polynesian feature not often seen on non-Polynesian crania. Judging from an inspection of numerous photographs (but without the advantage of a metric database), Ainu crania seem to show this same combination of features, as well as a concave nasal profile like that of Kennewick Man (Montandon 1927). Polynesians have concave nasal profiles as well. On Easter Island, within a sample of 96 adults with well-preserved nasal bones, 37.7 percent of males and 58.1 percent of females show a concave profile like that of Kennewick Man. A concavo-convex profile is slightly more common among males (56.6 percent), but not among the females (37.2 percent concavo-convex). In this tendency to reveal a concave nasal profile the Polynesians differ from most late American Indian groups. A sample of 38 adults from late prehistoric West Mexico shows no concave profiles. This trait has not been well quantified among North American groups but appears to be at a low frequency there too, especially for males. Peru appears to be a bit more like Polynesia on this particular trait with 30.7 percent of males ($n = 39$) showing the concave profile (Baker 1981). To more fully interpret the meaning of Kennewick Man's concave nasal profile it is clear that more data from modern and early *Homo sapiens* populations are needed.

Kennewick Man's paucity of discrete trait occurrences common to Polynesians—frontal grooves, malar foramina, and extra sutural bones—is noteworthy. In these respects Kennewick Man is unlike most Polynesians and, in fact, shows fewer discrete trait occurrences than individuals from most *Homo sapiens* populations.

Subtrochanteric femoral traits of Kennewick Man are of interest in that the thick anterior-posterior diameter of the proximal shaft (eurymeric form) clearly separates this individual from most later prehistoric American Indians. In this respect his femur form is more like that of other Paleoamericans, American Whites, and Europeans. The importance of this trait in establishing genetic relatedness has been questioned by Wescott (2006), who favors biomechanical stresses as an explanation for these shape differences. A study by Miller (see Gill 2001a) comparing femora of American Whites and Plains Indians of various ages (infancy through older adult categories), even though revealing both genetic and environmental factors, suggests that genetic factors are more important than environmental ones. A study of femoral platymeria among *Papio hamadryas* (Kesterke 2008), based upon pedigree data, reveals an important genetic factor in this trait. The genetic factor is about equal to the combined effects from environmental influences.

The large head diameters of Kennewick Man's femora likewise place him closer to the American Whites and Europeans. However, the torsion height measurement is more in line with those yielded by late prehistoric American Indians (i.e., high torsion). Clow (1997) reports a platymeric femur form for 208 adult Easter Islanders (99.04 percent plot outside the White sector shown in Figure 27.5), so in this characteristic, Kennewick Man departs from Polynesians as well as late American Indians.

Regarding the results of the comparison of the original Kennewick Man cranial bones to the stereolithographic model, it does seem that a pattern emerges. For most of the large cranial dimensions the model appears to be quite accurate. The few exceptions, involving mostly cranial height measurements, may be a result of a less faithful reproduction of parts of the cranial base. This is suggested not only by the fact that both auricular height and porion-bregma height are inflated by 3 millimeters on the model but also by the distorted proportions of the mastoid processes. This interpretation is further supported by a lack of definition to the anterior condylar canals and a failure of the model to show the pharyngeal fossa of the basilar occipital. Other shortcomings of the model seem to be related more to a problem of not faithfully reproducing fine details of the original, such as the form of the nasal sill, the root of the nasal bridge (as reflected by the low maxillofrontal subtense), and the failure to show certain suture lines and small foramina. Hopefully, the information presented here will assist future scientists and technicians working with stereolithographic model production toward improvement of this important technology.

Consideration of all of this osteological information leads to the question, what does this all mean? Before the present study of Kennewick Man's skeleton, a theoretical framework for explaining these contrasts between certain Paleoamerican and Archaic skeletons on the one hand, and the later more Mongoloid-looking American Indians on the other, was presented in two papers (Gill 2001b, 2008). Now, after close examination of the original Kennewick Man skeleton, plus DNA work that supports this hypothesis, even more confidence in this interpretation is justified.

In short, this view holds that the similarities among the Europeans, East Polynesians, the isolated Ainu population, and the Paleoamericans indicate that around 12,000 years ago much of Eurasia was inhabited by an extensive network of related peoples of a generalized and basic Caucasoid appearance. Beyond Europe these people were particularly prevalent in coastal and southern Asia and were typified by high, long skulls, prominent chins and noses, modest cheekbones, and parabolic palates. Judging from the appearance of contact-period Ainu and East Polynesians, some were also light-skinned with wavy hair and heavy male beards like the living Europeans.

Living far to the north and in a colder, inland environment were people possessing the Mongoloid skeletal trait complex. These inland hunters, according to Brace (1995), were adapted physically to processing hides and leather with their anterior teeth, thus the need for sturdy shovel-shaped anterior teeth and the robust skeletal structures, such as heavy cheekbones and mandibles, to power these strong anterior teeth. Straight black hair and dark eyes would seem to be among the soft tissue characteristics of this cold-adapted inland population. By two separate routes these two populations entered the New World, one by the Bering land bridge following megafauna (traditional view) and the other by a sea route following the coastline from north Asia to the Americas. A map illustrating these two major movements, and the relationship of the Asia-to-America sea route to the sea routes into Polynesia, is presented in Gill (2001b:453). An mtDNA study by Perego et al. (2009) provides maps showing the same two coastal and inland routes. Their evidence for two separate routes into the Americas from Asia is based upon the divergent spatial distribution within the Americas of two recently discovered mtDNA sub-haplogroups D4h3 and X2a. The D4h3 variant occurs in modern and ancient population samples from both North America and South America. This variant shows a higher frequency along the Pacific coast and to the west of the Andean Mountains. By contrast, X2a is detected only in North America and largely east of the Rocky Mountains. Lepper (2009) provides a concise summary of this work with comment by another DNA researcher. The potential significance of this work, when it supports fully the findings from the skeletal evidence, is hard to ignore.

CONCLUSION

It is clear from this evaluation of a number of discrete and continuous morphological traits of Kennewick Man's skull and femur that this skeleton is unique and does not fit within any major human population of today. For instance, the low number of accessory foramina and sutural bones is noteworthy, as are the unusual combinations of certain features like a eurymeric form to the proximal femur in conjunction with high torsion, and a Polynesian-like midface with a European-like mandible. On the other hand, regarding some of the best indicators of ancestry, for example, interorbital projection metrics, features of the nasal bridge, and characteristics of the palate, Kennewick Man comes quite close to modern Polynesians. The prominent leptorrhine nose in combination with a dull nasal sill, and a nasal root height just slightly below that for Europeans and American Whites is the classic Polynesian pattern, best described among the East Polynesian Easter Islanders. Some other Paleoamerican examples, such as Spirit Cave Man and some Archaic period skeletons (e.g., Harlan County Lake, Huntley, and Willson), show these same Polynesian parallels. Some of these same ancient Americans share with Kennewick Man the combination of a Polynesian-like midface and skull with a European-like mandible.

From this examination of specific morphological features of value in assessment of ancestry in modern forensic contexts, it appears that Kennewick Man's closest traceable affinities are with certain other Paleoamericans and the few Archaic period North Americans who continued for thousands of years to retain those early trait patterns. Among the most recent populations of the world it would seem to be the Polynesians who are the closest to Kennewick Man.

The lack of similarity of Kennewick Man to late prehistoric and modern American Indians is striking. The features that most typify late American Indian populations, such as platymeric femora and cranial features like wide heavy cheekbones, heavy mandibles with blunt rounded chins, strongly elliptic palates with straight palatine sutures, medium noses with medium nasal spines and sills, numerous accessory cranial foramina and extra sutural bones, are simply not to be found on Kennewick Man's cranium. This so-called Mongoloid skeletal trait complex, well-developed on late prehistoric American Indians, and even more so on late prehistoric and modern East Asians, appears to be almost totally lacking on Kennewick Man's skeleton. Morphological parallels between Kennewick Man and Europeans and other Caucasoids are much closer than they are with the so-called Mongoloid peoples. This is undoubtedly why the first few physical anthropologists to examine this skeleton concluded that it probably represented a frontier White and was not representative of a historic American Indian tribal group. Their skeletal evaluations were done adequately, but they missed the mark because of how ancient this unique skeleton eventually turned out to be.

This focused, forensic-style examination of specific traits of Kennewick Man's skeleton has produced results that rely on physical traits different from those utilized in parallel craniometric studies presented in Chapters 24 and 25. The traits highlighted here are noted specifically for their utility in ancestry assessment. The results complement conclusions derived from multivariate studies of Kennewick Man's cranial measurements. All of this coupled with recent findings from DNA research are pointing the same way, that is, to two separate migrations of biologically divergent populations into the Americas more than 10,000 years ago. The future of Paleoamerican origins research is promising if scientists are allowed to study these ancient skeletons. Meaningful answers to longstanding questions appear to be within grasp, and Kennewick Man's skeleton is helping to provide those answers.

REFERENCES

References listed below under the heading "COE" are documents from the administrative records filed with the district court by the US Army Corps of Engineers ("COE").

Baker, Scott J. 1981. "Non-metric Trait Comparisons Between Late Prehistoric Peruvians and Easter Islanders." Data on file at the Department of Anthropology, University of Wyoming, Laramie.

———. 1989. Osteological analysis. In *Archaeological and Osteological Analysis of Two Burial Sites Along Harlan County Lake, Nebraska: Chronological and Evolutionary Implications*, edited by William L. Tibesar, 69–87. Report to the U.S. Army Corps of Engineers, Kansas City District. Prepared by Larson-Tibesar Associates, Laramie.

Baker, Scott J., George W. Gill, and David A. Kieffer. 1990. Race and sex determination from the intercondylar notch of the distal femur. In *Skeletal Attribution of Race: Methods for Forensic Anthropology*, edited by George W. Gill and Stanley Rhine, 91–95. Maxwell Museum of Anthropology Paper 4. Maxwell Museum of Anthropology, Albuquerque.

Bass, William M. 2005. *Human Osteology: A Laboratory and Field Manual*. 5th edition. Missouri Archaeological Society Special Publication 2. Missouri Archaeological Society, Columbia.

Brace, C. Loring. 1995. *The Stages of Human Evolution*. 5th edition. Prentice Hall, Englewood Cliffs, NJ.

Brues, Alice M. 1990. The once and future diagnosis of race. In *Skeletal Attribution of Race: Methods for Forensic Anthropology,* edited by George W. Gill and Stanley Rhine, 1–7. Maxwell Museum of Anthropology Paper 4. Maxwell Museum of Anthropology, Albuquerque.

Byers, Steven N. 2011. *Introduction to Forensic Anthropology*. 4th edition. Pearson, Boston.

Chapman, Patrick M. 1991. "The Palate: Race and Sex." Manuscript on file at the Department of Anthropology, University of Wyoming, Laramie.

Chapman, Patrick M., and George W. Gill. 1993. "Use of the Palate and Palatine Suture in Race Identification." Paper presented at the 45th annual meeting of the American Academy of Forensic Sciences, Boston.

Clow, Charles M. 1997. "An Osteometric Description of the Pre-contact Easter Island Population." Master's thesis, Department of Anthropology, University of Wyoming, Laramie.

COE 9396. Letter from A. Miller, prosecuting attorney for Benton County, Washington, to L. Kirts, USACE, with enclosed letter regarding findings on examination of human remains, September 3, 1996.

Furgeson, Thomas A. 2003. "Regional Variation of Discrete Cranial Traits in a Prehistoric Easter Island Skeletal Sample." Master's thesis, University of Wyoming, Laramie.

Furgeson, Thomas A., and George W. Gill. 2005. Regional variation of discrete cranial traits in a prehistoric Easter Island skeletal sample. In *The Reñaca Papers: VI International Conference on Rapa Nui and the Pacific*, edited by C. M. Stevenson, J. M. Ramirez, F. J. Morin, and N. Barbacci, 191–200. The Easter Island Foundation and University of Valparaiso, Chile.

Gill, George W. 1971. "The Prehistoric Inhabitants of Northern Coastal Nayarit: Skeletal Analysis and Description of Burials." Ph.D. dissertation, University of Kansas, Lawrence.

———. 1990. Easter Island rocker jaws. *Rapa Nui Journal* 4(2):21.

———. 1998. Craniofacial criteria in forensic race identification. In *Forensic Osteology*. 2nd Edition. Edited by Kathleen J. Reichs, 293–317. Charles C. Thomas, Springfield, IL.

———. 2001a. Racial variation in the proximal and distal femur: Heritability and forensic utility. *Journal of Forensic Sciences* 46(4):791–799.

———. 2001b. Basic skeletal morphology of Easter Island and East Polynesia, with Paleoindian parallels and contrasts. In *Pacific 2000: Proceedings of the Fifth International Conference on Easter Island and the Pacific*, edited by Christopher M. Stevenson, Georgia Lee, and F. J. Morin, 447–456. Easter Island Foundation, Los Osos, CA.

———. 2004. "Population Variability in the Proximal Articulation Surfaces of the Human Femur and Humerus." Paper presented at the 56th Annual Meeting of the American Academy of Forensic Sciences, February 19, 2004, Dallas.

———. 2005. Appearance of the "Mongoloid Skeletal Trait Complex" in the northwestern Great Plains: Migration, selection, or both? In *Paleoamerican Origins: Beyond Clovis*, edited by Robson Bonnichsen, Bradley T. Lepper, Dennis Stanford, and Michael R. Waters, 257–266. Texas A&M Press, College Station.

———. 2008. Northwestern Plains Archaic skeletons with Paleoamerican characteristics. In *Skeletal Biology and Bioarchaeology of the Northwestern Plains*, edited by George W. Gill and Rick L. Weathermon, 241–262. University of Utah Press, Salt Lake City.

Gill, George W., and B. Miles Gilbert. 1990. Race identification from the midfacial skeleton: American blacks and whites. In *Skeletal Attribution of Race: Methods for Forensic Anthropology*, edited by George W. Gill and Stanley Rhine, 47–53. Maxwell Museum of Anthropology, Albuquerque.

Gill, George W., and Stanley Rhine. 1990. Appendix A to "A metric technique for identifying American Indian femora," by Randi Gilbert and George W. Gill. In *Skeletal Attribution of Race: Methods for Forensic Anthropology*, edited by George W. Gill and Stanley Rhine, 99. Maxwell Museum of Anthropology Paper 4. Maxwell Museum of Anthropology, Albuquerque.

Gill, George W., and Carlos J. Jimenez. 2004. Paleoamerican skeletal features surviving into the Late Plains Archaic. In *New Perspectives on the First Americans,* edited by Bradley T. Lepper and Robson Bonnichsen, 137–141. Texas A&M Press, College Station.

Gill, George W., and Cresta V. Deeds. 2008. Temporal changes in nasal sill development within the Northwestern Plains region. In *Skeletal Biology and Bioarchaeology of the Northwestern Plains*, edited by George W. Gill and Rick L. Weathermon, 263–270. University of Utah Press, Salt Lake City.

Gill, George W., and Henry W. Case. 2011. "Skeletal Biology and Population Affinities of the Prehistoric People of the Marismas Nacionales." Manuscript in possession of George W. Gill.

Gill, George W., Susan S. Hughes, Suzanne M. Bennett, and B. Miles Gilbert. 1988. Racial identification from the midfacial skeleton with special reference to American Indians and whites. *Journal of Forensic Sciences* 33(1):92–99.

Holborow, Amy A. 2002. "The Zygomaticomaxillary Suture: A Study of the Frequencies of Suture Shapes Within Various Major Populations of *Homo sapiens*." Master's thesis, University of Wyoming, Laramie.

Howells, William W. 1973. *Cranial Variation in Man: A Study by Multivariate Analysis of Patterns of Difference Among Human Populations.* Papers of the Peabody Museum of Archaeology and Ethnology 67. Harvard University, Cambridge, MA.

Jantz, Richard L. 1970. "Change and Variation in Skeletal Populations of Arikara Indians." Ph.D. dissertation, University of Kansas, Lawrence.

Jantz, Richard L., and Douglas W. Owsley. 1997. Pathology, taphonomy, and cranial morphometrics of the Spirit Cave Mummy. *Nevada Historical Society Quarterly* 40(1):62–84.

———. 2001. Variation among early North American crania. *American Journal of Physical Anthropology* 114(2):146–155.

Kesterke, Matthew J. 2008. "Heritability of Subtrochanteric Femur Shape (Platymeric Index) in *Papio hamadryas* and Implications for Hominid and Modern Human Postcranial Variation and Evolution." Master's thesis. University of Wyoming, Laramie.

Lepper, Bradley T. 2009. Genetics study: Two Paleoindian migration routes into the Americas. *Mammoth Trumpet* 24(4):9–11, 20.

Montandon, George. 1927. *Au Pays des Aïnou: Exploration Anthropologique*. Masson & Cie, Paris.

Olivier, Georges. 1969. *Practical Anthropology*. Charles C. Thomas, Springfield, IL.

Perego, Ugo A., Alessandro Achilli, Norman Angerhofer, Matteo Accetturo, Maria Pala, Anna Olivieri, Baharak Hooshiar Kashani, Kathleen H. Ritchie, Rosaria Scozzari, Qing-Peng Kong, Natalie M. Myres, Antonio Salas, Ornella Semino, Hans-Jürgen Bandelt, Scott R. Woodward, and Antonio Torroni. 2009. Distinctive Paleo-Indian migration routes from Beringia marked by two rare mtDNA haplogroups. *Current Biology* 19(1):1–8.

Powell, Joseph F., and Jerome C. Rose. 1999. Chapter 2. Report on the Osteological Assessment of the "Kennewick Man" Skeleton (CENWW.97. Kennewick). In *Report on the Non-Destructive Examination, Description, and Analysis of the Human Remains from Columbia Park, Kennewick, Washington* [October 1999]. National Park Service, http://www.nps.gov/archeology/kennewick/powell_rose.htm.

Rawlings, Kristen J. 2002. "Racial Variation in the Palate Shape and Form of the Transverse Palatine Suture." Master's thesis, University of Wyoming, Laramie.

Rogers, Martha. 2008. An analysis of discrete cranial traits of northwestern Plains Indians. In *Skeletal Biology and Bioarchaeology of the Northwestern Plains,* edited by George W. Gill and Rick L. Weathermon, 177–199. University of Utah Press, Salt Lake City.

Royer, Denise, George W. Gill, and Rick L. Weathermon. 1997. "Differences in Femoral Torsion Between American Indians and Whites." Paper presented at the annual meeting of the American Academy of Forensic Sciences, New York, NY.

Stefan, Vincent H. 2000. "Craniometric Variation and Biological Affinity of the Prehistoric Rapanui (Easter Islanders): Their Origin, Evolution and Place in Polynesian Prehistory." Ph.D. dissertation, University of New Mexico, Albuquerque.

Stewart, T. Dale. 1979. *Essentials of Forensic Anthropology.* Charles C. Thomas, Springfield, IL.

Weathermon, Rick L., and Mark E. Miller. 2008. History and development of the human remains repository at the University of Wyoming. In *Skeletal Biology and Bioarchaeology of the Northwestern Plains,* edited by George W. Gill and Rick L. Weathermon, 42–74. University of Utah Press, Salt Lake City.

Wescott, Daniel J. 2006. Ontogeny of femur subtrochanteric shape in Native Americans and American blacks and whites. *Journal of Forensic Sciences* 51(6):1240–1245.

Identity Through Science and Art

28

Karin S. Bruwelheide and Douglas W. Owsley

This chapter describes and illustrates the process and finished form of a facial reconstruction of Kennewick Man. Unlike previous facial reconstructions (Chatters 2000, 2001; Parfit 2000), this re-creation benefits from the information and analyses described in this volume, particularly craniometric comparisons to world population data. This information, not fully developed in earlier discussions of Paleoamerican remains, lends a deeper understanding of who this individual was and what he may have looked like. In this regard, the reconstruction is more resolved than in previous models, but it is not intended to end the discussion about the appearance of this man or his biocultural ties to those of the past as well as the present. Rather, this reconstruction is a component of the comprehensive investigation of the skeleton; in essence it is a visual bridge for translating current scientific findings to a broader audience, thus fostering dialogue on the complicated and incompletely understood process that led to the peopling of the Americas.

FACIAL RECONSTRUCTION

The process of facial reconstruction has been defined as "the scientific art of building the face onto the skull for the purposes of individual identification" (Wilkinson 2004:39). In two-dimensional reconstruction, the likeness of an individual is drawn over and in relationship to a skull image. In three-dimensional form, a facial reconstruction is sculpted over a skull or its model using a medium such as clay, or it is computer-generated. All methods rely on the relationship between the underlying framework, or architecture, of the skull and the hard (cartilage) and soft (muscle, fat, and skin) tissues of the face. Differences in skull shape, even minute ones, alter this relationship and individualize appearance.

Research on the relationship between facial bones and soft tissues for the purpose of reconstructing appearance in life began in Europe during the late nineteenth century. Anatomists collected tissue thickness data at designated landmarks on the skull from cadavers. With the assistance of sculptors, they used averaged tissue depth measurements and knowledge of facial anatomy to produce three-dimensional reconstructions on skulls of known and unknown individuals (His 1895; Kollmann and Büchly 1898; Welcker 1883). For skulls whose identity was known or suspected, the reconstructions were compared to portraits or busts of the individual, often with favorable results. Representations of more ancient remains for whom no comparative image existed could not be evaluated for accuracy, but the method initiated a scientific approach for discerning the faces of archaeological ancestors for the first time. Even ancient hominids became subjects for early facial reconstructions (McGregor 1926).

Throughout the mid- and late twentieth century, reconstructions in both forensic case work and osteoarchaeology increased. Russian anthropologist Mikhail Gerasimov (1971) was a pioneer in facial reconstruction techniques in Europe. Anthropologist Wilton Krogman and others in the field tested the accuracy of various methods of facial reconstruction in the United States (Krogman and McCue 1946; Krogman 1962; Snow et al. 1970). Method refinement and development led to more standardized approaches on both continents (Prag and Neave 1997; Taylor 2001; Wilkinson 2004), while a growing body of facial tissue-depth data, in addition to improved methods of skeletal interpretation, increased the effectiveness of facial reproductions. In a discussion of joint Smithsonian Institution and Federal Bureau of Investigation (FBI) forensic anthropology casework, Dr. Douglas Ubelaker (2000) noted a steady increase in facial reconstructions from the 1970s, when methods were just being tested, through the 1990s. In 1995, 60 percent of his FBI cases involved either facial reconstruction or photographic superimposition. Ubelaker's summary demonstrates that in contemporary forensic casework the method has become an accepted and widely used procedure toward establishing positive identification of unknown remains.

Museums and universities have incorporated greater numbers of facial reconstructions into public exhibitions and research programs. At the Smithsonian's Natural History Museum, for example, forensic facial reconstruction techniques were used to re-create life-size likenesses of early colonists from their skeletal remains in the exhibition *Written in Bone: Forensic Files of the 17th-Century Chesapeake* (http://anthropology.si.edu/writteninbone). Busts and full-body reconstructions of ancient hominids were included in the permanent exhibition of the Hall of Human Origins (http://humanorigins.si.edu). The popularity of producing facial reconstructions of mummified or ancient skeletonized remains—as drawings, computer reproductions, or three-dimensional sculptures—is based on recognition of the public's ability to connect more intimately to the past through these figures. Paleoamerican skeletal remains for which facial reconstructions have been done include the Peñón III woman from Mexico (radiocarbon date of 10,755 RC yr. BP, González et al. 2006; Sagrado 2004), the Wilson-Leonard II female (Lysek 1996) in Texas, and adult males from Horn Shelter in Texas (Bosque Museum 2006), and Spirit Cave (Bonnichsen and Schneider 2001) and Wizards Beach (Dixon 1999) in Nevada. A two-dimensional reconstruction was made of remains identified as Acha-2 (Arriaza et al. 1993:61), a nearly 9,000 year old skeleton from Chile.

Beginning in the 1990s computer programs were developed for facial reconstruction (Evenhouse et al. 1992; Evison et al. 1999; Ubelaker and O'Donnell 1992; Vanezis et al. 2000). The computer programs morph facial features or add facial components onto scanned or imaged skulls from limited libraries of data that match the biological profile of the skeleton. Because much of the debate on the effectiveness of reconstructions in forensic cases can be traced to the skill of the artist and the participation of a trained forensic anthropologist, computer programs aim to reduce the variability in style and artistic talent seen in manual reconstructions while decreasing the time that it takes to create a reconstruction. Programs are yet to be developed that address subtle variations in morphology that contribute to individual appearance or that produce an image that is as lifelike as some three-dimensional manual reconstructions.

RECONSTRUCTING KENNEWICK MAN'S FACE

Whether the final form of a facial reconstruction is a two-dimensional illustration, three-dimensional computer-generated image, or three-dimensional sculpture influences decisions related to its production. Given the potential for critical review, as well as possible exhibition use due to public interest, this attempt portrays Kennewick Man using a three-dimensional format. A life-size rendition of the face can be viewed from various angles, adding realism lacking in a drawing or a computer-generated reconstruction. Therefore, the desired product was a sculpted bust showing the head and neck.

An interdisciplinary collaborative effort was undertaken to produce this bust. It involved multiple individuals with different kinds of expertise—two skeletal anatomists, model makers, forensic and figurative sculptors, a photographic researcher, and a painter. Collectively, they formed a team whose objective was to align the figure's anatomy with the recent skeletal data. This included consideration of craniofacial morphology, which provided the baseline template for the reconstruction, and deliberation of subjective features such as head and facial hair, age-progression, skin color, and the expression of a frontal bone injury.

The multistep process of facial reconstruction, outlined elsewhere in varying degrees of complexity (Macleod and Hill 2001; Taylor 2001; Wilkinson 2004), began with detailed scientific assessment of the skeleton including determinations of age, sex, health, robusticity, and cranial morphology, specifically examining population relationships based on worldwide morphometric comparisons. With all reconstructions, this information, along with the identification of cranial anomalies, injuries, or evidence of disease, influences the outcome of the finished face. The Kennewick Man skeleton represents a male about 40 years of age, 5 feet 7 inches in stature, with moderate to marked development of muscle attachment areas on the postcranial bones, reflecting a physically active, rigorous lifeway. Morphometric results from the cranial analyses found greater similarities between

Kennewick Man and modern Polynesians and the Ainu of Japan, than to later Native Americans (Chapters 24, 25, and 26). These findings aided the second stage of the reconstruction, providing a relevant population framework from which to select historic photographs and facial features useful for finalizing the figure. Cranial observations specific to Kennewick Man include a small, healed depression fracture in the left midfrontal bone and slight deviation of the incomplete nasal bones toward the left side.

First Stage Reconstruction

Two models of the cranium and mandible were used in the reconstruction process. One cast served as the physical template for the reconstruction, and the other was used as a frequent reference guide for checking conformity of the progressing sculpture. These models, along with supporting biological information on the skeleton, were presented to forensic sculptor Amanda Danning. Anthropologists discussed how aspects of age, ancestry, body build, and distinctive features noted on the skull would be manifest in the sculpted clay form with Danning, whose experience includes the facial reconstruction of the Horn Shelter No. 2 skeleton and numerous historic personages, primarily from military battlefield contexts.

Using the tissue depths provided in Table 28.1, Danning sculpted the face, head, and neck using oil-based clay (Figure 28.1). While all reconstructions rely on

TABLE 28.1 Tissue depths (mm) for two reference samples used in the Kennewick Man reconstruction.

Measurement	Southwest American Indians (Asian-derived) Taylor 2001:352		American Caucasoids (European-derived) Taylor 2001:351		Kennewick Man Averaged Values
	Slender	Normal	Slender	Normal	
Midline					
Supraglabella	5.75	5.00	2.25	4.25	4.31
Glabella	5.75	5.75	2.5	5.25	4.81
Nasion	5.75	6.86	4.25	6.50	5.84
End of nasals	2.75	3.50	2.50	3.00	2.94
Mid-philtrum	7.50	9.75	6.25	10.00	8.38
Upper lip margin	8.25	9.75	9.75b	9.75	9.38
Lower lip margin	9.25	11.00	9.50b	11.00	10.19
Chin-lip fold	8.50	11.50	8.75	10.75	9.88
Mental eminence	8.00	12.00	7.00	11.25	9.56
Beneath chin	5.25	8.00	4.50	7.25	6.25
Bilateral					
Frontal eminence	4.75	4.25	3.00	4.25	4.06
Supraorbital	6.75	9.00	6.25	8.25	7.56
Suborbital	3.75	7.50	2.75	5.75	4.94
Inferior malar	10.00	14.00	8.50	13.25	11.44
Lateral orbit	8.00	12.50	5.00	10.00	8.88
Zygomatic arch, halfway	6.00	7.50	3.00	7.25	5.94
Supragleniod	5.75	8.50	4.25	8.50	6.75
Gonion	7.75	13.25	4.50	11.50	9.25
Supra M2	14.25	21.50	12.00	19.50	16.81
Occlusal line	15.50	20.75	12.00	18.25	16.63
Sub M_2	12.50	19.25	10.00	16.00	14.44

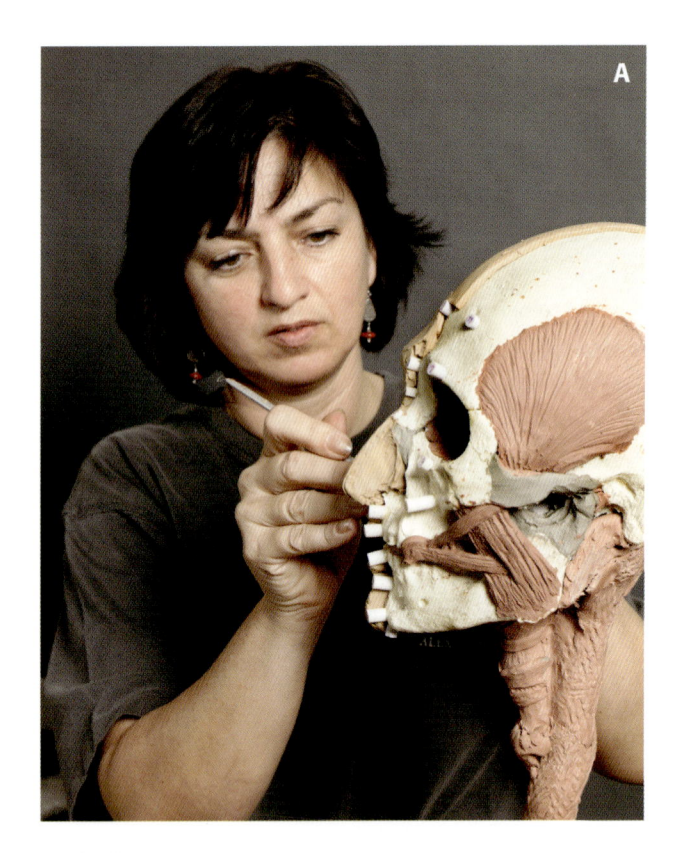

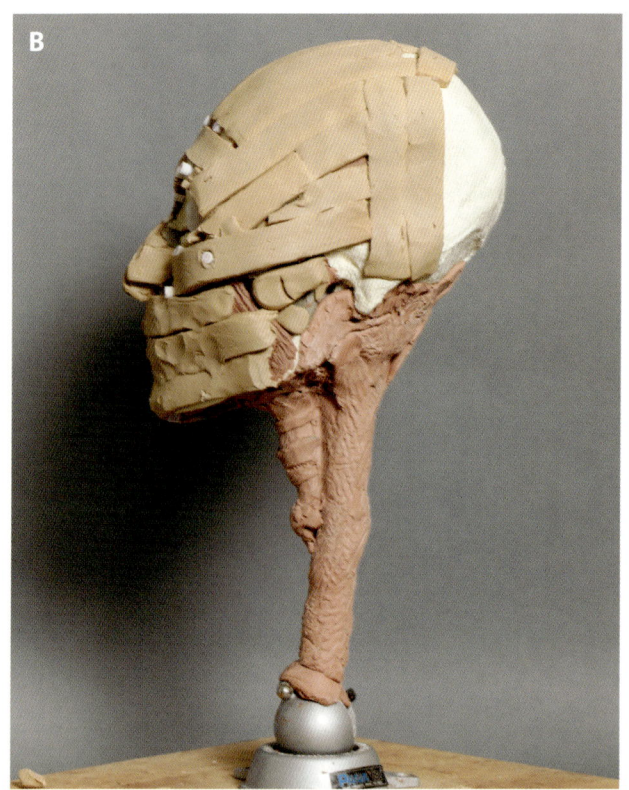

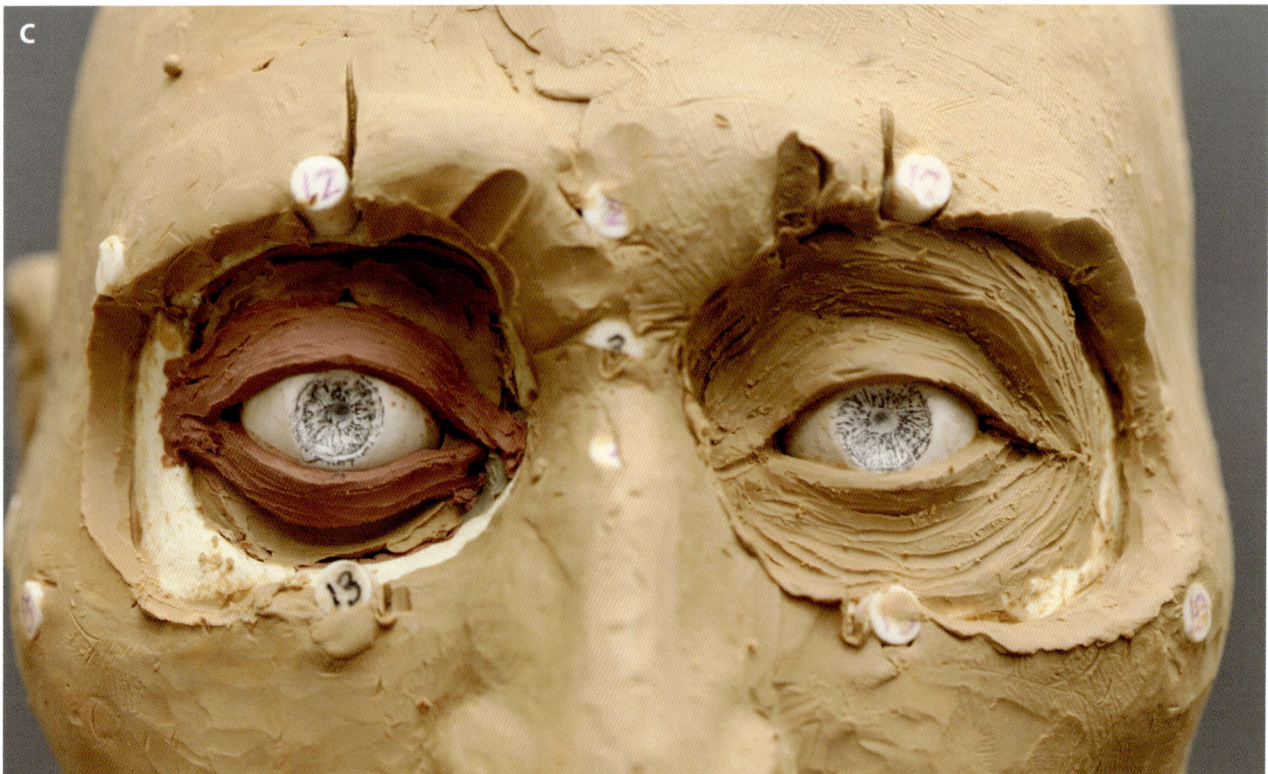

FIGURE 28.1 Progressive steps in the first-stage reconstruction process of Kennewick Man, sculpted by Amanda Danning: (A) initial sculpting of the face involving placement of tissue-depth markers on cranial landmarks and the development of facial musculature (temporalis and masseter); (B) left lateral view of the reconstruction showing the use of clay strips to cover sculpted muscles and fill spaces between tissue-depth markers; (C) close-up of sculpted eye muscle (orbicularis oculi) and tissue-depth markers defining the socket region of the face; (D) oblique view of sculpted head and unfinished neck with tissue depth markers still visible through clay; (E) frontal view of head and neck showing placement of sculpted clavicles and sternocleidomastoid muscles.

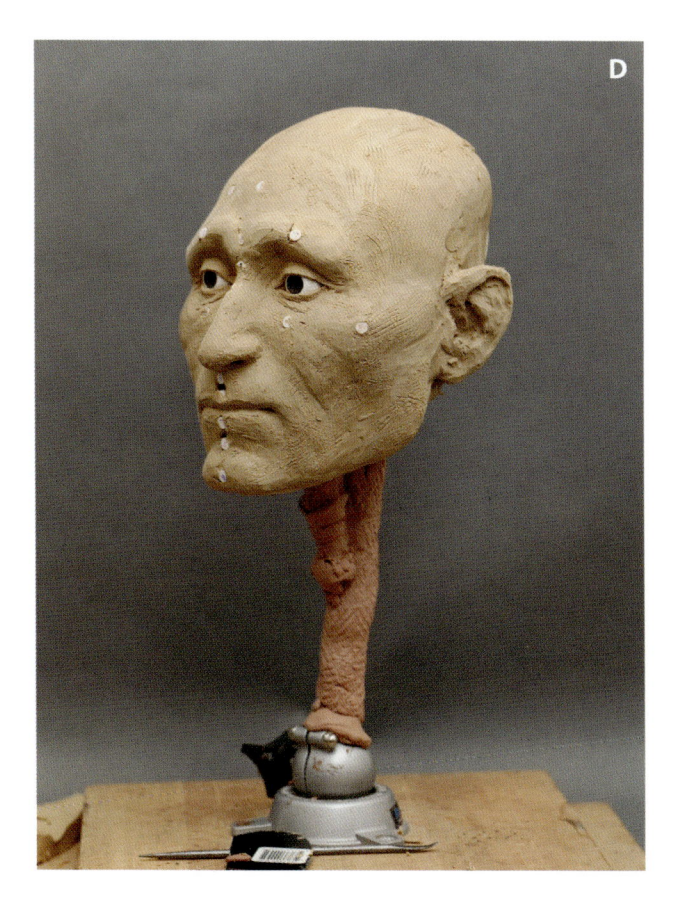

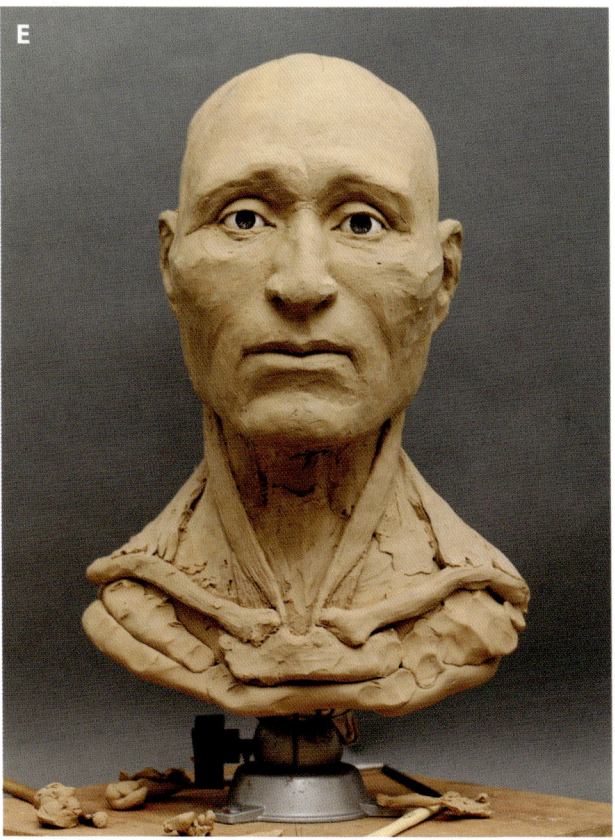

basically the same anatomical guidelines for placing the eyes and basic development of the mouth and nose, the process of building the layers of the face can differ between sculptors. The Russian technique, or anatomical method, relies on the artist's development of the musculature on the skull and neck. In contrast, the American technique, or tissue-depth method, utilizes average tissue depths at designated landmarks on the skull. Strips of clay are used to fill in the areas between these markers to re-create the face. Both methods have advantages, and many forensic sculptors, including Danning, use a combination of the two techniques, re-creating some of the facial musculature while guided by prescribed tissue thicknesses corresponding to the sex and ancestry of their subject (Taylor 2001). Hair and subtle details were not added at this stage. The guiding principle was that anatomical morphology and forensic experience should dictate structure (i.e., size, height, and width of the face, shape of the forehead, and brow ridge development) as well as placement of soft tissue features (i.e., locations of the eyes, ears, and mouth) during the initial stage of reconstruction.

The absence of tissue thickness data for Paleoamericans, Polynesians, and the Ainu was addressed by compiling a generic (i.e., somewhat population neutral) approximation of tissue-depth thicknesses using data from two series, Southwest American Indian males and American Caucasoid males. The tissue thickness data were derived from commonly referenced depth standards defined by ancestry, sex, and body weight compiled by Dr. Stanley Rhine and colleagues for contemporary forensic facial reconstructions (Taylor 2001). Tissue depths for slender and normal body weights from the two selected groups were combined and averaged based on the realistic assertion that Kennewick Man's age, diet, and active life as a hunter and fisherman would have resulted in a muscular, lean physique. Despite the recognized importance of using correct tissue-depth data for the age, sex, and ancestry of the subject skull, studies show that the underlying bone structure has greater influence on directing facial features. Using the wrong depth data will alter the reconstruction, but not dramatically (Wilkinson et al. 2002; Wilkinson 2004).

Second-Stage Reconstruction

The first-stage clay reconstruction was brought to StudioEIS in Brooklyn, New York, for finishing. StudioEIS specializes in the production of artworks and lifelike museum figures of contemporary, historical, and anthropological significance in a variety of mediums for museums and institutions worldwide. The company has sculpted reconstructions of some of America's notable figures including George Washington for George Washington's Mount Vernon Estate and Gardens in Virginia; Abraham Lincoln for displays at Gettysburg National Military Park in Pennsylvania and Lincoln's Cottage in Washington, DC, and Thomas Jefferson for Monticello in Virginia. The objectives achieved by bringing the Kennewick reconstruction to StudioEIS were to humanize the reconstruction and finalize the figure for display.

Refining the first stage of the facial reconstruction to achieve a realism deliberately not conveyed in the initial clay figure began with a remolding of the forensic sculptor's basic form. A copy of the original was produced using water-based clay (Figure 28.2). This clay press allowed technical freedom in finalizing the figure by pre-

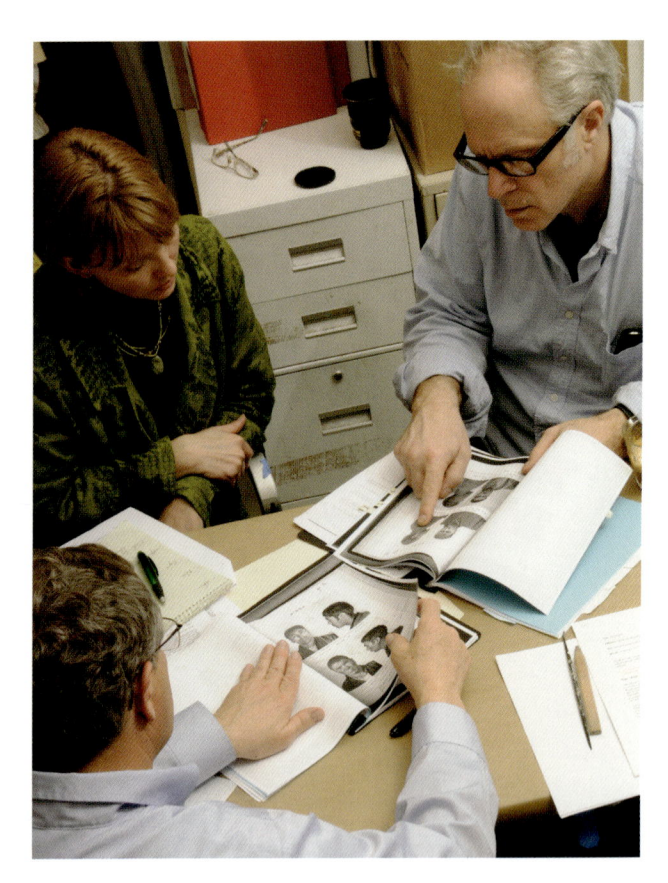

FIGURE 28.3 StudioEIS (Schwartz) and Smithsonian team members (Owsley and Bruwelheide) select facial features from historic reference images that correspond to features expressed in the Kennewick Man skull and first-stage reconstruction.

serving the forensic sculptor's original work for reference while subjective features and elements reflective of age and environmental exposure were added.

Reference images that would lend guidance to the finishing process were brought to the attention of the StudioEIS team under the direction of Ivan Schwartz (Figure 28.3). The selection of images was based on nearest neighbor and other statistical morphometric comparisons that strongly connect Kennewick Man's ancestry to circumpacific populations rather than with eastern and northeastern Asian populations historically classified as Mongoloid (Chapters 24 and 25; Owsley and Jantz 2001). General similarity to Polynesians and the Ainu implies descent from a shared or common ancestral population during the Late Pleistocene and Early Holocene. However, it is not realistic to assert that Kennewick Man was, for example, an

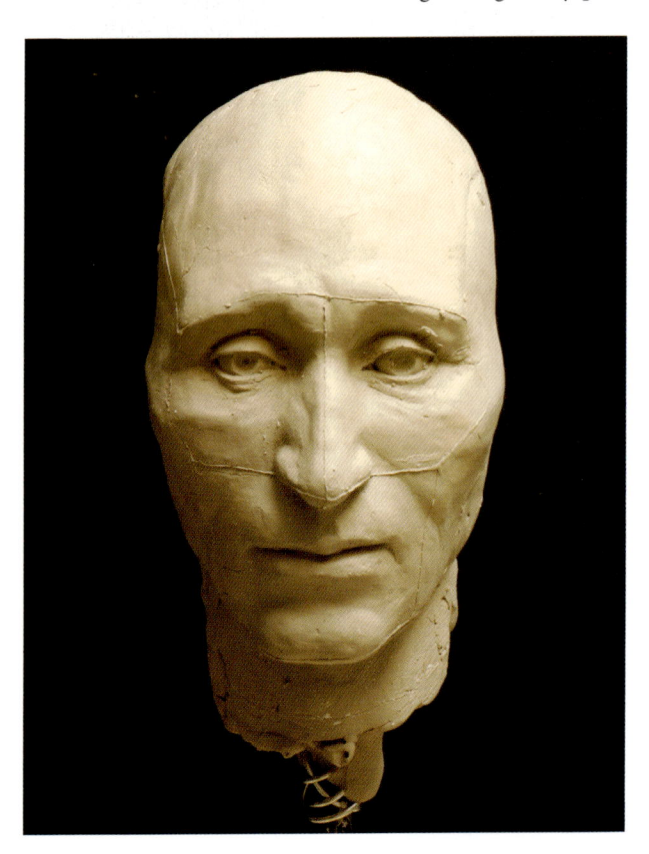

FIGURE 28.2 Clay press of completed first-stage reconstruction of Kennewick Man. Mold lines are visible on the reference model made using water-based clay.

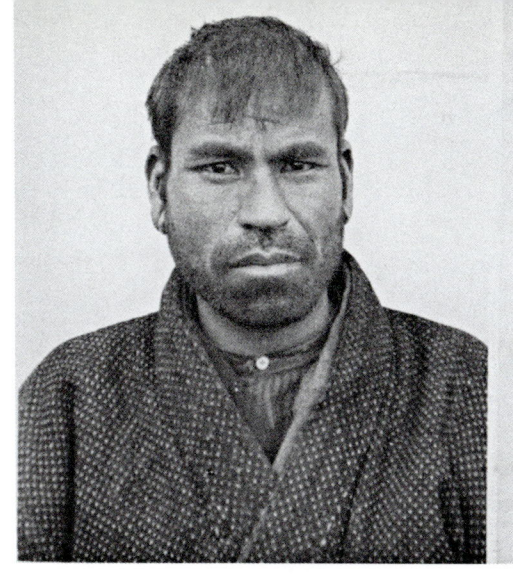

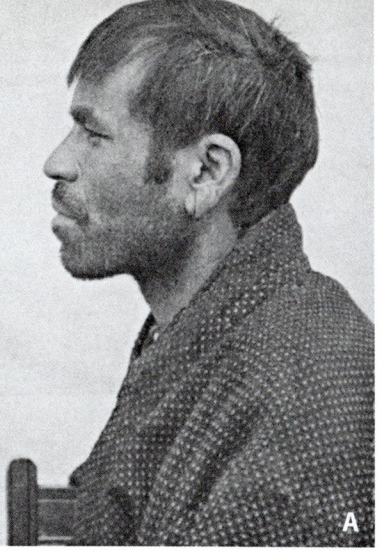

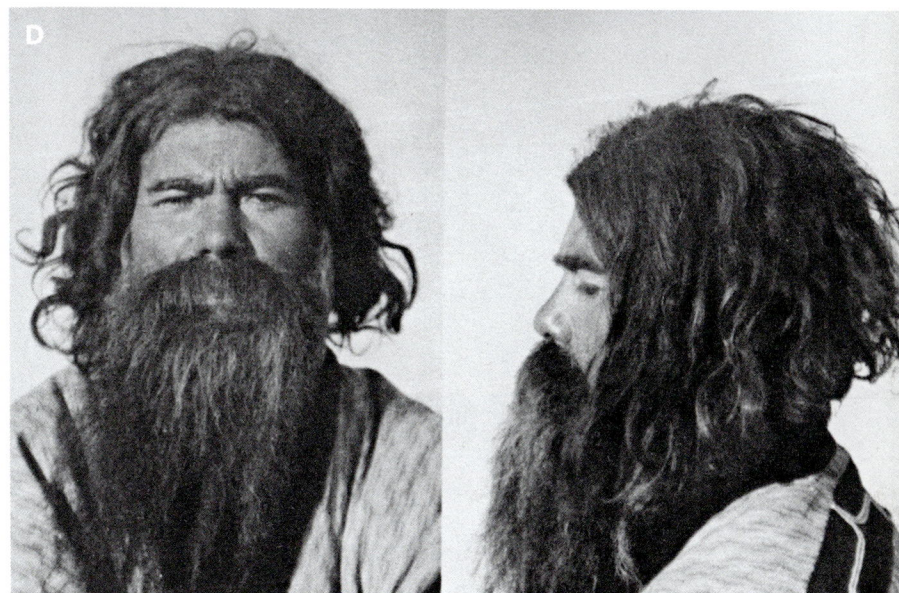

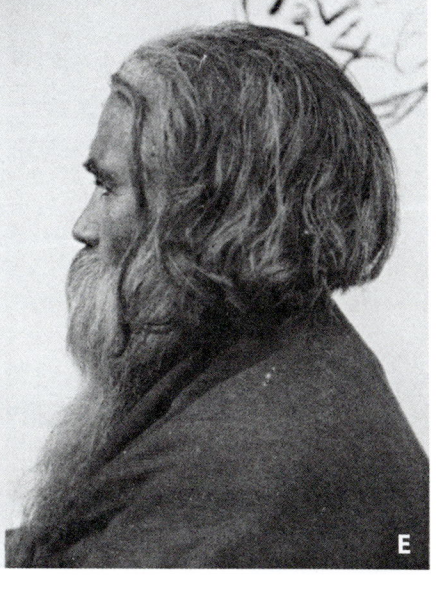

FIGURE 28.4 Historic photographs of Ainu adult males selected as reference images for the Kennewick Man reconstruction: (A) aged 42 years (Montandon 1927:pl. 4, figs. 13–14); (B) in 1908; (C) in 1909 or earlier; (D) Tcharakourè, age 56 years (Montandon 1927:pl. 10, figs. 39–40); (E) Ikonok, age 56 years (Montandon 1927:pl. 13, figs. 51–52); (F) Urap, in 1890.

FIGURE 28.5 Adding detail to the second-stage reconstruction of Kennewick Man using a reference model of the skull, the first-stage reconstruction, and select historic images. (A) StudioEIS sculptor Jiwoong Cheh; (B) Owsley and Bruwelheide conferring with Cheh on the expression of specific facial features.

FIGURE 28.6 In-progress second-stage reconstruction of Kennewick Man: (A) frontal and (B) right oblique views.

microevolutionary processes over the millennia have undoubtedly affected the craniofacial morphology of all groups. Yet, early photographs of the Ainu, along with groups like the Polynesian Moriori, can help with the portrayal of facial features and other characteristics. Historic images of late nineteenth to early twentieth-century Ainu (Figure 28.4) provided guidance for rendering physical features that tend to be patterned within populations, such as the form of the nose and eyes, skin color, and hirsutism, while also providing insight as to the range in individual variation. The males in these images show moderately pronounced brow ridges and nasal forms not unlike that expressed in the features of the Kennewick cranium. Several of these males provide references for appropriate age-related features, such as creases in the forehead and around the eyes. The images also give guidance in modifying the gaze or expression of the reconstruction, which was to be purposeful, with a naturalism and intensity expressed in the reference images. Without exception, the older males in the reference images and some younger ones as well show large amounts of dark facial and head hair. Wavy hair covers a majority of the men's faces, as in the Ainu male shown in Figure 28.4D. The hair, which is natural in appearance, having wave and texture that could be stylized in clay, was used as a guide for finalizing the Kennewick Man reconstruction.

Working with the three forms of reference materials including historic images of the Ainu and Moriori, along with a model of the Kennewick skull and the first-stage facial reconstruction by Danning, StudioEIS added details to the second-stage clay reconstruction (Figure 28.5). Consultation with team members continued throughout this second-stage process until a fairly advanced form of the head was complete (Figure 28.6). A purposeful gaze was incorporated into the reconstruction, the end of the nose and the nostrils were defined, and a lifetime of outdoor exposure to the sun and wind became expressed through creases etched into the forehead and around the eyes. The neck was given greater definition and hair was added. Based on the reference images of Ainu and Moriori males, which show greater amounts of facial hair than most East Asian groups, medium-length hair and a full beard and mustache were sculpted onto the Kennewick head. Although hair covers portions of the face, and its style and amount were at the discretion of the reconstruction team, the consensus was that more, rather than less, hair was likely to have been present. No head covering was used.

Painting

Upon completion, the second-stage clay bust (Figure 28.7) was molded, cast, and finished with a painting style based on variation in light and dark tones. Just as the materials used to create the piece needed careful consideration, painting and finishing techniques that add color and texture also required reflection. Painting adds another layer of interpretation, yet a figure devoid of color can appear ghostly. In contrast, the addition of realistic color can make the form too lifelike, allowing little room for personal interpretation on the part of the viewer. The style of painting selected for the Kennewick reconstruction lends realism without full interpretation of the subject's hair and skin color. Monochromatic tones with suggestions of color (darker skin and darker hair) resist presenting the figure with too much realism, but suggest enough of a realistic appearance that the figure is humanized.

StudioEIS artist Rebecca Spivack (Figure 28.8) described the coloration of the figure as "nuanced" with subtle suggestions of skin and hair color layered into the reconstruction—a hybrid form of tonal and realistic coloration. The colors applied are more vivid than true tonal painting, with highlights of color added to a tonal background. Since Kennewick Man would have experienced a lifetime outdoors with exposure to the elements, the weathered dark skin and naturalized hair of the figure were produced using warm, neutral tones.

Completed Reconstruction

This life-size facial reconstruction of Kennewick Man (Figure 28.9) differs from previous facial reconstructions in its final appearance and in the multilayered process used to achieve this outcome. Information not available in previous attempts was incorporated into this figure. An effort was made to avoid mannerisms of individual

FIGURE 28.7 Final second-stage reconstruction of Kennewick Man prior to painting: (A) frontal; (B) left oblique; (C) right lateral; and (D) posterior views.

FIGURE 28.8 Final Kennewick Man facial reconstruction: (A) frontal; (B) left oblique; (C) right oblique; (D) close-up, left oblique views.

FIGURE 28.9 Artist Rebecca Spivack adding color to the final form of the second-stage reconstruction of Kennewick Man.

artists by working with both a forensic sculptor and StudioEIS, along with direct and sustained consultation with physical anthropologists. At all times the reconstruction maintained the integrity of the underlying form that was dictated by the bone and created in the initial facial reconstruction by Danning. Features subject to artistic interpretation were developed only after extensive referral to the original skull and reference images. The decision to achieve a level of representation more resolved than earlier facial reconstructions allowed previously unassigned features, like hair and skin color, to be incorporated into the bust. The objective was to attain a level of naturalism and reality based on data generated by studying the skeleton.

SUMMARY

This volume brings together research findings from the scientific study of the Kennewick Man skeleton. The facial reconstruction is a product of this synthesis, incorporating information from the study of the bones into a lifelike sculpture that can be viewed and evaluated (Figure 28.10). As such, the facial reconstruction is "a three-dimensional report on the skull" (Prag and Neave 1997:10), another piece of Kennewick Man's story, and of the prehistoric puzzle encompassing research focused on the peopling of the New World.

FIGURE 28.10 Cast of Kennewick Man's skull with the completed reconstruction: (A) right oblique views; (B) frontal (next page).

This story is far from complete. Questions concerning the identity, life, and death of Kennewick Man and other early Americans remain unresolved. Debate on who Kennewick Man was and what he looked like will continue. New investigations will offer insights, but also challenges to scientists and the public, as the facial reconstruction of Kennewick Man and the story of what his bones reveal demonstrates. The evidence may not coincide with previously held perceptions of ancient America and what people from a particular time and place were thought to have looked like. Scientists need to be prepared to translate the new results of different types of research to the public in meaningful ways. Through this complex process, a fuller understanding of a skeleton and who it represents will be achieved.

This attempt at putting a face on Kennewick Man may be questioned, but the work was prompted by the query, "Was there another view of Kennewick man that would shed new light on his identity?" In raising and then trying to answer this question, scientists move toward a more complete understanding of the human experience of migration and dispersion into the Americas and the many ways it can be explored.

B

REFERENCES

Arriaza, Bernardo, Arthur Aufderhiede, and Iván Muñoz. 1993. Analisis antropologico fisico de la inhumacion de acha-2. In *Acha-2 Y Los Origenes Del Poblamiento Humano En Arica,* edited by Iván Muñoz, Bernardo Arriaza, and Arthur Aufderhiede, 47–61. Departmento de Arqueologia y Museologia, Universidad de Tarapaca, Arica, Chile.

Bonnichsen, Robson, and Alan L. Schneider. 2001. The case for a Pre-Clovis people. *American Archaeology* 5(4):35–39.

Bosque Museum. 2006. *The Horn Shelter Exhibit.* Catalog presented by the Bosque Museum, Clifton, TX.

Chatters, James C. 2000. Meet Kennewick Man. NOVA Online, Mystery of the First Americans. http://www.pbs.org/wgbh/nova/first/kennewick.html.

———. 2001. *Ancient Encounters: Kennewick Man and the First Americans.* Simon & Schuster, New York.

Dixon, E. James. 1999. *Bones, Boats, and Bison: Archaeology and the First Colonization of Western North America.* University of New Mexico Press, Albuquerque.

Evenhouse, R., M. Rasmussen, and L. Sadler. 1992. Computer-aided forensic facial reconstruction. *Journal of Biological Chemistry* 19(2):22–28.

Evison, M. P., O. M. Finegan, and T. C. Blythe. 1999. Computerized 3-D facial reconstruction: Research update. *Assemblage* 4: http://www.shef.ac.uk/assem/4/evison.html.

Gerasimov, M. M. 1971. *The Face Finder.* Hutchinson, New York.

González, Silvia, José Concepción Jiménez López, Robert Hedges, José Antonio Pompa y Padilla, and David Huddart. 2006. Early humans in Mexico: New chronological data. In *El Hombre Temprano en América y Sus Implicaciones en el Poblamiento de la Cuence de México,* edited by José Concepción Jiménez López, Silvia González, José Antonio Pompa y Padilla, and Francisco Ortiz Pedraza, 67–76. Instituto Nacional de Antropología e Historia, México.

His, W. 1895. Anatomische Forschungen über Johann Sebastian Bach Gebeine und Antlitz nebst Bemerkungen über dessen Bilder. *Abhandlungen der mathematisch-physikalischen Klasse der Königlichen Sachsischen Gesellschaft der Wissenschaften* 22:379–420.

Kollmann, J., and W. Büchly. 1898. Die Persistenz der Rassen und die Reconstruction der Physiognomie prähistorischer Schädel. *Archiv für Anthropologie* 25:329–359.

Krogman, Wilton Marion. 1962. *The Human Skeleton in Forensic Medicine.* Charles C. Thomas Publisher, Springfield, IL.

Krogman, W. M., and M. J. McCue. 1946. The reconstruction of the living head from the skull. *FBI Law Enforcement Bulletin* 17:7–12.

Lysek, Carol Ann. 1996. Putting a face on ancient people. *Mammoth Trumpet* 11(4):8–11.

Macleod, Iain, and Brian Hill. 2001. *Heads and Tales: Reconstructing Faces.* National Museum of Scotland Publishing Limited, Edinburgh.

McGregor, J. H. 1926. Restoring Neanderthal man. *National History* 26:288–293.

Montandon, George. 1927. *Au Pays des Aïnou: Exploration Anthropologique.* Masson & Cie, Paris.

National Museum of Natural History. 2010. Reconstructions of early human faces. http://humanorigins.si.edu/exhibit/gurche.

Owsley, Douglas W., and Richard L. Jantz. 2001. America's ancestors: Bioarchaeological investigations of ancient populations in the New World. In *International Colloquium on the First Americans,* Newsletter Special Issue (March), 8–9. Grant-in-Aid for Scientific Research on Priority Area (A) (1997–2000), Ministry of Education, Culture, Sports, Science and Technology, Monbukagakusho, Japan.

Parfit, Michael. 2000. Hunt for the first Americans. *National Geographic.* December: 41–67. Washington, DC.

Prag, John, and Richard Neave. 1997. *Making Faces Using Forensic and Archaeological Evidence.* British Museum Press, London.

Sagrado, María Villanueva. 2004. Reconstrucción facial escultórica de cráneos prehispánicos. *Arqueología* 11(65):48–53.

Snow, Clyde C., Betty P. Gatliff, and Kenneth R. McWilliams. 1970. Reconstruction of facial features from the skull: An evaluation of its usefulness in forensic anthropology. *American Journal of Physical Anthropology* 33(2):221–228.

Taylor, Karen T. 2001. *Forensic Art and Illustration.* CRC Press LLC, Boca Raton, FL.

Ubelaker, Douglas H. 2000. "A History of Smithsonian-FBI Collaboration in Forensic Anthropology, Especially in Regard to Facial Imagery." Paper presented at the 9th biennial meeting of the International Association for Craniofacial Identification, FBI, Washington, DC, July 24, 2000.

Ubelaker, D. H., and G. O'Donnell. 1992. Computer-assisted facial reproduction. *Journal of Forensic Science* 37(1):155–162.

Vanezis, P., M. Vanezis, G. McCombe, and T. Niblett. 2000. Facial reconstruction using 3-D computer graphics. *Forensic Science International* 108(2):81–95.

Welcker, H. 1883. *Schiller's Schädel und Todenmaske nebst Mittheilungen über Schädel und Todenmaske Kants*. Fr. Vieweg und Sohn, Braunschweig.

Wilkinson, Caroline M. 2004. *Forensic Facial Reconstruction*. University Press, Cambridge.

Wilkinson, C. M., R. A. H. Neave, and D. S. Smith. 2002. How important to facial reconstruction are the correct ethnic group tissue depths? *Proceedings of the 10th Meeting of the International Association of Craniofacial Identification, Bari, Italy*, 111–121.

LEARNING FROM EARLY HOLOCENE CONTEMPORARIES

The core of this volume is Kennewick Man's narrative. While much can be learned from this intensive study of a single human skeleton, it is comparative skeletal studies, geoarchaeology, and environmental and faunal information that ultimately provide the framework for interpreting Kennewick Man. So, this section reports on the recovery of ancient human remains from two carefully documented, but also highly unique sites. Studies of the archaeological evidence and physical remains are instructive on several levels, contributing to a better understanding of Kennewick Man and his place in the Paleoamerican world.

Chapter 29 reports on the 10,200-year-old disarticulated bones of a young adult male, named Shuká Káa by the Tlingit people, from On Your Knees Cave in southeast Alaska. This archaeological and paleontological investigation benefitted from working partnerships among scientists, government agencies, and the native community, with positive outcomes for all. Lithic material found at the cave could be tracked to distant quarries, and even genetic information could be obtained from a tooth. As a man from the north Pacific coast, Shuká Káa's isotope data provide a frame of reference for interpreting Kennewick Man's values; they are remarkably similar.

Next is a bioarchaeological reinterpretation of an 11,100-year-old burial of a middle-aged man and a child from Horn Shelter No. 2 in central Texas. We first saw these striking remains in 1998, a burial scene that needed additional interpretation benefitting from advances in technology and archaeological science, ethnographic insight, and further consideration of impressions gleaned from skeletal morphology.

Reports describe a boy and a man buried together with a flint knapper's tool kit at the Horn Shelter No. 2 site. Our first reaction was that the boy was more likely a girl and that the man's arm bones registered developmental/functional features indicative of activity-induced bone morphology that seemed out of the ordinary. Whereas Kennewick Man's skeletal features reflect a preoccupation with hunting, this man presents a different profile. Archaeological context, artifacts, and skeletal morphology suggest that the Horn Shelter male filled a completely different role in his social group. This new interpretation of the Horn Shelter No. 2 burial demonstrates the need for curation and continuing access to ancient skeletal remains. The bone biographies of these two individuals have yet to be fully written; there is potential to answer new questions about who they were and what their lives were like.

Evidence of Maritime Adaptation and Coastal Migration from Southeast Alaska

29

E. James Dixon, Timothy H. Heaton, Craig M. Lee, Terence E. Fifield, Joan Brenner Coltrain, Brian M. Kemp, Douglas W. Owsley, Eric Parrish, Christy G. Turner II, Heather J. H. Edgar, Rosita Kaahâni Worl, David Glenn Smith, and G. Lang Farmer

The oldest known human remains from southeast Alaska—dated to ca. 10,200 calendar years ago—were discovered at On Your Knees Cave (OYKC) located on Prince of Wales Island. The man, who died in his early twenties, was named Shuká Káa, or "Man Ahead of Us," by the Tlingit people of southeast Alaska. Analyses of the skeleton and artifacts provide new insights about the archaeology of the Northwest Coast of North America, including strong evidence that people obtained their food primarily from the sea and that they used watercraft to reach surrounding islands and the mainland.

SITE DESCRIPTION

Archaeological and paleontological testing and systematic excavations were conducted at OYKC (49-PET-408) between 1995 and 2004. The cave is located on Protection Head, a peninsula at the northwest end of Prince of Wales Island about 1 km from the coast and 125 m above current sea level (Figure 29.1). This multicomponent, stratified archaeological and paleontological site has two entrances, the Main Entrance where the human remains and most stone artifacts were found, and the smaller Ed's Dilemma Entrance to the east (Figure 29.2). The fossil record demonstrates that coastal southeast Alaska had unglaciated refugia during the last glacial maximum. OYKC was continuously used by mammals beginning about 45,000 RC yr. BP years ago, except for an interval between 20,300 and 17,200 CAL yr. BP (Heaton and Grady 2003), when the cave was probably over-ridden by glacial ice. Humans may have first used the cave as early as 10,300 ± 50 RC yr. BP (CAMS-42381), or 11,961–12,233 CAL yr. BP, based on a single ^{14}C determination from a bone tool recovered from the cave. At this time, Prince of Wales Island was inhabited by mammals including arctic fox (*Alopex lagopus*), red fox (*Vulpes vulpes*), caribou (*Rangifer tarandus*), bears (*Ursus arctos* and *Ursus americanus*), while ringed seal (*Phoca hispida*), harbor seal (*Phoca vitulina*), and Steller's sea lion (*Eumetopias jubatus*) swam in the adjacent ocean. A wide variety of fish and sea bird remains also have been recovered from OYKC and other caves throughout the region (Heaton and Grady 2003).

SPATIAL AND TAPHONOMIC ANALYSES OF THE REMAINS

Bone preservation in OYKC is good, and the amount of diagenetic carbon contamination is substantially less than in warmer latitudes. Shuká Káa's remains include: a mandible with partial dentition recovered in two pieces and subsequently rearticulated; a partial right pelvis and several small articulating and non-articulating fragments; one cervical, one lumbar, and four thoracic vertebrae; three ribs (one recovered in two fragments); three loose incisors; and one canine (Figures 29.3 and 29.4). No duplicate elements are present and the remains are from a single adult with pelvic morphology clearly indicative of a male. The remains provide the first opportunity to relate the genetic and physical attributes and associated artifacts with the earliest known inhabitants of the Northwest Coast of North America.

The spatial distribution and postmortem modifications of the bones demonstrate that this was not a deliberate burial. Shuká Káa's remains were disarticulated and scattered throughout Bear Passage, the primary passage extending north from the Main Entrance (Figure 29.4). Substantial breakage and the absence of limb bones imply postmortem transport of individual elements. The pelvis and mandible show carnivore gnawing indicating that scavengers—bears and foxes—modified and transported the bones within the cave (Chapter 17).

In addition to damage caused by scavenging, the bones were also affected by processes within the cave.

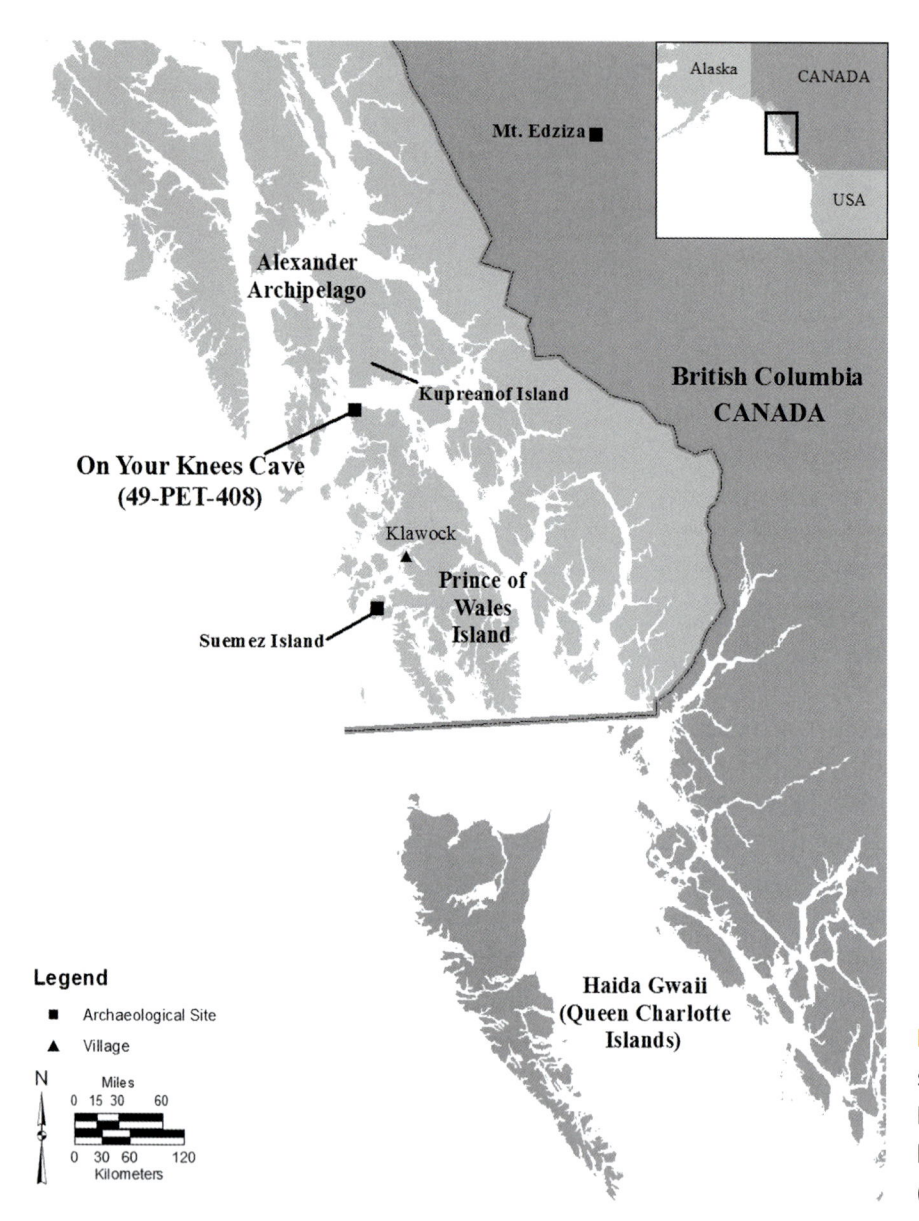

FIGURE 29.1 Map of the southern Northwest Coast of North America depicting the location of On Your Knees Cave (49-PET-408).

Scarring by cave rubble suggests occasional rock falls. Periodic seepage from small intermittent springs within the cave (Figure 29.2) may have provided sufficient water to transport bones during times of increased precipitation.

DATING

Microsamples extracted from the human pelvis and mandible were used in radiocarbon dating, dietary, and ecological analyses. Electric drills with small bits were used to avoid unnecessary damage or destruction. Molds and casts of the bones were made prior to sampling and have been retained for future study.

Purified bone collagen from the pelvis was dated to 9,880 ± 50 RC yr. BP (CAMS-32038) (calibrated two-sigma range 10,422–10,166 BP; $\delta^{13}C = -12.1‰$); mandibular bone collagen was dated to 9,730 ± 60 RC yr. BP (CAMS-29873) (calibrated two-sigma range 10,486–10,199 BP; $\delta^{13}C$ –12.5‰) (Dixon et al. 1997). The dates were calibrated by Calib 6.0 using the Mixed Marine and Northern Hemisphere Atmosphere calibration curve, which weights the correction for marine reservoir effect relative to the percentage of marine foods in the diet (McNeely et al. 2006). A local marine reservoir correction of $\Delta R = 495 ± 24$ was also made based on the weighted mean of six radiocarbon readings on pre-atomic bomb bivalves and a gastropod collected in the Alexander Archipelago (McNeely et al. 2006). The two dates yielded a pooled mean radiocarbon age[1] of death of 9,816 ± 42 RC yr. BP ($p < 0.05$, T = 3.1514, Xi^2 = 3.84,

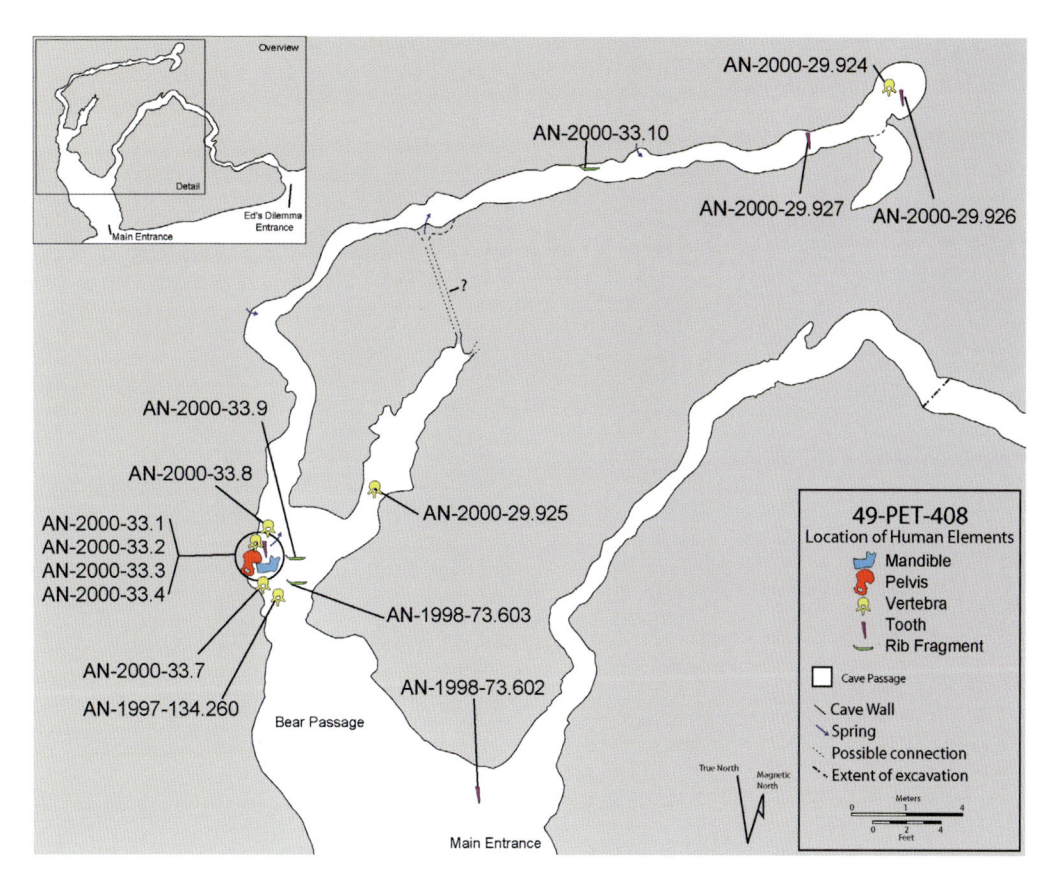

FIGURE 29.2 Spatial distribution of skeletal elements within On Your Knees Cave.

df = 1), which calibrates to 10,340–10,115 CAL yr. BP with a confidence interval median of 10,207 yr. BP.

A lower left second incisor (AN-1998-73.602) that articulates with the mandible was recovered from a stratigraphic level containing microblades and stone tools, making the mandible and tools from this stratigraphic level contemporaneous. The occupation range for this stratigraphic level is dated by AMS determinations from two pieces of wood charcoal: 9430 ± 140 RC yr. BP (AA-32866) (*Tsuga sp.*, hemlock), and 9170 ± 85 RC yr. BP (AA-32868) (probably *Thuja plicata*, western red cedar). These two dates provide a pooled mean age for this occupation of 10,411.5 ± 168.5 CAL yr. BP with a calibrated two sigma range between 10,242–10,580 CAL yr. BP.

DIET, INDIVIDUAL CHARACTERISTICS, AND CLIMATE
Isotope Analyses

Shuká Káa subsisted on a long-term diet of high trophic level marine foods such as seals and sea lions (pinnipeds) with very little terrestrial input. Two samples of bone collagen, one gelatinized and the other XAD extracted (Stafford et al. 1988), yielded a mean $\delta^{13}C$ value of −12.8 ± 1.6‰, in good accord with $\delta^{13}C$ values cited above, and a mean $\delta^{15}N$ value of 20.7 ± 0.2‰. Collagen values from prehistoric eastern Aleutian marine foragers ($n = 80$), with a mean $\delta^{13}C$ value of −12.5‰ and $\delta^{15}N$ of 19.9‰ (Byers et al. 2011; Coltrain et al. 2006), support this interpretation. The Aleut population subsisted nearly entirely on marine foods but with significant addition of littoral resources such as shoreline grasses and shell fish which are depleted in $\delta^{15}N$ relative to marine prey such as pinnipeds, likely accounting for the Aleut's very slightly depleted $\delta^{15}N$ value relative to that of Shuká Káa. This finding is also strongly supported by stable carbon isotope data on southeast Alaskan fauna reported by Heaton and Grady (2003:Table 2.3). Late Pleistocene fish-eating sea animals from southeast Alaska, such marine piscivores as ringed and harbor seals, exhibited mean $\delta^{13}C$ values in the −16.1‰ to −10.6‰ range, while sea birds ranged between −15.7‰ and −12.2‰. Shuká Káa's mean bone collagen $\delta^{13}C$ value of −12.8‰ falls at the upper end of both ranges. Allowing for ca. +1‰ fractionation between animal tissue and bone collagen, he was clearly targeting marine mammals and/or the same suite of prey as marine piscivores and piscivorous sea birds. In contrast, mixed feeders such as brown bear and the canids, Arctic fox and red fox, expressed mean $\delta^{13}C$ values of −17.5‰

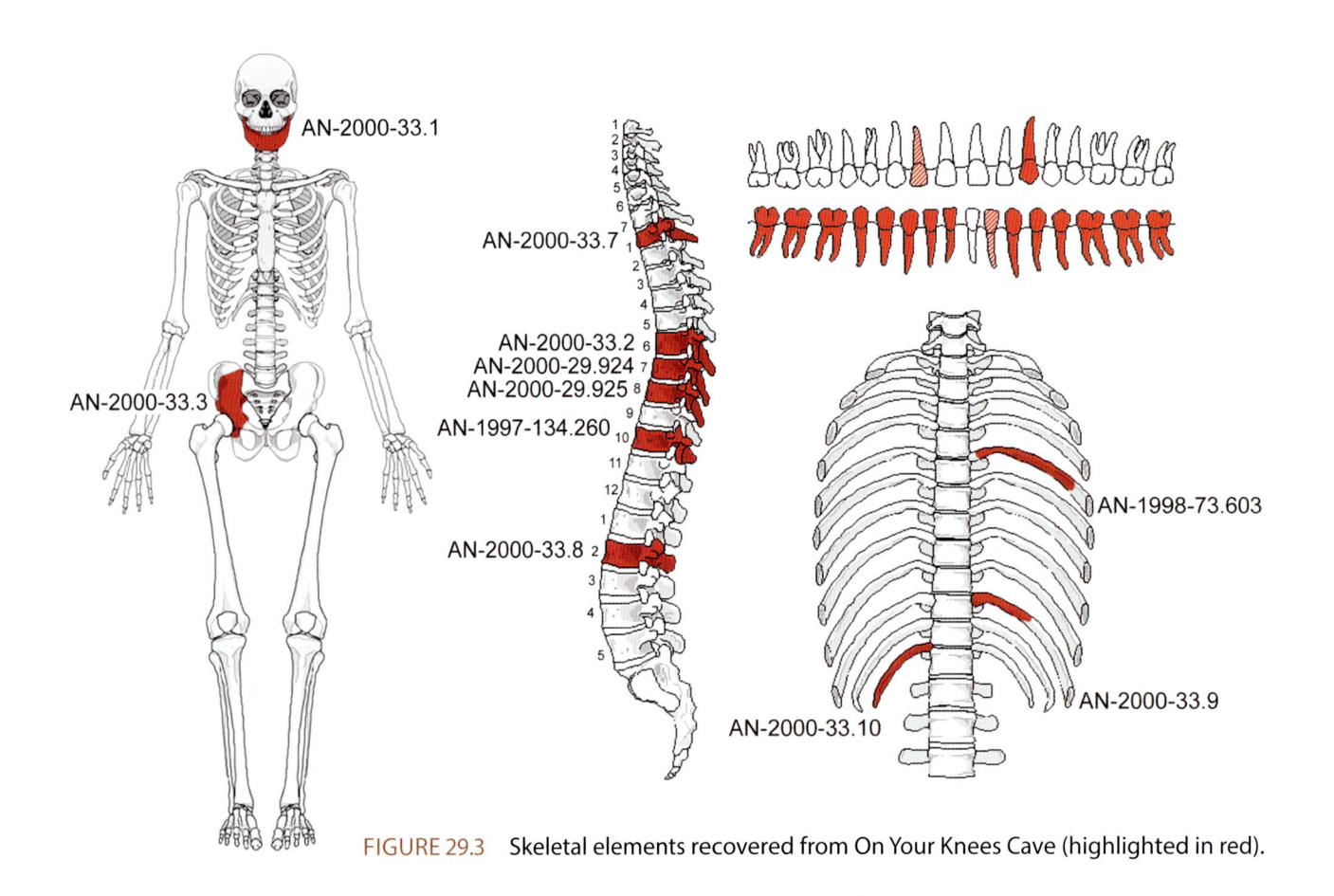

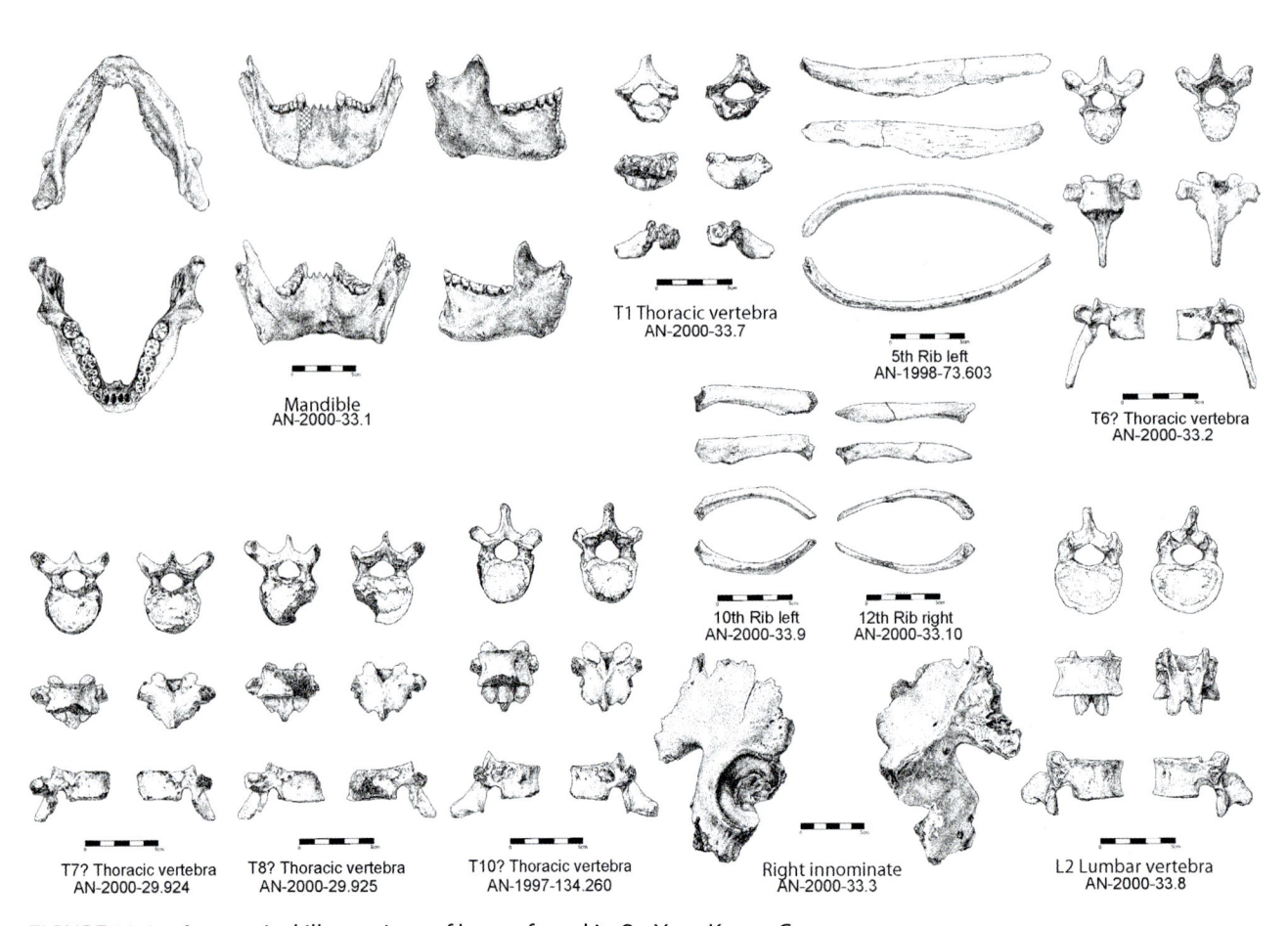

FIGURE 29.3 Skeletal elements recovered from On Your Knees Cave (highlighted in red).

Mandible
AN-2000-33.1

T1 Thoracic vertebra
AN-2000-33.7

5th Rib left
AN-1998-73.603

T6? Thoracic vertebra
AN-2000-33.2

T7? Thoracic vertebra
AN-2000-29.924

T8? Thoracic vertebra
AN-2000-29.925

T10? Thoracic vertebra
AN-1997-134.260

10th Rib left
AN-2000-33.9

12th Rib right
AN-2000-33.10

Right innominate
AN-2000-33.3

L2 Lumbar vertebra
AN-2000-33.8

FIGURE 29.4 Anatomical illustrations of bones found in On Your Knees Cave.

and −15.3‰ respectively, whereas terrestrial taxa, small mammals, the mustelids and artiodactyls, exhibited mean $\delta^{13}C$ values of −19.0 to 25.7‰.

Tooth enamel bioapatite from an incisor was also analyzed and corroborates these findings yielding a $\delta^{13}C$ value of −11.3‰, and $\delta^{18}O_{vPDB}$ value of −7.6‰ ($\delta^{18}O_{vSMOW}$ = 23.1‰). Stable carbon apatite-collagen spacing was merely +1.5‰ indicative of a diet high in animal protein (Hedges 2003; Passey et al. 2005). Aleut marine foragers analyzed for bone apatite $\delta^{13}C$ (n = 59) exhibited a virtually identical mean apatite $\delta^{13}C$ value of −11.4 ± 0.7‰, with mean apatite-collagen spacing of +1.3 ± 0.5‰.

The time averaged $\delta^{18}O$ signal in human bone and tooth enamel hydroxyapatite derives primarily from drinking water, and covaries with the $\delta^{18}O$ value of local meteoric water, which is fractionated by declining temperature, increases in elevation, and distance from a coastline, expressing further heavy fractionation in human tissue (Hoppe 2006; Koch 1998). A $\delta^{18}O$ value of 23.1‰$_{vSMOW}$ (−7.6‰$_{vPDB}$, Buzon et al. 2011) measured on the same enamel carbonate used to measure apatite $\delta^{13}C$ suggests that during Shuká Káa's lifetime a coastal climate slightly warmer than today prevailed in southeast Alaska, given a calculated, annual precipitation $\delta^{18}O$ value of ca. −10.5 to −12‰ (Iacumin et al. 1996; Keenleyside et al. 2011). In contrast, modern, weighted annual $\delta^{18}O$ values for southeast Alaska precipitation range from −18 to −22‰$_{vSMOW}$ (IAEA 2001). Shuká Káa's $\delta^{18}O$ value not only suggests a climate slightly warmer than today, but also indicates precipitation patterns similar to that of the eastern Aleutians, given a virtually identical mean $\delta^{18}O$‰ value for Aleut marine foragers (n = 59) dating to the pre- and proto-historic periods (Byers et al. 2011; Coltrain et al. 2006; Coltrain unpublished data).

Modern annual precipitation values for the eastern Aleutians range from −14‰ to −10‰$_{vSMOW}$ (IAEA 2001), fully in keeping with precipitation $\delta^{18}O$ calculated from Shuká Káa's enamel carbonate. Comparative $\delta^{18}O$ values on protohistoric Greenland Norse and Inuit remains range from a mean of 26.2‰$_{vSMOW}$ (17.2 ± 0.9‰ on enamel phosphate) during the Medieval Warm Period to 20.5‰$_{vSMOW}$ (11.6 ± 0.6‰ on enamel phosphate) with Neo-Boreal cooling (Fricke et al. 1995), placing Shuká Káa's enamel carbonate value within the context of similar data from other high latitude populations in the western hemisphere.

The strontium (Sr) isotopic composition of tooth enamel from Shuká Káa corresponds to a δSr value of −81.6, considerably lower than seawater Sr isotopic compositions over the past 6 million years (δSr = 0 to −30) (Capo and Depaolo 1990). This result precludes the possibility that enamel Sr was derived from seawater, via an exclusively marine diet. Instead, it is likely that Shuká Káa's enamel reflects the incorporation of Sr from local terrestrial sources, notably fresh drinking water.

Prince of Wales Island is comprised of Neoproterozoic to Late Triassic sedimentary and igneous rocks that constitute the allochthonous Alexander terrain found throughout much of southeast Alaska and adjacent portions of Canada (Churkin and Eberlein 1977; Gehrels and Saleeby 1987). Only limited Sr isotopic data from these rocks are available, but most of the intrusive and extrusive igneous rocks of the Alexander terrain have $^{87}Sr/^{86}Sr$ ratios between 0.704 and 0.708, while siliciclastic sedimentary rocks have higher $^{87}Sr/^{86}Sr$ ranging from 0.7056 to 0.710 (Boghossian and Gehrels 2000; Samson et al. 1989). Rock Sr isotopic compositions overlap the values determined for tooth enamel, suggesting that the Sr preserved in the tooth enamel was terrestrial in origin.

ARTIFACTS

Most of the artifacts from OYKC were recovered from excavations near the entrances to the cave and within the light zone. Cultural deposits at the Main Entrance did not continue beyond the first chamber of the Bear Passage, which suggests that human use of the cave was confined primarily to this chamber. In addition to hundreds of stone chips left from making and using stone tools, the artifacts include projectile points and/or knives (Dixon 2008), microblades, scrapers, choppers, flake cores, retouched flakes, and abrading stones. A few small pieces of red and orange ochre were found and were probably used as pigment (Mrzlack 2003). Artifacts found at the level of Shuká Káa's incisor and dating to the time of his death are stone and appear to be utilitarian.

Many of the artifacts were manufactured from stone procured from other islands or the mainland. Regional sea level history (Baichtal and Carlson 2010; Fedje and Josenhans 2000) indicates that Prince of Wales Island was also an island 10,200 calendar years ago, with a sea level approximately two meters higher than it is today.

Visual characterization and energy dispersive X-ray fluorescence (edXRF) demonstrate that the obsidian used in stone tool production is from Mt. Edziza, located on the British Columbia mainland, and from Suemez Island, located southeast of Prince of Wales Island (Figure 29.1) (Lee 2001). Obsidian data from OYKC and other early sites in the Alexander Archipelago (Lee 2007) indicate that people were coastal navigators who either procured stone directly from these sources or had established trade networks for obsidian and other types of exotic toolstone such as quartz crystal which was probably derived from nearby Kupreanof Island (Figure 29.1).

ORIGINS
Dentition

The mandibular dentition is complete except for the left central incisor. The left canine and all premolars and molars are in place in the alveoli of the mandible. The canines through first molars exhibit slight to moderate wear; the second and left third molars have only slight wear. Wear on the canines is distally oriented, and there are small, incisive edge notches, likely a result of task activities. The enamel of the mesiolingual cusp is missing due to an antemortem chip fracture. Radiography shows that both third molars are completely developed and the left third molar is slightly worn, indicating that it has been in occlusion for a short period (Figure 29.5). Crown attrition and the fact that third molars (wisdom teeth) erupt in the late teens or early twenties indicate that Shuká Káa died prior to reaching his midtwenties.

The facial surfaces of the two mandibular canines have depressed hypoplastic enamel defects indicating periods of developmental stress. On the right tooth, one inferior line is located 2.1 mm from the cemento-enamel junction (CEJ) and another line is 2.9 mm from the CEJ. On the left canine, the area of defect is bounded by two irregular lines and extends from 0.8 to 3.2 mm from the CEJ. The mandibular first premolars have single hypoplastic defects, measuring 2.7 mm from the CEJ.

Dental analysis was used to address relationships to early Asian and North American populations. Three buccolingual and three mesiodistal measurements were made from casts of the mandible and loose mandibular incisors. The average of these three measurements was calculated, the most outlying measure was removed, and the average recalculated. These measurements were compared with Asian and New World samples described in several sources (Doyle and Johnston 1977; Goaz and Miller 1966; Harris and Nweeia 1980; Kieser et al. 1985; Lukacs and Hemphill 1991; Matsumara 2001; Matsumara and Hudson 2005; Moorrees 1957). Comparing buccolingual measurements of the first molar, this individual's tooth size (10.8 mm) is the same as Chalcolithic Period III Mehrgarh (in modern Pakistan) (Lukacs and Hemphill 1991) and Sumatrans (Matsumara and Hudson 2005) (Table 29.1).

The mandibular first molars have two roots, which is more common in Sundadont dentitions. However, the mandibular second molar has six cusps, which is more commonly found in Sinodonts. The maxillary first molars are absent, so whether they had enamel extensions is not known. The mandibular first molars have grade one enamel extensions, and the second molar enamel extensions are grade two. This expression may provide insight into the condition of the maxillary molars, though

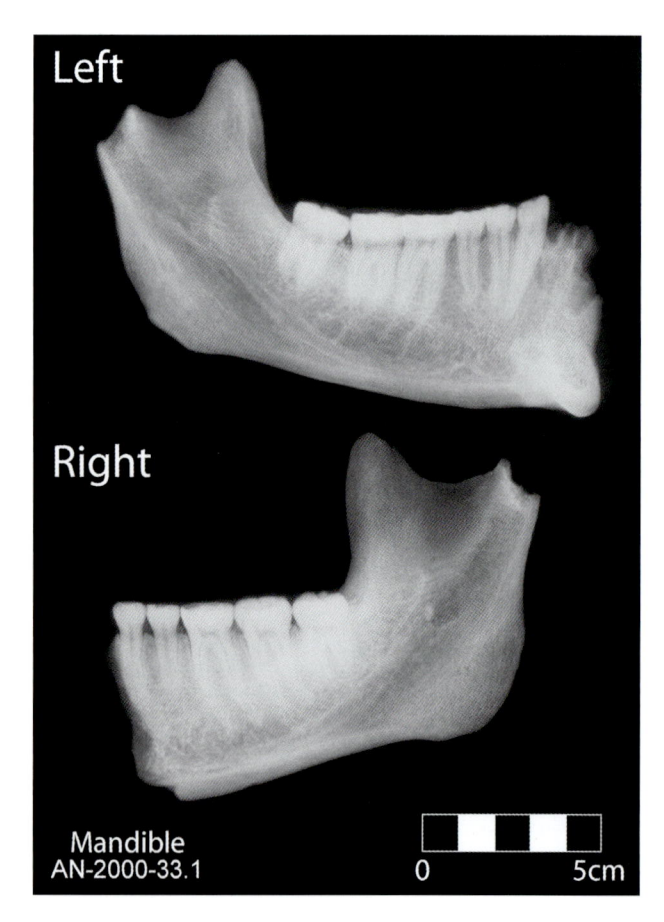

FIGURE 29.5 Radiograph of Shuká Káa's mandible (AN-200-33.1).

TABLE 29.1 Dental metrics for Shuká Káa and sample means for reference populations.

Population	Geographic Location	Mesiodistal Measures										Buccolingual Measures							Reference
		MDLI1	MDLI2	MDLC	MDLP ant	MDLP post	MDLM1	MDLM2	MDLM3	BLLI1	BLLI2	BLLC	BLLP ant	BLLP2 post	BLLM1	BLLM2	BLLM3	
Shuká Káa	Alaska	4.70	5.55	6.64	6.98	7.43	11.26	12.44	12.10	5.52	6.22	6.93	7.18	7.79	10.77	10.65	10.17	Edgar (measured from casts of teeth)
Shuká Káa	Alaska			7.05	7.65	7.70	11.95	12.85	12.80	.	.	7.60	7.75	8.05	11.30	11.60	11.05	Owsley (measured from teeth)
Aleut	Alaska	5.23	6.09	7.20	7.01	7.17	11.56	11.19	11.13	.	.	7.93	7.82	8.40	10.56	10.58	10.15	Moorrees 1957
Alaskan Eskimo	Alaska	.	.	.	7.42	7.85	10.86	10.45	10.30	.	.	.	6.80	6.67	11.55	10.95	11.05	Doyle and Johnston 1977
American Indian	Arizona Pueblos	.	.	.	7.02	7.54	10.47	10.04	9.89	.	.	.	6.27	6.87	10.72	10.55	9.90	Doyle and Johnston 1977
Ticuna Indians	Brazil	5.33	6.36	7.24	7.17	7.24	11.32	10.68		5.87	6.23	7.45	7.86	8.19	10.51	10.32		Harris and Nweeia 1980
Lengua Indians	Paraguay	5.71	6.56	7.68	7.68	7.62	11.98	11.32	10.95	.	.	.	8.10	8.34	11.04	10.57	10.58	Kieser et al. 1985
Peruvian Indians	Peru	5.62	6.38	.	7.35	7.45	11.92	11.03	11.19	.	.	.	8.18	8.61	11.03	10.76	11.03	Goaz and Miller 1966
Chalcolithic Period III (Mehrgarh)	Pakistan	5.53	6.07	6.89	7.00	7.29	11.39	11.05	10.92	5.93	6.34	7.60	7.95	8.42	10.75	10.54	10.27	Lukacs and Hemphill 1991
Neolithic Period I (Mehrgarh)	Pakistan	5.50	6.13	6.84	7.04	7.17	11.54	10.87	11.00	6.06	6.40	7.68	7.98	8.44	11.12	10.49	10.29	Lukacs and Hemphill 1991
Negritos	Southeast Asia	5.27	5.94	6.76	6.92	7.05	11.34	10.31		6.05	6.54	7.74	7.86	8.09	10.45	9.84		Matsumara and Hudson 2005
Andaman Islanders	Andaman Island	5.03	5.34	7.02	7.20	7.29	11.60	11.05		5.92	6.25	7.64	8.21	8.72	11.11	10.51		Matsumara and Hudson 2005
Non Nok Tha	Thailand	5.02	6.01	6.84	6.74	6.57	11.82	10.12		5.50	6.15	7.69	7.89	8.34	10.99	10.12		Matsumara and Hudson 2005
Ban Kao	Thailand	5.44	6.06	7.15	7.23	7.24	11.85	10.74		5.92	6.19	7.74	8.15	8.41	11.05	10.52		Matsumara and Hudson 2005
Laotians	Laos	5.29	6.18	7.13	6.92	7.04	11.48	10.40		6.22	5.96	7.37	7.98	8.29	10.77	10.02		Matsumara and Hudson 2005
Early Holocene Laotians	Laos	5.53	6.40	7.28	7.26	7.56	12.11	11.00		6.00	6.58	7.89	8.30	8.57	11.18	10.61		Matsumara and Hudson 2005

(continued)

TABLE 29.1 (*continued*) Dental metrics for Shuká Káa and sample means for reference populations.

Population	Geographic Location	Mesiodistal Measures								Buccolingual Measures							Reference
		MDLI1	MDLI2	MDLC	MDLP ant	MDLP post	MDLM1	MDLM2	MDLM3	BLLI1	BLLI2	BLLC	BLLP ant	BLLP2 post	BLLM1	BLLM2	BLLM3	
Bac Son	Vietnam	5.20	5.92	7.10	7.55	7.39	11.85	11.23		5.80	66.65	8.27	8.71	9.13	11.56	10.98		Matsumara and Hudson 2005
Philippines	Philippines	5.51	6.29	6.79	7.46	7.08	12.06	11.07		5.59	6.53	7.90	8.57	8.18	10.89	10.49		Matsumara and Hudson 2005
Dayak	Kalimantan	5.85	6.32	7.12	7.35	7.22	11.74	11.03		6.16	6.36	7.85	8.16	8.38	10.94	10.47		Matsumara and Hudson 2005
Lesser Sunda and Java Islanders	Lesser Sunda and Java Islands	5.61	6.48	7.32	7.31	7.20	11.78	10.58		5.97	6.44	8.16	8.15	8.41	10.78	10.23		Matsumara and Hudson 2005
Malay	Malay Peninsula	5.86	6.22	7.21	7.47	7.37	11.84	10.92		5.87	6.46	8.04	8.31	8.68	11.05	10.38		Matsumara and Hudson 2005
Gua Cha	Malay Peninsula	5.07	5.89	6.86	7.22	7.65	11.49	11.06		6.49	6.66	8.29	8.68	8.84	11.34	10.81		Matsumara and Hudson 2005
Guar Kepah	Malay Peninsula	5.87	6.86	7.34	7.85	7.89	12.56	12.19		6.82	7.32	8.75	9.09	9.25	11.76	11.18		Matsumara and Hudson 2005
Sumatra Islanders	Sumatra	5.48	6.11	7.00	7.35	7.22	11.53	10.81		5.81	6.29	7.81	8.16	8.50	10.75	10.29		Matsumara and Hudson 2005
Australian Aborigines	Australia	5.64	6.41	7.22	7.41	7.50	12.33	12.12		6.26	6.47	8.24	8.73	8.95	11.75	11.40		Matsumara and Hudson 2005
Loyalty Islanders	Loyalty Island	5.24	5.86	7.25	7.46	7.72	12.20	11.84		6.13	6.52	8.19	8.69	8.93	11.20	10.93		Matsumara and Hudson 2005
Amami-Okinawa Islanders	Japan	5.36	5.89	6.89	7.12	7.14	11.65	10.87		5.83	6.49	7.79	8.22	8.57	11.12	10.52		Matsumara and Hudson 2005
Native Jōmon	Japan	5.27	5.72	6.73	6.91	6.94	11.61	10.80		5.93	6.20	7.44	7.79	8.33	11.23	10.47		Matsumura 2001
Immigrant Yayoi	Japan	5.43	6.21	7.25	7.37	7.47	11.82	11.38		6.01	6.45	8.12	8.33	8.75	11.30	10.76		Matsumura 2001

TABLE 29.2 Summary list of dental traits in the Sundadont and Sinodont patterns, and traits observed in Shuká Káa.

Tooth	Trait	Sundadont	Sinodont	Shuká Káa
Maxillary central incisor	shoveling	0–2	3–6	
Maxillary central incisor	double shoveling	0–1	2–6	
Maxillary anterior premolar	single root	absent	present	
Maxillary first molar	enamel extension	0–1	2–3	*
Maxillary third molar	peg/reduced/congenital absence	absent	present	
Mandibular first molar	deflecting wrinkle	0–1	2–3	too worn
Mandibular first molar	three roots	absent	present	two roots
Mandibular second molar	cusp number	four cusps	more than four cusps	six cusps

* Mandibular M1 = 1; Mandibular M2 = 2

mandibular molars have been shown to have enamel extension more frequently than maxillary teeth (Risnes 1974). Whether the dental pattern was Sinodont or Sundadont (Turner 1990) cannot be determined due to wear, absence of maxillary teeth, and equivocal associations of observed characteristics (Table 29.2).

aDNA

Three attempts to extract DNA from bones (vertebrae and rib) were unsuccessful; however, results were obtained using material from the pulp cavity of a molar (Kemp et al. 2007). Analysis of mtDNA established Shuká Káa's maternal heritage as belonging to mitochondrial sub-haplogroup D4h3a, a minor founding Native American lineage that occurs primarily, but not exclusively (e.g., the precontact Hopewell sample C35-27 from the Pete Klunk Mound Group, Illinois (Bolnick and Smith 2007)) in extant and prehistoric populations of western North and South America (Kemp et al. 2007; Perego et al. 2009). This geographic pattern has been interpreted as the signature of an ancient migration of humans who entered the Americas along the coast (Erlandson et al. 2007; Perego et al. 2009). To date, the sister Asian branch to D4h3a, called D4h3b, has been observed in only one Han Chinese individual from Qingdao, Shangdon (Kemp et al 2007; Perego et al. 2009; Yao et al. 2002). Molecular sex determination confirmed Shuká Káa as male, and his Y-chromosome lineage was determined to belong to Q-M3* (Kemp et al. 2007), a Y-chromosome type that originated in Beringia and the most widespread in contemporary Native Americans (Kemp and Schurr 2010).

DISCUSSION

At a time when the analysis of ancient human remains is often contentious between researchers and Native Americans, the OYKC project helped to forge positive working partnerships between Alaska Native tribes, the United States Forest Service (USFS), and scientists. This cooperative relationship has been documented in the educational video *Kuwóot yas.éin: His Spirit is Looking Out From the Cave* (Worl et al. 2005). In 2007 Shuká Káa's remains were returned to the Native people of southeast Alaska who buried them in 2008 with a ceremony held in the village of Klawock (Figure 29.1) by tribes who invited scientists and USFS personnel to participate. Among the guests was the Secretary of the U.S. Department of Agriculture.

The human remains represent a single male in his early twenties. Dental enamel hypoplasia indicates that Shuká Káa survived at least two periods of developmental stress likely resulting from times of food scarcity during childhood. Carbon and nitrogen isotope analyses provide strong evidence that Shuká Káa subsisted on a diet of marine mammals, fish, and possibly shellfish. Acquiring lithic source material for the manufacture of the stone tools associated with Shuká Káa would have necessitated using watercraft to cross bodies of water between islands and the mainland. These independent sets of data demonstrate maritime subsistence and lithic procurement requiring the use of watercraft.

No other human remains of comparable age have been analyzed from this part of the world. Shuká Káa's remains provide a point of reference from which to compare archaeological evidence of similar age found elsewhere in North America and Northeast Asia. Shuká Káa's antiquity, location, and the geographic distribution of his mitochondrial type (D4h3a) suggest that he may be a descendant of a population of maritime hunter-gatherers that colonized the west coast of the Americas during or near the end of the late Pleistocene.

Presuming Shuká Káa was a local or regional resident, his remains and the artifacts dating to the same time period provide important insights into early lifeways along the Northwest Coast. This evidence indicates that people living in the Alexander Archipelago were coastal navigators with well-developed trade networks or procurement strategies for obtaining lithic material from sources that required the use of watercraft. Shuká Káa's isotope values suggest that early residents of the Northwest Coast were adapted to making their living primarily from the sea, but also used inland sites to harvest terrestrial resources.

Collectively, these lines of evidence document a well-developed maritime adaptation on the Northwest Coast of North America during the early Holocene. This adaptive pattern must have been established earlier than 10,200 years ago, in order for people to have discovered and exploited distant lithic sources and hibernacula such as OYKC. This evidence strengthens the Northwest Coast migration hypothesis. This hypothesis proposes that humans first entered the Americas (probably between 16,000 and 12,000 years ago) using watercraft along the Northwest Coast of North America to colonize refugia and deglaciated areas of the continental shelf exposed by lower sea levels.

NOTES

1. The calibration program uses the term "pooled mean radiocarbon age" to identify the results of a calculation which averages two or more radiocarbon readings (and their associated sigmas) that are relatively close in time.

REFERENCES

Baichtal, J. F., and Risa Carlson. 2010. "Latest developments in the Paleoshoreline: Predictive model for southwestern, southeast Alaska." Paper presented at the 37th annual meeting of the Alaskan Anthropology Association, Anchorage, AK.

Boghossian, Nevine D., and George E. Gehrels. 2000. Nd isotopic signature of metasedimentary pendants in the Coast Mountains between Prince Rupert and Bella Coola, British Columbia. In *Tectonics of the Coast Mountains, Southeastern Alaska and British Columbia*, edited by Harold H. Stowell and William C. McClelland, 77–87. Geological Society of America, Boulder, CO.

Bolnick, Deborah A, and David Glenn Smith. 2007. Migration and social structure among the Hopewell: Evidence from ancient DNA. *American Antiquity* 72(4): 627–644.

Buzon, M. R., C. A. Conlee, and G. J. Bowen. 2011. Refining oxygen isotope analysis in the Nasca region of Peru: An investigation of water sources and archaeological samples. *International Journal of Osteoarchaeology* 21(4):446–455.

Byers, David A., David R. Yesner, Jack M. Broughton, and Joan Brenner Coltrain. 2011. Stable isotope chemistry, population histories and late prehistoric subsistence change in the Aleutian Islands. *Journal of Archaeological Science* 38(1):183–196.

Capo, R. C., and D. J. Depaolo. 1990. Seawater strontium isotopic variations from 2.5 million years ago to the present. *Science* 249(4964):51–55.

Churkin, Michael, and G. Donald Eberlein. 1977. Ancient borderland terranes of the North American Cordillera: Correlation and microplate tectonics. *Geological Society of America Bulletin* 88(6):769–786.

Coltrain, Joan Brenner, M. Geoffrey Hayes, and Dennis H. O'Rourke. 2006. Hrdlička's Aleutian population-replacement hypothesis: A radiometric evaluation. *Current Anthropology* 47(3):537–548.

Dixon, E. James. 2008. Bifaces from On Your Knees Cave, Southeast Alaska. In *Projectile Point Sequences in Northwestern North America*, edited by Roy L. Carlson and Martin P. R. Magne, 11–18. Archaeology Press, Simon Fraser University, Burnaby, BC.

Dixon, E. James, Timothy H. Heaton, Terrence E. Fifield, Thomas D. Hamilton, David E. Putnam, and Frederick Grady. 1997. Late quaternary regional geoarchaeology of southeast Alaska karst: A progress report. *Geoarchaeology* 12(6):689–712.

Doyle, William J., and Olivia Johnston. 1977. On the meaning of increased fluctuating dental asymmetry: A cross populational study. *American Journal of Physical Anthropology* 46(1):127–134.

Erlandson, Jon M., Michael H. Graham, Bruce J. Bourque, Debra Corbett, James A. Estes, and Robert S. Steneck. 2007. The Kelp Highway Hypothesis: Marine ecology, the coastal migration theory, and the peopling of the Americas. *Journal of Island and Coastal Archaeology* 2(2):161–174.

Fedje, Daryl W., and Heiner Josenhans. 2000. Drowned forests and archaeology on the continental shelf of British Columbia, Canada. *Geology* 28(2):99–102.

Fricke, Henry C., James R. O'Neil, and Niels Lynnerup. 1995. Special Report: Oxygen isotope composition of human tooth enamel from medieval Greenland: Linking climate and society. *Geology* 23:869–872.

Gehrels, George E., and Jason B. Saleeby. 1987. Geologic framework, tectonic evolution, and displacement history of the Alexander Terrane. *Tectonics* 6(2):151–173.

Goaz, Paul W., and M. Clinton Miller. 1966. A preliminary description of the dental morphology of the Peruvian Indian. *Journal of Dental Research* 45(1):106–119.

Harris, Edward F., and Martin T. Nweeia. 1980. Tooth size of Ticuna Indians, Colombia, with phenetic comparisons to other Amerindian. *American Journal of Physical Anthropology* 53(1): 81–91.

Heaton, Timothy H., and Frederick Grady. 2003. The late Wisconsin vertebrate history of Prince of Wales Island, Southeast Alaska. In *Ice Age Cave Faunas of North America*, edited by Blaine W. Schubert, Jim I. Mead, and Russell W. Graham, 17–53. Indiana University Press, Bloomington.

Hedges, R. E. M. 2003. On bone collagen—apatite-carbonate isotopic relationships. *International Journal of Osteoarchaeology* 13(1–2):66–79.

Hoppe, Kathryn A. 2006. Correlation between the oxygen isotope ratio of North American bison teeth and local waters: Implication for paleoclimatic reconstructions. *Earth and Planetary Science Letters* 244(1–2):408–417.

IAEA. 2001. GNIP Maps and Animations. International Atomic Energy Agency, http://www-naweb.iaea.org/napc/ih/documents/userupdate/Waterloo/index.html.

Iacumin, P., H. Bocherens, A. Mariotti, and A. Longinelli. 1996. Oxygen isotope analyses of co existing carbonate and phosphate in biogenic apatite: A way to monitor diagenetic alteration of bone phosphate? *Earth and Planetary Science Letters* 142(1–2):1–6.

Keenleyside, Anne, Henry P. Schwarcz, Kristina Panayotova. 2011. Oxygen isotopic evidence of residence and migration in a Greek colonial population on the Black Sea. *Journal of Archaeological Science* 38(10):2658–2666.

Kemp, Brian M., Ripan S. Malhi, John McDonough, Deborah A. Bolnick, Jason A. Eshleman, Olga Rickards, Cristina Martinez-Labarga, John R. Johnson, Joseph G. Lorenz, E. James Dixon, Terence E. Fifield, Timothy H. Heaton, Rosita Worl, and David Glenn Smith. 2007. Genetic analysis of early Holocene skeletal remains from Alaska and its implications for the settlement of the Americas. *American Journal of Physical Anthropology* 132:605–621.

Kemp, Brian M., and Theodore G. Schurr. 2010. Ancient and modern genetic variation in the Americas. In *Human Variation in the Americas: The Integration of Archaeology and Biological Anthropology*, edited by Benjamin M. Auerbach, 12–50. Southern Illinois University, Carbondale.

Kieser, J. A., H. T. Groeneveld, and C. B. Preston. 1985. An odontometric analysis of the Lengua Indian dentition. *Human Biology* 57(4):611–620.

Koch, Paul L. 1998. Isotopic paleoecology and land vertebrates: Individuals, species, and ecosystems. PALAIOS 13(4):309–310.

Lee, Craig M. 2001. "Microblade Morphology and Trace Element Analysis: An Examination of Obsidian Artifacts from Archaeological Site 49-PET-408, Prince of Wales Island, Alaska." Master's thesis, University of Wyoming, Laramie.

———. 2007. "Origin and Function of Early Holocene Microblade Technology in Southeast Alaska, United States of America." Ph.D. dissertation, University of Colorado, Boulder.

Lukacs, John R., and Brian E. Hemphill. 1991. The dental anthropology of prehistoric Baluchistan: A morphometric approach to the peopling of South Asia. In *Advances in Dental Anthropology*, edited by Marc A. Kelley and Clark Spencer Larsen, 77–119. Wiley-Liss, New York.

Matsumura, H. 2001. Differentials of Yayoi immigration to Japan as derived from dental metrics. *HOMO– Journal of Comparative Human Biology* 52(2):135–156.

Matsumura, Hirofumi, and Mark J. Hudson. 2005. Dental perspectives on the population history of southeast Asia. *American Journal of Physical Anthropology* 127(2):182–209.

McNeely, R., A. S. Dyke, and J. R. Southon. 2006. Canadian marine reservoir ages, preliminary data assessment. Geological Survey of Canada, Open File 5049. http://geogratis.gc.ca/api/en/nrcan-rncan/ess-sst/81215fc0-7db0-5889-b311-26470c9f9193.html.

Moorrees, C. F. A. 1957. *The Aleut Dentition: A Correlative Study of Dental Characteristics in an Eskimoid People.* Harvard University Press, Cambridge, MA.

Mrzlack, Heather. 2003. "Ochre at 49-PET-408." Ph.D. dissertation. University of Colorado, Denver.

Passey, Benjamin H., Todd F. Robinson, Linda K. Ayliffe, Thure E. Cerling, Matt Sponheimer, M. Denise Dearing, Beverly L. Roeder, and James R. Ehleringer. 2005. Carbon isotope fractionation between diet, breath CO_2, and bioapatite in different mammals. *Journal of Archaeological Science* 32(10):1459–1470.

Perego, Ugo A., Alessandro Achilli, Norman Angerhofer, Matteo Accetturo, Maria Pala, Anna Olivieri, Baharak Hooshiar Kashani, Kathleen H. Ritchie, Rosaria Scozzari, Qing-Peng Kong, Natalie M. Myers, Antonio Salas, Ornella Semino, Hans-Jürgen Bandelt, Scott R. Woodward, and Antonio Torroni. 2009. Distinctive Paleo-Indian migration routes from Beringia marked by two rare mtDNA haplogroups. *Current Biology* 19(1):1–8.

Risnes, Steinar. 1974. The prevalence and distribution of cervical enamel projections reaching into the bifurcation on human molars. *European Journal of Oral Sciences* 82(6):413–419.

Samson, Scott D., William C. McClelland, P. Jonathan Patchett, George E. Gehrels, and Robert G. Anderson. 1989. Evidence from neodymium isotopes for mantle contributions to Phanerozoic crustal genesis in the Canadian Cordillera. *Nature* 337:705–709.

Stafford, Thomas W., Jr., Klaus Brendel, and Raymond C. Duhamel. 1998. Radiocarbon, ^{13}C and ^{15}N analysis of fossil bone: Removal of humates with XAD-2 resin. *Geochimica et Cosmochimica Acta* 52(9):2257–2267.

Turner, Christy G., II. 1990. "Major features of Sundadonty and Sinodonty, including suggestions about East Asian microevolution, population history, and late Pleistocene relationships with Australian Aboriginals. *American Journal of Physical Anthropology* 82(3):295–317.

Worl, Rosita, E. James Dixon, Terrence Fifield, Chuck Smythe, and Ted Timrick. 2005. Kuwóot Yas.éin: His Spirit Was Looking Out from the Cave. Sealaska Heritage Institute, Juneau. Digital Video Disc (DVD), 30 min.

Yao, Yong-Gang, Qing-Peng Kong, Hans-Jürgen Bandelt, Toomas Kivisild, and Ya-Ping Zhang. 2002. Phylogeographic differentiation of mitochondrial DNA in Han Chinese. *American Journal of Human Genetics* 70(3):635–651.

A New Look at the Double Burial from Horn Shelter No. 2

30

Margaret A. Jodry and Douglas W. Owsley

Nearly 11,100 years ago, a forty-year-old man and a ten- to-eleven-year-old girl were buried in a single, primary interment in a rock shelter located on the west bank of the Brazos River in Bosque County, Texas, 30 kilometers upstream from Waco (Figure 30.1). This chapter explores the forms and possible functions of a remarkable assemblage of items recovered primarily with the adult. Due to the near absence of durable items associated with her, the girl is addressed only briefly. As an opportunity to learn more about the lives of Paleoamericans, we explore the following questions: What do these items suggest about the recurring activities of this man? What insights can be gleaned from features of his anatomy that also bear on this subject?

INTRODUCTION AND SITE SETTING

Horn Shelter No. 2 (41BQ46) is a deeply stratified rock shelter situated at the juncture of the Grand Prairie and the Eastern Cross Timbers, a grassland-woodland ecotone lying 146 meters above mean sea level. The shelter, overlooking a prominent bend on the river, faces east and the rising sun (Figure 30.2). The overhang, which formed as an erosional recess in Cretaceous limestone, runs nearly 46 meters along the river with a maximum depth of about nine meters (Redder 1985). The shelter is naturally divided into two sections by a rock ledge (Figure 30.3). The stratigraphy in the north (Forrester 1985) and south (Redder 1985) ends of the shelter vary, indicating somewhat different depositional and cultural histories. The double burial was found in the larger south end, which is approximately 13.7 meters long, with an interior space extending about nine meters deep (Redder and Dial 2010) (Figure 30.4).

Knowledgeable avocational archaeologist Al Redder began excavations in the south end in 1966 and continued for 23 years. Here, almost five meters of natural and cultural deposits preserved a record of human occupation spanning some 13,200 years, from at least Clovis times through the early twentieth century (Redder 1985).

On August 15, 1970, beneath approximately two meters of overburden, a human cranium was partially unearthed. Further excavation in late October revealed the cranium of a second individual as well as a layer of nineteen large limestone slabs placed over the two bodies. This stone cap extended from the area of their necks to their feet, but did not cover their heads (Redder 1985). Although the top of the burial pit was not detected during excavation, examination of the burial fill provides secure evidence that the grave was dug from a surface within stratigraphic substratum 5G. This deposit formed from the decaying limestone of the ceiling and walls (Prochnow 2001) intermixed with the remains of intensive human habitation. This substratum contains dark, charcoal-stained sediment, hearths, abundant stone and

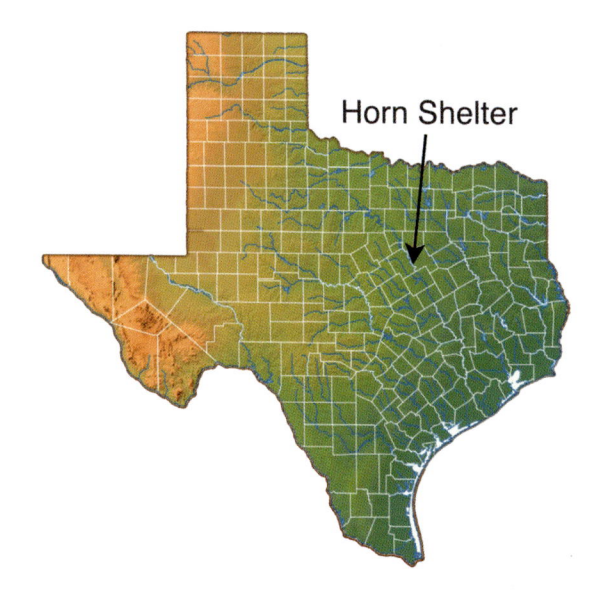

FIGURE 30.1 Location of Horn Shelter No. 2 (41BQ46), Bosque County, Texas.

FIGURE 30.2 A) View west toward the shelter site, B) Aerial view looking south/southwest, arrow points to the location of Horn Shelter No. 2.

bone tools, extensive faunal remains, and some preservation of botanical remains, in addition to the double burial itself (Redder 1985; Redder and Fox 1988; Redder and Dial 2010; Story 1990).

San Patrice projectile points recovered in 5G provide a cultural affiliation for the burial. Radiocarbon ages for this substratum range from 9980 to 9500 yr. BP (11,540 to 10,360 CAL yr. BP) (Bousman et al. 2004; Bousman and Oksanen 2012; Redder 1985). Ages measured directly from bone from the male and female in the double burial are 9710 ± 40 RC yr. BP (CAMS-60681) (11,158 ± 39 CAL yr. BP) and 9690 ± 50 RC yr. BP (CAMS-51794) (11,049 ± 134 CAL yr. BP), respectively (Owsley et al. 2010; Cal-Pal 2007). These latter dates provide some of the most reliable age determinations for the San Patrice period currently available (Owsley et al. 2010; Bousman and Oksanen 2012).

The cultural deposits in 5G represent a significant in-

FIGURE 30.3 Redder stands below the rock ledge during the 1988 excavation of the south end of Horn Shelter No. 2. View to south.

FIGURE 30.4 Redder (left) and Owsley standing in the south end of Horn Shelter No. 2. View to north.

FIGURE 30.5 Placement of individuals in the double burial; the girl is on the left side.

crease in the number and variety of artifacts and activities when compared to those recovered in deeper Clovis and Folsom strata (Redder 1985; Redder and Dial 2010; Story 1990). Redder reported that San Patrice (formerly referred to as Brazos Fishtail) groups lived in the shelter for extended periods, perhaps even year round (Redder and Fox 1988). This perceived difference in the use of the shelter during San Patrice times may coincide with the end, and certainly with the period just after, the Preboreal Oscillation, a 330-year cooling event spanning the interval from 11,550 to 11,220 CAL yr. BP (Bousman and Vierra 2012; Fisher et al. 2002; Rasmussen et al. 2007). San Patrice projectile points also occur in stratigraphic deposit 5F that underlies 5G sediments and the double burial. The cultural materials from 5F and 5G have not yet been fully cataloged, or analyzed in detail (Story 1990). Future work, including dating of substratum 5F, may clarify the temporal relationship between the San Patrice interval and the Preboreal Oscillation.

BURIAL DESCRIPTION

The adult and juvenile were buried facing west, the direction of the setting sun, and the back wall of the shelter. They were placed in semi-flexed positions on their left sides, with their heads to the south and feet to the north. The girl rested directly behind the man, facing his back. The arms of both individuals were flexed, with their hands near their faces (Redder and Fox 1988; Young 1988; Young et al. 1987) (Figure 30.5). The social and genetic relationships between the male and female are currently unknown. Neither individual shows evidence of trauma that could have caused death. What is clear is that the two individuals were buried at the same time with deliberate care.

The adult's head rested on ten closely grouped items. Covering his face was a turtle carapace, only a small portion of which was recovered intact (Redder 1985). Near the shell and his face were non-perforated badger claws and Swainson's Hawk talons. In the area of his head, neck, and chest were shell beads and four perforated coyote canine teeth (Figure 30.6). Beneath his pelvis was a relatively well-preserved turtle carapace that cradled an antler tool. In the space between the man's flexed arms and legs was a well-used deer antler implement. This tool extended out from his hip, with the perforated end of the tool lying near his lumbar vertebrae, perhaps suggesting

FIGURE 30.6   Shell beads and two of the four perforated coyote teeth. Most of these items were recovered near the man's head, neck, and torso.

it was attached to, or held within, something at his waist. Some of the shell beads may have been associated with the girl. The only other belonging thought to have been associated with her was an eyed bone needle, the tip of which was broken before burial. Wood, leather, herbs, textiles, thorns, fur, and/or feathers may have been present, but were not preserved.

Based on detailed study of the belongings buried with the man, augmented by ethnographic research and archaeological comparisons, we consider the idea that the adult buried in Horn Shelter No. 2 was a healer, perhaps a medicine man or shaman. Analysis of his anatomy, particularly the forearms and hands, provides intriguing information that may be in keeping with this idea.

Several features of this burial suggest that the man was distinguished within his social group: 1) the unusual location and form of the burial; 2) the ten items beneath his head, which may be a personal bundle; 3) the many articles worn around his neck and placed near his face and chest, which appear to be personal medicine and perhaps include clan references; and 4) unusually well-developed muscle attachments on his forearms, and robust development of his hands.

GRAVE LOCATION AND FORM

The burial at Horn Shelter No. 2 is unique in many aspects. To provide a comparative perspective, Table 30.1 identifies a sample of sixteen individuals from fifteen early Holocene rock shelter burials.

Of the fifteen shelter burials, the young female and adult male from Horn Shelter No. 2 are the only example of a double burial. The practice of single interments predominates among early Holocene burials in the Midcontinent, both in rock shelters and in open-air sites (Walthall 1999). Only one other early Holocene double burial is known; it is from a multicomponent, open-air habitation context at the Lawrence site in western Kentucky (Mocas 1977). Here, eleven early Holocene flexed burials were found on the edge of a sinkhole, in the up-

TABLE 30.1 Comparison of late Pleistocene/early Holocene rock shelter burials.

Site	State	Age and Sex	Burial Position	Burial Location	Stones Placed Over Burial	Belongings	RC yr. BP	Reference
Horn Shelter No. 2	Texas	Adult (40–45 years) Male	Flexed	5 meters distant from back wall, central location	19 large limestone rocks—do not cover heads	Abundant, including healing bundle	9710 ± 40	Redder 1985; Owsley et al. 2010; this volume
		Juvenile (10–11 years) Female				Eyed needle	9690 ± 50	
Spirit Cave, burial 1	Nevada	Adult Female?	Flexed	Against back wall	Single layer of rocks—covered entire burial	Textile mat covering	9270 ± 60	Wheeler and Wheeler 1969; Wheeler 1997; Owsley and Jantz 1999
Spirit Cave, burial 2	Nevada	Adult (45 years) Male	Flexed	Against back wall	Double layer of rocks—evidently covered entire burial	1 twined and 2 plaited textile coverings, fur blanket, leather moccasins, breechcloth	9454 ± 25	Wheeler and Wheeler 1969; Wheeler 1997; Owsley and Jantz 1999
Spirit Cave, cremation	Nevada	Adult (18–22 years) Female	N/A	Against back wall	No	2 twined and 1 plaited textile bags	9040 ± 50	Wheeler 1997; Tuohy and Dansie 1997; Owsley and Jantz 1999
Modoc Rock Shelter, burial 27	Illinois	Adult Female	Fully flexed	Near back wall	No	None	ca. 8000	Neumann 1967; Anderson 1991
Modoc Rock Shelter, burial 28	Illinois	Adult Female	Fully flexed	Near back wall	No	None	ca. 8000	Neumann 1967; Anderson 1991
Modoc Rock Shelter, burial 29	Illinois	Adult Male	Fully flexed	Near back wall	No	None	ca. 8000	Neumann 1967; Anderson 1991
Modoc Rock Shelter	Illinois	Infant	N/A	Near back wall	No	None	ca. 8000	Neumann 1967; Anderson 1991
Dust Cave, burial 5	Alabama	Adult Female	Tightly flexed	Near back wall	Small slabs of limestone	None	ca. 8000	Hogue 1994
Dust Cave, burial 15	Alabama	Child (5 years)	Tightly flexed	Near back wall	Small slabs of limestone	None	ca. 8000	Hogue 1994
Ashworth Shelter, burial 4	Kentucky	Adult Female	Flexed	Near back wall	Large limestone slabs	None (Kirk points within the body)	Early Holocene	DiBlasi 1981
Graham Cave	Missouri	Adult	Tightly flexed	Near eastern wall, one side of cave mouth	Limestone rock placed over crania area	Perforated canid canine	Early Holocene	Logan 1952
Arnold Cave	Missouri	Adult	Flexed	Near back wall	No	2 eyed needles, 3 chert bifaces	Early Holocene	Shippee 1966
Red Rock Shelter, burial 1	Arkansas	Adult	Tightly flexed	Near back wall	No	None	Early Holocene (Dalton)	McGimsey 1963; Walthall 1999
Red Rock Shelter, burial 2	Arkansas	Sub adult	Flexed	Near back wall	No	None	Early Holocene (Dalton)	McGimsey 1963; Walthall 1999

lands of the Cumberland River. Most of these individuals appear to have been adults. Mortuary items were encountered in only two of these burials. A grooved sandstone abrader and two stone bifaces accompanied an adult in Feature 73, while the grave containing the greatest number and diversity of items is a double burial found 1.5 meters away (Feature 72). This interment held two adult males, lying in flexed positions (Mocas 1985; Walthall 1999).

The double burial from Horn Shelter No. 2 has a central location within the utilized space of the shelter. The grave was dug nearly five and a half meters from the back wall, about midway between the shelter mouth and the rear wall (Figure 30.7). Of the fifteen shelter burials, it is the only one located a significant distance from the rear (or side) walls of the shelter. The dominant pattern, regardless of cultural affiliation, was to place early Holocene graves along the periphery of the shelter, either near or against the rear (86.6%) or side (6.7%) wall. Several of these burials were placed in shelters in which people did not live, yet the graves were still located near the rear wall (Table 30.1).

Ethnoarchaeological studies of rock shelter use among hunter-gatherers indicate that spaces near or against the rear wall most commonly served as sleeping areas during residential occupations (Gorecki 1991; Parkington and Mills 1991; Walthall 1998). Walthall (1999:19) suggests "the location of graves in such areas may have been symbolic, with the dead being placed in a sleeping posture, flexed and lying on the side, in areas normally reserved for such activity." Fourteen of the sixteen individuals were buried in a flexed position (Table 30.1). The burial position of an infant from Modoc Rock Shelter could not be determined (Anderson 1991), and this aspect is not applicable to the cremated individual from Spirit Cave (Owsley and Jantz 1999; Wheeler 1997).

Durable grave offerings accompanied three of fifteen burials (Table 30.1). Of the four people interred in these three graves, the man at Horn Shelter No. 2 stands out as the most richly-bestowed individual. The clothing and textile wrappings with the individuals at Spirit Cave are not considered durable grave offerings. Presumably, other persons were also dressed and possibly wrapped in fiber mats or animal skin coverings; the exceptional conditions of preservation in the dry Spirit Cave shelter are not comparable with other sites in the sample.

At least seven of the fifteen shelter burials were capped with stones (Table 30.1). While the presence of stones over early Holocene rock shelter graves is not unusual based on this sample, there is variation in their placement. The particular form of the capping over the Horn Shelter No. 2 double burial may have symbolic significance. Nineteen large slabs of limestone, from the eroded ceiling and wall were laid in a single layer so that it likely appeared as a slightly raised earthen mound with a visible, durable shell. This covering did not extend over their heads, but began at their shoulders and extended toward their feet (Figure 30.8). This arrangement, with heads protruding beyond a hard, domed exterior, suggests that the form of the grave may have been a turtle effigy. Direct association with five turtle carapaces, including one placed over the man's face implies a strong connection with turtles (Owsley et al. 2010).

The Horn Shelter double burial stands out relative to

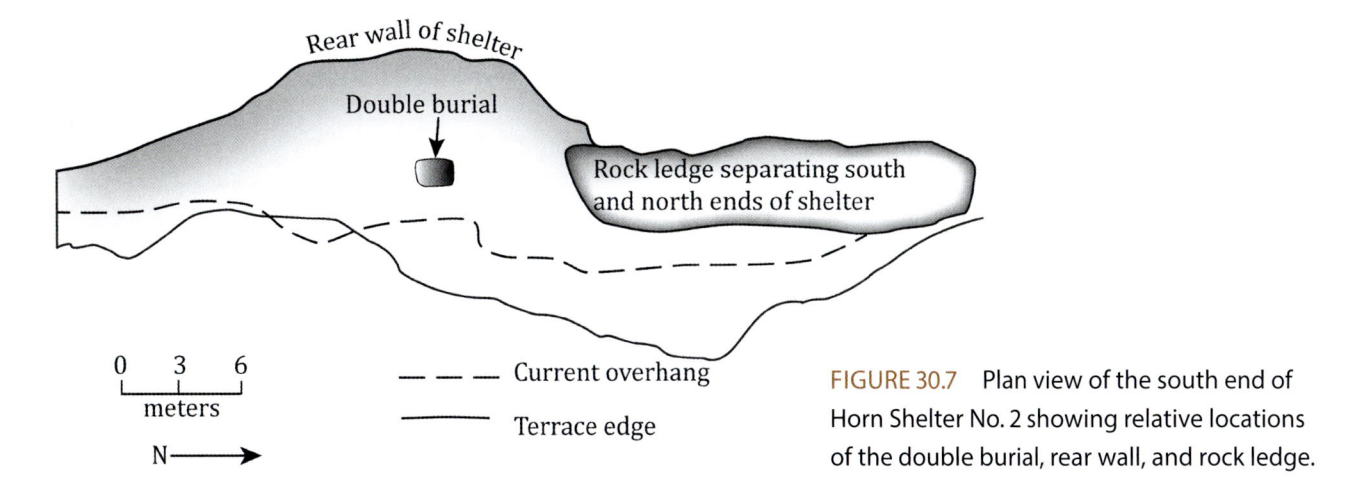

FIGURE 30.7 Plan view of the south end of Horn Shelter No. 2 showing relative locations of the double burial, rear wall, and rock ledge.

FIGURE 30.8 Limestone rocks found covering the pair from their shoulders to their feet. No rocks covered their heads.

FIGURE 30.9 In situ arrangement of items found beneath the man's head: a) 729-5G; b) 726-5G; c) 722-5G; d) 723-5G; e) 725-5G; f) 727-5G; g) 724-5G; h) 728-5G.

other early Holocene rock shelter burials with regard to the presence of a young female, unique placement near the center of the shelter, the turtle-like form of the grave covering, and the unusual richness of mortuary items buried with the male.

ITEMS PLACED BENEATH THE MAN'S HEAD

Many belongings are associated with the male, including a tightly clustered group of items on which his head rested (Table 30.2). These were likely originally held together in a perishable container (Figure 30.9). This grouping was removed in a matrix block that was later excavated and mapped by Redder under controlled circumstances. Ten objects are present: a slender deer bone tool (722-5G), an Edwards chert biface (723-5G), two tabular pieces of sandstone (724-5G, 725-5G), an ochre nodule (726-5G), two deer antler tools (727-5G, 728-5G), and three nested turtle carapaces (729-5G) (Figure 30.10).

For nearly thirty years, some of these items have been reported as flint knapping tools (Bousman and Oksanen 2012; Owsley et al. 2010; Redder and Dial 2010; Redder and Fox 1988; Redder 1985; Story 1990). The presence together of what appeared to be antler billets, abraders, and a stone biface led to this impression. In 2012, these items became available for close examination at the Smithsonian Institution's National Museum of Natural History and a revised understanding of this material emerged as a result. The interpretation of these objects as a flint knapper's tool kit was problematic, particularly because of the absence of stone projectile points, preforms, flake blanks, and a hammer stone in the bundle or elsewhere in the grave. While their absence is not definitive, flint knappers would be expected to have at least some of these items in their tool kit. In addition, the antler "billets" retained burrs at their bases and neither these burrs, nor the bases themselves, show evidence of percussive tool use (Maigrot 2005:Figure 8:3). In fact, to produce a functional soft hammer or billet, prehistoric flint knappers removed these burrs, a practice also adopted by modern

TABLE 30.2 Descriptive attributes and measurements of funerary items.

Catalog Number	Item	Material	Genus, Species, or Stone type	Placement	Use-wear	Maximum Length (mm)	Maximum Width (mm)	Maximum Thickness (mm)	Weight (gm[1])
716-5G	Multipurpose Tool	Antler	*Odocoileus virginianus* white-tailed deer	Near man's hip	Present	182.5	33.9	17.7	44.1
717-5G	Needle	Bone		Girl's abdominal area	Present	43			
718-5G	Perforated ornament	Tooth	*Canis latrans* coyote	Near man's neck	Present	38.2	9.3	6.1	1.5
719-5G	Perforated ornament	Tooth	*Canis latrans*	Near man's neck	Present	38.2	9.1	6.0	1.5
720-5G	Perforated ornament	Tooth	*Canis latrans*	Near man's neck					
721-5G	Perforated ornament	Tooth	*Canis latrans*	Near man's neck (Adheres to posterior cranium)	Present				
722-5G	Probable stylus	Bone	*Odocoileus virginianus*	Within bundle	Present	162.4	12.6	6.1	6.5
723-5G	Biface	Stone	Edwards chert	Within bundle	Edge wear absent	66.1	52.8	18.8	63.1
724-5G	Abrader	Stone	Sandstone	Within bundle	Present	75.7	65.0	12.4	93.7
725-5G	Abrader	Stone	Sandstone	Within bundle	Present	79.1	56.5	15.7	97.2
726-5G	Raw Material (pigment)	Hematite	Iron oxide concretion (Fe_2O_3)	Within bundle	Present	53.9	54.5	50.9	141.0
727-5G	*Odocoileus virginianus*	Antler	*Odocoileus virginianus*	Within bundle	Present	116.6	40.5	30.8	90.0
728-5G	Pestle	Antler	*Odocoileus virginianus*	Within bundle	Present	110.0	29.4	27.9	54.1
729-5G	Three carapaces	Bone	*Terrapene ornata* ornate box turtle	Within bundle	Present	147.7	128.7	ca 60	249.7
734-5G	Probable tool	Antler	*Odocoileus virginianus*	Beneath man's pelvis, inside carapace	Present	ca. 90	ca. 28		
735-5G	Carapace bowl	Bone	*Terrapene ornata*	Beneath man's pelvis	Present				71.2[2]
737-5G	83 Perforated beads	Shell	*Neritina reclivata* olive nerite	Several near man, majority displaced	Present				
738-5G	Perforated ornament	Shell	*Oliva sayana* Lettered olive	Within man's crania, displaced	Present	42.3	18.4	17.25	6.9
None	Claws	Bone	*Taxidea taxus* badger	Near man's cranium	Absent				
None	Talons	Bone	*Buteo swainsoni* Swainson's hawk	Near man's neck and cranium	Absent				
None	Carapace	Bone	*Terrapene ornata*?	Covering man's face					

[1]Some objects are glued and weight includes this aspect. [2]Combined weight of shell and antler.

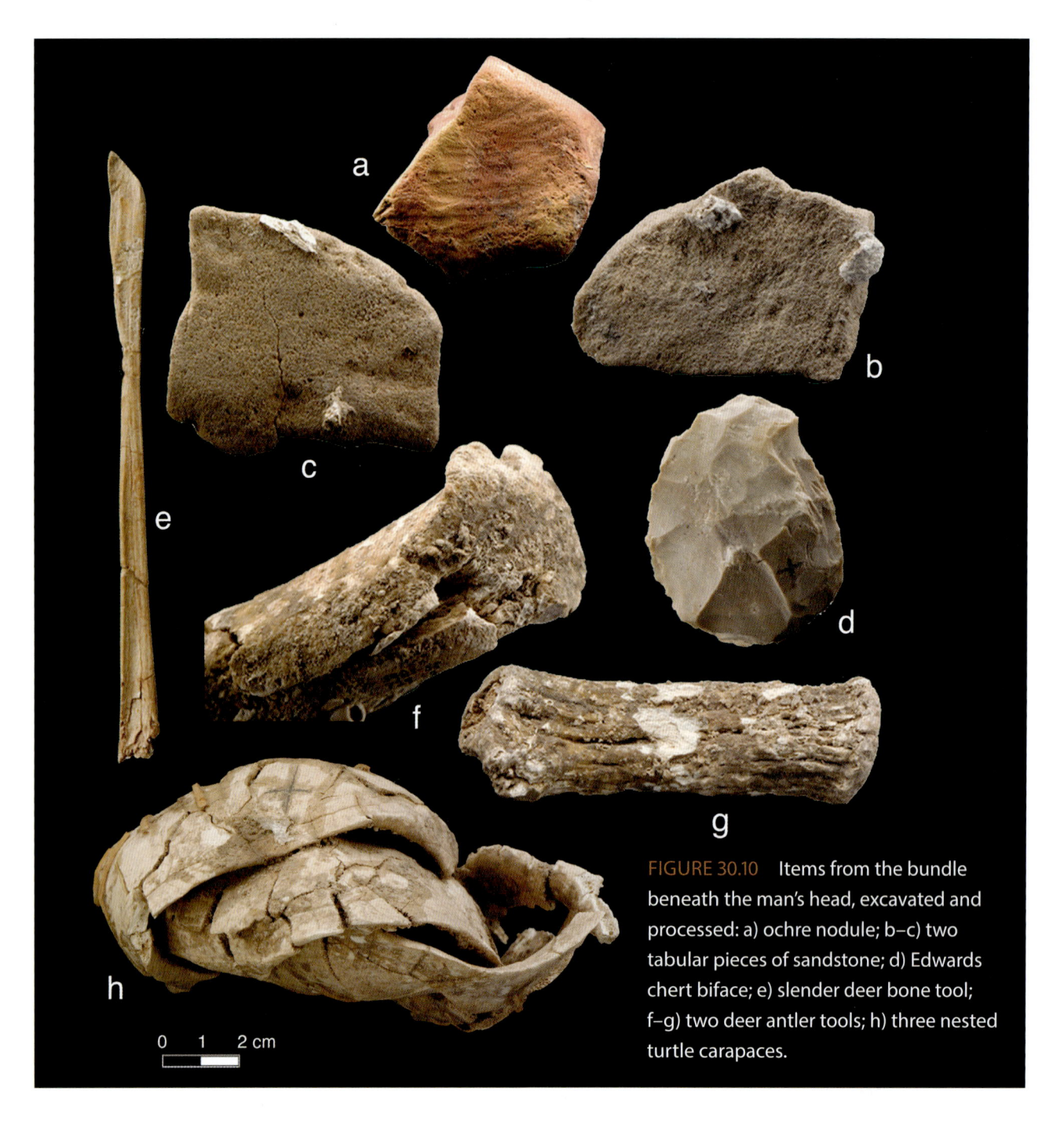

FIGURE 30.10 Items from the bundle beneath the man's head, excavated and processed: a) ochre nodule; b–c) two tabular pieces of sandstone; d) Edwards chert biface; e) slender deer bone tool; f–g) two deer antler tools; h) three nested turtle carapaces.

knappers. Experienced flint knapper and archaeologist Bruce Bradley concurs that in their present form, these particular antlers are unsuitable as billets (2013, personal communication). They would need additional shaping of brow tine remnants and burrs in order to be used as soft hammers. These objects could be raw material for future billets, but this study suggests that they were something else entirely.

The size of the biface (Figure 30.10) was also puzzling. If this stone was intended as a source of raw material for the production of flake blanks for the manufacture of San Patrice projectile points, it is surprisingly small, especially considering that the region has abundant sources of large chert nodules. The larger of two complete San Patrice points (364-5G) recovered in substratum 5G has maximum dimensions of 63.2 millimeters (length) × 33.4 millimeters (width) × .08 millimeters (thickness) (Figure 30.11). The maximum dimensions of the biface (723-5G) are 66.1 millimeters (length) × 52.8 millimeters (width) × 18.8 millimeters (thickness). No flake blanks can be removed from this biface that would be of suitable size for making San Patrice projectile points, and

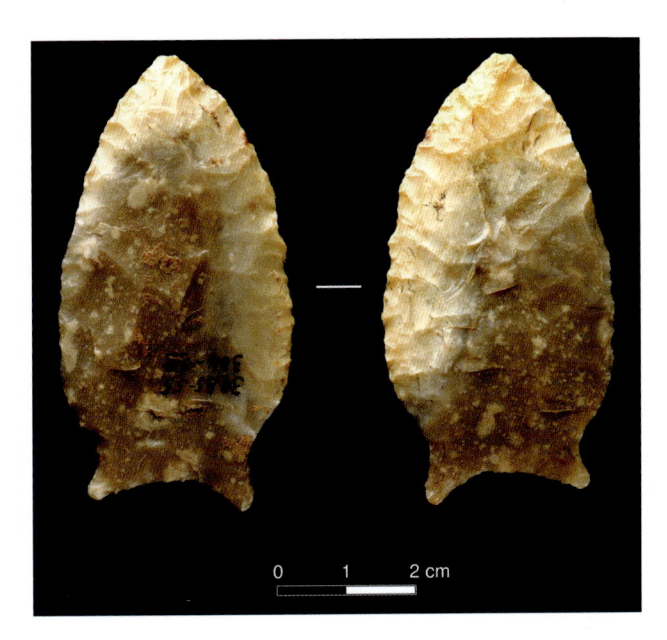

FIGURE 30.11 San Patrice projectile point (364-5G) found in substratum 5G, but not within the burial.

the biface itself is too thick and short to serve as a San Patrice preform.

In addition, recent examination of the biface edges with a scanning electron microscope suggests that it was not used as a cutting or scraping tool. The three largest, and most recent, flake scars indicate removal, by hard hammer percussion, of thin flakes ranging in length from nearly 27 to 42 millimeters, and width from about 18 to 25 millimeters. If the sharp-edged flakes removed from the biface were the intended tools, rather than the biface itself, then what activities might utilize such expedient, sharp cutting implements?

Antler Tools (727-5G, 728-5G)

Four deer antler tools are present. Two were in the bundle beneath the man's head (727-5G, 728-5G), a third was placed in the turtle carapace under his pelvis (735-5G), and a fourth was in an open area between his right radius-ulna and femur (Redder and Fox 1988:5). The first two antlers are described here.

The antlers in the bundle originated from different white-tailed deer (*Odocoileus virginianus*), as evidenced by variances in beam circumference, relative robustness, and development and position of the brow tines. The more gracile of the two (728-5G) is 23.7 millimeters in diameter as measured just above the burrs; the more robust antler (727-5G) measures 32.5 millimeters at this location. They were acquired as shed antlers and show no signs of having been cut from pedicles. A discontinuous veneer of calcite intermixed with tiny angular grains of quartz, chert, and occasionally, microscopically-visible bits of charcoal and hematite obscures much of the base and beam of each antler.

The greatest visibility of original working surface is present on antler 728-5G. Its flat to slightly convex base retains an angle of nearly 45 degrees (Figure 30.12b). Burr margins are distinctly rounded and polished. There is continuous rounding and polish of the burrs along the entire tool circumference and also in a perpendicular plane (Figure 30.12c). Repeated rubbing and friction of the burrs against curved surfaces is indicated.

Figure 30.12e shows the location and orientation of a striation that parallels the circumference of antler 728-5G and crosscuts the polished and rounded burr edges. Evidently, a small, abrasive particle scratched the surface while the tool was being used in a circular motion; likely one of its last uses as the scratch is not obliterated or overprinted (Adams 2013). This mark skips across intervening low areas between individual burrs and exhibits abrupt start and stop points. The rounding, polish, and striations along the sides of this antler indicate use with rocking and rotary motions against a curved surface.

This tool served as a pestle to pulverize and crush material within a concave mortar or bowl. Adams (2002) notes that pestle size is most often related to intended function, and smaller, lighter pestles, such as 728-5G and 727-5G, are expected to have been used to crush, grind, and stir. Pestles need not be of stone and wooden pestles are not uncommon.

Antler 728-5G also features the uncut remnant of a lightly smoothed brow tine bud located 60 millimeters from its base. Damage to this area overprints much of the original cortical surface. Definitive smoothing and high-gloss polish extend nearly 26 millimeters along two ridges of the antler beam creating a slight depression in one of them (Figure 30.13). This wear likely results from handling abrasion during use. Holding this antler in a working position as a pestle, the right thumb rests naturally and comfortably at this location. Comparative experimental information is needed to further assess this observation.

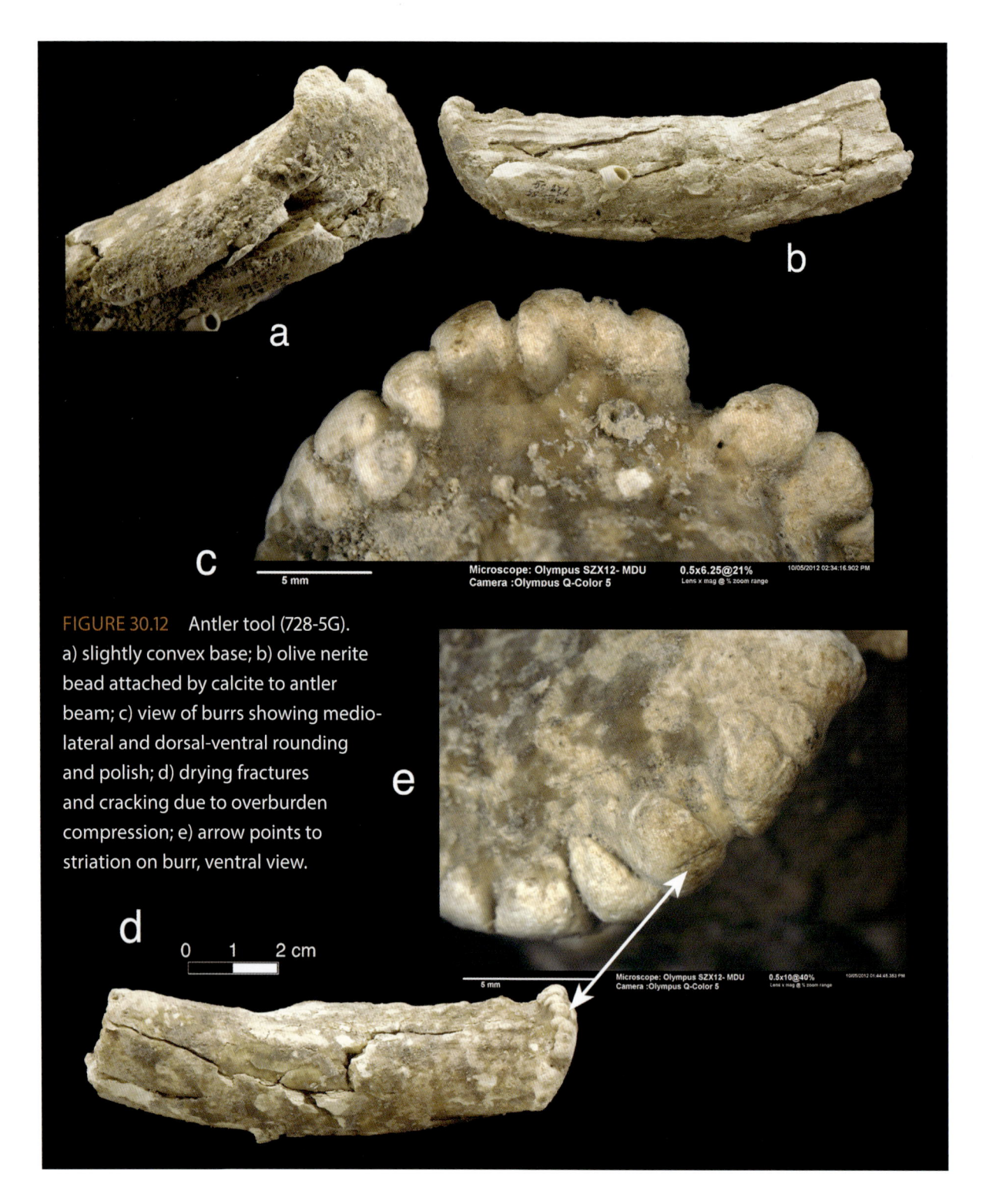

FIGURE 30.12 Antler tool (728-5G). a) slightly convex base; b) olive nerite bead attached by calcite to antler beam; c) view of burrs showing medio-lateral and dorsal-ventral rounding and polish; d) drying fractures and cracking due to overburden compression; e) arrow points to striation on burr, ventral view.

The weight of overlying sediments flattened and distorted the tool's distal end (Figure 30.14). Groups of multiple, parallel striations are present on the better preserved, groundward cortical surface. Groups 1 and 2 are oriented roughly parallel to the long axis of the beam. Groups 3 and 4 are oriented at different acute angles. Striations in all four groups are lightly covered by calcite.

The proximal and distal ends of the second antler, 727-5G, are largely covered by calcite, and have no observed use-wear striations. Where the original burr surfaces are visible, they are unmistakably rounded and polished (Figure 30.15). The burrs, like 728-5G, have a continuously shaped margin around the circumference and are rounded when viewed in a perpendicular plane.

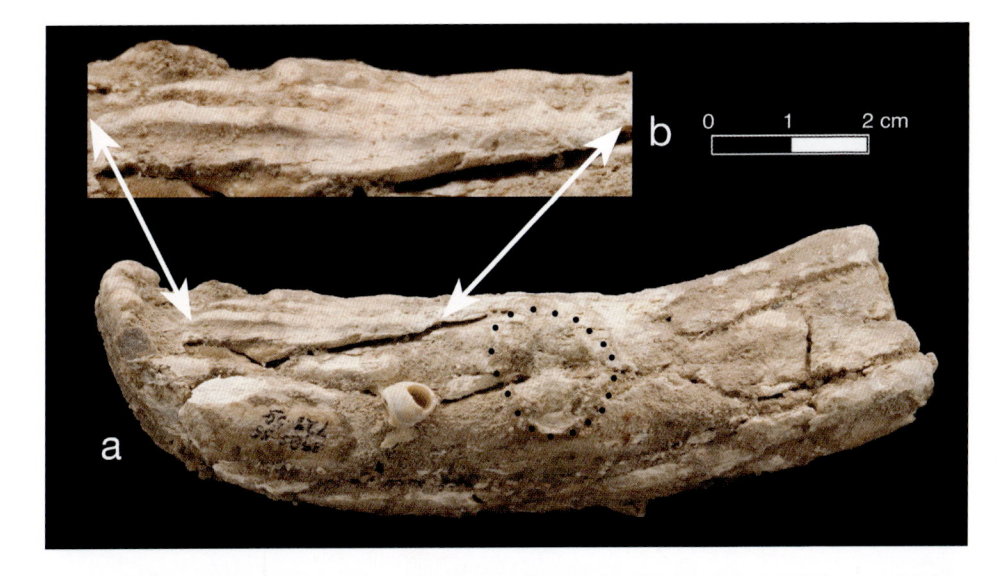

FIGURE 30.13 Antler tool (728-5G). a) dotted line encircles emerging brow tine; b) smoothing and polishing on ridges due to handling during pestle use.

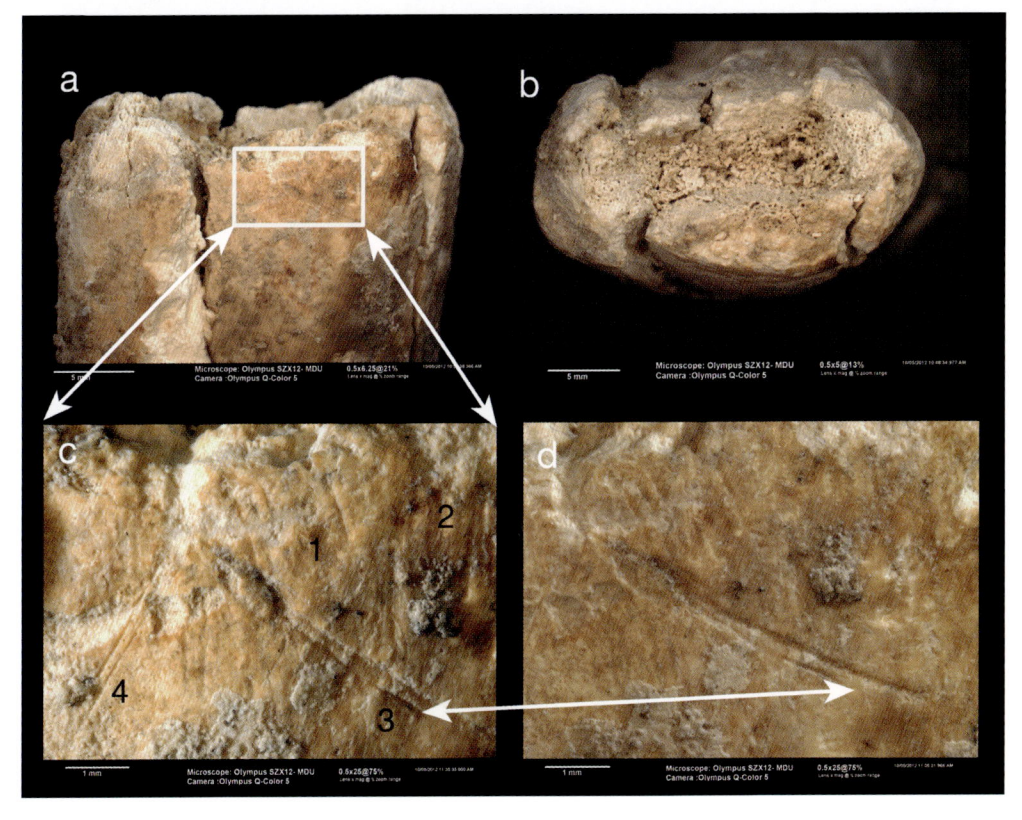

FIGURE 30.14 Distal end of antler tool (728-5G). a) groundward surface, area of striations indicated; b) transverse view showing crushing and compression of the antler beam; c) stereo-zoom micrograph showing four groups of parallel striations; d) magnified view of striation group 3.

The distal end of 727-5G was cut at a nearly 90 degree angle, just above the brow tine. The brow tine, located 80 millimeters above the base, appears to have been grooved and snapped as evidenced by a partial cut adjoining the protruding portion of the tine core.

In summary, surface modifications attributed to use-wear on the proximal ends of both antlers are similar. Symmetrical and continuous polish and rounding of the burrs on two planes indicate that this wear was not produced naturally, as when a deer rubs his antlers against a tree to remove velvet. Rounded and polished burr margins, symmetrically curved circumferences of the working ends, transverse striation on the burr of 728-5G, and the focal smoothing and polish on its beam provide compelling evidence that these tools were used as pestles within a hemispherical basin. The curved working surfaces of both antlers are good morphological fits for the curved interior surfaces of ornate box turtle (*Terrapene ornata*) carapaces, which are also present in the bundle. The fact that a third, similarly-sized antler tool was in-

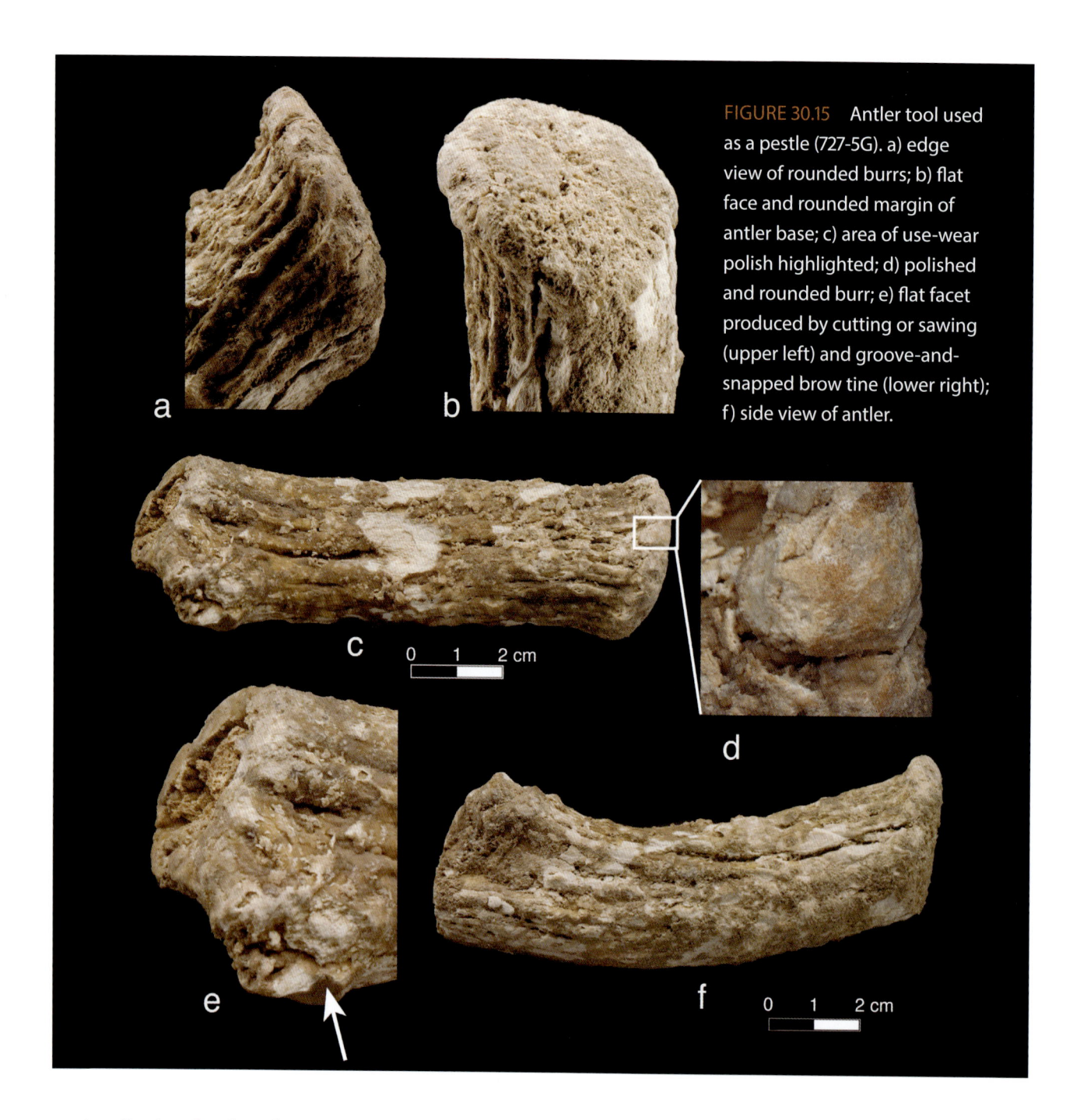

FIGURE 30.15 Antler tool used as a pestle (727-5G). a) edge view of rounded burrs; b) flat face and rounded margin of antler base; c) area of use-wear polish highlighted; d) polished and rounded burr; e) flat facet produced by cutting or sawing (upper left) and groove-and-snapped brow tine (lower right); f) side view of antler.

tentionally placed within a box turtle carapace at the time of burial reinforces the idea that antler tools and turtle shells were functionally related as pestles and bowls (Figure 30.16).

Antler (734-5G) and Carapace Bowl (735-5G) Beneath Man's Pelvis

A third antler (734-5G) was lying inside the turtle carapace (735-5G) recovered beneath the man's pelvis. The open side faced his lower back. Over time, calcite cemented small portions of his vertebrae to the antler, and the antler to the shell (Figure 30.16a). This antler is the least well-preserved and is more weathered than the carapace in which it was found.

In width, thickness, and surface texture, antler 734-5G, resembles the smaller antler (728-5G) in the bundle and appears to have come from a similarly sized deer, possibly the same cervid. The medial beam section measures 90 millimeters in length and 28 millimeters in width. One end is truncated (Figure 30.16a, left)

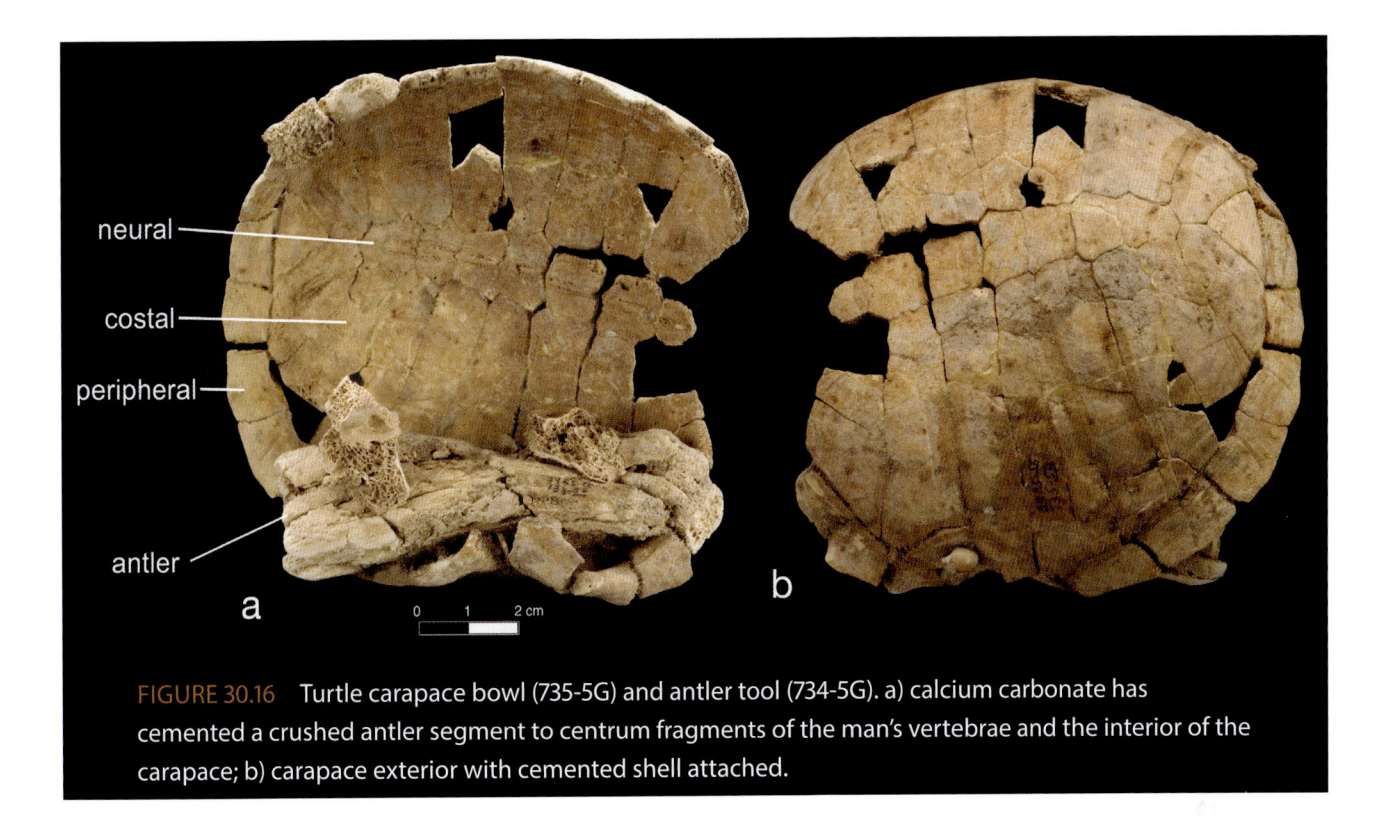

FIGURE 30.16 Turtle carapace bowl (735-5G) and antler tool (734-5G). a) calcium carbonate has cemented a crushed antler segment to centrum fragments of the man's vertebrae and the interior of the carapace; b) carapace exterior with cemented shell attached.

and its cross-section resembles the distal end of 728-5G (Figure 30.14). The opposite end is beveled with slightly rounded margins (Figure 30.16a, right). This tool was shaped by manufacture and use.

The associated carapace (735-5G) derives from a young adult box turtle with a solid shell, although the bony segments had not yet fully fused. The interior face of this carapace was protected in the burial and contains more than 80 preserved use-wear striations on interior neural and costal segments (Figure 30.17). They do not extend onto the ventral faces of peripheral segments, nor are they present on the carapace exterior. Vertebrae remnants are present on all seven neural segments. These exhibit use-wear polish and are infrequently cut by striations. Most striations on the neural segments parallel the spine and do not appear to have resulted from attempts to remove the vertebrae.

Striations on the interior of 735-5G are typically oriented in a single direction roughly parallel to the long axis of the carapace (Figure 30.17). The majority are elongated, U-shaped grooves with internal striations. V-shaped striations are also present, as are shorter, irregular marks oriented in multiple directions. These marks fall well within established criteria for identification of stone tool cut marks on bone (Lyman 1994; Shipman 1981; Shipman and Rose 1983).

The carapace interior developed a use-wear polish on the neural and costal surfaces that preceded the creation of striations. In many instances, one sees that a U-shaped groove removed previous polish and the base of this superimposed groove is unpolished. The superimposition of striations over polish and cross-cutting of different generations of striations show that the carapace was used repeatedly as a bowl.

Munro (2001) observed a similar distribution of striations in spur-thighed tortoise (*Testudo graeca*) carapace segments from Hayonim Cave, Israel, where Early and Late Natufian (12,800 BP–10,200 BP) people used modified tortoise carapaces as tools. The striations were formed when the carapaces were still articulated and are thought to result from the use of these shells as containers or palettes. Significantly, similar use-wear striations are absent from the interiors of at least 71 tortoise shells buried with a 12,000-year-old shaman at the Early Natufian site of Hilazon Tachtit, despite the fact that there is clear evidence for systematic butchery and roasting of tortoises (Munro and Grosman 2010; Munro 2013, personal communication). Striations like those on 735-5G do not

FIGURE 30.17 Stereo-zoom micrograph showing use-wear marks on the interior of a carapace (735-5G). a) U- and V-shaped striations on c5 to c3 (left to right); b) close-up from center of Figure 17a; c) cross-cutting striations on c8 to c4 and n7 to n5 (left to right), vertebrae are ground down (lower center); d) striations on c7 and c8 do not extend onto p10 and p11; (c = costal; n = neural; p = peripheral).

seem to be caused by butchery. Haury (2008) reports a similar combination of abrasions and polish on interior carapaces of box turtles (*Terrapene sp.*) in protohistoric Wichita assemblages at the Lower Walnut Settlement in Kansas. These modified turtle carapaces are also thought to have been used as bowls and containers.

Use-wear modifications on 735-5G evidently resulted from a combination of handling and abrasion by organic tools that produced polish, followed by scraping with a stone tool, perhaps to stir and remove substances within the bowl.

Three Nested Carapace Bowls (729-5G)

Three carapace bowls (729-5G) were placed together in the bundle. The lowest shell lay with its dorsal surface toward the earth, its bowl open skyward (upward), and the anterior portion eastward. The middle shell lay dorsal side skyward, open bowl facing the lower shell, anterior portion eastward. The upper shell lay dorsal side skyward, open bowl cupped over the middle shell, and the anterior end westward (Figure 30.18a). The man faced west with his cranium directly on (and at right angles to) the turtle shells.

The shapes of the carapace bowls are distorted by compression, but well preserved. Partial fusions of neural, costal, and peripheral bones suggest that these were young adult turtles. Redder removed some sediment from their interiors and stabilized them with glue and wood splints. Computed tomography scans confirm that no additional cultural objects lay within their interior spaces. The bowls remain nested together and visibility

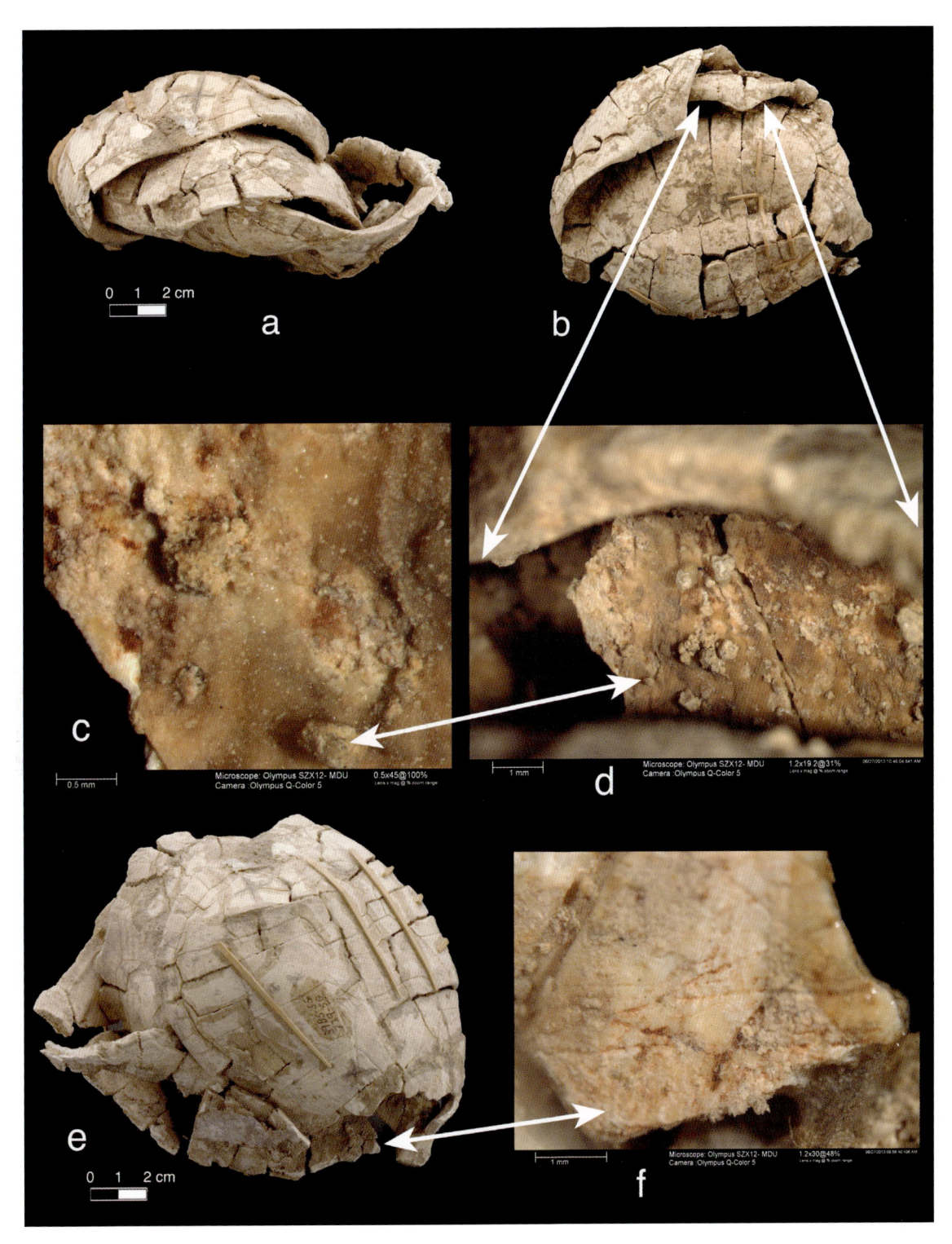

FIGURE 30.18 Nested turtle shell bowls (729-5G). a) cross-section that faced east in the burial; b) skyward (up) surface; c) micrograph of calcite overlying hematite, hematite exposed beneath quartz grain, arrow indicates area enlarged; d) hematite visible beneath veneer of calcite on costal bones of the middle turtle shell; e) groundward (down) surface; f) micrograph showing striations filled with red pigment on groundward turtle shell.

of ventral surfaces consists of one-third or less of each carapace. This includes: portions of the nuchal, left p1, right p1–p3 and adjoining portions of c1–c3, the suprapygal, pygal, and both p11 on the bottom shell; the outer margins of the right peripherals and adjacent portions of c2 to c4 on the middle shell; and the outer edges of the left peripherals, adjacent portions of left c1 to c3 and c8, and portions of the anterior neurals and adjacent c2 of the top shell (Figure 30.19).

Red residues are present on the middle and bottom bowls. X-ray diffraction analyses (XRD) confirm that the red residue on the ventral face of the right c3 of the middle bowl (Figure 30.18d) has the crystalline structure of hematite (Fe_2O_3, a form of iron oxide); the overlying minerals are calcite ($CaCO_3$, a form of calcium carbonate) and quartz (SiO_2) (Little 2013). Figure 30.18c is a composite micrograph showing where a quartz grain popped off the calcite surface exposing a fresh surface of hematite on the middle bowl.

Red residue and striations also occur on the suprapygal and pygal bones of the bottom bowl. Figure 30.18f is a micrograph of a small bone fragment glued to a costal segment. While close examination reveals that this is not an exact refit, hematite-filled, V-shaped striations located on the nearby suprapygal bone suggest that the illustrated fragment originated from the posterior end of this shell. The anterior portion of this bowl has smoothing and light polish along a prominent ridge on the nuchal bone interior. This wear resulted from repeated handling and use of the bowl. Similarly placed use-wear is present on a Cheyenne woman's medicine bowl used in childbirth, which was attended with much ceremony (NMNH E152811) (Figure 30.20).

Turtles hold special places in indigenous ideologies worldwide and are often perceived as totems and representations of longevity and fecundity. In the Americas, turtles are closely associated with the earth. The box turtle is known as the earth-bearer among Algonquian and Iroquoian speakers. The genus *Terrapene* (box turtles) is a name derived from an Algonquian word for turtle (Dodd 2001), a language that does not distinguish between turtles and tortoises (Ives Goddard, Curator and Senior Linguist, National Museum of Natural History, Smithsonian Institution, 2012, personal communication).

"The belief in the turtle as the upholder of the earth was common to all the Algonquian tribes, to which belong the Arapaho and Cheyenne, and to the northern Iroquoian tribes. Earthquakes were caused by his shifting his position from time to time. In their pictographs the turtle is frequently the symbol of the earth, and in their prayers it was sometimes addressed as mother. The most honored clan was the Turtle Clan; the most sacred spot in the Algonquian territory was Mackinaw, the 'Island of the Great Turtle'; the favorite medicine bowl of their doctors is the shell of a turtle" (Mooney 1973:976).

The life span of ornate box turtles approximates that of humans. From an indigenous perspective, they straddle worlds in that they live on land and enter water; they live above ground in warm seasons and hibernate below ground in cold seasons. They reappear at the onset of spring, when warmth and ground-softening rains facilitate emergence (Ernst and Lovich 2009). Soon after, they

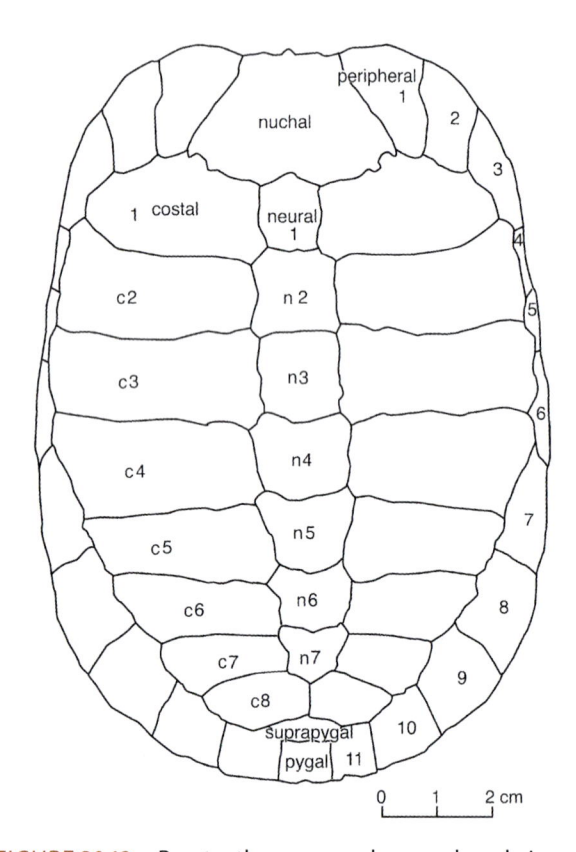

FIGURE 30.19 Box turtle carapace bones, dorsal view (c = costal, n = neural, p = peripheral). Adapted from Dodd (2001).

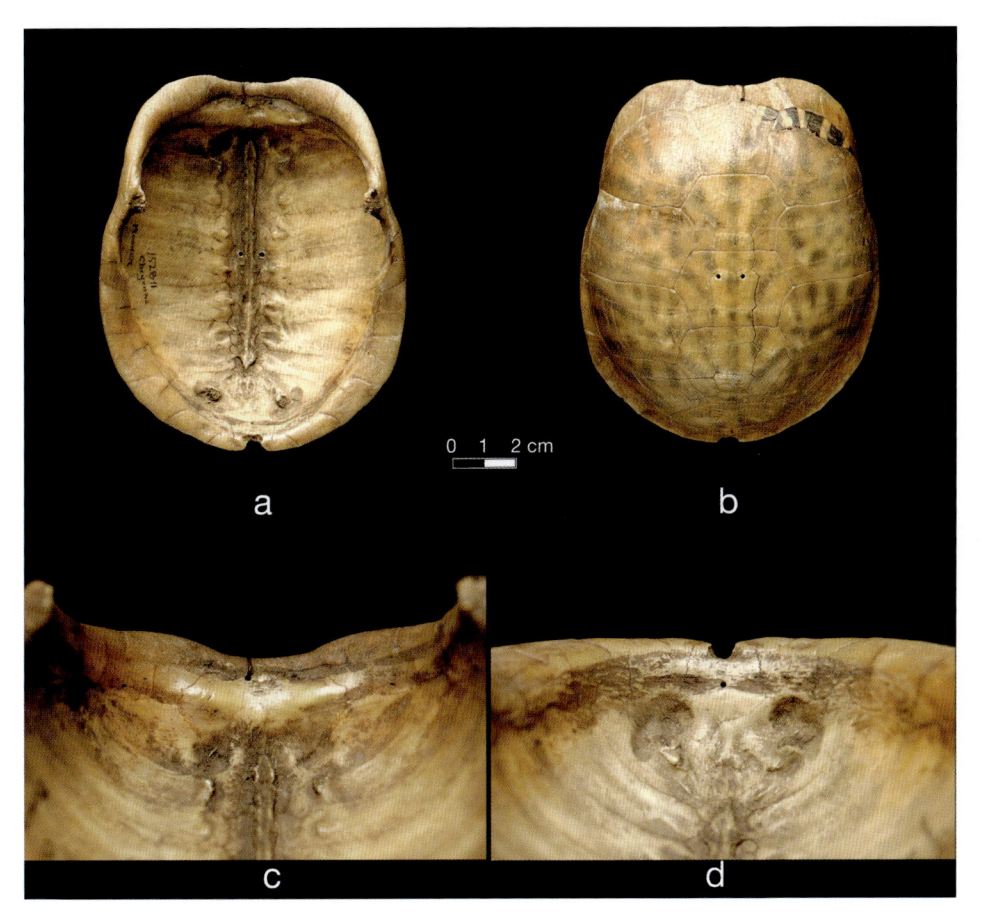

FIGURE 30.20 Cheyenne medicine bowl (E-152811). a) interior view; b) exterior view; c) use-wear polish on ventral nuchal ridge and adjoining peripherals, anterior end; d) use-wear polish on ventral suprapygal, pygal, and adjoining peripherals, posterior end.

begin to mate. Thus, turtle ecology assures that these creatures are naturally associated with the fertility of spring and the possibility of long life.

Turtle shells are valued and used in ceremonies in the Eastern Woodlands and Plains and are sometimes passed down through the generations. Most common in the Smithsonian Institution's ethnographic collections are leg rattles made of box turtles and hand-held rattles made of both box and snapping turtles. The archaeological remains of rattles can be distinguished from bowls by the presence in the former of both carapace and plastrons and the sound-makers that fill their interiors (i.e., small pebbles, beads, and other materials). Most rattles have drilled manufacture holes. The Seneca continue to make a box turtle rattle without a handle or holes, but the distinctive presence of carapace, plastrons, and sound-makers remains true.

Unlike rattles, turtle shell bowls are not common in the Smithsonian's ethnographic collections from the Eastern Woodlands and Plains. Only two are known to the authors, including the Cheyenne bowl previously noted. Turtle shell bowls from archeological contexts in these regions are documented (Blakeslee and Hawley 2006; Haury 2008). An example, of particular interest here, is a turtle carapace bowl used to hold red paint (#71, J. Stauffer Collection, Wichita State University) illustrated by Blakeslee (2008:96c) from the Great Bend Aspect (AD 1425 to early eighteenth century) in McPherson County, Kansas.

Ochre (726-5G)

Directly beneath the turtle bowls (729-5G) (Figure 30.9) lay a large piece of fine-grained, relatively soft, red and reddish-yellow colored ochre with use-wear markings (Figure 30.21). Evidence for repeated use prior to burial includes rounded and polished edges from prehensile and transport abrasion as well as overlapping patterns of grinding striations.

The two broad faces of the ochre differ in color and use-wear modifications. Face 1 (Figure 30.21a) is a moderate to dusky red color (5R 4/6 to 5R 3/4, Munsell Rock Color 2009). XRD analysis identifies this mineral as

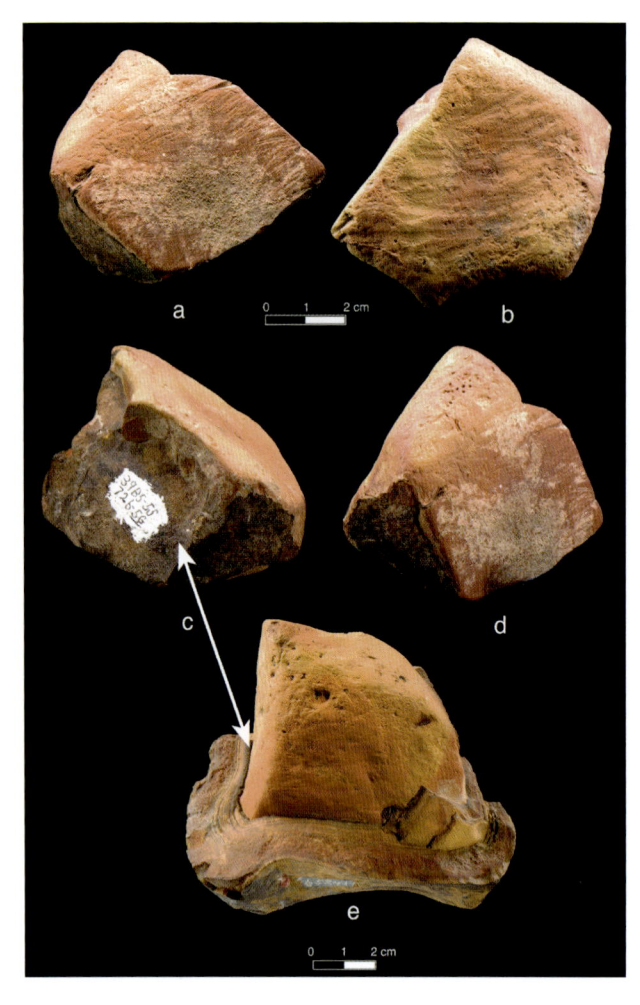

FIGURE 30.21 Comparison of the red ochre nodule (726-5G) with a natural iron oxide concretion (USNM 65241, bottom). a) Face 1 with grinding striations and smoothing; b) Face 2 with U-shaped grooves; c) hard, dark layer of Fe^2O^3 concretion; d) rounded and polished surface adjoining Face 1; e) iron oxide concretion from Robertson County, Texas. Arrow points to similar innermost concretion layer (Fe_2O_3) on both specimens.

hematite (Little 2013). The red face is nearly flat with rounded and polished edges. Overlapping groups of unidirectional striations are oriented perpendicular and at oblique angles to the long axis. Striations are typically less than one millimeter wide, with U- and V-shaped profiles. Most originate and/or terminate at the edges of the nodule (Figure 30.22a). These characteristics: a flat surface with unidirectional, parallel striations or groups of parallel striations at angles to each other; U- and V-shaped profiles that start and finish at the edges; and straight, un-frayed terminations have been produced experimentally by grinding dry red ochre on a sandstone abrader (Hodgskiss 2010:Figure 2d; see also Henshilwood et al. 2001).

A dull luster characterizes the valleys of these striations, whereas a shiny, metallic luster is present on plateaus between striations. This type of use-wear polish and smoothing has been observed on experimentally ground ochre pieces that were rubbed against human skin (Hodgskiss 2010:Figure 3e). A small area of smoothing on one corner of Face 1 (Figure 30.22a, arrow) has a dull luster and overprints prehistoric striations. Hodgkiss (2010:Figure 3f) suggests that such smoothing, which eradicates grinding striations, occurs when wet ochre is rubbed on animal hide, or possibly even human skin.

Thin patches of calcite overlie the grinding striations. Erratically-oriented scratches, in turn, superimpose and crosscut both the calcite and prehistoric use-wear marks, indicating post-depositional origins such as trampling and/or modern handling. These post-burial scuff marks and gouges are easily distinguished visually from pre-burial striations by their intrusive nature, distinctive reddish-orange color, powdery, matte texture, and random distribution (Figure 30.22f; see also Hodgkiss 2010: Figure 5b).

Face 2 graduates in color from a moderate red (5YR 4/6) to a moderate yellow brown (10YR 5/4, Munsell Rock Chart, 2009) (Figure 30.21b). Its surface topography gently undulates and prehistoric use-wear marks consist of relatively broad, U-shaped grooves or scrapes nearly 1.5 to 2.5 millimeters in width (Figure 30.21b). These sub-parallel marks are oriented at oblique angles to the long axis of the piece and do not extend to the edges. They are less evenly aligned than the striations produced by grinding (see Henshilwood 2001:Figure 8; Watts 2010). Hodgkiss (2010) produced such marks experimentally by scoring ochre to produce powder.

Grooves created experimentally by scraping with a stone flake produce use-wear marks that have internal microstriations. Microstriations do not characterize the prehistoric grooves on Face 2 of 726-5G, suggesting that a nonlithic tool produced these marks. Bone and wood tools used experimentally to score ochre did not leave internal striations on the ochre itself, but did leave bone or wood residues that, themselves, had an internally striated pattern. If the organic residues do not preserve,

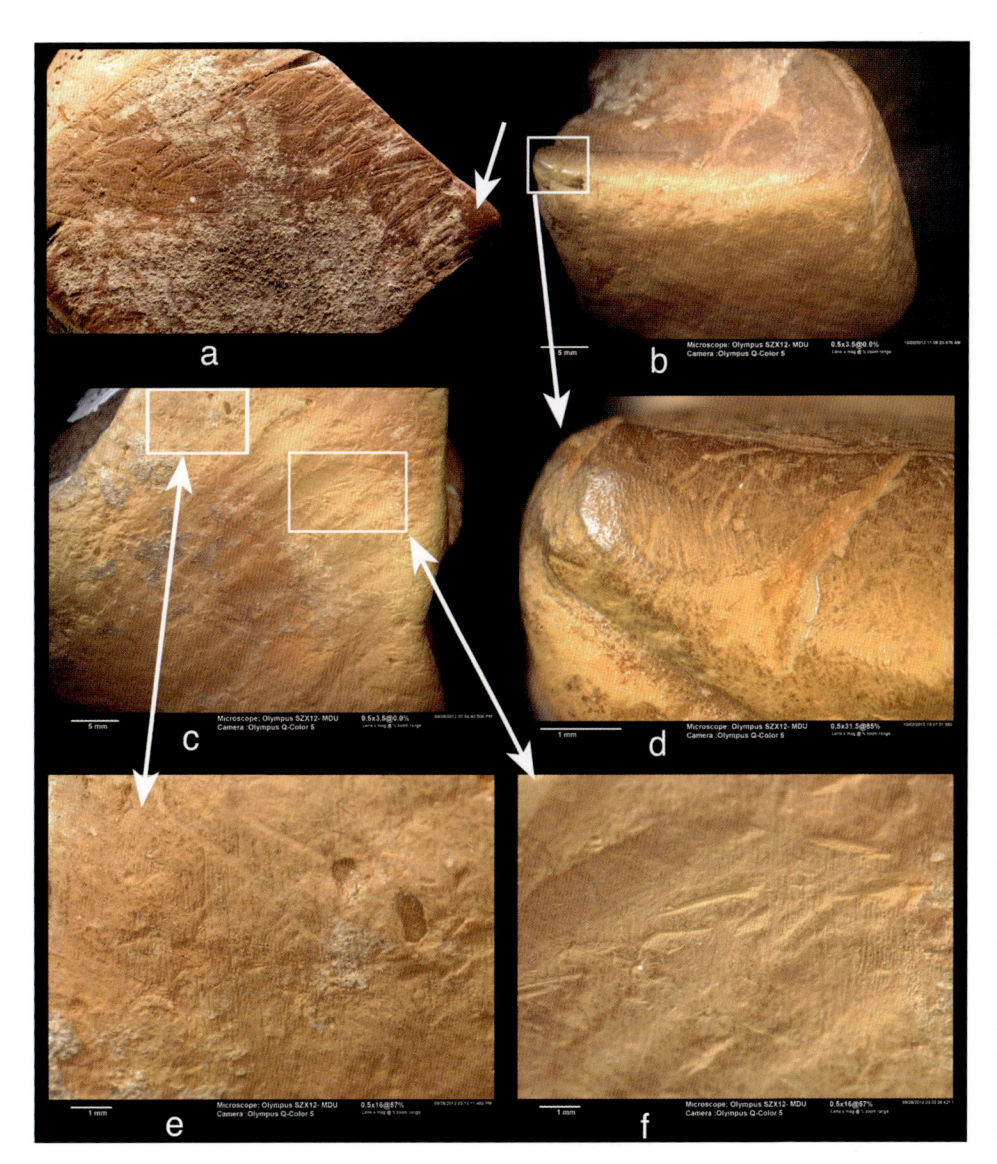

FIGURE 30.22 Stereozoom micrographs showing use-wear marks on red ochre (726-5G). a) grinding and smoothing use-wear, Face 1; b) rounded and polished edges; c) U-shaped grooves, Face 2; d) polish and parallel striations on margins; e–f) imprint of possible plant material used to wrap the ochre nodule that is superimposed on the use-wear, Face 2.

neither do their microstriations. A bone implement (722-5G) recovered beneath the ochre in the bundle has a stylus-like working end with dimensions consistent with those of the gouges. Use-wear modifications (discussed below) suggest that this tool was used to scrape and apply red pigment.

Overlying the prehistoric use-wear marks on Face 2 is a distinctive imprint revealed by microscopy. The pattern consists of symmetrical, longitudinally oriented, parallel microfurrows intersected at intervals by tiny circular nodes (Figure 30.22e–30.22f). The imprint is superimposed over the plateaus and valleys of the use-wear marks and postdates them. The fact that the imprint drapes smoothly over the surface topography of the piece suggests that a pliable material and a single event was involved. Where there are slight forks, the microfurrows appear to be pointing in the same direction, a kind of regularity that might suggest biogenic origin, such as a monocot leaf impression from plants such as grasses and agaves having strap-like leaves with parallel veins (William A. DiMichele, Curator of Fossil Plants, National Museum of Natural History, Smithsonian Institution, 2013, personal communication). The imprints do not branch out and reconnect (non-anastomosing) as in the vein skeleton of an oak leaf. The imprint appears to be inconsistent with woven plant material, in that there is only "warp," but no "weft" visible on Face 2. This imprint may have been left by organic material used to wrap the ochre nodule within the bundle, a common indigenous practice (see Sturtevant 1954 for the Seminole).

Nearly all margins of the nodule show smoothing, rounding, and well-developed polish. Intrusive to this polish are shallow, parallel microstriations with non-polished valleys (Figure 30.22d). Collectively, these marks are consistent with a combination of prehistoric handling and transport. The ochre margins that show the least rounding are those protected by a hard, iron-oxide concretion layer (Figure 30.21c), which is itself lightly rounded and polished.

The source of the ochre is unknown, but may lie in northeast and east central Texas where northeast-to-southwest trending geologic deposits produce iron oxide concretions. An iron oxide concretion collected during a U.S. Geological Survey in the Wheelock Prairie locality in Robertson County, Texas, is strikingly similar visually and in elemental composition to 726-5G. This comparative specimen (USM 65241) from the NMNH Department of Mineral Sciences was collected approximately 130 kilometers downstream from Horn Shelter No. 2. It consists of a multi-layered, ferruginous concretion that formed around a fine-grained ochre core (Figure 30.21e).

The ochre core of USM 65241 is a good match in terms of colors, texture and geochemistry for 726-5G. Likewise, the hard, innermost concretion layer of the former (Figure 30.21e) resembles the dark concretion surfaces on 726-5G (Figure 30.21, arrow). Non-invasive portable X-ray Fluorescence Spectroscopy (pXRF) demonstrates a sufficiently similar geochemical composition for both samples that the Robertson County locality warrants investigation as a possible source. In addition to the expected large iron (Fe) peak, both ochres share the presence of calcium (Ca), manganese (Mn), strontium (Sr), arsenic (As), lead (Pb), zinc (Zn), copper (Cu), zirconium (Zr), niobium (Nb), and tin (Sn) (Drake 2013, personal communication). Specific differences show a greater content of manganese and strontium in 726-5G (Coolidge 2013; Lee Drake, Bruker, 2013, personal communication).

The use of red ochre at Horn Shelter No. 2 is as an expression of an ancient and widespread practice. Its archaeological presence is coeval in time with *Homo sapiens* and is also known among Neanderthals (Riel-Salvatore et al. 2013; Watts 2010). Red ochre is still highly valued and widely used by indigenous peoples (Campbell 2007; Guilliford 2000; Hayden 2003; Lewis-Williams 2001; Stafford et al. 2003).

Red ochre is, and has been, used as an internal and external medicine (Clarke 2008; Peile 1979; Velo 1984; Wilcox 1911); a sacred substance in ceremonies and offerings (Bement et al. 1999; Gulliford 2000; Kohntopp 2010; Lewis-Williams 2001; Mooney 1973; Ricklis 1994; Salazar et al. 2011; Stafford et al. 2003; Stoffle et al. 1994) including burials (Bowler and Thorne 1976; Breternitz et al. 1971; Goebel and Slobodin 1999; Jenks 1937; McGee 1898; Muniz 2004; Owsley and Hunt 2001; Owsley et al. 2010; Potter et al. 2011); a pigment for body painting and tattooing (Campbell 2007; Colton 1948; Deter-Wolf 2013a; Krutak 2007, 2013; Le Moyne 1564; McGee 1898; Mooney 1973; Nutt 1947; Peterson and Lampert 1985); a coloring agent for rock art (Boyd 2003; Campbell 2007; Clottes and Lewis-Williams 1998; Lewis-Williams 1994, 1995, Lewis-Williams and Pearce 2004) and mobilary art (Campbell 2007; Eerkens et al. 2012; Henshilwood et al. 2009; Soffer 1985); an item of gifting and exchange (Dellenbaugh 1933; Mooney 1973; Osgood 1940; Peterson and Lampert 1985); an adhesive (Allain and Rigaud 1989; Wadley 2005; Wadley et al. 2004); a possible polishing agent for the hafted portion of Paleoamerican points (Titmus and Woods 1991); an insect repellent (Jenness 1955); and a substance in skin processing (Adams 2002, Audoin and Plisson 1982; Roper 1991; Semenov 1964; Rašková Zelinková 2010).

Evidence for repeated use of a large ochre nodule and the presence of red pigment on five of the ten bundle items (two turtle shell bowls, and two sandstone abraders and a bone stylus yet to be discussed) testify to its significance in the bundle and may suggest ways in which the man buried in Horn Shelter No. 2 used these items in service to his community and himself.

Bone Stylus (722-5G)

A slender tool fashioned from the rear leg of a white-tailed deer (*Odocoileus virginianus*) was lying partially buried directly beneath the ochre nodule (Figure 30.9). Postdepositional processes broke the tool into several pieces that Redder refit. This longitudinally split portion of the posterior-medial diaphysis of a right metatarsal extends 160 millimeters from just below the proximal articular surface toward the spatula-like, diagonal end of the tool, which was formed from the distal posterior shaft at a point where the bone naturally flattens.

The working end of the tool forms an angle of nearly 45 degrees. Its apex is slightly blunted from use and a glossy polish extends over a beveled facet on the interior face and across adjoining cortical edges (Figure 30.23b). The polish is more extensive on topographic high points. The cross-section of the diagonal surface is concave and rough in texture suggesting it was snapped, not cut (Figure 30.23c). Topographic lows on this concave surface are unpolished.

A pronounced focus of tool use was the tip, which resembles a stylus (Figure 30.23). The diagonal edge was also used. For a distance of at least 16 millimeters from the tip, the exterior surface shows fine striations that parallel the diagonal edge (Figure 30.23f). Polish becomes progressively developed toward this edge, covering the plateaus between striations, but absent within striations. Use-wear in the form of microchipping is present along the interior margin of the diagonal edge (Figure 30.23e). Longitudinal striations, with V- and U-shaped profiles, are present along the length of the tool on both interior and exterior surfaces. The latter appear to result from scraping with a stone tool. Well-developed prehensile polish and cross-cutting striations along the tool shaft imply extensive and repeated use.

The opposite (proximal) end of the tool was in direct contact with the ochre nodule upon recovery. This segment has red residue along its longitudinal margins

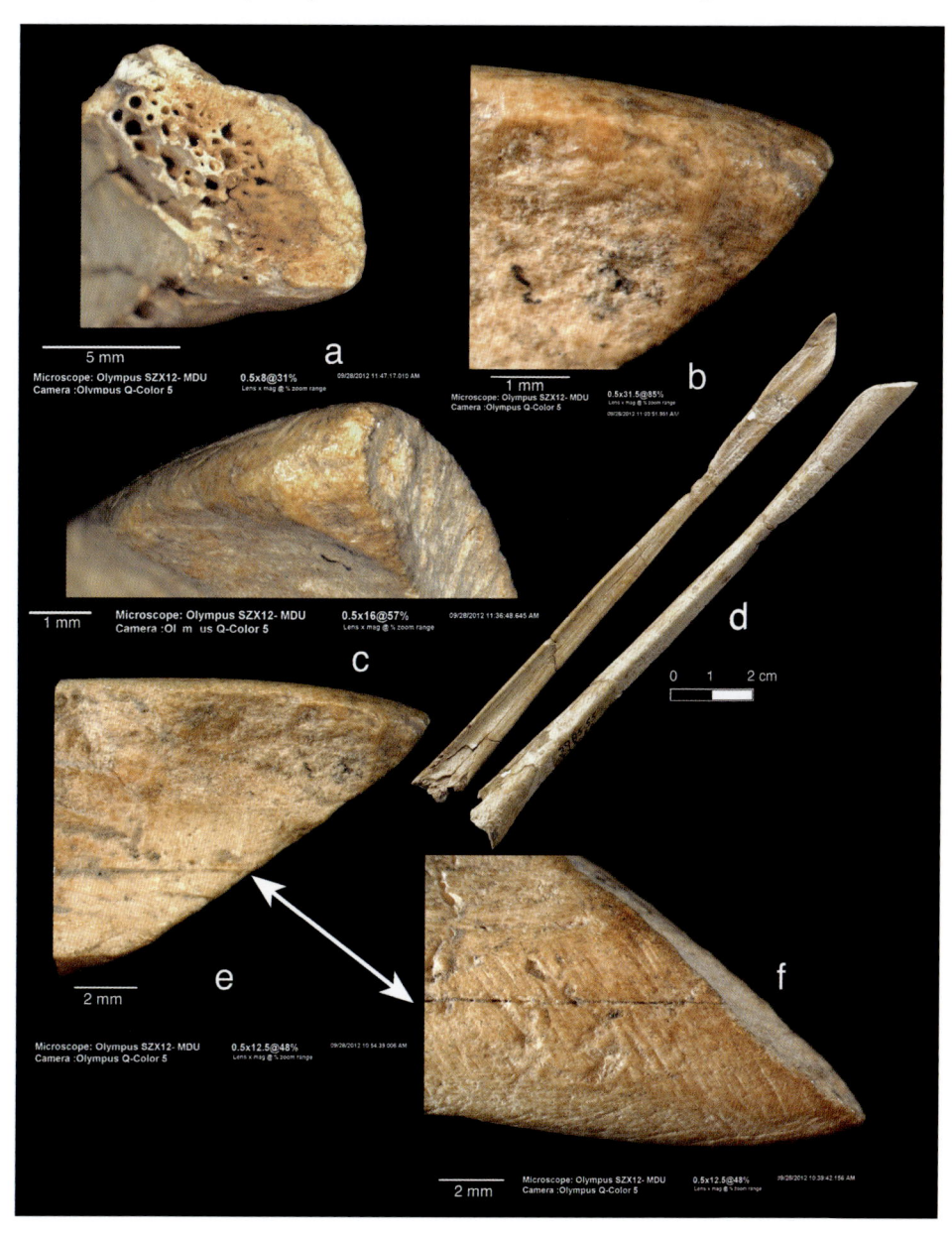

FIGURE 30.23 Deer bone stylus (722-5G). a) composite micrograph of red residue in trabecular bone of proximal metatarsal; b) composite micrograph of use-wear polish on the distal tip; c) composite micrograph of polish on the distal tip and absence of polish on the snap break (right); d) skyward (top) and groundward (bottom) views of stylus; e) micrograph of the interior face of the distal tip; f) micrograph of the exterior face of the distal tip showing parallel striations.

and within the trabecular structure (Figure 30.23a) that appear post-depositional in origin. The close spatial association in the bundle of the stylus and ochre suggest a functional relationship, an implication supported by ethnological and archaeological information on tool kits used in the manufacture and application of red pigments.

Two bone implements share similarities with the Horn Shelter No. 2 stylus. Each has been reported as a tool used in the preparation or application of red pigment. One was buried with a Late Mesolithic shaman from the site of Bad Dürrenberg in central Germany. This burial of a 25- to 35-year-old woman and a four- to six-month-old infant, dated between 9080 and 8230 CAL yr. BP, is one of the richest isolated Mesolithic graves in Europe (Porr and Alt 2006). The Bad Dürrenberg tool is the same length (162 millimeters) as the Horn Shelter No. 2 stylus and was similarly fashioned from a longitudinally split, posterior roe deer metatarsal (*Capreolus capreolus*) (Figure 30.24a). Trabecular bone on its distal tip preserves hematite residue. Porr and Alt (2006) identify it as a painting tool used to mix or apply red pigment on soft materials such as skin or leather. The absence of use-wear marks on its tip, and a general lack of polish, suggest that the tool had not been handled over a long period of time and perhaps was used during the burial ceremony itself.[1]

Back further in time, a Middle Stone Age bone tool (TK1-B1) with a diagonally-shaped end (Figure 30.24b) was found in Blombos Cave, South Africa (Henshilwood et al. 2011:Figure 3) in a workshop that produced two tool kits used to process red ochre. This tool, a broken canid ulna with red ochre residue was evidently used to stir the ochre-rich compound stored in an associated abalone shell and to "transfer some of the compound out of the shell" (Henshilwood et al. 2011:222). This bone implement was recovered directly beneath the abalone shell. This tool kit also included a quartzite cobble used to break and grind ochre and bone, a quartzite slab used to grind ochre, seal and bovid bone fragments, a quartz flake used as a grinder with a microchipped edge and ochre residues, and two quartzite microchips with red ochre residues.

A second abalone shell (TK-2), located 16 centimeters away, contained a similar red pigment mixture. Use-wear scratches on this abalone shell were likely produced "when quartz or ochre grains were gently moved across the nacre surface during mixing" (Henshilwood 2011:221). An associated quartzite stone used to grind pigment was lying on the shell. Elements of the TK-2 tool kit were reused, indicating that red pigment production was not a one-time event. These 100,000-year-old tool kits demonstrate deep antiquity for human manufacture, storage, and application of red pigment in quantities inconsistent with hide processing.

Abraders (724-5G, 725-5G)

The bundle also includes two pieces of tabular sandstone suited in both size and shape for handheld use as abraders. Both were in direct contact with each of the antler pestles (Figure 30.9); fragments of these antlers that detached during excavation remain cemented to the sandstone surfaces.

Abrader 724-5G was handled more extensively than 725-5G, as evidenced by greater edge rounding and smoothing from prehensile abrasion (Figure 30.25a). Its broad, relatively flat faces are ground smooth and converge in a wedge-shaped cross-section that tapers toward a long, curved edge. Thicknesses range from 5.5

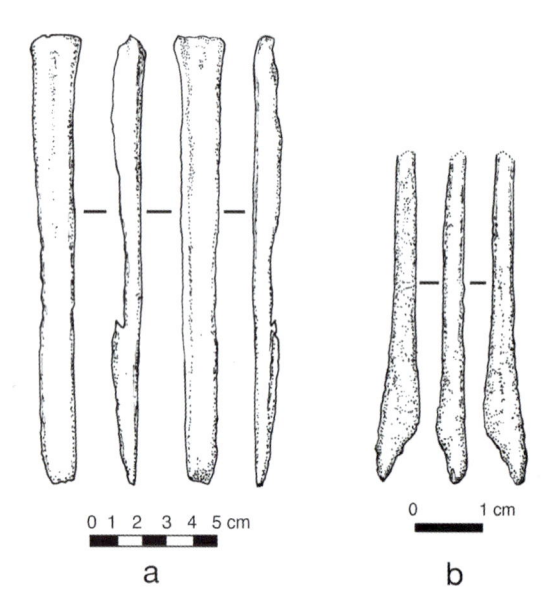

FIGURE 30.24 Prehistoric bone tools used to mix and apply pigment. a) Late Mesolithic split roe deer metatarsal with hematite residue, Bad Dürrenberg site, Germany, adapted from Porr and Alt 2006:Figure 5; b) Middle Stone Age broken canid ulna (TK1-B1) with red pigment residue, Blombos Cave, South Africa, adapted from Henshilwood et al. 2011:Figure 3.

to 12.4 millimeters. Both faces show use-wear markings. Microstriations parallel the long axis on the groundward face, which also retains red ochre residue near an outer edge (Figure 30.25a). Greater surface topography and natural vacuities are present on this face. The skyward surface exhibits greater abrasion, with microstriations oriented perpendicular to the abrader's long axis (Figure 30.25b). This face is roughened in several places by small pits and troughs.

The second abrader, 725-5G, is thicker overall, with less edge rounding, and greater surface topography (Figure 30.25). The working surfaces are roughened and pitted in several places, interspersed with smoothing. The wedge-shaped cross-section varies in thickness from 9 to 15.7 millimeters, with one very thin area (3.2 millimeters) at the rounded end. Tiny areas of ochre residue, overprinted with polish, are present along one edge (Figure 30.25c). This same area has a larger unidentified residue adhering to the surface (rather than crushed within quartz grains like the pigment residue). This compound is pinkish-tan in color with black specks that may be charcoal.

To aid identification and interpretation of residues, the abraders in the bundle were compared with sandstone abraders (of nearly the same size) bearing red pigment residues from Folsom components at Lindenmeier (F165), (Wilmsen and Roberts 1978: Figure115c), and Stewart's Cattle Guard (1981 D112-A2-19), (Jodry 1999: Figure 108d) sites in Colorado. The Lindenmeier abrader is flat with moderate to heavy coatings of red pigment on both faces. The Stewart's Cattle Guard tool is basin-shaped on one face with sparse traces of red pigment within the basin that increase to moderate densities toward the margins. The opposite face has irregular surface topography with multiple grinding facets including a deep, central groove used to abrade bone and/or stone. A convex surface near the tool margin of this face bears sparse pigment residues.

The two Folsom abraders were used on both faces to grind red pigment and pigment is preserved in spaces between quartz grains. With increasing amounts of residue, the quartz grains gain a red colored film and are eventually coated with solid particles. Pigment residues become de-

FIGURE 30.25 Abraders (724-5G and 725-5G). a) groundward surface of 724-5G with smoothing, parallel longitudinal microstriations, and red pigment residue; b) skyward surface of 724-5G with smoothing and parallel transverse microstriations overlain by pitting; c) 725-5G, groundward surface with red pigment and other residue; d) 725-5G, skyward surface with smoothing overlain with pitting.

tached when quartz grains and cement pop off the surface during subsequent tool resurfacing and/or non-pigment related tool use. Overprinting is evident on all four abraders with the Lindenmeier tool showing the least and the Horn Shelter No. 2 implements showing the greatest effects of this process. The spaces between quartz grains tend to best preserve the last remnants of pigment; outer margins preserve greater traces of red pigment than tool centers when subsequent tool use overprints the pigment.

In sum, there is evidence that red pigment was ground on the Horn Shelter No. 2 abraders, but ochre does not appear to have been the final material processed with either tool. Resurfacing and subsequent use of the tools with other materials evidently overprinted most traces of previous ochre grinding.

Biface (723-5G)

An Edwards chert biface was recovered in the lower portion of the bundle in contact with abrader 725-5G and antler pestle 728-5G (Figure 30.9). It was fashioned by hard hammer percussion and retains a small remnant of limestone cortex (Figure 30.26). Ray (1998: Figure 8.33) illustrates similar bifaces from a lithic workshop at the Big Eddy site in southwest Missouri. He reported these as rejected and discarded secondary bifaces, of probable San Patrice manufacture, made on ellipsoidal stream cobbles.

The size and irregular surface topography of the Horn Shelter No. 2 biface negates its suitability as a preform for San Patrice projectile point manufacture. Preliminary environmental scanning electron microscope (ESEM) examination of its edges suggests an absence of use-wear traces attributed to cutting or scraping. Limited micro-chipping on an edge beneath a series of stacked step-fractures (Figure 30.26b) appears to have resulted during manufacture (Figure 30.26c). Further examination of use-wear striations and polish awaits residue analysis and cleaning (see Kay 1998; Waters et al. 2011).

This biface may have been used as a core for the production of small flakes. Polish on some, but not all, high arrises may represent bag wear between episodes of flake removal. Figure 30.26 highlights the locations of two large flakes that were most recently removed. These flakes were of sufficient size to serve as handheld cutting implements. In keeping with other lines of contextual evidence, these flakes may have been used in practices involving incision and bloodletting, both widely practiced in aboriginal North America (Lyon 1996) including the Eastern Woodlands (Cushman 1899; Hilger 1951:88; Speck 1907, 1934) and Plains (Fowler and Flannery 2001; Grinnell 1923).

Bloodletting is one of the oldest healing practices and involves releasing small quantities of blood in order to cure illness, promote good health, and prevent disease. Historically it was the "most common, most versatile, and most trusted of surgical procedures" used in western medicine for almost two thousand years, until the nineteenth century (Kuriyama 1995:13). Small, thin, scalpel-like implements were among the tools employed by western practitioners for this purpose.

Among indigenous people of North America, Anishinabe (Chippewa) in the Northern Woodlands made incisions during healing practices "with sharp-edged flint stones (*Biwanak*), with pieces of broken porcelain sharpened to an edge, or with the points of knives" (Hilger 1951:95). The Wailaki in California used a flint knife to make small incisions in the skin at the location of the disease, which was then sucked out (Loeb 1934). Yellow Bear, a nineteenth-century Hopi shaman, used a three-inch, leaf-shaped stone knife to make an incision during a healing ceremony (Parsons 1939). Swanton (1931) related that Choctaw healers used a piece of flint to make eight or ten closely spaced incisions in the skin before they sucked out the disease. Esaw (Catawba) curers used a comb of rattlesnake teeth to draw blood (Lawson 1709). The Catawba (Speck 1934), Creek (Swanton 1946), and Choctaw (Swanton 1931) all practiced scarification. In addition to incision during bloodletting and scarification, tattooing was nearly ubiquitous in the Eastern Woodlands and Plains.

These traditional practices were widespread among indigenous peoples globally and continue into modern times. In 1970, for instance, Hayden (2003:88) witnessed a man transform into his remote ancestor in the Great Desert of Australia. As part of the process, the individual opened a vein in his arm with a sliver of glass and filled a small bowl with his blood. It was used as a substance to affix plant fluff to his body during the ceremony. Ethnographic and linguistic accounts from the Pacific establish the use of small, sharp flakes for piercing and cutting human skin during tattooing and scarification. One tech-

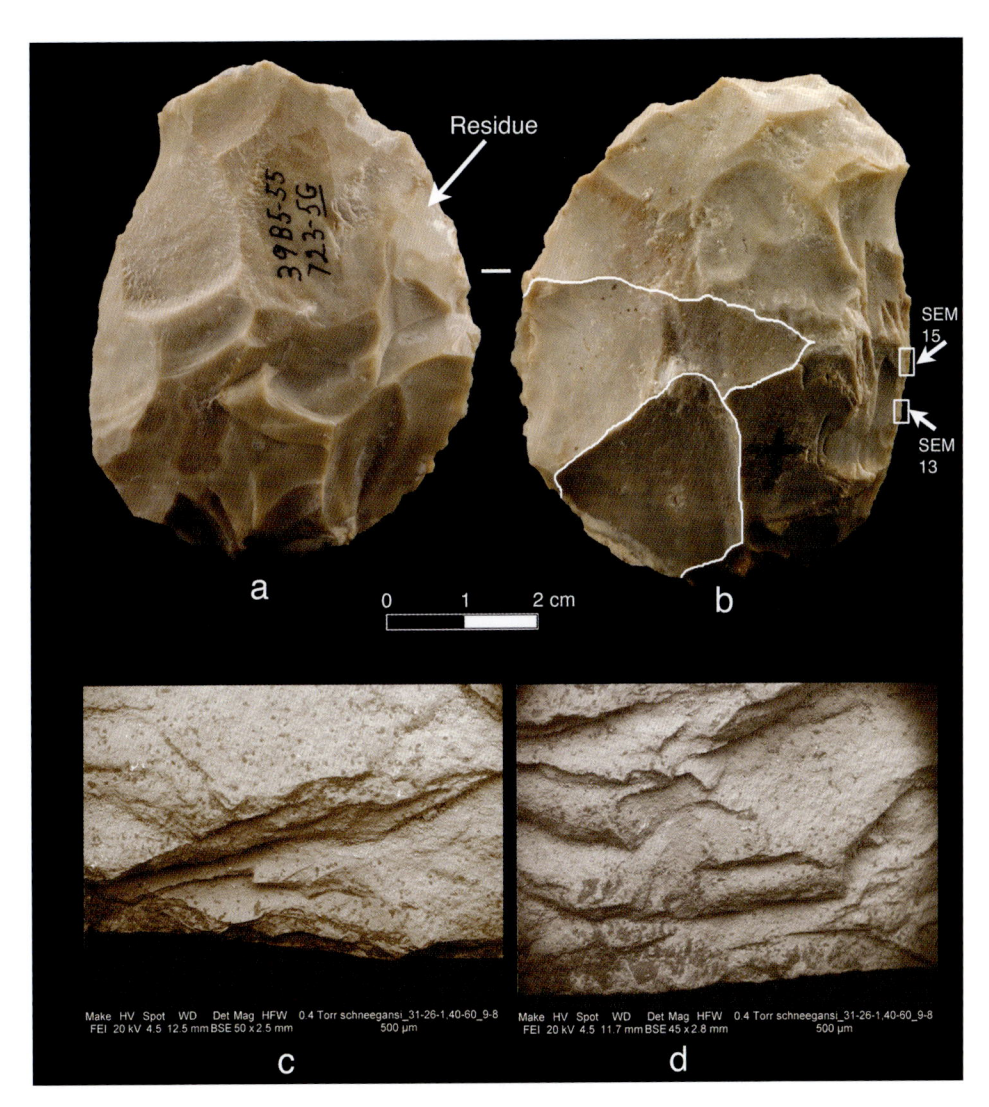

FIGURE 30.26 Biface (723-5G). a) skyward side of biface; b) groundward side of biface showing locations of micrographs and highlighted flake scars; (c–d) ESEM micrographs showing absence of use-wear from cutting or scraping.

nique involves making a series of deep cuts, two to four millimeters long, close to one another, executed in a row to create a line. These incisions were traditionally made using sharp obsidian, quartz, chert, or a sliver of bamboo (e.g., Ambrose 2012; Barton 1918). An old Ainu term for tattoo was *anchi-piri* (*anchi*, obsidian; *piri*, cut) (Batchelor 1901:21 as reported in Krutak 2007).

Ongoing investigations of small, utilized flakes and flake tools from archaeological sites in Melanesia integrate use-wear and residue (blood/pigment) analyses with experimental research. These data provide strong evidence that the use of small flakes in tattooing, scarification, and medical treatment of the human body extends back at least 5500 years (Kononenko and Torrence 2009; Kononenko 2011, 2012).

The two highlighted flakes (Figure 30.26) detached from biface 723-5G at Horn Shelter No. 2 lie within the morphological variation of these Melanesian flake tools (Kononenko and Torrence 2009:Figure 3) suggesting their suitability for incising human skin. Whether they were indeed used in this way remains an open question.

Studies aimed at identifying potential incision, scarification, and tattooing tools in North American archaeological sites are just beginning (e.g., Deter-Wolf 2010, 2013a; Deter-Wolf and Peres 2013). From ethnohistorical and ethnographic accounts of peoples of the Eastern Woodlands, Deter-Wolf (2013a:Table 2.1) compiled a list of implements used to cut and pierce human skin. In addition to a majority of organic tools (e.g., bone needles and awls, porcupine quills, piercing thorns, garfish teeth, fish spines, sharpened cane, and split bird feathers), flint points were used by the Omaha (Fletcher and La Flesche 1911) and Caddo (Vega 1993 [1605]), sharp flint stones were used by the Tuscarora (Heckewelder 1876 [1818]), and a gunflint was employed by the Ojibway (Anishinabe) (Long 1791). Experimental tattooing and incising indicates that flakes and gravers were highly effective for linear slicing, while serrations on a bifacially-flaked point resulted in an irregular wound (Deter-Wolf 2010; Kononenko 2011).

"At the present time there is not a sufficient library of use-wear data with which to distinguish patterns left on stone tools by tattooing human skin from those created by processing other soft hides" (Deter-Wolf and Peres 2013). Context is critical. Tools used for incision of human skin (and for applying pigment to it) are expected to be part of a larger tool kit, as is documented worldwide in the ethnographic literature regarding tattooing (Deter-Wolf 2013a; Krutak 2013) and healing practices.

OTHER BURIAL ITEMS

Mortuary items that were external to the bundle include two types of perforated marine shells, perforated coyote teeth, unperforated badger claws and Swainson's hawk talons, a perforated antler tool, an eyed bone needle, a box turtle carapace containing a possible pestle, and another covering the man's face.

Olive Nerite Beads (Lot 737-5G)

Eighty olive nerite (*Neritina reclivata*) beads from the burial were examined. These range in diameter from 5.7 to 13.6 millimeters and in height from 4 to 8.5 millimeters. All have deliberate perforations located at their posterior ends opposite the natural apertures (Figure 30.27).

The perforations were made from the outside, as the interior whorl obstructs tool access through the aperture. The holes have irregular oval shapes that range in maximum dimension from 2.5 to 4.8 millimeters. Their perimeters are typically jagged, with polished micro-indentations (Figure 30.27e). d'Errico et al. (2005:Figure 7) produced perforations of similar appearance experimentally by punching holes with a bone awl or stone tool.

The beads had been worn for some time prior to burial as evidenced by rounding and polish on at least 40 percent of the perforations and 65 percent of the apertures. Most of the former are encircled by flat facets of variable width that often show polish (Figure 30.27). d'Errico et al. (2005:16) determined that such use-wear patterns were "consistent with friction from rubbing against thread, skin, or other beads." All Horn Shelter No. 2 beads that retain remnants of original surfaces (80 percent) show pronounced polish on exterior body whorls.

Seventeen of the beads were mapped in situ; the rest were recovered while screening burial fill; therefore a detailed reconstruction of their layout at the time of burial is not possible. Story (1990:205) observed from their general distribution and size that the shells "may have been sewn onto the clothing of the adult."

Snail beads are among the earliest reported expressions of human adornment, dating from 82,000 years ago at Grotte des Pigeons in Taforalt, Morocco (Bouzouggar et al. 2007) and 75,000 years ago in Middle Stone Age levels at Blombos Cave, South Africa (d'Errico et al. 2005; Henshilwood et al. 2004). These ornaments, along with inscribed red ochre, are among ancient expressions of symbolism thought to be indicative of modern human behavior. The perforated olive nerite (*Neritina reclivata*) and lettered olive (*Oliva sayana*) shells from Horn Shelter No. 2 are the oldest reported shell beads from North America (Jodry 2010). Olive nerite snails inhabit the shallow coastal waters of the Gulf of Mexico (Russell

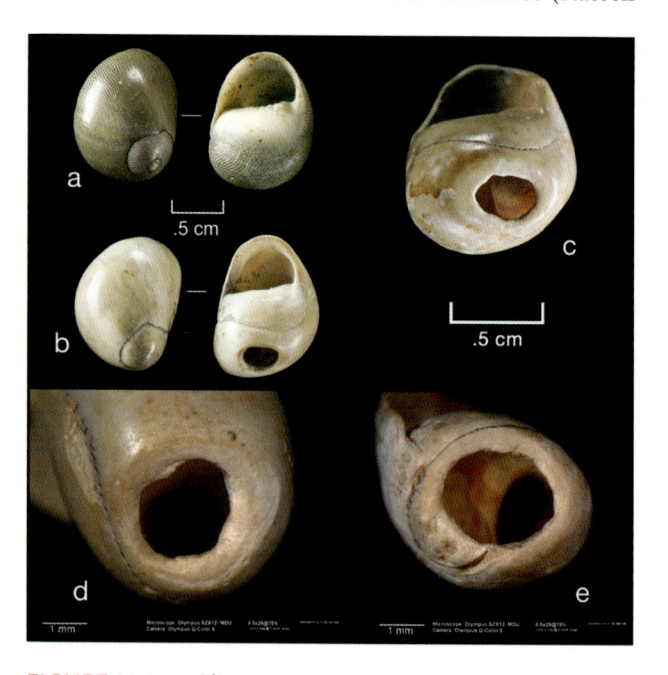

FIGURE 30.27 Olive nerite (*Neritina reclivata*) shells. a) modern comparative specimen, USNM 27236, Berlandier, Texas; b) bead, 737a-5G; c) bead 737d-5G shows micro-indentations on perforation's perimeter encircled by a flat, polished facet; composite micrograph of bead 737c-5G (d) and bead 737d-5G (e).

1941). Geographically, the closest sources to Horn Shelter No. 2 were the marshes and estuaries at the mouth of the Brazos River 300 kilometers downstream.

Lettered Olive Bead or Pendant (738-5G)

A single perforated lettered olive (*Oliva sayana*) shell was recovered inside the man's cranium, close to olive nerite beads and badger, Swainson's Hawk, and coyote items. It was not a fresh shell when collected prehistorically, but was an older, water worn, and preyed-upon shell (Figure 30.28). This may imply that it was valued for properties unrelated to the ornamental aesthetics of a colorful patterned exterior.

The shell is 43.4 millimeters long and 17.5 millimeters wide. Nearly two to three millimeters at the apex of its spire were removed to create a symmetrical perforation nearly three and a half millimeters in diameter (Figure 30.28c). The lip and upper interior of this hole are polished, with parallel, longitudinal striations visible on the lower interior (Figure 30.28e). The body whorl is polished with pronounced rounding on high points along the broken aperture margin. The shell had been worn for some time before burial.

This marine snail species ranges from North Carolina to Florida and the Gulf states of North America and Mexico. This gastropod inhabits near-shore waters on shallow sand flats near inlets. The closest geographic source was presumably the Brazos River mouth. Lettered olive, the lightning welk (*Busycon perversem*), and the minute dwarf olive (*Olivella minuta*) were extensively traded prehistorically and often included in burials as ornaments and/or talismans in south Texas and Mexico (Ricklis 1994).

Coyote Teeth (718-5G to 721-5G)

Four perforated coyote (*Canis latrans*) canines were near the man's head and neck. Specimens 719-5G (Figure 30.29) and 718-5G (Figure 30.30) were examined most closely, while 720-5G is on long-term loan and unavailable for study (see Redder 1985:Figure 4F). A mandibular canine (721-5G) from a mature animal is cemented by calcite to the right rear occipital bone of the man's cranium.

All were biconically drilled within 6 millimeters of the root tip (Redder and Fox 1988). The perforations taper toward the center, creating hourglass profiles. The outer perforation diameter of 718-5G is nearly 4.7 millimeters; the interior diameter is approximately 2.3 millimeters. The outer perforation diameter of 719-5G is nearly 5.3 millimeters with an interior diameter of approximately 3.1 millimeters. Encircling striations visible on the perforation interior of 718-5G suggest manufacture with a slender drill (Figure 30.30d). Rounding and polish on the perforation margins nearest the root tips of

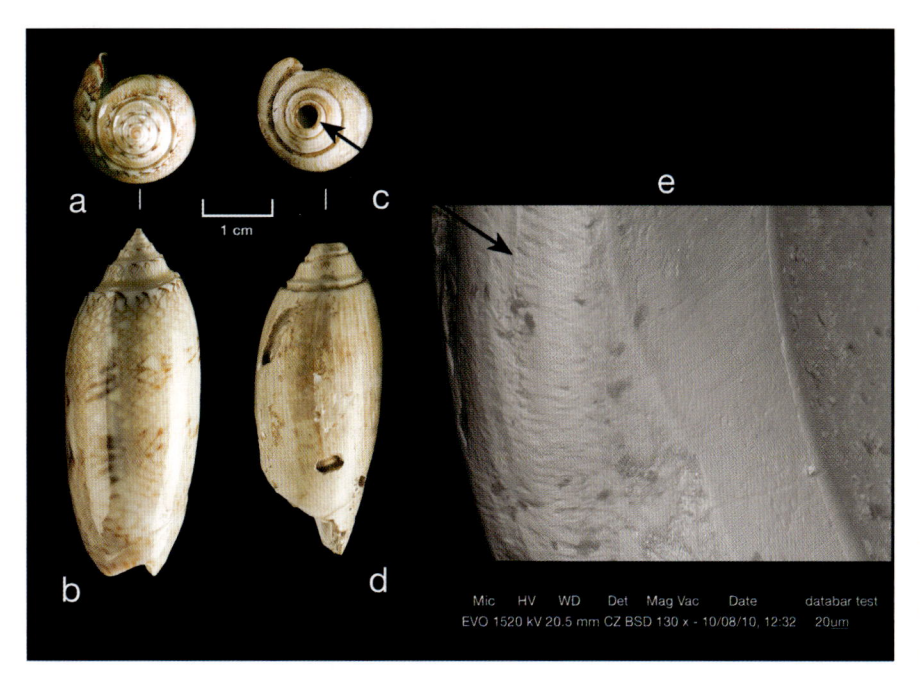

FIGURE 30.28 Lettered olive (*Oliva sayana*) shells. a–b) modern comparative specimen, USNM 404365, Long Boat Key, Florida; c) 738-5G, Horn Shelter No. 2, showing spire removed to create perforation; d) side view, 738-5G, showing natural attrition and predator scars; e) ESEM micrograph showing rounding and polish on the upper rim of the perforation and parallel, longitudinal striations on the perforation's lower wall.

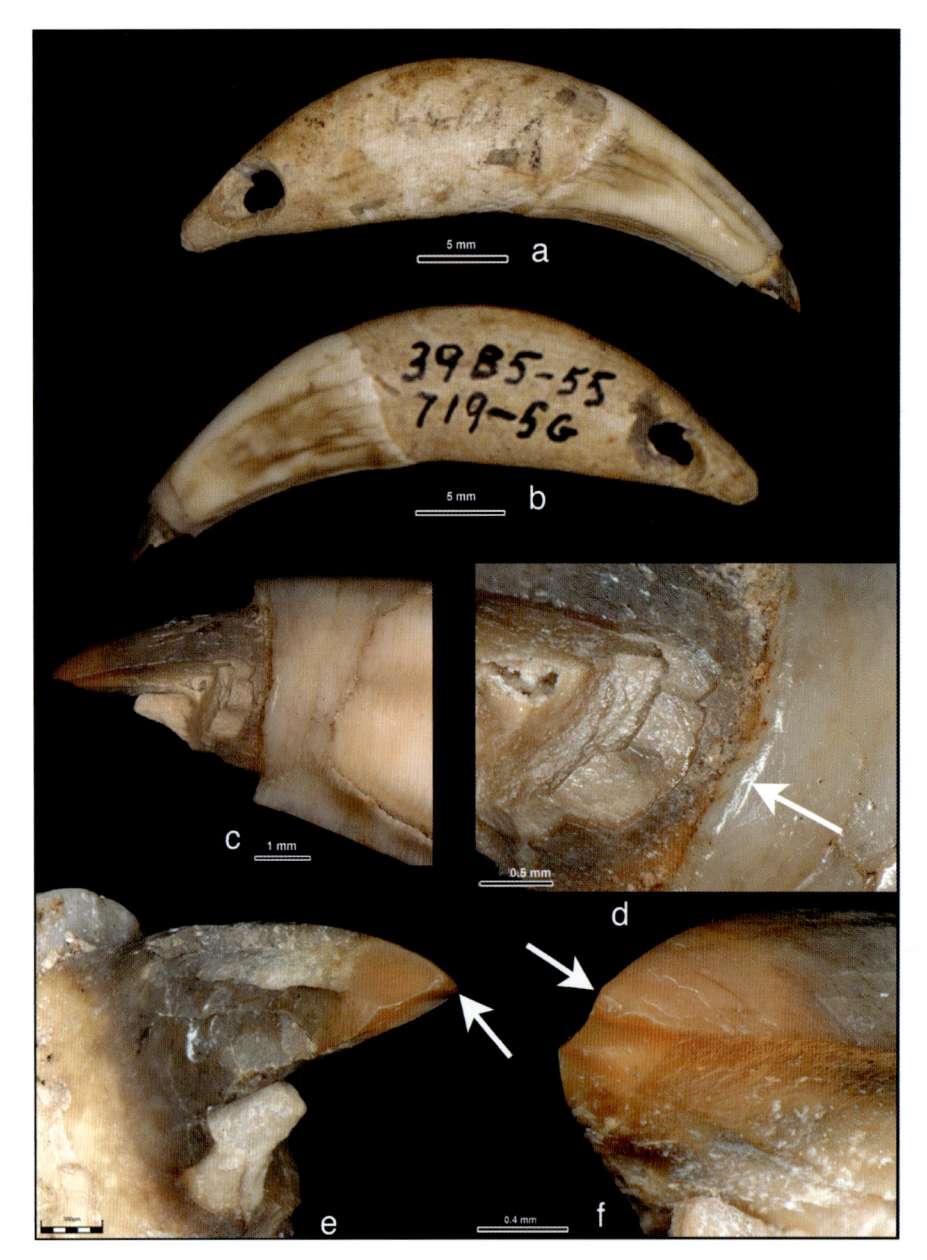

FIGURE 30.29 Perforated coyote canine shaped for tool use, possibly as a scarifier (719-5G). Composite micrographs: a–b) opposite views of perforated root and intentional removal of cusp-tip enamel; c) underlying dentin with enamel removed; dentin tip is only partially present as the lingual aspect is broken off; d) rounded and smoothed enamel margin; e) arrow points to use-wear rounding and polish on the very tip; f) use-wear rounding and polish on tip (arrow) compared with crisp, unpolished post-depositional fracture edges.

718-5G and 719-5G show that the canines had been worn for some time. Their recovered distribution suggests that they were around the man's neck at the time of burial.

The cusps of three of the canines have not been modified for tool use. The cusp of 720-5G is rounded and unbroken (Redder 1985:Figure 4F), as is the cusp of 721-5G. The cusp of 718-5G is longitudinally split by a 19.7 millimeter desiccation crack in the enamel and root (Figure 30.29). Half of the cusp is intact and shows natural enamel and dentin attrition that developed during the animal's life (Figure 30.29c, top); the lingual aspect of the cusp is missing a 2.5 millimeter-long fragment (Figure 30.29c, bottom). Fresh fracture edges and newly-exposed dentin indicate that this is a post-depositional break.

The cusp tip of 719-5G had been deliberately shaped and was used as a tool (Figure 30.30). Cusp enamel was removed from the tip around the entire circumference (Figure 30.30c), which exposed about 3.5 millimeters of the underlying dentin. Localized rounding and polish on the enamel margin (Figure 30.30d, arrow) reflects the enamel removal process and subsequent abrasion (possible tool use). The dentin tip is only partially present, as the lingual aspect is broken off (Figure 30.30c). A rem-

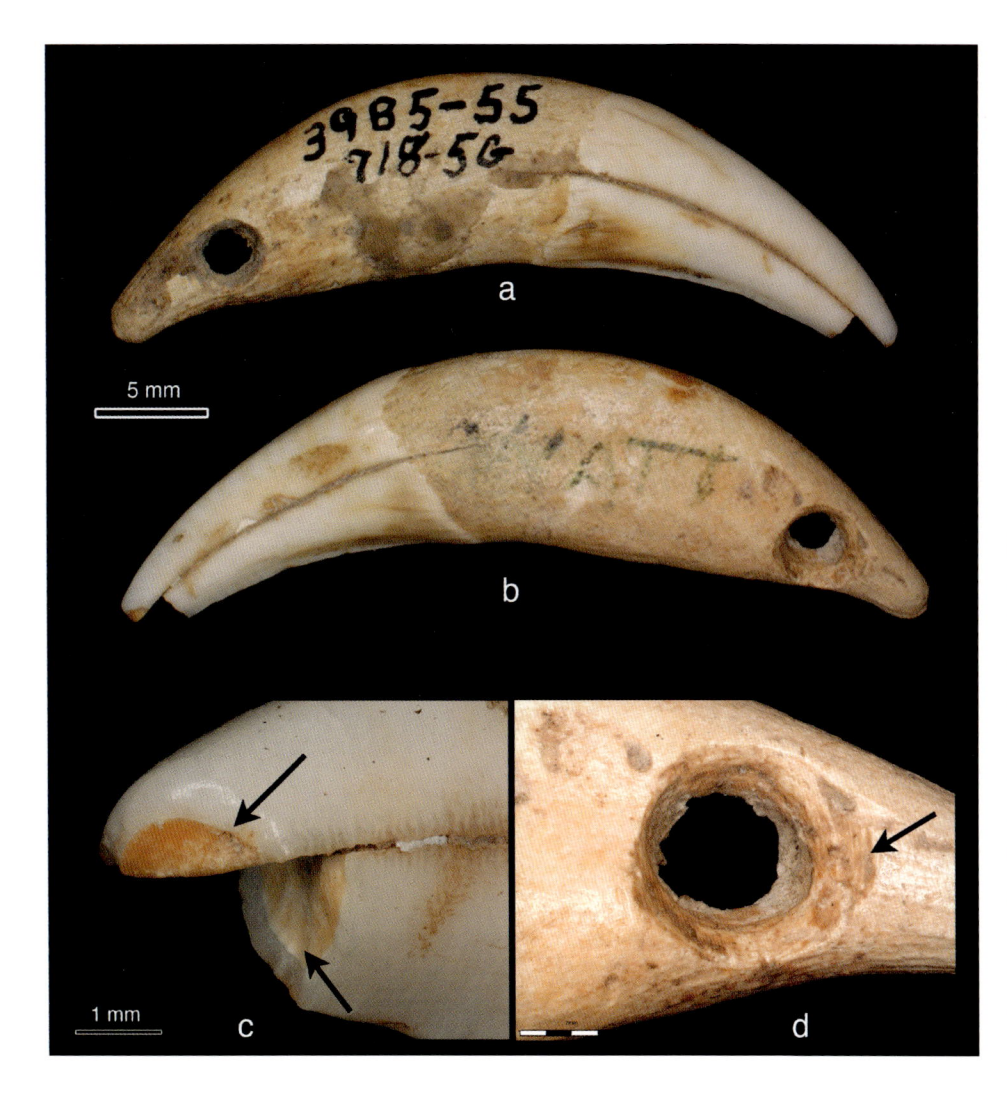

FIGURE 30.30 Perforated coyote canine (718-5G). Composite micrographs: a–b) opposite view of tip breakage and a linear desiccation crack; c) arrows point to natural enamel and dentin attrition during the life of the animal (above), post-depositional enamel and dentin fracture surfaces (below); d) arrow points to striations and polish in the perforation.

nant of the working end of the pointed tool is preserved. This tip is polished and slightly rounded (Figure 30.30e, arrow) from use.

The broken portion of the dentin tip shows multiple episodes of post-depositional breakage. The largest fracture split the tip longitudinally and exposed the pulp cavity (Figure 30.30c). The gray-brown discoloration seen on the fracture surface and adjoining tip remnant may suggest considerable age for this break. Additionally, there are microflake scars directed from two sides of the dentin tip. These post-depositional flake scars exhibit sharp, crisp fracture edges that are distinctly different from the adjoining edge of the tool tip (Figure 30.30), which is polished and slightly rounded.

For comparison, we examined published illustrations of perforated coyote canines from other archaeological sites. Of fourteen canines recovered in south Texas burials at the late prehistoric sites of Floyd Morris (41CF2) (Collins et al. 1969:Figure 9) and Ayala (41HG1) (Hester and Ruecking 1969: Figure 3), at least three show cusps chipped along desiccation cracks like 718-5G. Of seven with unbroken cusps, none are modified in a fashion similar to 719-5G. The perforated canines from Ayala and Floyd Morris were associated with perforated lettered olive shell beads and tinklers. The semi-circular distribution of shell ornaments and perforated canines from Floyd Morris suggested a necklace (Collins et al. 1969).

The dentin tip of 719-5G may have been used as a scarifier, i.e., a pointed tool intended to pierce human skin. Ethnographic and ethnohistorical accounts of incising and scarifying are numerous worldwide and minimally include bloodletting during initiations and other ceremonies, tattooing (both medicinal and decorative), and

Jodry and Owsley

scratching human skin during healing practices to more effectively apply infusions of medicinal plants (Mooney and Olbrechts 1932:70).

Identification of a probable scarifier (719-5G) at Horn Shelter No. 2 may indicate that bloodletting for ceremonial and/or medicinal purposes, perhaps tattooing, was practiced among San Patrice people 11,100 years ago. A 10,000-year-old scarifier has been reported from Hinds Cave in Val Verde County, Texas, nearly 450 kilometers southwest of Horn Shelter No. 2 (Andrews and Adovasio 1980: Figure 25; Deter-Wolf 2013a:Figure 2.5).

The Hinds Cave scarifier (Figure 30.31, right) consists of a sharpened antler tine (probably *Odocoileus virginianus*) set with resin or sap into "a complete, miniature, cylindrical, bi-mouthed coiled basket" (Andrews and Adovasio 1980:60). The tine is nearly 34.5 millimeters long and 9.1 millimeters in diameter at its base. The basket is nearly flush with the base of the tine, which was cut at right angles to its long axis. The basket, which seems to have served as a "finger-grip," is heavily worn and covers 26.6 millimeters of the total tine length. The tool tip is heavily polished and extends three to four millimeters beyond its basketry collar.

The Hinds Cave scarifier is reminiscent of 719-5G in terms of the lengths of the tool tips and the presence of encircling collars. The textile margin on the Hinds Cave

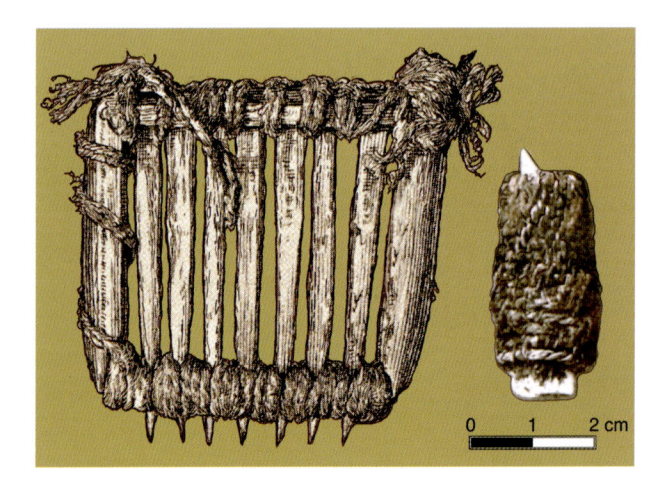

FIGURE 30.31 Scarifiers. Left) Cherokee *Ḵ'anv'Ga* scarifier made of sharpened turkey bones held in a turkey quill frame, adapted from Mooney and Olbrechts 1932:Plate 7; Right) 10,000-year-old antler tine scarifier encased in basketry from Hind's Cave, Texas, adapted from Deter-Wolf 2013a:Figure 2.5.

tool and the enamel border on the coyote tooth may have served as intentional guides that prevented the scarifier from penetrating too deeply into skin. This hypothesis is drawn from an ethnographic observation by Mooney, who worked with indigenous medicine people in the Southeast and Plains for nearly thirty years (1887–1918). Mooney and Olbrechts (1932:69) described a Cherokee scarification instrument (*Ḵ'anv'Ga*) (Figure 30.31, left) in the following way:

> It is made of seven splinters of bone of a turkey leg, set into a frame of a turkey quill . . . the seven splinters (about 5 cm. long and 3 mm. broad at the top; sharpened to a keen point at the bottom) are then struck through the bottom part of the frame, 1 or 2 millimeters of their cutting extremity piercing the frame at the bottom. With these seven sharp points the scratches are inflicted; and the ingenious way in which they are mounted prevents them from piercing so deep into the flesh as to inflict serious wounds.

Scarifier tip lengths (1 to 4 millimeters) may be functionally related to structural aspects of human skin, which includes a shallow epidermis (0.05 to 0.01 millimeters) underlain by dermis that varies in thickness from 0.5 to 5 millimeters. The uppermost blood vessels and nerve endings are encountered in the dermis. Experimental tattooing indicates that an effective tool allows the user to control the depth of penetration to avoid piercing the entire dermis. Should the latter occur, there is greater risk of infection and blurry tattoos can result (Deter-Wolf 2010).

Badger Claws (no catalog number)

Five unperforated badger (*Taxidea taxus*) claws were recovered from the double burial (Figure 30.32). Four are unpolished, devoid of calcite, and appear to be from the same animal. A fifth claw is coated in calcite. These items were not initially distinguished from raptor talons (Redder 1985; Redder and Fox 1988) and were later identified at the Smithsonian. One claw has the identifying label of "mouth cavity." Its provenience is confirmed in a color slide taken when Redder removed sediment from the man's mouth cavity and found it near an olive nerite shell bead and Swainson's hawk talons.

FIGURE 30.32 Badger claws (*Taxidea taxus*). a–d) appear to be from same animal; e) coated in calcite reflecting a difference in post-depositional history.

These items were likely displaced from the vicinity of the man's face and neck.

Among some North American indigenous groups, badgers are understood to have a special relationship with healing plants, whose roots lie underground where these nocturnal animals dwell. Among Hopi, for instance, badger clan members are thought to have an innate ability to cure and are ascribed supernatural power through their birth into this clan (Cushing 1883; Stephen 1936; Titiev 1972). Badger was the tutelary spirit of their *poswimkya* curing society whose members specialized in removing illness by means of sucking (Levy 1994).

Swainson's Hawk Talons (no catalog number)

Swainson's hawks (*Buteo swainsoni*) have one of the longest migration patterns of all North American raptors. They inhabit North America in the spring and summer and migrate in other seasons, primarily to South America. A few are known to winter along the Texas coast. Their appearance is visually striking with a dark reddish neck and upper chest, light colored feathers on their underside, and dark flight feathers.

Several hawk foot bones were found while excavating and screening burial fill near the man's head and neck (Figure 30.33). These items are well-preserved and lightly encrusted in calcite. Included among them are a left and a right first digit that appear to be from the same bird (Carla Dove, Program Manager, Division of Birds, National Museum of Natural History, Smithsonian Institution, 2006, personal communication) and a second and third digit, found in the man's mouth cavity, that articulate (Figure 30.33b–c).

Like the badger claws, the hawk talons are unperforated and unworn. The lack of polish (including proximal portions not covered by keratin) suggests that these talons and claws had a different history of cultural use and handling than the coyote teeth and marine shells. The former may have been offerings placed in the grave, while the latter were personal belongings worn by the man.

Perforated Antler Tool (716-5G)

A tool made from the distal portion of a left white-tailed deer antler was positioned in the burial with its perforation toward the man's right hip and its polished tip directed away from his torso (Figure 30.34). The tool is 182.5 millimeters long and encompasses portions of three tines. All intact margins (uncovered by calcite) show a light polish. Calcite occurs in sparse patches on the skyward surface; the groundward surface is more heavily coated (Figure 30.34b).

The slightly convex surface of the distal tine exhibits pronounced use-wear. Smoothing and high gloss polish cover the entire distal tool margin and extend onto the beam for nearly 12 millimeters (Figure 30.34f). Parallel oblique striations, oriented at nearly 45 degrees relative to the beam, crosscut the polish on the lateral margin near the tip (Figure 30.34e). While evidence for deliber-

FIGURE 30.33 Swainson's hawk talons (*Buteo swainsoni*). a) right first digit; b–c) articulated second and third digit; d–g) third digits.

ate use of the distal tine is clear, the function of either end of the tool remains uncertain.

A second tine, located near the tool's midpoint, was deliberately removed by a groove-and-snap technique. The encircling groove was not cut deep enough to permit a clean break and a remnant of the snapped tine remains (Figure 30.34g). Trabeculae exposed on this remnant are smoothed and polished, presumably due to actions related to tool use, handling, and/or transport.

At the junction of the third tine and beam is a perforation with a diameter of nearly 16 millimeters. Its interior surface is encrusted in calcite. The symmetry of this perforation suggests it was drilled (see Redder and Fox 1988). The perimeter of the perforation, if once a complete circle, is now hemispherical with fracture surfaces exposing cancellous tissue cemented with calcite (Figure 30.34c).

Redder and Fox (1988:5) and Story (1990:204) suggested that this perforated antler served as a shaft straightener. By inserting a shaft at an angle through such a hole and applying pressure on the inside of its curvature, a shaft can be straightened, while deliberate beveling and/or rounding of the hole's edges prevent denting the side of the shaft. For instance, the exterior margin of the perforation in the Clovis shaft wrench from Murray Springs, Arizona, is rounded and shows pronounced, purposeful beveling at its top and bottom (Haynes and Hemmings 1968; Haynes and Huckell 2007: Figure 5.17). In contrast, the perforation margin of 716-5G is not beveled or rounded. Rather, its interior and exterior surfaces adjoin at nearly right angles (Figure 30.34d). We do not concur that the perforation was used as a shaft straightener; function is uncertain.

Bone Needle (717-5G)

An eyed bone needle was found near the girl's waist (between her lower ribs and flexed femur). The needle is smooth and polished with longitudinal striations from manufacture and/or use. Its length upon recovery was 44.4 millimeters, with a thickness ranging from 1.7 to 2.5 millimeters. The diameter of the eye perforation was one millimeter (Redder and Fox 1988; Redder 1985:Figure 4d). This perforation later fragmented. Encircling striations (Figure 30.35a) suggest that the perforation was made using a rotary action with a stone tool. Experiments in bone needle manufacture, conducted in support of the analysis of Folsom needles from the Agate Basin site, Wyoming, indicate that eyes drilled with a "stone flake that was first given a pressure retouch along one edge and then snapped at an angle to form a sharp point" (Frison and Craig 1982:168) worked well and produced perforations similar to that on 717-5G. The needle's dis-

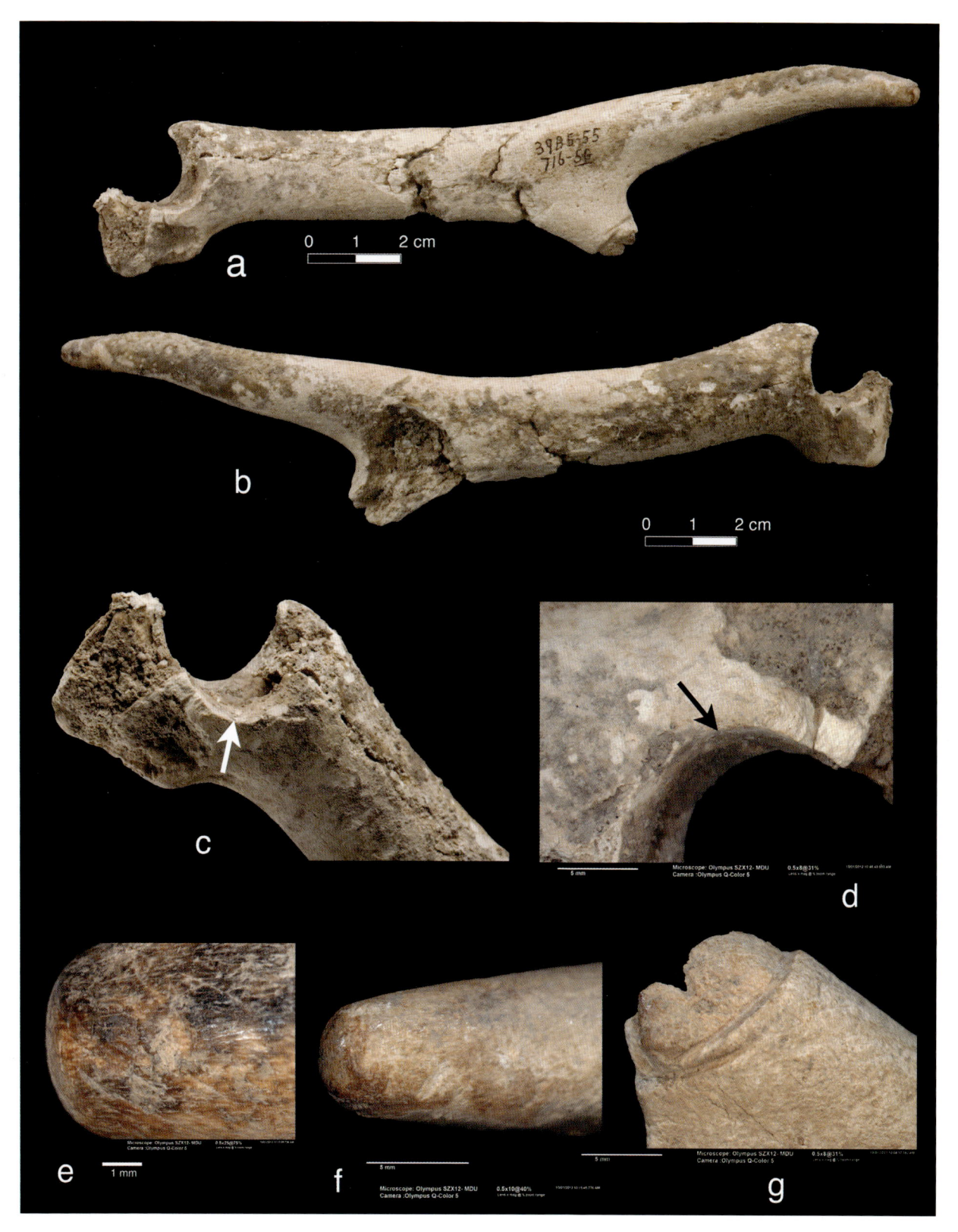

FIGURE 30.34 Perforated antler tool (716-5G). a–b) opposite views showing a perforated antler beam with a heavily polished and rounded tine; c) arrow points to area of perforation enlarged in d; d) composite micrograph showing unrounded, unpolished perforation margin with an edge angle of nearly 90 degrees; e) composite micrograph showing parallel oblique use-wear striations that cross-cut the polished side of the antler tine near the tip; f) micrograph showing smoothing and high-gloss polish on tip of the antler tine; g) composite micrograph showing cut marks made to remove the tine using a groove-and-snap technique; above this is a rounded and polished tine remnant that remained after snapping.

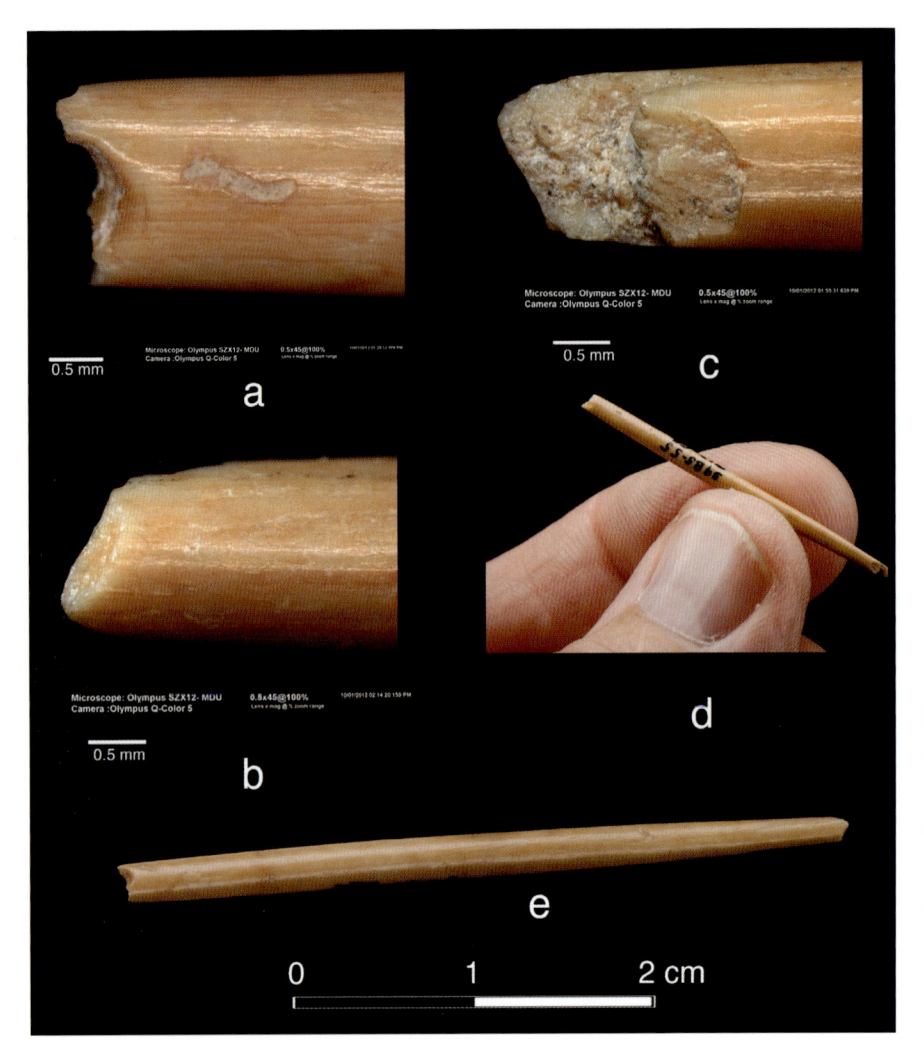

FIGURE 30.35 Eyed bone needle (717-5G). a) micrograph showing encircling striations within the perforation; b–c) composite micrographs showing green bone fractures on opposite sides of the needle tip; d) needle being held; e) side view of the smoothed and polished needle exterior.

tal tip was broken on opposite sides with green bone fractures (Figure 30.34b, c). These fractures evidently occurred prior to, or at the time of, burial.

FUNCTIONAL AND BEHAVIORAL ATTRIBUTES OF THE ADULT SKELETON

The human skeletons represent a middle-aged man and a girl aged 10 to 11 years. The skeletons are in good condition, despite their antiquity. Descriptions of the bones with illustrations were first presented in Young et al. (1987) and Young (1988). This addendum focuses on the adult male, specifically observations pertaining to health and arm bone morphology. This emphasis utilizes the biomechanical model presented by Wescott (Chapter 12), i.e., that bone development and architecture are shaped by loading forces and functional demands.

Evaluation of the auricular surfaces of the innominates and bone cortices and the lack of arthritic changes suggest an age of 37 to 44 years for this male. The auricular surfaces of the pelvic bones are uniformly dense with some striae, have only trace amounts of microporosity, and have no apical rim changes. The density of the cancellous bones is consistent with middle age. A few sternal rib ends are present and there is only slight cupping. There is trace development of intertrochanteric spicules on the right femur. Comparison of relative rib cortical area (cortical area divided by the total subperiosteal area) to an archaeological reference series is consistent with an age in the thirties (Sam D. Stout, Professor, The Ohio State University, 2006, personal communication).

As for his dentition, the man's bite pattern was edge-to-edge. Most of his teeth show near-complete loss of their crowns from attrition. The wear planes of the anterior teeth are fairly uniform in height with buccal-lingual rounding suggesting use in task activities (Young et al. 1987). The posterior molars are not quite as worn; masticatory function was maintained through adjusting wear planes among opposing teeth. Secondary dentin is evident in the infilled pulp chambers of most of the teeth.

Dental pathology was the most life-threatening health risk identified in the adult. Extreme wear in the maxillary dentition resulted in the loss of the right first incisor and abscessing of five additional teeth—the right molars and second incisor and the left first molar. The right third molar has a large, mesial interproximal cavity that cut into the root and adjacent enamel. In the mandible, the right first incisor, both first molars and the left second premolar were in the process of abscessing, and the left second molar had abscessed out and its socket was filling in. The left third molar has a carious lesion in its mesial interproximal surface, which began at the cementoenamel junction and was cutting into the enamel.

The skeleton exhibits little pathology, notably activity-related minor arthritic lipping in the elbow joints and slight margin lipping of the right hip socket (acetabulum). Multiple transverse growth-arrest lines are evident in radiographs of the femora and tibiae, indicating periods of illness or nutritional deficiency during childhood (Young 1988). The left fifth metatarsal has a healed, overriding, midshaft fracture, which was likely a source of discomfort.

The postcranial skeleton is moderately robust, with a medium build displaying well-developed muscle attachments. Young et al. (1987:281) described upper body development:

> Both humeri have pronounced deltoid crests and well-defined, roughened ridges for the attachment of the pectoralis major muscles. The radii exhibit very pronounced interosseous crests, with the left being extremely pronounced, suggesting strong, well-used arms. The interosseous borders of the ulnae are also well-defined, with the left, as in the radius, being more pronounced than the right. The right clavicle, however, is visibly more robust than the left, with a more pronounced conoid tubercle.

Several features indicate right-handedness. The right glenoid cavity (scapula) is slightly larger than the left, and its posterior margin is more rounded. The axillary ridge below the glenoid is thicker on the right side; the right ridge thickness 20 millimeters below the glenoid margin is 6.4 millimeters across while the left side measures 5 millimeters. Cross-section measurements of the right clavicle are larger than those of the left. The antero-lateral border of the right clavicle is thicker than the left, reflecting greater use of the deltoid muscle. The moderately developed deltoid tuberosity on the right humerus is larger than the left. The right ulna also displays slight arthritic lipping and microporosity of the semilunar joint surface, suggesting more strenuous use of the right elbow joint.

The scapulae have elongated coracoid processes—the origins of the coracobrachialis muscle and the short head of the biceps brachii. Coracobrachialis flexes and weakly adducts the shoulder while biceps brachii flexes the elbow, supinates the forearm, and weakly flexes the shoulder. The lateral border of the left scapula has a well-defined origin for the long head of the triceps brachii which extends the elbow; this attachment is only slightly defined on the right scapula. The acromion process of the left scapula has a well-defined origin for the deltoid muscle and a moderately defined insertion for the trapezius; the right scapula has only a moderately-defined origin for the deltoid and relatively no definition for trapezius.

The humeri are straight and round with moderately defined muscle attachments (Figure 30.36). Both left and right have a moderate crest of the greater tubercle, the insertion of the pectoralis major muscle, which adducts and medially rotates the upper arm. Both also have a defined lateral epicondylar ridge, origin for the brachialis, brachioradialis, and extensor carpi radialis longus muscles. Actions associated with these muscles are elbow flexion (brachialis and brachioradialis) and extension of the wrist (extensor carpi radialis longus). The left humerus has a rugged medial epicondyle, a common origin for muscles of elbow and wrist flexion. The left humerus also has a rugged lesser tubercle, the insertion for the subscapularis muscle of the rotator cuff which stabilizes the shoulder joint, and a prominent crest of the lesser tubercle, the insertion for the latissimus dorsi, which extends, adducts, and medially rotates the upper arm. The right humerus has a defined deltoid tuberosity, the insertion for the deltoid muscle, which abducts and can medially or laterally rotate the upper arm.

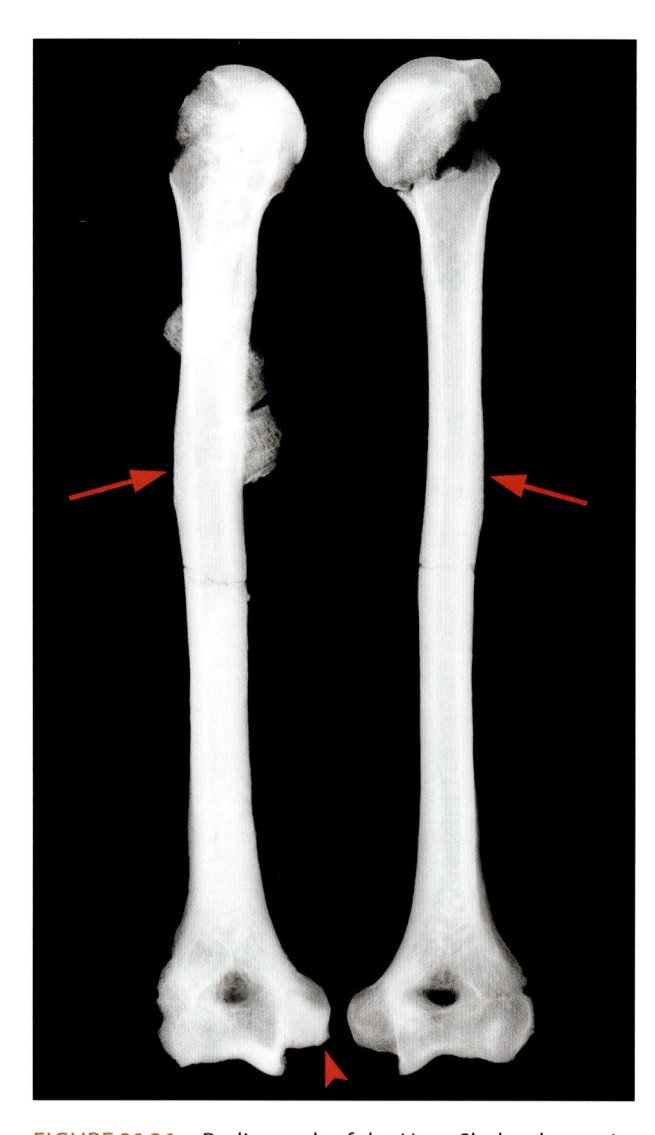

FIGURE 30.36 Radiograph of the Horn Shelter humeri. Development of the deltoid tuberosities (arrows) is greater on the larger right humerus (left). The left humerus has a septal aperture (perforated olecranon fossa). Compared to Kennewick Man (Figures 7.19, 7.21, and 7.22), these humeri are smaller, rounder, straighter, less asymmetrical, and show less development of the right ulnar collateral ligament (arrowhead). Rectangular shadows show vertebral centra cemented to the right humerus by carbonates.

The radii show bilateral hypertrophy of the interosseous crests. The proximal middle third of the left ulna was also affected. The crest is more pronounced on the left forearm than on the right (Figures 30.37 and 30.38). Two primary muscles are related to this hypertrophy. The muscle attachment located on the anterior midshaft of the radius represents the origin of flexor pollicis longus, which flexes the first digit (thumb). The other pronounced attachment located on the posterior midshaft of the ulna signifies the origin of abductor pollicis longus, which moves the first digit laterally. Other well-defined attachments on the ulnae and radii include the triceps brachii, anconeous, brachialis, flexor digitorum superficialis, abductor pollicis longus, and extensor indicis muscles. These muscles correspond with flexion and extension of the elbow, flexion of the wrist and hand, and extension/abduction of the first digit. Also, attachments for the supinator, pronator quadratus, and pronator teres muscles are well defined and correspond to rotation of the forearm. The triceps brachii, brachialis, and supinator muscles are relatively more well defined on the right arm and are associated with the actions of extending and flexing the elbow and laterally rotating the forearm. The flexor pollicis longus, abductor pollicis longus, extensor indicis, and pronator quadratus muscles are more well defined on the left forearm and are associated with the actions of flexing the first digit and moving it laterally, extending the index finger, and rotating the forearm medially.

To summarize, definition of upper extremity muscle attachments on both sides of the body show that this individual was performing actions of shoulder flexion and adduction, medial rotation of the arm, elbow flexion, wrist extension, forearm supination, as well as thumb flexion and abduction. The latter two motions are associated with the hypertrophied interosseous crests. Opposite forces are also indicated. The left upper extremity shows defined muscle attachments associated with the actions of shoulder extension, elbow extension, and wrist flexion. The right upper extremity shows defined muscle attachments associated with the actions of shoulder abduction, lateral rotation of the arm, forearm pronation, and index finger extension.

The proximal ends of the second and third metacarpals are fairly rugged on both hands. This location is consistent with the insertions of the flexor carpi radialis muscle (anteriorly) and the extensor carpi radialis longus and brevis muscles (posteriorly). These muscles produce flexion and extension, respectively, of the elbow and wrist. The phalanges have slight marginal palmar ridge development, another indication of active use.

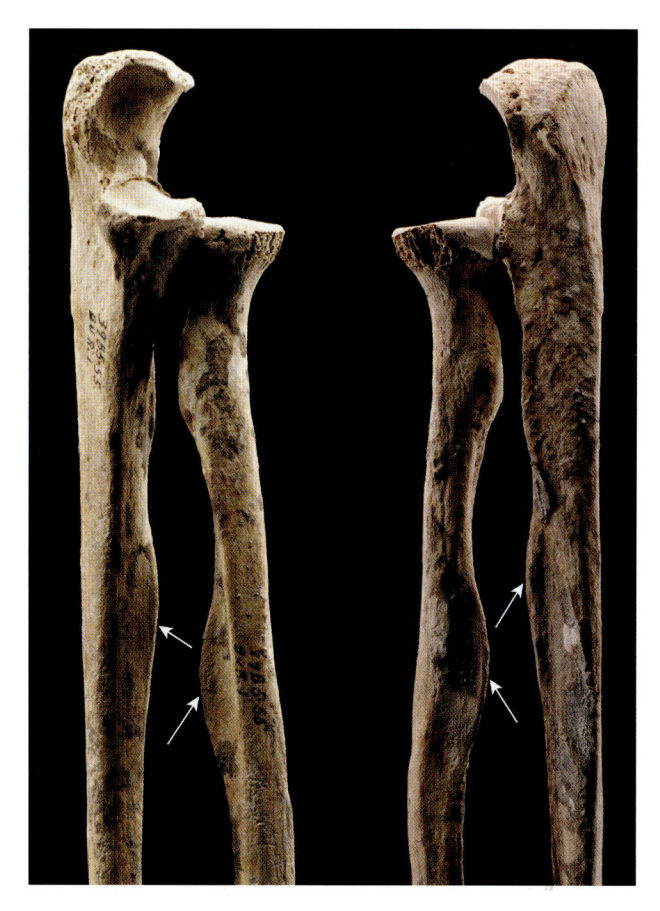

FIGURE 30.37 Anterior (left) and posterior views of the Horn Shelter left ulna and radius showing expansion and thickening of the interosseus crests (arrows).

Kennewick Man was larger, in terms of both stature and mass, than the male from Horn Shelter. The Horn Shelter man's height of 164.5 centimeters (approximately 5 feet 5 inches) was about 7 centimeters (almost 3 inches) shorter than Kennewick Man, and he weighed four and a half kilograms (nearly 10 pounds) less (Chapter 11). While Kennewick Man's hand bones are 11 percent longer (Chapter 13), the hands of the man from Horn Shelter were more robust relative to their length based on right first and second metacarpal length/width indices. First and second digit base robusticity relative to length is indicative of repetitive use of a precision grip.

Their morphological profiles also differ from one another. Kennewick Man's strongly developed right humerus, strengthened in response to unilateral torsion and medial-lateral bending, characterizes him as a hunter (Table 12.4). The Horn Shelter male's developed and sculpted upper limb bones are indicative of diverse muscle use in strenuous, sustained, and repetitive activities. The Horn Shelter male had strong arms with rounder humeri and less bilateral asymmetry (Figure 30.36). The cross-sectional geometry of his distal humerus shows slightly greater anterior-posterior bending strength than medial-lateral, whereas the opposite was true for Kennewick Man. But the development of the forearm musculature is of special interest and strongly suggests that the Horn Shelter male was preoccupied with activities other than hunting.

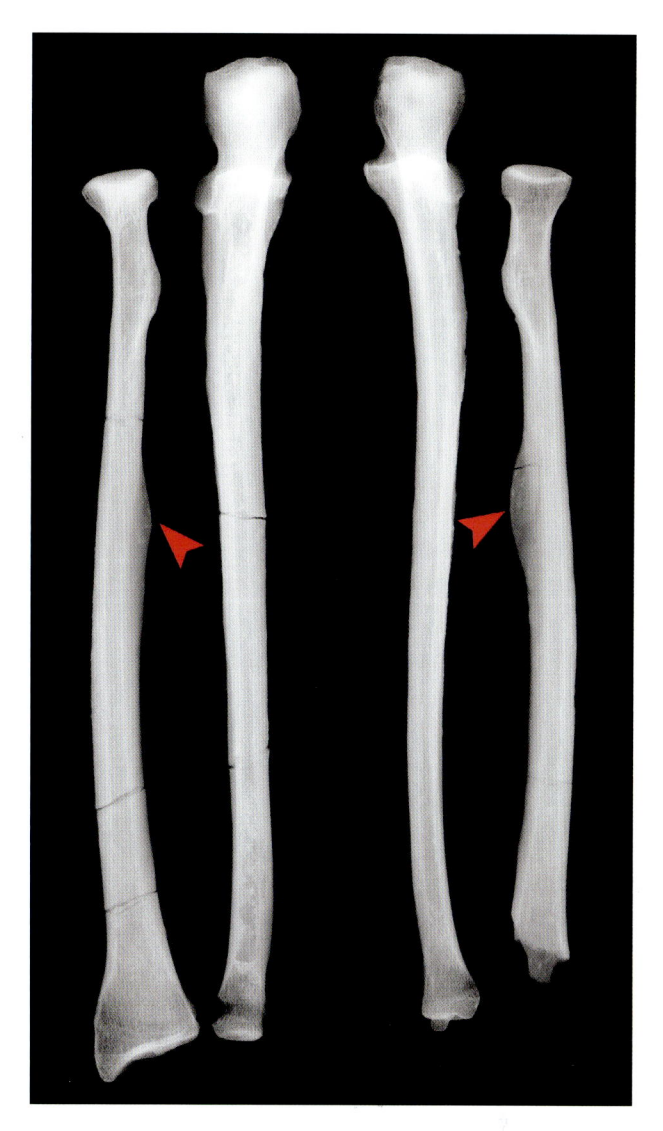

FIGURE 30.38 Radiograph of the radii and ulnae of the Horn Shelter male. This placement illustrates interosseus crests in the radii and shows greater development on the left radius. Kennewick Man lacks pronounced development of these crests (Figure 7.27).

DISCUSSION

Recent study of the Horn Shelter double burial provides new insights into technological, social, and ideological practices of San Patrice people who lived along the Eastern Woodland and Southern Plains peripheries nearly 11,100 years ago. The form and placement of the grave, pairing of two individuals of different age, sex, and status, a marked disparity in their belongings, and anatomical attributes of the adult collectively lend themselves to a level of interpretation that is beyond the reach of most Paleoamerican research.

The San Patrice heartland lies in Texas and Louisiana, with activities reported over a wider area to the north and west (Figure 30.39), including hunting buffalo (Hurst et al. 2010; Willey et al. 1978), acquiring stone (Jennings 2008; Ray 1998), and interacting with groups such as Dalton and Cody (Bousman et al. 2002, 2004; Bousman and Oksanen 2012; Johnson 1989; Owsley et al. 2010; Ray 1998; Ray et al. 1998). Located within this heartland is Horn Shelter No. 2, a prominent place on the banks of the Brazos River, the longest watercourse in Texas. This important corridor of travel, social interaction, and information flow connected the Gulf Coast and Southern Plains. San Patrice projectile points are abundant on the Upper Texas coast (Stright et al. 1999) and occur near the ancestral Brazos River headwaters in eastern New Mexico, near Portales (Hurst et al. 2009). Also important was the presence in Central Texas of Edwards chert, a highly-prized stone.

The double burial was placed in a central location in the south end of the shelter and was prominently marked by a turtle shell-like covering of 19 stones. These distinctions seem an apt foreshadowing of what lie beneath: a man with a personal bundle beneath his head and likely talismans near his head and chest. Lying behind him is a girl with a bone needle, shell beads, and perhaps perishable items that remain archaeologically invisible.

FIGURE 30.39 Map showing distribution of San Patrice sites. A dashed line depicts an approximate periphery of San Patrice site distribution; light green highlights the core area, which includes Horn Shelter No. 2.

Both individuals were facing westward. The cause(s) of their deaths are uncertain. Story reasoned (1990:205),

> the apparent simultaneous, or nearly simultaneous, death of two individuals suggest either that some sort of calamity had occurred or that one of the individuals was sacrificed. Given the richness of the offerings associated with the adult male, the latter seems possible. Regardless, the offerings strongly suggest that the status of the adult was different and that his loss was more deeply felt by the society. This would particularly be expected if the adolescent were a female and, hence, possibly the young wife of the adult.

Hunter-gatherers are conventionally conceived as egalitarian groups. Yet Hayden (1993a 1993b, 1995, 1998, 2003; Owens and Hayden 1997) argues for the recognition of transegalitarian hunter-gatherers (i.e., intermediate between egalitarian and class-structured societies) in Paleolithic Europe (see also Soffer 1985) and in the Northwest Coast of North America. Similarly, Speth et al. (2013) stress the importance of identifying social and political underpinnings of Paleoamerican lifeways. This study explored the context and content of mortuary items associated with a San Patrice adult male and integrated this information with osteological observations to infer recurring activities in his life and their sociocultural implications.

Possible Status and Role of the Adult Male

The Horn Shelter adult is the person most richly bestowed with funerary items in a comparative sample of early Holocene rock shelter burials. Study of tools found near his head and neck suggests that he served as a medicine person, perhaps a shaman, who was engaged in healing and other ceremonial practices that involved body painting and incision. The term medicine person follows Mooney and refers to individuals who treat the sick and cure diseases, but also serve in various capacities (depending on individual gifts, training, and lineage) as shamans, magicians, and/or divinators.

> To have an adequate idea of the social status of the medicine man we should bear in mind that in his person we find cumulated such professions and pursuits which in our society would correspond to those of the clergy, the educators, the philosophers and the historians, the members of the medical profession in its widest sense, i.e., physicians, surgeons, and chemists; and finally, to a certain degree, even to those of the politicians and the press. (Mooney and Olbrechts 1932:92).

Medicine people are universal among traditional indigenous people and shamanism exists among most hunter-gatherers (Eliade 1964; Harner 1980; Hayden 2003; Winkelman 1990). The antiquity of shamanism is established for the Upper Paleolithic (Hayden 2003; Winkelman 1990) in both rock art (Clottes and Lewis-Davis 1998) and burials (Grosman et al. 2008; Oliva 2000; Porr and Alt 2006). In Texas, shamanism is revealed in four-thousand-year-old rock art in the Lower Pecos Region (Boyd 1996, 2003) and in protohistoric burials from the Upper Texas Coast (Ricklis 1994, 1996).

Hultkrantz (1973:34) defines a shaman as a social functionary who, with the help of guardian spirits, attains an altered state of consciousness in order to create rapport with the supernatural realm on behalf of group members. "In order to understand shamanism, one must understand the traditional concept of the world as a sacred place imbued with vital forces that can be tapped into for one's benefit or that can erupt unexpectedly" (Hayden 2003:87). Animals, minerals, plants, elements of earth, air, fire and water, geographic places (particularly springs, waterfalls, and caves), celestial bodies, and ancestors hold life force and power (Moyes 2012; Vitebsky 2001).

Shamanic healing is based in a holistic view that integrates physical, mental, spiritual, social, and ecological aspects of well-being. In this view, imbalance and breakdown arise with some breaching of the interconnectedness of life and its sacred natural laws. To reestablish harmony and health, shamans enter an altered state of consciousness, travel to spirit realms, and consult with ancestors and other helping spirits to discover a therapeutic course of action. As shamans depend on the assistance and protection of spiritual allies, they cultivate good relations with them. Material items play a role in all of this.

The Material Culture of Scarification, Body Painting, and Tattooing

Ethnographic accounts reveal that scarification, body painting, and tattooing were embedded in systems of ceremonial and ritual action that expressed metaphysical and social conceptions of the world. Body modification was undertaken to mark significant life achievements and group identities, to enhance beauty, as medicinal remedies, and to embody the recipient with supernatural properties that promoted fertility, conveyed physical and spiritual protection, and attracted mates and prey animals (Krutak 1999, 2009, 2012). Marks and incisions that were intended to enhance one's access to, and support from, spirit realms (such as those received in rites of passage) were often applied by, or in conjunction with, a shaman or priest (Krutak 2012, 2013). Researchers have recently attempted to identify the material record of tattooing.

Deter-Wolf (2013a:Table 2.1) compiled a list of 48 indigenous tattooing implements and 37 descriptions of tattoo pigments used in the Eastern Woodlands and Plains and proposed an identifiable tattoo kit comprised of the following: incision and piercing tools, materials for creating pigment, pigment containers, pigment processing implements, utensils for mixing pigment to create paint and for its application to the skin, an enclosing bundle to contain these items, as well as medicinal materials, personal adornments and other ceremonial accoutrements. However, Deter-Wolf (2013a:67) explains that:

> Of the entire tool kit, only the actual needles and pigment containers (depending on material type), some vestiges of the pigments themselves, and bone or stone materials used for pigment processing, tool repair, or as ritual accoutrements may survive and eventually be recovered through excavation.

Table 30.3 compares Deter-Wolf's expectations with items in the double burial. All expectations are met, with the exception of a bundle cover, which was likely made of perishable materials. There is strong evidence that Horn Shelter tools were used in paint production, body painting, and scarification. In the absence of multiple tattoo needles in the bundle, definitive evidence of tattooing is less certain.

Additional early assemblages with ochre-related tool kits are the Magdalenian habitation site of Mas d'Azil, France (Figure 30.40) and the Bad Dürrenberg double burial in Germany (Porr and Alt 2006). Numerous ornaments, talismans, and tools in the latter accompanied an adult female shaman including turtle shell containers, grinding equipment, two needles, bone spatula with ochre residue, stone flakes, and a crane bone containing 31 microliths (possible scarifiers) buried in a 30-centimeter-thick layer of red pigment.

Table 30.4 places these and other sites in a timeline based on archaeological evidence for pigment processing and body modification. The earliest suggestions of body painting (d'Errico et al. 2009; Watts 2002, 2009) and tattooing (Deter-Wolf 2013a, 2013b) are inferred from an ochre processing workshop, ochre nodules shaped as crayons, and pointed bone tools at Blombos Cave, South Africa. Terminal Pleistocene and early Holocene body modification tools are represented at Mas d'Azil, Horn Shelter No. 2, Hinds Cave, and Bad Dürrenberg. The earliest direct evidence of tattooing is an eight-thousand-year-old adult male mummy from the Chinchorro culture, South America, with a tattoo on his upper lip (Allison 1996). Therapeutic tattoos on the Tyrolean Ötzi mummy indicate that a form of healing similar to Chinese acupuncture was being practiced 5.2 thousand years ago (Dorfer 1998, 1999) and a one thousand year old Chiribaya mummy from Southern Peru bears both therapeutic and ornamental tattoos differentiated by dissimilar forms, locations, and plant-based residues (Pabst et al. 2010).

Indigenous peoples in Texas extensively practiced tattooing and scarification (Newcomb 1961; Berlandier 1969). An account by Jean-Baptiste Talon in 1687 describes a scarification tool used by a shaman-curer on the Upper Texas Coast,

> Beside the perfect knowledge they have of the properties of medicinal herbs, as has been said, they draw blood and practice the sucking of wounds. To draw blood, they have little combs, made with the teeth of a kind of large rat found in that country, with which they scrape and tear the skin, [at] the very spot where one feels pain. When they want to cure a headache, they prick the skin in several places, depending on

TABLE 30.3 Comparison of items in indigenous tattoo tool kits with items associated with Horn Shelter No. 2 adult.

Items in Indigenous Tattoo Tool Kit (Deter-Wolf 2013a)	Items Associated with Horn Shelter Male	Catalog Number	Comments
Incising and piercing tools	Coyote canine scarifier Edwards chert biface as possible source of incising flakes?	719-5G 723-5G	Sharpened coyote canine Biface as possible source of incision flakes?
Raw material for creating pigment	Hematite nodule	726-5G	Utilized hematite nodule with impressions that may suggest it was wrapped in plant material
Container for holding pigment	Turtle shell bowls	729-5G	3 box turtle carapace bowls, 2 of which contain hematite residue
Implements for processing pigment	Sandstone abraders Antler pestles	724-5G; 725-5G 727-5G; 728-5G	Abraders with use-wear and pigment residue Probable use of pestles to process hematite
Utensils for mixing pigment with a liquid and applying pigment to skin	Stylus	722-5G	Bone spatula with use-wear on tip
Enclosing bundle to encase these items			Tight grouping of items under the man's head suggests presence of a container; impression on hematite may suggest it was wrapped in plant material
Medicinal materials	Hematite nodule	726-5G	Iron compounds in hematite create a powerful astringent effect, arrest hemorrhage, have antiseptic properties, and promote healing (Wilcox 1911 in Velo 1984); Gugadja, northwest Australia, used ochre as medicine, red preferred (Peile 1979)
Personal adornments worn by tattooist	Perforated coyote canines and marine shell beads and pendant	718-5G to 721-5G; 737-5G; 738-5G	The perforated and polished coyote canines and marine pendant appear to have been worn around his neck; the smaller shell beads may have adorned clothing
Ceremonial accoutrements		Not assigned	Badger claws and Swainson's hawk talons?

where the pain is, and then they suck the blood, which spurts out with great force. These practices are successful. (Bell 1987:225).

A tool of this description was found in a Late Prehistoric shaman burial (440 ± 70 RC yr. BP) at the Mitchell Ridge site, Galveston Island, Texas. It consisted of a single row of 18 sharpened hispid cotton rat incisors inset into an asphaltum handle (Ricklis 1994:Figure 5.29).

Two shark's teeth were placed in this healer's left hand. A nearly identical rat tooth tool was found in an eighteenth century burial at the Spanish mission of Espíritu Santo in Goliad, Texas, with an individual who also had two shark teeth resting in the left hand (Ricklis 1994). These scarifiers made of sharpened animal incisors are similar to the turkey bone scarifier used by Cherokee healers in the nineteenth century (Figure 30.31). The identification of a San Patrice scarifier in this study extends the practice

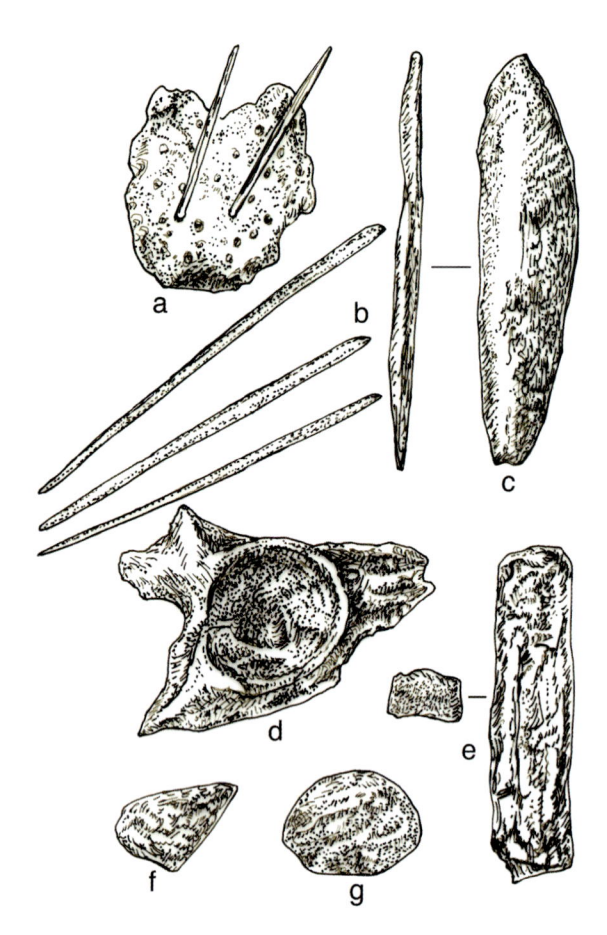

FIGURE 30.40 Upper Paleolithic modified red ochre and bone tools with ochre residue from Mas d'Azil, France (adapted from Péquart and Péquart 1962:Figure 157). a) red ochre and clay palette perforated with small holes; b) possible bipointed tattoo implements and *bâtonnets* with tapered tips and flattened ends; c) polished and stained spatula; d) reindeer innominate used to hold and process pigment; e) tool used to crush and grind pigment; f) red ochre shaped into "crayon"; g) polished red ochre nodule with use-wear from scraping or grinding.

of bloodletting with scarifier made of sharpened animal teeth to 11,100 years ago.

Shamanic Realms of Animal Familiars and Drumming

Many traditional people conceive of a cosmos comprised of Upper and Lower worlds in addition to the Middle world where humans live. Shamans are specialists in journeying to the Upper and Lower realms and are the ones with the necessary ability, spiritual allies, and responsibility to do so (Keeney 1999, 2000a, 2000b, 2003).

> The shaman's activity is based on ideas of space, and although the everyday world is permeated by spirits there are also other, separate realms to which shaman must travel. If one assumes that spirits exist, and that they exist in a different realm from ours and reach out to affect our health and our food supply, then it follows that when these things are disrupted someone must travel into the realm of the spirits to persuade them to behave differently. (Vitebsky 2001:15)

Animals aid shamans in their spiritual journeys. Birds, in particular, are associated with shamanic flight and carrying messages to the upper world (Davenport and Jochim 1988; Hatcher 1927–28; Hayden 2003; Vitebsky 2001; Winkelman 2000:62). Therefore, items from birds are frequently part of shamanic clothing, tools, and other accoutrements. Shamans also visit animal spirits residing in the underworld to obtain help needed to cure disease. Material items connected with particular underworld allies are also found among a shaman's belongings.

> Pawnee narratives tell how doctors obtained curing power by visiting councils of animal spirits in the underworld. At Waconda Spring, Kansas, the spirits in the council included beavers, otters, wolves, dogs, bison, bears, jackrabbits, muskrats, and mudpuppies (salamanders), along with species associated with the sky world, such as eagles and hawks (Murie [and Parks] 1981) (Blakeslee 2012:356).

Swainson's hawk talons and badger claws found near one another in the double burial were unperforated and both groups of bones shared a similar handling history that differed from the perforated items. Hawk and badger seem to be an ideologicly-related pair of animals in the context of this burial that may represent the sky and earth, the Upper and Lower worlds. The Osage, for instance, were divided into moieties symbolic of sky and earth (La Flesche 1916) and these symbols form an essential part of the design of their tattooing bundle covers (Krutak 2012). The association of badger with earth,

TABLE 30.4 Archaeological evidence of tattooed mummies and possible body painting, scarification, and/or tattooing tools.

Approximate Age (CAL yr. BP)	Culture or Era	Site or Area	Inferred Activities and Tattooing	Archaeological Evidence	References
84 to 100,000	Middle Stone Age	Blombos Cave, South Africa	Body painting and/or tattooing	2 tool kits for pigment manufacture, bone spatula with red pigment residue, abalone shell containers, pointed bone tools	Henshilwood et al. 2011; Deter-Wolf 2013a, 2013b
12,000	Upper Paleolithic Magdalenian	Mas d'Azil, Southern France	Body painting and/or tattooing	Modified red ochre nodules, perforated ochre and clay palette, reindeer bone container with ochre residue, bone spatula with ochre residue, pointed bone tools with ochre residue, pestle used to crush pigment	Péquart and Péquart 1962
11,100	San Patrice	Horn Shelter No. 2, Texas	Body painting, incision, and/or tattooing	Modified red ochre nodule, 3 turtle shell bowls (2 with red ochre reside), bone spatula, 2 abraders with red pigment residue, 2 antler pestles, and a chert biface; all contained in a probable bundle	this volume
10,000	Early Holocene	Hinds Cave, Texas	Scarification	Scarifier with use-wear made of sharpened antler tine encased in basketry "finger-grip"	Andrews and Adovasio 1980; Deter-Wolf 2013a
9,000 to 8,200	Late Mesolithic	Bad Dürrenberg, Central Germany	Body painting, incision, and/or tattooing	Turtle shell containers, bone spatula with ochre residue, grinding equipment, 2 bone needles, a crane bone containing 31 microliths, additional stone flakes	Porr and Alt 2006
8,000	Chinchorro	Northern Chile /Southern Peru, South America	Tattooing	Mummy of adult male with thin pencil moustache on upper lip	Allison 1996
5200	Chalcolithic	Ötzal Alps Austrian-Italian border	Tattooing	Mummy of adult male (Ötzi) with 15 groups of tattoo lines on the back and legs, opinions of 3 acupuncture societies indicate that 9 of the tattoos are located within 6 mm of traditional acupuncture points and represent therapeutic tattoos implying a form of medical therapeutics similar to Chinese acupuncture	Dorfer et al. 1998; Dorfer et al. 1999; Pabst et al. 2009
3800	Middle Holocene, Pre-Lapita	West New Britain, Papua, New Guinea	Incision during healing and/or tattooing	Obsidian unretouched and retouched flake tools with blood residues, red and black pigment residues, and use-wear striations like those produced experimentally during skin incision	Kononenko and Torrence 2009; Kononenko 2011, 2012
2500	Scythian	Altai Mountains	Tattooing	Mummy of adult male with decorative tattoos on arms and upper back and non-decorative, therapeutic tattoos along the spinal column	Dorfer et al. 1999; Pabst et al. 2009
1,000	Chiribaya	Chiribaya Alta, Southern Peru	Tattooing	Female mummy with decorative tattoos of animals and symbols with carbon residues on body extremities; tattooed circles with plant residues at acupuncture points on the neck that are interpreted as therapeutic tattoos	Dorfer et al. 1999; Pabst et al. 2010

healing, and the underworld follows naturally from its ecology as a nocturnal animal that dwells underground near the roots of healing plants.

Figure 30.41 honors and depicts animals associated with the adult. These items (with the possible exception of the antler tool) likely served as markers of clan and other social affiliations and as spiritual amulets connecting the man with tutelary animals, ancestors, elements, and/or places. Social and spiritual are intertwined in traditional society. The presence of five turtle carapaces in the burial and a possible turtle effigy created by stones capping the grave implies a particularly close relationship with this creature for which North America drew its name Turtle Island from Northeastern Woodland Peoples (Johansen and Mann 2000). Turtle shell bowls were used by Cheyenne (Figure 30.20) and Choctaw healers (Mooney and Olbrechts 1932).

Characteristic features of shamanistic healing practices include the use of drumming, other percussion, and singing, chanting, and dancing to induce trance. The Horn Shelter male exhibits robust development in his forearms and hands that set him apart. Repetitive, sustained drumming (using a handheld frame drum and beater) may be responsible for these osteological changes.

In virtually every region where shamanism is found, "the drum is the shamanic instrument par excellence" (Vitebsky 2001:78–79). "The drum is perhaps the most powerful item in the entire shamanic repertoire" and much of its power comes from the universal effects of its sound (Hayden 2003:59). Rhythmic auditory stimulation imposes a pattern on the brain that produces changes in the central nervous system that entrains both alpha and theta waves and creates visual sensations of color, pattern, and movement (Neher 1961, 1962; Winkelman 2000). Shamans say that it also entrains the human heart with the heartbeat of the earth. "Drum beats are mainly of low frequency, which means that more energy can be transmitted to the brain by a drum beat than from a sound stimulus of higher frequency" such as a rattle (Harner 1990:52).

The earliest evidence of shamanic drumming is a possible drum beater from the Upper Paleolithic Gravettian burial Brno II, Moravia, Czech Republic (Oliva 2000). This piece of reindeer antler is polished on the ends and was found with an adult male who was positioned between the bones of mammoths and woolly rhinoceroses. Six hundred fossil dentalium shells encircled his head and associated shamanic tools included stone rings, discs, and an ivory marionette sculpture with moveable head and arms (the only male sculpture known from the Upper Paleolithic) (Cook 2013). The drum itself was likely made of perishable leather, sinew, and wood.

Due to the nondurable materials used in hand drum construction these instruments are nearly invisible in the archaeological record. Ethnographic and other historic accounts from the Southeast and Plains document their use. A drum made by stretching a coyote hide over a wooden hoop was used by Coahuiltecans in South Texas during religious ceremonies (Newcomb 1961) and Choctaw used a small drum made of a wood cylinder with a leather head (Galloway and Kidwell 2004).

Skeletal evidence for the Horn Shelter male indicates that he was involved in strenuous, sustained, and repetitive activities involving use of his forearms that resulted in unusually well-developed muscle attachments (i.e., bilateral hypertrophy of his interosseous crests). Further, his habitual activities resulted in robusticity of his right thumb and first finger indicating he made repetitive use of a precision grip. These modifications may relate to shamanic drumming (i.e., sustained drumming for therapeutic effects that is undertaken repetitively over many years). His precision grip may relate to repetitive healing tasks including body painting. Future studies that investigate how the human skeleton responds to habitual activities are needed to confirm or reject the idea that shamanic drumming produces changes in skeletal morphology such as that observed on the forearms and hands of the Horn Shelter male.

As guardians of the mental and ecological equilibrium of a group and intermediaries between seen and unseen worlds, shamans were respected and sometimes feared. Traditional people today confer upon them spiritual, social, and political distinctions (Balzer 1996; Harner 1990; Krutak 2012; Vitebsky 2001) and did so in the past (Harner 1972; Winkelman 1990) as indicated by rich funerary accoutrements in shaman burials from the Upper Paleolithic and Late Mesolithic in Europe (Oliva 2000; Porr and Alt 2006) and an Early Natufian shaman burial in Israel (Grosman et al. 2008). The man in the San Patrice double burial appears to belong in this group.

FIGURE 30.41 Creatures represented in the double burial include coyote, badger, Swainson's hawk, lettered olive and olive nerite marine snails, ornate box turtles, and white-tailed deer.

NOTES

1. The woman and child were buried with two turtle carapace containers, grinding equipment, two bone needles, a crane bone containing 31 flaked microliths, additional stone flakes, beaver, and red deer bones, decorative plaques from boar tusk, a perforated boar tongue bone, 16 red deer incisors, two matching roe deer skull fragments with antlers, awls, and numerous freshwater mussel shells. Many of these items show wear and polish indicating an extensive period of use and are assumed to be the woman's personal belongings (Porr and Alt 2006). Porr and Alt (2006) reason that developmental craniovertebral anomalies (irregularly shaped foramen magnum and missing and underdeveloped posterior arches on the atlas vertebrae) resulted in neurological capabilities that predisposed her to shamanic practice. The richness and particular composition of the belongings buried with her provide a strong, independent line of converging evidence indicating that this distinguished woman was a shaman.

REFERENCES

Adams, Jenny L. 2013. Ground stone use-wear analysis: A review of terminology and experimental methods. *Journal of Archaeological Science*. http://dx.doi.org/10.1016/j.jas.2013.01.030.

———. 2002. *Ground Stone Analysis: A Technological Approach*. The University of Utah Press, Salt Lake City.

Allain, J., and A. Rigaud. 1989. Colles et mastics au Magdalenian. In Actes du Colloque International de Nemours 1987, sous la direction de Monique Olive et Yvette Taborin, 221–223. *Memoirs du Musée de Préhistoire d'Ile de France, No. 2*. Nemours.

Allison, Marvin J. 1996. Early mummies from Coastal Peru and Chile. In *Human Mummies: A Global Study of Their Status and the Techniques of Conservation (The Man in Ice)*, vol. 3, edited by Konrad Spindler, Harald Wilfring, Elisabeth Rastbichler-Zissernig, Dieter zur Nedden, and Hans Nothdurfter, 125–130. Springer-Verlag, Vienna.

Ambrose, Wal. 2012. Oceanic tattooing and the implied Lapita ceramic connection. *Journal of Pacific Archaeology* 3(1):1–21.

Anderson, Eve. 1991. "New Perspectives on Archaic Mortuary Behavior at Modoc Rock Shelter." Paper presented at the 56th annual meeting of the Society for American Archaeology, New Orleans, LA.

Andrews, Rhonda L., and James M. Adovasio. 1980. *Perishable Industries from Hinds Cave, Val Verde County, Texas*. Ethnology Monographs 5. Department of Anthropology, University of Pittsburgh.

Audoin, F., and H. Plisson. 1982. Les ocres et leurs témoins au Paléolithique en France: Enquête et expériences sur leur validité archéologique. *Cahiers du centre de recherches préhistoriques* 8:33–80.

Bailey, Garrick. 2004. Continuity and change in Mississippian civilization. In *Hero, Hawk, and Open Hand: American Indian Art of the Ancient Midwest and South*, edited by Richard F. Townsend and Robert V. Sharp, 83–91. Yale University Press, New Haven.

Balzer, Marjorie Mandelstam. 1996. Flights of the sacred, symbolism and theory in Siberian shamanism. *American Anthropologist* 98(2):305–318.

Barker, Pat, Cynthia Ellis, and Stephanie Damadio. 2000. Determination of Cultural Affiliation of Ancient Human Remains from Spirit Cave, Nevada. Bureau of Land Management, Nevada State Office. www.blm.gov/pgdata/etc/medialib/blm/nv/cultural/spirit_cave.Par.57656.File.dat/SC_finbal_July 26.pdf.

Barton, Francis R. 1918. Tattooing in south eastern New Guinea. *Journal of the Royal Anthropological Institute of Great Britain and Ireland* 48:22–79.

Batchelor, John. 1901. "Notes on the Ainu." *Transactions of the Asiatic Society of Japan* X(II):206–219.

Bell, Ann L., trans. 1987. Voyage to the Mississippi through the Gulf of Mexico, 1687. In *LaSalle, the Mississippi, and the Gulf: Three Primary Documents*, edited by Robert S. Weddle, 225–258. Texas A&M University Press, College Station.

Bement, Leland C., Brian J. Carter, Solveig A. Turpin. 1999. *Bison Hunting at Cooper Site: Where Lightning Bolt Drew Thundering Herds*. University of Oklahoma Press, Norman.

Berlandier, Jean Louis. 1969. *The Indians of Texas in 1830*. Edited by John Ewers. Smithsonian Institution Press, Washington DC.

Blakeslee, Donald J. 2008. Artifacts of Kansas, Great Bend Aspect. Wichita State University, Kansas. http://www.ksartifacts.info/GreatBend/96c%20turtle%20shell%20paint%20dish.jpg.

———. 2012. Caves and related sites in the Great Plains of North America. In *Sacred Darkness: A Global Perspective on the Ritual Use of Caves*, edited by Holley Moyes, 353–362. University of Colorado Press, Boulder.

Blakeslee, Donald J., and Marlin F. Hawley. 2006. The Great Bend Aspect. In *Kansas Archaeology*, edited by Robert J. Hoard and William E. Banks, 165–179. University of Kansas Press, Lawrence.

Bousman, C. Britt, Michael B. Collins, Paul Goldberg, Thomas Stafford, Jan Guy, Barry W. Baker, D. Gentry Steele, Marvin Kay, Anne Kerr, Glen Fredlund, Phil Dering, Vance Holliday, Diane Wilson, Wulf Gose, Susan Dial, Paul Takac, Robin Balinsky, Marilyn Masson, and Joseph F. Powell. 2002. The Paleoindian-Archaic transition in North America: New evidence from Texas. *Antiquity* 76:980–990.

Bousman, C. Britt, Barry W. Baker, and Anne C. Kerr. 2004. Paleoindian archaeology in Texas. In *The Prehistory of Texas*, edited by T. Perttula, 15–97. Texas A&M University Press, College Station.

Bousman, C. Britt, and Eric Oksanen. 2012. The Protoarchaic in central Texas and surrounding areas. In *From the Pleistocene to the Holocene: Human Organization and Cultural Transformations in Prehistoric North America*, edited by C. Britt Bousman and Bradley J. Vierra, 197–232. Texas A&M University Press, College Station.

Bousman, C. Britt, and Bradley J. Vierra. 2012. Chronology, environmental setting, and views of the terminal Pleistocene and early Holocene cultural transitions in North America. In *From the Pleistocene to the Holocene: Human Organization and Cultural Transformations in Prehistoric North America*, edited by C. Britt Bousman and Bradley J. Vierra, 1–15. Texas A&M University Press, College Station.

Bouzouggar, Abdeljalil, Nick Barton, Marian Vanhaeren, Francesco d'Errico, Simon Collcutt, Tom Higham, Edward Hodge, Simon Parfitt, Edward Rhodes, Jean-Luc Schwenninger, Chris Stringer, Elaine Turner, Steven Ward, Abdelkrim Moutmir, and Strambouli Abdelhamid. 2007. 82,000-year-old shell beads from North Africa and implications for the origins of modern human behavior. *Proceedings of the National Academy of Sciences U.S.A.* 104(24):9964–9969.

Bowler, James, and Allen G. Thorne. 1976. Human remains from Lake Mungo: Discovery and excavation of Lake Mungo III. In *The Origins of the Australians*, edited by R. L. Kirk and A. G. Thorne, 127–138. Australian Institute of Aboriginal Studies, Canberra.

Boyd, Carolyn. 1996. Shamanic journeys into the Otherworld of the Archaic Chichimec. *Latin American Antiquity* 7(2):152–164.

———. 2003. *Rock Art of the Lower Pecos*. Texas A&M University Press, College Station.

Breternitz, David A., Alan C. Swedlund, and Duane C. Anderson. 1971. An early burial from Gordon Creek, Colorado. *American Antiquity* 36(2):170–182.

CalPal. 2007. "Cologne Radiocarbon Calibration and Paleoclimate Research Package Online CalPal." quickcal 2007 ver. 1.5. http://www.calpal_online.de.

Campbell, Paul D. 2007. *Earth Pigments and Paint of California Indians: Meaning and Technology*. Sunbelt Press, Los Angeles.

Clarke, Philip. 2008. Aboriginal healing practices and Australian bush medicine. *Journal of the Anthropological Society of South Australia* 33:3–38.

Collins, Michael B., Thomas R. Hester, and Frank A. Weir. 1969. Two prehistoric sites in the Lower Rio Grande Valley of Texas. *Bulletin of the Texas Archeological Society* 40:119–146.

Colton, Harold S. 1948. Indian life—Past and present. In *The Inverted Mountains: Canyons of the West*, edited by Roderick Peattie, 111–129. Vanguard Press, New York.

Clottes, Jean, and David Lewis-Williams. 1998. *The Shamans of Prehistory: Trance and Magic in the Painted Caves*. Abrams Press, New York.

Cook, Jill. 2013. *Ice Age Art: The Arrival of the Modern Mind*. The British Museum Press, London.

Coolidge, Rhonda. 2013. "pXRF Analysis of Horn Shelter (726-5G) and USM65241." Ms. on file at Paleoindian/Paleoecology Program, Department of Anthropology, National Museum of Natural History, Washington, DC.

Cushing, Frank Hamilton. 1883. Zuni fetishes. In *Second Annual Report of the Bureau of [American] Ethnology*, 9–43. Smithsonian Institution, Washington, DC.

Cushman, Horatio B. (1899) 1962. *History of the Choctaw, Chickasaw, and Natchez Indians*. Edited by Angie Debo. Reprint, Redlands Press, Stillwater, OK.

Davenport, Demorest, and Michael A. Jochim. 1988. The scene in the shaft at Lascaux. *Antiquity* 62:558–562.

Dellenbaugh, Frederick S. 1933. Indian red paint. *Masterkey* 7:85–87.

d'Errico, Francesco, Christopher Henshilwood, Marian Vanhaerend, Karen van Niekerke. 2005. *Nassarius kraussianus* shell beads from Blombos Cave: Evidence for symbolic behaviour in the Middle Stone Age. *Journal of Human Evolution* 48:3–24.

d'Errico, Francesco, Marian Vanhaerend, Christopher Henshilwood, Graeme Lawson, Bruno Maureille, Henry D. Gambier, Anne-Marie Tillier, Marie Soressi, and Karen van Niekerke. 2009. From the origin of language to the diversification of languages: What can archaeology and palaeoanthropology say? In *Becoming Eloquent: Advances in the Emergence of Language, Human Cognition, and Modern Cultures*, edited by Francesco d'Errico and J. M. Hombert, 13–68. John Benjamins Publishing Company, Amsterdam.

Deter-Wolf, Aaron. 2010. "Flint, bone, and thorns: Using ethnohistorical data, experimental archaeology, and use-wear analysis to examine prehistoric tattooing in the Eastern United States." Paper presented at the 16th annual meeting of the European Association of Archaeologists, The Hague.

———. 2013a. Needle in a haystack: Examining the archaeological record for prehistoric tattooing. In *Drawing With Great Needles: Ancient Tattoo Traditions of North America*, edited by Aaron Deter-Wolf and Carol Diaz-Granados, 43–72. University of Texas Press, Austin.

———. 2013b. The material culture and Middle Stone Age origins of ancient tattooing. In *Tattoos and Body Modifications in Antiquity: Proceedings of the Sessions at the Annual Meetings of the European Association of Archaeologists in the Hague and Oslo, 2010–2011*, edited by Philippe Della Casa and Constanse Witt. Studies in Archaeology 9. University of Zurich.

Deter-Wolf, Aaron, and Tanya M. Peres. 2013. Flint, bone, and thorns: Using ethnohistorical data, experimental archaeology, and microscopy to examine ancient tattooing in eastern North America. In *Tattoos and Body Modifications in Antiquity: Proceedings of the Sessions at the Annual Meetings of the European Association of Archaeologists in the Hague and Oslo, 2010–2011*, edited by Philippe Della Casa and Constanze Witt. Studies in Archaeology 9. University of Zurich.

DiBlasi, Philip J. 1981. "A New Assessment of the Archaeological Significance of the Ashworth Site (11Bu236): A Study in the Dynamics of Archaeological Investigations in Cultural Resource Management." Master's thesis, University of Louisville, KY.

Dodd, C. Kenneth., Jr. 2001. *North American Box Turtles: A Natural History*. University of Oklahoma Press, Norman.

Dorfer, Leopold, Maximilian Moser, F. Bahr, E. Egarter-Vigl, and G. Dorr. 1998. 5200-year-old acupuncture in central Europe? *Science* 282:242–243.

Dorfer, Leopold, Maximilian Moser, F. Bahr, Konrad Spindler, E. Egarter-Vigl, S. Giullén, G. Dorr, and T. Kenner. 1999. A medical report from the stone age? *Lancet* 354:1023–25.

Eerkens, Jelmer W., Amy J. Gilreath, and Brian Joy. 2012. Chemical composition, mineralogy, and physical structure of pigments on arrow and dart fragments from Gypsum Cave, Nevada. *Journal of California and Great Basin Anthropology* 32(1):47–64.

Eliade, Mircea. 1964. *Shamanism: Archaic Techniques of Ecstasy*. Bollingen Series 76. Pantheon, New York.

Ernst, Carl H., and Jeffrey E. Lovich. 2009. *Turtles of the United States and Canada*. 2nd ed. The Johns Hopkins University Press, Baltimore.

Fisher, Timothy G., Derald G. Smith, and John T. Andrews. 2002. Preboreal oscillation caused by a glacial Lake Agassiz flood. *Quaternary Science Reviews* 21:873–878.

Fletcher, Alice C., and Francis La Flesche. 1911. The Omaha tribe. In *Twenty-Seventh Annual Report of the Bureau of American Ethnology, 1905–1906*, 17–672. Smithsonian Institution, Washington, DC.

Forrester, R. E. 1985. Horn Shelter Number 2: The north end, a preliminary report. *Central Texas Archaeologist* 10:21–35.

Fowler, Loretta, and Regina Flannery. 2001. Gros Ventre. In *Handbook of North American Indians*, vol. 2: *Plains*, edited by Raymond J. DeMallie, 677–694. Smithsonian Institution Press, Washington, DC.

Frison, George C., and Carolyn Craig. 1982. Bone, antler, and ivory artifacts and manufacture technology. In *The Agate Basin Site: A Record of the Paleoindian Occupation of the Northwestern High Plains*, edited by George C. Frison and Dennis J. Stanford, 157–173. Academic Press, New York.

Galloway, Patricia, and Clara Sue Kidwell. 2004. Choctaw in the east. In *Handbook of North American Indians*, vol. 14: *Southeast*, edited by Raymond D. Fogelson, 499–519. Smithsonian Institution Press, Washington DC.

Goebel, Ted, and Sergei B. Slobodin. 1999. The colonization of Western Beringia technology, ecology, and adaptations. In *Ice Age Peoples of North America: Environments, Origins, and Adaptations of the First Americans*, edited by Robson Bonnichsen and Karen L. Turnmire, 104–155. Oregon State University Press, Corvallis.

Gorecki, P. P. 1991. Horticulturalists as hunter-gatherers: Rock Shelter usage in Papua New Guinea. In *Ethnoarchaeological Approaches to Mobile Campsites: Hunter-Gatherer and Pastoral Case Studies, edited by Clive S. Gamble and William A. Boismier, 237–262. International Monographs in Prehistory, Ann Arbor.

Grinnell, George Bird. 1923. *The Cheyenne Indians: Their History and Ways of Life*, vol. 2. Yale University Press, New Haven.

Grosman, Leore, Natalie D. Munro, and Anna Belfer-Cohen. 2008. A 12,000-year-old shaman burial from the southern Levant (Israel). *Proceedings of the National Academy of Sciences* 105(46):17665–17669.

Gulliford, Andrew. 2000. *Sacred Objects and Sacred Places: Preserving Tribal Traditions*. University Press of Colorado, Boulder.

Harner, Michael. 1972. *The Jívaro: People of the Sacred Waterfalls*. Doubleday/Natural History Press, Garden City.

———. 1980. *The Way of the Shaman: A Guide to Power and Healing*. Harper and Row, San Francisco.

———. 1990. *The Way of the Shaman: A Guide to Power and Healing*. 10th anniversary ed. Harper and Row, San Francisco.

Hatcher, Mattie A., trans. 1927–1928. Descriptions of the Tejas or Asinai Indians, Part 4. *Southwestern Historical Quarterly* 30(4):283–304.

Haury, Chérie E. 2008. Bison and box turtles: Faunal remains from the Lower Walnut Settlement. *Plains Anthropologist* 53(208):503–530.

Hayden, Brian. 1993a. *Archaeology: The Science of Once and Future Things*. W. H. Freeman, New York.

———. 1993b. The cultural capacities of Neanderthals: A review and reevaluation. *Journal of Human Evolution* 24:113–146.

———. 1995. Pathways to power: Principles for creating socioeconomic inequalities. In *Foundations of Social Inequality*, edited by T. D. Price and G. Feinman, 15–85. Plenum Press, New York.

———. 1998. Practical and prestige technologies: The evolution of material systems. *Journal of Archaeological Method and Theory* 5(1):1–55.

———. 2003. *Shamans, Sorcerers, and Saints*. Smithsonian Books, Washington, DC.

Haynes, C. Vance, and E. Thomas Hemmings. 1968. Mammoth-bone shaft wrench from Murray Springs, Arizona. *Science* 159(3811):186–187.

Haynes, C. Vance, and Bruce B. Huckell. 2007. *Murray Springs: A Clovis Site with Multiple Activity Areas in the San Pedro Valley, Arizona*. Anthropological Papers 71. University of Arizona Press, Tucson.

Heckewelder, John Gotttlieb Ernestus. (1818)1876. *History, Manners, and Customs of the Indian Nations Who Once Inhabited Pennsylvania and the Neighboring States*. Memoir 12. Historical Society of Pennsylvania, Philadelphia.

Henshilwood, C. S., J. C. Sealy, R. Yates, K. Cruz-Uribe, P. Goldberg, F. E. Grine, R. G. Klein, C. Poggenpoel, K. van Niekerk, and I. Watts. 2001. Blombos Cave, Southern Cape, South Africa: Preliminary report on the 1992–1999 excavations of the Middle Stone Age levels. *Journal of Archaeological Science* 28(4):421–448.

Henshilwood, Christopher S., Francesco d'Errico, Karen L. van Niekerk, Yvan Coquinot, Zenobia Jacobs, Stein-Erik Lauritzen, Michel Menu, and Renata García-Moreno. 2011. A 100,000-year-old ochre-processing workshop at Blombos Cave, South Africa. *Science* 334:219–222.

Henshilwood, Christopher, Francesco d'Errico, Marian Vanhaeren, Karen van Niekerk, and Zenobia Jacobs. 2004. Middle Stone Age shell beads from South Africa. *Science* 304(5669):404.

Henshilwood, Christopher, Francesco d'Errico, and Ian Watts. 2009. Engraved ochres from the Middle Stone Age levels at Blombos Cave, South Africa. *Journal of Human Evolution* 57:27–47.

Hester, Thomas R., and Frederick Ruecking, Jr. 1969. Additional materials from the Ayala site, a prehistoric cemetery site in Hidalgo County, Texas. *Bulletin of the Texas Archeological Society* 40:147–166.

Hilger, Sister M. Inez. 1951. *Chippewa Child Life and Its Cultural Background*. Bureau of American Ethnology Bulletin 146. Smithsonian Institution, Washington, DC.

Hodgkiss, Tammy. 2010. Identifying grinding, scoring, and rubbing use-wear on experimental ochre pieces. *Journal of Archaeological Science* 37:3344–3358.

Hogue, S. Homes. 1994. The human skeletal remains from Dust Cave. *Journal of Alabama Archaeology* 40:173–191

Hultkrantz, Ake. 1973. A definition of shamanism. *Temenos* 9:25–37.

Hurst, Stance, Brian J. Carter, and Nancy B. Athfield. 2010. Investigations of a 10,214-year-old late Paleoindian bison kill at the Howard Gully site in southwestern Oklahoma. *Plains Anthropologist* 55:25–37.

Hurst, Stance, Jim Warnica, and Eileen Johnson. 2009. San Patrice points from the Hardwick site, a Paleoindian locality on the Southern High Plains of New Mexico. *Current Research in the Pleistocene* 26:70–73.

Jenks, Albert E. 1937. *Minnesota's Browns Valley Man and Associated Burial Objects*. Memoir 49. American Anthropological Association, Menasha.

Jenness, Diamond. 1955. *The Indians of Canada*. Bulletin 65. Anthropological Series 15. National Museum of Canada, Ottawa.

Jennings, Thomas A. 2008. San Patrice: An example of Late Paleoindian adaptive versatility in South-Central North America. *American Antiquity* 73(3):539–559.

Jodry, Margaret A. 1999. "Folsom Technological and Socioeconomic Strategies: Views from Stewart's Cattle Guard and the Upper Rio Grande Basin, Colorado." Ph.D. dissertation, The American University, Washington DC.

———. 2010. Walking in beauty: 11,000-year-old beads and ornaments from North America. *The Bead Forum* 57:1, 6–9.

Johansen, Bruce E., and Barbara A. Mann, eds. 2000. *Encyclopedia of the Haudenosaunee (Iroquois Confederacy)*. Greenwood Press: Westport.

Johnson, LeRoy, Jr. 1989. *Great Plains Interlopers in the Eastern Woodlands during Late Paleoindian Times: The Evidence from Oklahoma, Texas, and Areas Close By*. Report Series 36. Office of the State Archeologist, Texas Historical Commission, Austin.

Kay, Marvin. 1998. Scratchin' the surface: Stone artifact microwear evaluation. In *Wilson-Leonard: An 11,000-Year Archeology Record of Hunter-Gatherers in Central Texas*, vol. 3: *Artifacts and Special Artifacts Studies*, edited by Michael B. Collins, 743–794. Studies in Archeology 31. Texas Archeological Research Laboratory, University of Texas, Austin.

Keeney, Bradford. 1999. *Kalahari Bushmen Healers*. Profiles in Healing, Volume 3. Ringing Rocks Press, Philadelphia.

———. 2000a. *Gary Holy Bull, Lakota Yuwipi Man*. Profiles in Healing, Volume 1. Ringing Rocks Press, Philadelphia.

———. 2000b. *Guarani Shaman of the Forest*. Profiles in Healing, Volume 4. Ringing Rocks Press, Philadelphia.

———. 2003. *Ropes to God: Experiencing the Bushman Spiritual Universe*. Profiles in Healing, Volume 8. Ringing Rocks Press, Philadelphia.

Kohntopp, Steve W. 2010. *The Simon Clovis Cache, One of Oldest Archaeological Sites in Idaho*. Center for the Study of the First Americans, Texas A&M University, College Station.

Kononenko, Nina. 2011. Experimental and archaeological studies of use-wear and residues on obsidian artefacts from Papua New Guinea. Technical Reports of the Australian Museum, Online 21:1–244. http://dx.doi.org/10.3853/j.1835-4211.21.2011.1559.

———. 2012. Middle and late Holocene skin-working tools in Melanesia: Tattooing and scarification? *Archaeology in Oceania* 47(1):14–28.

Kononenko, Nina, and Robin Torrence. 2009. Tattooing in Melanesia: Local invention or Lapita introduction? *Antiquity* 83(320). http://antiquity.ac.uk/projgall/kononenko/.

Krutak, Lars. 1999. Saint Lawrence Island joint-tattooing: Spiritual/medicinal functions and intercontinental possibilities. *Etudes/Inuit/Studies* 23(1–2):229–252.

———. 2007. *The Tattooing Arts of Tribal Women*. Bennett & Bloom/Desert Hearts, London.

———. 2009. Of human skin and ivory spirits: Tattooing and carving in Bering Strait. In *Gifts From the Ancestors: Ancient Ivories of Bering Strait*, edited by William W. Fitzhugh, Julie Hallowell, and Aron Crowell, 190–203. Yale University Press, New Haven.

———. 2012. *Spiritual Skin: Magical Tattoos and Scarification. Wisdom. Healing. Shamanic Power. Protection*. Edition Reuss GMBH, Germany.

———. 2013. The art of enchantment: Corporeal marking and tattooing bundles of the Great Plains. In *Drawing With Great Needles: Ancient Tattoo Traditions of North America*, edited by Aaron Deter-Wolf and Carol Diaz-Granados, 131–174. University of Texas Press, Austin.

Kuriyama, Shigehisa. 1995. Interpreting the history of bloodletting. *Journal of the History of Medicine and Allied Sciences* 50(1):11–46.

La Flesche, Francis. 1916. Right and left in Osage ceremonies. In *Holmes Anniversary Volume: Anthropological Essays Presented to William Henry Holmes in Honor of His Seventieth Birthday, December 1, 1916*, 278–287. J. W. Bryan Press, Washington DC.

Lawson, John. 1709. *A New Voyage to Carolina: Containing the Exact Description and Natural History of That Country: Together with the Present State Thereof. And a Journal of a Thousand Miles, Travel'd thro' Several Nations of Indians Giving a Particular Account of Their Customs, Manners &c*. W. Taylor and F. Baker, London.

Le Moyne, Jacques. 1564. *The New World: The First Pictures of America*. Duell, Sloan, and Pearce, New York.

Levy, Jerrold E. 1994. Hopi shamanism: A reappraisal. In *North American Indian Anthropology: Essays on Society and Culture*, edited by Raymond J. DeMallie and Alfonso Ortiz, 307–327. University of Oklahoma Press, Norman.

Lewis-Williams, J. D. 1994. Rock art and ritual: Southern Africa and beyond. *Complutum* 5:277–289.

———. 1995. Modeling the production and consumption of rock art. *South African Archaeological Bulletin* 50:143–154.

———. 2001. Southern African shamanic rock art in its social and cognitive contexts. In *The Archaeology of Shamanism*, edited by Neil Price, 17–39. Routledge, London.

Lewis-Williams, J. D., and D. G. Pearce. 2004. *San Spirituality: Roots, Expressions, and Social Consequences*. Alta Mira Press, Walnut Creek, CA.

Little, Nicole C. 2013. "Open Lab XRD report to Margaret Jodry, Catalog Nos. 726-5G, 729-5G (red), and 729-5G (white)." Analysis date 7-23-2013. Museum Conservation Institute, Smithsonian Institution, Washington DC.

Loeb, E. M. 1934. The Western Kuksu cult. *University of California Publications in American Archaeology and Ethnology* 23(1):1–138.

Logan, Wilfred D. 1952. *Graham Cave: An Archaic Site in Montgomery County, Missouri*. Memoir 2. Missouri Archaeological Society, Columbia.

Long, John. 1791. *Voyages and Travels of an Indian Interpreter and Trader Describing the Manners and Customs of the North American Indians*. Robson, London.

Lyman, R. Lee. 1994. *Vertebrate Taphonomy*. Cambridge Manuals in Archaeology. Cambridge University Press.

Lyon, William S. 1996. *Encyclopedia of Native American Healing*. W. W. Norton & Company Ltd., New York.

Maigrot, Yolaine. 2005. Ivory, bone, and antler tools production systems at Chalain 4 (Jura, France): Late Neolithic site, 3rd millennium. In *From Hooves to Horns, from Mollusc to Mammoth: Manufacture and Use of Bone Artefacts from Prehistoric Times to the Present*, edited by Heidi Luik, Alice M. Choyke, Colleen E. Batey, and Lembi Lougas, 113–126. Proceedings of the 4th Meeting of the ICAZ Worked Bone Research Group, Tallinn, 26th-31st of August 2003. Tallinn Book Printers, Tallinn.

McGee, W. J. 1898. The Seri Indians. In *Annual Report of the Bureau of American Ethnology 1895–1896*, 17:1–344. U.S. Government Printing Office, Washington, DC.

McGimsey, Charles R. 1963. Two open sites and a shelter in Beaver Reservoir. *The Arkansas Archaeologist* 4(10):9–11.

Mocas, Stephen. T. 1977. *Excavations at the Lawrence Site, 15Tr3, Trigg County, Kentucky*. University of Louisville Archaeological Survey, Louisville.

———. 1985. An instance of Middle Archaic mortuary activity in western Kentucky. *Tennessee Anthropologist* 10:76–91.

Mooney, James. 1891. Catalog card for accession record 25718; ethnological catalog number E152811-0. Department of Anthropology, Ethnological Collections, Museum Support Center, Suitland, Maryland.

———. 1973. *The Ghost Dance Religion and Wounded Knee*. Dover Publications, Inc, New York.

Mooney, James, and Frans M. Olbrechts. 1932. *The Swimmer Manuscript: Cherokee Sacred Formulas and Medicinal Prescriptions*. Bureau of American Ethnology Bulletin 99. Smithsonian Institution, U.S. Government Printing Office, Washington, DC.

Moyes, Holley. 2012. *Sacred Darkness: A Global Perspective on the Ritual Use of Caves*. University Press of Colorado, Boulder.

Muniz, Mark P. 2004. Exploring technological organization and burial practices at the Paleoindian Gordon Creek site (5LR99), Colorado. *Plains Anthropologist* 49(191):253–279.

Munro, Natalie D. 2001. "A Prelude to Agriculture: Game Use and Occupation Intensity During the Natufian Period in the Southern Levant." Ph.D. dissertation, University of Arizona, Tucson.

Munro, Natalie D., and Leore Grosman. 2010. Early evidence (12,000 B.P.) for feasting in a burial cave in Israel. *Proceedings of the National Academy of Sciences* 107:15362–15366.

Murie, James R. 1981. *Ceremonies of the Pawnee, Part 1: The Siri*. Edited by Douglas R. Parks. Smithsonian Contributions to Anthropology, Volume 27(1). Washington DC.

Neher, A. 1961. Auditory driving observed with scalp electrodes in normal subjects. *Electroencephalography and Clinical Neurophysiology* 13:449–51.

———. 1962. A physiological explanation of unusual behavior in ceremonies involving drums. *Human Biology* 34:151–160.

Neumann, Holm Wolfram. 1967. *The Paleopathology of the Archaic Modoc Rock Shelter Inhabitants*. Report of Investigations 11. Illinois State Museum, Springfield.

Newcomb, William W. 1961. *The Indians of Texas from Prehistoric to Modern Times*. University of Texas Press, Austin.

Nutt, Rush. 1947. Nutt's trip to the Chickasaw country. *Journal of Mississippi History* 9(1):34–61.

Oliva, Martin. 2000. The Brno II Upper Paleolithic burial.

In *Hunter of the Golden Age: The Mid Upper Paleolithic of Eurasia 30,000–20,000 BP*, edited by Wil Roebroeks, Margherita Mussi, Jiří Svoboda and Kelly Fennema, 143–152. University of Leiden.

Osgood, Cornelius. 1940. *Ingalik Material Culture*. Publications in Anthropology 22. Yale University Press, New Haven.

Owens, D'ann, and Brian Hayden. 1997. Prehistoric rites of passage: A comparative study of transegalitarian hunter-gatherers. *Journal of Anthropological Archaeology* 16(2):121-161.

Owsley Douglas W., and David Hunt. 2001. Clovis and early Archaic crania from the Anzick site (24PA506) Park County, Montana. *Plains Anthropologist* 46(176):115–124.

Owsley, Douglas W., and Richard L. Jantz. 1999. Databases for Paleo-American skeletal biology research. In *Who Were the First Americans? Proceedings of the 58th Annual Biology Colloquium, Oregon State University*, edited by Robson Bonnichsen, 79–96. Texas A&M University Press, College Station.

Owsley, Douglas W., Margaret A. Jodry, Thomas W. Stafford, Jr., C. Vance Haynes, Jr., and Dennis J. Stanford. 2010. *Arch Lake Woman: Physical Anthropology and Geoarchaeology*. Texas A&M University Press, College Station.

Pabst, Maria Anna, Ilse Letofsky-Pabst, Elisabeth Brock, Maximilian Moser, Leopold Dorfer, L. Egarter-Vigl, and Ferdinand Hofer. 2009. The tattoos of the Tyrolean Iceman: A light microscopical, ultrastructural and element analytical study. *Journal of Archaeological Science* 36:2335–2341.

Pabst, Maria Anna, Ilse Letofsky-Pabst, Maximilian Moser, Konrad Spindler, Elisabeth Bock, Peter Wilhelm, Leopold Dorfer, Jochen B. Geigl, Martina Auer, Michael R. Speicher, and Ferdinand Hofer. 2010. Different staining substances were used in decorative and therapeutic tattoos in a 1000-year-old Peruvian mummy. *Journal of Archaeological Science* 37:3256–3262.

Parkington, J., and G. Mills. 1991. From space to place: The architecture and social organization of southern African mobile communities. In *Ethnoarchaeological Approaches to Mobile Campsites: Hunter-Gatherer and Pastoral Case Studies*, edited by Clive S. Gamble and William A. Boismier, 355–370. International Monographs in Prehistory, Ann Arbor.

Parsons, Elsie Clews. 1939. *Pueblo Indian Religion*. University of Chicago Press, Chicago.

Peile, A. R. 1979. Colours that Cure. *Hemisphere* 23:214–217.

Péquart, Marthe, and Saint-Just Péquart. 1962. Grotte du Mas d'Azil (Ariége), une Nouvelle Galerie Magdalénienne. *Annales de Paléotologie* 48:197–243.

Peterson, Nicolas, and Ronald Lampert. 1985. A central Australian ochre mine. *Records of the Australian Museum* 37(1):1–9.

Porr, M., and K. W. Alt. 2006. The burial of Bad Dürrenberg, Central Germany: Osteopathology and osteoarchaeology of a Latte Mesolithic shaman's grave. *International Journal of Osteoarchaeology* 16(5):395–406.

Potter, Ben. A., Joel D. Irish, Joshua D. Reuther, Carol Gelvin-Reymiller, and Vance T. Holliday. 2011. A terminal Pleistocene child cremation and residential structure from Eastern Beringia. *Science* 331(6020):1058–1062.

Prochnow, Shane J. 2001. "Paleohydrology and Geoarcheology of the South End of Horn Shelter, Number 2 along the Brazos River, Central Texas." Master's thesis, Baylor University, Waco.

Rašková Zelinková, Michaela. 2010. Reconstructing the "Chaîne Opératoire" of skin processing in Pavlovian bone artifacts from Dolní Věstonice I, Czech Republic. In *Ancient and Modern Bone Artefacts from America to Russia, Cultural, Technological and Functional Signature*, edited by Alexandra Legrand-Pineau, Isabelle Sidéra, Natacha Buc, Eva David and Vivian Scheinsohn, 191–200. BAR International Series 2136. Hadrian Books Ltd., Oxford.

Rasmussen, Sune O., Bo M. Vinther, Henrik B. Clausen, and Katrine K. Andersen. 2007. Early Holocene climate oscillations recorded in three Greenland ice cores. *Quaternary Science Reviews* 26:1907–1914.

Ray, Jack H. 1998. Cultural components. In *The 1997 Excavations at the Big Eddy Site (23CE426) in Southwest Missouri*, edited by Neal H. Lopinot, Jack H. Ray, and Michael D. Conner, 111–220. Special Publication 2. Center for Archaeological Research, Southwest Missouri State University, Springfield.

Ray, Jack H., Neal. H. Lopinot, Edwin R. Hajic, Rolfe D. Mandel. 1998. The Big Eddy Site: A multicomponent Paleoindian site on the Ozark border, Southwest Missouri. *Plains Anthropologist* 43(163):73–81.

Redder, Albert J. 1985. Horn Shelter Number 2: The south end, a preliminary report. *Central Texas Archaeologist* 10:37–65.

Redder, Albert J., and Susan Dial. 2010. Horn Shelter. www.texasbeyondhistory.net/horn/.

Redder, Albert J., and John W. Fox. 1988. Excavation and

positioning of the Horn Shelter's burial and grave goods. *Central Texas Archaeologist* 11:1–15.

Ricklis, Robert A. 1994. *Aboriginal Life and Culture on the Upper Texas Coast: Archaeology at the Mitchell Ridge Site, 41GV66, Galveston Island*. Coastal Archaeological Research, Inc., Corpus Christi.

———. 1996. *The Karankawa Indians of the Texas Coast: An Ecological Study of Cultural Tradition and Change*. University of Texas Press, Austin.

Riel-Salvatore, Julien, Ingrid C. Ludeke, Fabio Negrino, and Brigitte M. Holt. 2013. A spatial analysis of the late Mousterian levels of Riparo Bombrini (Balzi Rossi, Italy). *Canadian Journal of Archeology* 37(1):70–92.

Roper, Donna C. 1991. A comparison of contexts of red ochre use in Paleoindian and Upper Paleolithic sites. *North American Archaeologist* 12(4):289–301.

Russell, H. D. 1941. The recent mollusks of the Family Neritidae. *Bulletin of the Museum of Comparative Zoology* 88(4):347–404.

Salazar, Diego, D. Jackson, J. L. Guendon, H. Salinas, D. Morata, V. Figueroa, G. Manriquez and V. Castro. 2011. Early evidence (ca. 12,000 BP) for iron oxide mining on the Pacific coast of South America. *Current Anthropology* 52(3):463–475.

Semenov, S. A. 1964. *Prehistoric Technology: An Experimental Study of the Oldest Tools and Artefacts from Traces of Manufacture and Wear*. Adams and Dart, Bath.

Shippee, J. Mett. 1966. The archaeology of Arnold Research Cave, Calloway County, Missouri. *The Missouri Archaeologist* 28:1–107.

Shipman, Patricia. 1981. Applications of scanning electron microscopy to taphonomic problems. In *The Research Potential of Anthropological Museum Collections*, edited by Anne-Marie Cantwell, James B. Griffin, and Nan A. Rothschild, 357–385. Annals of the New York Academy of Science.

Shipman, Pat, and Jennie J. Rose. 1983. Early hominid hunting, butchering and carcass-processing behaviors: Approaches to the fossil record. *Journal of Anthropological Archaeology* 2:57–98.

Soffer, Olga. 1985. *The Upper Paleolithic of the Central Russian Plain*. Academic Press, New York.

Speck, Frank G. 1907. Some outlines of aboriginal culture in the southeastern states. *American Anthropologist* 9(2):287–295.

———. 1934. Catawba texts. *Columbia University Contributions to Anthropology* 24. Columbia Press, New York.

Speth, John D., Khori Newlander, Andrew A. White, Ashley K. Lemke, and Lars E. Anderson. 2013. Early Paleoindian big-game hunting in North America: Provisioning or Politics? *Quaternary International* 285:111–139.

Stafford, Michael D., George C. Frison, Dennis Stanford, and George Zeimans. 2003. Digging for the color of life: Paleoindian red ochre mining at the Powars II site, Plate County, Wyoming, U.S.A. *Geoarchaeology* 18(1):71–90.

Stephen, Alexander M. 1936. *Hopi Journal of Alexander M. Stephen*. Contributions to Anthropology 23. Columbia University, New York.

Stoffle, Richard W., David B. Halmo, Michael J. Evans, and Diane E. Austin. 1994. *Piapaxa 'Uipi (Big River Canyon): Southern Paiute Ethnographic Resource Inventory and Assessment for the Colorado River Corridor, Glen Canyon National Recreation Area, Utah and Arizona, and Grand Canyon National Park, Arizona*. University of Arizona Bureau of Applied Research in Anthropology, Tucson.

Story, Dee Ann. 1990. Cultural history of the Native Americans. In *The Archeology and Bioarcheology of the Gulf Coastal Plain*, vol. 1, edited by D. A. Story, J. A. Guy, B. A. Burnett, M. D. Freeman, J.C. Rose, D. G. Steele, B. W. Olive, and K. J. Reinhard, 163–366. Research Series 38. Arkansas Archeological Survey, Fayetteville.

Stright, Melanie J., Eileen M. Lear, and James F. Bennett. 1999. *Spatial Data Analysis of Artifacts Redeposited by Coastal Erosion: A Case Study of McFaddin Beach, Texas, Volume 1*. US Department of the Interior, Minerals Management Service, Herndon, Virginia.

Sturtevant, William C. 1954. Medicine bundles and busks of the Florida Seminole. *Florida Anthropologist* 7(2):31–70.

Swanton, John R. 1931. *Source Material for the Social and Ceremonial Life of the Choctaw Indians*. Bureau of American Ethnology Bulletin 103. Smithsonian Institution, Washington, DC.

———. 1946. *The Indians of the Southeastern United States*. Bureau of American Ethnology Bulletin 137. U.S. Government Printing Office, Washington DC.

Titiev, Mischa. 1972. *The Hopi Indians of Old Oraibi: Change and Continuity*. University of Michigan Press, Ann Arbor.

Titmus, Gene L., and J. C. Woods. 1991. A closer look at margin grinding on Folsom and Clovis points. *Journal of California and Great Basin Anthropology* 13:194–203.

Tuohy, D. R., and Amy J. Dansie. 1997. New information regarding early Holocene manifestations in the western Great Basin. *Nevada Historical Society Quarterly* 40:24–53.

Vega, Garcilaso de la. (1605) 1993. La Florida. Translated by Charmion Shelby and edited by David Bost. In *The de Soto Chronicles: The Expedition of Hernando de Soto to North America in 1539–1543*, vol. 2, edited by Lawrence A. Clayton, Vernon James Knight, Jr., and Edward C. Moore, 25–660. University of Alabama Press, Tuscaloosa.

Velo, Joseph. 1984. Ochre as medicine: A suggestion for the interpretation of the archaeological record. *Current Anthropology* 25(5):674.

Vitebsky, Piers. 2001. *Shamanism*. University of Oklahoma Press, Norman.

Wadley, L. 2005. Putting ochre to the test: Replication studies of adhesives that may have been used for hafting tools in the Middle Stone Age. *Journal of Human Evolution* 49:587–601.

Wadley, L., Williamson, B. S., and Lombard, M. 2004. Ochre in hafting in Middle Stone Age southern Africa: A practical role. *Antiquity* 78:661–675.

Walthall, John A. 1998. Rockshelters and hunter-gatherer adaptation to the Pleistocene/Holocene transition. *American Antiquity* 63(2):223–238.

———. 1999. Mortuary behavior and early Holocene land use in the North American Midcontinent. *North American Archaeologist* 20:1–30.

Waters, Michael R., Charlotte D. Pevny, and David L. Carlson. 2011. *Clovis Lithic Technology, Investigation of a Stratified Workshop at the Gault Site, Texas*. Texas A&M Press, College Station.

Watts, Ian. 2002. Ochre in the Middle Stone Age of southern Africa: Ritualized display or hide preservative? *South African Archaeological Bulletin* 57:64–74.

———. 2009. Red ochre, body painting, and language: Interpreting the Blombos ochre. In *The Cradle of Language*, edited by Rudolf *Botha* and Chris Knight, 62–92. Oxford University Press, Oxford.

———. 2010. The pigments from Pinnacle Point Cave 13B, Western Cape, South Africa. *Journal of Human Evolution* 59:392–411.

Wheeler, S. N. 1997. Cave burials near Fallon, Nevada. *Nevada Historical Society Quarterly* 40:15–23.

Wheeler, Sydney M., and Georgia N. Wheeler. 1969. Cave burials near Fallon, Churchill County, Nevada. Anthropological Papers 14:70–78. Nevada State Museum, Carson City.

Willey, P. S., B. R. Harrison, and Jack T. Hughes. 1978. The Rex Rogers site. In *Archeology at MacKenzie Reservoir*, edited by Jack T. Hughes and P. S. Wiley, 51–68. Archeological Survey Report 24. Office of the State Archeologist, Texas Historical Commission, Austin.

Wilcox, R. W. 1911. *Pharacology and Therapeutics*. Blackston, Philadelphia.

Wilmsen, Edwin N., and Frank H. H. Roberts. 1978. Lindenmeier, 1934–1974, concluding report on investigations. *Smithsonian Contributions to Anthropology* 24. Smithsonian Institution Press, Washington, DC.

Winkelman, Michael. 1990. Shamans and other "magico-religious" healers: A cross-cultural study of their origins, nature, and social transformations. *Ethos* 18:308–352.

———. 2000. *Shamanism: The Neural Ecology of Conscious Healing*. Bergin and Garvey, London.

Young, Diane. 1988. An osteological analysis of the Paleoindian double burial from Horn Shelter, Number 2. *Central Texas Archaeologist* 11:13–115.

Young, Diane E., Suzanne Patrick, and D. Gentry Steele. 1987. An analysis of the Paleoindian double burial from Horn Shelter No. 2, in central Texas. *Plains Anthropologist* 32(117):275–298.

ANTICIPATING KENNEWICK MAN'S FUTURE

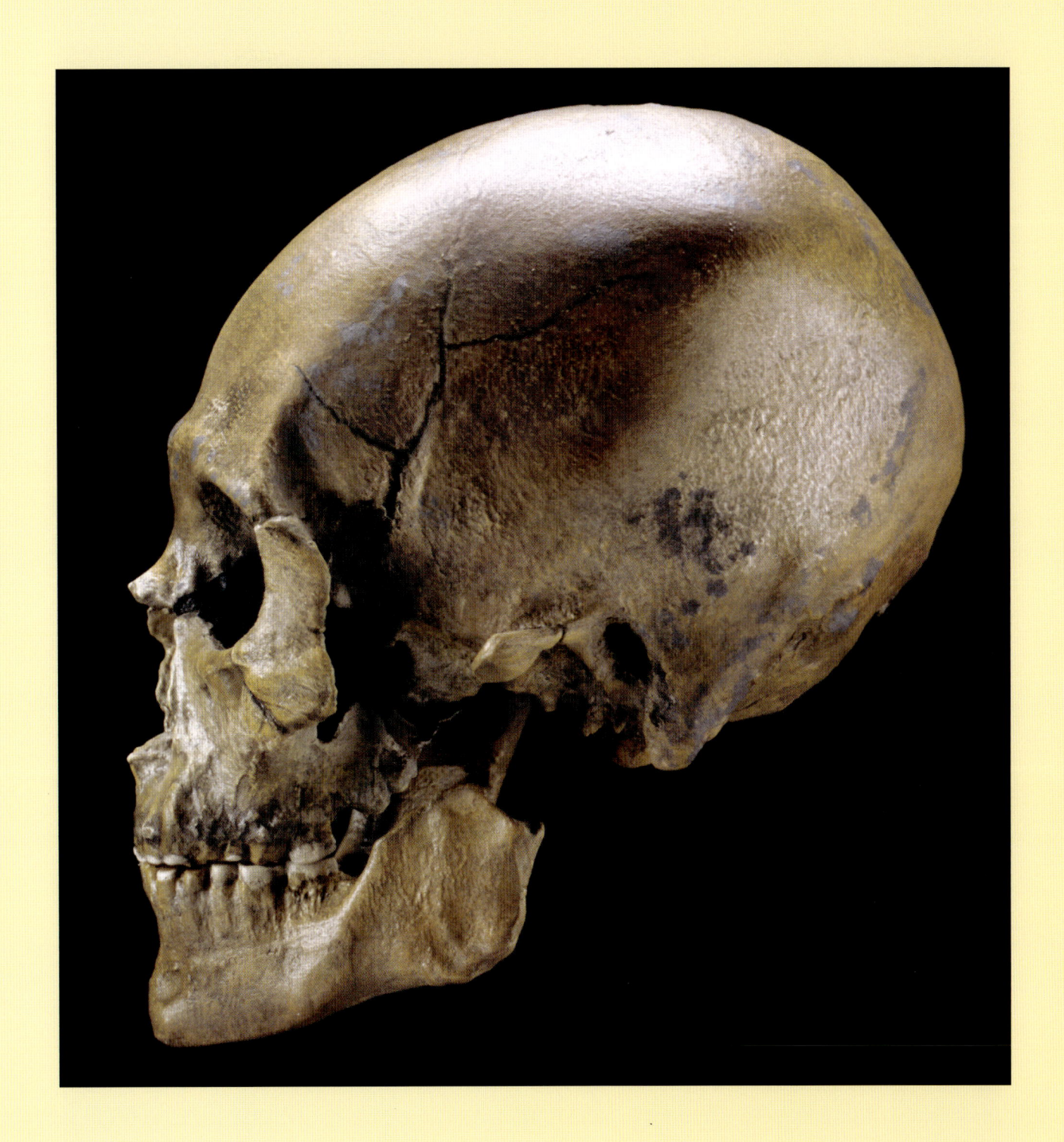

31 Storage and Care at the Burke Museum

Cleone H. Hawkinson

The US Army Corps of Engineers (Corps) has responsibility for the administrative control (curation) and the preservation of the Kennewick Man skeleton. In the fall of 1998, the Corps selected the Burke Museum at the University of Washington in Seattle to act as a neutral curation facility. In the agreement for services with the museum, the Corps defined their requirements to keep the bones not only secure but also in a stable museum environment where temperature and humidity change little. Twice yearly, the Corps has reported to the Federal District Court in Portland, Oregon, on the condition of the skeleton and related curation activities. To assist with this responsibility, the Corps has contracts for services with two conservators, Vicki Cassman and Nancy Odegaard. Since the transfer of the skeleton to the Burke Museum in 1998, the Corps has made improvements in the security and storage of the remains. Individual skeletal fragments are housed by element in custom boxes and kept in a museum-approved Delta® cabinet in a locked room in the museum basement. A data logger records, but does not control, the relative humidity and temperature inside the cabinet.

The Corps's "biannual" (*sic*; actually semiannual) condition reports are the primary source of the information used to evaluate the care of the Kennewick Man skeleton at the Burke Museum. During the court case, the Corps dismissed the plaintiff scientists' concerns about the stability of the environmental conditions at the Burke museum, despite three years of data logger evidence to the contrary. Subsequent data, recorded from late fall 1998 through 2008, were examined to evaluate the stability of the storage environment and to identify any patterns of significant or recurring fluctuation in temperature and humidity. Also of interest were possible periods of very low humidity and relatively stable temperature that suggest drying conditions.

STORAGE ENVIRONMENT FOR MUSEUM COLLECTIONS

Museum conservators agree on the need for a stable storage environment. Cassman and Odegaard (2007) list 10 agents of deterioration, citing Michalski (1990) and Waller (1994). Two of those agents were evaluated with the data collected (Cassman and Odegaard 2007:105):

Incorrect Temperature. Extremes of temperature and rapid, repetitive fluctuations are most problematic. Comfortable temperature for the living is likewise recommended for human remains.

Incorrect Humidity. Too high or too low or rapid and repetitive fluctuating humidity are all highly problematic for human remains.

Widely accepted museum standards for stable conditions generally fall within a range of 67–70 degrees Fahrenheit and 45–50 percent relative humidity (Centre de conservation du Québec et al. 1995; Fisher 1998; Thompson 1994). These conditions are stated in the agreement between the Corps and the Burke Museum (COE 5203). Sudden changes or repeated fluctuations in temperature and humidity, which significantly deviate from optimum museum conditions, contribute to deterioration (Buck and Gilmore 1998; Cassman et al. 2007; Centre de conservation du Québec et al. 1995; Fisher 1998; National Park Service Museum Handbook 1999).

Although temperature is easily measured, relative humidity is defined as a relationship rather than an independent variable. Relative humidity is the ratio of the partial pressure of water vapor in a mixture to the saturated vapor pressure of water at a prescribed temperature. When a change in temperature, pressure, or both occurs, relative humidity is affected. Such shifts produce predictable results.

- When temperature increases and pressure is stable, relative humidity goes down in a predictable way.
- When the temperature is at or above 67–70 degrees F and the relative humidity is low, evaporation occurs, causing extremely dry conditions that lead to drying, cracking, and flaking.
- If fluctuating temperature with higher moisture content follows very dry periods, the drying reverses and the increased moisture causes expansion.
- A high temperature combined with a high relative humidity gives molds and decay an ideal environment.

To move from these generalizations to understanding the effects of fluctuations on specific organic artifacts, such as ancient human bone, is a challenge. For example, the American Association of Museum Standards does not specifically mention how fluctuations affect bone when continued for many years (Fisher 1998). However, experience suggests it is reasonable to assume fluctuating conditions are not benign, especially on ancient bone with fragile surface layers. Studies of the effect of environmental changes on hygroscopic materials (those that take up water) offer some insight into the possible effects on bone. The *National Park Service Museum Handbook* (1999:15) states:

> ... (hygroscopic materials) swell or contract, constantly adjusting to the environment until the rate or magnitude of change is too great and deterioration occurs. Deterioration may occur in imperceptible increments, and therefore go unnoticed for a long time.... The damage may also occur suddenly (for example, cracking of wood). Materials particularly at high risk due to fluctuations are laminate and composite materials...

Research offers insight into the response rates of cracks in wood to fluctuating relative humidity (Knight and Thickett 2007). Sudden fluctuations affect surface layers, while changes taking place on a longer time scale affect deeper layers, which may never reach equilibrium. Sudden releases of accumulated stress in wood were recorded in jumps that occurred with no corresponding shift in relative humidity. Changes of this size would normally be expected to result from relative humidity changes of at least 10 percent. They emphasize that over time small fluctuations matter.

Even the long-term viability of Parafilm®, used to stabilize Kennewick postcranial fragments, depends on "normal environmental conditions," which are stated on the product website to be 45 to 90 degrees F at 50 percent humidity. The shelf life is three years under normal conditions. When contacted, the company had no data for conditions that fall outside their recommendations.

In discussing mechanical damage to organic artifacts, the American Society of Heating, Refrigerating, and Air-Conditioning Engineers (ASHRAE) Handbook, in chapter 21, on Museums Galleries, Archives, and Libraries (2007:21.3) clearly cautions: "Very low humidity or temperature also increases the stiffness of organic materials, making them more vulnerable to fracture."

Investigating specific environmental changes is critical to understanding ways to minimize (or delay) undesirable effects in storage environments. The ASHRAE Handbook describes a modeling technique called the Preservation Index (PI) with a free, downloadable tool called a Preservation Calculator. The index expresses the preservation quality of a storage environment, which aids in understanding the effects of temperature and relative humidity on natural aging of organic collections. As an example, this index has been applied to organic materials similar to archival record paper (e.g., photographs). Small deviations from the upper range of the generally accepted standard may be more important than previously recognized: " ... decreasing temperature from 70 to 65°F and relative humidity from 50 to 45% significantly increases expected lifetime ... Comparing the former 'ideal' set points (70°F and 50% rh) with current recommended parameters (65°F and 45% rh) the PI index increases from 39 to 64 years, or 64%" (ASHRAE Handbook 2007:21.3).

Such information makes it increasingly clear that the control of museum environments must be taken seriously if artifacts are to be preserved. Only a few museums, such as the Smithsonian Museum Support Center, in Suitland, Maryland, can provide a controlled environment for extraordinarily rare and precious artifacts. Nevertheless, museums have an obligation to provide a reasonably stable environment when they promise one in an agreement for services. This is particularly important if public funds are used to pay for museum services.

BURKE MUSEUM STORAGE

The Burke Museum's basement storage is secure. The director of the Burke Museum controls the key to the storage room and the Corps controls the key to the storage cabinet. ADT Securities Systems monitors an electronic alarm on the room door to control the date, time, and identity of the individual entering whenever the alarm is disabled. A motion sensor, located inside the storage room, is connected to the alarm (DOI 2976). The Corps notifies the Federal District Court in Portland, Oregon, whenever the storage room door is opened for any reason. The storage room contains only the cabinet holding the skeleton along with the paper and image archives related to the Kennewick Man skeleton. Access to the storage room has been limited to semiannual condition assessments and approved studies. The available record shows the storage room has been opened once for an unusual circumstance, when overhead water pipes broke in an adjoining room (Court Docket 273). The water did not flood the storage room. The locked storage room has a return air vent.

Multiple layers of protection are desirable to buffer environmental fluctuations in a storage environment (Cassman et al. 2007). With Burke Museum environmental controls in place and limited access, the basement closet should offer reasonably stable storage for the Kennewick Man collection. The museum-quality Delta steel storage cabinet, in which the boxes of bones are stored, offers the next layer of protection. Inside the closed corrugated blueboard alkaline-buffered boxes are layers of buffering to protect the bones. The conservators describe the custom boxes in detail (COE 1232):

> Most of the boxes were designed to hold the fragments in cavity-cut openings that would keep the fragments in alignment but with some separation to prevent abrasion. The padded system included a layer of polyfelt to absorb vibrations in movement during handling, a layer of Tyvek cloth to cover the polyester felt and prevent snagging, and a layer of 1/8- or 1/4-inch Volara foam for some lateral cushioning and stabilization. Individual cavity-cut openings for each bone were made in the Volara, closed-cell polyethylene foam . . . the use of a top pillow made of polyester batting covered with Tyvek for each box (to) place slight pressure on the bone during transport and prevent upward movements.

They identified the housing materials as Polyfelt (Gaylord #YAPF1440), Tyvek (Gaylord #YATR3050), Volara (University Products #170-4275 and 018-125), and PTFE polytetrafluoroethylene (Plastomer Products). They reported using the PTFE as a barrier layer on some of the risers and in the specialized cavity constructions.

A Dickson TL 120 data logger records, but does not control, the temperature and relative humidity inside the storage cabinet. According to the product specification, the temperature readings of this data logger are accurate to ±2.25 degrees F; measurements of relative humidity above 25 percent are accurate to ±5 percent at 77 degrees F (COE 626). Any relative humidity readings below 25 percent ±5 percent may be less reliable. Fluctuations recorded within the data logger specifications offer reasonably reliable insights about the stability of the environment.

Data have been recorded hourly, 24 hours a day, since late 1998. The data were downloaded twice yearly, and a summary was included with the condition assessments conducted by the Corps and the Burke Museum staff. The available reports do not mention recalibrating the data logger after downloading the data, although this is usually a standard curation procedure.

DOCUMENTING CHANGE DURING SEMIANNUAL CONDITION ASSESSMENTS

From 1999 to 2008 the Corps curation team, with the assistance of the Burke Museum staff, conducted semiannual condition assessments of the remains and archives. In the early years of these inspections, the conservators documented the change in the skeleton's condition using a subjective scale of 0–5 and added brief comments to describe soil deposits and bone changes. The 2005 reports show additional categories 2+ and 2++ on the scale.

Objective standards for the categories are not defined in the reports.

0 No significant observational changes since last condition assessment
1 Minor soil deposition
2 Soil deposition
2+ Off our original scale for soil deposition
2++ Significantly greater soil deposition than even the 2+ designation
3 Minute bone fragments deposited in container
4 Bone fragments in separate container such as a zip-lock bag
5 Increased cracking
6 Comments (this might include breaks, sampling or other changes)

A useful report appears in the record in 1999. A table gives measurements of changes in the cranial cracks for the period from August 1996 to November 1999 (COE 663). This table lists 12 areas of the cranium with cracks. The table, if updated, has not been included with the reports submitted after 1999. In the reports since 1999, the assessments are vague, so changes in condition are difficult to evaluate. The conservators report that the cracks "appear" to be lengthening. They also state that the stain on the malar "appears" to be spreading.

Individual cutouts in the custom boxes provide a snug fit to prevent fragments from touching or shifting. In 2004, 2005, and 2006 the plaintiffs' scientific team raised concerns about the tight fit for some fragments with Corps staff. During the prestudy assessment in December 2004, C. Wayne Smith commented to the curation team that some cutouts were too small, causing damage to the broken edges when fragments were removed from the box and returned. Smith expressed this concern to the conservators again during the studies in July 2005. In February 2006, Troy Case expressed similar concern about his difficulty in removing the foot bones from the box without damaging them. The available condition reports do not mention the scientists' concerns or action taken to modify the cutouts to prevent edge deterioration. Over the years, the reports consistently attribute handling during study as the cause of the deterioration.

The curation team also relies on photographs to document changes. In 2001 Chip Clark, senior scientific photographer for the Smithsonian's National Museum of Natural History, expressed concern for the need to apply consistent and standard scientific positioning, appropriate lighting, and use of scale (Chapter 6). The most recent condition photos were not available for this chapter, so their quality for objective or quantitative comparison is unknown. Access to the archives for an independent review of condition requires permission from the Corps.

EVALUATING ENVIRONMENTAL CONDITIONS

Presumably the purpose of the data logger is to monitor the storage environment so departures from stable conditions can be identified and corrected. Each Corps condition report includes a partial summary of the statistics (minimum, maximum, mean, standard deviation) for relative humidity and temperature. The reports include the lowest and the highest temperature reading. Relying on one high and one low value for the six-month period and averaging the data for the period obscures the evidence for a potentially fluctuating environment within the period. The data logger software generates graphs scaled in 10-degree increments. Although the graphs did not include target ranges for museum temperature (67–70 degrees F) or relative humidity (45 percent–50 percent), they clearly reveal that the storage environment regularly fell outside these ranges.

During the litigation in 1999 and 2000, the Corps (defendants) quarterly status to the court reported double-digit fluctuations in relative humidity. Although the plaintiffs expressed concern to the court, Michael Trimble, the Corps curatorial specialist, discounted environmental concerns in his status report to the court of May 2001 (Docket 428). The condition assessment of November 2001, six months later, continued to report no concern:

> . . . environmental conditions surrounding the remains at the Burke Museum were relatively stable since the last assessment beginning April 5, 2001. The relative humidity during the six months monitored here has varied from a low of 26.2% to a high of 51.5%. For the most part these changes were gradual, as expected with the transition from spring to summer and from summer to fall (Court Docket 490).

Museum standards do not mention exceptions for "expected seasonal variations" or sanction fluctuations of this magnitude. The report of the summary data includes no further explanation.

To gain a clearer understanding of the conditions in the storage environment, the plaintiff scientists filed a Freedom of Information Act request (FOIA) to the Corps in 2003. This request yielded hundreds of printed pages of raw logger data from 1998–2003. To expand the data set for this chapter, the Corps provided electronic files of the data from 2004–2008 without the need for an additional FOIA request.

Reevaluating the raw data offers the opportunity to see if information has been obscured, missed, or ignored in the individual summary reports. The data set has roughly 87,600 data points recorded by the data logger placed inside the Delta storage cabinet. With nearly half of the data in paper form, the data were visually inspected. To discern meaning without dampening critical information, observations for this chapter were limited to noting the monthly highs and lows for temperature and relative humidity. Unfortunately, this too dampens daily and weekly fluctuations that may be meaningful. The chapter on Collections Management Preventive Care in the American Association Museum Standards recommends limiting temperature fluctuation to 2 to 3 degrees within 24 hours as optimal (Fisher 1998:105). However, this evaluation does not analyze the Corps data in such detail. The raw data are available from the Corps for further analysis.

The Corps six-month reporting periods were from midmonth (December and June or early July). For this evaluation, the data were visually searched for minimum and maximum readings by month from late 1998 to 2008. This level of detail reveals environmental fluctuations beyond the statements reported in the semiannual inspections (Table 31.1). Figures 31.1 and 31.2 graphically illustrate how far from standard museum ranges the temperature and relative humidity actually fluctuated.

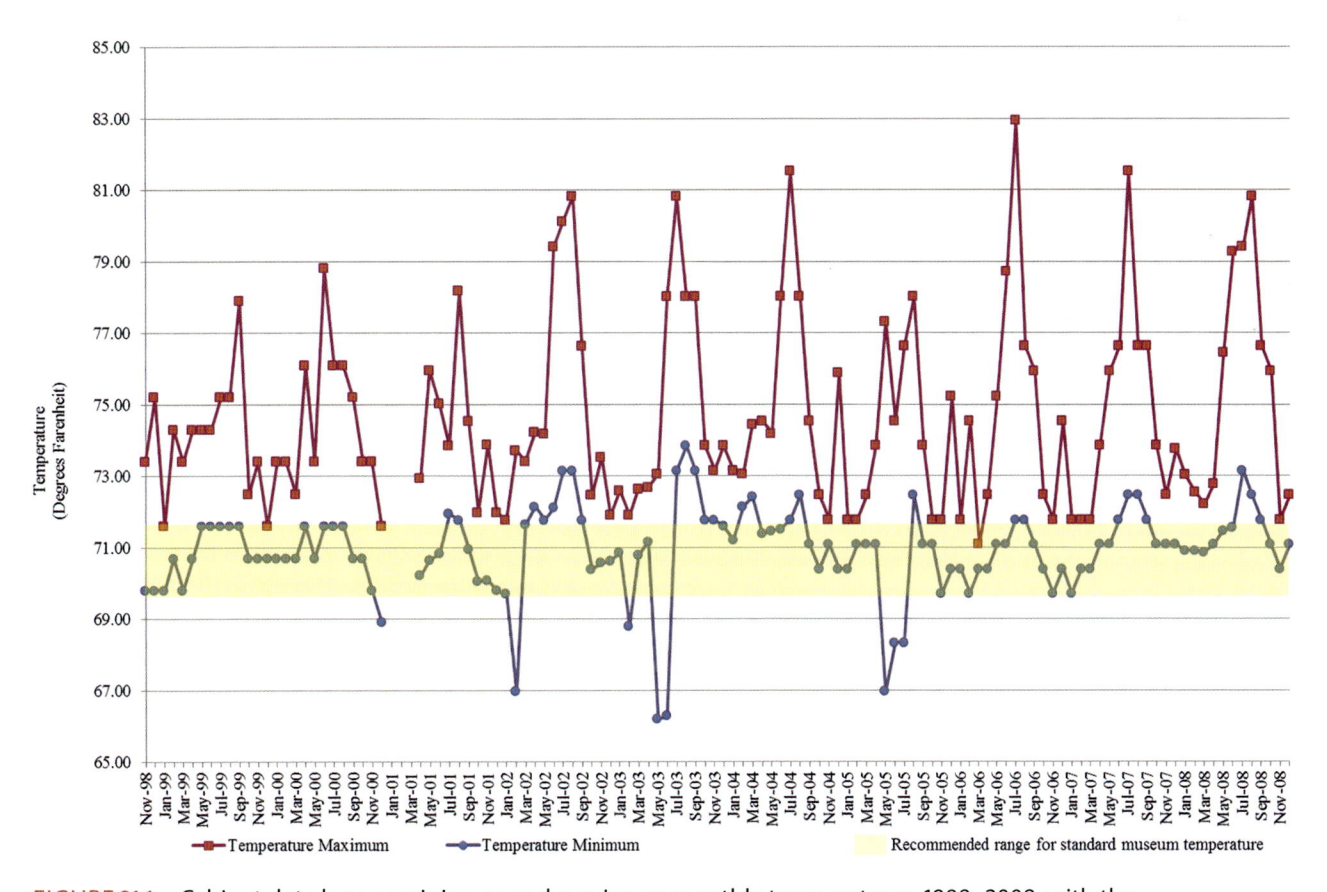

FIGURE 31.1   Cabinet data logger minimum and maximum monthly temperatures, 1998–2008, with the recommended range for standard museum temperature highlighted in yellow.

TABLE 31.1 Monthly relative humidity and temperature data, 1998–2008, showing departures from accepted museum standards. Measurements in bold are the lowest and highest recorded in each year.

	RH Min	RH Max	Range	Below 45% RH	Above 50% RH	Temp Min	Temp Max	Range	Below 67°F	Above 70°F
Nov. 1998	36.6%	45.8%	9.2%	8.4%		69.80	73.40	3.60		3.40
Dec. 1998	24.5%	33.0%	8.5%	20.5%		69.80	75.20	5.40		5.20
1998	**24.5%**	**45.8%**	**21.3%**			**69.80**	**75.20**	**5.40**		
Jan. 1999	31.6%	39.6%	8.0%	13.4%		69.80	71.60	1.80		1.60
Feb.1999	24.3%	40.8%	16.5%	20.7%		70.70	74.30	3.60		4.30
March 1999	28.0%	33.0%	5.0%	17.0%		69.80	73.40	3.60		3.40
April 1999	29.4%	31.1%	1.7%	15.6%		70.70	74.30	3.60		4.30
May 1999	30.8%	35.7%	4.9%	14.2%		71.60	74.30	2.70		4.30
June 1999	34.6%	53.4%	18.8%	10.4%	3.4%	71.60	74.30	2.70		4.30
July 1999	42.5%	47.1%	4.6%	2.5%		71.60	75.20	3.60		5.20
Aug. 1999	46.7%	52.7%	6.0%		2.7%	71.60	75.20	3.60		5.20
Sept. 1999	45.4%	52.0%	6.6%		2.0%	71.60	77.90	6.30		7.90
Oct. 1999	44.0%	47.1%	3.1%	1.0%		70.70	72.50	1.80		2.50
Nov. 1999	37.2%	51.5%	14.3%	7.8%	1.5%	70.70	73.40	2.70		3.40
Dec. 1999	35.9%	38.8%	2.9%	9.1%		70.70	71.60	0.90		1.60
1999	**24.3%**	**53.4%**	**29.1%**			**69.80**	**77.90**	**8.10**		
Jan. 2000	32.3%	36.3%	4.0%	12.7%		70.70	73.40	2.70		3.40
Feb. 2000	24.9%	33.0%	8.1%	20.1%		70.70	73.40	2.70		3.40
March 2000	30.3%	31.6%	1.3%	14.7%		70.70	72.50	1.80		2.50
April 2000	26.3%	34.5%	8.2%	18.7%		71.60	76.10	4.50		6.10
May 2000	32.8%	52.4%	19.6%	12.2%	2.4%	70.70	73.40	2.70		3.40
June 2000	41.2%	46.2%	5.0%	3.8%		71.60	78.80	7.20		8.80
July 2000	44.3%	51.0%	6.7%	0.7%	1.0%	71.60	76.10	4.50		6.10
Aug. 2000	49.7%	52.2%	2.5%		2.2%	71.60	76.10	4.50		6.10
Sept. 2000	48.6%	54.3%	5.7%		4.3%	70.70	75.20	4.50		5.20
Oct. 2000	45.4%	50.4%	5.0%		0.4%	70.70	73.40	2.70		3.40
Nov. 2000	32.7%	45.5%	12.8%	12.3%		69.80	73.40	3.60		3.40
Dec. 2000	31.0%	33.8%	2.8%	14.0%		68.90	71.60	2.70		1.60
2000	**24.9%**	**54.3%**	**29.4%**			**68.90**	**78.80**	**9.90**		
Jan. 2001	no data					no data				
Feb. 2001	no data					no data				
March 2001	no data					no data				
April 2001	26.2%	31.8%	5.6%	18.8%		70.22	72.93	2.71		2.93
May 2001	30.4%	34.3%	3.9%	14.6%		70.65	75.96	5.31		5.96
June 2001	34.3%	42.3%	8.0%	10.7%		70.82	75.01	4.19		5.01
July 2001	37.7%	43.8%	6.1%	7.3%		71.94	73.84	1.90		3.84
Aug. 2001	43.3%	50.0%	6.7%	1.7%		71.77	78.18	6.41		8.18
Sept. 2001	45.3%	48.9%	3.6%			70.95	74.53	3.58		4.53
Oct. 2001	41.8%	47.4%	5.6%	3.2%		70.05	71.98	1.93		1.98
Nov. 2001	39.7%	51.5%	11.8%	5.3%	1.5%	70.09	73.88	3.79		3.88
Dec. 2001	32.2%	40.7%	8.5%	12.8%		69.79	71.98	2.19		1.98
2001	**26.2%**	**51.5%**	**25.3%**			**69.79**	**78.18**	**8.39**		

(*continued*)

TABLE 31.1 (*continued*)   Monthly relative humidity and temperature data, 1998–2008, showing departures from accepted museum standards. Measurements in bold are the lowest and highest recorded in each year.

	RH Min	RH Max	Range	Below 45% RH	Above 50% RH	Temp Min	Temp Max	Range	Below 67°F	Above 70°F
Jan. 2002	29.9%	35.7%	5.8%	15.1%		69.70	71.77	2.07		1.77
Feb. 2002	28.0%	30.9%	2.9%	17.0%		66.96	73.71	6.75	0.04	3.71
March 2002	25.2%	28.0%	2.8%	19.8%		71.64	73.41	1.77		3.41
April 2002	26.2%	30.4%	4.2%	18.8%		72.15	74.23	2.08		4.23
May 2002	28.0%	35.7%	7.7%	17.0%		71.77	74.18	2.41		4.18
June 2002	32.3%	42.8%	10.5%	12.7%		72.11	79.41	7.30		9.41
July 2002	39.2%	42.8%	3.6%	5.8%		73.15	80.12	6.97		10.12
Aug. 2002	39.7%	42.3%	2.6%	5.3%		73.15	80.82	7.67		10.82
Sept. 2002	41.2%	43.3%	2.1%	3.8%		71.77	76.62	4.85		6.62
Oct. 2002	35.7%	42.3%	6.6%	9.3%		70.39	72.46	2.07		2.46
Nov. 2002	32.3%	41.8%	9.5%	12.7%		70.56	73.53	2.97		3.53
Dec. 2002	30.4%	36.2%	5.8%	14.6%		70.61	71.90	1.29		1.90
2002	**25.2%**	**43.3%**	**18.1%**			**66.96**	**80.82**	**13.86**		
Jan. 2003	29.4%	32.8%	3.4%	15.6%		70.86	72.59	1.73		2.59
Feb. 2003	27.5%	32.8%	5.3%	17.5%		68.80	71.90	3.10		1.90
March 2003	27.5%	30.4%	2.9%	17.5%		70.78	72.63	1.85		2.63
April 2003	29.4%	30.4%	1.0%	15.6%		71.16	72.67	1.51		2.67
May 2003	29.9%	48.4%	18.5%	15.1%		66.19	73.06	6.87	0.81	3.06
June 2003	38.2%	40.7%	2.5%	6.8%		66.28	78.01	11.73	0.72	8.01
July 2003	38.7%	41.8%	3.1%	6.3%		73.15	80.82	7.67		10.82
Aug. 2003	40.2%	43.8%	3.6%	4.8%		73.84	78.01	4.17		8.01
Sept. 2003	40.2%	42.8%	2.6%	4.8%		73.15	78.01	4.86		8.01
Oct. 2003	39.2%	45.8%	6.6%	5.8%		71.77	73.84	2.07		3.84
Nov. 2003	29.9%	39.2%	9.3%	15.1%		71.77	73.15	1.38		3.15
Dec. 2003	20.8%	30.4%	9.6%	24.2%		71.59	73.84	2.25		3.84
2003	**20.8%**	**48.4%**	**27.6%**			**66.19**	**80.82**	**14.63**		
Jan. 2004	21.7%	28.0%	6.3%	23.3%		71.21	73.15	1.94		3.15
Feb. 2004	24.8%	26.6%	1.8%	20.2%		72.15	73.06	0.91		3.06
March 2004	25.7%	27.5%	1.8%	19.3%		72.41	74.44	2.03		4.44
April 2004	26.6%	29.0%	2.4%	18.4%		71.38	74.53	3.15		4.53
May 2004	28.0%	34.7%	6.7%	17.0%		71.47	74.18	2.71		4.18
June 2004	29.4%	36.7%	7.3%	15.6%		71.51	78.01	6.50		8.01
July 2004	35.7%	42.8%	7.1%	9.3%		71.77	81.53	9.76		11.53
Aug. 2004	41.8%	49.5%	7.7%	3.2%		72.46	78.01	5.55		8.01
Sept. 2004	46.4%	48.9%	2.5%			71.08	74.53	3.45		4.53
Oct. 2004	41.8%	47.9%	6.1%	3.2%		70.39	72.46	2.07		2.46
Nov. 2004	35.7%	41.8%	6.1%	9.3%		71.08	71.77	0.69		1.77
Dec. 2004	30.9%	35.7%	4.8%	14.1%		70.39	75.87	5.48		5.87
2004	**21.7%**	**49.5%**	**27.8%**			**70.39**	**81.53**	**11.14**		

(*continued*)

TABLE 31.1 (continued)   Monthly relative humidity and temperature data, 1998–2008, showing departures from accepted museum standards. Measurements in bold are the lowest and highest recorded in each year.

	RH Min	RH Max	Range	Below 45% RH	Above 50% RH	Temp Min	Temp Max	Range	Below 67°F	Above 70°F
Jan. 2005	25.7%	32.8%	7.1%	19.3%		70.39	71.77	1.38		1.77
Feb. 2005	25.7%	31.3%	5.6%	19.3%		71.08	71.77	0.69		1.77
March 2005	26.6%	30.4%	3.8%	18.4%		71.08	72.46	1.38		2.46
April 2005	27.5%	31.8%	4.3%	17.5%		71.08	73.84	2.76		3.84
May 2005	31.3%	40.7%	9.4%	13.7%		66.96	77.31	10.35	0.04	7.31
June 2005	38.2%	43.3%	5.1%	6.8%		68.33	74.53	6.20		4.53
July 2005	37.2%	64.5%	27.3%	7.8%		68.33	76.62	8.29		6.62
Aug. 2005	45.3%	47.9%	2.6%			72.46	78.01	5.55		8.01
Sept. 2005	43.3%	47.4%	4.1%	1.7%		71.08	73.84	2.76		3.84
Oct. 2005	41.8%	44.3%	2.5%	3.2%		71.08	71.77	0.69		1.77
Nov. 2005	33.8%	42.3%	8.5%	11.2%		69.71	71.77	2.06		1.77
Dec. 2005	23.4%	34.3%	10.9%	21.6%		70.39	75.22	4.83		5.22
2005	**23.4%**	**64.5%**	**41.1%**			**66.96**	**78.01**	**11.05**		
Jan. 2006	28.4%	30.4%	2.0%	16.6%		70.39	71.77	1.38		1.77
Feb. 2006	5.6%	33.3%	27.7%	39.4%		69.71	74.53	4.82		4.53
March 2006	19.5%	23.9%	4.4%	25.5%		70.39	71.08	0.69		1.08
April 2006	23.9%	27.1%	3.2%	21.1%		70.39	72.46	2.07		2.46
May 2006	25.7%	32.8%	7.1%	19.3%		71.08	75.22	4.14		5.22
June 2006	32.3%	40.2%	7.9%	12.7%		71.08	78.71	7.63		8.71
July 2006	36.2%	41.2%	5.0%	8.8%		71.77	82.95	11.18		12.95
Aug. 2006	39.2%	42.8%	3.6%	5.8%		71.77	76.62	4.85		6.62
Sept. 2006	39.7%	43.8%	4.1%	5.3%		71.08	75.92	4.84		5.92
Oct. 2006	37.2%	43.3%	6.1%	7.8%		70.39	72.46	2.07		2.46
Nov. 2006	31.3%	40.7%	9.4%	13.7%		69.71	71.77	2.06		1.77
Dec. 2006	28.0%	32.8%	4.8%	17.0%		70.39	74.53	4.14		4.53
2006	**5.6%**	**43.8%**	**38.2%**			**69.71**	**82.95**	**13.24**		
Jan. 2007	24.8%	29.0%	4.2%	20.2%		69.71	71.77	2.06		1.77
Feb. 2007	24.3%	28.5%	4.2%	20.7%		70.39	71.77	1.38		1.77
March 2007	25.7%	30.4%	4.7%	19.3%		70.39	71.77	1.38		1.77
April 2007	28.0%	30.4%	2.4%	17.0%		71.08	73.84	2.76		3.84
May 2007	29.0%	32.3%	3.3%	16.0%		71.08	75.92	4.84		5.92
June 2007	31.8%	36.2%	4.4%	13.2%		71.77	76.62	4.85		6.62
July 2007	34.7%	46.9%	12.2%	10.3%		72.46	81.53	9.07		11.53
Aug. 2007	41.2%	45.8%	4.6%	3.8%		72.46	76.62	4.16		6.62
Sept. 2007	41.8%	45.8%	4.0%	3.2%		71.77	76.62	4.85		6.62
Oct. 2007	36.2%	42.8%	6.6%	8.8%		71.08	73.84	2.76		3.84
Nov. 2007	30.4%	37.2%	6.8%	14.6%		71.08	72.46	1.38		2.46
Dec. 2007	26.6%	31.8%	5.2%	18.4%		71.08	73.75	2.67		3.75
2007	**24.3%**	**46.9%**	**22.6%**			**69.71**	**81.53**	**11.82**		

(continued)

TABLE 31.1 *(continued)* Monthly relative humidity and temperature data, 1998–2008, showing departures from accepted museum standards. Measurements in bold are the lowest and highest recorded in each year.

	RH Min	RH Max	Range	Below 45% RH	Above 50% RH	Temp Min	Temp Max	Range	Below 67°F	Above 70°F
Jan. 2008	21.7%	27.5%	5.8%	23.3%		70.91	73.02	2.11		3.02
Feb. 2008	23.0%	26.6%	3.6%	22.0%		70.91	72.54	1.63		2.54
March 2008	24.3%	27.1%	2.8%	20.7%		70.86	72.20	1.34		2.20
April 2008	23.9%	26.2%	2.3%	21.1%		71.08	72.76	1.68		2.76
May 2008	24.8%	34.7%	9.9%	20.2%		71.47	76.44	4.97		6.44
June 2008	32.8%	36.2%	3.4%	12.2%		71.55	79.28	7.73		9.28
July 2008	36.2%	40.7%	4.5%	8.8%		73.15	79.41	6.26		9.41
Aug. 2008	38.2%	44.8%	6.6%	6.8%		72.46	80.82	8.36		10.82
Sept. 2008	41.2%	42.8%	1.6%	3.8%		71.77	76.62	4.85		6.62
Oct. 2008	35.7%	43.8%	8.1%	9.3%		71.08	75.92	4.84		5.92
Nov. 2008	34.7%	39.2%	4.5%	10.3%		70.39	71.77	1.38		1.77
Dec. 2008	31.8%	36.7%	4.9%	13.2%		71.08	72.46	1.38		2.46
2008	**21.7%**	**44.8%**	23.1%			**70.39**	**80.82**	10.43		

Temperature

Monthly high and low temperatures show repeated fluctuation beyond the desired range of 67–70 degrees F. Based on the raw data, monthly lows generally fell within the range (Figure 31.1). Monthly highs were consistently above 70 degrees F and, for some months, by as much as 8 to 10 degrees. Months with the highest temperatures also showed the highest within-month fluctuations. Federal museum curation sources state 70 degrees F as the upper limit for storage. (Smithsonian Museum Conservation Institute n.d.).

Recorded temperature fluctuations within six-month periods were not the smooth, gradual increase with a slow decline that conservators suggested in their reports. A summer pattern of widely ranging temperatures recurred annually. For example, in June 2002 the within-month range was 72.1 to 79.4 degrees F and fluctuated 7.3 degrees; in July the range was 73.1 to 80.1 degrees F (7 degrees fluctuation); and in August the range was 73.1 to 80.8 degrees F (7.7 degrees fluctuation). While these ranges are similar, they all start above the ideal upper limit of 70 degrees F and show daily and weekly fluctuations within each month. Temperature spikes, especially during July and August, are predictable and measures could have been taken to dampen effects across the warm months.

The following May (2003), the minimum temperature was 66.2 degrees F and the high temperature was 73.1 degrees F (a range of 6.9 degrees). In June 2003, the data logger recorded a return to a minimum temperature of 66.3 degrees F but the maximum climbed to 78.0 degrees F (a range of 11.7 degrees). The following month, the minimum temperature was 73.2 degrees F and the maximum was 80.8 degrees F (a range of 7.6 degrees) with intervening ups and downs. In 2005, before the plaintiffs' scientific studies in July, the within-month fluctuation in May was 10.3 degrees and in June, 6.2 degrees. The two biggest within-month summer temperature ranges occurred in July 2003 (fluctuated 11.7 degrees) and July 2006 (fluctuated 11.2 degrees). Scientific studies were not conducted during either month.

Temperature data show a predictable yearly cycle of cooling, heating, and then cooling again. The data also show continual temperature peaks and dips well beyond the target range within the summer months, with highs in July and August or September. These repeated seasonal and daily variations raise questions about the effect of perpetually shifting temperature on the condition of the bones and suggest that changes may have occurred while the bones were stored away. In their semiannual condition reports, The Corps claims that the deterioration

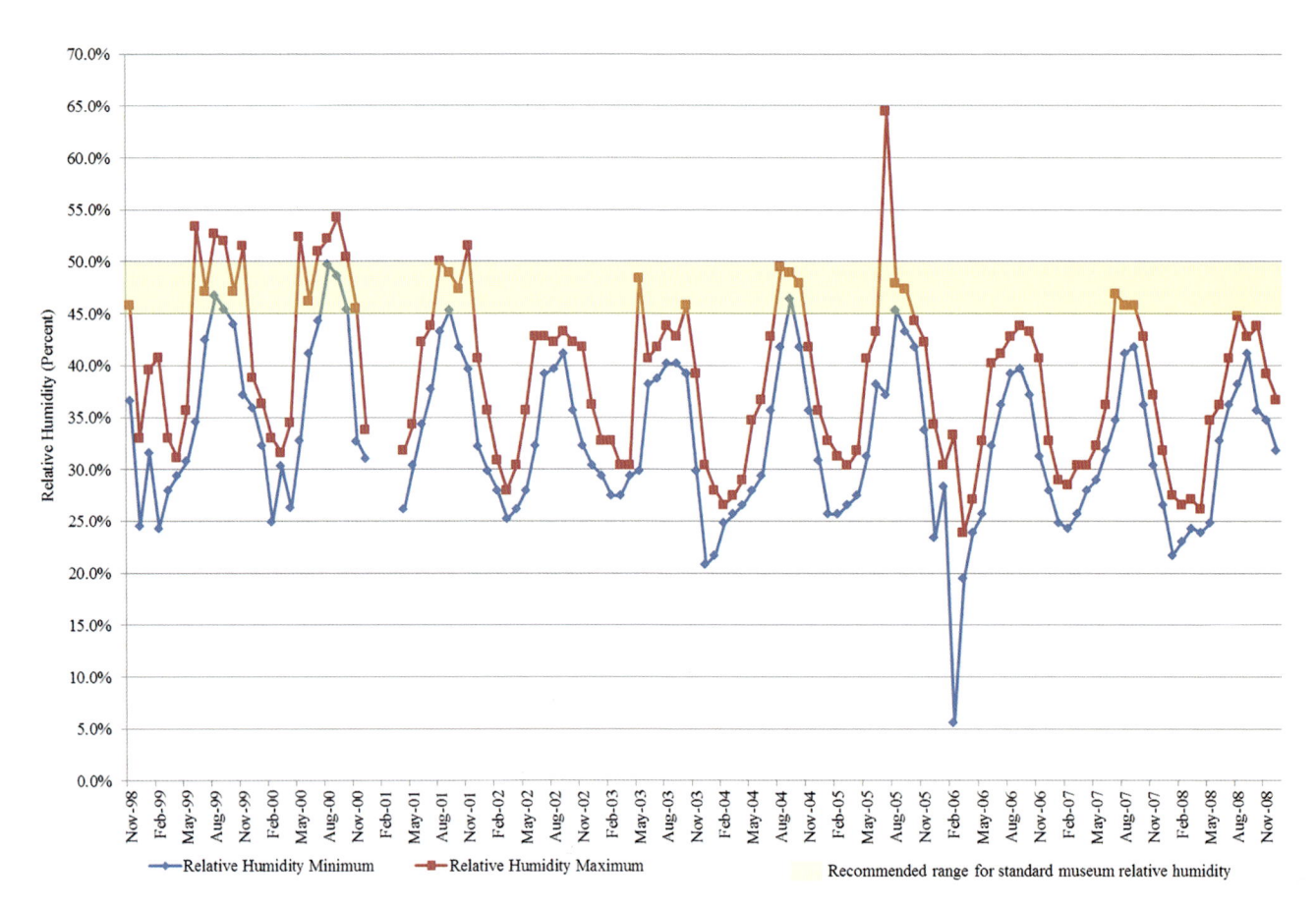

FIGURE 31.2 Cabinet data logger fluctuations in the minimum and maximum monthly relative humidity, 1998–2008, with the recommended range for standard museum relative humidity highlighted in yellow.

was caused by the scientists' handling of the bones; this is problematic.

Relative Humidity

The Corps agreement states that the Burke Museum is to maintain normal museum standards for relative humidity between 45 and 50 percent. After October 2000, the relative humidity fell consistently below the desired range, except during a few summer months and November 2001 (Figure 31.2). The agreement between the Burke Museum and the Corps made no mention of a 25 percent fluctuation in relative humidity due to seasonal shift.

Since 2001, between late November and May, data show yearly cycles of very low humidity with significant fluctuation (Table 31.1). Minimum relative humidity ranged from 20 to 30 percent. These readings may not be reliable because they are below the data logger specification. The range for monthly maximum humidity readings for the five- to six-month period is, in general, 30 to 35 percent relative humidity. This combination of low relative humidity with corresponding temperatures at or above 70 degrees F indicates dry conditions. Repeating this seasonal variability over the years raises concern for drying, cracking, or flaking of the surface of the bones and teeth.

In 2006 the Corps reported placing additional buffer materials around the bones to reduce the effects of varying relative humidity. A second data logger was placed inside a box to measure the effectiveness of the additional buffering. The summary statistics for the two data loggers (one inside the cabinet, the second inside one box containing bones) show similar results for the first four months in 2006. The relative humidity fluctuated inside the cabinet by 23.2 percent. Inside the box in the cabinet the relative humidity varied by 21 percent. In a stable environment, fluctuation in relative humidity is ideally about ±5 percent.

The June 2006 inspection report included two possible explanations for differences between readings in

TABLE 31.2 Comparison of data from a data logger inside the cabinet to data from a data logger inside a box holding remains in the cabinet.

	%RH Min / Max	Mean / S.D.	Temp °F. Min / Max	Mean / S.D.
February–June 2006				
Inside cabinet	17.0% / 40.2% range 23.2%	26.8% / 5.5	70.4° / 75.2° range 4.8°	71.4° / 0.8
Inside a box in the cabinet	23.5% / 45.0% range 21.5%	30.7% / 5.4	70.4° / 75.9° range 5.5°	71.2° / 0.7
June–December 2006				
Inside cabinet	30.4% / 43.8% range 13.4%	39.0% / 3.2	69.7° / 83.0° range 13.3°	72.8° / 2.2
Inside a box in the cabinet	34.1% / 49.5% range 15.4%	44.1% / 3.2	69.7° / 80.1° range 10.4°	72.7° / 1.75
July 2007–December 2007				
Inside cabinet	27.5% / 46.9% range 19.4%	39.3%/5.0	71.1° / 81.5° range 10.4°	73.1° /1.7
Inside a box in the cabinet	30.2% / 51.2% range 21%	44%/5.7	71.1° / 74° range 2.6°	73° /1.5
December 2007–July 2008				
Inside cabinet	21.7%/ 40.2% range 18.5%	27.9%/4.1	70.9° / 79.3° range 8.5°	72.1° /1.2
Inside a box in the cabinet	26.8% /38.1% range 11.3%	29.9%/2.4	71.1° / 78° range 6.3°	71.2° / 0.7

the two data loggers in that period—either disparities in instrument calibration or the role of organic materials with or in the remains. The report does not comment on the relative humidity itself or on the reliability of the data outside the data logger specifications. The December 2006 inspection report included this comment:

> The slight variance is most likely due to the buffering quality of the storage materials in the box surrounding the remains, which kept the relative humidity from dipping to the lowest point and these materials also held on to the humidity at the upper end. . . . (the second data logger) . . . confirm the utility of the buffering materials (Corps of Engineers June 2006 Condition Assessment p. 5).

The raw data from the second data logger was not available for monthly comparison. The summary values from the two data loggers given in the reports provided, between February 2006 and July 2008, show mixed results (Table 31.2). The data do not support the Corps' early claim that the buffering materials inside the box provide meaningful protection of the remains. In 2006 and 2007 the variance was greater in the buffered environment although the boilerplate explanation to the contrary is repeated in the reports. The later reports repeat this statement:

> The environmental data . . . showed that the environmental relative humidity with the storage box and cabinet did not vary greatly. The slight variance is most likely due to the buffering quality of the storage material in the box surrounding the remains, which kept the relative humidity from dipping to the lowest point and these materials also held on to the humidity at the upper end. Though this second data logger . . . does confirm the utility of the buffering materials (Corps of Engineers December 2006 Condition Assessment p. 4).

Similar statements appear in reports of December 2007:4 and July 2008:3.

The regular condition reports do not address why the curation team is satisfied with the dry, double-digit fluctuating ranges in relative humidity in the basement of the Burke Museum. Over time the data also show temperature rises well above maximum range for standard museum environments. When data are assessed monthly, it strongly suggests an unstable, potentially detrimental environment for storing ancient bone.

Plaintiffs' Studies in 2005 and 2006

The Corps attributes the presence of scientists during the study periods as the cause of fluctuating relative humidity and temperature. A statement in the Corps' Decem-

TABLE 31.3 Temperature and relative humidity during the 2005 and 2006 plaintiff studies.

Study period	Temp °F. Minimum	Temp °F. Maximum	Range	% RH Minimum	% RH Maximum	Range
July 2005, classroom	68.3°	76.6°	8.3°	37.2%	66.5%	29.3%
February 2006, basement laboratory	69.7°	74.5°	4.8°	5.6%	33.3%	27.7%

Note: Standard museum conditions are 67–70 degrees Fahrenheit and 45–50% relative humidity.

ber 2005 report characterized conditions during the 2005 studies in the classroom as "drastic." The curation team also expressed concerns about controlling environmental conditions for the February 2006 studies. A data logger recorded conditions in the classroom and basement laboratory during the two study periods (Table 31.3).

The data show that the minimum and maximum temperature readings during both studies fell within or above standard museum conditions. In contrast, the minimum and maximum readings for relative humidity were quite different during the two study periods. The July 2005 studies were held in a large classroom on the main floor of the museum with more than a dozen people present at any one time. The four days of studies in February 2006 were held in the smaller basement laboratory with a limit of about seven people in the room at a time.

During the 2005 study in the classroom, the relative humidity bracketed standard museum conditions, with the minimum and maximum relative humidity of 37.3 percent and 66.5 percent, respectively. With a dehumidifier to control relative humidity during the 2006 study in the basement laboratory, the minimum and maximum readings were 5.6 percent and 33.3 percent, respectively. Although the absolute fluctuations in relative humidity were similar (29.3 percent and 27.2 percent), the 2006 relative humidity appears to fluctuate significantly below the minimum relative humidity recorded during 2005 classroom studies. These data generally fall outside the data logger's specifications for reliability.

Prior to opening the storage unit for the July studies, the data for the second half of June show the relative humidity in the storage cabinet was on the rise with the maximum at 36.2 percent. After the studies and into August, the relative humidity in the storage unit was within the range of an optimal museum environment (46 percent to 49 percent).

In January 2006 the relative humidity inside the storage cabinet ranged from 28.4 percent to 30.4 percent. After the studies, the March relative humidity inside the cabinet ranged from 19.5 percent to 23.9 percent. The upper relative humidity readings during the February study were similar, although the minimum readings fell well below the January and March minimum ranges (Table 31.1). Although readings fall below the lower limits of the data logger specification, they suggest the bones were exposed to dry conditions before, during, and after the 2006 studies.

The hourly data during the two weeks of July 2005 studies provide insight into the conservators' claim that a rise in relative humidity was directly related to the number of people in the room during studies. Between the evening of Sunday, July 10 (7:15 p.m.) and Monday morning (7:15 a.m.), the bones were exposed to the peak relative humidity recorded for the entire two-week period. No one but the security guard had access to the room during that time. On Sunday, an evening reading at 6:15 p.m. of 56.7 percent jumped to 66.5 percent in one hour. Relative humidity remained between 65 and 66 percent all night, finally settling below 60 percent on Monday afternoon. The museum's environment control system, rather than the number of people in the laboratory, appears the source of the peak readings for relative humidity in the classroom.

An inspection report by the Corps contract conservators following the 2005 study (COE, Plaintiffs' Study of the Kennewick Remains February 2006:3) includes no reason for the dramatic departure from the widely accepted museum standard for relative humidity. The conservators' explanation, similar to those offered over the years, is not consistent with Corps data:

The recorded environmental data illustrate the significant changes that are expected with multi-persons, diverse activities, several exterior walls, constant collection movement, with long hours in an intense study session. The recorded environmental data also illustrate the gradual and minor seasonal changes that are to be expected during the storage period between inspections.

DISCUSSION

From 2001 to 2008, the conservators' reports include the summary environmental data and repeatedly attribute fluctuations to expected seasonal variation. Graphs, when included, lack target ranges, but clearly show a departure from acceptable conditions. The reports summarize changes in vague terms and provide a subjective code to document changes. More specific details may be archived, but a meaningful understanding of the changes is not possible from the reports available for review.

The first impression of the data suggests a problem. A curator might expect a standard museum-quality Delta cabinet to provide more protection from the fluctuating environmental conditions in a museum. If the Delta cabinet is providing the expected protection, significant swings in temperature and relative humidity in the museum would seem to be the problem. Another possible explanation is an improperly calibrated data logger. The available reports do not mention regular calibration of the data logger, although such a practice is routine. Without proper calibration, the data recorded for 10 years may be spurious. However, the regular patterning of the monthly data suggests fluctuating environments even if the relative position on the scale is off.

The Corps team appears to have largely misinterpreted or ignored the rich dataset they have collected. Twice a year they report data that falls outside standard museum conditions as exceptions and justify them as expected seasonal variation. No report mentions within- or between-month fluctuations, although the graphs generated with the summary statistics clearly show a closer look is warranted. If an understanding of the data across the span of a year (or years) has been sought, the reports do not mention this. The 10 years of data show predictable fluctuations in relative humidity and temperature so that preventive, stabilizing action could be taken.

In their reports the conservators continued to attribute changes in condition to handling during studies. However, they have reported soil deposition, expanding cracks and continued flaking during the years when no studies were conducted. For nearly five years after the government studies, the only access to the storage cabinet had been for semiannual condition assessments. The same is true after February 2006 when the plaintiff studies were completed. Douglas Owsley cautioned in his "Report of the Inventory of the Kennewick Skeleton" following his 1998 inventory, "On many fragments dirt is adhering to the bone cortex. The cortex will flake away with the dirt unless careful measures are followed" (COE 4323). Chapter 4 of *Human Remains: A Guide for Museums and Academic Institutions* describes the results of various physical stresses including "cracking caused by fluctuating relative humidity (teeth are particularly susceptible)" (Cassman et al. 2007:33). Owsley described each tooth in his inventory of the skeleton conducted in October 1998 (COE 4323). He described specific postmortem cracks, which were starting to break individual teeth apart, and noted the fragile remaining enamel.

Owsley's recommendations were not followed. Fluctuating environmental conditions cause drying and expanding, which in turn cause flaking and expanding cracks in fragile bone and teeth. The unstable environment is likely contributing to the expansion of the cranial cracks and other deterioration. The Corp has not acknowledged that environmental conditions may be a factor. Their reports indicate that they have overlooked the persistent and fundamental problem of the lack of a stable museum environment. The curatorial list of the agents of deterioration concluded with this remark: "The tenth agent of deterioration is curatorial neglect, which may include any or all of the other nine agents" (Cassman et al. 2007:105).

It is not clear why the Corps curation team never mentioned in their reports that standard museum environmental conditions were not being met nor why the Corps has neglected to hold the Burke Museum accountable to provide the environmental standards stated in their contract. These concerns are not new. These issues were covered in detail in the scientists' opening brief to the Oregon Federal District Court in 2001:

> Despite their obligations under ARPA and the Court's Remand, and contrary to their assurances that all

needed remedial actions had been taken, ... COE 6703-6704, 6731-6732, defendants allowed the skeleton to remain for months in substandard, unsafe conditions. Even after the relocation to Seattle, defendants' conduct still gives cause for concern.

The discrepancy between the Corps' idealized presentation and the data documenting the actual environment needs to be addressed in the preservation plan required by the 1998 Memorandum of Agreement.

PRESERVING KENNEWICK MAN

The Corps committed in their Memorandum of Agreement in September 1998 to develop a long-term preservation plan, not simply a storage (curation) plan, for the Kennewick Man skeleton (COE 4787). From the earliest days of conservator recommendations, Michael Trimble, Director of the Army Corps of Engineers' Mandatory Center of Expertise for the Curation and Management of Archaeological Collections of the Saint Louis, Missouri, District, has stated that the Corps' curation responsibilities include developing a long-term preservation plan. However, after the government studies in 1999, no such plan was developed. Trimble eventually conceded to the District Court that a plan would be developed after the litigation controversies were settled (Court Docket 428). The December 2008 inspection report indicates that the topic of long-term curation was currently under discussion but a preservation plan was not mentioned. This obligation of the Corps has yet to be fulfilled.

The Corps' preservation plan must provide environmental conditions appropriate for preserving the Kennewick Man skeleton. Ideally, this plan will establish a stable storage environment.

Since the discovery of Kennewick Man in 1996, both the safety and the security of the skeleton have improved. The skeletal fragments are now housed in custom boxes kept in a secure storage cabinet monitored for temperature and humidity. Access to the bones is controlled. All these factors represent a marked improvement over early curation efforts. However, even these improvements have not resulted in acceptable storage conditions.

Although the Corps curation staff has made some effort to protect the skeleton from the unstable environment, deterioration of the Kennewick Man's skeleton has already occurred. Specially designed boxes help to protect the remains from potentially destructive contaminants and from damage when shifted or moved, but they have caused concern by researchers. Some unnecessary wear of specific fragments occurs when elements are removed and replaced due to the constrictive size of the cutouts. More significantly, while the data logger records relative humidity and temperature, it does not control these variables. This unique dataset offers potential for gaining further insight into the effects of long-term, fluctuating temperature and relative humidity on ancient bone.

Since 2001 the skeleton has been subjected to widely fluctuating environmental storage conditions, which cause concern for Kennewick Man's long-term preservation. Claiming to have provided an appropriate storage environment despite clear evidence to the contrary, accepting seasonal variation as expected, and attributing deterioration solely to handling during studies are all unacceptable. The Corps has repeated these statements semiannually for 10 years, but has made no meaningful progress to address the fundamental issue of preserving the skeleton.

Neither the Corps nor the Burke Museum has taken precautionary steps to provide the environment that would slow deterioration of the skeleton. Corps environmental data show that significant, nongradual fluctuations in temperature and relative humidity have occurred in cycles over the years. Environmental conditions do not meet the stable museum standards that the Burke Museum is paid to provide. As of 2013, the Corps has presented no plan to ensure that normal museum standards are met. The Smithsonian Museum Support Center, a state of the art facility, was designed to maintain stable storage conditions for national treasures such as the Kennewick Man skeleton. The Corps, the agency responsible for preserving these remains, needs to reevaluate the decision for an appropriate curation facility.

REFERENCES

References listed below under the headings "COE" and "DOI" (with a number rather than a date) are documents from the administrative records filed with the district court by the Army Corps of Engineers ("COE") and the Department of the Interior ("DOI"). References listed under the heading "Court Docket" (with a number rather than a date) are other documents filed with the district court in the Kennewick Man lawsuit. Army Corps of Engineers biannual inspection and condition reports prepared after 2001 are on file in the St. Louis, Missouri, District Office.

American Society of Heating, Refrigerating, and Air-Conditioning Engineers, Inc. 2007. Chapter 21: Museum, Galleries, Archives, and Libraries. In *ASHRAE Handbook: Heating, Ventilating, and Air-Conditioning Applications*, edited by M. S. Owen. American Society of Heating, Refrigerating, and Air-Conditioning Engineers.

Buck, Rebecca A., and Jean Allman Gilmore, eds. 1998. *The New Museum Registration Methods.* American Association of Museums, Washington, DC.

Cassman, Vicki, Nancy Odegaard, and Joseph Powell, eds. 2007. *Human Remains: Guide for Museums and Academic Institutions.* Alta Mira Press, Lanham, MD.

Centre de conservation du Québec, Canadian Conservation Institute, and Université du Québec à Montréal. 1995. *Preventive Conservation in Museums*. Université du Québec à Montréal.

COE

626 Dickson Year 2000 Compliance Certificate and specification for the TL120 Temperature/Humidity Data Logger.

663 Timeline of reported conditions of cracks in the skull 11/99.

1232 Conservation Report on the Rehousing of the Kennewick Remains at the Burke Museum, Seattle, Washington, February 1999.

4323 Inventory Observations on the Kennewick Man Collection 10/28–29, 1998. [Author Douglas W. Owsley]

4787 Revised Joint Memorandum of Agreement Regarding the Transfer of the Remains to the Burke Museum, October 1, 1998.

5203 Federal Defendants Fourth Quarterly Status Report. July 1, 1998. Repository Summary Thomas Burke Memorial Washington State Museum, University of Washington, Seattle.

6703 Federal Defendants' Supplemental Status Report November 10, 1997.

6731 Federal Defendants' Response to Bonnichsen Plaintiffs' Status Report, October 24, 1997. Biannual Inspection Report: Plaintiffs' Study of the Kennewick Remains February 2006. On file in the St. Louis, Missouri, District Office.

Court Docket

273 Notice of pipe break at the Burke Museum on 5/18/00.

428 Status Report for May 2001 and Declaration of Michael K. Trimble, Ph.D. (entered 5/3/2001).

490 Federal Defendants' January 2002 Curation Status Report (entered 2/8/2002).

DOI

2976 Memorandum of Agreement for Cultural Services between the Army Corps of Engineers and the Thomas Burke Memorial Washington State Museum Attachment C: Collection Management Protocol included Special Procedures.

Fisher, Genevieve. 1998. Collections management: Preventive care. In *The New Museum Registration Methods*, edited by Rebecca A. Buck, and Jean Allman Gilmore, 104–105. American Association of Museums, Washington, DC.

Knight, Barry, and David Thickett. 2007. Determination of response rates of wooden objects to fluctuating relative humidity in historic properties. In *Museum Microclimates*, edited by Tim Padfield and Karen Borchersen, 85–88. National Museum of Denmark, Copenhagen.

Michalski, Stefan. 1990. An overall framework for preventive conservation and remedial conservation. In *Preprints of the International Council of Museums Committee for Conservation 9th Triennial Meeting, Dresden, 26–31 August 1990*, 589–591. ICOM Committee for Conservation, Los Angeles.

National Park Service. 1999. Chapter 4: Museum Collections Environment. In *National Park Service Museum Handbook, Part I: Museum Collections.* http://www.nps.gov/museum//publications/MHI/mushbkI.html.

Smithsonian Museum Conservation Institute. n.d. Department of Defense Guidelines and Standard Operating Procedures for Curating Archaeological Collections, 45.

Thompson, Garry. 1994. *The Museum Environment.* 2nd edition. Butterworth-Heinemann, Oxford, U.K.

Waller, Robert. 1994. Conservation risk assessment: A Strategy for Managing Resources for Preventive Conservation. In *Preventive Conservation, Practice, Theory and Research: Preprints of the Contribution to the Ottawa Congress 12–16 September 1994*, edited by Ashok Roy and Perry Smith, 12–16. International Institute for Conservation of Historic and Artistic Works, London.

Who Was Kennewick Man?

Douglas W. Owsley and Richard L. Jantz

Throughout this investigation of the skeleton of Kennewick Man, we have systematically sought answers to key questions about this unique individual who provides a rare glimpse of Paleoamerican life. We wanted to learn who Kennewick Man was, precisely how long ago he lived, what his daily life was like, the kinds of activities he pursued, the foods he consumed and, perhaps most of all, what land he might have called home. These fascinating yet complicated issues have taken time to answer and document. A synthesis of what we have come to know about Kennewick Man, including how he relates to other Paleoamericans, is presented here. In many ways, this volume provides a foundation for future analyses.

Kennewick Man, who lived nearly 9,000 years ago, was tall for his time, at 5 feet 7 inches. He was not long and lean, but generally wide-bodied and massive, at a weight of about 160 pounds. He easily moved his well-muscled body through the steppe-like land of his environment, where sagebrush and clumps of large-seeded grasses might have otherwise made walking the terrain difficult. No trees blocked his view; most of them grew much higher up on the hills. The river basin supported a few game animals, such as elk, deer, pronghorn antelope, and bighorn sheep, as well as multiple species of rabbit. The rushing Columbia River and two of its tributaries, the Yakima and Snake Rivers, would have provided another potential food source: millions of kilograms of spawning salmonids, along with other fish such as suckers, cyprinids, burbot, and white sturgeon. Based on stone tool technology, the archaeological record shows that at least two groups of people used the basin's resources along with Kennewick Man. One group, the people of the Western Stemmed tradition, ranged widely throughout the area, while a group identified by the Old Cordilleran tradition stayed close to the river corridors.

Kennewick Man's arm and leg bones provide details on how he interacted with his surroundings. He favored his right arm and hand in most activities. The bones of his right hand suggest he habitually used his fingers and thumb in a pinch (precision) grip holding objects tightly between his thumb and fingers, as in flint knapping. He mostly used his left hand to powerfully grip objects against the palm. The muscle attachment sites in his right shoulder, combined with his strong right arm, suggest Kennewick Man often held an object in front of him vertically while forcibly raising and lowering it, a motion that has been related to poling a boat or dipping a fish net. He routinely raised his arm with his elbow bent to hold something above his head—almost certainly an atlatl for propelling darts with forceful overhand motions. The vigorous and rapid use of his right shoulder in this way caused a glenoid rim fracture, a condition that was painful and undoubtedly affected his throwing ability.

As a result of absorbing the forward motion of his body after a throw with his right arm, his left leg became stronger than his right. The shapes of his leg bones also tell that he seldom ran long distances, but rather that he sprinted for short distances, making sudden turns and stops. Additionally, his leg bones suggest he routinely walked in shallow but fast moving water, probably to catch fish and other marine resources. Related to such an activity is Kennewick Man's slight ear condition—which did not greatly affect his hearing—that likely was the result of frequent contact with cold water. Although he had abnormal union of ankle bones in both feet, the condition was not severe and probably did not cause enough pain to break his stride. Both knees show joint surface damage due to stress on tightly flexed joints from habitual squatting.

Kennewick Man's bones tell of two major injuries. Even though not all of his ribs were recovered, he clearly suffered severe blunt force trauma to the chest. Years before his death, he broke six ribs; five on the right side healed improperly. Their pinched-together fracture edges show that the ribs remained disconnected as they healed. Kennewick Man's vigorous lifestyle did not allow sufficient time for recuperation, and his broken ribs failed to reunite.

A stone point deeply embedded in Kennewick Man's pelvis fractured the iliac crest with such force that the dart was probably thrown using an atlatl. The base of the point is enveloped by healed, remodeled bone, indicating that Kennewick Man was struck as a young adult. The body of the point, partially encapsulated by now-decayed fibrous tissue as part of the healing process, can be seen through gaps in both the internal and external surfaces of his right ilium. Kennewick Man was undoubtedly knocked down when his hip was pierced by the dart, but the point did not cut major blood vessels nor enter his abdomen. In fact, the impact location was amazingly fortuitous, resulting in basically a superficial trunk injury. Afterwards, damage to skin and tissue took time to heal completely. The painful injury limited his mobility and upper body bending in the short term, and possibly also caused a slight limp that disappeared over time.

The restricted view of the point through windows in the ilium makes it difficult to determine the exact cultural tradition of the point's manufacture. Computed tomography (CT) scanning and stereolithographic reproduction, however, reveal the shape and size of the broken projectile point, which is 53 mm in length with a maximum width of 20 mm. Nearly 3 mm of the tip shattered upon impact. The edges were serrated, which is common on Cascade points, but not limited to that style. Because of its long tapering haft element and thick bulbous tip, it is quite possibly a reworked Haskett point. Given the wide distribution of Haskett points in the Great Basin, learning where the stone was quarried could potentially provide information on where Kennewick Man was when he was hit, help assess whether he was an intruder subject to attack by local inhabitants, or suggest the home territory of his attacker.

Kennewick Man also had minor injuries to his head and neck. Two small depression fractures in his cranium are similar to injuries observed in the skeletons of other Paleoamericans who are thought to have caught birds using bolas. This hunting tool consisted of medium-sized stones tied together with twine or leather, twirled above the head to build momentum, and then released. A bola stone gone astray could have given the hunter a good-sized bump and a headache, but had no long-lasting effects. It is also possible that these depression fractures resulted from adversarial encounters, as these kinds of injuries have also been observed on the crania of other Paleoamerican males, usually on the frontal bone. Lesions in Kennewick Man's cervical vertebrae are possibly also related to the use of a bola. They indicate excessive and repeated flexing of his neck, as in looking skyward.

Tasks performed using his teeth and the methods Kennewick Man used to prepare his food ground away his tooth crowns. Only millimeters of enamel remain and secondary dentin fills visible pulp chambers. An analysis of enamel microabrasions shows that this significant wear resulted primarily from extremely fine grit in his food derived, in large part, from accidental inclusions such as windblown particles. As a youth he had a pleasant smile, but as he aged his worn teeth were likely a source of pain. Due to the advanced wear of his front teeth, our orthodontist has characterized the appearance of his facial expression later in life as that of a "gummy" smile. There is no evidence of cavities or periodontal disease and he lost just one tooth in life.

Task-related activities have resulted in rounded occlusal surfaces on Kennewick Man's front teeth and also prevented his first premolars from meeting when his jaws were closed. An unusually rounded first molar suggests he habitually held cordage or a similar material between his teeth on the right side, providing further evidence that he was right-handed.

Kennewick Man, who was approximately 40 years old at his death, was not frail; his bones were strong and robust. Despite his active life, he never broke his limb bones, or even the small bones in his hands and feet. While his skeleton gives no sign of his cause of death, it does tell how he was buried.

Since no one saw Kennewick Man's skeleton in situ, clues to his burial position were gleaned from his well-preserved bones. His skeleton was not subjected to highly destructive fluctuating wet and dry conditions because the Columbia Basin has always been in the rain shadow of the Cascade Mountain Range and because Kennewick

Man was buried on a terrace above the river channel. The little rain that did fall trickled through the soil, slowly dissolving calcium carbonate and, through evaporation, redepositing it on the bones almost six thousand years after Kennewick Man was buried. These concretions build from the bottom of an object to its top and serve as one line of evidence for the position of the articulated skeleton in the grave. In addition, the pattern of postmortem breaks in Kennewick Man's bones reinforces the positioning of the skeleton and details the sequence in which his bones fell onto the shoreline.

The completeness of the skeleton and lack of animal scavenging indicate that Kennewick Man was buried in an intentional, primary interment. He was buried on his back with his hands palm down at his sides. His face was up with his chin tucked to his neck; his legs were straight and parallel. His left foot rolled onto its side during decomposition, while the sole of his right foot settled downwards. His head was at the east, upstream, and his feet at the west with his left side closest to the river. Taphonomic and sediment analyses indicate that Kennewick Man was buried 70 to 90 cm below the current ground level with his head slightly higher than his feet.

Kennewick Man's skeleton was found soon after it eroded from his grave. Analysis of visible green algal stains on his bones showed an exposure time of one to four weeks before they were recovered. Official levels for Lake Wallula recorded high water periods that would have carried the bones away from the grave site if they had eroded out before this short window. Unfortunately, it is not possible to more thoroughly investigate the burial site or the riverbank terrace. In 1998, the US Army Corps of Engineers (Corps) buried it under tons of stones in a stated effort to reduce erosion in the area.

The skeleton of Kennewick Man was first dated as 8410 ± 60 RC yr. BP. While this radiocarbon date established that Kennewick Man did not die within the past few centuries, government-sponsored dating by additional laboratories returned numbers differing by 2,660 years (5750 ± 100 and 8410 ± 40 RC yr. BP). Redating the skeleton using left-over bone residue from this earlier work resulted in the more definitive date of 8358 ± 21 RC yr. BP (8690–8400 CAL yr. BP) by dating multiple chemical fractions from different skeletal elements, which contain collagen and have collagenous amino acid compositions. This case exemplifies the difficulty of measuring accurate geologic age on ancient skeletal remains and how errors in results compound research questions in archaeology and physical anthropology. It should also be viewed as cautionary when interpreting other skeletons that have similar dating histories and where a single age measurement controls interpretation.

WHERE DID KENNEWICK MAN LIVE?

Carbon, nitrogen, and oxygen stable isotope values measured from Kennewick Man's bone collagen tell of the foods that he ate and the water that he drank as an adult. They are therefore critical to solving the mystery of whether Kennewick Man lived in the Columbia River Valley or was a traveler from another land.

In Chapter 16, Kennewick Man's isotope values are evaluated under the assumption that he lived much his of adult life in the geographical region of his burial. Kennewick Man's carbon ($\delta^{13}C$ = –13.95‰) and nitrogen ($\delta^{15}N$ = 21.26‰) isotope values indicate a lack of herbivorous terrestrial species in his diet. Schwarcz et al. interpret this to mean that he primarily consumed salmonids from the river, as well as other animals that fed on these fish, such as bears or birds of prey. The Columbia River environment and the animals living in it experienced massive seasonal influxes of nitrogen and carbon as a result of the immense quantities of decaying carcasses of anadromous salmonids. No data exist to determine precisely how this would have altered food web isotope cycles and potentially Kennewick Man's values.

An alternative explanation for Kennewick Man's carbon and nitrogen isotope values is a diet based heavily on high-trophic marine mammals, which were not available in abundance in the Columbia River Basin. This suggests that he was a migrant from the Pacific coast and not local to the area near Kennewick, Washington, a scenario Schwarcz et al. also considered to be possible. Evidence supporting this argument is drawn from zooarchaeological studies of fauna found at early Holocene sites and species-specific isotope data for terrestrial and marine mammals, including human populations from the Northwest Coast and interior Plateau.

Cybulski (2010) has compared stable carbon isotope distributions for 249 individuals from the North Pacific Coast to 104 individuals from the interior Plateau, including previously published information on Kennewick Man. People living on the coast have higher, more positive, values characteristic of a marine signal. For example, Chisholm et al. (1982, 1983) reported a mean $\delta^{13}C$ value of −13.3‰ for 37 individuals from the coast of British Columbia and estimated that 90 percent of their dietary protein came from marine sources. Coastal groups are quite homogeneous because of their dependence on marine-derived protein, primarily salmon, but also other kinds of fish, sea mammals, and migratory seabirds and shorebirds (Cybulski 2010).

In contrast, carbon isotope values for human remains from the interior are more negative and more variable. Interior data have a median value of −20.0‰ with intra-area heterogeneity patterned according to relative consumption of C_3-derived terrestrial proteins. Plateau inhabitants "were more dependent on large game animals and plant foods although the degree of that dependence varied according to habitat and proximity to major river drainages" (Cybulski 2010:86). Isotope values decreased further inland because of reduced access to anadromous salmon (Chapter 16; Cybulski 2010; Lovell et al. 1986). Kennewick Man is the only individual in Cybulski's (2010) interior series with a carbon value similar to that of coastal consumers.

Big game was an important food resource on the Columbia Plateau (see Chapter 2 for a detailed description of the eastern Washington state environment 8500–8000 RC yr. BP). Faunal assemblages from roughly contemporaneous archaeological sites near Kennewick, namely the Marmes Rockshelter and the Lind Coulee site, contain high proportions of butchered bison, elk, deer, and pronghorn antelope bones (Daugherty 1956; Gustafson 1972; Irwin and Moody 1978). Other paleozoological studies verify the presence of these herbivores in the Columbia River Basin during the early Holocene (Ames et al. 1998; Butler and Campbell 2006; Chatters 1998; Chatters and Pokotylo 1998; Dixon and Lyman 1996; Lyman 2004, 2007; Lyman and Livingston 1983; McCorquodale 1985). Bears were present, but rare (Lyman 1986). Multiple species of birds were recovered from Early Period contexts at The Dalles Roadcut site along the Columbia River in Oregon (Cressman et al. 1960); four in particular were targeted. Gulls (opportunistic omnivores and scavengers of mostly fish and marine invertebrates) and cormorants (hunters of small fish and crustaceans via pursuit-diving) were consumed as supplemental foods (Hansel-Kuehn 2003). Bald eagles (scavenging and hunting raptors consuming primarily fish) and carrion-feeding California condors were hunted for feathers and talons (Hansel-Kuehn 2003).

Within this regional context, carbon and nitrogen isotope values for terrestrial food sources are plotted in Figure 32.1 along with Kennewick Man's data for comparison. Taking into account applicable trophic-level offsets in carbon and nitrogen isotope values of human consumers relative to animal protein in their diet (1‰ and 3–4‰, respectively), cervids (deer), bison, mountain sheep, and pronghorn antelope have too negative $\delta^{13}C$ values and too low $\delta^{15}N$ values to have been a significant component in Kennewick Man's diet.

Isotope values of representative bears, an omnivorous species, vary considerably based on local ecology. For example, brown bears from archaeological sites along the Northwest Coast (Fedje et al. 2011; Szpak et al. 2009) have higher nitrogen values and somewhat more positive carbon values than black bears and herbivores. Their values indicate a mixed diet comprised of both terrestrial and marine resources; conversely the isotope signal of the black bears suggests that they were consuming primarily terrestrial foods (Fedje et al. 2011). Based on his isotope value, it is highly unlikely that Kennewick Man made his living hunting bears—a perilous practice as indicated by the taphonomic study of human remains from On Your Knees Cave (Chapter 17).

Dependence on any of the aforementioned species for an extended length of time would have resulted in a different isotope signature than that recorded for Kennewick Man. He was not a broad-spectrum, terrestrial, hunter-forager.

Bird consumption was also likely inconsequential. Of the avifauna known to have been hunted during the early Holocene at The Dalles Roadcut site (Hansel-Kuehn 2003), isotope data available for higher trophic level raptors provide a frame of reference (Figure 32.2). Not included in the plot are archaeological gulls (*Larus* spp.) from Vancouver Island, for which only an average collagen carbon value of −13.6‰ was reported (Hobson 1987). This value reflects a coastal diet high in marine foods. Like bears, the diets of these omnivorous species

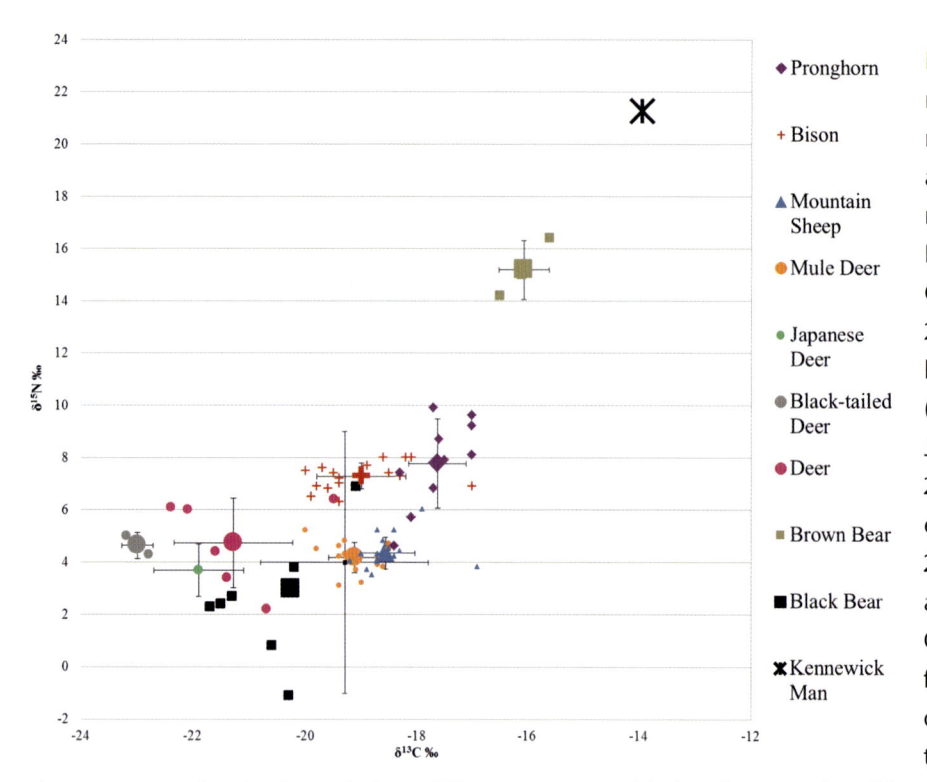

FIGURE 32.1 Carbon and nitrogen isotope values measured from collagen from adult terrestrial animal bones recovered from sites in the North Pacific and Northwest Coast. Black bear (Fedje et al. 2011; Szpak et al. 2009), brown bear (Fedje et al. 2011), bison (Coltrain and Leavitt 2002), Japanese deer (Yoneda et al. 2001), black-tailed deer (Topalov et al. 2013), deer (Fedje et al. 2011), mule deer, pronghorn, and mountain sheep (Ugan and Coltrain 2012) data were compiled from published literature. All data are archaeological, with the exception of the black-tailed deer; to correct for the Oceanic Suess Effect, 1.5‰ was added to these modern $\delta^{13}C$ values (Tieszen and Fagre 1993). Bars show one standard deviation (calculated) from mean carbon and nitrogen isotope values. The larger symbol represents the mean or weighted mean carbon and nitrogen isotope values for each animal group.

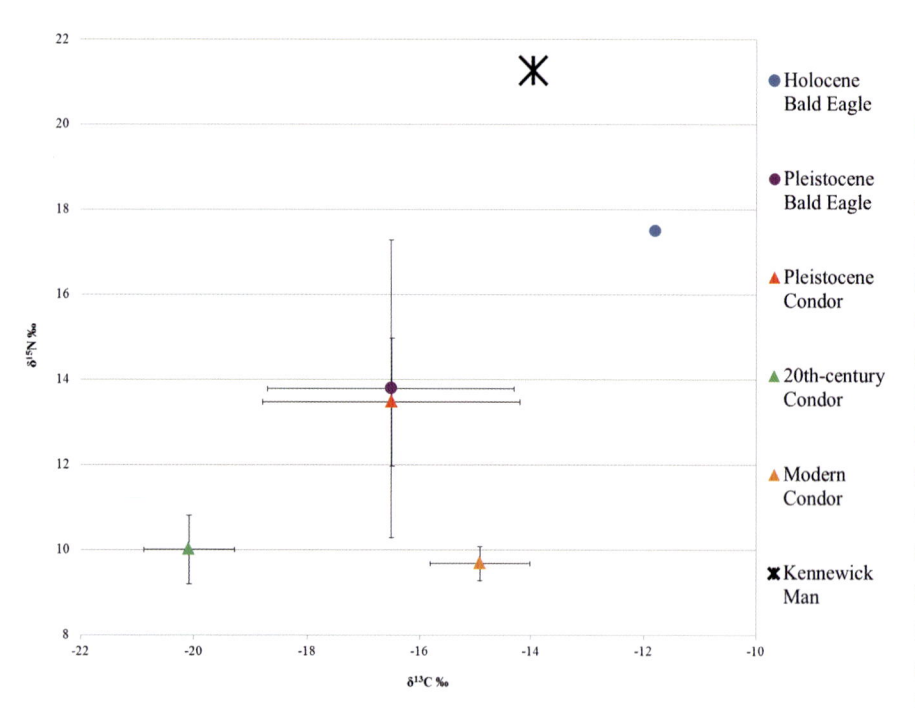

FIGURE 32.2 Carbon and nitrogen isotope values measured from collagen from modern and archaeological bird bones recovered from sites in the North Pacific and Northwest Coast. Pleistocene condor and bald eagle (extrapolated from Fox-Dobbs et al. 2006:Figure 2), twentieth-century and modern condor (extrapolated from Chamberlain et al. 2005:Figure 1), and Holocene bald eagle (Szpak et al. 2009) data were compiled from published literature. To correct for the Oceanic Suess Effect, 1.5‰ was added to all modern $\delta^{13}C$ values (Tieszen and Fagre 1993). Bars show one standard deviation (calculated) from mean carbon and nitrogen isotope values. The larger symbol represents the mean or weighted mean carbon and nitrogen isotope values for each animal group. (Modern condor values are dramatically different from their prehistoric ancestors, a consequence of artificial feeding programs to aid survival of the species (Chamberlain et al. 2005).)

vary; however, their nitrogen values are significantly below those of Kennewick Man.

Riverine and marine resources should also be considered as potential food sources for Kennewick Man. Resident fish of the Columbia River included suckers, cyprinids, burbot, and white sturgeon, in addition to spawning chinook, coho, and sockeye salmon, and steelhead trout. Salmonids would have been present for at least eight months of the year (Chapter 2; Chatters 1998), and the many thousands of salmon bones recovered from The Dalles Roadcut site provide evidence for intensive salmon fishing during the early Holocene (Butler 1993; Butler and O'Connor 2004; Cressman et al. 1960).

In addition to fish, Pacific lamprey (*Entosphenus tridentatus*), a parasitic anadromous species, were also present in the Columbia River (Chatters 1998; Hunn et al. 1998). In the ocean, lampreys feed on a wide range of fish, including, but not limited to, rockfish, Pacific cod, chinook and coho salmon, lingcod, and sable fish (Beamish 1980). While it has been reported that Pacific lampreys attach to marine mammals (Clemens et al. 2010; Close et al. 2002), it is more common for lampreys to instead be preyed upon by these species (Beamish 1980; Close et al. 1995; Williamson and Hillemeier 2001) and even for the diets of marine mammals to be comprised primarily of lampreys (Jameson and Kenyon 1977; Roffe and Mate 1984). While their ecology has been documented, an extensive literature search found no isotope data for adult Pacific lampreys. Since historic times, lampreys have been culturally important to the mid-Columbia Plateau tribes (e.g., Byram 2002; Close et al. 1995; Hunn 1990), and they are active in efforts to conserve the declining population (Close et al. 2002). Lampreys migrate upriver to spawn in the spring and early summer (Brostrom et al. 2010; Byram 2002; Hunn 1990; Kroeber and Barrett 1960; Miller and Seaburg 1990), and are typically caught near rapids or falls (Close et al. 1995; Close et al. 2002).

Harbor seals followed salmonids and lamprey up the Columbia River, to, but not beyond, Celilo Falls, a distance of about 324 kilometers from the Pacific Ocean (Lyman et al. 2002; Lyman 2011). Harbor seal bones have limited presence in faunal assemblages from archaeological sites along the Lower Columbia River; the Dalles Roadcut site, located approximately 10 kilometers downriver from Celilo Falls, is the farthest location upriver (Cressman et al. 1960; Lyman et al. 2002). A minimum number of six phocids (interpreted by Lyman et al. (2002) as harbor seals) have been recovered from early Holocene strata at this site (Cressman et al. 1960). The absence of pinniped remains beyond Celilo Falls indicates that they could not ascend these rapids (Lyman et al. 2002; Lyman 2011). Kennewick Man was discovered about 205 kilometers upriver from Celilo Falls.

Figure 32.3 compares carbon and nitrogen stable isotope values measured from Kennewick Man's bone collagen to those of pinnipeds (phocids and otariids), mustelids, and salmon recovered from archaeological contexts in the North Pacific and Northwest Coast (for a similar comparison, see Figure 16.1). Although there are differences, all of these species have marine carbon isotope values with salmon tending to be more negative and sea otters (mustelids) more positive. Sea otters, although having higher sample means than salmon, have comparatively low nitrogen isotope values. Nitrogen signatures in the salmonids are higher than the ungulates plotted in Figure 32.1, but far below that required as a major component of Kennewick Man's diet. Among prey groups in the northern Pacific, pinnipeds display the greatest enrichment in $\delta^{15}N$ and occupy the highest trophic level. Mean values for fur seals and Stellar sea lions (otariids, $\delta^{13}C = -14‰$ and $\delta^{15}N = 17.1‰$) and bearded, ringed, and harbor seals (phocids, $\delta^{13}C = -12.1‰$ and $\delta^{15}N = 18‰$) provide the best fit as the mainstay of Kennewick Man's diet, not just an occasional meal. Based on his isotope values, Kennewick Man's inferred diet reflects a coastal origin where pinnipeds were readily available.

Data for humans from along the Pacific Rim and the interior Plateau provide an additional basis for comparison (Figure 32.4). Represented groups include prehistoric Jōmon and Historic Ainu samples from Hokkaido and Honshu, Japan, and Sakhalin Island, Russia; Paleo- and Neo-Aleuts from the Aleutian archipelago; coastal villagers from the Alaska Peninsula, and early remains from On Your Knees Cave, as well as interior sites in Idaho—Buhl, Clarks Fork, and DeMoss. Kennewick Man has one of the highest nitrogen isotope values observed in this series, of which Aleuts and the man from On Your Knees Cave are most similar. Although very similar, Paleo-Aleuts relied slightly more on near-shore

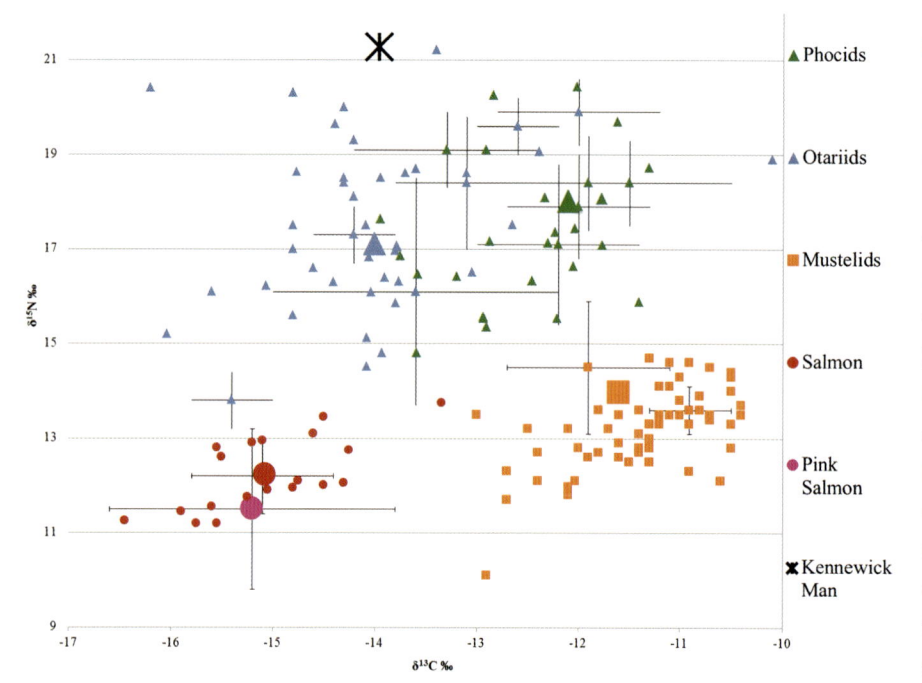

FIGURE 32.3 Carbon and nitrogen isotope values measured from collagen from bones of adult pinniped, mustelid, and salmon recovered from archaeological sites in the North Pacific and Northwest Coast. Mustelid (Misarti et al. 2009; Szpak et al. 2012), pinniped (Gifford-Gonzalez 2011; Misarti et al. 2009; Moss et al. 2006; Newsome et al. 2007; Szpak et al. 2009), and salmon (Misarti et al. 2009; Orchard 2011; Szpak et al. 2009) data were compiled from published literature. Bars show one standard deviation (calculated) from mean carbon and nitrogen isotope values. The larger symbol represents the mean or weighted mean carbon and nitrogen isotope values for each animal group.

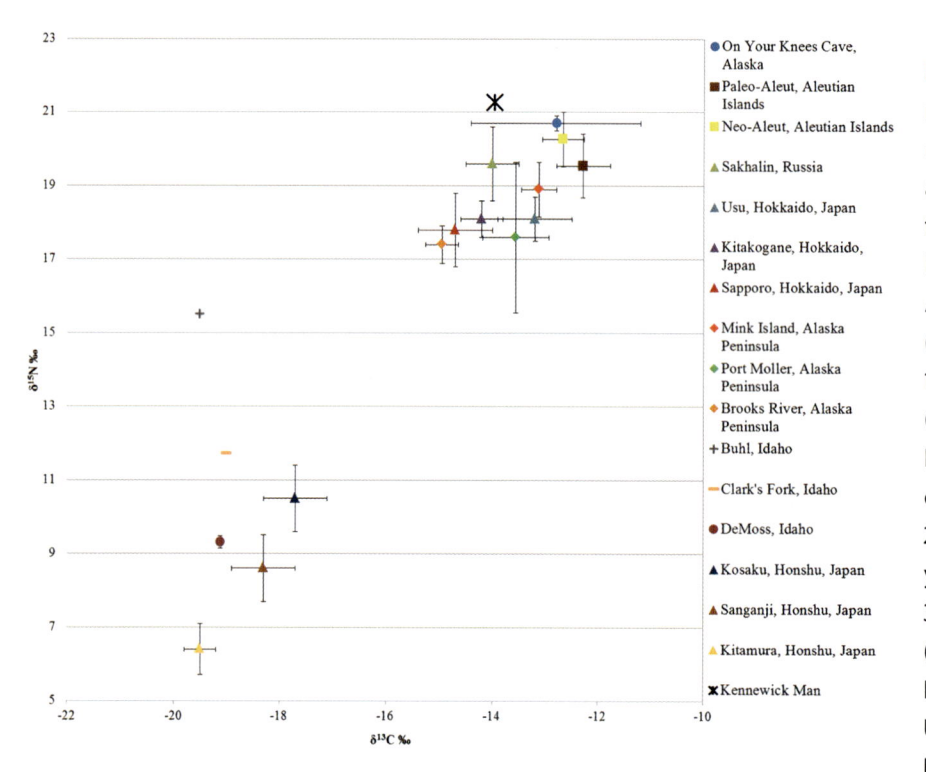

FIGURE 32.4 Carbon and nitrogen isotope values measured from collagen from adult human bones recovered from archaeological sites in the North Pacific and interior North America. On Your Knees Cave (Chapter 29); Port Moller (3547–1388 CAL yr. BP), Brooks River (1484–381 CAL yr. BP), and Mink Island (666–292 CAL yr. BP) sites on the Alaska Peninsula (Coltrain 2010); Paleo-Aleut (3434–407 CAL yr. BP) and Neo-Aleut (816–386 CAL yr. BP), Aleutian Islands (Byers et al. 2011); Kitakogane, Hokkaido, Japan (6000 BP); Usu, Hokkaido, Japan (2000 BP); Sapporo, Hokkaido, Japan (A.D. 1550, historic); Sakhalin, Russia (A.D. 1550, historic); Sanganji, Honshu (5000–2500 BP); Kosaku, Honshu (4100 BP); Kitamura, Honshu (4000 BP) (Minagawa and Akazawa 1992), and Buhl, Idaho (10,675 ± 95 RC yr. BP) (Green et al. 1998) data were compiled from published literature. Isotope values for the individuals from the sites of Clark's Fork (5930 ± 100 RC yr. BP) (Pennefather-O'Brien and Strezweski 2002) and DeMoss (5965 ± 60) (Green et al. 1986) are from Owsley's unpublished data. Each symbol represents the mean or weighted mean carbon and nitrogen isotope values. When applicable, bars show one standard deviation (calculated) from mean carbon and nitrogen isotope values.

resources while Neo-Aleuts gave slightly greater attention to high-trophic-level resources (Byers et al. 2011).

Individuals from sites on the Alaska Peninsula had less enriched nitrogen values with varied subsistence strategies centered on marine foods (Coltrain 2010). On Mink Island, marine foods represented about 85 percent of the diet including pinnipeds, fish, and invertebrates. The Port Moller faunal assemblage includes salmon, walrus, and caribou bones with approximately 25 percent of the human diet derived from terrestrial and low-trophic-level marine foods. At Brooks River, the diet consisted of 35 to 40 percent inland terrestrial foraging and hunting of caribou combined with low-trophic-level marine foods and salmon (Coltrain 2010). From the other side of the Pacific Rim, historic Ainu and prehistoric Jōmon hunter-fisher-gatherer economies varied according to geographic location. Groups from Honshu relied on terrestrial animals as much as, if not more, than marine resources. Northern groups from Hokkaido and Sakhalin relied extensively on marine foods (Minagawa and Akazawa 1992). As noted by Cybulski (2010), Plateau carbon isotope values are more negative and nitrogen values more variable and less enriched (lower) than coastal groups. The woman from Buhl consumed a diet comprised of high-trophic terrestrial foods with some access to anadromous fish (Green et al. 1998).

Rather than a reliance on salmon, Kennewick Man's isotope values reflect a diet comprised primarily of high-trophic-level marine pinnipeds. As only one species swam inland along the Columbia River, up to, but not beyond The Dalles, Oregon (Lyman et al. 2002), Kennewick Man could not have been a long-time resident of the area where he was found, but instead lived most of his adult life somewhere along the Northwest and North Pacific coast where marine mammals were readily available. Reinforcing this, Kennewick Man's collagen isotope values are most similar to Pacific coastal human populations with diets based primarily on marine mammals. Relative to other high latitude hunter-foragers, extremely similar values have been reported for the Sadlermiut of Hudson Bay, a group that depended almost entirely on seals and marine sea birds (see Coltrain et al. 2004).

With a $\delta^{18}O$ isotope value of 5.3 ‰ (VSMOW) obtained from phosphate, Kennewick Man's drinking water was calculated to have a $\delta^{18}O(w)$ of –25.5‰ (Chapter 16). As reported by Schwarcz et al., this value is depleted compared to contemporary values reported for the Columbia River at nearby Richland, Washington, to contemporary local precipitation, and to values measured from 6,000-year-old mollusk shells from the river. The average of ten $\delta^{18}O$ readings for the Columbia River at Richland, a few miles from Kennewick, is –17.1‰ (Coplen and Kendall 2000). This value is more positive than that of Kennewick Man. River samples taken from locations further inland are similar or more positive. Values near the Washington coast are even more positive, but gradually decrease going northward. For example, the Willapa River near Willapa mean is –8.5‰; the average for the Elwha River near Port Angeles on the Olympic Peninsula is –12.6‰; and the Skagit River average near Mount Vernon is –13.8‰ (Coplen and Kendall 2000). Using contemporary data as a guide, Kennewick Man's origin was even further north.

At a higher latitude, the modern, weighted $\delta^{18}O$ average for January precipitation in north central Alaska ranges from –22 to –26‰. Spring and summer $\delta^{18}O$ values for the Yukon River at Pilot Station, Alaska, about 200 kilometers from the coast of the Bering Strait, average –20.2‰, and $\delta^{18}O$ values for the Copper River at Chitina, Alaska, about 150 kilometers from the Gulf of Alaska, average –21.7‰ (Coplen and Kendall 2000).

Kennewick Man's depleted $\delta^{18}O$ value indicates that he drank cold river water originating in high elevation snow or glacier melt. For example, in AD 1736, Mount Logan in the Yukon Territory had a $\delta^{18}O$ of –28‰ and the Eclipse Icefield in the St. Elias Mountains, Yukon Territory has a long term, but more recent, average ice core $\delta^{18}O$ value of –25‰ (NOAA Paleoclimatology, Ice Core Gateway 2013) . Mount Logan is located about 100 kilometers inland from the Gulf of Alaska, and the St. Elias Mountains are even closer, just east of the southeast Alaska border. A –25‰ value for glacial melt mixed with precipitation may not be unreasonable for nine thousand years ago in the coastal northwest, and would likely characterize glacial melt water in southeast Alaska (Joan Coltrain, August 27, 2013 personal communication): "Given Kennewick Man's $\delta^{18}O$ value, he does not appear to be local to the Kennewick area, inland or coastal. If all we knew about him was his oxygen isotope chemistry, we might argue that he originated in inland, northern Alaska. But he dates to the early Holocene and precipitation $\delta^{18}O$ values should have been more negative than they are today. That has to be taken into consideration."

In conclusion, available isotope data for Kennewick Man suggest that he lived most of his life in coastal locations north of the state of Washington. In keeping with this finding are Kennewick Man's dental wear, reminiscent of that caused by working hides and similar to wear seen in early Eskimo dentitions, and his wide body breadth, a cold climate adaptation. Kennewick Man's bilateral asymmetry, shape, and strength of his humeri and left femur are not only consistent with proficient use of an atlatl, but also with the method of hunting seals shown in Figure 32.5. Was movement on ice and snow-covered ground a contributing factor to the tibia morphology described in Chapter 12?

WHO WERE KENNEWICK MAN'S RELATIVES?

The investigation into his origin is essentially a worldwide search for population relationships and genetic connections to Kennewick Man, information that contributes toward an understanding of his role in the peopling of the Americas. Several chapters in this volume have approached this topic through statistical analyses using craniometric data, and dental and nonmetric observations.

One way to address this question is to utilize cranial morphometric data. Though the original skull bones were not allowed to be pieced together by our research team, a reasonably accurate reproduction of Kennewick Man's skull could be created through noninvasive CT and 3D printing. Scans provided virtual representations of his skull components that could be analyzed in different ways and were also the basis for physical resin replicas that could be handled and reconstructed. Repeatedly comparing the stereolithographic reproductions with the original bone measurements resulted in as accurate a reconstruction as human skill and technological precision could obtain, and allowed for the collection of cranial measurements.

Although found in territory claimed by several tribes, Kennewick Man's large cranium and craniofacial morphology do not resemble present-day American Indians in the region, or even those from several thousand years ago. In fact, the lack of similarity between Kennewick Man and late prehistoric and modern Indians is striking. Like other Paleoamericans, his craniofacial differences make him distinctive. His skull more closely resembles the circumpacific populations of Polynesia. Yet, he was not from those populations. Instead, those peoples, along with Kennewick Man and other ancient Americans, have roots in coastal Asian groups, of which the prehistoric Jōmon culture is among the better known. Kennewick Man and other ancient individuals differ from modern American Indians in systematic ways that cannot be attributed to their subsistence strategy or to any other component of their environment. Rather, their difference is mainly genetic and as such carries information about their history and biological affinities.

FIGURE 32.5 Inuit seal hunter among ice floes, Alaska (ca. 1903–1915). Photographed by Lomen Brothers, Nome, Alaska. Glenbow Archives (detail of ND-1-1140a).

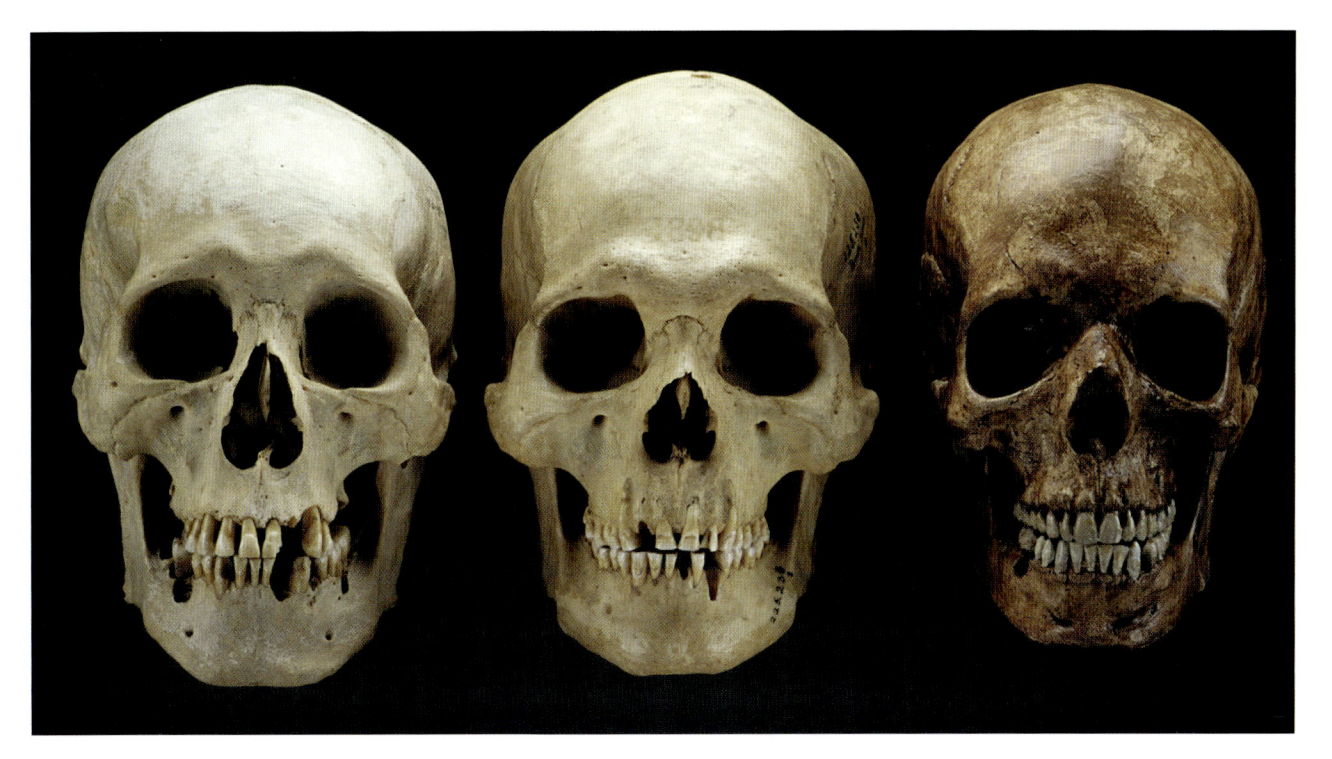

FIGURE 32.6 Frontal comparison of a Mongolian male (NMNH 278776, left), an American Indian from the Northern Plains (NMNH 225238), and an American White Male (W. Weir, right).

Similar craniofacial dimensions provide a basis for tracing biological relationships through time. Such analyses begin in forensic anthropological identifications and bioarchaeological research through the comparison of measurements for unknown crania to population reference samples. These multivariate statistical comparisons essentially ask which group(s) the unknown is most like and how good is the fit, (i.e., does it fall within the range of variation of a particular reference sample?). Visually, it is easy to appreciate the shape morphology shared by Mongolian and Northern Plains Indian male skulls when contrasted with the skull of an American White (Figure 32.6). The robust, wide faces with flaring zygomatics, and relatively short, broad crania of the Asian and American Indian skulls reflect a common ancestry in past millennia.

MORPHOMETRICS OF EARLY HOLOCENE CRANIA

Jantz and Spradley (Chapter 25) have drawn attention to the common, but mistaken, belief that early Holocene crania from North America are highly variable and exhibit few common features (e.g., Waguespack 2007). In fact, available crania show a number of similar characteristics which can be characterized as follows:

1. A large vault size is the most consistent feature; nearly all early Holocene crania are large compared to modern populations, being equaled only by modern Polynesians. Modern Polynesians average around 11.5 percent larger than modern American Indians, a difference that is easy to appreciate visually when regularly working with crania representing different world populations. For example, identically scaled frontal and lateral views (Figure 32.7) of a historic period Chatham Islands Moriori male cranium and a late prehistoric Chumash male cranium from Santa Cruz Island, California, clearly show the Moriori cranium is substantially larger. The lateral view also shows a forward projection of the face in the Moriori, a feature also seen in Kennewick Man.

2. Vaults have a strong tendency to be long and narrow. Spirit Cave is the most extreme in this regard, followed by Browns Valley, Stick Man, and Wizards Beach. In this feature they contrast markedly with most modern American Indians, which tend to be

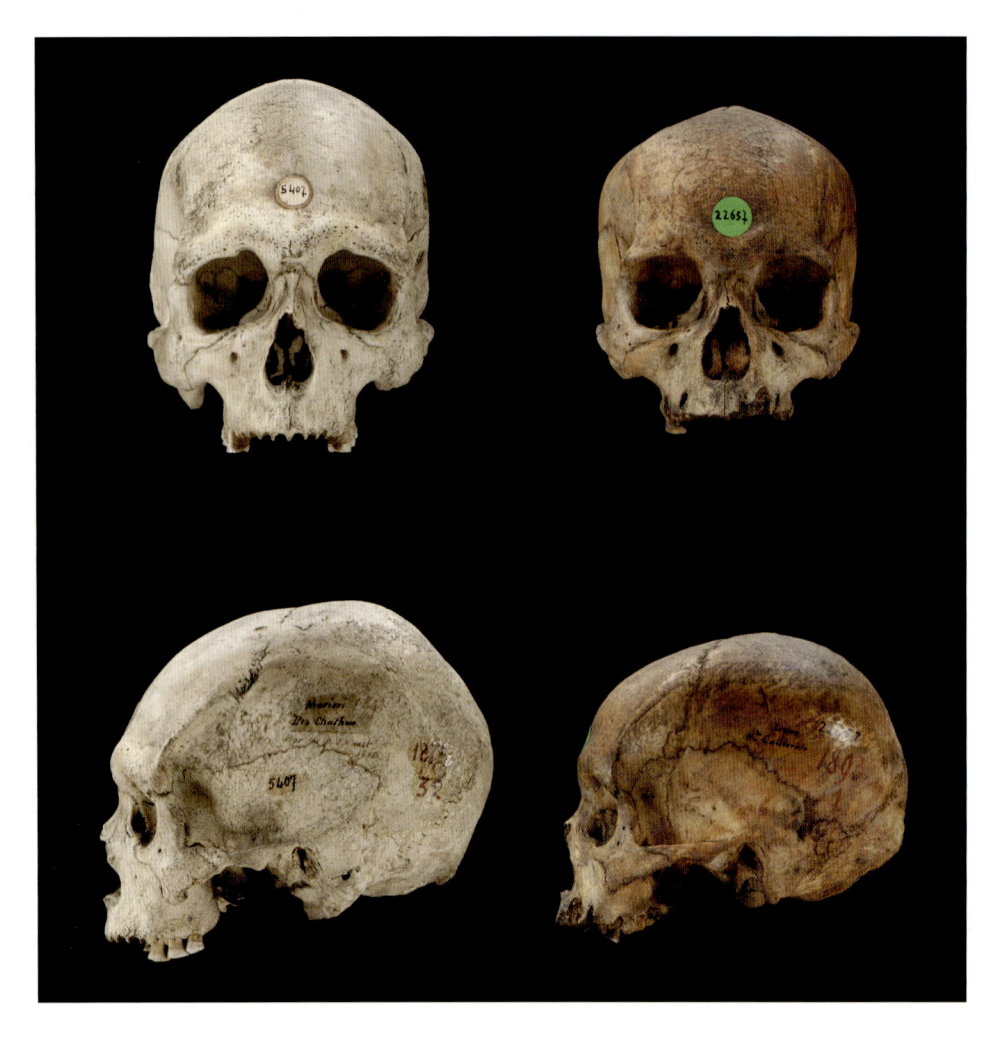

FIGURE 32.7 Identically scaled frontal and lateral views of the cranium of a Chatham Islands Moriori male (Musee de l'Homme, Paris, France 5407, left) with that of a Chumash male from California (Musee de l'Homme, Paris, France 22657). Musée de l'Homme, Collections d'anthropologie biologique, Paris.

short and broad. Only Arch Lake and Gordon Creek can be described as having a mesocranic shape.

3. Kennewick Man's frontal bone is extremely flat and in this way is similar to early Holocene human remains recovered from Browns Valley, Minnesota, Wizards Beach, Nevada, and Buhl, Idaho. Other early Holocene crania are less extreme but still do not approach the curved frontal morphology seen in European and African populations. Modern American Indian and Polynesians also tend toward flat frontals.

4. Low faces are common, but this trait is variable. Midfacial height, as measured between the landmarks of nasion and prosthion and correlated dimensions, shows considerable variation among ancient Americans. Some early crania can be characterized as high faced; Kennewick Man and Wizards Beach are the most extreme. Some are extremely low faced. We called attention to the low face seen in Arch Lake Woman (Owsley et al. 2010) and others who share in that configuration are Browns Valley, Gordon Creek, and Buhl.

Kennewick Man's craniofacial morphology is shared with other ancient Americans, such as San Miguel Man (Tuqan Man) recovered from the Channel Islands of California (Figure 32.8) and the cranial vault identified as Stick Man, also from Washington State (Chatters et al. 2000). Metric relationships of Kennewick Man's and San Miguel Man's crania relative to American Indian and Polynesian groups are shown in Figure 32.9. While this principal coordinates plot shows the extreme position of Kennewick Man, it also demonstrates that both Kennewick Man and San Miguel Man have morphological dimensions aligned with Polynesians, a reflection of some type of shared ancestral affinity. In terms

FIGURE 32.8 Frontal view comparison of Kennewick Man (left) and San Miguel Man. Similar features include the nasal aperture shape with vertical lateral margins, brow ridge configurations, the absence of flaring malars, and overall size and shape, although Kennewick Man is slightly higher headed.

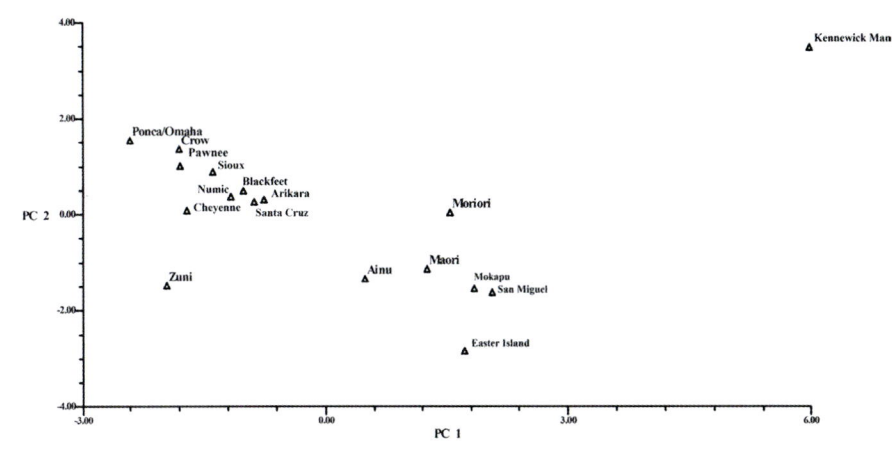

FIGURE 32.9 This bivariate principal coordinates (PC) plot shows the relationships of Kennewick Man and San Miguel Man to population samples representing Polynesians and American Indians. The plot is based on 20 components extracted from a covariance matrix of 34 metric variables. Modern American Indians and Polynesians, along with the Ainu, are differentiated by their cranial morphology, and both sets show internal cohesiveness. The American Indian groups are from the Plains (Blackfeet, Crow, Sioux, Arikara, Pawnee, Cheyenne, and the historically related Ponca and Omaha), Southwest (Zuni), Great Basin (Numic speakers), and California (Chumash from Santa Cruz Island). The first axis serves as the primary separator between American Indians and Polynesians. Kennewick Man occupies an extreme position on this axis, which one might characterize as "hyperpolynesian." The axis reflects cranial length, and especially facial forwardness in the sense described by Howells (1973) for Polynesians. Also important on this axis are midfacial size (which is smaller in Kennewick Man and San Miguel Man compared to American Indians) and nasal bone prominence (which is less prominent in Kennewick Man and San Miguel Man). Kennewick Man is also extreme on the second axis, which mainly reflects his wide posterior vault, high face and extreme frontal flatness. San Miguel Man (which occupies the space next to Mokapu Hawaiians) on the other hand, could be considered as having a morphology that is typical of Polynesians.

of nearest-neighbor statistics, the crania of Kennewick Man and San Miguel Man are most like the Moriori (Figure 32.10) of the Chatham Islands (Chapter 25).

As described in the morphometric chapters, early crania show markedly different features than those characterizing American Indian crania. Photographic comparisons visually illustrate some of the differences detected by these statistical analyses. Figures 32.11, 32.12, and 32.13 present Frankfort Horizontal images comparing the skull of Kennewick Man to those of Indians from California, the Northern Plains, and the Great Basin. Figure 32.14 compares the ancient skull of San Miguel Man to that of a Middle Period Chumash male (950 to 1150 AD). These two skulls represent morphologically distinct groups that occupied the same territory at different time periods.

Previously, we have identified metric similarity between the late Pleistocene Upper Cave Choukoutien 101 skull (near Peking, China) and early Americans, particularly the partially mummified remains from Spirit Cave in Nevada (Jantz and Owsley 2005). In our analysis of Spirit Cave craniometrics (Jantz and Owsley 1997) we further identified metric similarity of both crania to the historic Ainu, involving, in particular, the vault profile. At the same time, we found no affinity to any American Indian group. Brown (1999) characterizes Upper Cave 101 in parallel terms, noting that this male falls outside the range of modern East Asians, bears little likeness to them, and is an unlikely ancestor. Brown's assessment of the male from Upper Cave 101 is that, like the Spirit Cave mummy, he has Ainu affinities. That similarity is expressed in their juxtaposition in Figure 32.15, as is their differentiation from modern American Indians and greater similarity to Polynesians and the Ainu (Jantz and Owsley 1997).

A visual comparison of Spirit Cave and Upper Cave 101 (Figure 32.16) demonstrates their similarity. Lateral view facial profiles are very similar, as are vault heights and lengths, while the frontal view shows comparable nasal morphology and facial height proportionality. The primary difference between these two crania is size, with Upper Cave 101 being larger. Both exhibit similarities to the Ainu, Polynesians, and in some ways Europeans, and both are atypical of any modern population (Cunningham and Jantz 2003). Another example, the Upper Cave

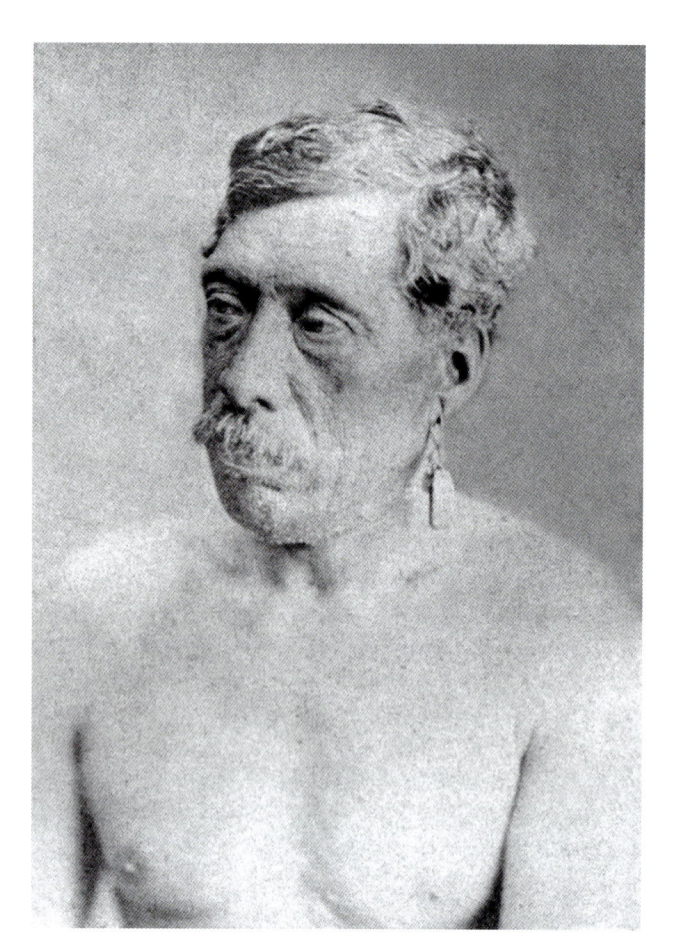

FIGURE 32.10 Moriori male from the Chatham Islands. "Portrait of a Moriori," January 1879, F. Dossetter, Christchurch, New Zealand (Potts 1882). Photo courtesy of the Rare Book Division of the Library of Congress.

103 female, shows some similarity to Southwest Pacific populations (Cunningham and Jantz 2003), but is even more atypical of modern populations.

DENTAL TRAITS

In addition to craniometrics, genetic relatedness can also be assessed using size and morphology of the teeth. Because Kennewick Man's teeth are so heavily worn, it is not possible to classify them as either Sinodont or Sundadont, the two major dental divisions in the Asian "Mongoloid dental complex" (Chapter 8; Scott and Turner 1997). Data on dental traits of adult Paleoamericans are hard to assemble because of extensive tooth wear. The dentitions of juveniles and young adults, however, suggest that dental morphology may distinguish Paleoamericans from recent American Indians

FIGURE 32.11 Frontal, right lateral, and posterior images comparing the skull of Kennewick Man (left) to that of an American Indian male from California (KER-60/KER-116, NMNH 372272).

FIGURE 32.12 Frontal and lateral views comparing the skull of Kennewick Man (right) to that of an American Indian male from the Northern Plains (NMNH 225238).

FIGURE 32.13 Frontal and lateral views comparing the skull of Kennewick Man (left) to that of an American Indian male from the western Great Basin (Nevada State Museum 735).

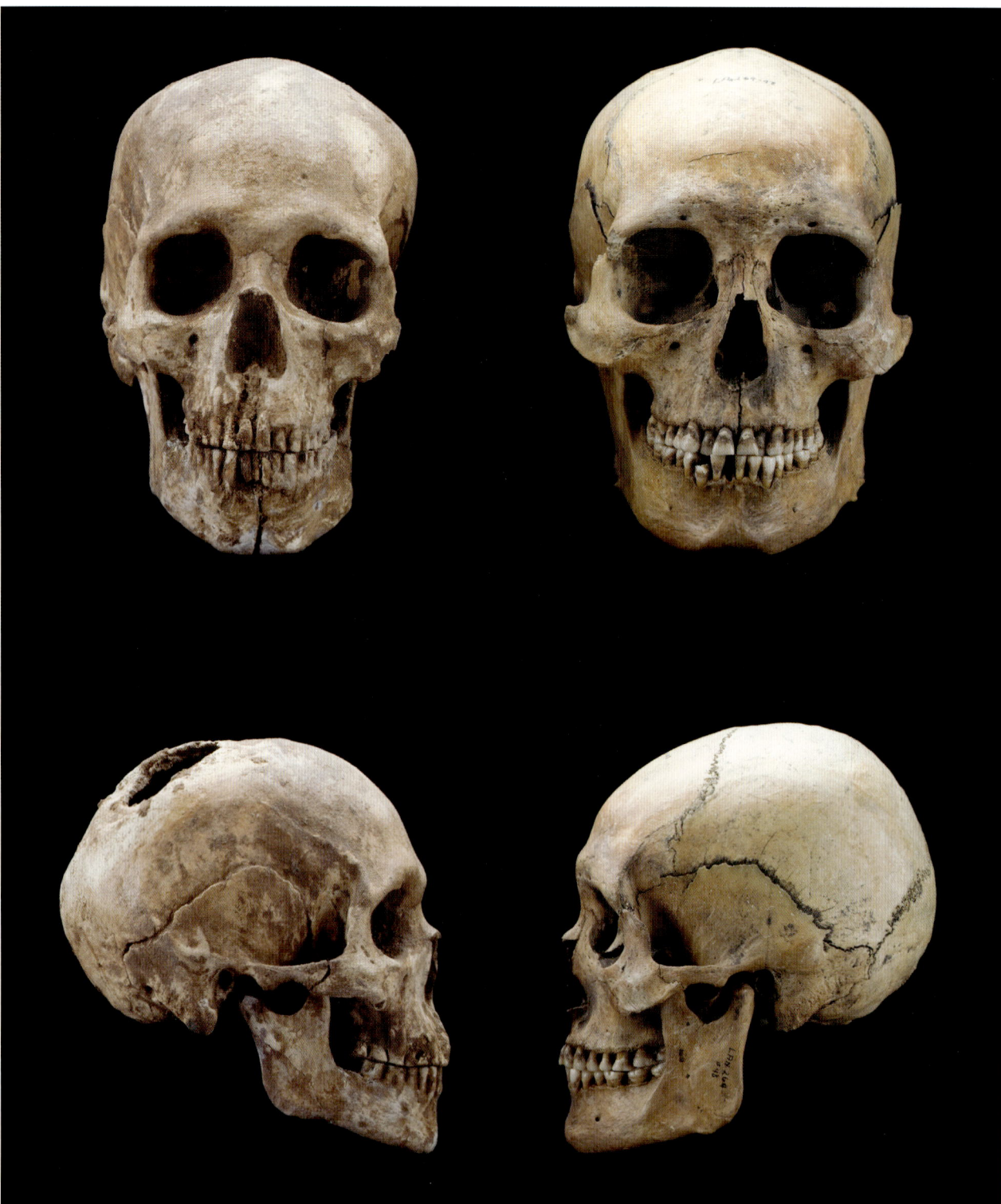

FIGURE 32.14　Frontal and lateral views comparing the ancient skull of San Miguel Man (left) to that of a Middle Period Chumash male (LAN 264-48). Both lived in the same area but at widely different times.

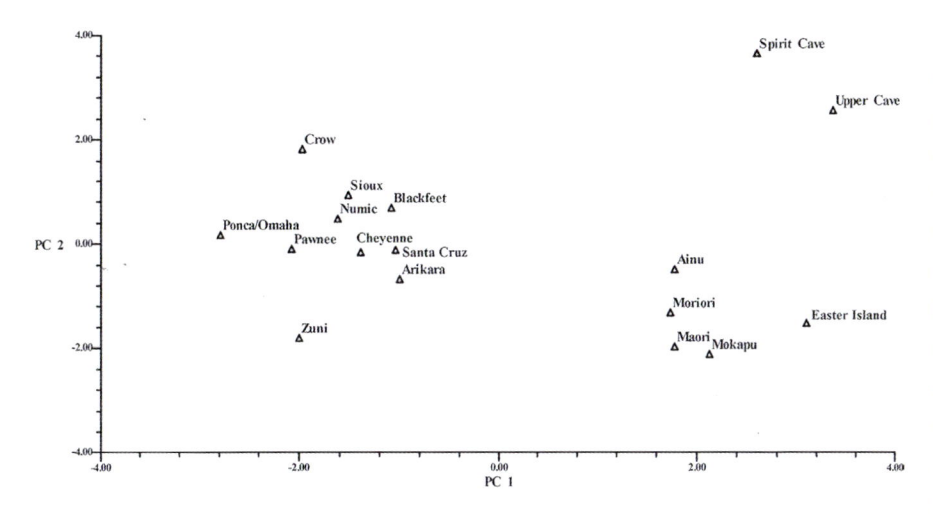

FIGURE 32.15 This bivariate principal coordinates (PC) plot compares the male from Upper Cave 101, China and the mummy from Spirit Cave, Nevada to modern American Indians and Polynesians. This plot is based on 25 principal components extracted from a covariance matrix of 40 variables. The comparative groups are the same as in Figure 4. Polynesians are again separated from American Indians on the first axis, and both early crania align with Polynesians. The first axis is mainly an expression of cranial length, but with a large contribution from posterior projection. Vault height is also reflected, separating the low-vaulted American Indians from the high-vaulted Polynesians, features that also characterize Spirit Cave and Upper Cave. Spirit Cave and Upper Cave are somewhat distinctive on the second axis, which is a reflection of the forward position of nasion and dacryon and prominent nasal bones.

(Owsley et al. 2010:80). Some ancient dentitions show less pronounced shovel-shaping of the maxillary central incisors, absence of winging and double shoveling, and the occurrence of Carabelli's trait, which occurs in low frequencies among recent American Indians (Scott and Turner 1997) (Figure 32.17). Despite these differences, arguments for the repatriation of ancient American skeletons have been made on the basis of presumed similarities in dental patterns, even when extreme dental wear precludes an accurate assessment. Goodman and Martin (1999), for example, linked the 9415 RC yr. BP Spirit Cave mummy from Nevada with contemporary American Indians in the Great Basin by categorizing his dentition as Sinodont. Initial work that categorized Kennewick Man's pattern as Sundadont is a similar kind of assessment (see Chapter 8 for further explanation). In both cases, however, the teeth are worn and the pattern is not evident (Figure 32.18).

NON-METRIC VARIATION

Kennewick Man presents an unusual array of discrete and continuous morphological traits that are not an easy match with any contemporaneous population (Chapter 27). Of the better markers of ancestry, Kennewick Man's midface, nasal features and palate characteristics, clearly different from American Indians, are most like Polynesians. Yet, his mandible lacks convex curvature of the lower border, a trait referred to as rocker jaw, a characteristic often associated with Polynesians. His mandible has a more obtuse gonial angle and a prominent square chin that resembles the European form. It also has a well-developed subcondylar tubercle (Figure 32.19). This non-pathological prominence was first noted by Poll (1902) in a Morrori mandible from the Chatham Islands, and occurs more frequently in Polynesians and Melanesians than in modern Asian Indians or Europeans (Dennison et al. 2005). The trait is not well known and has not been featured in anatomy texts based on European cadavers. The condylar process is the insertion point for the lateral ligament of the temporomandibular joint, and the presence of a developed tubercle potentially allows for greater lateral movement of the mandible, facilitating mastication of a prehistoric diet (Griffin 1977).

POSTCRANIAL MEASUREMENTS

If the morphology of the cranium provides a basis for tracing relationships, can similar information be derived from the rest of the skeleton? Unlike the cranium, influences on the morphology of the postcranial skeleton are not entirely known. Climate and activity are factors (Pearson 2000), but the degree to which postcrania

FIGURE 32.16 Frontal and lateral comparisons of casts of the Spirit Cave (left) and Upper Cave 101 skulls showing their similarity.

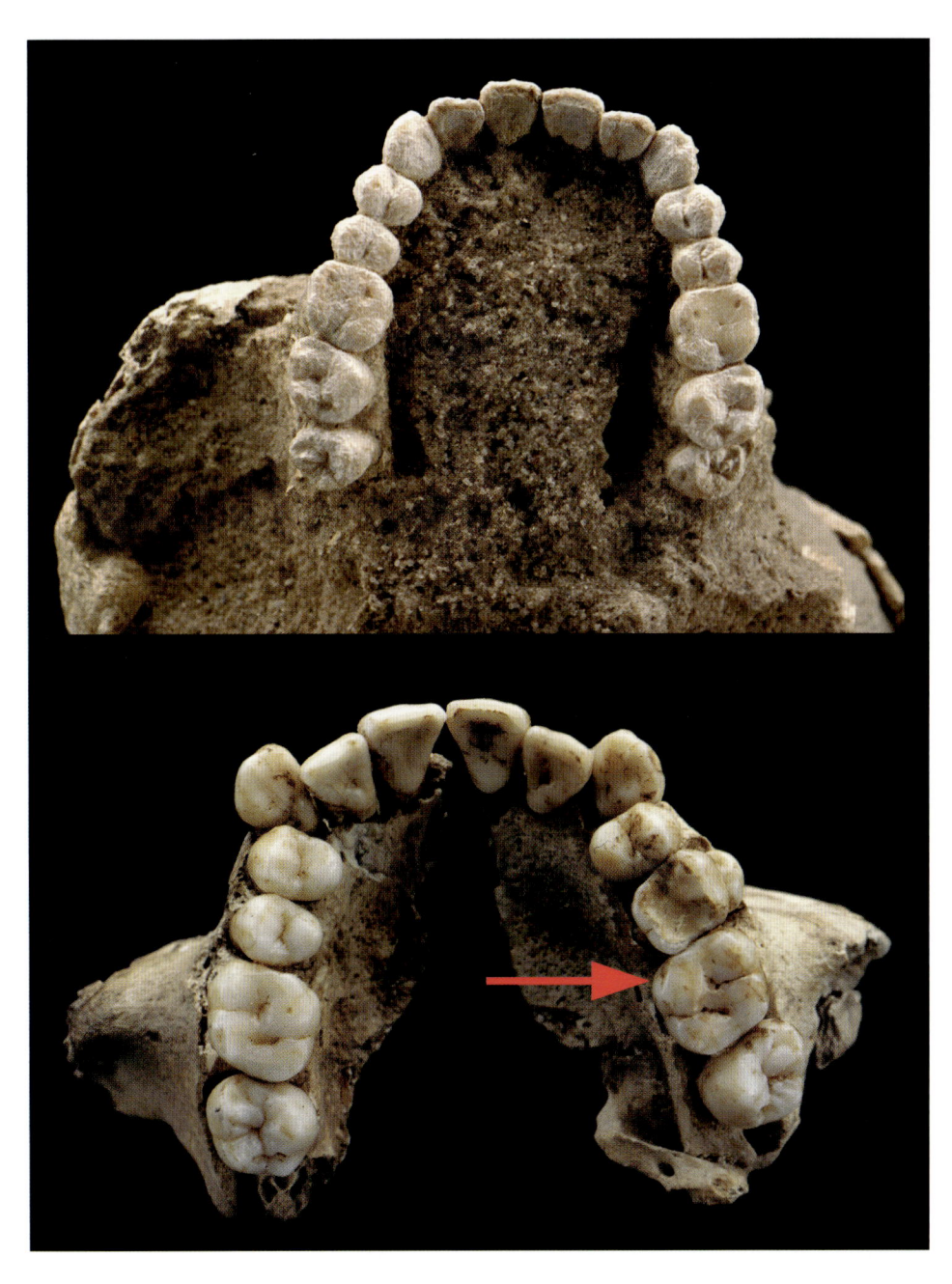

FIGURE 32.17 Occlusal view of the maxillary dentitions of Arch Lake Woman (top), aged 17–19 years, and the child from Horn Shelter No. 2 (bottom), aged 10 years. Visible dental traits include minor shovel-shaping of the central incisors, and absence of both double shoveling and incisor winging. The left maxillary deciduous second molar shows attrition of occlusal surface enamel. The adjacent first molar has a Carabelli's cusp (arrow).

reflect genetic relatedness has been little investigated. In one study of this type, Ishida (1993) has shown that Hawaiians differ from modern Japanese, but are similar to the Jōmon and Ainu, a relationship that has been argued from craniometric evidence (Chapter 24). Kennewick Man's cranial affinity to Polynesian populations may also be reflected in his postcranial skeleton.

In a previous study, Kennewick Man was included in a postcranial comparison of archaeological remains from two adjacent culture areas of northwestern North America, the Northwest Coast, and the inland Plateau (Cybulski 2010). This study concluded that Kennewick Man was Plateau adapted, tall with strong lower limbs, in contrast to recent Northwest Coast people with long and strong upper limbs and relatively short legs. Cybulski (2010:93–94) argues that the people of the Plateau were suited for hunting and gathering while those of the Northwest Coast were adapted to marine subsistence involving canoe travel.

To explore this generalization further, we compared the five indices (humero-femoral, humerus robusticity, platymeria, femur midshaft shape, and femur midshaft

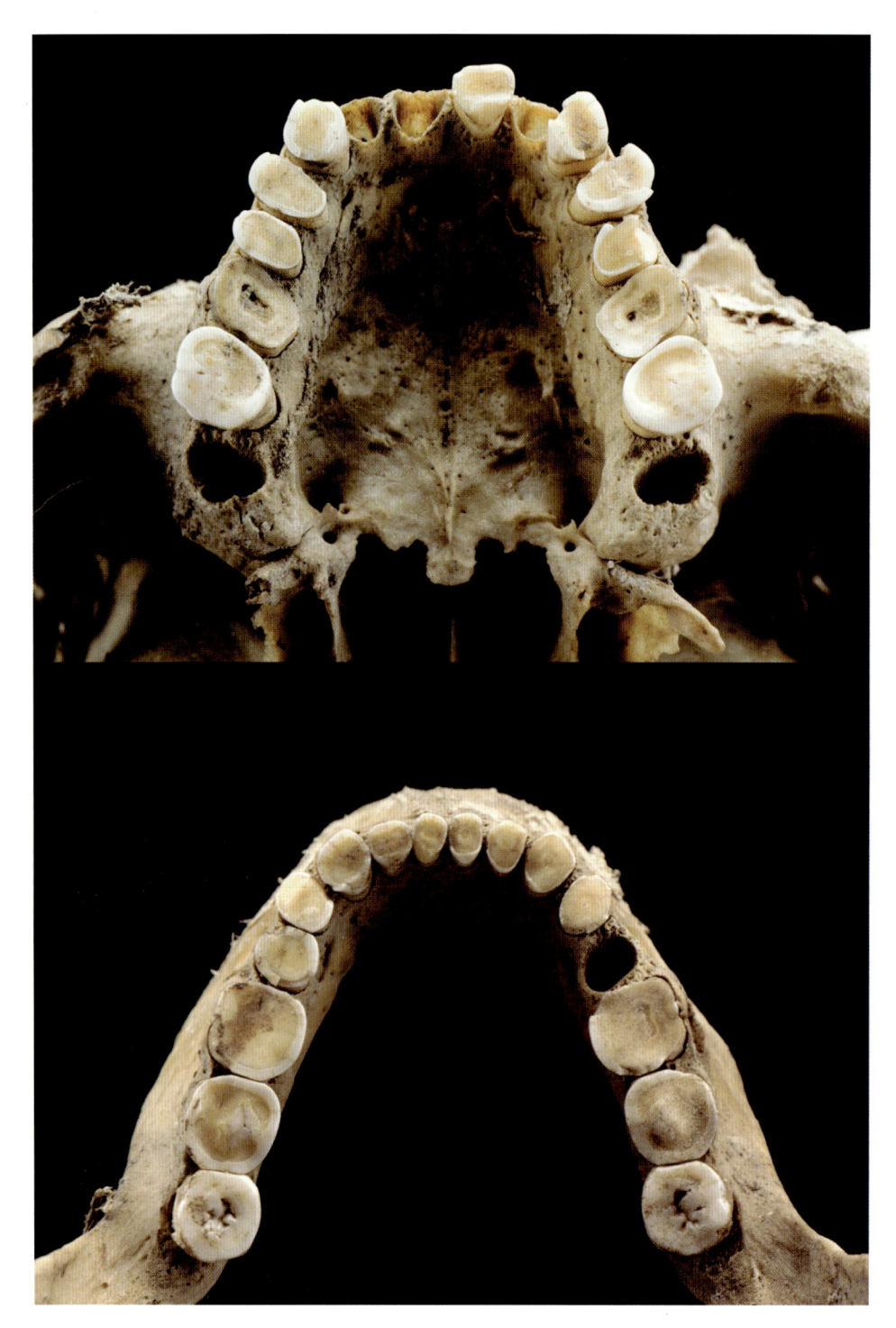

FIGURE 32.18 The heavily worn dentition of the Spirit Cave male.

robusticity) presented in Cybulski (2010:Table 4-2) for the Northwest Coast and Plateau to data for groups from Polynesia (Hawaiians) and Micronesia (Chamorro) (metric data from Ishida 1993), populations that exploited marine environments and traveled by canoe. Data were standardized using standard deviations in Cybulski (2010), and a simple Euclidean distance matrix was calculated. The distances are plotted on two principal coordinates in Figure 32.20. Plateau, Hawaiian, and Chamorro peoples occupy similar positions on the first principal coordinate (PC1), with the Chamorro differentiated on the second principal coordinate (PC 2). Variables influencing the first principal coordinate are humerus robusticity, femur midshaft shape and femur

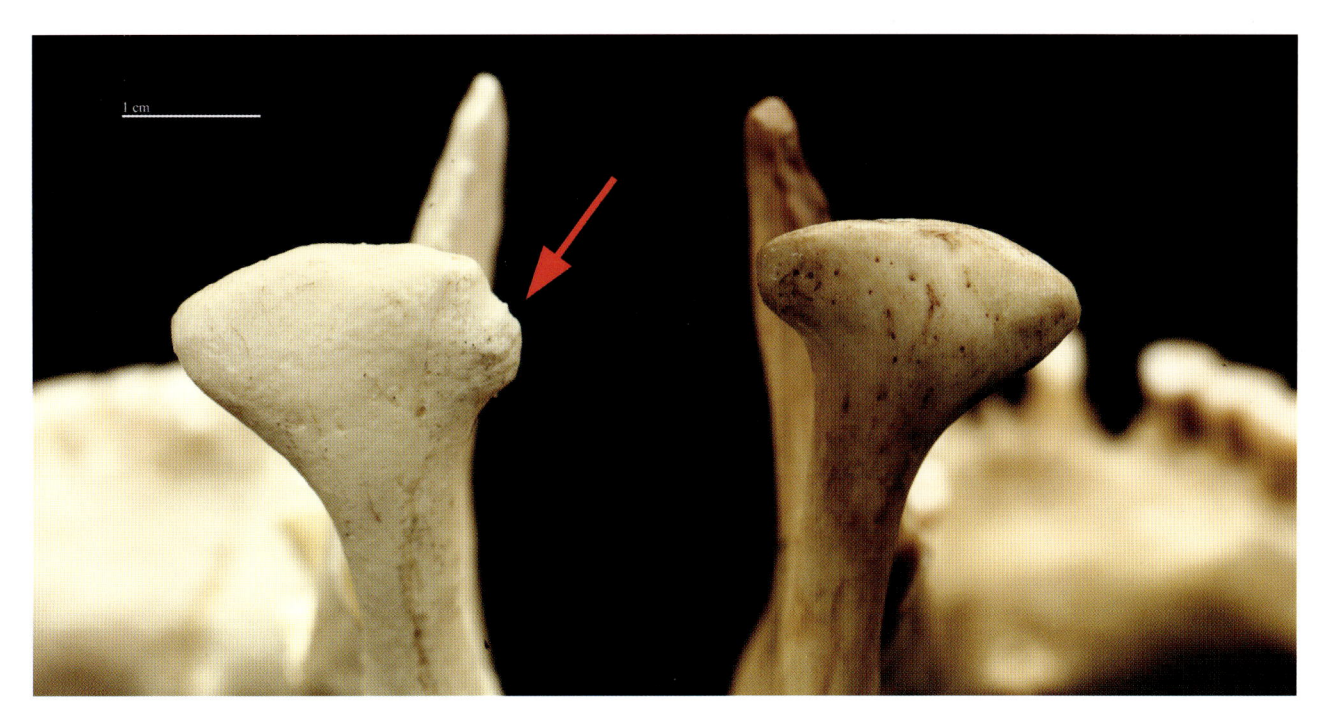

FIGURE 32.19 Posterior aspect of two mandibles showing a prominent lateral subcondylar tubercle (arrow) on the right condylar process of Kennewick Man, and trait absence in an American Indian from the eastern Great Basin (NMNH 225090). By definition, the protuberance is separated from the articular surface by a narrow groove (the paraglenoid sulcus).

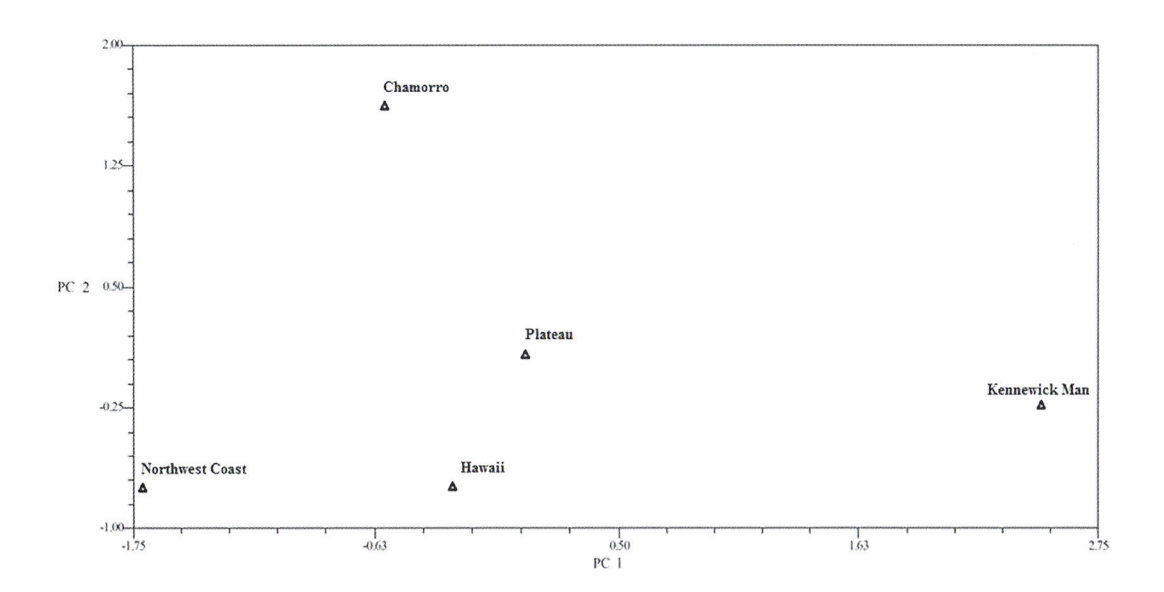

FIGURE 32.20 Principal coordinates plot comparing American Indians from the Northwest Coast and Plateau, Hawaiians, the Chamorro from Guam, and Kennewick Man using five postcranial indices. Kennewick Man is similar to both Hawiian and Plateau groups. Northwest Coast and Plateau data are from Cybulski (2010), Hawaiians and Chamorro from Ishida (1993), and Kennewick Man's measurements are from the present study.

robusticity. The second coordinate reflects platymeria and the humero-femoral index.

Kennewick Man is extreme on PC 1, a reflection of his low humeral robusticity, high femoral robusticity, and anterior-posterior elongation of the femur midshaft. These features separate him from the recent Northwest Coast population, as noted in Cybulski (2010). Kennewick Man is about equidistant from the Plateau and Hawaiian samples; it would be equally valid to argue that Kennewick Man was marine adapted because of his metric similarity to Hawaiians. Of the modern human populations that have been measured, Kennewick Man's height, mass, and bi-iliac body breadth are most like reported averages for males in the Lau Island Group of Fiji, though he was wider than their average body breadth (Auerbach, personal communication 2012).

KENNEWICK MAN'S FUTURE

Kennewick Man's bones are now placed in boxes of cut-to-fit foam and stored at the Burke Museum in Seattle, Washington. A careful examination of the court-required temperature and humidity records from 1998 to 2008 indicates wide fluctuations in both conditions; these are likely to promote mold and cause both flaking of bone layers and cracking of tooth enamel. In 1998, the Corps committed to a preservation plan, not simply a storage plan, for these ancient bones. To date, the authors know of no steps that have been taken to prevent deterioration.

Our understanding of Kennewick Man, however, is advancing. Scientific knowledge generally moves forward in increments, and answering one question often leads to more. This has been true for this investigation. Some unresolved issues require specific tests in order to tell Kennewick Man's story more completely. Several chapters in this volume have emphasized the need for additional focused research or have demonstrated the effective application of these advanced tests at other ancient sites. In the case of Kennewick Man's skeleton, excellent chemical and physical preservation will facilitate fine-tuned histological and molecular-level analyses—and we recommend that these be undertaken. Previously submitted, but not approved, proposals to the Corps have sought permission to evaluate Kennewick Man's chronological age and to determine the composition of the projectile point embedded in Kennewick Man's right ilium. To be clear, the following tests still need to be completed.

Precision in Determining Kennewick Man's Age

Estimates of Kennewick Man's chronological age have ranged from 35 to 55 years. In middle to older adult skeletons, age is difficult to determine from external criteria due to individual variation in the aging process. Degenerative features, such as tooth wear or arthritic changes, are affected by cultural factors or lifestyle and therefore cannot always be reliable indicators of age. Kennewick Man's life was far more rigorous than that of the individuals in the twentieth-century forensic and anatomical collections that have served as the reference bases for developing skeletal aging standards. Histological analysis of bone offers an independent, complementary means of more precisely determining Kennewick Man's age at death. Measured variables include intact osteon density, fragmentary osteon density, mean osteonal cross-sectional area, osteon population density, total subperiosteal area, cortical area, endosteal area, and relative cortical area (Stout personal communication 2005; Stout and Paine 1992). Haversian canals within Kennewick Man's bones are well preserved and will allow for this kind of analysis (Figure 3.6).

Ancient DNA

Genetic phylogeography research tracks DNA trees as a means of tracing migrations and inferring lineages. Obtaining sequenced genomes is now possible and is changing our perspective on human history on everything from Neanderthal-modern human relationships (Wall et al. 2013) to the peopling of the Americas (Raghavan et al. 2013; Rasmussen et al. 2014). Work in progress on Kennewick Man's DNA using residual samples may be informative but are not ideal (Figure 3.5). Handling during previous government-sponsored testing may have resulted in contamination and, in general, thick cortical bone and teeth produce the best results (Miloš et al. 2007; Nelson and Melton 2007). We recommend that Kennewick Man's remains be reviewed with the intent of obtaining a high-quality sample; a tooth with an intact pulp chamber may have the greatest chance of success. Such a sample might enable sequencing of both mitochrondrial and nuclear DNA, providing greater insight into his relationships with populations from Asian source regions and the New World.

The importance of obtaining the best DNA sample possible and analyzing it extensively is illustrated by Rasmussen et al.'s (2014) analysis of the Anzick child from Montana. Their results place him closer to the populations of South and Middle America than to those of North America. This is not a unique finding. Kemp et al. (2007) reported a subhaplotype in the individual from On Your Knees Cave that was widely distributed in North and South America, and most commonly found in the Cayapo of Brazil. Caution in generalizing too much from the findings on the single individual from the Anzick site is advised because of the absence of comparative samples from the lower 48 states and from Polynesia. Rasmussen et al. (2014) illustrate the complexity of the colonization of the Americas, which in turn demands that the genetics of Kennewick Man be examined thoroughly and carefully.

Isotope Analysis of a Tooth

The search for Kennewick Man's homeland can be advanced with isotope values from a tooth, possibly one of the detached third molars. Childhood dietary and geographic information can be obtained by measuring δC^{13} and δN^{15} values in dentin collagen and the δO^{18} value in enamel bioapatite. Unlike bones, teeth are not subject to remodeling during an individual's lifetime, and reflect the plants, animals, and drinking water consumed during childhood and adolescence.

XRF Analysis of the Projectile Point

The type of rock used to manufacture the projectile point embedded in Kennewick Man's ilium is still unknown. Identifying it as dacite, andesite, basalt, or some other vitrophyre is the essential first step in determining the source of the raw material. Columbia River and Snake River Plain basalts show elemental differences that vary from west to east (Stefano 2010). Fine-grained volcanic black or dark gray stone from northwestern North America has been classified and even sourced to specific quarries (see Bakewell 2005), suggesting that it would be possible to identify both the rock type of the projectile point and its origin.

As science progresses, additional studies beyond those that have been suggested here will become possible. Advances in molecular biology, instrumentation (high performance liquid chromatography and mass spectrometry), and the rapidly developing field of proteomics show, for example, that extracellular matrix proteins (i.e., peptide markers and protein sequences) can be identified. A recent proteomic analysis of a femur of a 43,000-year-old woolly mammoth, for example, identified 126 unique proteins (Cappellini et al. 2012). Even disease-related proteins can remain in bone for thousands of years (Schmidt-Schultz and Schultz 2004; Schultz et al. 2007) and may offer a potential way to determine the cause of Kennewick Man's death.

FINAL COMMENT

Ironically, our extensive skeletal study began with the Corps' refusal to allow any analysis whatsoever. That decision led to the formation of a plaintiff group, a court challenge, and now this volume, which presents the most complete analysis of any Paleoamerican skeleton to date.

Through this study, new interpretations have been developed, previous ones have been reinforced or updated, and others continue to be debated. Hopefully, Kennewick Man's story will not end here. Scientists seeking to address other topics or to validate our conclusions are encouraged to study this skeleton and those of other Paleoamericans. Examinations of the actual remains will continue to have enormous value; original bones can never be replaced by digital data or stereolithographic reproductions, regardless of their accuracy. A plastic model offers no data to future scientists interested in molecular or biochemical studies.

As we marvel at how rapidly our ability to learn from the human skeleton is advancing, and how rarely discoveries of Kennewick Man's magnitude occur, it is our responsibility to look to the future and emphasize that some skeletal collections and individual skeletons are particularly important in terms of what they can reveal about past populations. We, today, do not know the questions that will be significant in the future, but the remains of the earliest Americans, if respectfully cared for and curated, can be available to answer new questions. These human remains are America's ambassadors to our ancient past.

REFERENCES

Ames, Kenneth M., Don E. Dumond, Jerry R. Galm, and Rick Minor. 1998. Prehistory of the Southern Plateau. In *Handbook of North American Indians,* vol. 12: *Plateau,* edited by Deward E. Walker, 103–108. Smithsonian Institution Press, Washington DC.

Bakewell, Edward F. 2005. "The Archaeopetrology of Vitrophyric Toolstones, with Applications to Archaeology in the Pacific Northwest." Ph.D. dissertation, University of Washington, Seattle.

Beamish, Richard J. 1980. Adult biology of the River Lamprey (*Lampetra ayresi*) and the Pacific Lamprey (*Lampetra tridentata*) from the Pacific Coast of Canada. *Canadian Journal of Fisheries and Aquatic Sciences* 37:1906–1923.

Brostrom, Jody K., Christina Wang Luzier, and Katherine Thompson. 2010. Best Management Practices to Minimize Adverse Effects to Pacific Lamprey (*Entosphenus tridentatus*). U.S. Fish and Wildlife Service.

Brown, Peter. 1999. The first modern East Asians?: Another look at Upper Cave 101, Liujiang and Minatogawa 1. In *Interdisciplinary Perspectives on the Origins of the Japanese,* edited by K. Omoto, 105–130. International Research Center for Japanese Studies, Kyoto.

Butler, Virginia L. 1993. Natural versus cultural salmonid remains: Origin of The Dalles Roadcut bones, Columbia River, Oregon, U.S.A. *Journal of Archaeological Science* 20(1):1–24.

Butler, Virginia L., and Sarah K. Campbell. 2006. Northwest Coast and Plateau Animals. In *Handbook of North American Indians*, vol. 3: *Environment, Origins, and Population*, edited by Douglas H. Ubelaker, 263–273. Smithsonian Institution Press, Washington, DC.

Butler, Virginia L., and Jim E. O'Connor. 2004. 9000 years of salmon fishing of the Columbia River, North America. *Quaternary Research* 62(1):1–8.

Byram, Robert Scott. 2002. "Brush Fences and Basket Traps: The Archaeology and Ethnohistory of Tidewater Weir Fishing on the Oregon Coast." Ph.D. dissertation, University of Oregon, Eugene.

Byers, David A., David R. Yesner, Jack M. Broughton, and Joan Brenner Coltrain. 2011. Stable isotope chemistry, population histories and late Prehistoric subsistence change in the Aleutian Islands. *Journal of Archaeological Science* 38:183–196.

Cappellini, Enrico, Lars J. Jensen, Damian Szlarczyk, Aurélien Ginolhac, Rute A. R. da Fonseca, Thomas W. Stafford, Jr., Steven R. Holen, Matthew J. Collins, Ludovic Orlando, Eske Willerslev, M. Thomas P. Gilbert, and Jesper V. Olsen. 2012. Proteomic analysis of a Pleistocene mammoth femur reveals more than one hundred ancient bone proteins. *Journal of Proteome Research* 11(2):917–926.

Chamberlain, C. P., J. R. Waldbauer, K. Fox-Dobbs, S. D. Newsome, P. L. Koch, D. R. Smith, M. E. Church, S. D. Chamberlain, K. L. Sorenson, and R. Risebrough. 2005. Pleistocene to recent dietary shifts in California condors. *Proceedings of the National Academy of Sciences* 102(46):16707–16711.

Chatters, James C. 1998. Environment. In *Handbook of North American Indians,* vol. 12: *Plateau*, edited by Deward E. Walker, 29–48. Smithsonian Institution Press, Washington DC.

Chatters, James C., Steven Hackenberger, Alan J. Busacca, Linda Scott Cummings, Richard L. Jantz, Thomas W. Stafford, Jr., and R. Ervin Taylor. 2000. A possible second early-Holocene skull from Central Washington. *Current Research in the Pleistocene* 17:93–95.

Chatters, James C., and David Pokotylo. 1998. Prehistory: Introduction. In *Handbook of North American Indians,* vol. 12: *Plateau*, edited by Deward E. Walker, 73–80. Smithsonian Institution Press, Washington DC.

Chisholm, Brian S., D. Erle Nelson, and Henry P. Schwarcz. 1982. Stable-carbon isotope ratios as a measure of marine versus terrestrial protein in ancient diets. *Science* 216:1131–1132.

———. 1983. Marine and terrestrial protein in prehistoric diets on the British Columbia coast. *Current Anthropology* 24:396–398.

Clemens, Benjamin J., Thomas R. Binder, Margaret F. Docker, Mary L. Moser, and Stacia A. Sower. 2010. Similarities, differences, and unknowns in biology and management of three parasitic lampreys of North America. *Fisheries* 32(12):580–594.

Close, David A., Martin Fitzpatrick, Hiram Li, Blaine Parker, Douglas Hatch, and Gary James. 1995. "Status Report of the Pacific Lamprey (*Lampetra Trzdentata*) in the Columbia River Basin." Prepared by Oregon Cooperative Fishery Research Unit, Department of Fisheries and Wildlife, Oregon State University, Corvallis, Oregon; Columbia River Inter-Tribal Fish Commission, Portland, Oregon; Department of Natural Resources, Fisheries Program,

Confederated Tribes of the Umatilla Indian Reservation, Pendleton, Oregon. Prepared for U. S. Department of Energy, Bonneville Power Administration, Environment, Fish, and Wildlife, P.O. Box 3621, Portland, Oregon 97208-3621. Project Number 94-026. Contract Number 95BI39067.

Close, David A., Martin S. Fitzpatrick, and Hiram W. Li. 2002. The ecological and cultural importance of a species at risk of extinction, Pacific lamprey. *Fisheries* 27(7):19–25.

Coltrain, Joan Brenner. 2010. Alaska Peninsula stable isotope and radioisotope chemistry: A study in temporal and adaptive diversity. *Human Biology* 82(5–6):613–627.

Coltrain, Joan Brenner, M. Geoffrey Hayes, and Dennis H. O'Rourke. 2004. Sealing, whaling and caribou: The skeletal isotope chemistry of Eastern Arctic foragers. *Journal of Archaeological Science* 31:39–57.

Coltrain, Joan Brenner, and Steven W. Leavitt. 2002. Climate and diet in Fremont Prehistory: Economic variability and abandonment of maize agriculture in the Great Salt Lake Basin. *American Antiquity* 67(3):453–485.

Coplen, Tyler B., and Carol Kendall. 2000. Stable Hydrogen and Oxgen Isotope Ratios for Selected Sites in the U.S. Geological Survey's NASQAN and Benchmark Surface-water Networks. U.S. Geological Survey Open-File Report 00-160. USGS, Reston, VA.

Cressman, L. S., David L. Cole, Wilbur A. Davis, Thomas M. Newman, and Daniel J. Scheans. 1960. Cultural sequences at the Dalles, Oregon: A contribution to Pacific Northwest Prehistory. *Transactions of the American Philosophical Society* 50(10):1–108.

Cunningham, Deborah L., and Richard L. Jantz. 2003. The morphometric relationship of Upper Cave 101 and 103 to modern *Homo sapiens*. Journal of Human Evolution 45(1):1–18.

Cybulski, Jerome. 2010. Human skeletal variation and environmental diversity in northwestern North America. In *Human Variation in the Americas: The Integration of Archaeology and Biological Anthropology*, edited by Benjamin M. Auerbach, 77–112. Occasional Paper 38. Center for Archaeological Investigations, Southern Illinois University, Carbondale.

Daugherty, Richard D. 1956. Archaeology of the Lind Coulee Site, Washington. *Proceedings of the American Philosophical Society* 100(3):223–278.

Dennison, John, Jules Kieser, and Peter Herbison. 2005. The incidence and expression of the subcondylar tubercle of the mandible in early Polynesians, modern Indians, and modern Europeans. *Anthropologischer Anzeiger* 63(2):129–140.

Dixon, Susan L., and R. Lee Lyman. 1996. On the Holocene history of elk (*Cervus elaphus*) in eastern Washington. *Northwest Science* 70(3):262–272.

Fedje, Daryl, Quentin Mackie, Terri Lacourse, and Duncan McLaren. 2011. Younger Dryas environments and archaeology on the Northwest Coast of North America. *Quaternary International* 242:452–462.

Fox-Dobbs, Kena, Thomas A. Stidham, Gabriel J. Bowen, Steven D. Emslie, and Paul L. Koch. 2006. Dietary controls on extinction versus survival among avian megafauna in the late Pleistocene. *Geology* 34:685–688.

Gifford-Gonzalez, Diane. 2011. Holocene Monterey Bay fur seals: Distribution, dates, and ecological implications. In *Human Impacts on Seals, Sea Lions, and Sea Otters: Integrating Archaeology and Ecology in the Northeast Pacific*, edited by Todd J. Braje and Torben C. Rick, 221–242. University of California Press, Berkeley.

Goodman, Alan H., and Debra L. Martin. 1999. "Biological Analysis of the Spirit Cave Human Remains (AHUR 2064): Implications for Cultural Affiliation." Report submitted to BLM by Fallon Paiute-Shoshone Tribe.

Green, Thomas J., Bruce Cochran, Todd W. Fenton, James C. Woods, Gene L. Titmus, Larry Tieszen, Mary Anne Davis, and Susanne J. Miller. 1998. The Buhl burial: A Paleoindian woman from southern Idaho. *American Antiquity* 63(3):437–456.

Green, Thomas J., Max G. Pavesic, James C. Woods, and Gene Titmus. 1986. The DeMoss burial locality: Preliminary observations. *Idaho Archaeologist* 9:31–40.

Griffin, C. J. 1977. The prevalence of the lateral subcondylar tubercle of the mandible in fossil and recent man with particular reference to Anglo-Saxons. *Archives of Oral Biology* 22:633–639.

Gustafson, Carl Eugene. 1972. "Faunal Remains from the Marmes Rockshelter and Related Archaeological Sites in the Columbia Basin." Ph.D. dissertation, Washington State University, Pullman.

Hansel-Kuehn, Victoria June. 2003. "The Dalles Roadcut (Fivemile Rapids) Avifauna: Evidence for a Cultural Origin." Master's thesis, Washington State University, Pullman.

Hobson, Keith A. 1987. Use of stable-carbon isotope analysis to estimate marine and terrestrial protein content in gull diets. *Canadian Journal of Zoology* 65:1210–1213.

Howells, William W. 1973. *Cranial Variation in Man: A Study by Multivariate Analysis of Patterns of Difference Among Human Populations.* Papers of the Peabody Museum of Archaeology and Ethnology 67. Harvard University, Cambridge, MA.

Hunn, Eugene S. 1990. *Nch'i-Wána "The Big River": Mid-Columbia Indians and Their Land.* University of Washington Press, Seattle.

Hunn, Eugene S., Nancy J. Turner, and David H. French. 1998. Ethnobiology and subsistence. In *Handbook of North American Indians,* vol. 12: *Plateau,* edited by Deward E. Walker, 525–540. Smithsonian Institution Press, Washington DC.

Irwin, Ann M., and Ula Moody. 1978. The Lind Coulee Site (45GR97). Project Report 56. Submitted to The Bureau of Reclamation. Contract No. 14-06-100-8705. Washington Archaeological Research Center, Washington State University, Pullman.

Ishida, Hajime. 1993. Limb bone characteristics in the Hawaiian and Chamorro peoples. *Japan Review* 4:45–57.

Jameson, Ronald J., and Karl W. Kenyon. 1977. Prey of sea lions in the Rogue River, Oregon. *Journal of Mammalogy* 58(4):672.

Jantz, Richard L., and Douglas W. Owsley. 1997. Pathology, taphonomy, and cranial morphometrics of the Spirit Cave Mummy. *Nevada Historical Society Quarterly* 40(1):62–82.

———. 2005. Circumpacific populations and the peopling of the New World: Evidence from Cranial morphometrics. In *Paleoamerican Origins: Beyond Clovis,* edited by Robson Bonnichsen, Bradley T. Lepper, Dennis Stanford, and Michael R. Waters, 267–276. Texas A&M University Press, College Station.

Kemp, Brian M., Ripan S. Malhi, John McDonough, Deborah A. Bolnick, Jason A. Eshleman, Olga Rickards, Cristina Martinez-Labarga, John R. Johnson, Joseph G. Lorenz, E. James Dixon, Terence E. Fifield, Timothy H. Heaton, Rosita Worl, and David Glenn Smith. 2007. Genetic analysis of early Holocene skeletal remains from Alaska and its implications for the settlement of the Americas. *American Journal Physical Anthropology* 132: 605–621.

Kroeber, A. L., and S. A. Barrett. 1960. Fishing among the Indians of Northwestern California. *University of California Anthropological Records* 21(1).

Lovell, Nancy C., Brian S. Chisholm, D. Erle Nelson, and Henry P. Schwarcz. 1986. Prehistoric salmon consumption in interior British Columbia. *Canadian Journal of Archaeology* 10:99–106.

Lyman, R. Lee. 1986. On the Holocene history of *Ursus* in Eastern Washington. *Northwest Science* 60(2):67–72.

———. 2004. Late-Quaternary diminution and abundance of prehistoric bison (*Bison* sp.) in eastern Washington state, USA. *Quaternary Research* 62(1):76–85.

———. 2007. The Holocene history of pronghorn (*Antilocapra americana*) in eastern Washington State. *Northwest Science* 81(2):104–111.

———. 2011. A history of paleoecological research on sea otters and pinnipeds of the eastern Pacific Rim. In *Human Impacts on Seals, Sea Lions, and Sea Otters: Integrating Archaeology and Ecology in the Northeast Pacific*, edited by Todd J. Braje and Torben C. Rick, 19–40. University of California Press, Berkeley.

Lyman, R. Lee, Judith L. Harpole, Christyann Darwent, and Robert Church. 2002. Prehistoric occurrence of pinnipeds in the lower Columbia River. *Northwest Naturalist* 83(1):1–6.

Lyman, R. Lee, and Stephanie D. Livingston. 1983. Late Quaternary mammalian zoogeography of eastern Washington. *Quaternary Research* 20(3):360–373.

McCorquodale, Scott M. 1985. Archaeological evidence of elk in the Columbia Basin. *Northwest Science* 59(3):192–197.

Miller, Jay, and William R. Seaburg. 1990. Athapaskans of Southwestern Oregon. In *Handbook of North American Indians,* vol. 7: *Northwest Coast,* edited by Deward E. Walker, 580–588. Smithsonian Institution Press, Washington DC.

Miloš, Ana, Arijana Selmanović, Lejla Smajlović, René L.M. Huel, Cheryl Katzmarzyk, Adi Rizvić, and Thomas J. Parsons. 2007. Success rates of nuclear short tandem repeat typing from different skeletal elements. *Croatian Medical Journal* 48(4):486-493.

Minagawa, Masao, and Takeru Akazawa. 1992. Dietary patterns of Japanese Jōmon hunter-gatherers: Stable nitrogen and carbon isotope analyses of human bones. In *Pacific Northeast Asia in Prehistory,* edited by C. Melvin Aikens and Song Nai Rhee, 59–68. Washington State University Press, Pullman.

Misarti, Nicole, Bruce Finney, Herbert Maschner, and Matthew J. Wooller. 2009. Changes in northeast Pacific marine ecosystems over the last 4500 years: Evidence from stable isotope analysis of bone collagen from archaeological middens. *The Holocene* 19(8):1139–1151.

Moss, Madonna L., Dongya Y. Yang, Seth D. Newsome, Camilla F. Speller, Iain McKechnie, Alan D. McMillan, Robert J. Losey, and Paul L. Koch. 2006. Historical ecology and biogeography of North Pacific pinnipeds: Isotopes and ancient DNA from three archaeological assemblages. *Journal of Island & Coastal Archaeology* 1:165–190.

Nelson, Kimberlyn, and Terry Melton. 2007. Forensic mitochrondrial DNA analysis of 116 casework skeletal samples. *Journal of Forensic Sciences* 52(3):557–561

Newsome, Seth D., Michael A. Etnier, Diane Gifford-Gonzalez, Donald L. Phillips, Marcel van Tuinen, Elizabeth A. Hadly, Daniel P. Costa, Douglas J. Kennett, Tom P. Guilderson, and Paul L. Koch. 2007. The shifting baseline of northern fur seal ecology in the Northeast Pacific Ocean. *Proceedings of the National Academy of Sciences of the United States of America* 104(23):9709–9714.

NOAA Paleoclimatology. 2013. Ice Core Gateway. Eclipse Ice Field Ice Core Data. "Eclipse Ice Field 1996 Core Annual Accumulation, Stable Isotope, and Major Ion Data." Accessed August 29. www.ncdc.noaa.gov/paleo/icecore/trop/eclipse/eclipse.html.

Orchard, Trevor J. 2011. Late Holocene fisheries in Gwaii Haanas: Species composition, trends in abundances, and environmental or cultural explanations. In *The Archaeology of North Pacific Fisheries*, edited by Madonna L. Moss and Aubrey Cannon, 111–128. University of Alaska Press, Fairbanks.

Owsley, Douglas W., Margaret A. Jodry, Thomas W. Stafford, Jr., C. Vance Haynes, and Dennis J. Stanford. 2010. *Arch Lake Woman: Physical Anthropology and Geoarchaeology*. Texas A&M University Press, College Station.

Pearson, Osbjorn M. 2000. Activity, climate, and postcranial robusticity. *Current Anthropology* 41(4):569–607.

Pennefather-O'Brien, Elizabeth E., and Michael Strezewski. 2002. An initial description of the archaeology and morphology of the Clark's Fork skeletal material, Bonner County, Idaho. *North American Archaeologist* 23(2):101–115.

Poll, Heinrich. 1902. Über Schädel und Skelette der Bewohner der Chatham-Inseln. [On skulls and skeletons of the inhabitants of the Chatham Islands]. Translated by K. J. Dennison. *Zeitschrift für Morphologie und Anthropologie* 1:1–134.

Potts, Thomas Henry. 1882. *Out in the Open: A Budget of Scraps of Natural History, Gathered in New Zealand*. Lyttleton Times, Christchurch, New Zealand.

Raghavan, Maanasa, Pontus Skoglund, Kelly E. Graf, Mait Metspalu, Anders Albrechtsen, Ida Moltke, Simon Rasmussen, Thomas W. Stafford, Jr., Ludovic Orlando, Ene Metspalu, Monika Karmin, Kristiina Tambets, Siiri Rootsi, Reedik Mägi, Paula F. Campos, Elena Balanovska, Oleg Balanovsky, Elza Khusnutdinova, Sergey Litvinov, Ludmila P. Ospiova, Sardana A. Fedorova, Mikhail I. Voevoda, Michael DeGiorgio, Thomas Sicheritz-Ponten, Søren Brunak, Svetlana Demeshchenko, Toomas Kivisild, Richard Villems, Rasmus Nielsen, Mattias Jakobsson, and Eske Willerslev. 2013. Upper Palaeolithic Siberian genome reveals dual ancestry of Native Americans. *Nature* doi: 10.1038/nature12736.

Rasmussen, Morton, Sarah Anzick, Michael R. Waters, Pontus Skoglund, Michael DeGiorgio, Thomas W. Stafford, Jr., Simon Rasmussen, Ida Moltke, Anders Albrechtsen, Shane M. Doyle, G. David Poznik, Valbourg Gudmundsdottir, Rachita Yadav, Anna-Sapfo Malaspinas, Samuel S. White V., Morton E. Allentoft, Omar E. Cornejo, Kristiina Tambets, Anders Eriksson, Peter D. Heintzman, Monika Karmin, Thorfin Sand Korneliussen, David J. Meltzer, Tracey L. Pierre, Jesper Stenderup, Lauri Saag, Vera M. Warmuth, Margarida C. Lopes, Ripan S. Malhi, Søren Brunak, Thomas Sicheritz-Ponten, Ian Barnes, Matthew Collins, Ludovic Orlando, Francois Balloux, Andrea Manica, Ramneek Gupta, Mait Metspalu, Carlos D. Bustamante, Mattias Jakobsson, Rasmus Nielsen, and Eske Willerslev. 2014. The genome of a late Pleistocene human from a Clovis burial site in western Montana. *Nature* 506: 225–229. doi:10.1038/nature13025.

Roffe, Thomas J., and Bruce R. Mate. 1984. Abundance and feeding habits of pinnipeds in the Rogue River, Oregon. *The Journal of Wildlife Management* 48(4):1262–1274.

Schultz, Michael, Hermann Parzinger, Dmitrij V. Posdnjakov, Tatjana A. Chikisheva, and Tyede H. Schmidt-Schultz. 2007. Oldest known case of metastasizing prostate carcinoma diagnosed in the skeleton of a 2,700-year-old Scythian King from Arzhan (Siberia, Russia). *International Journal of Cancer* 121(12):2591–2595.

Schmidt-Schultz, Tyede H., and Michael Schultz. 2004. Bone protects proteins over thousands of years: Extraction, analysis, and interpretation of extracellular matrix proteins in archaeological skeletal remains. *American Journal of Physical Anthropology* 123:30–39.

Scott, G. Richard, and Christy G. Turner II. 1997. *The Anthropology of Modern Human Teeth: Dental Morphology and its Variation in Recent Human Populations.* Cambridge University Press.

Stefano, Christopher J. 2010. "Volatiles, Major Oxide, Trace Element and Isotope Geochemistry in the Snake River Plain and Columbia River Flood Basalts: Implications for the Evolution of a Continental Hotspot." Ph.D. dissertation, University of Michigan, Ann Arbor.

Stout, Sam D., and Robert R. Paine. 1992. Brief communication: Histological age estimation using rib and clavicle. *American Journal of Physical Anthropology* 87:111–115.

Szpak, Paul, Trevor J. Orchard, and Darren R. Gröcke. 2009. A late Holocene vertebrate food web from southern Haida Gwaii (Queen Charlotte Islands, British Columbia). *Journal of Archaeological Science* 36:2734–2741.

Szpak, Paul, Trevor J. Orchard, Iain McKechnie, and Darren R. Gröcke. 2012. Historical ecology of late Holocene sea otters (*Enhydralutris*) from northern British Columbia: Isotopic and zooarchaeological perspectives. *Journal of Archaeological Science* 39:1553–1571.

Tieszen, Larry L., and Tim Fagre. 1993. Carbon isotopic variability in modern and archaeological maize. *Journal of Archaeological Science* 20(1):25–40.

Topalov, Katarina, Arndt Schimmelmann, P. David Polly, Peter E. Sauer, and Mark Lowry. 2013. Environmental, trophic, and ecological factors influencing bone collagen $\delta^2 H$. *Geochimica et Cosmochimica Acta* 111:88–104.

Ugan, Andrew, and Joan Coltrain. 2012. Stable isotopes, diet, and taphonomy: A look at using isotope-based dietary reconstruction to infer differential survivorship in zooarchaeological assemblages. *Journal of Archaeological Science* 39:1401–1411.

Waguespack, Nicole M. 2007. Why we're still arguing about the Pleistocene occupation of the Americas. *Evolutionary Anthropology* 16(2):63–74.

Wall, Jeffrey D., Melinda A. Yang, Flora Jay, Sung K. Kim, Eric Y. Durand, Laurie S. Stevison, Chrstoperh Gignoux, August Woerner, Michael F. Hammer, and Montgomery Slatkin. 2013. Higher levels of Neanderthal ancestry in East Asians than in Europeans. *Genetics* 194: 199–209.

Williamson, Kathleen, and Dave Hillemeier. 2001. "An Assessment of Pinniped Predation Upon Fall-run Chinook Salmon in the Klamath River Estuary, CA, 1998." Prepared by Yurok Tribal Fisheries Program, 15900 Highway 101 North, Klamath, CA 95548.

Yoneda, Minoru, Masashi Hirota, Masao Uchida, Kazuhiro Uzawa, Atsushi Tanaka, Yasuyuki Shibata, and Masatoshi Morita. 2001. Marine radiocarbon reservoir effect in the western North Pacific observed in archaeological fauna. *Radiocarbon* 43(2A):465–471.

Acknowledgments

The editors acknowledge and thank those individuals and organizations whose support and help throughout the study of the Kennewick Man skeleton made this volume possible. James Chatters recognized the significance of the discovery and acted responsibly as a dedicated advocate for science. His persistent efforts, including thorough fieldwork resulting in the comprehensive recovery of the fragmented skeleton, made it possible for others to study the remains. In providing legal counsel, attorneys Alan Schneider and Paula Barran recognized the importance of the case to science and to future studies in Paleoamerican bioarchaeology. Their guidance, knowledge of the law, and legal strategy were unparalleled and essential for gaining access to the remains. We gratefully acknowledge the thoughtful deliberations and carefully constructed legal ruling of Magistrate Judge John Jelderks.

We thank our fellow plaintiffs for their perseverance. They were not motivated to study this skeleton for personal gain, but rather to ensure access for younger scholars interested in America's ancient prehistory. Their efforts recognize that future scientists will have methodological capabilities not yet developed to address research questions not yet formulated.

Thank you to the individuals who provided financial, managerial, and technical support, including affidavits, during the course of the litigation. The many contributors to Friends of America's Past helped with the expenses related to the legal case and to the editing of this volume. Whitney and Elizabeth MacMillan generously contributed to the bust that replicated Kennewick Man's appearance, allowing us to more clearly and deeply envision the man the bones represent. Nathan P. Myhrvold financially supported the application of industrial computed tomography that opened new avenues for revealing Kennewick Man's story. The Radiography Department at the University of Washington Medical Center provided technical support in the imaging of the skeleton. This was accomplished by Mario Ramos (Supervisor/CT scanner), Jody Bundrock (Technician), Belai Mewgesha (Technician), Erica Howard (Technician), Bisrat Baieh (Technician), Kim McKenzie (MA), and Mary Howe (Radiography Technician). Stephen Rouse aided in developing and formatting graphics for this volume. C. Wayne Smith, Nautical Archaeology Program, Texas A&M University, provided a conservation assessment of the Kennewick Man skeleton in 2004 and assisted with data collection during the 2005 study session. Phillip L. Walker graciously provided access to the San Miguel skeleton. Nicholas Herrmann digitized the San Miguel remains and determined the measurements of the Buhl cranium from photographs kindly provided by Todd Fenton. Archaeologist David G. Rice generously shared his knowledge of Northwest Coast pinnipeds. Geologist Bart Cannon provided information on igneous rocks of the Northwest and methods for their identification. Thomas Mann of the Library of Congress and his league of extraordinary gentlemen offered background and intellectual support. David Carlson of Texas A&M University helped with the transition following the untimely death of Rob Bonnichsen. Chris Pulliam, archaeologist with the US Army Corps of Engineers Mandatory Center of Expertise for the Curation and Management of Archaeological Collections, fostered cooperation during the skeletal analysis, serving as an effective liaison between the scientists and the Corps. During the Kennewick Man study sessions, the Burke Museum collections staff graciously provided working space and assisted with access to the skeleton.

Scientific and support staff at the Smithsonian Institution's National Museum of Natural History (NMNH) included photographers Donald E. Hurlbert, James Di Loreto, and Brittney Tatchell; photo assistant Debra Clark; digital technician Kristen N. Quarles; intern photographer Brittany M. Hance; X-ray technician Sandra Raredon; imaging specialist Scott Whittaker; and student interns Gabriela Lapera, Julia Franklin, and Keri Wilhelm. Maggie Dittemore of the Anthropology Library at NMNH helped obtain literature sources necessary for this volume. The work of NMNH senior scientific photographer Chip Clark is an unparalleled blending of biomedical and artistic imagery. Unless otherwise specified, the photographs in this volume reflect his expertise.

The editors gratefully acknowledge the meticulous efforts given by the contributors to this volume, which helps fulfill our obligation to Kennewick Man and to the American public. Diane Della-Loggia served as an editor for most of the chapters. We appreciate the constructive comments provided by the reviewers.

We gratefully thank Frances Miller Seay, Friends of America's Past and their supporters, and the Center for the Study of the First Americans for underwriting the production costs of this book.

The Rice Endowment for Forensic Anthropology made the completion of this volume possible.

Owsley thanks the Columbia Plateau Tribes for their invitation to meet at Central Washington University and at the Wanapum Heritage Center to discuss the results of the investigation in advance of this publication, with special acknowledgment for the gracious hospitality by Rex Buck, Sr. and Armand Minthorn.

This volume would not have been possible without the exceptional work of Kathryn Barca, Karin Bruwelheide, Sandra Schlachtmeyer, Vicki Simon, Catherine Taylor, and Aleithea Williams. No finer or more capable team could be assembled. Owsley also thanks Susanne, his wife of four decades, for her patience and encouragement as innumerable hours ticked away working on this project.

CHAPTER ACKNOWLEDGEMENTS
"The People Who Peopled America"

I am greatly indebted to the late Robson Bonnichsen for serving as a mentor and friend for more than two decades. I regret that he was unable to contribute to this volume, but his influence is reflected in my overview of the people who peopled America. I hope that he would have been pleased with the result.

I thank Doug Owsley for his invitation to contribute to this overview. The efforts of Rob, Doug, and the other plaintiffs in the landmark *Bonnichsen v. the United States* court decision made it possible for Kennewick Man/the Ancient One to more fully and completely convey to the modern world the story of his life and times.

The opinions expressed in this paper are those of the author and do not necessarily reflect those of the Ohio Historical Society, with whom he is employed as a curator of archaeology.

"Geography, Paleoecology, and Archaeology"

This chapter benefited from comments by Jenny Elf, Anna Prentiss, Thomas Stafford, and Philippe LeTourneau, who read earlier versions of the manuscript. Linda Scott Cummings of PaleoResearch Institute conducted the phytolith analysis and prepared the graphs presented in Figure 5. Claire Chatters created the artifact drawings presented as Figure 6. Thanks, too, to Doug Twehus, who led the author to the site where he found a Windust point more than 40 years ago. Mary Keith of the Seattle District, US Army Corps of Engineers was particularly helpful, providing digital copies of Corps maps and aerial photographs of the pre-reservoir Columbia River at Kennewick.

"Chronology of the Kennewick Man Skeleton"

Several people and institutions contributed funds and resources for the chronological study of Kennewick Man. Transportation and lodging during the Kennewick Man study sessions at the Burke Museum were provided by Friends of America's Past. Laura Phillips, Archaeology Collections Manager, and Megon Noble, Assistant Collections Manager at the Burke Museum provided invaluable support during the study sessions and facilitated subsequent loans to complete geochemical analyses. Christopher Pulliam and Michael Trimble, US Army Corps of Engineers, St. Louis District enabled study of Kennewick Man and approved additional loans for confirmatory chemical and biochemical analyses. Henry Schwarcz, McMaster University, Hamilton, Ontario, Canada measured stable isotope values. John Southon, W. M. Keck Carbon Cycle Accelerator Mass Spectrometry Laboratory, University of California-Irvine, California oversaw all AMS ^{14}C measurements and discussed interpretations concerning precisions and reservoir effects. Margaret Condron, Biopolymer Laboratory, David Geffen School of Medicine at University of California Los Angeles performed the quantitative amino acid analyses. Douglas

Owsley and Dennis Stanford, Department of Anthropology, Smithsonian Institution, Washington, DC, provided completion funding. Eske Willerslev, Centre for GeoGenetics, University of Copenhagen, Denmark funded final biogeochemical analyses that will lead to future research.

I am especially indebted to Wayne Smith, Department of Anthropology, Texas A&M University, College Station, Texas for making possible the major funding that supported the study and all analytical measurements. These funds were provided as a grant to the Texas A&M Research Foundation from the Hillcrest Foundation, founded by Mrs. W. W. Caruth, Sr. The chronological, isotopic, and biogeochemical research on Kennewick Man would have not been completed without this fundamental support from the Hillcrest Foundation.

"Reflections of a Former Army Corps of Engineers Archaeologist"

The photograph in Figure 1 is courtesy of James Chatters.

"Skeletal Inventory, Morphology, and Pathology"

Brian Spatola, Skeletal Collections Manager at the National Museum of Health and Medicine, provided background information on Army Medical Museum bones with gunshot wounds. Figure 17 was provided by Joseph H. Bailey / National Geographic Stock. Donald E. Hurlbert took the photographs shown in Figures 30, 31a, and 31b. Stephen Rouse created the images in Figures 37a–38b.

"Dental Microwear"

This work was supported in part by the National Science Foundation, the L.S.B. Leakey Foundation, and the Ruggles-Gates Fund for Biological Anthropology (the Royal Anthropological Institute of Great Britain and Ireland). We also thank James Brink, Linda Gordon, Zoe Henderson, Louise Humphrey, Dave Hunt, Rob Kruszynski, Giorgio Manzi, Alan Morris, David Morris, Ken Mowbray, Chris Stringer, Ian Tattersall, and Peter Ungar for allowing access to specimens in their care.

"Orthodontics"

Pamela Kay Rogers illustrated the images in Figures 1 and 2.

"Body Mass, Stature, and Proportions of the Skeleton"

I thank the curators and staff whose help at the Burke Museum made data collection efficient. Chris Ruff, Erik Trinkaus, Valerie DeLeon, Marta Lahr, and Mark Teaford all provided valuable insights to this research, and contributed useful feedback on previous versions of the analyses. Pamela Kay Rogers illustrated the image in Figure 1.

"Reconstructing Habitual Activities by Biomechanical Analysis of Long Bones"

I thank Dr. Bruno Frohlich of the Smithsonian Institution for assistance with CT images and Dr. Deborah Cunningham for valuable editorial comments.

"Bones of the Hands and Feet"

I thank the College of Humanities and Social Sciences of North Carolina State University for providing the Faculty Research and Professional Development Award that funded the travel for this project.

"Occupational Stress Markings and Patterns of Injury"

Photographs in Figures 1–5 were taken by Thomas W. Stafford, Jr., Stafford Research Laboratories, Inc.

"Stable Isotopic Evidence for Diet and Origin"

We acknowledge the US Army Corps of Engineers for making fragments of Kennewick Man's skeleton available for this study. Funding for these analyses was provided by the Natural Sciences and Engineering Research Council of Canada. Longstaffe thanks Emily Webb and Grace Yau for their assistance in the laboratory, Christine White for discussions, and the Natural Sciences and Engineering Research Council of Canada, the Canada Foundation for Innovation, and the Ontario Research Fund for financial support. In addition, this research was undertaken, in part, thanks to funding from the Canada Research Chairs program.

"Taphonomic Indicators of Burial Context"

The Corps provided water level data. Photographs in Figures 9, 40, 41, 42, and 43 were taken by Thomas W. Stafford, Jr., who also provided the geological cross-section in Figure 36. Marcia Bakry finalized the graphic illustration of this figure. Figures 13, 14, and 28 were prepared by Rebecca Snyder. Alice Tangerini (NMNH Botanical Illustrator) used an original photograph of the skeleton by Chip Clark to illustrate bone to bone adhesions (Figure 10) and macroabrasions and corrasion (Figure 27).

"Benthic Aquatic Algae: Indicators of Recent Taphonomic History"

Microscope slides of the algae were prepared for digital photomicrography with assistance of Katina E. Bucher (NMNH Botany) and Robert H. Sims (NMNH Botany), who also prepared the figures using Adobe Photoshop 3 on a MAC G5 computer. We thank James Chatters for use of habitat photographs (Figures 6–8). Alice Tangerini used Chip Clark's photograph of the skeleton to illustrate algal presence (Figure 1). Katina Bucher carefully reviewed the manuscript.

"Computed Tomography, Visualization and 3D Modeling"

The author thanks the following for their invaluable contributions: Chip Clark for his wonderful and detailed photography; Art Andersen of Virtual Surfaces for his overall work on the project and specifically for digitally separating the cranial fragments that had been glued together; Brian Wilcox of Point Data Marketing, Inc. for planning and coordinating the scanning and rapid prototyping contractors; Barry Smith, Chuck Smith and Forrest Martin of BIR, Inc. for ensuring the best possible results during our limited timeframe.

"Molding and Casting Methods"

Photographs in Figures 1–7 and in Figure 13 were taken by Steven J. Jabo, NMNH, Smithsonian Institution.

"The Point of the Story"

The following individuals offered advice and discussion on point typology: James Chatters, Bruce Bradley, Dale Croes, and Margaret Jodry. Alison Stenger provided the photo of the serrated Haskett projectile point from McMinnville, Oregon. Sandy Schlachtmeyer helped with library research and manuscript organization. Marjolein Admiraal provided information on Western Stem Tradition site locations and radiocarbon dates. I thank Marcia Bakry for her work on the illustrations, James Di Loreto for photography, and Stephen Rouse for creating the image in Figure 5.

"Two-Dimensional Geometric Morphometrics"

Michelle Hamilton provided assistance with Photoshop in the preparation of figures.

"Morphological Features that Reflect Population Affinities"

Graphics illustrating midfacial projection and femoral subtrochanteric proportions were prepared by Rick L. Weathermon, Senior Research Scientist, Department of Anthropology, University of Wyoming. The interorbital index calculations on the expanded sample of Easter Islanders included in Table 8 were provided by Dr. Vincent H. Stefan, Chair of the Department of Anthropology, Lehman College–CUNY, New York.

"Identity Through Science and Art"

Sculptors Amanda Danning and Jiwoong Cheh, painter Rebecca Spivak, and StudioEIS collaborated in the replication of Kennewick Man's appearance. Photographs in Figure 1 were taken by Donald E. Hurlbert. StudioEIS provided the photographs in Figures 2, 3, and 5–9. Brittney Tatchell photographed the images used in Figure 10, and assisted with editing the figures in this chapter. Historic photographs of Ainu adult males were obtained through the Smithsonian Institution National Anthropological Archives (Figures 4b (INV 07034800), 4c (INV 04734600), and 4f (INV 07034500)).

"Evidence of Maritime Adaptation and Coastal Migration from Southeast Alaska"

The Klawock Cooperative Association (Tribe) and the Craig Community Association (Tribe) supported and contributed to this study over the 12-year duration of the work. Sealaska Heritage Institute and Sealaska Corporation supported the work through ongoing consultation and funding of internships. The archaeological field work and preliminary analyses have been supported by the National Science Foundation (OPP-99-04258, 97-22858), the Tongass National Forest, the National Geographic Society, Denver Museum of Nature and Science, University of Colorado at Boulder, Institute of Arctic and Alpine Research (INSTAAR), and the Maxwell Museum of Anthropology, University of New Mexico. Thomas M. Bogan, D.D.S. provided valuable assistance in extracting molars for DNA analysis. The research team thanks Yarrow Vaara, Patrick Olsen, Frederick Grady, Linda Blankenship, and the many individuals who participated in the excavations of OYKC, Mim Dixon for editorial review, Mark R. Williams for technical support, and Kelly Monteleone for graphic assistance.

"A New Look at the Double Burial from Horn Shelter No. 2"

We gratefully acknowledge the long-term investigation of Horn Shelter No. 2 by Al Redder and his team. Mr. Redder assisted this study and generously donated materials and documentation to the Department of Anthropology, National Museum of Natural History, Smithsonian Institution.

We appreciate technical expertise from colleagues at the National Museum of Natural History including: Chip Clark, senior scientific photographer; Don Hurlbert, Jim DiLoreto, and Kristen Quarles in the Center for Scientific Imaging and Photography for additional photography and digitization; Scott Whittaker, SEM Lab, for microphotography; Marcia Bakry, Department of Anthropology, for the beautiful illustrations. Carla Dove and George Zug, Department of Vertebrate Zoology, assisted with bird and turtle identifications, respectively. Sorena Sorenson consulted on hematite, and Cathleen Brown facilitated loan and study of a comparative iron oxide sample (TM# 2065559) from the Mineral Sciences collections. Katie Barca's editorial assistance strengthened the prose. Brittney Tatchell's knowledge of skeletal anatomy and functional morphology were invaluable for interpreting muscle development and associated behavior.

David Rosenthal, Jim Krakker, and Felicia Pickering helped with collection access and care at the Museum Support Center, where Nicole Little provided XRD analysis of hematite from the burial, and Rhonda Coolidge and Eric Hollinger, Repatriation Office, assisted with XRF analyses of comparative and archaeological red ochre.

Special thanks to Lars Krutak, NMNH Repatriation Office, for sharing manuscripts and his extensive knowledge of body modification and healing. Adrian Deter-Wolf provided valuable consultation and manuscripts on the archaeological record of tattooing in Southeastern North America and replication and experimental use of tattooing tools.

Dennis Stanford, Darrin Lowery, Mike Frank, and Joe Gingerich of the Paleoindian Program, and visiting scholars Jim Adovasio and Mike Collins served as sounding boards. Bruce Bradley consulted on flint knapping tools, and San Patrice points. Bruce's expertise, support for new ideas, and friendship are deeply valued. Judy Knight shared ethnographic photographs and knowledge of pigment use and scarification in healing traditions in Gabon. Ted Timrek provided videotaped interviews with Al Redder. Jodry extends special gratitude to Frances Seay for financial, intellectual, and spiritual support.

To Dennis, who shares with me adventures in archaeology and life, provided enthusiasm, support, knowledge, and insight, and honored my efforts toward harmonizing seen and unseen worlds within current scientific paradigms, I extend my love and deep appreciation. I thank Angaangaq Lyberth, Jose Lucero, John Milton, and Others.

Deep appreciation and sincere gratitude are offered to the Elder and girl buried at Horn Shelter No. 2 for this respected opportunity to serve as a runner, to learn from and honor your belongings, and for assistance provided. May it be in a good way.

"Storage and Care at the Burke Museum"

NMNH conservator Catharine Hawks graciously provided a review of this manuscript.

"Who Was Kennewick Man?"

The skeleton of Tuqan Man from San Miguel Island and the Chumash skull from the Malibu site were examined courtesy of Phillip Walker, University of California, Santa Barbara. Amy Dansie facilitated the examination of the Great Basin skull and Spirit Cave Mummy at the Nevada State Museum. Moriori and Santa Cruz Island crania shown in Figure 2 were documented at the Musée de l'Homme Collections d'anthropologie biologique, Paris, courtesy of Alain Froment and Philippe Mennecier. The cast of Kennewick Man was provided courtesy of James Chatters and the National Geographic Society. Photographs in Figures 2 and 13 were taken by Donald E. Hurlbert. Stable isotope specialists Joan Brenner Coltrain, University of Utah, and Christine France, Smithsonian Institution's Museum Conservation Institute, provided thoughtful insights on interpretation of the data.

Index

Note: Figure/caption and table page numbers are shown in *italics* and for maps in **bold**.

aboriginal occupation standards, 97
abraders, 572–74
Acha-2 skeleton, 520
ACTIS analysis program, 422
Adams, Jenny L., 559
adhesions between bones, 270, 274, 338
Administrative Procedure Act of 1946, 94, 99–100
Afontova Gora II site, 15
age, determination of, 181–84, 279–80, 644. *See also* radiocarbon dating
Ainu peoples: ancestry, 463, 467; cranial morphology, 463–67, 478, 486, 634; dentition, 189, 463, 467; diet, 629; language spoken by, 468; nasal skeleton of, 513; R-matrices for, 464–67
Aka pygmies, 283
Aleuts: biomechanical analysis of long bones, 235; dentition, 188, 189, 200, *201*, 202; diet, 198, 539, 541; femoral to humeral strength proportions, 241; humeral strength and shape, 242; tibial and femoral strength and shape, 237
algae staining and colonization, 382–92; assessments of, 382, *383–85*, 386; on bones, 272, 275, 364, 374; erosion process and, 374, 388; interpreting historical events using, 390–91; taphonomic studies of, 388, *389*, 390; taxonomic identification of, 386–88
allometric analyses, 494–95, 497
al-Omaoui, I., 292
Alt, K. W., 572
Ambrose, Stanley H., 313
American Association of Museum Standards, 608, 611
American Homotype, 472, 483
American Indians: cranial morphology of, 631–32, 634, *635–37*; dental morphological traits of, 190, *191*; diversity among, 483; feature morphology of, 504–10, 512–14; inability to establish connection with Kennewick skeleton, 93–94, 95; joint diseases common among, 286; on repatriation of human remains, 5, 21, 22, 91; Tribal Claimants, 91, 92, 93–94, 102*n* 20, 117
American Museum of Natural History (New York), 198, 235
American Society of Heating, Refrigerating, and Air-Conditioning Engineers (ASHRAE) Handbook, 608
American technique for facial reconstruction, 523
Ames, Kenneth M., 38
amino acid analyses, 68, *70–71, 78, 79*, 84–85
AMS ^{14}C measurements, 68, *69, 79, 80*, 84, 85. *See also* radiocarbon dating
Amur River, 193
anatomical method: for estimating stature, 212, *213*; for facial reconstruction, 523
Ancient One. *See* Kennewick Man
Ancón skeletons, comparison of, 221–25
Andaman Islanders: dental microwear analyses of, 200, *201*, 202; diet, 197; femoral to humeral strength proportions, 241
Anderson, Art, 424, 427
Anderson, Duane, 16, 498
Angara River, 191
animals: damage to skeletons caused by, 324, 326–27, *327–29*, 330, 369, *370*; in shamanism, 592, 594, *595*
anisotropy, 199
anthroposcopic characteristics. *See* feature morphology
antler tools, 559–63, 581–82, *583*
Anzick: DNA analysis of, 645; funerary offerings at, 16, *17, 19*; ochre at, 330

Archaeological Resources Protection Act of 1979 (ARPA), 51, 90, 94, 98–99, 116
Arch Lake: cranial morphology, 485, 632; funerary offerings identified with, 20; ochre at, 330
Arctic Sadlermiut, osteometric study of, 222–24
Arikara: biomechanical analysis of long bones, 234–35; calcaneonavicular coalitions in, 256; dental microwear analyses of, 200, *201*, 202; diet, 198; femoral to humeral strength proportions, 241; pattern profile analyses of, 266–68, 274; tibial and femoral strength and shape, 237, 241
Arizona State University Dental Anthropology System (ASU DAS), 187, 189
Arlington Springs, 16
arm bones. *See* humeri; radii; ulnae
Armelagos, George J., 459–60
ARPA (Archaeological Resources Protection Act of 1979), 51, 90, 94, 98–99, 116
Arrow Game (film), 283
Asatru Folk Assembly, 118, 121
Ash Cave, 41
ASHRAE (American Society of Heating, Refrigerating, and Air-Conditioning Engineers) Handbook, 608
Ashworth Shelter, 18, 20
ASU DAS (Arizona State University Dental Anthropology System), 187, 189
atlatls, 156, *157*, 242, 305, 456
auditory exostoses, 150, 280–81
Auerbach, Benjamin M., 131, 213, 217, 218, 221, 481

Babbitt, Bruce, 94, 126
Bad Dürrenberg site, 572, 590
badger claws, 580–81, 592
Baker, Scott J., 510
Ballard, Joe N., 120

Banks, Larry D., 108–9
Bantu, 256
Barela, Ana M. F., 240
Barran, Paula, 110
Batak hunter-gatherers, tooth wear of, 188
Battelle Memorial Institute, 116–17
beads as funerary artifacts, 576–77
Beech Creek site, 39
Bellwood, Peter, 460
bending strength/rigidity of bone, 232–33, 235
Beneath World, 20
benthic aquatic algae, 364, 387–88, 391. *See also* algae staining and colonization
Benton County Coroner, 49, 115
Beringian Arctic, 190–91, 226, 467
Berryman, Hugh, 110, 111, *114,* 128, 130, 362
BIB (bi-iliac breadth) measurements, 217–18, 219
bifaces, 574–76
bi-iliac breadth (BIB) measurements, 217–18, 219
Bio-Imaging Research, Inc. (BIR), 418–19
biomechanical analysis of long bones: bone dimensions and derived properties, 234; comparative samples, 234–35; cross-sectional geometric properties, 232–34; femoral to humeral strength proportions, 241–42; humeral strength, shape, and asymmetry, 242–44; lower limb strength, shape, and asymmetry, 237–41; materials and methods, 232–35; results of, 235–37
Biopolymer Laboratory, 68
BIR (Bio-Imaging Research, Inc.), 418–19
BIR Custom MicroCT UHR/MCT Fast Scanner, 422, *423*
bird-bone tubes, 17
bison, as food source, 33, 35
Bitmap format, 419
Blangero, John, 483
block molds, *449,* 450–51
bloodletting, 574, 579–80, 592
Blum, Max, 192
Boas, Franz, 459–60
body mass: cranial size, relationship with, 481–82; estimation of, 217–19, 481–82

body painting, 590, *593*
body proportions, 219–20
bola stones, 38, 39, 290, 305
Bolton Standards study, *208,* 209
bone: adhesions between, 270, 274, 338; amino acid analyses on, 68, *70–71, 78, 79,* 84–85; bending strength/rigidity of, 232–33, 235; torsional strength/rigidity of, 233, 235. *See also* fractures; skeletal evidence; skeletal inventory; *specific bones*
bone-to-bone adhesions, 270, 274, 338
bone tools: needles, 582, 584; pins, 17; shaft wrench, 18; stylus, 570–72
Bonneville Power Administration (BPA), 109
Bonnichsen et al. v. United States et al. (1996), 2, 59, 91–94
Bonnichsen, Robson, 91, 92, 133
BPA (Bonneville Power Administration), 109
Brace, C. Loring, 91, 131, 487, 515
brachycephalization, 461
Bradley, Bruce, 558
Brazil, dentition in, 191, 466
breakage. *See* fractures; postmortem fractures
Breternitz, David A., 482
British Museum of Natural History, 197
Brown, Peter, 634
Browns Valley: body size, 482; cranial morphology, 479, 485, 631, 632; ochre at, 20
Bruwelheide, Karin, 110, *112, 113,* 114
Buhl Woman: comparison of craniofacial measurements, 465; cranial morphology, 482, 485, 631–32; diet and dietary stress identified in, 19, 629; geometric morphometrics, 495; stature estimation for, 481–82
Bull Brook site, 16
bull trout, as food source, 38
bunchgrass, 32
burial context, Kennewick Man's: algae colonization on bones, 272, 275, 364, 374; anatomical positioning of body, 324, 347–49, 362, 364, 365; animal damage to skeletons, 324, 326–27, *327–29,* 330, 369, *370;* bone colors and staining, 323–24, 330–31, *332–36,* 337;

corrasive damage to bone surfaces, 365, *366–68;* deliberate vs. natural, 324, 326–27, *328–29,* 330–31, *332–36,* 337; erosion process, timing of, 374, *375,* 376, *377–78,* 388; lake levels, 376, *377–78;* macroabrasions on bone surfaces, 365, *366,* 368, 369, 371; natural vs. deliberate interment, 324, 326–27, *328–29,* 330–31, *332–36,* 337; skeletal orientation relative to water, 364–65, *366–68,* 369, *370,* 371; stratigraphy of, 45–46, 371–72, *373. See also* burials; carbonate-cemented sediment concretions; discovery site; postmortem fractures
burials: artifacts, 17–18; caves as sites for, 20; dentition and, 187; double, 15, 19, 549–55, 580, 588–89, 594; flexed and semi-flexed, 17; impact on human skeletal remains, 7, 15; rock shelters as sites for, 553, *554,* 555–56; stratigraphic placement of, 45–46, 371–72, *373;* types of, 15–16, 17. *See also* burial context; *specific burial sites*
Burke Museum of Natural History and Culture (Seattle), 607–21; evaluation of environmental conditions, 610–11, *612–15,* 616–19; inventory of skeleton, 121–22; relative humidity considerations in, 607–8, 616–18; repository of skeleton, 64, 92, 101*n* 12; security of skeleton, 110, 609, 618, 620; storage environment for, 122, 607–9; taphonomic study conducted at, 110–14; temperature considerations in, 607–8, 615–16, 617–18; transfer to, 121; USACE agreement for skeleton storage, 607, 616
Burnett, S. E., 256
Busacca, Alan, 45
Butler, Virginia L., 237

CAC (Center of Alveolar Crest), 209–10, 211
calcaneonavicular coalitions, 252–53, *254,* 255–56, 273
calcaneus, guidelines for measurement of, 275
Caldera, Louis, 108, 126
California Channel Islands, 197, 280, 632

calipers, digital, 212, 429
Canada, Paleoamerican remains found in, 15
Canadian Museum of Man, 456
cannibalism, 18–19
Cannon, Bart, 131
canoes, as mode of transportation, 241–42, 467–68
capitate, guidelines for measurement of, 276–77
Carabelli's trait, 639, *641*
carapace bowls, 562, 563–64, *565*, 566–67, 594
carbonate-cemented sediment concretions: on clavicles, 347, 351; on cranium, 347, 350, *351*; on dentition, 347, 351; on femora, 349, 356–57, *358*; formation of, 337–38; on hand and foot bones, 269–70, *271–73*, 274, 348–49, 356, 361; on humeri, 348, 353, *354, 355*; on innominates, 348, 353; overall distribution on skeleton, 338, *339–46*, 347–50; on radii, 348, *349*, 355; on ribs, 353; on scapulae, 347–48, 351; on sternum, 351; on tibiae, 349, 357, 359–61; on ulnae, 348, *349*, 355–56; on vertebrae, 351, *352*, 353
carbonates: AMS ^{14}C measurements on, 68, *69*, 79, 84, 85; deposition on hand and foot bones, 269–70, *271–73*, 274; preparation for radiocarbon dating, 68, *76–77*
Carlson, David S., 460, 498
carp, as food source, 33
Cascade projectile points, 93, 452, 453, *454*
Cascade Range: glaciations, 33; Old Cordilleran Tradition sites in, 41; rainshadow effect, 31, 337; Western Stemmed Tradition sites in, 38, 39
Case, D. Troy, 131, *133*, 256, 258, 610
Cassman, Vicki, 123–24, 607
casting methods, 451
Catawba curers, 574
caves, as burial sites, 20
Center for the Study of First Americans, 127
Center of Alveolar Crest (CAC), 209–10, 211
Central Washington University, 115

Centre for GeoGenetics, University of Copenhagen, 75
cephalic index, 459–60, 461
cervical vertebrae: age-related changes in, 182; algae colonization on, 382; carbonate-cemented sediment concretions on, 351, *352*; morphology and pathology, 168, 286
Chalcolithic Period III Mehrgarh, 542
Chancellor's site, double burial at, 19
Chapman, Patrick M., 512, 513
Chatters, James C.: on bone colors and staining, 324; on chronological age of Kennewick Man, 182; collection of remains from discovery site, 115–16; cranial analyses by, 131, 149, 192, 280, 474, 475; on dental morphology, 191; on food sources, 237; on humeri pathology, 160, 283, 284; on natural vs. deliberate interment, 324; on projectile wound, 172, 176, 241, 282; on rib pathology, 165; on scapular pathology, 155, 284; skeletal inventory by, 90–91, 92, 117, 120–21, 139; on stratigraphic positioning of skeleton, 63, 372; on vertebral pathology, 286
Cheh, Jiwoong, 526
Chile, dentition in, 191
Chisholm, Brian S., 625
chloridoid grasses, 47
chronological age, determination of Kennewick Man's, 181–84, 279–80, 644
Chumash: cranial morphology, 631, *632, 634, 638*; dental microwear analyses of, 200, *201*; diet, 197
Clark, Chip, 110, 111, *113*, 114, *115*, 126, 130, 133, 610
Clark Fork River, 33
clavicles: bone colors and staining, 330–31; carbonate-cemented sediment concretions on, 347, 351; corrasive damage to, 365; morphology and pathology, 154–55, 284
climate: body breadth influenced by, 224; of discovery site, 31–33; limb proportions influenced by, 219; lower limb strength and, 241; for museum collections, 122, 607–9
Clover Island, 376

Clovis points, as funerary offerings, 16, 17, 19–20
Clow, Charles M., 514
cobble choppers, 39
collagen: amino acid analyses on, 68, *70–71*, 78, 79, 84–85; AMS ^{14}C measurements on, 68, *69*, 79, *80*, 84, 85; chemical purification of, 80, *81–83*, 84; isotopic analyses of, *72*, 80, 310; preservation among skeletal evidence, 61; pretreatment for radiocarbon dating, 64, 68
collar bones. *See* clavicles
Collections Management Preventive Care, 610
Columbia Basin region: climate and ecology, 31–33; cultural traditions in, 37–42; geochronology, 33–34; geography, 30, **31**; hydrology, 30–31; paleoecology, 34–35, *36*, 37
Columbia River, 31, 32, 34, 42, **43**
Colville Reservation, 91, 117
computed tomography (CT) scanning: applications of, 417; of cross-sectional geometric properties, 232–34; generation and format of data, 417–19, 422–24; limitations of, 417; machine configuration, 419–20; procedure used for skeletal evidence, 420, *421–22*, 422–23; projectile point model generated from, 418, 424–25, *426, 427, 428*; for prototypes, 430–31
concretions. *See* carbonate-cemented sediment concretions
Condron, Margaret, 68
conservation. *See* curation and conservation
Cook, Della, 131, *132*, 154, 165
cooking methods. *See* food processing technologies
Corps of Engineers. *See* United States Army Corps of Engineers
corrasive damage to bone surfaces, 365, *366–68*
COS (Curve of Spee), 209
cottonwood trees, 32
Coues, Elliott, 177, *178–79*
court case. See *Bonnichsen et al. v. United States et al.* (1996)
coyote teeth, 577–80

cranes, sandhill, 33
cranial index, 459–60, 461
cranial morphology: analysis of, 475–79; body size in relation to cranial size, 481–82; cranial size comparisons, 480–82; explanations of changes in, 498–99; genetic relatedness and, 639, 641–42, *643,* 644; in Holocene, 479–85, 486–87, 631–32, *633,* 634, *635–38;* measurements and interobserver variation, 472–73, *474;* modern populations, comparison to Kennewick Man, 473–79; population structure and, 482–85; statistical procedures for assessing, 475. *See also* feature morphology; geometric morphometric analyses
cranial plasticity, 459–61
cranium: carbonate-cemented sediment concretions on, 347, 350, *351;* corrasive damage to, 365, *367;* gender differences in size and shape of, 496, *497;* healing patterns in, 303; male sex features on, 147, 149–50, *151,* 280; morphology and pathology, 147, 149–50, *151,* 280; morphometric analysis of, 508–9; postmortem distortions of, 437, *439,* 440, *441;* prototype reconstruction of, 440, *442, 443, 444, 445–46,* 510–11. *See also* cranial morphology
Crater Lake, 34
cremation, 17, 18, 21, 39, 41
CT scanning. *See* computed tomography scanning
cuboid, guidelines for measurement of, 275
Cueva del Tecolote, 480
cultural affiliation, 5, 96–97
Cultural Resources Management Program (USACE), 108–9
Cummings, Linda Scott, 47
cuneiform: coalitions, 252, 253, 256, 273; guidelines for measurement of, 275–76
Cunningham, Deborah L., 241
curation and conservation: chain of custody, 116–17; in discovery and recovery process, 115–16; femur fragments, loss and recovery of, 119, 126–27; housing and access issues, 117–19, 120–21; image record evaluation, 126; inventory of skeleton, 116, 117–19, 121–22; in National Park Service studies, 122–26; in plaintiff team study, 128–30; in specialized studies, 130–31, *132–33;* taphonomic study, 110–14; USACE responsibilities for, 114–15, 116–18. *See also* Burke Museum of Natural History and Culture
Curtis, Donald, Jr., 118
Curve of Spee (COS), 209
Cutler Ridge, interpersonal violence identified at, 19
Cybulski, Jerome S., 235, 236, 238, 241, 624–25, 629, 641
cyprinids, 33

Dalton points, as funerary offerings, 19
dams, impact of, 30–33
Danning, Amanda, 521, 523
Dart Collection, 256
dating. *See* radiocarbon dating
Daux, Valérie, 317
David Geffen School of Medicine, 68
Deacy, Dave, 50, 452
Dead Birds (film), 283
Deadman Slough site, 16
deer, as food source, 38, 41, 198, 313, 319, 625
Delta® storage cabinets, 115, 122, 607, 609, 611, 619
dental microwear analyses: comparative samples, 197–99, 200, *201,* 202; factors impacting, 195–96; limitations of, 195; macroscopic examination, 196; materials and methods for, 196–97, 199–200; meat-eating, impact on, 197–99, 200, *201,* 202; microscopic examination, 197; molded impressions for, 199; objectives of, 195; results of, 200, *201,* 202; texture analysis used in, 199–200
dentition: age-related changes in, 184; carbonate-cemented sediment concretions on, 347, 351; crown and roots, 189, *190,* 191; cultural treatment, 189; enamel hypoplastic defects, 19; isotopic analyses teeth, 645; microwear analysis, 195–202; morphological traits, 189–92, 634, 639; observations, 187; oral pathology, 150–51, 188–89; Sinodont vs. Sundadont features, 189–92; tooth wear, 152, *153,* 187–88, 281–82, *463, 464. See also* orthodontics
Department of the Interior, U.S. (DOI), 92, 93–94
d'Errico, Francesco, 576
Designcraft, 425
Deter-Wolf, Aaron, 575, 590
DFA (discriminant function analysis), 225, 226
Diamond, Jared, 460
diatoms. *See* algae staining and colonization
DICOM format, 418, 419
diet and nutrition: cannibalism, 18–19; dental microwear impacted by, 195–96; food processing technologies and, 37–38, 39–40, 41, 195–96; isotopic evidence for, 59, 85, 310–19, 624–30, *626, 628;* limb proportions influenced by, 219
dietary stress, identification in Paleoamerican remains, 19, 21
digital calipers, 212, 429
Dillehay, Thomas D., 7, 15
discovery site: artifacts found near, 48–49; burial of, 92–93, 108, *109,* 110; chronology of skeletal evidence collection, 49–51, 90; climate and ecology, 31–33; geoarchaeology, 43–45; geochronology, 33–34; geography, 30, **31–32,** 42, **43;** geohydrology, 42; history of cultural traditions near, 37–42; hydrology, 30–31; paleoecology, 34–35, *36,* 37, 46–48; stratigraphy, 45–46, 371–72, *373. See also* burial context
discriminant function analysis (DFA), 225, 226
Dixon, E. James, 18
DNA analyses, 86, 545, 644–45
Dobe !Kung, 283
Dolní Věstonice, 19
Dolphin software, 208
Dongoske, Kurt E., 19
Doran, Glen H., 16
double burials, 15, 19, 549–55, 580, 588–89, 594. *See also* Horn Shelter No. 2
Dremel drills, 122, 123, 124
drumming, in shamanism, 594
Duarte, Marcos, 240

Duncan, Larry, 117
Dupont Tyvek pads, 122
Dutour, O., 297

Early Holocene Alluvium (Terrace), 34, 43–45, 377
Easter Islanders. *See* Polynesians
East Wenatchee site, funerary offerings at, 16
ecology of discovery site, 31–33
Ecuador, dentition in, 191
Edax Eagle II, 455
Edwards chert bifaces, 574–76
Egyptian mummy, musculoskeletal stress markings on, 292
elbow injuries, 283–84
Ellis Landing site, osteometric study of, 222–24, 226
El Zaatari, Sireen, 198, 200, 202
enamel hypoplastic defects, 19
environmental scanning electron microscopes (ESEMs), 574
environment for museum collections, 122, 607–9. *See also* climate
erosion process, timing of, 374, 375, 376, 377–78, 388
Esaw curers, 574
ESEMs (environmental scanning electron microscopes), 574
Eskimos: cranial morphology, 477; dentition, 188, 189, 196, 463; R-matrices for, 467
Eurasian Upper Paleolithic burial practices, 19
Europe, human skeletal remains in, 7

facial bones, morphology and pathology, 150, *151*
facial reconstruction: completed model, 527, 530, *531–32*; computer programs for, 520; first stage of, 521, *522–23*, 523; history of, 519–20; painting process, 527, *528–30*; public exhibitions and research programs on, 520; second stage of, 523–24, *525–26*, 527, *528*
Fagan, John L., 122–23
Fang, Madeline, 118, 119
feature morphology: of cranium, 508–9; interorbital projection metrics, 504–5, 511–12; of mandible, 506–8, 513, 639;

methods for, 503–4; of nasal skeleton, 505–6, 512, 513–14; of palate characteristics, 506, *507*, 512–13; population affinities and, 514–15; results of, 504–11; stereolithographic cranial model, accuracy and utility of, 510–11, 514
Federal Bureau of Investigation (FBI), 126–27, 519
femora: age-related changes in, 183; algae colonization on, 374; carbonate-cemented sediment concretions on, 349, 356–57, *358*; humeral strength proportions to, 241–42; macroabrasions on surface of, 369; methods for estimating body mass using, 217, 218–19; morphology and pathology, 179–80; osteochondritis at, 285, *286*; postmortem fractures of, 403–5; strength, shape, and asymmetry, 237–41; subtrochanteric traits, 509–10, 514
fibulae: macroabrasions on surface of, 365, *368*, 369; morphology and pathology, 181
Fiedel, Stuart, 483
field of view (FOV), 417–18
fingers. *See* hand and foot bones
First American human remains. *See* Paleoamerican human remains
Five Mile Rapids, 37, 38, 41
flail chest syndrome, 165, *166*, 286
Florida, Paleoamerican remains found in, 15
FOIA (Freedom of Information Act) request, 117, 611
food processing technologies, 37–38, 39–40, 41, 195–96
foot bones. *See* hand and foot bones
founder's effect, 190–91, 251
FOV (field of view), 417–18
Fox, John W., 582
fractures: cranial, 149; humeri, 159, 284; radii, 160; ribs, 162, *163–64*, 165–68, 285–86; scapular, 155, 156; secondary, 395. *See also* postmortem fractures
Freedom of Information Act (FOIA) request, 117, 611
Frison, George C., 283
Fryxell, Roald, 18
Fuegians: dental microwear analyses of, 200, *201*, 202; diet, 197–98

Fully method for stature estimation, 213, 215, 216
funerary offerings, 16, 17–18, 19–20

Gardenbroek, T. J., 166
GE DICOM viewing software, 146
Geffen, David, 68
GE Light Speed VCT/64 Slice Detector, 146
gender differences: in cranial size and shape, 496, 497; in limb proportions, 220; in musculoskeletal stress markings, 291
Generalized Procrustes analysis, 494
Genovés, Santiago, 213, 234
geoarchaeology of discovery site, 43–45
geochronology of discovery site, 33–34
geographic provenance, isotopic evidence for, 310, 317–18, 319, 624–30
geography of discovery site, 30, **31–32**, 42, **43**
geohydrology of discovery site, 42
geometric morphometric analyses: of allometry, 494–95, 497; benefits of, 492; comparative samples and methodology, 493–95; of dietary differences, 499; landmark data used in, 492, 493–94; results of, 495–99; of shape, 494, 495–96, 498; of size, 494–95, 496
geopetal structures, 338
Gerasimov, Mikhail, 519
Gill, George W., 91, 131, *132*, 503, 513
Gleser, Goldine C., 213
Goldendale site, 38
Goodman, Alan H., 639
Goodwin, A. J. H., 198
Gordon Creek: completeness of remains, 187; cranial morphology, 485, 632; ochre at, 330; stature estimation for, 481–82
Gore Creek: biomechanical analysis of long bones, 235, 236; osteometric study of, 220, 221–22, 226; tibial and femoral strength and shape, 237, 238
Gorto site, 16
Gower, J. C., 475
Grady, Frederick, 539
Graham Cave, 20
Gramly, Richard Michael, 20
Grand Coulee Dam, 33

graphitization, 68
grave offerings. *See* funerary offerings
Gravlee, Clarence C., 460
Green, Thomas J., 482
Gregory-Eaves, Irene, 315
Gresh, Ted, 33, 314
Grimes Shelter, interpersonal violence identified at, 18
Grine, F. E., 219, 221, 481
Gruss, Laura Tobias, 239
gunshot wounds, infections resulting from, 177, *178*

habitual activities: femoral to humeral strength proportions and, 241–42; humeri and, 242–44; lower limbs and, 237–41; scapulae and, 156, *157*
Hackenberry, Steve, 115
Haglund, William D., 369
Hall of Human Origins (exhibition), National Museum of Natural History, 520
hamate, guidelines for measurement of, 277
hand and foot bones: adhesions between, 270, 274; algae colonization on, 272, 275; asymmetry in, 256, *257*, 258, *259*, 260–62, 274; carbonate-cemented sediment concretions on, 269–70, *271–73*, 274, 348–49, 356, 361; comparison to Horn Shelter skeleton, 262–64, *265*, 266–69, 274; congenital defects in, 251–53, *254*, 255–56, 273–74; inventory of, 249–51; measurement guidelines for, 275–77; pathology, 181, 256; patina and color loss in, 270, 272, 275; pattern profile analyses of, 266–69, 274; postmortem fractures in, 401, *402*; taphonomic study of, 269–70, *271*, *272*, *273*, 274. *See also specific bones by name*
handedness: arm strength as indicator of, 155, 242; asymmetry of hand bones as indicator of, 256, 260, 261–62, 274; functional and developmental basis of, 159; molar wear as indicator of, 188
Hanford Reach National Monument, 35
harbor seals, as food source, 627
Harlan County Lake skeleton, 513
Harris lines, as sign of dietary stress, 19, 290

Harvest Technologies, 429
Haskett projectile points, 48, 49, 454–55, *456–57*
Hatwai site, projectile points discovered at, 48
Haury, Chérie E., 564
Haversian canals, *77*, 84
Hawikku, pattern profile analyses of, 266–68, 274
Hawkey, Diane E., 292
Hawkinson, Cleone, 110–11, 126, 131
Hawkins site, 16
hawk talons, 581, *582*, 592
Hayden, Brian, 574, 589
Haynes, C. Vance, 91, 92, 487
Hayonim Cave, 563
healing patterns, 299–304
Heaton, Timothy H., 539
Hebrews, changes in cephalic index, 460
height. *See* stature, estimation of
hematite. *See* ochre
Henneberg, M., 480, 481
Hicks, Brent A., 18
Hilazon Tachtit, 563
Hill, Kim, 18
Hinds Cave scarifier, 580
hip bone. *See* os coxae (innominates); projectile point wound
Historic Sites Act of 1935, 108
Hiwi society, violence in, 18
Hodgkiss, Tammy, 568
Hokkaido, 467
Holliday, Trenton W., 219, 220, 224, 227n 2
Holocene: cranial morphology in, 479–85, 486–87, 631–32, *633*, 634, 635–38; geochronology during, 34; Old Cordilleran Tradition during, 39; osteometric comparison of skeletons from, 220–27; paleoecological conditions during, 34, 35; population structure in, 483; subsistence activities during, 236–37; Western Stemmed Tradition during, 37
Holt, Brigitte M., 235
Hooton, Earnest A., 483
Hornby Creek, 38
Horn Shelter No. 2: ASU DAS study of, 187; biomechanical analysis of long bones, 235; burial description, 552–53,

589; comparison with Kennewick Man, 262–64, *265*, 266–69, 274, 587; cranial morphology, 480, 485; dentition, 207, *208*, 584–85, *641*; dietary stress identified at, 19; osteometric study of, 220, 221–22, 224, 226; site description, 549–50, *551*, 552, 588; skeletal attributes, *208*, 209–10, 584–87; stature estimation for, 481–82; status and role of male skeleton at, 589–92, 594; tibial and femoral strength and shape, 237
Horse Heaven Hills, 30, 31, 34
Hounsfield Units (HUs), 424, 430
Howells, William W., 472, 473–74, 475–76
Hrdlička, Aleš, 198, 472, 483
Huari Empire peoples, osteometric study of, 222–24
Huckleberry, Gary, 44–46, 92, 122, 324, 371–72
Hultkrantz, Ake, 589
Human Remains: A Guide for Museums and Academic Institutions (Cassman), 619
humeri: age-related changes in, 183; algae colonization on, 364, 382, 386; bone colors and staining, 331; carbonate-cemented sediment concretions on, 348, 353, *354*, 355; corrasive damage to, 365, 368; femoral strength proportions to, 241–42; macroabrasions on surface of, 369; medullary cavities of, 68, *75*, 122, *123*; morphology and pathology, 157–60, 283–84; postmortem fractures of, 410–12; reconstruction of, 146, *147*; strength, shape, and asymmetry, 242–44
humidity requirements for museum collections, 607–8, 616–18
Hunt, David, 110, 111, 129–30, 473
HUs (Hounsfield Units), 424, 430
hydrology of discovery site, 30–31
hypoplasia, 19

Idaho State University, 115
image record evaluation, 126
implements. *See* tools
Indian Knoll: femoral to humeral strength proportions, 241; osteometric

study of, 222–24, 225; tibial and femoral strength and shape, 237
Indians. *See* American Indians
injury and infection: auditory exostoses, 150, 280–81; chronological age and, 279–80; cranial injuries, 149, 280; elbow injuries, 283–84; healing patterns, 299–304; knee injuries, 285, *286*; oral infection, 281–82; osteochondritis, 283–84, 285, *286*; throwing injuries, 155–56; to vertebrae, 286. *See also* fractures; projectile point wound
innominates. *See* os coxae
in situ positioning of body, 324, 347–49, 362, 364, 365
Interior Department, U.S., 92, 93–94, 324
Interior Salish peoples, osteometric study of, 222–24, 226
interlandmark distances, 492, 494
interment. *See* burials
interorbital projection metrics, 504–5, 511–12
interpersonal violence, identification in Paleoamerican remains, 18–19, 21
Inuits, humeral strength and shape in, 242
inventory. *See* skeletal inventory
Irish, Joel D., 460
Ishida, Hajime, 641
isotopic analyses: of collagen, *72,* 80, 310; diet, determination through, 59, 85, 310–19, 624–30, *626, 628*; methods of, 310–11, *312*; origin, determination through, 310, 317–18, 319, 624–30; results of, 311, *312*; of teeth, 645
Italy, Paleolithic remains found in, 15–16

jackrabbits, as food source, 33, 198
Jantz, Richard L., 16–17, 91, 130, 440, 460, 473, 631
Japan, dentition in, 189
jawbone. *See* mandible
Jelderks, John, 127
Jerger mortuary ceremony, 20
jeweler's wax, 443, *444*
Johns Hopkins University, 131
Johnson, Floyd, 115
Johnson, Susan P., 313
joint surface changes, 297, 299, *300*
Jōmon peoples: craniofacial measurements, 465; dentition, 189, 467; diet, 629; humeral strength and shape, 242; in New World, 467–68; in Oceania, 468; origins of, 463, 467
JPEG2000 format, 419
juniper, 32

Kamchatka, 467
Kaskaskia Mine site, dietary stress identified at, 19
Keeley, Lawrence H., 18
Kelly, Robert L., 236
Kemp, Brian M., 645
Kendall's shape space, 492, 494
Kennewick (Washington), 30, **31–32**. *See also* discovery site
Kennewick Man: chronological age, determination of, 181–84, 279–80, 644; comparison with Horn Shelter, 262–64, *265,* 266–69, 274, 587; craniofacial measurements of, 463–67; diet and food sources, 59, 85, 310–19, 624–30; future research challenges, 644–45; non-metric variation in morphological traits, 639; origins of, 629–30; overall profile, 622–24; population relationships and genetic connections, 630–31; significance of discovery, 1, 5, 15. *See also* burial context; curation and conservation; discovery site; legal considerations; radiocarbon dating; skeletal evidence; skeletal inventory
Key, Patrick J., 475
Khoe-San: dental microwear analyses of, 200, *201*; diet, 198–99
Kintigh, Keith W., 22
Kirts, Linda, 120
Klawock, 545
Kline, Thomas C., Jr., 315
kneebone. *See* patellae
Kolman, Connie J., 86
Krantz, Grover S., 18, 116
Krogman, Wilton, 519
Kruskal-Wallis test, 223
Kumai, T., 253
Kuwóot yas.éin: His Spirit is Looking Out From the Cave (film), 545

Lahr, Marta Mirazón, 191
La Jolla W-12 skull: dental and skeletal measurements, *208,* 209, 210, 211; tool attrition in, 207
lake levels and erosion process, 376, *377–78*
Lake Wallula, 42, 323, 376, *377–78*
lampreys, as food source, 627
landmark data, 492, 493–94
Larsen, Clark, 125, 498
Lawrence site, double burial at, 553, 555
lawsuit. *See Bonnichsen et al. v. United States et al.* (1996)
Leavitt, Darrell G., 302
legal considerations: aboriginal occupation standards, 97; for administrative procedures, 99–100; for archaeological studies, 98–99; cultural affiliation standards, 5, 96–97; custody of skeleton, 59, 91–94; for discovery sites, 99; geological investigations of site and, 92; mishandling of skeleton and discovery site, 92–93; Native American, meaning of, 95–96, 100; relationship test, 95–96; repatriation claims, 5, 21, 22, 91; standing test, 97–98, 100–101, 103n 46. *See also specific legislation*
leg bones. *See* femora; fibulae; tibiae
lettered olive beads, 577
limb proportions, 219–20
Lind Coulee: faunal assemblages from, 625; projectile points discovered at, 48; as seasonal camp, 38
Lippert, Dorothy, 21
Little Salt Spring site, 20
Logan, Michael H., 460
long bones, biomechanical analysis of: bone dimensions and derived properties, 234; comparative samples, 234–35; cross-sectional geometric properties, 232–34; femoral to humeral strength proportions, 241–42; humeral strength, shape, and asymmetry, 242–44; lower limb strength, shape, and asymmetry, 237–41; materials and methods, 232–35; results of, 235–37
Longenecker, Julie, 117–18
Lovell, Nancy C., 315–16
Lower Walnut Settlement, 564
Lueck, Rhonda, 126
lumbar vertebrae: age-related changes in, 182; morphology and pathology, 170

lunate, guidelines for measurement of, 276

MacDonald, George, 456–57
MacMillian, Catherine, 115
macroabrasions on bone surfaces, 365, *366, 368, 369,* 371
Mahalanobis distances: and cranial morphology, 468, 476–77, 479–80, 484–85; in geometric morphometric analyses, 494, 495–96; procedures for obtaining, 475
male sex features on cranium, 147
Mal'ta site: dietary stress identified at, 19; double burial at, 15, 19; funerary offerings at, 19
Man Ahead of Us. See Shuká Káa skeleton
A Man Called "Bee": Studying the Yanomamo (film), 283
Mandatory Center of Expertise for the Curation and Management of Archaeological Collections (USACE), 118, 620
mandible: morphology and pathology, 150, *151*; morphometric analysis of, 506–8, 513, 639; prototype reconstruction of, 443
Maori: cranial morphology, 479; dentition, 196, 463
Marchi, Damiano, 239
maritime adaptations, evidence of, 537–46
Markowitz, Michael A., 460
Marks, Jonathan, 21
Marmes Rockshelter: cranial fractures at, 303; cremation evidence at, 39, 41; faunal assemblages from, 625; food processing technologies at, 38; interpersonal violence identified at, 18; projectile points discovered at, 49; as seasonal camp, 38
Martin, Debra L., 639
Mary Rose seamen, injuries to, 284, *285*
Marzke, M. W., 260, 261
Mas d'Azil site, 590, *592*
mathematical methods for estimating stature, 212–13
McClelland, Thomas, 51
McManamon, Francis, 92, 123–24
McNary Dam, 42, 379

Meadowcroft Rockshelter, interpersonal violence identified at, 19
meat-eating, dental microwear impacted by, 195–96
mechanical methods for estimating body mass, 217
medicine persons, 589
medullary cavities, 68, *75*, 122, *123*, 179
Meiklejohn, Christopher, 484
Memorandum of Agreement, 110, 619–20
Memorandum of Understanding for Curatorial Services, 122
Merbs, Charles F., 292
metacarpals: carbonate-cemented sediment concretions on, 356; guidelines for measurement of, 277
metatarsals: carbonate-cemented sediment concretions on, 361; guidelines for measurement of, 276; postmortem fractures in, 401, *402*
Mexico, Paleoamerican remains found in, 15
microwear analyses. See dental microwear analyses
migration: hypotheses, 483, 498; to North America, 1, 192; seasonal patterns of, 38, 41
Militello, Teresa, 119
Mini-Osteometric board, 275, 276
Minnesota Woman, 15
Mitutoyo digital calipers, 212, 429
Mobridge site, 198, 266
Modoc Rock Shelter, 555
molding methods, 447–48, *449,* 450–51
Mooney, James, 580
Moriori, cranial morphology of, 477–79, 631, *632,* 634
morphology. See cranial morphology; feature morphology
morphometric methods for estimating body mass, 217–19
mortuary ponds, 20
Mount Mazama, 34, 44, 46, 63, 371
MSM (musculoskeletal stress markings), 291–92, *293,* 294–97, *298*
multiple migration hypothesis, 498
Munro, Natalie D., 563
Munsell colors, 125, 162, 325, 326, *332–36*
musculoskeletal stress markings (MSM), 291–92, *293,* 294–97, *298*

museum collections, storage environment for, 122, 607–9
Museum of Anthropology (Rome), 198
mussels, as food source, 33, 35, 41, 313, 314

NAGPRA. *See* Native American Graves Protection and Repatriation Act of 1990
nasal skeleton, morphometric analysis of, 505–6, 512, 513–14
National Environmental Policy Act of 1970 (NEPA), 108
National Geodetic Vertical Datum, 376
National Historic Preservation Act of 1966 (NHPA), 94, 99
National Museum of Natural History (NMNH), 64, 110, 116, 198, 386, 447. *See also* Smithsonian Institution
National Park Service (NPS): examination of skeleton, 122–23; radiocarbon dating sponsored by, 59, 63, 123–26; taphonomic study by, 124–25
National Park Service Museum Handbook, 608
Native American Graves Protection and Repatriation Act of 1990 (NAGPRA): on aboriginal occupation, 97; on cultural affiliation, 5, 96–97; cultural items identified under, 95, 102–3*n* 34; effects on study rights, 94–95, 98; exclusion of remains from, 22; on legal standing requirements, 97–98, 100–101, 116; on meaning of Native American, 95–96, 100
Native American, legal meaning of, 95–96, 100. *See also* American Indians
navicular, guidelines for measurement of, 275
Neanderthals: DNA analysis on remains of, 86; humeral strength and asymmetry in, 242, *243*
nearest-neighbor analyses, 476–78, 480
needle-and-thread grass, 32
Nelson, Russell, 131, 132
Nemenyi post-hoc test, 323
NEPA (National Environmental Policy Act of 1970), 108
Neumann, Georg K., 483
neutral theory, 467

Nevada State Museum, 220–21
Neves, Walter A., 192, 498
Nez Perce Nation, 91, 94, 117
NHPA (National Historic Preservation Act of 1966), 94, 99
Nikander, R., 239
NMNH (National Museum of Natural History), 64, 110, 116, 198, 386, 447. *See also* Smithsonian Institution
Norr, Lynette, 313
North America, migration to, 1, 192
Norton, Gale, 91
Novoselovo VI site, 15
NPS. *See* National Park Service
Nubians, changes in cephalic index, 460
nutrition. *See* diet and nutrition

Oakhurst Rock Shelter, 198
occupational stress markers: healing patterns from traumatic injuries, 299–304; interpretations of, 304–6; joint surface changes, 297, 299, *300*; musculoskeletal stress markings (MSM), 291–92, *293*, 294–97, *298*; subsistence activities influencing, 290
ochre, 20, 330, 567–70. *See also* red ochre staining on bones
O'Connor, Jim E., 237
OCT (Old Cordilleran Tradition), 39–42, 452
Odegaard, Nancy, 607
Okanogan Highlands, 31, 33
Olbrechts, Frans M., 580
Old Cordilleran Tradition (OCT), 39–42, 452
olive beads, 576–77
Olivier method, 509
On Your Knees Cave (OYKC): artifacts at, 541–42; climate, 541; cooperation between researchers and Native Americans, 545; dietary stress identified at, 19; DNA analysis of, 645; site description, 537, *538–39*. *See also* Shuká Káa skeleton
oral infection, 281–82
origin, isotopic evidence for, 310, 317–18, 319, 624–30
orthodontics, 207–11; dental and skeletal measurements, *208*, 209–10, *211*; occlusion, 208–9; tooth attrition, 207–8. *See also* dentition
os coxae (innominates): age-related changes in, 182, *183*; carbonate-cemented sediment concretions on, 348, 353; morphology and pathology, 171–72, *173–75*, 176–79; prototype reconstruction of, 443. *See also* projectile point wound
osteochondritis, 283–84, 285, *286*
osteometry: body mass estimates, 217–19, 481–82; comparative samples, 220–22; limb and body proportions, 219–20; postcranial measurements, 146, *148*; skeletal element measurements, *208, 209*, 210, *211*, 214–15, *216*; stature estimates, 212–17, 234, 481–82; variation in morphological traits, 222–27
Ouzman, Sven, 21
Owsley, Douglas W.: on CT scans of Kennewick Man, 418–19; image record evaluation by, 126; on inattention to Paleoamerican skeletal remains, 16–17; skeletal inventory by, 90, 91, 92, 121, 139; taphonomic assessment by, 110
OYKC. *See* On Your Knees Cave

Pacific Northwest National Laboratory (PNNL), 115, 116–18
painting: in facial reconstruction, 527, *528–30*; of plaster casts, 451
Paisley Cave, 454
palate characteristics, morphometric analysis of, 506, *507*, 512–13
Paleoamerican human remains: age, sex, and social status of, 17; American Indian cooperation in study of, 21, 22; dietary stress identified in, 19, 21; funerary offerings with, 16, 17–18, 19–20; interpersonal violence identified in, 18–19, 21; lack of generalizability in, 21, 22; places of burial, 20; recovery of, 7, *8–14*, 15–17; repatriation of, 21, 22; time period of existence, 15; types of burials, 15–16, 17
paleoecology of discovery site, 34–35, *36, 37*, 46–48
PaleoResearch Institute, 46
Paleo-Tech Concepts, 258
Parafilm, 129, 147, 326, 608

parry fractures, 18
patellae: injuries to, 285, *286*; joint surface changes in, 299; morphology and pathology, 180
patina loss on hand and foot bones, 147, 162, 270, 272, 275, 330, 331
pattern profile analyses, 266–69, 274
patterns of healing, 299–304
Pearson, Osbjorn M., 234
Pecos Pueblo, cranial morphology of, 474–75
Pelican Rapids skeleton, 15
pelvis. *See* os coxae (innominates); sacrum
Perez, S. Ivan, 499
peroneal spasms, 253, 255, 256
Peru, dentition in, 191
Pfeiffer, Susan K., 241, 243
Pfitzner, W., 256
phalanges: carbonate-cemented sediment concretions on, 349, 356; guidelines for measurement of, 276, *277*; positioning and siding techniques, 249
photomacroscopy, 80–84
phytolith analyses, 47–48
phytoplankton. *See* algae staining and colonization
PI (Preservation Index), 608
pikeminnow, as food source, 36, 37, 38, 314
Pinhasi, Ron, 460
planning depth, 37
Pleistocene: geochronology during, 33, 34; glaciation during, 7, 33; subsistence activities during, 236; Western Stemmed Tradition during, 37
PNNL (Pacific Northwest National Laboratory), 115, 116–18
Point Data Marketing, Inc., 427
pollen analyses, 46–47, 48
Poll, Heinrich, 639
Polynesians: cranial morphology, 476, 477–79, 481, 486, 499, 632; dentition, 191; femur traits of, 514; interorbital projection metrics of, 512; mandibular morphology, 513, 639; nasal skeleton of, 512, 513–14; palate characteristics of, 512, 513; R-matrices for, 465, 467
ponds, mortuary, 20
Pope site, 16

population structure, cranial morphology and, 482–85
Porr, M., 572
postmortem fractures: anatomical positioning of body and, 393, 401, 405–6, 408–9, 410, 412; categories of, 362; descriptions and characteristics of, 393, *394, 396–99*; of femora, 403–5; in hand and foot bones, 401, *402*; of humeri, 410–12; mechanisms responsible for, 361–62, 393, 395; morphology of, 362, *363*, 395, 400; of radii, 407–8, 409–10; of scapulae, 406; sequential movement of skeleton into water resulting in, 412, *413*; taphonomic interpretations of, 400–401; of tibiae, 401–3; of ulnae, 406, *407*, 409
Powell, Joseph: on algae colonization on bones, 382; on animal damage to skeletons, 324; on bone colors and staining, 125, 323; on chronological age of Kennewick Man, 182, 279; on clavicular pathology, 154, 284; cranial analyses by, 149, 280, 473, 475, 498; on dental morphology, 191; on diversity of Native Americans, 483; on humeri pathology, 159–60, 283; interorbital projection metrics obtained by, 511–12; on lumbar vertebrae morphology and pathology, 170; on natural vs. deliberate interment, 324; on os coxae morphology and pathology, 179; on projectile wound, 172, 176, 282; reconstruction of bone fragments by, 123; on rib pathology, 165, 301; taphonomic assessment by, 124
power grip, 260, 261, 274
precision grip, 260–61, 274, 594
Preservation Index (PI), 608
projectile points: Cascade, 93, 452, 453, *454*; construction of, 456; Haskett, 48, 49, 454–55, *456–57*; of Old Cordilleran Tradition, 39, 452; San Patrice, 550, 552, 558–59, 588; of Western Stemmed Tradition, 37, 48–49, 454
projectile point wound: age at injury, 179, 282; cause of injury, 283; classification of projectile, 452, 453–55; CT generated model of, 424–25, *426, 427, 428*, 452, *454–55*; dimensions and characteristics of, 171, 452–53; future research needs, 457–58; healing patterns in, 290–91, 302–3; orientation and direction of entry, 172, *175–76*, 282, 456–57; pathological consequences of, 176–79, 282–83; visual inspection of, 171, *172–74, 452, 453*; XRF study of, 455, 645

Prospect Man: biomechanical analysis of long bones, 235; tibial and femoral strength and shape, 237, 240
prototype reconstruction: casting methods, 451; of cranium, 440, *442*, 443, 444, *445–46*, 510–11; equipment used for, 429; of innominates, 443; limitations of, 443–44, 447; of mandible, 443; molding methods, 447–48, *449*, 450–51; painting, 451; postmortem cranial distortions affecting, 437, *439*, 440, *441*; quality assurance in, 429–31, *432–36*, 437, *438*
pseudarthrosis in ribs, 165, 166, 167

Q-M3* chromosome type, 545

radii: age-related changes in, 183; algae colonization on, 382; bone colors and staining, 331; carbonate-cemented sediment concretions on, 348, *349*, 355; healing patterns in, 304; morphology and pathology, 160, *161*; postmortem fractures of, 407–8, 409–10
radiocarbon dating: advances in, 85–86; AMS ^{14}C measurements, 68, *69*, 79, *80*, 84, 85; initial measurements, 59, *60*, 61; by Interior Department, 93; by National Park Service, 59, 63, 123–26; objectives of, 63, 84; results and discussion, 68, *69–72*, 79–80, *81–83*, 84–85; sample selection and methods, 64, *65–67*, 68, *72–77*; by Waterways Experiment Station, 59, 61, *62*, 63
rainshadow effect, 31, 337
Rasmussen, Morton, 645
Rawlings, Kristen J., 513
Raxter, Michelle H., 213, 215–16, 217
Ray, Jack H., 574
reconstruction. *See* facial reconstruction; prototype reconstruction
Redder, Al, 549, 556, 564, 570, 580, 582

red ochre staining on bones, 39, 323, 324, 330, *331*
Regan, Marcia H., 253
Reid, Ken, 115
Relationship-Matrix technique, 464–67
relationship test, 95–96
relative humidity, 607–8, 616–18
Relethford, John H., 483
repatriation claims, 5, 21, 22, 91
Rhine, Stanley, 523
ribs: age-related changes in, 182; algae colonization on, 382; carbonate-cemented sediment concretions on, 353; healing patterns in, 300–302; morphology and pathology, 162–63, *164*, 165–68, 285–86; reassembly of fragments, 146, *147*
ricegrass, 32
Richman, Jennifer, 110
right-handedness: arm strength as indicator of, 155, 242; asymmetry of hand bones as indicator of, 256, 260, 261–62, 274; functional and developmental basis of, 159; molar wear as indicator of, 188
Ringold Formation, 33
Road Cut site: food sources at, 38–39, 625, 627; projectile points discovered at, 49; as seasonal camp, 38
rocker jaw, 508, 513, 639
rock shelter burials, 553, *554*, 555–56
Rose, Jennie, 369
Rose, Jerome C.: on algae colonization on bones, 382; on animal damage to skeletons, 324; on bone colors and staining, 323; on chronological age of Kennewick Man, 182, 279; on clavicular pathology, 154, 284; cranial analyses by, 149, 280, 473, 475; on dental morphology, 191; on humeri pathology, 159–60, 283; interorbital projection metrics obtained by, 511–12; on lumbar vertebrae morphology and pathology, 170; on natural vs. deliberate interment, 324; on os coxae morphology and pathology, 179; on projectile wound, 122–23, 172, 176, 282; on rib pathology, 165, 301; taphonomic assessment by, 124
Roseman, Charles C., 499

Ross, Ann H., 258
Rothschild, Bruce M., 284
routing effect, 313
Royer, Denise, 510
Ruff, Christopher B., 213, 217, 218, 224, 226, 239, 481
Rühli, F. J., 256
Russian technique for facial reconstruction, 523

sacrum: age-related changes in, 183; morphology and pathology, 170–71
Sadlermiut skeleton, 221–25
sagebrush, 31
salmon, as food source, 59, 237, 281, 290, 313–17, 319
saltgrasses, 32, 47
San Miguel Man (Tuqan Man): cranial morphology, 480, 485, 632, *633*, *638*; dietary stress identified in, 19
San Patrice projectile points, 550, 552, 558–59, 588
Sarna-Wojcicki, Andrei, 61
scale of maximum complexity, 199
scaphoid, guidelines for measurement of, 276
scapulae: age-related changes in, 183; bone colors and staining, 331; carbonate-cemented sediment concretions on, 347–48, 351; healing patterns in, 303–4; joint surface changes in, 299, *300*; macroabrasions on surface of, 369; morphology and pathology, 155–57, 284; postmortem fractures of, 406
scarification, 574–76, 579–80, 590–92, *593*
Schindler, Daniel E., 313
Schmitt, Daniel, 243
Schmucker, Betty J., 188
Schneider, Alan, 110, 128
Schwarcz, Henry, 624, 629
Schwartz, Ivan, 524
Sciulli, Paul W., 213
Scott, Robert S., 199
Sealaska Heritage Institute, 22
seasonal patterns of migration, 38, 41
secondary fractures, 395
security of skeleton at Burke museum, 110, 609, 618, 620

Seguchi, Noriko, 131
Sensofar Plu microscopes, 199
Sentinel Gap, projectile points discovered at, 48
sex, identification of, 17
shamanism, 589, 590–92, *593*, 594
Shaw, Collin N., 238, 239, 243–44
Shipman, Pat, 369
Shiwiar foragers, 283
Shoshone-Bannock tribes, 22
shoulder blades. *See* scapulae
shoulder girdle injuries, 284. *See also* clavicles; scapulae
Shuká Káa skeleton: ancestry of, 546; dentition, 542, *543–44*, 545; diet, 539, 541; DNA analysis of, 545; isotopic analyses of, 539, 541; radiocarbon dating of, 538–39; spatial and taphonomic analyses of remains, 327, *328*, 537–38, *540*. *See also* On Your Knees Cave (OYKC)
Sicilians, changes in cephalic index, 460
simometers, 503
Simpson, Ellie Kristina, 499
single migration hypothesis, 483, 498
Sinodonty, 189–92
skeletal evidence: arrangement of, 111, 114; chronology of discovery and collection, 49–51, 90; collagen preservation among, 61; examination of, 110–11, *112–15*; mishandling of, 92; stratigraphic origin of, 45–46, 371–72, *373*. *See also* biomechanical analysis of long bones; burial context; curation and conservation; discovery site; injury and infection; legal considerations; occupational stress markers; osteometry; radiocarbon dating; skeletal inventory
skeletal inventory: bone fragment identifications and catalog numbers, 139, *140–46*, 249–50; clavicles, 154–55; cranium and mandible, 147, 149–50, *151*, 280; curation and conservation efforts in, 116, 117–19, 121–22; dentition, 150–52, *153*; femora, 179–80; humeri, 157–60; os coxae, 171–72, *173–75*, 176–79; patellae, 180; postcranial measurements, 146, *148*; radii and ulnae, 160–61; ribs, 162–63, *164*, 165–68; sacrum, 170–71; scapulae, 155–57; sternum, 162; tibiae and fibulae, 180–81; vertebrae, 168–70. *See also* hand and foot bones; *specific related topics*
skull. *See* cranium
Sloan site: burial locations, 20; funerary offerings at, 17, 19–20
Smith, B. Holly, 152
Smith, C. Wayne, 110, 111, 129, 610
Smith, D., 125–26
Smithsonian Institution, 101*n* 12, 121, 472, 556, 608–9. *See also* National Museum of Natural History (NMNH)
snail beads, 576–77
Snake River, 30, 31, 32, 34, 38, 41
Snyder, Rebecca, 427
social status, identification of, 17
Solarmap Universal software, 199
Sparks, Corey S., 460
Speth, John D., 589
spine. *See* vertebrae
spiny hopsage, 32
Spirit Cave: biomechanical analysis of long bones, 235; completeness of remains at, 90, 212; cranial morphology, 631, 634, *640*; dentition, 639, *642*; fracture to frontal bone, 18, 303; mandibular morphology, 513; osteometric study of, 220, 221–22, 224, 226, 227 *n*4; stature estimation for, 481–82; tibial and femoral strength and shape, 237, 240
Spivack, Rebecca, 527, *530*
sports medicine, on throwing motions, 155–56, 297
Spradley, M. Katherine, 440, 473, 631
Stafford, Thomas W., Jr., 64, 92, 110, 111, 128–29, 130, 372
staining on bones, 323–24, 330–31, *332–36*, 337
standing test, 97–98, 100–101, 103*n* 46
Stanford, Dennis, 91, 130
stature, estimation of, 212–17, 234, 481–82
Steele, D. Gentry, 91
Stefan, Vincent H., 512
Stein, Julie K., 45–46, 122, 371–72
stereolithography, 425–27, 429, 510–11, 514
sternum: carbonate-cemented sediment concretions on, 351; morphology and pathology, 162

Stick Man: cranial morphology, 480, 631, 632; diet, 290; fracture to frontal bone, 303; red ochre on, 39
Stirland, A., 284
Stirling, Matthew, 198
Stock, Jay T., 238, 239, 241, 243–44
Stolf, Sandro F., 240
stone tools, 452–58, 574–76. *See also* bifaces; projectile points; projectile point wound
Story, Dee Ann, 576, 582, 589
stratigraphy of discovery site, 45–46, 371–72, *373*
stress markers. *See* occupational stress markers
StudioEIS, 523–24, 527, 530
sturgeon, as food source, 33
subsistence activities: occupational stress markers influenced by, 290; of Old Cordilleran Tradition, 41; of Western Stemmed Tradition, 38–39, 290. *See also* diet and nutrition
Sumatrans, 542
Sundadonty, 189–92
Sunghir mortuary ceremony, 19
surface complexity, 199
Swainson's hawk talons, 581, *582*, 592
Swanton, John R., 574
Swedlund, Alan, 16, 498
Swindler, Daris R., 192

talar beaking, 255, 256
Talon, Jean-Baptiste, 590–91
talus, guidelines for measurement of, 275
taphonomic studies: of algae staining and colonization, 388, *389,* 390; arrangement of skeleton, 111, 114; examination of skeleton, 110–11, *112–15;* of hand and foot bones, 269–70, *271, 272, 273,* 274; by National Park Service, 124–25; participants, 110, *111;* of postmortem fractures, 400–401; purpose of, 323; recording observations in, 324–26, *327;* safety and security measures for, 110. *See also* burial context
tattooing, 574–76, 579–80, 590, *593*
Taylor, Kathy, 124
Taylor, R. E., 59, 61, 85, 125–26, 310
Teaford, Mark, 131

teeth: attrition, 207–8; coyote, 577–80; isotopic analyses of, 645; wearing of, 152, *153,* 187–88, 281–82, 463, *464. See also* dentition; orthodontics
temperature requirements for museum collections, 607–8, 615–16, 617–18
Tepexpan Man, 279
terrestrial logistic mobility, 236, 237–38
Terry Anatomical Collection, 216, 256, 266, 267–68, 274
Texas A&M University, 127
thermogravimetric analyses, 45, 371, 372
thigh bones. *See* femora
thin-walled molds, 447–49
Thomas, David H., 459
Thomas, Will, 50, 452
thoracic vertebrae: age-related changes in, 182; algae colonization on, 382; carbonate-cemented sediment concretions on, 351, *352,* 353; morphology and pathology, 168–70
thorax. *See* clavicles; ribs; scapulae; sternum; vertebrae
3-D modeling, 425–27, 429
3Skull software, 472
three-migration hypothesis, 192
throwing exostosis, 155–56, 299, 305
tibiae: age-related changes in, 183; bone colors and staining, 330, 331; carbonate-cemented sediment concretions on, 349, 357, 359–61; morphology and pathology, 180–81; osteochondritis at, 285, *286;* postmortem fractures of, 401–3; strength, shape, and asymmetry, 237–40
TIFF format for data, 419
Tigara culture, osteometric study of, 222–24
Tikigagmiut: dental microwear analyses of, 200, *201,* 202; diet, 198
tissue-depth method for facial reconstruction, 523
Tlapacoya, 480
Tlingit, 22
Todd, Lawrence C., 236
toes. *See* hand and foot bones
tools: antler, 559–63; bone, 17, 18, 37, 39, 570–72, 582, 584; handling behavior, 260–61; of Old Cordilleran Tradition,

39; stone, 452–58, 574–76; of Western Stemmed Tradition, 37
toothpick grooves, 189
torsional strength/rigidity of bone, 233, 235
Tracy, Ray, 117, 119
Tribal Claimants, 91, 92, 93–94, 102*n* 20, 117
tribal lands, 95, 96, 103*n* 39
Tri-Cities Archaeological District, 48
Trimble, Michael, 118–19, 120, 123–24, 125, 610, 620
trophic level effect, 313
Trotter, Mildred, 213
Tukey multiple comparison test, 494, 496
Tuqan Man. *See* San Miguel Man
Turner, Christy, 131, *132,* 460, 467
Tuross, Noreen, 86
turtle carapaces, 562, 563–64, *565,* 566–67, 594
Twehus, Doug, 49
two-part block molds, *449,* 450–51
Tyrolean Iceman: biomechanical analysis of long bones, 235; tibial and femoral strength and shape, 239
Tyvek pads, 122, 609

Ubelaker, Douglas, 519–20
ulnae: age-related changes in, 183; algae colonization on, 382; bone colors and staining, 331; carbonate-cemented sediment concretions on, 348, *349,* 355–56; macroabrasions on surface of, 365; morphology and pathology, 160–61; postmortem fractures of, 406, *407,* 409
Umatilla Indian Reservation, 91, 94, 116, 117
United States Army Corps of Engineers (USACE): agreement with Burke Museum for skeleton storage, 607, 616; bone fragment identifications and catalog numbers assigned by, 139, *140–46,* 249–50; burial of discovery site by, 92–93, 108, *109,* 110; confiscation of skeleton, 90–91, 101*n* 3, 116; Cultural Resources Management Program, 108–9; curation and conservation responsibilities of, 114–15, 116–18; management

of Columbia River, 42; Mandatory Center of Expertise for the Curation and Management of Archaeological Collections, 118, 620; preservation plan for skeleton, 620; on safety and security measures for taphonomic study, 110; site geology investigation by, 92; transfer of custody, 116–17, 118, 121; Waterways Experiment Station (WES), 59, 61, *62*, 63

United States Coast and Geodetic Survey, 376

United States Forest Service (USFS), 545

United States Geological Survey (USGS), 371

United States National Herbarium, 386

University of Arkansas, 199

University of California, 59, 90, 91, 115–16, 311

University of Tennessee/Smithsonian Institution (UT/SI) database, 235, 472

University of Washington, 123, 130, 232, 418. *See also* Burke Museum of Natural History and Culture

University of Wyoming Human Osteology Laboratory protocol, 503–4, 510–11

UNIX platform, 422

Upper Cave 101, 634, *639–40*

USACE. *See* United States Army Corps of Engineers

USFS (United States Forest Service), 545

USGS (United States Geological Survey), 371

UT/SI (University of Tennessee/Smithsonian Institution) database, 235, 472

Vaara, Yarrow, 22

Van Gerven, Dennis P., 459–60, 498

Van Pelt, Jeff, 117–18

Varian Medical Systems, Inc., 419

vertebrae: age-related changes in, 182–83; algae colonization on, 382; carbonate-cemented sediment concretions on, 351, *352*, 353; morphology and pathology, 168–70, 286. *See also* cervical vertebrae; lumbar vertebrae; thoracic vertebrae

Vine site, food processing technologies at, 38

violence, interpersonal, 18–19, 21

virtual modeling, 425–27, 429

Virtual Surfaces, Inc., 427

Von Cramon-Taubadel, Noreen, 460

Vosberg site, cuneiform and calcaneonavicular coalitions in, 256

Wakeley, Lillian D., 371

Walker, Philip: on algae colonization on bones, 382; on animal damage to skeletons, 324, 369; on bone colors and staining, 125, 323–24; on cranial pathology, 149, 280; on humeri pathology, 160, 284; on projectile wound, 176, 282; on rib pathology, 165, 301–2; on in situ positioning of body, 324

Walthall, John A., 555

Wanapum Band, 91, 101*n* 8, 117

Warm Mineral Springs 3, dietary stress identified at, 19

Washington State Patrol Crime Laboratory, 131, 455

Washington State University, 116, 324, 372

Waters Alliance, 68

Waterways Experiment Station (WES), 59, 61, *62*, 63

Weaver, Timothy D., 241

weight. *See* body mass

Weiss, Elizabeth, 241–42, 292

WES (Waterways Experiment Station), 59, 61, *62*, 63

Wescott, Daniel J., 239, 241, 514

Western Idaho Burial complex, 41

Western Stemmed Tradition (WST), 37–39, *40*, 41–42, 48–49, 290, 454

white sturgeon, as food source, 33

Wilcox, Brian, 427

Wildcat Canyon site, 39

Williams, Aleithea, 110–11

Willson cranium, 513

Windover site: burial locations, 20; funerary offerings at, 17–18; interpersonal violence identified at, 18; osteometric study of, 222–24, 225, 226

Windust Caves, projectile points discovered at, 49

Wizards Beach: biomechanical analysis of long bones, 235; cranial morphology, 631, 632; mandibular morphology, 513; osteometric study of, 220–22, 224, 226; stature estimation for, 481–82; tibial and femoral strength and shape, 237

Wolff's Law, 291, 299

Worl, Rosita, 22

Wright, Richard, 477

Written in Bone: Forensic Files of the 17th-Century Chesapeake (exhibition), National Museum of Natural History, 520

WST (Western Stemmed Tradition), 37–39, *40*, 41–42, 48–49, 290, 454

X-ray diffraction (XRD), 45, 372, 566, 567–68

X-ray fluorescence (XRF), 455, 542, 570, 645

Yakama Reservation, 91, 94, 117

Yakima River, 30–31

Yellow Bear (shaman), 574

Young, Diane, 482, 585

Zeiss Universal microscope, 386, 387

Zumwalt, A. C., 292

Zuni Pueblo, pattern profile analyses of, 266–68, 274